Combinatorial Mathematics

This long-awaited textbook is the most comprehensive introduction to a broad swath of combinatorial and discrete mathematics. The text covers enumeration, graphs, sets, and methods, and it includes both classical results and more recent developments. Assuming no prior exposure to combinatorics, it explains the basic material for graduate-level students in mathematics and computer science. Optional more advanced material makes it also valuable as a research reference.

Suitable for a one-year course or a one-semester introduction, this textbook prepares students to move on to more advanced material. It is organized to emphasize connections among the topics and facilitate instruction, self-study, and research, with more than 2200 exercises (many accompanied by hints) at various levels of difficulty. Consistent notation and terminology are used throughout, allowing for a discussion of diverse topics in a unified language. The thorough bibliography, containing thousands of citations, makes this a valuable source for students and researchers alike.

Douglas B. West is Professor of Mathematics at Zhejiang Normal University and Professor Emeritus at the University of Illinois, where he won a campus-wide teaching award in 2002. Professor West has written more than 250 research articles on diverse topics in combinatorics and has advised 38 doctoral students. His earlier books include *Introduction to Graph Theory* (1996, 2001), a popular textbook adopted around the world for courses. He is Editor-in-Chief of *Discrete Mathematics* (since 2007) and Associate Editor of the *American Mathematical Monthly* (editing solutions for the Problems section since 1986). He also maintains web pages at https://faculty.math.illinois.edu/~west/ giving links to conferences in discrete mathematics and advice on writing mathematics.

Combinatorial Mathematics

DOUGLAS B. WEST

Zhejiang Normal University, China and University of Illinois, Urbana-Champaign

CAMBRIDGE
UNIVERSITY PRESS

CAMBRIDGE
UNIVERSITY PRESS

University Printing House, Cambridge CB2 8BS, United Kingdom

One Liberty Plaza, 20th Floor, New York, NY 10006, USA

477 Williamstown Road, Port Melbourne, VIC 3207, Australia

314–321, 3rd Floor, Plot 3, Splendor Forum, Jasola District Centre, New Delhi – 110025, India

79 Anson Road, #06–04/06, Singapore 079906

Cambridge University Press is part of the University of Cambridge.

It furthers the University's mission by disseminating knowledge in the pursuit of education, learning, and research at the highest international levels of excellence.

www.cambridge.org
Information on this title: www.cambridge.org/9781107058583
DOI: 10.1017/9781107415829

First published 2021

Printed in the United Kingdom by TJ International Ltd, Padstow Cornwall

A catalogue record for this publication is available from the British Library.

ISBN 978-1-107-05858-3 Hardback

Additional resources for this publication at www.cambridge.org/west

For my dear wife Ching

Contents

Part I — Enumeration

Chapter 3 – Generating Functions.   93

Chapter 4 – Further Topics.   153

Part II — Graphs

Chapter 5 – First Concepts for Graphs.   209

Part III — Sets

Part IV — Methods

Preface

Combinatorics is now a mature discipline. Although some see it as a maelstrom of isolated problems, it has central themes, techniques, and results that make it a surprisingly coherent subject. Meanwhile, it still rewards its students with endless discovery and delight.

This book introduces the reader to a substantial portion of combinatorics. It is not exhaustive in topics, results, or bibliography. However, it is thorough enough to equip the reader with the tools needed to read or do research in combinatorics or to apply combinatorics in other areas of mathematics and computer science. It assumes the maturity and sophistication of graduate students without assuming prior exposure to combinatorics. It assumes basic undergraduate mathematics, such as elementary set theory, induction, equivalence relations, limits, elementary calculus, and some linear algebra.

More advanced or specialized material is planned to appear in *The Art of Combinatorics*, a four-volume series of texts intended for researchers and for advanced graduate courses in combinatorics. Nevertheless, there is enough here to reward substantial study and investigation.

History and Rationale

Despite its fundamental nature and its explosive growth in recent decades, combinatorics still is not a standard part of mathematics instruction. Curricula (and mathematicians) are slow to change.

Combinatorial ideas appear in courses on elementary discrete mathematics, but such courses can be insubstantial. Serious undergraduate courses in combinatorics are seldom required for math majors. Graduate programs do not require combinatorics. Nevertheless, it is an elegant and valuable subject.

In the mid-1980s, I began to teach graduate courses in combinatorics at the University of Illinois. Excellent books existed for many topics, but every general textbook omitted substantial areas. Gathering material for such a textbook, I succumbed to the overabundance of riches before me. With so much beautiful material in combinatorics, the project grew to become four rotating courses taught from four books, now called *The Art of Combinatorics*.

In 1996, I realized that this structure served only students already committed to focusing on combinatorics. For others, an overview of the subject could have great value. An educated mathematical scientist should know some algebra and analysis, and also such a person should be acquainted with fundamental combinatorics and its relationships to other areas. Furthermore, disparities in preparation of entering students make a core course worthwhile to establish a common background before studying advanced material in combinatorics.

In 1997, I started a one-semester overview course to serve these goals. I extracted the fundamental material from *The Art of Combinatorics* and organized it to emphasize connections among topics. This book is the result. However, with so much beautiful combinatorics to choose from, I could not bring myself to cut the book down to one semester. It can support a two-semester sequence, analogous to fundamental two-semester sequences in classical areas of mathematics. It can also support various one-semester courses, as discussed later.

Since the scope is large, I have also sought to make the book useful as a research reference, rewarding further study after the courses are over. This leads to a fair amount of optional material allowing the reader to probe farther into the subject, plus remarks that provide statements and pointers to further results. Nevertheless, I still aim to keep the material accessible to graduate students.

Organization

One can organize combinatorial mathematics in many ways: by structures discussed, types of questions, methods used, etc. In a broad overview, the connections among topics are as important as the groupings within topics.

Most presentations of elementary combinatorics begin with enumeration or with graph theory; the former is the more classical approach. Natural enumerative questions arise in elementary graph theory, and many graph-theoretic arguments use basic counting techniques, so each informs the other. Here the basic notions of trees, cycles, and isomorphism are stated in Chapter 0 so that enumerative problems about graphs can be used as examples in Part I.

Part I presents the basics of bijective arguments, recurrence relations, generating functions, and inclusion-exclusion, with enhancements. Young tableaux and the elementary aspects of Pólya–Redfield counting appear here from a combinatorial point of view. Deeper algebraic aspects of enumeration are omitted.

Part II pursues central themes of elementary graph theory while reaching important and classical results, particularly those having broad applications. Graph theory is now a huge subject, so selecting fundamental core material is difficult. Many large topics are mentioned here at most in passing or in exercises; these include automorphism groups, Cayley graphs, graph representations, reconstruction, domination, decomposition, packings, genus, minors, nowhere-zero flows, Tutte polynomials, graph labelings, and structured families of graphs.

Part III explores our most general structural object: families of sets, generalizing graphs to hypergraphs. Four aspects of set systems are studied: Ramsey theory, extremal set theory, the structure of partially ordered sets and matroids, and combinatorial designs. Many aspects of posets and enumeration known as algebraic combinatorics are omitted (but Möbius inversion is in Chapter 15).

Part IV develops methods from probability, algebra, and geometry/topology and applies them to questions concerning graphs and sets. Also included are some applications of combinatorics to geometric questions. When discussing methods and connections, it helps to have the terminology and basic results of graph theory and enumeration available. Thus the material in the latter half of the book does depend on the earlier half.

Some topics omitted here are explored in *The Art of Combinatorics* or in my earlier *Introduction to Graph Theory*, which is a more patient and less sophisticated introduction to elementary graph theory. Some important topics in applied discrete mathematics are largely omitted here, partly because they often already have their own well-established courses; prominent among these are coding theory and linear programming.

The nearly 2200 exercises here apply ideas from the text and/or explore further concepts. Many have not appeared in texts before. More than 300 have hints with the problem statements, and another 380 have hints in the back of the book. Solutions are available to instructors via the book's website at

http://www.cambridge.org/west .

I have tried to indicate difficulty by marking easier problems with (−) and harder problems with (+). Problems of intermediate difficulty that are particularly interesting or instructive are marked with (◇). There is much ambiguity (and taste) in these designations, partly because the difficulty of finding a solution is not proportional to its length or its complexity. Thus these labels should be taken lightly.

Usage

Most schools have few regular graduate courses in combinatorics. At such schools this book is appropriate for a two-semester sequence to give a thorough introduction. Instead of separating graph theory from other topics to make two courses, this text integrates the topics into a coherent whole.

This approach enables students from other mathematical areas to acquire the fundamental material about enumeration, graphs, and sets in the first semester without continuing to the second. Also, topics that are best appreciated after knowing the fundamentals in several areas of combinatorics are omitted from courses that study only one part of combinatorics. Examples include the existence (Chapter 14) or construction (Chapter 10) of graphs with large chromatic number and girth, powerful techniques like the Regularity Lemma and entropy (Chapter 11), the application of projective planes to extremal graph problems (Chapter 13), the understanding of optimization via matroids (Chapter 11), and combinatorial applications of linear algebra and topology (Chapters 15 and 16). With the approach here, such applications enliven the second semester.

Nevertheless, the text can also be used separately for courses in graph theory and in "other" combinatorics, as described later.

In a two-semester sequence where most students take the full year and prior combinatorics is not expected, one can focus on Parts I and II in the first semester, III and IV in the second. Parts I and II concisely present the basic material taught in most undergraduate courses on combinatorics and graph theory, with a deeper point of view for graduate students. Students with prior exposure to the subject also benefit from this discussion. Classical topics in graph theory reside in Part II, but interactions between graphs and other topics and techniques appear in other chapters. Later topics are more independent, but the order of presentation here works well. Part III can be viewed as a third introductory area; it considers basic questions about sets and order relations. The methods of Part IV then apply to questions about the combinatorial contexts introduced in Parts I–III.

When the second semester is optional, with the first being the exposure that students from other areas will have to combinatorics, the first semester should be broader. Such a course has goals like those in a one-semester core course leading to multiple advanced courses in combinatorics. With this in mind, I have designated some sections and subsections as "optional", and in others the items marked "*" are optional. Such material is more technical or advanced and can be skipped at first reading without loss of continuity. In such a course, optional or more difficult topics should be skipped in Chapters 1–9 in order to present highlights of early portions of Chapters 10 and 12–14. I used this approach in a one-semester introduction at the University of Illinois that served as a departmental graduate exam course in combinatorics and prepared combinatorics students for four independent advanced courses.

That was a fast-moving course. I spent 16 lectures on enumeration (this can be less), 12 on graphs, and several each on Ramsey theory, posets, probabilistic methods, and designs, aiming in the latter topics mostly to introduce the basic ideas. Thus I was quite brief about signed involutions, the pattern inventory, and the magical properties of Young tableaux. Most students already had some acquaintance with graph theory (the clientele included many computer science students), so from Chapter 5 I presented just a few highlights and left the rest as background reading. In Chapters 6–9 I covered mostly fundamentals. Graduate students should see a bit extra, so in Chapter 6 I proved Plesník's Theorem instead of Petersen's Theorem, in Chapter 7 I gave the lower bound on connectivity of graph products, in Chapter 8 I approached Brooks' Theorem through list coloring (an important theme in modern graph theory), and in Chapter 9 I explained discharging. I skipped the subsections marked optional and also the proof of the Perfect Graph Theorem. In Section 9.3, one can present just enough about discharging to convey the idea.

In Chapter 10, I then presented a few choice pigeonhole applications and the main Ramsey theorem with several applications. In Chapter 12, the main goals were Dilworth's Theorem and LYM Orders for their connections with graph theory. Chapter 14 came before Chapter 13 to provide more time for homework problems, reaching the Local Lemma and threshold functions. In Chapter 13 the goal was the connection between latin squares and projective planes and the application of projective planes to extremal problems in graph theory.

Graph Theory vs. Combinatorics

Many institutions still have separate courses in graph theory and "other" parts of combinatorics, partly due to faculty interests.

A course in enumerative and set-theoretic combinatorics can use Part I and Part III. From graph theory one needs only the definitions from Chapter 0 to present Cayley's Formula (Section 1.3) and count isomorphism classes (Section 4.2). The discussion of chromatic polynomials in Section 4.1 using inclusion-exclusion can be skipped. Part II can be skipped completely. In Part III, one can skip graph-theoretic applications of the Pigeonhole Principle in Section 10.1, graph Ramsey Theory in 10.2, and all of 11.1. The material of Chapters 12 and 13 (except for symmetric chain decomposition of LYM orders and graph-theoretic applications of projective planes) is mostly accessible without graph theory.

I have used Part II for a graph theory course for masters students at Zhejiang Normal University in China. Moving more slowly to accommodate language difference, I did not go much beyond Part II. One would cover Part II except for some optional material. It does not require Part I except for binomial coefficients and simple bijective arguments (counting two ways). In Part III one can use graph-theoretic pigeonhole examples and graph Ramsey theory, possibly adding Section 11.1. The material on matroids in Section 11.3 generalizes results in graph theory, but it takes substantial time to develop the properties. It is more beneficial to include the basic material of Chapter 14, since probabilistic techniques are so effective and important and easily illustrated with graphs.

When a school has separate graduate courses in basic enumeration and basic graph theory, most likely most of the material from Parts III and IV will not be included, depending on the needs of the students and choices by the instructor. A subsequent course requiring the two basic courses can then cover topics from the last two Parts. Although the text often mentions connections between chapters, the chapters after Part II are relatively independent except for the background of language from the early parts.

It is worth noting that a two-semester sequence at U. Nebraska has for about 10 years used Part I combined with supplementary material on coding and information theory in the first semester, and Part II combined with selections from Chapters 10–15 in the second semester.

Highlights

Indeed, the connections between topics are among the features of this book. One aspect is pedagogical: we solve several fundamental problems repeatedly to show the usefulness of various techniques.

For example, Cayley's Formula to count labeled trees is obtained bijectively in Chapter 1, inductively in exercises, via generating functions and Lagrange Inversion in Chapter 3, and in Chapter 15 via the Matrix Tree Theorem and via eigenvalues. Derangements are counted via recurrence, generating functions, and inclusion-exclusion; Catalan numbers are also obtained repeatedly, including from Young tableaux. Turán's Theorem on the maximum number of edges in a graph containing no $(r + 1)$-vertex complete graph is proved inductively in Chapter 5, via extremality in Chapter 11, algebraically in exercises in Chapter 11, and probabilistically in Chapter 14. Planar graphs are characterized inductively in Chapter 9, via matroids in Chapter 11, and via dimension of partial orders in Chapter 16. The König–Egerváry max/min relation of Chapter 6 involving matchings and vertex covers in bipartite graphs is shown to be equivalent to Dilworth's Theorem on posets in Chapter 12 and is a special case of the Matroid Intersection Theorem in Chapter 11.

Other connections arise when techniques from one context are used to solve problems from other contexts. For example, common systems of distinct representatives (characterized in Chapter 7 using Menger's Theorem) are used to obtain symmetric chain decompositions of LYM orders in Chapter 12. Extremal problems for diameter in Chapter 5 and for graphs without 4-cycles in Chapter 11 are attacked using projective planes in Chapter 13. The problem in Chapter 8 of finding small triangle-free graphs with large chromatic number is discussed using Ramsey theory in Chapter 10. Probabilistic methods from Chapter 14 are used to obtained good bounds on crossing numbers in Chapter 16. Bounds on the graph connectivity parameter from Chapter 7 are obtained via eigenvalues in Chapter 15. Bounds on the list chromatic number, introduced for a richer study of graph coloring in Chapter 8, are obtained using the Discharging Method in Chapter 9 and the Combinatorial Nullstellensatz in Chapter 15. Pym's Theorem from Chapter 7 is used to prove the Planar Separator Theorem in Chapter 9.

Finally, since this is a graduate textbook, it covers the standard material of an elementary introduction efficiently in order to go beyond and offer the reader more. This permits the inclusion of many jewels that a standard elementary introduction at the undergraduate level cannot reach. Some of these were mentioned earlier in describing my overview course. Here are more of them.

In enumeration, we become familiar with the Delannoy numbers and the Eulerian numbers as additional basic counting models. We explain techniques such as Wilf's Snake Oil technique for evaluating sums, the Exponential Formula for obtaining generating functions, and Lagrange Inversion for extracting coefficients. We explore the combinatorial aspects of Young tableaux, obtaining not only the Hook-Length Formula and the Robinson–Schensted–Knuth Correspondence, but also Greene's Theorem about the largest union of k increasing subsequences in a permutation.

In graph theory, Chapter 5 provides unusual applications of the number of vertices of odd degree being even. Orientations with small outdegree, a beautiful application of Hall's Theorem by Hakimi in Chapter 6, are later applied to list coloring in Chapter 15. Optional material for advanced courses includes the proof of Tutte's f-Factor Theorem and the fast matching algorithm of Hopcroft and Karp in Chapter 6, and in Chapter 7 the Nash-Williams Orientation Theorem characterizing k-connected orientations (generalizing Robbins' Theorem on strong orientations). Also in Chapter 7, the standard sufficient conditions for spanning cycles in graphs are studied extended to long-cycle versions. List coloring provides a modern approach to coloring in Chapter 8, and the full form of Vizing's Theorem for multigraphs is proved. Chapter 9 includes a thorough introduction to the Discharging Method and an accessible proof of the Planar Separator Theorem.

In Part III on sets, we present many beautiful results that are not easy to find in general textbooks. Again some will typically be options for further reading. Chapter 10 includes the proof of the Stanley–Wilf Conjecture on pattern-avoiding permutations and the construction of graphs with large girth and chromatic number. The application of Ramsey's Theorem to table storage in computer science is unusual and makes use of allowing many colors. In Chapter 11, the Regularity Lemma of Szemerédi is presented, applied, and proved. Also, the notion of discrete entropy is developed to obtain extremal bounds on set counting problems. The basic structural aspects of posets are given in Chapter 12, along with important results about poset dimension, and the treatment of lattices leads to a rigorous discussion of correlational inequalities. Beyond the basic results about designs and projective planes, Chapter 13 includes the Multiplier Theorem for difference sets and the disproof of the Euler Conjecture.

Part IV on methods provides tools to attack many problems. Chapter 14 on the Probabilistic Method has a scope similar to the popular textbook of Alon and Spencer, including the basic methods, Dependent Random Choice, the Local Lemma, threshold functions, and concentration inequalities. The scope of Chapter 15 is similar to the well-known notes of Babai and Frankl on *Linear Algebra Methods in Combinatorics*; we also include Kastelyn's use of permanents to count perfect matchings in planar graphs and a discussion of Möbius inversion on posets. The Combinatorial Nullstellensatz is pursued as far as the recent strengthening of Thomassen's famous 5-choosability of planar graphs by Grytczuk and Zhu, show that every planar graph contains a matching whose edge-deletion yields a 4-choosable subgraph. Chapter 16 discusses geometric embeddings of graphs, applications of the Borsuk–Ulam Theorem and its relatives from combinatorial topology, and geometric aspects of partially ordered sets.

I hope this brief sampling whets the appetite for the delights ahead.

Acknowledgments

When C.L. Liu heard in the mid-1980s that I was accumulating text material on combinatorics, he showed me the lecture notes he had published as *Topics in Combinatorial Mathematics* (Math. Assoc. of America, 1972). These came from a summer seminar at Williams College in 1972 and were used in the combinatorics graduate course at the University of Illinois that I inherited from him. He proposed that we work them into a polished textbook; thus began *The Art of Combinatorics*. As described earlier, that project grew beyond the confines of a single volume, and the present text is closer to what he had in mind (but still more than twice as big). I thank him for the suggestion that started the process.

Also worthy of mention is Liu's earlier book *Introduction to Combinatorial Mathematics* (McGraw-Hill, 1968), which in 1972 introduced me to combinatorics. This book established the overall shape and subject matter for modern courses in combinatorics. Before it (at least in the U.S.) there was not much more than a compilation of chapters from eminent researchers who delivered a short course for engineers at UCLA (*Applied Combinatorial Mathematics*, 1964). Courses in elementary graph theory were initially shaped by the seminal textbooks of Berge (1962), Harary (1969), and Bondy and Murty (1976).

This text has benefitted by comments from many users and reviewers. Those who used pre-publication versions of the text several times include Garth Isaak (Lehigh U.), Art Benjamin (Harvey Mudd), Stephen Hartke, Christine Kelley, Xavier Pérez-Giménez, and Jamie Radcliffe (Nebraska), Jozsef Balogh (Illinois), and Mark Kayll and Cory Palmer (Montana). John Ganci and Leen Droogendijk each gave the book an extremely thorough reading, catching many glitches.

Other reviewers contributing insightful comments on early versions included Martin Aigner, Mike Albertson, Lowell Beineke, Miklós Bóna, Graham

Brightwell, Lynne Butler, Ira Gessel, Jay Goldman, Jerry Griggs, Mike Jacobson, Jenő Lehel, Herbert Maier, Michael Molloy, Chris Rodger, Bruce Rothschild, László Szekely, and Wal Wallis. Comments on particular chapters in later versions came from Noga Alon, Louis DeBiasio, Stefan Felsner, David Gunderson, Hemanshu Kaul, Sasha Kostochka, Cory Palmer, Pawel Prałat, Joel Spencer, Tom Trotter, Peter Winkler, and Günter Ziegler.

I also thank generations of students who labored with slowly evolving iterations of this material. Those who found numerous typos include Shivi Bansal, Alfio Giarlotta, Farzad Hassanzadeh, Bill Kinnersley, Darren Narayan, Radhika Ramamurthi, Michael Santana, Prasun Sinha, and Reza Zamani. I apologize to many others I have forgottten to mention over the long years of development.

At Cambridge University Press I thank my editor David Tranah for his patience through years of delays as I slowly refined the text. He accurately concluded that I view the book as a "work of art", which is part of why it took so long (25 or 35 years, depending on how you count). Clare Dennison shepherded the book through production, and Sarah Routledge gave it an incredibly thorough proofreading. All remaining errors, which I am sure exist, are solely my responsibility, especially since I continued to squeeze in exercises and make other refinements for months after she finished her job.

This book has been typeset using TeX. The scientific community owes a vast debt to its creator, Donald E. Knuth. With brilliance, foresight, and generosity, he has provided a common language for the publication and communication of technical material that is now used all over the world. Besides its versatility and free availability, its incredible genius is that it runs amazingly fast.

Chris Hartman taught me perl, which I used to convert earlier groff files to TeX. The "millenial" fonts were developed by Stephen Hartke, who helpfully made it possible to use them with plain TeX instead of LaTeX so that I could have greater control over spacing and placement of material. The references and indexes were assembled through herculean effort by Thomas Mahoney, who wrote scripts to handle most of the processing, built an effective computing environment for me to use in both China and the U.S., and patiently helped me resolve all my internet difficulties. Finally, I thank my wife, Ching Muyot, for her assistance, patience, and understanding with the crunch of each year's edition, no matter how many times I declared the book "essentially finished".

Feedback

I eagerly welcome comments on all aspects of this book. This includes selection and presentation of topics, errors made in mathematics or attribution or typography, items missing from the index, suggestions of additional hints, material that should be added if there is ever another edition, etc. Please send comments to dwest@illinois.edu. Errata will be listed at

http://www.math.uiuc.edu/~west/coerr.html .

Enjoy!

Douglas B. West
dwest@illinois.edu
Urbana, IL, and Jinhua, China

Chapter 0

Introduction

Combinatorial problems and arguments have a long history in mathematics, but only in the last half of the 20th century did they become a coherent subject. The discipline was long viewed as a collection of isolated tricks, but now the methods are more systematic, and the connections and applications between combinatorics and other areas of mathematics (in both directions) are being studied.

In this book we explore some of these connections and many fundamental results of combinatorics. We do not assume any prior exposure to combinatorics, but we assume mathematical maturity and basic undergraduate mathematics, including elementary set theory, induction, equivalence relations, limits, calculus, linear algebra, etc.

One can classify mathematical problems by the type of question, the object being studied, the method used, etc. These various aspects make it hopeless to impose a linear order of development in the study of mathematics. We emphasize different aspects at different times.

Our questions are of three general types. Given constraints specified for an object, does it exist? If such objects exist, how many are there? With respect to some criterion, which one is the best? These are the problems of Existence, Enumeration, and Extremality. We emphasize enumerative problems in Part I and problems of existence and extremality in most of the rest of the text.

We study objects that are discrete structures of various types. The simplest is a set. More complicated structures arise by imposing constraints or relations on sets or families of sets. We study various arrangements in Part I and graphs in Part II. In Part III we study structures such as hypergraphs, partially ordered sets, combinatorial designs, and matroids.

Finally, we also study methods of combinatorics. Many techniques arise in conjunction with particular structures, but some are used in many contexts and are worthy of study in their own right. Our focus on techniques is clearest in Part IV, where we discuss the probabilistic method, algebraic methods, and connections with geometry, but many other methods appear in earlier parts.

In this brief introduction, we review definitions from elementary mathematics, introduce elementary concepts about graphs for use in Part I, introduce elementary notions of probability as background for questions throughout the text, describe some additional discrete structures, and mention the basic notions of complexity. This is background material to be consulted as needed.

SETS, FUNCTIONS, AND RELATIONS

Our most fundamental object is a set. We build other structures from sets and relations. We use $\mathbb{N}_0$ for the set of nonnegative integers and $\mathbb{N}$ for the set of positive integers, also called the natural numbers. We let $[n]$ (pronounced "bracket n") denote the set $\{1, \ldots, n\}$ of the first n natural numbers, with $[0] = \varnothing$. We take as given the number systems $\mathbb{N}, \mathbb{Z}, \mathbb{Q}, \mathbb{R}, \mathbb{C}$ (natural numbers, integers, rational numbers, real numbers, complex numbers) and their elementary arithmetic and order properties. Similarly we assume the elementary operations and notation of sets, such as membership, containment, union, intersection, complement, and difference. We write the difference of sets A and B as $A - B$, not $A \setminus B$.

A **function** f from a set A to a set B, written $f \colon A \to B$, assigns each $x \in A$ an element $f(x) \in B$. The set A is the **domain**; f is *defined on* A. The **image** of $x \in A$ is $f(x)$, and $\{f(x) \colon x \in A\}$ is the **image** of f. The function is **injective** if each $y \in B$ is the image of at most one element of A. It is **surjective** if each $y \in B$ is the image of at least one element of A. It is a **bijection** if it is injective and surjective, and then it provides a **one-to-one correspondence** between A and B.

A set S is **finite** if a bijection from S to $[n]$ exists for some $n \in \mathbb{N}_0$; the value n is then the **size** of S, written $|S|$. *Counting* a finite set means computing its size. Two sets (finite or infinite) have the same **cardinality** if there is a bijection from one to the other. For finite sets, "cardinality" is a synonym for "size". A set with the same cardinality as $\mathbb{N}$ is **countable**.

A **sequence** is a function with domain $\mathbb{N}$ (or $\mathbb{N}_0$); we write a_n for the image of n, the nth **term** of the sequence. Usually one letter (such as f) denotes a function; similarly we use $\langle a \rangle$ to denote the sequence with terms of the form a_n.

A **list** is a function defined on $[n]$ for some $n \in \mathbb{N}_0$; this is the finite analogue of a sequence. We write a list a of *length* n as an n-**tuple** $(a_1, \ldots, a_n)$. (Many authors use "sequence" for an n-tuple; we try to use "sequence" only for functions on $\mathbb{N}$.) A **binary** n-**tuple** or $0, 1$-**list** is a list with entries in $\{0, 1\}$. Similarly, a $0, 1$-**matrix** or **binary matrix** has entries in $\{0, 1\}$. A **ternary list** has entries in $\{0, 1, 2\}$. An n-**ary list** takes values from a specified set of size n. An **arithmetic progression** is a list of equally spaced integers.

In contrast to a set, the order of elements in a list matters, and elements in lists may repeat. A **multiset** differs from a set by allowing repeated elements, but order remains unimportant. We can specify a multiset by specifying the set of distinct elements and their multiplicities. Since the order is unimportant but repetition is allowed, some authors refer to multisets as "unordered lists".

A **permutation** of a finite set S is a bijection from S to itself. Since a function on $[n]$ is a list, we may view a permutation σ of $[n]$ as a function from $[n]$ to $[n]$ or as a listing of $[n]$ in some order $a_1, \ldots, a_n$, with a_i denoting $\sigma(i)$. The latter is the **word form** of the permutation. Both viewpoints will be useful.

We often discuss sets whose elements are also sets. To avoid confusion, we use **class** and **family** as synonyms for "set". Instead of saying "a set in a set of sets", we say "a member of a family of sets".

The **cartesian product** of sets S and T is the set $S \times T$ of ordered pairs $\{(s, t) \colon s \in S, t \in T\}$. A **(binary) relation** between S and T is a subset of the cartesian product $S \times T$. When $S = T$, we call this a *relation on* S. We say that the pairs in a relation *satisfy* the relation.

When S is a family of sets, the **containment relation** on S is the set of pairs $(A, B) \in S \times S$ such that $A \subseteq B$. The pair (A, B) satisfies the **disjointness relation** when $A \cap B = \varnothing$. As a relation, disjointness is a property of pairs. Hence "a family of disjoint sets" technically has no meaning; nevertheless, by convention the word "disjoint" means "pairwise disjoint" when applied to sets in a family.

An **equivalence relation** on S is a relation R with these properties:
1) **reflexive** — $(x, x) \in R$ for all $x \in S$.
2) **symmetric** — $(x, y) \in R$ if and only if $(y, x) \in R$.
3) **transitive** — if $(x, y), (y, z) \in R$, then $(x, z) \in R$.

Containment is reflexive and transitive but not symmetric and hence not an equivalence relation. "Having the same cardinality" defines an equivalence relation, since the identity function is a bijection from a set to itself, the inverse of a bijection is a bijection, and the composition of bijections is a bijection.

An **equivalence class** of an equivalence relation R on S is a maximal subset T of S such that all pairs of elements of T satisfy R. Elements x and y are in the same equivalence class if and only if $(x, y) \in R$, and each element of S belongs to exactly one such class. Hence the equivalence classes of an equivalence relation on S form a partition of S, where a **partition** of a set S is a family of disjoint nonempty sets whose union is S. The sets in a partition are its **blocks**. Conversely, given a partition of S, putting $(x, y) \in R$ if x and y are in the same block yields an equivalence relation R on S. Hence partitions of S and equivalence relations on S are essentially the same notion.

0.1. Example. *Examples of equivalence relations.*

Congruence modulo n. Integers x and y are *congruent modulo n* if $x - y$ is a multiple of n. The equivalence classes are the subsets of $\mathbb{Z}$ having a fixed remainder upon division by n. These are the **congruence classes** modulo n, and the family of congruence classes modulo n is denoted by $\mathbb{Z}_n$.

Orbits under a permutation. Viewing a permutation σ of $[n]$ as a bijection from $[n]$ to $[n]$, we use σ^k to denote the bijection obtained by applying σ successively k times. The relation R on $[n]$ that puts $(i, j) \in R$ if $\sigma^k(i) = j$ for some $k \in \mathbb{N}_0$ is reflexive and transitive. Since bijections have inverses, also R is symmetric. The equivalence class containing i in this equivalence relation is the set of elements obtained as σ is repeatedly applied, called its **orbit**. ∎

0.2. Example. When S is a family of subsets of a "ground set" X, the **incidence relation** between X and S is the relation R consisting of the ordered pairs $(x, A) \in X \times S$ such that $x \in A$. When $X = \{x_1, \ldots, x_n\}$ and $S = \{S_1, \ldots, S_m\}$, we encode R as a $0, 1$-matrix with position (i, j) being 1 if $x_i \in S_j$ and 0 otherwise. This is the **incidence matrix** for R (with respect to the given indexing of X and S). The jth column of the incidence matrix is the **incidence vector** of the set S_j; this $0, 1$-vector records for each element of X whether it belongs to S_j. ∎

GRAPHS

Many natural relations are symmetric and irreflexive, where **irreflexive** means that for each x the pair (x, x) does *not* satisfy the relation (disjointness on a family of nonempty sets, for example). Such relations are modeled by "graphs".

A **graph** G is a pair consisting of a set $V(G)$ of **vertices** and a set $E(G)$ of **edges**, where each edge is a set of two vertices. The **order** of G is $|V(G)|$, and its **size** is $|E(G)|$. Vertices u and v forming an edge are **adjacent**, and the **neighbors** of v are the vertices adjacent to it. Note that the adjacency relation is symmetric and irreflexive. The two vertices in an edge are its **endpoints**, and we say that the edge **joins** its endpoints.

An edge and its endpoints are **incident**. The **incidence relation** of a graph G is the set of pairs (v, e) in $V(G) \times E(G)$ such that v and e are incident to each other (v is an endpoint of e). When discussing graphs we drop set brackets for edges and write the edge $\{u, v\}$ as uv (or vu). To display a graph, we represent each vertex by a point in the plane and each edge by a curve whose endpoints are the points assigned to the vertices of the edge; this is a **drawing** of the graph.

Structural properties of graphs do not depend on the names of the vertices. An **isomorphism** from a graph G to a graph H is a bijection $f \colon V(G) \to V(H)$ that preserves the adjacency relation: $uv \in E(G)$ if and only if $f(u)f(v) \in E(H)$. The resulting **isomorphism relation** is an equivalence relation on any set of graphs. Every graph isomorphic to G has the same structural properties as G, so we treat isomorphic graphs as "equal" when not given explicit vertex names.

Containment and union for graphs follow those notions for sets. A graph H is a **subgraph** of a graph G if $V(H) \subseteq V(G)$ and $E(H) \subseteq E(G)$. Similarly, the **union** of graphs G and H is the graph $G \cup H$ with vertex set $V(G) \cup V(H)$ and edge set $E(G) \cup E(H)$.

Many interesting enumerative questions can be asked about graphs. To permit discussion of such enumerative questions in Part I, we develop some elementary examples and properties of graphs here.

Many useful graphs are defined by their structural properties. A **path** is a graph whose vertices can be linearly ordered so that two vertices are adjacent if and only if they are consecutive in the ordering; the **endpoints** of the path are the first and last vertices in such an order. A **cycle** is a graph with an equal number of vertices and edges whose vertices can be placed around a circle so that two vertices are adjacent if and only if they appear consecutively along the circle.

A graph G is **connected** if for all $u, v \in V(G)$, it contains a path with endpoints u and v. A **component** of G is a maximal connected subgraph of G, where a **maximal** object of a given type is one that is not contained in another object of that type (**minimal** is defined similarly). The graph below has three components.

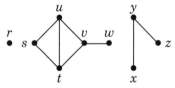

A u, v-**path** is a path with endpoints u and v. We say that u and v are *connected in* G if G contains a u, v-path. The **connection relation** on $V(G)$ is the set of pairs (u, v) such that G has a u, v-path; it is reflexive and symmetric. Our first proposition implies that it also is transitive and hence is an equivalence relation. The point is that although a u, v-path and a v, w-path together need not form a u, w-path, their union contains one. The equivalence classes of the connection relation are the vertex sets of the components of G.

The technique of **extremality** involves choosing an object that is extremal in some respect (a maximal connected subgraph, for example). This often requires finiteness, and our structures will be finite unless explicitly stated otherwise.

0.3. Proposition. If P is a u, v-path and P' is a v, w-path, then $P \cup P'$ contains a u, w-path.

Proof: We use extremality. Let x be the first vertex along P from u to v that also lies in P' (x exists, since both sets contain v). The union of the u, x-path in P and the x, w-path in P' is a u, w-path, since these subgraphs share only x. ∎

A **spanning subgraph** of G is a subgraph of G with vertex set $V(G)$. A **tree** is a graph that is connected and contains no cycles. A **spanning tree** in G is a spanning subgraph of G that is a tree.

0.4. Proposition. Every connected graph contains a spanning tree.

Proof: Since every vertex is itself a tree, every connected graph G contains at least one tree. Let T be a maximal tree contained in G.

If $V(T) \neq V(G)$, then a path from a vertex outside $V(T)$ to a vertex in $V(T)$ yields an edge e with endpoints in $V(T)$ and $V(G) - V(T)$. Let T' be the subgraph of G obtained from T by adding edge e and its endpoint v outside T. By the transitivity of the connection relation, T' is connected.

Also every cycle in T' appears in T, since v is in only one edge in T'. Thus T' has no cycle and is a tree. This contradicts our extremal choice of T. We conclude that $V(T) = V(G)$, and T is a spanning tree. ∎

The **degree** of a vertex in a graph is the number of edges incident to it. A **leaf** is a vertex of degree 1. A vertex of degree 0 is an **isolated vertex**.

A maximal path in a graph G is a path in G that is not a subgraph of another path in G. Thus every vertex of G adjacent to an endpoint of a maximal path in G must belong to the path.

0.5. Proposition. If all vertices in a finite graph G have degree at least 2, then G contains a cycle.

Proof: Since $V(G)$ is finite, G has a maximal path P. Let v be an endpoint of P. Since $d(v) \geq 2$, there is an edge vu not in P. Since P is maximal, u lies on P, and vu completes a cycle with the u, v-path in P. ∎

0.6. Proposition. Every tree with at least two vertices has at least two leaves. Deleting a leaf from a tree yields a tree with one less vertex.

Proof: A connected graph with at least two vertices has no isolated vertices. A tree has no cycle, so the endpoints of a maximal path in a tree with at least two vertices have degree 1.

Given a leaf x in a tree G, obtain G' from G by deleting x and its incident edge. Since deleting a vertex creates no new subgraphs, G' has no cycles. Hence it suffices to show that G' is connected. For distinct vertices $u, v \in V(G')$, there a u, v-path P in G. Since internal vertices along a path have degree at least 2, P does not contain x. Hence P is also a u, v-path in G'. ∎

0.7. Proposition. Every tree with n vertices has $n - 1$ edges. Furthermore, every graph with $n - 1$ edges that arises from n isolated vertices by iteratively adding an edge joining two components is a tree.

Proof: For the first statement, use induction on n. A 1-vertex tree has no edges. For $n > 1$, Proposition 0.6 provides a leaf whose deletion yields a tree with $n - 1$ vertices. Since it has $n - 2$ edges, the original tree has $n - 1$ edges.

For the second statement, note first that each such edge addition creates no cycles. Such a cycle would contain the new edge uv and another u, v-path from the previous graph, which does not exist since u and v were in different components in that graph. Also, each such edge addition reduces the number of components by 1. Hence $n - 1$ additions reduce the number of components to 1. Thus the resulting graph is acyclic and connected and is a tree. ∎

We assume familiarity with these results in order to study counting problems about trees in Part I. We discuss the structure of graphs more fully in Part II.

DISCRETE PROBABILITY

Many enumerative questions are easily motivated using discrete probability. In such questions, there is a set U (the "universe") of possible outcomes of some process. These outcomes are assumed to be equally likely; this is the meaning of phrases like "a random element" and "chosen uniformly at random". The probability of a desired property is then defined to be $|A|/|U|$, where A is the subset of U consisting of all outcomes having the desired property.

In Part III we will also consider countable spaces, where an outcome may be any natural number. Here we review the definitions of probability spaces.

0.8. Definition. A **discrete probability space** is a finite or countable set S with a function $\mathbb{P}$ defined on the subsets of S (called **events**) such that
a) If $A \subseteq S$, then $0 \le \mathbb{P}(A) \le 1$,
b) $\mathbb{P}(S) = 1$, and
c) If $A_1, A_2, \ldots$ are pairwise disjoint subsets of S, then
$$\mathbb{P}(\bigcup A_i) = \sum_{i=1}^{\infty} \mathbb{P}(A_i).$$

0.9. Remark. For a finite probability space, assuming $\mathbb{P}(A \cup B) = \mathbb{P}(A) + \mathbb{P}(B)$ when $A \cap B = \emptyset$ and applying induction on k yields $\mathbb{P}(A) = \sum_{i=1}^{k} \mathbb{P}(B_i)$ when $B_1, \ldots, B_k$ is a partition of A. This follows from (c) above but does not imply it, so we require the more general condition.

More general definitions of probability space allow the probability function to be defined only on subsets of S with certain properties, but the simple definition above suffices for our purposes. On the rare occasions where we mention continuous probability spaces, we will be informal (that is, non-rigorous). For example, when choosing a point at random from a region in the plane, we adopt the intuitive notion that the probability it lies in a particular subregion is proportional to the area, with "regions" simple enough not to worry about measurability.

Immediate consequences of Definition 0.8 include

    a) $\mathbb{P}(\varnothing) = 0$,

    b) $\mathbb{P}(\overline{A}) = 1 - \mathbb{P}(A)$, and

    c) $\mathbb{P}(A \cup B) = \mathbb{P}(A) + \mathbb{P}(B) - \mathbb{P}(A \cap B)$.

Furthermore, $\mathbb{P}(A) = \sum_{a \in A} \mathbb{P}(a)$, writing $\mathbb{P}(a)$ for $\mathbb{P}(\{a\})$ when $a \in S$. Elements of a probability space S are **sample points** or **outcomes** (of an "experiment"). ∎

0.10. Definition. Events A and B in a probability space are **independent** if $\mathbb{P}(A \cap B) = \mathbb{P}(A)\mathbb{P}(B)$. For events A and B with $\mathbb{P}(B) \neq 0$, the **conditional probability** *of A given B*, written $\mathbb{P}(A \mid B)$, is defined to be $\mathbb{P}(A \cap B)/\mathbb{P}(B)$.

0.11. Remark. Saying that a probability space is generated by making choices "independently" means that the space is a cartesian product. The probability of an outcome is the product of the probabilities of its coordinates in the factors. For example, when flipping a coin n times independently, each flip has outcome head or tail, each with probability $1/2$. The probability of a given list is 2^{-n}.

Making a choice **uniformly at random** means that the possible outcomes are equally likely. When flipping a coin, the flips are generated uniformly at random (probability $1/2$ of each outcome) and independently.

Studying conditional probability has the effect of normalizing or restricting the space to the points within the event B. The intuitive idea is that the conditional probability (given that B occurs) is the fraction of B where A also occurs.

The probability of a joint event can be computed as a product of conditional probabilities. For example,

$$\mathbb{P}(\textstyle\bigcap_{k=1}^{n} A_i) = \mathbb{P}(A_n \mid \bigcap_{k=1}^{n-1} A_i)\, \mathbb{P}(A_{n-1} \mid \bigcap_{k=1}^{n-2} A_i) \cdots \mathbb{P}(A_2 \mid A_1)\, \mathbb{P}(A_1). \qquad ∎$$

In the last half of the 20th century, advanced techniques for studying probability spaces found many applications to difficult combinatorial problems. Combinatorial techniques show that a "good enough" object exists by constructing it, but probabilistic methods are nonconstructive. An object with a desired property must exist when the probability of that property is nonzero in an appropriate probability space. Similarly, one can show the existence of an object with a large value of a parameter X by showing that the expected value of X is that large when the objects are randomly generated.

0.12. Definition. A **random variable** on a discrete probability space S is a function $X\colon S \to \mathbb{R}$. It is **discrete** when the range is finite or countable, often $\mathbb{N}_0$. Let $X = k$ denote the event $\{a \in S\colon X(a) = k\}$, and write $\mathbb{P}(X = k)$ for its probability. The **expectation** or **expected value** $\mathbb{E}(X)$ of X is $\sum_{a \in S} X(a)\mathbb{P}(a)$, when this sum converges. When X is a discrete random variable, we write this as $\mathbb{E}(X) = \sum_{k=0}^{\infty} k \cdot \mathbb{P}(X = k)$. The **pigeonhole property** of the expectation is the statement that there is an element of the probability space for which the value of X is as large as (or as small as) $\mathbb{E}(X)$.

Using the pigeonhole property requires a value or bound for $\mathbb{E}(X)$. Often the computation applies the "linearity of expectation" to an expression for X as a sum of simpler random variables. We restrict our attention to sums of finitely many random variables on discrete probability spaces. Analogous results hold in continuous probability spaces.

0.13. Lemma. (**Linearity of Expectation**) If X and $X_1, \ldots, X_k$ are random variables on the same space such that $X = \sum X_i$, then $\mathbb{E}(X) = \sum \mathbb{E}(X_i)$. Also $\mathbb{E}(cX) = c\mathbb{E}(X)$ for any constant c.

Proof: In a discrete probability space, each sample point contributes the same amount to each side of each of these equations. ∎

One reason for the influence of probabilistic methods is that exact counts are both too difficult and unnecessary in large structures. Somehow the probabilistic methods capture the most important aspects or most dominant terms. Probabilistic methods are especially effective for extremal problems when it turns out that "most" instances are near the optimum.

OTHER DISCRETE STRUCTURES

In addition to subsets, permutations, and graphs, many other structures are used to model combinatorial problems. We briefly describe several studied in this book in order to suggest the scope of the text. The reader should treat this section lightly; precise definitions and examples will be given later.

0.14. Example. *Digraphs and multigraphs.* General binary relations are modeled using **directed graphs** (**digraphs**); these differ from graphs in that the edges are *ordered* pairs of vertices. In an edge *from x to y*, the first vertex is the **tail** and the second is the **head**. We illustrate directed graphs by drawings that use arrows (curves with direction) for the edges.

We sometimes allow edges in digraphs to be pairs of the form (x, x), called **loops**. For example, the **functional digraph** of a function $f \colon A \to A$ has vertex set A; its edges are the pairs $(x, f(x))$ for $x \in A$. The fixed points of f become loops in the functional digraph. A permutation is a bijection $\sigma \colon A \to A$; its functional digraph consists of disjoint (directed) cycles corresponding to its orbits.

We may modify a graph or digraph both by allowing loops (one-vertex edges) and by allowing more than one edge with the same endpoints. Here $E(G)$ becomes a multiset. The resulting model is a **multigraph** or **multidigraph**. ∎

0.15. Example. *Order relations.* The containment relation on a family of sets is a reflexive relation that is transitive but not symmetric and hence is not an equivalence relation. It is a fundamental example of another important type of relation. A relation R is **antisymmetric** if $(x, y) \in R$ and $(y, x) \in R$ imply $x = y$. A relation is an **order relation** if it is reflexive, antisymmetric, and transitive.

Besides containment, other order relations include the divisibility relation on the set of positive integers and the componentwise order on $\mathbb{R}^n$, which contains the pair (u, v) when $u_i \leq v_i$ for all i. Other examples arise in scheduling: events occupy some interval in time, and we say that A "precedes" B if A ends before B begins. This defines an order relation on any set of intervals. We study order relations in Part III. ∎

0.16. Example. *Hypergraphs.* Another generalization of graphs allows edges of arbitrary size. A **hypergraph** G consists of a set $V(G)$ of vertices and a set $E(G)$ of edges, where an edge can be any subset of $V(G)$. A **k-uniform hypergraph** is a hypergraph whose edges all have size k; graphs are simply 2-uniform hypergraphs. A hypergraph is **regular** if every vertex is in the same number of edges. Since the edge set of a hypergraph with vertex set X is a family of subsets of X, hypergraphs can be studied from set-theoretic, order-theoretic, and graph-theoretic viewpoints. They will be helpful in Parts III and IV.   ∎

0.17. Example. *Designs and projective planes.* In Chapter 13 we study a special type of regular uniform hypergraph that has applications to design of experiments and to extremal combinatorial problems. A **block design** is a regular uniform hypergraph in which any two points (vertices) appear together in the same number of blocks (edges). Equivalently, any two rows in the incidence matrix for the membership relation of elements in blocks have the same dot product.

When any two points lie in exactly one common block, we may request also that any two blocks have exactly one common point. The resulting configurations are **projective planes**, in which the blocks are called **lines**. Projective planes can be obtained from finite fields and correspond to special families of latin squares. A **Latin square** of order n is an arrangement of symbols of n types in an n-by-n matrix so that each symbol appears exactly once in each row and in each column. A classical application is the assignment of types of fertilizers and seeds to regions in an agricultural plot to reduce the effect of soil differences.   ∎

0.18. Example. *Matroids.* In Chapter 11 we study another special structure. Matroids can be viewed as special families of subsets of a set. This interprets them as a special type of hypergraph. We postpone the precise definition and observe merely that the matroid context permits common generalizations of fundamental results in graph theory, linear algebra, and the theory of ordered sets. For example, matroids permit a natural generalization of the result that a spanning tree of minimum total weight in a connected graph with weighted edges can be found by iteratively including the cheapest edge that does not form a cycle with edges already chosen. Much of the elementary theory of dimension in linear algebra also arises as a special case of matroid properties.   ∎

COMPLEXITY

The growth of combinatorics has been stimulated by computer science, which studies the computational aspects of discrete mathematics. We will comment occasionally on the computational complexity of problems. A simple measure of the performance of an algorithm is its worst-case running time, as a function of the size of the input. A problem is *efficiently solved* if it has a solution algorithm whose running time is bounded by a polynomial in the size of the input.

The *size* of the input is its length in bits in some encoding of the problem. For our purposes, natural parameters such as the order of a matrix or the number of vertices suffice to measure size. A polynomial in n is bounded by a polynomial in

n^2 or n^3, so the manner of encoding input is unimportant unless the problem has exponentially large numbers as input data.

Complexity considers asymptotic growth rates. The set of functions whose magnitude is bounded above by a constant multiple of f (for sufficiently large arguments) is called $O(f)$. Several pertinent sets of functions arise when comparing growth rates to f, as listed below.

$$o(f) = \{g \colon |g(x)|/|f(x)| \to 0\}$$
$$O(f) = \{g \colon \exists c, a \in \mathbb{R} \text{ such that } |g(x)| \le |cf(x)| \text{ for } x > a\}$$
$$\Omega(f) = \{g \colon \exists c, a \in \mathbb{R} \text{ such that } |g(x)| \ge |cf(x)| \text{ for } x > a\}$$
$$\omega(f) = \{g \colon |g(x)|/|f(x)| \to \infty\}$$
$$\Theta(f) = O(f) \cap \Omega(f)$$

Derbyshire [2003] attributed "Big Oh notation" to Landau [1909], but Landau [1909, p. 883] borrowed it from Bachmann [1894].

Properly speaking, $o(f)$, $O(f)$, $\Omega(f)$, $\omega(f)$, and $\Theta(f)$ are sets, and it is correct to write $g \in O(f)$ when describing the growth rate of g. Mathematicians and computer scientists routinely write $g(n) = O(f(n))$ to mean $g \in O(f)$ (see Knuth [1976a]). Some avoid this by writing "$g(n)$ is $O(f(n))$", treating $O(f)$ as an adjective. We can compute with members of $O(f)$ somewhat as we do with congruence classes, but we have no symbol like "$\equiv$" for doing this. Thus in this book we sometimes use expressions like $f(n) = n^2 + O(n^{3/2})$ (meaning $f(n) - n^2 \in O(n^{3/2})$), but where the grammar is appropriate we use the membership symbol. The expression $g(n) \sim f(n)$ means that g is **asymptotic** to f, which can be written as $\lim_{n\to\infty} \frac{g(n)}{f(n)} = 1$ or $g(n) = f(n)(1 + o(1))$.

Complexity classes are studied using **decision problems** that have yes/no answers, such as "does the input graph have a spanning cycle?" Optimization problems (such as "what is the maximum length of a cycle in this graph?") can be solved using successive decision problems, such as "does this graph have a cycle of length at least k?", where k is part of the input. The class of decision problems solvable by a worst-case polynomial-time algorithm (polynomial in the size of the input) is called "P".

Many decision problems have no known polynomial-time solution algorithm but have a polynomial-time algorithm for verifying a YES answer. For example, existence of a spanning path can be verified by giving the order of vertices on such a path and checking that successive vertices are adjacent. When checking all possible permutations in parallel, each computation path is short. It is verifying a NO answer that is difficult.

A **deterministic algorithm** follows only one computation path on a given input. A **nondeterministic algorithm** follows multiple computation paths simultaneously. In the example above, given an input graph, such an algorithm checks all possible vertex orderings simultaneously to seek a spanning path; each ordering can be checked in polynomial time. A **nondeterministic polynomial-time** algorithm follows one computation path for each way of specifying a polynomial-length stream of bits, with each such computation running in polynomial time. The bit stream is not the input to the problem; the bits specify an option to consider, which in the example here is a vertex ordering.

A nondeterministic algorithm *solves* a decision problem if for every input I, the answer to the problem on I is YES if and only if the algorithm applied to I has at least one computation path that returns YES. The class of decision problems having nondeterministic polynomial-time solution algorithms is called "NP". Because a machine having the power to follow many computation paths in parallel can also follow one, $P \subseteq NP$.

Most computer scientists believe $P \neq NP$. In this context, a problem B is *as hard as* a problem A if a polynomial-time algorithm for B would yield a polynomial-time algorithm for A. This may involve using B as a subroutine in an algorithm to solve A or providing a polynomial-time transformation that converts an arbitrary instance of A into an instance of B such that the answer in B is YES if and only if the answer in A was YES. In either case, an algorithm for B yields an algorithm for A; we call this a **reduction** of A to B. A problem is **NP-hard** if it is as hard as every problem in NP in this sense (every problem in NP reduces to it). A problem *belonging to NP* is **NP-complete** if it is **NP-hard**.

Cook [1971] devised a generic transformation to reduce any problem in NP to that of deciding whether an input logical formula is true under some truth assignment for its variables. This problem is called **SATISFIABILITY** or **SAT**. Cook's result made SAT the first known NP-complete problem. To prove that a problem B is NP-hard, we can reduce SAT to B. Every problem proved NP-complete can then be used like SAT in this way. In practice, the known NP-complete problem in most NP-completeness proofs is one of a few fundamental NP-complete problems, such as those in Karp [1972].

A polynomial-time algorithm for any NP-complete problem could be used to construct a polynomial-time algorithm for each problem in NP, yielding P=NP. The conjecture that $P \neq NP$ is supported by the failure to find a polynomial-time algorithm for any problem in the large class of NP-complete problems, despite years of search.

Garey–Johnson [1979] gives a thorough and readable introduction to NP-completeness. Nowadays, an NP-completeness proof is only a beginning. One seeks to refine the boundary between P and NP by finding polynomial-time solution algorithms for large classes of inputs or by finding restricted classes of inputs where the problem remains NP-hard.

Mostly we use NP-completeness as motivation. NP-completeness justifies the study of heuristics that run quickly but do not obtain the optimal solution. It also motivates extremal problems: we study bounds on optimization problems in terms of various parameters of the input. A constructive proof of a bound for graphs in a particular class yields an algorithm that may approximate the true value of the parameter.

We hope that these few basic concepts suggest some of the underlying structure of this large subject. More thorough definitions and examples are given where the concepts are explored in depth. With this as background, we are ready to begin.

Chapter 1

Combinatorial Arguments

Enumerative questions are among the most natural in mathematics and have been studied for eons. A simple way to count a set is to list its elements, but this is inelegant and often impractical. We may view the study of enumeration as the search for ways to avoid exhaustive listing.

In this chapter we use combinatorial arguments. We give no precise definition of this term, but generally it refers to explicit counting arguments, "counting two ways", or the use of bijections to show that two sets have the same size.

Later chapters develop more sophisticated techniques. Combinatorial arguments are elementary but can be hard to find. Studying them is worthwhile, because elementary counting problems often serve as models or building blocks in harder problems, and combinatorial arguments can yield more information.

Because we assume no prior experience in combinatorics, many results and examples here appear also in undergraduate courses. Often we go farther, and our treatment is more concise and assumes more mathematical maturity.

To avoid excessive formality, we may omit explicit declaration of variables when the context is clear. The universe is the set of values where the resulting objects make sense. This may happen with the set $\mathbb{N}$ of natural numbers or the set $\mathbb{N}_0$ of nonnegative integers. Examples include "for all n-vertex graphs" and "for all odd n". A similar convention sometimes omits the limits on summations.

What does it mean to solve a counting problem? The problem may be expressed in terms of a parameter (variable), and we want to solve it simultaneously for each value of the parameter. With one parameter n, this yields a sequence $\langle a \rangle$ consisting of the values $\{a_n\}_{n=0}^{\infty}$. We may seek various expressions for the answer.

A *formula* for a_n expresses it as a function of the parameter(s). Sometimes we accept a *finite sum* as the answer. This is common when we use the *inclusion-exclusion principle* (Section 4.1), a technique that alternately overcounts and undercounts the desired set until each desired element is counted exactly once. An *asymptotic formula* (Section 2.3) for a_n is a formula $g(n)$ such that $g(n)/a_n \to 1$ as $n \to \infty$. Computational considerations make asymptotic analysis important, and asymptotic formulas may be preferable when exact solutions are too complicated.

An effective and fast procedure for computing individual terms or asymptotic behavior may be a good answer. A *recurrence relation* (Chapter 2) for $\langle a \rangle$ expresses a_n as a function of $a_0, \ldots, a_{n-1}$; appropriate initial values are needed to specify $\langle a \rangle$ completely. *Solving* a recurrence means obtaining a formula for a_n.

A *generating function* (Chapter 3) is a "formal" power series that encodes the sequence $\langle a \rangle$ by using the terms as coefficients. The simplest form is the *ordinary generating function* $\sum_{n=0}^{\infty} a_n x^n$. By "formal", we mean that the powers of x are placeholders for the terms in the sequence; we generally do not treat x as a number. Manipulating expressions for the generating function can lead to an explicit or asymptotic formula for a_n.

1.1. Classical Models

In this section we use combinatorial arguments to develop elementary models used in solving many counting problems. First we present several elementary techniques as "Principles"; the trivial examples here are just a warmup to illuminate a way of thinking.

ELEMENTARY PRINCIPLES

Our first two Principles are sometimes called "ad hoc" counting techniques. Nothing sophisticated is involved; we merely organize the set to simplify counting it, breaking the problem into smaller pieces.

1.1.1. Sum Principle: *If a finite set A is partitioned into sets $B_1, \ldots, B_k$, then* $|A| = \sum_{i=1}^{k} |B_i|$.
Recall that a partition of A is a family of disjoint nonempty sets whose union is A. This principle applies when a counting problem is broken into cases and is also called *Counting by Cases*. ∎

1.1.2. Product Principle: *If elements of a set A are built by successive choices and the number of options for the ith choice is independent of the outcomes of earlier choices, then $|A|$ is the product of the numbers of options for the choices.* This principle is also called *Counting by Stages*. The actual options at the ith stage may depend on earlier choices, but the number of options does not. The formula $|S \times T| = |S||T|$ for counting a cartesian product is a simple application of this principle. ∎

1.1.3. Example. *There are $\prod_{i=0}^{n-1}(2n - 1 - 2i)$ ways to pair $2n$ people.* The first person can be paired with another in $2n - 1$ ways. No matter how this choice is made, we can take the least indexed unpaired person and choose a partner for that person from the remaining people in $2n - 3$ ways. Continuing through n stages in this way produces $\prod_{i=0}^{n-1}(2n - 1 - 2i)$ distinct pairings, and every pairing is produced in this way. The 15 pairings for $n = 3$ appear below. ∎

$12 \cdot 34 \cdot 56$	$13 \cdot 24 \cdot 56$	$14 \cdot 23 \cdot 56$	$15 \cdot 23 \cdot 46$	$16 \cdot 23 \cdot 45$
$12 \cdot 35 \cdot 46$	$13 \cdot 25 \cdot 46$	$14 \cdot 25 \cdot 36$	$15 \cdot 24 \cdot 36$	$16 \cdot 24 \cdot 35$
$12 \cdot 36 \cdot 45$	$13 \cdot 26 \cdot 45$	$14 \cdot 26 \cdot 35$	$15 \cdot 26 \cdot 34$	$16 \cdot 25 \cdot 34$

1.1.4. Principle of Counting Two Ways: *When two formulas count the same set, their values are equal.*

Hermann Weyl once described the property of getting the same answer no matter how we count a finite set as one of the deepest theorems in mathematics. The depth is not in its truth but rather in its myriad applications. The point is not that we need to count a set twice; it is that we can prove equality of two formulas by *devising* an appropriate set whose size is given by both formulas, counted in different ways.   ∎

1.1.5. Example. $\sum_{i=1}^{n-1} i = n(n-1)/2$. This formula is easily proved by induction on n, but showing that both sides count the pairs of elements in $[n]$ provides more insight (recall from Chapter 0 that $[n] = \{1, \ldots, n\}$). The left side groups the pairs by the larger element, applying the Sum Principle (i pairs have larger element $i+1$). The right side applies the Product Principle by picking two elements successively and then dividing by 2 since order is irrelevant and we have counted each pair exactly twice.

Interchanging the order of summation. This operation can be viewed as an instance of counting two ways. Think of the value $f(i, j)$ as the entry in row i and column j of a matrix. The two sides of the identity $\sum_{i=1}^{m} \sum_{j=1}^{n} f(i, j) = \sum_{j=1}^{n} \sum_{i=1}^{m} f(i, j)$ sum the entries in the matrix, by rows or by columns.   ∎

1.1.6. Bijection Principle: *If there is a bijection from one set to another, then they have the same size.*

Recall that a **bijection** is a function $f \colon A \to B$ such that for all $b \in B$ exactly one $x \in A$ satisfies $f(x) = b$. The Bijection Principle is just the definition of equal size, but we use it as a counting technique. Establishing a bijection f from a set A of unknown size to a set B of known size computes $|A|$.

We use bijections when modeling part of a new problem as an instance of a classical problem. A bijection maps instances of the new problem into instances of the previously solved problem. When there is a bijection from A to B, we say that objects of A "correspond" to objects of B. A *bijection* is a function from one set to another, but a *one-to-one correspondence* between two sets puts them in pairs and can be followed in either direction. The latter term is less formal.   ∎

1.1.7. Example. *Within the set $[n]$, the number of subsets with k elements equals the number of subsets with $n - k$ elements.* Complementation of subsets within $[n]$ provides a bijection from one family to the other.

The number of binary lists of length n equals the number of subsets of $[n]$. Map any subset A of $[n]$ to the list a defined by $a_i = 1$ if $i \in A$ and $a_i = 0$ if $i \notin A$. The list a is the **incidence vector** of A. Any binary list a is the image of exactly one subset of $[n]$, namely the set $\{i \colon a_i = 1\}$. Hence the map is a bijection. This bijection enables us to view subsets as incidence vectors and vice versa.   ∎

Proofs using counting two ways or bijections are examples of combinatorial arguments. Asking for a **combinatorial proof** or **bijective proof** usually means asking for a proof using one of these techniques.

1.1.8. Pigeonhole Principle: *The maximum in a set of numbers is at least as large as the average (and the minimum is at least as small).*

In particular, placing more than kn objects into n boxes puts more than k objects into some box. This is not directly a counting technique, but it often helps in applications of counting formulas. It is another simple statement with subtle applications (for a precise proof, consider the contrapositive). We discuss applications and deep generalizations of the Pigeonhole Principle in Chapter 10, but the simple form will often be useful before that.   ∎

1.1.9. Example. *If A and B are finite and $f\colon A \to B$ and $g\colon B \to A$ are injections, then $|A| = |B|$, and f and g are bijections.* If $|B| < |A|$, then f puts more than $|B|$ objects into $|B|$ boxes, but such functions are not injective. Hence $|B| \geq |A|$. Similarly, g yields $|A| \geq |B|$, so $|A| = |B|$.

If some $x \in B$ lies outside the image of f, then f has at most $|A| - 1$ elements in its image, since we already have $|B| = |A|$. Again the Pigeonhole Principle contradicts the injectivity of f. Thus f (and similarly g) is surjective.   ∎

When A and B are infinite, and $f\colon A \to B$ and $g\colon B \to A$ are injections, it no longer follows that f and g are bijections, but still A and B must have the same cardinality (Exercise 47).

1.1.10. Polynomial Principle: *If two polynomials in one variable are equal at all positive integer values, then they are the same polynomial, meaning they are equal for all values of the variables and have the same coefficients.*

A polynomial of degree d is 0 at no more than d points. Hence polynomials of degree at most d that agree at $d + 1$ points also agree everywhere (consider their difference). A generalization for polynomials in more variables holds by induction on the number of variables; see Exercise 33. As in the one-variable case, equality at infinitely many points is not needed, but we usually invoke the principle when giving an argument that proves equality at all choices of positive integers.

Our first application of this principle (using two variables) is Proposition 1.1.18; see also Exercises 34–35.   ∎

1.1.11. Remark. Counting problems may arise from questions of probability. In a finite "sample space" U of *equally likely* outcomes, an **event** is a subset A of U. The **probability** of A, written $\mathbb{P}(A)$, is then defined as $|A|\,/\,|U|$. The probability of any property is the probability of the event consisting of all outcomes with that property (see Chapter 0 for further probabilistic concepts). In Chapter 14 we consider more sophisticated applications of probability, but meanwhile we occasionally use probabilistic questions to motivate counting problems.   ∎

WORDS, SETS, AND MULTISETS

In using bijective proofs to count sets, we need canonical sets of known sizes. Modeling a piece of a counting problem as an instance of one of these classical problems is a combinatorial argument.

1.1.12. Definition. A *k***-word** or **word** of **length** k is a list of k elements from a given set (the **alphabet**). A **simple word** is a word whose letters are distinct. A *k***-set** is a set with k elements; a k-set in a set S is a subset of S with k elements. We use $\binom{n}{k}$, read "*n choose k*", to denote the number of k-sets in an n-set. A **multiset** from a set S is a selection from S with repetition allowed.

Ways to have k items from an n-set

	no repetition	repetition allowed
ordered (arrangements)	simple k-words (k-permutations)	k-words/lists
unordered (selections)	subsets of size k (k-combinations)	multisets of size k

Words are simply lists; the term "word" is natural when we emphasize a particular finite alphabet from which they are formed. In contrast to sets, entries in a list have specified positions. Also sets (and simple words) have no repeated elements, while lists and multisets allow repetitions. The table above also lists older equivalent terms parenthetically.

The easiest of these four counting problems is for k-words. A list of length k can be viewed as a function f defined on $[k]$. The function maps $[k]$ into the set of entries allowed in the list, with $f(i)$ being the entry in position i.

1.1.13. Proposition. There are n^k words of length k from an alphabet S of size n. Equivalently, there are n^k functions from $[k]$ to S.

Proof: For the entry in each position there are n choices, regardless of the earlier choices made. The Product Principle applies. ∎

1.1.14. Remark. *Subsets as functions and lists.* Subsets of $[k]$ correspond to functions from $[k]$ to $\{0, 1\}$; there are 2^k. Given $A \subseteq [k]$, let $f_A(i) = 1$ if $i \in A$ and $f_A(i) = 0$ if $i \notin A$. The function f_A is the **characteristic function** of A (or, informally, the **membership function**). Since the domain of f_A is $[k]$, we can write f_A as a list of length k with entries in $\{0, 1\}$, as described in Example 1.1.7. ∎

Before solving the counting problems with repetition forbidden, we introduce notation for certain products.

1.1.15. Definition. For $n, k \in \mathbb{N}_0$, we define the following products:
  n **factorial**: $n! = \prod_{i=0}^{n-1}(n - i)$.
  falling factorial: $n_{(k)} = \prod_{i=0}^{k-1}(n - i)$.
  rising factorial: $n^{(k)} = \prod_{i=0}^{k-1}(n + i)$.

By convention, the result of an operation over an empty set is the identity element for that operation, so $0! = 1$ and $n_{(0)} = n^{(0)} = 1$. Since outer parentheses are not needed around a superscript or an exponent, their presence should have a special meaning as in our definition here. Many conflicting notations have been used for rising and falling factorials, as discussed at "MathWorld" (http://mathworld.wolfram.com/) and "PlanetMath" (http://planetmath.org/). Our notation was used long ago in published notes of a summer lecture course by C. L. Liu [1972].

A simple word of length n from the set $[n]$ writes the elements of $[n]$ in some order. Hence it is a permutation of $[n]$. Later we will discuss other ways of representing a permutation; this way is the **word form**. More generally, we may use only k elements.

1.1.16. Proposition. There are $n_{(k)}$ simple k-words from an alphabet of size n.

Proof: We form a simple k-word from a set of size n one position at a time. Since repetitions are forbidden, there are always $n+1-i$ ways to specify the ith element, no matter how the earlier elements were specified. By the Product Principle, the word can be formed in $\prod_{i=1}^{k}(n+1-i)$ ways. ∎

For the number of k-subsets, notations used in the past instead of $\binom{n}{k}$ include $C(n,k)$, $C_{n,k}$, $_kC_n$, $_nC_k$, and C_n^k; these are obsolete and now have other meanings.

1.1.17. Proposition. The number $\binom{n}{k}$ of k-sets in an n-set satisfies
$$\binom{n}{k} = \frac{n!}{k!(n-k)!} = \frac{n_{(k)}}{k!}.$$

Proof: In addition to the construction of simple k-words described in Proposition 1.1.16, we can also form a simple k-word by choosing a k-set and then placing the elements in order. For each k-set, there are $k!$ ways to permute the elements. By counting the simple k-words in these two ways, we have $k!\binom{n}{k} = n_{(k)}$, which yields the claimed formula. ∎

The numbers $\binom{n}{k}$ are called **binomial coefficients** due to their role in the Binomial Theorem. Our combinatorial definition of the symbol $\binom{n}{k}$ makes it 0 unless $0 \le k \le n$. Hence we need not worry about limits on the summation in the Binomial Theorem and many similar sums.

1.1.18. Proposition. (Binomial Theorem) For $n \in \mathbb{N}_0$,
$$(x+y)^n = \sum_k \binom{n}{k} x^k y^{n-k}.$$

Proof 1: Expanding $(x+y) \cdots (x+y)$ yields a sum of many terms. Each corresponds to a choice of x or y from each of the n factors. Terms of the form $x^k y^{n-k}$ arise by choosing exactly k of the n distinct factors to contribute x.

Proof 2: When x and y are integers, $(x+y)^n$ counts the n-words from an alphabet S of size $x+y$ (Proposition 1.1.13). We also count them by the number of positions using the first x letters in S. There are $\binom{n}{k} x^k y^{n-k}$ ways to choose k such positions, fill them, and fill the remaining positions (Product Principle). Summing over k counts all the words (Sum Principle). Hence we have counted a set in two ways, making the two sides equal whenever x and y are integers. By the Polynomial Principle, the two polynomials are equal. ∎

Choosing k objects from n types with no limit on repetitions produces a k-element *multiset* from $[n]$ (Definition 1.1.12). We care only how many objects are chosen of each type; the objects have no "positions". Thus each multiset corresponds to a vector of multiplicities for each type of object. To count multisets, we count these vectors of nonnegative integers.

1.1.19. Example. *The integer-sum problem.* Let $x_1, \ldots, x_n$ be variables, with x_i representing the number of copies of i chosen in a k-element multiset from $[n]$. The number of such multisets is the same as the number of solutions to $\sum_{i=1}^{n} x_i = k$ in nonnegative integers.

The solution vectors for $(n, k) = (3, 4)$ appear below.   ∎

(4,0,0)	(3,1,0)	(1,3,0)	(2,2,0)	(2,1,1)
(0,4,0)	(0,3,1)	(0,1,3)	(0,2,2)	(1,2,1)
(0,0,4)	(1,0,3)	(3,0,1)	(2,0,2)	(1,1,2)

1.1.20. Theorem. The number of k-element multisets from $[n]$, or equivalently the number of solutions to $\sum_{i=1}^{n} x_i = k$ in nonnegative integers, is $\binom{k+n-1}{n-1}$, which also equals $\binom{k+n-1}{k}$.

Proof: We model the integer-sum problem using subsets. For a nonnegative-integer solution, form a list of k dots and $n - 1$ bars, using x_1 dots, then one bar, then x_2 dots, then one bar, etc. The integer variables can be retrieved from such a list by letting x_i count the dots between the $(i - 1)$th and the ith bars, where we imagine a 0th bar before the list and an nth bar after it.

$$5 \qquad 2 \quad 0 \quad 3$$
$$\bullet \; \bullet \; \bullet \; \bullet \; \bullet \; | \; \bullet \; \bullet \; | \; | \; \bullet \; \bullet \; \bullet$$

Since we can retrieve uniquely from each arrangement of dots and bars the list of integers that yields it, we have established a bijection from the set of selections with repetition to the set of lists of k dots and $n - 1$ bars. The number of solutions is thus $\binom{k+n-1}{n-1}$ (or $\binom{k+n-1}{k}$), because the lists of dots and bars are specified by choosing the positions for the bars (or the dots).   ∎

The alternative formulas $\binom{k+n-1}{n-1}$ and $\binom{k+n-1}{k}$ can be confusing, since the model may arise with different (or reversed!) variables counting the objects and the types. Thus it is more valuable to understand the argument than to memorize the formula. The first formula emphasizes that the number of barriers between types is one less than the number of types. The second emphasizes that k is the size of the multiset. What happens when we require $x_1, \ldots, x_n$ to be positive?

1.1.21. Definition. A **composition** of an integer k is a list of positive integers summing to k. The entries in the list are the **parts** of the composition.

The last column of the listing in Example 1.1.19 shows the three compositions of 4 with three parts.

1.1.22. Corollary. There are $\binom{k-1}{n-1}$ compositions of k with n parts.

Proof 1: We count the solutions of $\sum_{i=1}^{n} x_i = k$ in positive integers by reducing it to a problem of counting nonnegative integer solutions. Solutions of $\sum_{i=1}^{n} x_i = k$ in positive integers correspond to solutions of $\sum_{i=1}^{n} y_i = k - n$ in nonnegative integers. Thus the number of solutions is $\binom{k-n+n-1}{n-1}$, which simplifies to $\binom{k-1}{n-1}$.

$$
\begin{array}{ccc}
(1,0,0) & (0,1,0) & (0,0,1) \\
(2,1,1) & (1,2,1) & (1,1,2) \\
\bullet\bullet\mid\bullet\mid\bullet & \bullet\mid\bullet\bullet\mid\bullet & \bullet\mid\bullet\mid\bullet\bullet
\end{array}
$$

Proof 2: Consider the model of dots and bars. Bars split the row of k dots into segments, which become parts. To make each segment have at least one dot, no two bars can be consecutive. Thus we select distinct places for the $n-1$ bars from the $k-1$ spaces between dots.   ∎

Discovering Proof 2 above after simplifying the computation in Proof 1 can be considered our first illustration of the **Erdős Principle**: If there is a simple answer, then there should be a simple proof!

EXERCISES 1.1

Problems marked "(−)" are easier or shorter than most (given the material of the text), while those marked "(+)" are harder than most. Problems marked "(◊)" are particularly valuable or instructive, not necessarily harder.

1.1.1. (−) When rolling n dice, what is the probability that the sum is even?

1.1.2. (−) Count all the rectangles with positive area formed by segments in a grid of m horizontal lines and n vertical lines.

1.1.3. (−) The roman alphabet has 21 consonants and 5 vowels. How many strings can be formed using r consonants and s vowels?

1.1.4. (−) Count the possible outcomes of an election with 30 voters and four candidates. Count those in which no candidate receives more than half of the votes.

1.1.5. (−) Prove that $(n^5 - 5n^3 + 4n)/120$ is an integer for every $n \in \mathbb{N}$.

1.1.6. (−) Count the orderings of a standard 52-card deck such that the 13 cards in the spade suit occur consecutively.

1.1.7. (−) Compute the probability that a random set of five cards from a standard 52-card deck has at least three cards with the same rank.

1.1.8. (−) Count the sets of six cards from a standard deck of 52 cards that have at least one card in every suit.

1.1.9. (−) Count the integers from 0 through 99,999 such that each digit occurs at most twice; leading zeros are counted as appearances of 0.

1.1.10. (−) In how many distinguishable ways can the letters in "Mississippi" be ordered?

1.1.11. (−) Given a bag with many marbles in each of four colors, count the ways to select 12 marbles. Count the distinguishable ways to put 12 of them in a row.

1.1.12. (−) New York City has about 7 million residents; suppose that each has 100 coins in a jar. The coins come in five types (pennies, nickels, dimes, quarters, half-dollars). Two jars are *equivalent* if they have the same number of coins of each type. Is it possible that no two people have equivalent jars?

1.1.13. (−) Count the compositions of k in which every part is even.

1.1.14. (◊) *Families of subsets.*
 (a) Count the subsets of $[n]$ that contain at least one odd number.
 (b) Count the k-sets in $[n]$ having no two consecutive integers.
 (c) Count the lists of subsets $A_0, A_1 \ldots A_n$ of $[n]$ such that $A_0 \subset A_1 \subset \cdots \subset A_n$. Count the lists such that $A_0 \subseteq A_1 \subseteq \cdots \subseteq A_n$.

1.1.15. (◊) Prove that the exponent on a prime p in the prime factorization of $\binom{2n}{n}$ is the number of powers p^k of p such that $\lfloor 2n/p^k \rfloor$ is odd. Use this to determine which primes divide $\binom{18}{9}$ and which divide $\binom{20}{10}$. For example, for $n = 3$, $\binom{6}{3} = 20$ and $20 = 2^2 \cdot 5$. The claim holds here, because $\lfloor 6/2 \rfloor$ and $\lfloor 6/4 \rfloor$ are odd, $\lfloor 6/3 \rfloor$ is even, $\lfloor 6/5 \rfloor$ is odd, and $\lfloor 6/p^k \rfloor = 0$ for all other prime powers p^k.

1.1.16. Given positive integers a and b, let $v(a, b) = \left(\binom{a}{b-1}, \binom{a}{b}, \binom{a}{b+1} \right)$. Prove that $v(c, d)$ cannot be a multiple of $v(a, b)$ when (a, b) and (c, d) are distinct pairs of positive integers.

1.1.17. (◊) Count the lists of m 1s and n 0s that have exactly k runs of 1s, where a **run** is a maximal set of consecutive entries with the same value.

1.1.18. (◊) For $S \subseteq [n]$, a **run** in S is a maximal set of consecutive integers in S (corresponding to a run of 1s in the incidence vector). For example, $\{1, 2, 4, 5, 6, 9\}$ is a subset of $[9]$ with three runs; every element is in one run.
 (a) Count the subsets of $[n]$ having k runs.
 (b) Count the t-element subsets of $[n]$ having k runs.
 (c) Let $s_1, \ldots, s_m$ be distinct positive integers. Among the t-element subsets of $[n]$ having k runs, count those having exactly r_i runs with length s_i, for $1 \le i \le m$. (Comment: Note that $k = \sum_{i=1}^{m} r_i$ and $t = \sum_{i=1}^{m} r_i s_i$.)

1.1.19. Determine the number of binary strings of length n in which the number of copies of 00 is the same as the number of copies of 11. For example, 00011011 has two copies of each. (Deutsch [2009])

1.1.20. Give a summation for the number of elements of $[3]^n$ with k odd entries that do not have a 1 next to a 3.

1.1.21. The **chords** of a convex n-gon are the segments that join two corners. Count the pairs of chords that cross inside the n-gon.

1.1.22. Count all triangles (not only the empty triangular regions) in the picture formed by drawing all $\binom{n}{2}$ chords of a convex n-gon, given that no three chords have a common internal point. When n is 4, 5, or 6, the answers are 8, 35, and 111.

1.1.23. We have six dice. Each has three red faces, two green faces, and one blue face. Determine the probability that three red faces, two green faces, and one blue face appear when the six dice are rolled.

1.1.24. Given a standard deck of playing cards with 52 cards in four suits of 13 values, which is more likely in a random set of five cards: having all five cards in the same suit, or having cards of five consecutive values?

1.1.25. A *trapezoid* is a quadrilateral with at least one pair of parallel sides. For $n \geq 4$, how many sets of four distinct points forming the vertices of a trapezoid exist among the vertices of a regular n-gon? (Nicoaescu [1986])

1.1.26. ($\diamond$) For a permutation π of $[n]$, the **displacement** of π is $\sum_{i=1}^{n} |i - \pi(i)|$. Prove that the largest displacement of a permutation of $[n]$ is $\lfloor n^2/2 \rfloor$. (Comment: The extremal permutation is not unique when n is odd and at least 3.)

1.1.27. ($\diamond$) Let A_n be the set of permutations of $[n]$. Let B_n be the set of n-tuples $(b_1, \ldots, b_n)$ such that $1 \leq b_i \leq i$ for each $i \in [n]$. Construct a bijection from A_n to B_n. Below we illustrate a possible correspondence for $n = 3$.

A_3	321	231	213	312	132	123
B_3	111	112	113	121	122	123

1.1.28. Using interchanges of two students, a professor wants to rearrange the n students sitting in a row so that no two students originally adjacent remain adjacent. Determine the minimum number of interchanges needed, given $n \geq 6$. (W. So)

1.1.29. Count the 0, 1-matrices with n^2 rows and n^2 columns such that (1) each row and column has exactly one 1, and (2) when the matrix is partitioned into n^2 blocks of n consecutive rows and n consecutive columns, each block contains exactly one 1. For example, when $n = 2$ the answer is 16. (Pratt [2011])

1.1.30. ($\diamond$) Count the permutations π of $[n]$ with $\pi(i + 1) \leq \pi(i) + 1$ for $1 \leq i \leq n - 1$. Give both a proof by induction and a bijective proof. (Deutsch [2001a])

1.1.31. (+) A permutation is **graceful** if the absolute differences between successive elements are distinct. Prove that if the set of elements in even-indexed positions of a graceful permutation of $[2n]$ is $[n]$, then the first and last elements differ by n. For example, 54637281 is such a permutation. (Comment: The converse also holds.) (Klove [1995])

1.1.32. A **necklace** is a circular arrangement that can rotate and flip without being considered different. Thus only one necklace can be made from three beads. A **crown**, in contrast, can rotate but not flip.
  (a) Count the necklaces with n beads that can be made from n distinct beads.
  (b) Count the crowns with n beads that can be made from a supply of k types of beads, given that n is prime (the k types need not all be used).

1.1.33. *Polynomial Principle in several variables.* Let p be a polynomial in $x_1, \ldots, x_k$. For $1 \leq i \leq k$, suppose that the degree of p as a polynomial in x_i is at most d_i, and let S_i be a set of $d_i + 1$ values. Prove that if p is 0 at all points in $\prod_{i=1}^{k} S_i$ (that is, on $\{(x_1, \ldots, x_k): x_i \in S_i \text{ for } 1 \leq i \leq k\}$), then p is zero everywhere. Conclude that if two polynomials in several variables are equal at all positive integer values, then they are equal at all values. (Hint: Assume the well-known statement that a polynomial of degree at most d in one variable is identically zero if it is zero at more than d values.)

1.1.34. ($\diamond$) Give a combinatorial proof that the identity below for falling factorials holds for all $x, y \in \mathbb{R}$.

$$(x + y)_{(n)} = \sum_{k} \binom{n}{k} x_{(k)} y_{(n-k)}.$$

1.1.35. ($\diamond$) *Flags on poles.*

(a) Obtain a simple formula for the number of ways to put m distinct flags on a row of r flagpoles. Poles may be empty, and changing the order of flags on a pole changes the arrangement. The formula must only use one "m" and one "r". (The answer is 6 for $m = r = 2$, as shown below.)

(b) Prove that the identity below for rising factorials holds for all $x, y \in \mathbb{R}$.

$$(x + y)^{(n)} = \sum_{k} \binom{n}{k} x^{(k)} y^{(n-k)}.$$

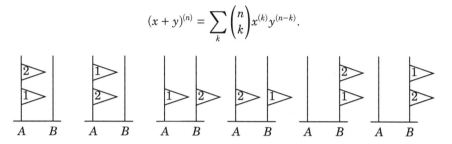

1.1.36. Given n distinct flags ($n \geq 1$) and many identical flagpoles, we want to put all the flags on equally spaced flagpoles on a rotating circular platform. We may use any number of poles, but each pole used must have at least one flag, and order of flags on a pole matters. Arrangements that can be rotated into each other are not distinguishable. Count the distinguishable ways to arrange the n flags. (Comment: There is a very short solution.)

1.1.37. For p prime and $n \in \mathbb{N}$, give a combinatorial proof that n divides $\binom{n+p-1}{p} - \binom{n}{p}$.

1.1.38. A spinner has a pointer that is equally likely to point to each of n regions numbered $1, 2, \ldots, n$. When we spin it three times, what is the probability that the sum of the selected numbers is n?

1.1.39. Prove the identity below by showing that both sides count the same set of ternary lists. (Karaivanov–Vassilev in Holland [2014]; identity (3.121) in Gould [1972])

$$\sum_{s=k}^{\lfloor n/2 \rfloor} \binom{n+1}{2s+1} \binom{s}{k} = \binom{n-k}{k} 2^{n-2k}$$

1.1.40. ($\diamond$) *Compositions of integers.*

(a) Count the solutions in positive integers to $\sum_{i=1}^{n} x_i \leq k$. Give a direct combinatorial argument for the resulting simple formula.

(b) Give a direct combinatorial argument to count all compositions of k (with any number of parts).

(c) For $k \geq 2$, prove that the number of compositions of k having an even number of parts equals the number having an odd number of parts.

(d) For $k \geq 2$, prove that the number of compositions of k with an even number of even parts equals the number with an odd number of even parts.

1.1.41. ($\diamond$) *Compositions of integers.*

(a) Determine the total number of parts over all compositions of k. (Hint: Consider part (b) of Exercise 1.1.40.)

(b) For $1 \leq m < k$, prove that all the compositions of k together have $(k - m + 3)2^{k-m-2}$ parts equal to m. For example, 1 occurs twice in the composition $1 + 2 + 1$ of 4, and all the compositions of 4 together have twelve 1s, five 2s, and two 3s. (Hint: Relate the 1s in compositions of k to compositions of $k - 1$.)

(c) Use parts (a) and (b) to evaluate $1 + \sum_{m=1}^{k-1} (k - m + 3)2^{k-m-2}$.

1.1.42. *The Weights Problem.* A *balance scale* tests whether the total weight put on the left and right sides is the same. If one can specify k known integer weights, what should they be to maximize the value n such that every unknown object with weight in $[n]$ can be correctly weighed? What is this maximum n? (Example: when $k = 2$ and the known weights are $\{1, 4\}$, one can weigh objects with weights in $\{1, 3, 4, 5\}$ but not 2. Using $\{1, 3\}$, one can balance every weight in $[4]$.)

1.1.43. Consider a balance scale and positive integer weights $w_1 \leq \cdots \leq w_k$. Let $S_0 = 0$, and let $S_j = \sum_{i=1}^{j} w_i$ for $1 \leq j \leq k$. Prove that it is possible to balance every integer weight from 1 to S_k if and only if $w_j \leq 2S_{j-1} + 1$ for $1 \leq j \leq k$.

1.1.44. Choose points x_1, x_2, x_3 on a circle. For each i, draw k_i chords from x_i to the arc joining the other two points. What is the maximum number of regions inside the circle? (Yaglom–Yaglom [1964])

1.1.45. Choose k distinct numbers at random from $[n]$, with all k-sets equally likely. Let X be the sum of the chosen numbers. Prove that if n is divisible by r, and $\gcd(k, r) = 1$, then X is divisible by r with probability $1/r$. (Kuczma [2000])

1.1.46. (+) For $k, m, n \in \mathbb{N}$ with $k \leq n$, prove the identity below by expressing the sum as the probability of some event certain to happen. (Xiao [2012])

$$\sum_{j=0}^{m} \frac{k\binom{n}{k}\binom{m}{j}}{(k+j)\binom{n+m}{k+j}} = 1.$$

1.1.47. **Schröder–Bernstein Theorem**. Let A and B be sets. Prove that if $f: A \to B$ and $g: B \to A$ are injections, then there is a bijection $h: A \to B$. (Hint: Consider the "chains" of elements formed by alternately using f to move from A to B and g to move from B to A, and define h on these families.)

1.2. Identities

Using the Sum and Product Principles in a counting problem may produce sums involving binomial coefficients. Standard formulas called **identities** can help evaluate such sums.

Proofs of identities using induction or factorial formulas may involve tedious manipulation. Combinatorial arguments can provide deeper understanding and more information. Algebraic arguments (such as manipulating an identity) may be easier to find; Graham–Knuth–Patashnik [1989] presents many. We focus first on combinatorial arguments and introduce other techniques later.

LATTICE PATHS AND PASCAL'S TRIANGLE

An argument by *counting two ways* shows that both sides of an identity count the same set. The difficulty is devising an appropriate set. Arguments for sums involving n^k often use the set of n-ary k-tuples (words of length k from an alphabet of size n). There are several equivalent models for a set counted by $\binom{n}{k}$; an argument using one can be translated into the others. We have discussed models involving subsets and binary lists; next we introduce the lattice path model.

1.2.1. Definition. A **lattice point** is a vector with integer coordinates. A **lattice step** changes one coordinate by 1. A **lattice walk** from a lattice point (often the origin) moves by lattice steps. A **lattice path** is a lattice walk in which each step increases one coordinate.

1.2.2. Example. *Lattice paths and binary lists.* A lattice path from $(0,0)$ to $(k, n - k)$ has length n, with k horizontal steps. Recording 1 for each horizontal step and 0 for each vertical step produces a binary n-tuple with k 1s. The path is determined by where the horizontal steps occur in the list, so exactly one path yields each such n-tuple. This establishes a bijection from the set of these lattice paths to the set of binary n-tuples with k 1s. ∎

$$\text{[lattice path figure]} \quad \leftrightarrow \quad 101101 \quad \leftrightarrow \quad \{1, 3, 4, 6\} \subseteq [6]$$

Many authors use the term "lattice path" more generally, allowing steps of other forms, but the restriction is convenient for us. A lattice path may end with a horizontal step or a vertical step. This proves the formula $\binom{n}{k} = \binom{n-1}{k} + \binom{n-1}{k-1}$, called **Pascal's Formula** in honor of Blaise Pascal (1623–1662). The array of numbers with $\binom{n}{0}, \dots, \binom{n}{n}$ in the nth row is called **Pascal's Triangle**, though it was known to Chinese mathematicians much earlier.

$$
\begin{array}{ccccccccccccc}
 & & & & & & 1 & & & & & & \\
 & & & & & 1 & & 1 & & & & & \\
 & & & & 1 & & 2 & & 1 & & & & \\
 & & & 1 & & 3 & & 3 & & 1 & & & \\
 & & 1 & & 4 & & 6 & & 4 & & 1 & & \\
 & 1 & & 5 & & 10 & & 10 & & 5 & & 1 & \\
1 & & 6 & & 15 & & 20 & & 15 & & 6 & & 1 \\
\end{array}
$$

Next we illustrate the use of various models in proving elementary identities.

1.2.3. Theorem. (Elementary Identities)

Elementary form	Generalization
(1) $\binom{n}{k} = \binom{n}{n-k}$	(see Remark 1.3.11)
(2) $\binom{n}{k} = \binom{n-1}{k} + \binom{n-1}{k-1}$	(see Remark 1.3.11)
(3) $k\binom{n}{k} = n\binom{n-1}{k-1}$	$\binom{k}{l}\binom{n}{k} = \binom{n}{l}\binom{n-l}{k-l}$
(4) $\sum_k \binom{n}{k} = 2^n$	$\sum_k r^k \binom{n}{k} = (r+1)^n$
(5) $\sum_{k=0}^{n} \binom{k}{r} = \binom{n+1}{r+1}$	$\sum_{k=-m}^{n} \binom{m+k}{r}\binom{n-k}{s} = \binom{m+n+1}{r+s+1}$
(6) $\sum_k \binom{n}{k}^2 = \binom{2n}{n}$	$\sum_k \binom{m}{k}\binom{n}{r-k} = \binom{m+n}{r}$

Proof: We give combinatorial proofs for these identities under the assumption that the parameters are all nonnegative integers.

(1) Choosing k positions for 1s or $n - k$ positions for 0s yields the same lists.

(2) Binary n-tuples with k 1s end with a 0 or with a 1; equivalently, a lattice path reaches $(k, n - k)$ from $(k - 1, n - k)$ or $(k, n - k - 1)$.

(3) From n people, we form committees of size k with a chair. On the left we choose the committee and then select the chair; on the right the chair is chosen first and then the rest of the committee. We call this the **Committee-Chair Identity**. The generalization is the **Subcommittee Identity**: each side counts committees of size k with a subcommittee of size l. These are also the lists of length n with l entries equal to A, $k - l$ equal to B, and $n - k$ equal to C. Because the positions for letters of each type can be chosen in various orders, there are many products of two binomial coefficients that count this set.

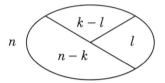

(4) In the special case, both sides count all subsets of n elements. The generalization is the Binomial Theorem with x set to r and y set to 1; alternatively, both sides count $(r + 1)$-ary lists of length n.

(5) For the special case, consider the $\binom{n+1}{r+1}$ binary $(n+1)$-tuples with 1s in $r+1$ positions. The left side counts them by the location of the rightmost 1; the kth term counts those where it is in position $k + 1$. Alternatively, the left side counts the lattice paths from $(0, 0)$ to $(r + 1, n - r)$ by the height at which the last horizontal step occurs. We call this the **Summation Identity**. The generalization counts binary $(m + n + 1)$-tuples with 1s in $r + s + 1$ positions by the position of the $(r + 1)$th 1; the kth term counts those where it is in position $m + k + 1$. The special case arises when $m = s = 0$.

(6) The general form is the **Vandermonde Convolution**; $m = n = r$ yields the special case. The left side counts r-sets from an $(m + n)$-set by how many elements come from the first k. Using lattice paths, the argument is that every lattice path from $(0, 0)$ to $(r, m + n - r)$ reaches the line $x + y = m$ at some point $(k, m - k)$. ∎

Writing $\binom{n}{k}$ as $\frac{1}{k!} n_{(k)}$ treats it as a polynomial of degree k in n. By the Polynomial Principle, Identities (4)–(6) in Theorem 1.2.3 are valid as polynomials in m and n. From this viewpoint, the general form of (6) is called **Vandermonde's Theorem**, in honor of Vandermonde [1772], but the convolution was known to Chu Shih-Chieh (now spelled Zhu Shijie) in 1303. It will be helpful to study the extension of binomial coefficients to non-integer arguments more formally.

1.2.4. Definition. For $k \in \mathbb{N}_0$, the **extended binomial coefficient** $\binom{u}{k}$ is the polynomial in u of degree k defined by

$$\binom{u}{k} = \frac{1}{k!} \prod_{i=0}^{k-1} (u - i) .$$

For $u \in \mathbb{N}_0$, the formula agrees with the combinatorial definition of $\binom{u}{k}$ in Definition 1.1.12; the value is 0 unless $0 \le k \le u$. The polynomial viewpoint leads to an extension of the Binomial Theorem that was proved by Isaac Newton (1643–1727) for rational exponents.

1.2.5. Theorem. (Extended Binomial Theorem) For $u, x \in \mathbb{R}$ with $|x| < 1$,

$$(1 + x)^u = \sum_{k \ge 0} \binom{u}{k} x^k .$$

Proof: By Taylor's Theorem (of calculus), the coefficient of $x^k/k!$ in the expansion of $(1 + x)^u$ as a power series in x is the kth derivative of $(1 + x)^u$ with respect to x, evaluated at $x = 0$. This value is $u_{(k)}$. ∎

1.2.6. Remark. *Negative binomial and multisets.* In Section 3.1 we will give a combinatorial interpretation of Theorem 1.2.5 in the case where u is a negative integer. Invoking Definition 1.2.4 as a polynomial yields

$$\binom{-n}{k} = (-1)^k \binom{n + k - 1}{k} .$$

Thus the expansion of $(1 + x)^{-n}$ is closely related to selection of k elements (with repetition) from n types of elements. ∎

There are many methods for evaluating sums. Pascal's Formula (Theorem 1.2.3(2)) may allow us to prove binomial coefficient identities by induction (as in Exercise 11). However, induction does not work unless we know the answer in advance; other methods may both discover and prove the formula. Bijective proofs may give more "refined" results (see Theorem 1.3.26 and Example 1.3.27, for example). Instead of devising new combinatorial arguments, we may also use identities already proved; indeed, this is the motivation for proving identities.

1.2.7. Application. *Initial values of a polynomial.* For a polynomial p, we can compute $\sum_{i=1}^n p(i)$ by writing the powers of i as integer combinations of binomial coefficients and applying Theorem 1.2.3(5). The transformation uses identities such as $i^2 = 2\binom{i}{2} + \binom{i}{1}$ and $i^3 = 6\binom{i}{3} + 6\binom{i}{2} + \binom{i}{1}$. To understand the identity for i^k, observe that the left side counts k-tuples with entries in $[i]$ by filling the positions independently, while the right side counts the same set according to the number of distinct values used. Up to cubes, this argument yields the formulas below. Exercise 8 requests the details. ∎

$$\sum_{i=1}^n i = \frac{n(n + 1)}{2}, \quad \sum_{i=1}^n i^2 = \frac{n(n + 1)(2n + 1)}{6}, \quad \sum_{i=1}^n i^3 = \frac{n^2(n + 1)^2}{4}.$$

The coefficient of $\binom{i}{r}$ in the expression for i^k counts the groupings of the positions in $[k]$ into r labeled boxes. We will obtain a general formula for this in Section 3.3. For now, it suffices that such coefficients exist, because $\binom{x}{j}$ is a polynomial in x of degree j, and hence $\binom{x}{0}, \ldots, \binom{x}{r}$ is a basis for the vector space of polynomials in x of degree at most r.

DELANNOY NUMBERS

Next we consider paths in the plane more general than lattice paths.

1.2.8. Definition. A **Delannoy path** is a path from $(0,0)$ to (m,n) with each step in $\{(1,0),(0,1),(1,1)\}$. The **Delannoy number** $d_{m,n}$ is the number of such paths. The numbers of the form $d_{n,n}$ are the **central Delannoy numbers**.

1.2.9. Example. Below we show the seven paths from $(0,0)$ to $(2,2)$ that arise in addition to the $\binom{4}{2}$ lattice paths; altogether, $d_{2,2} = 13$.

The Delannoy numbers were introduced by Henri Auguste Delannoy (1833–1915). The two formulas for $d_{m,n}$ that we discuss below appear in Delannoy [1889]; Banderier–Schwer [2005] discusses the history. Sulanke [2003] presents 29 problems for which the central Delannoy numbers provide the solution. ∎

Delannoy paths are arrangements of three types of steps, and $d_{m,n}$ counts such arrangements. When proving that a sum counts a particular set, we cut the set into pieces counted by the terms in the sum. A key step is to recognize what the index of summation means in describing the pieces. Doing that can make it easy to show that the summand counts the corresponding piece (see Exercise 49, for example).

1.2.10. Proposition. The Delannoy number $d_{m,n}$ is given by

$$d_{m,n} = \sum_j \binom{m}{j}\binom{n+m-j}{m} = \sum_k \binom{m}{k}\binom{n+k}{m}.$$

Proof: Since we know there are $\binom{n+m}{m}$ paths from $(0,0)$ to (m,n) using no diagonal steps, it makes sense to group the Delannoy paths by the number of diagonal steps, j. To reach (m,n) there must also be $m-j$ horizontal steps and $n-j$ vertical steps, in total $n+m-j$ steps. Given the number of steps of each type, the paths correspond to the orderings of the steps: the words with j copies of D, $m-j$ of H, and $n-j$ of V. Forming the words in two stages yields $\binom{n+m-j}{m}\binom{m}{m-j}$ as the number of words.

Summing over the number j of diagonal steps completes the count. Grouping instead by the number of horizontal steps, k, yields the second expression (also obtained from the first by substituting $k = m - j$). ∎

When no limits are given for the index parameter in a sum, as in the statement of Proposition 1.2.10, all terms where the summand is nonzero are included. When the summand involves binomial coefficients, the index values yielding nonzero terms are usually clear.

Our next objects are closely related to Delannoy paths.

1.2.11. Definition. The **lattice ball** of radius m in n dimensions is the set of lattice points in $\mathbb{Z}^n$ reachable from the origin by lattice walks with at most m steps, shown below for $m = n = 2$.

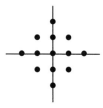

A more combinatorial description for the lattice ball is the set of all integer n-tuples such that the absolute values of the entries sum to at most m. This phrasing makes it easy to apply the integer equation model to count the points.

1.2.12. Proposition. (Golomb–Welch [1970], see also Vassilev–Atanassov [1994]) The size $l_{m,n}$ of the lattice ball of radius m in $\mathbb{Z}^n$ is given by

$$l_{m,n} = \sum_k \binom{n}{k}\binom{m}{k}2^k.$$

Proof: We count the integer n-tuples whose absolute values sum to at most m. Group them by the number of nonzero coordinates, k. We can pick these positions in $\binom{n}{k}$ ways and give signs to the nonzero entries in 2^k ways. To finish building a desired n-tuple, we give magnitudes to the k nonzero entries by solving $x_1 + \cdots + x_k \leq m$ in positive integers.

Taking partial sums of such a list yields a k-element subset of $[m]$, with the jth element being $\sum_{i=1}^{j} x_i$. To see that each k-set S in $[m]$ arises exactly once, write S as $a_1, \ldots, a_k$ in increasing order and let $x_i = a_i - a_{i-1}$ (with $a_0 = 0$). Now $x_1, \ldots, x_k$ is a solution to the inequality that yields S under the original map, and it is the only solution that yields S. Hence the map is a bijection, and there are $\binom{m}{k}$ solutions to the inequality. ∎

Surprisingly, this formula for $l_{m,n}$ is symmetric in m and n. The combinatorial Definition 1.2.8 for $d_{m,n}$ is also symmetric, but the formula in Proposition 1.2.10 is not. The real surprise, suggested by $d_{2,2} = 13 = l_{2,2}$, is that the two formulas are equal! We present a bijection found in 2006 by a student (Pavithra Prabhakar) after the second lecture of a combinatorics course; Sulanke [2003, Example 20] used a similar idea.

1.2.13. Theorem. (Delannoy [1889]) For $m, n \in \mathbb{N}_0$,

$$\sum_k \binom{m}{k}\binom{n+k}{m} = \sum_k \binom{n}{k}\binom{m}{k}2^k.$$

Proof: Let A be the set of Delannoy paths to (m, n), and let B be the lattice ball of radius m in $\mathbb{Z}^n$. By Propositions 1.2.10–1.2.12, the left side is $|A|$ and the right is $|B|$. We prove equality by a bijection $f \colon A \to B$.

From a Delannoy path P to (m, n), we form an integer list $(b_1, \ldots, b_n)$ with $\sum_{i=1}^{n} |b_i| \leq m$. Let $|b_i|$ be the number of horizontal or diagonal steps from the point where P first reaches vertical coordinate $i - 1$ to the point where P first reaches vertical coordinate i. The sign of b_i is positive if the final step is vertical, negative if it is diagonal. Let $f(P) = b$.

For every n-tuple b in the lattice ball, we show that $f(P) = b$ for exactly one Delannoy path P. Replace a nonnegative b_i with b_i horizontal steps and then one vertical step. Replace a negative b_i with $-b_i - 1$ horizontal steps and then one diagonal step. This produces a Delannoy path from $(0, 0)$ to (r, n), where r is the sum of the absolute values in b. Complete P by adding $m - r$ horizontal steps. Since the portion of P between the lines $y = i - 1$ and $y = i$ must yield b_i, this is the only way to form a Delannoy path that maps to b under f.   ∎

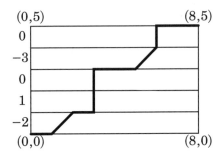

Using the Delannoy path and lattice ball models, Remark 2.1.11 gives an inductive proof of Theorem 1.2.13 using recurrences. The equality can be proved directly without these models by inventing a set counted by both sides. When devising a set to be counted by sum of products, often one defines a set of pairs; they can be counted by a two-step process, grouping the pairs according to one coordinate. Exercise 39 requests a proof of Theorem 1.2.13 by this method.

1.2.14.* Remark. The bijection in Theorem 1.2.13 restricts to various statements we have already proved. Forbidding diagonal steps in the path forbids negative entries in the n-tuple. The n-tuples become the nonnegative integer solutions to $\sum_{i=1}^{n} x_i \leq m$, or $\sum_{i=1}^{n+1} x_i = m$. These correspond to multisets of size m from $n + 1$ types; by Theorem 1.1.20, there are $\binom{m+n+1-1}{m}$. As expected, the resulting value $\binom{m+n}{n}$ counts the lattice paths from $(0, 0)$ to (m, n).

Forbidding vertical steps in the path forbids nonnegative entries in the n-tuple. Negating the n-tuple yields n positive integers summing to at most m. Again adding x_{n+1}, we obtain compositions of $m + 1$ with $n + 1$ parts, of which there are $\binom{m}{n}$ (Corollary 1.1.22). The corresponding paths arise by ordering n diagonal steps and $m - n$ horizontal steps in $\binom{m}{n}$ ways.   ∎

We have given only a few identities; others appear in the exercises. H. Gould [1972] listed more than 550 identities for binomial coefficients. Combinatorial argument is not our only tool. Other techniques developed in the next several chapters will enable us to prove identities or evaluate sums more systematically (and maybe with less cleverness!).

EXERCISES 1.2

1.2.1. (−) Give combinatorial proofs of $\binom{n}{k+1} = \frac{n-k}{k+1}\binom{n}{k}$ and $\binom{m+n}{m+k}\binom{m+k}{k} = \binom{m+n}{m}\binom{n}{k}$.

1.2.2. (−) Evaluate $\sum_{k=0}^{n}\binom{m+k}{k}$.

1.2.3. (−) Prove $\binom{n}{k} - \binom{n-2}{k} = \binom{n-1}{k-1} + \binom{n-2}{k-1}$.

1.2.4. (−) Prove that Pascal's Formula holds for the extended binomial coefficient.

1.2.5. (−) Use the Vandermonde Convolution to prove $\sum_{k}\binom{r}{m+k}\binom{s}{n-k} = \binom{r+s}{m+n}$.

1.2.6. (−) Prove $\sum_{k=0}^{n}\binom{m+k-1}{k} = \sum_{k=0}^{m}\binom{n+k-1}{k}$ and $\sum_{k}\binom{n}{k}\binom{m}{r+k} = \binom{n+m}{n+r}$.

1.2.7. (−) Evaluate $\binom{-1}{k}$.

1.2.8. (−) Complete the proofs of the formulas in Application 1.2.7 for $\sum_{i=1}^{n} i$, $\sum_{i=1}^{n} i^2$, and $\sum_{i=1}^{n} i^3$, using the expressions for powers of i in terms of binomial coefficients. Use these formulas to evaluate $\sum_{i=1}^{n}(2i^3 + 3i^2 - 5i)$.

1.2.9. (−) We put m white and n black marbles into a row of boxes. From the beginning of the row, no box can be left empty until all the marbles are used, and each box has at most one marble of each color. Prove that the number of ways to place the marbles in the boxes is the Delannoy number $d_{m,n}$. (Schwer [2002])

1.2.10. Prove $\binom{n-1}{k-1}\binom{n}{k+1}\binom{n+1}{k} = \binom{n-1}{k}\binom{n+1}{k+1}\binom{n}{k-1}$ without expanding into factorials (use known identities).

1.2.11. (◊) Use Pascal's Formula to prove the following identities by induction.

(a) $\binom{n}{k} = \frac{n!}{k!(n-k)!}$.

(b) $\sum_{i=0}^{n}\binom{i}{k} = \binom{n+1}{k+1}$ for $k, n \in \mathbb{N}_0$.

(c) $(x + y)^n = \sum_{k=0}^{n}\binom{n}{k}x^k y^{n-k}$.

(d) $\sum_{i=0}^{k}(-1)^i\binom{n}{k-i} = \binom{n-1}{k}$.

1.2.12. (◊) When flipping 100 fair coins, is it more likely that the numbers of heads and tails are equal or that they differ by two? Give an algebraic proof by manipulating formulas and a combinatorial proof by defining an injection.

1.2.13. (◊) Prove each identity below by counting a set in two ways. (One can count 3-tuples with special properties or various geometric arrangements.) Then use the two identities to obtain simple formulas that evaluate $\sum_{i=1}^{n}\sum_{j=1}^{n}\min\{i, j\}$ and $\sum_{i=1}^{n}\sum_{j=1}^{n}\max\{i, j\}$.

(a) $\sum_{i=1}^{n}\sum_{j=1}^{n}\min\{i, j\} = \sum_{k=1}^{n} k^2$.

(b) $\sum_{k=1}^{n} k^2 = 2\binom{n+1}{3} + \binom{n+1}{2}$.

1.2.14. For $m \in \mathbb{N}$, let $f(m) = \sum_{j=1}^{m}(m - j)2^{j-1}$. Obtain a simple formula for $f(m)$ inductively. Also give a combinatorial proof by counting a set in two ways.

1.2.15. (◊) Prove each identity below by counting a set in two ways. In each case, give a single direct argument without manipulating the formulas.

(a) $\binom{2n}{n} = 2\binom{2n-1}{n-1}$.

(b) $\sum_{k}\binom{k}{l}\binom{n}{k} = \binom{n}{l}2^{n-l}$.

(c) $\sum_{k=1}^{n} q^{k-1} = \frac{q^n-1}{q-1}$ for $q, n \in \mathbb{N}$ with $q \neq 1$.

(d) $\sum_{i=1}^{n} i(n - i) = \sum_{i=1}^{n}\binom{i}{2}$.

1.2.16. (+) Prove **Strehl's Identity** below (algebraically or combinatorially).

$$\sum_{k}\binom{n}{k}^2\binom{2k}{n} = \sum_{k}\binom{n}{k}^3 \qquad \text{(Strehl [1994, eq. 29])}$$

1.2.17. ($\diamond$) Use known identities to evaluate $\sum_{k=1}^{n} \sum_{i=1}^{k-1} (i-1)(k-i-1)$. Give a combinatorial proof of the resulting identity.

1.2.18. (+) For $m, r \in \mathbb{N}$, use known identities to prove $\sum_{k \geq 1} k\binom{m+1}{r+k+1} = \sum_{i=1}^{m} i 2^{i-1}\binom{m-i}{r}$, and give a combinatorial proof by devising a set that both sides count.

1.2.19. Use known identities to evaluate the sums below.

(a) $\sum_{k \geq 0} \frac{1}{k+1}\binom{n}{k}$.

(b) $\sum_{k=0}^{n} (-1)^k \binom{n}{k} \frac{1}{n+1-k}$.

1.2.20. ($\diamond$) Prove the following variation of Theorem 1.2.3(6):

$$\sum_{k=0}^{n} \binom{\lfloor n/2 \rfloor}{\lfloor k/2 \rfloor}\binom{\lceil n/2 \rceil}{\lceil k/2 \rceil} = \binom{n+1}{\lceil n/2 \rceil}.$$

1.2.21. Let $H_n = \sum_{i=1}^{n} \frac{1}{i}$ (so $H_0 = 0$). Suppose $0 \leq k \leq n$. (Based on Sondow [2003].)

(a) Prove $\sum_{j=k+1}^{n} (-1)^{j-k-1}\binom{n}{j}\frac{1}{j-k} = \binom{n}{k}(H_n - H_k)$. (Hint: Use induction on n.)

(b) Prove $\sum_{j=0}^{k-1} (-1)^{j-k-1}\binom{n}{j}\frac{1}{k-j} = \binom{n}{k}(H_n - H_{n-k})$.

1.2.22. ($\diamond$) *Reciprocal powers in sums.*

(a) Given $b_n = \sum_{k=1}^{n} a_k \binom{n}{k}$ for $n \geq 1$, prove $\sum_{k=1}^{n} \frac{b_k}{k} = \sum_{k=1}^{n} \frac{a_k}{k}\binom{n}{k}$.

(b) Prove $\sum \frac{1}{k_1 \cdots k_m} = \sum_{k=1}^{n} \frac{(-1)^{k-1}}{k^m}\binom{n}{k}$, where the sum on the left is over all $(k_1, \ldots, k_m)$ such that $1 \leq k_1 \leq \cdots \leq k_m \leq n$. (Dilcher [1995], 't Woord [1999])

1.2.23. For $n, m, j \in \mathbb{N}$, prove the identities below. (Generalizes Díaz-Barrero [2005]; see also Bang [1995] and 't Woord [1999].)

(a) $\sum_{k=j}^{n} (-1)^k \binom{n}{k} = (-1)^j \binom{n-1}{j-1}$.

(b) $\sum_{k=1}^{n} (-1)^{k+1}\binom{n}{k} \sum_{1 \leq i_1 \leq \cdots \leq i_m \leq k} \frac{1}{i_1 \cdots i_m} = \frac{1}{n^m}$.

1.2.24. Let S_n be the hexagonal arrangement consisting of n rings of dots, as illustrated below for $n \in \{1, 2, 3\}$. Let a_n be the number of dots in S_n. Compute a_n. Compute $\sum_{k=1}^{n} a_k$. The formula for $\sum_{k=1}^{n} a_k$ is rather simple; is there a direct combinatorial proof?

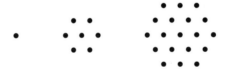

1.2.25. ($\diamond$) Prove $\sum_{k=1}^{n} k \cdot k! = (n+1)! - 1$ by induction and by counting two ways.

1.2.26. By counting a set in two ways, prove $\sum_{A \subseteq [n]} \sum_{B \subseteq [n]} |A \cap B| = n4^{n-1}$.

1.2.27. ($\diamond$) Evaluate $\sum_{S \subseteq [n]} \prod_{i \in S} \frac{1}{i}$ and $\sum_{S \subseteq [n]} (-1)^{|S|} \prod_{i \in S} \frac{1}{i}$.

1.2.28. Give both an algebraic proof (using known identities) and a combinatorial proof (counting a set in two ways) for the identity $(n-r)\binom{n+r-1}{r}\binom{n}{r} = n\binom{n+r-1}{2r}\binom{2r}{r}$.

1.2.29. ($\diamond$) Prove $\sum \binom{n}{k}\binom{n-k}{(m-k)/2} 2^k = \binom{2n}{m}$, summed over k with $m - k$ even. (Hint: Count m-subsets of $\{x_1, \ldots, x_n, y_1, \ldots, y_n\}$.) Conclude $\sum_{k=0}^{n} \binom{n}{k}\binom{n-k}{\lfloor (m-k)/2 \rfloor} 2^k = \binom{2n+1}{m}$. (R. Stanley)

1.2.30. Count the integer lists $a_{-m}, a_{-m+1} \ldots a_n$ such that $-p \leq a_{-m} \leq a_{-m+1} \leq \cdots \leq a_n \leq q$ and $a_{-1} \leq 0 \leq a_1$. (Djokovic [1991])

1.2.31. (+) For the identity below,

(a) Give a combinatorial proof (construct a set that both sides count).

(b) Use the Binomial Theorem to prove that each side is the coefficient of x^n in $(1 + 3x + x^2)^n$. Explain the sense in which the two proofs are the same. (Beckwith [2008])

$$\sum_{k \geq 0} \binom{n}{k}\binom{2k}{k} = \sum_{k \geq 0} \binom{n}{2k}\binom{2k}{k}3^{n-2k}$$

1.2.32. For the identity below, give a combinatorial proof, obtain the Summation Identity (elementary form of Theorem 1.2.3(5)) as a special case, and apply a polynomial identity from the text to express the general case as an instance of Vandermonde's Theorem (the polynomial version of Theorem 1.2.3(6)),

$$\sum_{i=0}^{k} \binom{a+i-1}{i}\binom{b+k-i-1}{k-i} = \binom{a+b+k-1}{k}.$$

1.2.33. ($\diamond$) For nonnegative integers m, n, r, and s, prove the identity below. (Ohtsuka–Tauraso [2018])

$$\sum_{k=0}^{s} \binom{m+r}{n-k}\binom{r+k}{k}\binom{s}{k} = \sum_{k=0}^{r} \binom{m+s}{n-k}\binom{s+k}{k}\binom{r}{k}$$

1.2.34. ($\diamond$) For $r \geq m+n$, prove $\sum_k \binom{r}{m+k}\binom{s}{n+k} = \binom{r+s}{r-m+n}$, and use it to evaluate $\sum_k k\binom{a}{k}\binom{b}{k}$. (Graham–Knuth–Patashnik [1989, p. 181])

1.2.35. Prove the identity below for $n \in \mathbb{N}$. (Sorel [2016])

$$\sum_{k=0}^{n} \binom{2n}{k}\binom{2n+1}{k} + \sum_{k=n+1}^{2n+1} \binom{2n+1}{k}\binom{2n}{k-1} = \binom{4n+1}{2n} + \binom{2n}{n}^2$$

1.2.36. For $n, m \in \mathbb{N}_0$ with $n \geq m$, prove $\sum_{k=0}^{m} \frac{\binom{m}{k}}{\binom{n}{k}} = \frac{n+1}{n-m+1}$. (Whitworth [1897])

1.2.37. For $a, b \in \mathbb{R}$ with $b > a+1$, prove $\sum_{k \geq 0} \frac{a^{(k)}}{b^{(k)}} = \frac{b-1}{b-1-a}$. Use it to evaluate $\sum_{k \geq 0} \binom{n+k}{n}^{-1}$. (Hint: First compute $\frac{b-1}{b-1-a} - \sum_{k=0}^{n} \frac{a^{(k)}}{b^{(k)}}$.)

1.2.38. Let $c_n = \binom{n}{\lfloor n/2 \rfloor}$. Prove that $\sum_{k=0}^{n} \binom{n}{k}c_k c_{n-k} = c_n c_{n+1}$. (Bloom [2002]) (Comment: A combinatorial proof counts "balanced" teams of sizes n and $n+1$ from n male/female pairs and one extra male. An algebraic proof uses factorials and Exercise 1.2.20.)

1.2.39. ($\diamond$) By Theorem 1.2.13, $\sum_k \binom{m}{k}\binom{n+k}{m} = \sum_j \binom{n}{j}\binom{m}{j}2^j$. Ignoring that the two sides count Delannoy paths and the lattice ball, give a direct combinatorial proof of the identity by showing that both sides count the following set S. Let M and N be sets with sizes m and n. Let S be the family of ordered pairs (A, B) such that A is a subset of M and B is an m-subset of $N \cup A$, as shown below.

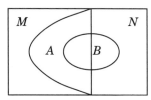

1.2.40. ($\Diamond$) For $m, n \in \mathbb{N}_0$, generalize the argument of Exercise 1.2.39 to prove the following polynomial identity by the Polynomial Principle:

$$\sum_k \binom{m}{k}\binom{n+k}{m}x^k = \sum_j \binom{m}{j}\binom{n}{j}x^{m-j}(1+x)^j .$$

1.2.41. Given $2m + 2n$ people named $x_1, \ldots, x_{m+n}$ and $y_1, \ldots, y_{m+n}$, we want to form n pairs using only pairs of the form $\{x_i, y_i\}$ for $1 \leq i \leq m+n$ or $\{x_i, x_{i+1}\}$ for $1 \leq i \leq m+n-1$ (each x_i can be involved in only one pair). Establish a bijection to show that the number of ways to form n pairs is the Delannoy number $d_{m,n}$. (Deutsch; see Sulanke [2003])

1.2.42. (+) The **augmented Aztec diamond** of order n consists of $2n+1$ centered rows of squares with lengths $2, 4 \ldots, 2n-2, 2n, 2n, 2n, 2n-2, \ldots, 4, 2$, shown below for $n = 3$. Prove that the number of ways to tile (partition) it using dominos (1-by-2 or 2-by-1 rectangles) is the central Delannoy number $d_{n,n}$. (Sachs–Zernitz [1994]) (Hint: The figure below suggests a way to obtain a path from a tiling. The easiest way to complete the proof is to prove a more general result. Comment: The ordinary Aztec diamond lacks the central row of length $2n$. It has exactly $2^{n(n+1)/2}$ tilings, a difficult result proved in Elkies–Kuperberg–Larsen–Propp [1992], Kuperberg [2002], Kuo [2004], and Brualdi–Kirkland [2005].)

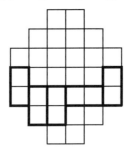

1.2.43. ($\Diamond$) Let $A = \{a_1, \ldots, a_n\} \subset \mathbb{R}$, with $a_1 < \cdots < a_n$. For $i \in [n]$, let S_i be the family of i-element subsets of A, and let $\sigma_i = \sum_{B \in S_i} \max(B)$. For $k \in [n]$, prove

$$a_k = \sum_{r=1}^n (-1)^{k+r}\binom{r-1}{k-1}\sigma_r. \tag{Ash [2010]}$$

1.2.44. (+) Let $\mathbb{S}_n$ be the set of permutations of $[n]$. For positive real $d_1, \ldots, d_n$, prove

$$\sum_{\pi \in \mathbb{S}_n} \frac{1}{d_{\pi(1)}(d_{\pi(1)} + d_{\pi(2)})\cdots(d_{\pi(1)} + \cdots + d_{\pi(n)})} = \frac{1}{d_1 \cdots d_n}.$$

Apply this to prove $\sum_{v \in S}(v_1 \cdots v_n)^{-1} = \sum_{w \in T}(w_1 \cdots w_n)^{-1}$, where $S = \{v \in [N]^n : \sum v_i \leq N\}$ and $T = \{w \in [N]^n : w_1, \ldots, w_n \text{ are distinct}\}$. (Hill [2004], solution to Lubell [2003])

1.2.45. Given $p, q \in \mathbb{N}$, let $d = \gcd(p, q)$. By counting a geometric arrangement of points, prove that $2\sum_{i=1}^{q-1}\lfloor ip/q \rfloor = (p-1)(q-1) + d - 1$.

1.2.46. Give a combinatorial proof that $\sum_{i=0}^m \lceil \frac{in}{m} \rceil = \sum_{j=0}^n \lceil \frac{jm}{n} \rceil$. (Hint: Use a geometric arrangement of points, as in Exercise 1.2.45.)

1.2.47. ($\Diamond$) *Binomial coefficients as polynomials.* Fix $k \in \mathbb{N}$.
   (a) Let $f_k(n) = k!\binom{n}{k} - n^k + \binom{k}{2}n^{k-1}$. Note that f_2 is identically 0. For $k \geq 3$, prove that f_k is a polynomial of degree $k - 2$.
   (b) Prove that $\sum_{i=0}^n i^k$ is a polynomial in n with leading terms $\frac{1}{k+1}n^{k+1} + \frac{1}{2}n^k$.

1.2.48. (◇) Let I be the set of polynomials with rational coefficients that have integer values at all integers.

(a) Prove that every polynomial f of degree k with rational coefficients can be expressed in exactly one way as $f(x) = \sum_{j=0}^{k} b_j \binom{x}{j}$ such that each b_j is rational.

(b) Prove that $f \in I$ if and only if the coefficients $b_0, \ldots, b_k$ in part (a) are integers. (Hint: Evaluate f at $\{0, \ldots, k\}$. Note that $\binom{0}{0} = 1$.)

1.2.49. By considering lists made from $n-2r$ copies of U (up) and $n+2r$ copies of D (down), prove the identity below. (Hint: To explain the left side, view each list as a list of pairs of successive elements; there are four types of pairs.)

$$\sum_k \binom{n}{2k}\binom{2k}{k-r} 2^{n-2k} = \binom{2n}{n-2r}. \qquad \text{(Callan [2003a])}$$

(Comment: A similar argument yields $\sum_k \binom{n}{2k+1}\binom{2k+1}{k-r} 2^{n-2k-1} = \binom{2n}{n-2r-1}$.)

1.2.50. (+) Prove the following identity by devising a set counted by both sides.

$$\sum_{k\geq 0} 2^k \binom{2m-k}{m+n} = 4^m - \sum_{j=1}^{n} \binom{2m+1}{m+j}. \qquad \text{(Knuth [2007])}$$

1.2.51. (+) Prove the following identity combinatorially. (Karaivanov–Vassilev [2017])

$$\sum_{s=k}^{\lfloor n/2 \rfloor} \binom{n+1}{2s+1}\binom{s}{k} = \binom{n-k}{k} 2^{n-2k}$$

1.3. Applications

Proving identities provides plenty of opportunity for clever combinatorial arguments, but often we want to count a new set rather than explain a given formula. In this section we study several problems of this type.

GRAPHS AND TREES

A special case of Proposition 1.1.13 is that every k-element set has 2^k subsets. We apply this to count graphs (see Chapter 0 for basic definitions about graphs).

1.3.1. Corollary. There are $2^{\binom{n}{2}}$ graphs with vertex set $[n]$.

Proof: Each edge is an unordered pair of vertices, and a graph with a specified vertex set is determined by choosing a family of vertex pairs as the edge set. There are $\binom{n}{2}$ pairs of vertices to choose from. ∎

We next count the trees (connected graphs without cycles) with vertex set $[n]$. Such trees have $n-1$ edges and arise from a single vertex by iteratively adding a new leaf with one old neighbor (see Chapter 0).

1.3.2. Example. *Small trees.* When $n \in \{1, 2\}$, there is one tree with vertex set $[n]$. With vertex set $[3]$, there are three trees, pairwise isomorphic. They are determined by choosing the central vertex, as shown below. With vertex set $[4]$, there are four "stars" and twelve paths; 16 trees in total. Case analysis yields 125 trees with vertex set $[5]$.   ■

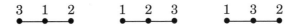

Note that $1, 3, 16, 125$ fits the sequence $\langle a \rangle$ with $a_n = n^{n-2}$. To prove that this is the formula, we seek a correspondence between the trees and the n^{n-2} lists of length $n - 2$ with entries in $[n]$. We first view the lists as n-tuples by adding a first entry 1 and last entry n. We then treat an n-tuple a with entries in $[n]$ as a function $f_a \colon [n] \to [n]$ defined by $f_a(i) = a_i$ for $1 \leq i \leq n$. A function from a set to itself has a useful graphical representation, which we use to define edges.

1.3.3. Definition. The **functional digraph** of a function $f \colon S \to S$ is the directed graph with vertex set S having an edge from x to $f(x)$ for each $x \in S$. Each vertex is the tail of exactly one edge.

1.3.4. Theorem. (**Cayley's Formula**; Cayley [1889]) There are n^{n-2} trees with vertex set $[n]$.

Proof: (Eğecioğlu–Remmel [1986]) There are n^n functions f from $[n]$ to $[n]$. Among these, n^{n-2} satisfy $f(1) = 1$ and $f(n) = n$. We establish a bijection from this set to the set of trees with vertex set $[n]$.

The functional digraph of f has vertex set $[n]$ and n edges. Since each vertex is the tail of exactly one edge, the digraph consists of cycles with directed trees appended (iteration of f takes an element eventually to a cycle). Loops at 1 and n are two of these cycles.

We modify the functional digraph of f to obtain a tree $\sigma(f)$. First place the vertices of the cycles in f on a horizontal line, with the least label in each cycle appearing first (followed in order by its successive images). List the cycles in decreasing order of their least elements. When $n = 15$, the n-tuple $(1, 5, 4, 3, 10, 12, 5, 15, 12, 7, 1, 7, 15, 3, 15)$ yields the diagram on the left.

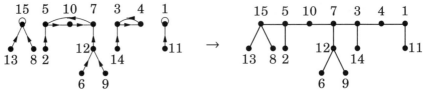

For each cycle, cut the edge from the last vertex to the first. Replace each such edge (except for the last cycle) with an edge to the first vertex of the next cycle. This destroys all cycles, reduces the number of edges by one, and connects the pieces. Ignoring the directions on the edges yields a connected graph with no cycles: a tree $\sigma(f)$ with vertex set $[n]$; see the outcome on the right above.

The vertices of the cycles in the functional digraph form a path P in $\sigma(f)$ from n to 1 along the horizontal line. A vertex on P began a cycle in f if and only if its label is less than all the labels that precede it on P. This is the key idea.

For a tree T with vertex set $[n]$, let P be the unique path from n to 1. Draw T with P along a line. Orient edges not in P toward P, and orient P from n toward 1. For each vertex along P whose label is less than all preceding it, make it the start of a new cycle by cutting the edge of P that reaches it and instead complete the current cycle. Since n is first and 1 is last, these two become cycles of length 1. We obtain a functional digraph f of the desired form, so σ is surjective.

A vertex on the path from n to 1 in $\sigma(f)$ is the least element of a cycle in f if and only if it is less than all earlier labels. Hence f is the only function with $\sigma(f) = T$. Thus σ is also injective, and there are n^{n-2} trees with vertex set $[n]$. ∎

The idea used to arrange the cycles of this functional digraph will be used again for the canonical cycle decomposition of a permutation in Section 3.1.

1.3.5.* Remark. Cayley's Formula has many proofs: Cayley [1889], Kirchhoff [1847] (Exercise 15.2.4), Prüfer [1918] (Exercise 12), Pólya [1937] (Theorem 3.3.29), Moon [1967], Rényi [1970] (Theorem 15.2.9), Joyal [1981], Pitman [1999], and others (Exercise 17). Equivalent results appeared earlier (Borchardt [1860], Sylvester [1857]), but Cayley expressed the problem in terms of graph theory and invented the term "tree". Proofs also appear in Chapter 3 of Biggs–Lloyd–Wilson [1976] and in Chapter 24 of Aigner–Ziegler [1999]. Moon [1970] wrote a book on enumerating classes of trees. ∎

MULTINOMIAL COEFFICIENTS

Cayley approached the tree-counting problem algebraically, using a generating function to enumerate labeled trees by their vertex degrees. The Eğecioğlu–Remmel bijection also provides this information, using our next counting formula (a special case was used in Proposition 1.2.10).

1.3.6. Proposition. There are $m!/\prod_{i=1}^{n} k_i!$ words of length m having exactly k_i letters of type i, where $\sum_{i=1}^{n} k_i = m$.

Proof 1: We can produce the arrangement by choosing positions for the letters of the first type, then choosing positions for the second type, and so on. The resulting product simplifies to the desired formula.

$$\binom{m}{k_1}\binom{m-k_1}{k_2}\binom{m-k_1-k_2}{k_3}\cdots = \frac{m!}{k_1!(m-k_1)!}\frac{(m-k_1)!}{k_2!(m-k_1-k_2)!}\cdots = \frac{m!}{\Pi k_i!}$$

Proof 2: The symmetry in the formula suggests a more direct argument. We temporarily assign subscripts to the letters of each type to distinguish them; now there are $m!$ arrangements. Each desired arrangement corresponds to exactly $\prod k_i!$ of these, because there are $\prod k_i!$ ways to assign subscripts once the positions of the letters are fixed. ∎

1.3.7. Corollary. The number of trees with vertex set $[n]$ in which vertices $1, \ldots, n$ have degrees $d_1, \ldots, d_n$, respectively, is $(n-2)!/\prod_{i=1}^{n}(d_i - 1)!$.

Proof: In the functional digraph f for the list $(1, a_2, \ldots, a_{n-1}, n)$, one edge leaves each i, and the number of edges entering i is the number of copies of i in the list. For each $i \notin \{1, n\}$, there is one incident edge in $\sigma(f)$ for each edge entering or leaving i in f. However, turning the cycles into a path loses the edge entering n and the edge leaving 1.

After deleting the first and last term in the list, it becomes true for all i that the degree of i in the tree $\sigma(f)$ is one more than the number of copies of i in the list $a_2, \ldots, a_{n-1}$. Therefore, we count trees with each i having degree d_i by counting lists of length $n - 2$ with $d_i - 1$ copies of i for each i. By Proposition 1.3.6, the formula is as claimed.   ∎

1.3.8. Example. *Trees with fixed degrees.* Consider 7-vertex trees whose vertices have degrees $3, 1, 2, 1, 3, 1, 1$. Here $\frac{(n-2)!}{\Pi(d_i-1)!} = 30$. Each tree is of one of the three types shown below. There are six ways to complete the first tree (pick from the remaining four vertices the two adjacent to vertex 1) and twelve ways to complete each of the others (pick the neighbor of vertex 3 from the remaining four, and then pick the neighbor of the central vertex from the remaining three).   ∎

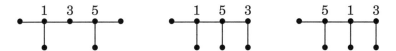

The number $k!/\prod_{i=1}^{n} k_i!$ is written as $\binom{k}{k_1,\ldots,k_n}$ or $P(k; k_1, \ldots, k_n)$. When $n = 2$, arranging the letters is the same as choosing positions for the first type, so $P(k_1 + k_2; k_1, k_2) = \binom{k_1+k_2}{k_1}$. Due to the next result, the numbers $\binom{k}{k_1,\ldots,k_n}$ are called **multinomial coefficients**.

1.3.9. Proposition. (Multinomial Theorem) For $n, k \in \mathbb{N}_0$,

$$\left(\sum_{i=1}^{n} x_i\right)^k = \sum \binom{k}{k_1, \ldots, k_n} \prod_{i=1}^{n} x_i^{k_i},$$

with the sum on the right over $(k_1, \ldots, k_n) \in \mathbb{N}_0^n$ with sum k.

Proof: In expanding $(\sum_{i=1}^{n} x_i)^k$, the contributions to the coefficient of $\prod_{i=1}^{n} x_i^{k_i}$ count the k-letter words with k_i letters of type i for each i.   ∎

1.3.10. Application. (Fermat's Little Theorem) If $n \in \mathbb{Z}$ and p is prime, then $n^p \equiv n \pmod{p}$. Furthermore, $n^{p-1} \equiv 1 \pmod{p}$ when p does not divide n.

Proof: (Leibniz) Multiplication with congruence classes is well defined, so we may assume that $n \in \mathbb{N}$. We rewrite n^p as $(\sum_{i=1}^{n} 1)^p$, which is obtained from $(\sum_{i=1}^{n} x_i)^p$ by setting each $x_i = 1$.

By the Multinomial Theorem, the coefficient of $\prod_{i=1}^{n} x_i^{k_i}$ in the expansion of $(\sum_{i=1}^{n} x_i)^p$ is $p!/\prod_{i=1}^{n} k_i!$. For each term of the form x_i^p, the coefficient is 1; there are n such terms. All other coefficients are divisible by p, since the denominator in the formula for those multinomial coefficients has no divisor of p. Hence $n^p = (1 + \cdots + 1)^p \equiv n \pmod{p}$. (See Exercise 22 for another proof.)

The second statement is the usual statement of Fermat's Little Theorem. It follows from the first by the existence of multiplicative inverses modulo p for numbers that are not multiples of p. ∎

1.3.11. Remark. The multinomial coefficients provide extensions of identities (1) and (2) from Theorem 1.2.3. Since $\binom{k}{k_1,\ldots,k_n}$ counts words with specified multiplicities of each letter, permuting $k_1,\ldots,k_n$ does not change the value. The words can also be interpreted as n-dimensional lattice paths. Grouping the paths by the direction of the last step extends Pascal's Formula:

$$\binom{n}{k_1,\ldots,k_t} = \binom{n-1}{k_1-1,k_2,\ldots,k_t} + \cdots + \binom{n-1}{k_1,\ldots,k_{t-1},k_t-1}.$$ ∎

THE BALLOT PROBLEM

A natural problem in combinatorial probability leads to a fundamental counting sequence for special sets of lattice paths or binary lists.

1.3.12. Example. *Bertrand's* **Ballot Problem.** Candidates A and B receive a and b votes, respectively, with $a \geq b$. When votes are counted in random order, what is the probability that A never trails? The $a+b$ votes can be removed from the box in $(a+b)!$ equally likely orders. When we record the votes as a list of As and Bs, each possible list arises from $a!\, b!$ orderings of the votes. This makes the A, B-lists equally likely, and it suffices to determine the fraction of these lists in which every initial segment has at least as many As as Bs. ∎

1.3.13. Theorem. (Bertrand [1887]) Among the lists formed from a copies of A and b copies of B, where $a \geq b$, there are $\binom{a+b}{a} - \binom{a+b}{a+1}$ lists in which every initial segment has at least as many As as Bs.

Proof: Lists consisting of a copies of A and b copies of B correspond to lattice paths from the origin to (a, b). The total number of paths is $\binom{a+b}{a}$, and the desired lists correspond to the paths that never move above the line $y = x$. We count these by subtracting the others from the total.

A path steps above the diagonal if it reaches $(k, k+1)$ for some k. Consider the first such occurrence. Modify the list after this by changing As to Bs and Bs to As. Now there are $b - k - 1$ additional As and $a - k$ additional Bs, so the total for the modified list is $(b - 1, a + 1)$.

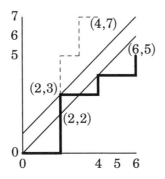

The switch reflects the part of the path after $(k, k + 1)$ through the line $y = x + 1$. The new path ends at $(b - 1, a + 1)$. Since $a + 1 > b - 1$, each path ending at $(b - 1, a + 1)$ rises above the line $y = x$. Reflecting the part of such a path after it first visits a point of the form $(k, k + 1)$ generates a path to (a, b). The second map inverts the first, so the bad paths to (a, b) correspond bijectively to all paths reaching $(b - 1, a + 1)$. Hence there are $\binom{a+b}{a+1}$ bad paths.   ∎

For the probability in Example 1.3.12, $(a + 1)\binom{a+b}{a+1} = b\binom{a+b}{a}$ yields

$$\frac{\binom{a+b}{a} - \binom{a+b}{a+1}}{\binom{a+b}{a}} = 1 - \frac{b}{a + 1} = \frac{a - b + 1}{a + 1}.$$

The switching argument in Theorem 1.3.13 is called the **reflection principle**, attributed to Antoine Désiré André [1887] (see Renault [2008] for what André actually did). According to Feller [1968, p. 72], "The reflection principle is used frequently in various disguises, but without the geometrical interpretation it appears as an ingenious but incomprehensible trick." That is, the argument works as well using 0, 1-lists; the reflection corresponds to complementing the final part of the list. For the analogous problem with $a > b$ and A being always ahead, the reflection is through the line $y = x$ and the formula is nicer (Exercise 5).

Theorem 1.3.13 leads to a short combinatorial proof of a beautiful identity for the central binomial coefficients. It appears as early as a text by Whitworth [1897] called *Choice and Chance*. In the 1930s P. Veress requested a combinatorial proof, found by G. Hajós (see Sved [1984], and Kleitman [1975]). Example 3.2.13 presents another proof.

1.3.14. Lemma. The central binomial coefficient $\binom{2m}{m}$ counts the following types of lattice paths of length $2m$ that start at $(0, 0)$.
(A) Those ending at (m, m).
(B) Those never rising above the line $y = x$.
(C) Those never returning to the line $y = x$.

Proof: Type A is the familiar lattice path model. The number of paths of Type B with length l (possibly l is odd) that end at (a, b) with $a + b = l$ and $a \geq b$ is $\binom{l}{a} - \binom{l}{a+1}$, by Theorem 1.3.13. Hence for the answer we sum over a and observe that the sum telescopes. The computation is

$$\sum_{a=\lceil l/2 \rceil}^{l} \left[\binom{l}{a} - \binom{l}{a+1} \right] = \binom{l}{\lceil l/2 \rceil}.$$

Paths of Type C first step to the right or up. If right, then after $(1, 0)$ the rest has length $2m - 1$ and does not rise above $y = x - 1$. Setting $l = 2m - 1$, the number of these paths is $\binom{2m-1}{m}$. The same number first step up, so the total number of Type C paths is $2\binom{2m-1}{m}$. The Committee-Chair and Complementation Identities yield $m\binom{2m}{m} = 2m\binom{2m-1}{m-1} = 2m\binom{2m-1}{m}$, so $2\binom{2m-1}{m} = \binom{2m}{m}$.   ∎

1.3.15. Theorem. For $n \in \mathbb{N}_0$,

$$\sum_{k=0}^{n} \binom{2k}{k}\binom{2n - 2k}{n - k} = 4^n.$$

Proof: There are 4^n binary lists of length $2n$; view them as lattice paths from the origin. Since $\binom{2k}{k}$ is the number of lattice paths from the origin to (k, k), we group the paths according to the point (k, k) where they last touch the line $y = x$.

Within such a path, the initial portion can combine with any path of $2n - 2k$ steps after (k, k) that does not return to $y = x$. By Lemma 1.3.14, there are $\binom{2n-2k}{n-k}$ such paths. Hence the sum counts all paths with $2n$ steps, grouped by the position of the last visit to the line $y = x$.   ∎

CATALAN NUMBERS

In the special case $a = b = n$ of Theorem 1.3.13, the number of good paths is $\frac{1}{n+1}\binom{2n}{n}$. These numbers turn up surprisingly often.

1.3.16. Definition. The **Catalan sequence** is defined by $C_n = \frac{1}{n+1}\binom{2n}{n}$ for $n \geq 0$. The number C_n is called the nth **Catalan number**.

A **ballot path** of length $2n$ is a lattice path from $(0, 0)$ to (n, n) that never rises above $y = x$. A **ballot list** of length $2n$ is a list of n 1s and n 0s such that each initial segment has at least as many 1s as 0s.

Netto [1901] named Catalan numbers in honor of Eugène Charles Catalan (1814–1894), (see Banderier–Schwer [2005]). Catalan [1838] studied them but called them "Segner numbers" after Segner [1759], who obtained a recurrence. For further discussion of the history, see I. Pak's appendix to Stanley [2015a].

The correspondence between paths and lists in Example 1.2.2 restricts to a correspondence between ballot paths and ballot lists. In Theorem 1.3.13, we proved that the nth Catalan number C_n counts both the ballot paths of length $2n$ and the ballot lists of length $2n$. Our next proof of this yields more general results (see Hilton–Pederson [1991] and Graham–Knuth–Patashnik [1989, 359–363]).

1.3.17. Theorem. If p and q are relatively prime, then the number of lattice paths from $(0, 0)$ to (p, q) that do not rise above the line $py = qx$ is $\frac{1}{p+q}\binom{p+q}{p}$.

Proof: Exactly $\binom{p+q}{p}$ words consist of p 1s and q 0s. Put them in groups; grouped with A are all words obtained by reading A cyclically from some starting point. If the rotations of A are not distinct, then A is periodic, making p and q both divisible by the number of periods. Since $\gcd(p, q) = 1$, each group has size $p + q$.

There are thus $\frac{1}{p+q}\binom{p+q}{p}$ sets of words. Since the words correspond bijectively to the lattice paths to (p, q), it suffices to show that each set contains exactly one word that corresponds to a good path.

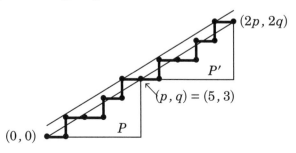

Given a lattice path P from $(0,0)$ to (p,q), let P' be the translation of P starting from (p,q). The subpaths of $P \cup P'$ having length $p+q$ that start at integer points along P correspond to words in the same set.

Let L be the line $\{(x,y): py - qx = 0\}$. Since $\gcd(p,q) = 1$, each line parallel to L contains at most one integer point of P. (Otherwise, the difference (a,b) satisfies $a/b = p/q$ with $a < p$, but p/q is already in lowest terms.)

Each line parallel to L has equation $py - qx = c$ for some c. The path $P \cup P'$ lies below the line parallel to L that has the highest value c among those hitting P, and it does not lie below any other line in this family. This selects a unique starting position among the cyclic rotations of P so that when the starting point is translated to the origin the path stays below the line $py = qx$. ∎

1.3.18. Definition. A q**-ballot list** is a list of n 0s and qn 1s where every initial segment has at least q times as many 1s as 0s ($q = 1$ yields ballot lists).

1.3.19. Corollary. There are $\frac{1}{qn+1}\binom{(q+1)n}{n}$ q-ballot lists of length $(q+1)n$.

Proof: Such lists correspond to lattice paths from $(0,0)$ to (qn, n) not rising above $qy = x$, but $\gcd(qn, n) \neq 1$. Instead, consider paths from $(-1, 0)$ to (qn, n) not rising above $(qn+1)y = n(x+1)$. Such a path starts horizontal and then follows a good path, since an integer point (a,b) above $qy = x$ but not $(qn+1)y = n(x+1)$ requires $bq \geq a+1$ and $b(qn+1) \leq n(a+1)$, a contradiction.

We therefore count paths from $(-1, 0)$ to (qn, n) not rising above $(qn+1)y = n(x+1)$. Moved one step rightward, these correspond to paths from $(0,0)$ to $(qn+1, n)$ not rising above $(qn+1)y = nx$. Since $\gcd(qn+1, n) = 1$, by Theorem 1.3.17 there are $\frac{1}{(q+1)n+1}\binom{(q+1)n+1}{qn+1}$ such paths. The formula simplifies by the Committee-Chair and Complementation Identities to the given expression. ∎

The numbers $\frac{1}{qn+1}\binom{(q+1)n}{n}$ are the **generalized Catalan numbers** or **Fuss–Catalan numbers** (first studied by Fuss [1791]). See Goulden–Serrano [2003] for other instances, also Exercises 43–45.

A related argument yields the same generalizations and others. We state the result here and leave the proof and applications to Exercises 37–39.

1.3.20. Theorem. (Cycle Lemma; Dvoretzky–Motzkin [1947]**)** For $n, m, k \in \mathbb{N}_0$ with $m \geq kn$, every arrangement of m 1s and n 0s in a circle has exactly $m - kn$ positions such that every clockwise segment starting there has more than k times as many 1s as 0s. ∎

$$
\begin{array}{ccccc}
 & & 0 \quad 0 & & \\
1 & & & & 1 \\
1 & & & & 1 \\
1 & & & & 0 \\
 & 1 & \underline{1} & &
\end{array}
$$

The example above has $(m,n,k) = (7,3,2)$, with the good position marked. Theorem 1.3.20 provides a bijection from q-ballot lists (with an initial 1 added) to cyclic arrangements with $qn + 1$ 1s and n 0s. Since the number of such cyclic arrangements is $\frac{1}{qn+1}\binom{(q+1)n}{n}$, Corollary 1.3.19 follows (see Peck [1989]).

For the Catalan numbers themselves, Exercise 6.19 of Volume 2 of *Enumerative Combinatorics* by Stanley [1999] lists 66 counting problems they solve. In a later book, Stanley [2015a] expanded the list to 214 combinatorial interpretations of the Catalan numbers. See also www-math.mit.edu/~rstan/ec/catadd.pdf. Below are a few classical examples.

> Ballot lists / ballot paths of length $2n$ (Corollary 1.3.19)
> Binary trees of $n + 1$ leaves (grouping $n + 1$ terms) (Theorem 1.3.23)
> Ordered trees with n edges (Example 1.3.25)
> Triangulations of a convex $(n + 2)$-gon (Example 1.3.24)
> Noncrossing pairings of $2n$ points on a circle (Exercise 31)
> Noncrossing partitions of $[n]$ (Exercise 41)
> Stack-sortable permutations of $[n]$ (Exercise 2.1.48)

Two of these examples use modified notions of trees.

1.3.21. Definition. A **rooted graph** has one vertex distinguished as a **root**. In a tree with root r, the neighbor of a vertex v on the path from v to r is the **parent** of v, and the other neighbors of v are its **children**. A **leaf** in a rooted tree is a vertex with no children. An **ordered tree** is a rooted tree in which the children of each vertex are given a fixed (left-to-right) linear order. A **binary tree** is an ordered tree in which every vertex has zero or two children.

The root in a rooted tree has no parent; it is considered a leaf when it has no children. Ordered trees have also been called **rooted plane trees**; the word "plane" suggests the fixed ordering on the children when the tree is drawn in the plane. In a binary tree, the **left subtree** and **right subtree** are the subgraphs obtained by deleting the root r; they are rooted at the left and right children of r, respectively. In some contexts in computer science, a vertex in a binary tree may have exactly one child, designated "left" or "right".

1.3.22. Example. A binary tree iteratively groups or "parenthesizes" its leaves. Each matched pair of parentheses corresponds to a non-leaf non-root vertex of the tree where the sets of leaves for two subtrees are "combined". We omit the outer parentheses corresponding to the combining operation at the root.

Below are the five binary trees with four leaves, the corresponding five ballot lists of length 6 under the bijection in Theorem 1.3.23, and the expressions of the trees as binary groupings of factors at the leaves.

A binary tree can be grown from the root by iteratively giving a leaf two children. The numbers of leaves, non-leaves, and edges grow by 1, 1, and 2, respectively. Thus a binary tree with $n + 1$ leaves has n non-leaf vertices and $2n$ edges (the root is a leaf when $n = 0$). ∎

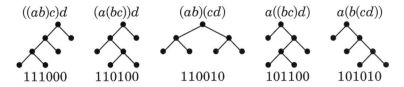

1.3.23. Theorem. The number of binary trees with $n+1$ leaves is the nth Catalan number, $\frac{1}{n+1}\binom{2n}{n}$.

Proof: Using induction on n, we establish a bijection f_n from the set A_n of binary trees with $n+1$ leaves to the set B_n of ballot lists of length $2n$ (we know the latter has size C_n). For $n = 0$, the 1-vertex tree corresponds to the empty ballot list.

Consider $n > 0$. A tree T in A_n has left and right subtrees T_l and T_r with k and $n+1-k$ leaves, for some $k \in [n]$. Under the bijections f_{k-1} and f_{n-k} provided by the induction hypothesis, these subtrees yield ballot lists of lengths $2k-2$ and $2n-2k$. Let $f_n(T) = (1, f_{k-1}(T_l), 0, f_{n-k}(T_r))$. The result is a ballot list of length $2n$, since the sublists are ballot lists and the added 1 precedes the added 0.

To prove that f_n is a bijection, we argue that for a ballot list b of length $2n$ exactly one binary tree T satisfies $f_n(T) = b$. This requires the initial 1 in b to be paired with a 0 that occurs immediately after the ballot list corresponding to T_l and before the list corresponding to T_r. By the induction hypothesis, there is exactly one choice for T_l and one for T_r. Hence T is determined uniquely. ∎

The key step for inductively producing the inverse of a bijection to the set of ballot lists is to decompose the list using the unique 0 that "matches" the initial 1 by the first equality in the bits of each type (first return to the diagonal). In Theorem 1.3.23, this discovers two smaller ballot lists to be used to find the inverse image (this is how one distinguishes the images of left and right subtrees in a list like 10101010).

Bijections involving binary trees are especially easy to verify inductively. This yields the next classical instance of the Catalan numbers.

1.3.24. Example. *There are $\frac{1}{n+1}\binom{2n}{n}$ triangulations of a convex $(n+2)$-gon.* This is the counting problem solved by Catalan. The result is surprising, but the bijection to a problem we already solved is easy to verify.

A parenthesized string of $n+1$ letters corresponds to a binary tree with $n+1$ leaves. With one root edge fixed, a triangulation of an $(n+2)$-gon yields such a tree in a natural way as suggested by the figure below. Exercise 30 requests the proof that the resulting map is a bijection. ∎

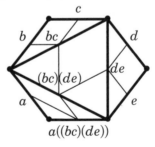

$$a((bc)(de))$$

An argument like that of Theorem 1.3.23 counts ordered trees with n edges.

1.3.25. Example. *There are $\frac{1}{n+1}\binom{2n}{n}$ ordered trees with n edges.* To define an inductive bijection to the set of ballot lists, cut the edge in T from the root to its leftmost child, forming ordered subtrees with $k-1$ and $n-k$ edges. By the induction hypothesis, these have corresponding ballot lists b and b' of lengths $2k-2$ and $2n-2k$; let $f(T) = (1, b, 0, b')$.

This bijection motivates our claim that bijective arguments give more information than inductive proofs. We record 1 when an edge is traveled to a subtree, then the subtree is traversed, then a 0 is recorded when following the edge back up, and finally we continue to the next subtree. Each edge contributes a 1 and later a 0. A 1 is followed immediately by its matching 0 if and only if the subtree has only one vertex and is a leaf. Thus the number of leaves in the ordered tree is the number of times that 0 follows 1 in the corresponding ballot list. Our bijection is "refined" by describing the images of subsets of interest. ∎

In a binary list, a **run** is a maximal sublist of consecutive equal entries.

1.3.26.* Theorem. The number of ordered trees with n edges and k leaves, called the **Narayana number** $N_{n,k}$, is $\frac{1}{k}\binom{n}{k-1}\binom{n-1}{k-1}$.

Proof: As observed in Example 1.3.25, it suffices to count the ballot lists of length $2n$ that have k runs of 1s (and k runs of 0). Appending a 0 to the end of the list yields a list whose corresponding path reaches $(n, n+1)$, still with k runs of each bit. As in Corollary 1.3.19, these are the paths reaching $(n, n+1)$ that do not step above the line $ny = (n+1)x$, and there is exactly one place to cut the cyclic arrangement of any list with n 1s and $n+1$ 0s to obtain the list for such a path.

Hence it suffices to count the cyclic arrangements of these bits with k runs of each type. Consider starting at a run of 1s. The 1s form a composition of n with k parts, and the 0s form a composition of $n+1$ with k parts in the spaces following runs of 1s. By Corollary 1.1.22, there are $\binom{n-1}{k-1}\binom{n}{k-1}$ ways to form these compositions. Starting at any one of the k runs of 1s yields a cyclic rotation of the same arrangement, so we divide by k. Because n and $n+1$ are relatively prime, there is no periodicity, and each cyclic arrangement arises in k ways. ∎

The Narayana numbers are named after Narayana [1955], though they were studied earlier (in more generality) by MacMahon [1916]; they arise again in Exercise 41 and in our last example.

1.3.27.* Example. Ballot paths have an alternative classical formulation. A **Dyck n-path** is a path in the plane from $(0, 0)$ to $(2n, 0)$ that takes n upsteps by $(1, 1)$ and n downsteps by $(1, -1)$ and never falls below the axis $y = 0$. A ballot path of length $2n$ becomes a Dyck n-path by turning horizontal steps (away from the diagonal) into upsteps and vertical steps (toward the diagonal) into downsteps. Each Dyck n-path thus arises from a unique ballot path, so there are C_n Dyck n-paths. Dyck paths are named for Walther von Dyck (1856–1934).

The Dyck model is more natural for some concepts. A **peak** in a Dyck path is an up-down subpath, while a **valley** is a down-up subpath. Left turns and right turns in a ballot path become peaks and valleys in a Dyck path, which are natural concepts. The Narayana number $N_{n,k}$ counts the Dyck n-paths having exactly k peaks. The **height** of a peak or valley is the vertical coordinate of its midpoint. See Exercise 32. ∎

We have noted that the Catalan numbers arise in many counting problems, such as in these exercises. Also, the formula for C_n can be found in many ways. Indeed, a theme of this text is the flexibility of using various techniques to solve a combinatorial problem. We therefore solve some problems repeatedly, such as deriving the Catalan formula (see Section 2.1 and Section 2.2), computing $\sum_{i=1}^{n} i^2$, studying the Delannoy numbers and the "derangement" problem, and enumerating various types of permutations and partitions.

EXERCISES 1.3

1.3.1. (–) A box has k_i letters of type i, for $1 \leq i \leq n$. Count the words (using all letters in the box) that have no two letters of type n adjacent.

1.3.2. (–) Given positive integers $k_1, \ldots, k_m$, how many ways are there to partition a set of n distinct objects so that there are k_i blocks of size i, for $1 \leq i \leq m$?

1.3.3. (–) Prove bijectively that the Catalan number C_n counts these sets:
(a) Nondecreasing functions $f \colon [n] \to [n]$ such that $f(i) \leq i$ for all i.
(b) Nonnegative integer sequences of length $2n + 1$ starting and ending at 0 such that consecutive entries differ by 1.
(c) Arrangements of $2n$ people in two rows of n such that heights increase in each row and in each column.

1.3.4. (–) Generate a random list $b_1, \ldots, b_n$ in the following way. For $1 \leq k \leq n$, choose b_k to be any of the k values in $[k]$, equally likely. Obtain a simple formula for the probability that the resulting list is nondecreasing.

1.3.5. (–) In Bertrand's Ballot Problem (Example 1.3.12) with outcome (a, b) and votes counted in random order, suppose that $a > b$. What is the probability that A is always ahead of B (after the start)?

1.3.6. (–) A fair coin is flipped $2n$ times, consecutively. Compute the probability that the lead changes, given that the final total is n heads and n tails.

1.3.7. (–) Prove the identities below by counting in two ways. (Whitworth [1897])
(a) $\displaystyle\sum_{k} \frac{1}{k+1}\binom{2k}{k}\binom{2n-2k}{n-k} = \binom{2n+1}{n}$.   (b) $\displaystyle\sum_{k=1}^{n} \frac{1}{k}\binom{2k-2}{k-1}\binom{2n-2k+1}{n-k} = \binom{2n}{n-1}$.

1.3.8. (◊) Prove bijectively that the number of graphs with vertex set $[n]$ in which all vertices have even degree is $2^{\binom{n-1}{2}}$.

1.3.9. Among the trees with vertex set $[n]$, use Corollary 1.3.7 to count those having $n-2$ leaves and those having two leaves.

1.3.10. (◊) Let $f(d_1, \ldots, d_n)$ be the number of trees with vertex set $[n]$ in which, for each i, the degree of i is d_i. Use induction on n to prove directly that $f(d_1, \ldots, d_n) = \binom{n-2}{d_1-1,\ldots,d_n-1}$. Use this to obtain Cayley's Formula n^{n-2} for the number of trees with vertex set $[n]$.

1.3.11. Let G be a graph with n vertices and $\binom{n}{2} - 1$ edges. Use Cayley's Formula to prove that G has exactly $(n-2)n^{n-3}$ spanning trees.

1.3.12. (◊) For a tree T with vertex set $[n]$, the **Prüfer code** $g(T)$ is formed by repeatedly deleting the least leaf in the remaining tree and recording its neighbor, $n-2$ times.
(a) Prove that g is a bijection from the set of trees with vertex set $[n]$ to the set $[n]^{n-2}$. (Comment: Prüfer's argument appears in Biggs–Lloyd–Wilson [1976].)
(b) Prove that a tree T with vertex set $[n]$ has $\{n-1, n\}$ as an edge if and only if the last entry in $g(T)$ is $n-1$ or n. Use this to solve Exercise 1.3.11.

1.3.13. For $n \geq 4$, construct a tree T with vertex set $[n]$ whose Prüfer code $g(T)$ (Exercise 1.3.12) differs from the list $f(2), \ldots, f(n-1)$ corresponding to it in Theorem 1.3.4.

1.3.14. ($\diamond$) Prove that $t_n = \sum_{k=1}^{n-1} k\binom{n-2}{k-1} t_k t_{n-k}$, where t_n is the number of trees with vertex set $[n]$. (Dziobek [1917])

1.3.15. ($\diamond$) Let t_n be the number of trees with vertex set $[n]$. Explain the identity below in terms of t_n. Prove it. Conclude that $t_n = \frac{n}{2} \sum_{k=1}^{n-1} \binom{n-2}{k-1} t_k t_{n-k}$. (Comment: The result appears in Lovász [1979]; the combinatorial argument is due to L. Smiley.)

$$2(n-1)n^{n-2} = \sum_{k=1}^{n-1} \binom{n}{k} k^{k-1}(n-k)^{n-k-1}$$

1.3.16. Let S be the set of pairs consisting of a rooted tree with vertex set $[n]$ and a marked vertex of that tree. Count S in two ways to prove that

$$n^n = \sum_{k=0}^{n-1} \binom{n}{k} k^k (n-k)^{n-k-1}. \qquad \text{(Rey [1997])}$$

(Comment: Using this on forests, Smiley proved $n(n+1)^{n-1} = \sum_{k=0}^{n-1} \binom{n}{k}(k+1)^k(n-k)^{n-k-1}$, which appears in Riordan [1968, p. 116].)

1.3.17. (+) *Cayley's Formula generalized.* A **forest** is a graph whose components are trees. Let $\mathcal{F}(n,k)$ be the set of rooted forests with vertex set $[n]$ in which the roots of the components are the vertices $1, \ldots, k$, and let $a_{n,k} = |\mathcal{F}(n,k)|$. This exercise develops three distinct proofs that $a_{n,k} = kn^{n-k-1}$. The formula appears in Cayley [1889].
(a) Use induction on n, with basis $a_{n,n} = 1$. (Hint: Delete the roots.) (David [2007])
(b) Prove $(k+1)!a_{n,k} = k!nka_{n,k+1}$ and use $a_{n,n} = 1$. (Hint: The operation obtaining members of $\mathcal{F}(n, k+1)$ from members of $\mathcal{F}(n, k)$ will need to be symmetrized by permuting $[k+1]$.) (Lovász [1979, problem 4.14]).
(c) Generalize Exercise 1.3.12(a). For $F \in \mathcal{F}(n, k)$, form a list by iteratively deleting the largest (nonroot) leaf and recording its neighbor, until only the roots remain. Prove that this establishes a bijection from $\mathcal{F}(n, k)$ to $[n]^{n-k-1} \times [k]$.

1.3.18. ($\diamond$) Let F be a forest with vertex set $[n]$ and t components having $n_1, \ldots, n_t$ vertices. Prove that exactly $n^{t-2} \prod_{i=1}^{t} n_i$ trees with vertex set $[n]$ contain F. (Moon [1967])

1.3.19. (+) Let $\mathbf{M}$ be the set of pairings of $[2n]$, and let $\mathbf{T}$ be the set of "unordered" binary trees with leaf set $[n+1]$. Each tree has n unlabeled nonleaf vertices and $2n$ edges; nonleaf vertices have exactly two children. "Unordered" means that the tree is unchanged when we exchange left and right subtrees at a vertex.
(a) Show that $|\mathbf{M}| = (2n)!/(n!2^n)$.
(b) Prove bijectively that $|\mathbf{T}| = |\mathbf{M}|$. (Hint: First extend the labeling of the leaves of such a tree to the full vertex set. When $1, \ldots, r$ have been used, assign $r+1$ to the unlabeled vertex that, among those with two labeled children, has the child with the smallest label.) (Schröder [1870], Erdős–Székely [1989])

1.3.20. Let p be a prime, and let the p-ary expansions of positive integers n and k be given by $n = \sum a_i p^i$ and $k = \sum b_i p^i$.
(a) Use $(x+1)^p \equiv (x^p + 1) \pmod{p}$ to prove $\binom{n}{k} \equiv \prod \binom{a_i}{b_i} \pmod{p}$. (Lucas [1878])
(b) Use part (a) to determine when $\binom{n}{k}$ is odd. Determine also for which n the binomial coefficients $\binom{n}{0}, \ldots, \binom{n}{n}$ are all odd.

1.3.21. Let p be a prime. Part (a) of Exercise 1.3.20 implies $\binom{pm}{pl} \equiv \binom{m}{l} \pmod{p}$. Strengthen this by giving a combinatorial proof that $\binom{pm}{pl} \equiv \binom{m}{l} \pmod{p^2}$. (Hint: Choose points from an m-by-p grid.) (Stanley [1986, p. 53])

1.3.22. For $a \in \mathbb{N}$, use cyclic rotations of p-tuples to prove that p divides $a^p - a$ when p is prime. (Comment: This yields a combinatorial proof of Fermat's Little Theorem.)

1.3.23. Let p be a prime, and let $n_1, \ldots, n_m$ be integers summing to n. Consider their p-ary expansions: $n = \sum_{j=0}^{k} a_j p^j$ and $n_i = \sum_{j=0}^{k} a_{i,j} p^j$.
   (a) Prove that the multinomial coefficient $\binom{n}{n_1, \ldots, n_m}$ is not divisible by p if and only if $a_j = \sum_{i=1}^{m} a_{i,j}$ for $0 \le j \le k$. (Dickson [1902])
   (b) Prove that the number of terms in the expansion of $(x_1 + x_2 + \cdots + x_m)^n$ whose coefficients are not divisible by p is $\prod_j \binom{a_j + m - 1}{m-1}$. (Howard [1974])

1.3.24. By comparing sets of words, determine which coefficients in the expansion of $(x_1 + \cdots + x_k)^n$ are the largest.

1.3.25. ($\diamond$) Prove bijectively that n pairwise-intersecting lines in $\mathbb{R}^2$ (no three at a point) cut the plane into $1 + n + \binom{n}{2}$ regions. (Steiner [1826])

1.3.26. Generalize the formula of Exercise 1.3.25 by finding the number of regions in $\mathbb{R}^d$ formed by n hyperplanes in general position. That is, no $d + 1$ of them have a common point, and for $k \le d$ any k of them intersect in a $(d-k)$-dimensional plane. (Schläfli [1852])

1.3.27. In an election where A receives a votes and B receives b votes, determine the probability that A never trails B by more than k votes during the counting, assuming that all orderings of the votes are equally likely.

1.3.28. ($\diamond$) Establish a direct bijection from the binary trees with $n + 1$ leaves to the ordered trees with $n+1$ vertices (Definition 1.3.21). Below are the instances of each for $n = 2$. (Comment: By Theorem 1.3.23, the Catalan number C_n thus counts each set.) (Bernhart)

1.3.29. ($\diamond$) The bijections in Exercise 1.3.28 and Example 1.3.25 show that there are $\frac{1}{n+1}\binom{2n}{n}$ ordered trees with $n + 1$ vertices. Drawing them all yields a total of $\binom{2n}{n}$ vertices. Prove that exactly half of these vertices are leaves in their trees. (Shapiro [1999])

1.3.30. Prove that the map suggested in Example 1.3.24 is a bijection from the set of triangulations of a convex $(n + 2)$-gon to the set of binary trees with $n + 1$ leaves, thereby showing that there are C_n such triangulations.

1.3.31. ($\diamond$) Place $2n$ points on a circle. Prove bijectively that the number of ways to pair the points by drawing noncrossing chords equals the number of ballot lists of length $2n$. The possibilities for $n = 3$ appear below.

1.3.32. ($\diamond$) Let E_n and O_n, respectively, be the number of Dyck n-paths (Example 1.3.27) having an even or an odd number of peaks at even height. Compute E_n and O_n in terms of the Catalan numbers. (Deutsch [2005])

1.3.33. ($\diamond$) Prove that the number of ballot lists of length $2n$ that are unchanged when they are reversed and complemented is $\binom{n}{\lceil n/2 \rceil}$.

1.3.34. ($\diamond$) *Lattice walks in the plane.* (See Definition 1.2.1; lattice walks can move in any direction, while lattice paths only increase coordinates.)

(a) A **positive lattice walk** is a lattice walk that starts at the origin and never falls below the horizontal axis. Prove that the number of positive lattice walks of length k is $\sum_{j=0}^{k} \binom{k}{j}\binom{j}{\lfloor j/2 \rfloor} 2^{k-j}$. (Hint: Consider Lemma 1.3.14. Comment: This sum was shown combinatorially to equal $\binom{2k+1}{k}$ in Exercise 1.2.29 and will be evaluated by other means in both Exercise 2.1.52 and Exercise 3.2.46.)

(b) In three dimensions, determine the number of walks of length n that start at the origin and don't fall below the horizontal plane. (Deutsch [2000])

1.3.35. ($\diamond$) *Generalization of ballot paths.*

(a) Prove that the number of lattice paths from $(0,0)$ to $(n, n+k)$ that never pass above the line $y = x + k$ in the plane is $\frac{k+1}{n+k+1}\binom{2n+k}{n}$.

(b) A population grows from one individual of type k in generation 0. For $i \geq 0$, each individual of type i in generation t produces one individual with each type $1, \ldots, i+1$ in generation $t+1$. For $k \geq 0$ and $n \geq 1$, prove that the total number of individuals in generation n is the same as the answer in part (a). (Hint: Trace the ancestry of each individual.) (See Kupka [1990], Beckwith [2006b])

1.3.36. For $q \in \mathbb{N}$, determine the number of lattice paths from $(0,0)$ to (n, qn) not rising above the line $y = qx$.

1.3.37. ($\diamond$) Prove the Cycle Lemma (Theorem 1.3.20). (Hint: Use induction.)

1.3.38. Apply the Cycle Lemma (Theorem 1.3.20) to solve Bertrand's Ballot Problem (Example 1.3.12).

1.3.39. For $n \in \mathbb{N}$, count the lists $(a_1, \ldots, a_n)$ of positive integers with $a_1 = 1$ such that $a_i - a_{i-1}$ is odd and at most 1 for $i > 1$. For example, for $n = 5$ the lists are $\{12345, 12343, 12341, 12323, 12321, 12123, 12121\}$. (Deutsch [1999])

1.3.40. (+) Valid parenthesizations of $x_0 \div x_1 \div \cdots \div x_n$ correspond to the C_n binary trees with $n+1$ leaves. Each such expression evaluates to a fraction with some variables in the numerator (including x_0) and the others (including x_1) in the denominator.

(a) For $n \in \mathbb{N}$, determine which fractions occur most often.

(b) Prove that the number of occurrences of the most frequent fractions is the number of nondecreasing nonnegative integer lists $b_0, \ldots, b_{n-1}$ such that $b_i \leq i/2$ for $0 \leq i \leq n-1$.

(c) Count the lists described in part (b). (Callan [2004])

1.3.41. ($\diamond$) *Noncrossing partitions of* $[n]$. A partition of $[n]$ is **noncrossing** if there are no a, b, c, d with $a < b < c < d$ such that a and c are in one block and b and d are in another. Thus $(14|25|3)$ is crossing and $(15|234)$ is noncrossing.

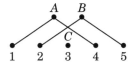   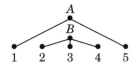

Establish a bijection from the set of noncrossing partitions of $[n]$ to the set of ballot lists of length $2n$. From the bijection, argue that the number of noncrossing partitions of $[n]$ with k blocks equals the number of ballot lists of length $2n$ with k runs of 1. Thus the value is the Narayana number $N_{n,k}$ of Theorem 1.3.26, counting the ordered trees with n edges and k leaves. (Prodinger [1983])

1.3.42. For $n, t \in \mathbb{N}$, count the nondecreasing integer lists $(a_1, \ldots, a_n)$ such that $1 \leq a_i \leq ti$ for $1 \leq i \leq n$. (Deutsch [2004a])

1.3.43. Establish a bijection from the set of q-ballot lists of length $(q + 1)n$ to the set of ordered $(q + 1)$-ary trees with $qn + 1$ leaves. The trees correspond to groupings of $qn + 1$ items, combining $q + 1$ at a time, shown below for $q = n = 2$. (Sands [1978])

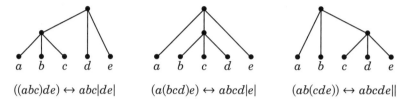

$$((abc)de) \leftrightarrow abc|de| \qquad (a(bcd)e) \leftrightarrow abcd|e| \qquad (ab(cde)) \leftrightarrow abcde||$$

1.3.44. Consider an election where A receives kn votes and B receives n votes. Given that the vote order is random, let p be the probability that throughout the counting the number of votes recorded for A is always at most k times the number of votes recorded for B. Let q be the analogous probability with "at most" replaced by "at least". Prove $p = q$.

1.3.45. ($\diamond$) *Counting q-ballot paths by the number of turns.*
(a) Let r and s be relatively prime positive integers. Prove that the number of lattice paths from $(0,0)$ to (r, s) that do not rise above the line $ry = sx$ and turn exactly $2k - 1$ times is $\frac{1}{k}\binom{r-1}{k-1}\binom{s-1}{k-1}$.
(b) For $q \in \mathbb{N}$, prove that the number of lattice paths from $(0,0)$ to (n, qn) that do not rise above the line $y = qx$ and turn exactly $2k - 1$ times is $\frac{1}{k}\binom{n-1}{k-1}\binom{qn}{k-1}$. (Cigler [1987]; see also Krattenthaler [1997])

1.3.46. ($\diamond$) Let $x_1, \ldots, x_n$ be a cyclic arrangement of integers with sum 1.
(a) Prove that for all $r \in [n]$, there is exactly one place to break the cycle into a linear arrangement with exactly r positive partial sums. Let p_j be the position from which j partial sums are positive. Show that if $r > s$, then the positions ending positive partial sums from position p_r include all the positions ending positive partial sums from p_s. For example, for $[x_1, \ldots, x_5] = [3, -4, 1, 2, -1]$, we have $(p_1, p_2, p_3, p_4, p_5) = (2, 5, 1, 4, 3)$. For the ending positions of the positive partial sums from each p_i, we have $p_1 = 2$: $\{1\}$; $p_2 = 5$: $\{1, 4\}$; $p_3 = 1$: $\{1, 4, 5\}$; $p_4 = 4$: $\{1, 3, 4, 5\}$; $p_5 = 3$: $\{1, 2, 3, 4, 5\}$.
(b) Let all lists of n As and n Bs be equally likely. Let X be the random variable counting the values i such that the ith A precedes the ith B. Apply part (a) to prove that $\text{Prob}(X = l) = 1/(n + 1)$ for each $l \in \{0, \ldots, n\}$. (Chung–Feller [1949])
(Comment: Many results similar or equivalent to parts of (a) have appeared: see Spitzer [1956], Raney [1960], Kierstead–Trotter [1988], Montágh [1991], Snevily–West [1998], and Chapter 3 of Mohanty [1979].)

1.3.47. ($\diamond$) Consider $f: [n] \to [n]$. Drivers $1, \ldots, n$ in order enter a street with parking spots $1, \ldots, n$. Driver i starts looking for parking at spot $f(i)$ and takes the first open spot. If none is open from $f(i)$ to n, then Driver i fails. The functions f such that all drivers succeed are **parking functions**. For example, if $f([9]) = (6, 4, 4, 3, 6, 8, 3, 1, 1)$ in order, then all nine drivers succeed, parking in order in spots $(6, 4, 5, 3, 7, 8, 9, 1, 2)$. Prove that f is a parking function if and only if there is a permutation σ of $[n]$ such that $f(i) \leq \sigma(i)$ for all i. (Riordan; see Knuth [1973, p. 545]) (Comment: More than 200 papers have been written about parking functions and related topics; see Yan [2015] for a survey.)

1.3.48. ($\diamond$) Prove that the number of parking functions on $[n]$ is $(n + 1)^{n-1}$ (Pyke [1959], Konheim–Weiss [1966]) (Hint: Add a position $n + 1$, which can be in the image of f, and move cars cyclically until they park; this proof is due to Pollak in the 1970s. Comment: Foata–Riordan [1974] gave a correspondence with labeled trees.)

Chapter 2

Recurrence Relations

Often the solutions to a counting problem can be built from solutions to a smaller problem of the same type. For example, let a_n count the 0, 1-lists of length n. A list of length $n-1$ can extend by acquiring a 0 or a 1. The resulting lists are distinct, and every list of length n arises in this way. Therefore, $a_n = 2a_{n-1}$ for $n \geq 1$. With $a_0 = 1$ (there is one way to "do nothing"), induction yields $a_n = 2^n$.

This example shows the steps in recursive solution of counting problems. Combinatorial arguments express a_n in terms of earlier values and a function of n. Along with initial values, this "recurrence" determines the sequence. "Solving" the recurrence means obtaining an exact or asymptotic formula for a_n.

A solution formula for a recurrence can be verified by induction. When no formula is apparent, solutions may arise by other techniques: the "characteristic equation method" for special recurrences, a more general "generating function method", substitution techniques, special methods for asymptotic solutions, etc.

We first concentrate on recurrences with one (integer) parameter. We may refer to the sequence of solutions to a counting problem as a **counting sequence**.

2.0.1. Definition. A **recurrence relation** or *recurrence* for a sequence $a_0, a_1, \ldots$ is an expression of the form $a_n = g(n, a_0, \ldots, a_{n-1})$. It has **order** k or **degree** k if the formula for a_n depends only on n and $a_{n-1} \ldots a_{n-k}$. The recurrence $a_n = g_1(n)a_{n-1} + \cdots + g_k(n)a_{n-k} + f(n)$ is **linear** if each g_i (and f) is independent of $\langle a \rangle$. Here $f(n)$ is the **inhomogeneous term**. If f is identically 0, then the relation is **homogeneous**; otherwise, it is **inhomogeneous**.

2.0.2. Example. Below we illustrate these definitions using several recurrences that we will derive for combinatorial problems. ∎

$a_n = a_{n-1} + n$	inhomogeneous, linear, order 1
$a_n = a_{n-1} + a_{n-2}$	homogeneous, linear, order 2
$a_n = (n-1)(a_{n-1} + a_{n-2})$	homogeneous, linear, order 2
$a_n = \sum_{k=1}^{n} a_{k-1}a_{n-k}$	homogeneous, nonlinear, no finite order

2.0.3. Remark. A recurrence relation does not by itself specify a sequence; for example, every constant sequence satisfies $a_n = a_{n-1}$. A recursive definition must include initial values that enable the recursive computation to proceed. A recurrence of order k requires k initial values to specify a sequence. We say that the recursive formula is *valid* for larger values. Initial values correspond to the basis step in inductive proofs; verifying that an explicit formula for a_n gives the initial values is the basis step in an inductive proof that the formula is correct for all n.

In some contexts $g(a_0, \ldots, a_n) = 0$ is a more natural form for recurrences. Here a recurrence is an operator, and one finds the sequences annihilated by it (its "nullspace"). Linear recurrences of order k then become $\sum_{i=0}^{k} g_i(n)a_{n-i} = f(n)$; the signs in the coefficients have changed. This convention is useful in the theory of solving recurrences, but in this text we emphasize the combinatorial aspects of describing a set using "earlier" instances. Thus we choose the "$a_n =$" form. ∎

2.0.4.* Remark. When studying integer sequences, one should know about the On-Line Encyclopedia of Integer Sequences (OEIS), developed by Neil J. A. Sloane and located at http://oeis.org. It contains enormous amounts of information about nearly every counting sequence in this book, including the Fibonacci numbers (sequence A000045), Catalan numbers (A000108), Derangement numbers (A000166), central Delannoy numbers (A001850), and many others. Readers can consult the OEIS for more information about special sequences and contribute interesting sequences not yet among the more than 100,000 sequences in the database. The welcome page describes the database and gives instructions for many ways of interacting with it. ∎

2.1. Obtaining Recurrences

Obtaining a recurrence for a counting problem usually involves bijective arguments. Some creativity may be needed; there is no algorithm. We give classical examples to illustrate typical arguments.

CLASSICAL EXAMPLES

In the simplest situation, our problem has one parameter, and we seek a single recurrence for the resulting counting sequence.

2.1.1. Example. *Regions in the plane.*
$$a_n = a_{n-1} + n \text{ for } n \geq 1, \text{ with } a_0 = 1.$$

Let an *n-configuration* be a set of n lines in the plane such that every two lines have one common point but no three lines have a common point. Let a_n be the number of regions formed by an n-configuration. It is not obvious that every n-configuration forms the same number of regions; this follows inductively when we establish a recurrence for a_n.

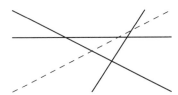

With no lines, $a_0 = 1$. To prove $a_n = a_{n-1} + n$ for $n \geq 1$ (and that all n-configurations form the same number of regions), consider an n-configuration with $n \geq 1$, and let L be one line. The other lines form an $(n-1)$-configuration; by the induction hypothesis, they create a_{n-1} regions. The intersections of L with the other $n-1$ lines cut L into n parts; each splits a region. Thus adding L increases the number of regions by n, and all n-configurations have $a_{n-1} + n$ regions. ∎

2.1.2. Example. *Fibonacci numbers.*

$$F_n = F_{n-1} + F_{n-2} \text{ for } n \geq 2, \text{ with } F_0 = 0 \text{ and } F_1 = 1.$$
$$\hat{F}_n = \hat{F}_{n-1} + \hat{F}_{n-2} \text{ for } n \geq 2, \text{ with } \hat{F}_0 = \hat{F}_1 = 1.$$

Imagine a linear parking lot with n spaces filled by two types of cars. Rabbits (sold by Volkswagen in the United States and Canada beginning in 1974) are small cars taking one space. Cadillacs (manufactured since 1902 and by General Motors since 1909) are large cars taking two spaces. Let $\hat{F}_n$ count the distinguishable ways to fill the parking lot. These correspond to the $1, 2$-lists with sum n (also called compositions of n using parts in $\{1, 2\}$); we emphasize this combinatorial model. The resulting sequence solves many problems; for example, $\hat{F}_n$ also counts the binary $(n-1)$-tuples with no consecutive 1s (Exercise 14).

In a $1, 2$-list with sum n, the last element may be 1 or 2. There are $\hat{F}_{n-1}$ ways to fill the earlier part if the list is 1 and $\hat{F}_{n-2}$ ways if it is 2. Thus $\hat{F}_n = \hat{F}_{n-1} + \hat{F}_{n-2}$ for $n \geq 2$. Only the empty list has sum 0, and one list has sum 1, so $\hat{F}_0 = \hat{F}_1 = 1$.

Changing the initial conditions to $F_0 = 0$ and $F_1 = 1$ yields another sequence $\langle F \rangle$. Since $F_1 = \hat{F}_0$ and $F_2 = \hat{F}_1$, inductively $F_n = \hat{F}_{n-1}$ for $n \geq 1$. The sequence $\langle F \rangle$ is the classical **Fibonacci sequence** introduced in 1202 by Leonardo of Pisa (ca. 1170–1250), known as Fibonacci. The Fibonacci recurrence $F_n = F_{n-1} + F_{n-2}$ is a natural model for growth from two previous stages of a process and occurs frequently in nature. The journal *The Fibonacci Quarterly* is devoted to the Fibonacci numbers and related topics. Extensive material on their history, mathematics, and applications appears in Koshy [2001]; we study them in Exercises 14–34.

Since $F_n = \hat{F}_{n-1}$, both sequences use the same numbers, and we call $\langle \hat{F} \rangle$ the **adjusted Fibonacci numbers**. Benjamin–Quinn [2003] also uses special notation to distinguish the adjusted and classical indexings, writing f_n for our $\hat{F}_n$.

The initial conditions $F_0 = 0$ and $F_1 = 1$ are more common in the literature and are appropriate in number theory due to divisibility properties (such as Exercise 22). Other initial conditions have been studied; the **Lucas numbers** arise from the Fibonacci recurrence with initial conditions $L_1 = 1$ and $L_2 = 3$; see Exercise 27. (The term "Fibonacci numbers" was popularized by Lucas.) ∎

Many identities can be proved easily by induction, but combinatorial arguments that present a set counted by both formulas can be more illuminating or provide more detailed results (see Theorem 1.3.26, for example). We illustrate this notion with Fibonacci numbers.

2.1.3. Application. *Combinatorial arguments for Fibonacci identities.* The initial conditions we use for $\langle \hat{F} \rangle$ facilitate combinatorial arguments using the model of 1, 2-lists with sum n. For example, consider the identity $\sum_{i=0}^{n} \hat{F}_i^2 = \hat{F}_n \hat{F}_{n+1}$ for $n \geq 0$. One can easily prove this by induction after replacing $\hat{F}_{n+1}$ with $\hat{F}_n + \hat{F}_{n-1}$.

Combinatorially, the product $\hat{F}_n \hat{F}_{n+1}$ counts the ways to choose two 1, 2-lists, with sums n and $n+1$ (as in a parking lot with two rows). In a 1, 2-list as a tiling with segments of lengths 1 and 2, we view each partial sum as a "breakpoint". In $\hat{F}_n^2$ of the pairs, both lists have a breakpoint at n; these are the pairs where the second list ends with 1. Among those where the second list ends with 2, there are $\hat{F}_{n-1}^2$ where both lists have a breakpoint at $n-1$, and so on.

When the last common breakpoint is known, there is only one way to complete the lists after that; it uses one 1 and the rest 2s. Thus there are exactly $\hat{F}_i^2$ pairs where the last common breakpoint is at i, which proves the identity.   ■

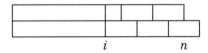

Next we turn to the famous **Hat-Check Problem**. If n people check their hats and retrieve them at random, what is the probability that each one gets someone else's hat? The new assignment of hats gives a permutation of the index set $[n]$, expressed as a bijection from $[n]$ to $[n]$. In the functional digraph (Definition 1.3.3), each number has one successor and one predecessor. Hence the functional digraph consists of disjoint cycles. Applying the function repeatedly sends an element x around its cycle. The successive elements form the **orbit** of x.

2.1.4. Definition. The **cycles** of a permutation are the orbits of elements under iteration of the permutation; these are the (ordered) vertex sets of the cycles in the functional digraph. A **fixed point** is a cycle of length 1. A **derangement** is a permutation with no fixed points. The number of derangements of an n-set is denoted by D_n.

2.1.5. Example. Viewing 641253798 as a permutation of $[9]$, the cycles are (163), (24), (89), (5), and (7), with 5 and 7 being fixed points.   ■

In the Hat-Check Problem, all $n!$ permutations are equally likely, and we need to count those with no fixed points. The word "derangement" is used because it is a variation on "arrangement" and because "deranged" means "crazy" or "mixed-up": that is, nothing is in its place.

2.1.6. Example. *Derangements of n objects.*
$$D_n = n! - \sum_{k=1}^{n} \binom{n}{k} D_{n-k} \text{ for } n \geq 1, \text{ with } D_0 = 1.$$
$$D_n = (n-1)(D_{n-1} + D_{n-2}) \text{ for } n \geq 2, \text{ with } D_0 = 1 \text{ and } D_1 = 0.$$

One way to form the $n!$ permutations of $[n]$ is to choose the fixed points and then derange the other elements. If we fix all the points, then there is one way to derange the remaining 0 elements, so it is sensible to set $D_0 = 1$. Now counting two ways yields $n! = \sum_{k=0}^{n} \binom{n}{k} D_{n-k}$, which is the first recurrence above.

This recurrence needs more computation for each successive value; fixed order is preferable. With $D_0 = 1$ and $D_1 = 0$ by inspection, we derive a second-order recurrence valid for $n \geq 2$. We count the ways to partition $[n]$ into cycles of length at least 2. There are two cases.

If the cycle containing the element n has length 2, then there are $n - 1$ ways to pick its other element, and there are D_{n-2} ways to derange the remaining elements (that is, to put them into cycles of length at least 2). Hence there are $(n-1)D_{n-2}$ derangements of this type.

If the cycle containing n is longer, then skipping it in its cycle still leaves cycles of length at least 2 and produces a derangement of $[n-1]$. On the other hand, every derangement of $[n]$ with n in such a cycle arises from a derangement of $[n-1]$ by inserting n immediately following some $x \in [n-1]$ on the cycle containing x. Hence there are $(n-1)D_{n-1}$ derangements of this type. ∎

Derangements are studied in Exercises 40–46. The recurrences above are linear (Definition 2.0.1); our next recurrence is not.

2.1.7. Example. *Catalan numbers and ballot paths.*

$$C_n = \sum_{k=1}^{n} C_{k-1} C_{n-k} \text{ for } n \geq 1, \text{ with } C_0 = 1.$$

In Section 1.3 we counted the ballot paths: lattice paths from $(0,0)$ to (n,n) that don't rise above the diagonal. We defined C_n to be the number of these paths and proved that $C_n = \frac{1}{n+1}\binom{2n}{n}$.

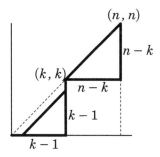

Every ballot path to (n,n) has some first return to the diagonal, say at (k,k). The part up to (k,k) begins rightward and then does not rise above $y = x - 1$ until it reaches $(k, k-1)$. Hence the possible initial portions of the path correspond bijectively to the ballot paths of length $2k - 2$. The portion from (k,k) to (n,n) is a translation of a ballot path of length $2n - 2k$. Hence exactly $C_{k-1}C_{n-k}$ ballot paths of length $2n$ first return to the diagonal at (k,k). Summing over the choices for k yields $C_n = \sum_{k=1}^{n} C_{k-1}C_{n-k}$ for $n \geq 1$, with initial condition $C_0 = 1$. This is the **Catalan recurrence**. The sequence begins $1, 1, 2, 5, 14, \ldots$. Only the specification of C_0 is needed as an initial condition. ∎

2.1.8. Remark. *Catalan numbers.* Exercises 47–56 study the Catalan numbers. In Section 1.3 we used bijections to show that this sequence solves many counting problems. Another method, often easier, is to prove that a new problem satisfies the Catalan recurrence and initial condition; the counting sequence $\langle a \rangle$ must then also be the Catalan sequence. As in combinatorial arguments for sums, the key step in proving $a_n = \sum_{k=1}^{n} a_{k-1}a_{n-k}$ is interpreting the index of summation so that the resulting subset of objects is counted by the desired summand.

For example, let a_n count the triangulations of a convex $(n+2)$-sided polygon with vertices $v_0, \ldots, v_{n+1}$ in order. Consider $n \geq 1$. In every triangulation, edge $v_{n+1}v_0$ belongs to some triangle; let v_k be its third corner. To complete the triangulation, we must triangulate the polygons formed by $v_0, \ldots, v_k$ and by $v_k, \ldots, v_{n+1}$, with $k+1$ and $n-k+2$ sides, respectively. They can be triangulated in a_{k-1} and a_{n-k} ways. Summing over k yields $a_n = \sum_{k=1}^{n} a_{k-1}a_{n-k}$ for $n \geq 1$, with $a_0 = 1$ (a 2-sided polygon has one (empty) triangulation). Hence $a_n = C_n$ for all n.

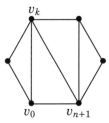

Showing $a_0 = 1$ and $a_n = \sum_{k=1}^{n} a_{k-1}a_{n-k}$ for $n > 0$ gives more than $a_n = C_n$. We also obtain a bijection from the given problem to any sequence of sets counted by the Catalan numbers. The bijection applies recursively to the objects with index $k-1$ and $n-k$ combined to form an object with index n. Thus it also matches the sets for the kth term in the recurrences for the two problems. For example, in Theorem 1.3.23 we used the position of the 0 that matches the initial 1 in a ballot list to recursively invert the bijection that maps binary trees to ballot lists. ∎

VARIATIONS

Problems having more than one parameter may lead to recurrences with more than one index. As with derangements and Fibonacci numbers, often one can derive recurrences for problems involving distinct objects or positions by studying how the last object or position is used.

2.1.9. Example. *Arrangements with n distinct available objects.*

$C(n,k) = C(n-1,k) + C(n-1,k-1)$	$C(n,k) = $ # k-subsets of $[n]$
$P(n,k) = P(n-1,k) + kP(n-1,k-1)$	$P(n,k) = $ # simple k-words from $[n]$
$S(n,k) = kS(n-1,k) + S(n-1,k-1)$	$S(n,k) = $ # partitions of $[n]$ with k blocks
$c(n,k) = (n-1)c(n-1,k) + c(n-1,k-1)$	$c(n,k) = $ # permutations of $[n]$ with k cycles

$C(n,k)$: When we select k from n elements, we may use or not use the nth element. This is Pascal's Formula for the binomial coefficients.

$P(n,k)$: When we list k from n elements, we may use or not use the nth element, and if we use it there are k locations it can occupy.

$S(n,k)$: When we partition $[n]$ into k blocks, the element n may or may not form a block itself. There are $S(n-1, k-1)$ partitions where it is alone. Otherwise, it may join any block in a partition of $[n-1]$ into k blocks; being nonempty, these are distinguishable by their members.

$c(n, k)$: When we permute $[n]$ using k cycles, the element n may or may not form a cycle by itself. There are $c(n - 1, k - 1)$ permutations where it is alone. Otherwise, there are $n - 1$ elements it can follow to turn a permutation of $[n-1]$ with k cycles into a permutation of $[n]$ with k cycles. (Compare this with the argument in Example 2.1.6.)

Each of these recurrences is valid for $n \geq 1$ and computes the value for (n, k) using values for $(n - 1, k)$ and $(n - 1, k - 1)$. Hence it suffices to specify initial values for $\{(0, k): k \in \mathbb{N}_0\}$. In each case, there is one construction when $k = n = 0$ and none when $k > n = 0$. Thus inductively the values are 0 whenever $k > n$. ∎

Next we find a recurrence for the size of the lattice ball (Definition 1.2.11). The solution $\sum \binom{m}{k}\binom{n}{k}2^k$ was found bijectively in Proposition 1.2.12. The first proof illustrates a useful technique: sometimes a summation can be eliminated from a recurrence by taking the difference of two instances of the recurrence. The second more directly obtains the desired recurrence.

2.1.10. Proposition. The number of lattice points within m lattice steps from the origin in $\mathbb{R}^n$ satisfies the recurrence $l_{m,n} = l_{m,n-1} + l_{m-1,n-1} + l_{m-1,n}$ for $m, n > 0$, with $l_{m,0} = l_{0,n} = 1$.

Proof: Here $l_{0,n}$ counts the all-0 vector and $l_{m,0}$ counts the vector with no entries. Let $B_{m,n} = \{(a_1, \ldots, a_n) \in \mathbb{Z}^n : \sum_{i=1}^n |a_i| \leq m\}$.

Proof 1 Group $B_{m,n}$ by the last coordinate. There are $l_{m-|k|,n-1}$ points in $B_{m,n}$ with $a_n = k$ and $l_{m-|k|,n-1}$ with $a_n = -k$. Thus $l_{m,n} = l_{m,n-1} + 2\sum_{k=1}^m l_{m-k,n-1}$. The same argument yields $l_{m-1,n} = l_{m-1,n-1} + 2\sum_{k=1}^{m-1} l_{m-1-k,n-1}$. Subtracting the second equation from the first yields $l_{m,n} - l_{m-1,n} = l_{m,n-1} + l_{m-1,n-1}$.

Proof 2 Among the groups in Proof 1, there are $l_{m,n-1}$ vectors with $a_n = 0$ and $l_{m-1,n-1}$ with $a_n = 1$. We claim that $l_{m-1,n}$ vectors remain. For each remaining vector, bringing its last coordinate one unit closer to 0 yields a vector in $B_{m-1,n}$. This produces each vector in $B_{m-1,n}$ exactly once; for example the vectors in $B_{m-1,n}$ with last coordinate 0 correspond to the vectors in $B_{m,n}$ with last coordinate -1. Thus there are $l_{m-1,n}$ vectors remaining, and we obtain $l_{m,n} = l_{m-1,n} + l_{m,n-1} + l_{m-1,n-1}$. ∎

2.1.11. Remark. *Delannoy paths and lattice ball.* Recall that the Delannoy number $d_{m,n}$ is the number of paths from $(0,0)$ to (m, n) that move by one of $\{(1, 0), (0, 1), (1, 1)\}$ at each step (Definition 1.2.8). The analogue of Pascal's Formula for Delannoy numbers is the recurrence $d_{m,n} = d_{m,n-1} + d_{m-1,n} + d_{m-1,n-1}$, valid for $m, n \geq 1$. Also $d_{m,0} = d_{0,n} = 1$.

Proposition 2.1.10 obtains the same recurrence and initial conditions for $l_{m,n}$. This gives a short proof (without finding a bijection) that the Delannoy numbers also count the points in the lattice ball, or a short proof of Theorem 1.2.13 that $\sum_k \binom{m}{k}\binom{n+k}{m} = \sum_k \binom{m}{k}\binom{n}{k}2^k$ if we instead counted the two sets separately as in Proposition 1.2.10 and Proposition 1.2.12. ∎

When a single recurrence for a sequence is not apparent, it may help to derive a system of recurrence relations for several sequences. Manipulating the system may then yield a recurrence for the original problem.

2.1.12. Example. *Systems of recurrences.*

$$\{a_n = b_n + c_n, \ b_n = a_{n-1}, \ c_n = b_{n-1}\} \text{ for } n \geq 1, \text{ with } a_0 = b_0 = 1.$$

Let a_n count the binary n-tuples with no consecutive 1s. Let c_n count those ending in 1, and let b_n count the rest. By definition, $a_n = b_n + c_n$. The lists ending in 0 arise by appending 0 to a shorter good list, but those ending in 1 can append 1 only to lists ending in 0. This yields the expressions for b_n and c_n. Substituting to eliminate b and c yields a single recurrence for a. It is the Fibonacci recurrence $a_n = a_{n-1} + a_{n-2}$ (obtained directly in Exercise 14). ∎

Recurrences may be indexed by objects that are not numbers. Each object must be evaluated in terms of objects evaluated earlier; finitely many computations must suffice to evaluate any object from the initial conditions. We consider a recurrence on graphs that evaluates each in terms of graphs with fewer edges.

2.1.13.* Example. A **spanning tree** in a graph G is a subgraph that is a tree and contains all the vertices; let $\tau(G)$ be the number of spanning trees. Cayley's Formula gives $\tau(K_n) = n^{n-2}$ (Theorem 1.3.4). For generalized graphs (loops and repeated edges allowed), we develop a recurrence.

For $e \in E(G)$, the graph $G - e$ is obtained from G by deleting e from the edge set. The graph $G \cdot e$ is obtained by deleting e and replacing its endpoints with a single vertex incident to all edges formerly incident to e. This is called **contracting** e; visually, e shrinks to a point. This may introduce loops and repeated edges and reduces the number of edges by 1.

$$\begin{aligned} \tau(G) &= \tau(G - e) + \tau(G \cdot e) && \text{when } e \text{ is not a loop in } G, \\ \tau(G) &= 0 && \text{when } G \text{ is disconnected,} \\ \tau(G) &= 1 && \text{when } G \text{ has one vertex.} \end{aligned}$$

The right side counts the spanning trees T by whether e is used. Those not containing e are the spanning trees of $G - e$. If T contains e, then the edges of $T - e$ form a spanning tree in $G \cdot e$. Conversely, adding e to the edge set of a spanning tree T' in $G \cdot e$ forms a spanning tree in G, since T' cannot include a path in G connecting the endpoints of e. Thus the number of spanning trees in G that contain e is $\tau(G \cdot e)$.

Initial conditions are needed when all edges are loops. Here we have one spanning tree if G has one vertex and none if G has more vertices.

Below, the (multi)graphs on the right each have four spanning trees, and the graph on the left has eight. The recurrence is useful in some special classes, but generally it requires exponentially many computations (in the number of edges). In Chapter 15 we develop a much faster method using determinants. Exercises 59–62 study $\tau(G)$ for special graphs. ∎

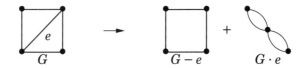

EXERCISES 2.1

2.1.1. (−) Obtain a recurrence relation for the number of ways to tile a 2-by-n checkerboard with n identical dominoes.

2.1.2. (−) Obtain a recurrence for the number of pairings of $2n$ people.

2.1.3. (−) Fix $r \in \mathbb{N}$. Let a_n be the number of regions formed by n lines in the plane, given that r of them are parallel, the other $n - r$ lines each intersect $n - 1$ lines, and no three lines have a common point. Obtain a recurrence for $\langle a \rangle$.

2.1.4. (−) Determine the number of binary n-tuples in which every run has odd length, where a **run** is a maximal string of consecutive equal entries. (Beckwith [2014])

2.1.5. (−) Count the symmetric 1, 2-lists with sum n (like 1221). (Alladi–Hoggatt [1975])

2.1.6. (−) A club grows as follows: at time 0 it has one man and no woman. At time t each man selects one woman to join, and each woman selects one man and one woman to join. What is the size of the club at time n, assuming that people never die?

2.1.7. (−) Let $\langle a \rangle$ satisfy $a_n = a_{n-1} + a_{n-2} + a_{n-3}$ for $n > 3$. Prove that $a_n \leq 2^{n-2}$ for $n \geq 2$ if $a_i = 1$ for $i \in \{1, 2, 3\}$, and $a_n < 2^n$ if $a_i = i$ for $i \in \{1, 2, 3\}$.

2.1.8. (−) Let $a_{n,k} = \binom{2n}{2k}/\binom{n}{k}$. Use the formula for the binomial coefficients to compute $a_{n,k}$ recursively in terms of $a_{n,k-1}$ and in terms of $a_{n-1,k}$, including initial conditions.

2.1.9. (−) Let D_n be the number of derangements of $[n]$, and let E_n be the number of permutations of $[n]$ having exactly one fixed point. Compute $D_n - E_n$ as a function of n.

2.1.10. An *arc diagram* consists of a binary n-tuple plus arcs joining some pairs of positions. The arcs can only join positions with opposite values, and each position is the left element of at most one arc. Below is an arc diagram of length 9. Determine the number of arc diagrams of length n. (Callan [2008])

$$0 \quad 1 \quad 1 \quad 0 \quad 1 \quad 0 \quad 1 \quad 1 \quad 0$$

2.1.11. Fix $r \in \mathbb{N}$. Let a_n be the number of outcomes of n coin flips (in order) that contain exactly r heads. Obtain a recurrence relation (in one variable!) for $\langle a \rangle$.

2.1.12. Let a_n count the binary n-tuples not having 0, 1, 1 consecutively. Find a system of recurrences for a_n (see Example 2.1.12), and reduce it to a single recurrence for $\langle a \rangle$.

2.1.13. (◊) A **Schröder n-path** is a path from $(0, 0)$ to $(2n, 0)$ in the first quadrant by steps in $\{(1, 1), (2, 0), (1, -1)\}$. An *uprun* is a maximal upward segment; the example below has two. One n-path has no uprun, and one has n. Let a_n and b_n count the Schröder n-paths having one uprun and $n - 1$ upruns, respectively ($\langle a \rangle$ begins $0, 1, 4, 11, \ldots$, and $\langle b \rangle$ begins $0, 1, 4, 9, \ldots$). Derive first-order recurrence relations for $\langle a \rangle$ and for $\langle b \rangle$. (Deutsch [2004b])

2.1.14. For $n \geq 1$, let a_n be the number of binary $(n - 1)$-tuples with no consecutive 1s. Give two proofs that $a_n = \hat{F}_n$, as follows.
  (a) Show that $\langle a \rangle$ satisfies the same recurrence and initial conditions as $\hat{F}_n$.
  (b) Establish a bijection from the set of binary $(n - 1)$-tuples with no consecutive 1s to the set of 1, 2-lists with sum n.

2.1.15. Two adjacent panes of glass produce multiple images because some light reflects each time a surface is encountered. Let a_n be the number of ways an image can reflect n times *from inner surfaces* before emerging. As shown below, $a_0 = 1$, $a_1 = 2$, $a_2 = 3$. Obtain a recurrence relation for a_n. (Moser [1963])

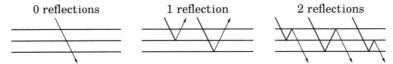

0 reflections　　　　　　　　1 reflection　　　　　　　2 reflections

2.1.16. For $n \geq 5$, consider the digraph formed from n points around a circle by adding an edge from each point to the next point and to the point after that. Count the cycles (as subgraphs) in the digraph (cycles must follow arrows). In particular, when $n = 5$ the answer is 12. (Juvan–Mohar–Škrekovski [1998b])

2.1.17. Prove the following by induction and by combinatorial argument.

$$(a)\ \hat{F}_n = \sum_{i=0}^{n} \binom{n-i}{i}. \qquad (b)\ 1 + \sum_{i=0}^{n} \hat{F}_i = \hat{F}_{n+2}.$$

2.1.18. Give inductive and combinatorial proofs of (a). Use it to prove (b).

$$(a)\ \sum_{i=0}^{n} \hat{F}_{2i} = \hat{F}_{2n+1}. \qquad (b)\ \sum_{i=0}^{2n-1} (-1)^i \hat{F}_{2n-i} = \hat{F}_{2n-1}.$$

2.1.19. ($\diamond$) Prove that the classical Fibonacci number F_n counts the compositions of n using odd parts and the compositions of $n + 1$ using parts greater than 1.

2.1.20. ($\diamond$) **Cassini's Identity**. Prove $\hat{F}_n^2 = \hat{F}_{n-1}\hat{F}_{n+1} + (-1)^n$ for $n \geq 1$, both inductively and combinatorially. (Comment: For classical Fibonacci numbers, this is $F_m^2 = F_{m-1}F_{m+1} + (-1)^{m-1}$, proved by Cassini in 1680). Use the identity to explain Lewis Carroll's "proof" of 64 = 65 (and larger analogues). (Weaver [1938])

2.1.21. ($\diamond$) For $m, n \in \mathbb{N}_0$ with $m \geq n$, prove $\hat{F}_n\hat{F}_m - \hat{F}_{n-1}\hat{F}_{m+1} = (-1)^n \hat{F}_{m-n}$ inductively and combinatorially. (Comment: This identity generalizes Cassini's Identity. When reindexed and written using classical Fibonacci numbers as $F_{n+1}F_m - F_nF_{m+1} = (-1)^n F_{m-n}$, it is known as **d'Ocagne's Identity**.)

2.1.22. ($\diamond$) Prove bijectively that $\hat{F}_{m+n} = \hat{F}_m\hat{F}_n + \hat{F}_{m-1}\hat{F}_{n-1}$. Conclude for $k \in \mathbb{N}$ that $\hat{F}_{n-1}$ divides $\hat{F}_{kn-1}$, and hence F_n divides F_{kn}.

2.1.23. For classical Fibonacci numbers, Exercise 2.1.22 says $F_{m+n+1} = F_{m+1}F_{n+1} + F_mF_n$. Use this and Cassini's Identity (Exercise 2.1.20) to prove **Catalan's Identity** for $a < b$: $F_b^2 - F_{b-a}F_{b+a} = (-1)^{b-a}F_a^2$. (Comment: See Melham–Shannon [1995] and Rao [1953].)

2.1.24. ($\diamond$) **Vajda's Identity**. Consider $r, s, t \in \mathbb{N}$ with $r < s$.
(a) Give a combinatorial proof of $\hat{F}_r\hat{F}_{s-1} - \hat{F}_{r-1}\hat{F}_s = (-1)^r\hat{F}_{s-r-1}$.
(b) Give a combinatorial proof of $\hat{F}_{r+t}\hat{F}_s - \hat{F}_r\hat{F}_{s+t} = (-1)^{r-1}\hat{F}_{t-1}\hat{F}_{s-r-1}$.
(c) For classical Fibonacci numbers, conclude Vajda's Identity for $i, j, n \in \mathbb{N}$:
$$F_{n+i}F_{n+j} - F_nF_{n+i+j} = (-1)^n F_i F_j.$$
(Comment: This is a common generalization of d'Ocagne's Identity and Catalan's Identity; see Basin–Hoggatt [1963]. It was attributed by Dickson to Tagiuri [1900] and was rediscovered in Everman–Danese–Venkannayah [1960].)

2.1.25. Extend classical Fibonacci numbers to negative indices via the recurrence. Prove $F_{-n} = (-1)^{n-1}F_n$ for $n \in \mathbb{N}$. (Graham–Knuth–Patashnik [1989, p. 279])

2.1.26. ($\diamond$) Prove that every positive integer has a unique expression as the sum of a set of nonconsecutive Fibonacci numbers (1 and 2 cannot both be used). (Lekkerkerker [1952], Zeckendorf [1972])

2.1.27. The Lucas numbers $\langle L \rangle$ satisfy the Fibonacci recurrence with $L_1 = 1$ and $L_2 = 3$. Prove $L_n = F_{n-1} + F_{n+1}$. Conclude $L_n = F_{2n}/F_n$.

2.1.28. Let a_n count the circular lists of n bits with no consecutive 1s (rotations are distinct, so $a_2 = 3$ and $a_3 = 4$). Compute a_n using Fibonacci numbers.

2.1.29. For $n, d \in \mathbb{N}$ with $d \le n$, determine the number of partitions of $[n]$ into arithmetic progressions all having constant difference d and having length at least 1. Do the same for length at least 2. (Getz–Jones [2003])

2.1.30. For $n \in \mathbb{N}$, let $a_n = \sum_{k=1}^n \frac{1}{k2^{n-k}}$, let $b_n = \sum_{k=1}^n [k\binom{n}{k}]^{-1}$, and let $c_n = 2^{-n+1}\sum_{k \text{ odd}} \binom{n}{k}\frac{1}{k}$, Note that $a_0 = b_0 = c_0 = 0$. Prove $a_n = b_n = c_n$ for all n, by showing that each sequence satisfies $2nx_n = nx_{n-1} + 2$ for $n \ge 1$ (see Galperin–Gauchman [2004]).

2.1.31. ($\diamond$) Let G be the square grid on n^2 points in horizontal and vertical paths.
(a) Partition the edges of G into $2n - 2$ zig-zag paths as shown for $n = 4$. Let a_n count the edge sets with no two consecutive edges on any such path. Prove $a_n = \prod_{j=1}^{n-1} \hat{F}_{2j+1}^2$.

(b) Let b_n count the partitions of the vertices into up/right lattice paths ($b_2 = 9$, shown below). Prove $b_n = \prod_{j=1}^{n-1} \hat{F}_{2j+1}^2$. (Stanley [1992] gives a more general result.)

2.1.32. (+) Consider cards labeled 1 through n in some order. If the top card is m, we reverse the order of the first m cards. The process stops only when card 1 is at the top. Prove that the process must stop. Let a_n be the maximum (over all initial orderings) of the number of steps in the process. Prove $a_n < \hat{F}_n$. (Hint: Prove that if k distinct cards appear at the top during the process, then there are at most $\hat{F}_k - 1$ steps.) (Knuth)

2.1.33. *Generalized Fibonacci.* Let $a_{n+1} = a_n + a_{n-1}$ for $n \geq 3$, with a_1 and a_2 fixed.

(a) Determine T so that $a_{n+2}a_n + (-1)^n T = a_{n+1}^2$ for all $n \in \mathbb{N}$. (Comment: T is the only x such that $a_{n+2}a_n + (-1)^n x$ is always square.) (Horadam [1961], Kantrowitz [1986])

(b) Prove $a_{n+1} = F_{n-1}a_1 + F_n a_2$ for $n \geq 2$, using the classical Fibonacci numbers.

2.1.34. The Fibonacci numbers of order t are defined by $F_n^{(t)} = 2^{n-1}$ if $1 \leq n \leq t$, and $F_n^{(t)} = \sum_{j=1}^t F_{n-j}^{(t)}$. The binomial coefficient of order t, $C_{n,k}^{(t)}$, counts the n-tuples with entries in $\{0, \ldots, t-1\}$ and total sum k. Prove the statements below.

(a) $F_n^{(t)}$ counts binary $(n-1)$-tuples with no t consecutive 1s.

(b) $C_{n,k}^{(t)} = C_{n,(t-1)n-k}^{(t)}$ and $C_{n,k}^{(t)} = \sum_{i=0}^k C_{n-1,k-i}^{(t)}$, if $n > 0$ and $0 \leq k \leq n(t-1)$.

(c) Use the recurrences to prove $F_n^{(t)} = \sum_{r \geq 0} C_{n-r,r}^{(t)}$.

(d) Give a combinatorial proof of (c).

2.1.35. ($\diamond$) Let $a_n = 2a_{n-1} + a_{n-2}$ for $n \geq 2$, with $a_0 = 1$ and $a_1 = 2$. Use Delannoy paths to prove that $a_n = \sum \frac{(i+j+k)!}{i!j!k!}$, summed over nonnegative integer triples (i, j, k) such that $i + j + 2k = n$. (Comment: This sequence is the **Pell sequence**.) (Deutsch [1998])

2.1.36. Given $c_1, \ldots, c_k \in \mathbb{N}$, define $\langle a \rangle$ by $a_n = \sum_{i=1}^k a_{n-c_i}$ for $n > 0$, with $a_0 = 1$ and $a_n = 0$ for $n < 0$. Prove $a_n = \sum (\sum_{i=1}^k m_i)! / \prod_{i=1}^k m_i!$, where the sum runs over all nonnegative $m_1, \ldots, m_k$ such that $\sum_{i=1}^k c_i m_i = n$. (Barra [2000])

2.1.37. Fix $k \in \mathbb{N}$ and let $b_{n,j} = n^j \binom{k}{j}$. Prove $b_{n,j} = \sum_{i=0}^j \binom{k-i}{j-i} b_{n-1,i}$. Give both a combinatorial proof and a proof using identities.

2.1.38. ($\diamond$) Let k be a natural number. Create an array of k rows with row 1 being the natural numbers. To obtain row $j+1$ from row j, cross out every $(k+1-j)$th entry in row j and take partial sums of the sequence that remains (leave blanks under the crossed-out entries). Prove that row k consists of the numbers n^k. (Hint: Guess and prove a formula for the numbers on the longest rising diagonals of the triangular wedges, using a recurrence.)

1	2	3	4	5	6	7	8	9	10	11	12
1	3	6		11	17	24		33	43	54	
1	4			15	32			65	108		
1				16				81			

(Comment: The case $k = 2$ states $\sum_{i=1}^n 2i - 1 = n^2$. The procedure is called **Moessner's Process**, after Moessner [1951]. Proofs appear in Perron [1951], Long [1966, 1982], and Bender–Kochman–West [1990], generalized in Paasche [1956] and Long [1986].)

2.1.39. (+) Let $a_{0,0} = 1$, and let $a_{m,n} = 0$ when $m < 0$ or $n < 0$. For $m, n \in \mathbb{N}$, let

$$a_{m,n} = \begin{cases} a_{m,n-1} + a_{m-1,n} & \text{if } m+n \text{ is even,} \\ a_{m,n-1} + 2a_{m-1,n} & \text{if } m+n \text{ is odd.} \end{cases}$$

By a combinatorial argument, obtain a summation for $a_{n,n}$ and prove that it equals the central Delannoy number $d_{n,n}$. (Sulanke [2001])

2.1.40. Use the formula $n! = \sum_{k=0}^n \binom{n}{k} D_{n-k}$ of Example 2.1.6 to obtain the recurrence $D_n = \sum_{k=1}^n (k-1)\binom{n}{k} D_{n-k}$ for the derangement numbers.

2.1.41. ($\diamond$) Let $d(n, k)$ count the derangements of $[n]$ with k cycles. Derive a recurrence.

2.1.42. Let $D_o(n)$ and $D_e(n)$ be the numbers of derangements of $[n]$ having an odd or an even number of cycles, respectively. Compute $D_o(n) - D_e(n)$. (Stathopoulos [2012])

2.1.43. Determine the number of permutations π of $[n]$ that are derangements satisfying the increase condition $\pi(i+1) \leq \pi(i) + 1$ for $1 \leq i \leq n-1$. (Deutsch [2001a])

2.1.44. ($\Diamond$) Let A_n be the set of permutations of $[n]$ such that position i does not have element $i+1$, for $1 \le i \le n-1$. Let B_n be the set of those in which no element is followed immediately by the next larger element. Thus $A_3 = \{123, 312, 321\}$ and $B_3 = \{132, 213, 321\}$.

(a) Prove $|A_n| = D_n + D_{n-1}$ for all n, where D_n is the number of derangements of $[n]$. (Hint: Consider the position of the element 1.)

(b) Obtain a second-order recurrence for $|B_n|$, and use it to conclude $|B_n| = D_n + D_{n-1}$. (Hint: Consider how the element n is used.)

(Comment: Thus $|A_n| = |B_n|$; a bijective proof is requested in Exercise 3.1.27.)

2.1.45. Let $D_{n,k}$ be the number of permutations of $[n+k]$ having no fixed points in $[n]$. Generalize the argument of Example 2.1.6 to obtain a recurrence for $D_{n,k}$.

2.1.46. Prove the identity below involving the derangement numbers.

$$\sum_{j=0}^{k} \binom{k}{j} D_{k+n-j} = k! \sum_{j=0}^{\min\{n,k\}} \binom{k}{j}\binom{k+n-j}{k} D_{n-j} \qquad \text{(Callan [1998])}$$

2.1.47. ($\Diamond$) Prove recursively (no bijections needed) that in each case below, a_n is the nth Catalan number. (Comment: In each case, one can also prove the claim bijectively by establishing a bijection from the set being counted to the set of ballot lists of length $2n$.)

(a) Let a_n be the number of ordered trees with $n+1$ vertices (see Definition 1.3.21). The ordered trees with four vertices appear below.

(b) Let a_n be the number of noncrossing pairings of $2n$ points on a circle. The noncrossing pairings of six points appear below.

(c) Let a_n be the number of configurations of pennies on a base row of n pennies, where pennies can be added so that each one not in the base rests on two in the row immediately below it. An example for $n = 6$ appears below. (Propp)

2.1.48. ($\Diamond$) A **stack** stores numbers with only the most recent number (the *top element*) accessible. A stack processes an input list as follows. Let x be the next input number (if any), and let y be the top element of the stack (if any). Move x to the stack when x exists and y does not or when $x < y$; otherwise move y to the output. Eventually all input elements move input to the stack and later to the output. A list is **stack-sortable** if the output is in increasing order. For example, $(1, 5, 4, 2, 3)$ is stack-sortable, but the output for $(3, 5, 1, 2, 4)$ is $(3, 1, 2, 4, 5)$. Let a_n be the number of stack-sortable permutations of $[n]$.

(a) Prove recursively that a_n is the Catalan number C_n.

(b) Prove that a list $x_1, \ldots, x_n$ of distinct numbers is stack-sortable if and only if there do not exist i, j, k with $i < j < k$ such that $x_k < x_i < x_j$.

2.1.49. Prove that the Catalan numbers satisfy the recurrence below.

$$C_n = \binom{2n}{n} - \frac{1}{2} \sum_{k=0}^{n-1} C_k \binom{2n-2k}{n-k}.$$

2.1.50. ($\diamond$) *Noncrossing partitions of* $[n]$. A partition of $[n]$ is **noncrossing** if there are no a, b, c, d with $a < b < c < d$ such that a and c are in one block and b and d are in another. Thus $(14|25|3)$ is crossing and $(15|234)$ is noncrossing. Let a_n be the number of these.

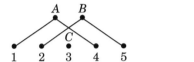 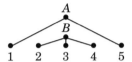

(a) Prove recursively that a_n is the Catalan number C_n.

(b) Use the recurrence to obtain a bijection from noncrossing partitions of $[n]$ to ordered trees with n edges (Definition 1.3.21). (Prodinger [1983])

(c) From the bijection in part (b), observe that the number of noncrossing partitions of $[n]$ with l blocks equals the number of ordered trees with n edges and l leaves. (Comment: The sizes of these subsets are the Narayana numbers; see Theorem 1.3.26.)

2.1.51. ($\diamond$) Let a_n count the partitions of $3n$ points on a circle into n nonintersecting triangles. Let b_n count the lists of $2n$ 1s and n 0s such that every prefix has at least twice as many 1s as 0s (2-ballot lists; Definition 1.3.18). Prove $a_n = b_n$ by showing that the satisfy the same recurrence. (Comment: By Corollary 1.3.19, $b_n = \frac{1}{2n+1}\binom{3n}{n}$.)

2.1.52. ($\diamond$) A **positive lattice walk** in $\mathbb{Z}^n$ starts at the origin, at each step changes one coordinate by 1, and visits no point with last coordinate negative. Vertices and edges may repeat. Let a_n be the number of such walks of length n in $\mathbb{Z}^2$ ending on the horizontal axis.

(a) Obtain a recurrence for $\langle a \rangle$. Show that a_n is the Catalan number C_{n+1}.

(b) Let w_n be the total number of positive lattice walks of length n in $\mathbb{Z}^2$. Use part (a) to obtain a recurrence for $\langle w \rangle$, and conclude $w_n = \binom{2n+1}{n}$. (Guy [2000])

(c) Use part (b) to show that the total number of positive lattice walks of length n in three dimensions is $\sum_{k=0}^{n} \binom{n}{k}\binom{2k+1}{k}2^{n-k}$.

(Comment: Exercise 1.3.34 obtains another formula for w_n. It and part (b) provide a combinatorial proof of $\sum_{k=0}^{n} \binom{n}{k}\binom{k}{\lceil k/2 \rceil}2^{n-k} = \binom{2n+1}{n}$. For the corresponding problem in three dimensions, part (c) and Exercise 1.3.34 yield $\sum_{k=0}^{n} \binom{n}{k}\binom{k}{\lceil k/2 \rceil}4^{n-k} = \sum_{k=0}^{n} \binom{n}{k}\binom{2k+1}{k}2^{n-k}$.)

2.1.53. (+) Let S_n count the paths from $(0,0)$ to (n,n) that never rise above $y = x$ and use steps in $\{(0,1),(1,0),(1,1)\}$. Note that $(S_0, S_1, S_2, S_3) = (1, 2, 6, 22)$.

(a) Prove that S_n is divisible by 3 when n is positive and even. (Shapiro–Rogers [1989])

(b) Prove $a_n = 6a_{n-1} - a_{n-2} - 2S_{n-1}$ for $n \geq 2$, where a_n is the nth central Delannoy number $d_{n,n}$ (see Definition 1.2.8). (Peart–Woan [2002])

(Comment: S_n is the nth "large" **Schröder number** (there are also "small" ones), named for Ernst Schröder (1841–1902) after Schröder [1870]. These numbers satisfy $(n+1)S_n = 3(2n-1)S_{n-1} - (n-2)S_{n-2}$ for $n \geq 2$; Sulanke [1998] gave a bijective proof. Also $na_n = 3(2n-1)a_{n-1} - (n-1)a_{n-2}$ for the Delannoy numbers.)

2.1.54. ($\diamond$) **Shapiro** n**-paths** are lattice paths from $(0,0)$ to $(2n,2n)$ avoiding the odd points $(2i-1, 2i-1)$ on the diagonal; let S_n be the number of these. Prove $S_n = C_{2n}$.

2.1.55. ($\diamond$) A **Dyck** n**-path** is a path from $(0,0)$ to $(2n,0)$ using steps that move by $(1,1)$ or $(1,-1)$ without falling below the x-axis. Show that the number of Dyck $(2n)$-paths that avoid $\{(4k,0): 1 \leq k \leq n-1\}$ is twice the number of Dyck $(2n-1)$-paths. (Callan [2003b])

2.1.56. Two teams play until one wins n games. Team A has probability p of winning any game, independently. Let $q = 1 - p$. Let a_n be the expected number of games played.

(a) Prove $a_n = n \sum_{k=0}^{n-1} \binom{n+k}{k}(p^n q^k + p^k q^n)$.

(b) Prove $a_n = n \sum_{k=0}^{n-1} C_k (pq)^k$, where $C_k = \frac{1}{k+1}\binom{2k}{k}$. (Shapiro–Hamilton [1993])

2.1.57. ($\diamond$) An **up-down permutation** alternates ascents and descents, an ascent first. Let a_n be the number of up-down permutations of $[n]$; note $(a_0, \ldots, a_4) = (1, 1, 1, 2, 5)$. Let b_n be the number of equivalence classes of permutations of $[n]$ under "low-flip" operations that reverse $x_1, \ldots, x_k$ if $k = n$ or if $x_{k+1} > \max\{x_1, \ldots, x_k\}$; two permutations are equivalent if low-flips can turn one into the other. Prove $a_n = b_n$ for $n \geq 0$. (Knuth [2009]) (Comment: These are called **secant-and-tangent numbers**; see Lehmer [1935], Knuth–Buckholtz [1967], Riordan [1968], Comtet [1970, 1974], Gould [1972], etc.)

2.1.58. For $n \in \mathbb{N}$, let $T_k(n)$ be the coefficient of $(-1)^k x^{n-k}$ in the expansion of $\prod_{i=0}^{n-1}(x - i)$.

(a) Give a combinatorial proof that $T_k(n) = T_k(n-1) + (n-1)T_{k-1}(n-1)$.

(b) Prove $T_k(n) = \sum_{i=k}^{n-1} i T_{k-1}(i)$ for $k \geq 1$.

(c) Compute simple formulas for $T_k(n)$ when $k \in \{0, 1, 2\}$.

(d) Prove that T_k is a polynomial of degree $2k$.

2.1.59. Let the graph G_n be the path $\langle x_1, \ldots, x_n \rangle$ plus one vertex adjacent to all x_i (shown below for $n = 5$). Let a_n be the number of spanning trees in G_n.

(a) Obtain a many-term recurrence for $\langle a \rangle$.

(b) Obtain a second-order recurrence for $\langle a \rangle$.

(c) Prove $a_n = F_{2n}$, where F_k is the kth classical Fibonacci number.

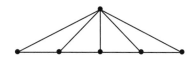

(Comment: Relations for Fibonacci/Lucas numbers and counts of spanning trees in special graphs have been studied in Sedláček [1970], Hilton [1972], Fielder [1974], Myers [1975], Shannon [1978], Mikola [1980], and Exercises 60–61.)

2.1.60. A k-**tree** is a graph obtained from a complete graph with k vertices by iteratively adding a new vertex whose neighbors induce a complete graph with k vertices. For example, ordinary trees are 1-trees, and the graphs in Exercise 2.1.59 are 2-trees. A vertex with degree k in a k-tree is called a **simplicial vertex**. Prove that all n-vertex 2-trees having exactly two simplicial vertices have the same number of spanning trees (this includes the graphs in Exercise 2.1.59). (Comment: In fact, these 2-trees are the n-vertex 2-trees with the most spanning trees.) (Xiao–Zhao [2013])

2.1.61. Form a graph G_n from two n-vertex paths by making corresponding vertices adjacent as shown. Obtain a second-order recurrence for the number of spanning trees.

2.1.62. Let G be a graph with m spanning trees. Let G' be the graph obtained by replacing each edge of G with k copies of that edge. Let G'' be the graph obtained by replacing each edge $uv \in E(G)$ with a u, v-path of length k through $k - 1$ new vertices. Determine $\tau(G')$ and $\tau(G'')$ in terms of m, k, and $\tau(G)$.

2.1.63. Let a_n be the number of domino tilings of a 4-by-n rectangle. Obtain a bounded-order recurrence for $\langle a \rangle$. (Rymer [1979], Zhou [2005])

2.1.64. ($\diamond$) *A Summation Identity for permutations.* As in Example 2.1.9, let $c(n, k)$ be the number of permutations of $[n]$ with k cycles. For $m \geq k$, prove $\sum_{j=0}^{m} \frac{c(j,k)}{j!} = \frac{c(m+1,k+1)}{m!}$. (Hint: After multiplying by $m!$, there is both a direct bijective proof and an inductive proof using a known identity.

2.1.65. *Gambler's Ruin.* Player A starts with \$$r$, B with \$$s$. They flip a fair coin until one goes broke. On a head, A pays B \$1; on a tail, B pays A \$1. (Kraitchik [1942])
    (a) Prove that the probability of A winning is $\frac{r}{r+s}$.
    (b) Prove that the expected number of flips until the game ends is rs.

2.1.66. (+) A random walk starts at 0. On each step, it moves up or down by 1, each with probability .5 (independent of earlier steps). Compute the expected number of steps until the first time the walk is k steps below the highest value it has reached. (Hint: Find a recurrence for the expected number.) (Palacios [2012])

2.1.67. (+) The gambler and the devil play with n red balls and $n + 1$ blue balls. The gambler starts with one unit of money (infinitely divisible). At each round, the gambler bets some money (maybe 0), and the devil picks a remaining ball. When the ball is red, the gambler loses the bet; when blue, the gambler gains that amount. The selected ball is discarded. The gambler wants to maximize his money at the end, while the devil wants to minimize it. If both play optimally, how much does the gambler end with? (Hint: Solve the more general problem with r red balls and b blue balls.) (Pudaite [2000])

2.1.68. Compute $\det M$, where M is the n-by-n matrix whose first row and column is all 1, and otherwise $m_{i,j} = m_{i-1,j} + m_{i,j-1} + x m_{i-1,j-1}$. (Bacher [2001])

2.1.69. Let m and n be nonnegative integers. Prove

$$\sum_{i=0}^{\lfloor m/2 \rfloor} \sum_{j=0}^{\lfloor n/2 \rfloor} \binom{i+j}{j}^2 \binom{m+n-2i-2j}{n-2j} = \frac{\lceil \frac{n+m}{2} \rceil! \lceil \frac{n+m+1}{2} \rceil!}{\lfloor n/2 \rfloor! \lfloor m/2 \rfloor! \lceil n/2 \rceil! \lceil m/2 \rceil!}.$$

(Hint: Writing each side as $a_{m,n}$, compute $a_{m,n} - a_{m,n-1} - a_{m-1,n}$.) (Andrews–Paule [1992])

2.1.70. *Destructive testing.* When dropped from the bth floor of a building, beanbags break. From higher floors they also break, but from lower floors they survive and can be re-used. Let $f(t, k)$ be the largest value b that can be determined with certainty when k beanbags are available and up to t tests are permitted. Prove that $f(t, 1) = t$ and that $f(t, k) = f(t-1, k-1) + f(t-1, k) + 1$ for $k > 1$. (G. Peterson)

2.2. Elementary Solution Methods

Finding a recurrence for a counting sequence can be a step toward a solution formula. We next study methods for obtaining such formulas from recurrences. A simple formula for a_n obtained from a recurrence may have an elegant direct combinatorial proof. Although we may then discard the recursive approach in presenting the final solution, solving the recurrence remains a useful tool.

THE CHARACTERISTIC EQUATION METHOD

Linear recurrence relations with constant coefficients can be solved by the *characteristic equation method* or the more versatile *generating function method* discussed later. When it applies, the first method is faster, abbreviating standard computations in the second method. It also closely parallels standard methods for elementary differential equations. In this discussion all recurrence relations are linear with constant coefficients, though we may neglect to say so.

2.2.1. Definition. For the linear constant-coefficient recurrence relation

$$a_n = c_1 a_{n-1} + \cdots + c_k a_{n-k} + f(n),$$

the associated **characteristic polynomial** is the polynomial ϕ defined by $\phi(x) = x^k - c_1 x^{k-1} - \cdots - c_k x^0$, the **characteristic equation** is the equation $\phi(x) = 0$, and its solutions are the **characteristic roots**.

2.2.2. Remark. *Solutions to homogeneous recurrences.* For the homogeneous recurrence $a_n = \alpha a_{n-1}$, the characteristic equation is $x = \alpha$. Successive substitution shows that the solution is $\alpha^n a_0$. In other words, the solution has the form $A\alpha^n$, and the constant A is determined by the initial condition. This behavior generalizes to higher order.

Consider the homogeneous recurrence $a_n = c_1 a_{n-1} + \cdots + c_k a_{n-k}$. When α is a characteristic root, substituting α in the characteristic equation and multiplying by α^{n-k} yields $\alpha^n = c_1 \alpha^{n-1} + \cdots + c_k \alpha^{n-k}$. Multiplying by a constant A does not disturb the equality, so $a_n = A\alpha^n$ is a solution.

When both α and β are characteristic roots, the equations showing that $A\alpha^n$ and $B\beta^n$ satisfy the recurrence can be summed to show that setting $a_n = A\alpha^n + B\beta^n$ also satisfies the recurrence.

A specific sequence is determined by the initial conditions. For $k = 2$ with roots α and β, we choose values for A and B in $A\alpha^n + B\beta^n$ so that the formula for a_n reduces to the values for a_0 and a_1 when $n \in \{0, 1\}$. ∎

2.2.3. Example. For the homogeneous recurrence $a_n = 3a_{n-1} - 2a_{n-2}$, the characteristic equation is $x^2 = 3x - 2$, with roots 1 and 2. Hence every sequence of the form $a_n = A + B2^n$ satisfies the recurrence.

Setting $n = 0$ yields $a_0 = A + B$, and $n = 1$ yields $a_1 = A + 2B$. Given the values for a_0 and a_1 as initial conditions, the system of two linear equations determines A and B to complete the solution. ∎

Irrational or complex numbers may arise as characteristic roots. Nevertheless, when the initial conditions and coefficients of the recurrence are integers, each term in the resulting sequence must be an integer.

2.2.4. Example. The Fibonacci recurrence $\hat{F}_n = \hat{F}_{n-1} + \hat{F}_{n-2}$ has the characteristic equation $x^2 - x - 1 = 0$, with roots $\alpha_1 = (1 + \sqrt{5})/2$ and $\alpha_2 = (1 - \sqrt{5})/2$. Requiring $\hat{F}_0 = \hat{F}_1 = 1$ in the general solution $\hat{F}_n = A\alpha_1^n + B\alpha_2^n$ yields $1 = A + B$ and $1 = A\alpha_1 + B\alpha_2$. The solution is

$$\hat{F}_n = \frac{1}{\sqrt{5}} \left(\frac{1 + \sqrt{5}}{2} \right)^{n+1} - \frac{1}{\sqrt{5}} \left(\frac{1 - \sqrt{5}}{2} \right)^{n+1} .$$

For the classical Fibonacci numbers ($F_0 = 0$; $F_1 = 1$), we obtain the solution $F_n = (\alpha_1^n - \alpha_2^n)/\sqrt{5}$. This is **Binet's Formula**, proved by de Moivre [1730] using generating functions and rediscovered by Binet [1843] and Lamé [1844]. The characteristic root $(1 + \sqrt{5})/2$ is the **Golden Ratio**.   ∎

Since sums and constant multiples of sequences satisfying a homogeneous recurrence also satisfy it, the solutions form a linear space. Since k initial values are required to specify a sequence satisfying a recurrence of order k, the space of solutions has dimension k. The next lemma makes this precise.

2.2.5. Lemma. The solutions to a homogeneous constant-coefficient linear recurrence of order k form a k-dimensional subspace of the space of all sequences.

Proof: For completeness, consider sequences from the field $\mathbb{C}$ of complex numbers; they form an infinite-dimensional vector space. Being closed under addition of sequences and under multiplication by constants, the set U of solutions to a homogeneous linear recurrence with constant coefficients is a subspace.

Specify a solution sequence whose 0th through $(k-1)$th terms are all 0 except that the jth term is 1. These sequences are linearly independent. Also, every solution sequence is a linear combination of these, using the values in the initial conditions as coefficients. Thus the k special sequences form a basis for the solution space, so it has dimension k.   ∎

Lemma 2.2.5 suggests that the method used above finds all the solutions, *when the characteristic roots are distinct.*

2.2.6. Example. When $a_n = 2a_{n-1} - a_{n-2}$, the characteristic polynomial is $x^2 - 2x + 1$, which factors as $(x-1)^2$. For any constants A_1 and A_2, the expression $A_1 1^n + A_2 1^n$ specifies a constant sequence. Nevertheless, the initial conditions may have $a_0 \neq a_1$. For example, when $a_0 = 1$ and $a_1 = 2$, it is easy to check by induction that $a_n = n + 1$ is the solution.   ∎

The next theorem finds the missing solutions. "General Solution" means that all sequences satisfying the recurrence are described; a specific sequence arises by choosing coefficients in the polynomials P_i to satisfy the initial conditions.

2.2.7. Theorem. (**General Solution**) Let $\langle a \rangle$ satisfy a homogeneous linear recurrence relation of order k with constant coefficients. If $\alpha_1, \ldots, \alpha_r$ with multiplicities $d_1, \ldots, d_r$ are the distinct characteristic roots, then $a_n = \sum_i P_i(n)\alpha_i^n$, where each P_i is a polynomial of degree less than d_i.

Proof: The characteristic polynomial factors as $\phi(x) = \prod_{i=1}^{r}(x - \alpha_i)^{d_i}$. Multiplying by x^{n-k} yields

$$x^n - c_1 x^{n-1} - \cdots - c_k x^{n-k} = x^{n-k} \prod_{i=1}^{r} (x - \alpha_i)^{d_i} .$$

Using the product rule to differentiate the right side retains $(x - \alpha_i)^{d_i-1}$ as a factor in each term. More generally, differentiating j times when $j < d_i$ retains $(x - \alpha_i)^{d_i-j}$ as a factor in each term on the right; so the value at $x = \alpha_i$ remains 0. Differentiating j times on the left, multiplying both sides by x^j, and then setting $x = \alpha_i$ yields

$$n_{(j)}\alpha_i^n - c_1(n-1)_{(j)}\alpha_i^{n-1} - \cdots - c_k(n-k)_{(j)}\alpha_i^{n-k} = 0 \ .$$

Therefore, setting $a_n = n_{(j)}\alpha_i^n$ is a solution. Note that the falling factorial $n_{(j)}$ is a polynomial in n of degree j. Every polynomial with degree less than d_i is a linear combination of $\{n_{(0)}, \dots, n_{(d_i-1)}\}$. Hence $a_n = P_i(n)\alpha_i^n$ is a solution whenever P_i is a polynomial of degree less than d_i (P_i may have complex coefficients).

By linearity, every linear combination of such solutions from distinct roots is also a solution. By Lemma 2.2.5, the set of sequences satisfying the recurrence is a vector space of dimension k. In the expression for the solution, there are k parameters (the coefficients in the polynomials). For a finite set of distinct pairs (α, r), Exercise 20 shows that the sequences defined by setting $a_n = n^r\alpha^n$ are independent. Hence we have described a k-dimensional space of solutions, which consists of all solutions of the recurrence.   ∎

A direct proof that the sequences of the form $n^r\alpha^n$ are independent involves the Vandermonde determinant, which we have not derived. An alternative proof using only the dimension of vector spaces will follow from Theorem 2.2.20.

Specifying k initial conditions determines one sequence and hence the coefficients of the polynomials $P_1, \dots, P_r$ in the general solution. To find it, we use the initial conditions to impose k linear equations on the k unknown coefficients. For $0 \le i < k$, set $n = i$ in the formula for a_n and equate this to the given value a_i. The system is uniquely solvable because the space of sequences (1) satisfying the recurrence and (2) specified by the general solution are the same.

2.2.8. Example. Consider the recurrence $a_n = 2a_{n-1} - 2a_{n-2} + 2a_{n-3} - a_{n-4}$ for $n \ge 4$, with $(a_0, a_1, a_2, a_3) = (1, 3, 5, 5)$. The characteristic equation is $x^4 - 2x^3 + 2x^2 - 2x + 1 = 0$, which factors as $(x^2 + 1)(x - 1)^2 = 0$. The characteristic roots are $i, -i, 1, 1$. Thus the general solution is

$$a_n = A_1 i^n + A_2(-i)^n + A_3 1^n + A_4 n 1^n.$$

To determine these coefficients, we invoke the initial conditions:

$$
\begin{array}{llrcl}
n = 0: & \quad & 1 &=& A_1 + A_2 + A_3 \\
n = 1: & \quad & 3 &=& iA_1 - iA_2 + A_3 + A_4 \\
n = 2: & \quad & 5 &=& -A_1 - A_2 + A_3 + 2A_4 \\
n = 3: & \quad & 5 &=& -iA_1 + iA_2 + A_3 + 3A_4
\end{array}
$$

Summing the equations in pairs eliminates A_1, A_2 to solve for A_3, A_4, and then the first two yield A_1 and A_2. The solution is $(A_1, A_2, A_3, A_4) = (-1/2, -1/2, 2, 1)$, which yields $a_n = n + 2 - (i^n + (-i)^n)/2$.   ∎

Next we consider inhomogeneous linear recurrences. The method is motivated by the observation that the solutions form an *affine subspace* of the space of all sequences. The all-0 sequence is no longer a solution.

2.2.9. Example. *The* **Tower of Hanoi** *problem.*

$$a_n = 2a_{n-1} + 1 \text{ for } n \geq 1, \text{ with } a_0 = 0 .$$

In this problem, n rings on a peg must move to another peg, with a third peg available as workspace. No ring can be placed upon a smaller ring. Let a_n be the minimum number of steps needed to move the pile. The bottom ring can move to the desired peg only when we have moved the rings above it to the third peg, and then after moving the bottom ring we must move the other rings on top of it. Thus $a_n = 2a_{n-1} + 1$. It takes no steps to move no rings. The value more than doubles with each increase in n, so the growth is exponential. This is a linear first-order recurrence.

Lucas popularized this problem using the legend of the Tower of Benares: monks were moving a pile of 64 heavy golden disks according to these rules, and the world would end when they finished. In fact, there was no such legend; Lucas made it up! (Singmaster [1998, p. 581]). An early solution is in Ball [1892].

Dozens of papers have studied variations on this problem, such as starting from an arbitrary configuration of the disks, moving disks only cyclically, finding the minimum number of moves when p pegs are available, etc. A discussion of work on the p-peg problem appears in Klavžar–Milutinović–Petr [2002].

In the 3-peg recurrence, each value roughly twice the previous value, the solution should be close to 2^n. In fact, with $a_0 = 0$, it follows by induction that setting $a_n = 2^n - 1$ solves the recurrence. Proof by induction works only when we already know the answer. If not given the formula $2^n - 1$, how can we find it? ∎

2.2.10. Remark. *Solutions to inhomogeneous recurrences.* To solve such a recurrence, we first seek a single sequence $\langle p \rangle$ satisfying the recurrence, *without regard to the initial conditions*. We call this a **particular solution**. We then obtain the **general solution** by adding $\langle p \rangle$ to the general solution of the homogeneous recurrence obtained by dropping the inhomogeneous term. For each solution $\langle h \rangle$ to the homogeneous recurrence, linearity (a special case of Exercise 10) implies that setting $a_n = p_n + h_n$ solves the original inhomogeneous recurrence.

$$p_n = \sum_{i=1}^{k} c_i p_{n-i} + f(n)$$
$$h_n = \sum_{i=1}^{k} c_i h_{n-i}$$
$$p_n + h_n = \sum_{i=1}^{k} c_i (p_{n-i} + h_{n-i}) + f(n)$$

As in Theorem 2.2.7, *general solution* means that all solutions of the recurrence have been described. The general solution again is parametrized by k constants; these are the constants in the general solution of the homogeneous part of the recurrence. To complete the solution, we solve linear equations in these constants to make the general formula satisfy the initial conditions. ∎

When the inhomogeneous term is nonzero, many books suggest "guessing" the form of a particular solution. Usually it is not a guess at all; the next theorem guarantees the form of a particular solution for certain kinds of recurrences.

For example, this theorem guarantees that the recurrence $a_n = 2a_{n-1} + 1$ in Example 2.2.9 has a particular solution that is a constant sequence. If some constant b works, then b must satisfy $b = 2b + 1$. This yields $b = -1$. The characteristic root is 2, so the general solution is $a_n = A2^n - 1$. The initial condition $a_0 = 0$ requires $A = 1$, so $a_n = 2^n - 1$, with no inductive proof needed.

2.2.11. Theorem. Let R denote the recurrence $a_n = \left(\sum_{i=1}^k c_i a_{n-i} \right) + F(n)\alpha^n$ for $n \geq k$, where F is a polynomial of degree d. If α has multiplicity r as a characteristic root of the homogeneous recurrence obtained by setting $F(n) = 0$ (r may be 0), then the recurrence has a solution of the form $a_n = P(n)n^r\alpha^n$, where P is a polynomial of degree at most d.

Proof: Since R generates a sequence from any initial conditions, particular solutions exist. The key is to find a particular solution to R among the solutions of a homogeneous recurrence S of higher order.

Let H denote the homogeneous recurrence obtained from R by eliminating the inhomogeneous term, and let ϕ be the characteristic polynomial of H. Let S be the homogeneous recurrence whose characteristic polynomial in x is $(x - \alpha)^{d+1}\phi(x)$.

Step 1: *Every solution to R is a solution to S.* From the expression for a_n given by R, subtract α times the expression for a_{n-1} given by R:

$$a_n - \alpha a_{n-1} = \sum_{i=1}^k (c_i a_{n-i} - \alpha c_i a_{n-i-1}) + [F(n) - F(n-1)]\,\alpha^n.$$

This new recurrence is still satisfied by $\langle a \rangle$, since we simply subtracted equalities satisfied by $\langle a \rangle$. The recurrence has order $k + 1$, since it relates $a_n, \ldots, a_{n-k-1}$. Its characteristic equation is

$$x^{k+1} - \alpha x^k = \sum_{i=1}^k (c_i x^{k+1-i} - \alpha c_i x^{k-i})$$

which simplifies to $(x - \alpha)\phi(x) = 0$. Most importantly, $F(n) - F(n-1)$ is a polynomial in n of degree $d - 1$.

Applying this operation d times, the resulting recurrence has characteristic polynomial $(x - \alpha)^d \phi(x)$ and inhomogeneous term $G(n)\alpha^n$, where G is a polynomial of degree 0; that is, $G(n)$ is a constant. One more application of the differencing operation reduces the inhomogeneous term to $0 \cdot \alpha^n$, so the recurrence is now homogeneous and has characteristic polynomial $(x - \alpha)^{d+1}\phi(x)$. That is, it is the recurrence S. Furthermore, throughout each step, $\langle a \rangle$ remains a solution.

Step 2: *Choosing a particular solution.* Since S is a homogeneous recurrence, its solutions are as described in Theorem 2.2.7. Since α is a root of multiplicity $d + r + 1$ for $(x - \alpha)^{d+1}\phi(x)$, the solutions to S have the form $Q(n)\alpha^n + h(n)$, where Q is a polynomial of degree at most $d + r$ and h is a solution to H in terms of the characteristic roots other than α.

By Step 1, R has a solution of this form. The sum of this solution and any solution to H is another solution to R, so we have one of the form $Q(n)\alpha^n$. Furthermore, α^n times any polynomial of degree less than r is a solution to H, so we may also add such a sequence to our solution to cancel the coefficients of Q for powers less than r. We have now shown that there is a solution of the form $P(n)n^r\alpha^n$, where P is a polynomial of degree at most d.

Having proved that there is a particular solution of this form, we can use the recurrence R to find one. Substitute $P(n)n^r\alpha^n$ into R. With α known, cancel α^{n-k} from both sides of R to obtain

$$P(n)n^r\alpha^k = \sum_{i=1}^{k} c_i P(n-i)(n-i)^r \alpha^{k-i} + F(n)\alpha^k.$$

We choose the coefficients in $P(n)$ to make the polynomials in n obtained on both sides be the same polynomial. Equating coefficients or evaluating at $d+1$ values of n yields $d+1$ equations in $d+1$ unknowns to determine the coefficients.   ∎

Theorem 2.2.11 can also be proved by the generating function method (Exercise 19). As remarked in the proof, we can force two polynomials to be equal by equating the coefficients of corresponding powers or by requiring the values to be equal at more points than the degree. The next example illustrates both methods.

2.2.12. Example. *Regions among lines, again (Example 2.1.1).* Let $a_n = a_{n-1} + n$ for $n \geq 1$, with $a_0 = 1$. Using iteration, a recurrence of the form $a_n = a_{n-1} + f(n)$ has the solution $a_n = a_0 + \sum_{i=1}^{n} f(i)$. Since $\sum_{i=1}^{n} i = \binom{n+1}{2}$, we obtain $a_n = 1 + \binom{n+1}{2}$.

Alternatively, we can use the characteristic equation method, which we do to illustrate the method. Here $\phi(x) = x - 1$, with root 1. The inhomogeneous term is $n \cdot 1^n$, which has the form $F(n)c^n$ with F having degree 1 and c being a characteristic root of multiplicity 1. Theorem 2.2.11 guarantees a particular solution of the form $a_n = (B_1 n + B_2)n1^n$.

Substituting the sequence $B_1 n^2 + B_2 n$ into the recurrence yields the polynomial equation $B_1 n^2 + B_2 n = B_1(n-1)^2 + B_2(n-1) + n$ to determine B_1 and B_2.

When equating corresponding coefficients, the coefficients of n^2 yield $B_1 = B_1$. This gives no information, which is good, since corresponding powers give three equations but we have only two parameters. The coefficients of n yield $B_2 = -2B_1 + B_2 + 1$, or $B_1 = 1/2$. The constant term yields $0 = B_1 - B_2$, and hence $B_2 = B_1 = 1/2$. The general solution is $a_n = A1^n + \binom{n+1}{2}$. Satisfying the initial condition $a_0 = 1$ yields $A = 1$.

When evaluating at fixed values, $n = 0$ yields $0 = B_1 - B_2$, and $n = 1$ yields $B_1 + B_2 = 1$. Again $B_1 = B_2 = 1/2$, and we obtain the same solution. This method avoids the work of extracting the coefficients of powers of n.

A simple solution formula to a recurrence may suggest a direct counting argument. Exercise 13 requests it for this problem.   ∎

2.2.13. Remark. The form of the inhomogeneous term in Theorem 2.2.11 seems quite special, but the method extends to handle sums of such terms. When the inhomogeneous term is $f(n) + g(n)$, a particular solution can be found by summing solutions obtained when the inhomogeneous term is $f(n)$ or $g(n)$ alone. This property is called the **superposition principle** (see Exercise 10 and Exercise 14).   ∎

THE GENERATING FUNCTION METHOD

Here we introduce generating functions as a tool to solve recurrences and to explain the characteristic equation method. We postpone the combinatorial aspects of generating functions to Chapter 3.

2.2.14. Definition. The **generating function** for a sequence $\langle a \rangle$ of complex numbers is the *formal power series* $\sum_{n=0}^{\infty} a_n x^n$; the term is also used for any expression $A(x)$ with power series expansion $\sum_{n=0}^{\infty} a_n x^n$. The **coefficient operator** $[x^k]$ extracts the coefficient of x^k in a (formal) power series in x, so $[x^k]A(x) = a_k$ when $A(x) = \sum_{n=0}^{\infty} a_n x^n$.

The word "formal" in "formal power series" indicates that x^n serves not as a number but rather as a placeholder for the coefficient a_n. At present we ignore the distinction between power series and formal power series, because a Taylor series expansion of a function (around 0) will also be its formal power series (we return to this issue in Section 3.2).

2.2.15. Algorithm. The **generating function method** uses the following steps to solve a recurrence for $\langle a \rangle$ (see the example below).
(1) Sum the recurrence over its "region of validity" (the values of the parameter where the recurrence holds) to introduce the generating function $A(x)$ and obtain an equation that $A(x)$ satisfies.
(2) Solve this equation to express $A(x)$ in terms of x.
(3) Find the formal power series expansion and set $a_n = [x^n]A(x)$. ∎

2.2.16. Example. *Regions among lines, yet again.* Let $a_n = a_{n-1} + n$ for $n \geq 1$, with $a_0 = 1$. Multiplying by x^n and summing over $n \geq 1$ yields

$$\sum_{n \geq 1} a_n x^n = \sum_{n \geq 1} a_{n-1} x^n + \sum_{n \geq 1} n x^n.$$

With $A(x) = \sum_{n \geq 0} a_n x^n$, the first sum is missing the first term in the generating function and becomes $A(x) - 1$. The second needs x factored out to make the subscript and exponent agree; it is thus $x A(x)$.

Since the geometric series yields $\sum_{n=0}^{\infty} x^n = (1-x)^{-1}$, termwise differentiation expresses the third sum as $x \frac{d}{dx}(1-x)^{-1}$. Hence our equation for $A(x)$ becomes $A(x) - 1 = x A(x) + \frac{x}{(1-x)^2}$, which simplifies to

$$A(x) = \frac{1}{1-x} + \frac{x}{(1-x)^3}.$$

Now $a_n = [x^n]\left((1-x)^{-1} + x(1-x)^{-3}\right)$. The first term contributes 1. For the second term, we seek the coefficient of x^{n-1} in the expansion of $(1-x)^{-3}$. We will see in Lemma 2.2.17 that $(1-x)^{-3} = \sum_{n=0}^{\infty} \binom{n+2}{2} x^n$. Hence the answer is $a_n = 1 + \binom{n+1}{2}$, as in Example 2.2.12. ∎

In Example 2.2.16, the generating function method automatically incorporates the inhomogeneous term and the initial conditions. It applies more generally than the characteristic equation method, but the latter is usually more efficient when both apply.

As in Example 2.2.16, completing the generating function method often requires the series for expressions such as $(1 - cx)^{-k}$. We will give a combinatorial proof avoiding Taylor's Theorem in Section 3.1. (When the summand is written as $\binom{n+k-1}{n}$, the expression is valid for all real k.)

2.2.17. Lemma. For $k \in \mathbb{N}$, the power series expansion of $(1 - cx)^{-k}$ is

$$\frac{1}{(1 - cx)^k} = \sum_{n=0}^{\infty} \binom{n + k - 1}{k - 1} c^n x^n.$$

Proof: In Theorem 1.2.5, Taylor's Theorem of calculus yields the power series $(1 + y)^u = \sum_{n \geq 0} \binom{u}{n} y^n$, using the extended binomial coefficient. With $y = -cx$ and $u = -k$, we obtain $(1 - cx)^{-k} = \sum_{n \geq 0} (-1)^n \binom{-k}{n} c^n x^n$, called the **negative binomial expansion**. By Definition 1.2.4, $\binom{-k}{n} = \frac{1}{n!}(-k)_{(n)}$. As in Remark 1.2.6, $(-1)^n \binom{-k}{n} = \binom{k+n-1}{n} = \binom{n+k-1}{k-1}$. ∎

2.2.18. Application. We describe the general steps for applying the generating function method to constant-coefficient linear recurrences of order k; it may be helful to read the example below along with this.

Suppose $a_n = \sum_{i=1}^{k} c_i a_{n-i} + f(n)$ for $n \geq k$. To apply the generating function method, we multiply by x^n and sum over n with $n \geq k$. After factoring out $c_i x^i$, the term involving a_{n-i} becomes a multiple of $A(x)$ with initial terms missing.

Moving multiples of $A(x)$ to the left and the expressions for missing initial terms to the right yields $Q(x)A(x) = P(x) + R(x)$, where $Q(x) = 1 - \sum_{i=1}^{k} c_i x^k$ (a polynomial), P is a polynomial arising from the initial terms, and $R(x)$ is the formal power series for the inhomogeneous terms. Note that $Q(x) = \prod_{i=1}^{k}(1 - \alpha_i x)$, where $\alpha_1, \ldots, \alpha_k$ are the roots of the characteristic polynomial, since $Q(x) = x^k \phi(1/x)$.

How do we find $R(x)$? Consider the inhomogeneous term $F(n)c^n$ of Theorem 2.2.11 (with $\deg(F) = d$), so $R(x) = \sum_{n \geq k} F(n) c^n x^n$. Since $\binom{n+j-1}{j-1}$ is a polynomial in n of degree $j - 1$, we can express $R(x)$ as a linear combination of expressions of the form $(1 - cx)^{-j}$ for $1 \leq j \leq d + 1$.

We have written $A(x)$ as $\hat{P}(x)/\hat{Q}(x)$, where $\hat{P}$ and $\hat{Q}$ are polynomials, with $\hat{Q}(x)$ inheriting a higher power of $(1 - cx)$ from the denominator of $R(x)$. A ratio of two polynomials is called a **rational function**.

To expand a generating function $A(x)$ that is a rational function, we first use partial fractions to write the expression for $A(x)$ as a linear combination of expressions of the form $(1 - \alpha x)^{-j}$. We then extract $[x^n]A(x)$ by expanding these expressions using Lemma 2.2.17.

The effect of the inhomogeneous term here explains (indeed, proves) the claim of Theorem 2.2.11. The extra powers of $(1 - cx)$ in the denominator produce the same effect on the expansion (introducing c^n times a higher-degree polynomial) as the particular solution claimed in Theorem 2.2.11 (see Exercise 19).

The initial conditions determine P. The generating function method automatically picks the one solution to the recurrence that satisfies the initial conditions. To check for computational errors, it may be a good idea to show that the resulting formula for a_n satisfies the recurrence. ∎

The only part of the method we have not explained in detail is the partial fraction expansion. This we illustrate before proving that it works.

2.2.19. Example. *Partial fractions and the Heaviside method.* Consider $a_n = 2a_{n-1} - a_{n-2} + 2^n$ for $n \geq 2$, with $a_0 = 1$ and $a_1 = 3$. Multiply by x^n and sum over $n \geq 2$. The inhomogeneous term 2^n turns into $(2x)^2/(1-2x)$. The generating function $A(x)$ satisfies

$$A(x) - 1 - 3x = 2x[A(x) - 1] - x^2 A(x) + \sum_{n \geq 2}(2x)^n,$$

which becomes $[1 - 2x + x^2]A(x) = -2x + 3x + 1 + (2x)^2/(1-2x)$. Hence

$$A(x) = \frac{1+x}{(1-x)^2} + \frac{4x^2}{(1-2x)(1-x)^2} = \frac{1-x+2x^2}{(1-2x)(1-x)^2}.$$

To obtain $[x^n]A(x)$, we need the numerators in the partial fraction expansion $A(x) = \frac{B}{1-x} + \frac{C}{(1-x)^2} + \frac{D}{1-2x}$. Clearing fractions yields

$$1 - x + 2x^2 = B(1-x)(1-2x) + C(1-2x) + D(1-x)^2. \qquad (*)$$

As in Example 2.2.12, there are two methods to compute the constants. Equating coefficients yields three linear equations for $\{B, C, D\}$.

Alternatively, the **Heaviside method** (named for Oliver Heaviside) saves work. It uses the fact that equal polynomials are equal at all values. To express $f(x)/\prod(1 - \alpha_i x)$ as $\sum b_i/(1 - \alpha_i x)$ when the roots are distinct, first clear fractions. Letting $x = 1/\alpha_i$ sets all but one term on the right to 0 and yields $f(1/\alpha_i) = b_i \prod_{j \neq i}(1 - \alpha_j/\alpha_i)$. This computes b_i directly. With repeated roots, the method still computes the coefficients of the highest-power terms, and then evaluating at other appropriate values simplifies computation of the other coefficients.

For example, setting $x = 1$ and $x = 1/2$ in $(*)$ yields $2 = -C$ and $1 = D/4$. Setting $x = 0$ yields $1 = B + C + D$, and hence $B = -1$. Finally,

$$a_n = [x^n]\left(\frac{-1}{1-x} + \frac{-2}{(1-x)^2} + \frac{4}{1-2x}\right) = -1 - 2(n+1) + 4 \cdot 2^n.$$

After such a computation, it is a good idea to check that the resulting formula agrees with the initial conditions! ∎

We have justified the generating function method for recurrences solved by the characteristic equation method, except for proving that partial fraction expansion works. The next theorem does that, without saying how to find a partial fraction expansion. It implies that the linear equations for partial fraction expansion always have a unique solution.

For simplicity, we state the theorem only in the homogeneous setting. It extends to inhomogeneous terms of the form $F(n)c^n$ because the generating functions for these are again rational functions, as discussed in Application 2.2.18. The theorem then implies that the particular solutions arise as solutions to homogeneous recurrences of higher order.

2.2.20. Theorem. Let $c_1, \ldots, c_k$ be complex numbers with $c_k \neq 0$, and let $Q(x) = 1 - c_1 x - \cdots - c_k x^k$. If $Q(x) = \prod_{i=1}^{r}(1 - \alpha_i x)^{d_i}$ for distinct $\alpha_1, \ldots, \alpha_r$, then the following are equivalent for a sequence $\langle a \rangle$.

(A) $\langle a \rangle$ satisfies $a_n = c_1 a_{n-1} + \cdots + c_k a_{n-k}$ for $n \geq k$.
(B) $\langle a \rangle$ has generating function $P(x)/Q(x)$, where P is
     a polynomial of degree less than k.
(C) $\langle a \rangle$ has generating function that is a linear combination of
     expressions of the form $(1 - \alpha_i x)^{-j}$, where $1 \leq j \leq d_i$.
(D) $a_n = \sum_{i=1}^{r} P_i(n) \alpha_i^n$ for $n \geq 0$, where each P_i is
     a polynomial of degree less than d_i.

Proof: (Stanley [1986, pp. 202–203]) Let V_A, V_B, V_C, V_D be the sets of sequences satisfying conditions A, B, C, D above. These sets are closed under addition and under multiplication by constants. Hence each is a linear subspace of the space of sequences in $\mathbb{C}$. Each is specified by k independent parameters (initial conditions $a_0, \ldots, a_{k-1}$, coefficients of P, coefficients of k functions of the form $(1 - \alpha_i x)^{-j}$, or d_i coefficients in each P_i). Hence all have dimension at most k.

By Lemma 2.2.5, $\dim(V_A) = k$. When a vector space is contained in another of the same finite dimension, they are equal. It therefore suffices to show $V_A \subseteq V_B \subseteq V_C \subseteq V_D$. We show this by A $\Rightarrow$ B $\Rightarrow$ C $\Rightarrow$ D (equivalent to justifying the generating function method). The proof just reviews earlier discussion. Let $A(x)$ be the generating function for $\langle a \rangle$.

A $\Rightarrow$ B. Multiply the recurrence by x^n and sum over n at least k. The term $c_i a_{n-i}$ becomes $c_i x^i [A(x) - G_i(x)]$, where $G_i(x) = \sum_{j=0}^{k-i} a_j x^j$. Moving the terms with $A(x)$ to the left yields $Q(x)A(x) = P(x)$ for a polynomial P of degree less than k.

B $\Rightarrow$ C. This step is partial fraction expansion. Consider $Q(x) = f(x)g(x)$. If $f(x)$ and $g(x)$ have no common zeros, then there exist polynomials C and D such that $C(x)f(x) + D(x)g(x) = 1$ (see Exercise 21). Multiplying by $P(x)/Q(x)$ yields an expansion; this generalizes for r distinct roots. With P having degree less than k, in the resulting expansion $\sum_{i=1}^{r} p_i(x)/(1 - \alpha_i x)^{d_i}$ the degree of p_i is less than d_i. Expanding $p(x)/(1 - ax)^d$ into $\sum_{j=1}^{d} b_j/(1 - ax)^j$ follows inductively by clearing fractions and matching coefficients.

C $\Rightarrow$ D. By Lemma 2.2.17, $\frac{b_j}{(1 - \alpha_i x)^j} = \sum_{n \geq 0} b_j \binom{n+j-1}{j-1} \alpha_i^n x^n$. For fixed j, the expression $\binom{n+j-1}{j-1}$ is a polynomial in n of degree $j - 1$. A linear combination of these expressions for $1 \leq j \leq d_i$ (using coefficients $b_1, \ldots, b_{d_i}$) is a polynomial of degree less than d_i. Doing this for all i yields $a_n = \sum_{i=1}^{r} P_i(n) \alpha_i^n$, where each P_i is some polynomial of degree less than d_i. ∎

2.2.21. Remark. Stanley treated Theorem 2.2.20 as a statement in the theory of rational generating functions, part of the calculus of finite differences, whose origins have been ascribed to Taylor in 1717, Stirling in 1730, and Euler in 1755.

Theorem 2.2.20 also justifies the characteristic equation method, because $Q(x) = \prod_{i=1}^{r}(1 - \alpha_i x)^{d_i}$ if and only if each α_i is a characteristic root of multiplicity d_i. This follows from $Q(x) = x^k \phi(1/x)$.

Finding the numerators in the partial fraction expansion of the generating function is equivalent to finding coefficients to satisfy the initial conditions in the characteristic equation method. This "trivial" step may take the most effort by hand but is easily done by computer. ∎

The generating function method also works on some nonlinear recurrences where the equation for $A(x)$ is quadratic or even differential. It may still be possible to extract coefficients and thus solve the recurrence.

We apply this to the Catalan recurrence, obtaining once again the Catalan numbers. Here we use products of generating functions for the first time. When multiplying formal power series, each term in the first is multiplied by each term in the second, and the coefficient of x^r in the product accumulates all contributions involving r factors of x. That is,

$$\Big(\sum_{m\geq 0} b_m x^m\Big)\Big(\sum_{m\geq 0} c_m x^m\Big) = \sum_{r\geq 0}\Big(\sum_{k=0}^{r} b_k c_{r-k}\Big)x^r.$$

2.2.22. Theorem. If $a_0 = 1$, and $a_n = \sum_{k=1}^{n} a_{k-1}a_{n-k}$ for $n \geq 1$, then

$$\sum_{n\geq 0} a_n x^n = \frac{1 - \sqrt{1 - 4x}}{2x}, \quad \text{and} \quad a_n = \frac{1}{n+1}\binom{2n}{n}.$$

Proof: Shift the index to rewrite the recurrence as $a_n = \sum_{k=0}^{n-1} a_k a_{n-1-k}$ for $n \geq 1$. To apply the generating function method, multiply by x^n and sum over $n \geq 1$. The left side becomes $A(x) - 1$, where $A(x) = \sum_{m\geq 0} a_m x^m$.

The right side becomes $x \sum_{n=1}^{\infty}\big(\sum_{k=0}^{n-1} a_k a_{n-1-k}\big)x^{n-1}$. Letting $r = n-1$ writes this as $x[A(x)]^2$ using the formula for products of power series.

The resulting equation for $A(x)$ is $A(x) - 1 = x[A(x)]^2$. By the quadratic formula, $A(x) = \big(1 \pm \sqrt{1-4x}\big)/(2x)$. We claim that only the minus sign gives a valid solution. By definition, the coefficient of x^{-1} in $A(x)$ must be 0, so the constant terms in the numerator must cancel. By the Extended Binomial Theorem (Theorem 1.2.5), $\sqrt{1-4x} = \sum_{k\geq 0}\binom{1/2}{k}(-4x)^k$. The term for $k = 0$ in this sum is 1; hence we must choose the negative sign to make the constants cancel.

The coefficient of x^{-1} is the only term affected by the 1 in the numerator. Thus to compute a_n for $n \geq 0$, we seek $[x^n]\frac{-1}{2}(1-4x)^{1/2}x^{-1}$, which equals $\frac{-1}{2}[x^{n+1}](1-4x)^{1/2}$. We now have the tools to compute

$$a_n = -\frac{1}{2}\binom{1/2}{n+1}(-4)^{n+1} = \frac{(-1)2^{n+1}}{2(n+1)!}(-2)^{n+1}\prod_{i=0}^{n}\Big(\frac{1}{2}-i\Big)$$

$$= \frac{(-1)2^n\, n!}{(n+1)!\, n!}\prod_{i=0}^{n}(2i-1) = \frac{\prod_{i=1}^{n}(2i)\prod_{i=1}^{n}(2i-1)}{(n+1)n!n!} = \frac{1}{n+1}\binom{2n}{n}. \quad\blacksquare$$

The generating function method is valid also with multiple indices. Fixing one index defines a sequence of generating functions, say $A_n(x) = \sum_{k\geq 0} a_{n,k}x^k$. The recurrence for the coefficients yields a recurrence for this sequence. The generating function method in one variable can then yield a generating function for the sequence of generating functions, say as $B(x,y) = \sum_{n\geq 0} A_n(x)y^n$.

2.2.23.* Example. *Binomial coefficients.* Let $a_{n,k} = a_{n-1,k-1} + a_{n-1,k}$ for $n \geq 1$, with $a_{n,0} = 1$ and $a_{0,k} = 0$ for $n, k \geq 1$. Failing to notice that this is the recurrence given by Pascal's Formula, we solve for $a_{n,k}$.

For $n \in \mathbb{N}_0$, let $A_n(x) = \sum a_{n,k} x^k$. Multiply the original recurrence by x^k and sum over $k \geq 1$ to get $A_n(x) - 1 = x A_{n-1}(x) + A_{n-1}(x) - 1$ for $n \geq 1$. This simplifies to $A_n(x) = (x+1)A_{n-1}(x)$, with $A_0(x) = 1$.

We now solve the recurrence for $\langle A \rangle$. Failing to notice that $A_n(x) = (x+1)^n$, we define $B(x,y) = \sum A_n(x) y^n$. Multiplying the recurrence for A_n by y^n and summing over $n \geq 1$ yields $B(x,y) - 1 = (1+x)y B(x,y)$, and hence $B(x,y) = [1 - (1+x)y]^{-1}$.

Now $A_n(x)$, the coefficient of y^n in the expansion of $B(x,y)$, is $(1+x)^n$. Expanding this in turn to obtain the coefficient of x^k yields $\binom{n}{k}$ as the solution to the original recurrence in two indices. ∎

The generating function method also works on systems of recurrences. We multiply the recurrences by x^n and sum over n to obtain a system of equations for the generating functions.

2.2.24.* Example. *Systems of recurrence relations.* Suppose that $a_n = a_{n-1} + b_{n-1}$, $b_n = b_{n-1} + c_{n-1}$, and $c_n = a_{n-1} - b_{n-1}$, all valid for $n \geq 1$. The resulting system below can be solved for A, B, and C as functions of x. ∎

$$\begin{aligned} A(x) - a_0 &= xA(x) + xB(x) \\ B(x) - b_0 &= xB(x) + xC(x) \quad \text{or} \\ C(x) - c_0 &= xA(x) - xB(x) \end{aligned} \qquad \begin{aligned} (1-x)A - xB &= a_0 \\ (1-x)B - xC &= b_0 \\ -xA + xB - C &= c_0 \end{aligned}$$

EXERCISES 2.2

To prove a formula by induction, one must already know the formula. Solutions found non-inductively can be checked (for mistakes) using induction.

2.2.1. (–) Let $\langle a \rangle$ satisfy $a_n = 2a_{n-1} + 3a_{n-2}$ for $n \geq 2$.
(a) Prove that a_n is odd for all $n \geq 0$ when a_0 and a_1 are odd.
(b) Solve for a_n in terms of a_0 and a_1.

2.2.2. (–) Let $\langle a \rangle$ satisfy $(n-3)a_n = na_{n-1} - n^2 - n$ for $n \geq 4$, with $a_3 = 10$. Prove $a_n = \binom{n}{3} + 2\binom{n}{2} + \binom{n}{1}$ for $n \geq 3$. Is there a simpler expression for a_n?

2.2.3. (–) Let $\langle a \rangle$ satisfy $a_n = a_{n-4}$ for $n \geq 4$. Use the characteristic equation method to develop a single formula for a_n in terms of n and the initial conditions a_0, a_1, a_2, a_3.

2.2.4. (–) Let $a_n = 3a_{n-1} - 2a_{n-2} + 1$ for $n \geq 2$, with $a_0 = 2$ and $a_1 = 4$. Obtain a simple formula for a_n.

2.2.5. (–) At the start of each year, \$100 is added to a savings account. At the end of the year, 5% interest is added. Obtain and solve a recurrence for the account value at the end of the nth year. The account is empty before the first year.

2.2.6. (–) A mortgage loan begins with \$50,000 owed to the bank. During each year, 5% of the unpaid loan amount is added as interest. At the end of each year, the borrower pays \$5,000. Obtain a recurrence for the unpaid amount at the end of the nth year. Use a calculator to determine the number of years needed to pay off the mortgage. What happens if the interest rate is 10% instead of 5%?

2.2.7. (−) Consider a set of n circles in the plane such that each intersects every other (twice) and no three circles meet at a point. Obtain and solve a recurrence for the number of regions into which the circles cut the plane.

2.2.8. (−) Let $\alpha_1, \ldots, \alpha_k$ be integers, and let $\lambda_1, \ldots, \lambda_k$ be real numbers. Let $a_n = \sum_{i=1}^{k} \lambda_i \alpha_i^n$. Prove that if $a_0, \ldots, a_{k-1}$ are integers, then so is a_n for $n \geq 0$.

2.2.9. (−) Let $\langle a \rangle$ satisfy $a_n = c_1 a_{n-1} + c_2 a_{n-2}$, with initial values a_0, a_1. Express the generating function for a as a ratio of two polynomials.

2.2.10. (−) Let $\langle b \rangle$ and $\langle d \rangle$ satisfy inhomogeneous linear kth-order recurrences given by $a_n = \left(\sum_{i=1}^{k} h_i(n) a_{n-i} \right) + f(n)$ and $a_n = \left(\sum_{i=1}^{k} h_i(n) a_{n-i} \right) + g(n)$, respectively. Prove that $\langle b \rangle +$ $\langle d \rangle$ satisfies $a_n = \left(\sum_{i=1}^{k} h_i(n) a_{n-i} \right) + f(n) + g(n)$.

2.2.11. (◊) Let a_n be the number of n-tuples in $[4]^n$ that have at least one 1 and have no 2 appearing before the first 1 (note that $\langle a \rangle$ begins $0, 1, 6, \ldots$). Obtain and solve a recurrence for $\langle a \rangle$. Give a direct counting argument (without using summations) to prove the resulting simple formula.

2.2.12. For $m, n \in \mathbb{N}$, let $a_{m,n}$ be the number of squares (of all sizes, with positive area) formed by a grid of m horizontal and n vertical lines, given that the two sets of lines are equally spaced by the same amount. Obtain and solve a recurrence for $a_{m,n}$. Give a short alternative argument for a different solution formula.

2.2.13. Example 2.2.12 solves a recurrence to show that the number of regions formed by n pairwise intersecting lines in the plane is $1 + n + \binom{n}{2}$. Give a bijective counting argument to prove this formula.

2.2.14. Solve the recurrences below, given $a_0 = a_1 = 1$.
   (a) $a_n = 4a_{n-1} - 4a_{n-2} - n + 6$.
   (b) $a_n = 5a_{n-1} - 6a_{n-2} + 2n - 1 + 2^n$.

2.2.15. The chords of a convex n-gon are the segments joining pairs of vertices. Let a_n denote the number of pairs of chords that cross.
   (a) Compute $a_n - a_{n-1}$.
   (b) Solve the recurrence relation in part (a) to obtain a formula for a_n. (Comment: Exercise 1.1.21 requests a direct combinatorial argument for the answer.)

2.2.16. (◊) Choose n points on a circle so that no point inside the circle lies on three of the $\binom{n}{2}$ chords. Let a_n be the number of regions formed inside the circle by the chords.
   (a) Obtain the recurrence relation $a_n = a_{n-1} + f(n)$ for $n \geq 1$, where $f(n) = n - 1 + \sum_{i=1}^{n-1} (i-1)(n-1-i)$, with initial condition $a_0 = 1$.
   (b) Solve the recurrence of part (a) to obtain a formula for a_n as a sum of three binomial coefficients. (Hint: Avoid the characteristic equation method.)

2.2.17. Let $a_n = n^3$. Find a constant-coefficient first-order linear recurrence relation satisfied by $\langle a \rangle$. Does there exist a homogeneous constant-coefficient higher-order linear recurrence relation satisfied by $\langle a \rangle$?

2.2.18. (◊) Suppose that $\langle a \rangle$ satisfies $a_n = \left(\sum_{i=1}^{k} c_i a_{n-i} \right) + F(n) \alpha^n$, where F is a polynomial of degree d. Prove that $\langle a \rangle$ satisfies the homogeneous recurrence whose characteristic polynomial is $(x - \alpha)^{d+1} \phi(x)$, where $\phi(x)$ is the characteristic polynomial of the recurrence that arises by setting $F(n) = 0$.

2.2.19. Use the generating function method to prove Theorem 2.2.11. That is, given the recurrence $a_n = \left(\sum_{i=1}^{k} c_i a_{n-i} \right) + F(n) c^n$, where F is a polynomial of degree d and c has multiplicity r as a characteristic root, prove that there is a solution of the form $a_n = P(n) n^r c^n$, where P is a polynomial of degree at most d.

2.2.20. Given (j, α) with $j \in \mathbb{N}_0$ and $\alpha \in \mathbb{C}$, let $a_n = n^j \alpha^n$. The proof of Theorem 2.2.7 can be completed by showing that for any finite set of distinct such pairs (j, α), the resulting sequences are linearly independent in the vector space of sequences over $\mathbb{C}$, meaning that no linear combination of these sequences gives the all-0 sequence.

(a) Show that in an equation of dependence, attention can be restricted to the pairs with $|\alpha|$ maximized, and among these to the pairs with j maximized.

(b) For $\beta_1, \ldots, \beta_m$, the **Vandermonde matrix** is the m-by-m matrix with β_r^{s-1} in position (r, s). Its determinant is $\prod_{i<j}(\beta_j - \beta_i)$. Use this to complete the proof.

2.2.21. Let $\mathbb{C}[x]$ denote the family of polynomials in x with complex coefficients. A set in $\mathbb{C}[x]$ is an **ideal** if it is closed under addition and under multiplication by any polynomial.

(a) Prove that every ideal in $\mathbb{C}[x]$ has the form $\{pg: p \in \mathbb{C}[x]\}$ for some polynomial g.

(b) Prove that if f and g have no common zeros, then there exist $c, d \in \mathbb{C}[x]$ such that $cf + dg = 1$. (Comment: This completes the proof of Theorem 2.2.20. Assume the Fundamental Theorem of Algebra: every member of $\mathbb{C}[x]$ is a product of linear factors.)

2.2.22. Prove that the Tower of Hanoi Problem with n rings (Example 2.2.9) is solved in $2^n - 1$ moves as follows: On odd-indexed moves, move the smallest ring one peg rightward (cyclically). On even-indexed moves, move a non-smallest ring. (Buneman–Levy [1980], Walsh [1983]) (Comment: The procedure is essentially unique, although there are other descriptions, such as Hayes [1977], Wood [1981], Cull–Ecklund [1982], and Lavallée [1982]. In Walsh [1982], if we never move the same ring twice in a row, then it suffices to never let two rings with indices of the same parity touch.)

2.2.23. ($\diamond$) For each recurrence below, valid for $n \geq 2$ with $a_0 = a_1 = 1$, solve for a_n using both the characteristic equation method and the generating function method. What changes if the initial conditions are $a_0 = 1$ and $a_1 = 0$?

(a) $a_n = 3a_{n-1} - 2a_{n-2} + 2^n$.

(b) $a_n = 4(a_{n-1} - a_{n-2}) + 2^n + 1$.

2.2.24. Let $a_n = 4a_{n-1} - 5a_{n-2} + 2a_{n-3} + 2^{n-2}$ for $n \geq 3$, with $(a_0, a_1, a_2) = (1, 2, 7)$. Obtain a simple formula for a_n.

2.2.25. For $m \in \mathbb{N}_0$, let $a_m = 2^{-m} \sum_{k=0}^m \binom{2m+1}{2k} 3^k$.

(a) Prove $a_m = \frac{1}{2}(1 + \sqrt{3})(2 + \sqrt{3})^m + \frac{1}{2}(1 - \sqrt{3})(2 - \sqrt{3})^m$.

(b) Use part (a) to prove $a_m = 4a_{m-1} - a_{m-2}$ for $m \geq 2$.

(c) Conclude that a_m is an odd integer for all m. (Alkan [1995])

2.2.26. ($\diamond$) Let a_n be the number of words of length n on the alphabet $\{0, 1, 2\}$ such that 1 and 2 are never adjacent.

(a) Obtain a second-order recurrence relation for $\langle a \rangle$.

(b) Solve for a_n using both the characteristic equation method and the generating function method.

2.2.27. *The k-Fibonacci numbers.* Fix $k \in \mathbb{N}$. Let $F_{k,n} = kF_{k,n-1} + F_{k,n-2}$ for $n \geq 2$, with $F_{k,0} = 0$ and $F_{k,1} = 1$.

(a) Solve the recurrence to obtain a formula for $F_{k,n}$.

(b) Use the formula for $F_{k,n}$ to obtain a simple expression for $\sum_{i=0}^n \binom{2n+1}{i} F_{k,2n+1-2i}$. Note that $(F_{1,0}, F_{1,1}, F_{1,2}) = (1, 5, 25)$. (Plaza–Falcón [2016])

2.2.28. The Fibonacci recurrence can be encoded in matrices as $\binom{F_{n+1}}{F_n} = \binom{1\ 1}{1\ 0}\binom{F_n}{F_{n-1}}$. Use this expression and matrix eigenvalues to prove $F_n = \frac{1}{\sqrt{5}}(\alpha_1^n - \alpha_2^n)$, where $\alpha_1 = (1 + \sqrt{5})/2$ and $\alpha_2 = (1 - \sqrt{5})/2$. (The initial conditions are $F_0 = 0$ and $F_1 = 1$.)

2.2.29. ($\diamond$) *Characteristic roots from eigenvalues.* Let $\langle a \rangle$ satisfy $a_n = \sum_{i=1}^{k} c_i a_{n-i}$.

(a) Express $(a_n, \ldots, a_{n-k+1})^T$ in terms of $(a_{n-1}, \ldots, a_{n-k})^T$ using a matrix A.

(b) Derive the relationship between the roots of the characteristic polynomial of the recurrence and the eigenvalues of A (with proof).

(c) Use A to determine $\lim_{n\to\infty} a_n/a_{n-1}$.

2.2.30. ($\diamond$) Given $a_n = 2a_{n-1} + a_{n-2}$ for $n \geq 2$, with $a_0 = 1$ and $a_1 = 2$, use the generating function method to prove that $a_n = \sum \frac{(i+j+k)!}{i!j!k!}$, summed over all nonnegative integer triples (i, j, k) such that $i + j + 2k = n$. (Comment: This is again the **Pell sequence** obtained combinatorially in Exercise 2.1.35.)

2.2.31. Let $a_{n,k}$ be the number of compositions of n with k parts. Derive a recurrence and use the generating function method to obtain $\sum_{n,k\geq 0} a_{n,k} x^n y^k$ and compute $a_{n,k}$.

2.2.32. ($\diamond$) Let $b_{n,k}$ be the number of k-subsets of $[n]$ having no two consecutive integers.

(a) Obtain a recurrence relation in two indices for these numbers.

(b) Use the generating function method in two variables to compute $b_{n,k}$. (Hint: Follow Example 2.2.23. Let $B_k(x) = \sum_{n\geq 0} b_{n,k} x^n$, and use the recurrence of part (a) to obtain a recurrence for this sequence of generating functions.)

2.2.33. ($\diamond$) Let $a_{n,k}$ be the number of ways to partition $[n]$ into k sets so that no two consecutive numbers are in the same set. ("Partition" means that the k sets are nonempty and unlabeled; for example, $a_{n,2} = 1$.) Define $b_{n,k}$ similarly, except that for $b_{n,k}$ also n is forbidden to be in the same set with 1; for example $b_{n,2}$ is 1 if n is even, but 0 if n is odd.

(a) Obtain a recurrence relation for $a_{n,k}$.

(b) Let $A_k(x) = \sum_{n=0}^{\infty} a_{n,k} x^n$. Use the generating function method to express $A_k(x)$ as a ratio of polynomials.

(c) Obtain a recurrence for $b_{n,k}$ involving the number $a_{n,k}$. Use this and part (b) to express $\sum_{n=0}^{\infty} b_{n,k} x^n$ as a ratio of polynomials. (Knuth [2005])

2.2.34. ($\diamond$) The Delannoy numbers satisfy $d_{m,n} = d_{m,n-1} + d_{m-1,n-1} + d_{m-1,n}$ for $m, n \geq 1$, with $d_{m,0} = d_{0,n} = 1$ for $m, n \geq 0$. Find the generating function $\sum_{m\geq 0} \sum_{n\geq 0} d_{m,n} x^m y^n$. Using a clever factorization of the denominator, prove that $d_{m,n} = \sum_k 2^k \binom{m}{k}\binom{n}{k}$. (Hint: Let $(1-x)$ and $(1-y)$ be factors of the denominator. Comment: The combination of Proposition 1.2.10 and Theorem 1.2.13 yields this formula combinatorially.) (Callan [2003c])

2.2.35. ($\diamond$) A row of n lightbulbs, initially all off, must be turned on. Bulb 1 can be turned on or off at any time. For $i > 1$, bulb i can be turned on or off only when bulb $i-1$ is on and all earlier bulbs are off. Let a_n be the number of steps needed to turn all on; note that $\langle a \rangle$ begins $(0, 1, 2, 5, \ldots)$. Let b_n be the number of steps to turn on bulb n for the first time.

(a) Find a recurrence for $\langle b \rangle$ and solve it.

(b) Use $\langle b \rangle$ to find a recurrence for $\langle a \rangle$. (Tucker)

(c) Solve the recurrence for $\langle a \rangle$.

2.2.36. Let a_n be the number of domino tilings of $[0, 2n] \times [0, 4]$ in which dominos do not cross the $n-1$ vertical cuts from $(2i, 2)$ to $(2i, 4)$ shown as heavy lines below.

(a) Obtain and solve a second-order recurrence for $\langle a \rangle$. (Hint: It may be easiest to first obtain a system of recurrences for $\langle a \rangle$ and another sequence $\langle b \rangle$.)

(b) Let $C(m, j)$ be the set of nonnegative integer vectors $(x_0, \ldots, x_j)$ with sum m. Prove $a_n = \sum_{k=0}^{n} 2^{n-k} \sum_{x \in C(n-k,k)} \prod_{i=0}^{k} \hat{F}_{2x_i}$. (Tauraso [2006])

2.2.37. (+) Let a_n count the ways to add noncrossing chords to a fixed convex $(n+2)$-gon. The empty set is allowed, so $(a_0, a_1, a_2, a_3) = (1, 1, 3, 11)$. Find a recurrence and obtain the generating function for $\langle a \rangle$. (Schröder [1870], Stanley [1971])

2.2.38. ($\diamond$) **Touchard's Formula** (Shapiro [1976]). Given n points on a circle, let t_n count the ways to pair some by noncrossing chords and color the others red or blue. (Given n politicians at a circular table, those not in the red party or blue party are shaking hands in a noncrossing pairing; n need not be even.)

(a) Show that $t_n = \sum_{k \geq 0} \binom{n}{2k} 2^{n-2k} C_k$, where $C_n = \frac{1}{n+1}\binom{2n}{n}$.

(b) Find a recurrence for t_n, and use it to obtain $\sum t_n x^n$.

(c) From parts (a) and (b), prove that $C_{n+1} = \sum_{k \geq 0} \binom{n}{2k} 2^{n-2k} C_k$.

2.2.39. Let L be the L-shaped region of area 3 obtained from $[0,2] \times [0,2]$ by deleting $[1,2] \times [1,2]$. Let a_n count the ways to cut a 4-by-n rectangle into unit squares and unrotated translations of L. Express $\sum a_n x^n$ as a ratio of two polynomials. (Deutsch [2001])

2.2.40. (+) Let $\phi = (1+\sqrt{5})/2$; note that $\phi^2 = \phi + 1$. Let $f(1) = 1$, and for $n \geq 1$ let

$$f(n+1) = \begin{cases} f(n) + 2 & \text{if } f(f(n) - n + 1) = n \\ f(n) + 1 & \text{otherwise} \end{cases}$$

Prove $f(n) = \lfloor \phi n \rfloor$, and determine when $f(f(n) - n + 1) \neq n$. (Doster [1990])

2.3. Further Topics

We consider asymptotic solution of recurrences and a technique using recurrences to evaluate summations.

THE SUBSTITUTION METHOD

The **substitution method** reduces a recurrence to a simpler recurrence for a related sequence. In $a_n^2 = a_{n-1}^2 + 2$, for example, we can substitute $b_n = a_n^2$, solve for b_n, and take its square root. In general, we let b_n be a suitable function of a_n and rewrite the recurrence in terms of $\langle b \rangle$.

2.3.1. Example. $a_n = na_{n-1} + n!$ *for* $n \geq 1$, *with* $a_0 = 1$. The recurrence is close to a recurrence for $n!$. This suggests letting $b_n = a_n/n!$ to obtain

$$n!b_n = n(n-1)!b_{n-1} + n!,$$

which simplifies to $b_n = b_{n-1} + 1$. Since $a_0 = 1$, we have $b_0 = 1$, and hence the solution to the new recurrence is $b_n = n+1$. We obtain $a_n = (n+1)!$ as the solution to the original recurrence. ∎

2.3.2. Proposition. The number D_n of derangements of $[n]$ is given by

$$D_n = n! \sum_{k=0}^{n} \frac{(-1)^k}{k!}.$$

Proof: In Example 2.1.6 we showed $D_n = (n-1)(D_{n-1} + D_{n-2})$ for $n \geq 2$, with $D_0 = 1$ and $D_1 = 0$. (Even without knowing D_0, given only D_1 and D_2, we can define D_0 to extend the region of validity of the recurrence.)

We use the substitution method. To eliminate the factor $n - 1$, introduce an auxiliary sequence $\langle a \rangle$ defined by $a_n = D_n/n!$. Note that a_n is then the probability that a random permutation of $[n]$ is a derangement. From $D_0 = 1$ and $D_1 = 0$ we obtain $a_0 = 1$ and $a_1 = 0$. Substitution yields

$$n! a_n = (n - 1)(n - 1)! a_{n-1} + (n - 1)! a_{n-2},$$

and then dividing by $n!$ yields $a_n = (1 - 1/n) a_{n-1} + (1/n) a_{n-2}$.

Now rewrite the recurrence for $\langle a \rangle$ as

$$a_n - a_{n-1} = -\tfrac{1}{n}(a_{n-1} - a_{n-2}).$$

The natural substitution $b_n = a_n - a_{n-1}$ for $n \geq 2$ leads to $b_n = -\tfrac{1}{n} b_{n-1}$. This substitution *reduces the order* of the recurrence, analogous to substitutions reducing the order of differential equations. To recover $\langle a \rangle$ from $\langle b \rangle$, use $\sum_{k=1}^{n} b_k = \sum_{k=1}^{n}(a_k - a_{k-1}) = a_n - a_0$; the sum "telescopes".

The initial condition for $\langle b \rangle$ is $b_1 = a_1 - a_0 = -1$. The recurrence then yields $b_n = (-1)^n/n!$. Using the value of a_0 and the solution for b_k,

$$a_n = 1 + \sum_{k=1}^{n} \frac{(-1)^k}{k!} = \sum_{k=0}^{n} \frac{(-1)^k}{k!} .$$

Finally, $D_n = n! \sum_{k=0}^{n} (-1)^k/k!$.   ∎

The sum $\sum_{k=0}^{n} (-1)^k/k!$ is a partial sum of the series $\sum_{k \geq 0} x^k/k!$ at $x = -1$. As n grows, the probability of avoiding fixed points rapidly approaches the constant e^{-1}, where e denotes the base of the natural logarithm. Other derivations of the formula appear in Exercise 4, Corollary 3.3.17, and Theorem 4.1.13.

Substitutions can involve some guesswork. The principle of reducing to a simpler recurrence can suggest helpful substitutions.

2.3.3. Example. Define $\langle a \rangle$ by $n a_n = (n + 1) a_{n-1}$, with $a_0 = 1$. Multiplying the left side by $n + 1$ and the right side by n would permit canceling $n(n + 1)$. To achieve this, let $a_n = (n + 1) b_n$. (Also suggested by dividing the original recurrence by $n(n + 1)$.) We obtain $b_n = b_{n-1}$, with $b_0 = 1$. Thus $b_n = 1$ and $a_n = n + 1$, which would be quite easy to check if we guessed it initially.   ∎

2.3.4. Example. *Eliminating summations.* Subtracting successive instances of a recurrence may eliminate a summation. For example, consider $a_n = \sum_{i=0}^{n-1} a_i$ for $n \geq 1$, with $a_0 = 1$. We also write $a_{n-1} = \sum_{i=0}^{n-2} a_i$. Subtraction yields $a_n - a_{n-1} = a_{n-1}$, and hence $a_n = 2 a_{n-1}$. The solution to this first order recurrence is $a_n = 2^n$.

Unfortunately, this is wrong, since $2^n \neq \sum_{i=0}^{n-1} 2^i$. Since our original recurrence is valid only for $n \geq 1$, the expression $a_{n-1} = \sum_{i=0}^{n-2} a_i$ is valid only for $n \geq 2$. Thus we can subtract this expression to obtain $a_n - a_{n-1} = a_{n-1}$ only when $n \geq 2$. We use the original recurrence to obtain $a_1 = 1$, and now we have $a_n = 2^{n-1}$ for $n \geq 1$, with $a_0 = 1$. This does check: $2^{n-1} = 1 + \sum_{i=1}^{n-1} 2^{i-1}$.

The subtraction trick can be simpler or quicker than eliminating summations by substitution. In this example, substituting $b_n = \sum_{i=0}^{n} a_i$ into $a_n = \sum_{i=0}^{n-1} a_i$ yields $b_n - b_{n-1} = b_{n-1}$ for $n \geq 1$, with $b_0 = 1$. Thus $b_n = 2^n$, and $a_n = b_n - b_{n-1} = 2^{n-1}$ for $n \geq 1$ as before.   ∎

2.3.5. Example. *Binary search.*

$$a_n = a_{\lceil (n-1)/2 \rceil} + 1 \text{ with } a_0 = 0.$$

We seek a number x in a sorted list of length n. Let a_n be the number of probes we may need. When we probe location j and find y, we compare y with x. If $y \neq x$, then we consider only the earlier or later part of the list, depending on whether $x < y$ or $x > y$. We may always be forced to search the larger part. We minimize the larger size by probing the middle location; the resulting algorithm is **binary search**. After the first probe, we may need to solve a problem of size $\lceil \frac{n-1}{2} \rceil$. Hence $a_n = 1 + a_{\lceil (n-1)/2 \rceil}$. Also $a_0 = 0$, since no table of size 0 contains x.

For the growth rate of a_n, consider n when $n + 1$ is a power of 2. That is, define a new sequence by $b_k = a_{2^k-1}$ for $k \geq 0$. Substitution in the recurrence for $\langle a \rangle$ yields $b_k = b_{k-1}+1$. Also $a_0 = 0$ yields $b_0 = 0$, and hence $b_k = k$ for $k \geq 0$. Thus $a_n = \log_2(n + 1)$ when $n + 1$ is a power of 2. The full solution $a_n = \lceil \log_2(n + 1) \rceil$ follows by induction on n; it is the statement "$a_n = k$ for $2^{k-1} \leq n < 2^k$".   ∎

2.3.6.* Remark. The technique of splitting a problem into smaller subproblems and combining the solutions is called **divide-and-conquer**. Solutions to the recurrences that result from this process can sometimes be found by substitutions. Consider $a_n = ca_{\lceil n/d \rceil} + e$. To study this for powers of d, let $b_k = a_{d^k}$, so $b_k = cb_{k-1} + e$. When $c \neq 1$, we obtain $b_k = Ac^k + \frac{e}{1-c}$, or $a_n = An^{\log_d c} + \frac{e}{1-c}$.

For a linear inhomogeneous term en, the behavior depends on the ratio c/d. The substitution yields $b_k = cb_{k-1}+ed^k$. When $c \neq d$, we obtain $b_k = Ac^k + \frac{ed}{d-c}d^k$, yielding $a_n = An^{\log_d c} + \frac{ed}{d-c}n$ (with A determined by the initial condition). When $c = d$, the solution is $b_k = Ad^k + ekd^k$, which becomes $a_n = An + en\log_d n$.   ∎

ASYMPTOTIC ANALYSIS

Exact solutions may be unnecessary or too complicated. We may be content with a simple formula that approximates or is asymptotic to a_n. For example, in a linear recurrence with constant coefficients it may be enough to know the exponential growth that arises from the characteristic root of largest magnitude.

The substitution method can provide asymptotic information. We "peel off the controlling behavior" by substitutions, yielding a simpler recurrence. If substitutions yield a constant sequence, then we have explained everything. Our first example is easy to solve exactly.

2.3.7. Example. *The Tower of Hanoi, again:* $a_n = 2a_{n-1} + 1$ with $a_0 = 0$. We expect a_n to be about 2^n, so we set $a_n = 2^n b_n$ (setting $a_n = 2^n + b_n$ does not help!). Substituting and dividing by 2^n yields $b_n = b_{n-1} + 2^{-n}$, with $b_0 = 0$. Iterating yields $b_n = \sum_{i=1}^n 2^{-i} = 1 - 2^{-n}$, and hence $a_n = 2^n - 1$.   ∎

Asymptotic analysis first seeks the most extreme behavior. This may be factorial growth (or worse). Next is exponential behavior and then perhaps polynomial. Finding substitutions to extract the extreme behavior may take some cleverness. To avoid guessing, we make substitutions with parameters and then determine the values of the parameters that simplify the unexplained behavior.

2.3.8. Application. *Stirling's Formula.* Consider $a_n = na_{n-1}$ for $n \geq 1$, with $a_0 = 1$. The solution is $a_n = n!$. Computation with $n!$ is difficult for large n, so we seek an approximation using simpler functions. Since $\alpha^n < n! < n^n$ for fixed α when n is large, we set $b_n = a_n/n^n$ and study b_n. Substituting into $a_n = na_{n-1}$ yields $b_n = (1 - 1/n)^{n-1}b_{n-1}$. Since $(1 - 1/n)^{n-1} \to e^{-1}$, we have the "asymptotic recurrence" $b_n \sim e^{-1}b_{n-1}$. This suggests that the behavior of $\langle b \rangle$ is exponential, which is "simpler". (We could have tried the more general $b_n = a_n/n^{\beta n}$, but then we find that setting $\beta = 1$ eliminates the superexponential behavior from $\langle b \rangle$.)

We expect b_n to be close to e^{-n}, so we set $c_n = e^n b_n$. Substituting into the recurrence for $\langle b \rangle$ yields $c_n = e(1 - 1/n)^{n-1}c_{n-1}$. The ratio c_n/c_{n-1} tends to 1, which is necessary (but not sufficient) for the behavior of $\langle c \rangle$ to be "algebraic" (bounded by a polynomial). A careful look at c_n/c_{n-1} will lead us to the right exponent for the leading term. Suppose $c_n \sim \beta n^\alpha$ for some constants β and α. This yields $c_n/c_{n-1} \sim n^\alpha/(n-1)^\alpha = (1 - 1/n)^{-\alpha}$. By the Extended Binomial Theorem (Theorem 1.2.5), the expansion is $1 + \frac{\alpha}{n} + \frac{a(a+1)}{2n^2} + \cdots$; only the linear term is accurate since we parametrized only the leading behavior of c_n.

To determine α, we compare with our exact expression for the ratio: $c_n/c_{n-1} = e(1 - 1/n)^{n-1}$. Letting $W = c_n/c_{n-1}$, we obtain

$$\ln W = \ln e + (n - 1)\ln(1 - 1/n)$$
$$= 1 + (n - 1)(-\tfrac{1}{n} - \tfrac{1}{2n^2} - \tfrac{1}{3n^3} - \tfrac{1}{4n^4} - \cdots)$$
$$= \tfrac{1}{(1\cdot2)n} + \tfrac{1}{(2\cdot3)n^2} + \tfrac{1}{(3\cdot4)n^3} + \cdots.$$

Using the exponential series on $e^{\ln W}$ yields a series for W in powers of $1/n$. We obtain $W = 1 + \frac{1}{2n} + \frac{7}{24n^2} + \frac{3}{16n^3} + \cdots$. This yields $\alpha = \frac{1}{2}$ in the expansion for c_n/c_{n-1}.

To further refine the analysis, define d_n by $c_n = n^{1/2}d_n$. Substituting into the recurrence for $\langle c \rangle$ yields the exact recurrence $d_n = e(1 - 1/n)^{n-1/2}d_{n-1}$. Now we hope that d_n tends to a constant C. If so, then undoing the substitutions so far yields $n! \sim C(n/e)^n\sqrt{n}$.

To prove this, let $f_n = \ln d_n$; we show that f_n tends to a limit. By the Monotone Convergence Theorem, it suffices to show that $\langle f \rangle$ is decreasing and bounded below. The recurrence for d_n yields $f_n = f_{n-1} + 1 + (n - 1/2)\ln(1 - 1/n)$. Note that $b_1 = a_1 = 1$, $d_1 = c_1 = e$, and $f_1 = 1$. Expanding the series for $\ln(1 - 1/n)$ shows that the product $(n - 1/2)\ln(1 - 1/n)$ is less than -1; thus $\langle f \rangle$ is decreasing, and $f_n < 1$ for $n > 1$. To obtain a lower bound, we compute

$$f_n = 1 + \sum_{k=2}^{n}\left[1 + \left(k - \frac{1}{2}\right)\ln\left(1 - \frac{1}{k}\right)\right] = 1 + \sum_{k=2}^{n}\left[1 - \left(k - \frac{1}{2}\right)\left(\frac{1}{k} + \frac{1}{2k^2} + \frac{1}{3k^3} + \cdots\right)\right]$$

$$= 1 - \sum_{k=2}^{n}\sum_{t\geq2}\left(\frac{1}{t+1} - \frac{1}{2t}\right)k^{-t} = 1 - \sum_{t\geq2}\frac{t-1}{2t(t+1)}\sum_{k=2}^{n}k^{-t}$$

$$> 1 - \sum_{t\geq2}\frac{t-1}{2t(t+1)}\int_1^{\infty}k^{-t}dk = 1 - \frac{1}{2}\sum_{t\geq2}\frac{t-1}{t(t+1)}\frac{-1}{t-1}k^{-t+1}\Big|_{k=1}^{\infty}$$

$$= 1 - \frac{1}{2}\sum_{t\geq2}\frac{1}{t(t+1)} = 1 - \frac{1}{2}\sum_{t\geq2}\left(\frac{1}{t} - \frac{1}{t+1}\right) = \frac{3}{4}.$$

Since $\langle f \rangle$ is decreasing and bounded below, it has a limit, by the Monotone Convergence Theorem. Undoing the substitutions yields $n! = a_n = rn^n e^{-n} n^{1/2}$, where $e^{3/4} < r < e$. Furthermore, we have proved that $n!/(n^n e^{-n} n^{1/2})$ tends to a constant C. This asymptotic approach cannot compute C, although we have confined it between 2.1170 and 2.7183. We will in fact prove $C = \sqrt{2\pi} \approx 2.5066$.

Continuing the asymptotic analysis expands d_n in powers of $1/n$. For example, set $d_n = e_n(1 + \frac{\gamma}{n})$. We choose γ to make e_n/e_{n-1} as close to 1 as possible. Computing e_n/e_{n-1} by substituting into d_n/d_{n-1} yields $e_n/e_{n-1} = 1 + O(n^{-3})$ when $\gamma = \frac{1}{12}$ (Exercise 12). The process continues forever, producing what is called an **asymptotic series**; each substitution computes another coefficient. Using the overall constant $\sqrt{2\pi}$ computed below, the resulting asymptotic expansion is

$$n! = \sqrt{2\pi n}\left(\frac{n}{e}\right)^n \left[1 + \frac{1}{12n} + \frac{1}{288n^2} - \frac{139}{51840n^3} - \cdots\right]. \qquad \blacksquare$$

2.3.9. Theorem. (Stirling's Formula) $n! \sim n^n e^{-n}\sqrt{2\pi n}$.

Proof: Let $s_n = \frac{n!}{n^n e^{-n} n^{1/2}}$. The outcome of the asymptotic analysis in Application 2.3.8 is a proof that s_n tends to a constant C. To compute the constant, we use the **Wallis product**, which expresses $\pi/2$ as an infinite product: $\frac{\pi}{2} = \prod_{k=1}^{\infty} \frac{2k}{2k-1}\frac{2k}{2k+1}$ (derived in Exercise 13). We combine the exact expression $n! = s_n(n/e)^n\sqrt{n}$ and the fact that s_n tends to a limit value C to compute

$$\frac{\pi}{2} = \lim_{n\to\infty}\prod_{k=1}^{n}\frac{2k}{2k-1}\frac{2k}{2k+1} = \lim_{n\to\infty}\frac{4^{2n}(n!)^4}{(2n+1)(2n)!^2}$$

$$= \lim_{n\to\infty}\frac{4^{2n}(s_n(n/e)^n\sqrt{n})^4}{(2n+1)(s_{2n}(2n/e)^{2n}\sqrt{2n})^2} = \lim_{n\to\infty}\frac{s_n^4 n^2}{(2n+1)s_{2n}^2 \cdot 2n} = \frac{C^4}{4C^2} = \frac{C^2}{4}.$$

From this we conclude $C = \sqrt{2\pi}$. $\qquad \blacksquare$

2.3.10. Example. *Coin-flipping.* When a fair coin is flipped n times (and n is even), what is the probability of having the same number of heads and tails? There are 2^n equally likely sequences, and exactly $\binom{n}{n/2}$ have the same number of heads and tails. Stirling's Formula yields

$$\binom{n}{n/2} = \frac{n!}{(n/2)!^2} \approx \frac{n^n e^{-n}\sqrt{2\pi n}}{[(n/2)^{n/2}e^{-n/2}\sqrt{2\pi(n/2)}]^2} = \frac{2^n}{\sqrt{\pi n/2}}.$$

Dividing by 2^n tells us that the probability of equal numbers of heads and tails is about $c/\sqrt{n}$, where $c = (2/\pi)^{1/2}$. For $\binom{n}{an}$, see Exercise 14.

Without knowing Stirling's Formula, we can still obtain the leading (exponential) behavior of $\binom{n}{n/2}$. We have observed that $\lim x_{n+1}/x_n = L$ yields exponential behavior of the form L^n. More precisely, $\lim x_{n+1}/x_n = L$ implies $\lim x_n^{1/n} = L$ (Exercise 15). Taking the ratio of successive middle binomial coefficients yields 2^n as the leading behavior of $\binom{n}{\lfloor n/2 \rfloor}$. $\qquad \blacksquare$

2.3.11. Example. *Inversion of functions.* A natural question in many contexts is how big n must be so that $\binom{n}{n/2} \geq t$. Example 2.3.10 yields an approximate answer. Roughly, we are given $t = c2^n/\sqrt{n}$, where c is a constant, and we want to invert the expression of t in terms of n to obtain n in terms of t. The process is similar to asymptotic analysis of recurrences: we approximate the leading behavior and use substitutions to modify it to a more accurate solution.

We first take logarithms, where $\lg t$ denotes the base-2 logarithm: $\lg t \approx n - \frac{1}{2}\lg n$, dropping the additive constant. Setting n to $\lg t$ leaves the right side too small. Since the right side increases with n, we must enlarge n. With experience, it is apparent that n should be about $\lg t + \frac{1}{2}\lg\lg t$, but where does this come from?

From the leading behavior, substitute by setting $n = (\lg t) + x$, with x a function of t. Now the right side is $\lg t + x - \frac{1}{2}\lg(x + \lg t)$. If we correctly found the leading behavior, then $x \in o(\lg t)$, and we rewrite the last term as $\frac{1}{2}\lg[(\lg t)(1 + x/\lg t)]$, which equals $\frac{1}{2}[\lg\lg t + \lg(1 + x/\lg t)]$. When $x/\lg t \to 0$, our overall expression "behaves like" $\lg t + x - \frac{1}{2}\lg\lg t$, which simplifies to $\lg t$ when $x = \frac{1}{2}\lg\lg t$.

Essentially, the lower order term explains the effect of the lower order factor $\sqrt{n}$ in the original function. The point is that when we set n to be $\lg t + \frac{1}{2}\lg\lg t$, the difference between the two sides is of lower order than when we set n to be $\lg t$. Thus we can write $n = \lg t + \frac{1}{2}\lg\lg t + o(\lg\lg t)$. With further substitutions, we could reduce the error even further.   ∎

THE WZ METHOD (optional)

Many combinatorial identities evaluate sums. Combinatorial proofs are hard to find, so mathematicians have sought automated methods to compute sums, analogous to symbolic computation for integration. Here we describe a technique that enables computers to evaluate many sums that arise in practice.

The technique relates instances of a formula over different values of a parameter, like a recurrence does. Also, recurrences are used in the method (albeit in a step we will not explore in depth). This makes the method more an application of recurrences rather than a derivation or solution of recurrences. Like other subsections marked "optional", this is a topic not used in the rest of the book and may be skipped without loss of continuity.

Developed and popularized by H. Wilf and D. Zeilberger, this method is known as the **WZ method**. It is sketched in Wilf [1990, pp. 120–126] and presented in detail in Petkovšek–Wilf–Zeilberger [1996], a book with the simple title "**A = B**".

Consider the identity $\sum_k U(n, k) = H(n)$ for $n \geq 0$. Dividing by $H(n)$ yields an equivalent identity $\sum_k F(n, k) = 1$. This transformation discards the bijective approach; we are no longer summing integers. Nevertheless, since the right side is now constant, it suffices to verify the equality at $n = 0$ and prove $\sum_k [F(n + 1, k) - F(n, k)] = 0$ for $n \geq 0$.

The crux of the WZ method is finding an expression $G(n, k)$ such that $F(n + 1, k) - F(n, k) = G(n, k + 1) - G(n, k)$. This turns the needed sum into a telescoping sum.

2.3.12. Proposition. If F and G satisfy

$$F(n+1,k) - F(n,k) = G(n,k+1) - G(n,k)$$

and $\lim_{k \to \pm\infty} G(n,k) = 0$, then $\sum_k F(n,k)$ is independent of n.

Proof: Given such an expression, we write the difference in $\sum_k F(n,k)$ for two successive values of n in terms of G:

$$\sum_{k=A}^{B} [F(n+1,k) - F(n,k)] = \sum_{k=A}^{B} [G(n,k+1) - G(n,k)] .$$

The sum over G telescopes to $G(n,B+1) - G(n,A)$. Summing over all k yields the desired result when $\lim_{B \to \infty} G(n,B) = \lim_{A \to -\infty} G(n,A) = 0$.   ∎

A pair (F,G) satisfying the conditions of Proposition 2.3.12 is a **WZ pair** and certifies that $\sum_k F(n,k)$ is independent of n.

2.3.13. Example. $\sum_k \binom{n}{k} = 2^n$. Of course we know this, but let us see how the WZ method proves it. Let $F(n,k) = \binom{n}{k}/2^n$. If we set $G(n,k) = -\binom{n}{k-1}/2^{n+1}$, then the behavior of the binomial coefficient yields $\lim_{k \to \pm\infty} G(n,k) = 0$, and we need only verify the difference condition. The check is

$$\frac{\binom{n+1}{k}}{2^{n+1}} - \frac{2\binom{n}{k}}{2^{n+1}} = \frac{-\binom{n}{k}}{2^{n+1}} - \frac{-\binom{n}{k-1}}{2^{n+1}} .$$

By Proposition 2.3.12, $\sum_k F(n,k)$ is independent of n. To show that the value is 1, we set $n = 0$ in the sum, obtaining $\sum_k \binom{0}{k} 2^{-0} = 1$.   ∎

The difficulty is finding $G(n,k)$. When $\frac{F(n+1,k)}{F(n,k)}$ and $\frac{F(n,k-1)}{F(n,k)}$ both are rational functions of n and k, and $\sum_k F(n,k) = 1$, usually there exists such a function G of the form $G(n,k) = R(n,k)F(n,k-1)$, where $R(n,k)$ is a rational function of n and k. Moreover, there is an algorithm that finds such G when it exists or proves that it does not exist. This is the advantage of computerizing the process, along with the fact that computers can quckly perform manipulations that humans do not even want to try.

The computation in the example above shows that $R(n,k) = -\frac{1}{2}$ works when $F(n,k) = \binom{n}{k}/2^n$. To prove $\sum_k (-1)^k \binom{n}{k}\binom{2k}{k} 4^{n-k} = \binom{2n}{n}$, the magic expression is $R(n,k) = \frac{2k-1}{2n+1}$ (see Exercise 19). The next one looks more complicated.

2.3.14. Example. Dixon's Identity. The identity is

$$\sum_{k=-\infty}^{\infty} (-1)^k \binom{a+b}{a+k}\binom{a+c}{c+k}\binom{b+c}{b+k} = \frac{(a+b+c)!}{a!b!c!} .$$

When we set $a = n$, the magic expression is $R(n,k) = \frac{(c+1-k)(b+1-k)}{2(n+k)(n+b+c+1)}$.   ∎

To seek the expression $G(n, k)$ to mate with $F(n, k)$, we treat n as a parameter, defining $f(k) = F(n + 1, k) - F(n, k)$ for fixed n and seeking $g(k)$ such that $g(k + 1) - g(k) = f(k)$ to become $G(n, k)$. Since summation is the inverse to difference, we want $g(k) = \sum_{j=0}^{k} f(j)$ if the values are defined for $k \geq 0$. However, we also want $g(k)$ in "closed form".

2.3.15. Definition. An expression g in terms of k is in **closed form** if $g(k+1)/g(k)$ is a rational function of k. Such an expression g is also called a **hypergeometric term**. A **hypergeometric series** is a power series where the ratio of successive coefficients is a rational function of the summation index.

In a geometric series, the **term ratio** (ratio of successive terms) is constant. In a more general hypergeometric series, the term ratio is a ratio of polynomials. Although $k!$ and most ratios of binomial coefficients are in closed form, k^k is not. Closed form may involve factorials, polynomials, and exponentials in the index.

For $f(k)$ in closed form, an algorithm of Gosper [1978] tests whether there is a hypergeometric term $g(k)$ such that $f(k) = g(k + 1) - g(k)$, constructing it if so. Graham–Knuth–Patashnik [1989] presents the algorithm (see also Petkovšek–Wilf–Zeilberger [1996]); it is available in some symbolic computation software packages. We sketch the steps.

2.3.16. Algorithm. (sketch of **Gosper's Algorithm**). Given a hypergeometric term $f(k)$, we seek g such that $f(k) = g(k + 1) - g(k)$.
  (1) Let $r(k) = f(k + 1)/f(k)$.
  (2) Express $r(k)$ as $\frac{a(k)}{b(k)} \frac{c(k+1)}{c(k)}$, where a and b are polynomials such that $a(k)$ and $b(k + j)$ are relatively prime for all $j \in \mathbb{N}_0$.
  (3) Find a nonzero polynomial h solving $a(k)h(k + 1) - b(k - 1)h(k) = c(k)$, if such a polynomial exists.
  (4) If (3) fails, say g does not exist. Otherwise, let $g(k) = \frac{b(k-1)h(k)}{c(k)} f(k)$. ∎

As discussed in Petkovšek–Wilf–Zeilberger [1996] (Section 5.3), every rational function r has a factorization as requested in Step 2, and there is an algorithm to find it. We then seek h in Step 3 by equating corresponding coefficients.

When f has a particular form, we can state the factorization in Step 2 explicitly. If there are polynomials p and q and constants m_i such that

$$f(k) = \alpha^k \frac{p(k)(m_1 + k)!(m_2 - k)!}{q(k)(m_3 + k)!(m_4 - k)!} ,$$

then first simplify by cancellation (for example, write $(m_1 + k)!/(m_3 + k)!$ as $(m_1 + k)_{(m_1-m_3)}$ when $m_1 \geq m_3$). Dropping factors that are then absent, we can choose the polynomials a, b, and c roughly as below

$$a(k) = \alpha(m_1 + k + 1)(m_4 - k)q(k) ,$$
$$b(k) = q(k + 1)(m_3 + k + 1)(m_2 - k) ,$$
$$c(k) = p(k) .$$

2.3.17. Example. We apply Gosper's Algorithm to find the magic expression in Example 2.3.13. We are given $F(n, k) = \binom{n}{k}2^{-n}$, so

$$f(k) = \binom{n+1}{k}2^{-n-1} - \binom{n}{k}2^{-n} = \frac{1}{2^{n+1}}\binom{n}{k}\frac{2k-n-1}{n+1-k}.$$

Dividing $f(k+1)$ by $f(k)$ yields $r(k) = \frac{2k-n+1}{k+1}\frac{n+1-k}{2k-n-1}$. We set $a(k) = n+1-k$, $b(k) = k+1$, and $c(k) = 2k-n-1$. For Step 3, we seek a solution to

$$(n+1-k)h(k+1) - kh(k) = 2k-n-1.$$

Choosing $h(k) = -1$ satisfies the equation. The algorithm now outputs

$$G(n, k) = \frac{k(-1)}{2k-n-1}\frac{1}{2^{n+1}}\binom{n}{k}\frac{2k-n-1}{n+1-k} = \frac{-k}{n+1-k}\binom{n}{k}\frac{1}{2^{n+1}} = \frac{-1}{2}\binom{n}{k-1}.$$

With $F(n, k) = \binom{n}{k}s^{-n}$, indeed $G(n, k) = -\frac{1}{2}F(n, k-1)$, as in Example 2.3.13. In general, as stated in Step 4, we are setting

$$G(n, k) = \frac{b(k-1)h(k)}{c(k)}(F(n+1, k) - F(n, k)).\qquad\blacksquare$$

2.3.18. Proposition. If Gosper's Algorithm returns successful output $g(k)$ from input $f(k)$, then $g(k+1) - g(k) = f(k)$.

Proof: We compute

$$g(k+1) - g(k) = \frac{b(k)h(k+1)}{c(k+1)}f(k+1) - \frac{b(k-1)h(k)}{c(k)}f(k)$$

$$= f(k)\left[\frac{b(k)h(k+1)}{c(k+1)}r(k) - \frac{b(k-1)h(k)}{c(k)}\right]$$

$$= f(k)\left[\frac{a(k)h(k+1)}{c(k)} - \frac{b(k-1)h(k)}{c(k)}\right] = f(k).\qquad\blacksquare$$

Before concluding that the WZ method solves everything, note that our discussion of the method requires advance knowledge of the value of the sum. Otherwise, we wouldn't even know the expression $F(n, k)$ that starts the process!

Fortunately, there are also methods to automate finding the sum. These stem from an algorithm in the thesis of Sister Mary Celine Fasenmyer [1946]. The basic idea is to find a recurrence satisfied by the summand and then sum it to find a recurrence satisfied by the sum. Fasenmyer's algorithm automates the search for the first recurrence when the summand belongs to an appropriate class.

Zeilberger [1982, 1990, 1991] developed faster and more general algorithms. Petkovšek [1991] automated the search for solutions of the resulting recurrences. The algorithms find sums of hypergeometric terms or show that the sum has no formula of a certain type. Again, see Petkovšek–Wilf–Zeilberger [1996].

EXERCISES 2.3

2.3.1. (−) Use the substitution method to solve the recurrences below.
(a) $a_n = (1 - \frac{1}{n+1})a_{n-1}$ for $n \geq 1$, with $a_0 = 1$.
(b) $a_n = \frac{2}{3}(1 + \frac{2}{3^n+1})a_{n-1}$ for $n \geq 1$, with $a_0 = 1$.

2.3.2. (−) Give the exact solution formula (for all n) for the recurrences below.
(a) $a_n = a_{\lceil n/2 \rceil} + 1$ for $n \geq 2$, with $a_1 = 0$.
(b) $a_n = a_{\lfloor (n-1)/2 \rfloor} + 1$ for $n \geq 2$, with $a_1 = 1$. (Comment: This is the recurrence for the worst-case number of probes in binary search for an element in a sorted n-element list.)

2.3.3. (−) Let $H_n = \sum_{k=1}^{n} \frac{1}{k}$. Prove for $n \in \mathbb{N}$ that $\sum_{i=1}^{n-1} H_i = n(H_n - 1)$.

2.3.4. (−) Use the recurrence $D_n = (n-1)(D_{n-1} + D_{n-2})$ of Example 2.1.6 (with $D_0 = 1$ and $D_1 = 0$) to prove $D_n = nD_{n-1} + (-1)^n$. Use the new recurrence to prove $D_n = n! \sum_{k=0}^{n} (-1)^k/k!$.

2.3.5. Find the general solution to $a_n = a_{n/2} + 2a_{n/4} + 3n/4$ when $n = 2^k$.

2.3.6. (◇) Solve the recurrence $a_n = 2a_{n-1}^2$ with $a_0 = 1$.

2.3.7. (◇) *Finding the max and min in a set S.* When $|S| = 1$, the one element is both answers. When $|S| = 2$, compare the two numbers and return the result. When $|S| > 2$, split S into sets T and T' of sizes $\lfloor n/2 \rfloor$ and $\lceil n/2 \rceil$, apply the algorithm to T and T', and compare the answers for the two subsets. Let a_n be the number of comparisons used when $|S| = n$. Derive a recurrence for a_n. Use the substitution method to obtain a formula for a_n when n is a power of 2.

2.3.8. (◇) In the game "High/low", a number is chosen uniformly at random from $[n]$, where n is known. The player makes a guess, winning the prize if correct. Otherwise, she learns whether the guess is high or low. When each guess is made randomly from the remaining possible values, the expected numbers of guesses when $n \in \{0, 1, 2, 3\}$ are $0, 1, 3/2, 17/9$, respectively.
The **harmonic number** H_n is defined by $H_n = \sum_{k=1}^{n} 1/k$. Obtain a formula for the expected number of guesses made to win the prize, using one harmonic number in the formula. (Hint: Obtain a recurrence with many terms. Simplify via natural substitutions and reduce to a first-order recurrence.) (Taylor [1991])

2.3.9. Let $n(n-3)a_n = (n-1)(n^2 - 3n + 1)a_{n-1} - (n-2)^3 a_{n-2}$ for $n \geq 4$, with $a_2 = a_3 = 1$. Use substitutions to obtain a simple formula for a_n. (Wilf [1997])

2.3.10. Let $a_n = a_{n-1} + a_{n-2}/(n-1)$ for $n \geq 2$, with $a_0 = \alpha$ and $a_1 = \beta$. Find a formula for a_n (one summation may appear in the formula).

2.3.11. Let $a_{n+1} = (2n+3)a_n - 2na_{n-1} + 8n$ for $n \geq 1$, with $a_0 = 1$ and $a_1 = 3$. Use substitutions to obtain an asymptotic formula for a_n. (Doster [1994])

2.3.12. (◇) *Continuation of Application 2.3.8 for Stirling's Formula.* Given the recurrence $d_n = e(1 - 1/n)^{n-1/2}d_{n-1}$, let $d_n = e_n(1 + \gamma/n)$ and choose γ so that $e_n/e_{n-1} = 1 + O(n^{-3})$.

2.3.13. (◇) Let $a_n = \int_0^{\pi/2} (\sin x)^n dx$ for $n \in \mathbb{N}_0$.
(a) Use integration by parts to prove $a_n = \frac{n-1}{n} a_{n-2}$ for $n \geq 2$.
(b) Use the sequence $\langle a \rangle$ to prove the **Wallis product**: $\prod_{n=0}^{\infty} \frac{2n}{2n-1} \frac{2n}{2n+1} = \frac{\pi}{2}$.

2.3.14. *Approximations for binomial coefficients.* For k fixed and n growing, $\binom{n}{k} \sim \frac{n^k}{k!}$.

(a) Using Stirling's Formula, prove $\binom{n}{\alpha n} \approx \dfrac{\left(\alpha^\alpha(1-\alpha)^{1-\alpha}\right)^{-n}}{\sqrt{\alpha(1-\alpha)2\pi n}}$.

(b) An ordinary deck of cards consists of 13 cards each in four suits. A bridge hand distributes these randomly to four players around a table, each receiving 13 cards. Opposite players are partners. To the nearest power of 10, approximate the probability that some partnership will have all 13 cards of one suit.

2.3.15. Consider limits taken as $n \to \infty$.

(a) Prove that if $\lim(x_{n+1} - x_n) = L$, then $\lim x_n/n = L$.

(b) Apply part (a) to prove that if $\lim(a_{n+1}/a_n) = M$, then $\lim(a_n^{1/n}) = M$.

(c) Conclude $\lim\left(\binom{n}{\lfloor n/2 \rfloor}\right)^{1/n} = 2$.

2.3.16. Balls $1, \ldots, n$ will go in boxes $1, \ldots, n$. Put ball 1 into a randomly chosen box. Thereafter, for j from 2 to n, put ball j in box j if box j is empty; otherwise, put it in a random empty box. Let μ_n be the expected number of j such that ball j is in box j.

(a) Prove $\mu_n = 1 + \dfrac{1}{n}\sum_{k=2}^{n}\left(k - 2 + \mu_{n-k+1} - \dfrac{1}{n-k+1}\right)$.

(b) Conclude $\mu_n = n - H_{n-1}$, where $H_m = \sum_{j=1}^{m} 1/j$. (Deshpande–Deshpande [2010])

2.3.17. Suppose $a_n = 2a_{n-1} + a_{n-2}$ for $n \geq 2$, with $a_0 = 1$ and $a_1 = 2$. Let $c_{n,i,k} = \frac{(n-k)!}{i!(n-i-2k)!k!}$ when i, k, and $n - i - 2k$ are all nonnegative; otherwise $c_{n,i,k} = 0$.

(a) Prove $c_{n,i,k} = c_{n-1,i,k} + c_{n-1,i-1,k} + c_{n-2,i,k-1}$ for $n \geq 2$ and all i and k.

(b) Use part (a) to prove $a_n = \sum \frac{(i+j+k)!}{i!j!k!}$, summed over nonnegative integer (i, j, k) with $i + j + 2k = n$. (Comment: This is again the Pell sequence.) (Stockmeyer [2000])

2.3.18. (+) An usher seats n people in a row of n narrow seats, choosing a seat for each arrival in turn. In order to seat a new arrival, the people neighboring that seat in each direction up to the first empty seat must stand so they can sit together at one time. For example, if seat 4 is next filled when seats $1, 3, 5, 6, 8$ are already filled, then those in seats $3, 5, 6$ must stand so that $3, 4, 5, 6$ all sit together. The usher wants to minimize the total number of seatings, counting 1 each time any person sits. Let $f(n)$ be the optimal value; for $1 \leq n \leq 5$, the answers are $1, 3, 5, 8, 11$. In general, filling the seats in order from one end yields $(n^2 + n)/2$ seatings (the worst!), but the optimum is asymptotic to $n \log_2 n$.

(a) Let $g(n) = (n+1)k - 2^k + 1$, where $k = \lceil \log_2(n+1) \rceil$. Prove $g(n) = g(\lfloor (n-1)/2 \rfloor) + g(\lceil (n-1)/2 \rceil) + n$.

(b) Prove that $g(n) - g(n-1)$ is a nondecreasing function of n.

(c) Prove $f(n) = g(n)$ for $n \in \mathbb{N}$. (Fredman–Knuth [1974], Vanden Eynden [1991])

2.3.19. Use the WZ method with $R(n, k) = \frac{2k-1}{2n+1}$ to prove

$$\sum_k (-1)^k \binom{n}{k}\binom{2k}{k} 4^{n-k} = \binom{2n}{n}.$$

2.3.20. Use the WZ Method with $G(n, k) = \frac{-k}{n} F(n, k)$ to prove that

$$\sum_{k=0}^{t}(-1)^k \frac{1}{n+k}\binom{t}{k} = \frac{1}{n}\binom{t+n}{t}^{-1}. \tag{Wilf}$$

(Comment: Here $F(n, k)$ is the summand divided by the right side; $\sum_k F(n, k) = 1$ can also be proved using Vandermonde's Theorem and other familiar identities.)

Chapter 3

Generating Functions

In Section 2.2, we called $\sum_{n=0}^{\infty} a_n x^n$ the **generating function** for a sequence $\langle a \rangle$ of complex numbers. There it was just a tool for solving recurrence relations, but generating functions have deeper combinatorial aspects. They yield systematic solutions for many counting problems. Often we obtain an expression for the generating function and a formula for a_n directly, without using recurrences.

There are many types of generating functions (see Stanley [1986]). We have only mentioned the *ordinary generating function* (OGF), useful in studying multi-sets and selection problems. A variation called the *exponential generating function* (EGF) is useful for enumeration of "labeled" structures. We will also apply OGFs to study partitions of integers.

3.1. Ordinary Generating Functions

Let a_n be the solution to a counting problem with n as a parameter. A problem may have many parameters; at present we treat all but one as fixed and use the varying parameter as the index of a sequence. We associate a generating function in one variable with this counting sequence.

3.1.1. Definition. A **formal power series** is an expression of the form $\sum_{n=0}^{\infty} a_n x^n$. The **(ordinary) generating function** (OGF) for a sequence $\langle a \rangle$ of complex numbers is the formal power series $\sum_{n=0}^{\infty} a_n x^n$ or any expression having this formal power series expansion (also called **generating series** in the literature on formal power series).

When objects in a universe U are assigned nonnegative integer values by a weight function w, with $a_n = |\{u \in U : w(u) = n\}|$, we say that the generating function for $\langle a \rangle$ is the **enumerator by** w for U or is **indexed by** w.

3.1.2. Example. Let a_n be the number of binary lists of length n. The generating function for $\langle a \rangle$ is $\sum_{n=0}^{\infty} 2^n x^n$; it is the enumerator of binary lists, indexed by length. We say "indexed by length" because the values of the weight function index the terms in the sequence $\langle a \rangle$.

Next consider k-subsets of an n-set. We fix one parameter to form a generating function, writing $a_k = \binom{n}{k}$. In $A(x) = \sum_{k=0}^{n} \binom{n}{k} x^k$, the coefficient of x^k is the number of k-subsets of $[n]$. Thus $A(x)$, or more properly $A_n(x)$, is the enumerator of subsets of $[n]$, indexed by size. Since the coefficient of x^k in the expansion of $(1+x)^n$ is $\binom{n}{k}$, we also say that $(1+x)^n$ is the generating function for $\langle a \rangle$.

In a problem with several parameters, it may not be obvious which is most useful as the index of summation in a generating function. We can also form a generating function in two variables: $\sum_{n=0}^{\infty} \sum_{k=0}^{\infty} a_{n,k} x^n y^k$.   ∎

3.1.3.* Remark. The word "function" in "generating function" indicates only the notational form; x^n is just a placeholder for the term a_n in the sequence, not a number. In algebraic combinatorics the term **generating series** is used instead of generating function; the series "generates" the sequence of coefficients.

Since x is not numerical, we call x a "formal variable". Our first proof of $(1+x)^n = \sum_{k=0}^{n} \binom{n}{k} x^k$ manipulated x as a symbol, not a number. This viewpoint also excuses the notation "$A(x)$" for a generating function. One should not write "$f(x)$" for a numerical function f, since $f(x)$ is a number. We write "$A(x)$" for a generating function because "argument" and "value" have no numerical meaning; the notation indicates that we are encoding a sequence using powers of x.

The question of convergence when the formal variable takes a numerical value is mostly irrelevant, but numerical convergence can be helpful in extracting asymptotic formulas for coefficients (Section 3.4).   ∎

MODELING COUNTING PROBLEMS

The importance of generating functions arises from the way they model counting arguments using the Sum and Product Principles, based on the behavior of sum and product of formal power series. This allows generating functions to model ad hoc counting for a sequence of counting problems all at once.

3.1.4. Definition. The **sum** and **product** of formal power series $\sum_{n=0}^{\infty} a_n x^n$ and $\sum_{n=0}^{\infty} b_n x^n$ are defined as follows:

$$\text{sum}: \quad \sum_{n=0}^{\infty} (a_n + b_n) x^n \qquad \text{product}: \quad \sum_{n=0}^{\infty} \left(\sum_{j=0}^{n} a_j b_{n-j} \right) x^n.$$

The **convolution** of two sequences $\langle a \rangle$ and $\langle b \rangle$ is the sequence of coefficients in the product of their generating functions. The term *convolution* is also used for a sum of the form $\sum_{j=0}^{n} a_j b_{n-j}$.

3.1.5. Example. *Multisets from two types of objects.* The formal power series $\sum_{k=0}^{\infty} x^k$ is the enumerator by size for multisets from one type of object. There is exactly one multiset consisting of k objects of the same type. Marbles of the same color are indistinguishable, so there is only one way to have 10 white marbles.

With two types, such as black and white marbles, there are $k+1$ ways to form a multiset of size k, since the first type can contribute anywhere from 0 to k objects. Thus the generating function is $\sum_{k=0}^{\infty} (k+1) x^k$. Alternatively, to reach size k, the first type contributes j and the second contributes $k-j$, for some j. Hence by the definition of product the generating function also equals $\left(\sum_{k=0}^{\infty} x^k \right)^2$.   ∎

3.1.6. Example. *Two types of coins.* With k coins on a table, each may show heads or tails. The number of heads can be 0 through i, so there are $i+1$ distinguishable ways to have i coins. Thus the generating function for selections of coins, indexed by the number of coins, is $\sum_{k=0}^{\infty}(k+1)x^k$.

Now consider coins of two values, such as nickels and dimes. A selection with k coins uses j nickels and $k-j$ dimes, for some j. If we also note how many of each show heads or tails, then we have $j+1$ ways to have j nickels and $k-j+1$ ways to have $k-j$ dimes, and each choice for the nickels can be paired with each choice for the dimes. Thus the generating function for distinguishable selections of nickels and dimes, indexed by the number of coins, is $\sum_{k\geq 0}\left[\sum_{i=0}^{k}(i+1)(k-i+1)\right]x^k$. By the rule for multiplying formal power series, this is $\left[\sum_{k=0}^{\infty}(k+1)x^k\right]^2$.

We described this example in this way to illustrate that the coefficients in a factor may be other than 0 or 1. In fact, we have just discussed the enumerator for multisets from four types: head-nickels, tail-nickels, head-dimes, and tail-dimes. The generating function is $\left(\sum_{k=0}^{\infty}x^k\right)^4$. ∎

Products of generating functions model the enumeration of objects formed as ordered pairs when the weight functions are "additive".

3.1.7. Lemma. Let a_k, b_k, c_k count the elements with weight k in sets A, B, C, respectively. The OGF for $\langle c \rangle$ is the product of the OGFs for $\langle a \rangle$ and $\langle b \rangle$ if and only if, for all k, the elements in C with weight k correspond to the pairs $(\alpha, \beta) \in A \times B$ such that k is the sum of the weights of α and β.

Proof: The correspondence specified is precisely the condition for $c_k = \sum_{j=0}^{k} a_j b_{k-j}$, which by Definition 3.1.4 expresses the OGF for $\langle c \rangle$ as the product of the OGFs for $\langle a \rangle$ and $\langle b \rangle$. ∎

The product behavior generalizes to objects built in n stages. If objects in C correspond to n-tuples in $A_1 \times \cdots \times A_n$, and the weight of an n-tuple is the sum of the weights of the entries, then the OGFs satisfy $C(x) = \prod_{i=1}^{n} A_i(x)$, using Lemma 3.1.7 and induction on n.

3.1.8. Example. *Multisets (selections with repetition).* Let a_k be the number of multisets of size k from n types of objects, where n is fixed. Thus $\sum_{k\geq 0} a_k x^k$ is the enumerator by size for multisets from n types of objects.

To specify a multiset, we list the multiplicity of each type of object; the size is the sum of these multiplicities. Since the generating function for multisets from one type is $\sum_{k=0}^{\infty}x^k$, we obtain $\sum_{k\geq 0} a_k x^k = \left(\sum_{k=0}^{\infty}x^k\right)^n$.

Each contribution to the expansion of the product is obtained by choosing a term from each factor, independently. Within the jth factor, the different terms represent options for the jth type in forming the compound object. The sum of the exponents is the index (size) of the resulting compound object. This is why we say that sum and product of generating functions model the use of the Sum and Product Principles to solve a whole sequence of counting problems at once.

Consider red, white, and blue marbles. A multiset with three red, one white, and two blue marbles corresponds to choosing x^3 from the first factor, x from the second, and x^2 from the third. This contributes 1 to the coefficient of x^6 in the product. Each choice of a term from each factor generates one multiset and contributes 1 to the coefficient of x^n, where n is the total size of the multiset.

In fact, we already know the coefficients in the generating function for multisets from n types. In Theorem 1.1.20, we proved $a_k = \binom{k+n-1}{n-1}$. Equating the two expressions for the OGF yields a combinatorial proof of an algebraic identity (for each n) about formal power series:

$$\left(1 + x + x^2 + \cdots\right)^n = \sum_{k=0}^{\infty} \binom{k+n-1}{n-1} x^k.$$

Coefficients in factors need not always equal 1 (see Example 3.1.6).   ∎

Formal power series are specified by their coefficients; two formal power series are equal if and only if corresponding coefficients are equal. We need convenient notation for a specified coefficient. We repeat the definition from Chapter 2.

3.1.9. Definition. The **coefficient operator** $[x^k]$, applied to a (formal) power series in x, returns the coefficient of x^k in the series. That is, if $A(x) = \sum_{n=0}^{\infty} a_n x^n$, then $[x^n]A(x) = a_n$.

When the scope of the coefficient operator is clear, we usually do not place parentheses around its argument.

The definition of sum (and multiplication by constants) makes the set of formal power series in x an infinite-dimensional vector space. The definition of product satisfies the additional axioms needed to make it an "algebra". By the definition of product, the formal power series $1x^0 + 0x^1 + 0x^2 + \cdots$ (or simply 1) is an identity element for multiplication. We *define* $A(x)^{-1}$ to be $B(x)$ when $A(x)B(x) = 1$ (note that multiplication of formal power series is commutative). The condition for $A(x)$ having a multiplicative inverse is $[x^0]A(x) \neq 0$ (Exercise 6).

We obtained the power series for $(1 - x)^{-n}$ in Lemma 2.2.17. Here we obtain the *same expression* as a formal power series expansion. We will consider why these expressions are the same in the next section.

3.1.10. Theorem. For $n \in \mathbb{N}$, the formal power series expansion of the generating function $(1 - x)^{-n}$ is

$$(1 - x)^{-n} = \sum_{k=0}^{\infty} \binom{k+n-1}{n-1} x^k.$$

Proof: The product of $(1-x)$ and $\sum_{j=0}^{\infty} x^j$ is 1, so $(1-x)^n \left(\sum_{j=0}^{\infty} x^j\right)^n = 1^n = 1$, and therefore $(1-x)^{-n} = \left(\sum_{j=0}^{\infty} x^j\right)^n$. Since $[x^k]\left(\sum_{j=0}^{\infty} x^j\right)^n = \binom{k+n-1}{n-1}$ (Example 3.1.8), also $[x^k](1-x)^{-n} = \binom{k+n-1}{n-1}$.   ∎

3.1.11. Example. *Multisets with restricted multiplicities.* When multiplicity is unrestricted, the factor for each type of element in the OGF is $1 + x + x^2 + \cdots$, which equals $(1-x)^{-1}$. Restrictions on usage affect the choices listed in a factor. When S is the set of multiplicities allowed for a type of object, the generating function for that factor is $\sum_{k \in S} x^k$.

For ordinary subsets, the choice for each type is omission or inclusion ($S = \{0, 1\}$), and the generating function is $(1 + x)^n$.

When each type *must* be used, we cannot select none. The enumerator is then $(x + x^2 + x^3 + \cdots)^n$, which equals $(\frac{x}{1-x})^n$. Thus

$$[x^k]\left(\frac{x}{1-x}\right)^n = [x^{k-n}](1-x)^{-n} = \binom{k-n+n-1}{n-1} = \binom{k-1}{n-1},$$

as in Corollary 1.1.22. The x in the numerator of each factor corresponds to requiring a positive integer in the integer-sum model.

When a type must have even usage, its factor is $1 + x^2 + x^4 + \cdots$, which equals $(1 - x^2)^{-1}$. Any restriction on multiplicities can be encoded. For example, in how many ways can we pick 20 coins that are pennies, nickels, or dimes, with at least three nickels and at most four dimes (pennies unrestricted)? With $\sum_{k=0}^{r} x^k = \frac{1-x^{r+1}}{1-x}$, we compute

$$[x^{20}]\frac{1}{1-x}\frac{x^3}{1-x}\frac{1-x^5}{1-x} = [x^{20}]\frac{x^3-x^8}{(1-x)^3}$$

$$= [x^{17}]\sum_k\binom{k+3-1}{3-1}x^k - [x^{12}]\sum_k\binom{k+3-1}{3-1}x^k = \binom{19}{2} - \binom{14}{2}.$$

Example 4.1.6 addresses restricted multiplicities in another way.   ∎

3.1.12. Remark. By definition, a generating function systematically solves a *sequence* of problems. The automatic bookkeeping performed by modeling with sums and products of generating functions avoids case analysis, and the expression for the full generating function is often simpler than more specific formulas. For this reason, we do not artificially end factors at x^t even when we seek only the coefficient of x^t.

Also, an expression with formal power series expansion $\sum_{k=0}^{\infty} a_k x^k$ cannot depend on k, just as the value of $\sum_{k=0}^{n} f(k)$ cannot depend on k.

By "combinatorial argument" to obtain a generating function, we mean using generating functions to model the sum (case) and product (stage) choices that are made when building the objects being enumerated, in such a way that the exponent on the formal variable accumulates the contributions to the index parameter. In Example 3.1.11, a multiset is formed by choosing some permitted number of copies of each type, and the size of each resulting multiset is the sum of the multiplicities chosen from each type. Accumulating the multiplicities in the exponent models this additivity of size (the index parameter).   ∎

The notion of generating function extends naturally for problems with m parameters: the coefficient of the monomial $x_1^{k_1} \cdots x_m^{k_m}$ is the number of objects in which the value of the ith parameter is k_i, for all i. For example, when flipping nickels and dimes in Example 3.1.6, we could define a generating function in which the coefficient of $x^r y^s$ is the number of outcomes with r coins, of which s show heads (see Exercise 16).

3.1.13. Example. Let $A(x, y) = \sum_{n=0}^{\infty} \sum_{k=0}^{\infty} a_{n,k} x^k y^n$ with $a_{n,k} = \binom{n}{k}$. Now

$$A(x, y) = \sum_{n=0}^{\infty} \sum_{k=0}^{\infty} \binom{n}{k} x^k y^n = \sum_{n=0}^{\infty} (1 + x)^n y^n = \frac{1}{1 - y - xy}.$$

Thus $(1 - y - xy)A(x, y) = 1$. For $n \geq 1$, we read this as $a_{n,k} - a_{n-1,k} - a_{n-1,k-1} = 0$, which is the binomial recurrence (Pascal's Formula).

This example reverses the process of Example 2.2.23; generating functions can be used to obtain recurrence relations.   ∎

PERMUTATION STATISTICS

There are $n!$ permutations of $[n]$, but we can group them by the value of some parameter. Such parameters are **permutation statistics**. Natural statistics for permutations include number of fixed points, number of cycles, etc. These parameters have important roles in algebraic combinatorics and in the theory of special functions, but here we study only the counting problems.

We use $\mathbb{S}_n$ for the set of permutations of $[n]$. This "blackboard" font is reserved for important sets with special algebraic structure, such as $\mathbb{N}, \mathbb{Z}, \mathbb{Q}, \mathbb{R}$ for number systems and $\mathbb{F}_q$ for the finite field of order q. The permutations of $[n]$ form a "group" under composition (see Section 4.2), called the **symmetric group** (on n elements). Its fundamental role makes a special font appropriate; we use $\mathbb{S}$ for "symmetric". The historical notation is S in Fraktur font, which is hard to read.

It may be hard to count subsets of $\mathbb{S}_n$ explicitly. Instead, we seek a generating function indexed by the relevant parameter. Our first example illustrates the combinatorial approach of finding a generating function by describing the choices made in building the objects.

3.1.14. Definition. An **inversion** in $\sigma \in \mathbb{S}_n$ is a pair (σ_i, σ_j) with $i < j$ and $\sigma_i > \sigma_j$.

3.1.15. Proposition. The enumerator of $\mathbb{S}_n$ by number of inversions is

$$(1 + x)(1 + x + x^2) \cdots (1 + x + x^2 + \cdots + x^{n-1}).$$

Proof: We build a permutation σ (viewed as an ordering of $[n]$) by successively inserting $1, \ldots, n$. Elements r and s with $r < s$ form an inversion in σ if and only if s is inserted leftward of r; their relative order cannot change later. The number of inversions in which s is the higher element is the number of elements in positions after s when s is inserted.

We have one factor in the generating function for each insertion. Inserting s creates between 0 and $s - 1$ insertions. For each placement value, the choices for the number of new inversions when inserting later values are the same.

After inserting all elements, we should contribute 1 to the coefficient of x^t, where t is the total number of inversions. The factor $(1 + x + \cdots + x^{s-1})$ lists the choices when we insert s.   ∎

Next we enumerate $\mathbb{S}_n$ by number of cycles.

3.1.16. Definition. The **2-line form** or **tabular representation** of a permutation records the domain on top and the corresponding images below. The **word form** of a permutation of $[n]$ is simply the second row of the 2-line form, listing the image of i in position i.

A **cycle representation** of a permutation is a list of its cycles, grouped within parentheses. The **canonical cycle representation** puts the least element first in each cycle and puts cycles in decreasing order of least elements.

3.1.17. Example. The permutation with tabular representation $\binom{123456789}{461752398}$ has word form 461752398, and the cycles are (4731), (62), (89), and (5). The canonical cycle representation is (89)(5)(26)(1473).

Dropping the parentheses in this representation yields 895261473, the word form of another permutation. We show that this process defines a bijection. ∎

3.1.18. Lemma. The canonical cycle representation $f(\sigma)$ of a permutation σ (without parentheses) defines a bijection from $\mathbb{S}_n$ (given as words) to itself.

Proof: In the canonical cycle representation, the next cycle begins when a new left-to-right minimum is reached. By starting with the word form of a permutation τ and inserting parentheses before the left-to-right minima, we obtain the canonical cycle representation of a permutation σ such that $f(\sigma) = \tau$. Since a surjection from a finite set to itself is a bijection, f is thus a bijection. ∎

We used this idea in the enumeration of trees in Theorem 1.3.4. Knuth [1968] described the bijection as "An unusual correspondence"; it also appeared in Rényi [1962] and Foata–Schützenberger [1970]. It yields short proofs for many enumerative and probabilistic problems; see Exercises 23–31. We give one example.

3.1.19. Proposition. When permutations of $[n]$ are equally likely, the distribution of the length of the cycle containing a fixed element k is uniform, with probability $1/n$ for each length.

Proof: By symmetry, it suffices to prove the claim when $k = 1$. The length of the cycle containing 1 in a permutation π is $n + 1$ minus the position of 1 in the canonical cycle representation of π. By Lemma 3.1.18, this position is uniformly distributed over all n positions. ∎

Next we use canonical cycle representation to enumerate $\mathbb{S}_n$ by number of cycles. Note that setting $x = 1$ combines the $n!$ permutations into one class.

3.1.20. Theorem. Let $c(n, k)$ be the number of elements of $\mathbb{S}_n$ with k cycles. The enumerator of $\mathbb{S}_n$ by number of cycles, $C_n(x)$, is given by

$$C_n(x) = \sum_{k=1}^{n} c(n, k)x^k = x^{(n)}.$$

Proof: By definition, each permutation with k cycles contributes 1 to $[x^k]C_n(x)$. Recall that $x^{(n)} = \prod_{i=1}^{n}(x+i-1)$. To show that $C_n(x) = x^{(n)}$, we interpret $(x+i-1)$ as encoding the options when inserting element i to build the canonical cycle representation of a permutation. We insert elements in increasing order as in Proposition 3.1.15, but now the choices have different effects on the index parameter.

Again the factor for inserting i is independent of the choices made for earlier elements, so the generating function is the product of these factors.

Choosing "x" from $(x + i - 1)$ means starting a new (leftmost) cycle in the representation with i as least element. This increases the number of cycles, so the objects built with this choice contribute to a term with higher exponent. The number $i - 1$ counts the ways to make i follow a smaller number j in its cycle. For example, the permutation of Example 3.1.17 is built as below. Here (i, x) indicates an increase in the number of cycles by inserting i as the left, while (i, j) indicates the insertion of i following the smaller number j.

$(1, x)$:	(1)	$(4, 1)$:	(2)(143)	$(7, 4)$:	(5)(26)(1473)
$(2, x)$:	(2)(1)	$(5, x)$:	(5)(2)(143)	$(8, x)$:	(8)(5)(26)(1473)
$(3, 1)$:	(2)(13)	$(6, 2)$:	(5)(26)(143)	$(9, 8)$:	(89)(5)(26)(1473)

Since i is largest when inserted, it does not change left-to-right minima unless it is put first. When expanding the product, the exponent on x in a term is the number of times the choice started a new cycle.

Each permutation is counted exactly once and with the correct exponent, because there is exactly one way to reconstruct the choices from the canonical cycle representation of a permutation. To reconstruct the situation when i was inserted, we discard all elements of the canonical cycle representation that are larger than i; their insertion does not affect the cycle structure on the first i numbers. Now i is first if and only if the choice was made to start a cycle with i, and otherwise i immediately follows the element it was chosen to follow. ∎

Theorem 3.1.20 can be interpreted as a polynomial identity. Another technique for polynomial identities, different from building generating functions, is the Polynomial Principle (Remark 1.1.10). To illustrate the difference, we reprove Theorem 3.1.20 using the Polynomial Principle.

3.1.21. Remark. *Enumerating* $\mathbb{S}_n$ *by cycles, again.* (Stanton–White [1986, p. 79]) By the Polynomial Principle, it suffices to show for $x \in \mathbb{N}$ that both sides of $\sum_{k=1}^{n} c(n, k)x^k = x^{(n)}$ count the same set. The right side counts placements of n distinct flags on x flagpoles in a row; order on flagpoles matters (Exercise 1.1.35).

To show that the left side also counts these arrangements, begin with the cycle representation of a permutation of $[n]$. For each cycle, assign the elements to one of the x flagpoles. A permutation with k cycles has x^k such assignments. Put all the flags in all the cycles assigned to a given pole on that pole, in the order given by the canonical cycle representation. That is, each cycle has its least element first, and the cycles are entered in decreasing order of least elements.

For each arrangement of flags, we retrieve the cycles (and the assignment of cycles to poles). On each pole, a new cycle begins with each new minimum among the flags listed so far. Hence the pairs consisting of a permutation of $[n]$ with k cycles and an x-ary word of length k correspond bijectively to the arrangements of n flags on x flagpoles. ∎

Next we introduce another important permutation statistic and the corresponding 2-parameter family of counting numbers.

3.1.22. Definition. For $\sigma \in \mathbb{S}_n$, a **run** is a maximal increasing string in the word form. There is a **descent** or **ascent** at i if $\sigma_i > \sigma_{i+1}$ or $\sigma_i < \sigma_{i+1}$, respectively. The **descent set** $D(\sigma)$ is the subset of $[n-1]$ at which descents occur. The **Eulerian number** $A(n, k)$ is the number of permutations of $[n]$ having k runs (and $k - 1$ descents).

3.1.23. Example. The permutation 791368452 has four runs (79, 1368, 45, and 2), three descents (at 2, 6, and 8), and five ascents. For $\sigma \in \mathbb{S}_n$, the ascent set of σ is $[n - 1] - D(\sigma)$. When σ has k runs, it has $k - 1$ descents and $n - k$ ascents, and its reverse has $k - 1$ ascents and $n - k$ descents.   ∎

The Eulerian numbers are the subject of the first chapter of Bóna [2004], and there is an entire book about them (Petersen [2015]). Relevant surveys include Carlitz [1958–59], Foata–Schützenberger [1970], Knuth [1973], and Charalambides [2002]. The most common notation now is $A(n, k)$, though $A_{n,k}$ is also used. Knuth [1968] used $\left\langle {n \atop k} \right\rangle$ for $A(n, k)$, but this notation meant $A(n, k + 1)$ in Graham–Knuth–Patashnik [1989], letting k count descents rather than runs.

The Eulerian numbers are studied in Exercises 34–43. Many of the exercises use a natural recurrence.

3.1.24. Proposition. With $A(0, k) = \delta_{0,k}$, for $n, k \in \mathbb{N}$ the Eulerian numbers satisfy $A(n, k) = kA(n - 1, k) + (n - k + 1)A(n - 1, k - 1)$.

Proof: When inserting n into a permutation of $[n - 1]$, the number of runs is unchanged if n goes at the end of a run; otherwise, it increases by 1. Hence permutations with k runs arise by inserting n into permutations of $[n - 1]$ with k runs in k ways and by inserting n into permutations of $[n - 1]$ with $k - 1$ runs in $(n - 1) - (k - 1) + 1$ ways. Since every permutation arises by successive insertions in a unique way, this counts each permutation of $[n]$ with k runs exactly once. ∎

We obtain a formula for $A(n, k)$ from the next result by using an inversion relationship for generating functions.

3.1.25. Theorem. (**Worpitzky's Identity**; Worpitzky [1883]) For $n \geq 1$, the Eulerian numbers $A(n, k)$ satisfy

$$x^n = \sum_{k=1}^{n} A(n, k) \binom{x - k + n}{n}.$$

Proof: (Gessel) When $x \in \mathbb{N}$, the left side counts functions from $[n]$ to $[x]$. To turn these into permutations, express them as distributions of $[n]$ into x successive bins, with each bin ended by a vertical bar. Write the elements of each bin in increasing order. This produces a permutation of $[n]$ with x bars inserted. There is a bar at the end of each run and possibly in additional locations (empty bins produce consecutive bars). Below we show two ways of distributing $[9]$ into 6 bins so that the resulting permutation is 791368452.

$$79 \mid 136 \mid 8 \parallel 45 \mid 2 \mid \qquad \text{and} \qquad \mid 79 \mid 13 \mid 68 \mid 45 \mid 2 \mid$$

We have represented the x^n distributions as **barred permutations** of $[n]$ with x bars; such an object consists of the word form of a permutation with x bars inserted so that there is at least one bar immediately following every run (including the last). We now count them by grouping them according to the number of runs in the permutation.

If the permutation has k runs, then k of the bars must appear at the ends of the runs. (Note that $\binom{x-k+n}{n} = 0$ if $k > x$.) There remain $x - k$ bars to put in $n + 1$ possible locations, with repetition allowed. Hence there are $\binom{x-k+n}{n}$ ways to choose positions for the extra bars. (Bars are identical, so it doesn't matter which k bars were placed initially.)

Thus the identity counts barred permutations of $[n]$ with x bars in two ways. Since the argument is valid for every positive integer x, the result is an identity in polynomials, by the Polynomial Principle. ∎

Barred permutations may have appeared first in the thesis of Gessel [1977]; see also Gessel–Stanley [1978]. Bóna [2004, p. 6] gives another combinatorial proof of Worpitzky's Identity. The identity is sometimes used to define the Eulerian numbers. It expresses the power x^n in terms of the Eulerian numbers; our goal now is to invert this relationship to obtain a formula for $A(n, k)$.

3.1.26. Theorem. The Eulerian number $A(n, k)$ (the number of permutations of $[n]$ with k runs) is given by

$$A(n, k) = \sum_{i=0}^{k}(-1)^i\binom{n + 1}{i}(k - i)^n.$$

Proof: Change x to r in Worpitzky's Identity to treat it as a statement about positive integers. Note that $A(n, k) = 0$ when $k = 0$ or $k > n$, and recall that every run has a bar; that is, $\binom{r-k+n}{n} = 0$ when $k > r$. Thus we change the summation limits without changing the sum to obtain

$$r^n = \sum_{k=0}^{r} A(n, k)\binom{r - k + n}{n}.$$

Let $A(x) = \sum_{k=0}^{n} A(n, k)x^k$ and $B(x) = \sum_{k\geq 0}\binom{k+n}{n}x^k$; note that n is fixed. Worpitzky's Identity says $C(x) = A(x)B(x)$, where $C(x) = \sum_{r\geq 0} r^n x^r$. Thus $A(x) = \frac{1}{B(x)}C(x)$. Since $B(x) = \frac{1}{(1-x)^{n+1}}$, we have $A(x) = (1 - x)^{n+1}C(x)$. By the product rule, $A(n, k) = [x^k]A(x) = \sum_{i=0}^{k}(-1)^i\binom{n+1}{i}(k - i)^n.$ ∎

Many inversion formulas can be proved by substituting the original relation into the desired expression and showing that unwanted contributions vanish. This may require various identities and be rather painful! It also requires knowing the formula for the inverse in advance. Exercise 34 proves Theorem 3.1.26 this way. The proof by inverting Worpitzky's Formula in Theorem 3.1.26 is easy and does not require knowing the formula for $A(n, k)$.

The generating function $\sum_{k=0}^{n} A(n, k)x^k$ is the nth **Eulerian polynomial**, $A_n(x)$. The proof above used $A_n(x) = (1 - x)^{n+1}\sum_{k\geq 0} k^n x^k$. Euler *defined* $A_n(x)$ this way. Curiously, this expression for a polynomial involves an infinite sum!

EXERCISES 3.1

3.1.1. (−) Use generating functions to count the distinguishable selections of six marbles from a pile consisting of three red marbles, four white marbles, and five blue marbles. Verify the answer by describing the selections explicitly.

3.1.2. (−) Let $r_1, \ldots, r_n$ and $s_1, \ldots, s_n$ be natural numbers with $r_i \leq s_i$ for all i. Let a_k be the number of multisets of size k from n types of objects such that the number e_i of objects of the ith type satisfies $r_i \leq e_i \leq s_i$. Express the generating function for $\langle a \rangle$ as $(1 - x)^{-n}$ times a product of polynomial factors.

3.1.3. (−) A child wants to buy candy. Four types of candy have prices two cents, one cent, two cents, and five cents per piece, respectively. Build the enumerator by total cost for the number of ways to buy candy.

3.1.4. (−) Let a_n be the number of nonnegative integer solutions to $4e_1 + 2e_2 + e_3 + 2e_4 = n$. Find the generating function for $\langle a \rangle$.

3.1.5. (−) Let $a_{n,k}$ be the number of permutations of $[n]$ having k inversions (Proposition 3.1.15). Obtain a recurrence relation for these numbers.

3.1.6. (−) From Worpitzky's Identity (Theorem 3.1.25), prove

$$x^n = \sum_{k=0}^{n} A(n, k)\binom{x + k - 1}{n}.$$

3.1.7. (−) Compute the Eulerian number $A(4, k)$ for all k from the definition and from Theorem 3.1.26. Compute the congruence class of $A(p - 1, k)$ modulo p when p is prime.

3.1.8. (◇) Let $b_{n,k}$ be the number of k-subsets of $[n]$ with no consecutive integers. Let $B_k(x) = \sum_{n \geq 0} b_{n,k} x^n$. Build $B_k(x)$ using combinatorial arguments. Use known generating function expansions to obtain a formula for $b_{n,k}$.

3.1.9. Sicherman dice. When rolling dice, on any die each face is equally likely to be the outcome. Two 4-sided dice labeled $(1, 2, 2, 3)$ and $(1, 3, 3, 5)$ have the same distribution of sums as two normal 4-sided dice labeled $(1, 2, 3, 4)$.

(a) Use generating functions to prove that no other pair of 4-sided dice in positive integers has the same sum distribution as the normal dice.

(b) (+) Determine the pairs of 6-sided dice with positive integer labels that have the same sum distribution as two normal 6-sided dice. (Broline [1979], Gallian–Rusin [1979])

3.1.10. (◇) A coin is flipped 14 times, and three flips are tails. Use a generating function to determine the probability that no five consecutive flips are heads.

3.1.11. In Example 3.1.6, the enumerator for distinguishable flippings of pennies, nickels, and dimes by number of coins is $\left(\sum_{k=0}^{\infty} (k + 1)x^k \right)^3$. Give both an algebraic argument and a direct combinatorial argument to show that the generating function equals $(1 - x)^{-6}$.

3.1.12. A meeting has five delegates from each of 100 countries. How many ways are there to form a committee of 25 delegates using at most one person from each country? What is the enumerator for committees with at most three people from each country, indexed by committee size? (Note: People are distinguishable.)

3.1.13. Ten pairs of socks are washed; the two socks in a pair are indistinguishable. Some are lost; use a generating function to count the distinguishable ways that exactly eight socks survive. Explain why these outcomes are not equally likely when all socks are equally likely to survive. (Comment: Extracting the desired coefficient from the generating function requires computation.)

3.1.14. ($\diamond$) Let a_n be the number of ways to choose $r \in \mathbb{N}_0$, roll one six-sided die r times, and obtain outcomes with sum n. Express the generating function for $\langle a \rangle$ as the ratio of two polynomials that each have at most three terms. Obtain a recurrence for $\langle a \rangle$ from the generating function. Give a direct combinatorial argument for the recurrence.

3.1.15. ($\diamond$) Given an alphabet consisting of a vowels and c consonants, let b_n be the number of words of length n not having consonants in consecutive positions. Obtain the generating function $B(x) = \sum b_n x^n$ as the ratio of two polynomials.

3.1.16. Given t types of coins, let $a_{n,k}$ be the number of distinguishable ways to have a multiset of n coins on a table with k showing heads. Find a closed-form expression (without summations) for the generating function $\sum_{n,k} a_{n,k} x^n y^k$.

3.1.17. Let $a_{n,k}$ be the number of ways to tile a 3-by-n rectangle using unit squares and k copies of L, the 2-by-2 square missing the top right 1-by-1 corner (no rotation allowed). Build the generating function $\sum_{n,k \geq 0} a_{n,k} x^n y^k$.

3.1.18. Let $a_{m,n,k}$ be the number of lattice paths from $(0,0)$ to (m,n) using steps in $\{(1,0),(0,1),(1,1)\}$ such that k diagonal steps are taken. Build the generating function $\sum_{m,n,k \geq 0} a_{m,n,k} x^m y^n z^k$.

3.1.19. ($\diamond$) Let $\hat{F}(x) = \sum_{n \geq 0} \hat{F}_n x^n$, where $\langle \hat{F} \rangle$ is the adjusted Fibonacci number.
    (a) Obtain $\hat{F}(x)$ from the Fibonacci recurrence.
    (b) Obtain $\hat{F}(x)$ by building it combinatorially (without the recurrence!), using the model that $\hat{F}_n$ is the number of 1,2-lists with sum n.
    (c) Expand the generating function to prove $\hat{F}_n = \sum_{k=0}^{n} \binom{k}{n-k}$.

3.1.20. A **k-colored graph** is a graph with its vertices partitioned into nonempty sets $V_1, \ldots, V_k$ such that adjacent vertices lie in different sets. Prove that the enumerator of k-colored graphs with vertex set $[n]$, by number of edges, is

$$\frac{1}{k!} \sum \binom{n}{n_1, \ldots, n_k} (1+x)^{\frac{1}{2}(n^2 - \sum n_i^2)},$$

where the sum is over choices of nonzero $n_1, \ldots, n_k$ with sum n.

3.1.21. Let $a_{n,k}$ be the number of directed graphs with vertex set $[n]$ that have exactly k sources (vertices of indegree 0) and no directed cycles. Note $a_{n,n} = 1$. For $n > k$, prove

$$a_{n,k} = \binom{n}{k} \sum_{j=1}^{n-k} a_{n-k,j} (2^k - 1)^j 2^{k(n-k-j)}.$$

Explain how to obtain from this the generating function indexed by number of edges.

3.1.22. Given a k-set $S \subseteq [n]$, let $\pi(S)$ be the permutation consisting of S in increasing order followed by $[n] - S$ in increasing order. Letting the k-sets be equally likely, determine the expected number of inversions in $\pi(S)$. (Comment: Also the variance is exactly $(n+1)/6$ times the expectation, for each k.) (Deshpande–Laghate [2003])

3.1.23. ($\diamond$) Compute the expected number of cycles in a random permutation of $[n]$,
    (a) by using the generating function in Theorem 3.1.20, and
    (b) by using Proposition 3.1.19 and linearity of expectation (the expectation of a sum is the sum of the expectations). (Lovász [1979, p. 25])

3.1.24. ($\diamond$) Compute the probability that a random permutation of $[n]$ has exactly two cycles, in two ways:
    (a) by using the generating function in Theorem 3.1.20, and
    (b) by considering random canonical cycle representations (see Lemma 3.1.18).

Exercises for Section 3.1

3.1.25. Given $S \subseteq [n]$, compute the probability that all elements of S are in the same cycle in a random permutation of $[n]$.

3.1.26. Let $m_0 = n$. Starting with $i = 0$, iteratively choose $x_i \in [m_i]$ at random, subtract x_i from m_i to obtain m_{i+1}, and continue (stopping if $m_{i+1} = 0$). Let Y_k be the number of times that k is the randomly chosen integer. Prove that the distribution of the random vector $(Y_1, \ldots, Y_n)$ is the same as the distribution of the random vector $(X_1, \ldots, X_n)$, where X_k is the number of k-cycles in a random permutation of $[n]$. (Bach [1994])

3.1.27. ($\diamond$) For $n \geq 1$, view $\mathbb{S}_n$ as the set of permutations of $[n]$ in word form. Let $A_n = \{\pi \in \mathbb{S}_n : \pi_i \neq i + 1 \text{ for } 1 \leq i \leq n-1\}$. Let $B_n = \{\sigma \in \mathbb{S}_n : \sigma_{j+1} \neq \sigma_j + 1 \text{ for } 1 \leq j \leq n-1\}$. Use the canonical cycle representation to prove bijectively that $|A_n| = |B_n|$. (Comment: The sizes of A_n and B_n were computed in Exercise 2.1.44.)

3.1.28. ($\diamond$) The keys to n boxes are put randomly in the boxes, one per box. The boxes are locked by closing them (a box may contain its own key). Find the probability that breaking open k random boxes will allow unlocking the remaining boxes. (Hint: Consider canonical cycle representations.) (Bognár–Mogyoródi–Prékopa–Rényi–Szász [1970, p. 56])

3.1.29. ($\diamond$) The **reverse canonical representation** of a permutation writes each cycle with largest element first and puts the cycles in increasing order of those first elements. Let $\psi_n \colon \mathbb{S}_n \to \mathbb{S}_n$ take each $\pi \in \mathbb{S}_n$ to the permutation obtained by dropping the parentheses in this representation of π. Characterize and count the fixed points of ψ_n. (Deutsch [2008])

3.1.30. Put elements of $\mathbb{S}_{2n}$ into E_{2n} or O_{2n} if their cycle lengths are all even or all odd, respectively ($|E_4| = |O_4| = 9$). Prove $|E_{2n}| = |O_{2n}|$, bijectively. (Schmidt [1994])

3.1.31. ($\diamond$) Let $C_n(x)$ be the enumerator of $\mathbb{S}_n$ by number of cycles, with $c(n, k) = [x^k]C_n(x)$. Theorem 3.1.20 proves combinatorially that $C_n(x) = x^{(n)}$.
  (a) Use canonical cycle representations to prove $c(n + 1, m + 1) = \sum_{k=m}^{n} c(n, k)\binom{k}{m}$.
  (b) Use part (a) to prove Theorem 3.1.20 by induction.
  (c) By Example 2.1.9, $c(n, k) = (n - 1)c(n - 1, k) + c(n - 1, k - 1)$ for $n, k \geq 1$. Use this and the generating function method to prove Theorem 3.1.20 again.

3.1.32. ($\diamond$) Create an array with row 0 being repeated copies of 1. To obtain row $j + 1$ from row j, cross out the first element of row j, then the second subsequent entry, then the third after that, and so on. Take partial sums of the remaining entries to form row $j + 1$, with blanks under the deleted entries. Prove that the first entry in row j is $j!$. (Hint: The diagonal of the jth wedge sums to $j!$. Explain those values using Exercise 3.1.31. Comment: Guy [1988] calls this an instance of **Moessner's Process**; see Exercise 2.1.38.)

1	1	1	1	1	1	1	1	1	1	1	1	1	1	1
1		2	3		4	5	6		7	8	9	10		
2			6	11			18	26	35					
6					24	50								
24														

3.1.33. Let $F_n(z) = \sum_{k \geq 0} c(k, n)z^k/k!$, where $c(k, n)$ is the number of permutations of $[k]$ with n cycles. Prove $F_n(z) = [-\ln(1 - z)]^n/n!$.

The remaining problems in this section involve the Eulerian numbers $A(n, k)$, where $A(n, k)$ denotes the number of permutations of $[n]$ with k runs.

3.1.34. Substitute the expression for $(k - r)^n$ given by using Worpitzky's Identity with $x = k - r$ into the sum $\sum_{r=0}^{k}(-1)^r\binom{n+1}{r}(k - r)^n$ to prove that the sum equals $A(n, k)$.

3.1.35. Prove $A(n, k) = \sum_{r=0}^{k} (-1)^r \binom{n+1}{r}(k - r)^n$ from Proposition 3.1.24.

3.1.36. *Eulerian numbers.* There are $A(n, k)$ permutations of $[n]$ with k runs.
    (a) Show that $\sum_{k=1}^{n} A(n, k) = n!$ and $A(n, k) = A(n, n + 1 - k)$.
    (b) Prove $\sum_{k=1}^{n} k A(n, k) = \frac{1}{2}(n + 1)!$ using the identities of part (a).
    (c) Prove $\sum_{k=1}^{n} k A(n, k) = \frac{1}{2}(n + 1)!$ by counting two ways.
(Comment: Thus the expected number of runs in a random permutation is $\frac{n+1}{2}$.)

3.1.37. ($\diamond$) There are $n!$ lists $(b_1, \ldots, b_n)$ in $\mathbb{N}^n$ such that $b_i \leq i$ for all i. Let $B(n, k)$ be the number that omit $k - 1$ elements of $[n]$. Prove $B(n, k) = A(n, k)$. (Bóna [2004, p. 31])

3.1.38. ($\diamond$) For $n \in \mathbb{N}$, prove that $\sum_{k \geq 0} k^n/2^k$ is an integer. (Hint: Consider the proof of Theorem 3.1.26.)

3.1.39. ($\diamond$) Two games begin with players A and B each having \$1. On each play, A or B wins \$1. When A has \$$a$ and B has \$$b$, the probability that A gets the next dollar is $a/(a + b)$ in Game 1, but in Game 2 it is $b/(a + b)$. For $1 \leq k \leq n$, determine in each game the probability that A has \$$k$ when the total is \$$(n + 1)$. (Comment: Game 1 has been called **Pólya's Urn**; see Johnson–Kotz [1977].)

3.1.40. ($\diamond$) An **increasing tree** is rooted with integer vertices, increasing along paths from the root. Let $\mathbf{T}_n$ be the set of increasing trees with vertices $\{0, \ldots, n\}$. For $\sigma \in \mathbb{S}_n$, form σ' by putting 0 before the word form of σ. For $i \in [n]$, create the edge ji, where j is the last element before i in σ' that is less than i. Let $f(\sigma)$ be the resulting graph on $\{0, \ldots, n\}$.
    (a) Prove that f is a bijection from the set of permutations of $[n]$ to $\mathbf{T}_n$.
    (b) Prove that $\mathbf{T}_n$ has $c(n, k)$ elements whose root has degree k, where $c(n, k)$ counts the permutations of $[n]$ with k cycles. (Hint: Consider canonical cycle representations.)
    (c) Prove that $\mathbf{T}_n$ has $A(n, k)$ elements with k (non-root) leaves.
(Comment: Elements of $\mathbf{T}_n$ correspond to inclusion arrangements of n square envelopes with distinct sizes. There are $n!$ arrangements, of which $c(n, k)$ have k envelopes contained in no others and $A(n, k)$ have k envelopes containing no others.)

3.1.41. ($\diamond$) For $\sigma \in \mathbb{S}_n$, there is an **excedance** at i if $\sigma(i) > i$. For example, the excedances in 791368452 are at $\{1, 2, 5, 6\}$. Given σ, let $\hat{\sigma}$ denote the permutation whose word form is obtained by dropping the parentheses in the canonical cycle representation of σ. Prove that σ has an excedance at i if and only if $\hat{\sigma}$ has an ascent at i. Thus the number of permutations of $[n]$ with k ascents is the number of permutations of $[n]$ with k excedances, and $A(n, k)$ permutations of $[n]$ have $n - k$ excedances. (Foata–Schützenberger [1970])

3.1.42. ($\diamond$) For $\sigma \in \mathbb{S}_n$, there is a **weak excedance** at i if $\sigma(i) \geq i$. Use Exercise 3.1.41 to prove that the number of permutations of $[n]$ having k weak excedances is $A(n, k)$. (Hint: Compare σ with its *reverse complement* σ^*, defined by $\sigma^*(i) = n + 1 - \sigma(n + 1 - i)$.) (Bóna [2004, p. 32])

3.1.43. (+) *Smith College Diploma Problem.* By tradition, the graduating students form a circle and diplomas are distributed at random. Students having their own diplomas leave; each remaining student passes the diploma she holds to the student on her right to complete the round. Maurer [1973] asked for the probability that the process takes k rounds. Don West [1974] used recurrence and induction; Gessel [2001] gave a bijective solution.
    (a) For $\sigma \in \mathbb{S}_n$, let $s_1, \ldots, s_m$ be the non-fixed elements in increasing order. Let $\sigma'(s_i) = \sigma(s_{i+1})$ if $i < m$ and $\sigma'(s_m) = \sigma(s_1)$. The fixed points of σ are also fixed points of σ'. Prove that if σ has k excedances (Exercise 3.1.41), then σ' has $k - 1$ excedances. For example, $32514 \to 52143 \to 12345$.
    (b) Use part (a) and Exercise 3.1.41 to prove that the number of permutations of the n diplomas that result in k rounds for sorting is $A(n, k)$.

3.2. Coefficients and Applications

Formal power series expansions of generating functions can be manipulated much like ordinary power series in calculus, with the additional freedom of ignoring issues of convergence. (Niven [1969] presents a readable introduction to the theory.) When a function has a power series expansion at 0, that is also its formal power series expansion. In Lemma 2.2.17 we expanded $(1-x)^{-n}$ using Taylor's Theorem. In Theorem 3.1.10, the same expression emerged as the formal power series expansion of $(1-x)^{-n}$ using the combinatorial interpretation of formal power series products. This is no coincidence.

Equality holds because operations on formal power series agree with corresponding operations on power series that converge at numerical values. Since this holds for sum and product, statements like $A(x)^m A(x)^n = A(x)^{m+n}$ have the same proofs for formal power series as for power series.

To extend the correspondence, we define differentiation and "evaluation" for formal power series to agree with these operations on power series. In calculus, Taylor series are termwise differentiable, using uniform convergence of series of functions. For formal power series, this is the *definition* of differentiation. We will skip formal verifications, focusing instead on the *use* of generating functions.

3.2.1. Definition. When $A(x)$ is the generating function with expansion $A(x) = \sum_{n=0}^{\infty} a_n x^n$, **evaluation at 0** is defined by setting $A(0) = a_0$. The **derivative** of a formal power series $\sum_{n=0}^{\infty} a_n x^n$ is the formal power series $\sum_{n=1}^{\infty} n a_n x^{n-1}$; we write $A'(x)$ for the derivative of a formal power series $A(x)$ in x.

Since the termwise behavior holds by definition, the sum and product rules for differentiation of formal power series are easy to verify (Exercise 30). Generally we have the following statement, which motivates writing $A'(x)$ for the derivative of a formal power series $A(x)$:

The derivative of the formal power series expansion of $A(x)$ is the formal power series expansion of the derivative of $A(x)$.

The occurrence of power series as formal power series is another excuse for our (ab)use of the word "function" in the term *generating function*. Some formal power series converge in a neighborhood of the origin when viewed as power series, but some do not. Evaluation of $A(x)$ at numerical values for x (such as $x = 1$) is allowed when the sum converges (such as when $A(x)$ is a polynomial in x).

OPERATIONS AND SUMMATIONS

Since the formal power series expansion of $(1-x)^{-1}$ is also the power series for $(1-x)^{-1}$, the correspondence between operations allows us to derive our favorite formal power series expansion by differentiation.

3.2.2. Example. $(1-x)^{-n} = \sum_{k \geq 0} \binom{k+n-1}{n-1} x^k$ *by formal differentiation.* In contrast to Example 3.1.8, we differentiate $(n-1$ times) both sides of $(1-x)^{-1} = \sum_{k \geq 0} x^k$. This yields $\frac{(n-1)!}{(1-x)^n} = \sum_{k \geq n-1} k(k-1) \cdots (k-n+2) x^{k-n+1}$. Now divide by $(n-1)!$ and "shift the index" by $n-1$.   ∎

3.2.3. Remark. *Shifting the index.* The negative binomial expansion often arises as $\sum_{k \geq 0} \binom{k}{r} x^k$. We find the generating function by

$$\sum_{k \geq 0} \binom{k}{r} x^k = \sum_{k \geq r} \binom{k}{r} x^k = \sum_{k \geq 0} \binom{k+r}{r} x^{k+r} = \frac{x^r}{(1-x)^{r+1}}.$$

Here we discard terms that are 0, then shift the index, then use the known expansion of $(1-x)^{-n}$. Lowering the starting value of the index (by r) is done by adding r to each appearance of the index in the summand. We often do this to introduce or remove a power of the formal variable.   ∎

3.2.4. Remark. *Extracting coefficients.* We often study the generating function $A(x)$ to seek a formula for a_n. In principle, we can differentiate n times, divide by $n!$, and evaluate at 0, since $a_n = \frac{1}{n!} A^{(n)}(0)$. In practice, this approach is useless.

Usually we obtain coefficients from a known series, such as $(1+x)^n$, $(1-x)^{-n}$, $\frac{1-x^{n+1}}{1-x}$, or e^x. We operate on the function and expansion simultaneously, maintaining equality. This works with sums, products, derivatives, and partial fractions. The other operations below are easy but useful manipulations. Property (3) was used in Remark 3.2.3. Property (4) is a common application for differentiation, where we also shift the index to restore equality of exponent and subscript.   ∎

3.2.5. Proposition. If A, B, C are the ordinary generating functions for $\langle a \rangle$, $\langle b \rangle$, and $\langle c \rangle$, respectively, then

(1) $c_n = a_n + b_n$ for all n $\Leftrightarrow$ $C(x) = A(x) + B(x)$.

(2) $c_n = \sum_{i=0}^{n} a_i b_{n-i}$ for all n $\Leftrightarrow$ $C(x) = A(x)B(x)$.

(3) $b_n = \begin{cases} a_{n-r} & \text{for } n \geq r \\ 0 & \text{for } n < r \end{cases}$ $\Leftrightarrow$ $B(x) = x^r A(x)$.

(4) $b_n = na_n$ $\Leftrightarrow$ $B(x) = xA'(x)$.

(5) $c_n = \sum_{i=0}^{n} a_i$ for all n $\Leftrightarrow$ $C(x) = \dfrac{A(x)}{1-x}$ (special case of (2))

(6a) $b_n = \begin{cases} a_n & \text{for } n \text{ even} \\ 0 & \text{for } n \text{ odd} \end{cases}$ $\Leftrightarrow$ $B(x) = \frac{1}{2}\left[A(x) + A(-x)\right]$.

(6b) $b_n = \begin{cases} a_n & \text{for } n \text{ odd} \\ 0 & \text{for } n \text{ even} \end{cases}$ $\Leftrightarrow$ $B(x) = \frac{1}{2}\left[A(x) - A(-x)\right]$.

(7) $b_n = \begin{cases} a_{n/m} & \text{when } m \mid n \\ 0 & \text{when } m \nmid n \end{cases}$ $\Leftrightarrow$ $B(x) = A(x^m)$.

Proof: (1,2) These are the definitions of arithmetic on formal power series (Definition 3.1.4). Convolution gives the coefficient of x^n in a product of series.

(3) Multiplying by powers of the formal variable has the effect of shifting indices in the sequence of coefficients.

(4) Differentiation of $A(x)$ introduces a linear factor (Definition 3.2.1).

(5) Multiplying $A(x)$ by $(1-x)^{-1}$ produces a new generating function where the coefficient of x^n is the sum of the coefficients up to x^n in $A(x)$.

(6) Odd or even terms in the expansion of $A(x)$ can be canceled out by forming the sum $[A(x) + A(-x)]/2$ or the sum $[A(x) - A(-x)]/2$.

(7) Substituting $y = x^m$ spreads out the terms.   ∎

These manipulations are useful in several ways. We can start with the expansion of a known generating function and manipulate it to obtain a desired sum, operating simultaneously on the generating function. We may also recognize the sum as a *coefficient* in a generating function. We call these *algebraic proofs* for evaluating summations. As with recurrence relations, discovering the value of the sum may suggest a short combinatorial proof that was not initially apparent.

3.2.6. Example. *Summing the coefficients.* The terms of a sequence can be summed (if the sum converges) by setting the argument of the generating function to 1. For example, $\sum k\binom{n}{k}$ arises from the binomial expansion $\sum \binom{n}{k}x^k$ by differentiating and then setting $x = 1$. Thus

$$\sum_{k \geq 0} k\binom{n}{k} = \frac{d}{dx}(1+x)^n\Big|_{x=1} = n2^{n-1}.$$

Combinatorially, both sides count chaired committees from n people.   ∎

3.2.7. Example. *Even coefficients.* To sum the even coefficients in the expansion of $(1+x)^n$, Proposition 3.2.5(6) first yields

$$\sum_{i \geq 0} \binom{n}{2i}x^{2i} = \frac{1}{2}\Big[(1+x)^n + (1-x)^n\Big].$$

Setting $x = 1$ yields 2^{n-1} when $n > 0$. When $n = 0$, both $(1+x)^n$ and $(1-x)^n$ equal 1, and setting $x = 1$ yields 1, not $1/2$.

Combinatorially, the sum counts the even subsets of an n-set. The number is 2^{n-1}, since for each subset of $[n-1]$, including or omitting n yields one even subset and one odd subset (when $n > 0$).   ∎

3.2.8. Example. *Even minus Odd.* Setting $x = -1$ in $A(x)$ gives the sum of the even coefficients minus the sum of the odd coefficients. If $A(x)$ enumerates a set having m_0 objects with even index and m_1 with odd index, then $A(-1) = m_0 - m_1$.

For example, $(1+x)^n$ enumerates subsets of $[n]$ by size. Setting $x = 1$ yields all 2^n subsets, and setting $x = -1$ proves equality between the numbers of subsets with even size and odd size (if $n > 0$).

The generating function $\prod_{j=0}^{n-1} \sum_{i=0}^{j} x^i$ enumerates permutations of $[n]$ by number of inversions (Proposition 3.1.15). Setting $x = 1$ yields all $n!$ permutations, and setting $x = -1$ proves equality between the numbers of permutations with even and odd numbers of inversions (if $n > 1$).

The generating function $\prod_{j=1}^{n}(x + j - 1)$ enumerates permutations of $[n]$ by number of cycles (Theorem 3.1.20). Setting $x = 1$ yields all $n!$ permutations, and setting $x = -1$ proves equality between the numbers of permutations with even and odd numbers of cycles (if $n > 1$).   ∎

3.2.9. Example. *Summation of polynomials, revisited.* We saw in Application 1.2.7 that every polynomial p of degree d in k is a linear combination of $\binom{k}{d}, \ldots, \binom{k}{0}$. This and $\sum_{k=0}^{n} \binom{k}{j} = \binom{n+1}{j+1}$ allowed us to compute $\sum_{k=0}^{n} p(k)$.

Here instead we use Proposition 3.2.5(3–5). In $\sum_{k\geq0} k^m x^k$, successive factors of k are introduced by differentiating and then multiplying by x to restore the exponent. Thus $\sum_{k\geq0} x^k = \frac{1}{1-x}$ and $\sum_{k\geq0} kx^k = \frac{x}{(1-x)^2}$ and $\sum_{k\geq0} k^2 x^k = \frac{x+x^2}{(1-x)^3}$. By (5), $\sum_{k=0}^{n} k^2 = [x^n]\frac{x+x^2}{(1-x)^4}$. Summing $[x^{n-1}](1-x)^{-4}$ and $[x^{n-2}](1-x)^{-4}$ yields

$$\sum_{k=0}^{n} k^2 = \binom{n-1+3}{3} + \binom{n-2+3}{3} = \frac{(n+1)n(2n+1)}{6}.$$

Exercise 13 suggests another generating function approach.   ∎

Evaluation of sums by convolutions (Proposition 3.2.5(2)) deserves special attention. We use the convenient abbreviation "**OGF**" for "ordinary" generating function (in the next section we study a different type of generating function). The value of $\sum_{i=0}^{n} a_i b_{n-i}$ is the coefficient of x^n in the product of the OGFs for $\langle a \rangle$ and $\langle b \rangle$. The coefficient operator conveniently encodes this phrase in notation. Theorem 1.2.3 gave combinatorial arguments for the first two examples below.

3.2.10. Example. *The Vandermonde Convolution* $\sum_{k=0}^{r} \binom{m}{k}\binom{n}{r-k}$. Recognizing the sum as a convolution with $a_k = \binom{m}{k}$ and $b_k = \binom{n}{k}$, we have

$$\sum_{k=0}^{r} \binom{m}{k}\binom{n}{r-k} = [x^r] \sum_k \binom{m}{k} x^k \sum_k \binom{n}{k} x^k$$

$$= [x^r](1+x)^m (1+x)^n = [x^r](1+x)^{m+n} = \binom{m+n}{r}.$$ ∎

3.2.11. Example. *The convolution* $\sum_{k=0}^{n} k(n-k)$. Let $a_k = b_k = k$ and $c_n = \sum_{k=0}^{n} a_k b_{n-k}$. The generating function $\sum_{k\geq0} kx^k$ is $x/(1-x)^2$ (see Example 3.2.9). We obtain $c_n = \binom{n+1}{3}$, because c_n is the coefficient of x^n in the computation below.

$$\sum_{k\geq0} kx^k \sum_{k\geq0} kx^k = \left(\frac{x}{(1-x)^2}\right)^2 = \frac{x^2}{(1-x)^4} = \sum_{k=0}^{\infty} \binom{k+4-1}{4-1} x^{k+2}.$$ ∎

3.2.12.* Example. *Evaluation of* $\sum_{k\geq0}(-1)^k \binom{n-k}{k}\binom{n-2k}{m-k}$ *for* $m, n \in \mathbb{N}_0$. With so many uses of k, it is unclear how to write this as a simple convolution. However, when $m + k \leq n$, the Subcommittee Identity yields $\binom{n-k}{k}\binom{n-2k}{m-k} = \binom{n-k}{m}\binom{m}{k}$, which rewrites the sum as $\sum_k (-1)^k \binom{m}{k}\binom{n-k}{m}$. Let $a_k = (-1)^k \binom{m}{k}$ and $b_k = \binom{k}{m}$. The OGF for $\langle b \rangle$ from Remark 3.2.3 yields

$$\sum_{k\geq0}(-1)^k \binom{n-k}{k}\binom{n-2k}{m-k} = \sum_{k\geq0}(-1)^k \binom{m}{k}\binom{n-k}{m} = \sum_{k\geq0} a_k b_{n-k}$$

$$= [x^n](1-x)^m \frac{x^m}{(1-x)^{m+1}} = \begin{cases} 1 & \text{if } n \geq m \\ 0 & \text{if } n < m \end{cases}.$$ ∎

Next we apply generating functions to convolutions in a reverse way. Instead of finding the unknown OGF for the convolution of sequences with known OGFs, we use the known OGF for the convolution to find the OGF for a factor.

3.2.13. Example. *Central binomial coefficients:* $\sum_{n\geq 0} \binom{2n}{n} x^k = (1-4x)^{-1/2}$. Let $A(x)$ be this generating function. The convolution $\sum_{k=0}^{n} \binom{2k}{k}\binom{2n-2k}{n-k}$ equals $[x^n]A(x)^2$. Knowing the coefficients in $A(x)$ is equivalent to knowing the sum. In Theorem 1.3.15, combinatorial argument gave $\sum_{k=0}^{n} \binom{2k}{k}\binom{2n-2k}{n-k} = 4^n$. Thus $A(x)^2 = (1-4x)^{-1}$, so $\sum_{n\geq 0} \binom{2n}{n} x^n = (1-4x)^{-1/2}$.

If we did not know the sum, we could still find it and $A(x)$ by using the Catalan generating function. Termwise differentiation yields $A(x) = \frac{d}{dx} xC(x)$, where $C(x) = \sum_{k\geq 0} \frac{1}{k+1}\binom{2k}{k} x^k$. In Theorem 2.2.22 we proved $C(x) = (1 - \sqrt{1-4x})/(2x)$. Applying $A(x) = \frac{d}{dx} xC(x)$ yields $A(x) = (1-4x)^{-1/2}$. Hence the OGF for the convolution is $(1-4x)^{-1}$, and the value of the sum is 4^n. ∎

3.2.14. Remark. The Inversion Principle. Let $C(x)$ be a generating function such that $C(0) \neq 0$, which is the condition for existence of the multiplicative inverse $1/C(x)$ (Exercise 6). If $\langle a \rangle$ and $\langle b \rangle$ are sequences whose OGFs $A(x)$ and $B(x)$ satisfy $B(x) = A(x)C(x)$, then also $A(x) = B(x) \cdot \frac{1}{C(x)}$. Thus if $C(x) = \sum_{k\geq 0} c_k x^k$ and $\frac{1}{C(x)} = \sum_{k\geq 0} c'_k x^k$, then the statements

$$b_k = \sum_{i=0}^{k} a_i c_{k-i} \text{ for all } k \qquad \text{and} \qquad a_k = \sum_{i=0}^{k} b_i c'_{k-i} \text{ for all } k$$

are equivalent (see Exercises 7–8). Our formula for the Eulerian numbers in Theorem 3.1.26 used the Inversion Principle, with $(1-x)^{-(n+1)}$ and $(1-x)^{n+1}$ playing the roles of $C(x)$ and $\frac{1}{C(x)}$. ∎

Proposition 3.2.5(4) has an important application in probability.

3.2.15. Application. *Expected value.* Let X be a discrete random variable (Definition 0.12). Write $\mathbb{P}(X = k)$ for the probability that X has value k. When X is restricted to $\mathbb{N}_0$, the **probability generating function** for X is the generating function $\sum \mathbb{P}(X = k)x^k$.

The **expectation** or **expected value** of X, written $\mathbb{E}(X)$, is the average of its values when weighted by their probabilities: that is, $\mathbb{E}(X) = \sum_k k\mathbb{P}(X = k)$. Computing this from the probability generating function $A(x)$ via $\mathbb{E}(X) = A'(1)$ is valid when $A(x)$ is analytic at $x = 1$.

Consider a sequence of coin flips. On each flip, independently, the outcome is a head with probability p. Let X be the index of the first head. Since the flips are independent, $\mathbb{P}(X = k) = p(1-p)^{k-1}$; this is the **geometric distribution**. The probability generating function is

$$A(x) = \sum_{k\geq 1} p(1-p)^{k-1}x^k = \frac{p}{1-p}\left(\frac{1}{1-(1-p)x} - 1\right).$$

Thus $\mathbb{E}(X) = A'(1) = \frac{p}{1-p} \cdot \frac{1-p}{(1-(1-p)1)^2} = \frac{1}{p}$. ∎

SNAKE OIL

When using convolution to evaluate a sum, we treat the sum as a coefficient in the product of two generating functions. Generating functions can help evaluate sums in other ways.

The **Snake Oil method** was named by Wilf [1990, pp. 108–120]. Although Snake Oil was originally an ointment for joint pain made from the Chinese Water Snake, the term came to refer to ineffectual elixirs peddled by traveling hoaxsters in 19th-century western America, marketed as relieving all kinds of ailments. Wilf used the term for this technique because it can produce amazingly simple "cures" to evaluate sums, seemingly by pure magic and without knowing the value in advance.

The idea is to view a sum with a parameter n as the coefficient of x^n in a generating function $A(x)$. When the order of summation in the expression for $A(x)$ is interchanged, it may be unexpectedly easy to perform the new inner sum on n explicitly, especially if n appears only once in the expression. Several of our examples come from Wilf [1990]; see also Exercises 36–49.

3.2.16. Example. *Evaluating* $\sum_{k \geq 0} \binom{k}{n-k}$. Let a_n be the desired sum, with $A(x) = \sum_{n \geq 0} a_n x^n$. Interchanging the order of summation allows us to perform the inner sum as follows:

$$\sum_{n \geq 0} \sum_{k \geq 0} \binom{k}{n-k} x^n = \sum_{k \geq 0} x^k \sum_n \binom{k}{n-k} x^{n-k} = \sum_{k \geq 0} x^k (1+x)^k = \frac{1}{1-x-x^2}.$$

The last expression enumerates 1,2-lists by their sum. Hence it is the OGF for adjusted Fibonacci numbers (see Exercise 3.1.19), and $a_n = \hat{F}_n$.

Had we known that we wanted $\hat{F}_n$, we could have set $i = n-k$ at the beginning to transform the sum to $\sum_i \binom{n-i}{i}$, which evaluates to $\hat{F}_n$ by a pleasant combinatorial argument. Snake Oil did not require knowing the value of the sum. ∎

Exercise 3.1.31 requests a bijective proof of the identity in the next example. Snake Oil evaluates the sum without first knowing the value.

3.2.17. Example. $\sum_{k=m}^{n} c(n,k)\binom{k}{m} = c(n+1, m+1)$. Here $c(n,k)$ is the number of permutations of $[n]$ with k cycles; by Theorem 3.1.20, $\sum_{k=0}^{n} c(n,k)x^k = x^{(n)}$. We find the sum as a coefficient in a generating function. Since $\binom{k}{m}x^m$ is easy to sum over m when k is fixed, we introduce a generating function with index m. After interchanging the order of summation, we use Theorem 3.1.20 to compute

$$\sum_m \sum_k c(n,k)\binom{k}{m} x^m = \sum_k c(n,k) \sum_m \binom{k}{m} x^m = \sum_k c(n,k)(1+x)^k = (1+x)^{(n)}.$$

To extract the coefficient of x^m cleanly, a small trick takes us to the generating function for permutations of $[n+1]$:

$$\sum_{k=m}^{n} c(n,k)\binom{k}{m} = [x^m](1+x)^{(n)} = [x^{m+1}]x^{(n+1)} = c(n+1, m+1). \quad ∎$$

Snake Oil can succeed when the summand has several factors. The method splits the summand to perform simpler sums over fewer factors. This is especially promising when a parameter appears only once in the summand. We use it next for a short proof of the equality of two formulas for the Delannoy numbers. Theorem 1.2.13 and Exercise 1.2.39 give combinatorial arguments; here we show more easily that both formulas give the coefficients in the same generating function.

3.2.18. Example. $\sum_k \binom{m}{k}\binom{n+k}{m} = \sum_k \binom{m}{k}\binom{n}{k}2^k$. Snake Oil proves the identity by showing that both sides have the same generating function, indexed by n. Multiply by x^n, sum over n, and interchange the order of summation. On both sides we use $\sum_{r\geq 0}\binom{r}{k}x^r = \frac{x^k}{(1-x)^{k+1}}$.

On the left we compute

$$\sum_k \binom{m}{k}x^{-k}\sum_{n\geq 0}\binom{n+k}{m}x^{n+k} = \sum_k \binom{m}{k}x^{-k}\frac{x^m}{(1-x)^{m+1}}$$

$$= \frac{x^m}{(1-x)^{m+1}}(1+x^{-1})^m = \frac{(1+x)^m}{(1-x)^{m+1}}.$$

In evaluating $\sum_{n\geq 0}\binom{n+k}{m}x^{n+k}$, we have used that the coefficient $\binom{m}{k}$ is 0 unless $k\leq m$, so all the terms needed to form $(1-x)^{m+1}$ are present.

On the right we compute

$$\sum_k \binom{m}{k}2^k\sum_{n\geq 0}\binom{n}{k}x^n = \frac{1}{1-x}\sum_k \binom{m}{k}2^k\left(\frac{x}{1-x}\right)^k$$

$$= \frac{1}{1-x}\left(1+\frac{2x}{1-x}\right)^m = \frac{(1+x)^m}{(1-x)^{m+1}}. \qquad \blacksquare$$

When a parameter in the sum appears more than once, Snake Oil may still work after introducing an extra free variable. The desired sum then becomes a special case of a more general sum.

3.2.19.* Example. $\sum_k(-1)^{n-k}\binom{2n}{k}^2 = \binom{2n}{n}$. With n appearing so often in $(-1)^{n-k}\binom{2n}{k}\binom{2n}{2n-k}$, it is hard to perform the inner sum after multiplying by x^n, summing over n, and interchanging the order of summation.

Instead, we evaluate the more general sum $\sum_k(-1)^{n-k}\binom{2n}{k}\binom{2n}{m-k}$ and then set $m = 2n$. Now the parameter m appears only once. Introduce the OGF indexed by m. After interchanging the order of summation,

$$\sum_k(-1)^{n-k}\binom{2n}{k}x^k\sum_{m\geq 0}\binom{2n}{m-k}x^{m-k} = \sum_k(-1)^{n-k}\binom{2n}{k}x^k(1+x)^{2n}$$

$$= (-1)^n(1-x)^{2n}(1+x)^{2n} = (-1)^n(1-x^2)^{2n}.$$

In this generating function we seek the coefficient of x^m. It is 0 when m is odd, and when m is even it is $[y^{m/2}](-1)^n(1-y)^{2n}$, which equals $(-1)^{n-m/2}\binom{2n}{m/2}$. For $m = 2n$, this is precisely $\binom{2n}{n}$, as desired. $\qquad \blacksquare$

EXERCISES 3.2

3.2.1. (−) Compute the coefficient of x^{10} in the following generating functions.
  (a) $(1+x)^3(1-x)^{-3}$.   (b) $(x^2+x^3+x^4)^4$.   (c) $(1-2x)^{-3}$.

3.2.2. (−) Let $a_k = \binom{n}{\lfloor k/2 \rfloor}$. Determine the generating function $\sum_{k \geq 0} a_k x^k$.

3.2.3. (−) For $n \in \mathbb{N}$, let $a_n = \sum_{k=1}^n \frac{1}{k 2^{n-k}}$. Find $\sum_{n \geq 1} a_n x^n$. (Comment: See Exercise 2.1.30 for alternative expressions of a_n.)

3.2.4. (−) Use generating functions to evaluate the sums below.

  (a) $\displaystyle\sum_{k=0}^r (-1)^k \binom{n}{k}\binom{n}{r-k}$ (b) $\displaystyle\sum_{k=1}^n (-1)^{k-1} k \binom{n}{k} 2^{n-k}$

3.2.5. (−) Use generating functions to evaluate the sums below for all $n \geq 0$, and give combinatorial proofs of the resulting identities.

  (a) $\displaystyle\sum_i 2i\binom{n}{2i}$ (b) $\displaystyle\sum_{k=0}^n k\binom{n}{k}^2$

3.2.6. (−) Prove that a formal power series $\sum_{n \geq 0} a_n x^n$ has a multiplicative inverse if and only if $a_0 \neq 0$. Prove that the inverse is unique when it exists.

3.2.7. (−) Prove that the following four statements (each over all $k \in \mathbb{N}$) are equivalent.

  (a) $b_k = \sum_{i=0}^k \binom{n}{i} a_{k-i}$ (c) $a_k = \sum_{i=0}^k (-1)^i \binom{i+n-1}{n-1} b_{k-i}$
  (b) $b_k = \sum_{i=0}^k \binom{n}{k-i} a_i$ (d) $a_k = \sum_{i=0}^k (-1)^{k-i} \binom{k-i+n-1}{n-1} b_i$

3.2.8. (−) Prove that the following four statements (each over all $k \in \mathbb{N}$) are equivalent.

  (a) $b_k = \sum_{i=0}^k \binom{i+n}{n} a_{k-i}$ (c) $a_k = \sum_{i=0}^k (-1)^i \binom{n+1}{i} b_{k-i}$
  (b) $b_k = \sum_{i=0}^k \binom{k-i+n}{n} a_i$ (d) $a_k = \sum_{i=0}^k (-1)^i \binom{n+1}{k-i} b_i$

3.2.9. Let $f(x)$ be the generating function for $\langle a \rangle$, and let $b_n = \sum_{k>n} a_k$. Prove that the generating function $g(x)$ for $\langle b \rangle$ is given by $g(x) = \frac{f(1)-f(x)}{1-x}$.

3.2.10. (◊) *Restricted multisets.* Let n be even.
  (a) Find the generating function (indexed by size) for multisets from $[n]$ having odd multiplicity of each odd number and even multiplicity of each even number. For example, $(1,1,1,2,2,3)$ is such a multiset of size 6 when $n = 4$.
  (b) Extract the coefficient of x^k in the generating function.

3.2.11. Let $\frac{3-3x}{1-x-2x^2}$ be the generating function for the sequence $a_0, a_1, a_2, \ldots$. Without obtaining a formula for a_k, compute $\sum_{k=0}^n a_k$ as a function of n.

3.2.12. (◊) Use generating functions to prove

$$\sum_i (-1)^i \binom{n}{i}\binom{k-2i+n-1}{n-1} = \binom{n}{k}.$$

3.2.13. *Alternative computation of $\sum_{k=1}^n k^2$.*
  (a) Express k^2 as a linear combination of polynomials of the form $\binom{k+j}{j}$.
  (b) Use part (a) to prove $\sum_{k=1}^n k^2 = n(n+1)(2n+1)/6$.

3.2.14. ($\diamond$) Compute $[x^n](1-x^2)^{-1/2}$.

3.2.15. ($\diamond$) Use generating functions to evaluate the sums below, and then prove the second inductively.

(a) $\displaystyle\sum_k \frac{1}{k+1}\binom{2k}{k}\frac{1}{n-k+1}\binom{2n-2k}{n-k}$

(b) $\displaystyle\sum_{i=0}^{k}(-1)^i\binom{n}{k-i}$

3.2.16. Let $b_n = \hat{F}_{2n}$ and $c_n = \hat{F}_{2n+1}$, where $\langle\hat{F}\rangle$ is the adjusted Fibonacci sequence. From the OGF for $\langle\hat{F}\rangle$, obtain the OGFs for $\langle b\rangle$ and $\langle c\rangle$, and use them to obtain recurrences for these sequences.

3.2.17. Let $a_n = \binom{2n-1}{n}$ for $n \geq 1$. Find the generating function $\sum_{n\geq 1} a_n x^n$.

3.2.18. ($\diamond$) Use generating functions to evaluate the sums below, and give combinatorial proofs of the resulting identities.

(a) $\displaystyle\sum_{j=0}^{k}\binom{n+k-j-1}{k-j}\binom{m+j-1}{j}$

(b) $\displaystyle\sum_{k=0}^{n}\frac{1}{k+1}\binom{2k}{k}\binom{2n-2k}{n-k}$

3.2.19. ($\diamond$) Count the lattice paths that have endpoints in $[n]\times[n]$ and take only rightward or upward unit steps (evaluate all sums). For $n \in \{1,2,3\}$, the values are 1, 10, 53.

3.2.20. ($\diamond$) Example 3.2.13 obtains the generating function $\sum_n\binom{2n}{n}x^n = (1-4x)^{-1/2}$. Use this in a combinatorial argument about lattice paths to obtain the generating function below, where $C(x)$ is the OGF for the Catalan numbers.

$$\sum_{n=0}^{\infty}\binom{2n+k}{n}x^n = \frac{C(x)^k}{\sqrt{1-4x}}$$

3.2.21. *Modified Catalan generating function.* Let $B(x) = \sum_{k\geq 0}C_k x^{k+1} = \sum_{k\geq 0}\binom{2k}{k}\frac{x^{k+1}}{k+1}$.
   (a) Prove $B(x)B'(x) = \frac{1}{2}(B'(x)-1)$, and use this to prove the identity below, where the sum is over $k, l \in \mathbb{N}_0$ with $k+l=n$.
   (b) Using lattice paths, give a bijective proof of the identity.

$$\sum\frac{\binom{2k}{k}\binom{2l+2}{l+1}}{k+1} = 2\binom{2n+2}{n}$$ (Dályay [2016])

3.2.22. ($\diamond$) Use OGFs to sum $\sum_{k=0}^{n-1}4^{n-k}\frac{1}{k+1}\binom{2k}{k}$. (Whitworth [1897])

3.2.23. Evaluate the sum $\sum_{k\geq 0}(-1)^k\binom{n-k}{k}x^k$. Conclude $\sum_{k\geq 0}\binom{n-k}{k} = \hat{F}_n$.

3.2.24. Let $m!! = \prod_{k=0}^{\lfloor(m-1)/2\rfloor}(m-2k)$. Evaluate $\sum_{i=0}^{n}\binom{n}{i}(2i-1)!!(2n-2i-1)!!$. (Note that $(-1)!! = 1$.) (Dzhumadiĺdaeva [2009])

3.2.25. ($\diamond$) Let $P(m,n,r) = \sum_{k=0}^{r}(-1)^k\binom{m-2k}{n}\binom{r}{k}$. Prove that if $0 \leq r \leq n \leq m$ and $n > (m+1)/2$, then $P(m,n,r) > 0$ and $\sum_{r=0}^{n}P(m,n,r) = \binom{m+2}{n}$. (Hint: Let $F_{n,r}(x) = \sum_{m\geq 0}P(m,n,r)x^m$.) (Deshpande–Welukar [2003])

3.2.26. Let $a_{m,n} = [x^m y^n](1-x-y+2xy)^{-1}$. Prove $a_{m,n} = \sum_{k\geq 0}(-1)^k\binom{m}{k}\binom{n}{k}$ and $a_{2j,2j+2} = (-1)^j\frac{1}{j+1}\binom{2j}{j}$. (Hint: $1-x-y+2xy = (1-x)(1-y)(1+\frac{xy}{(1-x)(1-y)})$.) (Gessel [1994])

3.2.27. (+) Let $f(n)$ be the number of binary words of length n in which the numbers of occurrences of consecutive 00 and consecutive 01 are the same. Prove

$$\sum_{n=0}^{\infty} f(n)t^n = \frac{1}{2}\left(\frac{1}{1-t} + \frac{1+2t}{\sqrt{(1-t)(1-2t)(1+t+2t^2)}}\right). \qquad \text{(Stanley [2011])}$$

3.2.28. A gambler insists on playing until he is ahead by one game. Assume that he has probability p of winning any single game, independently.
(a) Let α be the probability that the match ends. Prove $\alpha = 1$ for $p \geq 1/2$ and $\alpha < 1$ for $p < 1/2$. (Hint: Consider the Catalan generating function.)
(b) For $p > 1/2$, use generating function techniques to give an expression for the expected number of games in the match (do not compute the value).
(c) For $p > 1/2$, find an equation for the expected number of games in the match that computes it directly and simply without using generating functions.

3.2.29. (+) For nonnegative integers m and n, prove the identity below. (Hint: Both sides equal the coefficient of x^{m+n} in the same generating function. A direct combinatorial proof was requested in Exercise 1.2.50.) (Keselman [2008], solution to Knuth [2007])

$$\sum_{k=0}^{\infty} 2^k \binom{2m-k}{m+n} = 4^m - \sum_{j=1}^{n} \binom{2m+1}{m+j}.$$

3.2.30. Let $A(x)$ and $B(x)$ be formal power series in x.
(a) Prove that the derivative of $A(x) + B(x)$ is $A'(x) + B'(x)$.
(b) Prove that the derivative of $A(x)B(x)$ is $A'(x)B(x) + A(x)B'(x)$.
(c) Suppose that $B(x)$ is the multiplicative inverse of $A(x)$. By taking the derivative on both sides of $A(x)B(x) = 1$ and rearranging terms, show that $B'(x) = \frac{-B(x)A'(x)}{A(x)} = \frac{-A'(x)}{[A(x)]^2}$.
Thus the ordinary formula for the derivative of $[A(x)]^{-1}$ with respect to x holds also for differentiation of formal power series.

3.2.31. Prove $\sum_{k=0}^{\infty} \binom{k+n-1}{k} 2^{-k} = 2^n$, and apply this to evaluate $\sum_{k=1}^{\infty} k2^{-k}$. Verify the result by using another method to evaluate the latter sum.

3.2.32. For $m, n \in \mathbb{N}_0$, evaluate $\displaystyle\sum_{k\in\mathbb{Z}} (-1)^k \binom{m+n}{m+k}\binom{m+n}{n+k}$.

3.2.33. For $m, n \in \mathbb{N}_0$, evaluate $\displaystyle\sum_{k=0}^{\lfloor m/2 \rfloor} (-1)^k \binom{n}{k}\binom{m-2k+n-1}{n-1}$.

3.2.34. ($\diamond$) *Converting sums to integrals.*
(a) For $r, s \in \mathbb{N}$, prove $\int_0^1 x^r(1-x)^s dx = \frac{1}{r+s+1}\binom{r+s}{s}^{-1}$.
(b) For $n, m \in \mathbb{N}$, prove $\sum_{k=0}^{n} (-1)^k \binom{n}{k}\frac{1}{m+k} = \frac{1}{m}\binom{m+n}{n}^{-1}$.
(c) For $n \in \mathbb{N}$, prove $\sum_{k=1}^{n} (-1)^k \binom{n}{k}\frac{1}{k} = -H_n$, where $H_n = \sum_{i=1}^{n} \frac{1}{i}$.

3.2.35. Prove the identity below. (Comment: The sums can be related to r rolls of a die with b sides, where s and t are nonnegative integers with $r + s + t = rb$.) (Wardlaw [1989])

$$\sum_{k=0}^{\lfloor s/b \rfloor} (-1)^k \binom{r}{k}\binom{s+r-1-bk}{s-bk} = \sum_{k=0}^{\lfloor t/b \rfloor} (-1)^k \binom{r}{k}\binom{t+r-1-bk}{t-bk}$$

Exercises 36–49 use Snake Oil; those up to Exercise 39 are from Wilf [1990].

3.2.36. Let $a_n = \sum_{k \geq 0} \binom{k}{n-k} 2^k$. Find the formula for a_n in terms of n (no summations).

3.2.37. Evaluate $\sum_k \binom{n+k}{2k} 2^{n-k}$ for $n \in \mathbb{N}_0$.

3.2.38. ($\diamond$) For $m, n \in \mathbb{N}_0$, prove
$$\sum_k \binom{n+k}{m+2k}\binom{2k}{k}\frac{(-1)^k}{k+1} = \binom{n-1}{m-1}.$$

3.2.39. ($\diamond$) Let $A_n(y) = \sum_k \binom{n}{k}\binom{2k}{k} y^k$ for $n \in \mathbb{N}_0$. Let $A(x, y) = \sum_{n \geq 0} A_n(y) x^n$.
(a) Prove $A(x, y) = [(1 - x)(1 - x - 4xy)]^{-1/2}$.
(b) Use part (a) to prove $\sum_k \binom{n}{k}\binom{2k}{k}(-1/4)^k = 2^{-2n}\binom{2n}{n}$.
(c) Use part (a) to evaluate $\sum_k \binom{n}{k}\binom{2k}{k}(-2)^{-k}$.

3.2.40. ($\diamond$) Use Snake Oil to evaluate $\sum_k c(n + 1, k + 1)\binom{k}{m}(-1)^{k-m}$.

3.2.41. ($\diamond$) The central Delannoy number $d_{n,n}$ is the number of paths from $(0, 0)$ to (n, n) by steps in $\{(1, 0), (0, 1), (1, 1)\}$. Those using $n - k$ diagonal steps correspond to arrangements of these $n - k$ with k vertical and k horizontal steps. Hence $d_{n,n} = \sum_k \binom{n+k}{2k}\binom{2k}{k}$. Use Snake Oil to obtain the generating function $\sum_{n=0}^{\infty} d_{n,n} x^n$.

3.2.42. ($\diamond$) Use Snake Oil to obtain the generating function $\sum a_{m,k} x^m y^k$, where $a_{m,k} = \sum_r \binom{r}{k-r}\binom{m}{r}$, and then give a combinatorial proof that this is the generating function.

3.2.43. ($\diamond$) Evaluate $\sum_{k=0}^{n} (-1)^k \binom{n}{k}\binom{m+n-k}{n-k}$ twice: by convolutions and by Snake Oil.

3.2.44. Use Snake Oil to prove the identity below. (Hint: Compare with Example 3.2.18. Comment: Exercise 1.2.31 requested a combinatorial proof.)
$$\sum_{k \geq 0} \binom{n}{k}\binom{2k}{k} = \sum_{k \geq 0} \binom{n}{2k}\binom{2k}{k} 3^{n-2k}$$

3.2.45. For $n, k \in \mathbb{N}_0$, evaluate $\sum_{j=0}^{k-1} (-1)^j \binom{k-1}{j}\binom{n-j}{k-j}$.

3.2.46. ($\diamond$) A **positive lattice walk** is a lattice walk in the plane starting at $(0, 0)$ that does not fall below the horizontal axis (each step moves one unit horizontally or vertically). In Exercise 1.3.34, the number of positive lattice walks of length n is determined to be $\sum_{k=0}^{n} \binom{n}{k}\binom{k}{\lfloor k/2 \rfloor} 2^{n-k}$. Here we evaluate the sum.
(a) Let $b_k = \binom{k}{\lfloor k/2 \rfloor}$. Determine the OGF for $\langle b \rangle$. (Hint: Use Example 3.2.13.)
(b) Use Snake Oil to evaluate $\sum_{k=0}^{n} \binom{n}{k}\binom{k}{\lfloor k/2 \rfloor} 2^{n-k}$ (Comment: The computation looks messy, but the expressions simplify to a known generating function.)

3.2.47. ($\diamond$) Use Snake Oil to evaluate $\sum_{r=k}^{\lfloor n/2 \rfloor} \binom{n+1}{2r+1}\binom{r}{k}$ for $n \geq 2k$. (Comment: Exercise 1.1.39 requested a combinatorial proof.) (M. Wildon, in Holland [2014])

3.2.48. Use Snake Oil to evaluate the sum below, and give a combinatorial proof of the resulting identity. (Hint: Simplify $\sum_{m,n} a_{m,n} x^m y^n$, where $a_{m,n}$ is the expression below.)
$$\sum_{k \geq 0} \binom{m-k}{k}\binom{n+k}{2k} + \sum_{k \geq 0} \binom{m-k-1}{k}\binom{n+k}{2k+1}$$

3.2.49. (+) Use Snake Oil to evaluate $\sum_{0 \leq k \leq n/3} 2^k \frac{n}{n-k}\binom{n-k}{2k}$ for $n \geq 1$. (Hint: Use $\frac{n}{n-k} = 1 + \frac{k}{n-k}$ to simplify the dependence on n.) (Gessel [1995])

3.3. Exponential Generating Functions

We used OGFs in studying subset problems because products of formal power series have the proper effect on coefficients to model compound selection problems. For counting problems involving "labeled" objects, a different type of generating function plays the appropriate role.

3.3.1. Definition. The **exponential generating function (EGF)** for a sequence $\langle a \rangle$ is $\sum a_n x^n / n!$. An EGF is also called an **exponential enumerator**; an OGF is an **ordinary enumerator**.

3.3.2. Example. *k-ary words, indexed by length.* There are k^n words of length n consisting of letters from a set of size k. The EGF for k-ary words, enumerated by length, is $\sum_{n=0}^{\infty} k^n x^n / n!$. When x is a number, this is the power series expression for e^{kx}, where again e denotes the base of the natural logarithm. As discussed earlier for OGFs, the formal power series has the same behavior under addition and multiplication as the exponential function given by e^{kx}. Hence we say that e^{kx} *is the exponential enumerator by length for n-ary words.*

Words are the natural ordered analogue of multisets. An n-word uses a multiset of size n from $[k]$, but the elements are chosen in order.

When $k = 1$, we obtain e^x, so the EGF for k-ary words is the product of k copies of the EGF for 1-ary words; that is, $e^{kx} = (e^x)^k$. We need to understand the combinatorial meaning of the product to explain why the k-fold product enumerates the words from an alphabet of size k.   ∎

MODELING LABELED STRUCTURES

The condition on the coefficients that characterizes when an EGF is the product of two other EGFs arises from the definition of product for formal power series, just as it does for OGFs.

3.3.3. Lemma. The EGF for a sequence $\langle c \rangle$ is the product of the EGFs for sequences $\langle a \rangle$ and $\langle b \rangle$ if and only if $\langle c \rangle$ is the **binomial convolution** of $\langle a \rangle$ and $\langle b \rangle$, meaning that for $n \in \mathbb{N}_0$,

$$c_n = \sum_{j=0}^{n} \binom{n}{j} a_j b_{n-j}.$$

Proof:

$$\sum_{n=0}^{\infty} a_n \frac{x^n}{n!} \sum_{n=0}^{\infty} b_n \frac{x^n}{n!} = \sum_{n=0}^{\infty} \left(\sum_{j=0}^{n} \frac{a_j}{j!} \frac{b_{n-j}}{(n-j)!} \right) x^n = \sum_{n=0}^{\infty} \left(\sum_{j=0}^{n} \binom{n}{j} a_j b_{n-j} \right) \frac{x^n}{n!}. \qquad ∎$$

3.3.4. Example. *"Words" of length n.* The Roman alphabet consists of a set A of five vowels and a set B of 21 consonants. A word is any string from the alphabet. With 26^n words of length n, the exponential enumerator by length is e^{26x}.

Alternatively, every word is a merger of an all-vowel word and an all-consonant word. Every word of length n uses j letters from A and $n - j$ letters from B, for some j. After choosing j positions for vowels, we fill them with a word from A and the rest with a word from B. Thus there are $\sum_{j=0}^n \binom{n}{j} 5^j 21^{n-j}$ ways to form a word of length n. By the Binomial Theorem, the sum has value $(5 + 21)^n$.

We have confirmed that the counting sequence for words from an alphabet of size 26 is the binomial convolution of the sequences for words from alphabets of sizes 5 and 21. Hence Lemma 3.3.3 ensures that the EGF for the compound problem is the product of the EGFs for the two smaller problems. This provides a combinatorial proof of $e^{26x} = e^{5x}e^{21x}$.   ∎

Understanding the use of EGFs means understanding when the counting sequence for a compound problem is the binomial convolution of the sequences for the component problems. The answer lies in "labels".

3.3.5. Definition. A **labeled structure** is an object formed on a specified finite set of distinct labels. Given a family A of labeled structures, let A_S denote the subset of A in which S is the set of labels used. The family A is **symmetric** if $|A_S|$ depends only on $|S|$. Let $a_{|S|} = |A_S|$; in the EGF for the counting sequence $\langle a \rangle$, the index corresponds to the number of labels used.

For n-words (Example 3.3.4), the n positions are the labels. Letters may be used repeatedly, but each label is used once. Words from a fixed alphabet form a symmetric family. When putting a word from A into j positions and a word from B in the other $n - j$ positions, the number of ways depends only on j, not on *which* j positions (labels) are used. In allocating the labels to the two subproblems, binomial convolution and EGF product both model the counting process.

3.3.6. Lemma. Given symmetric families A, B, C with counting sequences $\langle a \rangle$, $\langle b \rangle$, $\langle c \rangle$, the EGF for $\langle c \rangle$ is the product of the EGFs for $\langle a \rangle$ and $\langle b \rangle$ if and only if, for all S, objects in C_S correspond to objects in A_T and B_{S-T} for some $T \subseteq S$.

Proof: The description of objects in C_S yields $c_{|S|} = \sum_{k=0}^{|S|} \binom{|S|}{k} a_k b_{n-k}$. This expresses $\langle c \rangle$ as the binomial convolution of $\langle a \rangle$ and $\langle b \rangle$, which by Lemma 3.3.3 is the condition for the EGFs to satisfy $C(x) = A(x)B(x)$.   ∎

Multiplying EGFs corresponds to forming labeled structures in stages. That is, Lemmas 3.3.3–3.3.6 extend by induction to products with k factors. This explains the observation in Example 3.3.2 that $e^{kx} = (e^x)^k$. The important point is the allocation of labels, which for k-ary words are the *positions*. We allocate the labels to k sets (a set may receive no labels) and use the ith letter in all positions of the ith set. Thus the EGF is the product of the EGFs for each letter, each of which is e^x. Exercise 12 formalizes the generalization to products of k EGFs.

To clarify the role of independent stages in labeled enumeration, we present an example where the labels do not correspond to positions.

3.3.7. Example. *Flags on flagpoles.* We have n distinct flags to put on r flagpoles; each flag is used once. Hence the flags are the labels.

When $r = 1$, there are $n!$ structures with label set $[n]$; we arrange the flags in some order. In the EGF, the coefficient of $x^n/n!$ is $n!$, so the EGF for flags on one pole, indexed by the number of flags, is $(1 - x)^{-1}$.

When $r = 2$, we must allocate the flags to two labeled substructures (arrangements on poles A and B). This yields the binomial convolution $\sum \binom{n}{j} j!(n - j)!$ for the coefficient of $x^n/n!$ in the EGF. By Lemma 3.3.6, the EGF for flag arrangements on two poles is $(1 - x)^{-2}$.

Using induction, the EGF for flag arrangements on r poles, indexed by the number of flags, is $(1 - x)^{-r}$. This is also the OGF for multisets from r types. The relevant multiset gives the multiplicity of usage (number of flags) for each pole. An arrangement allocates labels (flags) to poles and puts them in order; there are $n!$ arrangements for each multiset.

That is, the coefficient of $x^n/n!$ in the expansion of $(1 - x)^{-r}$ is $n!$ times the coefficient of x^n. We obtain the same answer $r^{(n)}$ as in Exercise 1.1.35:

$$(1 - x)^{-r} = \sum_{n=0}^{\infty} \binom{n + r - 1}{r - 1} x^n = \sum_{n=0}^{\infty} (n + r - 1)_{(n)} \frac{x^n}{n!} = \sum_{n=0}^{\infty} r^{(n)} \frac{x^n}{n!}. \qquad \blacksquare$$

In labeled enumeration, the labels are distinct and all used, and the index in the EGF is the number of labels. We use products of EGFs when we build the labeled structures in stages described by allocation of labels. With this in mind, we return to the discussion of words.

3.3.8. Example. *Words with restricted use of letters.* The EGF e^x models unrestricted multiplicity. As with OGFs for multiset problems, keeping terms for allowed multiplicities solves many word-counting problems. When we must use a particular letter, the factor for it is $e^x - 1$, since there is no allowed word of length 0 formed using that letter.

When a letter must be used at most once, the factor for it is the EGF $1 + x$ (note that $x/1! = x$). This describes simple words. The EGF for simple words from an alphabet of size k, indexed by length (the labels are the positions), is $(1 + x)^k$. To check this, note that

$$(1 + x)^k = \sum_{j=0}^{k} \binom{k}{j} x^j = \sum_{j=0}^{k} k_{(j)} \frac{x^j}{j!}.$$

The table below compares the analogous OGFs and EGFs for natural multiplicity conditions. The OGFs are enumerators by the total size of the multiset; the EGFs are by the length of the word. $\qquad \blacksquare$

Multisets and Words from k Types, Indexed by Size (Length)

Multiplicity	Multisets (OGF)	Words (EGF)
unrestricted	$(1 - x)^{-k}$	e^{kx}
≤ 1 of each type	$(1 + x)^k$	$(1 + x)^k$
≥ 1 of each type	$\left(\frac{x}{1-x}\right)^k$	$(e^x - 1)^k$

3.3.9. Example. *Coefficients of EGFs.* When the EGF can be expressed as a linear combination of powers of e^x, the coefficient of $x^n/n!$ is easy to extract. This can occur with restrictions on the usage of certain letters in forming k-ary words, which are the EGF analogue of multisets with restricted repetitions.

For example, when a letter must be used with even multiplicity, the factor associated with it in building the EGF for the resulting words is $\frac{1}{2}(e^x + e^{-x})$. Similarly, $\frac{1}{2}(e^x - e^{-x})$ when the multiplicity must be odd (see Proposition 3.2.5(6)). Hence the EGF for ternary words with an even number of 0s, odd number of 1s, and any number of 2s is $\frac{1}{2}(e^x + e^{-x})\frac{1}{2}(e^x - e^{-x})e^x$, which simplifies to $\frac{1}{4}(e^{3x} - e^{-x})$. The coefficient of $x^n/n!$ is $\frac{1}{4}(3^n - (-1)^n)$.

An additive multiple of e^0 affects only the constant term. For example, the EGF for binary lists with the multiplicities of 0 and 1 both even is $\frac{1}{4}(e^x + e^{-x})^2$, which equals $\frac{1}{4}(e^{2x} + 2 + e^{-2x})$. For $n > 0$, the coefficient of $x^n/n!$ is $2^{n-2} + (-2)^{n-2}$, but we must add $1/2$ when $n = 0$ to get 1. ∎

STIRLING AND DERANGEMENT APPLICATIONS

Enumerating words where each letter of the alphabet must be used leads to a classical application. Recall that a **partition** of a set A is a set of disjoint nonempty subsets with union A. The subsets are the **blocks** of the partition, and the blocks are neither ordered nor labeled.

3.3.10. Definition. The **Stirling number** $S(n, k)$ (or $S_{n,k}$) is the number of partitions of $[n]$ into k (nonempty) blocks. When the k blocks are numbered 1 through k, partitions become **ordered partitions**.

3.3.11. Theorem. $S(n, k) = \dfrac{1}{k!} \displaystyle\sum_{i=0}^{k} (-1)^i \binom{k}{i} (k-i)^n.$

Proof: Since nonempty blocks are distinguished by their members, each partition into k blocks can be ordered in $k!$ ways. Hence there are $k!S(n, k)$ ordered partitions of $[n]$. Putting objects into blocks assigns a block name from $[k]$ to each element of $[n]$. Ordered partitions are thus functions from $[n]$ to $[k]$, or words with $[k]$ as the alphabet and $[n]$ as the set of positions (labels).

The condition that each block is nonempty requires each element of $[k]$ to appear in the word. As in Example 3.3.8, the EGF (indexed by length) is $(e^x - 1)^k$, since the option of multiplicity 0 is prohibited for each type of letter. Thus

$$\sum_{n=0}^{\infty} k!S(n, k)\frac{x^n}{n!} = (e^x - 1)^k.$$

Applying the Binomial Theorem to the EGF yields $\sum_{i=0}^{k}(-1)^i\binom{k}{i}e^{x(k-i)}$, which expands to $\sum_{i=0}^{k}(-1)^i\binom{k}{i}\sum_{n\geq 0}(k-i)^n\frac{x^n}{n!}$. Interchanging the order of summation yields the sum in the claimed expression as the coefficient of $x^n/n!$. As we have noted, $S(n, k)$ is then obtained by dividing by $k!$. ∎

The trick of temporarily distinguishing blocks requires all blocks to be nonempty so that all partitions are counted $k!$ times. Section 4.1 proves Theorem 3.3.11 again using the Inclusion-Exclusion Principle. The Stirling numbers and the Eulerian numbers can be obtained from each other (Exercise 25).

Actually, $S(n, k)$ is the Stirling number *of the second kind*. The "first kind" involves permutations instead of partitions.

3.3.12. Definition. The **signless Stirling number** $c(n, k)$ is the number of permutations of $[n]$ with k cycles (see Theorem 3.1.20). The **Stirling number** $s(n, k)$ *(of the first kind)* is defined by $s(n, k) = (-1)^{n-k} c(n, k)$.

Knuth [1968] used $\left\{ {n \atop k} \right\}$ for $S(n, k)$ and $\left[{n \atop k} \right]$ for $c(n, k)$. The two kinds of Stirling numbers are related by forming inverse (infinite) matrices; that is, $\sum_{k \geq 0} S(n, k) s(k, m) = \delta_{n,m}$. We prove this by showing that the Stirling numbers transform between two bases for the vector space of polynomials. Like $\{x^n\}_{n \geq 0}$, the falling factorials $\{x_{(n)}\}_{n \geq 0}$ and the rising factorials $\{x^{(n)}\}_{n \geq 0}$ also form bases for the space of polynomials since there is one of each degree.

3.3.13. Theorem. For $n \in \mathbb{N}_0$, the Stirling numbers satisfy

$$\sum_{k=0}^{n} S(n, k) x_{(k)} = x^n \qquad \text{and} \qquad \sum_{k=0}^{n} s(n, k) x^k = x_{(n)}.$$

Proof: By the Polynomial Principle, it suffices to prove the first identity for $x \in \mathbb{N}$. The right side counts n-words from $[x]$. Alternatively, form words with k distinct letters by partitioning the n positions into k blocks (in $S(n, k)$ ways) and assigning the blocks to letters in $[x]$ (in $x_{(k)}$ ways). Finally, sum over k.

We already proved the nontrivial part of the second identity in Theorem 3.1.20, which is the middle equality below. We compute

$$\sum_{k=0}^{n} s(n, k) x^k = (-1)^n \sum_{k=0}^{n} c(n, k)(-x)^k = (-1)^n (-x)^{(n)} = x_{(n)}. \qquad \blacksquare$$

3.3.14. Corollary. $\sum_{k \geq 0} S(n, k) s(k, m) = \delta_{n,m}$ for $n, m \in \mathbb{N}_0$.

Proof: By definition, $\delta_{n,m}$ is 1 for $n = m$ and 0 for $n \neq m$. We compute

$$x^n = \sum_{k=0}^{n} S(n, k) x_{(k)} = \sum_{k=0}^{n} S(n, k) \sum_{m=0}^{k} s(k, m) x^m = \sum_{m=0}^{n} \left[\sum_{k=m}^{n} S(n, k) s(k, m) \right] x^m.$$

When expressing x^n as a polynomial, the coefficient of x^n must be 1, and the coefficients of other powers must be 0. $\qquad \blacksquare$

By Theorem 3.3.13, the values $j! S(k, j)$ are the coefficients used in writing m^k as a linear combination of $\binom{m}{k}, \ldots, \binom{m}{0}$ (recall Application 1.2.7). That is, $\sum_{j=0}^{k} S(k, j) x_{(j)} = x^k$ becomes $\sum_{j=0}^{k} j! S(k, j) \binom{x}{j} = x^k$. The Summation Identity (Theorem 1.2.3(5)) then implies that $\sum_{m=1}^{n-1} m^{k-1}$ is a polynomial in n of degree k. Exercise 2.1.38 shows that the leading coefficient of this polynomial is $1/k$, and the next coefficient is $-1/2$. The full story of the polynomial is told by the Bernoulli numbers, named for Jakob Bernoulli (1654–1705), who discovered the relationship. Exercise 24 derives the coefficients of the polynomial.

In general, the "connection coefficients" that relate two bases for the space of polynomials are quite interesting. These behave nicely when the bases are sequences of polynomials satisfying a general form of the Binomial Theorem.

3.3.15.* Definition. A sequence $\{p_n \colon n \geq 0\}$ of polynomials is of **binomial type** if $p_0 = 1$ and each p_n has degree n and satisfies

$$p_n(x+y) = \sum_{k=0}^{n} \binom{n}{k} p_k(x) p_{n-k}(y).$$

The ordinary powers, falling factorials, and rising factorials are sequences of binomial type (Exercises 1.1.34–35; see also Exercise 29). The elegant theory of such sequences gives many characterizations of them and associates a natural linear operator with each sequence (the operator corresponding to $\{x^n \colon n \geq 0\}$ is differentiation). We refer interested readers to Mullin–Rota [1970], Roman–Rota [1978], Aigner [1979, pp. 99–118], and Roman [1984].

We continue with applications of EGFs. Binomial convolution plays the role for EGFs that convolution plays for OGFs (there are also analogues of the other operations in Proposition 3.2.5). The sum $\sum_{i=0}^{n} \binom{n}{i} a_i b_{n-i}$ can be evaluated if we can find the EGFs for $\langle a \rangle$ and $\langle b \rangle$ and find the coefficients of their product.

3.3.16. Example. To evaluate $\sum_{k=0}^{n} \binom{n}{k} m^k$ without the Binomial Theorem, let $a_k = m^k$ and $b_{n-k} = 1$. The EGFs are e^{mx} and e^x, so

$$\sum_{k=0}^{n} \binom{n}{k} m^k = \left[x^n/n! \right] e^{mx} e^x = \left[x^n/n! \right] e^{(m+1)x} = (m+1)^n \; . \qquad \blacksquare$$

The next application is more important and echoes Remark 3.2.14. In the context of Example 3.3.16, it yields $m^n = \sum_{i=0}^{n} (-1)^k \binom{n}{k} (m+1)^{n-k}$.

3.3.17. Theorem. The EGF and formula for derangement numbers are

$$D(x) = \frac{e^{-x}}{1-x} \qquad \text{and} \qquad D_n = n! \sum_{k=0}^{n} \frac{(-1)^k}{k!}.$$

Proof: We know $n! = \sum_{k=0}^{n} \binom{n}{k} D_{n-k}$ (Example 2.1.6); permutations are formed by picking the fixed points and deranging the rest. The EGF for $\langle a \rangle$ when $a_n = n!$ is $1/(1-x)$. From the binomial convolution, we obtain $(1-x)^{-1} = e^x D(x)$, so $D(x) = e^{-x}/(1-x)$. To extract the coefficients, recall that multiplying by $(1-x)^{-1}$ sums the initial terms of a power series. Thus

$$D_n = \left[x^n/n! \right] \frac{e^{-x}}{1-x} = n! \sum_{k=0}^{n} \left[x^k \right] e^{-x} = n! \sum_{k=0}^{n} \frac{(-1)^k}{k!}. \qquad \blacksquare$$

We can also view this analysis of $n! = \sum \binom{n}{k} D_{n-k}$ as the EGF analogue of the generating function method for solving recurrences. Instead of x^n, we multiply by $x^n/n!$ before summing over n to introduce the EGF. To illustrate this approach, we apply it also to the second order recurrence for derangements.

3.3.18. Example. *Derangements another way.* We know $D_n = (n-1)(D_{n-1} + D_{n-2})$ for $n \geq 2$, with $D_0 = 1$ and $D_1 = 0$ (Example 2.1.6). Since permutations are labeled structures, we seek the EGF $D(x)$ instead of the OGF. Due to the coefficient $n-1$, we multiply by $x^{n-1}/(n-1)!$ instead of $x^n/n!$ before summing over $n \geq 2$:

$$\sum_{n\geq 2} D_n \frac{x^{n-1}}{(n-1)!} = \sum_{n\geq 2} D_{n-1} \frac{x^{n-1}}{(n-2)!} + \sum_{n\geq 2} D_{n-2} \frac{x^{n-1}}{(n-2)!}.$$

We can manipulate powers of x to make exponents agree with subscripts, but factorials are more stubborn. To shift a factorial, differentiate; that is, if $A(x) = \sum_{n\geq 0} a_n x^n/n!$, then $A'(x) = \sum_{n\geq 1} a_n x^{n-1}/(n-1)!$. Thus the generating function method with EGFs can lead to differential equations.

Here we obtain $D'(x) = xD'(x) + xD(x)$ (using $D_1 = 0$). From

$$\frac{D'(x)}{D(x)} = \frac{x}{1-x} = -1 + \frac{1}{1-x},$$

integration yields $\ln D(x) = -x - \ln(1-x) + C$ for some constant C, and thus $D(x) = e^{C-x}/(1-x)$. Since $D(0) = D_0 = 1$, we must have $C = 0$. ∎

The expression $n! = \sum_{k=0}^{n} \binom{n}{k} D_{n-k}$ can be inverted directly to compute D_n as an instance of a general inversion formula like that for OGFs in Remark 3.2.14.

3.3.19. Theorem. (Binomial Inversion Formula) For sequences $\langle a \rangle$ and $\langle b \rangle$, the following are equivalent
(A) $a_n = \sum_{k=0}^{n} \binom{n}{k} b_{n-k}$ for all $n \in \mathbb{N}_0$.
(B) $b_n = \sum_{k=0}^{n} (-1)^k \binom{n}{k} a_{n-k}$ for all $n \in \mathbb{N}_0$.
Proof: Let $A(x)$ and $B(x)$ be the EGFs for $\langle a \rangle$ and $\langle b \rangle$. Multiplying by $x^n/n!$ and summing over n converts statements (A) and (B) to $A(x) = e^x B(x)$ and $B(x) = e^{-x} A(x)$, respectively, which are equivalent. ∎

THE EXPONENTIAL FORMULA

EGFs are useful for labeled enumeration because the binomial convolution $\sum \binom{n}{k} a_k b_{n-k}$ introduces the factor $\binom{n}{k}$ that counts allocations of labels to subproblems. The Exponential Formula generalizes this idea. We begin with a classical application to motivate the general formula.

3.3.20. Example. *General and connected graphs.* There are $2^{\binom{n}{2}}$ graphs with a specified set of n vertices. Hence the EGF for graphs by number of vertices is given by $G(x) = \sum_{n\geq 0} 2^{\binom{n}{2}} x^n/n!$. By convention one graph has no vertices.

Let $C(x)$ be the EGF for (labeled) connected graphs, indexed by number of vertices. We relate $C(x)$ to $G(x)$. Each graph is formed from components. Consider first a graph with exactly two components. If we temporarily index them as first component and second component, then we form such a graph by partitioning the set of vertex labels into X and Y and placing a connected graph on label set X and another on label set Y.

With n labels, we can do this in $\sum_{j=1}^{n-1} \binom{n}{j} c_j c_{n-j}$ ways, where c_n is the coefficient of $x^n/n!$ in $C(x)$. We divide by 2, since either component could be the "first", but in graphs we do not number components. We can run the sum from 0 to n if $c_0 = 0$. Thus, we say that the one graph with no vertices is not connected. Now the EGF for two-component graphs is $\sum_{n \geq 0} \frac{1}{2} \left(\sum_{j=0}^{n} \binom{n}{j} c_j c_{n-j} \right) x^n/n!$, which equals $C(x)^2/2$.

For graphs with k components, the argument is similar. Partition the n labels into k nonempty sets, form a connected graph on each set, and divide by $k!$ since there are $k!$ ways to index the components of a graph with labeled vertices. By iterating the product rule for EGFs, we find that the EGF for graphs with k components is $C(x)^k/k!$. This formula works also when $k = 0$, since $C(x)^0 = 1$ and there is one graph with no vertices; by convention it has no components.

Summing over k to count every graph yields $G(x) = e^{C(x)}$. ∎

The formula $G(x) = e^{C(x)}$ expresses $G(x)$ as the composition of two formal power series. We pause to say precisely what composition means.

3.3.21. Definition. A sequence $A_1(x), A_2(x), \ldots$ of formal power series **converges** to a formal power series $\sum_{n \geq 0} a_n x^n$ if for each n there exists k_n such that $[x^n] A_k(x) = a_n$ when $k \geq k_n$.

3.3.22. Remark. For a formal power series $C(x)$, let $C_k(x) = \sum_{j=0}^{k} c_j x^j$. In the composition $A(B(x))$, we want to substitute $B(x)^n$ for y^n in $A(y) = \sum_{n \geq 0} a_n y^n$. If this yields a formal power series in x, then it should equal $\lim_{k \to \infty} A_k(B_k(x))$. When $B(0) = 0$, computing a coefficient in the composition is a finite process; $[x^n] A_k(B_k(x))$ is fixed for $k \geq n$. Hence the composition converges. This is why we require $C(0) = 0$ in Example 3.3.20.

The composition $A(B(x))$ also converges when B is a polynomial; again $[x^n] A_k(B_k(x))$ is fixed when k is large. Also, when $B(0) = 0$, the derivative agrees with the ordinary chain rule: the derivative of $A(B(x))$ is $A'(B(x))B'(x)$. ∎

Since $C(0) = 0$ in Example 3.3.20, the composition $e^{C(x)}$ makes sense. More generally, the relationship between EGFs $C(x)$ and $G(x)$ is valid whenever general structures are formed from component structures as in Example 3.3.20. Recall that *symmetric* in the statement below means that the number of structures with a given label set depends only on the number of labels.

3.3.23. Theorem. (The Exponential Formula) Let $G(x)$ and $C(x)$ be the EGFs for symmetric families of "general" and "component" labeled structures. Suppose that $G(0) = 1$ and $C(0) = 0$. If general structures are formed by partitioning the set of labels and placing a component structure on the labels in each block, then $G(x) = e^{C(x)}$.

Proof: By hypothesis, the one general structure with no elements has no components, and every component structure has at least one element.

By the argument of Example 3.3.20, the EGF for general structures with k components is $C(x)^k/k!$: partition the labels into set $B_1, \ldots, B_k$, choose a component with each label set, and divide by $k!$ to cancel the overcounting caused by indexing the components. A general structure may have any number of components, so $G(x) = \sum_{k \geq 0} C^k(x)/k! = e^{C(x)}$. The term $k = 0$ yields the specified value for structures with no labels. ∎

The terminology we have chosen reflects the application to graphs. Simpler examples arise from sets and from permutations.

3.3.24. Example. *Partitions of an n-set.* The set of labels is $[n]$; each must be used once. The "components" of a partition are its blocks. A set of labels forms one block in one way (no ways if the set is empty), so the EGF for component structures is $e^x - 1$. By the Exponential Formula, the EGF for set partitions, indexed by the size of the set, is $e^{e^x - 1}$. By convention, the one partition of $\varnothing$ has no blocks.

The total number of partitions of an n-set is the **Bell number** B_n, named for Eric Temple Bell (unfortunately, Bell and Bernoulli have the same initial). Since $B_n = \sum_{k=0}^n S(n, k)$, we have

$$\sum_{n=0}^{\infty} \sum_{k=0}^n S(n, k)\frac{x^n}{n!} = e^{e^x - 1}.$$ ∎

3.3.25. Example. *Permutations and involutions.* The "components" of a permutation are its cycles; we partition the labels and form a cycle on each block. There are $(j - 1)!$ distinct cycles that we can form from j labels, if $j \geq 1$. As the EGF for component structures, we thus have $C(x) = \sum_{j \geq 1} x^j/j$, which yields $C(x) = -\ln(1 - x)$. Hence the EGF for arbitrary permutations is given by $G(x) = e^{-\ln(1-x)} = (1 - x)^{-1}$, which tells us that there are $n!$ permutations of $[n]$.

An **involution** is a permutation whose square is the identity: all cycles have length 1 or 2. For such permutations, the enumerator $C(x)$ for component structures becomes $x + x^2/2$. By the Exponential Formula, the EGF for involutions, indexed by length, is $e^{x + x^2/2}$. ∎

Given the relation $G(x) = e^{C(x)}$, it is possible to compute the coefficients in $C(x)$ recursively from the coefficients of $G(x)$.

3.3.26. Theorem. Let $C(x) = \sum c_n x^n/n!$ and $G(x) = \sum g_n x^n/n!$. If $G(x) = e^{C(x)}$, with $g_0 = 1$ and $c_0 = 0$, then

$$c_n = g_n - \sum_{k=1}^{n-1} \binom{n-1}{k-1} c_k g_{n-k} \text{ for } n \geq 1.$$

Proof: The equivalence between binomial convolution and products of EGFs (Lemma 3.3.3) extends to the Exponential Formula. Since $G(x) = e^{C(x)}$ implies $G(x) = \sum_{k \geq 0} C(x)^k/k!$ (when $g_0 = 1$ and $c_0 = 0$), we have $g_n = \sum_{k \geq 0}[x^n/n!]C(x)^k/k!$.

We can therefore view $G(x)$ and $C(x)$ as EGFs for general and component structures, where we form a general structure by partitioning the labels, placing a component structure on each block of the partition, and dividing by $k!$ to eliminate the indexing of the blocks.

This combinatorial relationship leads to a recurrence. In a general structure with label set $[n]$ for $n > 1$, the label n appears in a component of some size k with $k \geq 1$. To complete the structure, we choose the labels belonging to the same component as n, choose a component on these k labels, and choose a general structure on the remaining $n - k$ labels. Thus $g_n = \sum_{k=1}^{n} \binom{n-1}{k-1} c_k g_{n-k}$. The term for $k = n$ counts the n-element components, since $g_0 = 1$. ∎

3.3.27. Example. *Connected and general (labeled) graphs.* Let $C(x)$ and $G(x)$ be the EGFs for connected and general graphs by number of vertices. Since there are $2^{\binom{n}{2}}$ graphs with vertex set $[n]$, we count connected graphs among them by $c_n = 2^{\binom{n}{2}} - \sum_{k=1}^{n-1} \binom{n-1}{k-1} 2^{\binom{n-k}{2}} c_k$. ∎

3.3.28.* Remark. The Exponential Formula generalizes in a way that gives a combinatorial explanation for composition of EGFs. In forming general structures, we again place a component structure on each of the k blocks in a partition of the labels. However, there may be h_k ways to form a general structure from the k components, instead of just one. Let $H(x)$ be the EGF for $\langle h \rangle$.

General structures with k components are counted as before, except that each contribution is multiplied by h_k. Hence the EGF for general structures with k components is $h_k C(x)^k / k!$. Summing over k yields $G(x) = H(C(x))$. This is the **Compositional Formula** for EGFs whose counting problems are related in the way we just described. We require $c_0 = 0$ and $g_0 = h_0$. The special case $H(x) = e^x$ with each $h_k = 1$ is the Exponential Formula. The restriction $c_0 = 0$ is needed so that the computation of coefficients in the composition is a finite process.

A formula for formal composition of EGFs is due to Faà di Bruno [1855, 1857]. The history of the Exponential Formula and the Compositional Formula appears in Stanley [1999, p. 65]. Early special cases of the Exponential Formula appear in Touchard [1939] and Riddell–Uhlenbeck [1953]. General combinatorial interpretations of exponentiation of EGFs were developed independently in Foata–Schützenberger [1970], Bender–Goldman [1971], and Doubilet–Rota–Stanley [1972]. Joyal [1981] gave a combinatorial interpretation of composition. Stanley [1978, 1986, 1999] further developed the theory. ∎

There are several beautiful applications of the Exponential Formula. Pólya applied it to prove Cayley's Formula n^{n-2} for the number of trees with vertex set $[n]$. This proof also uses the Lagrange Inversion Formula. We treat the formula as a tool for applications, stating a simple form of it here informally and postponing the proof until later.

Lagrange Inversion Formula: If $x = y/\phi(y)$, where $\phi(y)$ is a power series with $\phi(0) = 1$, then the coefficient of x^n in the expansion of y as a power series in x is $[y^{n-1}]\phi(y)^n/n$.

A **rooted labeled structure** is a labeled structure with one label distinguished as a "root." If there are g_n labeled structures of a given type on a set of n labels, then there are ng_n rooted labeled structures. If $G(x) = \sum g_n x^n / n!$, then the EGF for the rooted structures is $xG'(x)$.

3.3.29. Theorem. (**Cayley's Formula**) There are n^{n-2} trees with vertex set $[n]$.

Proof: (Pólya [1937]) Let t_n be the number of trees, and let $T(x)$ be the EGF for $\langle t \rangle$. There are nt_n rooted labeled trees with vertex set $[n]$. Letting y be the EGF for rooted labeled trees, we have $y(x) = xT'(x)$. We obtain an equation relating x and y and solve it for y using Lagrange inversion.

A **rooted forest** is a disjoint union of rooted trees. Let $U(x)$ be the exponential enumerator for rooted forests by number of vertices. By the Exponential Formula (Theorem 3.3.23), $U(x) = e^{y(x)}$.

Next we show $y = xU(x)$. A labeled rooted tree is formed by choosing a root and joining it to the roots of a labeled rooted forest on the remaining $n - 1$ vertices. Thus y enumerates labeled structures that consist of one labeled vertex and a rooted forest, and the rule for multiplying EGFs yields $y = xU(x) = xe^y$.

Rewriting this as $x = y/e^y$, we apply Lagrange inversion to obtain

$$\left[x^n\right] y(x) = \left[y^{n-1}\right] \frac{(e^y)^n}{n} = \left[y^{n-1}\right] \frac{e^{ny}}{n} = \frac{n^{n-1}}{n(n-1)!}.$$

Thus $y(x) = \sum_{n \geq 0} n^{n-1} x^n / n!$, and n^{n-1} rooted trees have vertex set $[n]$. ∎

Lagrange Inversion applies to power series, ignoring whether they arise as OGFs or EGFs. An application to enumeration of rooted trees, using OGFs because the vertices are not labeled, appears in Exercise 47. To emphasize this point we give another such application here. We use OGFs and Lagrange Inversion, *not* the Exponential Formula.

3.3.30. Example. A **noncrossing tree** consists of $n + 1$ specified points on a circle as vertices, with edges drawn as noncrossing chords. With three points, all trees are noncrossing, but not with four. Of the 16 trees on four points of a circle, four paths have crossings (shown below), leaving 12 noncrossing trees.

This value agrees with the general formula $\frac{1}{n}\binom{3n}{n-1}$ (equal to $\frac{1}{2n+1}\binom{3n}{n}$) that we derive in the next theorem. The formula may look familiar; these are the Fuss–Catalan numbers defined after Corollary 1.3.19 for the 2-ballot sequences. In Exercise 2.1.51 we obtained the same recurrence for them that we obtain next for the noncrossing trees; here we solve it using Lagrange Inversion. ∎

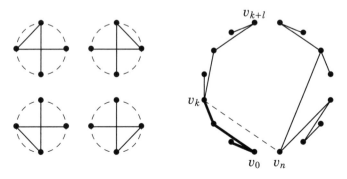

3.3.31. Theorem. The number t_n of noncrossing trees on a given set of $n + 1$ points on a circle is $\frac{1}{n}\binom{3n}{n-1}$.

Proof: (see Lossers–Pinkham [1989]) Name the points v_0 to v_n, clockwise. We derive a recurrence for t_n, valid for $n \geq 1$. From a noncrossing tree T, we extract three subtrees.

Let k be the least index of a neighbor of v_n in T. Since T is noncrossing, $v_0, \ldots, v_k$ induce a noncrossing subtree, shown in bold above. Let $k + l$ be the highest index of a vertex whose path to v_n in T arrives via v_k. The vertices $v_k, \ldots, v_{k+l}$ also induce a noncrossing subtree, and their paths to v_n in T all arrive via v_k. This leaves some number m of vertices other than v_n whose paths to v_n do not arrive via v_k. They are $v_{k+l+1}, \ldots, v_{n-1}$, and with v_n they induce a third noncrossing subtree.

The numbers of vertices in these three subtrees are $k + 1$, $l + 1$, and $m + 1$. Given k, l, m, the subtrees can be formed in $t_k t_l t_m$ ways. Since v_k appears in two of the subtrees, we have $(k + 1) + (l + 1) + (m + 1) = n + 2$, or $k + l + m = n - 1$. To obtain all the trees, we sum over all triples (k, l, m) such that $k + l + m = n - 1$, obtaining $t_n = \sum_{k+l+m=n-1} t_k t_l t_m$.

The vertices are fixed and there is no allocation of labels to form the subtrees. Hence we introduce an OGF, letting $z(x) = \sum_{n \geq 0} t_n x^n$. Note that $t_0 = 1$. Summing over the region of validity ($n \geq 1$), we obtain

$$z(x) - 1 = \sum_{n \geq 1} \left[\sum_{k+l+m=n-1} t_k t_l t_m \right] x^n = x[z(x)]^3.$$

Expressing x in terms of z yields $x = (z - 1)/z^3$. We want to invert this relationship via Lagrange Inversion, but the numerator is not z, and the denominator is not 1 at $z = 0$. We fix both difficulties by setting $y = z - 1$. Now $x = y/\phi(y)$, where $\phi(y) = (1 + y)^3$. By the Lagrange Inversion Formula,

$$\left[x^n \right] y(x) = \left[y^{n-1} \right] \frac{(1 + y)^{3n}}{n} = \frac{1}{n}\binom{3n}{n-1}.$$

Returning to z affects only the constant, so $t_n = [x^n]y(x) = \frac{1}{n}\binom{3n}{n-1}$ for $n \geq 1$. ∎

We usually say that OGFs are good for unlabeled enumeration and EGFs for labeled enumeration, but why is the noncrossing tree problem unlabeled when there are specified points on the circle? The key is the kind of convolution used when combining generating functions for parts of the problem. The number of ways to assemble three trees in obtaining the recurrence is $t_k t_l t_m$, not $\binom{n-2}{k,l,m} t_k t_l t_m$, because there was no allocation of labels to subproblems.

THE LAGRANGE INVERSION FORMULA (optional)

When x is expressed as a function of y by a formal power series (subject to some restrictions) and we want to express y as a function of x, the Lagrange Inversion Formula may enable us to invert the relationship. Wilf [1990, p. 139] presents a proof via complex analysis. In that proof, a coefficient in a power series is studied by translating it into a complex integral. A change of variables is performed, and then the translation is reversed to return to the context of series.

We present a direct and self-contained approach in the context of formal power series. In his survey paper, Gessel [2016] attributes this proof to Jacobi [1830]. Stanley [1999, pp. 38–39] presents a similar proof that he attributes to Lagrange [1770]. Stanley also presents two combinatorial proofs based on counting forests, due to Raney [1960] (simplified by Schützenberger [1971]) and to Labelle [1981]. For further history, see Stanley [1999, p. 67].

To find the coefficient of x^n in a power series in x, we can divide by x^{n+1} and then take the coefficient of x^{-1}. We need series having terms with negative exponents. The "residue" of such a series is the same notion as in complex analysis for Laurent series around an isolated singular point.

3.3.32. Definition. A **formal Laurent series** is an expression of the form $\sum_{n\in\mathbb{Z}} a_n x^n$ such that $a_n \neq 0$ for only finitely many negative indices (all $a_n \in \mathbb{C}$). When $f(x) = \sum_n a_n x^n$, the **derivative** is the formal Laurent series $f'(x)$ defined by $f'(x) = \sum_n n a_n x^{n-1}$. The **residue** of the formal Laurent series $\sum_n a_n x^n$ is a_{-1}.

When $f(y)$ is a formal Laurent series and $g(x)$ is a formal power series with constant term 0, the composition $f(g(x))$ is well-defined as a formal Laurent series in x, because there is a finite procedure to compute the coefficient of any power of x (see Remark 3.3.22). The equation $y = g(x)$ expresses a change of variables. We can study f as a formal Laurent series in y or in the variable x related to it by this change. The relationship between the residues behaves like a "chain rule" for change of variables; we will use it to prove the Lagrange Inversion Formula.

3.3.33. Lemma. (Jacobi) Let f be a formal Laurent series in y and g be a formal power series in x. If $[x^0]g(x) = 0$ and $[x^1]g(x) \neq 0$, then

$$[y^{-1}]f(y) = [x^{-1}]f(g(x))g'(x).$$

Proof: By linearity, it suffices to consider the case $f(y) = y^k$ for $k \in \mathbb{Z}$. If $k \neq -1$, then $[y^{-1}]f(y) = 0$. Since $f(y) = y^k$, we have

$$f(g(x))g'(x) = g(x)^k g'(x) = \frac{1}{k+1}\frac{d}{dx}g(x)^{k+1}.$$

Since g is a formal power series in x, so are g^{k+1} and its derivative, and hence the coefficient of x^{-1} is 0, as desired.

If $k = -1$, then $[y^{-1}]f(y) = 1$. In this case, $f(g(x))g'(x) = g'(x)/g(x)$. The hypotheses on g lead to $g(x) = xh(x)$, where $h(x)$ is a formal power series with nonzero constant term. Thus $h(x)$ has a multiplicative inverse, and we can write $h'(x)/h(x)$ as a formal power series.

By the product rule for differentiation, $g'(x) = h(x) + xh'(x)$. Thus

$$\frac{g'(x)}{g(x)} = \frac{1}{x} + \frac{h'(x)}{h(x)}.$$

Since $h'(x)/h(x)$ is a formal power series, $[x^{-1}]f(g(x))g'(x) = 1$. ∎

In the form of the Lagrange Inversion Formula applied in Theorem 3.3.29, x is expressed as a function of y via $x(y) = y/\phi(y)$. Since $\phi(0) = 1$, the expansion of $\phi(y)^{-1}$ begins with 1, so $x(y) = y + c_2 y^2 + \cdots$.

We want to express y as a function of x. Since $x = 0$ when $y = 0$, we have $[x^0]y(x) = 0$. Now the key point is that the composition is the identity: $x(y(x)) = x$. That is, $x = y(x) + c_2 y(x)^2 + \cdots$. Since the constant term in $y(x)$ is 0, equating coefficients tells us that $[x^1]y(x) = 1$.

3.3.34. Theorem. (Lagrange Inversion Formula, special case) Let $\phi(y)$ be a formal power series with $[y^0]\phi(y) = 1$. If $x = y/\phi(y)$, then for $n \geq 1$ the coefficient of x^n in the expansion of y in terms of x is given by

$$[x^n]y(x) = \frac{1}{n}[y^{n-1}]\phi(y)^n.$$

Proof: In order to use Lemma 3.3.33, we shift the series to make the desired coefficient the residue. We use $[z^{n-1}]z^n f(z) = [z^{-1}]f(z)$ twice.

$$
\begin{aligned}
\tfrac{1}{n}[y^{n-1}]\phi(y)^n &= \tfrac{1}{n}[y^{n-1}]\left(\frac{y}{x(y)}\right)^n && \text{definition of } \phi \\
&= \tfrac{1}{n}[y^{-1}]\frac{1}{(x(y))^n} && \text{shift of series} \\
&= \tfrac{1}{n}[x^{-1}]\frac{y'(x)}{x^n} && \text{Lemma 3.3.33 (details below)} \\
&= \tfrac{1}{n}[x^{n-1}]y'(x) && \text{shift of series} \\
&= \tfrac{1}{n}n[x^n]y(x) && \text{def. of differentiation} \\
&= [x^n]y(x)
\end{aligned}
$$

In the crucial step, Lemma 3.3.33 yields

$$[y^{-1}]f(y) = [x^{-1}]f(y(x))y'(x)$$

when $f(y)$ is a formal Laurent series in y and $y(x)$ is a formal power series with $[x^0]y(x) = 0$ and $[x^1]y(x) \neq 0$ (which holds in this case). Let $f(y) = 1/(x(y))^n$. Since $\phi(y)$ is a formal power series in y, also $x(y)$ is a formal power series in y, and $f(y)$ is a formal Laurent series. On the right side, $f(y(x))$ becomes $1/x(y(x))^n$, expressed as a formal Laurent series in x. Since the composition $x(y(x))$ is the identity, the right side becomes $[x^{-1}]y'(x)/x^n$, as desired. ∎

Applications of Theorem 3.3.34 require finding the coefficient of y^{n-1} in a formal power series in y. In Theorem 3.3.29, this caused no difficulty, because the power series in question was an exponential series.

A more general version of Theorem 3.3.34 allows composing an additional function h with $y(x)$, such as squaring $y(x)$ when we need to enumerate ordered pairs. See Exercise 49 for the proof and Exercises 50–54 for applications.

3.3.35. Theorem. (Lagrange Inversion Formula) Let $\phi(y)$ and $h(y)$ be formal power series with $[y^0]\phi(y) = 1$. If $x = y/\phi(y)$, then the coefficient of x^n in the expansion of $h(y)$ as a power series in x is given by

$$[x^n]h(y(x)) = \frac{1}{n}[y^{n-1}]h'(y)\phi(y)^n.$$

EXERCISES 3.3

3.3.1. (–) Let n be an even number. Find the exponential generating function $B_n(x)$ for lists from $[n]$ such that each odd number is used an odd number of times and each even number is used an even number of times.

3.3.2. (–) Use an EGF to determine the number of ways to distribute 10 people into three rooms so that each room has at least one person.

3.3.3. (–) Cards are dealt in a row from a standard 52-card deck, and the values are recorded (suits are ignored). Build an EGF for these lists, indexed by length.

3.3.4. (–) Build an EGF for the ways to put distinct objects into k distinct boxes with at least m objects in each box, indexed by the number of objects. How does this change if the boxes are not distinguishable?

3.3.5. (–) The people in a club arrive for a movie showing. A subset S gets in early with special coupons. Another subset T waits in a queue to get in. The remaining people give up and go home. Let a_n be the number of ways this can all happen when the club has n people. Obtain a simple expression for the EGF of $\langle a \rangle$.

3.3.6. (–) Use binomial convolution to evaluate $\sum_{k=0}^n k \binom{n}{k}$.

3.3.7. (–) Apply Binomial Inversion to the explicit formula for the Stirling number $S(n,k)$ to obtain the formula for k^n as a linear combination of the binomial coefficients $\{\binom{k}{i}: 0 \le i \le k\}$. Explain the resulting formula combinatorially.

3.3.8. (–) *Stirling Inversion.* Use Corollary 3.3.14 to prove that the statements below are equivalent.
  (a) $a_n = \sum_{k=1}^n s(n,k) b_k$ for all $n \in \mathbb{N}$.
  (b) $b_n = \sum_{k=1}^n S(n,k) a_k$ for all $n \in \mathbb{N}$.

3.3.9. (–) Let c_n be the number of ways that n children can be arranged in teams, with a captain for each team chosen from the team members. Prove that the EGF for $\langle c \rangle$ is e^{xe^x}. (Stanley [1978] states this problem using idempotent functions.)

3.3.10. (–) Use the generating function method with EGFs to solve the recurrence $a_n = na_{n-1} + n!$ for $n \ge 1$, with $a_0 = 1$.

3.3.11. (–) Compute $[x^3](1+x)^\alpha$ by composing the series for $e^{\alpha y}$ and $\ln(1+x)$.

3.3.12. Let $A^1, \dots, A^n$ and C be symmetric families of labeled structures such that objects in $C_{[k]}$ correspond bijectively to distributions of the label set $[k]$ into n sets $S_1, \dots, S_n$ and choices of elements from $A^i_{S_i}$ for $1 \le i \le k$. Let $a^i_k = |A^i_{[k]}|$ and $c_k = |C_{[k]}|$. Prove that the EGF for $\langle c \rangle$ is the product of the EGFs for $\langle a^1 \rangle, \dots, \langle a^n \rangle$.

3.3.13. (◊) Let a_n be the number of words of length n from the alphabet $\{w, x, y, z\}$ such that x appears an even number of times and y appears an odd number of times. Build the EGF for $\langle a \rangle$ and use it to obtain a formula for a_n. Give a direct combinatorial argument to explain the resulting formula.

3.3.14. (◊) Evaluate $\sum_{k=0}^m \binom{n}{k}\binom{n-k}{m-k}$ using binomial convolution, and give a bijective proof of the resulting identity.

3.3.15. (◊) Let $b_{r,n}$ be the number of ways to place n distinct flags on r distinct flagpoles, each pole having at least one flag (order of flags on poles matters). For example, $b_{2,4} = 72$.
  (a) Use generating function arguments to build the EGF $B_r(x) = \sum b_{r,n} x^n/n!$.
  (b) Use the EGF in part (a) to obtain a simple formula for $b_{r,n}$.
  (c) Give a direct combinatorial proof of the formula in part (b).

3.3.16. ($\diamond$) Let a_n be the number of involutions on an n-element set. Derive a recurrence for $\langle a \rangle$, and use it to obtain the EGF, which appears in Example 3.3.25.

3.3.17. ($\diamond$) A **ranking** of candidates in an election allows ties. Let a_n be the number of rankings of n distinct candidates. Note $a_2 = 3$ and $a_3 = 13$.
(a) Obtain the EGF for $\langle a \rangle$.
(b) Using (a), prove that $\sum_{k \geq 0} k^n / 2^k$ is an integer (also in Exercise 3.1.38).

3.3.18. Use the Stirling numbers of the second kind to count the trees with vertex set $[n]$ that have exactly k leaves. (Rényi [1959])

3.3.19. ($\diamond$) Prove that $S(n-1, k-1)$ is the number of ways to partition $[n]$ into k sets with no two consecutive values in the same set. (Hint: See Exercise 2.2.33.)

3.3.20. ($\diamond$) For $0 \leq k < n$, give three proofs to evaluate $\sum_{i=0}^{n} (-1)^i \binom{n}{i} i^k$.
(a) Use induction. (Hint: Use an identity to reduce the exponent on i.)
(b) Use OGFs.
(c) Use the Stirling numbers of the second kind.
(Comment: This yields $\sum_{i=0}^{n} (-1)^i \binom{n}{i} p(i)$ whenever p is a polynomial of degree less than n.)

3.3.21. Let $P(m, n)$ be the set of nonnegative integer vectors $a = (a_1, \ldots, a_{m+1})$ such that $\sum a_i = n$ and $\sum i a_i = m + n$.
(a) Find a direct combinatorial argument for

$$S(m + n, n) = \sum_{a \in P(m,n)} (n+1)(n+2)\cdots(n+m) \binom{n}{a_1 \cdots a_{m+1}} \prod_{i=2}^{m+1} \frac{1}{i!^{a_i}}.$$

(b) Prove that $(n+1)^{(m)}$ divides $\prod_{i=2}^{m+1} i^{\lfloor m/i \rfloor} S(m+n, n)$. (Knuth [1993])

3.3.22. Count the partitions of proper subsets of $[n]$ into k blocks.

3.3.23. ($\diamond$) *Identities for Stirling numbers.*
(a) Prove bijectively that $S(n+1, m+1) = \sum_k \binom{n}{k} S(k, m)$.
(b) Apply part (a) to prove $\binom{n}{m} = \sum_k S(n+1, k+1) s(k, m)$.

3.3.24. ($\diamond$) Define the **Bernoulli number** B_n by $B_0 = 1$ and $\sum_{j=0}^{m} \binom{m+1}{j} B_j = 0$ for $m \geq 1$.
(a) Prove that the EGF $B(x)$ for the Bernoulli numbers is $x/(e^x - 1)$.
(b) Evaluate the sum of the $(k-1)$th powers of the first $n-1$ positive integers in terms of the Bernoulli numbers by proving

$$\sum_{m=1}^{n-1} m^{k-1} = \frac{1}{k} \sum_{j=0}^{k-1} \binom{k}{j} B_j n^{k-j}.$$

3.3.25. ($\diamond$) *Stirling numbers and Eulerian numbers.*
(a) Prove bijectively that $k! S(n, k) = \sum_{i=0}^{k} \binom{n-i}{k-i} A(n, i)$. (Riordan [1964])
(b) Use part (a) to prove $\sum_{k=0}^{n} k! S(n, k) x^{n-k} = \sum_{i=0}^{n} A(n, i)(x+1)^{n-i}$. Conclude that there are $\sum_{j=1}^{n} 2^{j-1} A(n, j)$ ordered partitions of $[n]$ (it is 3 for $n = 2$ and 13 for $n = 3$).

3.3.26. *Another inversion formula.* Prove that the two statements below are equivalent. (Recall that $\binom{u}{0}$ is the constant polynomial 1 as a polynomial in the real variable u.)
(a) $b_k = \sum_{i=0}^{k} \binom{n-i}{k-i} a_i$ for $0 \leq k \leq n$.
(b) $a_k = \sum_{i=0}^{k} (-1)^{k-i} \binom{n-i}{k-i} b_i$ for $0 \leq k \leq n$.
Conclude $A(n, k) = \sum_{i=0}^{k} (-1)^{k-i} \binom{n-i}{k-i} i! S(n, i)$ from (a) of Exercise 3.3.25. (Bóna [2004, p. 13])

3.3.27. Given $n \geq m \geq 2$, prove $n^m S(n, m) \geq m^n \binom{n}{m}$. (Pité [2017], Nikšić [2017])

3.3.28. For positive integers m, n, N, give a bijective proof of

$$\sum_{j=1}^{\min(n,N)} \binom{N}{j} j! S(n,j)(N-j)^m = \sum_{i=1}^{\min(m,N)} \binom{N}{i} i! S(m,i)(N-i)^n .$$ (Khan [1991])

3.3.29. ($\diamond$) Define a sequence of polynomials by letting $p_n(x) = \sum_{j=0}^{n} S(n,j)x^j$ for $n \in \mathbb{N}_0$. (These have been called both the **Touchard polynomials** and the **exponential polynomials**; note $p_0(x) = 1$.) Give a combinatorial proof of the polynomial identity

$$p_n(x+y) = \sum_{k=0}^{n} \binom{n}{k} p_k(x) p_{n-k}(y).$$

3.3.30. ($\diamond$) A **Stirling permutation** linearly orders two copies of $[n]$ so that for all i, all entries between the two copies of i exceed i. Let a *skyline* be a Stirling permutation having the additional property that no strictly increasing triple has its last two entries consecutive in the arrangement. For example, 122133 is a skyline. Let a_n be the number of skylines of length $2n$, and let $A(x) = \sum_{n \geq 0} a_n x^n/n!$.
 (a) Prove $A'(x) = e^{2x} A(x)$, and conclude $A(x) = e^{(e^{2x}-1)/2}$.
 (b) Prove $a_n = \sum_{k=0}^{n} 2^{n-k} S(n,k)$ by using part (a).
 (c) Obtain part (b) by establishing a correspondence between skylines of length $2n$ and partitions of $[n]$ with some elements marked. (Callan [2011])

3.3.31. ($\diamond$) The **Bell number** B_n is the total number of partitions of $[n]$.
 (a) Prove $B_{n+1} = \sum_{k=0}^{n} \binom{n}{k} B_k$ for $n \geq 0$.
 (b) Use part (a) to prove that the EGF for the Bell numbers is e^{e^x-1}.

3.3.32. ($\diamond$) Let B_n be the nth Bell number.
 (a) Prove $\sum_{k=1}^{\infty} \frac{k^n}{k!} x^k = e^x \sum_{k=1}^{n} S(n,k)x^k$.
 (b) Conclude **Dobiński's Formula**: $B_n = e^{-1} \sum_{k \geq 1} k^n/k!$. (Dobiński [1877])
 (c) Conclude also that the difference between the number of partitions with an even and an odd number of blocks equals $e \sum_{k \geq 1} (-1)^k k^n/k!$.

3.3.33. *Bell numbers.* For $k \in [n]$, prove $B_n \geq k^{n-k}$. Conclude $\left(\frac{n}{\ln n}\right)^{n(1-1/\ln n)} \leq B_n \leq n^n$.

3.3.34. Let S be a set of n marbles, consisting of two indistinguishable white marbles and one each in $n-2$ other colors. Prove that the number of distinguishable partitions of S is $(B_n + B_{n-1} + B_{n-2})/2$, where B_m is the number of partitions of $[m]$. (When $n = 3$, there are four: $WWB, WW|B, W|BW, W|W|B$.) (G. Beck)

3.3.35. A **principal submatrix** is a submatrix obtained by extracting the same set of columns as rows. A symmetric matrix is **positive semidefinite** if all its principal submatrices have nonnegative determinant.
 (a) A **partial partition** of a set X is a partition of a nonempty subset of X. Show that the number of partial partitions of $[n]$ is $B_{n+1} - 1$.
 (b) Prove that the number of positive semidefinite $0, 1$-matrices of order n is the Bell number B_{n+1}. (Schmidt [1995])

3.3.36. (+) Use the formula for the derangement numbers to give a combinatorial proof of the identity below.

$$\sum_{l=0}^{n} (-1)^l \binom{n}{l} (2l)! \sum_{m=0}^{2l} \frac{(-1)^m}{m!} = \sum_{l=0}^{n} (-1)^l \binom{n}{l} 2^{n-l}(n+l)!$$ (Yu [1997])

3.3.37. ($\diamond$) *Derivation of the Eulerian polynomial from its EGF.* The nth Eulerian polynomial A_n, written with t as the formal variable, is defined by $A_n(t) = \sum_{k=1}^{n} A(n,k)t^k$, where $A(n,k)$ is the number of permutations of $[n]$ with k runs.

(a) Prove $A(n,k) = A(n-1,k-1) + \sum_{m=1}^{n-1} \sum_{j=1}^{k} \binom{n-1}{m} A(m,j)A(n-1-m,k-j)$ for $n, k \geq 1$.

(b) Use part (a) to obtain a recurrence for the Eulerian polynomials.

(c) Let $A(x) = \sum_{n \geq 0} A_n(t)x^n/n!$. Use part (b) to prove $A'(x) = (t-1)A(x) + A^2(x)$.

(d) Use part (c) to prove $A(x) = \frac{1-t}{1 - t e^{(1-t)x}}$.

(e) Use part (d) to prove $A_n(t) = (1-t)^{n+1} \sum_{i \geq 0} i^n t^i$. (See Theorem 3.1.26.)

3.3.38. Let $G(x)$ be the EGF for graphs, indexed by number of vertices. A graph is **even** if all vertices have even degree. Prove that $1 + \int_0^x G(t)dt$ is the EGF for even graphs, indexed by number of vertices, and that $\ln(1 + \int_0^x G(t)dt)$ is the EGF for connected even graphs.

3.3.39. Let b_n be the number of unordered binary trees with n labeled leaves (interchanging the left and right subtrees of a vertex does not change the tree). The non-leaf vertices have no labels (when $n = 1$ the root is a leaf). Set $b_0 = 0$. Let $B(x)$ be the EGF for $\langle b \rangle$.

(a) Prove $B(x) = x + \frac{1}{2}B(x)^2$.

(b) Use part (a) to prove $b_n = \prod_{i=1}^{n-1}(2i-1)$.

(Comment: Equality of b_n with the number of pairings of $[2n-2]$ is requested bijectively in Exercise 1.3.19.) (Schröder [1870]; see also Erdős–Székely [1989])

3.3.40. ($\diamond$) Let c_n be the number of ways to arrange n children in circles (holding hands) with one child standing alone inside each circle. Each circle has at least one child (not counting the one inside). Prove that the EGF for $\langle c \rangle$ is $(1-x)^{-x}$. (Stanley [1978])

3.3.41. Let a_n be the number of distinct matrices expressible as the sum of an n-by-n permutation matrix and its inverse. Prove $\sum a_n x^n/n! = (1-x)^{-1/2}e^{x/2+x^2/4}$. (Stanley [1978])

3.3.42. ($\diamond$) Let $G(x)$ be the EGF for graphs where every vertex has degree 2, indexed by number of vertices. Let $F(x)$ be the EGF for permutations with no fixed points or 2-cycles.

(a) Use the Exponential Formula to prove

$$G(x) = \frac{e^{-x/2 - x^2/4}}{\sqrt{1-x}} \qquad \text{and} \qquad F(x) = \frac{e^{-x - x^2/2}}{1-x}.$$

(b) Give a bijective proof of $F(x) = G(x)^2$. (Wilf [1990, p. 77])

3.3.43. ($\diamond$) Let a_n be the number of permutations of $[n]$ whose cycles all have odd length. Let b_n be the number of such permutations having an even number of cycles. Let $A(x)$ and $B(x)$ denote the EGFs for $\langle a \rangle$ and $\langle b \rangle$.

(a) Use the Exponential Formula to prove $A(x) = \left(\frac{1+x}{1-x}\right)^{1/2}$ and $B(x) = (1-x^2)^{-1/2}$.

(b) Let p be the probability that a random permutation of $[n]$ has an even number of cycles, all of odd length. Let q be the probability that a string of n coin tosses has exactly $n/2$ heads. By expanding $B(x)$, prove the startling fact that $p = q$. (Wilf [1990, p. 75])

3.3.44. ($\diamond$) A **2-colored bipartite graph** is a bipartite graph (Definition 5.1.9) with the first and second parts named. Let $F(x)$ and $G(x)$ respectively be the EGFs for 2-colored bipartite graphs and for bipartite graphs (parts unnamed), indexed by number of vertices.

(a) Use the Exponential Formula to prove $G(x) = \sqrt{F(x)}$.

(b) Give a bijective proof of $G(x)^2 = F(x)$. (Hint: Group the components of a 2-colored bipartite graph by which part contains the least vertex.) (Wilf)

3.3.45. ($\diamond$) Let a_n and b_n be the numbers of derangements of $[n]$ having an even number and an odd number of cycles, respectively. Use the Exponential Formula to determine $|a_n - b_n|$. (Comment: Solved using recurrences in Exercise 2.1.42.) (Stathopoulos [2012])

3.3.46. A **block** is a connected graph such that the subgraph obtained by deleting any one vertex is connected. The blocks of a connected graph are the maximal such subgraphs (Chapter 7). A 1-vertex graph is not a block, since we take the graph with no vertices to be disconnected. Let b_n and c_n count the blocks and the connected graphs with vertex set $[n]$. Let $B(x) = \sum_{k\geq 2} b_n x^n/n!$ and $C(x) = \sum_{k\geq 1} c_n x^n/n!$. A **rooted graph** marks one vertex.

(a) Let $R_k(x)$ be the EGF for rooted connected graphs whose root appears in k blocks. Prove $R_k(x)/x = (R_1(x)/x)^k/k!$. (Hint: For what is $R_k(x)/x$ the EGF?)

(b) Let $R(x)$ be the EGF for rooted connected graphs, so $R(x) = \sum_{k\geq 0} R_k(x)$. Prove $R_1(x)/x = \sum_{k\geq 2} b_k R(x)^{k-1}/(k-1)!$.

(c) Conclude $B'(xC'(x)) = \ln(C'(x))$, where $'$ is differentiation. (Riddell [1951], Ford–Uhlenbeck [1956]; see Harary–Palmer [1973, pp. 9–11], Stanley [1999, pp. 119–120].)

3.3.47. ($\diamond$) Fix $k \in \mathbb{N}$, and let a_n be the number of k-ary ordered trees with n vertices; k-ary means each vertex has 0 or k children (see Definition 1.3.21).

(a) Let $y(x)$ be the OGF for $\langle a \rangle$. Obtain the functional equation $y = x(1 + y^k)$.

(b) Use Lagrange Inversion to obtain a simple expression for a_n. (Hint: Note that a_n is nonzero only for some n.) (see Stanley [1999, p. 175])

(c) Use part (b) to prove the formula for the Catalan numbers (yet again).

3.3.48. Let f be a formal Laurent series in y, and let g be a formal power series in x such that $[x^0]g(x) = \cdots = [x^{m-1}]g(x) = 0$ and $[x^m]g(x) \neq 0$. Generalize Lemma 3.3.33:

$$[y^{-1}]f(y) = \frac{1}{m}[x^{-1}]f(g(x))g'(x).$$

3.3.49. ($\diamond$) Prove the general Lagrange Inversion Formula (Theorem 3.3.35): If $\phi(y)$ and $h(y)$ are formal power series in y with $\phi(0) = 1$, and $x = y/\phi(y)$, then $h(y)$ expands as a power series in x as given below. (Hint: Mimic the proof of Theorem 3.3.34.)

$$[x^n]h(y(x)) = \frac{1}{n}[y^{n-1}]h'(y)\phi(y)^n.$$

3.3.50. Find the power series expansion of e^{ay} in powers of x, given $y = xe^y$.

3.3.51. ($\diamond$) *Generalization of Cayley's Formula, again.* Let $a_{n,k}$ be the number of rooted forests with vertex set $[n]$ and root set $[k]$. Apply Theorem 3.3.35 to prove $a_{n,k} = kn^{n-k-1}$. (Hint: Express $a_{n,k}$ in terms of a coefficient in y^k, where $y(x)$ is the EGF for rooted trees (see Theorem 3.3.29).)

3.3.52. ($\diamond$) The Catalan recurrence yields the equation $C(x) - 1 = xC(x)^2$ for the generating function $C(x)$. Letting $y = xC(x)$, use the general form of Lagrange Inversion (Theorem 3.3.35) to prove the identity below. (Comment: Setting $k = 1$ yields another derivation of the formula for C_n.)

$$[x^n]C(x)^k = \frac{k}{2n+k}\binom{2n+k}{n}$$

3.3.53. Apply Lagrange Inversion (Theorem 3.3.35) to the equation $y = x\sqrt{1+y}$ to prove

$$\left(\frac{\sqrt{4+x^2}+x}{2}\right)^k = \sum_{n=0}^{\infty}\frac{k}{n+k}\binom{\frac{1}{2}(n+k)}{n}x^n \ .$$

3.3.54. (+) Given $a_n = \sum_k \binom{k}{n-k}b_k$ for $n \geq 0$, use Theorem 3.3.35 for $n \geq 1$ to prove

$$b_n = \frac{1}{n}\sum_k(-1)^{n-k}\binom{2n-k-1}{n-1}ka_k. \qquad \text{(Wilf [1990, pp. 140–141])}$$

3.4. Partitions of Integers

Corollary 1.1.22 counts lists of n positive integers with sum k; the order of the integers matters. Sometimes, we do not care about the order.

3.4.1. Example. *Distributions of bridge hands.* A card deck has 13 cards each in "spades", "hearts", "diamonds", and "clubs". A *bridge hand* consists of 13 cards. Strategy depends heavily on the distribution of a player's cards among suits. A hand having 5 spades, 4 hearts, 4 diamonds, and no clubs lies in an equivalence class with one having 4 spades, no hearts, 5 diamonds, and 4 clubs. A "5440 distribution" denotes a hand with five cards in one suit and four in each of two others. ∎

3.4.2. Definition. A **partition** of an integer n is a multiset of positive integers with sum n. A **composition** of n places those integers in a fixed order. The numbers used in a partition or composition are its **parts** (parts can be repeated). Since the order of parts in a partition of n is irrelevant, we canonically write them in nonincreasing order.

The possible bridge distributions are the partitions of the integer 13 into *at most* four parts (one or more suits may have no cards). Exercise 6 compares the probabilities of bridge distributions.

GENERATING FUNCTION METHODS

To enumerate partitions with specified properties, indexed by the sum, we seek an OGF where the coefficient of x^n is the number of such partitions of n (the OGF is appropriate because we are partitioning identical units). One way to specify a partition is by the number of times each integer is used as a part. This suggests building the generating function by using a factor for each type of number allowed to be a part.

3.4.3. Theorem. The OGFs for partitions using parts in $\{1, \ldots, k\}$, partitions with largest part k, and all partitions are, respectively,

$$\prod_{i=1}^{k} \frac{1}{1-x^i}, \qquad x^k \prod_{i=1}^{k} \frac{1}{1-x^i}, \qquad \prod_{i=1}^{\infty} \frac{1}{1-x^i}.$$

Proof: Let e_i be the number of copies of i in a partition. Partitions of n using parts in $[k]$ correspond to nonnegative integer solutions of $1e_1 + 2e_2 + \cdots + ke_k = n$. The index in the OGF is the sum; when j copies of i are used, the contribution to the exponent is $j \cdot i$. Hence the factor listing the options for i is $1 + x^i + x^{2i} + \cdots$, which equals $(1-x^i)^{-1}$.

Deleting x^0 from the factor for copies of k ensures that at least one k is used. The infinite product allows all positive integers as parts. ∎

The third generating function of Theorem 3.4.3 is the "master" generating function for partitions; it includes all partitions we might want to count. Restrictions on usage of particular parts can be incorporated into the corresponding factors. For example, the generating function for partitions of n in which all parts are odd, 1 is used at least three times, and 3 is used at most three times is

$$x^3(1-x)^{-1}(1+x^3+x^6+x^9)(1-x^5)^{-1}(1-x^7)^{-1}\cdots.$$

Generating function methods can sometimes be used to find the asymptotic behavior of a sequence of coefficients. Let $a_{n,k}$ be the number of partitions of n using parts in $[k]$. For small k, we can extract the exact formula for $a_{n,k}$. Note that $a_{n,1} = 1$ and $a_{n,2} = 1 + \lfloor n/2 \rfloor$.

3.4.4. Proposition. The number of partitions of n using parts in $\{1, 2, 3\}$ is the nearest integer to $\frac{1}{12}(n+3)^2$.

Proof: By Theorem 3.4.3, the generating function for partitions using parts in [3] is $(1-x)^{-1}(1-x^2)^{-1}(1-x^3)^{-1}$. Note that

$$\prod_{i=1}^{3}\frac{1}{1-x^i} = \frac{1}{(1-x)^3}\frac{1}{1+x}\frac{1}{1+x+x^2} = \frac{A}{(1-x)^3} + \text{other terms}.$$

Multiplying by the denominators leaves at least one factor of $(1-x)$ in every term except the first. Hence setting $x = 1$ (Heaviside method, Example 2.2.19) yields $1 = 6A$, so $A = 1/6$. Since $[x^n](1-x)^{-3} = \binom{n+2}{2}$, the leading behavior will be $n^2/12$ if the contributions from the other terms grow at most linearly in n.

To be more precise, we factor $1 + x + x^2$ over the complex numbers. With $\omega = e^{2\pi i/3}$, the denominator is $(1-x)^3(1+x)(1-\omega x)(1-\omega^2 x)$ in linear factors. Expanding by partial fractions yields

$$\prod_{i=1}^{3}\frac{1}{1-x^i} = \frac{1/6}{(1-x)^3} + \frac{1/4}{(1-x)^2} + \frac{17/72}{1-x} + \frac{1/8}{1+x} + \frac{1/9}{1-\omega x} + \frac{1/9}{1-\omega^2 x}.$$

Summing the coefficients of x^n in these terms yields

$$a_{n,3} = \frac{1}{6}\binom{n+2}{2} + \frac{1}{4}(n+1) + \frac{17}{72} + \frac{(-1)^n}{8} + \frac{1}{9}(\omega^n + \omega^{2n}).$$

Note that $\frac{1}{6}\binom{n+2}{2} + \frac{1}{4}(n+1) = \frac{1}{12}(n+3)^2 - \frac{1}{3}$. Moving $\frac{1}{12}(n+3)^2$ to the left side and applying the triangle inequality on the right yields, as desired,

$$\left| a_{n,3} - \frac{1}{12}(n+3)^2 \right| \le \frac{7}{72} + \frac{1}{8} + \frac{2}{9} < \frac{1}{2}.\qquad\blacksquare$$

Similar reasoning applies to $a_{n,k}$ for any fixed k. The dominant contribution comes from $(1-x)^{-k}$. We obtain the leading coefficient.

3.4.5. Theorem. The number of partitions of n using parts in $[k]$ is asymptotic to $\frac{1}{k!(k-1)!}n^{k-1}$.

Proof: The generating function is the first formula in Theorem 3.4.3. Since $1 - x^r = (1-x)(1 + x + \cdots x^{r-1})$, we have

$$\prod_{i=1}^{k}\frac{1}{1-x^r} = \frac{1}{(1-x)^k}\frac{1}{1+x}\frac{1}{1+x+x^2}\cdots\frac{1}{1+x+\cdots+x^{k-1}}.$$

To obtain the partial fraction expansion and complete the analysis, we must further factor the denominator into complex linear factors, as in Proposition 3.4.4. Note the use there of the primitive cube roots of unity. In general, $1 - x^r$ is the product of $1 - \theta x$ over all primitive dth roots of unity over all d that divide r. (For example, $1 - x^4 = (1-x)(1+x)(1-ix)(1+ix)$.) When we factor $\prod_{r=1}^{k}(1 - x^r)$, we obtain $1 - \theta x$ as a factor $\lfloor k/d \rfloor$ times if θ is a primitive dth root.

In the partial fraction expansion, we therefore have constant multiples of $\frac{1}{(1-\theta x)^j}$ for $1 \le j \le \lfloor k/d \rfloor$. The contribution of a fixed dth-root θ to the coefficient of x^n in the expansion of the generating function is thus $p_\theta(n)\theta^n$, where p_θ is a polynomial of degree at most $\lfloor k/d \rfloor - 1$.

The only one of these polynomials having a term with degree as large as $k - 1$ is the one associated with $(1 - x)^k$. To find its coefficient, we multiply by $\prod_{r=1}^{k}(1 - x^r)$ (the product of the denominators) and set $x = 1$ (again using the Heaviside method). For this operation we use the initial factorization given above, where setting $x = 1$ yields $1 = A \cdot k!$, so $A = 1/k!$. We obtain

$$[x^n]\frac{A}{(1-x)^k} = \frac{1}{k!}\binom{n+k-1}{k-1} \sim \frac{1}{k!(k-1)!}n^{k-1}.$$

Furthermore, the magnitude of each root is 1. Hence the magnitude of the leading behavior of the coefficient of x^n is bounded by $\frac{1}{k!(k-1)!}n^{k-1}$ plus the sum of polynomials in n with degree less than $k - 1$. We conclude $a_{n,k} \sim \frac{1}{k!(k-1)!}n^{k-1}$. ∎

The asymptotic behavior of the total number of partitions of n is more complicated. We will obtain an upper bound. In order to judge how good it is, we first build many partitions. Let $p(n)$ denote the total number of partitions of n, so $\sum_{n\ge0}p(n)x^n = \prod_{i\ge1}(1 - x^i)^{-1}$.

3.4.6. Remark. *A lower bound.* For fixed k, the number of partitions of n using parts in $[k]$ is asymptotic to $n^{k-1}/(k!(k-1)!)$ as $n \to \infty$. When k grows slowly enough with n, the formula should still be fairly accurate. With $k = \sqrt{n}$, by Stirling's Formula it becomes roughly $n^{\sqrt{n}}/(\sqrt{n}/e)^{2\sqrt{n}}$, which simplifies to $e^{2\sqrt{n}}$.

This lower bound seems reasonable; not many partitions can have very large parts. Partitions of n with largest part k correspond to partitions of $n - k$ into parts no bigger than k. When k is large, $p(n - k)$ is much smaller than $p(n)$, but there are many terms between $k = \sqrt{n}$ and $k = n$, so the outcome is unclear. ∎

The intuition of Remark 3.4.6 is not bad; the actual upper bound is less than $e^{2.565\sqrt{n}}$, so the order of growth of the logarithm is correct. For asymptotic analysis, we treat the generating function as a power series and worry about convergence. The short proof of this upper bound is due to van Lint [1974].

3.4.7. Theorem. The number $p(n)$ of partitions of n satisfies
$$p(n+1) < \frac{\pi}{\sqrt{6n}} e^{2\pi\sqrt{n/6}}.$$

Proof: Let $P(x) = \sum_{n\geq 0} p(n)x^n = \prod_{k\geq 1}(1-x^k)^{-1}$. For $0 < x < 1$, we obtain a numerical bound on $\ln P(x)$ and then use it to bound $\ln p(n+1)$. The idea behind this comparison is the following, using that $\langle p \rangle$ is increasing:

$$P(x) = \sum_{k\geq 0} p(k)x^k > \sum_{k>n} p(k)x^k > p(n+1)\sum_{k>n} x^k = p(n+1)\frac{x^{n+1}}{1-x}. \qquad (*)$$

Since $P(x)$ is a product, it is convenient to study $\ln P(x)$. The inner sum below is bounded by an absolutely convergent geometric series, so we can interchange the order of summation and evaluate the new inner geometric series $\sum_{k\geq 1}(x^j)^k$.

$$\ln P(x) = -\sum_{k=1}^{\infty} \ln(1-x^k) = \sum_{k=1}^{\infty}\sum_{j=1}^{\infty}\frac{x^{jk}}{j} = \sum_{j=1}^{\infty}\frac{1}{j}\frac{x^j}{1-x^j}.$$

Next we bound $\frac{x^j}{1-x^j}$. Since $jx^{j-1} < 1 + x + \cdots + x^{j-1} = \frac{1-x^j}{1-x}$ when $0 < x < 1$,

$$\frac{x^j}{1-x^j} = \frac{x}{j}\frac{jx^{j-1}}{1-x^j} < \frac{1}{j}\frac{x}{1-x}.$$

Since $\frac{x}{1-x}$ is independent of j, we have $\ln P(x) < \frac{x}{1-x}\sum_{j=1}^{\infty} j^{-2} = \frac{\pi^2}{6}\frac{x}{1-x}$.

Now combine $(*)$ with the bound on $\ln P(x)$ (also set $u = \frac{1-x}{x}$, so $x^{-1} = 1 + u$):

$$\ln p(n+1) < \ln\frac{P(x)(1-x)}{x^n \cdot x} < \frac{\pi^2}{6}\cdot\frac{x}{1-x} + n\ln\frac{1}{x} + \ln\frac{1-x}{x}$$
$$< \frac{\pi^2}{6}\cdot\frac{1}{u} + nu + \ln u.$$

Here we have also used $\ln(1+u) < u$. Choosing $u = \pi/\sqrt{6n}$ to make the bound small, we obtain $\ln p(n+1) < 2\pi\sqrt{n/6} + \ln u$, as desired. ∎

The upper bound in Theorem 3.4.7 is within a constant factor of the true behavior. Hardy–Ramanujan [1918] showed $p(n) \sim (4n\sqrt{3})^{-1}e^{2\pi\sqrt{n/6}}$. Rademacher [1937] gave more detail; see Chandrasekharan [1970] or Andrews [1976, Chap. 5].

3.4.8.* Remark. *Analytic methods.* Various methods from complex analysis shed light on the coefficients. If a formal power series $A(x)$ converges absolutely when the argument is set to some complex number in a neighborhood of the origin, then the function is analytic in that neighborhood. Now $[x^n]A(x)$ is the residue of $A(x)/x^{n+1}$, and the Cauchy Integral Formula yields $a_n = \frac{1}{2\pi i}\int_C \frac{A(x)}{x^{n+1}}dx$, where C is a simple closed curve around the origin. The partition function is analytic for $|x| < 1$ and has singularities at every complex root of unity on the unit circle. This leads to the asymptotic analysis of $p(n)$. These notions are explored in Andrews [1976, Chapter 6] and Wilf [1990, Chapter 5]. ∎

FERRERS DIAGRAMS

We have an OGF for the partitions of n with largest part k. What about the partitions of n with k parts? In fact, these sets have the same size. This and many combinatorial identities arise from a geometric view of partitions.

3.4.9. Definition. The **Ferrers diagram** of a partition λ of n is an array of n dots, with λ_i dots in row i, where λ_i is the ith largest part. The rows are left-justified, each at least as long as the row below it.

When drawn with boxes instead of dots, Ferrers diagrams are called **Young diagrams**. We can cut and reassemble a diagram to turn a partition of one type into a partition of another type. When this yields a bijection, the two types of partitions are equinumerous. For example, we can count dots by columns instead of by rows. This maps the partition 5,3,1,1 shown above into 4,2,2,1,1.

3.4.10. Definition. The **conjugate** λ^* of a partition λ is the partition whose Ferrers diagram is the transpose of the Ferrers diagram of λ.

3.4.11. Proposition. The number of partitions of n with largest part k equals the number of partitions of n into k parts.

Proof: Taking the conjugate converts every partition with k parts into a partition with largest part k, and conjugation is a bijection. ∎

By Proposition 3.4.11, the OGFs for partitions with largest part k and partitions into k parts are equal. Similarly, partitions with parts in $[k]$ correspond to partitions with at most k parts.

Euler [1748] proved the following identity algebraically (Exercise 13):

$$\prod_{i=1}^{\infty}(1 + x^i) = \prod_{i=1}^{\infty}(1 - x^{2i-1})^{-1}.$$

The left side is the OGF for partitions into distinct parts, and the right side is the OGF for partitions into odd parts. The natural bijective proof presented next shows that corresponding coefficients are equal. See Sylvester–Franklin [1882], Pak [2006], and Exercises 14–15 for extensions.

3.4.12. Proposition. The number of partitions of n into distinct parts equals the number of partitions of n into odd parts.

Proof: (Glaisher [1883]) Map an odd-parts partition into a distinct-parts partition by iteratively combining two identical parts until no identical parts remain.

Each part in the resulting partition is a power of 2 times an odd number, say $2^k(2j + 1)$, obtained by combining 2^k copies of $2j + 1$ from the original odd-parts partition. Since every positive integer is expressible as $2^k(2j + 1)$ in a unique way, for each distinct-parts partition there is exactly one odd-parts partition that maps to it in this way. Hence the map is a bijection. ∎

Next consider congruence classes of **integer triangles**; triangles whose sides have integer lengths. Two triangles are congruent if their side-lengths form the same integer partition. For perimeter m, we partition m into three parts a, b, c with $a \geq b \geq c$, but they must satisfy the strict triangle inequality $b + c > a$.

The methods we have discussed do not constrain one part in terms of others, so the triangle inequality is hard to enforce directly. Exercise 18 develops an OGF for the congruence classes of triangles, indexed by perimeter. Although somewhat longer than the proof here, that approach does not require prior knowledge of the answer. Here we use Ferrers diagrams to give a short proof.

3.4.13. Theorem. The OGF for congruence classes of triangles with integer-length sides, indexed by perimeter, is

$$\frac{x^3}{(1 - x^2)(1 - x^3)(1 - x^4)}.$$

Proof: We show that 3-part partitions satisfying the strict triangle inequality correspond to partitions with parts in $\{2, 3, 4\}$ and at least one 3.

For a partition of the first type, the triangle inequality ensures that the first row of the diagram exceeds the second row by less than the length of the third. Therefore, moving the excess to a fourth row yields a new Ferrers diagram. The new fourth row is strictly shorter than the third, so the conjugate partition has a 3. Since we moved all the excess, the conjugate has no 1; it uses only parts in $\{2, 3, 4\}$, with at least one 3.

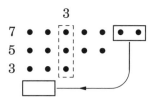

Given a partition with conjugate of the second type, this process reverses to yield the unique partition of the first type that maps to it. Since the conjugate has at least one 3, the excess of the largest resulting part over the second part is less than the third part, and hence the three parts satisfy the triangle inequality. Thus we have a bijection, and the two generating functions are the same. ∎

Many other bijections for partitions can be found in the survey by Pak [2006]; see also Sylvester–Franklin [1882]. We close with a classical such argument to prove another famous identity that Euler proved algebraically.

3.4.14. Theorem. (Euler's Identity)

$$\prod_{i=1}^{\infty}(1 + x^{2i-1}) = 1 + \sum_{k=1}^{\infty} \frac{x^{k^2}}{(1 - x^2)(1 - x^4)\cdots(1 - x^{2k})}$$

Proof: (possibly due to Durfee) We prove that both expressions enumerate **self-conjugate partitions** (those unchanged by transposing the diagram). To do this, we cut the diagram of a self-conjugate partition in two ways.

The coefficient of x^n in $\prod(1 + x^{2i-1})$ counts partitions of n into distinct odd parts. In a self-conjugate partition, grouping dots by the smaller coordinate cuts the Ferrers diagram into L-shapes as on the left below. Corresponding rows and columns have the same length and share a diagonal dot, so each piece has odd size. Also, the sizes are distinct. Conversely, a partition into distinct odd parts yields the self-conjugate partition that generates it by bending each part in the middle. Hence $\prod(1 + x^{2i-1})$ is the OGF for self-conjugate partitions.

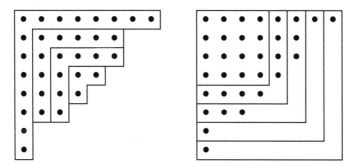

Each self-conjugate partition has some number k of dots on the diagonal. The largest square in the diagram (the "**Durfee square**") has k^2 dots. In the example above, $k = 4$. The number of dots at any distance to the right of the Durfee square is the same as the number below it, so grouping the remaining dots by distance from the square partitions them into even parts. These parts do not reach the diagonal, so they are at most $2k$. Thus the OGF for self-conjugate partitions with Durfee square of size k is the term for k on the right side of the identity ($k = 0$ yields the one partition of 0). Summing over k completes the proof. ∎

BULGARIAN SOLITAIRE (optional)

Ferrers diagrams facilitate solution of a curious problem that occurred to many people independently over the years. Martin Gardner [1983] gave the problem its name in his column in *Scientific American*.

3.4.15. Problem. (Bulgarian Solitaire) Given n identical coins distributed into piles, a *move* takes one coin from each pile and puts them into a new pile. What happens under repeated moves?

Gardner phrased the problem using cards (hence "solitaire"), with $n = 45$. Since the coins are identical and the piles are not distinguished, each configuration is a partition of n, and the specified move defines a function B (for "Bulgarian") from the set of partitions of n to itself. Since the set is finite, repeated moves must eventually reach a cycle.

When n is a triangular number (having the form $\binom{k+1}{2}$), it is easy to see that the partition $k, k-1, \ldots, 1$ is fixed under B, and such partitions are the only fixed points (Exercise 48). More surprising is that when $n = \binom{k+1}{2}$ this is the only cycle. Brandt [1982] proved this and described all the cycles for all values of n (we will count them in Application 4.2.22). Akin–Davis [1985] discussed the proof in terms of dynamical systems.

Similar processes studied in Akin–Davis [1985], Cannings–Haigh [1992], Yeh [1995], Servedio–Yeh [1995], Broline–Loeb [1995], and Griggs–Ho [1998] were called Austrian solitaire, Montreal solitaire, Carolina solitaire, etc.   ∎

3.4.16.* Remark. Let $d(\lambda)$ be the number of iterations of B needed to reach a partition on the final cycle when starting from λ. Brandt [1982] conjectured that $d(\lambda) \leq k(k-1)$ when $n = \binom{k+1}{2}$. This was proved by Igusa [1985], Bentz [1987], and Etienne [1991] (the latter submitted in 1984), and the bound is sharp (Exercise 50). Igusa and Etienne proved more generally that $d(\lambda) \leq k(k-1)$ when $n = \binom{k}{2} + r$ with $0 < r \leq k$.

Griggs–Ho [1998] proved $d(\lambda) \leq (k-1)^2$ when $\binom{k}{2} < n < \binom{k+1}{2}$, which is sharp when $n = \binom{k+1}{2} - 1$ with $k \geq 4$. They also conjectured the form of partitions maximizing $d(\lambda)$ over partitions of n and found necessary conditions (not sufficient) for partitions achieving the maximum. Partitions achieving the maximum are not unique, even when $n = \binom{k+1}{2}$.   ∎

Study of B is aided by Ferrers diagrams. Given the diagram T for a partition λ, the diagram for $B(\lambda)$ is obtained by removing the first column of T and inserting it instead as a new row.

3.4.17. Example. The key to understanding B is to form the diagram for $B(\lambda)$ in another way. A **slant** in the diagram is a set of positions (i, j) having the same value of $i + j$ (the top left position is $(1, 1)$). The kth slant, counting from the upper left, has k possible positions.

Consider $B(7, 6, 3, 3) = (6, 5, 4, 2, 2)$. When T has s rows, $B(T)$ gains s as a part. We can produce the new diagram by putting the dots taken from the start of each row as a new top row. We then move each subsequent column up by one row (see the middle diagram below), and s appears as a part in the correct position (indicated by the circled dot). Since dots moved from the left to the top, all dots in the first s columns that were in the kth slant remain in the kth slant, and dots in the later columns move to the $(k-1)$th slant. One can equivalently say that every slant rotates down by one position, except that when this leaves dots with a gap above them they move up into the previous slant.   ∎

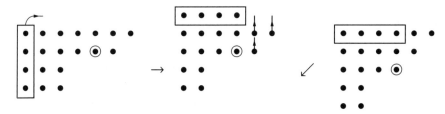

The kth slant is **complete** if it contains k dots, **incomplete** if it is nonempty but has fewer than k dots.

3.4.18. Theorem. (Brandt [1982]) A partition is in a cycle under Bulgarian solitaire if and only if its Ferrers diagram has at most one incomplete slant.

Proof: (Griggs–Ho [1998]) Under the description of B in Example 3.4.17, complete slants are preserved (dots rotate by one position). When T has at most one incomplete slant, $B(T)$ is obtained by one cyclic shift on the incomplete slant. No dots move upward, because the first column becomes the full top row.

For the converse, dots remain on the same slants except for those in the rightmost columns (as described in Example 3.4.17), which move to a shorter slant. Dots never move to a longer slant. Hence for a partition on a cycle, iteration of B can never move a dot to a shorter slant.

If T has more than one incomplete slant, then there is a value r such that T has a gap in slant r and a dot in slant $r + 1$. If T is on a cycle, then these slants shift cyclically with each application of B. However, since the sizes of slant r and slant $r + 1$ are relatively prime, the gap in slant r catches up to the dot in slant $r + 1$, reaching the same column. This array is not a Ferrers diagram, and hence such a diagram cannot be on a cycle under B. ∎

DISTRIBUTION MODELS (summary)

Many of the classical counting problems can be phrased in the context of distributions of objects into boxes. There are conditions on the objects, the boxes, and the allowed distributions. We use a uniform notation with k being the number of objects and n being the number of boxes.

A distribution of k distinct objects into n distinct boxes is a function $f: [k] \rightarrow [n]$, assigning a box to each object. We use all k objects, but boxes may be empty. It suffices to list $f(1), \ldots, f(k)$ in order. Thus the distributions correspond to words of length k from n letters.

When the order of letters is unimportant, words using the same multiset of letters are equivalent, and we count the equivalence classes (the multisets). The word model is now less natural than the distribution model using k indistinguishable (or "identical") objects.

Making boxes indistinguishable yields another equivalence relation: a distribution groups the objects without labels on the groups. This yields partitions of the set $[k]$ into (at most) n blocks. When the objects also are identical, a distribution just groups k identical dots and forms a partition of the integer k into (at most) n parts. The number of blocks or parts is exactly n when we restrict to distributions that use all the boxes. We can also restrict distributions by allowing only one object in each box.

We obtain twelve distribution problems, in an array called "The Twelvefold Way". Indexing the rows, distributions may be **U**nrestricted, **I**njective (at most one per box), or **S**urjective (no box empty). Indexing the columns, objects may be **D**istinct or **I**dentical, and boxes may be **D**istinct or **I**ndistinguishable, Stanley [1986] publicized the table, attributing the idea to G.-C. Rota and the term to Joel Spencer, suggested by the Eightfold Way of Buddhism. Where possible, we give a classical model, a formula, a generating function, and a recurrence (several of the recurrences appear in Example 2.1.9). We use subscripts (for example, $S_{k,n}$ instead of $S(k, n)$) for clarity. We break the table into two parts.

	k Distinct Objects n Distinct Boxes	k Identical Objects n Distinct Boxes
U	Words of length k from n types (repetition okay) $\overline{P}_{n,k} = n^k \qquad \overline{P}_{n,k} = n\overline{P}_{n,k-1}$ $\sum_{k\geq 0} \overline{P}_{n,k}x^k = \frac{1}{1-nx}$ $\sum_{k\geq 0} \overline{P}_{n,k}x^k/k! = e^{nx}$ $\overline{P}_{n,k} = \overline{P}_{n-1,k} + \sum_i \binom{n}{i}\overline{P}_{n-1,k-i}$	Multisets of size k from n types (repetition okay) $\overline{C}_{n,k} = \binom{k+n-1}{n-1} = (-1)^k\binom{-n}{k}$ $\sum_{k\geq 0} \overline{C}_{n,k}x^k = (\frac{1}{1-x})^n$ $\overline{C}_{n,k} = \overline{C}_{n-1,k} + \overline{C}_{n,k-1}$
I ≤ 1	Simple words of length k from n types (no repetition) $P_{n,k} = \frac{n!}{(n-k)!}$ $\sum_{k\geq 0} P_{n,k}x^k/k! = (1+x)^n$ $P_{n,k} = P_{n-1,k} + kP_{n-1,k-1}$ $P_{n,k} = nP_{n-1,k-1}$	Subsets of size k from n elements (no repetition) $C_{n,k} = \binom{n}{k} = \frac{n!}{k!(n-k)!}$ $\sum_{k\geq 0} C_{n,k}x^k = (1+x)^n$ $C_{n,k} = C_{n-1,k} + C_{n-1,k-1}$
S ≥ 1	Partitions of k objects into n labeled blocks $\overline{S}_{k,n} = n!S_{k,n}$ $\overline{S}_{k,n} = \sum_{i=0}^n (-1)^i\binom{n}{i}(n-i)^k$ $\sum_{k\geq 0} \overline{S}_{k,n}x^k/k! = (e^x - 1)^n$ $\overline{S}_{k,n} = n(\overline{S}_{k-1,n-1} + \overline{S}_{k-1,n})$	Compositions of integer k into n positive parts (solns to $\sum_{i=1}^n e_i = k;\ e_i \geq 1$) $q_{n,k} = \binom{k-1}{n-1}$ $\sum_{k\geq 0} q_{n,k}x^k = (\frac{x}{1-x})^n$ $q_{n,k} = q_{n,k-1} + q_{n-1,k-1}$

References

UDD: Proposition 1.1.13, Example 3.3.8 IDD: Proposition 1.1.16 SDD: Theorem 3.3.11	UID: Theorem 1.1.20 Example 3.1.8, Theorem 3.1.10 IID: Proposition 1.1.17 SID: Corollary 1.1.22

In the second part of the table (below), our model here of distributing k objects into n boxes exchanges the meanings of n and k that were used in the text for surjective distributions or partitions of integers. For Stirling numbers of the second kind, we thus write $S_{k,n}$ here; written in second position, n is the number of blocks. For partitions of an integer with a restricted number of parts, the first argument is always the integer being partitioned.

	k Distinct Objects n Indistinguishable Boxes	k Identical Objects n Indistinguishable Boxes
U	Partitions of k-set into at most n blocks $\sum_{i=0}^{n} S_{k,i}$ For $n = k$, is the **Bell number** B_k $\sum_{k\geq 0} B_k x^k/k! = e^{e^x - 1}$ $B_k = \sum_{i=0}^{k-1} \binom{k-1}{i} B_i$	Partitions of integer k into at most n parts $\overline{p}_{k,n} = \sum_{i=0}^{n} p_{k,i}$ $\sum_{k\geq 0} \overline{p}_{k,n} x^k = \dfrac{1}{\Pi_{i=1}^{n}(1 - x^i)}$ $\overline{p}_{k,n} = \overline{p}_{k-n,n} + \overline{p}_{k,n-1}$
I ≤ 1	1 way if $k \leq n$ 0 ways if $k > n$	1 way if $k \leq n$ 0 ways if $k > n$
S ≥ 1	Partitions of k-set into n blocks $S_{k,n} = \frac{1}{n!} \sum_{i=0}^{n}(-1)^i \binom{n}{i}(n - i)^k$ $\sum_{k\geq 0} S_{k,n} x^k/k! = \frac{1}{n!}(e^x - 1)^n$ $\sum_{k\geq 0} S_{k,n} x^k = \dfrac{x^n}{\Pi_{i=1}^{n}(1 - ix)}$ $S_{k,n} = S_{k-1,n-1} + n S_{k-1,n}$	Partitions of integer k into n parts $\sum_{k\geq 0} p_{k,n} x^k = \dfrac{x^n}{\Pi_{i=1}^{n}(1 - x^i)}$ $p_{k,n} = p_{k-n,n} + p_{k-1,n-1}$

References

UDI: Exercise 3.3.31	UII Theorem 3.4.3
IDI: trivial	III: trivial
SDI: Theorem 3.3.11	SII: Theorem 3.4.3

Bogart [1990] expanded the table to 18 cells by also considering the order of distinct elements within a box. Arrangements of k flags on n flagpoles, counted by the rising factorial $n^{(k)}$, are distributions of k distinct objects into n distinct boxes (cell UDD), where in addition order of objects within the boxes matters.

Similarly, permutations of $[k]$ with n cycles arise by putting k distinct objects surjectively into n identical boxes (cell SDI) and then cyclically arranging within each box. Hence $c(k, n)$ is not in our table.

Permutations of $[k]$ with n runs (Theorem 3.1.26) can be viewed as surjective distributions of k objects into n boxes, but the largest object in one box must exceed the smallest object in the next. When constraints involve more than one box, the distribution language is not very useful.

Proctor [2006] suggested increasing the number of cells to 24 and then 30. His first augmentation allowed an arbitrary number of boxes, thus introducing the Bell numbers and the total number of partitions of an integer. His other augmentation, suggested by Brylawski, specified multiplicities within boxes, introducing multinomial coefficients.

A summary of Proctor's table with links to the relevant counting sequences appears in the index of Sloane's On-Line Encyclopedia of Integer Sequences at http://oeis.org. We consider only the Twelvefold Way because we seek only an overview of basic results to show some of the relationships among them.

EXERCISES 3.4

3.4.1. (−) Count the partitions of 30 into 1s and 3s using an odd number of 3s. Express the answer using generating functions. Verify the answer by describing these partitions explicitly.

3.4.2. (−) In how many ways can a roll of five ordinary six-sided dice of different colors sum to 20? How many distinguishable ways if the dice are identical? (It suffices to give answers using generating functions.)

3.4.3. (−) Build a generating function to count the positive integer solutions to $\sum_{i=1}^{n} e_i = k$ such that $e_1 \geq \cdots \geq e_n$. (Hint: Choose the index appropriately.)

3.4.4. (−) Use generating functions to show that every nonnegative integer has a unique binary expansion.

3.4.5. (−) Prove that the number of partitions of $n+1$ having no 1 is at most the number of partitions of n in which the number of copies of 1 is positive and is less than the size of the smallest other part (if such a part exists). For example, when $n = 3$ the partitions of the first type are 4 and 22, while those of the second type are 111 and 21. (Merca [2014b])

3.4.6. (−) List the possible distributions of bridge hands with no void suit (Example 3.4.1; a *void* is a suit with no cards). Which distribution is more likely, 4333 or 5431?

3.4.7. (◇) Let a_n count the distinguishable ways to toss n indistinguishable dice and obtain an even sum. Prove $\sum a_n x^n = \frac{1}{2}(1-x)^{-3}[(1-x)^{-3} + (1+x)^{-3}]$.

3.4.8. (◇) The left and right pillars of an arch must have the same total height. They will be built using blocks of height 1 or 2, but blocks of height 2 may not sit on blocks of height 1. Let a_n be the number of ways to do this using a total of n blocks. Note that $a_0 = 1$, $a_1 = 0$, $a_2 = 2$, $a_3 = 2$, etc.
　　(a) Build the OGF for $\langle a \rangle$ by showing that a_n equals the number of partitions of n into red 2s, blue 2s, red 3s, and blue 3s that use at most one color of 3s.
　　(b) Determine the asymptotic behavior of a_n. (Beckwith [2005])

3.4.9. Let $p_{n,k}$ be the number of partitions of n into k parts. Give bijective proofs for the following facts.
　　(a) $p_{2r+k,r+k}$ is independent of k.
　　(b) $p_{r+k,k}$ counts the partitions of r into parts of size $\leq k$.
　　(c) $p_{r+k,k}$ counts the partitions of $r + k(k+1)/2$ into k distinct parts.

3.4.10. Let $p_{n,k}$ be the number of partitions of n into k parts, and $A_k(x) = \sum_n p_{n,k} x^n$.
　　(a) Prove bijectively that $p_{n,k} = p_{n-1,k-1} + p_{n-k,k}$. When is the equation valid?
　　(b) Obtain a recurrence for A_k. Solve it and explain the result combinatorially.

3.4.11. Let $p_{j,k}(n)$ be the number of partitions of n that have j parts and have k as the largest part. Compute $\sum_{n=0}^{\infty} p_{j,k}(n)$.

3.4.12. Prove that the number of partitions (of any integer) whose Ferrers diagram fits in an i-by-$(n-i)$ rectangle is one more than the number of permutations of $[n]$ having exactly one descent, at position i.

3.4.13. Prove Proposition 3.4.12 by direct algebraic manipulation of the generating functions. That is, give an algebraic proof of $\prod_{i=1}^{\infty}(1 + x^i) = \prod_{i=1}^{\infty}(1 - x^{2i-1})^{-1}$. (Euler [1748])

3.4.14. (◇) *Generalization of Proposition 3.4.12.* For $n, k \in \mathbb{N}_0$, use generating functions to prove that the number of partitions of n having k even parts is the same as the number of partitions of n in which k is the largest repeated part ($k = 0$ when there is no repeated part). (Comment: Proposition 3.4.12 is the case $k = 0$.) (Andrews–Deutsch [2016])

3.4.15. ($\Diamond$) *Further generalization of Proposition 3.4.12.*

(a) (**Glaisher's Theorem**) Prove for $d \in \mathbb{N}$ that the number of partitions of n with no part divisible by d equals the number of partitions of n with no part occuring at least d times. (Comment: Proposition 3.4.12 is the case $d = 2$.) (Glaisher [1883])

(b) Prove that the number of partitions of n having exactly k parts divisible by d is the same as the number of partitions of n in which k is the largest part that occurs at least d times. For example, 6 has one partition with three even parts $(2 + 2 + 2)$ and one in which the largest repeated part is 3 $(3 + 3)$. (Comment: Setting $k = 0$ yields part (a); setting $d = 2$ yields Exercise 3.4.14.) (Smoot [2018])

3.4.16. Prove that for $n \geq 2$, exactly half of the partitions of n into powers of 2 have an even number of parts. For example, when $n = 2$ the partitions are 2 and 11, when $n = 3$ they are 21 and 111, and when $n = 4$ they are 4, 22, 211, 1111.

3.4.17. Let a_n be the number of partitions of n in which each part is at least as big as the sum of all subsequent parts. Let b_n be the number of partitions of n into powers of 2. Prove $a_n = b_n$ for $n \geq 1$. (Beckwith [2009])

3.4.18. ($\Diamond$) For $n \in \mathbb{N}_0$, let a_n be the number of congruence classes of triangles with integer-length sides and perimeter n. Equivalently, a_n is the number of partitions of n into three parts less than $n/2$. This reproves Theorem 3.4.13, which is forbidden from use here.

(a) For $k \in \mathbb{N}_0$, prove that a_{2k} counts the partitions of k into three parts.

(b) For $k \geq 2$, prove $a_{2k-3} = a_{2k}$.

(c) Use parts (a) and (b) and manipulation of generating functions to prove that the OGF for $\langle a \rangle$ is $x^3/[(1-x^2)(1-x^3)(1-x^4)]$.

3.4.19. ($\Diamond$) Let a_n be the number of partitions of n into three parts less than $n/2$, corresponding to congruence classes of integer triangles.

(a) For $n \geq 3$, prove
$$a_n = a_{n-3} + \begin{cases} 0 & \text{if } n \text{ is even,} \\ \lfloor \frac{n+1}{4} \rfloor & \text{if } n \text{ is odd.} \end{cases}$$

(b) For $n \geq 3$, prove $a_n = a_{n-3} + \frac{1-(-1)^n}{8}(n + i^{n+1})$, where $i = \sqrt{-1}$.

(c) Obtain the generating function for $\langle a \rangle$ from the recurrence.

3.4.20. With a_n as in Exercise 3.4.19, prove $a_n = a_{n-2} + \lfloor n/3 \rfloor - \lfloor n/4 \rfloor$ for $n \geq 2$. Find the generating function $\sum_{n \geq 0} \lfloor n/k \rfloor x^n$, and use it to find the generating function $\sum_{n \geq 0} a_n x^n$.

3.4.21. Prove the identity below. (MacMahon [1916], Beckwith [2000])
$$\sum_{n=1}^{\infty} \frac{x^{n(n+1)/2}}{1-x^n} = \sum_{n=1}^{\infty} \frac{x^n}{1-x^{2n}}$$

3.4.22. Let $A(x) = \sum_{n=0}^{\infty} a_n x^n/n!$ and $B(x) = \sum_{n=1}^{\infty} b_n x^n/n!$. Given a partition λ of the integer n, let e_i be the number of parts equal to i in λ. Prove
$$A(B(x)) = a_0 + \sum_{n=1}^{\infty} \left(\sum_{k=1}^{n} a_k \cdot \sum \frac{n! \prod b_{\lambda_j}}{\prod_i (i!)^{e_i} (e_i!)} \right) \frac{x^n}{n!},$$
where the inner sum is over partitions λ of n with k parts.

3.4.23. ($\Diamond$) Let a_n be the total number of 2s over all partitions of n. Let b_n be the total number of nonrepeated parts over all partitions of $n - 1$. For $n \geq 1$, both sequences begin $(0, 1, 1, 3, 4, \ldots)$. Prove $a_n = b_n$ for all n by showing that for each sequence the OGF is $\frac{x^2}{1-x^2} \prod_{i=1}^{\infty} \frac{1}{1-x^i}$. (Deutsch [2006])

3.4.24. ($\diamond$) Use partitions to prove $\sum_{k=0}^{\infty} x^{2^{k+1}-1} \prod_{j=k+1}^{\infty}(1 + x^{2^j-1}) = \frac{x}{1-x}$. (GCHQ Problem Solving Group [2018])

3.4.25. ($\diamond$) Prove $\sum \prod_{j\geq 1} \binom{\lambda_j}{\lambda_{j+1}} = 2^{n-1}$, where the sum is over all partitions λ of n (when λ has k parts, by convention $\lambda_{k+1} = 0$). (Beckwith [2011])

3.4.26. For a partition λ, let $m(\lambda)$ denote the total number of parts, and let $m_i(\lambda)$ denote the number of parts equal to i. Prove

$$\sum \frac{m(\lambda)!}{\prod_i m_i(\lambda)!} = 2^n - F_n,$$

where the sum is over all partitions of $n+1$ having at least one 1 and F_n is the nth classical Fibonacci number. (Merca [2014a])

3.4.27. ($\diamond$) Let $p(n,m)$ be the number of partitions of n with m parts, and let $H(x,t) = \sum_{m,n} p(n,m)x^n t^m$.
    (a) Prove $H(x,t) = \prod_{k=1}^{\infty} \frac{1}{1-x^k t}$.
    (b) Let $f(n)$ be the sum, over all partitions λ of n, of the number of parts in λ. Let $F(x) = \sum_{n\geq 0} f(n)x^n$. Prove $F(x) = \sum_{k\geq 1} \frac{x^k}{1-x^k}P(x)$, where $P(x) = \sum_{n\geq 0} p(n)x^n$ with $p(n)$ being the number of partitions of n. (Stanley [2014])

3.4.28. ($\diamond$) A **double-partition** of the integer n consists of two partitions λ and μ such that $\min\{\lambda_i, \mu_i\} \geq \max\{\lambda_{i+1}, \mu_{i+1}\}$ for all i (trailing zeros added as needed) and $\sum \lambda_i + \sum \mu_i = n$. Note that the numbers of nonzero parts in λ and μ must differ by at most 1. Interchanging λ and μ yields a different double-partition when λ and μ are distinct. Let $a_{k,n}$ be the number of double-partitions of n such that λ and μ each have at most k parts. Let $A_k(x) = \sum_{n=0}^{\infty} a_{k,n}x^n$. Prove $A_k(x) = \prod_{i=1}^{k}(1 + x^{2i-1})/\prod_{i=1}^{2k}(1 - x^i)$. (Hint: Use induction on k.) (Corteel–Savage)

3.4.29. ($\diamond$) Let λ and μ be integer partitions, and let λ^* and μ^* be their conjugates. By counting a set in two ways, prove $\sum_{i,j} \min\{\lambda_i, \mu_j\} = \sum_k \lambda_k^* \mu_k^*$.

3.4.30. (+) Let n be a positive integer. For each positive integer k, let $q_k = \lfloor n/k \rfloor$, let $r_k = n - kq_k$, and let $f(k) = \binom{q_k}{2}k + r_k q_k$. Prove combinatorially that $f(k) \geq f(k+1)$ for all $k \in \mathbb{N}$. (Hint: For each k, create a Ferrers diagram for a partition of n in which $f(k)$ counts something. Transform the diagram for k into the diagram for $k+1$ by small changes that don't increase f.) (Griggs)

3.4.31. Prove the variation of Euler's Identity given below.

$$\prod_{i=1}^{\infty}(1 + x^{2i}) = 1 + \sum_{k\geq 1} \frac{x^{k(k+1)}}{(1-x^2)(1-x^4)\cdots(1-x^{2k})}$$

3.4.32. ($\diamond$) Prove the variation of Euler's Identity for all partitions.

$$\prod_{i=1}^{\infty} \frac{1}{1-x^i} = 1 + \sum_{k\geq 1} \frac{x^{k^2}}{(1-x)^2(1-x^2)^2\cdots(1-x^k)^2}$$

3.4.33. ($\diamond$) Prove the variation of Euler's Identity for partitions into distinct parts.

$$\prod_{i=1}^{\infty}(1 + x^i y) = 1 + \sum_{k\geq 1} \frac{x^{k(k+1)/2}y^k}{(1-x)(1-x^2)\cdots(1-x^k)}$$

3.4.34. With m_i being the number of copies of i in a partition λ, let $f(\lambda) = 1/\prod_{i \geq 1} m_i! i^{m_i}$. Prove for $n \in \mathbb{N}$ that the sum of $f(\lambda)$ over partitions of n equals 1. (Stanley [1986])

3.4.35. Rogers–Ramanujan partitions are partitions into distinct parts that do not use two consecutive integers as parts. For $n > 1$, prove that at least half of the Rogers–Ramanujan partitions of n do not use 1 as a part. (Andrews [1997a])

3.4.36. ($\diamond$) For $n \in \mathbb{N}$, let $P(n) = \sum_{i=0}^{n-1} p(i)$, where $p(i)$ counts all partitions of i.
   (a) Consider a set S of marbles consisting of one black marble and $n-1$ white marbles. White marbles are indistinguishable from one another. Prove that $P(n)$ is the number of distinguishable partitions of S.
   (b) Prove that $P(n)$ is the sum, over all partitions of n, of the number of distinct parts in the partition. The count is one for 4, $2+2$, and $1+1+1+1$, and it is two for $3+1$ and $2+1+1$, so the sum is 7 when $n = 4$. (G. Beck)

3.4.37. (+) In a Dyck n-path (Example 1.3.27), a **peak** is a consecutive pair UD; its **height** is the y-coordinate reached between its U and D. Counting the peak heights with multiplicity yields the peak multiset: for example, the peak multiset of $UDUDUD$ is $\{1, 1, 1\}$. Prove that the number of distinct multisets that occur as peak multisets of Dyck n-paths is $2^n - 1 - \sum_{k=1}^{n-1} p(k)$, where $p(k)$ is the number of partitions of k. (Callan–Deutsch [2012])

3.4.38. ($\diamond$) **Euler's Pentagonal Number Theorem**.
   (a) Let $p_e(n)$ and $p_o(n)$ count the partitions of n into an even number and an odd number of distinct parts, respectively. For example, $p_e(9)$ counts $\{8+1, 7+2, 6+3, 5+4\}$, while $p_o(9)$ counts $\{9, 6+2+1, 5+3+1, 4+3+2\}$. Use the left diagram below to show that $p_e(n)$ equals $p_o(n)$ except when n has the form $\sum_{i=1}^{m}(m+i)$ or $\sum_{i=0}^{m-1}(m+i)$, in which case they differ by 1. (Franklin [1881])

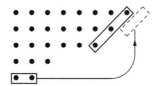

   (b) Let $\omega(m) = (3m^2 - m)/2$. Use part (a) to prove

$$\prod_{k \geq 1}(1 - x^k) = 1 + \sum_{m \geq 1}(-1)^m \left(x^{\omega(m)} + x^{\omega(-m)} \right). \qquad \text{(Euler [1748])}$$

(Comment: "Pentagonal Number Theorem" arises from writing $\omega(m)$ as $\sum_{j=0}^{m-1}(3j+1)$. The sum counts dots arranged pentagonally as on the right above, so $\omega(m)$ is a *pentagonal number*. Similarly, $\sum_{j=0}^{m-1}(j+1)$ is a *triangular number* and $\sum_{j=0}^{m-1}(2j+1)$ is a *square number*.)

3.4.39. *Pentagonal numbers, continued.*
   (a) With $p(n)$ being the number of partitions of n and $\omega(m) = (3m^2 - m)/2$, use Exercise 3.4.38b to prove the identity below for $n \geq 1$.

$$p(n) = \sum_{m \geq 1}(-1)^{m-1}\left[p\big(n - \omega(m)\big) + p\big(n - \omega(-m)\big) \right] \qquad \text{(Euler [1748])}$$

(Hint: For $n \geq 0$, let a_n be $p(n)$ minus the right side of the identity. Prove $\sum_{n \geq 0} a_n x^n = 1$.)
   (b) Prove $\sum_{j \geq 1}(-1)^j q_j(n - \omega(j)) = 0$, where $q_j(k)$ is the number of partitions of k not having j or $2j$ as a part. (For $n = 9$, we compute $-q_1(9) + q_2(6) - q_2(0) = -4 + 5 - 1 = 0$.) (Andrews [1997b])

3.4.40. ($\diamond$) The **Gaussian polynomial** $\left[\begin{smallmatrix} m+n \\ m \end{smallmatrix}\right]_q$ is defined by $\left[\begin{smallmatrix} m+n \\ m \end{smallmatrix}\right]_q = \prod_{j=1}^{m} \frac{1-q^{n+j}}{1-q^j}$.

(a) Let $p_{m,n;k}$ be the number of partitions of k into at most m parts of size at most n. Prove $p_{m,n;k} = p_{m-1,n;k} + p_{m,n-1;k-m}$.

(b) For fixed m and n, prove $\sum_k p_{m,n;k} q^k = \left[\begin{smallmatrix} m+n \\ m \end{smallmatrix}\right]_q$. Conclude that $\left[\begin{smallmatrix} m+n \\ m \end{smallmatrix}\right]_q$ is a polynomial and is symmetric in m and n.

3.4.41. Letting m_i be the number of parts equal to i in a partition λ, prove for $n \in \mathbb{N}$ that $\prod_\lambda \left(\prod_{k \geq 1} k^{m_k} \right) = \prod_\lambda \left(\prod_{j \geq 1} m_j! \right)$, where each outer product is over all partitions of n. For example, the values of both sides for $1 \leq n \leq 4$ are $1, 2, 6, 96$, respectively. (Hint: Show that the logarithms of both sides are equal.) (Kirdar–Skyrme [1982], Hoare [1986])

3.4.42. ($\diamond$) A *corner* of a partition is a dot in its Ferrers diagram that is the last dot in its row and in its column. Let $\gamma(\lambda)$ denote the number of corners in a partition λ, and let $\mathcal{P}(n)$ denote the set of all partitions of n.

(a) Prove $\sum_{\lambda \in \mathcal{P}(n)} \gamma(\lambda) = |\mathcal{P}(n-1)| + \sum_{\lambda \in \mathcal{P}(n-1)} \gamma(\lambda)$.

(b) Conclude $\sum_{n \geq 0} \sum_{\lambda \in \mathcal{P}(n)} \gamma(\lambda) x^n = \frac{x}{1-x} \prod_{i \geq 1} \frac{1}{1-x^i}$. (Pak [2006])

3.4.43. ($\diamond$) Let λ and μ be conjugate partitions, with trailing zeros added. Prove that the infinite multisets $\{\lambda_i + i\colon i \geq 1\}$ and $\{\mu_i + i\colon i \geq 1\}$ are the same. (Knuth [2018])

3.4.44. (+) Prove that the number of partitions of n equals the number of partitions (of all integers) into distinct parts whose odd-indexed parts sum to n. For example, the partitions of 4 are 4, $3+1$, $2+2$, $2+1+1$, and $1+1+1+1$, and the partitions into distinct parts whose odd-indexed parts sum to 4 are 4, $4+3$, $4+2$, $4+1$, and $3+2+1$. (Schmidt [1997])

3.4.45. (+) Let $t(n)$ be the number of unordered factorizations of n into divisors greater than 1 (for example, $t(12) = 4$). Prove $\sum_{n=2}^{\infty} t(n)/n^2 = 1$. (Beckwith [1998])

3.4.46. Let A be an m-by-n integer matrix. Given $\lambda_1, \dots, \lambda_m$ with $n \geq \lambda_1 \geq \cdots \geq \lambda_m \geq 0$, let $f_A(\lambda) = \sum_{i=1}^{m} \sum_{j=1}^{\lambda_i} a_{i,j}$. A *move* subtracts 1 from some entry and adds 1 to the next entry rightward or below or to no entry. Prove that some sequence of moves reduces A to the zero matrix if and only if $f_A(\lambda) \geq 0$ for all such λ. (Bennett [1992])

3.4.47. Let f be a function defined on the subsets of $[n]$ by letting $f(S) = S \cup \{1\}$ if $1 \notin S$ and $f(S) = [n] - \{k-1\colon k \in S\}$ if $1 \in S$. Show that f^2 has a unique fixed point S^* reached by iterating f from any initial set S. (Huang–Scully [2003])

3.4.48. ($\diamond$) Determine all the partitions that are fixed points in Bulgarian solitaire.

3.4.49. ($\diamond$) Determine the number of steps in Bulgarian solitaire needed to turn the position consisting of a single pile of size n into a position on a cycle. For example, $5 \to 41 \to 32 \to 221 \to 311 \to 32$, so the answer is 2 when $n = 5$. (Cranston–West [2013])

3.4.50. For $k \geq 3$, let λ be the partition of $\binom{k+1}{2}$ whose parts are two copies of $k-1$, two copies of 1, and one copy each of 2 through $k-2$. Prove that Bulgarian solitaire takes $k(k-1)$ steps to reach the "staircase" partition from λ. (Griggs–Ho [1998])

3.4.51. For a partition λ, let $d(\lambda)$ be the number of moves of Bulgarian solitaire needed to reach a partition that lies on a cycle when starting from λ.

(a) Prove that if the Ferrers diagram for λ is contained in the Ferrers diagram for μ, then $d(\lambda) \leq d(\mu)$. (Akin–Davis [1985])

(b) As stated in Remark 3.4.16, $d(\lambda) \leq k(k-1)$ when λ is a partition of $\binom{k+1}{2}$ (Exercise 3.4.50 shows that this is sharp). Conclude that $d(\lambda) \leq k(k-1)$ when λ is a partition of n with $\binom{k}{2} < n < \binom{k+1}{2}$. (Igusa [1985], Bentz [1987], Etienne [1991], Griggs–Ho [1998])

Chapter 4

Further Topics

The topics in this chapter exhibit common themes of indirect counting techniques and applications to permutations and lattice paths. The Inclusion-Exclusion Principle gives a combinatorial explanation for many alternating sums, while the Pólya–Redfield method counts equivalence classes of objects made indistinguishable by symmetry operators. Young tableaux, which arise in the theory of representations of symmetric groups, have surprising connections to permutations and sorting.

4.1. The Inclusion-Exclusion Principle

"Sieve methods" are counting methods that within a universe allow only a desired set to survive a process of overcounting and undercounting. The Inclusion-Exclusion Principle is the most common sieve and is our focus. Knuth attributed it to *Doctrine of Chances* by de Moivre [1718]. Riordan [1958] wrote

> The logical identity on which it rests is very old; Dickson's *History of the Theory of Numbers* (vol. I, p. 119) mentions its appearance in a work by Daniel da Silva in 1854, but Montmort's solution in 1713 of a famous problem, known generally by its French name, "le problème des rencontres" . . . effectively uses it and it may have been known to the Bernoullis.

Montmort corresponded with Nicolaus Bernoulli on such topics during 1710-12.

THE BASIC PRINCIPLE

Many elementary texts describe inclusion-exclusion as a way to count the elements in a union of sets. Summing the sizes of two sets counts the intersection twice, so $|A_1 \cup A_2| = |A_1| + |A_2| - |A_1 \cap A_2|$. For three sets, the computation is

$$|A_1 \cup A_2 \cup A_3| = \begin{array}{l} +|A_1| \quad -|A_1 \cap A_2| \\ +|A_2| \quad -|A_2 \cap A_3| \quad +|A_1 \cap A_2 \cap A_3|. \\ +|A_3| \quad -|A_3 \cap A_1| \end{array}$$

The first column overcounts each element in more than one A_i. An element in any two sets is cancelled by subtracting the intersection, but this completely cancels elements in all three sets. We correct by adding back the number of elements in all three sets. Now every element of the union is counted exactly once.

4.1.1. Example. *The classical model.* In most generalizations and applications with n sets, we want instead to count the elements of an appropriate universe U that are *outside* the union of specified sets. We can do this by subtracting the formula for the size of the union from $|U|$.

$$\left|\overline{A_1 \cup A_2}\right| = |U| - |A_1| - |A_2| + |A_1 \cap A_2|$$

$$\left|\overline{A_1 \cup A_2 \cup A_3}\right| = |U| \begin{array}{l} -|A_1| \\ -|A_2| \\ -|A_3| \end{array} \begin{array}{l} +|A_1 \cap A_2| \\ +|A_2 \cap A_3| \\ +|A_3 \cap A_1| \end{array} - |A_1 \cap A_2 \cap A_3|.$$

This model is natural for counting intersections instead of unions, since $\overline{\bigcup_{i=1}^n A_i} = \bigcap_{i=1}^n \overline{A_i}$. Imagine n constraints on objects in U. If A_i is the set of elements violating the ith constraint, then the desired objects (satisfying all the constraints) are those in $\bigcap_{i=1}^n \overline{A_i}$.

For example, the derangements of $[n]$ are the permutations of $[n]$ with no fixed points. If A_i is the set of permutations that fix element i, then $D_n = \left|\bigcap_{i=1}^n \overline{A_i}\right|$. Similarly, surjective distributions of objects into boxes are the distributions with no box empty, and the numbers in $[m]$ relatively prime to m are the numbers not divisible by any prime factor of m. We will discuss all of these.

Consider the Venn diagram of sets $A_1, \ldots, A_n$ in a universe U. In addition to counting the elements in the region outside all the circles, we may want to count the elements in any given cell. The cell corresponding to the index set T for $T \subseteq [n]$ consists of the elements that are in A_i for $i \in T$ and are not in A_i for $i \notin T$. We introduce precise notation for these concepts. ∎

4.1.2. Definition. Let $A_1, \ldots, A_n$ be subsets of a universe U. For $x \in U$, let $P(x)$ be the **property set** of x, defined by $P(x) = \{i \in [n]: x \in A_i\}$. For $S, T \subseteq [n]$, let $f(T) = |\{x \in U: P(x) = T\}|$ and $g(S) = \left|\bigcap_{i \in S} A_i\right|$. ∎

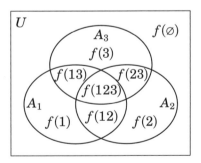

The notation here echoes the meaning. Often $A_1, \ldots, A_n$ are defined by specified properties for the elements, so $P(x)$ indexes the **P**roperties satisfied by x. Also, $f(T)$ counts the elements with a **f**ixed property set T, while $g(S)$ counts those whose property set is **g**reater than or equal to S. We drop set brackets and commas in the arguments of f and g when specifying a particular subset of $[n]$.

The relationship between g and f is $g(S) = \sum_{T \supseteq S} f(T)$. That is, $\bigcap_{i \in S} A_i$ consists of the elements of U belonging to A_i for $i \in S$ and possibly to other sets also. Thus $g(1) = |A_1| = f(1) + f(12) + f(13) + f(123)$ and $g(12) = |A_1 \cap A_2| = f(12) + f(123)$ when $n = 3$ and the sets are A_1, A_2, A_3. We use $g(S)$ and $\left|\bigcap_{i \in S} A_i\right|$ interchangeably.

In many settings, $\left|\bigcap_{i \in S} A_i\right|$ is easy to compute directly but $f(T)$ is not. For example, in the derangements problem $\left|\bigcap_{i \in S} A_i\right| = (n - |S|)!$, because after fixing the elements indexed by S we can permute the remainder arbitrarily, not caring whether other elements also become fixed points.

When $g(S)$ is easy to compute, we want to *invert* the relationship $g(S) = \sum_{T \supseteq S} f(T)$ to compute $f(T)$ using values of $g(S)$. This is like inversion principles in Chapter 3 for sequences $\langle a \rangle$ and $\langle b \rangle$; now values are indexed by subsets of $[n]$ instead of by $\mathbb{N}_0$. (Notation such as $N(S)$ and $N^*(S)$ or $N_=(S)$ and $N_\geq(S)$ has been used, but f and g better suggest the more general context of inversion.)

The next theorem provides a formula for the inversion. The proof uses the observation that a nonempty finite set has the same number of subsets of even size and odd size.

4.1.3. Theorem. (Inclusion-Exclusion Principle — PIE). Let $A_1, \ldots, A_n$ be sets in a universe U. For $T \subseteq [n]$, let $f(T)$ be the number of elements $x \in U$ such that $x \in A_i$ if and only if $i \in T$. The inclusion-exclusion formula for f is

$$f(T) = \sum_{S \supseteq T} (-1)^{|S|-|T|} \left|\bigcap_{i \in S} A_i\right|.$$

Proof: We prove that each element of U contributes equally to both sides. An element x contributes to the term for S in the sum if and only if x appears in $\bigcap_{i \in S} A_i$. Since $S \supseteq T$, only elements in $\bigcap_{i \in T} A_i$ contribute.

Recall that $P(x) = \{i \in [n] : x \in A_i\}$. Thus x appears in $\bigcap_{i \in S} A_i$ precisely when $T \subseteq S \subseteq P(x)$. When $|S| - |T|$ is even, the contribution is positive; when $|S| - |T|$ is odd, it is negative. Thus x contributes $+1$ for each even subset of $P(x) - T$ and -1 for each odd subset of $P(x) - T$.

Nonempty sets have equal numbers of even subsets and odd subsets, so the net count from x is 0 when $P(x) \neq T$, as desired. When $P(x) = T$, only the term $S = T$ contributes, and x counts $+1$ on both sides. ∎

Most applications need $f(\varnothing)$, given by $f(\varnothing) = \sum_{S \subseteq [n]} (-1)^{|S|} \left|\bigcap_{i \in S} A_i\right|$. We start with $|U|$ and next subtract the sizes of the sets A_i, letting $\left|\bigcap_{i \in \varnothing} A_i\right| = |U|$. Treating $\bigcap_{i \in \varnothing} A_i$ as U makes sense because intersection over no sets should be the "intersective identity" U; every element is in all of "none of the sets".

One of the earliest applications counts the elements of $[m]$ relatively prime to m (the number 1 is included). The resulting value $\varphi(m)$ is the **Euler totient**; the function is studied extensively in number theory (see Exercises 15–16).

4.1.4. Proposition. Let $p_1, \ldots, p_n$ be the distinct prime factors of m. The number $\varphi(m)$ of elements of $[m]$ relatively prime to m is given by

$$\varphi(m) = \sum_{S \subseteq [n]} (-1)^{|S|} \frac{m}{\prod_{i \in S} p_i}.$$

Proof: Within the universe $[m]$, let A_i be the set of multiples of p_i. The numbers relatively prime to m are those in none of $A_1, \ldots, A_n$. To apply the inclusion-exclusion formula, we compute $\left|\bigcap_{i \in S} A_i\right|$. The numbers divisible by all of $\{p_i : i \in S\}$ are those divisible by $\prod_{i \in S} p_i$, and thus $\left|\bigcap_{i \in S} A_i\right| = m / \prod_{i \in S} p_i$. ∎

Note that $\varphi(1) = 1$. As another example, $60 = 2^2 \cdot 3 \cdot 5$, so

$$\varphi(60) = 60 - \tfrac{60}{2} - \tfrac{60}{3} - \tfrac{60}{5} + \tfrac{60}{6} + \tfrac{60}{10} + \tfrac{60}{15} - \tfrac{60}{30} = 16.$$

4.1.5. Remark. (1) In the PIE formula, the sets with a given size all contribute with the same sign. Letting h_k be the sum of $\left|\bigcap_{i \in S} A_i\right|$ over all S with size k, we can write the formula as $f(\varnothing) = \sum_{k=0}^{n} (-1)^k h_k$.

(2) With h_k defined in this way, the value h_k *does not equal the number of elements lying in at least k of the sets.* The sum overcounts elements lying in more than k sets.

(3) In many applications (*not* Proposition 4.1.4), $\left|\bigcap_{i \in S} A_i\right|$ depends only on $|S|$. The sets of size k all make the same contribution, and the sum with 2^n terms becomes a sum with $n+1$ terms involving binomial coefficients. Letting g_k be the common value of $\left|\bigcap_{i \in S} A_i\right|$ when $|S| = k$ reduces the formula to

$$f(\varnothing) = \sum_{k=0}^{n} (-1)^k \binom{n}{k} g_k. \qquad ∎$$

4.1.6. Example. *Multisets with restricted usage.* A girl has red, white, and blue marbles and wants to bring 20 to school. In how many ways can she do this? When 20 marbles of each color are available, the multiset formula yields $\binom{20+2}{2}$ ways. If she has only seven of each, then we must discard multisets having at least eight of the same color.

Let $a_{m,n,r}$ be the number of multisets of size m from n types of elements using *fewer* than r of each type. These correspond to the integer solutions to $x_1 + \cdots + x_n = m$ such that $0 \le x_i < r$ for each i.

Let U be the set of all multisets of size m from n types: $|U| = \binom{m+n-1}{n-1}$. For $1 \le i \le n$, let A_i be the subset of U where type i violates its limit. These correspond to $x_i \ge r$ in the integer-sum model. Since the units being distributed are identical, we can count such solutions by reserving r units for x_i and distributing the rest arbitrarily. Thus $|A_i| = \binom{m-r+n-1}{n-1}$. This may include distributions where other constraints also are violated.

Violating k constraints leaves $m - kr$ units to distribute. Thus $\left|\bigcap_{i \in S} A_i\right| = \binom{m-|S|r+n-1}{n-1}$, which depends only on $|S|$. By inclusion-exclusion,

$$a_{m,n,r} = \sum_{k=0}^{n} (-1)^k \binom{n}{k} \binom{m - kr + n - 1}{n - 1}.$$

When $m = n(r-1)-1$, the answer must be n, since we just choose a marble of one type to leave home. Thus inclusion-exclusion provides a combinatorial proof that $\sum_{k=0}^{n} (-1)^k \binom{n}{k} \binom{(n-k)r-2}{n-1} = n$. If $m = n(r-1)+1$, then we have proved that $\sum_{k=0}^{n} (-1)^k \binom{n}{k} \binom{(n-k)r}{n-1} = 0$. ∎

4.1.7. Theorem. The formula for $S(n, k)$, the number of partitions of an n-element set into k blocks, is

$$S(n, k) = \frac{1}{k!} \sum_{i=0}^{k} (-1)^i \binom{k}{i} (k - i)^n.$$

Proof: In each partition, the blocks are distinguished by their elements, so there are $k!$ distinguishable ways to index the blocks. Hence we can compute $k!S(n, k)$ by counting surjective distributions of $[n]$ into k labeled blocks.

Altogether there are k^n distributions of $[n]$ into k labeled blocks with blocks allowed to be empty. Let A_j be the subset in which block j is empty. Given a set S of blocks required to be empty, we distribute $[n]$ arbitrarily into the remaining blocks. Hence $\left|\bigcap_{j \in S} A_j\right| = (k - i)^n$ whenever $|S| = i$. Combining the terms for sets of the same size yields, as desired,

$$k!S(n, k) = f(\varnothing) = \sum_{i=0}^{k} (-1)^i \binom{k}{i} (k - i)^n. \qquad \blacksquare$$

We proved Theorem 4.1.7 using generating functions in Theorem 3.3.11. Stirling numbers appear also in Exercises 47–50.

We have used PIE as an inversion formula: given sets $A_1, \ldots, A_n$ within a universe U, we count each $\bigcap_{i \in S} A_i$ and obtain $f(\varnothing)$, the number of elements outside all the sets. Next we use PIE to evaluate sums by interpreting them as inclusion-exclusion computations. This has the same flavor as evaluating convolutions using generating functions.

4.1.8. Example. *PIE evaluation of sums.* Consider the sum $\sum_{k=0}^{n} (-1)^k \binom{n}{k} c_k$. If the sum is the inclusion-exclusion formula for the number of elements outside some sets $A_1, \ldots, A_n$ in a universe U, then counting those elements in another way gives the value of the sum.

To fit this model, we need $|U| = c_0$ and $|A_i| = c_1$; we seek natural sets with these sizes. Some creativity may be needed. The sets must be chosen so that also $c_k = \left|\bigcap_{i \in S} A_i\right|$ whenever $|S| = k$.

For example, when $c_k = 2^{n-k}$, we seek a universe of size 2^n, such as the subsets of $[n]$. To obtain $|A_i| = 2^{n-1}$ for $1 \le i \le n$, let A_i consist of the subsets containing i. Now $\left|\bigcap_{i \in S} A_i\right| = 2^{n-|S|}$, as desired. Hence the sum counts the subsets of $[n]$ outside all of $A_1, \ldots, A_n$. These are precisely the subsets containing no elements. The only such set is the empty set, so $\sum_{k=0}^{n} (-1)^k \binom{n}{k} 2^{n-k} = 1$. (In the language of the Binomial Theorem, we have just proved that $(-1 + 2)^n = 1$.)

Similarly, in $\sum_{k=0}^{n} (-1)^k \binom{n}{k}$, each c_k is 1. The sum counts by inclusion-exclusion the items in a universe U of size 1 that belong to none of n sets each equal to U. There is no such item if $n > 0$ and one such if $n = 0$, so the sum is 0 if $n > 0$ and 1 if $n = 0$. We have counted a set both by PIE and directly. $\blacksquare$

4.1.9. Example. *Evaluation of* $\sum_{k=0}^{n} (-1)^k \binom{n}{k} \binom{m+n-k}{r-k}$. Here $c_0 = \binom{m+n}{r}$; we want a universe of that size. Let U be the family of r-sets in $[m + n]$. To model this as an inclusion-exclusion computation, we need n sets of size $\binom{m+n-1}{r-1}$. Let A_i consist of the r-sets in $[m + n]$ that use element i. When $|S| = k$, a member of $\bigcap_{i \in S} A_i$

has k required elements, so such members are completed by choosing $r - k$ of the remaining $m + n - k$ elements. Thus $\left|\bigcap_{i \in S} A_i\right| = \binom{m+n-k}{r-k}$. Since $c_k = \binom{m+n-k}{r-k}$, we conclude that the sum is the inclusion-exclusion sum to count $f(\varnothing)$.

The r-sets in none of $A_1, \ldots, A_n$ are the r-sets in $[m+n]$ that use none of the first n elements. Thus the value of the sum is $\binom{m}{r}$. We did not need to know the value in advance to compute the sum.

This example generalizes the sum in Exercise 3.2.42. ∎

Using inclusion-exclusion in this way is an argument by "counting two ways" (see also Exercises 29–34). Next we return to the setting of finding a formula to count a known set. We rederive the formula for the Eulerian numbers.

4.1.10.* Theorem. The Eulerian number $A(n, k)$ is given by

$$A(n, k) = \sum_{i=0}^{k} (-1)^i \binom{n + 1}{i}(k - i)^n.$$

Proof: (Stanley; see Bóna [2004, p. 8–9]) Let U be the universe of barred permutations of $[n]$ with k bars, defined in the proof of Theorem 3.1.25 to be permutations of $[n]$ in word form with k bars inserted, having at least one bar at the end of each run. We showed there that $|U| = k^n$; the elements correspond to distributions of $[n]$ into k boxes, since the numbers placed immediately before any bar must be written in increasing order. The $A(n, k)$ permutations with exactly k runs are the elements of U having one bar at the end of each run and no others.

A barred permutation with k bars but fewer than k runs may have excess bars in any of the $n + 1$ positions. For $1 \le j \le n + 1$, let A_j be the set of elements of U having *at least one* excess bar in the jth position. Since $f(\varnothing) = A(n, k)$, it suffices to show that the sum is the inclusion-exclusion computation for $f(\varnothing)$.

For $S \subseteq [n+1]$, we count the barred permutations having at least one excess bar in each position indexed by S. When $|S| = i$, form a barred permutation with $k - i$ bars and then add one bar at each position indexed by S. Bars are identical, so there is no distinction between the bars distributed in the two steps.

For $|S| \le k$, we have $\left|\bigcap_{j \in S} A_j\right| = (k - |S|)^n$, since there are $(k - i)^n$ barred permutations of $[n]$ with $k - i$ bars. When $|S| > k$, the value is 0, because no barred permutation with k bars has excess bars in more than k positions. Hence the sum stops at k even though we avoid $n + 1$ sets. ∎

Whitney [1932b] applied PIE to count proper colorings of graphs. A **coloring** of a graph is a function that assigns each vertex a label from a set of *colors*. A **proper coloring** gives distinct labels to adjacent vertices. Typically, we use $[k]$ for the set of colors. We are placing the n distinguishable vertices into k distinguishable boxes, with adjacent vertices forbidden to go in the same box.

4.1.11.* Application. For a set S of edges in a graph G, let $c(S)$ be the number of components in the subgraph with vertex set $V(G)$ and edge set S. For the number $\pi_G(k)$ of proper colorings of G using colors in $[k]$,

$$\pi_G(k) = \sum_{S \subseteq E(G)} (-1)^{|S|} k^{c(S)}.$$

Proof: The universe U of unrestricted colorings is the set of all functions from $V(G)$ to $[k]$; this has size k^n, where $n = |V(G)|$. Let A_i denote the set of colorings in which the vertices of the ith edge have the same color. We want to count the colorings belonging to none of these sets.

To apply inclusion-exclusion, we compute $|\bigcap_{i \in S} A_i|$. Let H be the subgraph of G with vertex set $V(G)$ and edge set S. A coloring in $\bigcap_{i \in S} A_i$ makes each edge of H monochromatic. If x and y are the ends of a path in H, then by transitivity x and y have the same color. Thus all vertices of a component of H (a maximal connected subgraph) have the same color. The color can be chosen in k ways, independently, on each component of H. Thus $|\bigcap_{i \in S} A_i| = k^{c(S)}$, as claimed. ∎

In a graph G with n vertices and m edges, subgraphs with vertex set $V(G)$ are **spanning subgraphs**. Spanning subgraphs with one edge have $n - 1$ components; those with more edges have fewer components. Thus Application 4.1.11 shows that $\pi_G(k)$ is a polynomial in k with leading terms $k^n - mk^{n-1}$; it is called the **chromatic polynomial** of G. Exercises 36–38 concern chromatic polynomials and their properties.

4.1.12.* Example. *A chromatic polynomial.* The computations in Application 4.1.11 begin $k^n - mk^{n-1} + \binom{m}{2}k^{n-2}$ for $|S| \le 2$. When $|S| = 3$, the number of components is again $n - 2$ if the three edges form a triangle; otherwise it is $n - 3$. The graph below has two triangles, and the other eight triples of edges yield one component each. All subgraphs with four or five edges have only one component. Hence the computation is

$$\pi_G(k) = k^4 - 5k^3 + 10k^2 - (2k^2 + 8k^1) + 5k - k = k^4 - 5k^3 + 8k^2 - 4k.$$

By ad hoc counting, one can also see that $\pi_G(k) = k(k-1)(k-2)(k-2)$. This is 0 when k is 1 or 2, but $\pi_G(3) = 6$. ∎

RESTRICTED PERMUTATIONS

For selection problems, we considered restricted multiplicities. For arrangement problems, we consider restricted positions. The first example is the derangement problem; derangements are permutations without fixed points. We counted these in Proposition 2.3.2 and Corollary 3.3.17; inclusion-exclusion yields the same formula almost immediately.

4.1.13. Theorem. The formula for D_n, the number of derangements of $[n]$ (permutations with no fixed points), is

$$D_n = n! \sum_{k=0}^{n} \frac{(-1)^k}{k!}.$$

Proof: Let $U = \mathbb{S}_n$. For $i \in [n]$, let A_i be the set of permutations that fix i; we seek $f(\varnothing)$. If $|S| = k$, then the elements of $\bigcap_{i \in S} A_i$ have k specified fixed points and permute the remaining elements of $[n]$ arbitrarily. Thus $\left|\bigcap_{i \in S} A_i\right| = (n-k)!$. Collecting terms with $|S| = k$ yields

$$f(\varnothing) = \sum_{k=0}^{n} (-1)^k \binom{n}{k}(n-k)! = n! \sum_{k=0}^{n} \frac{(-1)^k}{k!}. \qquad \blacksquare$$

4.1.14. Example. *Permutations with restrictions.* The derangement problem generalizes to forbid any values from any positions. The **permutation matrix** for a permutation σ is the $0,1$-matrix with 1 in entry (i,j) if and only if $\sigma(i) = j$. Forbidding value j from position i of the word form forbids 1 from entry (i,j) of the matrix. We view the forbidden locations for 1s in the matrix as squares on a chessboard, forming a **board of forbidden positions**. Below is a board with five forbidden positions for permutations of $[4]$; any value may appear in position 4.

The universe is the set of n-by-n permutation matrices. Let A_s consist of those matrices with 1 in square s of the forbidden board B. We count the desired permutations by inclusion-exclusion. There exist permutations with 1s in all positions indexed by S if and only if S has at most one position in every row and column; such a set of positions in a matrix (or on a board) is **independent**.

If the positions in S are independent in B, then $\left|\bigcap_{i \in S} A_i\right| = (n - |S|)!$. Since *the universe has only permutation matrices*, the number of such terms with $|S| = k$ is not $\binom{n}{k}$; it is the number of ways to choose k independent positions on B. Let $r_k(B)$ denote this number. For the board above, $(r_0, r_1, r_2, r_3, r_4) = (1,5,7,2,0)$. PIE then counts the permutation matrices with no 1s in forbidden entries. $\qquad \blacksquare$

4.1.15. Proposition. Given a board B of forbidden positions for the 1s in an n-by-n permutation matrix, the number of permutations of $[n]$ having no elements in forbidden positions is $\sum_{k=0}^{n}(-1)^k r_k(B)(n-k)!$, where $r_k(B)$ is the number of k-element subsets of B having no two elements in the same row or column.

Proof: The sum is the PIE computation for $f(\varnothing)$ within the universe of permutation matrices, where A_i consists of the permutation matrices with 1 in the ith square. We have $\left|\bigcap_{i \in S} A_i\right| = (n - |S|)!$ when S has no two positions in the same row or column, and otherwise $\left|\bigcap_{i \in S} A_i\right| = 0$. $\qquad \blacksquare$

In the derangement problem, the forbidden positions are those along the diagonal, and every subset of these is independent. In this case $r_k(B) = \binom{n}{k}$, and the computation reduces to Theorem 4.1.13. In general, we model independent positions by pairwise non-attacking rooks on a chessboard, since a **rook** is a piece that can attack (or move to) any position in its row or column (if there are no intervening pieces). This is why we call the set of forbidden positions a *board*.

4.1.16. Definition. For a board B of positions in a grid, the **rook polynomial** $R_B(x)$ is the generating function $\sum r_k(B)x^k$, where $r_k(B)$ is the number of placements of k pairwise non-attacking rooks on B.

4.1.17. Remark. The rook polynomial factors when a board B decomposes into sub-boards B_1 and B_2 occupying distinct rows and distinct columns. Since k rooks on B must consist of j on B_1 and $k-j$ on B_2, and rooks on the two sub-boards cannot attack each other, the rule for multiplying OGFs yields $R_B(x) = R_{B_1}(x)R_{B_2(x)}$. For example, we compute $R_B(x)$ for the board in Example 4.1.14 as $(1+3x+x^2)(1+2x)$, which equals $1 + 5x + 7x^2 + 2x^3$.

To compute $r_k(B)$ in general, group rook placements by whether they occupy a fixed square s in B. Deleting s from B yields a board $B - s$, and deleting the entire row and column containing s yields a board we write as $B \cdot s$. Counting the placements of k rooks by whether they use s yields $r_k(B) = r_k(B - s) + r_{k-1}(B \cdot s)$. Note that $r_0(B) = 1$ for every board and $r_k(\varnothing) = 0$ for $k > 0$. Multiplying by x^k and summing where the recurrence is valid $(k \geq 1)$ yields $R_B(x) - 1 = R_{B-s}(x) - 1 + xR_{B \cdot s}(x)$. Thus

$$R_B(x) = R_{B-s}(x) + xR_{B \cdot s}(x).$$

An extensive classical treatment of rook polynomials appears in Riordan [1958]. See also Exercises 50–54. ∎

How many permutations have exactly k elements in forbidden positions? For inclusion-exclusion in general, we develop a generating function to enumerate objects by "number of properties". Having property i is the same as being in set A_i in the formulation of PIE using subsets.

4.1.18. Theorem. For sets $A_1, \ldots, A_n$ in a universe U, let h_k be the sum of $\left|\bigcap_{i \in S} A_i\right|$ over all sets S of size k. If f_p is the number of elements of U in exactly p of the sets $A_1, \ldots, A_n$, then

$$\sum_{p=0}^{n} f_p x^p = \sum_{k=0}^{n} (x-1)^k h_k.$$

Proof: As in Theorem 4.1.3, we show that each element contributes the same to both sides. For $a \in U$, the contribution of a to the left side is x^p, where $p = |\{i \in [n]: a \in A_i\}|$. Since the number of k-subsets of $\{A_1, \ldots, A_n\}$ whose intersection contains a is $\binom{p}{k}$, element a contributes $\binom{p}{k}$ times to h_k. Hence a contributes $\sum_{k=0}^{n}(x-1)^k\binom{p}{k}$ to the right side. Since $\binom{p}{k} = 0$ for $k > p$, the sum is x^p by the Binomial Theorem. ∎

4.1.19.* Remark. Using the Binomial Theorem to expand $(x-1)^k$ in Theorem 4.1.18 yields $\sum_{k=0}^{n}\left(\sum_{p=0}^{k}\binom{k}{p}(-1)^{k-p}x^p h_k\right)$ on the right. Interchanging the order of summation and extracting the coefficient of x^p yields $f_p = \sum_{k=p}^{n}(-1)^{k-p}\binom{k}{p}h_k$. Some authors express Theorem 4.1.18 in this form.

Setting $x = 0$ reduces to the usual inclusion-exclusion formula $f(\varnothing) = \sum_{k=0}^{n}(-1)^k h_k$. The general formula of Theorem 4.1.18 can also be proved by summing the inclusion-exclusion formulas for each $f(T)$ (Exercise 25).

To apply Theorem 4.1.18 to permutations with restricted positions given by a board B, recall that $h_k = r_k(B)(n-k)!$. With f_p being the number of permutations having p elements in forbidden positions,

$$\sum_{p=0}^{n} f_p x^p = \sum_{k=0}^{n} (x-1)^k r_k(B)(n-k)!. \qquad \blacksquare$$

4.1.20. Example. *Expected number of fixed points* (Montmort [1708]). Consider an urn with balls labeled 1 through n. We extract the balls one by one, randomly. What is the expected number of times that the label on a ball equals the step on which it is drawn? These occurrences were called "rencontres", and this question was the "problème des rencontres".

In the language of forbidden positions, we are seeking the expected number of fixed points in a random permutation of $[n]$. Let $N(x)$ be the generating function for permutations by the number of positions with forbidden elements: $N(x) = \sum f_p x^p$. As in Application 3.2.15, we find the expected weight of a random object in a finite universe U by differentiating the weight enumerator: the expected weight is $N'(1)/|U|$.

To enumerate permutations by fixed points, we use the diagonal board of forbidden positions used to count derangements. We obtain

$$N(x) = \sum_{k=0}^{n} (x-1)^k \binom{n}{k}(n-k)! = n! \sum_{k=0}^{n} \frac{(x-1)^k}{k!}.$$

The answer is $N'(1)/n!$, which equals 1, a remarkably simple answer.

When a fancy method yields a simple answer, we seek a simpler explanation (we called this the "Erdős Principle"). Here, the number X of fixed points is $\sum X_i$, where X_i is 1 if i is a fixed point and 0 otherwise. By the linearity of expectation, $\mathbb{E}(X) = \sum_{i=1}^{n} \mathbb{P}(X_i = 1)$ (see Chapter 14 for further explanation and use of this concept). Since each element has probability $1/n$ of being fixed, $\mathbb{E}(X) = 1$. $\blacksquare$

We consider another classical example.

4.1.21. Example. *Problème des ménages* (Lucas [1891]). The hosts of a party with n male/female couples want to seat men and women alternately at a circular table with no spouses adjacent. Index the couples $1, \dots, n$ by the circular ordering of sex A. Let position i be the position between A_i and A_{i+1}. Neither B_i nor B_{i+1} is allowed to sit there. Acceptable seatings of $B_1, \dots, B_n$ now correspond to permutations of $[n]$ not having i or $i+1$ in position i for any i (modulo n).

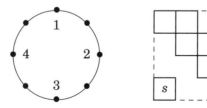

To count the allowed seatings, compute $r_k(B)$ with this board of forbidden positions. Applying the recurrence to the square s in the lower left yields two zig-zag boards; we must place k rooks on a board with $2n - 1$ positions or $k - 1$ rooks on a board with $2n - 3$ positions.

Instead of continuing the recurrence, note that l non-attacking rooks on a zig-zag board with m positions correspond to a $0, 1$-string of length m with l copies of 1, no two consecutive. These strings arise from an all-0 string of length $m - l$ by choosing l distinct places to insert a 1. There are $\binom{m-l+1}{l}$ such choices.

With (m, l) set to $(2n - 1, k)$ or $(2n - 3, k - 1)$, the computation yields $r_k(B) = \binom{2n-k}{k} + \binom{2n-k-1}{k-1} = \frac{2n}{2n-k}\binom{2n-k}{k}$. Having seated one sex, the enumerator of arrangements by number of consecutively-seated couples is

$$\sum_{k=0}^{n} \frac{2n}{2n-k}\binom{2n-k}{k}(n-k)!(x-1)^k.$$

Setting $x = 0$ counts the arrangements with no couples consecutive.   ∎

SIGNED INVOLUTIONS

The basic Inclusion-Exclusion Principle has many generalizations. It generalizes to the classical Möbius Inversion Formula of number theory and further to the setting of partially ordered sets (see Section 15.2).

Here we discuss a simple but flexible generalization. Inclusion-exclusion counts a set within a larger universe X by canceling unwanted elements. If we weight the elements ± 1 so that unwanted elements come in plus/minus pairs, and elements of the desired set F have positive weight, then $|F|$ will be the sum of the weights over X. We consider a more general setting.

4.1.22. Definition. An **involution** is a permutation whose square is the identity; the cycles have length 1 or 2. With respect to a partition of X into a *positive part* X^+ and a *negative part* X^-, a **signed involution** on X is an involution τ such that every 2-cycle pairs a positive element with a negative element. Let F_τ and G_τ denote the sets of fixed points within X^+ and X^- under a signed involution τ.

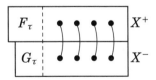

4.1.23. Example. *Parity of permutations.* A permutation σ is **even** [**odd**] when the number of inversions in σ is even [odd] (see Proposition 3.1.15). A more common definition is the parity of the number of transpositions needed to turn σ into the identity permutation; these are the same because every exchange of two elements in the word form changes the number of inversions by an odd amount.

even permutations:	123 231 312	
odd permutations:	213 321 132	

Interchanging the first two elements in the word form changes the parity. Repeating the transposition changes it back. Thus we have a signed involution with $X = \mathbb{S}_n$, where X^+ and X^- are the sets of even and odd permutations, respectively. There are no fixed points unless $n = 1$.   ∎

4.1.24. Proposition. If τ is a signed involution on a set X weighted by $w(x) = 1$ for $x \in X^+$ and $w(x) = -1$ for $x \in X^-$, then

$$\sum_{x \in X} w(x) = |F_\tau| - |G_\tau|.$$

Proof: The contributions to the sum from 2-cycles in τ cancel, leaving those from fixed points: $|F_\tau|$ with weight $+1$, $|G_\tau|$ with weight -1.   ∎

The inclusion-exclusion formula for $f(\varnothing)$ is a special case of Proposition 4.1.24. We embed U in a still larger set X and show that the alternating sum in the formula for $f(\varnothing)$ is the sum of the weights for a signed involution on X having $f(\varnothing)$ fixed points, all in X^+.

4.1.25. Example. *PIE by signed involution.* We have a universe U and sets $A_1, \ldots, A_n$. We seek $f(\varnothing)$. Recall the "property set": $P(x) = \{i \in [n]: x \in A_i\}$. In the inclusion-exclusion formula, each element x of U appears in $\bigcap_{i \in S} A_i$ for each S such that $S \subseteq P(x)$. Grouping the contributions to reflect this yields

$$f(\varnothing) = \sum_{S \subseteq [n]} (-1)^{|S|} \left| \bigcap_{i \in S} A_i \right| = \sum_{x \in U} \sum_{S \subseteq P(x)} (-1)^{|S|} = \sum_{(x,S) \in X} (-1)^{|S|},$$

where X is the set of ordered pairs (x, S) such that $x \in U$ and $S \subseteq P(x)$. The contribution is $+1$ for even $|S|$ and -1 for odd $|S|$. Partition X into X^+ and X^- by the parity of $|S|$. To apply Proposition 4.1.24, we need a signed involution with respect to (X^+, X^-) whose fixed points lie in X^+ and correspond to the elements of U outside all the sets (those with $P(x) = \varnothing$). Proposition 4.1.24 then proves the inclusion-exclusion formula.

Given a pair $(x, S) \in X$, view the elements of S as *marked* elements of $P(x)$. Let τ operate on elements of X by switching whether the largest element of $P(x)$ is marked, if any element of $P(x)$ exists. That is,

$$\tau(x, S) = \begin{cases} (x, S) & \text{if } P(x) = \varnothing, \\ (x, S - i) & \text{if } i = \max P(x) \in S, \\ (x, S \cup i) & \text{if } i = \max P(x) \notin S. \end{cases}$$

Applying τ again acts on the same element of $P(x)$ and returns us to (x, S). Hence τ is an involution. It is a signed involution relative to (X^+, X^-), because τ changes the parity of $|S|$ when it makes a change. The fixed points are $\{(x, \varnothing): P(x) = \varnothing\}$. These lie in X^+ and correspond bijectively to the elements of U outside all of $A_1, \ldots, A_n$.   ∎

Many applications of signed involution follow this pattern. We embed the desired set in a larger set X. For elements in X, we define a "switch" operation such that applying it twice retrieves the original element. The desired fixed points have no candidates for switching, so the map does not change them (see Exercise 57, for example).

By taking advantage of cancellation, signed involutions can make it possible to find short proofs that avoid needless detailed counting.

4.1.26. Proposition. (Deutsch [2005]) For a fixed convex n-gon, if e_n and o_n count the dissections by nonintersecting diagonals into an even or an odd number of bounded regions, then $e_n - o_n = (-1)^n$.

Proof: Let the parity of a dissection be the parity of the number of regions. If a signed involution τ pairs dissections with opposite parity, then $e_n - o_n$ is the number of even fixed points minus the number of odd fixed points. Adding or deleting one diagonal changes the parity; we want τ to do this and repeating τ to undo it by changing the same diagonal.

Let $v_1, \ldots, v_n$ be the vertices in order around the n-gon. If the diagonals incident to v_n are not all present, then let k be the index before the first non-neighbor of v_n (if any exists), and let l be the index of the next actual neighbor of v_n. (For example, when no diagonals are present, there is only one region and $(k, l) = (1, n - 1)$.) Define τ by inserting $v_k v_l$ if it is missing and removing $v_k v_l$ if it is present. When τ inserts an edge, no existing edge crosses it, so the image is indeed a dissection.

The dissection is unchanged by τ if and only if k and l are undefined, meaning that v_n is adjacent to all the other corners and there are $n - 2$ regions. When τ produces a change, the parity changes. Because τ does not change the set of edges incident to v_n, the values k and l for a dissection are the same as for its image under τ. Thus repeating τ changes the presence of the same diagonal, restoring the original dissection.

We have shown that τ pairs even and odd dissections except for the one fixed point, whose parity is the same as the parity of n. ∎

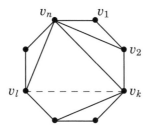

Exercises 56–67 concern signed involutions.

DETERMINANTS AND PATH SYSTEMS (optional)

Many counting problems about lattice paths are solved by determinants. Gessel–Viennot [1985] explained this phenomenon by using signed involutions. Lattice paths can be viewed as paths in a directed graph. Every determinant can be interpreted as a signed counting problem in a weighted bipartite digraph. The common generalization is a theorem by Lindström [1973] that includes the results on lattice paths. Aigner [2001] gave an elegant exposition of the topic with many striking applications. We begin with a simple example as motivation.

4.1.27. Example. We know how to count lattice paths using binomial coefficients. Consider disjoint pairs of lattice paths with specified endpoints. For example, with initial points defined by $x_1 = (0, 1)$ and $x_2 = (1, 0)$, and terminal points defined by $y_1 = (1, 2)$ and $y_2 = (3, 2)$, there are eight such pairs, shown below.

An x, y-path is a path from x to y. By computing the familiar binomial coefficients, the numbers of x_i, y_j-paths, arranged in a matrix for $i, j \in \{1, 2\}$, are $\left(\begin{smallmatrix} 2 & 4 \\ 1 & 6 \end{smallmatrix}\right)$. As it happens, the determinant of this matrix also equals 8. We will see that this is no coincidence.   ■

We defined the relevant concepts for directed graphs in the Introduction. Recall that each edge in a **directed graph** (**digraph**) G with vertex set $V(G)$ and edge set $E(G)$ is an ordered pair (u, v) of vertices, abbreviated to uv. An edge is *directed* from its first vertex (the **tail**) to its second vertex (the **head**). An x, y-**path** consists of vertices that can be ordered as $v_0, \ldots, v_k$ with $v_0 = x$ and $v_k = y$ and edges $\{v_i v_{i+1} \colon 0 \le i \le k - 1\}$.

For example, we used *functional digraphs* in Chapter 1; this is the special case where each vertex is the tail of exactly one edge. A path in the functional digraph of f follows repeated iteration of f.

4.1.28. Example. Lattice paths can be modeled as paths in a digraph with grid points as the vertices and allowed transitions as the edges. The ballot paths of length 6 are the x, y-paths in the digraph on the left below.

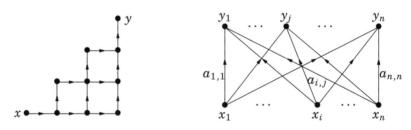

We introduce a weight $w(e)$ on each edge e. In the digraph G on the right above, the edges are $\{x_i y_j \colon i, j \in [n]\}$. Let $w(x_i y_j) = a_{i,j}$, where the numbers $a_{i,j}$ are the entries in an n-by-n matrix A. Using sets of paths in G, we will interpret the computation of $\det A$ combinatorially.   ■

4.1.29. Definition. In a weighted digraph G, the **weight** of a subgraph is the product of the weights of its edges. Given vertex subsets $X, Y \subseteq V(G)$, with $X = \{x_1, \ldots, x_n\}$ and $Y = \{y_1, \ldots, y_n\}$, the X, Y-**path matrix** has as its (i, j)-entry the sum of the weights of all paths from x_i to y_j. An X, Y-**path system** $\mathcal{P}$ consists of a permutation $\sigma_\mathcal{P}$ of $[n]$ and paths $P_1, \ldots, P_n$ such that P_i is a path from x_i to $y_{\sigma_\mathcal{P}(i)}$. It is a **disjoint-path system** if the paths are pairwise disjoint.

4.1.30. Example. *Determinants and path systems.* For the weighted digraph on the right in Example 4.1.28, the path matrix is the matrix A of order n that provided the weights; there is exactly one x_i, y_j-path for each pair (i, j), and its weight is $a_{i,j}$. All the path systems are disjoint-path systems, and there is one such system $\mathcal{P}_\sigma$ for each permutation σ.

In terms of permutations, $\det A = \sum_{\sigma \in \mathbb{S}_n}(\text{sign } \sigma) \prod_{i=1}^n a_{i,\sigma(i)}$, where the sign of a permutation σ is $+1$ or -1 depending on whether the number of inversions is even or odd (see Example 4.1.23). Since $w(\mathcal{P}_\sigma)$ in this example is precisely $\prod_{i=1}^n a_{i,\sigma(i)}$, we obtain $\sum_{\sigma \in \mathbb{S}_n}(\text{sign } \sigma)w(\mathcal{P}_\sigma) = \det A$.

Our aim is to extend this formula to all disjoint-path systems in all digraphs. Aigner wrote, "It is precisely this step to arbitrary graphs that makes the [Theorem] so widely applicable — and what's more, the proof is stupendously simple and elegant." ∎

To keep the sums finite, we consider only finite digraphs without cycles.

4.1.31. Theorem. (Lindström [1973], Gessel–Viennot [1985]) Let X and Y be n-sets of vertices in a finite acyclic digraph G with edges weighted by w. If A is the X, Y-path matrix, and $\mathbf{P}$ is the set of disjoint-X, Y-path systems, then

$$\sum_{\mathcal{P} \in \mathbf{P}}(\text{sign } \sigma_\mathcal{P})w(\mathcal{P}) = \det A$$

Proof: Let $\mathbf{Q}$ be the set of all X, Y-path systems, disjoint or not. We seek a signed involution on $\mathbf{Q}$ whose fixed points are the disjoint-path systems and whose systems paired in 2-cycles have the same weight and have associated permutations with opposite sign. This is a signed involution with X^+ and X^- being the sets of X, Y-path systems whose associated permutations are even or odd, respectively. The paired contributions cancel in $\sum_{\mathcal{P} \in \mathbf{Q}}(\text{sign } \sigma_\mathcal{P})w(\mathcal{P})$, leaving just $\sum_{\mathcal{P} \in \mathbf{P}}(\text{sign } \sigma_\mathcal{P})w(\mathcal{P})$. The final step is to interpret the sum over all of $\mathbf{Q}$ as a determinant, proving

$$\sum_{\mathcal{P} \in \mathbf{P}}(\text{sign } \sigma_\mathcal{P})w(\mathcal{P}) = \sum_{\mathcal{P} \in \mathbf{Q}}(\text{sign } \sigma_\mathcal{P})w(\mathcal{P}) = \det A.$$

Let P_i denote the x_i, $y_{\sigma(i)}$-path in a path system $\mathcal{P}$ with permutation σ. Let i^* be the least i such that P_i intersects another path in $\mathcal{P}$ (if such an index exists). Let z be the first vertex of P_{i^*} shared with another path, and let j^* be the smallest index other than i^* such that P_{j^*} also contains z. Let $\tau(\mathcal{P})$ be the path system obtained from $\mathcal{P}$ by switching the portions of P_{i^*} and P_{j^*} after z. Let σ' be the permutation associated with $\tau(\mathcal{P})$.

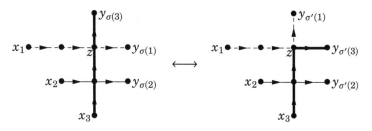

The switch transposes $\sigma(i^*)$ and $\sigma(j^*)$ in the word form of σ to obtain σ'. Transpositions reverse sign, so $\operatorname{sign} \sigma_{\tau(\mathcal{P})} = -\operatorname{sign} \sigma_{\mathcal{P}}$. Also $\tau(\tau(\mathcal{P})) = \mathcal{P}$, since paths remain unchanged before intersections, so τ selects the same intersection point and pair of paths in $\tau(\mathcal{P})$ and undoes the switch.

For a disjoint-path system, no intersection is found and τ makes no change. Always $w(\mathcal{P}) = w(\tau(\mathcal{P}))$, since both systems use the same edges with the same multiplicities. Thus τ is a signed involution having the desired properties, so $\sum_{\mathcal{P} \in \mathbf{Q}}(\operatorname{sign} \sigma_{\mathcal{P}})w(\mathcal{P}) = \sum_{\mathcal{P} \in \mathbf{P}}(\operatorname{sign} \sigma_{\mathcal{P}})w(\mathcal{P})$.

For the remaining step, recall that $\det A = \sum_\sigma(\operatorname{sign} \sigma)a_{1,\sigma(1)} \cdots a_{n,\sigma(n)}$, where $a_{i,j}$ is the sum of the weights of all x_i, y_j-paths. Each path system associated with σ chooses one x_i, $y_{\sigma(i)}$-path for each i, and its weight is the product of their weights. Hence $\prod_{i=1}^n a_{i,\sigma(i)}$ is the sum of the weights of all the path systems associated with σ. Summing over σ yields $\sum_{\mathcal{P} \in \mathbf{Q}}(\operatorname{sign} \sigma_{\mathcal{P}})w(\mathcal{P})$, as desired. ∎

Aigner noted that basic properties of determinants follow easily from Theorem 4.1.31. For example, $\det A^T = \det A$ follows by reversing the edges in the digraph of Example 4.1.28. The path system associated with σ is associated with σ^{-1} instead, but they have the same sign, so the computation of the determinant is the same. Exercise 62 uses Theorem 4.1.31 to prove that a matrix with linearly dependent rows has determinant 0.

Another application generalizes the product formula for determinants to products of non-square matrices. We will apply this result in Chapter 15. This proof is very easy; others require tedious manipulation or clever linear algebra.

4.1.32. Proposition. (**Cauchy–Binet Formula**) If A and B are n-by-p and p-by-n matrices, respectively, with $n \leq p$, then

$$\det(AB) = \sum_{S \subseteq \mathbf{S}}(\det A_S)(\det B_S),$$

where $\mathbf{S}$ is the family of all n-subsets of $[p]$, A_S consists of the columns of A indexed by S, and B_S consists of the rows of B indexed by S.

Proof: (Aigner [2001]) Form a digraph with vertex set $X \cup Y \cup Z$, where $X = \{x_1, \ldots, x_n\}$, $Y = \{y_1, \ldots, y_n\}$, and $Z = \{z_1, \ldots, z_p\}$. Let the edge set be $(X \times Z) \cup (Z \times Y)$, with $w(x_i z_j) = a_{i,j}$ and $w(z_j y_k) = b_{j,k}$. Entry (i, k) in AB is $\sum_j w(x_i z_j)w(z_j y_k)$, so AB is the X, Y-path matrix. By Theorem 4.1.31, $\det(AB)$ sums the weights of nonintersecting X, Y-path systems.

Such a path system $\mathcal{P}$ selects for each i a vertex $z_{\sigma'(i)}$ on a path from x_i to $y_{\sigma(i)}$; let S be the set of indices selected. Since $w(\mathcal{P})$ is the product of all edge weights, $w(\mathcal{P})$ is the product of weights of path systems in the digraph computations for $\det A_S$ and $\det B_S$. Also, $\operatorname{sign} \sigma$ is the product of the signs of the permutations of $[n]$ for those path systems. Therefore, the right side gives the sum of weights over all nonintersecting X, Y-path systems, grouped by the choice of S. ∎

When all disjoint-X, Y-path systems have the same associated permutation σ, a determinant counts them. For this discussion, we return to the context of lattice paths from Example 4.1.27 and name the initial points $p_1, \ldots, p_n$ and terminal points $q_1, \ldots, q_n$. The digraphs for lattice paths have edges oriented upward and rightward as in Example 4.1.28.

4.1.33. Corollary. (Gessel–Viennot [1985]) Let $h((a,b),(c,d)) = \binom{d-b+c-a}{c-a}$. If every lattice disjoint-path system from $\{p_1, \ldots, p_n\}$ to $\{q_1, \ldots, q_n\}$ matches p_i to q_i for all i, then the number of such disjoint-path systems is the determinant of the matrix with (i,j)-entry $h(p_i, q_j)$.

Proof: Form the digraph on $\mathbb{Z}^2$ with edges for upward and rightward steps. The number of paths from point p to point q is $h(p,q)$. By Theorem 4.1.31, the specified determinant sums the signed weights of all disjoint-path systems from the initial points to the terminal points. Under the given conditions, the only such systems use the identity permutation, which has positive sign. The weight of every path system is 1, since all the edge weights are 1, so the sum of the weights is the number of disjoint-path systems. ∎

4.1.34. Example. *Example 4.1.27, continued.* The hypothesis of Corollary 4.1.33 holds when $(p_1, \ldots, p_n)$ and $(q_1, \ldots, q_n)$ are nondecreasing in first coordinates and nonincreasing in second coordinates. This property holds in Example 4.1.27. By Corollary 4.1.33, the number of disjoint paths of paths from p_1 and p_2 to q_1 and q_2 is $\det \left| \begin{smallmatrix} 2 & 4 \\ 1 & 6 \end{smallmatrix} \right|$, which is 8.

We can also enumerate the disjoint-path systems in more detail by generating functions. When each edge on the line $y = s$ has weight x_s, for each s, the coefficient of $\prod x_s^{e_s}$ counts the disjoint-X, Y-path systems having exactly e_s horizontal steps on $y = s$. This allows us to enhance the definition of $h(p,q)$ via

$$h((a,b),(c,d)) = \sum_{b \le k_1 \le \cdots \le k_{c-a} \le d} \prod x_{k_i}.$$

Setting all $x_i = 1$ yields the previous value $\binom{d-b+c-a}{c-a}$ that counts each path once. In Example 4.1.27, the resulting generating function is

$$x_1^2 x_2 + x_1 x_2^2 + x_0 x_2^2 + 2x_0 x_1 x_2 + x_0 x_1^2 + x_0^2 x_2 + x_0^2 x_1. \qquad \blacksquare$$

Exercise 61 applies Corollary 4.1.33 to count paths *avoiding* a specified set of lattice points. Other applications of Theorem 4.1.31 appear in Lindström [1973] and Gessel–Viennot [1989].

As a striking application of the method, we present Aigner's proof of MacMahon's formula for the number of rhombic tilings of a regular hexagon of side-length n. Cut the hexagon into $6n^2$ equilateral triangles; each rhombus consists of two triangles with side-length 1 sharing an edge. Below we show the triangles in a hexagon of length 3 and one of its rhombic tilings.

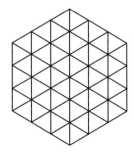
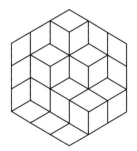

4.1.35. Theorem. (MacMahon [1916]) The number of rhombic tilings of a hexagon with side-length n is the determinant of the matrix with $\binom{2n}{n+i-j}$ in position (i, j).

Proof: (Aigner [2001]) Let L and R be the lower left and upper right edges of the diagram. We interpret rhombic tilings as disjoint-path systems joining points on L and R. Rhombi in a tiling have three possible orientations. Shade the rhombi whose short diagonal is parallel to L and R. Each unshaded rhombus has two edges parallel to L and R. The segment joining the centers of those edges can be viewed as a step in a path from L to R. These paths naturally partition the unshaded rhombi; the paths can be seen as climbing 3-dimensional piles of boxes fixed in one coordinate.

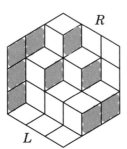 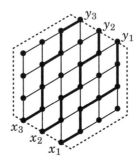

These edges belong to the graph on the right above; orient them from lower left to upper right to form a digraph G. The edge crossing an unshaded rhombus continues a path that enters it at the lower left, so the chosen edges form paths from L to R. Each rhombus has only one entrance and one exit, so the paths are disjoint. Thus the unshaded rhombi yield a disjoint-path system from $x_1, \ldots, x_n$ to $y_1, \ldots, y_n$ in G.

The map is a bijection; each edge in a disjoint-path system bisects one rhombus, which is the unshaded rhombus that generates it in the corresponding tiling. (The path system has n^2 steps up and n^2 steps rightward, so every tiling has n^2 rhombi with each orientation.)

Hence it suffices to count the disjoint-path systems in G. They all consist of paths from x_i to y_i for $i \in [n]$ and have positive sign. By Theorem 4.1.31, the number of disjoint-path systems is the determinant of the matrix whose (i, j)-entry is the number of paths from x_i to y_j. This is $\binom{2n}{n+i-j}$, the ordinary count of lattice paths from $(0, 0)$ to $(n + i - j, n + j - i)$.   ∎

4.1.36. Remark. *Evaluation of combinatorial determinants.* The reader may wonder whether the determinant in Theorem 4.1.35 evaluates nicely. In fact, it does: if the (i, j)-entry of A is $\binom{2n}{n+i-j}$, then $\det A = \prod_{i=0}^{n-1} \frac{(2n+i)_{(n)}}{(n+i)_{(n)}}$, where as usual the parenthesized subscript indicates the falling factorial. When $n = 3$, the number of tilings evaluates to 980.

Many methods for evaluating determinants arise combinatorially; Krattenthaler [1999, 2005] is a thorough two-part survey. Most of these methods are beyond the scope of this text, but a simple method called the **Condensation Method** works for some determinants including those in this application.

The method is often attributed to Dodgson [1866] (the clergyman known as Lewis Carroll), but it may have been known earlier. The identity relates several determinants, computing $\det A$ in terms of determinants of submatrices. For a square matrix A of order n, with $I, J \subseteq [n]$, let $\hat{A}_{I:J}$ denote the matrix obtained by deleting from A the rows indexed by I and the columns indexed by J. The identity is

$$\det A \det \hat{A}_{1,n:1,n} = \det \hat{A}_{n:n} \det \hat{A}_{1:1} - \det \hat{A}_{n:1} \det \hat{A}_{1:n}.$$

This yields inductive proofs of some determinant formulas. For $n, a, b \in \mathbb{N}$, let $M_n(a, b)$ denote the n-by-n matrix whose (i, j)-entry is $\binom{a+b}{a-i+j}$. The result is $\det M_n(a, b) = \prod_{j=1}^{a} \prod_{k=1}^{b} \frac{n+j+k-1}{j+k-1}$ (see Exercise 69). ∎

Further applications of Theorem 4.1.31 appear in Exercises 61–65.

EXERCISES 4.1

4.1.1. (−) How many ways are there to place 10 distinct people within three distinct rooms? How many ways are there to place 10 distinct people within three distinct rooms so that every room receives at least one person? How many decimal n-tuples contain at least one each of $\{1, 2, 3\}$ (leading 0s allowed)?

4.1.2. (−) Two distinct dice are rolled (together) n successive times. What is the probability that each roll of doubles (two 1s, ..., two 6s) appears?

4.1.3. (−) 13 cards are chosen at random from a standard deck of 52 cards. Compute the probability of each event below.
   (a) At least one card in each suit.
   (b) No cards in some suit.
   (c) Some four cards with the same value.

4.1.4. (−) Given the special case $f(\emptyset) = \sum_{S \subseteq [n]} (-1)^{|S|} g(S)$ of Theorem 4.1.3, prove the general formula for $f(T)$ by reducing it to the special case for $f(\emptyset)$.

4.1.5. (−) How many natural numbers in $[252]$ are relatively prime to 252?

4.1.6. (−) How many natural numbers in $[200]$ have no divisor in $\{6, 10, 15\}$?

4.1.7. (−) Given pennies, nickels, dimes, quarters, and half-dollars, how many ways can one select n coins so that no type of coin is selected more than four times?

4.1.8. (−) How many permutations of $[n]$ leave no odd number fixed?

4.1.9. (−) Use inclusion-exclusion to prove that $\sum (-1)^k \binom{n}{k}(n-k)^n = n!$.

4.1.10. (−) A 6-sided die is rolled repeatedly until the numbers 1 through 5 have appeared at least once each. What is the probability that this happens in the first n rolls? Use a calculator to find the least n where the probability exceeds .5.

4.1.11. (−) Compute the rook polynomial of the board below.

4.1.12. (−) Compute the rook polynomials for all boards with at most four squares.

4.1.13. (−) Using Theorem 4.1.31 and the example of lattice paths with initial points $\{(1,0),(0,1)\}$ and terminal points $\{(2,2),(3,3)\}$, explain the role of the assumption in Corollary 4.1.33 that disjoint-path systems occur only with the identity permutation. In general, what happens when that condition fails?

4.1.14. Given fixed positive integers $s_1,\ldots,s_n$, let $a_{n,k}$ be the number of integer solutions to $z_1+\cdots+z_n=k$ such that $0\le z_i<r_i$ for each i.
    (a) Obtain the generating function $\sum a_{n,k}x^k$ (as a ratio of polynomials).
    (b) Without using (a), compute $a_{n,k}$ as a finite sum.
    (c) Explain why the answers to (a) and (b) are equivalent.

4.1.15. Prove that $\sum_{k\mid n}\varphi(k)=n$, where φ is the Euler totient function. (Gauss)

4.1.16. (◇) Let φ denote the Euler totient function.
    (a) For $m\in\mathbb{N}$, prove that $\varphi(m)=m\prod_{p\in P(m)}(1-1/p)$, where $P(m)$ is the set of distinct prime factors of m.
    (b) For $n\in\mathbb{N}$ with $n>\min\{m/2,m-\varphi(m)\}$, prove that $\sum_{i=1}^n(-1)^i\binom{n}{i}i^m$ is divisible by m. (Hint: What does the sum count?) (Popescu [1999])

4.1.17. Let M be a multiset consisting of two each of n types of letters. When all arrangements of M into a row are equally likely, what is the probability that no two consecutive letters are the same?

4.1.18. Given n boys and n girls, use inclusion-exclusion to obtain sums for the number of ways to pair the $2n$ people as lab partners under the following criteria. (No simple closed formulas are available.)
    (a) For each i, the ith tallest boy is not matched to the ith tallest girl.
    (b) Same condition as (a) and also each pair has one person of each sex.

4.1.19. (◇) A mathematics department has n professors and $2n$ courses, each professor teaching two each semester. How many ways are there to assign the courses in the fall semester? How many ways are there to assign them in the spring so that no professor teaches the same two courses in the spring as in the fall?

4.1.20. (◇) At a circular table are n students taking an exam. The exam has four versions. Given that no two neighboring students have the same version and $n>1$, how many ways are there to assign the exams? Do not leave the answer as a sum.

4.1.21. (◇) Let a_n count the permutations of $[n]$ with no cycles of length 2.
    (a) Use PIE to obtain a formula for a_n (a summation).
    (b) Use the Exponential Formula to obtain the same answer as in part (a).
    (c) What is the asymptotic probability that a permutation has no 2-cycles?

4.1.22. (◇) Count the distinguishable ways to seat the people in n married couples at a rotating round table with $2n$ seats so that no person sits next to his or her spouse. The sexes need not alternate. (Leave the answer as a summation.)

4.1.23. (◇) Let $S=\{a_1,\ldots,a_k\}\subseteq[n-1]$, with $a_1\le\cdots\le a_k$. Count the permutations of $[n]$ whose descent set is contained in S. Use the resulting formula to count the permutations of $[n]$ whose descent set equals S.

4.1.24. Prove Theorem 4.1.3 by substituting $\left|\bigcap_{i\in S}A_i\right|=\sum_{Q\supseteq S}f(Q)$ into the formula and simplifying. Explain how this proof relates to that in the text.

4.1.25. Prove Theorem 4.1.18 from Theorem 4.1.3 by algebraic computation.

4.1.26. Determine the number of graphs with vertex set $[n]$ that have exactly m edges and exactly k isolated vertices.

4.1.27. A number $n \in \mathbb{N}$ is a *power* if it equals a^m for some $a, m \in \mathbb{N}$ with $m \geq 2$. For $S \subseteq \mathbb{N}$, let $l(S)$ be the least common multiple of the elements of S. For $x \geq 4$, prove that the number of powers in the interval $[1, x]$ is $\sum_S (-1)^{|S|+1} \lfloor x^{1/l(S)} \rfloor$, where the sum is over all nonempty subsets of the integer interval $[2, \lfloor \log_2 x \rfloor]$. (Nyblom [2004])

4.1.28. ($\diamond$) Let t_n be the number of triangles in the drawing of the equilateral triangular grid with side length n, shown below for $n = 3$. Note that $t_1 = 1$, $t_2 = 5$, and $t_3 = 13$. Use PIE to derive a third-order recurrence for $\langle t \rangle$. Solve it by the characteristic equation method. (Hint: Express the inhomogeneous term using exponentials.)

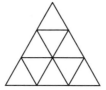

4.1.29. ($\diamond$) Given $m, n, r \in \mathbb{N}_0$, evaluate each sum below using generating functions, and give inclusion-exclusion proofs of the resulting identities. In part (c), $m \geq n$.

(a) $\displaystyle\sum_{k=0}^{n} (-1)^k \binom{n}{k}(n-k)$

(c) $\displaystyle\sum_{k=0}^{n} (-1)^k \binom{n}{k}\binom{m-k}{r}$

(b) $\displaystyle\sum_{k=0}^{m} (-1)^k \binom{n}{k}\binom{n-k}{m-k}$

(d) $\displaystyle\sum_{k=0}^{n} (-1)^k \binom{n}{k}\binom{r+n-k-1}{r}$

4.1.30. For $n \geq t$, apply Theorem 4.1.3 directly to prove the identity below. (Hint: Consider the universe of n-sets in $[2n]$, and do not shift the index of summation. Comment: Shifting the index and additional substitutions can turn this identity into part (c) of Exercise 4.1.29; instead we seek a direct proof.)

$$\sum_{r=t}^{n} (-1)^{r-t} \binom{n-t}{r-t}\binom{2n-r}{n-r} = \binom{n}{n-t}$$

4.1.31. ($\diamond$) For $n \in \mathbb{N}$, use inclusion-exclusion to give a combinatorial proof of the identity below. (Beckwith [2006a])

$$\sum_{k=0}^{n} (-1)^k \binom{n}{k}\binom{2n-2k}{n-1} = 0$$

4.1.32. ($\diamond$) For $n, k \in \mathbb{N}$, evaluate the sum below, both by using generating functions and by using inclusion-exclusion. (Cox–Thieu [2015])

$$\sum_{i=0}^{k} (-1)^i \binom{k}{i}\binom{kn-in}{k+1}$$

4.1.33. Evaluate the sum below. Use inclusion-exclusion to prove the resulting identity.

$$\sum_{k=0}^{r-1} (-1)^k \binom{r-1}{k}\binom{n-k}{r-k}$$

4.1.34. Let b, r, s, t be nonnegative integers such that $r + s + t = rb > 0$. Prove the identity below using constrained compositions of integers. (Wardlaw [1989])

$$\sum_{k=0}^{\lfloor s/b \rfloor}(-1)^k\binom{r}{k}\binom{s+r-1-bk}{s-bk}=\sum_{k=0}^{\lfloor t/b \rfloor}(-1)^k\binom{r}{k}\binom{t+r-1-bk}{t-bk}$$

4.1.35. Let $A_1,\ldots,A_k$ be finite sets. For $J\subseteq[k]$, let $N_J=\left|\bigcup_{j\in J}A_j\right|$, and let $S_m=\sum_{|J|=m}N_J$. Let T_m be the number of elements that belong to exactly m of the sets $A_1,\ldots,A_k$. Prove $T_m=\sum_{i=1}^{k}(-1)^{k+i+m+1}\binom{i}{k-m}S_i$. (Sofair [2014])

4.1.36. Every tree with n vertices has $n-1$ edges, and the cycle with n vertices has n edges. Use the inclusion-exclusion formula to derive the chromatic polynomials of trees with n vertices and cycles with n vertices.

4.1.37. In a proper coloring of a graph G, the vertices receiving a particular color must form an **independent set** (a set of pairwise nonadjacent vertices). Hence another formula for the chromatic polynomial is $\sum_{r=1}^{n}k_{(r)}p_r(G)$, where $p_r(G)$ denotes the number of partitions of $V(G)$ into r independent sets.
(a) Let T be an n-vertex tree. Prove that $p_r(T)=S(n-1,r-1)$, where $S(m,k)$ is the Stirling number of the second kind.
(b) Compute the chromatic polynomial of T from the information above. (Hint: Use a result about Stirling numbers from Chapter 3.) (Voloshin [2002])

4.1.38. Prove that a graph G with n vertices and m edges has at most $\frac13\binom{m}{2}$ cycles with three edges. Conclude that the coefficient of k^{n-2} in $\pi_G(k)$ is positive unless $|E(G)|\le 1$.

4.1.39. ($\diamond$) Let a_n be the number of permutations of $[n]$ such that no number i is followed immediately by $i+1$. Use PIE to derive a formula for a_n, and from it conclude $a_n=D_n+D_{n-1}$, where D_n counts the derangements of $[n]$. Explain how the first computation changes when we also forbid n to be followed by 1. (Comment: $a_n=D_n+D_{n-1}$ was proved using recurrences in Exercise 2.1.44.) (Brualdi [2010, Section 6.5])

4.1.40. ($\diamond$) Let a_n be the number of permutations of $[n]$ such that position i does not contain element $i+1$, for $1\le i\le n-1$. Use Proposition 4.1.15 to prove $a_n=D_n+D_{n-1}$, where D_n counts the derangements of $[n]$. Use rook placements to give a direct proof. (Comment: Exercise 3.1.27 compares this set and that in Exercise 4.1.39, having the same size.)

4.1.41. ($\diamond$) Without using derangement numbers, use Theorem 4.1.18 to count the permutations of $[n]$ having an even number of fixed points. From the resulting formula, find the approximate probability (when n is large) that a random permutation of $[n]$ has an even number of fixed points. Is the number of fixed points more likely to be even or odd?

4.1.42. ($\diamond$) For $n\ge2$, let f_n be the polynomial defined below. Interpret $[x^r]f_n(x)$ combinatorially; that is, find a set that it counts. Use this to determine the degree of f_n. (Hint: Consider the combinatorial meaning of $\prod_{j=0}^{n-1}(x+j)$.) (Caro–Pohoata [2011])

$$f_n(x)=\sum_{i=0}^{n}\binom{n}{i}(-x)^{n-i}\prod_{j=0}^{i-1}(x+j)$$

4.1.43. ($\diamond$) Count the k-subsets of $[n-1]$ having exactly r runs of consecutive elements. Use this to obtain a double summation for the number of permutations of $[n]$ in which no two numbers differing by 1 are adjacent (in word form).

4.1.44. (+) Squares on an n-by-n chessboard alternate black and white. How many sets of k white squares and $n-k$ black squares have no two in a row or column? Interpret the result in terms of permutations. (Andrews–Wang [1988])

4.1.45. (+) Let $\eta(j)$ denote the number of positions having 1 in the binary expansion of j. Prove that $\sum_{j=0}^{2^n-1}(-1)^{n-\eta(j)}j^r$ is 0 when $r < n$ and is $n!2^{n(n-1)/2}$ when $r = n$. (Callan [1999])

4.1.46. (◊) Prove the identity below, where $s(m, k)$ is the signed Stirling number of the first kind (recall that $s(m, k) = (-1)^{n-k}c(n, k)$, where $c(n, k)$ is the number of permutations of $[m]$ having k cycles). (Kauers [2011])

$$\sum_{k=0}^{n}(-1)^k\binom{2n}{n+k}s(n+k, k) = \prod_{i=1}^{n}(2i-1)$$

4.1.47. (◊) *Stirling numbers and derangements.*
    (a) Use known formulas or EGFs to prove the identity below.
    (b) Prove it also by counting a set in two ways. (Leuck [1987])

$$\sum_{k=0}^{n}k^r\binom{n}{k}D_{n-k} = n!\sum_{m=0}^{n}S(r, m)$$

4.1.48. Prove the identities below.

    (a) $\displaystyle\sum_{j=0}^{n}(-1)^j\frac{1}{j!}\frac{(n-j)^n}{(n-j)!} = 1$ (b) $\displaystyle\sum_{j=0}^{n-1}(-1)^j\frac{1}{j!}\frac{(n-1-j)^n}{(n-1-j)!} = \binom{n}{2}$

4.1.49. (+) For $n \in \mathbb{N}$, prove $\sum_{k=1}^{n}\frac{(-1)^{k-1}}{k}\binom{n}{k}(n-k)^n = n^n(\frac{1}{2} + \frac{1}{3} + \cdots + \frac{1}{n})$. (Ungar [1984])

4.1.50. (◊) Let T be the triangular board with n rows of lengths $1, \ldots, n$. For the coefficients of the rook polynomial, prove both bijectively and recursively that $r_k(T) = S(n+1, n+1-k)$, where $S(m, j)$ is the number of partitions of $[m]$ into j blocks. (Hint: Encode the positions of T as $\{(i, j): 1 \le i < j \le n+1\}$.)

4.1.51. Let $R_{m,n}(x)$ be the rook polynomial for a full m-by-n board. Give combinatorial arguments (bijections) to prove
    (a) $R_{m,n}(x) = R_{m,n-1}(x) + mxR_{m-1,n-1}(x)$.
    (b) $\frac{d}{dx}R_{m,n}(x) = mnR_{m-1,n-1}(x)$.

4.1.52. (◊) Let B be a board whose squares form the Ferrers diagram of a partition λ with at most n parts (append 0s as needed). Prove $\sum_{k=0}^{n}r_k(B)x_{(n-k)} = \prod_{i=1}^{n}(x+\lambda_{n+1-i}-i+1)$. For example, with $\lambda = (3, 1)$ and $n = 3$, both sides equal $x^2(x+1)$. (Comment: The resulting polynomial is called the **factorial polynomial** of B.) (Goldman–Joichi–White [1975])

4.1.53. (◊) Two boards are *equivalent* if their rook polynomials are equal. They are *complementary* if they can be expressed as a partition of a rectangular board.
    (a) For complementary boards B and B' in an m-by-n rectangle, prove

$$r_k(B') = \sum_{j=0}^{k}(-1)^j r_j(B)\binom{n-j}{k-j}\binom{m-j}{k-j}(k-j)! \ .$$

    (b) Prove that two boards are equivalent if they have equivalent complements in some rectangular board. Conclude that the boards below are equivalent. Using the rook polynomial, find a third board equivalent to them that cannot be obtained from them by rotations and/or permutations of rows and/or columns.

4.1.54. (+) For a board B, let $R_B^*(x) = \sum_k r_k(B)x_{(n-k)}$, where $x_{(n-k)}$ denotes the falling factorial. Let B and B' be complementary boards in an n-by-n square. Prove $R_{B'}^*(x) = (-1)^n R_B^*(-x-1)$. (Hint: For $x \in \mathbb{N}_0$, add x extra rows to the square.) (Chow [1996])

4.1.55. ($\diamond$) Give three proofs of the identity below: by identities, by Snake Oil, and by Theorem 4.1.18.

$$\sum_{k=p}^{n} (-1)^{k-p} \binom{k}{p}\binom{n}{k} 2^{n-k} = \binom{n}{p}.$$

4.1.56. A mathematics department has m students, of which n are male. All m students want to take the combinatorics course, limited to l students and graded pass/fail. Assume that $l, n \le m$. Define a signed involution on the set of all possible grade reports to prove the identity below. (Comment: There are also other proofs.) (West–Wiedemann [1988])

$$\sum_{i=0}^{l} \sum_{j=0}^{i} (-1)^j \binom{m-i}{m-l}\binom{n}{j}\binom{m-n}{i-j} = 2^l \binom{m-n}{l}.$$

4.1.57. ($\diamond$) Let X be the set of all partitions of n. For $i \in \{0, 1\}$, let $p_i(n)$ be the number of elements of X whose number of even parts is congruent to i modulo 2. Prove that the number of partitions of n into distinct odd parts equals $p_0(n) - p_1(n)$. (Gupta [1976])

4.1.58. ($\diamond$) For a permutation π of $[n]$, the **displacement** of π is $\sum_{i=1}^{n} |i - \pi(i)|$.
   (a) Prove that every permutation has even displacement.
   (b) Let $f(n, k)$ be the number of even permutations of $[n]$ with displacement $2k$ minus the number of odd permutations of $[n]$ with displacement $2k$. Prove $f(n, k) = (-1)^k \binom{n-1}{k}$. (Hint: Obtain a recurrence for $f(n, k)$ from a signed involution on the set of permutations of $[n]$ with displacement $2k$.) (Olmsted [1986])

4.1.59. ($\diamond$) Colored balls 1 through m are put in boxes 1 through k so that no box is empty and no box has two balls with the same color. The coloring of the balls is fixed, and they do not all have the same color. Let $P_m(k)$ be the number of such arrangements. Prove $\sum_{k=1}^{m} \frac{(-1)^k}{k} P_m(k) = 0$. (Schmuland [2013])

4.1.60. (+) Given $m, n, q \in \mathbb{N}$ such that $0 \le n - m < q \le n$, prove

$$\sum_{k \equiv q \,(\mathrm{mod}\,2)} \binom{(k+q)/2 - 1}{k-1}\binom{n - (k+q)/2}{m-k} \equiv \binom{n}{m} \,(\mathrm{mod}\,2).$$

(Hint: Define an involution on the set of arrangements of m unit squares and $n - m$ dominoes in a row of length $2n - m$.) (Knuth [1996])

4.1.61. ($\diamond$) Let T be a set of integer lattice points along a single lattice path from $(0, 0)$ to (r, s), with $(0, 0), (r, s) \notin T$. Let $f_{r,s}(T)$ be the number of lattice paths from $(0, 0)$ to (r, s) that avoid all of T.
   (a) Use Corollary 4.1.33 to compute $f_{r,s}(T)$.
   (b) Use inclusion-exclusion to compute $f_{r,s}(T)$.
(Comment: Together, (a) and (b) provide a combinatorial proof of an identity.)

4.1.62. ($\diamond$) Use Theorem 4.1.31 to prove that $\det A = 0$ when a matrix A has linearly dependent rows $u_1, \dots, u_n$. (Hint: Given $u_n = \sum_{i=1}^{n-1} c_i u_i$, form a digraph with vertex set $\{x_1, \dots, x_n\} \cup \{y_1, \dots, y_n\}$ having edges $x_i y_j$ for $(i, j) \in [n-1] \times [n]$ and edges $x_n x_i$ for $i \in [n-1]$. Define edge weights appropriately.) (Aigner [2001])

4.1.63. A *path diagram* is a Ferrers diagram with dots replaced by numbers such that the number in position (i, j) is the number of paths from the bottom of column j to the right end of row i, moving one box up or right at each step, always staying within the diagram. Let M be the largest square matrix containing the upper left box (in bold below).

22	10	3	2	1	1
6	3	1	1	1	
3	2	1			
1	1	1			

(a) Prove that for every path diagram, $\det M = 1$. (Hint: Use Theorem 4.1.31.) (Carlitz–Roselle–Scoville [1971]; Aigner [2001])

(b) Prove that the Catalan numbers form the only sequence $\langle a \rangle$ of real numbers such that for all $n \in \mathbb{N}_0$,

$$\det \begin{pmatrix} a_0 & a_1 & \cdots & a_n \\ a_1 & a_2 & \cdots & a_{n+1} \\ \vdots & \vdots & \ddots & \vdots \\ a_n & a_{n+1} & \cdots & a_{2n} \end{pmatrix} = \det \begin{pmatrix} a_1 & a_2 & \cdots & a_{n+1} \\ a_2 & a_3 & \cdots & a_{n+2} \\ \vdots & \vdots & \ddots & \vdots \\ a_{n+1} & a_{n+2} & \cdots & a_{2n+1} \end{pmatrix} = 1.$$

(Hint: Apply part (a). Comment: Matrices of this form are called **Hankel matrices**. Related work appears in Kellogg [1997] and Radoux [1997].)

4.1.64. ($\diamond$) For $m \geq 0$ and $n \geq 1$, prove the determinant identity below, where $S_{n,k}$ is the Stirling number of the second kind. (Hint: Use a weighted digraph like that shown on the right for the case $(m, n) = (3, 4)$. The diagonal edges have weight 1 and the horizontal edges at height k have weight k.) (Aigner [2001])

$$\det \begin{pmatrix} S_{m+1,1} & S_{m+1,2} & \cdots & S_{m+1,n} \\ S_{m+2,1} & S_{m+2,2} & \cdots & S_{m+2,n} \\ \vdots & \vdots & \ddots & \vdots \\ S_{m+n,1} & S_{m+n,2} & \cdots & S_{m+n,n} \end{pmatrix} = (n!)^m$$

4.1.65. (+) Let $S = \{a_1, \ldots, a_n\} \subseteq \mathbb{N}$. The **GCD matrix** A of S has (i, j)-entry $\gcd(a_i, a_j)$. Prove that if S is closed under taking divisors, then $\det A = \prod_{i=1}^{n} \varphi(a_i)$, where φ is the Euler totient function. (Hint: Form a digraph using three copies of S. For elements $q, r \in S$ with $q \mid r$, add edges from r_1 to q_2 and from q_2 to r_3, with $w(r_1 q_2) = \varphi(q)$ and $w(q_2 r_3) = 1$. Apply Theorem 4.1.31. Comment: The formula is due to Smith [1876]; this proof is from Altinisik–Sagan–Tuglu [2005], generalized in Exercise 15.2.35.)

4.1.66. ($\diamond$) *Monotone trees and alternating permutations.* Let M_n be the set of n-vertex ordered trees with vertex labels $[n]$ such that each vertex has at most two children and labels increase along paths from the root. For $T \in M_n$ with left subtree T_1, right subtree T_2, and root label a, recursively define $f(T)$ to be the concatenation $f(T_1), a, f(T_2)$.

(a) Prove that f is a bijection from M_n to the set of permutations of $[n]$.

(b) When $\pi = (a_1, \ldots, a_n)$, the element a_i is a *climb*, *slide*, *peak*, or *valley* if the values move up-up, down-down, up-down, or down-up, respectively, in the triple (a_{i-1}, a_i, a_{i+1}). Characterize each type in terms of whether a_i has left and/or right children in $f^{-1}(\pi)$. (Set $a_0 = a_{n+1} = 0$, so a_1 is a climb or a peak, and a_n is a slide or a peak.) (Stanley [1986])

(c) Let n be odd. In an *alternating permutation* of $[n]$, odd positions are peaks and even positions are valleys. Characterize the elements of M_n mapped to alternating permutations by f. Prove that $[n]$ has exactly $|d_n - e_n|$ alternating permutations, where $[n]$ has d_n permutations with an odd number of ascents and e_n with an even number ($d_3 - e_3 = 2$, as shown below). (Hint: Using part (b), define an appropriate involution.) (Foata–Schützenberger [1970, 83–84])

permutation	123	132	213	231	312	321
# of ascents	2	1	1	1	1	0
alternating?	no	no	yes	no	yes	no

4.1.67. ($\diamond$) Let $\mathbf{G}$ be the set of graphs with vertex set $[n]$ and m edges. A graph is *even* if every vertex has even degree. Let $w_{n,m}$ be the number of even graphs in $\mathbf{G}$. Let X be the set of pairs (G, S) such that $G \in \mathbf{G}$ and $S \subseteq V(G)$.

(a) Define a signed involution on X for which the set of fixed points consists of all pairs (G, S) in which G is an even graph.

(b) Let $a_{k,l}$ be the number of graphs in $\mathbf{G}$ such that l of the m edges join $[k]$ to $[n] - [k]$. Use the involution in part (a) to prove $w_{n,m} 2^n = \sum_{k=0}^{n} \binom{n}{k} \sum_{l=0}^{m} (-1)^l a_{k,l}$.

(c) Obtain the generating function for even graphs on $[n]$ by number of edges:

$$\sum w_{n,m} x^m = 2^{-n} (1 + x)^{\binom{n}{2}} \sum_{k=0}^{n} \binom{n}{k} \left(\frac{1 - x}{1 + x} \right)^{k(n-k)}.$$

4.1.68. Let A be a square matrix in which every entry not in the first row or column is the sum of the two entries immediately above it and to its left. Prove that if every entry in the first column of A is 1, then $\det(A) = 1$. (Comment: The special case for binomial coefficients is in Rupp [1930].) (Ionin [2014])

4.1.69. (+) *The Condensation Method for determinants* (Remark 4.1.36).

(a) For an n-by-n matrix A, let $\hat{A}_{I;J}$ denote the matrix obtained by deleting from A the rows indexed by I and the columns indexed by J. Prove

$$\det A \det \hat{A}_{1,n:1,n} = \det \hat{A}_{n:n} \det \hat{A}_{1:1} - \det \hat{A}_{n:1} \det \hat{A}_{1:n}. \qquad \text{(Dodgson [1866])}$$

(Comment: Zeilberger [1997] gave a bijective proof.)

(b) For $n, a, b \in \mathbb{N}$, let $M_n(a, b)$ be the n-by-n matrix with (i, j)-entry $\binom{a+b}{a-i+j}$. Use part (a) to prove $\det M_n(a, b) = \prod_{j=1}^{a} \prod_{k=1}^{b} \frac{n+j+k-1}{j+k-1}$ (see Zeilberger [1996]). (MacMahon [1916])

4.2. Pólya–Redfield Counting

When we count "distinguishable" objects, the objects are grouped into classes by an equivalence relation, and we count these classes. For example, we can form a string of n beads drawn from k types in k^n ways, but when we join the ends to make a necklace, the necklaces obtained from it by cyclic rotations and flips are indistinguishable from it.

A classical application counts the isomorphism classes of n-vertex graphs. There are $2^{\binom{n}{2}}$ graphs with vertex set $[n]$, but two graphs are isomorphic when a vertex permutation turns one into the other. Since bijections can be composed or inverted, isomorphism is an equivalence relation. To view this as a "distinguishable colorings" problem, we view vertex pairs as colored by "edge" or "non-edge".

Two simple examples introduce the technique.

4.2.1. Example. *Painted batons.* A baton is partitioned into n bands of equal length that can be painted with any of k colors; there seem to be k^n paintings. However, the baton can be flipped without changing it, so most equivalence classes of colorings are listed twice. Symmetric colorings appear only once. Therefore, before dividing by 2, we add the number of symmetric colorings to make each equivalence class be counted exactly twice. The number of distinguishable paintings is thus $(k^n + k^{\lceil n/2 \rceil})/2$.   ∎

4.2.2. Example. *Tri-cornered hats.* A clown with a three-cornered hat wants to pin a flower to each corner. Given k types of flowers, how many distinguishable patterns exist? There are k^3 colorings of the fixed hat, but most equivalence classes appear three times in the list, since the hat can rotate to turn one pattern into another. The k monochromatic patterns appear only once. We add $2k$ so that those classes also are counted thrice. The number of patterns is $(k^3 + 2k)/3$.   ∎

In these examples, the equivalence classes have different sizes, so we cannot just divide the total number of objects by the size of the classes. The idea in general is to adjust the counting so that each equivalence class contributes the same amount and then divide by that amount.

BURNSIDE'S LEMMA

Equivalence classes of colorings result from "symmetry operators" that permute the things being colored. When we color pieces of a physical object, a **symmetry** of the object is a rigid motion that leaves it occupying the same position. Typically, a symmetry permutes the pieces of the object being colored. For example, a regular n-gon has $2n$ symmetries in space; n rotations and n reflections (flips). These operations permute the vertices or edges.

Symmetries can be inverted and can be composed. Under composition, the symmetries of an object form a group.

4.2.3. Definition. A **group** is a set G together with a composition law satisfying the four properties below. Here we write composition as multiplication.
   (1) *identity*: there exists $e \in G$ such that $e\pi = \pi = \pi e$ for all $\pi \in G$.
   (2) *inverse*: for $\pi \in G$, there is a unique $\pi^{-1} \in G$ with $\pi\pi^{-1} = e = \pi^{-1}\pi$, where e is the identity element.
   (3) *closure*: $\pi\sigma \in G$ for all $\pi, \sigma \in G$.
   (4) *associativity*: $(\pi\sigma)\tau = \pi(\sigma\tau)$ for all $\pi, \sigma, \tau \in G$.

Two colorings of a set X are equivalent if we can move the underlying object to turn one coloring into the other. Such a motion π permutes X.

4.2.4. Definition. Given a set Y of *colors*, a **coloring** of X is a function $f \colon X \to Y$. Given a group G of permutations of X, colorings f and f' are **indistinguishable** or **equivalent** if there exists $\pi \in G$ such that $f'(\pi(x)) = f(x)$ for all $x \in X$. We then say that π *turns f into f'*.

4.2.5. Remark. Note that "π turns f into f'" means that f' assigns to $\pi(x)$ the same color that f assigns to x, for all $x \in X$. Using identity, inverse, and closure for G, indistinguishability is reflexive, symmetric, and transitive. Thus it is an equivalence relation on the colorings; we want to count the equivalence classes.

Symmetries of an object permute the set X being colored and thereby permute the set C of colorings. The group structure is the same: the permutation of C induced by composing π and σ is the composition of the permutations of C induced by π and σ. Thus we may view $\pi \in G$ as a symmetry operator, a permutation of X, or a permutation of C. We use the same notation π in each context.

When G is a group of permutations of a set C, and $u \in C$, the **orbit** of u is $\{\pi(u): \pi \in G\}$. The defining properties of a group imply that the orbits partition C, and "in the same orbit" is an equivalence relation. In our model, the orbit of a coloring is the set of colorings equivalent to it.

When all elements of G map a coloring f to distinct colorings, the orbit of f has size $|G|$, and dividing by $|G|$ counts it as one class. The orbit is smaller when two elements of G take f to the same f'. Since every element of G maps f to some element of its orbit, summing the number taking f to each element of its orbit again counts the orbit $|G|$ times, and then dividing the total by $|G|$ counts it once. A classical lemma of group theory will simplify the computation. ∎

4.2.6. Lemma. If G is a group of permutations of C, and $u, v \in C$ are in the same orbit, then $|\{\pi \in G: \pi(u) = v\}| = |\{\pi \in G: \pi(v) = v\}|$.

Proof: Let $A = \{\pi \in G: \pi(u) = v\}$ and $B = \{\pi \in G: \pi(v) = v\}$. We establish injections in both directions. Since u, v are in the same orbit, there is some $\sigma \in G$ such that $\sigma(u) = v$; we use this fixed σ to define the injections. For each $\pi \in B$, the composition $\pi\sigma$ takes u to v. For each $\pi' \in A$, the composition $\pi'\sigma^{-1}$ takes v to v. This defines functions from B to A and from A to B. Both are injective, because for τ fixed we can multiply $\pi_1\tau = \pi_2\tau$ on the right by τ^{-1} to obtain $\pi_1 = \pi_2$. ∎

4.2.7. Lemma. (Burnside's Lemma). Under the action of a group G of permutations of C, the number of orbits of C is $\frac{1}{|G|} \sum_{\pi \in G} \psi(\pi)$, where $\psi(\pi)$ is the number of elements of C left fixed by the action of π on C.

Proof: We count 1 for each orbit (equivalence class) by counting $|G|$ for each orbit and dividing the total by $|G|$. Choose a distinguished element u_E from each orbit E. Apply each π in G to u_E, and group the elements of G according to the image $\pi(u_E) \in E$. Thus $|G| = \sum_{v \in E} |\{\pi: \pi(u_E) = v\}|$. Using Lemma 4.2.6, the number of operators in G taking u_E to any element v equals the number of operators fixing v, which we write as $\phi(v)$. The computation of the number of orbits becomes

$$\sum_E 1 = \frac{1}{|G|} \sum_E |G| = \frac{1}{|G|} \sum_E \sum_{v \in E} |\{\pi: \pi(u_E) = v\}| = \frac{1}{|G|} \sum_{v \in C} \phi(v) = \frac{1}{|G|} \sum_{\pi \in G} \psi(\pi).$$

The sum $\sum_{v \in C} \phi(v)$ counts all pairs (v, π) such that $\pi(v) = v$, grouped by the colorings. The sum $\sum_{\pi \in G} \psi(\pi)$ counts the same set, grouped by the operators. We just interchanged the order of summation. ∎

The last step in Lemma 4.2.7 is crucial for computations. We study questions where the set X and group G acting on X are fixed, but the color set Y and thus C varies. The computations for the baton and hat in Examples 4.2.1–4.2.2 were those of the final step in Burnside's Lemma.

Although Lemma 4.2.7 is generally known as Burnside's Lemma, Wikipedia says: "William Burnside stated and proved this lemma, attributing it to Frobenius [1887] in his 1897 book on finite groups. But even prior to Frobenius, the formula was known to Cauchy in 1845. In fact, the lemma was apparently so well known that Burnside simply omitted to attribute it to Cauchy. Consequently, this lemma is sometimes referred to as the lemma that is not Burnside's."

When counting colorings left fixed by ψ, we use the observation that ψ fixes a coloring if and only if for all $x \in X$ it assigns the same color to x and πx. We impose this requirement one cycle at a time.

4.2.8. Lemma. In the set of all colorings of X from a set of k colors, the number $\psi(\pi)$ of colorings left fixed by the action of π on X is k^t, where t is the number of cycles created by π on X.

Proof: To be fixed by π, a coloring must give the same color to all elements in a cycle produced by π on X. There are k choices for that color, and the choice on each cycle is made independently of the other choices. ∎

4.2.9. Example. *4-bead necklaces.* We model the problem as coloring X (the corners of a square) from a set of k colors. The square has eight symmetries; four rotations and four reflections. Rotations of $0°$, $90°$, $180°$, and $270°$ have $4, 1, 2, 1$ cycles, respectively, as permutations of the corners. Non-diagonal flips have two cycles of length 2; diagonal flips have one cycle of length 2 and two of length 1.

Among the eight permutations, $1, 2, 3, 2$ have $4, 3, 2, 1$ cycles, respectively. By Burnside's Lemma, there are $\frac{1}{8}(k^4 + 2k^3 + 3k^2 + 2k)$ distinguishable colorings of X. When $k = 2$, the answer is 6. There are two monochromatic necklaces, two where one color occurs once, and two with two beads of each color.

We can add constraints by applying Burnside's Lemma to restricted sets of colorings. If adjacent beads must have distinct colors, then the only operators that fix legal colorings are the identity, the $180°$ rotation, and the diagonal flips, fixing two legal colorings each. Burnside's Lemma then yields $\frac{1}{8}(2 + 2 + 2 + 2)$ as the number of legal 4-bead necklaces. ∎

Dividing by $|G|$ when using Burnside's Lemma provides a computational check. The number of equivalence classes is an integer. If the computation produces a non-integer, then mostly likely some operator or some colorings fixed by an operator have been overlooked.

THE PATTERN INVENTORY

By recording how many times each color is used, we can count the distinguishable patterns using each color a specified number of times. For this we need the lengths of the cycles in the permutations of X. We represent a cycle of length j by the variable x_j.

4.2.10. Definition. The **cycle structure** of a permutation π on a set is $\prod x_j^{e_j}$, where π has e_j cycles of length j. The **cycle index** $Z_G(x_1, x_2, \cdots)$ of a group G of permutations is $|G|^{-1}$ times the generating function that enumerates G by cycle structure, meaning that the coefficient of $\prod x_j^{e_j}$ counts the elements of G having cycle structure $\prod x_j^{e_j}$ as permutations of the set being colored.

4.2.11. Example. We have applied Burnside's Lemma to count colorings of the vertices of a rotating triangle and a square in space (Examples 4.2.2–4.2.9). For the rotating triangle, the cycle index is $\frac{1}{3}(x_1^3 + 2x_3^1)$. For 4-bead necklaces, it is $\frac{1}{8}(x_1^4 + 2x_4 + 3x_2^2 + 2x_1^2 x_2)$. We just record, for each element of the group, the lengths of the cycles in the corresponding permutation of the objects being colored.

In the cycle structure $\prod x_j^{e_j}$ for a permutation of X, note that $\sum j e_j = |X|$, because each element of X appears in one cycle. ∎

For cycle indices, Burnside's Lemma states the following:

4.2.12. Corollary. If G is a group of permutations of X, then the number of distinguishable k-colorings of X under G is $Z_G(k, k, k, \cdots)$.

Proof: Setting $x_j = k$ for each j accumulates $\psi(\pi)$ for each $\pi \in G$; we then divide by $|G|$ to count one for each class. ∎

We now want to record how many times each color is used.

4.2.13. Definition. Let $Y = \{y_1, \ldots, y_k\}$ be a set of colors. The **pattern inventory** of colorings of X is the OGF in which the coefficient of $\prod y_i^{e_i}$ is the number of equivalence classes of colorings that for each i have color y_i on e_i elements of X.

The pattern inventory records the options instead of just the number of options. Instead of setting $x_j = k$ in the cycle index, we let x_j list the k options "j blacks *or* j reds *or* j greens . . ." To do this, we set $x_j = y_1^j + y_2^j + y_3^j + \cdots$, in the usual way of building generating functions to model the listing of alternatives. The choice from the expression for x_j records a contribution of j to the exponent of the chosen color. This represents using that color on all j vertices of the corresponding cycle.

4.2.14. Example. *Pattern inventory for 4-bead necklaces.* Setting $x_j = b^j + r^j$ in the cycle index of Example 4.2.11 yields the pattern inventory

$$\tfrac{1}{8}[(b + r)^4 + 2(b^4 + r^4) + 3(b^2 + r^2)^2 + 2(b + r)^2(b^2 + r^2)]$$

for distinguishable necklaces with blue and red beads. This simplifies to $b^4 + b^3 r + 2b^2 r^2 + br^3 + r^4$. With two beads of each color we have two patterns: alternating or non-alternating. ∎

4.2.15. Example. A cube has 24 "symmetries", meaning rigid rotations that leave it occupying the same space. Other than the identity, there are three rotations about the axis through a pair of opposite faces, one about the axis through a pair of opposite edges, and two about the axis through a pair of opposite vertices. The groups we consider when coloring the vertices, faces, or edges of a cube are isomorphic (each corresponds to the rigid motions of the cube), but the cycle structures are different, because the set X being colored is different. The cycle structures are listed below, with "multiplicity" being the number of motions of the type indicated.

Axis	Degrees	Mult.	Vertices	Faces	Edges
anywhere	0	1	x_1^8	x_1^6	x_1^{12}
opp. faces	90,270	6	x_4^2	$x_1^2 x_4$	x_4^3
opp. faces	180	3	x_2^4	$x_1^2 x_2^2$	x_2^6
opp. verts.	120,240	8	$x_1^2 x_3^2$	x_3^2	x_3^4
opp. edges	180	6	x_2^4	x_2^3	$x_1^2 x_2^5$

The resulting cycle indices are

$$\frac{1}{24}(x_1^8 + 6x_4^2 + 9x_2^4 + 8x_1^2 x_3^2) \qquad \text{for coloring vertices,}$$
$$\frac{1}{24}(x_1^6 + 6x_1^2 x_4 + 3x_1^2 x_2^2 + 8x_3^2 + 6x_2^3) \qquad \text{for coloring faces,}$$
$$\frac{1}{24}(x_1^{12} + 6x_4^3 + 3x_2^6 + 8x_3^4 + 6x_1^2 x_2^5) \qquad \text{for coloring edges.}$$

Setting each x_j to k counts the equivalence classes of colorings when k colors are available. Setting $x_j = \sum_{i=1}^{k} y_i^j$ forms the pattern inventory listing equivalence classes by how many times each color is used. ∎

Using Burnside's Lemma, we can count isomorphism classes of graphs with n vertices. With the pattern inventory, we can count them by the number of edges.

4.2.16. Application. *Isomorphism classes of n-vertex graphs.* Graphs G and H with the same vertices are *isomorphic* if some permutation π of the vertices turns G into H. That is, $uv \in E(G)$ if and only if $\pi(u)$ and $\pi(v)$ are adjacent in H.

A graph G is formed by deciding for each pair of vertices whether they are adjacent or not. This yields two colors: "edge" and "non-edge". A permutation of the vertices can turn one coloring into another by inducing a permutation of the vertex *pairs*. The cycle index for the counting of classes is always for a group of permutations *of the objects being colored*.

The group of permutations of the vertices is $\mathbb{S}_n$. These permutations induce $n!$ permutations of the set X of vertex pairs. The resulting group is the **pair group** $\mathbb{S}_n^{(2)}$. In Example 4.2.25 we describe it for general n. Meanwhile, the table below shows the computation of $Z_{\mathbb{S}_4^{(2)}}$.

cycle form of vertex perm.	# of this type	Induced edge perm. 12 13 14 23 24 34	Induced cycle struc.
(1)(2)(3)(4)	1	12 13 14 23 24 34	x_1^6
(1234)	6	23 24 21 34 31 41	$x_4 x_2$
(123)(4)	8	23 21 24 31 34 14	x_3^2
(12)(34)	3	21 24 23 14 13 43	$x_1^2 x_2^2$
(12)(3)(4)	6	21 23 24 13 14 34	$x_1^2 x_2^2$

The cycle index of $\mathbb{S}_4^{(2)}$ is $\frac{1}{24}(x_1^6 + 6x_2x_4 + 8x_3^2 + 9x_1^2x_2^2)$. Setting each $x_i = 2$ yields 11 as the number of isomorphism classes with four vertices.

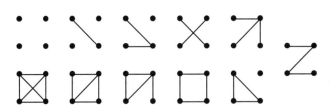

Setting x_j to $1+y^j$ models the choices in constructing graphs invariant under π so that the number of edges accumulates in the exponent. We can choose no edges (y^0) or j edges on a cycle consisting of j pairs. There is such a factor for each cycle of the permutation, and multiplying them yields a contribution of 1 to the coefficient of y^k for each graph with k edges that is unchanged by π.

Thus $\psi(\pi)$ is a generating function for the graphs left fixed by π, indexed by the number of edges. Summing over all π to count each isomorphism class once accumulates those with k edges in the coefficient of y^k.

For example, there are three isomorphism classes with three edges:

$$[y^3]\tfrac{1}{24}\left((1+y)^6 + 6(1+y^2)(1+y^4) + 8(1+y^3)^2 + 9(1+y)^2(1+y^2)^2\right)$$
$$= \tfrac{1}{24}\left(\binom{6}{3} + 6\cdot 0 + 8\binom{2}{1} + 9\binom{2}{1}\binom{2}{1}\right) = 3. \qquad\blacksquare$$

The argument of Application 4.2.16 works for all n.

4.2.17. Corollary. (Redfield [1927], Pólya (see Harary [1955b])) The number of isomorphism classes of n-vertex graphs with m edges is the coefficient of y^m in the OGF obtained by letting $x_j = 1 + y^j$ for each x_j in the cycle index $Z_{\mathbb{S}_n^{(2)}}$ of the pair group on n vertices. $\qquad\blacksquare$

Application 4.2.16 uses one color of weight 1 (edge) and one color of weight 0 (non-edge). We may have several colors of each weight.

4.2.18. Definition. Given a_i options with weight i, the generating function $\sum_{i\geq 0} a_i y^i$ is the **weight enumerator** $w(y)$. When options have various weights in r colors, use $w(y_1, \ldots, y_r)$: the coefficient on each monomial counts the options where the weights on each color are given by the exponents.

Weight enumerators yield a common generalization of the pattern inventory arguments we have given for necklaces and for isomorphism classes of graphs.

4.2.19. Example. Given three types of paint that cost one dollar per use and another that costs two dollars, the weight enumerator (by cost) is $w(y) = 3y + y^2$. To enumerate distinguishable colorings by total cost, set $x_j = w(y^j)$ in the cycle index for each j, because using a color on a j-cycle uses it j times.

Similarly, with two types of red paint and two types of blue paint, costing $2 or $4 per face, let $w(r, b) = r^2 + r^4 + b^2 + b^4$. The coefficient of $r^i b^j$ in the pattern inventory counts the distinguishable paintings using $i of red paint and $j of blue paint. To compute it, set $x_j = w(b^j, r^j)$ in the cycle index. $\qquad\blacksquare$

We summarize the general discussion in the following theorem. We have essentially explained the proof; it uses Burnside's Lemma to argue that each class of colorings has been counted and assigned to the correct weight.

4.2.20. Theorem. (**Pólya's Theorem**; Pólya [1937], Redfield [1927]) Let G be a group of permutations of X, and let w be a weight enumerator for colors. The OGF for equivalence classes of colorings of X under action of G, by weight in each color, is obtained from the cycle index Z_G by setting each x_j to the expression obtained from w by replacing each variable by its jth power.   ∎

CLASSICAL CYCLE INDICES

Various classical groups arise in using the Pólya–Redfield Theorem.

4.2.21. Example. *The cyclic group.* Counting distinguishable jeweled crowns uses the cycle index of the **cyclic group** C_n, the group of rotations of an n-gon. With elements $\mathbb{Z}_n$, the jth element π_j takes i to $(i + j) \pmod n$ and has d cycles of length n/d, where $d = \gcd(j, n)$. Its cycle structure is $x_{n/d}^d$. The values j with $\gcd(j, n) = d$ are multiples of d; we have $\gcd(id, n) = d$ when $\gcd(i, n/d) = 1$. Thus there are $\phi(n/d)$ permutations with cycle structure $x_{n/d}^d$, where $\phi(k)$ counts the numbers in $[k]$ that are relatively prime to k (the *Euler totient*—Proposition 4.1.4). For ease of application, we exchange d and n/d in the sum to write

$$Z_{C_n} = \frac{1}{n} \sum_{d|n} \phi(d) x_d^{n/d}. \qquad\qquad ∎$$

4.2.22. Application. *Crowns with r missing jewels.* A crown with k places for diamonds is missing r of them; how many distinguishable ways can this happen? We are asking for the number of distinguishable cyclic arrangements of r 1s and $k - r$ 0s under rotational symmetry.

The answer also counts the cycles in Bulgarian solitaire on partitions of n when $n = \binom{k}{2} + r$ (Theorem 3.4.18). The "move" in Bulgarian solitaire maps the set of Ferrers diagrams with n dots to itself. We showed that the Ferrers diagrams lying on cycles in the resulting functional digraph consist of $k - 1$ full slants plus r dots in the slant of size k. Each move shifts these dots one step down the slant, cyclically. Thus the number of cycles is again the number of cyclic arrangements of r 1s and $k - r$ 0s.

Applying Example 4.2.21, our choices for x_d in the pattern inventory are y^0 or y^d to accumulate contributions by the number of 1s. Since the cycle structures of permutations in C_n have the form $x_d^{n/d}$, we obtain contributions to the desired term only when d divides both k and r. Setting $x_d = (1 + y^d)^{k/d}$ for all d, contributions to the coefficient of y^r arise when we take r/d of the k/d cycles to have 1s. Using the Binomial Theorem, the number of cyclic arrangements is

$$\frac{1}{k} \sum_{d|\gcd(k,r)} \phi(d) \binom{k/d}{r/d}. \qquad\qquad ∎$$

4.2.23.* Example. *The dihedral group.* To count distinguishable necklaces, we need the cycle index for the **dihedral group** D_n, the group of symmetry operators on an n-gon in space. It has the same rotational operators as C_n plus n "flips." We divide the cycle index for C_n by 2 and add the contributions for the flips. If n is odd, then every flip uses an axis between a vertex and the opposite edge, so each consists of a fixed point and $(n-1)/2$ pairs forming 2-cycles. If n is even, then the $n/2$ flips around an axis through opposite edges consist solely of 2-cycles, while the $n/2$ flips around an axis through opposite vertices have two fixed points. Thus

$$Z_{D_n} = \begin{cases} \frac{1}{2}Z_{Cn} + \frac{1}{2}x_1 x_2^{(n-1)/2} & \text{when } n \text{ is odd} \\ \frac{1}{2}Z_{Cn} + \frac{1}{4}x_2^{n/2} + \frac{1}{4}x_1^2 x_2^{n/2-1} & \text{when } n \text{ is even.} \end{cases}$$ ∎

4.2.24.* Example. *Symmetric group.* The **symmetric group** $\mathbb{S}_n$ is the group of all permutations of $[n]$. A cycle structure $\prod_k x_k^{e_k}$ corresponds to the partition λ of n having e_k parts of size k. To count the permutations with this cycle structure, view the cycles linearly as

$$(-\,-\,-\,-\,-\,-\,-\,-\,-)(-\,-\,-\,-\,-)(-\,-)(-\,-),$$

in decreasing order of length, waiting to be filled. There are $n!$ ways to put $[n]$ in the blanks. Each permutation with this structure arises in $\prod k^{e_k} e_k!$ ways, because (1) a cycle of length k has k different starting points, and (2) cycles of the same length can appear in any order. Hence $n!/\prod k^{e_k} e_k!$ permutations have cycle structure $\prod_k x_k^{e_k}$. Letting $\lambda \vdash n$ indicate a sum over all partitions of n,

$$Z_{\mathbb{S}_n} = \sum_{\lambda \vdash n} \prod_k \frac{x_k^{e_k}}{k^{e_k} e_k!}.$$ ∎

4.2.25.* Example. *The pair group.* The **pair group** $\mathbb{S}_n^{(2)}$ is the group of permutations of the pairs in $[n]$ induced by $\mathbb{S}_n$. We used this in Application 4.2.16 and Corollary 4.2.17. The cycle structure of the permutation in $\mathbb{S}_n^{(2)}$ induced by $\pi \in \mathbb{S}_n$ depends only on the cycle structure for π. Pairs are of two types: those contained in one cycle of π and those consisting of elements from two cycles in π.

For two elements on the same cycle, π shifts each by one along the cycle, leaving them separated by the same distance along the cycle. Thus a cycle of length k in π generates $\lfloor (k-1)/2 \rfloor$ cycles of length k among the pairs contained in it. When k is even it also generates one cycle of length $k/2$; pairs with displacement $k/2$ return to themselves after $k/2$ steps.

A pair of elements from distinct cycles of lengths r and s repeats after $\mathrm{lcm}\,(r,s)$ iterations of π. The induced permutation has $\gcd(r,s)$ cycles of length $\mathrm{lcm}\,(r,s)$ on the rs pairs chosen from the two cycles.

Thus when π has cycle structure $\prod_k x_k^{e_k}$, the induced permutation in $\mathbb{S}_n^{(2)}$ has cycle structure

$$\prod_{\text{odd } k} x_k^{e_k(k-1)/2} \prod_{\text{even } k} x_k^{e_k(k/2-1)} x_{k/2}^{e_k} \prod_k x_k^{\binom{e_k}{2}k} \prod_{r \neq s} x_{\mathrm{lcm}\,(r,s)}^{e_r e_s \gcd(r,s)}.$$ ∎

EXERCISES 4.2

4.2.1. (−) A game is played on a table with 15 balls in a triangular array. The array can slide on the table and rotate. There is a supply of three types of balls. How many distinguishable ways are there to arrange balls in the array?

4.2.2. (−) The clown of Example 4.2.2 wants to put flowers also on the edges of the tri-cornered hat; the same k types are available, and we count distinguishable patterns under rotation. The number of patterns on edges alone is the same as the number on vertices alone. Why is the number of patterns with flowers on both the edges and the vertices not equal to the square of this?

4.2.3. (−) The n bands of a baton (Example 4.2.1) are to be painted under the additional requirement that adjacent bands must not have the same color. With k colors available, how many distinguishable ways are there to paint the baton?

4.2.4. (−) Given that n is prime, count the distinguishable n-bead necklaces that can be formed when k colors of beads are available.

4.2.5. (−) For $n = 6$, $n = 9$, and $n = 10$, count the n-bead necklaces that can be formed using three colors of beads.

4.2.6. (−) Determine the cycle indices for coloring the vertices, edges, or faces of a tetrahedron that can rotate in space.

4.2.7. (−) Determine the pattern inventory for colorings of the squares of a rotating 4-by-4 chessboard with two colors available.

4.2.8. (−) Let $c(x)$ be the OGF for isomorphism classes of connected graphs, indexed by number of vertices. Find the OGF for isomorphism classes of graphs with two components.

4.2.9. (−) A **double domino** is a solid 2-sided chip of wood such that each side consists of two squares sharing an edge. Each of the four squares is marked with a number of dots between 0 and $k-1$, inclusive. Determine the number of distinguishable double dominoes.

4.2.10. The squares of a 4-by-4 chessboard are to be painted (front only), with four colors available for each square. Determine the number of distinguishable paintings that have four squares of each color.

4.2.11. Determine the cycle index for coloring the vertices *and* the edges of an n-gon that can rotate or flip (movement in space around any axis).

4.2.12. (◊) **Caterpillars** are trees in which the deletion of all leaves leaves a path. Prove that the number of isomorphism classes of n-vertex caterpillars is $2^{n-4} + 2^{\lfloor n/2 \rfloor - 2}$ if $n \geq 3$. (Harary–Schwenk [1973], Kimble–Schwenk [1981])

4.2.13. Lemma 4.2.6 states that when a group G acts on a set S, the number of permutations that map x to y is the same for all y in the orbit of x. Use this to count the ways to pair up $2n$ people. (Hint: Let S be the set of pairings.)

4.2.14. (◊) Partitions of an integer n are obtained from compositions of n by ignoring the order of the parts. Use Burnside's Lemma and the symmetric group $\mathbb{S}_3$ to count the partitions of 9 into three parts. List the partitions explicitly to check the computation.

4.2.15. (◊) Let p, n, l be positive integers with p prime. Use Burnside's Lemma to prove that $n^{p^l} - n^{p^{l-1}}$ is divisible by p^l. (Chappell–Hartman)

4.2.16. With four colors of beads available, how many distinguishable 6-bead necklaces are there in which no two adjacent beads have the same color?

4.2.17. ($\diamond$) Consider distinguishable n-bead necklaces formed from three colors of beads. Use the methods of this section to answer the following questions.

(a) For $n = 4$, how many have no two consecutive beads of the same color?

(b) For $n = 6$, how many have no two of the three colors used equally often?

(c) For $n = 9$, how many use all three colors three times each?

4.2.18. With three colors of beads available, how many distinguishable n-bead necklaces are there with each color used at least once, for $n = 4$ and $n = 7$?

4.2.19. Determine the number of isomorphism classes of 4-vertex loopless multigraphs in which each pair of vertices occurs as the set of endpoints of at most two edges. Use the pattern inventory to compute the number of these classes that have six edges. Draw them explicitly to check the computation.

4.2.20. ($\diamond$) A financial company is designing a new symbol: a regular tetrahedron built from six metal bars of equal length. Given that the bars can be copper, silver, or gold, determine the number of distinguishable ways there are to do this (the tetrahedron can rotate in space but not reflect). How many ways are there using two bars of each type?

4.2.21. ($\diamond$) *Pattern inventories for cubes.*

(a) Count the distinguishable 17-dollar cubes that can be built using (as edges) \$1 lead bars, \$2 silver bars, and \$3 gold bars.

(b) Determine the number of distinguishable ways to paint the faces of a cube with three colors so that each color is used twice. Describe them geometrically.

4.2.22. Use Pólya's Theorem to count isomorphism classes of spanning subgraphs of $K_{3,3}$.

4.2.23. Use Pólya's Theorem to count the isomorphism classes of 5-vertex graphs with four edges.

4.2.24. Determine the number of terms in the pattern inventory for n-bead necklaces with k colors of beads available.

4.2.25. ($\diamond$) A rotating square table has a pocket at each corner. Each pocket may have 1, 2, or 3 balls. Compute the total number of distinguishable arrangements. Use Pólya's Theorem to compute the number of distinguishable arrangements with a total of 7 balls. Describe how the computation would change if we have two colors of balls, but the balls within a pocket must all have the same color.

4.2.26. An **even graph** is a graph whose vertex degrees are all even.

(a) Compute the weight enumerator for isomorphism classes of connected even graphs with at most five vertices, indexed by number of vertices.

(b) Use part (a) and Pólya's Theorem to enumerate, by number of vertices, the even graphs with three components in which every component has at most five vertices. Check your work by explicitly listing those with nine vertices.

4.2.27. Let e_n and o_n count the isomorphism classes of graphs on n vertices whose number of edges is even or odd, respectively. Prove that $e_n \geq o_n$ for all n and that equality holds if and only if n is congruent to 2 or 3 modulo 4. (Schmidt [1993])

4.2.28. ($\diamond$) *The* **Ordinary Exponential Formula**.

(a) By comparing coefficients, prove $\sum_{n \geq 0} Z_{\mathbb{S}_n}(x_1, x_2, \cdots) = \mathrm{e}^{\sum_{k \geq 1} x_k/k}$.

(b) Prove $g(x) = \mathrm{e}^{\sum_{k \geq 1} c(x^k)/k}$, where $g(x)$ and $c(x)$ are the OGFs by number of vertices for isomorphism classes of graphs and isomorphism classes of connected graphs. (Comment: This unlabeled version of the Exponential Formula applies whenever a general labeled structure consists of component structures and the symmetric group acts on the set of labels to produce isomorphisms.)

(c) For OGFs $g(x)$ and $c(x)$ related by $g(x) = \mathrm{e}^{\sum_{k \geq 1} c(x^k)/k}$, prove $c_n = g_n - \frac{1}{n} \sum_{k=1}^{n-1} k c_k g_{n-k}$.

4.2.29. Let $r(x)$ be the OGF by number of vertices for isomorphism classes of rooted trees.

(a) Use Exercise 4.2.28 to prove $r(x) = xe^{\Sigma_{k\geq 1} r(x^k)/k}$.

(b) Obtain a recurrence for the coefficients of $r(x)$.

4.2.30. Let $T_r(x)$ be the OGF by number of vertices for isomorphism classes of rooted trees such that every vertex is reachable from the root by a path with at most r edges.

(a) Find a formula for T_r in terms of T_{r-1}.

(b) In terms of these functions, find an expression for the OGF by number of vertices for isomorphism classes of unrooted trees in which the longest path has d edges.

4.3. Permutations and Tableaux

The subject of Young tableaux expands upon our discussions of permutations, partitions of integers, and bijections. Young tableaux correspond to representations of the symmetric group and are central in the study of symmetric functions (see Sagan [1991], Fulton [1997], and Stanley [1999]). We confine our attention to their enumeration and their connections to monotone sublists in permutations.

THE HOOK-LENGTH FORMULA

A generalization of Bertrand's Ballot Problem considers m candidates instead of two. Candidate i receives λ_i votes, with $\lambda_1 \geq \cdots \geq \lambda_m$ and $\sum \lambda_i = n$. What is the probability that the partial vote totals always maintain the same order while the votes are being counted?

Record j in the row for the candidate receiving the jth vote. At the end the elements of $[n]$ occupy the positions in the Ferrers diagram for a partition λ. By construction, the numbers increase in each row. Partial votes totals satisfy the stated condition if and only if each candidate reaches k votes before the next candidate does; thus the numbers must also increase in each column, as shown below. To solve our problem, we divide the number of such tableaux for λ by the multinomial coefficient $\binom{n}{\lambda_1,\ldots,\lambda_m}$, which counts all vote sequences.

$$1\ 3\ 5\ 7$$
$$2\ 4\ 8$$
$$6\ 9$$

4.3.1. Definition. A **reverse plane partition** consists of positive integers in the positions of a Ferrers diagram with rows and columns nondecreasing. It is a **standard Young tableau** if the entries are $1, \ldots, n$, distinct. A **generalized Young tableau** or **column-strict tableau** is a reverse plane partition with strictly increasing columns. Its **shape** is the list of row-lengths.

The word "partition" indicates that the entries form a partition of their total, "plane" suggests the geometric arrangement, and "reverse" refers to nondecreasing instead of nonincreasing rows and columns. (Standard) Young tableaux were introduced in Young [1901] to study matrix representations of permutations but actually were used earlier by Frobenius [1900]. They have many connections to algebraic combinatorics and to combinatorial aspects of permutations. Much of the subject (and the terminology) was developed by French mathematicians.

Let $f(\lambda)$ be the number of Young tableaux of shape λ (the value needed in the generalized ballot problem). The first formula for $f(\lambda)$ was found by Frobenius [1900] using group representations. The modern formula uses the sizes of particular subsets of the diagram.

4.3.2. Definition. In a tableau with shape λ, the **hook** $H_{i,j}$ consists of all positions (i, s) and (r, j) with $s \geq j$ or $r \geq i$. The **hook length** $h_{i,j}$ is the size of $H_{i,j}$. A **corner** is a position in the Ferrers diagram with no item to its right or below (an (i, λ_i) with $\lambda_i > \lambda_{i+1}$).

4.3.3. Lemma. Let position (α, β) be a corner of shape λ. If $i \leq \alpha$ and $j \leq \beta$, then $(h_{i\beta} - 1) + (h_{\alpha j} - 1) = h_{i,j} - 1$.

Proof: By shifting the sets labeled x and u in the figure, we conclude that $\big(H_{i\beta} - \{(\alpha, \beta)\}\big) \cup (H_{\alpha j} - \{(\alpha, \beta)\})$ has the same size as $H_{i,j} - \{(i, j)\}$. ∎

A tableau puts $[n]$ into a shape λ in one of $n!$ ways. The result is a standard Young tableau if and only if for each (i, j) the entry in position (i, j) is the least entry in $H_{i,j}$. When $H_{i,j}$ is filled randomly, position (i, j) is least with probability $1/h_{i,j}$. Knuth [1973] joked that we would like to multiply these values to obtain $1/\prod h_{i,j}$ as the probability of being a Young tableau, but the events are far from independent (consider $h_{2,1}$ and $h_{1,2}$ when $\lambda = (2, 2)$).

Nevertheless, this formula is correct! The original proof used a determinant for $f(\lambda)$ known to Frobenius [1900] and Young [1902] (Exercise 13). Another uses the fundamental theorem of algebra (Bandlow [2008]). Ours is probabilistic.

4.3.4. Theorem. (**Hook-Length Formula**; Frame–Robinson–Thrall [1954]) For a partition λ of n, the number $f(\lambda)$ of standard Young tableaux of shape λ is $n!/\prod h_{i,j}$, where the product is over all positions (i, j).

Proof: (Greene–Nijenhuis–Wilf [1979]). Let $F(\lambda) = n!/\prod h_{i,j}$; we want to prove $f(\lambda) = F(\lambda)$. We extend the domains of F and f by setting $F(\lambda) = f(\lambda) = 0$ when λ is a list that is not nonincreasing.

When a list λ is in fact a partition (that is, non-increasing), let $\hat{\lambda}_k$ denote the list of numbers obtained from λ by subtracting 1 from λ_k. Since the element n must occupy the last position in its row, we have the recurrence $f(\lambda) = \sum_{k=1}^{m} f(\hat{\lambda}_k)$, where m is the number of rows in the shape.

Since F and f agree when $f(\lambda) = 0$ and when $n = 1$, it suffices to prove $F(\lambda) = \sum_{k=1}^{m} F(\hat{\lambda}_k)$, or equivalently $1 = \sum F(\hat{\lambda}_k)/F(\lambda)$, which we interpret probabilistically. We define an experiment whose outcomes correspond to rows in the shape. It suffices to prove that the probability of the outcome in row α is $F(\hat{\lambda}_\alpha)/F(\lambda)$.

Each trial of the experiment is a path in the tableau. Begin at a random cell, each with probability $1/n$. Thereafter, when at position (i, j), move to another position in $H_{i,j}$, each with probability $1/(h_{i,j} - 1)$. This move increases i or j; repeat until a corner square is reached. The outcome of the trial is the corner square (α, β) that is reached.

To find the probability of outcome (α, β), we compute conditional probabilities from each starting location. We then sum over starting locations and multiply by $1/n$ to remove the conditioning. In computing the conditional probability, we do not consider each path from (a, b) to (α, β) separately. We combine paths by recording only the list A of rows and the list B of columns used in the path. For example, from (a, b) the paths $(a, b), (a, \beta), (\alpha, \beta)$ and $(a, b), (\alpha, b), (\alpha, \beta)$ both appear in the conditional event A, B with $A = (a, \alpha)$ and $B = (b, \beta)$. The grouped probabilities will correspond to terms in the formula $F(\hat{\lambda}_\alpha)/F(\lambda)$.

Let $p(A, B)$ be the probability that the path visits row set A and column set B, given the start at (a, b). The probability is 0 if A or B is empty; it is 1 if (a, b) is a corner and $A = \{a\}$, $B = \{b\}$. Otherwise, let $A = \{a, a' \ldots \alpha\}$ and $B = \{b, b' \ldots \beta\}$ in order. We can move to (a, b') and thereafter have $A, (B - \{b\})$, or we can move to (a', b) and thereafter have $(A - \{a\}), B$. Since the probability of moving from (a, b) to any location in the hook is $1/(h_{a,b} - 1)$, the recurrence is

$$p(A, B) = \frac{1}{h_{a,b} - 1} p(A - \{a\}, B) + \frac{1}{h_{a,b} - 1} p(A, B - \{b\})$$

We claim that $p(A, B) = Q(A, B)$, where

$$Q(A, B) = \prod_{i \in A - \alpha} \frac{1}{h_{i,\beta} - 1} \prod_{j \in B - \beta} \frac{1}{h_{\alpha,j} - 1}.$$

Observe that $p(A, B) = 1 = Q(A, B)$ when $|A| = |B| = 1$; we proceed by induction. Since $Q(A - \{a\}, B) = (h_{a,\beta} - 1)Q(A, B)$ and $Q(A, B - \{b\}) = (h_{\alpha,b} - 1)Q(A, B)$, the recurrence and the induction hypothesis yield

$$p(A, B) = \frac{1}{h_{a,b} - 1} \left[(h_{a,\beta} - 1)Q(A, B) + (h_{\alpha,b} - 1)Q(A, B) \right].$$

Lemma 4.3.3 yields $h_{a,\beta} - 1 + h_{\alpha,b} - 1 = h_{a,b} - 1$, and thus $p(A, B) = Q(A, B)$. The probability computation finishes with $\mathbb{P}(\alpha, \beta) = (1/n) \sum p(A, B)$, where the summation runs over all A, B such that $\alpha \in A$ and $\beta \in B$.

Now, consider the ratio of $F(\hat{\lambda}_\alpha)$ and $F(\lambda)$. Each has a factorial and a product of hook lengths. When the corner position $\alpha\beta$ is deleted, the factor $h_{\alpha\beta} = 1$ disappears, and the hooks in row α and column β decrease by one. Since $(n - 1)!/n! = 1/n$, this yields

$$\frac{F(\hat{\lambda}_\alpha)}{F(\lambda)} = \frac{1}{n} \prod_{i < \alpha} \frac{h_{i,\beta}}{h_{i,\beta} - 1} \prod_{j < \beta} \frac{h_{\alpha,j}}{h_{\alpha,j} - 1} = \frac{1}{n} \prod_{i < \alpha} \left(1 + \frac{1}{h_{i,\beta} - 1} \right) \prod_{j < \beta} \left(1 + \frac{1}{h_{\alpha,j} - 1} \right).$$

In expanding this product, every term corresponds to choosing a set A' of row indices less than α and a set B' of column indices less than β. Let $A = A' \cup \{\alpha\}$, and let $B = B' \cup \{\beta\}$. The corresponding term in the product is precisely $Q(A, B)$. The terms together generate all A, B that contain α, β. As desired, we obtain

$$\frac{F(\hat{\lambda}_\alpha)}{F(\lambda)} = \frac{1}{n} \sum Q(A, B) = \frac{1}{n} \sum p(A, B) = \mathbb{P}(\alpha, \beta). \qquad \blacksquare$$

In this experiment, the probability of each outcome is the fraction of tableaux whose largest element is in that corner. The proof is really a counting argument; probabilistic language does the bookkeeping.

4.3.5. Remark. The weighting of corner squares in this experiment lets us generate Young tableaux of a given shape with equal probability. First, run a trial and put n in the corner location specified by the outcome. Delete this position, run the experiment for the smaller shape, and place $n - 1$ in the resulting location. Repeat until the entire tableau is filled. At each step, the weights for placing k are as they should be, given the placement of $k + 1, \ldots, n$.

Also, since it is the size of a set, $n!/ \prod h_{i,j}$ is always an integer. $\blacksquare$

To count column-strict tableaux with shape λ and entries in $[m]$, see Exercise 14. Thrall [1952] introduced a variation called **shifted tableaux**, where the shape λ has distinct parts and the rows start at positions (i, i) instead of $(i, 1)$; there is a counting formula using generalized hooks (Exercise 15).

Researchers long sought a simple bijective argument for the Hook-Length Formula. There are $n!$ ways to fill the shape λ with $[n]$. One would like to prove the formula by grouping them into sets of size $\prod h_{i,j}$, each containing one standard Young tableau. Complicated bijective proofs were found by Franzblau–Zeilberger [1982], Remmel [1982], Zeilberger [1984], and Krattenthaler [1995]. Novelli–Pak–Stoyanovskii [1997] found a bijection that maps each permutation into a pair consisting of a standard Young tableaux of shape λ and an element of a set of size $\prod h_{i,j}$. Generating a random permutation and applying this bijection provides another way to generate a random standard tableau with shape λ.

4.3.6.* Remark. *Bijection for Hook-Length Formula* (Novelli–Pak–Stoyanovskii [1997]). Number the cells of the shape λ in reverse lexicographic (RL) order, first the last column from bottom to top, then the next, and so on, ending with n in the upper left corner. Enter the given permutation into the cells in this "RL" order.

The bijection involves moving elements to reach a standard tableau and recording data about the process. A **hook function** f assigns an integer to each position in the shape, with $-(\lambda_j^* - i) \le f(i, j) \le \lambda_i - j$, where λ^* is the conjugate of λ. Given λ, there are $\prod h_{i,j}$ hook functions.

Now consider an entry a with entries b to the right and c below (let entries outside the shape be infinite). Leave a in place if $a < \min\{b, c\}$. Otherwise, switch a with the smaller of b and c and continue from the new position of a. We call this movement "bubbling" by analogy with one-dimensional "bubble sort". Bubbling a until it stops produces a new tableau.

If cells 1 through l (in RL order) are increasing in rows and columns and we bubble the entry in cell $l + 1$, then cells 1 through $l + 1$ of the resulting tableau will be increasing in rows and columns. Hence successively bubbling the elements in cells 1 through n produces a standard Young tableau.

During this process, we develop a hook function, starting with $f(i, j) = 0$ for all (i, j). When a bubbles from position (i, j) in a tableau with hook function f, it comes to rest in a new position (i', j') with $i' \geq i$ and $j' \geq j$. We modify the hook function, changing values only in column j. The value in row i' of column j becomes $j' - j$. The values that were in rows $i + 1$ through i' (if any) move one position higher and decrease by 1. Since the higher cell has one more cell below it than the lower one, and position (i', j) has at least $j' - j$ positions to its right, the result is again a hook function.

Below we apply the bijection to the permutation 123456 in shape $(2, 2, 2)$. Bubbling 1 from the corner produces no change. In each figure, we underscore the element next to bubble. To complete the proof, we must prove that every pair consisting of a standard tableau and a hook function with that shape arises from exactly one permutation by this map. The probabilistic proof remains shorter, but the bijection is useful. ∎

$$
\rightarrow
\begin{array}{cc} 6 & 3 \\ 5 & \underline{2} \\ 4 & 1 \end{array}\;
\begin{array}{cc} 0 & 0 \\ 0 & 0 \\ 0 & 0 \end{array}
\rightarrow
\begin{array}{cc} 6 & \underline{3} \\ 5 & 1 \\ 4 & 2 \end{array}\;
\begin{array}{cc} 0 & 0 \\ 0 & -1 \\ 0 & 0 \end{array}
\rightarrow
\begin{array}{cc} 6 & 1 \\ 5 & 2 \\ \underline{4} & 3 \end{array}\;
\begin{array}{cc} 0 & -2 \\ 0 & -1 \\ 0 & 0 \end{array}
$$

$$
\rightarrow
\begin{array}{cc} 6 & 1 \\ \underline{5} & 2 \\ 3 & 4 \end{array}\;
\begin{array}{cc} 0 & -2 \\ 0 & -1 \\ 1 & 0 \end{array}
\rightarrow
\begin{array}{cc} \underline{6} & 1 \\ 2 & 4 \\ 3 & 5 \end{array}\;
\begin{array}{cc} 0 & -2 \\ 0 & -1 \\ 1 & 0 \end{array}
\rightarrow
\begin{array}{cc} 1 & 4 \\ 2 & 5 \\ 3 & 6 \end{array}\;
\begin{array}{cc} -1 & -2 \\ 0 & -1 \\ 1 & 0 \end{array}
$$

THE RSK CORRESPONDENCE

We next study a remarkable correspondence between permutations and pairs of same-shape Young tableaux. It enables us to count the n-element Young tableaux and to compute the maximum size of the union of k increasing sequences in a given permutation. Its properties are so astonishing that Knuth [1973] wrote, "The unusual nature of these coincidences might lead us to suspect that some sort of witchcraft is operating behind the scenes!" We call it the *RSK Correspondence* to honor Robinson [1938], Schensted [1961], and Knuth [1970].

Algebraists discovered using group representations that there are $n!$ pairs of same-shape Young tableaux with n elements. The RSK Correspondence proves this combinatorially. To count the tableaux with n elements, it then suffices to count permutations of $[n]$ whose images under the RSK Correspondence are pairs of identical tableaux.

Counting these permutations uses a remarkable property of the RSK Correspondence: if it maps σ to the pair (P, Q), then it maps σ^{-1} to (Q, P). Hence the permutations corresponding to pairs of identical tableaux are the *involutions* (permutations consisting of fixed points and 2-cycles). Involutions with i 2-cycles correspond to partitions of $[n]$ into $n - i$ blocks with i blocks of size 2. Summing over i yields the formula below to count both involutions and standard Young tableaux with n elements.

$$\sum_{i=0}^{\lfloor n/2 \rfloor} \frac{n!}{2^i i! (n - 2i)!}$$

4.3.7. Example. For $n = 4$, the sum yields 10 involutions. They appear below, using the two-line notation for permutations. Also shown are the 10 Young tableaux, using the correspondence we will develop. ∎

$$
\begin{array}{ccccc}
1\,2\,3\,4 & 1\,2\,3\,4 & 1\,2\,3\,4 & 1\,2\,3\,4 & 1\,2\,3\,4 \\
1\,2\,3\,4 & 2\,1\,3\,4 & 3\,2\,1\,4 & 4\,2\,3\,1 & 1\,3\,2\,4
\end{array}
$$

$$
\begin{array}{ccccc}
1\,2\,3\,4 & 1\,3\,4 & 1\,4 & 1\,3 & 1\,2\,4 \\
 & 2 & 2 & 2 & 3 \\
 & & 3 & 4 &
\end{array}
$$

$$
\begin{array}{ccccc}
1\,2\,3\,4 & 1\,2\,3\,4 & 1\,2\,3\,4 & 1\,2\,3\,4 & 1\,2\,3\,4 \\
1\,4\,3\,2 & 1\,2\,4\,3 & 2\,1\,4\,3 & 3\,4\,1\,2 & 4\,3\,2\,1
\end{array}
$$

$$
\begin{array}{ccccc}
1\,2 & 1\,2\,3 & 1\,3 & 1\,2 & 1 \\
3 & 4 & 2\,4 & 3\,4 & 2 \\
4 & & & & 3 \\
 & & & & 4
\end{array}
$$

A permutation $\sigma\colon [n] \to [n]$ can be expressed as a set of pairs $\{(i, \sigma_i)\}$. Knuth generalized the correspondence to allow the input σ to be any multiset of n integer pairs $\{(q_i, p_i)\}$. The output is then a pair P, Q of same-shape column-strict tableaux. We describe the insertion of a new pair recursively.

4.3.8. Algorithm. INSERT(P, Q, p, q) If p is as large as everything in row 1 of P, return the new pair P', Q' obtained by putting p at the end of row 1 of P and putting q at the end of row 1 of Q.

Otherwise, p *bumps* the first value p' in row 1 that exceeds p. Replace p' with p in row 1 of P and leave row 1 of Q unchanged; these become the top rows of the new pair P', Q'. For the remaining rows of P', Q', use INSERT$(\overline{P}, \overline{Q}, p', q)$, where $\overline{T}$ denotes the tableau consisting of all but the top row of T. ∎

4.3.9. Example. When processing a permutation, the pairs are inserted in increasing order of the values of q_i. If $\sigma = 617258934$, then the situation when we insert the next-to-last pair $(8, 3)$ is

$$
\begin{array}{llcll}
P & Q & & P' & Q' \\
1\,2\,5\,8\,9 & 1\,3\,5\,6\,7 & & 1\,2\,3\,8\,9 & 1\,3\,5\,6\,7 \\
6\,7 & 2\,4 & \rightarrow & 5\,7 & 2\,4 \\
 & & & 6 & 8
\end{array}
$$

Here 3 bumps 5, 5 bumps 6, and 6 is inserted at the end of row 3. Hence Q acquires 8 in a new position at the end of row 3. To complete $P(\sigma), Q(\sigma)$, we insert $(9, 4)$. Here 4 bumps 8 to row 2, and the ninth new position is at the end of row 2.

In the generalized input allowing any multiset of pairs as input, the insertion order is lexicographic on the pairs (q_i, p_i). For example, with $(\mathbf{q}, \mathbf{p}) = \binom{133444555}{211246133}$ (written with q_i above p_i), the last insertion is

P	Q		P'	Q'
1 1 1 3 6	1 3 4 4 4		1 1 1 3 3	1 3 4 4 4
2 2 4	3 5 5	$\rightarrow$	2 2 4 6	3 5 5 5	∎

This algorithm grows the common shape in one location; the end of the row where insertion succeeds without bumping. If (q, p) was the pair being inserted, then we add q to the current tableau Q in the new location to obtain the new Q'. This new position receiving q is the end of a row. Since we want nondecreasing rows, we INSERT input pairs $\{(q_i, p_i)\}$ in increasing order of q_i. For permutations, this reduces to inserting as p_i the values $\sigma_1 \ldots \sigma_n$ in order. We write q_i first in generalized input pairs to agree with the 2-line notation for permutations.

4.3.10. Definition. The **RSK Correspondence** is the map that takes a generalized input σ to the pair $(P(\sigma), Q(\sigma))$ of tableaux generated by successively INSERTing input pairs (q_i, p_i) in lexicographic order (nondecreasing order in q, and nondecreasing in p among pairs with the same q). In the output, $P(\sigma)$ is the **P-symbol** and $Q(\sigma)$ is the **Q-symbol** of the input σ.

The RSK Correspondence is also called the **Robinson–Schensted Correspondence**, **Schensted's Algorithm**, or the **bumping procedure**. It was implicit in Robinson [1938], explicit in Schensted [1961] (independently), and generalized beyond permutations in Knuth [1970].

The proof that $P(\sigma)$ and $Q(\sigma)$ are column-strict uses a technical lemma about the positions where tableaux successively grow.

4.3.11. Lemma. Let p_i and p_{i+1} be values successively INSERTed into a column-strict tableau P, creating positions A and B, respectively.

(a) The successive values bumped when inserting p_i form a strictly increasing sequence. The successive column indices of the bumped positions (and the new position) form a nonincreasing sequence.

(b) B is in a column to the right of A if and only if $p_{i+1} \geq p_i$.

Proof: (a) An element bumps only a larger element, so the values strictly increase. If in row i the element x in column j is bumped, then x cannot enter a column past j in row $i + 1$, because the columns of P are strictly increasing.

(b) We use induction on the number of rows. Whether it bumped or not, p_i is in row 1 when we start to INSERT p_{i+1}. If p_{i+1} enters without bumping, then $p_{i+1} \geq p_i$, a new rightmost position is created, and B is farther right than A. If p_i does not bump but p_{i+1} bumps, then $p_{i+1} < p_i$ and B is not farther right than A, since no later row reaches the column of A.

In the remaining case, p_i bumps some p'_i and p_{i+1} bumps some p'_{i+1}. Now $p_{i+1} \geq p_i$ if and only if $p'_{i+1} \geq p'_i$. Since A and B are the positions created when p'_i and p'_{i+1} are inserted in the tableau $\overline{P}$ consisting of the remaining rows, the induction hypothesis completes the proof.   ∎

4.3.12. Lemma. The tableaux $P(\sigma)$ and $Q(\sigma)$ produced by the RSK Correspondence are column-strict.

Proof: The processing order makes the rows of $Q(\sigma)$ nondecreasing, and the bumping procedure makes the rows of $P(\sigma)$ nondecreasing. Since dropping (q_n, p_n) from the lexicographic input order leaves the rest in lexicographic order, it suffices to show that the outputs of INSERT(P, Q, p, q) are column-strict when P and Q are column-strict and produced by pairs lexicographically earlier than (q, p).

The value q is as large as all entries in Q and appears at the bottom of its column. All copies of q appear as $(q, p_j), \ldots, (q, p_n)$ at the end of the input, with $p_j \leq \cdots \leq p_n$. By Lemma 4.3.11b, the columns of the new locations for these insertions are strictly increasing and thus distinct.

For P, let t be the column where p enters row 1. Since P is column-strict, all entries in column t of P are larger then p. Suppose that p bumps an element p'. By Lemma 4.3.11a, p' cannot survive past column t when inserted in row 2. If p' enters row 2 under element p'' of row 1, then $p'' \leq p < p'$. Continuing the INSERTion leaves every element in row 2 larger than the element above it. The result of inserting p' into $\overline{P}$ is also column-strict, since this is the P-symbol for the input consisting of elements bumped from the first row (in the same order). ∎

Next we show that the RSK Correspondence is a bijection. Since we insert the pairs in a canonical order, the input is specified by the multiplicities of distinct pairs. Encode the input as a matrix A whose (r, s)-entry is the multiplicity of (r, s) in the input. From A, we can recover the pairs and INSERT them in lexicographic order. If $\sigma \in \mathbb{S}_n$, then A is an n-by-n permutation matrix.

4.3.13. Theorem. The RSK Correspondence maps nonnegative integer matrices with total sum n bijectively into pairs of column-strict tableaux with n cells. It restricts to a bijection from $\mathbb{S}_n$ to the set of pairs of same-shape standard Young tableaux with entries $[n]$.

Proof: It suffices to invert the procedure. Given same-shape tableaux P and Q on n elements, we specify an inverse for the last INSERT operation.

DELETE(P, Q). Let q be the rightmost copy of the largest element in Q, and let p be the entry in that position in P; say that they are in row k. Delete q from Q to obtain the new Q'; also delete p from P. If $k = 1$, then return the new tableau pair and the deleted elements. If $k > 1$, then let p' be the rightmost value in row $k - 1$ that is strictly less than the "unbumped element" p. Replace p' with p in P and let p' become the unbumped element. Decrease k and repeat.

Since P and Q are column-strict, the deleted position was a corner. Column-strictness also makes the next unbumped element p' well-defined. The column indices of the "unbumps" are nondecreasing, and the values unbumped are strictly decreasing. The resulting P' and Q' are column-strict tableaux of the same shape. The modifications undo in reverse the changes performed by INSERT(P', p', Q', q) when (q, p') is the last pair in a lexicographically ordered sequence and the pair (P', Q') arises by applying RSK to the preceding portion of the sequence. Induction on the input length completes the argument. ∎

Using Lemma 4.3.11, we can obtain information about a permutation σ in terms of the positions of elements in $P(\sigma)$ and $Q(\sigma)$.

4.3.14. Definition. A permutation σ of $[n]$ (in word form) has an **ascent** at position i if $\sigma(i+1) > \sigma(i)$; otherwise there is a **descent** at i. It has an **advance** at value x if value $x+1$ is rightward of x; otherwise there is a **retreat** at x.

Reading the word form from left to right and writing $+$ or $-$ for ascents or descents produces the **up-down sequence**. Following the values from low to high and writing $+$ or $-$ for advances or retreats produces the **right-left sequence**.

4.3.15. Example. Let $\sigma = 617258934$. This permutation has advances at values $\{1, 2, 3, 6, 7, 8\}$, retreats at values $\{4, 5\}$, ascents at positions $\{2, 4, 5, 6, 8\}$, and descents at positions $\{1, 3, 7\}$. Hence the right-left sequence is $+ + + - - + + +$ and the up-down sequence is $- + - + + + - +$. These sequences can also be obtained by tracing the numbers through $P(\sigma)$ and $Q(\sigma)$, respectively. ■

$$
\begin{array}{cc}
P & Q \\
1\,2\,3\,4\,9 & 1\,3\,5\,6\,7 \\
5\,7\,8 & 2\,4\,9 \\
6 & 8
\end{array}
$$

4.3.16. Theorem. (Schützenberger [1963], Foulkes [1976]). A permutation σ has an ascent at position c if and only if $c+1$ is rightward from c in $Q(\sigma)$. It has an advance at c if and only if $c+1$ is rightward from c in $P(\sigma)$.

Proof: The row and column inequalities of tableaux imply that $c+1$ cannot appear in a position with both indices smaller or both indices larger than the position of c. Hence the only possible movements are rightward (larger column index) and downward (larger row index). By Lemma 4.3.11b, the bumping procedure places $c+1$ in a column to the right of c in $Q(\sigma)$ if and only if $\sigma(i+1) \geq \sigma(i)$.

Since the up-down and right-left sequences interchange when we replace σ with σ^{-1}, the proof of the second statement will follow from $P(\sigma) = Q(\sigma^{-1})$. ■

A tool used to prove $P(\sigma) = Q(\sigma^{-1})$ will allow us to obtain the maximum length of an increasing sublist of σ.

4.3.17. Definition. In the 2-line input $(\mathbf{q}, \mathbf{p})$, the pair (q_i, p_i) belongs to **class** t if p_i initially enters P in the tth position of the first row. We refer to p_i or q_i as a class t element or to (q_i, p_i) as a class t pair.

4.3.18. Example. *Nondecreasing sublist.* In our example $\sigma = 617258934$, element p_i is in class t if and only if the longest nondecreasing sublist up to p_i has t terms.

$$
\begin{array}{cl}
\mathbf{q} & 1\,2\,3\,4\,5\,6\,7\,8\,9 \\
\mathbf{p} & 6\,1\,7\,2\,5\,8\,9\,3\,4 \\
t & 1\,1\,2\,2\,3\,4\,5\,3\,4
\end{array}
$$
■

4.3.19. Lemma. In the class t pairs $(q_{i_1}, p_{i_1}), \ldots, (q_{i_m}, p_{i_m})$, the second coordinates are strictly decreasing, and the first coordinates are strictly increasing. Also, these pairs are those for which the longest nondecreasing sublist of p ending at this position has t elements.

Proof: Each p_i in class t bumps the previous element in class t, and thus in order they are strictly decreasing. The lexicographic input order then implies that the corresponding q_i's are strictly increasing.

The top left element is always the least yet seen; thus the longest nondecreasing sublist ending at a class 1 element is just that element. By induction on t and the number of elements seen, the current element in position t of row 1 is always the least element yet seen that ends a nondecreasing sublist of length t. ∎

4.3.20. Corollary. (Schensted's Theorem; Schensted [1961]) The maximum length of an increasing sublist of a permutation σ is the length of the first row of $P(\sigma)$. For a decreasing sublist it is the length of the first column. ∎

Schensted noted that the second statement follows from the first by reversing the permutation. Later we prove much stronger statements than Corollary 4.3.20. Note that the elements of the first row in $P(\sigma)$ need not themselves form an increasing subsequence of σ; in Example 4.3.18, the first row of $P(\sigma)$ is 12349, but the only increasing sequence of length 5 in σ is 12589.

SWITCHING P-SYMBOL AND Q-SYMBOL

To finish counting n-element standard Young tableaux, we still need $P(\sigma^{-1}) = Q(\sigma)$, equivalent to $Q(\sigma^{-1}) = P(\sigma)$. For the generalized input matrix A, interchanging $\mathbf{q}$ and $\mathbf{p}$ corresponds to transposing A. The next lemma and theorem obtain $P(A^T) = Q(A)$ and $Q(A^T) = P(A)$ directly from class t pairs. We then interpret the RSK Correspondence geometrically for an alternative proof.

4.3.21. Lemma. The pair (q_k, p_k) belongs to class t of $(\mathbf{q}, \mathbf{p})$ if and only if the pair (p_k, q_k) belongs to class t of $(\mathbf{p}, \mathbf{q})$.

Proof: If (q_k, p_k) is in class t of $(\mathbf{q}, \mathbf{p})$, then by Lemma 4.3.19 some nondecreasing subsequence $p_{i_1}, \ldots, p_{i_t}$ ends with p_k. From the lexicographic input order, $q_{i_1}, \ldots, q_{i_t}$ is also nondecreasing. When p and q are interchanged and re-ordered lexicographically, the indices of these pairs remain in the same order. Thus q_k will appear at the end of a nondecreasing subsequence of length at least t. By Lemma 4.3.19, (p_k, q_k) belongs to class t or higher in the $(\mathbf{p}, \mathbf{q})$ input. The same argument applies when the switch is repeated, so the pairs (q_k, p_k) from $(\mathbf{q}, \mathbf{p})$ and (p_k, q_k) from $(\mathbf{p}, \mathbf{q})$ belong to the same class in their respective processing. ∎

4.3.22. Theorem. (Schützenberger [1963], Knuth [1970]). Given a finite nonnegative integer matrix A, the P-symbols and Q-symbols satisfy $P(A^T) = Q(A)$ and $Q(A^T) = P(A)$. In particular, if σ is a permutation, then $P(\sigma^{-1}) = Q(\sigma)$ and $Q(\sigma^{-1}) = P(\sigma)$.

Proof: The algorithm ends with the last class t element in position t of row 1 in P. The value in position t of row 1 of Q was put there when the position was first created; it is the q_i from the first pair in class t. Lemma 4.3.21 says that the pairs in class t are the same (with coordinates reversed) when we transpose the input A; Lemma 4.3.19 says that they are processed in opposite order. Together these facts imply that transposing A interchanges the first rows of P and Q.

We have proved the basis step for induction on the number of rows. The rows below row 1 are formed by applying the same algorithm to a modified input. The modified input consists of the pairs (q_i, p_j) such that p_i bumps p_j from row 1. We need only show that the modified input creating the pair of tableaux below the first row for $(\mathbf{p}, \mathbf{q})$ is the transpose of the modified input that does so for $(\mathbf{q}, \mathbf{p})$.

If p_i bumps p_j from position t of row 1, then (q_j, p_j) is the class t pair before (q_i, p_i). By Lemmas 4.3.19–4.3.21, (p_i, q_i) and (p_j, q_j) are successive class t pairs for A^T, and q_j bumps q_i from position t when we insert (p_j, q_j). Thus (p_j, q_i) is in the modified input for A^T whenever (q_i, p_j) is in the modified input for A. ∎

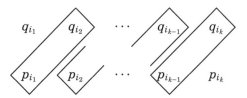

Since the P-symbol and Q-symbol are formed so differently, it is surprising that inverting a permutation exchanges them. The result becomes natural under a geometric interpretation called **planarization** developed by Viennot [1977]. It encodes bumping in a special diagram that treats $P(\sigma)$ and $Q(\sigma)$ symmetrically.

Given input σ with corresponding matrix A, put $A_{r,s}$ points at position (r, s) in the plane (the pair (r, s) occurs k times in σ). Let $T(\sigma)$ be the resulting arrangement of points with multiplicities. On this multiset we build what Viennot called *outstanding lines*. The tth outstanding line corresponds to the class t pairs in the bumping procedure. These lines yield row 1 of $P(A)$ and $Q(A)$ in a symmetric manner. They also yield the input for building the remaining rows, inductively.

4.3.23. Example. *Planarization, outstanding lines, and skeletons.* We illustrate the construction with the sample permutation $\sigma = \left(\begin{smallmatrix} 123456789 \\ 617258934 \end{smallmatrix}\right)$ used by Viennot. Its planarization appears below. In placing this diagram in the plane, we have temporarily adopted the "French notation", in which everything is indexed from the bottom left. (Indexing from the upper left is the "English convention".)

The first line is determined by the elements of the input $T(\sigma)$ that are minimal in the component-wise dominance ordering, defined by $(i, j) \leq (r, s)$ if and only if $i \leq r$ and $j \leq s$. The minimal points of $T(\sigma)$ are those that dominate no other points; here they are $(1, 6)$ and $(2, 1)$. (If there are repeated points in the input, then we just use one copy of each minimal point, but here for simplicity we only discuss permutations.)

Let M denote the set $\{(q_{i_1}, p_{i_1}) \ldots (q_{i_k}, p_{i_k})\}$ of minimal points of $T(\sigma)$. Since none dominates any other, we may index them so that $q_{i_1} < \cdots < q_{i_k}$ and $p_{i_1} > \cdots > p_{i_k}$. From M we form the **outstanding line** consisting of the vertical and horizontal zig-zag path linking the following points in order: (q_{i_1}, ∞), (q_{i_1}, p_{i_1}), (q_{i_2}, p_{i_1}), $(q_{i_2}, p_{i_2}) \ldots (q_{i_k}, p_{i_{k-1}})$, (q_{i_k}, p_{i_k}), (∞, p_{i_k}).

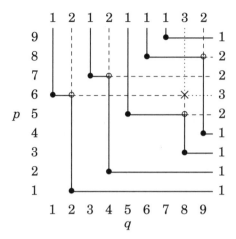

Geometrically, consider light shining from the left and below. The outstanding line is the outline of the shadow of M. Alternatively, it is the outline of the
set of points in the plane that dominate points of M.

The outstanding line changes direction at $2k-1$ points. Of these, k form M;
the other $k-1$ we place in the **skeleton** of $T(\sigma)$, denoted $S(\sigma)$. The full skeleton consists of these points plus the skeleton of $T(\sigma) - M$. Thus we find the first
outstanding line, put the inside turning points in the skeleton, delete the minimal points M from $T(\sigma)$, and iterate until no points remain. If $T(\sigma)$ has n points,
then $S(\sigma)$ has $n - \lambda_1$ points, where λ_1 is the number of outstanding lines. ∎

In the diagram of Example 4.3.23, we have iterated the process of finding the
skeleton. The solid points are $T(\sigma)$, with solid outstanding lines. The open points
are the skeleton $S(\sigma)$, with dashed outstanding lines. The skeleton of the skeleton
has only the one point denoted × and one outstanding (dotted) line. An outstanding line receives label k where it exits the diagram if it arises when forming the
skeleton the kth time. The creation of the skeleton has the same effect as bumping, and the labels of the lines give the rows of the elements in $P(\sigma)$ and $Q(\sigma)$!

4.3.24. Theorem. (Viennot [1977]). Given the permutation σ as input to the
RSK Correspondence,
    (1) The skeleton $S(\sigma)$ is the set of pairs produced by the bumping proce-
dure for computation of $P(\sigma)$ and $Q(\sigma)$ below row 1.
    (2) The kth row of $P(\sigma)$ consists of the rows labeled k in the planarization.
The kth row of $Q(\sigma)$ consists of the columns labeled k in the planarization.
    (3) $P(A^T) = Q(A)$ and $Q(A^T) = P(A)$.

Proof: The point (q_i, p_i) is a minimal point of σ in the plane if and only if p_i is a
left-to-right minimum in p, making it a class 1 pair for $(\mathbf{q}, \mathbf{p})$. Since the class t
pairs are the class 1 pairs for the input in which the earlier classes are deleted,
the points of σ on the tth outstanding line of σ are the class t pairs for $(\mathbf{q}, \mathbf{p})$.
    (1) If (q_j, p_j) and (q_i, p_i) are consecutive in class t, then (q_i, p_j) belongs to the
skeleton. Also, p_i bumps p_j when inserted (Lemma 4.3.19), and hence (q_i, p_j) belongs to the input below row 1.

(2) By (1), it suffices to prove this for $k = 1$ and apply induction on the number of rows. As observed in proving Theorem 4.3.22, the value in position t of row 1 of P is the p-coordinate of the last class t pair in the input, and the value in position t of the first row of Q is the q-coordinate of the first class t pair, placed when the position is created. Since the tth outstanding line visits the class t pairs in order, these two values are the horizontal and vertical lines on which the tth outstanding line exits the planarization diagram. Being an outstanding line of $T(\sigma)$, the label is 1, correctly placing these elements in row 1 of $P(\sigma)$ and $Q(\sigma)$. When the input pairs are not restricted to be distinct, several outstanding lines may exit at the same value.

(3) Transposing the input means reflecting the planarization diagram through the line $x = y$, which interchanges rows and columns.   ∎

JEU DE TAQUIN (optional)

By Schensted's Theorem, the length of the longest nondecreasing subsequence in a list is the length of the first row in its P-symbol. More generally, how can we find the largest size of the union of k nondecreasing sequences? Greene proved that this is the sum of the lengths of the first k rows in the P-symbol.

We need several ideas to prove this. The first generalizes the notion of "nondecreasing sequence" to planar arrangements and incorporates the properties of advances while ignoring consecutivity among the values.

4.3.25. Definition. A **partial tableau** consists of numbers in some positions in rows and columns, strictly increasing in columns and nondecreasing in rows. A **nondecreasing sequence** in a partial tableau T is a nondecreasing sequence of elements in positions with increasing column indices and nonincreasing row indices. A k-**sequence** in T is a union of k nondecreasing sequences in T. The **Greene number** $G_{k,l}(T)$ is the largest size of a k-sequence in the first l positions in T under lexicographic order on (value, column).

When the entries are distinct, the "first l positions" are those of the l smallest values. In the partial tableau $T(\sigma)$ obtained by writing the elements of a permutation σ diagonally from lower left to upper right, nondecreasing sequences are the same as increasing sequences in σ; this motivates Definition 4.3.25. To prove Greene's Theorem, we need

(1) $T(\sigma)$ and $P(\sigma)$ have the same Greene numbers, and
(2) in $P(\sigma)$ the first k rows form a largest k-sequence.

We will move from $T(\sigma)$ to $P(\sigma)$ using a class of partial tableaux containing both.

4.3.26. Example. *Three partial tableaux and their Greene numbers.* Let $\sigma = 617258934$. The diagonal tableau $T(\sigma)$ puts the values in order of the word form. As noted above, $G_{k,n}(T(\sigma))$ is the maximum size of the union of k increasing sequences in σ. The other tableaux below are obtained by sliding elements; all three have the same Greene numbers.

In R (in fact, it is $P(\sigma)$), the first two rows form a largest 2-sequence, as do 1234 and 6789. Also $T(\sigma)$ has 1234 and 6789 as a 2-sequence of size 8, but it cannot be found greedily; the only increasing sequence of length 5 is 12589, and deleting it leaves no increasing triple.

We will slide elements to move from $T(\sigma)$ to $P(\sigma)$.   ■

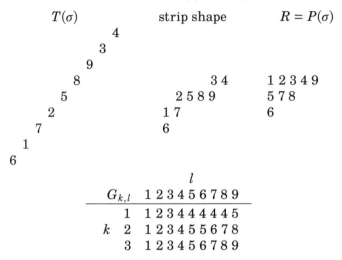

	l
$G_{k,l}$	1 2 3 4 5 6 7 8 9
	1	1 2 3 4 4 4 4 4 5
k	2	1 2 3 4 5 5 6 7 8
	3	1 2 3 4 5 6 7 8 9

The process of moving from $T(\sigma)$ to the "reduction" R is called **Jeu de Taquin**. In French, the term means "Game of Teasing", or brain-teaser. It is the French name for the English **15-puzzle**. In this puzzle, tiles numbered 1 through 15 slide in a 4-by-4 grid with one empty position; the aim is to put the numbers in order. We need an operation to slide numbers in a partial tableau.

4.3.27. Definition. Given two partitions λ and μ such that $\lambda_i \geq \mu_i$ for all i, a **skew tableau** of shape (λ, μ) is a partial tableau filling the positions in the diagram of λ that are outside the diagram of μ. A **hole** in such a tableau is an unfilled position within the skew tableau shape: outside μ but inside λ.

The tableau $T(\sigma)$ is a skew tableau, where λ is a triangular shape and μ is the next smaller triangle. The strip shape is a more complicated skew tableau, and in $P(\sigma)$ the partition μ is empty. Jeu de Taquin reduces the vacant portion μ to an empty shape by successively filling holes.

4.3.28. Definition. Given a hole in a skew tableau, with elements immediately below and to the right, a **SHIFT** modifies the tableau by sliding into the hole the smaller of those two elements (from below if they are equal); positions outside λ are assumed to hold $+\infty$.

Given a skew tableau of shape (λ, μ), a **FILL** operation deletes a corner of μ to form μ', which leaves a skew tableau of shape (λ, μ') with one hole, and then successively applies SHIFT to flush the hole out of the shape, leaving a skew tableau of shape (λ', μ'), where λ' is obtained from λ by deleting the corner reached by the hole.

In Example 4.3.26, the strip shape arises from $T(\sigma)$ by FILL operations that compare consecutive elements of σ. This leaves σ_{i+1} rightward of σ_i if $\sigma_{i+1} \geq \sigma_i$ (by a vertical SHIFT) and σ_{i+1} above σ_i if $\sigma_{i+1} < \sigma_i$ (by a horizontal SHIFT). After comparing σ_i and σ_{i+1}, a horizontal SHIFT is followed by SHIFTing the later elements left (by FILLing each later corner of the current μ), while a vertical SHIFT is followed by SHIFTing the earlier elements up (by FILLING each earlier corner of the current μ. The resulting strip shape encodes the up-down sequence of σ.

Further FILLing transforms the strip shape to a tableau R, which in Example 4.3.26 we see is $P(\sigma)$ from Example 4.3.15. **Jeu de Taquin** is the process of successively FILLing corners of μ, in some order. We will show Jeu de Taquin always produces a column-strict tableau from $T(\sigma)$ and that it is always $P(\sigma)$.

Jeu de Taquin was invented by Schützenberger [1977]. Our motivation here is the purely combinatorial result of Greene. Viennot called the tableau $R(\sigma)$ resulting from Jeu de Taquin the **redressé** of the original, meaning "straightened"; we use the English word **reduction**.

The plan. We show first that Jeu de Taquin can reduce the diagonal tableau $T(\sigma)$ to $P(\sigma)$ by modeling the bumping procedure with SHIFTs. Next, in a column-strict tableau with n elements, $G_{k,n}$ is just the sum of the lengths of the first k rows. The fact that every reduction $R(\sigma)$ equals $P(\sigma)$ follows because the Greene numbers determine exactly one column-strict tableau and each SHIFT in Jeu de Taquin preserves the Greene numbers. Greene's Theorem is then a corollary.

4.3.29. Lemma. For an input σ, some instance of Jeu de Taquin reduces the diagonal tableau $T(\sigma)$ to the P-symbol $P(\sigma)$.

Proof: We model the bumping procedure, by induction on the length of σ. No SHIFT is needed when σ has length 1. Otherwise, form σ' from σ by ignoring the last element x and inductively transform $T(\sigma')$ into $P(\sigma')$ using Jeu de Taquin. FILLing corners now can move $P(\sigma')$ up from below and x in from the right to obtain a partial tableau where μ consists of one part equal to the length of the first row in P', with x immediately to its right and $P(\sigma')$ below it.

We now model the bumping procedure to insert x into $P(\sigma')$. Apply SHIFT to the left of x until x slides left no farther. Now we have $P(\sigma')$ plus x sitting above the element x' that x bumps when inserted into $P(\sigma')$. SHIFTs into the other positions of row 1 (in decreasing order) now produce the first row of $P(\sigma)$, with x above x'. Now x' sits alone in row 2, ready to enter the tableau below the first row of $P(\sigma')$. In particular, since the list of column indices of bumped positions is nonincreasing (Lemma 4.3.11a), the newly bumped element y' is no farther left than where it should enter the next row. By the induction hypothesis, we can perform the rest of the bumping procedure with further SHIFTs.

We have not made these SHIFTs in the order of Jeu de Taquin, but they can be reordered to express the process (after the induction hypothesis) as FILLing row 1 from right to left. As x moves in, the rest of each FILL pulls up an element from row 2 behind it. After x reaches its resting place, the first SHIFT in each fill to its left pulls up the element from row 2. Instead of performing these immediately, we can postpone them to subsequent FILLs, since they are not disturbed by FILLING the hole to the left of x. The induction hypothesis similarly permits reordering the other SHIFTs to complete each FILL in the first row. ∎

4.3.30. Lemma. If T is a column-strict tableau (without holes), then the first k rows of T form a largest k-sequence in T.

Proof: For a tableau T of shape λ, let $f_k(T) = \sum_{i=1}^{k} \lambda_i$. Since the first k rows form a k-sequence, it suffices to show that no k-sequence is larger than $f_k(T)$. We use induction on the number of elements; the claim is trivial when T has one element. Let x in position (i, j) be the rightmost copy of the largest entry in T; delete x to obtain T'. Every k-sequence in T that omits x is a k-sequence in T'; by the induction hypothesis, we need only consider a k-sequence S containing x.

Since $S - \{x\}$ is a k-sequence in T', we have $|S| \le f_k(T') + 1$. If $i \le k$, then $f_k(T) = f_k(T') + 1$ and the bound holds. Hence we may assume $i > k$. Now every nondecreasing sequence in T containing x lies completely below the first k rows. Also it has size at most j, since the elements of a nondecreasing sequence occupy distinct columns. The remainder of S is a $(k - 1)$-sequence in T. Also $j \le \lambda_k$, since $i > k$. Thus $|S| \le f_{k-1}(T) + \lambda_k = f_k(T)$. ∎

4.3.31. Lemma. The shape of a column-strict tableau and the placement of its multiset of entries is determined by the Greene numbers.

Proof: Let T and T' be column-strict tableaux with the same elements. If $T \ne T'$, then let l be the least integer such that the subtableaux T_l and T'_l formed by the first l values in T and in T' differ (repeats of a single value are taken in order from left to right to form the first l values).

Note that T_l and T'_l differ only in where the last element is, say in row k in T_l and row k' in T'_l. We may assume $k < k'$ by symmetry. By Lemma 4.3.30, $G_{k,l}(T) = f_k(T_l) > f_k(T'_l) = G_{k,l}(T')$. Thus distinct tableaux with the same entries cannot have the same Greene numbers. ∎

The proofs of Lemmas 4.3.30–4.3.31 yield an algorithm to construct the column-strict tableau determined by the Greene numbers. It suffices to find for each l the row for the lth element; this is the least k such that $G_{k,l} > G_{k,l-1}$.

Finally, we show that each SHIFT in a FILL preserves the Greene numbers.

4.3.32. Lemma. If T is a skew tableau with one hole, and T' is obtained from T by a SHIFT into the hole, then their Greene numbers are the same: $G_{k,l}(T') = G_{k,l}(T)$ for all k and l.

Proof: A horizontal SHIFT does not change the sets of entries forming k-sequences, so here $G_{k,l}(T') = G_{k,l}(T)$ holds trivially. Vertical SHIFTs can change k-sequences. We prove $G_{k,l}(T') \ge G_{k,l}(T)$ for this case and leave the similar inequality $G_{k,l}(T) \ge G_{k,l}(T')$ as Exercise 17.

Let S of size $G_{k,l}(T)$ be a largest k-sequence among the first l elements in T. Suppose that T' arises by SHIFTing y from row j to row $j - 1$. The set S is still a k-sequence unless y lies in a sequence A in S with elements to its right in row j. Let $A = aybc$, where b is the part after y in row j.

Since T' has only one hole, all positions in row $j - 1$ above elements of b are filled; let b' be this list above b in row $j - 1$. Since $\max b' < \max b$, we have $\max b' < \min c$; also $y \le \min b'$ since y SHIFTs up. If no element of b' appears in S, then we can modify A by replacing b with b' to obtain $G_{k,l}(T') \ge G_{k,l}(T)$.

Otherwise, let z be the leftmost element of b' in S, and let B be the nondecreasing sequence in S containing z. Write $B = a'dd'$, where d is the part of B in row j after y (d' begins with z). If d is empty, then y can move from A into B.

If $d \neq \varnothing$, then break b into b_1 and b_2, where b_1 is the part to the left of the right end of d. Let b'_1 be the part of b' above b_1. Replace b_1 by b'_1 in A and switch the latter parts of A and B to form A' and B':

$$A = aybc \qquad A' = ayb'_1d'$$
$$B = a'dd' \qquad B' = a'db_2c$$

Note that $A' \cap B' = \varnothing$ and $|b'_1| + |b_2| = |b|$, so $|A'| + |B'| = |A| + |B|$. The elements of b'_1 are available; by the choice of z, none of them are in S.

The new sequences move strictly rightward in columns and weakly upward in rows. It suffices to check inequalities at the splices. If $b'_1 \neq \varnothing$, then $y \leq \min b'_1 \leq \max b'_1 < \max b_1 \leq \max d \leq \min d'$. If $b'_1 = \varnothing$, then the assumption $B \cap b' \neq \varnothing$ guarantees $y \leq \min d'$, since d' cannot avoid row $j - 1$. For B', it suffices that db_2 lies entirely in row j. New elements are bounded by others used, so all elements are still among the l smallest. ∎

Various cases in the proof of Lemma 4.3.32 change lists. Although $G_{k,l}$ does not change, the previous largest k-sequence may no longer be a k-sequence. We already saw that the top line of $P(\sigma)$ need not be a nondecreasing sequence in σ.

4.3.33. Theorem. The reduction R of a skew tableau T is independent of the order of FILLs. Also, the sum $f_k(R)$ of the first k row-lengths in R is the maximum size of a k-sequence in T.

Proof: Successive FILLs at corners of μ produce a column-strict tableau. By Lemma 4.3.32, the Greene numbers remain unchanged. By Lemma 4.3.31, the shape and placement of entries in a column-strict tableau with fixed entries and Greene numbers is fixed. Since also Lemma 4.3.29 shows that $P(\sigma)$ can be reached, the reduction R is independent of the sequence of FILLs, and $f_k(P(\sigma))$ is the maximum size of a k-sequence in $T(\sigma)$, which corresponds to the union of k increasing sequences in σ. ∎

As a corollary, we obtain Greene's Theorem. In order to state it for both increasing and decreasing sequences, we need one more observation.

4.3.34. Corollary. If σ^* denotes the reverse (word form) of a permutation σ, then $P(\sigma^*)$ is the transpose of $P(\sigma)$.

Proof: Initially, $T(\sigma^*)$ is the transpose of $T(\sigma)$. If a tableau T has distinct entries and T' is the transpose of T, then the result of SHIFT to fill position (j, i) in T' is the transpose of the result of SHIFTing to fill (i, j) in T. Hence for any sequence of SHIFTs reducing $T(\sigma)$, there is a corresponding sequence of SHIFTs that reduces $T(\sigma^*)$ and maintains the transpose relationship at every step. ∎

4.3.35. Corollary. (Greene's Theorem; Greene [1974]) The maximum size of the union of k increasing [decreasing] subsequences of a permutation σ is the total size of the first k rows [columns] of $P(\sigma)$.

Proof: Since Lemma 4.3.29 implies that the reduction is $P(\sigma)$ when T is the diagonal tableau of σ, the claim for increasing sequences is a special case of Theorem 4.3.33. For decreasing subsequences, combine this with Corollary 4.3.34. ∎

EXERCISES 4.3

4.3.1. (−) Find the P-symbol and Q-symbol for the permutation 596134278. Find the two-line generalized input σ such that the RSK Correspondence yields the results below.

$$P(\sigma) = \begin{matrix} 1\ 1\ 1\ 2\ 2 \\ 2\ 3\ 3 \end{matrix} \qquad Q(\sigma) = \begin{matrix} 1\ 1\ 2\ 3\ 3 \\ 2\ 2\ 3 \end{matrix}$$

4.3.2. (−) Use Young tableaux to prove the Erdős–Szekeres Theorem: every list of $n^2 + 1$ distinct numbers has a monotone subsequence of length $n + 1$.

4.3.3. (−) How many permutations σ of $[n]$ have element 2 in the first row of $P(\sigma)$?

4.3.4. *The Non-Messing-Up Theorem.* Consider a matrix such that in each row the entries appear in increasing order. Prove that after sorting the entries of each column into increasing order, the entries of the rows are still in increasing order.

4.3.5. (◊) Find the minimum and maximum of the number of comparisons between elements when applying the RSK Correspondence to a permutation of $[n]$.

4.3.6. A riffle shuffle of a permutation σ cuts σ into initial and final segments and then merges them. Let $c(\sigma)$ be the minimum number of riffle shuffles that can sort σ into order. Let $d(\sigma)$ be the number of descents in σ. Prove $c(\sigma) = \lceil \log_2(1 + d(\sigma)) \rceil$. (Schwenk [1986])

4.3.7. Show that the number of ways to label the vertices of an n-vertex ordered tree (Definition 1.3.21) with $[n]$ so that each label is less than those on its descendants is $n!$ divided by the numbers of vertices in each subtree rooted at a single vertex. (Knuth)

4.3.8. Prove combinatorially that $(n^2)!$ is divisible by $n^n \prod_{i=1}^{n-1}[i(2n-i)]^i$.

4.3.9. A tableau with shape λ is a drawing with $\sum \lambda_i$ squares in the plane. The corners of the squares have degrees 2, 3, or 4 in the drawing. In terms of λ_1, k, and the number of distinct parts in λ, count the corner points with odd degree. (Stanley [2014])

4.3.10. Obtain a simple formula for the number of permutations of $[rs]$ having no increasing sublist with $s+1$ terms and no decreasing sublist with $r+1$ terms. (Schensted [1961])

4.3.11. (◊) *Permutations with forbidden patterns.*
   (a) Establish a one-to-one correspondence between tableaux consisting of two rows of length n each and pairs of n-element same-shape tableaux with at most two rows.
   (b) Conclude that the number of permutations of $[n]$ having no decreasing sublist of length 3 is the Catalan number C_n.
   (c) Exercise 2.1.48 shows that C_n also counts the permutations of $[n]$ having no triple $i < j < k$ such that $\sigma(k) < \sigma(i) < \sigma(j)$. Conclude that for each of the six orderings of three elements, the set of permutations of $[n]$ whose triples avoid that pattern has size C_n.

4.3.12. (◊) Show that the number of fixed points in an involution is the number of columns having odd length in its P-symbol. (Schützenberger [1963])

4.3.13. ($\diamond$) *Alternative proof of Hook-Length Formula.* Let λ be a partition of n with m parts, and define $f(\lambda)$ and $F(\lambda)$ as in Theorem 4.3.4.

(a) Group terms in the Hook-Length Formula $F(\lambda)$ to prove

$$\frac{F(\lambda)}{n!} = \frac{\prod_{i<j}(h_{i,1} - h_{j,1})}{\prod_i h_{i,1}!} = \frac{\prod_{i<j}(\lambda_i - i - \lambda_j + j)}{\prod_i (\lambda_i - i + m)!}$$

(b) With $\mathbb{S}_m$ being the set of permutations of m, prove

$$f(\lambda) = \sum_{\sigma \in \mathbb{S}_m} (\text{sign}\,\sigma)\binom{n}{\lambda_1 + \sigma(1) - 1, \ldots, \lambda_m + \sigma(m) - m},$$

(c) Prove the Hook-Length Formula $f(\lambda) = F(\lambda)$ by proving that the expressions in parts (a) and (b) are equal. (Hint: Let $b_i = \lambda_i + m - i$. For any values $b_1, \ldots, b_m$, the product $\prod_{i<j}(b_i - b_j)$ is the **Vandermonde determinant** for the matrix with entry b_i^{m-j} in position (i, j). Comment: This formula for $f(\lambda)$ was known to Frobenius [1900] and Young [1902]; MacMahon [1916] found another proof. This proof is from Linial [1982].)

4.3.14. (+) Given a partition λ with r parts, and $m \geq r$, let $f(\lambda; m)$ count the column-strict tableaux with shape λ and elements in $[m]$. Let M be the m-by-m matrix with (i, j)-entry $\binom{\lambda_j + m - j}{m - i}$, where $\lambda_j = 0$ for $r < j \leq m$.

(a) Prove $\det M = V(\lambda_1 + m - 1, \lambda_2 + m - 2, \ldots, \lambda_m)/\prod_{i=1}^m (m - i)!$, where $V(b_1, \ldots, b_m)$ is the Vandermonde determinant with (i, j)-entry b_j^{m-i}, equal to $\prod_{i<j}(b_i - b_j)$.

(b) Prove $f(\lambda; m) = \det M$. (Littlewood [1940]; see also Bender–Knuth [1972])

4.3.15. (+) For a partition λ with distinct parts, the **shifted shape** is the set of locations $\{(i, j): i \leq j \leq i + \lambda_i - 1\}$; row i starts with the diagonal position (i, i). With distinct parts, no row extends past the previous row. A **standard shifted tableau** fills the positions in a shifted shape of size n with the elements of $[n]$, increasing in both rows and columns. Thrall [1952] proved that the number of standard shifted tableaux with shape λ is

$$n!\frac{\prod_{i<j}(\lambda_i - \lambda_j)}{\prod_i \lambda_i! \prod_{i<j}(\lambda_i + \lambda_j)},$$

(a) For a position (i, j) in the shifted shape, define the *shifted hook* $H_{i,j}^*$ by

$$H_{i,j}^* = \{(i, s): s \geq j\} \cup \{(r, j): r \geq i\} \cup \{(j + 1, s): s \geq j + 1\}.$$

Below we show the hook $H_{1,2}^*$ in the shape $(5, 3, 2)$. Use part (a) of Exercise 4.3.13 to show that Thrall's formula equals $n!/\prod_{i,j}|H_{i,j}^*|$ (for $(5, 3, 2)$, both formulas yield 54).

(b) Prove Thrall's formula for the number of shifted tableaux. (Hint: Prove that the formula and the number of shifted tableaux satisfy the same recurrence. Comment: Other proofs appear in Stanley [1972] and Sagan [1980]; the latter is probabilistic.)

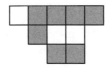

4.3.16. Consider a zig-zag skew shape with n boxes, two in each row (except one in the first row if n is odd), as shown below. Let a_n be the number of skew tableaux of this shape with entries $[n]$ (these correspond to permutations with up-down sequence $+ - + - \cdots$, called **alternating permutations**; see Stanley [2010] for a survey).

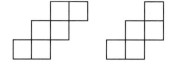

(a) Find a (nonlinear) recurrence for $\langle a \rangle$.

(b) Let $g(x) = \sum a_n x^n/n!$ and $h(x) = \frac{1}{2}[g(x) - g(-x)]$. Prove $h(x)g(x) = g'(x) - 1$.

(c) Solve the differential equation in (b) to obtain $g(x) = \sec x + \tan x$. (André [1881])

4.3.17. Let T be a skew tableau with one hole, and let T' be a partial tableau obtained from T by a vertical SHIFT filling the hole. Prove $G_{k,l}(T) \le G_{k,l}(T')$ for all k, l. (Comment: This completes the proof of Greene's Theorem.)

4.3.18. The **row-canonical permutation** of a standard Young tableau is obtained by reading the tableau row by row (left to right) starting at the bottom. Its **column-canonical permutation** reads it column by column (bottom to top) starting at the left. Prove that the P-symbol of a row-canonical or column-canonical permutation is the Young tableau from which it was obtained. Conclude also that the reverse of a row-canonical permutation is the column-canonical permutation for the transposed tableau.

4.3.19. ($\diamond$) In a permutation σ (in word form), a transposition of two consecutive values is **admissible** if some value between their values appears in a position next to one of them. (Comment: Such transpositions are also called **Knuth transformations**, with permutations being **Knuth-equivalent** if one can be obtained from the other via such operations. The definition and results extend also to generalize inputs $(\mathbf{q}, \mathbf{p})$.) (Knuth [1970])

(a) Prove $P(\sigma) = P(\sigma')$ when σ' is obtained from σ by an admissible transposition. (Comment: There is a proof by induction, but using Jeu de Taquin is much simpler.)

(b) Prove that every permutation transforms to a row-canonical permutation using admissible transpositions. (Hint: Model the bumping procedure.)

(c) Conclude from (b) that every permutation can be transformed to a column-canonical permutation using admissible transpositions.

4.3.20. For a tableau T, the **deletion operator** produces a tableau $d(T)$ by emptying the upper left corner and filling the hole by Jeu de Taquin. Define τ from σ by deleting the first occurrence of the least element. Prove $P(\tau) = d(P(\sigma))$. (Schützenberger [1963])

4.3.21. ($\diamond$) A **balanced tableau** of shape λ is a tableau in which for every position (i, j), the hook $H_{i,j}$ (Definition 4.3.2) has $\lambda_j^* - i$ numbers less than entry (i, j) and $\lambda_i - j$ numbers greater than entry (i, j), where λ^* is the conjugate of λ. For example, there are three balanced tableaux of shape $(3, 1)$ and 16 of shape $(3, 2, 1)$. An n-**staircase tableau** is a tableau with shape λ such that $\lambda_i = n - i$ for $1 \le i \le n - 1$. For the n-staircase shape, let $\mathbf{B}$ be the set of balanced tableaux, and let $\mathbf{Y}$ be the set of standard Young tableaux.

(a) Prove that an n-staircase tableau T is balanced if and only if for every position (i, j) and all k with $i < k \le n - j$, exactly one of the entries (k, j) and $(i, n + 1 - k)$ (in the hook $H_{i,j}$) is less than entry (i, j).

(b) One can pass from the identity permutation to its reverse via $\binom{n}{2}$ steps of transposing two adjacent elements to increase the number of inversions by 1. Let $\mathbf{P}$ be the set of such paths. Suppose that a path $P \in \mathbf{P}$ transposes adjacent elements a and b on step t. Let $\phi(P)$ be an n-staircase tableau defined by putting t in position $(n + 1 - b, a)$, for $1 \le t \le \binom{n}{2}$. Prove that ϕ is a bijection from $\mathbf{P}$ to $\mathbf{B}$.

(Comment: Edelman–Greene [1987] proved this and other results about balanced tableaux, including the amazing result that for *every shape* λ there are the same number of balanced tableaux and Young tableaux. The combination of those results provided a combinatorial proof of the algebraic result by Stanley [1984] that $|\mathbf{P}|$ and $|\mathbf{B}|$ are equal.)

Chapter 5

First Concepts for Graphs

The usefulness of graphs stems from their flexibility to model diverse situations: for example, in Theorem 4.1.35 we used paths in a directed graph to model tilings of hexagons by rhombi. In Part II we discuss graphs for their own sake, emphasizing proof techniques and classical topics. With this foundation, later we explore further connections between graph theory and other areas.

5.1. Definitions and Examples

We introduced some definitions about graphs and directed graphs in Chapter 0 to facilitate discussion of enumerative questions. We repeat these for convenience and provide others to be consulted as needed when reading later material. Although graph theory begins with an onslaught of definitions, most are natural and visual, and the terms become familiar with use.

GRAPHS AND SUBGRAPHS

5.1.1. Definition. A **graph** G is an ordered pair consisting of a **vertex set** $V(G)$ and an **edge set** $E(G)$, where each **edge** (element of $E(G)$) is a set of two **vertices** (elements of $V(G)$). The vertices of an edge are its **endpoints**. We write xy for an edge with endpoints x and y. The **order** of a graph G is $|V(G)|$; its **size** is $|E(G)|$. A graph with order n is an n-**vertex graph**.

A graph is defined "**on**" its vertex set by specifying the edges. It is **finite** if its vertex set and edge set are finite. Our graphs are finite unless stated otherwise.

5.1.2. Definition. A **path** with n vertices is a graph whose vertices can be named $v_1, \ldots, v_n$ so that the edges are $\{v_i v_{i+1} : 1 \le i \le n-1\}$. We write $\langle v_1, \ldots, v_n \rangle$ to specify a path having vertices $v_1, \ldots, v_n$ in order.

A **cycle** with n vertices is a graph whose vertices can be named $v_1, \ldots, v_n$ so that the edge set is $\{v_i v_{i+1} : 1 \le i \le n-1\} \cup \{v_n v_1\}$. We write $[v_1, \ldots, v_n]$ to specify this cycle. The **length** of a path or cycle is the number of edges. A cycle is an **odd cycle** [**even cycle**] if its length is odd [even].

The paths $\langle v_1, \ldots, v_n \rangle$ and $\langle v_n, \ldots, v_1 \rangle$ are the same graph. The square brackets used to denote a cycle suggest "closing" it. Since the starting point and direction do not matter, Definition 5.1.2 gives $2n$ ways to specify any n-vertex cycle.

5.1.3. Definition. A **subgraph** of a graph G is a graph H such that $V(H) \subseteq V(G)$ and $E(H) \subseteq E(G)$. A **spanning subgraph** of G is a subgraph H such that $V(H) = V(G)$. An **induced subgraph** of G, induced by vertex set $S \subseteq V(G)$, is the subgraph H with vertex set S such that $E(H) = \{uv \in E(G) \colon u, v \in S\}$; we write this as $H = G[S]$. A graph **contains** its subgraphs.

5.1.4. Example. Every cycle of length n contains n paths with n vertices; these are not induced subgraphs. However, a cycle of length n does contain $(n-1)$-vertex paths as induced subgraphs. The graph below has 49 subgraphs (we list three); 16 are induced subgraphs. Its only cycle is an odd cycle.

The only spanning induced subgraph of a graph G is G itself.   ∎

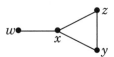

vertex set	edge set	spanning subgraph?	induced subgraph?
w, x, y, z	xy, yz	yes	no
x, y, z	xy, yz	no	no
x, y, z	xy, yz, xz	no	yes

A graph defines relations on its vertex set.

5.1.5. Definition. Vertices x and y in a graph G are **adjacent** (or **neighbors**) if $xy \in E(G)$. We may write $x \leftrightarrow y$ for "x is adjacent to y" and $x \nleftrightarrow y$ for "x is not adjacent to y", denoting the **adjacency** and **nonadjacency** **relations** on $V(G)$. The **neighborhood** $N_G(x)$ or $N(x)$ of a vertex x is $\{y \in V(G) \colon xy \in E(G)\}$. An **isolated vertex** is a vertex x with $N(x) = \varnothing$.

The **degree** of a vertex v in a graph G, written $d_G(v)$ or $d(v)$, is the number of edges containing v. The minimum vertex degree is $\delta(G)$; the maximum is $\Delta(G)$. A graph G is **regular** if $\Delta(G) = \delta(G)$; k-**regular** if that degree is k. Graphs that are 3-regular are also called **cubic graphs**.

5.1.6. Definition. A **complete graph** is a graph whose vertices are pairwise adjacent. A set of pairwise adjacent vertices is a **clique**; it is a k-**clique** when it has size k. A set of pairwise nonadjacent vertices is an **independent set**, also called a **stable set**. The **complement** of a graph G is the graph $\overline{G}$ with vertex set $V(G)$ defined by $uv \in E(\overline{G})$ if and only if $uv \notin E(G)$.

A complete graph with n vertices is $(n-1)$-regular; a cycle is 2-regular. The vertex set of a complete subgraph is a clique. In Example 5.1.4, the largest clique has three vertices, and the largest stable set has two. In the complement, $\langle y, w, z \rangle$ is a path and x is isolated, having no neighbor. The complement of an n-vertex graph G has $\binom{n}{2} - |E(G)|$ edges, and the complement of a regular graph is regular.

5.1.7. Definition. A graph G is **bipartite** if $V(G)$ is the union of two independent sets called its **parts**. An X, Y-**bigraph** is a bipartite graph G with parts X and Y, and (X, Y) is then a **bipartition** of G. An X, Y-bigraph with all of X adjacent to all of Y is a **complete bipartite graph**. A k-**partite graph** is a graph whose vertex set is the union of k independent sets.

5.1.8. Example. A graph is 1-partite if and only if it has no edges, such as the complement of a complete graph. Every graph is k-partite for some k, since individual vertices are independent sets. When $l \geq k$, every k-partite graph is also l-partite, since independent sets may be empty. The complement of a complete bipartite graph consists of two disjoint complete graphs.

The vertices of a path or cycle in a bipartite graph must alternate between the parts. Hence *every cycle in a bipartite graph is an even cycle*. Below are three drawings of a complete bipartite graph with three vertices in each part. The fourth drawing also has six vertices, all of degree 3, but it cannot be a drawing of a bipartite graph because it has an odd cycle.   ∎

We draw a graph on paper by assigning each edge a curve joining points assigned to the endpoints of the edge. We require that curves for distinct edges have only finitely many common points. Since the incidence relation associating each curve with its endpoints is the incidence relation of the graph, we think of the drawing *as* the graph. More precisely, it is a member of the isomorphism class of the graph, which brings us to the next topic.

ISOMORPHISM

To specify a graph, we must specify the vertices and edges. We can list edges or neighborhoods. Also matrix representations can be useful.

5.1.9. Definition. For an ordering $v_1, \ldots, v_n$ of $V(G)$, the **adjacency matrix** $A(G)$ is the $0, 1$-matrix having $A_{i,j} = 1$ if and only if $v_i \leftrightarrow v_j$.

Vertex v and edge e are **incident** if v is an endpoint of e. Also edges with a common endpoint are **incident**. Given a vertex ordering $v_1, \ldots, v_n$ and an edge ordering $e_1, \ldots, e_m$, the **incidence matrix** $M(G)$ is the $0, 1$-matrix defined by $M_{i,j} = 1$ if and only if $v_i \in e_j$.

5.1.10. Example. Below are two adjacency matrices and an incidence matrix for the 4-vertex path $\langle w, x, y, z \rangle$ with edges e_1, e_2, e_3 in order.   ∎

$$
\begin{array}{c|cccc}
 & w & x & y & z \\
\hline
w & 0 & 1 & 0 & 0 \\
x & 1 & 0 & 1 & 0 \\
y & 0 & 1 & 0 & 1 \\
z & 0 & 0 & 1 & 0
\end{array}
\qquad
\begin{array}{c|cccc}
 & w & y & z & x \\
\hline
w & 0 & 0 & 0 & 1 \\
y & 0 & 0 & 1 & 1 \\
z & 0 & 1 & 0 & 0 \\
x & 1 & 1 & 0 & 0
\end{array}
\qquad
\begin{array}{c|ccc}
 & e_1 & e_2 & e_3 \\
\hline
w & 1 & 0 & 0 \\
x & 1 & 1 & 0 \\
y & 0 & 1 & 1 \\
z & 0 & 0 & 1
\end{array}
$$

The adjacency matrix depends on the vertex ordering. However, structural properties do not depend on this ordering or on how we draw a graph or name its vertices. Isomorphism captures this notion.

5.1.11. Definition. An **isomorphism** from a graph G to a graph H is a bijection $f \colon V(G) \to V(H)$ such that $uv \in E(G)$ if and only if $f(u)f(v) \in E(H)$. When an isomorphism exists, we write $G \cong H$ and say "G *is isomorphic to* H". The **isomorphism relation** is the set of pairs (G, H) such that $G \cong H$.

5.1.12. Remark. *Isomorphism classes and "unlabeled" graphs.* Isomorphisms can be composed and inverted. Hence isomorphism is an equivalence relation on graphs, and the equivalence classes are called **isomorphism classes**. A drawing of a graph specifies a member of its isomorphism class, so we often omit vertex and edge labels in drawings.

The informal term **unlabeled graph** indicates an isomorphism class. When two graphs are isomorphic, we often use the same name for them, because the object of importance is the isomorphism class, and all members of it have the same structural properties. Thus we often write $G = H$ instead of $G \cong H$; asking whether a given graph *is* G means asking whether it is isomorphic to G. We say "graph" in discussing an isomorphism class to refer to each member of the class.

Exercises 15–27 study isomorphism. To prove $G \cong H$, one shows that a particular vertex bijection is an isomorphism or that G and H have the same structural description. To prove $G \ncong H$, one shows that some aspect of their structure that does not depend on vertex names differs. For example, a bipartite graph cannot be isomorphic to a graph having an odd cycle. ∎

5.1.13. Definition. *Special isomorphism classes.* We use C_n, P_n, K_n for the cycle, path, and complete graph of order n, respectively, viewed as isomorphism classes. We use $K_{r,s}$ for the complete bipartite graph with parts of sizes r and s. A **star** is a graph of the form $K_{1,t}$. The **claw** is $K_{1,3}$. The **triangle** is K_3.

We use "H is a subgraph of G" to mean that G has a subgraph isomorphic to H. Thus we say that C_k is a subgraph of G whenever G has a cycle with k vertices. A subgraph of G isomorphic to H is a **copy** of H in G.

5.1.14. Definition. An **automorphism** of G is an isomorphism from G to itself; G is **vertex-transitive** if for any $u, v \in V(G)$, some automorphism maps u to v. It is **edge-transitive** if for any $e, f \in E(G)$, some automorphism maps e (as a set of two vertices) to f.

The identity is an automorphism, composition of automorphisms is an automorphism, and they are invertible, so they form a group under composition.

5.1.15. Example. The cycle C_n has $2n$ automorphisms (n rotations and n reflections), corresponding to the $2n$ ways we can specify the cycle by listing its vertices (compare to Example 4.2.23).

The path P_n has two automorphisms, and K_n has $n!$. Although $K_{r,s}$ has $r!s!$ automorphisms when $r \neq s$, it has $2r!s!$ when $r = s$. Both C_n and K_n are vertex-transitive and edge-transitive; P_n is neither. Although $K_{r,s}$ is edge-transitive, it is not vertex-transitive when $r \neq s$.

THE PETERSEN GRAPH AND HYPERCUBES

The complete graph K_n has $n!$ automorphisms, but its structure is not very interesting, and it has too many edges for practical applications. The cycle C_n has only n edges, but also only $2n$ automorphisms. We next discuss a 10-vertex graph that is almost as sparse as C_{10} but has many more automorphisms and rich structure. It is an important example in many contexts, and there are many families in which this is the smallest or the only graph. An entire book was devoted to topics involving it (Holton–Sheehan [1993]).

5.1.16. Example. *The Petersen graph.* The graphs below are isomorphic (more precisely, any two of them are isomorphic). We refer to any graph isomorphic to these as the **Petersen graph**. A structural description of the Petersen graph names its vertices as the 2-element subsets of [5] (we write the vertex names without set brackets). Two vertices form an edge in this graph if and only if as 2-sets from [5] they are disjoint. Although the graph is named for Petersen [1891], it appeared earlier (using the rightmost drawing) in Kempe [1886, p. 11].   ∎

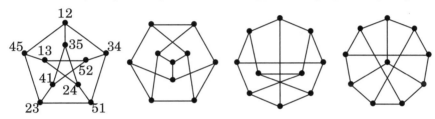

We derive several important properties of the Petersen graph.

5.1.17. Definition. A k-**cycle** is a cycle of length k. The **girth** of a graph with a cycle is the length of a shortest cycle in the graph (with no cycle, the girth is infinite). A **chord** of a cycle C is an edge not in C whose endpoints are in C.

5.1.18. Proposition. The Petersen graph G is 3-regular, and any two nonadjacent vertices in it have exactly one common neighbor. Also it has girth 5 but no spanning cycle.

Proof: View the vertices as 2-sets in [5], with adjacency being disjointness. There are three ways to select two elements of $[5] - \{a, b\}$, so every vertex has degree 3.

If ab and cd are nonadjacent, then they have a common element. Their union omits one pair in [5], which is their unique common neighbor.

A 3-cycle requires three pairwise disjoint pairs in [5], which don't exist. A 4-cycle thus requires nonadjacent vertices with two common neighbors, also forbidden. Hence every cycle has length at least 5, and $[12, 34, 51, 23, 45]$ is a 5-cycle.

Suppose that G contains a spanning cycle C. The remaining edges must be chords of C. Since G has girth 5, chords can only join vertices that are opposite or nearly-opposite on C. Making two consecutive vertices of C adjacent to the vertices opposite them creates a 4-cycle. By symmetry, we may therefore assume $v_0 v_4 \in E(G)$, where $C = [v_0, \dots, v_9]$. Now every way of giving v_9 another neighbor creates a cycle of length at most 4.   ∎

In fact, the Petersen graph is the only 3-regular graph with 10 vertices that has girth at least 5 and has no spanning cycle (Exercise 32). The structural description immediately makes it vertex-transitive and edge-transitive. Every permutation of [5] permutes the pairs in [5] and preserves the disjointness relation; this generates 120 automorphisms of the Petersen graph. In fact, these are all the automorphisms (Exercise 33).

Our next family of graphs has many applications in computer science.

5.1.19. Example. The **k-dimensional cube** or **hypercube** is the graph Q_k with vertex set $\{0,1\}^k$ such that two k-tuples are adjacent if and only if they differ in exactly one position. Below we draw Q_3. The number of ways to form two vertices differing in one coordinate is exactly $k2^{k-1}$, so $|E(Q_k)| = k2^{k-1}$. Each edge joins vertices whose names have an even and an odd number of 1s, so Q_k is bipartite.

Permuting coordinates and/or switching 0 and 1 within fixed coordinates generates $k!2^k$ automorphisms of Q_k; these are all the automorphisms (Exercise 42).

For $k \geq 1$, the graph Q_k arises naturally from two copies of Q_{k-1}. Append 0 to the name of each vertex in the first copy, and append 1 to the name of each vertex in the second copy. Adding one edge from each vertex of the first copy to the corresponding vertex of the second copy completes Q_k. Below we show Q_3 generated from Q_2 in this way, with the first copy outside and the second copy inside. The inductive description facilitates many proofs about Q_k by induction on k. ∎

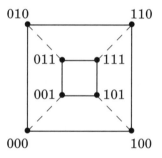

The inductive construction of Q_k is a special case of a general operation for combining two graphs.

5.1.20. Definition. The **cartesian product** of graphs G and H, written $G \square H$, is the graph with vertex set $V(G) \times V(H)$ such that (u,v) is adjacent to (u',v') if and only if (1) $u = u'$ and $vv' \in E(H)$, or (2) $v = v'$ and $uu' \in E(G)$.

5.1.21. Example. The cartesian product is symmetric; $G \square H \cong H \square G$. As examples, $C_3 \square K_2$ appears below, and $Q_k = Q_{k-1} \square K_2$ if $k \geq 1$. In general, $G \square H$ is an edge-disjoint union of copies of H for each vertex of G plus copies of G for each vertex of H. Thus $d_{G \square H}(u,v) = d_G(u) + d_H(v)$.

We use $\square$ instead of $\times$ to avoid confusion with other product operations, reserving $\times$ for the cartesian product of vertex sets. The symbol $\square$ visually represents the observation that $K_2 \square K_2$ is a cycle with four vertices. This notation was popularized by Nešetřil. ∎

We mention several additive combining operations on graphs.

5.1.22. Definition. The **union** of graphs G and H is the graph $G \cup H$ with vertex set $V(G) \cup V(H)$ and edge set $E(G) \cup E(H)$. We use $G + H$ for the **disjoint union** when $V(G)$ and $V(H)$ are disjoint; kG is the disjoint union of k copies of G. The **join** of G and H, denoted $G \oplus H$, is obtained from $G + H$ by adding edges to make all of $V(G)$ adjacent to all of $V(H)$.

5.1.23. Example. *Union and join.* Using the natural vertex names, $G \cup \overline{G}$ is a complete graph. The notation $G + H$ for disjoint union (usually for unlabeled graphs) is convenient for making "arithmetic of graphs" consistent with ordinary arithmetic: $mG = G + \cdots + G$. The symbol " $\oplus$ " illustrates the graph $K_2 \oplus K_2$, with the interior "+" reflecting the additivity of the sizes of the vertex sets. This is consistent with the notation " $\square$ " for cartesian product.

Using $G = K_{1,2}$ and $H = K_3$, we illustrate below their disjoint union, their join, and their union in the case where the vertices are named so that the endpoints of G are two of the vertices of H.

We can express $K_{r,s}$ using $K_{r,s} = \overline{K_r + K_s}$ or $K_{r,s} = (rK_1) \oplus (sK_1)$. More generally, $\overline{G + H} = \overline{G} \oplus \overline{H}$. Using the word "join" in "u and v are joined by an edge" is consistent with the definition of the join of two graphs. ∎

$G + H$ $\qquad\qquad$ $G \oplus H$ $\qquad\qquad$ $G \cup H$

5.1.24.* Remark. There is no standard notation for the order and size as graph parameters. Textbook authors have tried: $\nu(G), \varepsilon(G)$ (Bondy–Murty [1976]); $v(G), e(G)$ (Bondy–Murty [2008]); $|G|, e(G)$ (Bollobás [1978, 1998]); $n(G), e(G)$ (West [1996, 2001]); $|G|, \|G\|$ (Diestel [1997, etc.]). Conflicts arise with using v and e as a specific vertex and edge and using n as an integer variable. Reserving special characters has problems when there are several graphs under discussino: p, q (Harary [1969], Gould [1988]); v, e (Wallis [2000]); n, m (Chartrand–Lesniak [1986, etc.], Volkmann [1991, 1996], Buckley–Lewinter [2002]). Although $|G|, \|G\|$ avoids these problems, internet respondents disliked them. V. Strehl suggested $\#V(G), \#E(G)$; this avoids the difficulties but is not in common use.

We therefore introduce no notation for these operators, but we often say "let G be an n-vertex graph" or "let $n = |V(G)|$". Thus we encourage using n for the order of a particular graph, with no global definition (this is common in computer science). We similarly often let $m = |E(G)|$, reserving "e" for a specific edge. ∎

EXERCISES 5.1

5.1.1. (−) Count the automorphisms of P_n, C_n, and K_n.

5.1.2. (−) Using the disjointness definition of the Petersen graph, prove that the graphs of Example 5.1.16 are isomorphic.

5.1.3. (−) Prove by induction that $|E(Q_k)| = k2^{k-1}$.

5.1.4. (−) Prove that each graph below is isomorphic to $C_3 \,\square\, C_3$.

5.1.5. (−) Let G be a graph with rs vertices that decomposes into rK_s and sK_r. Prove $G \cong K_r \,\square\, K_s$. Construct infinitely many examples to prove that in general a graph with rs vertices may decompose into rH and sF, where $r = |V(F)|$ and $s = |V(H)|$, without being isomorphic to $F \,\square\, H$.

5.1.6. (−) Determine which pairs of graphs below are isomorphic.

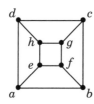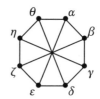

5.1.7. (−) Prove that the graph below is isomorphic to $\overline{Q}_3$.

5.1.8. (−) Let G be a graph with adjacency matrix A and incidence matrix M. What do the entries in position (i, j) of A^2 and MM^T say about G?

5.1.9. (−) In terms of a known enumerative problem, find the number of isomorphism classes of X, Y-bigraphs such that $|X| = m$ and $|Y| = n$ and every vertex of X has degree 1.

5.1.10. (−) In terms of the numbers of vertices and edges in a graph G, count the vertices and edges in the cartesian product $G \,\square\, \cdots \,\square\, G$ with k factors all isomorphic to G.

5.1.11. (−) Prove that if $\delta(G) \geq 3$, then some cycle in G has a chord. (Czipzer [1963])

5.1.12. Let G_n be the graph whose vertices are the permutations of $[n]$, with two permutations $a_1, \ldots, a_n$ and $b_1, \ldots, b_n$ adjacent if they differ by switching two entries. Prove that G_n is bipartite.

5.1.13. Determine the least number of vertices in a graph in which the minimum length of an even cycle is r and the minimum length of an odd cycle is s. (Harary–Kovacs [1982])

5.1.14. (+) Determine all r such that $C_r \,\square\, K_2$ can be expressed as the union of two cycles.

5.1.15. Determine whether the graphs below are isomorphic.

5.1.16. Prove that the two graphs below are isomorphic.

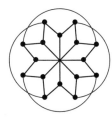 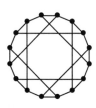

5.1.17. Determine which pairs of graphs below are isomorphic.

5.1.18. Determine which pairs of graphs below are isomorphic.

 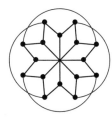 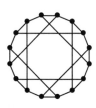

5.1.19. Determine which pairs of graphs below are isomorphic.

5.1.20. (◊) Determine which pairs of graphs below are isomorphic.

5.1.21. (◊) Determine which pairs of graphs below are isomorphic, presenting the proof by testing the fewest possible pairs.

   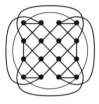 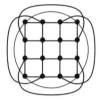

5.1.22. For each k at least 2, determine the smallest n such that
(a) There is a k-regular graph with n vertices.
(b) There exist nonisomorphic k-regular graphs with n vertices.

5.1.23. For $k \geq 2$, determine the smallest n such that there exist non-isomorphic k-regular bipartite graphs with n vertices.

5.1.24. Determine the number of isomorphism classes of 3-regular graphs with six vertices and with eight vertices.

5.1.25. (◊) A graph isomorphic to its complement is **self-complementary**. Prove that there is a self-complementary graph with n vertices if and only if n or $n-1$ is divisible by 4. Determine for which n there is a regular self-complementary n-vertex graph.

5.1.26. Let G be a self-complementary graph. Let σ be a permutation of $V(G)$ that is an isomorphism from G to $\overline{G}$. Prove that every cycle of σ other than fixed points has length divisible by 4. (Ringel [1963])

5.1.27. Let G be an n-vertex graph with $G \cong \overline{G}$ and $n \equiv 1 \pmod 4$. Prove that G has at least one vertex of degree $(n-1)/2$.

5.1.28. Suppose that n is congruent to 0 or 1 modulo 4. Construct an n-vertex graph G with $\frac{1}{2}\binom{n}{2}$ edges such that $\Delta(G) - \delta(G) \leq 1$.

5.1.29. Construct a 6-vertex graph having no nontrivial automorphism. Construct a graph having exactly three automorphisms.

5.1.30. *Edge-transitive versus vertex-transitive.*
(a) For $n \geq 4$, form a graph G by replacing each edge of K_n with a path of length two (the new vertices have degree 2). Prove that G is edge-transitive but not vertex-transitive.
(b) Prove that every edge-transitive graph that is not vertex-transitive is bipartite.

5.1.31. Prove that the Petersen graph has no cycle of length 7.

5.1.32. Prove that the Petersen graph is the only 3-regular graph with 10 vertices that has girth at least 5.

5.1.33. View the Petersen graph as the graph whose adjacency relation is the disjointness relation on the 2-element subsets of [5]. Prove that every automorphism is determined by a permutation σ of [5], with vertex ij mapped to the vertex $\sigma(i)\sigma(j)$ for all $i, j \in [5]$.

5.1.34. Count the independent sets of size 4 and the independent sets of size 3 in the Petersen graph. Find all smallest sets whose deletion eliminates all independent 4-sets.

5.1.35. Let e and e' be non-incident edges in the Petersen graph. Prove that e and e' lie in a 5-cycle or in a 6-cycle. Determine the number of pairs of nonincident edges that lie in a 5-cycle and the number that do not.

5.1.36. The **Odd graph** O_k is the disjointness graph on the k-subsets of $[2k+1]$ (O_2 is the Petersen graph). For $k > 2$, determine the girth of O_k.

5.1.37. For the **Heawood graph** below, determine all cycle lengths and find a second spanning cycle.

5.1.38. For even n, determine the minimum number of distinct cycle lengths in an n-vertex graph with minimum degree $n/2$.

5.1.39. Determine the smallest bipartite graph that is not a subgraph of the k-dimensional cube Q_k for any k.

5.1.40. Define a graph Q'_k with vertex set $\{0, 1\}^k$ by $uv \in E(Q'_k)$ if and only if u and v *agree* in exactly one coordinate. Prove that $Q'_k \cong Q_k$ if and only if k is even. (D.G. Hoffman)

5.1.41. For $k \in \mathbb{N}$, let G be the subgraph of Q_{2k+1} induced by the vertices in which the number of ones and zeros differs by 1. Prove that G is vertex-transitive and edge-transitive, and compute the order, size, and girth of G.

5.1.42. Prove that Q_k has exactly $k!2^k$ automorphisms.

5.1.43. A **subcube** of dimension r in Q_k is a subgraph induced by 2^r vertices that agree on some $k - r$ coordinates.
   (a) Prove that every cycle of length $2r$ in Q_k lies in a subcube of dimension at most r. Prove that this subcube is unique when $r = 2$ or $r = 3$ but need not be unique when $r = 4$.
   (b) Count the 4-cycles and 6-cycles in Q_k.

5.1.44. ($\diamond$) *Regular graphs with no short cycle.*
   (a) Prove that every k-regular graph with girth at least 4 has at least $2k$ vertices. Determine all such graphs where equality holds.
   (b) Prove that every k-regular graph with girth at least 5 has at least $k^2 + 1$ vertices. Find examples where equality holds when k is 2 or 3.

5.1.45. ($\diamond$) Prove that every k-regular graph with girth at least 6 has at least $2k^2 - 2k + 2$ vertices. (Comment: The Heawood graph of Exercise 5.1.37 achieves equality when $k = 3$.)

5.1.46. (+) Fix $n, k \in \mathbb{N}$ with $1 < k < n-1$. Let G be an n-vertex graph such that every k-vertex induced subgraph of G has m edges. For $l \geq k$, prove that every induced subgraph of G with l vertices has $m\binom{l}{k}/\binom{l-2}{k-2}$ edges. Use this to prove $G = K_n$ or $G = \overline{K}_n$.

5.1.47. (+) A (k, g)-**cage** is a smallest k-regular graph with girth g. For $k \geq 2$ and $g \geq 3$, prove that there exists a k-regular graph with girth g. (Hint: For $g \geq 5$, construct such a graph from a $(k-1)$-regular graph H with girth g and a graph with girth $\lceil g/2 \rceil$ that is regular with degree $|V(H)|$.) (Erdős–Sachs [1963])

5.2. Vertex Degrees

Vertex degrees give more information than the numbers of vertices and edges. Here we briefly consider enumerative, existential, and extremal problems about degrees. We also introduce corresponding terminology for directed graphs.

THE DEGREE-SUM FORMULA

A simple counting argument yields a fundamental tool counting edges in terms of vertex degrees. It has been called the "First Theorem of Graph Theory" and the **Handshaking Lemma** (modeling handshakes as edges in a graph).

5.2.1. Proposition. (Degree-Sum Formula) For a graph G with m edges,
$$m = \frac{1}{2} \sum_{v \in V(G)} d(v).$$

Proof: Grouped by edges, there are $2m$ pairs (v, e) such that e is an edge incident to vertex v. Grouped by vertices, there are $\sum_{v \in V(G)} d(v)$ pairs. ∎

Thus a k-regular n-vertex graph has $kn/2$ edges (Q_k has $k2^{k-1}$ edges).

5.2.2. Corollary. Every graph has an even number of vertices of odd degree, and thus no regular graph of odd degree has an odd number of vertices. ∎

We expressed the argument for the Degree-Sum Formula as "counting two ways" for the set of incidences of edges with vertices. More simply, summing the degrees counts every edge twice. Here is another example of multiple counting.

5.2.3. Example. *The Petersen graph has twelve 5-cycles.* The Petersen graph has 15 edges (it is 3-regular with ten vertices). We claim that every edge lies in four 5-cycles. If so, then $4 \cdot 15$ counts every edge in every 5-cycle; since a 5-cycle has five edges, there must thus be 12 of them.

A 5-cycle through edge e contains a copy of P_4 with central edge e. The endpoints of such a path are nonadjacent, since the girth is 5. Nonadjacent vertices have one common neighbor, so each copy of P_4 extends to one 5-cycle. Since e has two incident edges at each end, and there are no triangles, e is the central edge of exactly four copies of P_4. Thus e lies in four 5-cycles. ∎

The simple statements in Corollary 5.2.2 have nontrivial applications. If a graph has an odd-degree vertex, then it must have another. We present one application; see also Exercise 12, Exercise 13, Theorem 5.4.7, and Theorem 16.2.2.

5.2.4. Proposition. If a rectangular region R is decomposed into nonoverlapping rectangular subregions $R_1, \ldots, R_k$, and each R_i has a side of integer length, then the full region R has a side of integer length.

Proof: Translate one corner of R to the origin, as illustrated below. Form a bipartite graph G whose two parts are the subrectangles and the integer lattice points, with R_i adjacent to (r, s) if (r, s) is a corner of R_i.

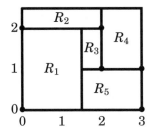
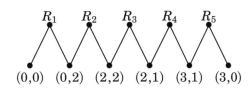

Since each subrectangle has a side of integer length, it has an even number of integer corners; hence each vertex R_i has even degree in G. An integer point is the corner of an even number of subrectangles if it is internal or on the side of R, but it is a corner of only one subrectangle if it is a corner of R. Hence any vertex of odd degree in G is an integer corner of R. Since the origin is one such corner, Corollary 5.2.2 yields another. Hence R has a side of integer length. ∎

Wagon [1987] published 14 proofs of Proposition 5.2.4, some using complex analysis. Another proof appeared in Winkler [2004, pp. 6–7].

DEGREE LISTS

We may ask whether some graph in a given class has specified values of certain parameters? For example, consider the vertex degrees. We can ask this about the class of all n-vertex graphs.

5.2.5. Definition. The **degree list** or **degree sequence** of a graph G is the list of its vertex degrees, usually written in nonincreasing order. A list of integers is **graphic** if it is the degree list of some graph G; then G **realizes** it.
   When $v \in V(G)$, **vertex deletion** yields the graph denoted $G - v$ by deleting v and all its incident edges from G.

For example, $(3, 3, 1, 1)$ is not graphic; when two vertices in a 4-vertex graph have degree 3, the other two vertices must have degree at least 2. Degree "sequence" is the traditional term, but a sequence is a function defined on $\mathbb{N}_0$, so the term "list" is more correct. We next determine which integer lists are graphic.

5.2.6. Theorem. (Havel [1955], Hakimi [1962]) For $n > 1$, a nonnegative integer list d of size n is graphic if and only if d' is graphic, where d' is the list of size $n - 1$ obtained from d by deleting its largest element k and subtracting 1 from its k next largest elements.

Proof: If $d = 3322211$, for example, then $d' = 211211$. Given a list d and a graph G' with degree list d', we add a new vertex adjacent to vertices in G' having degrees that were reduced in forming d' from d. The resulting graph G realizes d and proves that the condition is sufficient.

For necessity, we begin with a graph G realizing d and find G' realizing d'. Let $k = \Delta(G)$, and let w be a vertex of degree k in G. Let S be a set of k other vertices in G with largest degrees. If $N(w) = S$, then we let $G' = G - w$.

Otherwise, some vertex x of S is missing from $N(w)$, and w has a neighbor z outside S (since $d(w) = |S|$). Note that $d(x) \geq d(z)$, by the choice of S. Since $w \in N(z) - N(x)$, some vertex y outside $\{x, z, w\}$ must satisfy $y \in N(x) - N(z)$, as illustrated below. Replacing $\{wz, xy\}$ with $\{wx, yz\}$ in $E(G)$ yields a realization of d such that w has more neighbors in S. Since $|N(w) \cap S|$ can increase at most k times, repeating this procedure converts G into a graph G^* that realizes d and has $N(w) = S$. Now $G^* - w$ is the desired graph G' realizing d'. ∎

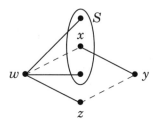

The operation we used has further applications.

5.2.7. Definition. Given $xy, zw \in E(G)$ and $yz, wx \in E(\overline{G})$, a **2-switch** is the replacement of $\{xy, zw\}$ with $\{wx, yz\}$ in G.

5.2.8. Theorem. (Fulkerson–Hoffman–McAndrew [1965], Berge [1970]) If G and H are graphs with vertex set V, then 2-switches can turn G into H if and only if $d_G(v) = d_H(v)$ for every $v \in V$.

Proof: Every 2-switch preserves vertex degrees, so the condition is necessary. For the converse, we use induction on $|V|$. When $|V| \leq 3$, each specification of degrees is realized by at most one graph.

For $|V| \geq 4$, let $k = \Delta(G)$, and let w be a vertex of degree k. Let S be a fixed set of k vertices other than w with the highest degrees. As in the proof of Theorem 5.2.6, we use 2-switches to turn G into a graph G^* with $N_{G^*}(w) = S$ and turn H into a graph H^* with $N_{H^*}(w) = S$.

Since $N_{G^*}(w) = N_{H^*}(w)$, deleting w leaves graphs $G' = G^* - w$ and $H' = H^* - w$ with $d_{G'}(v) = d_{H'}(v)$ for every vertex v. By the induction hypothesis, G' transforms to H' by 2-switches. Since these do not involve w, and w has the same neighbors in G^* and H^*, the same 2-switches transform G^* to H^*. Hence we can transform G to H by transforming G to G^*, then G^* to H^*, then (in reverse order) the transformation of H to H^* (the inverse of a 2-switch is a 2-switch). ∎

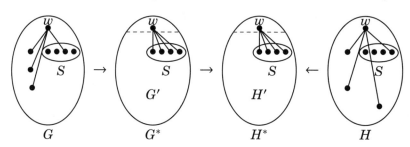

EXTREMALITY

An important technique in discrete mathematics is the selection of an extremal instance of some structure. The absence of a more extreme structure gives extra leverage to prove the desired conclusion. We give two examples here.

5.2.9. Theorem. Every graph G with m edges has a bipartite subgraph with at least $m/2$ edges.

Proof: Among all bipartite subgraphs of G, let H be one having the most edges. We may assume that H has bipartition X, Y with $X \cup Y = V(G)$, and every edge of G having endpoints in both X and Y belongs to H.

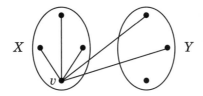

If some vertex v contributes fewer than half of its incident edges to H, then v has more neighbors in its own part of the bipartition than in the opposite part. Moving v to the other part puts more edges into H than it deletes from H, yielding a larger bipartite subgraph.

Since H is a bipartite subgraph with the most edges, we conclude that $d_H(v) \geq d_G(v)/2$ for every $v \in V(G)$. Summing over all vertices and applying the Degree-Sum Formula yields $|E(H)| \geq |E(G)|/2$. ∎

We could also start with a spanning bipartite subgraph and move vertices to gain edges until we reach the desired degree property. Choosing a largest bipartite subgraph skips directly to that point. Many proofs by extremality have equivalent inductive or algorithmic versions; using extremality is usually shorter. We next use extremality by choosing a vertex of largest degree.

5.2.10. Definition. A graph is **triangle-free** if it does not contain K_3. More generally, a graph G is H-**free** if G has no induced subgraph isomorphic to H. The **Turán graph** $T_{n,r}$ is the n-vertex complete r-partite graph whose parts differ in size by at most 1.

5.2.11. Theorem. (**Turán's Theorem**; Turán [1941]) The largest n-vertex K_{r+1}-free graph is $T_{n,r}$. For the triangle-free case $r = 2$, it has $\lfloor n^2/4 \rfloor$ edges.

Proof: (Erdős [1970]) Every r-partite graph is K_{r+1}-free, since any $r + 1$ vertices have two in the same part, which are not adjacent. A largest r-partite graph is a complete r-partite graph. If two parts differ in size by more than 1, then moving a vertex from the larger part to the smaller part gains more edges than it loses. Hence the graph described is the unique largest r-partite graph with n vertices.

It thus suffices to prove that when G is K_{r+1}-free, there is an r-partite graph H with $V(H) = V(G)$ such that $d_H(v) \geq d_G(v)$ for all $v \in V(G)$ (and therefore $|E(H)| \geq |E(G)|$). We prove this by induction on r, trivial for $r = 1$.

Suppose $r > 1$. Choose $x \in V(G)$ with $d_G(x) = \Delta(G)$. Since $K_{r+1} \not\subseteq G$, we have $K_r \not\subseteq G'$, where $G' = G[N(x)]$. By the induction hypothesis, there is an $(r-1)$-partite graph H' with $d_{H'}(v) \geq d_{G'}(v)$ for all $v \in N(x)$.

Let $S = V(G) - N(x)$. Form H from H' by making every vertex of S adjacent to every vertex of $N(x)$, leaving S an independent set. Note that H is r-partite. Now $d_H(v) = d_G(x) = \Delta(G)$ for $v \in S$. Hence $d_H(v) \geq d_G(v)$ for all $v \in S$, and H has all the required properties.

In the special case $r = 2$, the triangle-free graph with the most edges has parts of size $\lfloor n/2 \rfloor$ and $\lceil n/2 \rceil$, and the size is $\lfloor n^2/4 \rfloor$.

Uniqueness of the extremal graph also follows inductively.   ∎

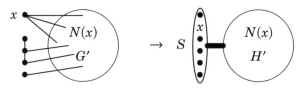

Mantel [1907] (jointly with others!) proved the case $r = 2$. We will revisit Turán's Theorem in Chapters 11 and 14 to study other techniques.

DIRECTED GRAPHS

Edges in a graph are unordered pairs of distinct vertices, so graphs model binary relations that are symmetric and irreflexive. A general relation on a set V is a subset of the cartesian product $V \times V$. We can model general relations using ordered pairs of vertices as edges.

5.2.12. Definition. A **directed graph** or **digraph** G is an ordered pair consisting of a **vertex set** $V(G)$ and an **edge set** $E(G)$, with $E(G) \subseteq V(G) \times V(G)$. An edge *exits* from its first vertex (the **tail**) and *enters* its second vertex (the **head**). We write xy instead of (x, y) for an edge from x to y. An edge from x to x is a **loop**. Given a digraph G, we may write $x \to y$ to mean $xy \in E(G)$.

The term *digraph* was introduced by Harary [1955a] at the request of Pólya (as told in Harary [1989]). It is natural to allow loops in digraphs, such as in the functional digraphs of permutations. For the analogue of vertex degrees, we count entering and exiting edges separately.

5.2.13. Definition. For a vertex x in a digraph G, the **indegree** $d_G^-(x)$ and **outdegree** $d_G^+(x)$ are the number of edges entering x or leaving x, respectively. The **out-neighborhood** $N_G^+(x)$ is $\{y: x \to y\}$; the **in-neighborhood** $N_G^-(x)$ is $\{w: w \to x\}$. Subscripts may be dropped. A vertex with indegree 0 is a **source**; one with outdegree 0 is a **sink**.

Given a vertex ordering $v_1, \ldots, v_n$, the **adjacency matrix** A of a digraph is the 0, 1-matrix defined by $A_{i,j} = 1$ if and only if $v_i \to v_j$. With edges $e_1, \ldots, e_m$, the **incidence matrix** M of a loopless digraph has $M_{i,j} = 1$ if v_i is the tail of e_j, $M_{i,j} = -1$ if v_i is the head of e_j, and $M_{i,j} = 0$ otherwise.

5.2.14. Example. Compare the matrices for the digraph below with those for P_4 in Example 5.1.10. ∎

$$A(G) = \begin{array}{c} \\ w \\ x \\ y \\ z \end{array} \begin{pmatrix} \begin{array}{cccc} w & x & y & z \end{array} \\ 0 & 1 & 0 & 0 \\ 0 & 0 & 1 & 0 \\ 0 & 1 & 0 & 0 \\ 0 & 0 & 1 & 0 \end{pmatrix}$$

$$M(G) = \begin{array}{c} \\ w \\ x \\ y \\ z \end{array} \begin{pmatrix} \begin{array}{cccc} a & b & c & d \end{array} \\ +1 & 0 & 0 & 0 \\ -1 & +1 & -1 & 0 \\ 0 & -1 & +1 & -1 \\ 0 & 0 & 0 & +1 \end{pmatrix}$$

A graph can be viewed as a special kind of digraph.

5.2.15. Definition. A digraph is **symmetric** if its adjacency matrix is symmetric (a graph is a symmetric loopless digraph). A digraph D is **antisymmetric** if xy and yx are not both edges when $x, y \in V(D)$. To form the **underlying graph** of an antisymmetric digraph, make the edges unordered. An **orientation** of a graph G is an antisymmetric digraph whose underlying graph is G. An **oriented graph** is a digraph that is an orientation of a graph.

In drawing a digraph, we view the curve for an edge as an "arrow" from its tail to its head. Basic properties of digraphs are like those of graphs, so we focus primarily on graphs. For example, definitions of sub(di)graphs and isomorphism are the same as those for graphs. Nevertheless, interesting problems for digraphs arise when we consider orientations of graphs.

5.2.16. Definition. A **tournament** is an orientation of a complete graph. In the directed sense, a **path** is a digraph whose vertices can be ordered as $v_1, \ldots, v_k$ such that $v_i \to v_j$ if and only if $j = i+1$. A **cycle** consists of a path $\langle v_1, \ldots, v_k \rangle$ plus the edge $v_k v_1$.

Thus, a path or cycle in a digraph is a "consistent" orientation of an undirected path or cycle. A tournament is a natural model for a competition in which each team plays every other, with one winner in each game.

5.2.17. Definition. In a digraph, a **king** is a vertex from which every vertex is reachable by a path of length at most 2. A **successor** of v is a vertex w such that vw is an edge. A **predecessor** of v is a vertex u such that uv is an edge.

5.2.18. Proposition. (Landau [1953]) Every tournament has a king.

Proof: Let x be a vertex of maximum outdegree in a tournament T. If x is not a king, then some vertex y is not reachable from x in at most two steps. Hence no successor of x is a predecessor of y. Since T is an orientation of a complete graph, every successor of x is thus a successor of y, as is x. Hence $d^+(y) > d^+(x)$, contradicting the choice of x. ∎

An algorithmic proof starts with an arbitrary vertex and moves to a vertex of larger degree if it is not a king. The extremal choice of a vertex with maximum degree goes directly to the key point. Exercise 43 generalizes Proposition 5.2.18.

EXERCISES 5.2

5.2.1. (–) Prove or disprove:
  (a) Deleting a vertex of maximum degree cannot increase the average degree.
  (b) Deleting a vertex of minimum degree cannot reduce the average degree.

5.2.2. (–) Count the copies of C_{2k} in $K_{r,s}$.

5.2.3. (–) Let G be a graph with maximum degree 3. Prove that $V(G)$ can be partitioned into two sets that each induce a subgraph consisting of isolated edges and vertices.

5.2.4. (–) For $k \in \mathbb{N}$, prove that every graph G has a k-partite subgraph H (Definition 5.1.7) with at least $(1 - 1/k)|E(G)|$ edges.

5.2.5. (–) Determine the minimum number of edges in an n-vertex graph having no independent set of size 3.

5.2.6. (–) Prove that some n-vertex tournament satisfies $d^+(v) = d^-(v)$ for every vertex v if and only if n is odd.

5.2.7. (–) An oriented graph G is **transitive** if $xy, yz \in E(G)$ implies $xz \in E(G)$. Prove that K_n has exactly $n!$ transitive orientations.

5.2.8. Count the cycles of lengths 6, 8, and 9 in the Petersen graph.

5.2.9. Use graphs and bijections (not algebra!) to prove the following:

  (a) $\binom{n}{2} = \binom{k}{2} + k(n-k) + \binom{n-k}{2}$ for $0 \le k \le n$.

  (b) If $\sum n_i = n$, then $\sum \binom{n_i}{2} \le \binom{n}{2}$.

5.2.10. ($\diamond$) Let G be a graph with average vertex degree a.
  (a) In terms of $d(x)$, determine when $G - x$ has average degree at least a. Conclude that if $a > 0$, then G has a subgraph with minimum degree greater than $a/2$.
  (b) For $\varepsilon > 0$, show that a subgraph with minimum degree greater than $(a/2) + \varepsilon$ cannot be guaranteed.

5.2.11. Let G be a graph with positive average degree a. Let $t(v)$ denote the average of the degrees of the neighbors of v. Prove that $t(v) \ge a$ for some $v \in V(G)$. Construct infinitely many graphs G such that $t(v) > a$ for all $v \in V(G)$. (Ajtai–Komlós–Szemerédi [1980])

5.2.12. Let F be a family of subsets of $[n]$, with $[n] \notin F$. Prove that the number of sets $S \subseteq [n]$ such that S contains an odd number of sets in F is even. (J. Edmonds)

5.2.13. ($\diamond$) A *mountain range* is the curve of a piecewise-linear function from $(a, 0)$ to $(b, 0)$ in the upper half-plane. Travelers A and B start at $(a, 0)$ and $(b, 0)$, respectively. Give a graph-theoretic proof (without induction) that they can meet by traveling on the mountain range so that at all times their heights above the horizontal axis are the same.

5.2.14. ($\diamond$) For $n \ge 2$, determine whether an n-vertex graph can have n distinct vertex degrees. In an n-vertex graph with $n-1$ distinct vertex degrees, what are the possibilities for the repeated and missing degrees?

5.2.15. (+) Let G be a graph such that $|N(x) \cap N(y)|$ is odd for all $x, y \in V(G)$. Prove that $|V(G)|$ is odd. (Engel [1998], Royle [2010], Silwal [2018])

5.2.16. ($\Diamond$) Let d be a list of n nonnegative integers with even sum that differ by at most 1 and all are at most $n - 1$. Prove that d is graphic.

5.2.17. Given a nonincreasing list d of nonnegative integers, let d' be obtained by deleting d_k and subtracting 1 from the d_k largest elements remaining in the list. Prove that d is graphic if and only if d' is graphic. (Wang–Kleitman [1973])

5.2.18. Let $a_1, \ldots, a_k$ be integers such that $0 < a_1 < \cdots < a_k$.
(a) Construct a graph with $a_k + 1$ vertices whose *set* of distinct vertex degrees is $a_1, \ldots, a_k$. (Kapoor–Polimeni–Wall [1977])
(b) Prove that when $a_1 = 1$, the construction is unique if and only if there is a unique graph on fewer than a_k vertices with degree set $\{a_2 - 1, \ldots, a_{k-1} - 1\}$.

5.2.19. Let G and H be graphs with $V(G) = V(H)$ and $d_G(v) = d_H(v)$ for all $v \in V(G)$. Use induction on number of edges belonging to exactly one of G and H to prove that G can be turned into H by using 2-switches. (Comment: This yields an alternative proof of Theorem 5.2.8.)

5.2.20. Let p and q be nonnegative integer lists, with $p = (p_1, \ldots, p_r)$ and $q = (q_1, \ldots, q_s)$. The pair (p, q) is **bigraphic** if some bipartite graph has $p_1, \ldots, p_r$ as the vertex degrees in one part and $q_1, \ldots, q_s$ as the degrees in the other. When $\sum p_i > 0$, prove that (p, q) is bigraphic if and only if (p', q') is bigraphic, where (p', q') is obtained from (p, q) by deleting the largest element k from p and subtracting 1 from the k largest elements of q.

5.2.21. Let A and B be r-by-s matrices with entries in $\{0, 1\}$, having the same vector of row sums and the same vector of column sums. Prove that A can be transformed into B via steps in which the 0s and 1s are interchanged in a 2-by-2 permutation submatrix (that is, submatrices $\binom{0\,1}{1\,0}$ and $\binom{1\,0}{0\,1}$ can be substituted for each other). (Ryser [1957])

5.2.22. An **expansion** of a graph replaces two edges with paths of length 2 through two new vertices joined by an edge. **Erasure** is the inverse. Prove that a 2-switch can be performed using expansions and erasures. Use this to show that every 3-regular graph arises from K_4 by expansions and erasures.

5.2.23. Given an X, Y-bigraph G and edges $x_1 y_1$ and $x_2 y_2$ with $x_1, x_2 \in X$ and $y_1, y_2 \in Y$, say that a *bridging operation* deletes $x_1 y_1$ and $x_2 y_2$ and adds two new vertices x_3 and y_3 with $N(x_3) = \{y_1, y_2, y_3\}$ and $N(y_3) = \{x_1, x_2, x_3\}$. Prove that every 3-regular bipartite graph can be obtained from a graph whose components are isomorphic to $K_{3,3}$ by repeated bridging operations. (Kotzig [1966a])

5.2.24. For $m > 0$, prove that every graph with m edges has a bipartite subgraph with strictly more than $m/2$ edges. Construct a sequence of graphs (with $m > 0$) whose largest bipartite subgraphs tend to half the number of edges.

5.2.25. Prove that the maximum number of edges in a bipartite subgraph of the Petersen graph is 12. (Comment: Bondy–Locke [1986] proved that every triangle-free graph with maximum degree 3 has a bipartite subgraph with at least 4/5 of its edges.)

5.2.26. Prove or disprove: Every graph has a vertex partition into two nonempty sets such that each vertex has at least half of its neighbors in its own set.

5.2.27. ($\Diamond$) For a graph G with no isolated vertices, prove that $V(G)$ can be partitioned into sets of size at least 2 whose induced subgraphs each have a spanning star. (Winkler)

5.2.28. A directed graph is **unipathic** if for all vertices x and y there is at most one (directed) x, y-path. Let T_n be the tournament with vertex set $[n]$ such that $i \to j$ if and only if $i < j$. What is the maximum number of edges in a unipathic subgraph of T_n? How many unipathic subgraphs have that size? (Maurer–Rabinovitch–Trotter [1980])

5.2.29. ($\diamond$) *Two inductive proofs of Mantel's Theorem* (the case $r = 2$ of Theorem 5.2.11).
(a) Given adjacent vertices x and y in an n-vertex graph, prove that xy lies in at least $d(x) + d(y) - n$ triangles. Use this to prove by induction on n that every n-vertex graph with more than $n^2/4$ edges contains a triangle.
(b) Prove that an n-vertex graph G with more than $n^2/4$ edges has a vertex v such that $G - v$ has more than $(n-1)^2/4$ edges, again yielding Mantel's Theorem by induction.

5.2.30. ($\diamond$) *Strengthening Mantel's Theorem* (the case $r = 2$ of Theorem 5.2.11).
(a) For a vertex v in an n-vertex graph G, let $f(v)$ be the maximum size of an independent set contained in $N(v)$. Prove $\sum_{v \in V(G)} f(v) \le n^2/2$, and determine which graphs achieve equality. (Hint: Consider a largest independent set in G.)
(b) Use part (a) to obtain the case $r = 2$ of Theorem 5.2.11. (Galvin [1999])

5.2.31. ($\diamond$) Let G be an n-vertex graph having an orientation in which every triangle is oriented cyclically. Prove that $|E(G)| \le n^2/3$ and that this is sharp. (Brown–Harary [1970])

5.2.32. The game "bridge" has two teams of two partners each. Consider a club in which four players cannot play if two of them have previously been partners that night. One night 15 members arrive, but one wants to study graph theory. The other 14 play until each has partnered with four others. After six more games (12 partnerships), they cannot find four players with no pair of previous partners. Prove that adding the graph theorist allows at least one more game to be played. (adapted from Bondy–Murty [1976, p. 111])

5.2.33. ($\diamond$) Let $t_r(n)$ be the number of edges in the n-vertex r-partite Turán graph, and let G be an n-vertex graph not containing K_{r+1}. Prove that if $t_r(n) - |E(G)| \le s$, then G can be made into an r-partite graph by deleting at most s edges. (Hint: Consider the proof of Turán's Theorem (Theorem 5.2.11).) (Füredi [2015])

5.2.34. ($\diamond$) Let G be a graph such that $\frac{|E(H)|}{|V(H)|} \le k$ for all $H \subseteq G$. By improving a bad orientation, prove that G has an orientation in which every vertex has outdegree at most k.

5.2.35. Determine the least n such that some two nonisomorphic n-vertex tournaments have the same list of outdegrees.

5.2.36. ($\diamond$) Let T be an n-vertex tournament.
(a) Prove that $\sum_{v \in V(T)} \binom{d^+(v)}{2} = \sum_{v \in V(T)} \binom{d^-(v)}{2}$ by counting a set in two ways.
(b) Prove that T has $\binom{n}{3} - \sum_{v \in V(T)} \binom{d^+(v)}{2}$ 3-cycles. (Kendall–Smith [1940])
(c) When n is odd, determine the maximum possible number of 3-cycles in T.

5.2.37. Prove that every tournament T with no source has at least three kings. For $n \ge 3$, construct an n-vertex tournament with no source and only three kings.

5.2.38. Prove that there is no 4-vertex tournament in which every vertex is a king. For $n \ge k \ge 1$, construct an n-vertex tournament with exactly k kings, except possibly when $k = 2$ and when $n = k = 4$. (Exercise 5.2.37 excludes $k = 2$.) (Maurer [1980])

5.2.39. ($\diamond$) Prove that $p_1, \ldots, p_n \in \mathbb{N}_0$ with $p_1 \le \cdots \le p_n$ are the outdegrees of some tournament if and only if $\sum_{i=1}^{k} p_i \ge \binom{k}{2}$ for $1 \le k < n$ and $\sum_{i=1}^{n} p_i = \binom{n}{2}$. (Landau [1953])

5.2.40. ($\diamond$) Let G and H be tournaments on a vertex set V. Prove that $d_G^+(v) = d_H^+(v)$ for all $v \in V$ if and only if G can be turned into H using direction-reversals on cycles of length 3. (Hint: Consider the graph of edges oriented differently in G and H.) (Ryser [1964])

5.2.41. Given an ordering $v_1, \ldots, v_n$ of the vertices of a tournament G, let S denote the sum of $j - i$ over edges $v_j v_i$ such that $j > i$. Prove that every vertex ordering of G that minimizes S puts the vertices in nonincreasing order of outdegree.

5.2.42. ($\diamond$) In an ordering L_0 of the vertices of a tournament G, a *reverse edge* is an edge $yx \in E(G)$ such that y immediately follows x in L_0. Switching the order of such x and y may create more reverse edges. Form $L_0, L_1, \cdots$ by iteratively switching vertices of reverse edges. Prove that this always reaches a list with no reverse edges. What is the maximum number of steps to termination? (Comment: When the vertices are numbers, and each edge points to the higher number, the result implies that iteratively switching any two wrongly ordered adjacent numbers always sorts a list.) (Locke [1995])

5.2.43. ($\diamond$) Prove that every loopless digraph D has an independent set S such that every vertex outside S is reached from S by a path of length at most 2. (Hint: Use induction on $|V(D)|$. Comment: This generalizes Proposition 5.2.18.) (Chvátal–Lovász [1974])

5.3. Connection and Decomposition

We have defined paths and cycles in graphs and digraphs. Here we discuss their use in connection and traversal. We also introduce multigraphs and the notion of graph decomposition.

COMPONENTS AND WALKS

5.3.1. Definition. A u, v-**path** is a path with first and last vertices u and v, called its **endpoints**; other vertices are **internal vertices**. A graph G is **connected** if it contains a u, v-path for all $u, v \in V(G)$. The **components** of G are its maximal connected subgraphs.

The **connection relation** on the vertex set of a graph G is the set of pairs $u, v \in V(G)$ such that G has a u, v-path; we say "u is connected to v" or "u and v are connected by a path". A **connected set** of vertices is a set S such that the induced subgraph $G[S]$ is connected.

To state the stronger property that u and v are adjacent, we avoid ambiguity by saying "u and v are joined by an edge", not "u and v are connected". For digraphs, we use a stronger notion of connection.

5.3.2. Definition. In a digraph, a **path** $\langle v_1, \ldots, v_k \rangle$ is a subdigraph such that $v_i \to v_j$ if and only if $j = i + 1$. A digraph is **strongly connected** or **strong** if it has a u, v-path for each ordered pair (u, v) of vertices. The **strong components** of a digraph are its maximal strongly connected subgraphs.

For example, the digraph in Example 5.2.14 has three strong components, with vertex sets $\{w\}$, $\{x, y\}$, and $\{z\}$.

We need more general notions of movement. The definitions and results hold for both graphs and digraphs, with edges in digraphs being ordered pairs.

5.3.3. Definition. A **walk** is a vertex list $\langle v_0, \ldots, v_k \rangle$ such that each $v_i v_{i+1}$ is an edge. The **length** of $\langle v_0, \ldots, v_k \rangle$ is k, the number of edges "traversed". A walk or cycle is **odd** or **even** when its length is odd or even. A walk with first vertex u and last vertex v is a u, v-**walk**; these are its **endpoints**. A walk is **closed** if its endpoints are the same.

5.3.4. Lemma. Given distinct vertices u and v in a graph or digraph, every u, v-walk contains the vertices of a u, v-path as a sublist.

Proof: "Contains" refers to the vertex list, not the subgraph relation. The vertices of the desired u, v-path must appear in order in the walk, but they need not appear consecutively.

If no vertex repeats, then a u, v-walk W just traverses a u, v-path. If some vertex x repeats, then deleting the portion of W after the first x up to and including its next appearance leaves a shorter u, v-walk contained in W. Hence a shortest u, v-walk contained in W has no repeated vertex and traverses a u, v-path. ∎

5.3.5. Corollary. The union of a u, v-path and a v, w-path contains a u, w-path. In particular, the connection relation is transitive.

Proof: Concatenating a u, v-path and a v, w-path yields a u, w-walk, and Lemma 5.3.4 applies. ∎

5.3.6. Remark. Proving connectedness for a graph can be nontrivial. For example, given numbers $d_1, \ldots, d_n$, let S be the set of graphs with vertices $v_1, \ldots, v_n$ in which the degree of v_i is d_i, for each i. Define a graph with vertex set S by making two vertices adjacent if one can be obtained from the other by a 2-switch. Theorem 5.2.8 proves that every such graph is connected. Exercise 20 is another problem of this type. For undirected graphs, Corollary 5.3.5 implies that for connectedness it suffices to show that there are paths from a single vertex to all others. Connected graphs are those having one component. ∎

5.3.7. Remark. By Corollary 5.3.5, the connection relation for a graph is an equivalence relation on its vertex set. This motivates the term "component" for a maximal connected subgraph. The equivalence classes of the connection relation are the vertex sets of the components; they are the maximal connected sets.

In a directed graph, the connection relation need not be symmetric (consider a path, for example). We can define vertices u and v to be **strongly connected** if both a u, v-path and a v, u-path exist. By Corollary 5.3.5 for digraphs, this is an equivalence relation, and its equivalence classes are the vertex sets of the strong components. ∎

Lemma 5.3.4 has an analogue for closed walks. Note that a single vertex forms a closed walk of length 0, but this is not a cycle.

5.3.8. Lemma. Every odd closed walk contains the vertices of an odd cycle (in order) as a sublist.

Proof: Given an odd closed walk W, let W' be a minimal odd closed walk contained in W (viewing W cyclically, without a fixed beginning vertex). If some vertex repeats in W', then it splits W' into portions of odd and even positive lengths, each of which is closed. The odd portion is a shorter odd closed walk contained in W. Hence no vertex repeats, and the vertices of W' in order traverse a cycle.   ∎

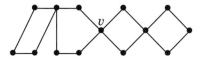

Lemma 5.3.8 yields an important characterization of bipartite graphs.

5.3.9. Theorem. (König) A graph is bipartite if and only if it has no odd cycle.

Proof: A walk in a bipartite graph must alternate between the parts. Hence every closed walk (including every cycle) has even length.

Conversely, when G has no odd cycle, we construct a bipartition of each component H of G. Choose any $v \in V(H)$. If for some vertex x there is both an even v, x-walk and an odd v, x-walk, then they combine to form a closed odd walk, which by Lemma 5.3.8 contains an odd cycle. Hence we can label each $x \in V(H)$ as "odd" or "even" by the parity of the v, x-walks in H. This is a bipartition of H, because appending the edge xy to a v, x-walk shows that x and y have opposite parity.   ∎

Theorem 5.3.9 guarantees short proofs of whether a graph *is* bipartite (exhibit a bipartition) or *is not* bipartite (exhibit an odd cycle).

CYCLES AND CUT-EDGES

Deleting parts of a graph can increase the number of components. Recall that $G - v$ denotes the subgraph of G obtained by deleting a vertex v. We denote edge deletion similarly; we need to know whether we are deleting a vertex or an edge.

5.3.10. Definition. The subgraph of G obtained by deleting an edge e is denoted $G - e$. A **cut-vertex** or **cut-edge** of G is a vertex or edge whose deletion leaves a subgraph with more components than G.

Cut-edges (also called **bridges**) can be characterized using cycles.

5.3.11. Proposition. An edge is a cut-edge if and only if it belongs to no cycle. Adding an edge to a graph combines the vertex sets of two components into one or leaves the vertex sets of components unchanged.

Proof: Consider an edge uv in a graph G. If u and v are in the same component in $G - uv$, then G and $G - uv$ have the same number of components. Hence the following are equivalent: (1) uv is a cut-edge of G, (2) u and v are in distinct components of $G - uv$, (3) $G - uv$ has no u, v-path, (4) G has no cycle containing uv.

Since deleting an edge can only increase the number of components by 1, adding an edge can only reduce the number of components by 1.   ∎

5.3.12. Corollary. Every graph with n vertices and k edges has at least $n - k$ components. Every connected n-vertex graph has at least $n - 1$ edges, which is sharp for paths.

Proof: With n vertices and no edges, there are n components. Each added edge reduces the number of components by at most 1.   ∎

Finite graphs have maximal (nonextendible) paths; choosing one is another use of extremality. A **nontrivial graph** is a graph with at least one edge.

5.3.13. Proposition. Every nontrivial graph has at least two vertices that are not cut-vertices.

Proof: Let u be an endpoint of a maximal (nonextendible) path P. Since $P - u$ is connected and $N(u) \subseteq V(P)$, the neighbors of u lie in one component of $G - u$. Hence u is not a cut-vertex.   ∎

 u

5.3.14. Lemma. A graph with minimum degree at least 2 has a cycle.

Proof: Again consider an endpoint u of a maximal path P. Since $d(u) \geq 2$, there is an edge incident to u other than its incident edge on P. By the maximality of P the other endpoint of this edge is also in $V(P)$, thereby completing a cycle.   ∎

We first apply this lemma to "decomposition" of graphs.

5.3.15. Definition. A **decomposition** of a graph G is a family of subgraphs that partition $E(G)$. An **even graph** is a graph where every vertex degree is even.

5.3.16. Theorem. Every even graph decomposes into cycles.

Proof: We use induction on the number of edges, m. A graph with no edges has a decomposition into 0 cycles. If $m > 0$, then G has a nontrivial component H, and $\delta(H) \geq 2$. By Lemma 5.3.14, H has a cycle C. Deleting the edges of C reduces the degree of its vertices by 2, leaving an even graph G'. By the induction hypothesis, G' decomposes into cycles, which with C complete a decomposition of G.   ∎

The reader deserves an apology about "even" and "odd". An odd cycle is an even graph, and a graph that is not an even graph is not called an odd graph.

Given a family $\mathcal{F}$, we can ask whether an input graph G decomposes into graphs in $\mathcal{F}$. Decomposing a complete graph into copies of a smaller complete graph is the central topic of Design Theory (Chapter 13). A complete graph decomposes into two copies of a graph F if and only if F is self-complementary.

Hajós conjectured that every n-vertex even graph decomposes into at most $\lfloor n/2 \rfloor$ cycles, and Gallai conjectured that every connected n-vertex graph decomposes into at most $\lceil n/2 \rceil$ paths. These have been open for many years; see Exercise 48 for complete graphs. Lovász [1968a] proved that every n-vertex graph decomposes into at most $\lfloor n/2 \rfloor$ subgraphs when both paths and cycles are allowed.

The exercises pose many decomposition problems. For example, does the Petersen graph decompose into copies of P_4? Does it decompose into copies of $K_{1,3}$? The **center** of a star with at least two edges is its vertex of maximum degree.

5.3.17. Proposition. A k-regular graph G decomposes into copies of the star $K_{1,k}$ if and only if G is bipartite.

Proof: *Sufficiency.* For an X, Y-bigraph G, form a decomposition using the stars with centers at vertices of X. They are copies of $K_{1,k}$ since G is k-regular, and they form a decomposition because every edge has exactly one endpoint in X.

Necessity. The claim is trivial for $k = 1$, so we may assume $k \geq 2$. Let X be the set of centers of the stars in the decomposition. Since G is k-regular, each star uses all the edges at its center; hence X is an independent set. Since the stars decomposes G, no edges join vertices outside X.   ■

EULERIAN CIRCUITS

5.3.18. Example. *The Königsberg Bridge Problem.* It is said that graph theory began in Königsberg. Located on the Pregel river at the island of Kneiphopf, Königsberg had seven bridges as shown on the left below. The citizens wondered whether it was possible to take a stroll crossing each bridge exactly once.   ■

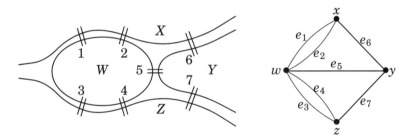

We model this problem by shrinking the land masses to single points to obtain the diagram of connections shown on the right above. We want to traverse the edges once each. However, the diagram has multiple edges joining vertices. This suggests a model more general than graphs.

5.3.19. Definition. A **multigraph** G is an ordered pair consisting of a vertex set $V(G)$ and an edge multiset $E(G)$ consisting of subsets of $V(G)$ with size 1 or 2. A **loop** is an edge consisting of one vertex. **Multiedges** are repeated copies of edges with the same endpoints. The **multiplicity** of an edge is its multiplicity as an element of $E(G)$. The **degree** of a vertex is the sum of the multiplicities of its incident edges, except that each loop contributes twice.

The convention that a loop contributes twice to the degree of its vertex extends the Degree-Sum Formula (and its consequences). A loop would arise in Example 5.3.18 from a bridge whose endpoints lie in the same land mass. We treat loops and paired multiedges in a multigraph as cycles of lengths 1 and 2.

5.3.20. Remark. *Graphs vs. multigraphs.* Many definitions and results for graphs hold also for multigraphs (with the same proofs). Exceptions are statements that need distinctness of the other endpoints of the edges incident to a given vertex. The **underlying graph** of a loopless multigraph is the graph having the same edges, but with multiplicity 1.

In the adjacency matrix of a multigraph, entry $A_{i,j}$ counts the edges with endpoints v_i and v_j. A loopless multigraph can be viewed as a complete graph with a nonnegative integer weight function giving edge multiplicities. An isomorphism is a vertex bijection that preserves edge multiplicities.

When moving from vertex to vertex in a multigraph, there may be a choice of edges. For technical precision, one should thus define a walk as an alternating list $v_0, e_1, v_2, \ldots, e_k, v_k$ of vertices and edges. In the absence of multiedges, listing only the vertices is precise. Extending elementary concepts to multigraphs is straightforward, so we will not be formal about notation for walks.

Graphs have no loops or multiedges. We may emphasize this by calling a graph a **simple graph**, particularly when also discussing multigraphs. When an object under discussion is a multigraph, the term "subgraph" refers to any multigraph contained in it; we avoid the awkward term "submultigraph".

Of course, the digraph model also extends to allow multiedges.   ∎

In the Königsberg problem, the people wanted a special walk in the multigraph of Example 5.3.18. Such walks are named in honor of Euler's paper about the problem and its generalizations.

5.3.21. Definition. A **trail** is a walk that traverses each edge at most once (vertices may repeat). A **circuit** is a closed trail, viewed cyclically regardless of the starting vertex (it is an equivalence class of closed trails with the same cyclic ordering of edges). An **Eulerian trail** in a multigraph G is a trail traversing all edges of G. An **Eulerian circuit** is a closed Eulerian trail.

The difference between a trail and the subgraph consisting of its edges is that the trail specifies an order on the edges.

The Königsberg multigraph has no Eulerian trail. Each visit of a trail to a vertex contributes twice to the degree by its entrance and exit. Only at the start and end can the degree be odd. Thus a graph with an Eulerian trail has at most two vertices of odd degree, and only even graphs can have Eulerian circuits. Euler [1736] observed that this condition is also sufficient for graphs with only one nontrivial component, but the first published full proof was by Hierholzer [1873].

5.3.22. Theorem. A multigraph G has an Eulerian circuit if and only if it has at most one nontrivial component and all vertex degrees are even.

Proof: *Necessity.* Each visit of a circuit to a vertex uses two incident edges.

Sufficiency. Since G is an even graph, it decomposes into cycles, by Theorem 5.3.16. Traversing any cycle is following a closed trail. Hence a decomposition of G into cycles is a set of edge-disjoint circuits partitioning $E(G)$.

Since such a set of circuits exists, we may take a smallest one, S. We claim that S consists of a single Eulerian circuit. If S has two distinct circuits, and no two circuits have a common vertex, then the circuits traverse distinct nontrivial components of G. This contradicts the hypothesis, so there must be two circuits sharing a vertex v. We can traverse one, starting and ending at v, and then traverse the other, thereby forming a single circuit and partitioning $E(G)$ into fewer circuits than S does.   ∎

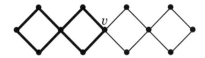

An **Eulerian [multi][di]graph** is a [multi][di]graph having an Eulerian circuit. For digraphs, the corresponding necessary degree condition is $d^+(v) = d^-(v)$ for all $v \in V(G)$ (Exercise 58).

With the characterization of Eulerian multigraphs, it is easy to solve the problem of optimal decomposition into trails, where "decomposition" means that each edge is traversed by exactly one of the trails. The transformation by adding edges to pair vertices of odd degree is a common technique that takes advantage of the generality of multigraphs.

5.3.23. Corollary. If G is a connected multigraph with $2k$ vertices of odd degree, where $k > 0$, then G decomposes into k trails and no fewer.

Proof: A trail contributes odd degree only to its endpoints, so a partition into trails has a trail ending at each odd-degree vertex. To prove that k trails suffice, add k edges joining pairs of odd-degree vertices. The resulting multigraph G' is Eulerian. Deleting the k added edges from an Eulerian circuit of G' cuts it into k edge-disjoint trails decomposing G.   ∎

EXERCISES 5.3

5.3.1. (−) Prove that the complement of any disconnected graph is connected.

5.3.2. (−) Let k, l, n be nonnegative integers with $k + l = n$. Find necessary and sufficient conditions on (k, l, n) such that there exists a connected n-vertex graph with k vertices of even degree and l vertices of odd degree.

5.3.3. (−) Prove or disprove the following statements about graphs.
(a) Every union of two distinct u, v-walks contains a cycle.
(b) Every union of two distinct u, v-paths contains a cycle.
(c) A circuit contains a cycle through each of its vertices.
(d) If $\delta(G) \geq 2$, then every vertex of G belongs to some cycle.

5.3.4. (−) Prove that every finite digraph with no vertex of outdegree 0 contains a cycle.

5.3.5. (−) Prove that the number of v_i, v_j-walks of length k in a graph is entry (i, j) in A^k, where A^k is the kth power (multiplication) of the adjacency matrix.

5.3.6. (−) Prove that a multigraph G is bipartite if and only if every subgraph H of G has an independent set consisting of at least half of $V(H)$.

5.3.7. (−) The **Möbius ladder** M_k is the graph formed by adding to a $2k$-cycle a set of k edges joining opposite vertices on the cycle. Determine all k such that M_k is bipartite. (Comment: Why is this graph called a "Möbius ladder"?)

5.3.8. (–) Let $D_1, \ldots, D_k$ be the strong components of a digraph D. Let D^* be the loopless digraph with vertices $v_1, \ldots, v_k$ such that $v_i \rightarrow v_j$ if and only if $i \neq j$ and D has an edge from D_i to D_j. Prove that D^* has no cycle. Conclude that in every digraph, some strong component has no entering edges, and some strong component has no exiting edges.

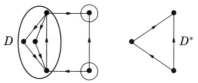

5.3.9. (–) Determine all k such that the hypercube Q_k decomposes into 4-cycles.

5.3.10. (–) Prove that in a connected graph that is not a complete graph, every vertex belongs to an induced copy of P_3. Does the same conclusion hold for every edge?

5.3.11. (–) Decompose Q_3 into copies of $K_{1,3}$ and into copies of P_4. Does the Petersen graph decompose into copies of P_4? Does it decompose into copies of P_6?

5.3.12. (–) Prove or disprove: A 3-regular graph with more than six vertices cannot decompose into three paths.

5.3.13. (–) Characterize the pairs of positive integers r, s such that $K_{r,s}$ decomposes into two isomorphic subgraphs.

5.3.14. (–) Let G be a disconnected graph with n vertices. Find the maximum possible value of $\delta(G)$. When $\delta(G) = k$, find the maximum possible value of $\Delta(G)$.

5.3.15. (–) Let G and H be two disconnected graphs with the same vertex set. Prove that some two vertices are in different components in both G and H.

5.3.16. (–) Prove or disprove the following statements.
   (a) No connected Eulerian graph has odd size and even order.
   (b) If e and f are incident edges in an Eulerian graph G, then e and f appear consecutively in some Eulerian circuit of G.

5.3.17. (–) Two Eulerian circuits are equivalent if they have the same pairs of consecutive edges, viewed cyclically. For example, a cycle has only one equivalence class of Eulerian circuits. How many equivalence classes of Eulerian circuits are there in $K_{2,r}$?

5.3.18. (–) For $k, r \in \mathbb{N}$, define G on the set of integer k-tuples by $uv \in E(G)$ if and only if $\sum |u_i - v_i| = r$. Prove that G is disconnected when r is even and bipartite when r is odd.

5.3.19. Let G be the graph defined on all binary k-tuples by making x and y adjacent if they differ in exactly two positions. Determine the number of components of G.

5.3.20. Let G_n be the graph whose vertices are the permutations of $[n]$ in word form, with permutations $a_1, \ldots, a_n$ and $b_1, \ldots, b_n$ adjacent if they differ by interchanging two consecutive entries. Prove that G_n is connected.

5.3.21. Let G be a connected graph with no induced subgraphs that are paths or cycles of length 3. Prove that G is a complete bipartite graph.

5.3.22. For $n \geq 6$, prove that the minimum number of edges that must be deleted from K_n to obtain a graph having a decomposition into two disconnected subgraphs without isolated vertices is 4. (Hint: Use induction. Comment: This result strengthens the easy fact that the complement of a disconnected graph is connected.) (Caro–Roditty [2004])

5.3.23. Given vertices u and v in a connected graph G, prove that there is a unique partition of $V(G)$ into sets A and B inducing connected subgraphs such that $u \in A$, $v \in B$, and every edge joining A and B is incident to v. (Liu [1985])

5.3.24. ($\diamond$) In a connected graph G, let P and Q be two paths chosen to maximize the sum of their lengths. Prove that P and Q have a common vertex.

5.3.25. ($\diamond$) Prove that a connected graph G with at least three vertices has two vertices x and y such that (1) $G - \{x, y\}$ is connected and (2) x and y are adjacent or have a common neighbor. (Hint: Consider a longest path.)

5.3.26. ($\diamond$) For integer weights on $E(K_n)$, not all even, prove the following equivalent.
    (A) Every cycle has even total weight.
    (B) Every triangle has even total weight.
    (C) The edges with odd weight form a spanning complete bipartite subgraph.

5.3.27. A **signed graph** assigns $+1$ or -1 to each edge. A signed graph is **balanced** if every cycle has an even number of negative edges. Prove that a signed graph G is balanced if and only if $V(G)$ can be expressed as the disjoint union of sets A and B such that an edge is negative if and only if it joins A and B. (Harary [1953], Cartwright–Harary [1956]) (Comment: Zaslavsky [1998] and Kaiser–Lukot'ka–Rollová [2017] survey signed graphs.)

5.3.28. ($\diamond$) Let D be a strongly connected orientation of a graph G. Prove that if G has an odd cycle, then D has an odd cycle.

5.3.29. (+) For a strong digraph D, let $f(D)$ be the length of the shortest closed walk visiting every vertex. Prove that the maximum of $f(D)$ over all n-vertex strong digraphs is $\lfloor (n+1)^2/4 \rfloor$ if $n \geq 2$. (Vizing–Goldberg [1969], Cull [1980])

5.3.30. Let G be an oriented graph with $|V(G)| \geq 3$. Prove that edges can be added to extend G to a strongly connected oriented graph if and only if $V(G)$ contains no nonempty proper subset S such that $xy \in E(G)$ for all $x \in S$ and $y \in \overline{S}$. (Boesch–Tindell [1980], Farzad–Mahdian–Mahmoodian–Saberi–Sadri [2006])

5.3.31. (+) For a graph G, let $c(G)$ denote the least k such that every edge lies in a cycle of length at most k (infinite when G has a cut-edge). Prove for $n \geq 3$ that the minimum of $|E(G)| + c(G)$, over n-vertex connected graphs, is $n + \lceil 2\sqrt{n-1} \rceil$. (Butler–Mao [2007])

5.3.32. Let W be a nontrivial closed walk that does not contain a cycle. Prove that some edge of W occurs twice in succession (once in each direction).

5.3.33. Let C be a closed walk in a graph G, and let H be the subgraph of G whose edges are those used an odd number of times in C. Prove that H is an even graph.

5.3.34. ($\diamond$) A **parity subgraph** of a graph G is a spanning subgraph H such that $d_G(v) \equiv d_H(v) \pmod 2$ for all $v \in V(G)$. Prove that a parity subgraph contains every cut-edge of G.

5.3.35. ($\diamond$) *Cut-edges.*
    (a) Prove that a k-regular bipartite graph has no cut-edge when $k \geq 2$.
    (b) Prove that every even graph has no cut-edge. For each odd k with $k \geq 3$, construct a k-regular connected graph having a cut-edge.

5.3.36. (+) Determine the maximum number of cut-edges in a connected 3-regular graph with n vertices. (O–West [2010] determined this for $(2r+1)$-regular n-vertex graphs.)

5.3.37. ($\diamond$) Let $d_1, \ldots, d_n$ be the vertex degrees of a graph G, with $d_1 \leq \cdots \leq d_n$. Prove that G is connected if $d_j \geq j$ whenever $j \leq n - 1 - \Delta(G)$.

5.3.38. ($\diamond$) Let G be a graph with minimum degree k and girth g.
    (a) For $k \geq 2$, prove that G has a cycle of length at least $k + 1$ and that this is sharp.
    (b) For $k \geq 3$, prove that G has an even cycle.
    (c) For $k \geq 2$, prove that G has a cycle of length at least $(g-2)(k-1) + 2$.

5.3.39. ($\diamond$) *Non-bipartite graphs have short odd cycles.* Let G be a non-bipartite triangle-free n-vertex graph, and let $k = \delta(G)$. Let C be a shortest odd cycle in G, with length l.

(a) Prove that every vertex not in $V(C)$ has at most two neighbors in $V(C)$.

(b) Prove that $l \leq 2n/k$. (Campbell–Staton [1991])

(c) Prove that the inequality of part (b) is best possible when k is even.

5.3.40. ($\diamond$) Prove that some loopless multigraph has degree list $d_1, \ldots, d_n$ if and only if $\sum d_i$ is even and no value is more than half the total. (Hakimi [1962])

5.3.41. Prove that there is a connected multigraph whose vertex degrees are the positive integers $d_1, \ldots, d_n$ if and only if $\sum d_i$ is even and at least $2n - 2$.

5.3.42. Prove that the degree list of any loopless multigraph has a realization whose underlying graph has no cycle or has one cycle, with length 3. (Will–Hulett [2004])

5.3.43. (+) Let G be a graph with $2r + 1$ vertices in which any r vertices have a common neighbor. Prove that some vertex is adjacent to all others. (Burungale [2006])

5.3.44. (+) Suppose that n wires run from the top to the bottom of a tall building through a tube. The correspondence between the tops and the bottoms is unknown. The inspector has a battery and a lamp. When a circuit is completed using battery, lamp, and wires, the lamp lights. Prove that the inspector can match up the ends using only one trip up and down the building. The inspector can link some pairs of wires at the bottom, go up and make similar links and tests, and finally return to the bottom to complete the analysis.

5.3.45. ($\diamond$) Prove that K_n decomposes into three isomorphic subgraphs if and only if $n + 1$ is not divisible by 3.

5.3.46. Decompose the Petersen graph into three isomorphic connected subgraphs.

5.3.47. Let G be a k-regular graph, where k is odd. Prove that in any decomposition of G into paths, the average length of the paths is at most k.

5.3.48. ($\diamond$) Prove that K_n decomposes into $\lceil n/2 \rceil$ paths. Prove that K_n decomposes into $\lfloor n/2 \rfloor$ cycles when n is odd. (Walecki, in Lucas [1892])

5.3.49. ($\diamond$) Prove that every connected graph has an orientation with at most one vertex of odd outdegree. Conclude that a connected graph with an even number of edges decomposes into paths with two edges. (Rotman [1991])

5.3.50. Use induction on k to prove that a connected graph with $2k$ edges decomposes into paths with two edges. Is this true without the hypothesis of connectedness?

5.3.51. ($\diamond$) Nash-Williams [1970] conjectured that when n is sufficiently large, an n-vertex even graph G with $|E(G)|$ divisible by 3 decomposes into triangles if $\delta(G) \geq \frac{3}{4}n$. Use the graph H_t to prove that the degree threshold is sharp, where H_t is formed from C_4 by expanding each vertex into K_t and each edge into $K_{t,t}$.

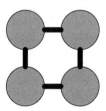

5.3.52. Let G be an n-vertex graph with $\delta(G) \geq n/4$, no cut-edge, and $|E(G)|$ divisible by 3. Barát–Thomassen [2006] proved that G decomposes into copies of $K_{1,3}$. Prove that this is sharp using a graph formed from $4K_{n/4}$ by one edge for each pair of components.

5.3.53. Prove that the edges of every n-vertex graph can be covered using at most $\lfloor n^2/4 \rfloor$ edges and triangles, with equality needed only for $K_{\lfloor n/2 \rfloor, \lceil n/2 \rceil}$. (Erdős–Goodman–Pósa [1966]) (Comment: Bollobás [1976a] generalized to coverings by edges and copies of K_r.)

5.3.54. (+) The **broom** $B_{n,r}$ is the n-vertex graph formed from the union of the path P_{n-r} and the star $K_{1,r}$ by letting the center of the star be an endpoint of the path. Prove that if K_n decomposes into copies of $B_{n,r}$, then n is even and $n \geq 4r - 2$. (Kovář–Kubesa–Meszka [2012]) (Comment: the condition is also sufficient, except for $(n, r) = (6, 2)$.)

5.3.55. (+) For a graph G, let $\omega(G) = \max\{r: K_r \subseteq G\}$. Prove $\sum_{i=1}^{k} \omega(G_i) \leq n + \binom{k}{2}$ for any decomposition $(G_1, \ldots, G_k)$ of K_n. Provide a construction to show that the bound is best possible when $n \geq \binom{k}{2}$. (Füredi–Kostochka–Škrekovski–Stiebitz–West [2005])

5.3.56. ($\Diamond$) Prove or disprove each statement below.

(a) Every graph G has an orientation D such that for every $v \in V(G)$, the numbers of edges entering v and leaving v in D differ by at most 1.

(b) Every graph G has an orientation D such that for every $S \subseteq V(G)$, the numbers of edges entering S and leaving S in D differ by at most 1.

5.3.57. ($\Diamond$) Let G be a connected graph having $|E(G)|$ even and an even number of vertices of each odd degree. Prove that G decomposes into two spanning subgraphs with the same degree list. Show that the last condition is needed. (Choi–Ozkahya–West [2010])

5.3.58. *Eulerian digraphs.* Let D be a digraph with no isolated vertex.

(a) Prove that D has an Eulerian circuit if and only if $d^+(v) = d^-(v)$ for each vertex v and the underlying graph is connected.

(b) Suppose that $d^-(v) = d^+(v)$ for every vertex v, except that $d^+(x) - d^-(x) = k = d^-(y) - d^+(y)$. Prove that D contains k edge-disjoint x, y-paths.

5.3.59. ($\Diamond$) *de Bruijn cycles.* For $k, l \in \mathbb{N}$, form a digraph G with $V(G) = [k]^l$ (called the **de Bruijn graph**) as follows. For $u, v \in [k]^l$, create an edge uv if the last $l - 1$ positions in u form the same $(l - 1)$-tuple as the first $l - 1$ positions in v.

(a) Prove that the de Bruijn graph is an Eulerian digraph. (de Bruijn [1946])

(b) Prove that there is a cyclic arrangement of k^{l+1} characters chosen from $[k]$ such that the k^{l+1} strings of length $l + 1$ are distinct. (Good [1946], Rees [1946])

5.3.60. (+) *Alternative characterization of Eulerian graphs.*

(a) For $uv \in E(G)$ when G is Eulerian, prove that an odd number of u, v-trails in $G - uv$ reach v only at the end, and that an even number of these are not paths. (Toida [1973])

(b) Let v be a vertex of odd degree in a graph G. Prove that for some edge incident to v, the number of cycles through it is even.

(c) Use parts (a) and (b) to show that a nontrivial connected graph is Eulerian if and only if every edge belongs to an odd number of cycles. (McKee [1984])

5.4. Trees and Distance

When a graph models a communication network, its spanning trees are the minimal subgraphs that permit communication among all vertices. We restate the definitions from the Introduction.

5.4.1. Definition. An **acyclic graph** is a graph with no cycles. A **tree** is a connected acyclic graph. A graph whose components are all trees is a **forest**. A **spanning tree** in a graph is a spanning subgraph that is a tree.

We studied enumeration of trees in Part I, counting those with a fixed vertex set in Chapter 1, isomorphism classes in Chapter 4, and spanning trees of a given graph (recursively) in Chapter 2 (see also Theorem 15.2.5). Here we consider graph-theoretic properties and applications of trees. We also study distances in graphs and their relation to trees.

PROPERTIES OF TREES

We expand on properties proved earlier. A **leaf** is a vertex of degree 1.

5.4.2. Proposition. Every nontrivial tree has at least two leaves. Deleting a leaf from a tree yields a tree.

Proof: Nontrivial trees have no isolated vertices. Since trees are acyclic, the endpoints of every maximal path are leaves.

Deleting a vertex cannot create a cycle. A leaf v in a graph G cannot be a cut-vertex of G, since v can have a neighbor only in one component of $G - v$. Thus if v is a leaf of a tree G, then $G - v$ is acyclic and connected and hence is a tree. ∎

Proposition 5.4.2 implies that each tree with $n + 1$ vertices arises from a tree with n vertices by adding an edge to a new vertex. This justifies inductive proofs about trees that perform the induction step by "growing a new leaf" from an arbitrary vertex of an arbitrary smaller tree.

The fact that the number of edges in a tree is one less than the number of vertices is one of the most fundamental properties of a tree.

5.4.3. Proposition. For a graph G with n vertices, any two of the following three properties implies the third (and hence characterize trees).
(a) G is acyclic.
(b) G is connected (that is, G has one component).
(c) G has $n - 1$ edges.

Proof: Adding an edge to a graph reduces the number of components (by 1) if and only if it joins vertices from distinct components and hence creates no cycle. Starting with no edges, we thus conclude that a graph with k edges has exactly $n - k$ components if and only if it has no cycles.

Therefore, an acyclic graph has one component if and only if it has $n-1$ edges, and a graph with $n - 1$ edges and one component must be acyclic. ∎

Various other properties also characterize the graphs that are trees. It is efficient to prove a cycle of implications.

5.4.4. Proposition. For each property below, a graph G is a tree if and only if it satisfies that property.
(a) G is connected and every edge is a cut-edge.
(b) G contains exactly one u, v-path whenever $u, v \in V(G)$.
(c) G has no cycle, and adding any edge creates exactly one cycle.

Proof: (G is a tree) $\Rightarrow$ (a): By Proposition 5.3.11, an edge is a cut-edge if and only if it is in no cycle. Since a tree is acyclic, every edge is a cut-edge.

(a) $\Rightarrow$ (b): Since G is connected, it contains a u, v-path. If G has more than one u, v-path, then let xy be an edge belonging to one but not all of them. By Corollary 5.3.5, $G - xy$ contains an x, y-path, contradicting xy being a cut-edge.

(b) $\Rightarrow$ (c): Two vertices on a cycle are connected by two paths along the cycle, so (b) implies that G has no cycle. For $u, v \in V(G)$, adding the edge uv creates exactly one cycle since G contains exactly one u, v-path.

(c) $\Rightarrow$ (G is a tree): Adding a missing edge uv creates a cycle, so G contains a u, v-path. Hence G is connected. With lack of cycles also given, G is a tree. ∎

5.4.5. Proposition. Every connected graph contains a spanning tree.

Proof: Since deleting an edge of a cycle does not disconnect a graph, iteratively deleting edges from cycles in a connected graph leads to a connected acyclic spanning subgraph of the original graph. ∎

The properties of trees allow us to switch edges in spanning trees, moving from one spanning tree to another (see also Exercises 64–65).

5.4.6. Theorem. If T, T' are spanning trees of a graph G and $e \in E(T) - E(T')$, then there exists $e' \in E(T') - E(T)$ such that both $T - e + e'$ and $T' + e - e'$ are spanning trees.

Proof: Since e is a cut-edge of T, we may let U and U' be the vertex sets of the two components of $T - e$ (in the 8-vertex graph below, T is bold and T' is solid). Let $e = uu'$, with $u \in U$ and $u' \in U'$. Because T' is a spanning tree, T' contains a unique u, u'-path. This path has an edge from U to U'; let e' be such an edge.

The only edge of T joining U and U' is e, so $e' \in E(T') - E(T)$. Since e' joins the components of $T - e$, the subgraph $T - e + e'$ is a spanning tree. Since e' is in the u, u'-path in T', it is in the unique cycle formed by adding e to T'. Hence $T' + e - e'$ also is a spanning tree. ∎

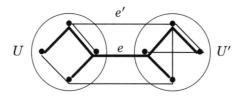

5.4.7.* Theorem. (Berman [1986]) Let T be a spanning tree in a graph G with at least three vertices. If $d_G(v) - d_T(v)$ is odd for every vertex that is not a leaf of T, then G has an even number of spanning trees T' (including T) such that $d_{T'}(v) = d_T(v)$ for all $v \in V(G)$.

Proof: (Cameron–Edmonds [1999]) We define an auxiliary graph H such that the vertices of odd degree in H correspond to the desired trees. Corollary 5.2.2 then implies that the number of them is even.

Specify one non-leaf vertex w in T. The vertices of H are the spanning trees in G whose vertex degrees agree with T (Type 1) or agree with T except for having degree $d_T(w) + 1$ at w and degree $d_T(u) - 1$ at some other vertex u (Type 2).

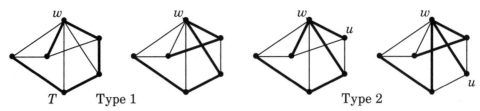

Spanning trees S and S' that are vertices of H are adjacent in H if and only if each tree has exactly one edge not belonging to the other. Thus a neighbor S' of S is obtained from S by adding one edge of S' and deleting one edge of S.

One endpoint of the added edge increases in degree; call it v. If S is Type 1, then $v = w$. If S is Type 2, then v must be the vertex u such that $d_S(u) = d_T(u) - 1$. Since u is not isolated in S, in both cases v is not a leaf in T. Hence the number of choices for the added edge is odd when S is Type 1 and is even when S is Type 2.

Let vx be the added edge, where $x \in N_G(v) - N_S(v)$. Adding vx creates a unique cycle with edges of S. The degree of x must not increase, because already degree is increasing at v. Hence the edge deleted to form S' must be the edge reaching x along the v, x-path in S. The resulting tree S' has vertex degrees putting it in $V(H)$; it is Type 1 if $y = w$ and Type 2 otherwise.

We have proved $d_H(S) = d_G(v) - d_S(v)$ and shown that the degree is odd when S is Type 1 and even otherwise. Hence the number of Type 1 trees is even.   ∎

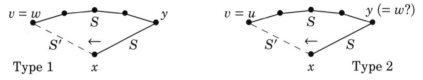

A famous result of Smith (see Tutte [1946]) states that in a 3-regular graph the number of spanning cycles containing a particular edge is even. A stronger version of this follows easily from Theorem 5.4.7.

5.4.8.* Corollary. (Thomason [1978]) If every vertex of G except possibly the endpoints of a particular edge e has odd degree, then the number of spanning cycles containing e is even.

Proof: (Cameron–Edmonds [1999]) Deleting e from a spanning cycle of G leaves a spanning tree T in which every vertex other than the endpoints of e has degree 2. Hence those vertices have odd degree outside T, and Theorem 5.4.7 applies. Each of the resulting trees becomes a spanning cycle by adding e.   ∎

DISTANCE AND DIAMETER

In a connected graph, each vertex can be reached from every other; how many steps are needed? The concept is valid for graphs or digraphs.

5.4.9. Definition. If x is connected to y in G, then the length of the shortest x, y-path is the **distance** from x to y, written $d_G(x, y)$ or simply $d(x, y)$. If G has no x, y-path, then $d(x, y) = \infty$.

The distance function of a graph is a metric on the vertex set: it is nonnegative, symmetric, positive for distinct vertices, and satisfies the triangle inequality $d(x, y) + d(y, z) \geq d(x, z)$ (Exercise 38). The distance function for a digraph is not symmetric unless the digraph is symmetric.

One measure of the performance of a communication network is the worst time it might take to reach one point from another.

5.4.10. Definition. The **diameter** of a graph G is $\max_{u,v \in V(G)} d(u, v)$. The **eccentricity** $\varepsilon(u)$ of a vertex u is $\max_{v \in V(G)} d(u, v)$. The **radius** of G equals $\min_{u \in V(G)} \varepsilon(u)$. The **center** of G is the set of vertices with least eccentricity.

The diameter is the largest vertex eccentricity. The term "diameter" comes from geometry, where diameter is the greatest distance between two vertices in a set. When G is not connected, the diameter, the radius, and each eccentricity are infinite. The graph below has three vertices in its center; we have labeled each vertex with its eccentricity.

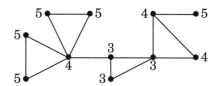

The uniqueness of paths in trees makes it easy to compute distances. Since every path is the shortest (and only!) path between its endpoints, the maximum of the vertex distances is the length of the longest path.

The next observation, about centers of trees, was used in Section 4.2.

5.4.11. Theorem. (Jordan [1869]) The center of a tree consists of one vertex or two adjacent vertices.

Proof: Let d be the diameter of a tree T. Let S be the central vertex or edge on a path P of length d in T. The vertices of S have eccentricity $\lceil d/2 \rceil$, since a vertex at distance greater than $\lceil d/2 \rceil$ from $v \in S$ would combine with half of P to form a path longer than P.

All vertices outside S have eccentricity larger than $\lceil d/2 \rceil$, by the uniqueness of paths in trees. From a vertex u, follow a path to P, and then add at least $\lceil d/2 \rceil$ edges to reach the farther end of P, with equality in the last step only if the first vertex reached on P is in S. ∎

Theorem 5.4.11 also has an easy inductive proof (Exercise 44) and a proof using the Pigeonhole Principle (Exercise 10.1.24).

Beyond trees, small diameter remains desirable for efficiency of communication. A complete graph has diameter 1, while the star $K_{1,n-1}$ has diameter 2 with only $n - 1$ edges, but physical constraints may limit the maximum degree. The problem of finding the smallest diameter among n-vertex graphs with maximum degree k can be solved by instead determining for all k and d the maximum number of vertices in a graph with diameter d and maximum degree at most k.

5.4.12. Proposition. (**Moore bound**) Graphs with maximum degree k and diameter d have at most $1 + k\frac{(k-1)^d-1}{k-2}$ vertices.

Proof: Let G be a graph with maximum degree k and diameter d. From a vertex v, at most $k(k-1)^{i-1}$ vertices have distance i in G, and all vertices are within distance d. Thus G has at most $k\sum_{i=1}^{d}(k-1)^{i-1}$ vertices other than v. Evaluating the geometric sum yields the desired bound. ∎

When $d(v) < k$ for some vertex v, starting the search at v multiplies the bound by the fraction $d(v)/k$. Thus graphs within a fraction $1/k$ of the bound must be k-regular. Equality requires adding edges among the leaves of the search tree to establish distance at most d for all vertex pairs. As shown above, the Petersen graph accomplishes this when $d = 2$ and $k = 3$ (further motivating study of the Petersen graph!).

Equality in the Moore bound requires the full tree grown from a vertex and hence girth at least $2d + 1$. This is best possible; every graph of diameter d that is not a tree has girth at most $2d + 1$ (Exercise 9).

5.4.13. Example. *Moore graphs.* A **Moore graph** is a k-regular graph (for $k \geq 3$) that has diameter d, girth $2d + 1$, and order $1 + k\frac{(k-1)^d-1}{k-2}$ (any two of these three restrictions imply the third; see Exercise 59).

Damerell [1973] and Bannai–Ito [1973] independently proved that no Moore graph exists with diameter at least 3. For diameter 2, Hoffman–Singleton [1960] proved that Moore graphs exist only for $k = 3$, $k = 7$, and possibly $k = 57$ (Theorem 15.3.26). For $k = 3$, we have the Petersen graph. For $k = 7$, the 50-vertex Hoffman–Singleton graph contains many copies of the Petersen graph (Exercise 59). It remains unknown whether a 57-regular Moore graph exists. ∎

For diameter 2, the least number of edges in an n-vertex graph with maximum degree k is known exactly when $k \geq (n+1)/2$ (Erdős–Rényi–Sós [1966]) and asymptotically when $\sqrt{n-1} \leq k \leq n/2$ (Bollobás [1971]).

Why $\sqrt{n-1}$? The Moore bound requires $n \leq 1 + k^2$ for diameter 2. Graphs with diameter 2 and maximum degree near $\sqrt{n}$ don't look much like trees. We can get close with a cartesian product.

5.4.14. Example. *Diameter 2 with degree as small as $2\sqrt{n}$.* Given $n = m^2$, let $G_m = K_m \square K_m$. Thus $V(G_m) = \{(i, j): 1 \leq i, j \leq m\}$, with (i, j) and (k, l) adjacent if $i = k$ or $j = l$. Each vertex has degree $2m - 2$, and non-adjacent vertices (i, j) and (k, l) have common neighbors (i, l) and (k, j).

However, G_m is regular of degree $2\sqrt{n} - 2$, not $\sqrt{n-1}$. Reducing the degree to approximately $\sqrt{n-1}$ requires finite projective planes, which can be obtained from finite fields; see Chapter 13. For values of n having the form q^2+q+1 (where q is a power of a prime), the result is a nearly-regular graph of diameter 2 with maximum degree $q + 1$, which equals $\lceil \sqrt{n-1} \rceil$ (some vertices have degree q). ∎

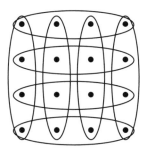

OPTIMIZATION ON WEIGHTED GRAPHS

The notion of length suggests a more general graph model: a **weighted graph** has real-valued edge weights. In a weighted graph, the length of a walk is the sum of the weights of its edges; the *unweighted case* is when each edge weight is 1. Classical problems in this setting are computing distances and finding a spanning tree of minimum total weight.

Algorithms to find minimum-weight spanning trees in connected weighted graphs rest on Theorem 5.4.6 or similar observations. Kruskal's Algorithm is a well-known example; it is a "greedy" algorithm. In Chapter 11 we study a more general setting where this algorithm works.

5.4.15. Theorem. (Kruskal's Algorithm; Kruskal [1956]) Given a connected n-vertex graph with edge weights, a minimum weight spanning tree is constructed by $n - 1$ repetitions of selecting a cheapest edge that completes no cycle with edges already selected.

Proof: Always there are edges that join components in the spanning subgraph of chosen edges, so the algorithm produces a spanning tree.

Let T be a tree found by the algorithm, and let T^* be a minimum-weight spanning tree sharing the most edges with T. If $T^* \neq T$, then because they have the same size there exists $e \in E(T) - E(T^*)$. By Theorem 5.4.6, there exists $e' \in E(T^*) - E(T)$ such that both $T - e + e'$ and $T^* + e - e'$ are spanning trees.

Since $T^* + e - e'$ shares more edges with T than does T^*, the choice of T^* implies that e has larger weight than e'. On the other hand, since adding e' to $T - e$ creates no cycles, both e and e' were eligible when the algorithm choose e, and thus e cannot have larger weight than e'. The contradiction yields $T^* = T$. ∎

5.4.16. Example. Choices in Kruskal's Algorithm depend only on the order of weights, not the values. Below we have used positive integers as weights to indicate the order of examination of edges. The four cheapest edges are selected, but then we cannot take the edges of weight 5 or 6. We can take the edge of weight 7, but then not those of weight 8 or 9. ∎

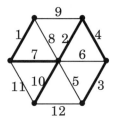

Other early algorithms were Borůvka [1926], Jarník [1930], and Prim [1957] (Exercise 67). Modern improvements use clever data structures to merge components quickly. Fast versions appear in Tarjan [1984] when the edges are presorted by cost and in Gabow–Galil–Spencer–Tarjan [1986] when they are not. For further discussion and references, see Ahuja–Magnanti–Orlin [1993, Chapter 13]. More recent developments appear in Karger–Klein–Tarjan [1995].

Next we consider finding shortest paths. Dijkstra's Algorithm (Dijkstra [1959] and Whiting–Hillier [1960]) solves this problem quickly, using the observation that the u,v-portion of a shortest u,z-path is a shortest u,v-path. It finds shortest paths from u to all other vertices z in increasing order of $d(u,z)$, creating a spanning tree from u. The **distance** $d(u,z)$ in a weighted graph is the minimum sum of the edge weights along a u,z-path (we assume nonnegative weights).

5.4.17. Algorithm. (**Dijkstra's Algorithm** for distances from one vertex.) Given a graph (or digraph) G with nonnegative edge weights and a starting vertex u, with weight $w(xy)$ on edge xy (and $w(xy) = \infty$ if $xy \notin E(G)$), we maintain a set S of vertices to which a shortest path from u is known and a tentative distance $t(z)$ from u to each $z \notin S$, being the length of the shortest u,z-path yet found.

The initial state is $S = \{u\}$, $t(u) = 0$, and $t(z) = w(uz)$ for $z \neq u$. At each iteration, select v outside S such that $t(v) = \min_{z \notin S} t(z)$. Add v to S. Explore edges from v to the rest of $V(G) - S$ to update tentative distances: for each edge vz with $z \notin S$, set $t(z)$ to $\min\{t(z), t(v) + w(vz)\}$.

The iteration continues until $S = V(G)$ or until $t(z) = \infty$ for every $z \notin S$. At the end, set $d(u,v) = t(v)$ for all v. ∎

5.4.18. Example. In the example below, shortest paths from u are found to the other vertices in the order a,b,c,d,e, with distances $1,3,5,6,8$, respectively. To obtain the paths, we only need the edge on which each reaches its destination, because the earlier portion of a shortest u,z-path that reaches z along vz is a shortest u,v-path. Here the final edges on the paths to a,b,c,d,e generated by the algorithm are ua,ub,ac,ad,de, respectively; these are the edges of the spanning tree generated from u. ∎

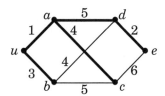

With the phrasing given in Algorithm 5.4.17, Dijkstra's Algorithm works also for digraphs, generating an out-tree rooted at u if every vertex is reachable from u. The proof works for graphs and for digraphs.

5.4.19. Theorem. Given a [di]graph G and a vertex $u \in V(G)$, Dijkstra's Algorithm computes $d(u, z)$ for every $z \in V(G)$.

Proof: We prove the stronger statement that at each step, (1) $t(z) = d(u, z)$ if $z \in S$, and (2) if $z \notin S$, then $t(z)$ is the minimum length of a u, z-path reaching z directly from S.

We use induction on $|S|$. Basis step: $|S| = 1$. From the initialization, $S = \{u\}$, $d(u, u) = t(u) = 0$, and the minimum length of a u, z-path reaching z directly from S is $w(uz)$, which is the initial value for $t(z)$ and is infinite when $uz \notin E(G)$.

Induction step: Suppose that (1) and (2) are true when $|S| = k$. Let v be a vertex among $z \notin S$ such that $t(z)$ is smallest. Let $S' = S \cup \{v\}$. We first show $d(u, v) = t(v)$. A shortest u, v-path must exit S before reaching v. By the induction hypothesis, the length of the shortest path going directly to v from S is $t(v)$. The induction hypothesis and choice of v imply that a path reaching v via any vertex outside S has length at least $t(v)$. Hence $d(u, v) = t(v)$, and (1) holds for S'.

To prove (2) for S', let z be a vertex outside S other than v. By hypothesis, the shortest u, z-path reaching z directly from S has length $t(z)$ (∞ if there is no such path). When we add v to S, we also consider paths reaching z from v. Since we have now computed $d(u, v) = t(v)$, the shortest such path has length $t(v) + w(vz)$, which we compare with the previous value of $t(z)$ to find the shortest path reaching z directly from S'.

Thus (1) and (2) hold for the new set S' of size $k + 1$.   ∎

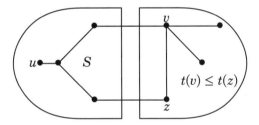

By maintaining $d(u, x) \le t(z)$ for all $x \in S$ and $z \notin S$, the algorithm computes distances from u in nondecreasing order. It yields $d(u, v) = \infty$ when v is unreachable from u. The special case for unweighted graphs is **Breadth-First Search** from u; here both the algorithm and the proof (Exercise 70) are simpler.

EXERCISES 5.4

5.4.1. (−) Prove that each property below characterizes the class of forests.
(a) Every induced subgraph has a vertex of degree at most 1.
(b) Every connected subgraph is an induced subgraph.
(c) If two paths have a common vertex, then their intersection is a path.

5.4.2. (−) Prove that a tree with $2k$ vertices of odd degree decomposes into k paths.

5.4.3. $(-)$ Prove that a tree with maximum degree k has at least k leaves. For $n > k > 1$, construct an n-vertex tree having maximum degree k and exactly k leaves.

5.4.4. $(-)$ Determine the number of vertices in a tree with average degree a.

5.4.5. $(-)$ Determine the possible numbers of vertices in a tree with each degree in $\{1, k\}$.

5.4.6. $(-)$ Prove that every nontrivial tree has at least two maximal independent sets, with equality only for stars. (Note: maximal $\neq$ maximum.)

5.4.7. $(-)$ Count the isomorphism classes of n-vertex trees with diameter 3.

5.4.8. $(-)$ Decompose the graph below into two isomorphic spanning trees.

5.4.9. $(-)$ Prove that a connected graph G with a cycle has girth at most $2\operatorname{diam}(G) + 1$. For $k \in \mathbb{N}$, exhibit a graph with diameter k and girth $2k + 1$.

5.4.10. $(-)$ Prove or disprove: If T is a tree in which every vertex adjacent to a leaf has degree at least 3, then T has two leaves with a common neighbor.

5.4.11. $(-)$ Given a graph G, let G' be the graph with vertex set $V(G)$ such that $xy \in E(G')$ if and only if $d_G(x, y) \in \{1, 2\}$. Prove $\operatorname{diam}(G') = \lceil \operatorname{diam}(G)/2 \rceil$.

5.4.12. $(-)$ Find the least radius among spanning trees of the k-dimensional hypercube.

5.4.13. $(-)$ In the graph $K_1 \oplus C_4$, assign weights $(1, 1, 2, 2, 3, 3, 4, 4)$ to the edges in two ways: one way so that the minimum-weight spanning tree is unique, and another way so that the minimum-weight spanning tree is not unique.

5.4.14. Let G be an n-vertex graph such that deleting any one vertex of G yields a tree. Determine the number of edges in G, and use this to determine G itself.

5.4.15. For $k > 2$, prove that if the vertices of a k-regular graph can be partitioned into sets A and B that each induce a tree, then $|A| = |B|$. (X. Lv)

5.4.16. $(\diamond)$ Let $d_1, \ldots, d_n$ and $c_1, \ldots, c_n$ be positive integers.
    (a) Prove that some n-vertex tree has degrees $d_1, \ldots, d_n$ if and only if $\sum d_i = 2n - 2$.
    (b) Prove that some $2n$-vertex tree has degrees $d_1, \ldots, d_n$ in one part of the bipartition and $c_1, \ldots, c_n$ in the other if and only if $\sum d_i = \sum c_i = 2n - 1$. (Hollingsworth [2013])

5.4.17. $(\diamond)$ Let T be a tree with k edges. Prove that if G is a graph with minimum degree at least k, then $T \subseteq G$. (Comment: By Exercise 5.2.10, the same conclusion follows when G has average degree at least $2k$. The Erdős-Sós Conjecture states that it suffices to have average degree more than $k - 1$, or equivalently more than $n(k - 1)/2$ edges.)

5.4.18. Prove for $n > k$ that every tree with k edges is a subgraph of every n-vertex graph with more than $n(k - 2) - \binom{k}{2}$ edges. (Comment: The Erdős-Sós Conjecture implies this.)

5.4.19. $(\diamond)$ Two n-vertex graphs G and H **pack** if $H \subseteq \overline{G}$.
    (a) Prove that if $|E(G)|\,|E(H)| < \binom{n}{2}$, then G and H pack. This is sharp, since K_n does not pack with $K_2 + (n - 2)K_1$. For even n, find another sharpness example. (Comment: Sauer–Spencer [1978] proved this and that G and H pack if $2\Delta(G)\Delta(H) < n$, which also is sharp. See also Kierstead–Kostochka [2009].)
    (b) Prove that if T is an n-vertex tree, and G is an n-vertex graph with $|E(\overline{G})| < n/2$, then $T \subseteq G$. Is the bound best possible? (Zhou [1984])

5.4.20. (+) Let G be a graph such that the degrees of any two vertices that are distance 2 apart sum to at least $2k$ (and there is at least one such pair). Prove that G contains every tree with k edges. (T. Jiang)

5.4.21. ($\diamond$) Prove that if K_n decomposes into k spanning connected subgraphs, then $n \geq 2k$. Show that this bound is sharp by decomposing K_{2k} into k spanning double-stars, where a **double-star** is a tree of diameter 3. (Hint: Use double-stars whose central edges share no vertices.) (Lovász [1966], Palumbíny [1973])

5.4.22. ($\diamond$) For $2 \leq d < n$, determine the maximum number of edges in an n-vertex graph having diameter d, and find all the extremal graphs. (Ore [1968], Qiao–Zhan [2019])

5.4.23. Let G be a tree. Prove that $V(G)$ can be split into two nonempty sets such that each vertex has at least half its neighbors in its own set if and only if G is not a star.

5.4.24. Let U be the set of nonleaf vertices in a forest G with k nontrivial components. Prove that G has $2k + \sum_{u \in U}[d(u) - 2]$ leaves. In terms of r alone, determine the number of vertices in a tree whose nonleaf vertices are one of degree i for each i with $2 \leq i \leq r$.

5.4.25. Let e be an edge in a connected multigraph G. Prove that e is a cut-edge of G if and only if e belongs to every spanning tree of G. Prove that e is a loop if and only if e belongs to no spanning tree of G.

5.4.26. ($\diamond$) Prove that a connected n-vertex multigraph has exactly one cycle if and only if it has exactly n edges.

5.4.27. ($\diamond$) Let G be an n-vertex graph with m edges, where $n \geq 4$ and $m \geq 2n - 3$. Prove that G has two cycles of the same length. (Comment: Chen–Lehel–Jacobson–Shreve [1998] proved that $m \geq n + \sqrt{2n - 4} - 1$ suffices and is asymptotically sharp.)

5.4.28. Let T be a tree of even order. Prove that T has exactly one spanning subgraph in which every vertex has odd degree.

5.4.29. ($\diamond$) A **parity subgraph** of a graph G is a graph H contained in G such that $d_H(v) \equiv d_G(v) \pmod 2$ for all $v \in V(G)$.

(a) Prove that every spanning tree of a connected graph G contains exactly one parity subgraph of G. (Itai–Rodeh [1978])

(b) Hajós conjectured that every even graph with n vertices decomposes into at most $\lfloor n/2 \rfloor$ cycles. Prove that if this conjecture is true, then every n-vertex graph decomposes into fewer than $3n/2$ cycles and edges. (Pyber [1984, 1992], Dean [1987])

5.4.30. ($\diamond$) Let $e_1, \ldots, e_k$ be distinct edges in a graph G having k edge-disjoint spanning trees. Prove that G has edge-disjoint spanning trees $T_1, \ldots, T_k$ with $e_i \in T_i$ for $1 \leq i \leq k$.

5.4.31. ($\diamond$) Let G be a tree with k leaves. Prove that G is the union of paths $P_1, \ldots, P_{\lceil k/2 \rceil}$ such that $P_i \cap P_j \neq \varnothing$ for all $i \neq j$. (Ando–Kaneko–Gervacio [1996])

5.4.32. Let G be an n-vertex graph having a decomposition into k spanning trees. Suppose also that $\Delta(G) = \delta(G) + 1$. For $2k \geq n$, show that this is impossible. For $2k < n$, determine the degree list of G in terms of n and k.

5.4.33. (+) *Spanning trees with fixed degrees.*

(a) Let G be a graph decomposing into trees S and T. Prove that the number of decompositions of G into trees S' and T' with $(d_{S'}(v), d_{T'}(v)) = (d_S(v), d_T(v))$ for all $v \in V(G)$ is even. (Berman [1986], Cameron–Edmonds [1999])

(b) Let G be 4-regular, with $e, e' \in E(G)$. Prove that G has an even number of decompositions into spanning cycles with e in the first and e' in the second. (Thomason [1978])

5.4.34. Let x and y be the endpoints of a maximal path P in a graph G, and let l be the length of P. Prove that $d(x, y) \leq \max\{2 + l - d(x) - d(y), 2\}$. (Tracy [2000])

5.4.35. *Decomposition into trees.* (Chung [1978a])

(a) Let G be a connected graph with at least three vertices. Prove that G has two vertices x, y such that $G - \{x, y\}$ is connected and $d(x, y) \leq 2$.

(b) (+) Prove that every connected n-vertex graph G decomposes into $\lceil n/2 \rceil$ trees. (Hint: Prove that G has a tree decomposition $T_1, \dots, T_{\lceil n/2 \rceil}$ and $V(G)$ has a partition into pairs $S_1, \dots, S_{\lceil n/2 \rceil}$ such that $S_i \subseteq V(T_i)$ for all i (except that $|S_1| = 1$ when n is odd).

(c) Let G be a graph with n vertices and k components. Prove that G decomposes into at most $\lfloor (n + k)/2 \rfloor$ trees.

5.4.36. ($\Diamond$) Let $g(n, k)$ be the maximum possible girth of an n-vertex graph with $n + k$ edges. Trivially $g(n, 0) = n$. Prove $g(n, 1) = \lfloor (2n + 2)/3 \rfloor$ and $g(n, 2) = \lfloor (n + 2)/2 \rfloor$. (Comment: Bollobás–Thomason [1997] proved $g(n, k) \leq \lfloor \frac{n+k}{k} \rfloor \log_2(2k)$, and Bollobás–Szemerédi [2002] improved this to $g(n, k) \leq \frac{2}{3}(n + k)(\log_2 k + \log_2 \log_2 k + 4)$.)

5.4.37. ($\Diamond$) *Diameter and complements.* Let G be a graph.

(a) Prove (diam $G \geq 3 \Rightarrow$ diam $\overline{G} \leq 3$) and (diam $G \geq 4 \Rightarrow$ diam $\overline{G} \leq 2$).

(b) Prove that if G is regular, then diam $G \geq 3$ implies diam $\overline{G} \leq 2$.

5.4.38. ($\Diamond$) *Diameter and radius.* Let G be a graph.

(a) Prove the triangle inequality for distance: $d_G(u, v) + d_G(v, w) \geq d_G(u, w)$.

(b) Use part (a) to prove that diam $G \leq 2\,\mathrm{rad}\,G$.

(c) Given positive integers r, d with $r \leq d \leq 2r$, construct a graph with radius r and diameter d. (Hint: Build a suitable graph with one cycle.)

5.4.39. Let S be a set of k values in $\mathbb{Z}_n$ that is unchanged under negation. Let G be the k-regular graph with vertex set $\mathbb{Z}_n$ whose vertices are adjacent if their difference modulo n is in S. Prove that diam $G > 3$ requires $k < \lfloor n/2 \rfloor$. (A. Pasotti)

5.4.40. Let G be a connected Eulerian graph with at least three vertices. A vertex v in G is *extendible* if every trail beginning at v can be extended to form an Eulerian circuit of G. For example, in the graphs below only the marked vertices are extendible. Prove the following statements about G (adapted from Chartrand–Lesniak [1986, p. 61]).

a) A vertex $v \in V(G)$ is extendible if and only if $G - v$ is a forest. (Ore [1951])

b) If v is extendible, then $d(v) = \Delta(G)$. (Bäbler [1953])

c) All vertices of G are extendible if and only if G is a cycle.

d) If G is not a cycle, then G has at most two extendible vertices.

 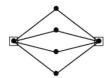

5.4.41. ($\Diamond$) Prove that the isomorphism class of a tree is determined by the distances among its leaves. That is, a tree S with leaves $\{x_1, \dots, x_k\}$ is determined by knowing for each i and j the value of $d_S(x_i, x_j)$. (Smolenskii [1962])

5.4.42. Prove that every automorphism of a tree T with $|V(T)|$ odd has a fixed point.

5.4.43. *Pathological behavior of eccentricity.*

a) Prove that the center of a graph can be disconnected and can have components arbitrarily far apart, by constructing a graph where the center consists of two vertices and the distance between these two vertices is k.

b) Prove that if G is a tree, then every vertex x outside the center of G has a neighbor with eccentricity $\varepsilon(x) - 1$.

c) For $r \geq 3$, construct a graph with radius r in which some vertex x has eccentricity $r + 2$ and has no neighbor with eccentricity $r + 1$.

5.4.44. Theorem 5.4.11 states that the center of a tree consists of one vertex or two adjacent vertices. Give two more proofs of this fact, as follows:
   (a) By induction on the number of vertices.
   (b) By showing directly that any two vertices in the center of a tree must be adjacent.

5.4.45. ($\diamond$) Let the *weight* of a vertex v in a tree T be the maximum order of a component of $T - v$. A **centroid** is a vertex of smallest weight. Prove that a tree has one centroid or two adjacent centroids, the latter only when each has weight $|V(T)|/2$. (Jordan [1869])

5.4.46. Prove that every tree T has a vertex v such that for all $e \in E(T)$, the component of $T - e$ containing v has at least $\lceil |V(T)|/2 \rceil$ vertices. Prove that v is unique or there are two adjacent such vertices.

5.4.47. ($\diamond$) For $v \in V(G)$, let $s(v) = \sum_{w \in V(G)} d(v, w)$. The set of vertices minimizing $s(v)$ is called the **barycenter** of G.
   (a) Prove that if G is a tree with edges xy and yz, then $2s(y) < s(x) + s(z)$. Use this to prove that the barycenter of a tree consists of one vertex or two adjacent vertices.
   (b) Let G be a tree of diameter d. Determine the maximum possible distance in G between the center and the barycenter.

5.4.48. Let T be a tree with n vertices, k leaves, and maximum degree k. Determine the maximum and minimum possible values of diam T.

5.4.49. ($\diamond$) The **Kneser graph** $K(n, k)$ (Kneser [1955]) is defined on the k-element subsets of $[n]$, with two k-sets adjacent if and only if they are disjoint. For example, the Petersen graph is $K(5, 2)$. Prove diam$(K(n, k)) = 2$ for $n \geq 3k - 1$. Prove diam$(K(2k + 1, k)) = k$. (Comment: In general, diam$(K(n, k)) = \lceil \frac{k-1}{n-2k} \rceil + 1$; see Valencia-Pabon & Vera [2005].)

5.4.50. ($\diamond$) For a connected n-vertex graph, the **average distance** is the average of all $\binom{n}{2}$ distances between pairs of vertices. Determine the average distance in P_n, and prove that every connected n-vertex graph other than P_n has smaller average distance.

5.4.51. Determine the average distance between points in the k-dimensional hypercube Q_k (averaged over all pairs of points).

5.4.52. Let G be a connected graph of order n and minimum degree k, with $2 \leq k \leq n - 3$. Prove diam $G \leq 3\frac{n-2}{k+1} - 1$. Prove that equality can hold whenever $k \geq 2$ and $\frac{n-2}{k+1}$ is an integer greater than 1. (Moon [1965a])

5.4.53. Let $F_1, \ldots, F_k$ be forests whose union is G. Prove that $k \geq \max_{H \subseteq G} \lceil \frac{|E(H)|}{|V(H)|-1} \rceil$. (Comment: Chapter 11 has a proof using matroids that equality holds for some decomposition into forests.) (Nash-Williams [1964], Edmonds [1965a])

5.4.54. Let G be a graph having k edge-disjoint spanning trees. Prove that for any partition of $V(G)$ into r parts, there are at least $k(r - 1)$ edges of G whose endpoints are in different parts of the partition. (Comment: Chapter 11 has a proof using matroids that the condition is also sufficient.) (Tutte [1961a], Nash-Williams [1961], Edmonds [1965b].)

5.4.55. ($\diamond$) Prove that a graph G is a forest if and only if for every set of pairwise intersecting paths in G, some vertex belongs to all the paths in the set.

5.4.56. (+) Prove that every n-vertex tree other than $K_{1,n-1}$ is contained in its complement. (Burns–Schuster [1978])

5.4.57. Let S be an n-element set, and let $A_1, \ldots, A_n$ be n distinct subsets of S. Prove that S has an element x such that the sets $A_1 \cup \{x\}, \ldots, A_n \cup \{x\}$ are distinct. (Hint: Define a graph G with vertices $a_1, \ldots, a_n$ such that $a_i a_j \in E(G)$ if and only if A_i and A_j differ by one element. Use that element as a label on the edge. Prove that some forest contains all the labels that occur on edges, and use it to obtain the desired element x.) (Bondy [1972a])

5.4.58. (+) Let G be a graph with fewest vertices among k-regular graphs with girth at least g, where $k \geq 2$ and $g \geq 3$. (By Exercise 5.1.47, such graphs exist.) Prove that G has diameter at most g. (Hint: If $d_G(x, y) > g$, then obtain from G a smaller k-regular graph with girth at least g.) (Erdős–Sachs [1963])

5.4.59. ($\Diamond$) *Moore graphs.*
   (a) For k-regular connected graphs (with $k \geq 3$), prove that any two of the three properties {diameter d, girth $2d + 1$, order $1 + \frac{k[(k-1)^d - 1]}{k-2}$} imply the third.
   (b) The **Hoffman–Singleton graph** with 50 vertices is the graph G formed by adding to ten disjoint 5-cycles $Q_0, \ldots, Q_4$ and $R_0, \ldots, R_4$ a 5-regular bipartite graph as follows. Let the vertex sets of the 5-cycles be copies of $\mathbb{Z}_5$ in order. Make $i \in V(Q_j)$ and $l \in V(R_k)$ adjacent if and only if $2l \equiv i + 2jk \pmod 5$. Use part (a) to prove that G has diameter 2.

5.4.60. For odd d, prove that the maximum number of vertices in a tree with diameter d and maximum degree k is $\sum_{i=1}^{(d+1)/2} (k-1)^{i-1} + \sum_{i=1}^{(d+1)/2} (k-1)^{i-1}$.

5.4.61. Let G be a graph with order n, size m, maximum degree k, and diameter d. Prove $m \geq \binom{n}{2} / \sum_{i=1}^{d} (k-1)^{i-1}$. (Hint: Count paths starting at each edge.) (Erdős–Rényi [1962])

5.4.62. Let G be a connected graph with distinct edge weights. Without Kruskal's Algorithm, prove that G has only one minimum-weight spanning tree.

5.4.63. Suppose that in the hypercube Q_k, each edge whose endpoints differ in coordinate i is given weight 2^i. Compute the minimum weight of a spanning tree.

5.4.64. ($\Diamond$) Let G be a connected n-vertex graph. Define a graph G' on the set of spanning trees of G by putting $TT' \in E(G')$ if and only if T and T' have exactly $n-2$ common edges. Prove that G' is connected. Obtain a formula for $d_{G'}(T, T')$.

5.4.65. ($\Diamond$) Let T' be a spanning tree in a graph G with minimum-weight spanning tree T. Prove that T' can transform into T by steps exchanging one edge of T' for one edge of T so that the edge set is always a spanning tree and the total weight never increases.

5.4.66. A **minimax** or **bottleneck** spanning tree is a spanning tree in which the maximum of the edge weights is as small as possible. Prove that every minimum-weight spanning tree is a minimax spanning tree.

5.4.67. ($\Diamond$) **Prim's Algorithm** grows a spanning tree from a given vertex of a connected weighted graph, iteratively adding the cheapest edge between a vertex already absorbed and a vertex not yet absorbed, finishing when all the vertices have been absorbed. (Ties are broken arbitrarily.) Prove that Prim's Algorithm produces a minimum-weight spanning tree. (Jarník [1930], Prim [1957], Dijkstra [1959], independently).

5.4.68. ($\Diamond$) In a connected weighted graph, iteratively delete a heaviest non-cut-edge until no cycle remains. Prove that what remains is a minimum-weight spanning tree.

5.4.69. A greedy algorithm for minimum-weight spanning path iteratively selects the edge of least weight so that the edges chosen so far form a disjoint union of paths. After $n-1$ steps, a spanning path is obtained. Prove that this algorithm always gives a minimum-weight spanning path, or give an infinite family of counterexamples where it fails.

5.4.70. *Breadth-First Search for distances from a vertex u_0 in an unweighted graph.* Place the initial vertex u_0 on a queue, and declare its distance from u_0 to be 0. While there remains a vertex in the queue, remove the oldest vertex v from the queue, declare its neighbors that have not yet been reached to have distance $d(u_0, v) + 1$ from u_0, and add them to the queue. When the queue is empty, say that any unreached vertices have infinite distance from u_0. Prove that this algorithm computes distances from u_0 to all vertices.

Chapter 6

Matchings

Can we fill jobs using qualified applicants from an applicant pool? Can we pair students as roommates with all pairs compatible? How do we find a strategy for playing Tic-Tac-Toe? These are all questions about matchings in graphs.

6.0.1. Definition. A **matching** in a graph G is a set of pairwise non-incident edges, also called **independent edges**. A matching **covers** (or **saturates**) the vertices in its edges. A matching that covers all of $V(G)$ is a **perfect matching** in G. The **matching number** $\alpha'(G)$ is the maximum size (number of edges) of a matching in G.

Study of matchings involves existence (conditions for a perfect matching), enumeration (how many perfect matchings), and optimization (finding a maximum-sized matching). The traditional term "saturates" is from optimization; it indicates that the usage of edges at a vertex has reached its bound. We use the simpler word "covers". A vertex not covered by a matching M is "missed" by M. For the history of matching theory, see Mulder [1992] and Plummer [1992].

6.1. Matching in Bipartite Graphs

Filling jobs with qualified applicants is a matching problem in a bipartite graph. The parts are the jobs and the applicants; the edges record which applicants can do which jobs. We ask whether some matching fills all the jobs.

6.1.1. Remark. We can restate the job-filling problem using sets. For the ith job, there is a set A_i of qualified applicants. The problem is to find a **system of distinct representatives (SDR)** of the sets $A_1, \ldots, A_n$, meaning distinct elements $z_1, \ldots, z_n$ such that $z_i \in A_i$ for all i.

The two problems are equivalent; an SDR of a family of sets corresponds to a matching covering X in a natural bipartite graph. The **incidence graph** of a family $A_1, \ldots, A_n$ is the X, Y-bigraph with $X = \{A_1, \ldots, A_n\}$ and $Y = \bigcup_{i \in [n]} A_i$ such that $A_i \in X$ is adjacent to $y \in Y$ if and only if $y \in A_i$. An SDR with z_i representing A_i for all i corresponds to the matching $\{A_i z_i : i \in [n]\}$; we use this correspondence freely.

In discussing matchings covering X, we have used n for $|X|$. When there are n sets and n elements, the matrix of the membership relation between the sets and the elements is a $(0, 1)$-matrix of order n. An SDR then corresponds to a permutation of $[n]$ that maps indices in one part into indices of the matched vertices in the other part. Hence in the context of bipartite matching with parts of equal size, it is convenient to let the total number of vertices be $2n$. ∎

HALL'S THEOREM

Philip Hall [1935] characterized when an SDR exists. There is an obvious necessary condition. For $J \subseteq [n]$, an SDR for a family $A_1, \ldots, A_n$ must have distinct elements representing the sets indexed by J, and thus $\left|\bigcup_{i \in J} A_i\right| \geq |J|$ is necessary. In terms of matchings in bipartite graphs, the condition is stated using neighborhoods.

6.1.2. Definition. The **neighborhood** of a set of vertices is the union of the neighborhoods of its members. In notation, $N(S) = \bigcup_{v \in S} N(v)$.

If an X, Y-bigraph has a matching of size $|X|$, then the elements of any $S \subseteq X$ have distinct neighbors in the matching, so $|N(S)| \geq |S|$. **Hall's Condition** is the statement that $|N(S)| \geq |S|$ for all $S \subseteq X$ (or, in the SDR setting, $\left|\bigcup_{i \in J} A_i\right| \geq |J|$ for all $J \subseteq [n]$). Hall proved that this necessary condition for a matching that covers X is also sufficient.

6.1.3. Theorem. (**Hall's Theorem**; P. Hall [1935]) In an X, Y-bigraph, some matching covers X if and only if $|N(S)| \geq |S|$ for all $S \subseteq X$.

Proof: (M. Hall [1948], Halmos–Vaughan [1950], Mann–Ryser [1953]). As remarked above, existence of such a matching requires Hall's Condition. We prove sufficiency by induction on $|X|$ for such a graph G. When $|X| = 1$, the claim is immediate. For $|X| > 1$, we consider two cases.

If $|N(S)| > |S|$ for every nonempty proper subset S of X, then we fix a vertex $x \in X$ and choose $y \in N(x)$, which exists since Hall's Condition prohibits isolated vertices in X. The graph $G - \{x, y\}$ satisfies Hall's Condition, because each subset of $X - \{x\}$ loses at most one neighbor, y. The induction hypothesis yields a matching covering $X - \{x\}$ in $G - \{x, y\}$, and adding the edge xy completes the desired matching in G.

In the remaining case (see figure below), there is a nonempty proper subset S of X such that $|N(S)| = |S|$. Let G_1 be the subgraph induced by $S \cup N(S)$, and let $G_2 = G - (S \cup N(S))$. Hall's Condition holds for G_1, since $N(T) \subseteq N(S)$ for $T \subseteq S$. Hall's Condition also holds for G_2 as follows: since $N_{G_2}(T) = N_G(T \cup S) - N_G(S)$ for $T \subseteq X - S$, we have

$$|N_{G_2}(T)| = |N_G(T \cup S)| - |N_G(S)| \geq |T \cup S| - |S| = |T|.$$

By the induction hypothesis, G_1 has a matching covering S and G_2 has a matching covering $X - S$. Their union is the desired matching in G. ∎

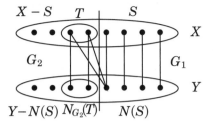

6.1.4.* Remark. A closer look at the argument of Theorem 6.1.3 yields a quantitative result (M. Hall [1948]): if G is an X, Y-bigraph satisfying Hall's Condition, then at least $\prod_{i=1}^{\min\{\delta,n\}}(\delta + 1 - i)$ matchings in G cover X, where $n = |X|$ and $\delta = \min_{x \in X} d(x)$ (Exercise 34). That is, when Hall's Condition holds there tend to be many matchings covering X. For example, $K_{r,s}$ has $\prod_{i=1}^{r}(s + 1 - i)$ matchings covering the part of size r. ∎

Hall's Theorem states that an obvious necessary condition for existence of a matching that covers X is also sufficient. The British mathematician Crispin St. James Alva Nash-Williams coined an acronym for such results: TONCAS (The Obvious Necessary Conditions are Also Sufficient). Such results usually guarantee a short verification of the answer. Here, one can prove that some matching covers X by exhibiting one. One can prove that none exists by exhibiting a set $S \subseteq X$ with $|N(S)| < |S|$. Always exactly one of these alternatives holds.

It should be noted, however, that the proof given here does not lead to a fast algorithm for checking Hall's Condition or showing that the guaranteed matching exists. The techniques in Section 6.3 are useful for that.

Hall's Theorem has many TONCAS-type applications. We seek an auxiliary X, Y-bigraph H such that the claimed sufficient condition implies Hall's Condition for H, and a matching in H covering X yields the desired conclusion. We present such an application; others appear in Exercises 19–23. A more detailed version of this result is in Exercise 18.

6.1.5. Corollary. (Hakimi [1965]) For $d \in \mathbb{N}$, a graph G has an orientation in which each vertex has outdegree at most d if and only if every subgraph H of G has at most $d\,|V(H)|$ edges.

Proof: *Necessity:* If some subgraph H has more than $d\,|V(H)|$ edges, then in any orientation some vertex of H is the tail of more than d edges.

Sufficiency: Form an X, Y-bigraph G' in which $X = E(G)$ and Y consists of d copies of $V(G)$. For each $v \in V(G)$, we have vertices $v_1, \ldots, v_d \in Y$. For each $uv \in X$, make uv adjacent to all copies of u and v in Y.

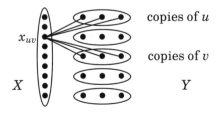

A matching M in G' that covers X selects an endpoint of each edge uv, by matching uv to a copy of u or a copy of v. Since there are d copies of each vertex, each vertex of G is selected at most d times. Therefore, orienting each edge so that the endpoint selected by M is the tail produces the desired orientation of G.

It thus suffices to show that G' has a matching that covers X. To do this, we verify Hall's Condition. For each $S \subseteq X$, there is a corresponding subgraph H of G whose edge set is S. Let U be the set of vertices incident to the edges of S in G. By the hypothesis, $|S| \leq d\,|U|$. However, $d\,|U| = |N_{G'}(S)|$, since the neighbors of S in Y are precisely the d copies of the vertices incident to edges of S. Thus Hall's Condition holds.   ∎

The application of Hall's Theorem to regular bipartite graphs is called the "Marriage Theorem". Multiedges can be important to achieve regularity. *Our statements about existence (but not enumeration) of matchings in bipartite graphs extend to bipartite multigraphs.* (For k-regular bipartite multigraphs, a sharp lower bound on the number of perfect matchings is $n!(k/n)^n$, weaker than Remark 6.1.4. The bound follows from the famous van der Waerden Conjecture, proved independently by Egorychev [1981] and Falikman [1981]; see Exercises 15.2.30-31.)

6.1.6. Corollary. (**Marriage Theorem**; König [1916], Frobenius [1917]) Every nontrivial regular bipartite multigraph has a perfect matching.

Proof: In a k-regular bipartite multigraph with parts X and Y, consider any $S \subseteq X$. Counting edges joining S and $N(S)$ yields $k\,|S| \leq k\,|N(S)|$, since $N(S)$ receives all the edges from S and maybe more. "Nontrivial" implies $k \geq 1$, so we can divide by k, and Hall's Condition holds.   ∎

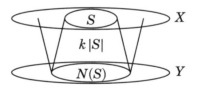

The **biadjacency matrix** or **reduced adjacency matrix** of an X, Y-bigraph G is the submatrix of $A(G)$ induced by the rows for X and the columns for Y. For a k-regular bipartite multigraph, this matrix is a nonnegative integer matrix whose rows and columns all sum to k. The positions for a perfect matching are those of the 1s in a permutation matrix: one 1 in each row and each column. Deleting a perfect matching leaves a $(k-1)$-regular subgraph and reduces each row and column by 1. By induction on k, every nonnegative integer matrix whose rows and columns all sum to k is a sum of permutation matrices.

6.1.7.* Definition. A **doubly stochastic matrix** is a nonnegative matrix whose rows and columns all have sum 1. A **convex combination** is a linear combination using nonnegative coefficients that sum to 1.

If the entries in a doubly stochastic matrix M are rational, then we can multiply M by the least common multiple of the denominators (call it k) to obtain a nonnegative integer matrix whose rows and columns all have sum k. As we have just observed, this is the biadjacency matix of a k-regular bipartite multigraph G, and the matrix can be expressed as a sum of permutation matrices. Let $P_1, \ldots, P_r$ be the permutation matrices used, and let c_i be the number of times P_i is used. By letting the coefficient λ_i of P_i be c_i/k, we have $M = \sum_{i=1}^{r} \lambda_i P_i$. This expresses M as a convex combination of permutation matrices, proving the special case of the next theorem in which the entries are all rational.

6.1.8.* Theorem. (**Birkhoff–von Neumann Theorem**; Birkhoff [1946], von Neumann [1953]) Every doubly stochastic matrix is a convex combination of permutation matrices.

In fact, the full strength of the theorem (for any real entries) can be proved by induction on the number of nonzero entries using Hall's Theorem (Exercise 25). One may wonder about the size of the coefficients. Is it possible to find a permutation matrix whose 1s correspond to entries in M that are not too small?

6.1.9.* Example. An island of area n is home to n married couples, each consisting of a farmer and a rancher. The Ministry of Agriculture splits the island into n farms of area 1. The Ministry of Meat independently splits it into n ranches of area 1. The Ministry of Marriage requires that each couple's farm and ranch overlap. Theorem 6.1.8 guarantees that such an assignment exists. Can we match the farms and ranches so that each couple's intersection is large?

In the example below, the ranches $(1, 2, 3)$ are horizontal strips, and the farms (a, b, c) extend vertically. The matrix records areas of intersection. Since 1 and 2 share area .5 only with a, one of them must match to a farm with area only .25. This example is the worst; when $n = 3$ there always are independent positions with values at least .25. We next find the worst case in general. ∎

$$\begin{array}{c} \\ 1 \\ 2 \\ 3 \end{array} \begin{array}{ccc} a & b & c \\ \left(\begin{array}{ccc} .5 & .25 & .25 \\ .5 & .25 & .25 \\ 0 & .5 & .5 \end{array}\right) \end{array}$$

6.1.10.* Theorem. (Marcus–Ree [1959], Floyd [1990]) Every doubly stochastic matrix of order n has n independent entries all with value at least ε, where $\varepsilon = \frac{4}{(n+1)^2}$ for odd n and $\varepsilon = \frac{4}{n(n+2)}$ for even n. Furthermore, no larger minimum can be guaranteed.

Proof: We use the language of Example 6.1.9. We first generalize the construction. Partition a square of area n into n equal strips as ranches. Let T be a set of t ranches. Let $t - 1$ farms intersect each ranch in T with area $\frac{1}{t}$. Split the remaining $\frac{1}{t}$ for each ranch in T equally among the remaining $n+1-t$ farms. Some ranch in T must match to a farm intersecting it in area $\frac{1}{n+1-t}\frac{1}{t}$. Set $t = \lceil \frac{n}{2} \rceil$.

To prove the guarantee of finding a matching with common area at least ε for each pair, define a bipartite graph G by making ranches and farms adjacent when they have common area at least ε. It suffices to prove that G satisfies Hall's Condition. Let $\mu(A)$ denote the area of A.

For Hall's Condition, consider a set S of k ranches, and let R be their union. We need k farms that each intersect some ranch in S with area at least ε. If $\mu(F \cap R) \geq \frac{1}{n+1-k}$ for some farm F, then F intersects some ranch in S with area at least $\frac{1}{k(n+1-k)}$ and hence at least ε. We show that there are at least k such farms.

Index the farms $F_1, \ldots, F_n$ so that $\mu(F_1 \cap R) \geq \cdots \geq \mu(F_n \cap R)$. Let $a = \mu(F_k \cap R)$. Since $\mu(F_i \cap R) \leq 1$ for $i < k$ and $\mu(F_i \cap R) \leq a$ for $i \geq k$,

$$k - 1 + a(n + 1 - k) \geq \sum_{i=1}^{n} \mu(F_i \cap R) = \mu(R) = k.$$

Simplifying yields $a \geq \frac{1}{n+1-k}$. We conclude that $N(S)$ contains $F_1, \ldots, F_k$, and Hall's Condition holds. ∎

MIN-MAX RELATIONS

Hall's Theorem leads to a formula for the maximum size of a matching in a bipartite graph. It turns out to equal a trivial upper bound.

6.1.11. Definition. Let $\alpha'(G)$ denote the maximum size of a matching in a graph G. For an X, Y-bigraph G, the **defect** $\mathrm{df}(S)$ of a set $S \subseteq X$ is $|S| - |N(S)|$.

One would like to call $\mathrm{df}(S)$ the "deficiency" and write $\mathrm{def}(S)$, but we use that term and notation later to discuss matching in general graphs.

6.1.12. Corollary. (Ore's Defect Formula; Ore [1955]). In an X, Y-bigraph G, a formula for the maximum size of a matching is given by

$$\alpha'(G) = \min_{S \subseteq X} \{|X| - \mathrm{df}(S)\} .$$

Proof: Every matching misses at least $\mathrm{df}(S)$ vertices of S, so $\alpha'(G) \leq |X| - \mathrm{df}(S)$ for every S. We construct a matching of size $|X| - d$, where $d = \max_{S \subseteq X} \mathrm{df}(S)$. Form G' by adding d vertices to Y, each adjacent to all of X. Now G' satisfies Hall's Condition, since each subset of X has received d more neighbors. Hence G' has a matching of size $|X|$. By the construction, at most d edges of this matching are in $G' - G$, so the remaining edges form a matching of the desired size in G. ∎

The size of a matching is bounded using a natural "dual" concept.

6.1.13. Definition. A **vertex cover** is a set of vertices having at least one endpoint of every edge; the stars centered at these vertices cover the edges. The minimum size of a vertex cover in a graph G is denoted $\beta(G)$.

6.1.14. Theorem. (König–Egerváry Theorem; König [1931], Egerváry [1931]). If G is bipartite, then max |matching| = min |vertex cover|. That is,

$$\alpha'(G) = \beta(G).$$

Proof: No vertex can cover more than one edge of a matching, so $\beta(G) \geq \alpha'(G)$. We seek a vertex cover of size $\alpha'(G)$. Let G be an X, Y-bigraph. By Corollary 6.1.12, there is a set $T \subseteq X$ such that $\alpha'(G) = |X| - |T| + |N(T)|$. Since $N(T) \cup (X - T)$ is a vertex cover, $\beta(G) \leq |N(T)| + |X - T| = \alpha'(G)$. ∎

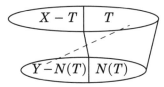

A statement of equality for the optima of minimization and maximization problems over a class of inputs (like Theorem 6.1.14) is a **min-max relation**. Theorem 6.1.14 ensures that in a bipartite graph we can always prove optimality of a maximum matching by finding a vertex cover of the same size. Exercise 42 requests another proof.

Matching and vertex cover are called **dual problems** because every vertex cover must have size at least as big as every matching, making $\beta(G) \geq \alpha'(G)$ trivial for every graph G. A similar pair of optimization problems involves independent sets of vertices. The **independence number** of a graph is the maximum size of an independent set of vertices. An X, Y-bigraph may have independent sets larger than $\max\{|X|, |Y|\}$, as shown below.

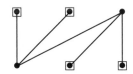

Just as no vertex covers two edges of a matching, no edge contains two vertices of an independent set. Again we have a dual covering problem.

6.1.15. Definition. An **edge cover** of G is a set of edges that cover the vertices of G (only graphs without isolated vertices have edge covers). We list notation for related optimization parameters:

maximum size of independent set	$\alpha(G)$
maximum size of matching	$\alpha'(G)$
minimum size of vertex cover	$\beta(G)$
minimum size of edge cover	$\beta'(G)$

We use $\beta(G)$ for minimum vertex cover due to its interaction with maximum matching. We put the "prime" on $\beta'(G)$ rather than on $\beta(G)$ because $\beta(G)$ counts a set of vertices and $\beta'(G)$ counts a set of edges.

Always $\beta'(G) \geq \alpha(G)$. We now prove the min-max relation.

6.1.16. Theorem. (**König's Other Theorem**, König [1916]) If G is a bipartite graph with no isolated vertices, then $\alpha(G) = \beta'(G)$.

Proof: Let Q be a minimum vertex cover in an X, Y-bigraph G. Let $R = X \cap Q$ and $S = Y \cap Q$. Since $\alpha'(G) = \beta(G)$ by the König–Egerváry Theorem, G has a matching M of size $|Q|$. Since Q covers every edge, M matches R into $Y - S$ and S into $X - R$, and each edge of M covers one vertex of $(Y - S) \cup (X - R)$. Adding one edge to cover each remaining vertex of $(Y - S) \cup (X - R)$ completes an edge cover of size $|Y - S| + |X - R|$. Furthermore $(Y - S) \cup (X - R)$ is an independent set, so we have produced an edge cover and an independent set of the same size; they are a smallest edge cover and largest independent set. $\blacksquare$

6.1.17. Remark. Every n-vertex graph G satisfies $\alpha(G) + \beta(G) = n$, since the complement of any independent set of vertices is a vertex cover. Surprisingly, the edge parameters satisfy the same equality: **Gallai's Theorem** (Gallai [1959]) states that $\alpha'(G) + \beta'(G) = n$ for every n-vertex graph G without isolated vertices.

For bipartite graphs, this follows from the proof of Theorem 6.1.16, where we produced an edge cover of size $n - \alpha'(G)$. It was a smallest edge cover since we produced an independent set of the same size. For non-bipartite graphs this is not enough: Exercise 50 requests a proof of the full statement. $\blacksquare$

EXERCISES 6.1

6.1.1. $(-)$ Prove $\alpha'(G) \geq \delta(G)/2$ for every graph G. Prove $\alpha'(G) \geq \delta(G)$ for every bipartite graph G. In each case, give infinitely many examples where equality holds.

6.1.2. $(-)$ A **cycle-factor** of a digraph is a set of (directed) cycles such that each vertex lies on exactly one of the cycles. Prove that a digraph D has a cycle-factor if and only if $|N^+(S)| \geq |S|$ for all $S \subseteq V(D)$. (Ore [1962])

6.1.3. $(-)$ Let T be an n-vertex tree with independence number k. Determine $\alpha'(T)$.

6.1.4. $(-)$ Prove that a graph G is bipartite if and only if $\alpha'(H) = \beta(H)$ for all $H \subseteq G$.

6.1.5. $(-)$ Prove that $\alpha'(G) \geq m/\Delta(G)$ when G is a bipartite graph with m edges.

6.1.6. $(-)$ Prove that $\alpha(G) \leq n/2$ when G is a nontrivial regular n-vertex graph. Prove that equality holds if and only if G is bipartite.

6.1.7. $(-)$ Prove $\frac{n}{\Delta(G)+1} \leq \alpha(G) \leq n - \delta(G)$ for every n-vertex graph G.

6.1.8. Prove that a bipartite graph G has a perfect matching if and only if $|N(S)| \geq |S|$ for all $S \subseteq V(G)$, and present an infinite class of examples to prove that this characterization does not hold for all graphs.

6.1.9. Let G be a bipartite graph having a perfect matching. For $r < |V(G)|/2$, prove that if every matching of size r extends to a perfect matching, then also every matching of size $r - 1$ extends to a perfect matching.

6.1.10. Given n red and n blue points in the plane, with no three on a line, prove that there is a matching of all red points to blue points by straight line segments so that no two of the segments cross.

6.1.11. Let M be a matching in the hypercube Q_k such that the vertices covered by M induce no edges other than M. Prove that M is contained in a perfect matching of Q_k. (Vandenbussche–West [2009])

6.1.12. The hypercube Q_k has vertex set $\{0, 1\}^k$; the *weight* of a vertex is the number of nonzero positions. Let M be a perfect matching in Q_k. For each i, compute the number of edges of M whose endpoints have weights i and $i + 1$.

6.1.13. For $k \geq 2$, prove that Q_k has at least $2^{2^{k-2}}$ perfect matchings.

6.1.14. Let A and B be disjoint independent sets in a graph G, with $|A| = \alpha(G)$. Prove that $G[A \cup B]$ contains a matching of size $|B|$.

6.1.15. ($\diamond$) Let G be an X, Y-bigraph with no isolated vertices and with $|X| \leq |Y|$. Prove that Hall's Condition ($|N(S)| \geq |S|$ for all $S \subseteq X$) is equivalent to the statement "$|T| - |N(T)| \leq |Y| - |X|$ for all $T \subseteq Y$".

6.1.16. ($\diamond$) Let G be a connected regular bipartite graph. Prove that any graph obtained from G by deleting one vertex from each part has a perfect matching. Conclude that every edge in G appears in some perfect matching. For $k \geq 2$, construct a k-regular bipartite graph having two nonincident edges that do not appear together in a perfect matching.

6.1.17. Let G be an X, Y-bigraph. Prove that there exists $x \in X$ such that every edge incident to x belongs to some maximum matching.

6.1.18. ($\diamond$) Prove that a multigraph G with m edges has an orientation with specified out-degree d_i at each vertex v_i if and only if $\sum_{v_i \in V(G)} d_i = m$ and $\sum_{v_i \in U} d_i \geq |E(G[U])|$ for every $U \subseteq V(G)$. (Hakimi [1965]) (Comment: The special case $G = K_n$ was proved in Landau [1953], characterizing "score sequences" of tournaments (Exercise 5.2.39). Hall's Theorem was used to prove that result in Bang–Sharp [1979].)

6.1.19. ($\diamond$) Consider a deck of mn cards with n suits and m values; each value appears on one card in each suit. The cards are placed randomly in a grid with n rows and m columns.

(a) Prove that it is possible to choose one card from each of the m columns so that the values are all distinct.

(b) Use part (a) to prove that it is possible to iteratively exchange positions of two cards of equal value so that eventually each suit appears in each column. (Enchev [1994])

6.1.20. ($\diamond$) Let T be a set of k permutations of $[n]$. Prove that if $k \leq n/2$, then some permutation of $[n]$ disagrees in every position with every member of T. Prove that if $k > n/2$, then there may be no such permutation. (Kézdy–Snevily; see Cameron–Wanless [2005])

6.1.21. A travel club is planning vacations. Trips $t_1, \ldots, t_n$ are available, but trip t_i has capacity c_i. Each person likes some trips and will take at most one. In terms of which people like which trips, derive a necessary and sufficient condition for being able to fill all trips (to capacity) with people who like them.

6.1.22. Let G be an X, Y-bigraph having a matching that covers X. Given $S, T \subseteq X$ such that $|N(S)| = |S|$ and $|N(T)| = |T|$, prove that $|N(S \cap T)| = |S \cap T|$.

6.1.23. ($\diamond$) Let G be an X, Y-bigraph with no isolated vertices, with $d(x) \geq d(y)$ whenever $xy \in E(G)$ with $x \in X$ and $y \in Y$. Prove that G has a matching covering X. (Hint: Consider a smallest set violating Hall's Condition.) (F. Galvin)

6.1.24. A **positional game** consists of a set X of positions and a family $W_1, \ldots, W_m$ of winning sets of positions (Tic-Tac-Toe has nine positions and eight winning sets). Two players alternately choose positions; the player who first occupies the positions of a winning set wins. Let a be the minimum size of a winning set, and let b be the maximum number of winning sets containing a given position. Prove that Player 2 can force a draw if $a \geq 2b$. (Comment: More generally, Player 2 can force a draw in d-dimensional Tic-Tac-Toe when the sides are long enough, given the appropriate definition of winning sets. For fixed side-length, Player 1 can win if the dimension is large enough; see Chapter 10.)

6.1.25. Prove the Birkhoff–von Neumann Theorem (Theorem 6.1.8) that every doubly stochastic matrix can be written as a convex combination of permutation matrices. (Hint: Use induction on the number of nonzero entries.)

6.1.26. Let M be a nonnegative integer matrix of order n in which every row and column has sum t. Determine the least value r such that if $t \geq r$, then M must contain a set of n entries, with one in each row and each column, that all exceed 2. (Stanley [2015b])

6.1.27. (+) Let G be an n-vertex graph with $\delta(G) \geq k$.
(a) Let F' and F be forests with the same number of components. Prove that if $F' \subseteq F$ and $|V(F)| \leq k + 1$, then every copy of F' in G is contained in a copy of F in G.
(b) Let F be a forest with k edges and at most n vertices. Prove that $F \subseteq G$. (Hint: When F has no isolated vertices, deleting a leaf from each component yields a k-vertex forest. Comment: The case where F is a tree is Exercise 5.4.17.) (Brandt [1994])

6.1.28. Use the König–Egerváry Theorem to prove both the Marriage Theorem and Hall's Theorem.

6.1.29. ($\diamond$) For an X, Y-bigraph G with $|X| = |Y| = n$, prove $\alpha'(G) \geq \min\{n, y\delta(G)\}$:
(a) Using Ore's Defect Formula.
(b) Using the König–Egerváry Theorem.

6.1.30. Let G be a connected X, Y-bigraph with vertex degrees $d_1, \dots, d_n$ in nonincreasing order. Prove $\alpha'(G) \geq \max_k \min\{|X| - k, |Y| - k, 2d_{n-k-1}\}$ (this strengthens Exercise 6.1.29). (Jahanbekam–West [2013])

6.1.31. A multigraph is r-semiregular if every vertex degree is r or $r - 1$.
(a) For $0 \leq r \leq d$, prove that every loopless d-semiregular multigraph has a spanning r-semiregular submultigraph. (Thomassen [1981c])
(b) Prove that every graph with maximum degree at most $s + t - 1$ is the union of graphs G_1 and G_2 with $\Delta(G_1) \leq s$ and $\Delta(G_2) \leq t$. (Lovász [1966])

6.1.32. Let $A_1, \dots, A_m$ be k-sets such that each element of the union lies in at most ℓ sets. Prove that there exist disjoint $B_1, \dots, B_m$ of size $\lfloor k/\ell \rfloor$ such $B_i \subseteq A_i$ for all i. (T. Jiang)

6.1.33. ($\diamond$) Let G be an X, Y-bigraph without $(m + 1)K_2$ as an induced subgraph.
(a) Prove that if G has no isolated vertices, then $N(S) = Y$ for some $S \subseteq X$ with $|S| \leq m$. (Liu–Zhou [1997])
(b) Prove that G has at most $m(\Delta(G))^2$ edges. (Faudree–Gyárfás–Schelp–Tuza [1989])

6.1.34. ($\diamond$) Let G be an X, Y-bigraph with $m = |X|$ and $\delta = \min_{x \in X} d(x)$. Prove that if G satisfies Hall's Condition for X, then at least $\prod_{i=1}^{\min\{\delta, m\}} (\delta + 1 - i)$ matchings in G cover X. (Hint: Follow the method of Theorem 6.1.3, proving this lower bound instead of mere existence. Multiedges are forbidden.) (M. Hall [1948])

6.1.35. Exhibit a perfect matching in the graph drawn below or give a short proof that it has none. (Lovász–Plummer [1986, p. 7])

6.1.36. Determine the maximum number of edges in a bipartite graph that has no matching with k edges and contains no star with l edges. (Isaak)

6.1.37. ($\diamond$) Prove that every subgraph of $K_{n,n}$ with more than $(k-1)n$ edges has a matching of size at least k.

6.1.38. Prove that in a bipartite graph G, a vertex belongs to some smallest vertex cover if and only if it is covered by every maximum matching.

6.1.39. For $k \le r \le s$, characterize the maximal subgraphs of $K_{r,s}$ that have no matching of size k, and determine which have the most edges.

6.1.40. ($\diamond$) Let G be an X, Y-bigraph with $\alpha'(G) = |X| = r$. Prove that at most $\binom{r}{2}$ edges of G lie in no maximum matching. Show that this is sharp for every r.

6.1.41. ($\diamond$) Determine the largest number b such that every maximal matching in every 3-regular graph G has size at least $b|V(G)|$.

6.1.42. (+) Let G be a bipartite graph, and let G' be a minimal spanning subgraph of G such that $\beta(G') = \beta(G)$. Prove that G' has no two incident edges, and use this to prove the König–Egerváry Theorem. (Lovász [1975])

6.1.43. ($\diamond$) Let G be an n-vertex graph with m edges. Prove $\beta(G) \le n/2 + m/6$, with equality only when every component of G is K_3 or K_4. (Hint: Consider cases depending on $\Delta(G)$.) (Henning–Yeo [2013a])

6.1.44. Let G be a nonbipartite graph having exactly one cycle, C. Prove that $\alpha(G) \ge (n-1)/2$, with equality if and only if $G - V(C)$ has a perfect matching.

6.1.45. ($\diamond$) Let G be a graph with vertex cover number k.
  (a) Build a vertex cover R greedily as follows: iteratively select a vertex of maximum degree in the remaining graph, add it to R, and delete it, until the remaining graph has no edges. Prove that the final set R may be as large as $k \ln k - O(k)$. (Hint: Construct an X, Y-bigraph with $|X| = k$ and $|Y| = \sum_{i=2}^{k} \lfloor k/i \rfloor$ so that the algorithm chooses all of Y.)
  (b) Another algorithm adds to R the vertices of a remaining edge and deletes them. Prove that this always produces a vertex cover of size at most $2k$.

6.1.46. ($\diamond$) Build an independent set S greedily as follows: iteratively select a vertex v of minimum degree in the remaining graph, add it to S, and delete $\{v\} \cup N(v)$. Prove that the final independent set S has size at least $\sum_{u \in V(G)} \frac{1}{d_G(u)+1}$. (Caro [1979], Wei [1981])

6.1.47. For a graph G, order $V(G)$ as $v_1, \dots, v_n$ by iteratively deleting a vertex of maximum degree in $G - \{v_1, \dots, v_{i-1}\}$ and calling it v_i. Let $k = \lceil \sum_{i=1}^{n} \frac{1}{1+d_G(v_i)} \rceil$. Prove that $\{v_{n-k+1}, \dots, v_n\}$ is an independent set in G. (Caro–Tuza [1991])

6.1.48. ($\diamond$) Let G be an n-vertex graph not having $K_{1,3}$ as an induced subgraph. Prove $\alpha(G) \le \frac{2n}{\delta(G)+2}$.

6.1.49. Let G be an n-vertex graph with no isolated vertex.
  (a) Prove $\alpha'(G) \ge \frac{n}{1+\Delta(G)}$. (Hint: Use induction on $|E(G)|$.) (Weinstein [1963])
  (b) Prove $\beta'(G) \le \frac{n\Delta(G)}{1+\Delta(G)}$, and show that both bounds are sharp infinitely often for each maximum degree.

6.1.50. ($\diamond$) Prove Gallai's Theorem: $\alpha'(G) + \beta'(G) = n$ whenever G is an n-vertex graph with no isolated vertices. (Comment: Since G need not be bipartite, the König–Egerváry Theorem and König's Other Theorem are not relevant.) (Gallai [1959])

6.1.51. *Consequences of Gallai's Theorem (Exercise 6.1.50).* Let M be a maximal matching and L be a minimal edge cover in a graph G having no isolated vertices.
  (a) Prove that $|M| = \alpha'(G)$ if and only if M is contained in a smallest edge cover.
  (b) Prove that $|L| = \beta'(G)$ if and only if L contains a largest matching. (Norman–Rabin [1959], Gallai [1959])

6.1.52. An edge e of a graph G is α-**critical** if $\alpha(G-e) > \alpha(G)$. Let xy and xz be α-critical edges in G. Prove that some induced subgraph of G is an odd cycle containing xy and xz. (Hint: Let Y and Z be largest independent sets in $G - xy$ and $G - xz$, respectively. Let $H = G[Y \triangle Z]$. Prove that every component of H has the same number of vertices from Y and from Z. Conclude that y and z lie in the same component of H.) (Berge [1970])

6.1.53. α-critical edges.
    (a) Prove that in a bipartite graph, every α-critical edge intersects every largest independent set of vertices.
    (b) Prove that in a tree, the α-critical edges form a matching. (Zito [1991])
    (c) Prove that both properties above may fail in general graphs.

6.2. Matching in General Graphs

The matching theorems that we proved for bipartite graphs do not hold for general graphs.

6.2.1. Example. *Odd cycles.* Note that $\alpha(C_{2k+1}) = \alpha'(C_{2k+1}) = k$, but $\beta(C_{2k+1}) = \beta'(C_{2k+1}) = k + 1$. That is, the min-max relations fail. Also Hall's Theorem fails: although C_{2k+1} satisfies "$|N(S)| \geq |S|$ for all $S \subseteq V(G)$" (see Exercise 6.1.8), it has no perfect matching. ∎

Nevertheless, we still seek a necessary and sufficient condition for the existence of a perfect matching in a general graph.

6.2.2. Definition. For $k \in \mathbb{N}$, a k-**factor** in a graph G is a k-regular spanning subgraph of G. In the context of factors, let an **odd component** of a graph be a component of odd order (odd number of vertices). We use $o(H)$ to denote the number of odd components of a graph H.

A perfect matching is a set of edges; a 1-factor is a subgraph whose edge set is a perfect matching. The terms are almost interchangeable.

TUTTE'S 1-FACTOR THEOREM

As with Hall's Theorem, we start with an obvious necessary condition. If a graph G has a 1-factor, and $S \subseteq V(G)$, then $G - S$ must have at most $|S|$ odd components, since each odd component of $G - S$ must have a vertex matched into S by the 1-factor. Tutte proved that this condition is also sufficient for a 1-factor.

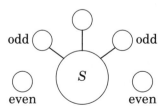

6.2.3. Theorem. (**Tutte's 1-Factor Theorem**; Tutte [1947]) A graph G has a 1-factor if and only if $o(G - S) \leq |S|$ for all $S \subseteq V$. ∎

Tutte's Theorem has many proofs, including Gallai [1950, 1963, 1964], Maunsell [1952], Kotzig [1959–60], Edmonds [1965c], Anderson [1971], Lovász [1975], Mader [1973]. Like Hall's Theorem, Tutte's Theorem leads to a min-max relation for the maximum size of a matching. Every matching must leave an uncovered vertex in at least $o(G - S) - |S|$ odd components of $G - S$. This yields an upper bound on $\alpha'(G)$; the Berge–Tutte Formula is the statement that equality holds. This can be proved from Tutte's Theorem by a proof like our proof of Ore's Defect Formula from Hall's Theorem (see Exercise 4). The forced uncovered amount $\max\{o(G-S)-|S|\}$ is analogous to the notion of "defect" in bipartite matching, and again we have a special term for it.

6.2.4. Definition. The **deficiency** def(S) of a vertex set S in a graph G is $o(G - S) - |S|$. **Tutte's Condition** is the statement "$o(G - S) \leq |S|$ for all $S \subseteq V(G)$". A set with positive deficiency is a **Tutte set**.

graph class	characterization	min-max relation
bipartite	Hall's Condition $\forall S \subseteq X,\ |N(S)| \geq |S|$	Ore's Defect Formula $\alpha'(G) = \min_{S \subseteq X}[|X| - \mathrm{df}(S)]$
general	Tutte's Condition $\forall S \subseteq V(G),\ o(G - S) \leq |S|$	Berge – Tutte Formula $\alpha'(G) = \min_{S \subseteq V(G)} \frac{1}{2}[n - \mathrm{def}(S)]$

In fact, the min-max relation and the characterization are equivalent. We have remarked that Tutte's Theorem implies the min-max relation (Exercise 4). Conversely, Tutte's Theorem is a special case of the Berge–Tutte Formula. Just as Hall's Condition is the statement "max df$(S) = 0$", so Tutte's Condition is the statement "max def$(S) = 0$". Thus Hall's Theorem and Tutte's Theorem follow as special cases of the min-max relations.

To emphasize this relationship, we prove the full statement of the Berge–Tutte Formula directly. We begin with two lemmas.

6.2.5. Lemma. (**Parity Lemma**) If S is a set of vertices in an n-vertex graph G, then $o(G - S) - |S| \equiv n \pmod 2$.

Proof: Counting all vertices shows that $o(G - S) + |S| \equiv n \pmod 2$. ∎

6.2.6. Remark. The Parity Lemma aids in proving Tutte's Condition when $|V(G)|$ is even. To prove $o(G - S) \leq |S|$, it then suffices to prove $o(G - S) < |S| + 2$, since two integers with the same parity cannot differ by 1. ∎

Many results about maximum matchings can be proved by studying vertex sets with maximum deficiency. When T is a maximal such set (contained in no other), the components of $G - T$ are very well behaved.

6.2.7. Lemma. Let T be a maximal subset of $V(G)$ among those having largest deficiency. If u is a vertex of an odd component C of $G - T$, then $C - u$ satisfies Tutte's Condition. Also, $G - T$ has no even components.

Proof: Subscripts on def denote the relevant graph. For $S \subseteq V(C - u)$, the odd components of $C - u - S$ are in $C - u$. Comparing $o(G - T - u - S)$ to $o(G - T)$, we lose one odd component (C) and gain odd components in $C - u - S$. Thus

$$\operatorname{def}_G(T \cup \{u\} \cup S) = o(G - T - u - S) - (|T| + 1 + |S|)$$
$$= o(G - T) - 1 + o(C - u - S) - |T| - 1 - |S|$$
$$= \operatorname{def}_G(T) - 2 + \operatorname{def}_{C-u}(S).$$

Since T is a maximal set of maximum deficiency and $T \cup \{u\} \cup S$ contains T, we have $\operatorname{def}_G(T \cup \{u\} \cup S) < \operatorname{def}(T)$. Thus $\operatorname{def}_{C-u}(S) < 2$. Since $C - u$ has even order, $\operatorname{def}_{C-u}(S)$ is even, by the Parity Lemma. We conclude that $\operatorname{def}_{C-u}(S) \leq 0$ for all $S \subseteq V(C - u)$, which is Tutte's Condition.

If $G - T$ has an even component C, then let $T' = T \cup \{v\}$, where v is a leaf of a spanning tree of C. We have $|T'| = |T| + 1$ and $o(G - T') = o(G - T) + 1$. Hence $\operatorname{def}_G(T') = \operatorname{def}_G(T)$, which contradicts the choice of T.   ∎

6.2.8. Theorem. (**Berge–Tutte Formula**; Berge [1958]). In an n-vertex graph G, the maximum number of vertices covered by a matching is $n - d$, where $d = \max_{S \subseteq V(G)} \operatorname{def}(S)$. That is,

$$\alpha'(G) = \min_{S \subseteq V(G)} \tfrac{1}{2}(n - \operatorname{def}(S)).$$

Proof: We know $\alpha'(G) \leq \frac{1}{2}(n - \operatorname{def}(S))$ for all $S \subseteq V(G)$. Hence it suffices to construct a matching leaving only d vertices uncovered. We use induction on n; the claim is trivial for $n = 0$. For $n > 0$, let T be a maximal set with deficiency d.

By Lemma 6.2.7, $G - T$ has no even components, and $C - u$ satisfies Tutte's Condition when u is a vertex in an odd component C of $G - T$. Since $C - u$ has fewer vertices than G, the induction hypothesis yields a perfect matching in $C - u$. Since this holds for any vertex u in any component C of $G - T$, and there are $|T| + d$ such components, it suffices to cover T using edges to $|T|$ distinct components of $G - T$. The resulting matching leaves only d uncovered vertices.

Let Y be the set of components of $G - T$. To match T into Y, we define an auxiliary T, Y-bigraph H. Let H have an edge ty for $t \in T$ and $y \in Y$ if and only if t is adjacent in G to some vertex of the component of $G - T$ corresponding to y. To obtain a matching that covers T, we apply Hall's Theorem to H.

For $S \subseteq T$, the vertices of Y outside $N_H(S)$ remain as odd components when only $T - S$ is deleted. Using also $\operatorname{def}(T - S) \leq d$ and $|Y| = |T| + d$, we compute

$$|N_H(S)| \geq |Y| - o(G - (T - S))$$
$$= |Y| - (\operatorname{def}(T - S) + |T - S|) \geq |Y| - d - |T| + |S| = |S|.$$

Since this is true for every subset S of T, Hall's Condition holds, and Hall's Theorem guarantees the needed matching in H.   ∎

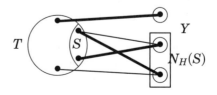

Like many other characterization theorems, Tutte's 1-Factor Theorem guarantees short proofs of both yes and no answers. To prove that G has a 1-factor, present one. To prove that G has no 1-factor, present a set with positive deficiency (a "Tutte set"). Always one of these two proofs is available. Note that if S is a Tutte set in a graph of even order, then $o(G - S) - |S| \geq 2$, by the Parity Lemma.

Applications of Tutte's Theorem usually involve showing that some other condition implies Tutte's Condition and hence guarantees existence of a 1-factor. There are 3-regular graphs without 1-factors; below we show the smallest. Nevertheless, long before Tutte's Theorem, Petersen proved a sufficient condition that now follows easily from Tutte's Theorem.

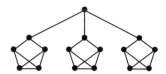

6.2.9. Corollary. (Petersen [1891]) Every 3-regular graph with no cut-edge has a 1-factor.

Proof: Given any vertex subset S, let m be the number of edges joining S to $G - S$. Since G is 3-regular, $m \leq 3\,|S|$. Since any odd component H of $G - S$ must have even degree-sum, H is joined to S by an odd number of edges. This number is at least 3 since G has no cut-edge. Hence $m \geq 3o(G - S)$, so $o(G - S) \leq |S|$, and Tutte's Theorem applies. ∎

A more careful look at the argument yields various strengthenings. For example, every edge lies in some 1-factor (Schönberger [1934]). Extensions to k-regular graphs were proved by Bäbler [1938] and Berge [1973] (see also Exercise 25 and Exercise 32). Most of these follow from Theorem 6.2.10 (even the hypothesis here can be weakened; see Exercise 29). The condition of even order in the statement is included to cover the case of even k.

6.2.10. Theorem. (Plesník [1972]) Let G be a k-regular loopless multigraph of even order that remains connected when at most $k - 2$ edges are deleted. If G' arises from G by deleting at most $k - 1$ edges, then G' has a 1-factor.

Proof: We prove that G' has no Tutte set. Consider $S \subseteq V(G')$, and let m be the number of edges of G' joining S to odd components of $G' - S$. Since $G' \subseteq G$ and G is k-regular, $m \leq k\,|S|$. If we can prove $m \geq ko(G' - S) - 2(k - 1)$, then dividing by k will yield $|S| > o(G' - S) - 2$. Since $|V(G')|$ is even, the Parity Lemma then requires $|S| \geq o(G' - S)$, as desired.

For the lower bound on m, consider an odd component H of $G' - S$. Let l be the number of edges in G leaving $V(H)$; by hypothesis, $l \geq k - 1$. Note that $\sum_{v \in V(H)} d_H(v) = k\,|V(H)| - l$. Since H is a graph, the sum is even, but $|V(H)|$ is odd, so k and l have the same parity. Thus $l \geq k$.

Over all odd components of $G' - S$, the total is at least $ko(G' - S)$, but this counts edges of G. Up to $k - 1$ such edges may be missing in G', counted from both ends if they join odd components of $G' - S$. Thus we subtract $2(k - 1)$ to ensure having a lower bound on m. That is, $m \geq ko(G' - S) - 2(k - 1)$, as desired. ∎

6.2.11. Corollary. (Berge [1973]) Let G be a k-regular loopless multigraph of even order. If deleting at most $k - 2$ edges leaves a connected subgraph, then every edge of G appears in some perfect matching.

Proof: Delete $k - 1$ of the k edges incident to one endpoint of the edge desired in the 1-factor, and apply Theorem 6.2.10.   ∎

Lemma 6.2.5 and Lemma 6.2.7 also lead to the **Gallai–Edmonds Structure Theorem** describing the structure of maximum matchings (Exercise 40).

GENERAL FACTORS OF GRAPHS

A **factor** is a spanning subgraph of G; we consider the vertex degrees. For surveys, see Akiyama–Kano [1985], Volkmann [1995], and Plummer [2007].

6.2.12. Definition. Given a graph G and $f: V(G) \to \mathbb{N}_0$, an f-**factor** of G is a spanning subgraph H such that $d_H(v) = f(v)$ for all $v \in V(G)$.

Recall that for $k \in \mathbb{N}$, a k-factor of G is a k-regular factor; this is the special case of f-factor when $f(v) = k$ for all v. Petersen found a sufficient condition for 2-factors, which we prove using only Eulerian circuits and bipartite matching. We observed in Chapter 5 that even graphs decompose into edge-disjoint cycles. For regular graphs, these cycles can be grouped into 2-factors.

6.2.13. Theorem. (**Petersen's 2-Factor Theorem**; Petersen [1891]) Every regular multigraph of positive even degree has a 2-factor.

Proof: Let G be a $2k$-regular graph with vertices $v_1, \ldots, v_n$. By considering each component, we may assume that G is connected. Hence G is Eulerian; let C be an Eulerian circuit. Define a U, W-bigraph H with $U = \{u_1, \ldots, u_n\}$ and $W = \{w_1, \ldots, w_n\}$ by making u_i and w_j adjacent if v_j immediately follows v_i somewhere on C. Since C leaves and enters each vertex k times, H is k-regular.

Hence the Marriage Theorem yields a 1-factor in H. It picks one edge "leaving" v_i (incident to u_i in H) and one edge "entering" v_i (incident to w_i in H). Thus the edges of a 1-factor in H form a 2-factor in G.   ∎

6.2.14. Example. *Construction of a 2-factor.* Consider the Eulerian circuit in K_5 that successively visits 1231425435. The corresponding bipartite graph H is on the right. For the 1-factor whose u, w-pairs are 12, 25, 54, 43, 31, the resulting 2-factor is the cycle $[1, 2, 5, 4, 3]$. The remaining edges form another 1-factor that yields the remaining cycle $[1, 4, 2, 3, 5]$.   ∎

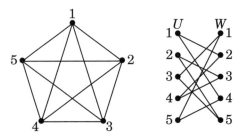

The general f-factor problem specifies the degree at each vertex. Multiedges are irrelevant for 1-factors but becomes important for f-factors. Tutte [1952] found a necessary and sufficient condition for a multigraph to have an f-factor (proved earlier by Belck [1950] for regular factors). Tutte's first proof was quite difficult; later [1954] he reduced the problem to checking for a 1-factor in a related graph. We describe this construction. It is a beautiful example of transforming a graph problem into a previously solved problem.

6.2.15. Example. *A graph transformation* (Tutte [1954]). When f satisfies the obvious necessary condition that $f(w) \leq d(w)$ for all w, we construct a graph H that has a 1-factor if and only if G has an f-factor. Let $e(w) = d(w) - f(w)$; this is the *excess degree* at w and is nonnegative.

To construct H, replace each vertex v with a copy of $K_{d(v),e(v)}$ having parts $A(v)$ and $B(v)$ of sizes $d(v)$ and $e(v)$, respectively. For each $vw \in E(G)$, add an edge joining one vertex of $A(v)$ to one vertex of $A(w)$. Each vertex of $A(v)$ participates in one such edge.

The figure below shows a multigraph G, vertex labels given by f, and the resulting graph H. The bold edges in H form a 1-factor that corresponds to an f-factor of G. The f-factor is not unique.   ∎

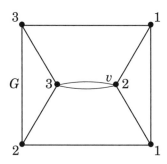 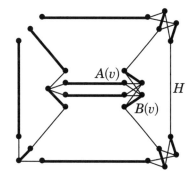

6.2.16. Theorem. A multigraph G has an f-factor if and only if the **blow-up graph** H constructed from G and f in Example 6.2.15 has a 1-factor.

Proof: *Necessity.* Start with the edges in H corresponding to an f-factor in G. They leave $e(v)$ vertices of $A(v)$ unmatched, and these match into $B(v)$ (for each v) to complete a 1-factor of H.

Sufficiency. From a 1-factor of H, deleting $B(v)$ and the vertices of $A(v)$ matched into $B(v)$ leaves $f(v)$ vertices of degree 1 corresponding to v. Doing this for each v and merging the remaining $f(v)$ vertices of each $A(v)$ yields a subgraph of G with degree $f(v)$ at each vertex v.   ∎

By Theorem 6.2.16, any algorithm to test for 1-factors can be used to test for an f-factor. Nevertheless, we still seek a necessary and sufficient condition in terms of the structure of G. Applications of the resulting Tutte f-Factor Theorem include a proof of the Erdős–Gallai [1960] characterization of degree lists of graphs (Exercise 43). The theorem can be proved by translating the statement of Tutte's 1-Factor Condition for the blow-up into a structural condition for G. First we show necessity of the condition directly.

6.2.17. Definition. Given disjoint sets Q, $T \subseteq V(G)$, let $\|Q, T\|$ denote the number of edges in G with one endpoint each in Q and T. When $f \colon V(G) \to \mathbb{N}_0$, let $f(S) = \sum_{v \in S} f(v)$. Similarly, $d_H(S) = \sum_{v \in S} d_H(v)$.

6.2.18. Example. *The f-factor condition.* Given disjoint vertex sets S and T in G, the condition states a bound on the contributions to $f(T)$ in a hypothetical f-factor F. Thus $f(T)$ cannot exceed that bound. The necessary condition is

$$f(T) \le f(S) + d_{G-S}(T) - q(S, T),$$

where $q(S, T)$ is the number of components of $G - S - T$ such that $f(Q) + \|Q, T\|$ is odd, where Q denotes the vertex set of the component.

Consider how the edges of an f-factor can contribute to $f(T)$. At most $f(S)$ can join S to T, each contributing once to $f(T)$. The remaining contribution at each vertex $v \in T$ is bounded by $d_{G-S}(v)$, so $f(T) \le f(S) + d_{G-S}(T)$. However, satisfying this is not sufficient (Chvátal gave a 13-vertex example of insufficiency).

We can tighten the bound further by subtracting 1 for some components $G[Q]$ of $G - S - T$. If the f-factor uses any edge from Q to S, then that edge reduces the contribution from $f(S)$ in the bound on $f(T)$ by 1. If no such edge is used, and the f-factor uses all edges joining T to Q, then these edges plus the edges used within Q form a subgraph H of F. Since no edge from Q to S is used, the quantity $f(Q) + \|Q, T\|$ counts each edge of H twice, and hence it must be even. If it is odd, then the edges joining T to Q cannot all be used, which reduces the contribution to the bound from $d_{G-S}(T)$. Hence when $f(Q) + \|Q, T\|$ is odd we get one way or another a reduction of at least 1 in the bound on $f(T)$. ∎

6.2.19. Theorem. (**Tutte's f-Factor Theorem**; Tutte [1952, 1954]). If G is a multigraph and $f \colon V(G) \to \mathbb{N}_0$, then G has an f-factor if and only if

$$f(T) \le f(S) + d_{G-S}(T) - q(S, T) \qquad (*)$$

for all disjoint $S, T \subseteq V(G)$, where $q(S, T)$ counts the sets Q such that $G[Q]$ is a component of $G - S - T$ and $f(Q) + \|Q, T\|$ is odd.

Proof*: Necessity was shown in Example 6.2.18; consider the converse. With $S = \varnothing$ and $T = \{w\}$, $(*)$ yields $f(w) \le d(w)$ for every vertex w; hence the blow-up graph H from G and f is well-defined. By Theorem 6.2.16, it suffices to prove that $(*)$ guarantees Tutte's 1-Factor Condition for H.

When Tutte's Condition fails, there is a Tutte set C (with $o(H - C) > |C|$). We use a *minimal* Tutte set to construct a partition of $V(G)$ (with $R = V(G) - S - T$) violating $(*)$. Let a *parasite* in a Tutte set C be a vertex with neighbors in at most two components of $H - C$. If x is a parasite in C, then

$$o(H - (C - x)) \ge o(H - C) - 1 > |C| - 1 = |C - x|,$$

so a minimal Tutte set C has no parasite.

Using C, we prove that each $v \in V(G)$ is in one of three types:

Type 1	Type 2	Type 3
$B_v \subseteq C$	$B_v \cap C = \varnothing$	$B_v \cap C = \varnothing$
$A_v \cap C = \varnothing$	$A_v \subseteq C$	$A_v \cap C = \varnothing$

Below we show edges of $H - C$ in bold. If B_v has a vertex x in C and a vertex y outside C (shown on the left), then the vertices of A_v not in C all lie in one component of $H - C$, and x is a parasite. Hence $B_v \subseteq C$ or $B_v \cap C = \varnothing$ (on the right). If $B_v \subseteq C$, then any $x \in A_v \cap C$ has at most one neighbor in $H - C$ and is a parasite, so $A_v \cap C = \varnothing$ (Type 1); this applies also when $B_v = \varnothing$ (that is, $e(v) = 0$).

For $B_v \neq \varnothing$, if $B_v \cap C = \varnothing$ and there exist $z \in A_v - C$ and $x \in A_v \cap C$, then B_v lies in one component of $H - C$. Now x has neighbors in at most two components of $H - C$ and is a parasite. Hence C contains all or none of A_v (Type 2 or 3).

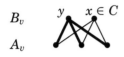

Given a minimal Tutte set C, let T, S, R be the subsets of $V(G)$ consisting of the Type 1, Type 2, and Type 3 vertices, respectively. For v of each type, B_v and A_v are distributed to C and $V(H) - C$ as listed below.

	size	$v \in T$	$v \in S$	$v \in R$
B_v	$e(v)$	C	$V(H) - C$	$V(H) - C$
A_v	$d(v)$	$V(H) - C$	C	$V(H) - C$

We claim that $o(H - C) > |C|$ violates $(*)$ for this choice of R, S, T. By construction, $|C| = \sum_{v \in S} d(v) + \sum_{v \in T} e(v) = d_G(S) + d_G(T) - f(T)$. In the figure below, edges of $H - C$ are in bold. Let $\mathbf{R}$ be the family of vertex sets of components of $G[R]$. The components of $H - C$ are of four types:

Description of component	Order	Number with odd order
$y \in B_v$ with $v \in S$	1	$\sum_{v \in S} e(v)$
yz with $y \in A_u$, $z \in A_v$, $uv \in E(G[T])$	2	none
$y \in A_v$ for each edge $uv \in [S, T]$	1	$\|T, S\|$
one for each $Q \in \mathbf{R}$	large	$q(S, T)$

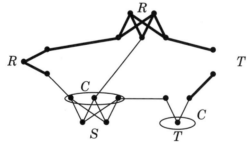

Consider the large component of $H - C$ corresponding to some connection class $Q \in \mathbf{R}$. It has $\sum_{v \in Q}(d(v) + e(v)) + \|Q, T\|$ vertices, since it gains one vertex for each edge from Q to T. Since $e(v) = d(v) - f(v)$, this quantity has the same parity as $f(Q) + \|Q, T\|$. By definition, the number of components of $G[R]$ for which $f(Q) + \|Q, T\|$ is odd is $q(S, T)$.

Now we rewrite $o(H - C) > |C|$ as

$$q(S, T) + \sum_{v \in S}[d(v) - f(v)] + \|T, S\| > d_G(S) + d_G(T) - f(T).$$

When we cancel $d_G(S)$ and rearrange what remains, we have

$$q(S, T) + f(T) > f(S) + (d_G(T) - \|T, S\|).$$

The final quantity in parentheses equals $d_{G-S}(T)$, so $(*)$ fails. ∎

EXERCISES 6.2

6.2.1. (−) Let M be a perfect matching in a k-regular graph G, where k is odd. Prove that M contains every cut-edge of G.

6.2.2. (−) Find maximum matchings in the graph G below and in the graph G' obtained by adding the edge xy. For each prove optimality using "duality".

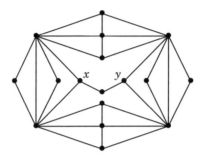

6.2.3. (−) Let G be a $(2r-1)$-regular graph. Prove that G decomposes into r factors whose components are paths or cycles. For example, the Petersen graph decomposes into a perfect matching and $2C_5$.

6.2.4. Prove the Berge–Tutte Formula (Theorem 6.2.8) from Tutte's 1-Factor Theorem (Theorem 6.2.3).

6.2.5. (◊) Prove that a tree T has a perfect matching if and only if $o(T − v) = 1$ for every $v \in V(T)$. (Chungphaisan)

6.2.6. (◊) *Minimal Tutte sets.*
(a) Let S be a minimal Tutte set in a graph G. Prove that every vertex in S has neighbors in at least three odd components of $G − S$.
(b) Let G be a connected graph with $|V(G)|$ even. Prove that if G does not contain $K_{1,3}$ as an induced subgraph, then G has a 1-factor. (Sumner [1974], Las Vergnas [1975])
(c) Use part (b) to prove that every connected graph with $|E(G)|$ even decomposes into 3-vertex paths. (Chartrand–Polimeni–Stewart [1973])

6.2.7. Example 5.2.3 shows that every edge in the Petersen graph lies in four 5-cycles. Use this to count the perfect matchings.

6.2.8. Derive a formula for the number of 1-factors in $K_{n,n}$ that do not match x_i to y_i for any i, and for the number of 1-factors in K_{2n} that do not match x_{2i-1} to x_{2i} for any i. No simple closed formulas are known.

6.2.9. For each $k \in \mathbb{N}$ except $k = 1$, construct a k-regular graph with no 1-factor.

6.2.10. For $k \in \mathbb{N}$ with $k \geq 2$, construct a $2k$-regular connected graph having no spannning subgraph in which at most $k − 2$ vertices have degree $2k$ and the rest have degree $2k − 1$.

6.2.11. (◊) Let G be a graph with $\alpha'(G) = k$.
(a) Determine the maximum possible value of $\beta(G)$.
(b) Determine the minimum possible size of a maximal matching in G.

6.2.12. Let G be a graph with $\alpha'(G) = k$. Prove that G has at most $k(k − 1)$ edges that belong to no maximum matching. Construct examples to show that this bound is best possible for every k. (F. Galvin)

6.2.13. Obtain sufficiency in Hall's Theorem (Theorem 6.1.3) from Tutte's Theorem. (Hint: Transform an X, Y-bigraph G into a graph G' that has a 1-factor if and only if G has a matching that covers X.)

6.2.14. Let T be a set of vertices in a graph G. Prove that G has a matching that covers T if and only if for all $S \subseteq V(G)$, the number of odd components of $G - S$ contained in $G[T]$ is at most $|S|$. (Lovász [1965], extended to a min-max relation in Bollobás [1978])

6.2.15. ($\diamond$) Let G be a graph of even order not having $K_{1,r+1}$ as an *induced* subgraph ($K_{1,r+1} \subseteq G$ is allowed). Prove that if $G - S$ is connected whenever $S \subseteq V(G)$ with $|S| < r$, then G has a 1-factor. (Sumner [1976])

6.2.16. Given graphs G and H, the **tensor product** $G * H$ is the graph with vertex set $V(G) \times V(H)$ having (u, v) adjacent to (u', v') if and only if $uu' \in E(G)$ and $vv' \in E(H)$. Prove that the following conditions are equivalent for G (George [1991]):
(A) For every graph H, the graph $G * H$ has no 1-factor.
(B) $G * K_2$ has no 1-factor.
(C) There exists $S \subseteq V(G)$ such that $G - S$ has more than $|S|$ isolated vertices.

6.2.17. ($\diamond$) For a 3-regular graph G with n vertices and c cut-edges, prove $\alpha'(G) \geq n/2 - c/3$. (Chartrand–Kapoor–Lesniak–Schuster [1984])

6.2.18. (+) Let $i(H)$ be the number of isolated vertices in a graph H; always $i(H) \leq o(H)$. For $m \geq 2$, prove that a graph G has a factor whose components are nontrivial stars with at most m edges if and only if $i(G - S) \leq m|S|$ for all $S \subseteq V(G)$. (Amahashi–Kano [1982])

6.2.19. ($\diamond$) *Extension of König–Egerváry to general graphs.* Given a graph G, let $S_1, \ldots, S_k$ and T be subsets of $V(G)$ such that each S_i has odd size. These sets form a **generalized cover** of G if every edge of G has at least one endpoint in T or has both endpoints in some S_i. The *weight* of a generalized cover is $|T| + \sum(|S_i| - 1)/2$. Let $\beta^*(G)$ be the minimum weight of a generalized cover. Prove $\alpha'(G) = \beta^*(G)$. (Hint: Use the Berge–Tutte Formula, Theorem 6.2.8. Comment: Since every vertex cover is a generalized cover, $\beta^*(G) \leq \beta(G)$.)

6.2.20. ($\diamond$) Use induction on n to determine the maximum number of edges in a $2n$-vertex graph having no 1-factor.

6.2.21. For even n, prove that the maximum number of edges in an n-vertex graph having a Tutte set of size k is $\binom{k}{2} + k(n - k) + \binom{n-2k-1}{2}$. Use this to find the maximum number of edges in an n-vertex graph with no 1-factor. (Erdős–Gallai [1961])

6.2.22. ($\diamond$) Prove $\alpha'(G) \geq \min\{\delta(G), \lfloor n/2 \rfloor\}$ when G has n vertices. (Erdős–Pósa [1962])

6.2.23. Let G have vertex degrees $d_1, \ldots, d_n$ in nonincreasing order. Strengthen Exercise 6.2.22 by proving $\alpha'(G) \geq \max_k \min\{d_{n-k}, \lfloor (n - k)/2 \rfloor\}$. (Jahanbekam–West [2013])

6.2.24. Prove that a 3-regular graph G has a 1-factor when all its cut-edges lie on one path (Petersen [1891]). Conclude that the minimum number of cut-edges in a 3-regular graph with no 1-factor is 3. (Comment: More generally, every $(2r + 1)$-regular graph with at most $2r$ cut-edges has a 2-factor, and this is sharp (Hanson–Loten–Toft [1998]).)

6.2.25. ($\diamond$) *Sharpness of the connectivity threshold for 1-factors in regular graphs.* Let k be an integer at least 3, and let l be the least odd number greater than k.
(a) Construct a k-regular graph G of even order that has no 1-factor and remains connected whenever fewer than $k - 2$ edges are deleted. (Hint: Use $(l + 1)k - 2$ vertices.)
(b) Prove that every answer to part (a) has at least $(l + 1)k - 2$ vertices. (Hint: For a Tutte set S, treat the large odd components and small odd components of $G - S$ differently.) (See Niessen–Randerath [1998] for a more general result.)

6.2.26. ($\diamond$) Let G be a 3-regular graph. Prove that G has a 1-factor if and only if G decomposes into copies of P_4. (Comment: Favaron–Genest–Kouider [2010] conjectured that every $(2k+1)$-regular graph with a perfect matching decomposes into paths of length $2k+1$.) (Kotzig [1957], Bouchet–Fouquet [1983])

6.2.27. The graph on the left below is the **Heawood graph**. Prove that every 2-factor in this graph is a spanning cycle. Is this true also for the other graph?

 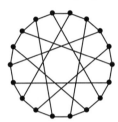

6.2.28. Prove Corollary 6.2.11 directly from Tutte's 1-Factor Theorem. That is, if xy is an edge in a k-regular graph G of even order such that G cannot be disconnected by deleting fewer than $k-1$ edges, then $G - \{x, y\}$ has no Tutte set.

6.2.29. Weaken the hypothesis of Theorem 6.2.10 by requiring only that vertex sets of odd size are not disconnected from the rest of G by deleting fewer than $k-2$ edges. Prove that any G' obtained from G by deleting at most $k-1$ edges has a 1-factor. (Cruse [1977])

6.2.30. *Petersen's proof of Petersen's 2-Factor Theorem (Theorem 6.2.13).*
(a) Prove that every 4-regular multigraph decomposes into two 2-factors.
(b) For $k > 2$, prove by induction on $|E(G)|$ that every multigraph G with $\Delta(G) \leq 2k$ decomposes into k factors with maximum degree at most 2. (Petersen [1891])

6.2.31. ($\diamond$) A d-regular multigraph is *primitive* if it does not decompose into two regular factors with degree less than d. By Petersen's Theorem, no primitive multigraphs have d even and $d > 2$. Let G be a d-regular n-vertex loopless multigraph with d odd.
(a) Construct an example where G is primitive with $n = 3d + 1$. (J.J. Sylvester)
(b) Prove $\alpha'(G) \geq n/3$.
(c) For $\alpha'(G) = n - s$, prove that G has an odd-regular factor with degree at most $2s+1$. Conclude that G cannot be primitive if $n \leq 3d - 3$. (Petersen [1891]; see Mulder [1992])

6.2.32. ($\diamond$) A *balloon* in a graph G is a maximal subgraph without cut-edges that is incident to exactly one cut-edge of G; let $b(G)$ be the number of balloons in G (Petersen [1891] called such subgraphs "leaves"). Let G be a connected n-vertex 3-regular graph.
(a) Prove $b(G) \leq (n + 2)/6$, with equality achievable when $n \equiv 4 \pmod 6$.
(b) Prove $\alpha'(G) \geq \frac{n}{2} - \frac{b(G)}{3}$. (Hint: Use the Berge–Tutte Formula.)
(b) Conclude $\alpha'(G) \geq (4n - 1)/9$. Show that equality can hold when $n \equiv 16 \pmod{18}$.
(Comment: For $r \in \mathbb{N}$, Henning–Yeo [2007] proved $\alpha'(G) \geq \frac{n}{2} - \frac{r}{2} \frac{(2r-1)n+2}{(2r+1)(2r^2+2r-1)}$ for any $(2r + 1)$-regular graph G with n vertices. O–West [2010] proved that equality holds if and only if G arises from a tree with degrees 1 and $2r + 1$ having all leaves on one side of the bipartition by attaching a copy of $\overline{P_3} + rK_2$ at each leaf; below is an example.)

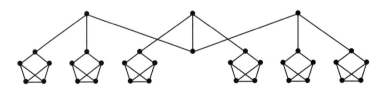

6.2.33. (+) For $k \geq 2$ and $g \geq 2$ with g even, construct a k-regular bipartite (multi)graph with girth g. (Hint: Construct it from a $(k-1)$-regular bipartite graph H with girth g and a bipartite graph with girth $\lceil g/2 \rceil$ that is $|V(H)|$-regular. Comment: The requirement of being bipartite strengthens the result of Erdős–Sachs [1963]; see Exercise 5.1.47.)

6.2.34. (+) Let G be a $2m$-regular graph with girth g, and let T be a tree with m edges and diameter d. Prove that if $d < g$, then G decomposes into copies of T. (Hint: Using Theorem 6.2.13, produce inductively such a decomposition in which each vertex of G is used once as an image of each vertex of T.) (Häggkvist)

6.2.35. (+) Let G be a $2n$-vertex graph. Let $P(c)$ be the statement "$|N(S)| \geq c|S|$ whenever S is a set of at most $2n/c$ vertices." (Anderson [1973])
  (a) Prove that $P(4/3)$ implies that G has a 1-factor. (Hint: In applying Tutte's 1-Factor Theorem, consider the cases $|S| \geq n/2$ and $|S| < n/2$ separately.)
  (b) Prove that $P(c)$ does not imply that G has a 1-factor when $1 \leq c < 4/3$.

6.2.36. ($\Diamond$) Let v be a vertex in a graph G. Prove $\operatorname{def}(G - v) \leq \operatorname{def}(G) + 1$.

6.2.37. ($\Diamond$) A graph is **factor-critical** if every subgraph obtained by deleting one vertex has a 1-factor. Prove the following statements.
  (a) Every factor-critical graph has odd order, is connected, and is not bipartite.
  (b) A connected graph is factor-critical if and only if $\varnothing$ is its only Tutte set.
  (c) A connected graph is factor-critical if and only if no vertex is in every maximum matching. (Hint: Consider a Tutte set of maximum deficiency.) (Gallai [1963b])

6.2.38. Let v be a vertex in a factor-critical graph G. Let G' be a connected graph obtained from G by splitting v into two vertices whose neighborhoods form a nontrivial partition of $N_G(v)$. Prove that G' has a 1-factor.

6.2.39. Let G be a connected graph such that every edge is in a triangle and all cycles are triangles. Suppose also that every vertex has degree 2 or 4.
  (a) Prove that G is factor-critical.
  (b) Prove that for every odd-sized set S of vertices having degree 2, $G-S$ has a unique perfect matching. (Lovász–Plummer [1986, p. 89]; see also Došlić–Rautenbach [2015])

6.2.40. ($\Diamond$) In a graph G, let B be the set of vertices covered by every maximum matching. Let A be the set of vertices in B having a neighbor outside B. Let $C = B - A$ and $D = V(G) - B$ (see figure below). Let $G_1, \ldots, G_k$ be the components of $G[D]$. The **Gallai–Edmonds Structure Theorem** (Gallai [1963b], Edmonds [1965c]) asserts the claims below about every maximum matching M in G; prove them. (Hint: Let T be a maximal set of maximum deficiency in G. Use Lemma 6.2.7 to find the sets A, C, D; see West [2011].)
  (a) M covers C and matches A into distinct components of $G[D]$.
  (b) Each G_i is factor-critical and has a near-perfect matching in M.
  (c) If $\varnothing \neq S \subseteq A$, then $N(S)$ intersects at least $|S| + 1$ of $G_1, \ldots, G_k$.
  (d) $\operatorname{def}(A) = \operatorname{def}(G) = k - |A|$.

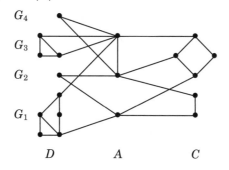

6.2.41. (+) Let G be a $2n$-vertex graph having exactly one 1-factor, M.

(a) Prove that if G has no cut-vertex, then for every two vertices $x, y \in V(G)$, the graph $G - \{x, y\}$ has a 1-factor. (Lovász [1972c])

(b) Prove that G has a cut-edge that belongs to M. (Kotzig [1959])

(c) Prove $\delta(G) \leq \lfloor \log_2(n+1) \rfloor$ and show that this bound is sharp.

(d) Prove that G has at most n^2 edges and that this bound is sharp.

(Comment: Part (c) is **Hetyei's Theorem** and appears in Lovász [1972c]. Generalizing (d), one can ask for the maximum number of edges in a $2n$-vertex graph having exactly p perfect matchings; see Dudek–Schmitt [2012] and Hartke–Stolee–West–Yancey [2013].)

6.2.42. (+) Let G be an even multigraph, and let f assign an odd integer to each vertex. Prove that G has an even number of f-factors. (Cameron–Edmonds [1999])

6.2.43. (+) *f-Factor Theorem and graphic lists.* The f-Factor Theorem of Tutte [1952, 1954] (Theorem 6.2.19) states that a graph G has an f-factor if and only if

$$f(T) \leq f(S) + d_{G-S}(T) - q(S, T)$$

for all disjoint sets $S, T \subseteq V(G)$, where $q(S, T)$ counts the components $G[Q]$ of $G - S - T$ such that $f(Q) + \|Q, T\|$ is odd.

(a) *The Parity Lemma.* Let $\delta(S, T) = f(T) - f(S) - d_{G-S}(T) + q(S, T)$. Prove that $\delta(S, T)$ has the same parity as $f(V(G))$ for disjoint sets $S, T \subseteq V(G)$.

(b) Let $f(v_i) = d_i$, where $\sum d_i$ is even and $d_1 \geq \cdots \geq d_n$. Use the f-factor condition and part (a) to prove that K_n has an f-factor if and only if $\sum_{i=1}^{k} d_i \leq (n-1-s)k + \sum_{i=n+1-s}^{n} d_i$ for all k, s with $k + s \leq n$.

(c) Show that $d_1, \ldots, d_n$ is graphic when $d_1 \geq \cdots \geq d_n$ if and only if $\sum d_i$ is even and $\sum_{i=1}^{k} d_i \leq k(k-1) + \sum_{i=k+1}^{n} \min\{k, d_i\}$ for $1 \leq k \leq n$. (Erdős–Gallai [1960])

6.2.44. Belck [1950] proved the special case of Tutte's f-Factor Theorem for regular factors, restricting f to assign the same integer to all vertices. Use this result to prove that if j, k, l are three odd integers with $j < k < l$ and G has a j-factor and an l-factor, then G also has a k-factor. (Katerinis [1985])

6.2.45. For a graph G and a function $f \colon V(G) \to \mathbb{N}_0$, say that G is f-**soluble** if there exists $w \colon E(G) \to \mathbb{N}_0$ such that $\sum_{uv \in E(G)} w(uv) = f(v)$ for all $v \in V(G)$.

(a) Prove that G has an f-factor if and only if the graph H obtained from G by subdividing each edge twice and defining f to be 1 on the new vertices is f-soluble. (This reduces testing for an f-factor to testing f-solubility.)

(b) Given G and an $f \colon E(G) \to \mathbb{N}_0$, construct a graph H (with proof) such that G is f-soluble if and only if H has a 1-factor. (Tutte [1954])

6.3. Algorithmic Aspects

Min-max relations guarantee short proofs of optimality by exhibiting solutions to dual problems. However, we still must *find* optimal solutions. The needed tool for maximum matching is augmenting paths. We will use them to give a fast algorithm for bipartite matching and then consider more general weighted versions of the dual matching and vertex cover problems.

AUGMENTING PATHS

Maximal matchings need not be maximum(-sized) matchings.

6.3.1. Example. *Maximal $\neq$ maximum.* In the drawing of P_6 below, the solid edges form a maximal matching; we can add no other. However, the three dashed edges form a larger matching. $\blacksquare$

6.3.2. Definition. Given a matching M, an *M***-alternating path** is a path that alternates between edges in M and edges not in M. An M-alternating path whose endpoints are missed by M is an *M***-augmenting path**.

If P is an M-augmenting path, then replacing $M \cap E(P)$ with $E(P) - M$ produces a new matching M' with one more edge than M. Thus a maximum matching admits no augmenting path. Furthermore, this characterizes maximum matchings. Petersen [1891] stated the characterization for the special case of regular graphs, without giving a proof; Berge [1957] provided a proof in general.

6.3.3. Definition. The **symmetric difference** $A \triangle B$ of sets A and B is $(A \cup B) - (A \cap B)$. The **symmetric difference** $G \triangle H$ of graphs G and H has vertex set $V(G) \cup V(H)$ and edge set $E(G) \triangle E(H)$.

6.3.4. Theorem. (Berge [1957]) A matching M in a graph G is a maximum matching in G if and only if G has no M-augmenting path.

Proof: We have noted that an M-augmenting path produces a larger matching. Conversely, when G has a matching M' larger than M, we construct an M-augmenting path. Let $F = M \triangle M'$. Since M and M' are matchings, every vertex has at most one incident edge from each of them. Hence $\Delta(F) \leq 2$ when F is viewed as a subgraph of G.

Since $\Delta(F) \leq 2$, the components of F are paths and cycles (as illustrated below). Since each component alternates between M and M', each cycle in F has even length. Since $|M'| > |M|$, some component has more edges of M' than of M. Such a component must be a path that starts and ends with an edge of M' and hence is an M-augmenting path. $\blacksquare$

Theorem 6.3.4 is useful in studying the properties of maximum matchings. For example, the thorough three-volume text by Schrijver [2003] on combinatorial optimization presents a direct proof of the Berge–Tutte Formula using induction and augmenting paths.

Most algorithms for finding maximum matchings are based on augmenting paths. A simple such algorithm for bipartite graphs also provides a constructive proof of the König–Egerváry Theorem.

6.3.5. Algorithm. *Augmenting path algorithm for an X, Y-bigraph G.* Given a matching M, let U be the set of vertices in X not covered by M. Grow a forest rooted at U by exploring M-alternating paths, letting S and T be the sets of vertices reached in X and Y, respectively. To do this, explore from vertices of S along edges not in M; when a vertex of Y covered by M is reached, add it to T, and add its neighbor via M to S (see the figure below).

Reaching a vertex of Y missed by M completes an M-augmenting path; augment the matching. Because the search always reaches X along edges of M and reaches Y along edges not in M, every edge is explored at most once, and recording the edge used to reach each vertex permits retrieval of the augmenting path.

If no vertex of Y missed by M is reached, then when all vertices in S have been explored the set $(X - S) \cup T$ forms a vertex cover of size $|M|$. This vertex cover proves that M is a maximum matching in G. ∎

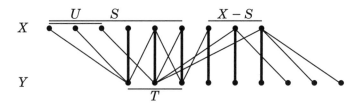

The augmenting path characterization of maximum matchings (Theorem 6.3.4) holds also for nonbipartite graphs, but odd cycles make the exploration of alternating paths more difficult. Edmonds [1965c] overcame this difficulty using the "Blossom Algorithm" in his famous paper "Paths, Trees, and Flowers".

6.3.6. Example. An M-alternating path can travel in either direction around an odd cycle in order to continue onward from any of its vertices, as suggested below with M in bold. A search for shortest M-augmenting paths from u reaches x via the edge ax outside M. If we do not also consider a longer path reaching x via dx in M, then we miss the augmenting path u, v, a, b, c, d, x, y. To postpone deciding which route to take, the algorithm shrinks such an odd cycle to a single vertex before continuing to seek an augmenting path. ∎

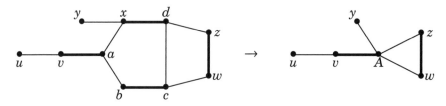

The running time of the algorithm is cubic in the number of vertices. With another approach, describing a polyhedron whose vertices correspond to matchings, Edmonds [1965d] extended the result to weighted graphs.

WEIGHTED BIPARTITE MATCHING

Like the shortest path problem (Section 5.4), maximum matching generalizes to weighted graphs. We assign weights to edges and seek the matching with maximum total weight. The maximum matching problem is the special case where edges of G have weight 1 and edges not in G have weight 0. The dual notion of vertex cover also extends. For bipartite graphs, we present a common solution to the weighted matching and cover problems using augmenting paths.

When $|X| = |Y| = n$, the problem reduces to seeking a maximum-weight perfect matching in $K_{n,n}$ with nonnegative edge weights. If $G \neq K_{n,n}$, then we can insert the missing edges with weight 0 without changing the answer; hence we may assume $G = K_{n,n}$. We may also assume that all weights are nonnegative, by changing negative weights to 0, solving the resulting problem, and deleting chosen edges with weight 0.

6.3.7. Example. *Weighted bipartite matching and its dual.* A farming company has n farms and n stores. Each farm can produce corn to supply one store. The profit from sending the output of farm i to store j is $w_{i,j}$. This yields a weighted X, Y-bigraph with $X = \{x_1, \ldots, x_n\}$ and $Y = \{y_1, \ldots, y_n\}$; the weight on edge $x_i y_j$ is $w_{i,j}$. The company seeks the matching with maximum total weight.

The government offers to pay subsidies to reduce corn production. The government will pay u_i if the company doesn't use farm i and v_j if it doesn't use store j. If $u_i + v_j < w_{i,j}$, then the company makes more by using the edge $x_i y_j$ than by accepting the payments u_i and v_j. In order to stop all production, the government must offer amounts such that $u_i + v_j \geq w_{i,j}$ for all (i, j). The government wants to find such values for $\{u_i\}$ and $\{v_j\}$ to minimize the cost $\sum u_i + \sum v_j$. ∎

6.3.8. Definition. A **transversal** of an n-by-n matrix consists of n positions, one in each row and each column. Finding a transversal with maximum sum is the **Assignment Problem**. This is the matrix formulation of **Maximum Weighted Matching**, where nonnegative weight $w_{i,j}$ is assigned to edge $x_i y_j$ of $K_{n,n}$ and we seek a perfect matching M to maximize the total weight $w(M)$.

With these weights, a **(weighted) cover** is a choice of labels $u_1, \ldots, u_n$ and $v_1, \ldots, v_n$ such that $u_i + v_j \geq w_{i,j}$ for all i, j. The **cost** $c(u, v)$ of a cover (u, v) is $\sum u_i + \sum v_j$. The **minimum weighted cover** problem is that of finding a cover of minimum cost.

A minimization and a maximization problem are **dual problems** when the value of every solution to the minimization problem is at least as large as the value of every solution to the maximization problem. Thus a solution to either problem establishes a bound for the other. (See Section 8.3 and Section 11.3 for further discussion of duality.) The next lemma shows that weighted matching and weighted cover are dual problems.

6.3.9. Lemma. (**Weighted matching/cover duality**) If M is a perfect matching in a weighted bipartite graph G and (u, v) is a cover, then $c(u, v) \geq w(M)$. Also, $c(u, v) = w(M)$ if and only if M consists of edges $x_i y_j$ with $u_i + v_j = w_{i,j}$, and then M and (u, v) are optimal.

Proof: Since M is a perfect matching, summing the constraints $u_i + v_j \geq w_{i,j}$ that arise from its edges yields $c(u, v) \geq w(M)$ for every cover (u, v). Furthermore, if $c(u, v) = w(M)$, then equality must hold in each of the n inequalities summed. Finally, since $c(u, v) \geq w(M)$ for every matching and every cover, $c(u, v) = w(M)$ implies that there is no matching with weight greater than $c(u, v)$ and no cover with cost less than $w(M)$. ■

The observation that a matching and cover can have equal weight and cost only by using edges covered with equality leads to an algorithm.

6.3.10. Definition. The **equality subgraph** $G_{u,v}$ for the cover (u, v) is the spanning subgraph of $K_{n,n}$ whose edges are $\{x_i y_j : u_i + v_j = w_{i,j}\}$. In the cover (u, v), the **excess** $e_{i,j}$ for the pair (i, j) is $u_i + v_j - w_{i,j}$.

If $G_{u,v}$ has a perfect matching, then its weight is $\sum u_i + \sum v_j$, and by Lemma 6.3.9 we have the optimal matching and cover. Otherwise, we use the equality subgraph and matrix of excesses to change the cover (u, v). The resulting algorithm was named the "Hungarian Algorithm" by Kuhn in honor of the work of König and Egerváry on which it is based.

6.3.11. Algorithm. (**Hungarian Algorithm**; Kuhn [1955], Munkres [1957]). Let X and Y be the parts in the bipartition of $K_{n,n}$. Given a matrix of weights on $E(K_{n,n})$ and a weighted cover (u, v) as in Definition 6.3.8, we first find a matching M and a vertex cover Q of the same size in the equality subgraph $G_{u,v}$ (by any method, such as the Augmenting Path Algorithm). Let $R = Q \cap X$ and $T = Q \cap Y$. Our matching of size $|Q|$ consists of $|R|$ edges from R to $Y - T$ and $|T|$ edges from T to $X - R$, as shown below.

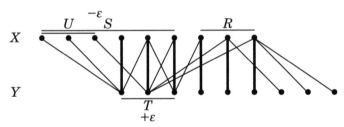

To seek an equality subgraph having a larger matching, we modify (u, v) to introduce an edge from $X - R$ to $Y - T$ while maintaining equality on all edges of M. Edges joining $X - R$ and $Y - T$ are not in $G_{u,v}$ and thus have positive excess. Let ε be the minimum among these; that is,

$$\varepsilon = \min\{e_{i,j} : x_i \in X - R,\ y_j \in Y - T\}.$$

Reducing u_i by ε for all $x_i \in X - R$ maintains the cover condition for these edges while bringing at least one into the equality subgraph. To maintain the cover condition for the edges from $X - R$ to T, increase v_j by ε for $y_j \in T$, as shown in the figure above. Recompute the equality subgraph.

The algorithm proceeds by iteratively adjusting the cover to obtain a cover whose equality subgraph contains a perfect matching. ■

We explained the adjustment step using the equality subgraph, but redrawing a changing equality subgraph is awkward. More efficient is using the matrix of excesses, called the **excess matrix** for the cover.

6.3.12. Example. *Solving the Assignment Problem.* The first matrix below lists the weights $(w_{i,j})$. The others display a cover (u, v) and the corresponding excess matrix. The 0s in an excess matrix correspond to edges in the equality subgraph $G_{u,v}$; an *independent family* of 0s (no two in a row or column) corresponds to a matching in $G_{u,v}$. We underscore independent 0s to mark a maximum matching, shown in bold in the equality subgraphs for the first two excess matrices.

A set of rows and columns covering the 0s in the excess matrix is a **covering set**; these form a vertex cover in $G_{u,v}$. A covering set of size less than n yields progress, since the next weighted cover costs less. For the 0s in the excess matrix, we find an independent family and a covering set of the same size. In a small matrix, we can do this by inspection.

$$
\begin{pmatrix}
6 & 3 & 4 & 3 & 4 \\
5 & 8 & 6 & 4 & 6 \\
6 & 2 & 3 & 4 & 1 \\
3 & 7 & 6 & 5 & 0 \\
4 & 5 & 8 & 2 & 3
\end{pmatrix}
\rightarrow
\begin{array}{c}
\begin{matrix} 0 & 0 & 0 & 0 & 0 \end{matrix} \\
\begin{matrix}
6 \\ 8 \\ 6 \\ 7 \\ 8
\end{matrix}
\begin{pmatrix}
0 & 3 & 2 & 3 & 2 \\
3 & 0 & 2 & 4 & 2 \\
\underline{0} & 4 & 3 & 2 & 5 \\
4 & \underline{0} & 1 & 2 & 7 \\
4 & 3 & \underline{0} & 6 & 5
\end{pmatrix} R \\
 \quad T \quad T
\end{array}
$$

We use "R" and "T" to label the rows and columns of the covering set and the corresponding vertex cover in $G_{u,v}$. At each iteration, we compute the minimum excess on the positions that are *not in a covered row or column* (in rows of $X - R$ and columns of $Y - T$). These uncovered positions have positive excess; that is, the corresponding edges are not in the equality subgraph. As suggested earlier, let ε be the minimum of these excesses. We reduce the label u_i by ε on rows not in R and increase the label v_j by ε on columns in T. The outlined positions in the diagram are the ones that change to reach the next matrix.

$$
\begin{array}{c}
\begin{matrix} 1 & 1 & 0 & 0 & 0 \end{matrix} \\
\begin{matrix}
5 \\ 7 \\ 5 \\ 6 \\ 8
\end{matrix}
\begin{pmatrix}
0 & 3 & 1 & 2 & 1 \\
3 & 0 & 1 & 3 & 1 \\
\underline{0} & 4 & 2 & 1 & 4 \\
4 & \underline{0} & 0 & 1 & 6 \\
5 & 4 & \underline{0} & 6 & 5
\end{pmatrix} \\
 \; T \quad T \quad T
\end{array}
\qquad
\begin{array}{c}
\begin{matrix} 2 & 2 & 1 & 0 & 0 \end{matrix} \\
\begin{matrix}
4 \\ 6 \\ 4 \\ 5 \\ 7
\end{matrix}
\begin{pmatrix}
0 & 3 & 1 & 1 & 0 \\
3 & \underline{0} & 1 & 2 & 0 \\
\underline{0} & 4 & 2 & 0 & 3 \\
4 & 0 & 0 & \underline{0} & 5 \\
5 & 4 & \underline{0} & 5 & 4
\end{pmatrix}
\end{array}
$$

In the example given, the first iteration reduces the cost of the cover but does not increase the size of the maximum matching in the equality subgraph. The second iteration produces a perfect matching. Using the first three columns as a covering set in the first iteration would augment the matching immediately.

The transversal of 0s at the end identifies a perfect matching whose total weight equals the cost of the final cover. Its edges have weights $6, 8, 8, 5, 4$ in the original data; they sum to 31. The labels $4, 6, 4, 5, 7$ and $2, 2, 1, 0, 0$ in the final cover satisfy these edges exactly and also sum to 31. The value of the optimal solution is unique, but the solution itself is not; this example has many maximum weight matchings and many minimum cost covers, all summing to 31. ∎

6.3.13. Theorem. The Hungarian Algorithm finds a maximum weight matching and a minimum cost cover.

Proof: The algorithm begins with a cover. It can terminate only when the equality subgraph has a perfect matching, which guarantees equal value for the current matching and cover. Suppose that (u, v) is the current cover and that the equality subgraph has no perfect matching. Let (u', v') denote the new lists of numbers assigned to the vertices. Because ε is the minimum of a nonempty finite set of positive numbers, $\varepsilon > 0$.

We verify first that (u', v') is a cover. The change of labels on vertices of $X - R$ and T yields $u_i' + v_j' = u_i + v_j$ for edges $x_i y_j$ from $X - R$ to T or from R to $Y - T$. If $x_i \in R$ and $y_j \in T$, then $u_i' + v_j' = u_i + v_j + \varepsilon$, and the weight remains covered. If $x_i \in X - R$ and $y_j \in Y - T$, then $u_i' + v_j'$ equals $u_i + v_j - \varepsilon$, which by the choice of ε is at least $w_{i,j}$.

The algorithm terminates only when the equality subgraph has a perfect matching, so it suffices to prove termination. If the weights $w_{i,j}$ are rational, then multiplying by their least common denominator yields an equivalent problem with integer weights, where the changes made in labels are always integers. By the König–Egerváry Theorem, $|T| < |X - R|$ when the equality subgraph has no perfect matching. The changes in labels reduce the cost of the cover by a multiple of $|X - R| - |T|$, which is an integer. Since the cost is bounded below by the weight of a perfect matching, after finitely many iterations we have equality.

For real-valued weights, see the next remark. ∎

6.3.14. Remark. The algorithm also works with real-valued weights if we obtain covers more carefully. Because M remains in the equality subgraph, the size of the matching never decreases. Since it can increase at most n times, it suffices to show that it must increase within n iterations.

If we find M using the Augmenting Path Algorithm, then the last iteration produces a vertex cover by exploring M-alternating paths from the subset U of uncovered vertices in X. With S and T being the sets of vertices reached in X and Y, the vertex cover is $R \cup T$, where $R = X - S$.

A step of the Hungarian Algorithm using this cover $R \cup T$ maintains equality on M and all edges in M-alternating paths from U. Edges from T to R disappear from the equality subgraph, but they don't appear in M-alternating paths from U. Introducing an edge from S to $Y - T$ either creates an M-augmenting path or increases T while leaving U unchanged. Since T can increase at most n times, we obtain a larger matching in the equality subgraph within n iterations. ∎

When running the Hungarian Algorithm with all inputs $w_{i,j} \in \{0, 1\}$, with initial cover assigning 1 for each u_i and 0 for each v_j, the excesses always remain 0 or 1. Hence the values in the weighted cover are all integers. The algorithm produces a minimum weighted cover, but it does not use integers larger than 1 (using nonnegative cover values). Thus in the special case of maximum matching, the dual cover problem reduces to finding $\beta(G)$.

We have not discussed weighted matching in general graphs. Edmonds [1965d] found an algorithm for this, which was implemented in time $O(n^3)$ by Gabow [1975] and by Lawler [1976]. Faster algorithms appear in Gabow [1990] and in Gabow–Tarjan [1989].

Weighted bipartite matching is an important special case of the general problem of optimizing a linear function of several variables under linear constraints and requiring variables to have integer values (like "use an edge or don't"). Linear optimization problems always have natural dual problems, and when the integrality constraints are dropped the optimal values of the two problems are always equal. We refer the reader to the many textbooks on Linear Programming.

FAST BIPARTITE MATCHING (optional)

In a bipartite graph with n vertices and m edges, exploring all edges for each augmentation could take time $O(nm)$ to find a maximum matching, since $\alpha'(G)$ may be linear in n. By looking first for short augmenting paths, Hopcroft–Karp [1973] reduced the running time to $O(\sqrt{n}m)$.

The idea is to explore M-alternating paths simultaneously from all M-uncovered vertices of X, seeking disjoint M-augmenting paths of the same length with one examination of the edge set. Subsequent augmentations must use longer paths, so the procedure splits into phases finding paths of the same length. The key is that not many phases are needed. The first remark follows from the proof of Theorem 6.3.4.

6.3.15. Remark. If M and M' are matchings, with $|M'| = s > r = |M|$, then $M \triangle M^*$ contains at least $s - r$ disjoint M-augmenting paths.   ∎

We next show that successive shortest augmenting paths get longer.

6.3.16. Lemma. If P is a shortest M-augmenting path and P' is $M \triangle P$-augmenting, then $|P'| \geq |P| + 2 |P \cap P'|$ (treating paths as edge sets).

Proof: The edge set $M \triangle P$ is the matching obtained by using P to augment M. Let N be the matching given by using P' to augment $M \triangle P$. Since $|N| = |M| + 2$, Remark 6.3.15 guarantees disjoint M-augmenting paths P_1 and P_2 in $M \triangle N$. Each is at least as long as P, since P is a shortest M-augmenting path.

Since N arises from M by switching on P and then switching on P', an edge belongs to exactly one of M and N if and only if it belongs to exactly one of P and P'. That is, $P \triangle P' = M \triangle N$. Hence $|P \triangle P'| = |M \triangle N| \geq |P_1| + |P_2| \geq 2|P|$, and

$$2|P| \leq |P \triangle P'| = |P| + |P'| - 2|P \cap P'|.$$

We conclude $|P'| \geq |P| + 2|P \cap P'|$.   ∎

6.3.17. Lemma. Augmenting paths of the same length in a list of successive shortest augmentations share no vertices.

Proof: Let $P_1, \ldots, P_t$ be successive shortest augmenting paths starting from some matching, in which P_k and P_l are two closest paths of the same length that are not disjoint. Let M be the matching that results from using the augmenting paths $P_1, \ldots, P_k$. Thus every vertex of P_k is covered by an edge of M in P_k.

By the choice of P_k and P_l, no two of $P_{k+1}, \ldots, P_l$ share a vertex. Hence P_l is an M-augmenting path. Since every vertex of P_k belongs to an edge of $P_k \cap M$, the M-augmenting path P_l cannot end at a vertex of P_k, and it cannot have an internal vertex on P_k without sharing an edge of M with P_k. However, by Lemma 6.3.16 P_k and P_l share no edge.

The contradiction implies that there is no such pair (k, l). ∎

6.3.18. Theorem. (Hopcroft–Karp [1973]) Maximum matching can run in $O(\sqrt{n}m)$ time on bipartite graphs with n vertices and m edges.

Proof: Lemmas 6.3.16–6.3.17 imply that searching simultaneously for shortest M-augmenting paths from all vertices of X not covered by M yields disjoint paths, and further augmenting paths are longer. The augmentations of one length are thus found by exploring the edges at most once.

List the resulting paths as $P_1, \ldots, P_s$ in order by length, with $s = \alpha'(G)$. Each P_{i+1} is an augmenting path for the matching M_i formed by using $P_1, \ldots, P_i$. It suffices to prove that these paths have at most $2\sqrt{s}$ distinct lengths.

Let $r = s - \lceil \sqrt{s} \rceil$. Because $|M_r| = r$ and the maximum matching has size s, Remark 6.3.15 yields $s - r$ disjoint M_r-augmenting paths. The shortest of these uses at most $\frac{r}{s-r}$ edges from M_r. Hence P_{r+1} has at most $2\frac{r}{s-r} + 1$ edges, and each of $P_1, \ldots, P_r$ has no more than that. Since augmenting paths have odd length, $P_1, \ldots, P_{r+1}$ have at most $\frac{r}{s-r} + 1$ lengths. Even if the remaining $\lceil \sqrt{s} \rceil - 1$ paths have distinct lengths, altogether we use at most $\frac{r}{s-r} + \lceil \sqrt{s} \rceil$ phases. Since $\frac{r}{s-r} < \frac{s}{\lceil \sqrt{s} \rceil} \le \sqrt{s}$, we have the desired upper bound of at most $2\lceil \sqrt{s} \rceil$ phases. ∎

On general graphs, Edmonds' Blossom Algorithm runs in time $O(n^3)$. There is a complicated faster algorithm with running time $O(\sqrt{n}m)$ (Micali–Vazirani [1980]); the proof of correctness is in Vazirani [1994].

STABLE MATCHINGS (optional)

Instead of total weight, we may consider other criteria for matchings. Given n men and n women, we want to establish n "stable" marriages. If man x and woman a are paired with other partners, but x prefers a to his current partner and a prefers x to her current partner, then they might leave their current partners and switch to each other. In this situation we say that the unmatched pair (x, a) is an **unstable pair**.

6.3.19. Definition. A perfect matching is a **stable matching** if it yields no unstable unmatched pair.

6.3.20. Example. Given men x, y, z, w, women a, b, c, d, and preferences listed below, the matching $\{xa, yb, zd, wc\}$ is stable. ∎

Men $\{x, y, z, w\}$	Women $\{a, b, c, d\}$
$x : a > b > c > d$	$a : z > x > y > w$
$y : a > c > b > d$	$b : y > w > x > z$
$z : c > d > a > b$	$c : w > x > y > z$
$w : c > b > a > d$	$d : x > y > z > w$

In their paper "College admissions and the stability of marriage", Gale and Shapley proved that a stable matching always exists and can be found using a relatively simple algorithm.

6.3.21. Algorithm. (**Gale–Shapley Proposal Algorithm**) Given preference rankings by each of n men and n women, iteratively perform the following. Each man proposes to the highest woman on his preference list who has not previously rejected him. If each woman receives exactly one proposal, then stop and use the resulting matching. Otherwise, every woman receiving more than one proposal rejects all except the one that is highest on her preference list. Every woman receiving a proposal says "maybe" to the most preferred proposal received. ∎

6.3.22. Theorem. (Gale–Shapley [1962]) The Proposal Algorithm produces a stable matching.

Proof: *Key Observation:* the sequence of proposals made by each man is non-increasing in his preference list, and the sequence of men to whom a woman says "maybe" is nondecreasing in her preference list, culminating in the man assigned. This holds because men propose repeatedly to the same woman until rejected, and women say "maybe" to the same man until a better offer arrives.

Once a woman has a proposal, she keeps receiving proposals and giving one "maybe" answer. Hence if $n-1$ women have rejected man x, then they are all holding proposals from $n-1$ distinct men who propose to them again on the next round. When x proposes to the nth woman, a matching is completed.

Hence the algorithm can only terminate with a matching. It does terminate, since at each iteration the total length of the remaining proposal lists decreases.

If the result is not stable, then some unstable unmatched pair (x, a) has x matched to b and y matched to a. By the key observation, x never proposed to a, since a received a mate less desirable than x. The key observation also implies that x would not have proposed to b without earlier proposing to a. This contradiction confirms the stability of the result. ∎

The proposal algorithm is asymmetric; which sex is happier? When the first choices of the men are distinct, they all get their first choice, and the women receive whoever proposed. In Example 6.3.20, when women propose the algorithm immediately yields the matching $\{xd, yb, za, wc\}$; all women receive their first choices. In fact, among all stable matchings, every man is happiest in the one produced by the male-proposal algorithm, and every woman is happiest under the female-proposal algorithm (Exercise 24). Societal conventions thus favor men.

The algorithm is used in another setting. Each year, the graduates of medical schools submit preference lists of hospitals where they want to be residents. The hospitals also have preferences; a hospital with multiple openings acts as several hospitals with the same preference list. Chaos in the market for residents (then called interns) forced hospitals to devise and implement the algorithm ten years before the Gale–Shapley paper defined and solved the problem! The result was the National Resident Matching Program, a non-profit corporation established in 1952 to provide a uniform appointment date and matching procedure.

Since the medical organizations ran the algorithm, it is not surprising that initially they did the proposing and were happier with the outcome. This is even clearer in another setting; students applying for jobs have preferences, but the employers make the proposals, called "job offers". Unhappiness with the NRMP caused the system to be changed in 1998 to a student-proposing algorithm. In 1998 the system processed 35,823 applicants for 22,451 positions. Additional details are at nrmp.aamc.org/nrmp/mainguid/ on the World Wide Web.

There may be stable matchings other than those found by the two versions of the proposal algorithm. To seek a "fair" stable matching, we could give each person a number of points with which to rate preferences. The weight for the pair xa is then the sum of the points that x gives to a and a gives to x. The Hungarian Algorithm would yield a matching of maximum total weight, but this might not be a stable matching (Exercise 23). Other approaches appear in the books Knuth [1976b] and Gusfield–Irving [1989] on stable marriages and related topics.

EXERCISES 6.3

6.3.1. (–) Given a graph G, suppose that $S \subseteq V(G)$ is covered by some matching. Prove or disprove each statement below:

(a) S is covered by some maximum matching.

(b) S is covered by every maximum matching.

6.3.2. (–) Use the symmetric difference operation to give an alternative proof that if x and y are vertices in a tree G, then G has exactly one x, y-path.

6.3.3. (–) Using nonnegative edge weights, construct a 4-vertex weighted graph in which the matching of maximum weight is not a matching of maximum size.

6.3.4. (–) Show how to use the Hungarian Algorithm to test for the existence of a perfect matching in a bipartite graph.

6.3.5. (–) Give an example of the stable matching problem with two men and two women in which there is more than one stable matching.

6.3.6. (–) Determine the stable matchings resulting from the Proposal Algorithm with men proposing and with women proposing, given the preference lists below.

$$
\begin{array}{ll}
\text{Men } \{u, v, w, x, y, z\} & \text{Women } \{a, b, c, d, e, f\} \\
u: a > b > d > c > f > e & a: z > x > y > u > v > w \\
v: a > b > c > f > e > d & b: y > z > w > x > v > u \\
w: c > b > d > a > f > e & c: v > x > w > y > u > z \\
x: c > a > d > b > e > f & d: w > y > u > x > z > v \\
y: c > d > a > b > f > e & e: u > v > x > w > y > z \\
z: d > e > f > c > b > a & f: u > w > x > v > z > y
\end{array}
$$

6.3.7. (◊) Let G be a graph with m edges and maximum degree k.

(a) Let M be a maximum matching in G. For $k \geq 3$, prove that the number of edges joining vertices covered by M to vertices not covered by M is at most $(k-1)|M|$. Use this to prove $\alpha'(G) \geq 2m/(3k-1)$. (Feng [2009])

(b) Let M be a maximal matching in G. Prove that M has size at least $m/(2k-1)$. For each k, construct infinitely many k-regular graphs having some maximal matching with $m/(2k-1)$ edges (thereby proving that the inequality is sharp). (Biedl–Demaine–Duncan–Fleischer–Kobourov [2004])

6.3.8. ($\diamond$) Two people play a game on a graph G by alternately selecting vertices $v_1, v_2, \cdots$ such that for $i > 1$, vertex v_i is adjacent to v_{i-1} and has not previously been chosen. The last player able to select a vertex wins. Prove that the second player has a winning strategy if G has a perfect matching and that the first player has a winning strategy if G has no perfect matching.

6.3.9. Prove that every nontrivial tree has a maximum matching that covers every non-leaf vertex plus at least one leaf. Determine which trees have a matching covering at least two leaves. (Lih–Lin–Tong [2006])

6.3.10. Let G be an X, Y-bigraph with $|X| \leq |Y|$. Prove that G has a maximal path with an endpoint in Y. (Kündgen–Ramamurthi [2002])

6.3.11. Let M and M' be matchings in an X, Y-bigraph G such that $S \subseteq X$ is covered by M and $T \subseteq Y$ is covered by M'. Prove that G contains a matching that covers $S \cup T$.

6.3.12. Let G be an X, Y-bigraph, not necessarily finite. Prove that if G has a matching covering X and a matching covering Y, then G has a perfect matching. (J. Zaks)

6.3.13. ($\diamond$) Given a function $f \colon V(G) \to \mathbb{N}_0$ on a graph G, an f-**matching** in G is a subset M of $E(G)$ having at most $f(v)$ edges incident to each vertex v. An M-**augmenting trail** is a trail that alternates between edges outside and in M, beginning and ending with edges outside M at vertices having "excess capacity". Use the transformation in Example 6.2.15 to prove that an f-matching M in G is a maximum-sized f-matching if and only if G has no M-augmenting trail. (Gondran–Minoux [1984])

6.3.14. (+) Let G be a connected X, Y-bigraph with girth at least $2s$. Prove that if the distance between any two vertices of X having degree 1 is at least $2s$, then every subset of X with size at most s is covered by some matching in G. (Hint: Consider a smallest subset of X that cannot be covered.) (Horák–Tuza)

6.3.15. Let u be a vertex missed by a matching M. Prove that if no M-augmenting path starts at u, then u is missed by some maximum matching.

6.3.16. ($\diamond$) *Proof of König–Egerváry Theorem by augmenting paths.* Let M be a matching in an X, Y-bigraph G. Let U be the subset of X missed by M. Let W be the set of vertices reachable from U by M-alternating paths (this includes U). Let $S = W \cap X$ and $T = W \cap Y$. Prove that if G has no M-augmenting path, then $T \cup (X - S)$ is a vertex cover of size $|M|$.

6.3.17. ($\diamond$) Find a transversal of maximum total sum (weight) in each matrix below. Prove that there is no larger weight transversal by exhibiting a solution to the dual problem. Explain why this proves that there is no larger transversal.

(a)	(b)	(c)
4 4 4 3 6	7 8 9 8 7	1 2 3 4 5
1 1 4 3 4	8 7 6 7 6	6 7 8 7 2
1 4 5 3 5	9 6 5 4 6	1 3 4 4 5
5 6 4 7 9	8 5 7 6 4	3 6 2 8 7
5 3 6 8 3	7 6 5 5 5	4 1 3 5 4

6.3.18. Given a connected graph G with nonnegative edge weights, the **Chinese Postman Problem** (posed in Guan [1962]) is the problem of finding the shortest closed walk that includes all the edges. Reduce this problem to a problem of finding a perfect matching of minimum total weight in a complete graph with weights on the edges.

6.3.19. Find a minimum-weight transversal in the matrix below, and use duality to prove that the solution is optimal. (Hint: Use a transformation of the problem.)

$$\begin{pmatrix} 4 & 5 & 8 & 10 & 11 \\ 7 & 6 & 5 & 7 & 4 \\ 8 & 5 & 12 & 9 & 6 \\ 6 & 6 & 13 & 10 & 7 \\ 4 & 5 & 7 & 9 & 8 \end{pmatrix}$$

6.3.20. ($\diamond$) *The Bus Driver Problem.* Suppose that bus drivers are paid overtime for the time by which their routes in a day exceed t. Let there be n bus drivers, n morning routes with durations $x_1, \ldots, x_n$, and n afternoon routes with durations $y_1, \ldots, y_n$. The objective is to assign one morning route and one afternoon route to each driver to minimize the total overtime. Express this as a weighted matching problem. Prove that overtime is minimized by giving the ith longest morning route and ith shortest afternoon route to the same driver, for each i. (R.B. Potts)

6.3.21. Let the entries in matrix A have the form $w_{i,j} = a_i b_j$, where $a_1, \ldots, a_n$ are numbers associated with the rows and $b_1, \ldots, b_n$ are numbers associated with the columns. Determine the maximum weight of a transversal of A. What happens when $w_{i,j} = a_i + b_j$? (Hint: In each case, guess the general pattern by examining the solution when $n = 2$.)

6.3.22. ($\diamond$) A mathematics department offers k seminars in different topics to its n students. Each student will take one seminar; the ith seminar will have k_i students, where $\sum k_i = n$. Each student submits a preference list ranking the k seminars. An assignment of the students to seminars is *stable* if no two students can both obtain more preferable seminars by switching their assignments. Show how to find a stable assignment using weighted bipartite matching. (Isaak)

6.3.23. Consider n men and n women, each assigning $n-i$ points to the ith person in his or her preference list. Let the weight of a pair be the sum of the points assigned by those two people. Construct an example where no maximum weight matching is a stable matching.

6.3.24. ($\diamond$) Prove that if man x is paired with woman a in some stable matching, then a never rejects x in the Proposal Algorithm with men proposing. Conclude that among all stable matchings, *every* man is happiest in the matching produced by this algorithm. (Hint: Consider the first occurrence of such a rejection.)

6.3.25. In the Stable Roommates Problem, each of $2n$ people has a preference ordering on the other $2n - 1$. A stable matching is a perfect matching such that no unmatched pair prefers each other to their current roommates. Prove that the preferences below do not permit a stable matching. (Gale–Shapley [1962])

$$a: b > c > d \qquad b: c > a > d \qquad c: a > b > d \qquad d: a > b > c$$

6.3.26. In the Stable Roommates Problem, let each individual declare a top portion of the preference list as "acceptable". Define the *acceptability graph* to be the graph whose vertices are the people and whose edges are the pairs who rank each other as acceptable. Prove that all sets of rankings with acceptability graph G lead to a stable matching if and only if G is bipartite. (Abeledo–Isaak [1991]).

Chapter 7

Connectivity and Cycles

Every connected graph contains a spanning tree, but a tree is just barely connected; deleting any edge or non-leaf vertex disconnects it. In a communication network, we want to preserve service by ensuring that the graph (or digraph) of possible transmissions remains connected even when some vertices or edges fail.

7.1. Connectivity Parameters

How difficult is it to disconnect a graph? We consider first deletion of vertices and then deletion of edges.

SEPARATING SETS

7.1.1. Definition. A **separating set** or **vertex cut** of a graph G is a set $S \subseteq V(G)$ such that $G - S$ has more than one component. A graph G is k-**connected** if it has more than k vertices and every vertex cut has size at least k. The **connectivity** of G, written $\kappa(G)$, is the maximum k such that G is k-connected.

A non-complete graph G has a separating set, and $\kappa(G)$ is the minimum size of such a set. In this case, the neighborhood of a vertex of minimum degree is a separating set, and hence $\kappa(G) \le \delta(G)$. Also, if a graph G is k-connected, and $k > 0$, then G is also $(k-1)$-connected, by the definition.

7.1.2. Example. *Connectivity of K_n and $K_{r,s}$.* A complete graph has no separating set. By requiring k-connected graphs to have more than k vertices, we obtain $\kappa(K_n) = n - 1$. This allows general connectivity results (such as $\kappa(G) \le \delta(G)$) to hold also for K_n. A graph with more than two vertices has connectivity 1 if it is connected and has a cut-vertex. A graph with more than one vertex has connectivity 0 if and only if it is disconnected. Unfortunately, K_1 is an anomaly; it is connected but has connectivity 0.

Every induced subgraph of $K_{r,s}$ having at least one vertex from each part is connected. Hence every vertex cut contains a full part, and $\kappa(K_{r,s}) = \min\{r, s\}$ (the convention for K_2 is consistent with this).   ∎

Since $\delta(G)$ is an upper bound on $\kappa(G)$, it is natural to ask when equality holds. Such graphs have been called **maximally connected**. Our first family with this property also solves the extremal problem for the minimum number of edges in a k-connected graph G with n vertices. Since $\delta(G) \geq k$, the Degree-Sum Formula requires at least $\lceil kn/2 \rceil$ edges; this is achieveable whenever $n > k$.

7.1.3. Example. *The k-connected Harary graph $H_{k,n}$.* Given $2 \leq k < n$, place vertices $1, \ldots, n$ around a circle, equally spaced. Make each vertex adjacent to the nearest $\lfloor k/2 \rfloor$ vertices in each direction around the circle. If k is odd, then also add the edges from i to $i + \lfloor n/2 \rfloor$ for $1 \leq i \leq \lceil n/2 \rceil$.

When kn is even, $H_{k,n}$ is k-regular. When kn is odd, vertex $(n+1)/2$ has degree $k + 1$ and the rest have degree k. See $H_{4,8}$, $H_{5,8}$, and $H_{5,9}$ below.   ∎

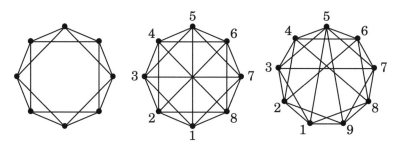

The first general method for proving a graph k-connected is the definition: consider a separating set S and prove $|S| \geq k$. For the Harary graphs, we will use another tool for the last case; we postpone it until after the main proof.

7.1.4. Theorem. (Harary [1962a]) $\kappa(H_{k,n}) = k$, and hence the minimum number of edges in a k-connected graph on n vertices is $\lceil kn/2 \rceil$.

Proof: Let $r = \lfloor k/2 \rfloor$ and $G = H_{k,n}$. Since $\delta(G) = k$, it suffices to prove $\kappa(G) \geq k$. Let S be a separating set; we prove $|S| \geq k$.

Choose $x, y \in V(G) - S$. Deleting x and y from the circular arrangement of vertices leaves two maximal segments A and B of consecutive vertices. In $G - S$, we have the potential for traveling from x to y in a clockwise or a counterclockwise direction, through A or B.

Since each vertex is adjacent to the next r vertices in each direction, there is an x, y-path in $G - S$ via A or B unless S contains r consecutive vertices both in A and in B. Thus $|S| \geq k$ unless k is odd and S consists exactly of r consecutive vertices in A and r consecutive vertices in B.

In this case, when n is even we find an x, y-path in $G - S$ using the diagonal edge at x or y. Let x' and y' be the neighbors of x and y along these edges, respectively. Label A and B so $|A| \geq |B|$; now $x', y' \in A$ (see figure below). Since S has r vertices between x and y in B, there are also r vertices between x' and y' in A. Therefore, deleting r consecutive vertices in A leaves intact the x', y-path in A or the x, y'-path in A; adding xx' or yy' completes an x, y-path in $G - S$.

For odd n, the lack of rotational symmetry leads to annoying technical details in such arguments. Instead, note that $H_{k,n}$ is obtained from $H_{k,n-1}$ by using the operation in Definition 7.1.5, which by Lemma 7.1.6 preserves k-connectedness.

The new vertices 1 and n each have degree k, and each vertex that was adjacent to the split vertex now has a neighbor in $\{1, n\}$.   ∎

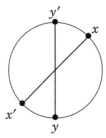

7.1.5. Definition. A **vertex k-split** forms a graph H from a graph G by deleting one vertex x and replacing it with adjacent vertices x_1 and x_2 such that $N_H(x_1) \cup N_H(x_2) = N_G(x) \cup \{x_1, x_2\}$ and $d_H(x_i) \geq k$.

7.1.6. Lemma. If a graph H arises from a k-connected graph G by a vertex k-split, then H is k-connected.

Proof: Form H by splitting x into x_1 and x_2 as in Definition 7.1.5. Let S be a separating set of H, and let $X = \{x_1, x_2\}$. If $S \cap X = \varnothing$, then also S separates G, so $|S| \geq k$. If $X \subseteq S$, then $H - S = G - T$, where $T = (S - X) \cup \{x\}$, so $|S| > |T| \geq k$.

The remaining case is $|S \cap X| = 1$, say $x_1 \in S$. Let $T = S - \{x_1\} \cup \{x\}$. If T separates G, then $|S| = |T| \geq k$, so assume $G - T$ is connected. Now $H - S$ is obtained from $G - T$ by adding x_2 and the edges joining x_2 to its neighbors. Hence $H - S$ is connected unless S contains $N_H(x_2)$, again yielding $|S| \geq k$.   ∎

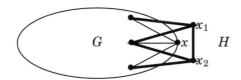

Theorem 7.1.4 implies that $\lceil kn/2 \rceil$ edges and minimum degree k are the smallest values that *allow* an n-vertex graph to be k-connected. The smallest value of the minimum degree that *forces* an n-vertex graph to be k-connected (for $1 \leq k < n$) is much larger: $\lceil (n + k - 2)/2 \rceil$ (Exercise 17).

Next we show that hypercubes are also maximally connected.

7.1.7. Example. $\kappa(Q_k) = k$. The hypercube Q_k is k-regular, so $\kappa(Q_k) \leq k$. We prove the lower bound by induction on k. Note that $\kappa(Q_k) = k$ for $k \leq 1$, since Q_0 and Q_1 are complete graphs.

For $k \geq 2$, express Q_k as two copies of Q_{k-1} (called G_1 and G_2) plus a matching joining corresponding vertices (each vertex of Q_{k-1} becomes two vertices in Q_k joined by an edge). By the induction hypothesis, G_1 and G_2 are $(k-1)$-connected.

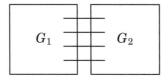

Let S be a separating set in Q_k. If S disconnects G_1 or G_2, then S must have at least $k - 1$ vertices in that copy of Q_{k-1}. Also S must have at least one vertex in the other copy, since otherwise any two vertices of $G - S$ would be connected by a path through the other copy. On the other hand, if $G_1 - S$ and $G_2 - S$ both are connected, then S must contain at least one vertex from every edge joining G_1 and G_2 to avoid $G - S$ being connected; this yields $|S| \geq |V(G_1)| = 2^{k-1} \geq k$. ∎

The argument in Example 7.1.7 also proves that whenever G is k-connected, the product $G \,\square\, K_2$ is $(k + 1)$-connected. The result generalizes further.

7.1.8. Theorem. (Niu–Zhu [1994], Chiue–Shieh [1999]) If G and H are connected graphs, then $\kappa(G \,\square\, H) \geq \kappa(G) + \kappa(H)$.

Proof: The cartesian product $G \,\square\, H$ (Definition 5.1.20) decomposes into a copy G_v of G for each $v \in V(H)$ and a copy H_u of H for each $u \in V(G)$. Let $n_G = |V(G)|$ and $n_H = |V(H)|$, and let S be a separating set of $G \,\square\, H$.

Case 1: S separates some G_v or H_u. By symmetry, let $G_v - S$ be disconnected, with (u, v) and (u', v) in different components. Thus $|S \cap V(G_v)| \geq \kappa(G)$. If $H_u - S$ or $H_{u'} - S$ is disconnected, then S also has $\kappa(H)$ vertices in H_u or $H_{u'}$, yielding $|S| \geq \kappa(G) + \kappa(H)$. Otherwise, with $H_u - S$ and $H_{u'} - S$ connected, we claim that S contains a vertex of each G_w with $w \neq v$. If not, then we can travel from (u, v) to (u, w) in $H_u - S$, from (u, w) to (u', w) in $G_w - S$ (since $S \cap V(G_w) = \varnothing$), and from (u', w) to (u', v) in $H_{u'} - S$, completing a walk from (u, v) to (u', v). We conclude $|S| \geq \kappa(G) + n_H - 1 \geq \kappa(G) + \kappa(H)$.

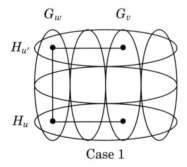 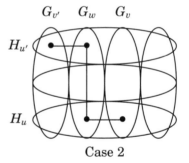

Case 1   Case 2

Case 2: S separates no G_v or H_u. Choose (u, v) and (u', v') in different components of $(G \,\square\, H) - S$; note that $u \neq u'$ and $v \neq v'$. For any $w \in V(H)$ with $w \notin \{v, v'\}$, we must have (u, w) or (u', w) in S, since otherwise we can travel from (u, v) to (u, w) in $H_u - S$, from (u, w) to (u', w) in $G_w - S$, and from (u', w) to (u', v') in $H_{u'} - S$, since S does not separate any copy of G or H. Similarly, S contains (x, v) or (x, v') for $x \notin \{u, u'\}$. Finally, S contains (u', v) and (u, v'). Thus

$$|S| \geq (n_G - 2) + (n_H - 2) + 2 = (n_G - 1) + (n_H - 1) \geq \kappa(G) + \kappa(H). \qquad \blacksquare$$

Equality holds when $\kappa(G) = \delta(G)$ and $\kappa(H) = \delta(G)$, since $\kappa(G \square H) \leq \delta(G \square H) = \delta(G) + \delta(H)$. However, $\kappa(G \,\square\, H)$ can be large when $\kappa(G) = \kappa(H) = 1$ (Exercise 27). In fact, always $\kappa(G \square H) = \min\{\delta(G) + \delta(H), |V(G)| \kappa(H), |V(H)| \kappa(G)\}$, stated without proof by Liouville [1978] and proved by Špacapan [2008] (Exercise 28).

EDGE CUTS

Edge deletion may be more relevant than vertex deletion. For example, perhaps transmitters (vertices) are secure, but communication links (edges) can be disrupted. In this setting, the redundancy of multiedges can be valuable.

7.1.9. Definition. A **disconnecting set** of edges in a multigraph G is a set $F \subseteq E(G)$ such that $G-F$ is disconnected; G is k-**edge-connected** if every disconnecting set has size at least k. The **edge-connectivity** $\kappa'(G)$ is $\max\{k: G$ is k-edge-connected$\}$. For $S, T \subseteq V(G)$, write $[S, T]$ for the set of edges joining S to T. An **edge cut** is a set of the form $[S, \overline{S}]$, where $\varnothing \neq S \subset V(G)$. A minimal set of edges whose deletion increases the number of components is a **bond**.

Deleting any edge cut of a graph G disconnects it: $G - [S, \overline{S}]$ has no path from S to $\overline{S}$. In fact, all minimal disconnecting sets of edges have this form.

7.1.10. Proposition. For a graph G, every minimal disconnecting set is an edge cut. If G is connected, then an edge cut $[S, \overline{S}]$ is a bond if and only if the induced subgraphs $G[S]$ and $G[\overline{S}]$ are both connected.

Proof: If $G - F$ is disconnected, then let T be the vertex set of a component of $G - F$. The edge cut $[T, \overline{T}]$ is contained in F, so F is not a minimal disconnecting set unless $F = [T, \overline{T}]$.

If G is connected, F is the edge cut $[S, \overline{S}]$, and both $G[S]$ and $G[\overline{S}]$ are connected, then no proper subset of $[S, \overline{S}]$ disconnects G, so F is a bond. On the other hand, if $G[S]$ is not connected, and T is the vertex set of a component of $G[S]$, then $[T, \overline{T}]$ is a proper subset of $[S, \overline{S}]$, as shown below, so F is not a bond.   ∎

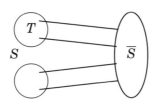

7.1.11. Example. *Edge-connectivity of K_n.* Every edge cut in K_n has size $k(n-k)$ for some k with $1 \leq k \leq n-1$. The minimum occurs at $k = 1$, so $\kappa'(K_n) = n - 1$. ∎

Deleting one endpoint of each edge in an edge cut F deletes every edge of F. Hence we expect that $\kappa(G) \leq \kappa'(G)$, but we must avoid leaving a connected subgraph by deleting all isolated vertices of $G - F$.

7.1.12. Theorem. (Whitney [1932a]) $\kappa(G) \leq \kappa'(G) \leq \delta(G)$.

Proof: The edges incident to a vertex v of minimum degree disconnect G; hence $\kappa'(G) \leq \delta(G)$. To prove $\kappa(G) \leq \kappa'(G)$, let $k = \kappa'(G)$, and let $[S, \overline{S}]$ be a smallest edge cut. If all of S is adjacent to all of $\overline{S}$, then $k = |S| |\overline{S}| \geq |V(G)| - 1 \geq \kappa(G)$.

Otherwise, there exist $x \in S$ and $y \in \overline{S}$ with $xy \notin E(G)$. For each edge of $[S, \overline{S}]$ incident to x, put the other endpoint into T. For other edges of $[S, \overline{S}]$, put the endpoints in S into T, as shown below. Now $|T| \leq k$, and x and y lie in different components of $G - T$. Hence $\kappa(G) \leq \kappa'(G)$.   ∎

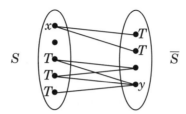

When $\kappa'(G) < \delta(G)$, no smallest edge cut isolates a vertex; in fact, both sides of a smallest edge cut $[S, \overline{S}]$ must then be larger than $\delta(G)$. This follows from a simple expression for the size of an edge cut.

7.1.13. Proposition. If $S \subseteq V(G)$, then $\left|[S, \overline{S}]\right| = \left[\sum_{v \in S} d(v)\right] - 2\,|E(G[S])|$.

Proof: The sum $\sum_{v \in S} d(v)$ counts edges in $G[S]$ twice and edges in $[S, \overline{S}]$ once. ∎

7.1.14. Corollary. If G is a graph and $\left|[S, \overline{S}]\right| < \delta(G)$ for some nonempty proper subset S of $V(G)$, then $|S| > \delta(G)$.

Proof: By Proposition 7.1.13, $\delta(G) > \sum_{v \in S} d(v) - 2\,|E(G[S])|$. Since $d(v) \geq \delta(G)$ and $2\,|E(G[S])| \leq |S|\,(|S|-1)$, we have $\delta(G) > |S|\,\delta(G) - |S|\,(|S|-1)$. This inequality requires $|S| > 1$, so rearranging and canceling $|S| - 1$ yields $|S| > \delta(G)$. ∎

We give a sufficient condition for equality in the trivial upper bound. Note that diameter 2 does *not* imply $\kappa(G) = \delta(G)$; consider $K_1 \oplus 2K_r$.

7.1.15. Theorem. (Plesník [1975]) If $\operatorname{diam} G = 2$, then $\kappa'(G) = \delta(G)$.

Proof: If $\left|[S, \overline{S}]\right| < \delta(G)$, then Corollary 7.1.14 yields $|S| > \delta(G)$ and $|\overline{S}| > \delta(G)$. If each vertex in S has a neighbor in $\overline{S}$, then $\left|[S, \overline{S}]\right| \geq |S| > \delta(G)$. Otherwise, some $x \in S$ has no neighbor in $\overline{S}$, and similarly some vertex $y \in \overline{S}$ has no neighbor in S. Now $d(x, y) > 2$, contradicting $\operatorname{diam} G = 2$. ∎

BLOCKS

Decomposition into components can reduce a problem to connected graphs. There is a similar decomposition into connected pieces without cut-vertices.

7.1.16. Definition. A **block** of a graph G is a maximal connected subgraph of G that has no cut-vertex.

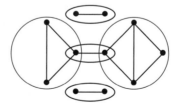

7.1.17. Example. If H is a block of G, then H as a graph has no cut-vertex, but H may contain cut-vertices of G. The graph above has five blocks, circled. ∎

7.1.18. Remark. *Properties of blocks.* An edge in a cycle is not a block, since it lies in a larger subgraph with no cut-vertex. An edge (with its endpoints) is a block if and only if it is a cut-edge; the blocks of a tree are its edges. Any block with more than two vertices is 2-connected. The blocks of a graph are its isolated vertices, its cut-edges, and its maximal 2-connected subgraphs.

If blocks B and B' share two vertices, then their union cannot be disconnected by deleting one vertex. Thus B and B' fail the maximality condition in Definition 7.1.16. Hence any two blocks in a graph share at most one vertex.

This implies that the blocks of a graph form a decomposition of it. A vertex shared by two blocks must be a cut-vertex of G. The interaction between blocks and cut-vertices is described by a special graph. The **block-cutpoint graph** of a graph G is the bipartite graph H whose parts are the cut-vertices and the blocks of G, with cut-vertex v adjacent to block B in H if and only if $v \in B$.

The block-cutpoint graph is a forest with a component for each component of G (Exercise 34). The leaves of this graph correspond to blocks of G containing only one cut-vertex of G; such blocks are **leaf blocks**.

EXERCISES 7.1

7.1.1. $(-)$ Prove that if G has more than k vertices and is not k-connected, then G has a separating set of size exactly $k - 1$.

7.1.2. $(-)$ Prove that G is k-connected if and only if $G \oplus K_r$ is $(k + r)$-connected.

7.1.3. $(-)$ For $k, l, m \in \mathbb{N}$ with $k \le l \le m$, construct a graph $G_{k,l,m}$ with $\kappa(G_{k,l,m}) = k$, $\kappa'(G_{k,l,m}) = l$, and $\delta(G_{k,l,m}) = m$. (Chartrand–Harary [1968])

7.1.4. $(-)$ Let M be a matching of size r in $K_{r,s}$, where $r \le s$. Prove that the graph $K_{r,s} - M$ is $(r - 1)$-connected unless $(r, s) = (2, 2)$.

7.1.5. $(-)$ A **cactus** is a connected graph whose blocks are all edges or cycles. Prove that a connected graph is a cactus if and only if no two vertices are joined by three internally disjoint paths. Conclude that a connected graph with no even cycle is a cactus.

7.1.6. $(-)$ Determine the smallest 3-regular graph with connectivity 1.

7.1.7. $(-)$ Construct a graph with degree list 5543333 that is 3-connected. Prove that it is 3-connected by using vertex splits.

7.1.8. $(-)$ For $n, k \in \mathbb{N}$ with $n - k$ odd and at least 3, show that the list with $k - 1$ copies of $n - 1$ and $n - k + 1$ copies of k is graphic but has no k-connected realization. (F. Jao)

7.1.9. $(-)$ Prove that every even graph has even edge-connectivity.

7.1.10. Determine $\kappa(G)$, $\kappa'(G)$, and $\delta(G)$ for each graph G below. (Hint: For the graph on the left, use Proposition 7.1.13 to establish the edge-connectivity.)

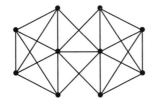 

7.1.11. A graph G is *k-expansive* if for every set $S \subseteq V(G)$ with $|S| \leq n-k$, there are at least $|S| + k$ vertices of G whose neighborhood intersects S. Prove that every k-expansive graph is k-connected. Provide infinitely many k-connected graphs that are not k-expansive.

7.1.12. For $k \geq 2$, prove that the only separating sets of size at most k in the hypercube Q_k are vertex neighborhoods. (Ramras [2004])

7.1.13. Let G be an n-vertex graph such that $\kappa(G) = \kappa'(G) < n - 1$. Prove that every smallest edge cut consists of one edge incident to each vertex of a smallest separating set.

7.1.14. Prove that a connected graph is k-edge-connected if and only if each of its blocks is k-edge-connected.

7.1.15. Use Proposition 7.1.13 to prove that the Petersen graph has an edge cut of size k if and only if $3 \leq k \leq 12$ (thus its edge-connectivity is 3).

7.1.16. Let G be an n-vertex triangle-free graph, with $\delta(G) \geq 3$. Prove that if $n \leq 11$, then G is 3-edge-connected. Prove sharpness by finding a 3-regular bipartite graph with 12 vertices that is not 3-edge-connected. (F. Galvin)

7.1.17. ($\Diamond$) Let G be an n-vertex graph. Prove that if $\delta(G) \geq (n + k - 2)/2$, where $1 \leq k \leq n - 1$, then G is k-connected. Prove sharpness by constructing an n-vertex graph with minimum degree $\lfloor (n + k - 3)/2 \rfloor$ that is not k-connected.

7.1.18. (+) Let G be an n-vertex graph with $n \geq k + l$ and $\delta(G) \geq \frac{n+l(k-2)}{l+1}$. Prove that if $G - S$ has more than l components, then $|S| \geq k$. Prove that the hypothesis on $\delta(G)$ is sharp for $n \geq k + l$ by constructing an appropriate n-vertex graph with minimum degree $\lfloor \frac{n+l(k-2)-1}{l+1} \rfloor$. (Comment: This generalizes Exercise 7.1.17.)

7.1.19. ($\Diamond$) *Generalization of Exercise 5.3.37 to $(k + 1)$-connected graphs.* (Bondy [1969])
  (a) Let G have degrees $d_1, \dots, d_n$ in nondecreasing order, and suppose $0 \leq k \leq n - 2$. Prove that if $d_j \geq j + k$ whenever $j \leq n - 1 - d_{n-k}$, then G is $(k + 1)$-connected.
  (b) Prove sharpness by constructing an example for each k to show that requiring $d_j \geq j + k$ whenever $j < n - 1 - d_{n-k}$ does not imply $\kappa(G) > k$.

7.1.20. ($\Diamond$) Let G be an n-vertex graph with m edges. For $k \geq 2$, prove that if $n \geq 2k-1 > 1$ and $m > (2k - 3)(n - k + 1)$, then G has a k-connected subgraph. Conclude that average vertex degree a guarantees a subgraph with connectivity at least $a/4$. (Mader [1972])

7.1.21. Let G be an n-vertex graph.
  (a) Prove that if $\delta(G) \geq n - 2$, then $\kappa(G) = \delta(G)$.
  (b) For $n \geq 4$, prove that part (a) is sharp by constructing an n-vertex graph with minimum degree $n - 3$ and connectivity less than $n - 3$.
  (c) Prove that if $\delta(G) = n - 3$ and $\overline{G}$ contains no 4-cycle, then $\kappa(G) = \delta(G)$.

7.1.22. ($\Diamond$) Let G be an n-vertex graph such that $d(x) + d(y) \geq n - 1$ whenever $xy \notin E(G)$. Prove that $\kappa'(G) = \delta(G)$. As a function of n, determine the least k such that $\delta(G) \geq k$ implies $\kappa'(G) = \delta(G)$.

7.1.23. ($\Diamond$) Prove that if G is a bipartite graph with diameter 3, then $\kappa'(G) = \delta(G)$. (Hint: Enhance the argument of Theorem 7.1.15.) (Plesník–Znám [1989])

7.1.24. ($\Diamond$) Let G be a graph with diameter 2.
  (a) Prove that $\kappa(G) = \delta(G)$ when G has girth at least 5.
  (b) Prove that part (a) is sharp by constructing an example with girth 4, diameter 2, and $\kappa(G) < \delta(G)$. (Hint: Create a 4-vertex cut S such that $G - S = 2K_{3,3}$. Comment: Soneoka–Nakada–Imase–Peyrat [1987] proved $\kappa(G) = \delta(G)$ for G with girth g and diameter at most $2\lceil g/2 \rceil - 3$. Diameter at most $2\lceil g/2 \rceil - 2$ guarantees $\kappa'(G) = \delta(G)$.)

7.1.25. (◇) Let G be an n-vertex bipartite graph. Prove the following implications and show that they are sharp.

(a) If $\delta(G) > n/4$, then $\kappa'(G) = \delta(G)$. (Volkmann [1988])

(b) If $\delta(G) \geq n/3$, then $\kappa(G) = \delta(G)$. For sharpness, construct for each odd integer k a bipartite graph G with $\delta(G) = k$ and $|V(G)| = 3k + 1$ that is not k-connected. (Kostochka)

7.1.26. (+) For $k \geq 2$, prove that every graphic list of integers with smallest value at least k is the degree list of some k-edge-connected graph. (Hint: Consider a realization whose smallest edge cut has size h. If $h < k$, then find a 2-switch that produces a graph with fewer edge cuts of size at most h.) (Edmonds [1964])

7.1.27. Theorem 7.1.8 states $\kappa(G \square H) \geq \kappa(G) + \kappa(H)$ for connected graphs G and H.

(a) Prove that the bound can be weak by computing $\kappa(G \square H)$ when $G = H = K_1 \oplus 2K_r$.

(b) Prove that $\kappa'(G \square H) \geq \kappa'(G) + \kappa'(H)$ when G and H are connected.

7.1.28. (+) Strengthen Theorem 7.1.8 by proving, for nontrivial graphs G and H,

$$\kappa(G \square H) = \min\{\delta(G) + \delta(H), |V(G)|\,\kappa(H), |V(H)|\,\kappa(G)\}.$$

(Hint: Prove that if S is a separating set with $|S| < \min\{|V(G)|\,\kappa(H), |V(H)|\,\kappa(G)\}$, then S has at least $\delta(G) + 1$ vertices in some two copies of G and $\delta(H) + 1$ vertices in some two copies of H.) (Špacapan [2008]; also proved in Govorčin–Škrekovski [2014])

7.1.29. (◇) Let H be a spanning subgraph of a connected graph G. Prove that H is a spanning tree if and only if $G - E(H)$ is a maximal subgraph containing no bond of G. (Comment: Note that H is a spanning tree if and only if H is a maximal subgraph containing no cycle. See Chapter 11 to relate the two statements.)

7.1.30. (◇) Prove that the edges of any edge cut can be partitioned into bonds.

7.1.31. (◇) *Edge cuts and cycles.*

(a) Let F be a nonempty set of edges in a graph G. Prove that F is an edge cut if and only if F has an even number of edges in every cycle in G. (Hint: For sufficiency, find an appropriate bipartition of the components of $G - F$.)

(b) A **signing** of a graph is a map $\sigma: E(G) \to \{+1, -1\}$. A signing is *positive* if each cycle has an even number of negative edges. Prove that the number of positive signings of any connected n-vertex graph is 2^{n-1}. (See also Exercise 5.3.27 about signed graphs.)

7.1.32. For $n \geq k$, prove that an n-vertex graph with no k-connected subgraph has at most $\binom{n}{2} - \frac{(n-k+1)^2-1}{3}$ edges. (Matula [1983])

7.1.33. Prove that G is an even graph if and only if every block of G is Eulerian.

7.1.34. (◇) Let H be the block-cutpoint graph (Remark 7.1.18) of a graph G that has a cut-vertex. (Harary–Prins [1966])

(a) Prove that H is a forest.

(b) Prove that at least two blocks of G contain only one cut-vertex of G.

(c) Prove that G has exactly $k + \sum_{v \in V(G)}(b(v) - 1)$ blocks, where k is the number of components of G and $b(v)$ is the number of blocks containing v.

(d) Prove that every graph has fewer cut-vertices than blocks.

7.1.35. Prove that a graph has no connected induced subgraph with three leaf blocks if and only if it is claw-free and net-free, where the claw is $K_{1,3}$ and the **net** is the graph formed from K_3 and $\overline{K}_3$ by adding a matching joining them. (Kelmans [2006])

7.1.36. The **cyclic edge-connectivity** of a graph is the least number of edges whose deletion leaves a disconnected graph with a cycle in each component. For $m \geq 6$, prove that $C_{m-2} \oplus 2K_1$ is 4-connected and has cyclic edge-connectivity m. (Plummer [1972])

7.2. Properties of k-Connected Graphs

Being k-connected is more restrictive than being connected. When G is connected and $x, y \in V(G)$, there is an x, y-path. When G has k such paths sharing no internal vertices, separating y from x requires deleting at least k vertices. We will show that this obviously sufficient condition for being k-connected is also necessary. When phrased appropriately, these concepts apply also to digraphs.

7.2.1. Definition. A digraph D is **strongly connected** (or **strong**) if it contains a path from each vertex to every other. It is **k-connected** if it has more than k vertices and $D - S$ is strong for any set $S \subseteq V(D)$ with $|S| < k$. The **connectivity** $\kappa(D)$ is the maximum k such that D is k-connected. For $\varnothing \neq S \subset V(D)$, the **edge cut** $[S, \overline{S}]$ is the set of edges from S to $\overline{S}$. A digraph D is **k-edge-connected** if every edge cut has size at least k. The **edge-connectivity** $\kappa'(D)$ is the minimum size of an edge cut.

Note that a graph or digraph G is k-edge-connected if and only if for every nonempty proper vertex subset S, at least k edges leave S.

MENGER'S THEOREM

When studying paths from one specified vertex to another, there is a natural local version of separation, valid for both graphs and digraphs.

7.2.2. Definition. For $xy \notin E(G)$, an x, y-**separating set** is a subset S of $V(G) - \{x, y\}$ such that $G - S$ has no x, y-path. We write $\kappa(x, y)$ for the minimum size of such a set. Paths from x to y are **independent** if they share no internal vertex. We write $\lambda(x, y)$ for the maximum size of a set of pairwise independent x, y-paths.

Every x, y-separating set is at least as large as every family of independent x, y-paths, so always $\kappa(x, y) \geq \lambda(x, y)$. Thus like vertex cover and matching in Chapter 6, separation and path-packing are dual optimization problems (see also Section 8.3 for further discussion of duality). Again, a guarantee of equality for the optima of a pair of dual problems is called a **min-max relation**, guaranteeing that an optimal solution to an instance of one problem can be proved optimal by exhibiting a solution to the other problem. Min-max relations also often lead to polynomial-time algorithms to find optimal solutions.

Menger [1927] proved $\kappa(x, y) = \lambda(x, y)$ for nonadjacent vertices x and y in a graph G. Separating G requires separating some two vertices, so $\kappa(G)$ equals $\min\{\kappa(x, y): xy \notin E(G)\}$, which yields a global duality. The global version and analogues for edge-connectivity and digraphs were observed by others, but all eight variants are called **Menger's Theorem**. We start with a similar result.

7.2.3. Definition. For $X, Y \subseteq V(G)$, an X, Y-**path** is a path from X to Y visiting X and Y only at its endpoints. An X, Y-**barrier** is a vertex set Z such that $G - Z$ has no X, Y-path. An X, Y-**link** is a set of pairwise disjoint X, Y-paths.

We use "barrier" instead of "cut" since an X, Y-barrier may intersect X and Y; in particular, X and Y themselves are X, Y-barriers.

7.2.4. Theorem. (**Pym's Theorem**; Pym [1969]) In a graph or digraph G, the minimum size of an X, Y-barrier is the maximum size of an X, Y-link.

Proof: We use induction on $|V(G)| + |E(G)|$; the claim is trivial for small graphs. Let k be the minimum size of an X, Y-barrier in G; we may assume $k \geq 1$. For the induction step, we consider two cases (illustrated on the left and right below).

Case 1. G has an X, Y-barrier Z of size k other than X or Y. Since $G - Z$ has no X, Y-path, every X, Y-path visits Z. Hence no X, Z-path and Z, Y-path share a vertex outside Z. Let G_1 and G_2 be the subgraphs of G induced by the vertices on X, Z-paths and on Z, Y-paths, respectively; we have $V(G_1) \cap V(G_2) = Z$. Since Z is not X or Y and $k \geq 1$, both G_1 and G_2 are smaller than G.

Since every X, Y-path visits Z, every X, Z-barrier is an X, Y-barrier. Hence G_1 has no X, Z-barrier of size $k - 1$. By the induction hypothesis, in G_1 there is an X, Z-link of size k. Similarly, G_2 has a Z, Y-link of size k. They meet at Z to form an X, Y-link of size k.

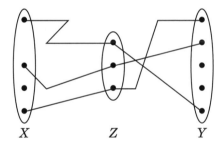 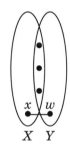

$X \qquad\qquad Z \qquad\qquad Y \qquad\qquad X \; Y$

Case 2. G has no X, Y-barrier of size k other than X and/or Y. By symmetry, we may assume $|X| = k$. If $X \subseteq Y$, then the link consists of paths of length 0. Otherwise, there exists $x \in X - Y$. Since X is a smallest X, Y-barrier, the graph $G - (X - \{x\})$ has an X, Y-path; it begins along some edge xw with $w \notin X$. Let Z' be a smallest X, Y-barrier in $G - xw$.

If $G - xw$ has an X, Y-link of size k, then it is an X, Y-link of size k in G. Otherwise, by the induction hypothesis, $|Z'| < k$. Now Z' is not an X, Y-barrier in G, so $G - Z'$ has an X, Y-path. Every such path must use the edge xw. Hence $Z' \cup \{x\}$ and $Z' \cup \{w\}$ are X, Y-barriers in G. Since these sets have size at most k, the hypothesis for Case 2 yields $Z' \cup \{x\} = X$ and $Z' \cup \{w\} = Y$. Now the 1-vertex paths in Z' together with the x, w-path of length 1 form the desired X, Y-link. ∎

The analogous duality holds for edge-connectivity. Given k pairwise edge-disjoint x, y-paths, we must delete at least k edges to make y unreachable from x. Appropriate notation and the "line graph" operation yields the min-max relations for edges from those for vertices as corollaries of Pym's Theorem.

7.2.5. Definition. Let $\kappa'(x, y)$ be the minimum number of edges whose deletion makes y unreachable from x, and let $\lambda'(x, y)$ be the maximum size of a set of pairwise edge-disjoint x, y-paths.

The **line graph** of a (multi)graph G, written $L(G)$, is the graph whose vertices are the edges of G, with $ef \in E(L(G))$ when $e, f \in E(G)$ with $e = uv$ and $f = vw$. The **line digraph** of a digraph is defined in the same way with edges being ordered pairs.

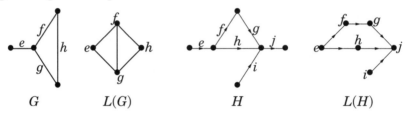

$$G \qquad\qquad L(G) \qquad\qquad H \qquad\qquad L(H)$$

The line graph operation motivates our use of the "prime" for related parameters involving vertices or edges. We have $\alpha'(G) = \alpha(L(G))$, and the next proof shows that $\kappa'(G)$ is closely related to $\kappa(L(G))$.

7.2.6. Theorem. (Menger [1927], Ford–Fulkerson [1956]) If x and y are vertices in a (multi)graph or digraph G, then $\kappa(x, y) = \lambda(x, y)$ when $xy \notin E(G)$, and $\kappa'(x, y) = \lambda'(x, y)$ always.

Proof: Pym's Theorem yields the first statement, as follows. For graphs, use $X = N(x)$ and $Y = N(y)$, and add edges from x to X and from Y to y to complete the desired paths. For digraphs, use $X = N^+(x)$ and $Y = N^-(y)$.

For the second statement, apply Pym's Theorem to the line graph of G. Let X be the set of edges incident to [leaving] x, and let Y be the set of edges incident to [entering] y. The X, Y-link of size $\kappa'(x, y)$ in $L(G)$ transforms into k pairwise edge-disjoint x, y-paths in G.

These edge statements extend without change to multi[di]graphs. Under the line graph operation, multiedges become distinct vertices, and the application of Pym's Theorem is unchanged. ∎

The resulting characterization of k-connected graphs (the "global" version of Menger's Theorem) was observed by Whitney [1932a]. The global versions for edges and digraphs appeared in Ford–Fulkerson [1956]. For vertices we need a lemma (valid also for digraphs; see Exercise 8).

7.2.7. Lemma. Deletion of an edge reduces connectivity by at most 1.

Proof: If S is a smallest separating set of $G - xy$, then $S \cup \{x\}$ or $S \cup \{y\}$ separates G (yielding $\kappa(G) \le \kappa(G - xy) + 1$), unless x and y are the only vertices of $G - S$. In that case $|S| = |V(G)| - 2 \ge \kappa(G) - 1$, so again $\kappa(G) \le \kappa(G - xy) + 1$. ∎

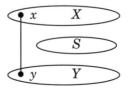

7.2.8. Theorem. (**Menger's Theorem**; Whitney [1932a], Ford–Fulkerson [1956])
For a graph or a digraph G,
(1) G is k-connected if and only if $\lambda(x,y) \geq k$ for all $x,y \in V(G)$, and
(2) G is k-edge-connected if and only if $\lambda'(x,y) \geq k$ for all $x,y \in V(G)$.

Proof: Since $\kappa'(G) = \min_{x,y \in V(G)} \kappa'(x,y)$, Theorem 7.2.6 yields (2).

For $\kappa(G)$, Theorem 7.2.6 yields $\kappa(x,y) = \lambda(x,y)$ for $xy \notin E(G)$, and $\kappa(G)$ is the least of these values. We need only show that $\lambda(x,y) \geq \kappa(G)$ for $xy \in E(G)$. Since xy forms an x,y-path and lies in no other x,y-path,

$$\lambda_G(x,y) = 1 + \lambda_{G-xy}(x,y) = 1 + \kappa_{G-xy}(x,y) \geq 1 + \kappa(G-xy) \geq \kappa(G). \qquad \blacksquare$$

APPLICATIONS OF MENGER'S THEOREM

Our first application is a short proof of a useful fact.

7.2.9. Corollary. $\kappa(G) = \kappa'(G)$ when $\Delta(G) = 3$.

Proof: Always $\kappa(G) \leq \kappa'(G)$, so it suffices to show for $x,y \in V(G)$ that G has $\kappa'(G)$ independent x,y-paths. By Menger's Theorem, G has $\kappa'(G)$ edge-disjoint x,y-paths. Since $\Delta(G) = 3$, two such paths share no internal vertices, so the $\kappa'(G)$ paths *are* independent. $\qquad \blacksquare$

The next lemma gives another way (besides vertex k-splits) to enlarge a graph while preserving k-connectedness.

7.2.10. Lemma. (**Expansion Lemma**) If G' is formed from a k-connected graph G by adding a vertex y with at least k neighbors in G, then G' is k-connected.

Proof: If $|S| < k$, then $G - S$ is connected, and a neighbor of y remains, so by transitivity $G' - S$ is connected. $\qquad \blacksquare$

Analogous conditions yield expansion lemmas for digraphs and for edge-connectivity of multigraphs; see Exercise 23.

7.2.11. Definition. For $x \in V(G)$ and $U \subseteq V(G)$, an x,U-**fan** of size k is a set of k paths from x to U that pairwise share only x and reach U only at their ends.

7.2.12. Lemma. (**Fan Lemma**; Dirac [1960]) A graph with more than k vertices is k-connected if and only if it has an x,U-fan of size k for each choice of x and U with $|U| \geq k$ and $x \notin U$.

Proof: Let G be k-connected, and construct G' from G by adding a new vertex y adjacent to all of U. Since G is k-connected, the Expansion Lemma implies that G' is k-connected, and by Menger's Theorem there is in G' a set of k independent x,y-paths. Stopping these paths where they reach U produces an x,U-fan of size k in G, as illustrated below.

If G is not k-connected, then G has a separating $(k-1)$-set S (Exercise 7.1.1). Choose x and y in distinct components of $G - S$, and let $U = S \cup \{y\}$. There is no x,U-fan, since every x,y-path intersects S. $\qquad \blacksquare$

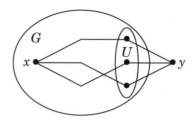

Exercises 28–29 generalize the Fan Lemma. We give an application.

7.2.13. Theorem. (Dirac [1960]) When $k \geq 2$, in any k-connected graph G each set of k vertices lies on some cycle.

Proof: We use induction on k. When $k = 2$, two independent x, y-paths guaranteed by Menger's Theorem form a cycle containing $\{x, y\}$.

For $k > 2$, observe that G is also $(k-1)$-connected, and choose $x \in S$. By the induction hypothesis, G has a cycle C through $S - \{x\}$. If $|V(C)| = k - 1$, then consider an $x, V(C)$-fan of size $k - 1$. Using two successive such paths to form a detour through x completes the desired cycle.

If $|V(C)| \geq k$ and $x \notin V(C)$, then use an $x, V(C)$-fan of size k. Partition $V(C)$ into $k-1$ segments, each starting at an element of $S - \{x\}$. Since the fan has size k, two paths reach C in the same segment (by the Pigeonhole Principle). Detour from C along them to visit x between two vertices of $S - \{x\}$. ∎

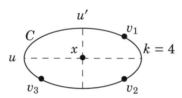

When $k \geq 3$, the converse does not hold, as shown when G is a cycle. Bondy–Lovász [1981] proved the stronger result that for $k \geq 3$ in a k-connected graph every set of k vertices lies on an even cycle.

To apply Menger's Theorem, model a problem by defining a graph or digraph whose paths yield the desired objects. Exercise 31 requests such proofs of Hall's Theorem and the König–Egerváry Theorem. (One can also prove Menger's Theorem from those results, as König did to fill a gap in Menger's original proof.)

Ford and Fulkerson solved a more general problem. Recall that an SDR for a family $A_1, \ldots, A_m$ of sets consists of distinct elements $z_1, \ldots, z_m$ with $z_i \in A_i$.

7.2.14. Definition. Let **A** and **B** be two families of sets, each of size m. A **common system of distinct representatives (CSDR)** is a set of m elements that form an SDR for **A** and also for **B**.

7.2.15. Theorem. (Ford–Fulkerson [1958]) Let **A** and **B** be families of m sets: $\mathbf{A} = \{A_1, \ldots, A_m\}$ and $\mathbf{B} = \{B_1, \ldots, B_m\}$. Using the notation $A(I) = \bigcup_{i \in I} A_i$ and $B(J) = \bigcup_{j \in J} B_j$, families **A** and **B** have a (CSDR) if and only if

$$\left| A(I) \cap B(J) \right| \geq |I| + |J| - m \quad \text{for each pair } I, J \subseteq [m].$$

Proof: We prove that the condition of satisfying the equivalent inequality

$$\left| A(I) \cap B(J) \right| + (m - |I|) + (m - |J|) \geq m \qquad (*)$$

for all $I, J \subseteq [m]$ is both necessary and sufficient for the existence of a CSDR.

Create a digraph G with special vertices s and t and vertex sets $\mathbf{A'}$, $\mathbf{B'}$, and X, where $\mathbf{A'} = \{a_i \colon A_i \in \mathbf{A}\}$, $\mathbf{B'} = \{b_j \colon B_j \in \mathbf{B}\}$, and $X = \left(\bigcup A_i \right) \cup \left(\bigcup B_j \right)$. As illustrated below, the edges are

$$\begin{array}{cc} \{sa_i \colon A_i \in \mathbf{A}\}, & \{a_i x \colon x \in A_i\}, \\ \{b_j t \colon B_j \in \mathbf{B}\}, & \{x b_j \colon x \in B_j\}. \end{array}$$

An s, t-path is $\langle s, a_i, x, b_j, t \rangle$ for $x \in A_i \cap B_j$. Thus $\mathbf{A}$ and $\mathbf{B}$ have a CSDR if and only if G has m independent s, t-paths. By Menger's Theorem, m independent s, t-paths exist if and only if every s, t-separating set has size at least m.

Let R be a minimal s, t-separating set. Let $I = \{i \in [m] \colon a_i \notin R\}$ and $J = \{j \in [m] \colon b_j \notin R\}$. Since the vertices of $\mathbf{A'}$ and $\mathbf{B'}$ indexed by I and J are not being deleted, R will be an s, t-cut if and only if $A(I) \cap B(J) \subseteq R$, and equality will hold when R is minimal. In that case,

$$|R| = \left| A(I) \cap B(J) \right| + (m - |I|) + (m - |J|).$$

Therefore, requiring $(*)$ for all $I, J \subseteq [m]$ is both necessary and sufficient for each s, t-separating set to have size at least m. ∎

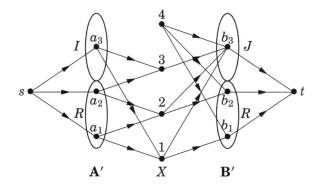

7.2.16. Example. *Digraph for CSDR.* In the figure above, the elements are $\{1, 2, 3, 4\}$, with $\mathbf{A} = \{12, 23, 31\}$ and $\mathbf{B} = \{14, 24, 1234\}$. When R contains $\{a_1, a_2, b_1, b_2\}$ but not a_3 or b_3, we set $I = J = \{3\}$. Now R is an s, t-cut if and only if it also contains $\{1, 3\}$, which is $\left(\bigcup_{i \in I} A_i \right) \cap \left(\bigcup_{j \in J} B_j \right)$. ∎

Like many min-max relations, Menger's Theorem follows from the Max-Flow Min-Cut Theorem of network flow theory. Conversely, Menger's Theorem implies both the flow theorem (for rational capacities) and most of its combinatorial applications. For example, Theorem 7.2.15 above is usually proved in the language of network flows, but Menger's Theorem suffices. We will apply Theorem 7.2.15 in Chapter 12.

2-CONNECTED AND 3-CONNECTED GRAPHS

Menger's Theorem characterizes k-connected graphs but does not say how to construct them. Construction or decomposition procedures for a class of graphs facilitate the development of iterative algorithms and inductive proofs. For graphs that are 2-connected or 3-connected, such procedures exist using vertex splits. First we introduce an inverse operation.

7.2.17. Definition. Given an edge e in a graph G, the **contraction** $G \cdot e$ is the graph obtained from G by replacing the endpoints of e with a single vertex whose neighbors are their neighbors in G outside e. We say that $G \cdot e$ is obtained from G by **contracting** e. A k-**contractible edge** is an edge whose contraction leaves a k-connected graph.

The endpoints of an edge e may have common neighbors. When discussing multigraphs, we may let these yield multiedges with the new vertex in $G \cdot e$. Here we consider only graphs and do not introduce multiedges. Thus contracting the edge $x_1 x_2$ introduced by splitting vertex x inverts the splitting operation.

An easy construction procedure for 3-connected graphs results from finding 3-contractible edges.

7.2.18. Lemma. (Contraction Lemma; Tutte [1961]) Every 3-connected graph other than K_4 has a 3-contractible edge.

Proof: (Thomassen [1980a]) Let G be a 3-connected graph with at least five vertices. If G has no contractible edge, then for each edge e in G the graph $G \cdot e$ has a separating 2-set. This 2-set must consist of the vertex formed by contracting e and some other vertex z, which we call a *companion* of e. If $e = xy$, then $\{x, y, z\}$ is a separating triple in G.

Among all the edges of G, choose xy and a companion z of xy so that the resulting disconnected graph $G - \{x, y, z\}$ has a component H with the largest possible number of vertices. Let H' be another component of $G - \{x, y, z\}$, as sketched below. Since $\{x, y, z\}$ is a minimal separating set, each element of $\{x, y, z\}$ has a neighbor in both H and H'.

Let u be a neighbor of z in H'. Since $uz \in E(G)$, we are guaranteed a separating set $\{z, u, v\}$ of G. The subgraph of G induced by $V(H) \cup \{x, y\}$ is connected. Deleting v from this subgraph, if it occurs there, cannot disconnect it, since $\{z, v\}$ would be a separating set of G. Therefore $G[V(H) \cup \{x, y\}] - v$ lies in a component of $G - \{z, u, v\}$ with more vertices than H, which contradicts the choice of $\{x, y, z\}$. Hence G does have a contractible edge.   ∎

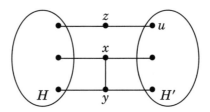

We use Lemma 7.2.18 in characterizing planar graphs in Chapter 9.

7.2.19. Theorem. A graph is 3-connected if and only if it can be obtained from K_4 by a sequence of vertex 3-splits.

Proof: By Lemma 7.1.6, every graph so constructed is 3-connected. By Lemma 7.2.18, every 3-connected graph can be obtained. ∎

For $k > 3$, finding k-contractible edges is harder. In fact, when k is even, the k-connected graph $H_{k,n}$ of Example 7.1.3 has no k-contractible edge (Exercise 33). Thomassen [1981b] proved that every triangle-free k-connected graph has a k-contractible edge (Exercise 34).

A more traditional and harder characterization of 3-connected graphs is due to Tutte [1961]. We use only disjoint 3-splits, where a **disjoint k-split** requires the neighborhoods of the new vertices to be disjoint. Inverting a disjoint 3-split means finding an edge *not on a triangle* to contract. Tutte proved that the 3-connected graphs are the graphs that arise by disjoint 3-splits and edge additions from graphs of the form $K_1 \oplus C_{n-1}$, called **wheels**.

Similarly, the 2-connected graphs can be obtained from C_3 by disjoint 2-splits and edge additions, but there is a more useful way to describe the construction procedure. We start with a special case of disjoint 2-split.

7.2.20. Definition. In a graph G, **subdivision** of an edge uv is the operation of replacing uv with a path u, w, v through a new vertex w.

If $d(v) \geq 2$, then subdivision of uv is equivalent to a vertex 2-split at v where one new vertex inherits only u and the other inherits all other neighbors of v. By Lemma 7.1.6, subdivision therefore preserves 2-connectedness. We also want to apply it to 2-edge-connected graphs.

7.2.21. Lemma. Edge-subdivision preserves 2-edge-connectedness.

Proof: Let G be a 2-edge-connected graph, and let H be obtained from G by adding w to subdivide uv. Since cut-edges are those in no cycle, every edge of G lies in a cycle. The same edge in H lies in the corresponding cycle, lengthened by the replacement of uv with uw and wv if it used uv. Similarly, uw and wv lie in a cycle corresponding to a cycle through uv.

Also, lengthening paths of G (if needed) shows that H is connected. Since H is connected and has no cut-edge, it is 2-edge-connected. ∎

The construction procedure for 2-connected or 2-edge-connected graphs assembles them from a decomposition into cycles and paths.

7.2.22. Definition. An **ear** in a graph G is a path whose internal vertices have degree 2 in G and whose endpoints have degree greater than 2. An **ear decomposition** of G is a decomposition $Q_k, \ldots, Q_0$ such that Q_0 is a cycle and Q_i for $i \geq 1$ is an ear in $Q_0 \cup \cdots \cup Q_i$.

A **closed ear** in a graph G is a cycle whose vertices have degree 2 in G

except for one of degree at least 4. A **weak ear decomposition** of G is a decomposition $Q_k, \ldots, Q_0$ such that Q_0 is a cycle and Q_i for $i \geq 1$ is either an ear or a closed ear in $Q_0 \cup \cdots \cup Q_i$.

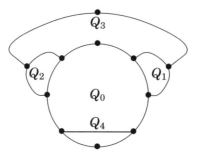

Weak ear decompositions are also called "closed-ear decompositions". We use "weak" to emphasize that the requirements for the decomposition are relaxed; each added Q_i may now be an ordinary ear or a closed ear.

7.2.23. Theorem. (Whitney [1932a]) A graph is 2-connected if and only if it has an ear decomposition. A multigraph is 2-edge-connected if and only if it has a weak ear decomposition. Furthermore, when such decompositions exist they can start with any cycle.

Proof: *Sufficiency.* Cycles are 2-connected, so it suffices to show that adding an ear or closed ear preserves the desired properties. Ear addition can be accomplished by adding an edge (or a new vertex with two neighbors) and then possibly subdividing. These operations preserve 2-connectedness and 2-edge-connectedness.

When adding a closed ear to a 2-edge-connected graph, the properties of being connected and having every edge on a cycle are maintained.

Necessity. Given a 2-connected graph G, we build an ear decomposition of G from any cycle C. Let $G_0 = C$. Consider a subgraph G_i built by adding ears. If $G_i \neq G$, then we may select an edge $uv \in E(G) - E(G_i)$ with $u \in V(G_i)$ (since G is connected). Since G is 2-connected, there is a path P in $G - u$ from v to $V(G_i)$ (it has length 0 if $v \in V(G_i)$). Now $\{uv\} \cup P$ is an ear that can be added to G_i. Repeating the argument absorbs all of G.

For a weak ear decomposition of a 2-edge-connected (multi)graph, the proof is the same, except that we find a path in $G - uv$ instead of $G - u$. If the resulting path ends at u, then we obtain a closed ear addition; otherwise we have an ear addition as before.   ∎

Weak ear decompositions yield a short solution of the "one-way street" problem. When can the streets of a road network all be made one-way and still permit each location to reach every other? A **strong orientation** of a graph is an orientation that is a strongly connected digraph (Definition 7.2.1).

7.2.24. Corollary. (Robbins [1939]) A multigraph has a strong orientation if and only if it is 2-edge-connected.

Proof: One cannot travel both directions across an edge cut having fewer than two edges. Hence the condition is necessary.

For sufficiency, use a weak ear decomposition. Orient the initial cycle consistently. Orient each subsequent ear or closed ear Q from one end (u) to the other (v). By transitivity, the new digraph is strong: old vertices can reach u, which can reach all new vertices, and new vertices can reach v, which can reach all old vertices. After all of G is added, we have a strong orientation.   ∎

Similarly, an obvious necessary condition for a k-edge-connected orientation of G is $\kappa'(G) \geq 2k$. We next generalize Corollary 7.2.24 to prove that this is sufficient. The proof of the generalization is much more difficult.

HIGHLY CONNECTED ORIENTATIONS (optional)

In a k-edge-connected orientation of G, every edge cut needs at least k edges in each direction, so G must be $2k$-edge-connected. Nash-Williams [1960] generalized Robbins' Theorem by proving this obvious necessary condition sufficient.

It is easy for Eulerian multigraphs. Orienting consistently along an Eulerian circuit crosses each cut the same number of times in each direction. If each edge cut has size at least $2k$, the orientation is then k-edge-connected. For the general result, Nash-Williams showed that the vertices of odd degree can be paired by paths so that an Eulerian circuit of the rest gives the desired orientation.

Lovász simplified the proof that $2k$-edge-connected multigraphs have k-edge-connected orientations. Our presentation follows the survey by Frank [1993].

7.2.25. Definition. (Lovász) In a multigraph or digraph G, with $X \subseteq V(G)$, let $\delta_G(X)$ or $\delta(X)$ denote the number of edges of G leaving X, so $\delta(X) = \left|[X, \overline{X}]\right|$.

Note that $\delta(X) < \sum_{v \in X} d(v)$ when X is not an independent set. The $\delta(X)$ notation makes it easy to compare the sizes of related edge cuts.

7.2.26. Proposition. Given a graph G and $X, Y \subseteq V(G)$,
  (a) $\delta(X) + \delta(Y) = \delta(X \cap Y) + \delta(X \cup Y) + 2\left|[X - Y, Y - X]\right|$.
  (b) $\delta(X) + \delta(Y) = \delta(X - Y) + \delta(Y - X) + 2\left|[X \cap Y, \overline{X \cup Y}]\right|$.

Proof: Each equation counts a set in two ways. In terms of the multiplicities shown in the figure below, the equations are

$$(b + e + f + c) + (a + e + f + d) = (a + b + e) + (c + e + d) + 2f$$
$$(b + e + f + c) + (a + e + f + d) = (a + c + f) + (b + d + f) + 2e$$ ∎

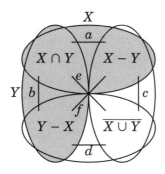

A function r on sets is **submodular** if $r(X \cap Y) + r(X \cup Y) \le r(X) + r(Y)$ for all sets X and Y (see Chapter 11). Proposition 7.2.26a implies that δ is submodular. Note that G is k-edge-connected if and only if $\delta(X) \ge k$ whenever $\varnothing \ne X \subset V(G)$.

7.2.27. Definition. A multigraph G is k-**edge-connected relative to** a vertex z if $\delta(X) \ge k$ for all $\varnothing \ne X \subset V(G) - \{z\}$. If G has edges uz and zv, then the u, v-**shortcut of** z is the graph $G' = G - uz - zv + uv$.

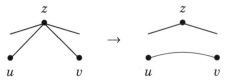

A u, v-shortcut of z may have uv as a multiedge. Multiedges cause no trouble when discussing edge-connectivity (we ignore loops). The content of "relative to" in this definition is to ignore the condition for the cut between z and $V(G) - \{z\}$.

Lovász proved that when z has even degree there is a shortcut of z that preserves the edge-connectivity relative to z. Mader [1978, 1982] proved stronger shortcut results. The proof of Lovász's weaker version is more difficult when k is odd. Fortunately, we only need the even case.

7.2.28. Lemma. (Shortcut Lemma; Lovász [1974, 1979]**)** If z is a vertex of even degree in a multigraph G that is k-edge-connected relative to z, then for all $u \in N(z)$ there exists $v \in N(z)$ such that the u, v-shortcut of z is also k-edge-connected relative to z.

Proof: (for even k only). Let u be a fixed neighbor of z, and let $V' = V(G) - \{z\}$. Let X be a nonempty proper subset of V', and consider $v \in N(z) - \{u\}$. Among the edges leaving X, the u, v-shortcut of z destroys uz or vz if X contains u or v, and it adds uv if X contains exactly one of $\{u, v\}$. Thus the shortcut violates $\delta(X) \ge k$ only if $u, v \in X$ and $\delta(X) \le k + 1$.

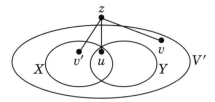

We call X **dangerous** if X is a nonempty proper subset of V' that contains u and satisfies $\delta(X) \le k + 1$. A u, v-shortcut of z is k-edge-connected relative to z if and only if every dangerous set omits v. Letting M be the union of all dangerous sets, it thus suffices to show that

$$M \text{ is dangerous and omits some neighbor of } z.$$

To show that M is dangerous, it suffices to show that the union of any two dangerous sets X and Y is dangerous. We may assume that $X - Y$ and $Y - X$ are nonempty. We prove first that $\delta(X \cup Y)$ is odd and then use this to prove that $X \cup Y$ is a proper subset of V' and satisfies $\delta(X \cup Y) \le k + 1$.

Since X and Y are dangerous, $\delta(X), \delta(Y) \le k + 1$. By the definition of dangerous, $u \in X \cap Y$, so $uz \in [X \cap Y, \overline{X \cup Y}]$. The hypothesis on relative edge-connectivity yields $\delta(X - Y) \ge k$ and $\delta(Y - X) \ge k$. We use Proposition 7.2.26b to rewrite $\delta(X) + \delta(Y)$ in terms of $X - Y$ and $Y - X$:

$$2k + 2 \ge \delta(X) + \delta(Y) = \delta(X - Y) + \delta(Y - X) + 2\big|[X \cap Y, \overline{X \cup Y}]\big| \ge 2k + 2.$$

Equality must hold throughout, and hence $\delta(X) = \delta(Y) = k + 1$ and $\delta(X - Y) = \delta(Y - X) = k$. Now, since $\delta(Y) + \delta(X - Y) \equiv \delta(X \cup Y) \pmod 2$, we conclude that $\delta(X \cup Y)$ is odd. Also, $\delta(X \cap Y) + \delta(X - Y) \equiv \delta(X) \pmod 2$, so $\delta(X \cap Y)$ is odd.

Since $\delta(X \cup Y)$ is odd and $\delta(V') = d(z)$, evenness of $d(z)$ yields $X \cup Y \ne V'$. Now relative edge-connectivity, grouping of the edges leaving X and leaving Y, and the dangerousness of X and Y yield

$$k + k \le \delta(X \cap Y) + \delta(X \cup Y) \le \delta(X) + \delta(Y) \le 2k + 2.$$

Since $\delta(X \cap Y)$ and $\delta(X \cup Y)$ are odd, they must both equal $k + 1$, and hence $X \cup Y$ is dangerous.

We have proved that M is dangerous. If there is no dangerous set, then every u, v-shortcut of z works, so we may assume $M \ne \varnothing$. Since M is a nonempty proper subset of V' and G is k-edge-connected relative to z, we have $\delta(V' - M) \ge k$.

If M contains all neighbors of z, then $\delta(M \cup \{z\}) = \delta(M) - d(z)$. Since $d(z)$ is a positive even number and M is dangerous, we then have

$$k \le \delta(V' - M) = \delta(M \cup \{z\}) = \delta(M) - d(z) \le \delta(M) - 2 \le k - 1.$$

The contradiction implies that M omits some $v \in N(z)$, and the u, v-shortcut of z is k-edge-connected relative to z. ∎

To prove the Orientation Theorem from the Shortcut Lemma inductively, we need a vertex of even degree. With $k = 2t$ and seeking a t-edge-connected orientation, we will discard edges from our k-edge-connected multigraph to obtain a **minimal k-edge-connected multigraph**, one where the deletion of any edge destroys k-edge-connectedness. Submodularity of the degree function now makes it fairly easy to obtain a vertex with degree k.

7.2.29. Lemma. (Mader [1971]) Every minimal k-edge-connected multigraph has a vertex of degree k.

Proof: If $\delta(X) > k$ for all $\varnothing \ne X \subset V(G)$, then deleting any edge leaves a k-edge-connected multigraph. Thus $\delta(X) = k$ for some set X.

Suppose that $G[X]$ has an edge xy. Since $G - xy$ is not k-edge-connected, there is a set $Z \subset V(G)$, with Z containing exactly one of the vertices x and y, such that $k - 1 \ge \delta_{G-xy}(Z) = \delta(Z) - 1$. Since $\delta(Z) \ge k$, equality holds.

Now k-edge-connectedness of G and submodularity of δ yield

$$k + k \le \delta(X \cap Z) + \delta(X \cup Z) \le \delta(X) + \delta(Z) = k + k.$$

Since G is k-edge-connected, we obtain $\delta(X \cap Z) = k$. Since Z contains exactly one of $\{x, y\}$, the set $X \cap Z$ is smaller than X.

Hence a minimal set X such that $\delta(X) = k$ must be an independent set. Since each vertex of X has at least k incident edges leaving X, we have $|X| = 1$, and this is the desired vertex of degree k. ∎

7.2.30. Theorem. (**Orientation Theorem**; Nash-Williams [1960]) An n-vertex multigraph has a t-edge-connected orientation if and only if it is $2t$-edge-connected.

Proof: (Lovász [1974, 1979], see Frank [1993]). *Necessity.* We need t edges each way across each cut. *Sufficiency.* We use induction on n. For $n = 2$, the two vertices are joined by at least $2t$ edges.

For $n > 2$, consider a $2t$-edge-connected multigraph G. We discard edges to obtain a minimal $2t$-edge-connected multigraph; we later orient the deleted edges arbitrarily. By Lemma 7.2.29, what remains has a vertex z of degree $2t$. Lovász's Shortcut Lemma iteratively finds shortcuts of z until we reduce the degree of z to 0. This process maintains $2t$-edge-connectedness relative to z. At the end, deleting z yields a $2t$-edge-connected multigraph G' with $n - 1$ vertices.

By the induction hypothesis, G' has a t-edge-connected orientation. Orient G by replacing each shortcut edge uv with $\{uz, zv\}$ or $\{vz, zu\}$, agreeing with the orientation of uv. For $X \neq \{z\}$, lifting uv preserves $\delta(X) \geq t$ in the orientation; the only edge lost is uv, and if uv leaves X, then uz or zv is a new edge leaving X, depending on whether $z \in X$. After t lifts, also $\delta(\{z\}) = t$, so now $\delta(X) \geq t$ whenever $\varnothing \neq X \subset V(G)$. ∎

Mader [1978] used his Shortcut Lemma to give another proof of the Strong Orientation Theorem. Other proofs of the Orientation Theorem use polyhedral combinatorics (see Frank [1980a], Frank–Tardos [1984]).

We close this section with the analogue of Lemma 7.2.29 for vertices. A **minimal k-connected graph** is a graph from which the deletion of any edge destroys k-connectedness. Halin [1969, 1971] proved that every such graph has a vertex of degree k. Mader strengthened Halin's result by showing that every cycle contains a vertex of degree k. We need a technical lemma.

7.2.31. Lemma. (Mader [1972]). Let ax and ay be edges in a minimal k-connected graph G, with $d_G(a) \geq k+1 \geq 3$. If S is a $(k-1)$-set separating $G - ax$, and T is a $(k-1)$-set separating $G - ay$, then the component of $G - ay - T$ containing y has fewer vertices than the component of $G - ax - S$ containing a.

Proof: The graph $G - ax - S$ has two components, with vertex sets A containing a and X containing x. Also $G - ay - T$ has two components, with vertex sets B containing a and Y containing y (see figure below). Note that $x \notin Y$ and $y \notin X$.

The claim is $|Y| < |A|$. Since Y is partitioned by its intersections with the sets A, S, X, and since $A \cap B$ is nonempty (it contains a), it suffices to show (1) $|Y \cap S| \leq |A \cap T|$, and (2) $Y \cap X = \varnothing$.

(1) Let $U = (S - Y) \cup (T - X)$. If $|A \cap T| < |Y \cap S|$, then $|U| < |S| = k - 1$. Since $d(a) \geq k+1$, the vertex a has a neighbor z outside $U \cup \{x, y\}$. All neighbors of a not in $U \cup \{x, y\}$ lie in $A \cap B$. Since a is the only vertex of $A \cap B$ having a neighbor (in G) in $X \cup Y$, the set $U \cup \{a\}$ separates z from $X \cup Y$. This contradicts the k-connectedness of G, since $|U \cup \{a\}| \leq k - 1$, and hence $|Y \cap S| \leq |A \cap T|$.

(2) Let $D = X \cap Y$, and suppose $D \neq \varnothing$. Let W be the set of vertices outside D having neighbors in D. Since $x \notin Y$ and $y \notin X$, no vertex of W is in A or in B. Hence $W \subseteq (S \cap Y) \cup (T - A)$. From (1), this yields $|W| \leq |T| = k - 1$. This is impossible, since W is a separating set and G is k-connected, so $D = \varnothing$. ∎

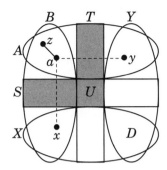

7.2.32. Theorem. (Mader [1972]). Every cycle in a minimal k-connected graph contains a vertex of degree k.

Proof: Let $[a_1, \ldots, a_l]$ be a cycle with no vertex of degree k, and let S_i be a separating $(k-1)$-set of $G - a_{i-1}a_i$, where a_0 denotes a_l. Let n_i be the order of the component of $G - a_{i-1}a_i - S_i$ containing a_i. By applying Lemma 7.2.31 with $a_i = a$, $S_i = S$, and $S_{i+1} = T$, we obtain $n_i > n_{i+1}$. Iteration yields the contradiction $n_1 > n_2 > \cdots > n_l = n_0 > n_1$. ∎

Via a simple counting argument, Theorem 7.2.32 implies that almost half the vertices of a minimal k-connected graph have degree k.

7.2.33. Corollary. (Bollobás [1978, p. 25]). Every minimal k-connected graph G with n vertices has at least $\frac{(k-1)n+2}{2k-1}$ vertices of degree k.

Proof: Let S be the set of vertices having degree k. By the Degree-Sum Formula, G has at least $\frac{1}{2}(kn + n - |S|)$ edges. By Theorem 7.2.32, $G - S$ is a forest, and hence $G - S$ has at most $n - |S| - 1$ edges. Since deleting S removes at most $k|S|$ edges, $\frac{1}{2}(kn + n - |S|) - k|S| \le n - |S| - 1$. This yields the desired inequality. ∎

Corollary 7.2.33 is almost the best possible bound. Mader [1979] (see Mader [1996], a survey) proved that a minimal k-connected graph with n vertices has at least $\frac{(k-1)n+2k}{2k-1}$ vertices of degree k. Equality holds for infinitely many examples, such as those in Exercise 59.

EXERCISES 7.2

7.2.1. (−) Prove or disprove: If P is a u, v-path in a 2-connected graph G, then G has a u, v-path Q sharing no internal vertices with P.

7.2.2. (−) Let x and y be nonadjacent vertices in a 2-connected graph G. Prove that if $G - x - y$ is connected, then y lies on a cycle in $G - x$. Is this true when $xy \in E(G)$?

7.2.3. (−) From a connected graph G, form G' by adding the edge xy whenever $d_G(x, y) = 2$. Prove that G' is 2-connected.

7.2.4. (−) Prove that a graph G is 2-connected if and only if G can be obtained from C_3 by edge additions and edge subdivisions.

7.2.5. (–) A **thread** in a graph is a path that is maximal subject to the condition that the internal vertices have degree 2. Prove or disprove: The last ear added in an ear decomposition of a 2-connected graph G that is not a cycle can be any thread in G.

7.2.6. (–) Let G be a graph with at least three vertices, none isolated. Prove that G is 2-connected if and only if any two edges appear in a common cycle.

7.2.7. (–) Prove that a graph with at least three vertices is 2-connected if and only if for every 3-tuple (x, y, z) of vertices, the graph has an x, z-path through y. (Chein [1968])

7.2.8. For an edge e in a digraph D, proved $\kappa(D - e) \geq \kappa(D) - 1$.

7.2.9. (–) Prove that a graph G with at least four vertices is 2-connected if and only if for every pair X, Y of disjoint vertex subsets with $|X|, |Y| \geq 2$, there are two disjoint paths from X to Y in G that have no internal vertex in X or Y.

7.2.10. (–) Let v be a vertex in a 3-connected 3-regular graph G. Form G' by expanding v into a triangle whose vertices are joined to the original neighbors of v by a matching. Thus G' is a 3-regular graph. Prove that G' is 3-connected.

7.2.11. (–) Show that Theorem 7.2.13 is best possible by constructing, for all $k \in \mathbb{N}$ with $k \geq 3$, a k-connected graph having $k + 1$ vertices that do not lie on a cycle.

7.2.12. (–) For $k \geq 2$, prove that a graph G with at least $k + 1$ vertices is k-connected if and only if for disjoint $S, T \subseteq V(G)$ with $|S| = k - 2$ and $|T| = 2$, there is a cycle in G that contains T and avoids S. (Lick [1973])

7.2.13. (–) Prove that a connected graph G with more than k vertices is k-connected if and only if $\kappa_G(x, y) \geq 2$ whenever $d(x, y) = 2$. (Li [1994], Naatz [2000])

7.2.14. (–) Use Theorem 7.2.19 to prove that the Petersen graph is 3-connected.

7.2.15. (–) Let H' be a k-connected graph obtained by contracting the edges of a matching in a graph H with $\delta(H) \geq k$. Prove that H is k-connected. (Savage–Zhang [1998])

7.2.16. (–) Let G be a 2-connected n-vertex graph with m edges. Prove that in every ear decomposition of G, the number of ears after the initial cycle is $m - n$.

7.2.17. (–) Let G be a 2-connected graph. Prove that if G has an ear decomposition with initial cycle length at least l where each added ear has length at least $l - 1$, then G has girth at least l. Prove that the converse is not true (provide a counterexample). (Kelmans)

7.2.18. (–) Győri [1978] and Lovász [1977] proved that a graph G with more than k vertices is k-connected if and only if for all distinct $v_1, \ldots, v_k \in V(G)$ and $n_1, \ldots, n_k \in \mathbb{N}$ summing to $|V(G)|$, there is a partition $V_1, \ldots, V_k$ of $V(G)$ such that each V_i has size n_i, contains v_i, and induces a connected subgraph. Show that the condition is sufficient.

7.2.19. Prove or disprove: If G is a 2-connected graph that is not a cycle, then G contains a cycle C such that $G - V(C)$ is connected.

7.2.20. For a graph G with $x, u, v \in V(G)$, prove $\lambda'_G(u, v) \geq \min\{\lambda'_G(x, u), \lambda'_G(x, v)\}$. Does the analogous inequality hold for $\lambda_G(u, v)$?

7.2.21. (◇) The **pinch** operation subdivides k edges of a digraph (each becomes a path of length 2 through a new vertex) and merges the k new vertices into a single vertex. Prove that pinching a k-edge-connected digraph always yields a k-edge-connected digraph.

7.2.22. (◇) Prove that if G is an Eulerian graph, then $\lambda'_G(x, y)$ is even for all $x, y \in V(G)$. (Comment: M. Ghorbani conjectured that the converse is true; this remains open.)

7.2.23. *Expansion Lemmas.*

(a) Let G' be obtained from a k-edge-connected multigraph G by adding a new vertex w with degree at least k (multiedges are allowed). Prove that G' is k-edge-connected.

(b) Prove analogues of part (a) and the Expansion Lemma (Lemma 7.2.10) for digraphs.

7.2.24. ($\diamond$) Let G be a k-edge-connected multigraph, with $k \geq 2$. Prove that G has a closed trail through any k specified vertices.

7.2.25. Fix $j, k \in \mathbb{N}$ with j even and $j \leq k$. Form a k-regular graph G from $2K_{k+1}$ by deleting a matching of size $j/2$ from each component and adding a matching of size j joining the components. Prove $\kappa'(G) = j$. (Kostochka)

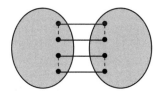

7.2.26. A u, v-**necklace** is a graph consisting of cycles $C_1, \ldots, C_k$ with $u \in C_1$ and $v \in C_k$ such that consecutive cycles share one vertex and non-consecutive cycles are disjoint. Prove that a graph G is k-edge-connected if and only if for all $u, v \in V(G)$ there are k pairwise edge-disjoint paths such that the union of any two is a u, v-necklace. (T. Jiang)

7.2.27. ($\diamond$) Let G be an n-vertex graph with connectivity k and diameter d. Prove both $n \geq k(d-1) + 2$ and $\alpha(G) \geq \lceil (1+d)/2 \rceil$. For $k, d \geq 2$, construct an example such that equality holds in both bounds. (Watkins [1967])

7.2.28. Let G be a k-connected graph. For sets $S, T \subset V(G)$ with size at least k, prove that G has k pairwise disjoint S, T-paths.

7.2.29. ($\diamond$) Let X and Y be disjoint sets of vertices in a k-connected graph G. Let w be a positive integer function on $X \cup Y$ such that $\sum_{x \in X} w(x) = \sum_{y \in Y} w(y) = k$. Prove that there are k independent X, Y-paths such that for each vertex $v \in X \cup Y$, the number of these paths having an endpoint at v is $w(v)$.

7.2.30. Assuming that $\kappa'(x, y) = \lambda'(x, y)$ and $\kappa(x, y) = \lambda(x, y)$ hold for digraphs (the latter when xy is not an edge), derive the same statements for graphs.

7.2.31. ($\diamond$) Prove the König–Egerváry Theorem and Hall's Theorem from Menger's Theorem. Prove Hall's Theorem from the Ford–Fulkerson CSDR Theorem.

7.2.32. ($\diamond$) Fix $k \geq 2$, and let G be a k-connected graph with n vertices.

(a) For $n \geq 2k$, prove that G has a cycle of length at least $2k$.

(b) For $n \geq 3k$, prove that G has a cycle of length at least $3k$ if $C_4 \not\subseteq G$. (Kostochka)

7.2.33. *k-contractible edges* (see Martinov [1982]).

(a) Prove that a k-connected k-regular graph in which every edge lies in a triangle has no k-contractible edge. (Comment: This includes $H_{k,n}$ of Example 7.1.3 for even k.)

(b) The Harary graph $H_{2,n}$ is also denoted C_n^2. For $n \geq 7$, prove that contracting some two edges in C_n^2 leaves C_{n-2}^2 (multiedges after contraction are turned into simple edges).

7.2.34. (+) Prove that every triangle-free k-connected graph has a k-contractible edge. (Thomassen [1981b]) (Comment: Thomassen used this to prove that every $(k+3)$-connected graph has a cycle such that deleting its vertices leaves a k-connected graph.)

7.2.35. Let S be a set of vertices in a graph G such that $\kappa(x, y) \geq k$ for all $x, y \in S$. Prove that if $|S| \leq k + 1$, then the vertices in S lie on a single path in G. Show this cannot be guaranteed when $|S| = k + 2$.

7.2.36. ($\Diamond$) A graph is k-**linked** if for every choice of distinct vertices $s_1, \ldots, s_k$ and $t_1, \ldots, t_k$, there exist disjoint paths $P_1, \ldots, P_k$ such that P_i is an s_i, t_i-path. Prove that every k-linked graph is $(2k - 1)$-connected. The graph below is 5-connected; prove that it is not 2-linked (Watkins [1968]). For general k, construct a $(3k - 3)$-connected graph that is not k-linked. (Comment: Thus $f(k) \geq 3k - 2$, where $f(k)$ is the least j such that every j-connected graph is k-linked. Thomassen [1980b] and Seymour [1980] proved $f(2) = 6$. The best general result is $f(k) \leq 10k$; in fact, $2k$-connected graphs with average degree at least $10k$ are k-linked (Thomas–Wollan [2005]).)

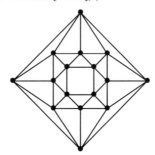

7.2.37. Let G be a k-edge-connected graph. Prove that the line graph $L(G)$ is k-connected and $(2k - 2)$-edge-connected.

7.2.38. Use Tutte's 1-Factor Theorem to prove that every connected line graph with an even number of vertices has a perfect matching. Conclude that every connected graph with an even number of edges decomposes into paths of length 2.

7.2.39. ($\Diamond$) Let $A_1, \ldots, A_m$ and $B_1, \ldots, B_m$ be two partitions of a set E such that all the sets A_i and B_j have the same size. Prove that the two set systems have a common system of distinct representatives. (Ryser [1963, p. 51])

7.2.40. Let G_1 and G_2 be disjoint k-connected graphs with $k \geq 2$. For $i \in \{1, 2\}$, choose $v_i \in V(G_i)$ and let $X_i = N_{G_i}(v_i)$. Let B be an X_1, X_2-bigraph that has no isolated vertex and has a matching of size at least k. Prove that $(G_1 - v_1) \cup (G_2 - v_2) \cup B$ is k-connected.

7.2.41. ($\Diamond$) Let $\beta(j)$ be the least k such that any two vertices in a k-connected graph are joined by a path P such that $G - V(P)$ is j-connected. Prove $\beta(1) = 3$. (Tutte [1961]) (Comment: Lovász conjectured that $\beta(j)$ exists; Chen–Gould–Yu [2003] proved $\beta(2) \leq 5$.)

7.2.42. Let u be a vertex in a 3-connected graph G. Prove that $G - u$ contains a cycle C such that $G - V(C)$ is connected.

7.2.43. Prove that applying the expansion operation of Exercise 5.2.22 to a 3-connected graph yields a 3-connected graph. Obtain the Petersen graph from K_4 by expansions. (Comment: Tutte [1966] proved that all 3-regular 3-connected graphs arise this way.)

7.2.44. Prove that if every edge in a connected graph G is the unique common edge in some pair of cycles, then G is 3-edge-connected. Prove this property for the Petersen graph.

7.2.45. Use induction on the distance between vertices to prove that a graph with at least three vertices is 2-connected if and only if for any two vertices x and y there exist two independent x, y-paths. (Whitney [1932a])

7.2.46. Let G be a graph having no induced subgraph that is C_k for $k > 3$ or K_4^- (five edges). Prove that every block of G is a complete graph. (Lehel)

7.2.47. ($\Diamond$) For a connected graph G with $|V(G)| \geq 3$, prove the following equivalent.
  (A) G is 2-edge-connected.
  (B) Every edge of G appears in a cycle.
  (C) G has a closed trail containing any specified pair of edges.
  (D) G has a closed trail containing any specified pair of vertices.

7.2.48. Let v be a vertex of a 2-connected graph G. Prove that v has a neighbor u such that $G - u - v$ is connected. (Chartrand–Lesniak [1986, p. 51])

7.2.49. Let G be an n-vertex connected graph having a cut-vertex. In terms of n, determine the maximum number of edges that may need to be added to change G into a 2-connected graph. Obtain all n-vertex graphs achieving the maximum.

7.2.50. ($\Diamond$) Let C be a cycle in a 2-connected graph G that is not a cycle. Prove that C contains an ear of G whose deletion leaves a 2-connected subgraph. That is, removal of ears in an ear decomposition of G can start by breaking any cycle.

7.2.51. ($\Diamond$) Let s and t be vertices in a 2-connected graph G. Prove that the vertices of G can be linearly ordered so that each vertex outside $\{s, t\}$ has a neighbor that is earlier in the order and a neighbor that is later in the order. (Hint: Use ear decompositions. Comment: This is called an s, t-**numbering** of G.)

7.2.52. ($\Diamond$) Let G be a 2-connected graph, and fix $r \in V(G)$. Prove that G has two spanning trees such that for every $v \in V(G)$, the r, v-paths in the two trees are independent. (Hint: Use ear decomposition to prove the stronger statement for each r that G has two trees and a labeling of $V(G) - \{r\}$ by real numbers such that in one tree the labels increase along paths from r and in the other they decrease along paths from r.)

  (Comment: Itai and Rodeh conjectured a k-connected graph always has k such trees. This was proved for $k = 3$ in Itai–Zehavi [1989] and for $k = 4$ in Curran–Lee–Yu [2006].)

7.2.53. (+) Alice and Bob play a game on a 2-connected n-vertex graph G. Alice picks vertices u and v. Next Bob orients up to $f(n)$ of the edges. Alice then orients the remaining edges and selects an edge e, which may have been oriented by her or by Bob. If the orientation contains a u, v-path through e, then Bob wins; otherwise, Alice wins. Prove that the least $f(n)$ such that Bob always has a winning strategy is $2n - 3$. (Kerimov [2009])

7.2.54. Prove or disprove: If G is a k-edge-connected graph with minimum degree greater than k, then some two vertices in G are joined by more than k edge-disjoint paths.

7.2.55. Let G be a minimal 2-connected n-vertex graph other than C_n.
  (a) Prove that if P is the last ear in an ear decomposition of G, then the graph G' formed before adding P is a minimal 2-connected graph. (Dirac [1967], Plummer [1968])
  (b) For $n \geq 4$, prove that G has at most $2n - 4$ edges, with equality holding only when $G = K_{2,n-2}$. (Comment: Mader [1972] proved that for $n \geq 3k - 2$, a minimal k-connected n-vertex graph has at most $k(n - k)$ edges, achieved only by $K_{k,n-k}$ when $n \geq 3k - 1$.)

7.2.56. ($\Diamond$) *Minimal 2-connected graphs.* (Dirac [1967], Plummer [1968])
  (a) For a 2-connected graph G, prove that $G - xy$ is 2-connected if and only if x and y lie on a cycle in $G - xy$. Conclude that a 2-connected graph is a minimal 2-connected graph if and only if no cycle has a chord.
  (b) Vertices x and y are **compatible** in a graph G if no x, y-path in G has a chord. For $t \geq 2$, let each of $G_1, \ldots, G_t$ be an edge or a minimal 2-connected graph such that x_i and y_i are compatible vertices in G_i. Form G by merging y_i with x_{i+1} for $1 \leq i < t$ and y_k with x_1. Prove that G is a minimal 2-connected graph, except when $k = 2$ and $K_2 \in \{G_1, G_2\}$. Conversely, prove that every minimal 2-connected graph can be constructed from edges and smaller 2-connected graphs in this way. (Hint: Use part (a).)

7.2.57. Given vertices x and y in a minimal 2-connected graph G such that no x, y-path has a chord, prove that every x, y-path has an internal vertex with degree 2 in G. Use this to improve Corollary 7.2.33 by showing that every minimal 2-connected graph with n vertices has at least $(n + 4)/3$ vertices of degree 2. (Dirac [1967], Plummer [1968])

7.2.58. ($\diamond$) *Applications of Mader's Theorem (Theorem 7.2.32).*

(a) Prove: a minimal k-connected graph has at least k vertices of degree k.

(b) Let S be any k-set in a minimal k-connected graph G, and let C be a component of $G - S$. Prove that C contains a vertex of degree k in G (see Mader [1972]).

(c) Prove that a minimal k-connected graph G has at least $k + 1$ vertices of degree k. (Comment: This improves Corollary 7.2.33 when $|V(G)| \leq 2k$.)

7.2.59. Construct a graph $G_{k,m}$ as follows. Begin with $mK_{k-1,k}$, having $m(k - 1)$ vertices $y_{r,j}$ and mk vertices $x_{r,i}$ such that $y_{r,j}$ and $x_{r,i}$ are adjacent for $i \in [k]$ and $j, r \in [m]$. Also make $x_{r,i}$ and $x_{r+1,i}$ adjacent for $i \in [k]$ and $r \in [m-1]$. Finally, add two vertices z and z' of degree k having neighbors $\{x_{1,i}\}$ and $\{x_{m,i}\}$, respectively. Prove that $G_{k,m}$ is a minimal k-connected graph with $\frac{(k-1)n+2k}{2k-1}$ vertices of degree k. (Bollobás [1978, p. 25], Mader [1979])

7.3. Spanning Cycles

A k-connected graph has a cycle through any k vertices (Theorem 7.2.13), but cycles through larger sets are not guaranteed (Exercise 7.2.11). However, when the minimum degree is large enough in terms of the number of vertices, there is a cycle through all the vertices. We first consider conditions for spanning cycles, and later we discuss "long cycle" versions of these results.

Spanning cycles were introduced by Kirkman in 1855 in a paper on polyhedral graphs. Nevertheless, they are called **Hamiltonian cycles** to honor Sir William Rowan Hamilton (1805–1865). In studying non-commutative algebra, he devised a game on the dodecahedron graph where one player would specify a 5-vertex path and the other would extend it to a spanning cycle. Hamilton sold "The Icosian Game" to a dealer in puzzles, for £25. In another version, "The Traveller's Dodecahedron", the vertices were named for 20 important cities. Neither was commercially successful, but Hamilton's name was associated with the concept.

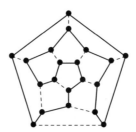

A graph with a spanning cycle is a **Hamiltonian graph**. Until the 1970s, interest in spanning cycles was related to the Four Color Problem (Chapter 9). Later, the **Traveling Salesman Problem (TSP)** became an important computational problem: it asks for the minimum weight of a spanning cycle in an edge-weighted input graph. Testing for a spanning cycle is a special case and is NP-complete, even for 3-regular triangle-free planar graphs or for line graphs.

Survey papers about Hamiltonian graphs and related topics include Bondy [1978a, 1995], Gould [1991, 2003, 2014], Faudree [1996, 2001], Lesniak [1996], Kawarabayashi [2001], Yamashita [2004], Li [2013], Kühn–Osthus [2014], and many others. We will not discuss analogues for directed graphs, surveyed in Bermond–Thomassen [1981] and Zhang–Song [1991].

PROPERTIES OF HAMILTONIAN GRAPHS

Every Hamiltonian graph is 2-connected, since deleting any vertex leaves a spanning path. Bipartite graphs suggest an extension of this condition.

7.3.1. Example. *Bipartite graphs.* A spanning cycle in a bipartite graph visits the two parts alternately, so such a cycle requires the parts to have the same size. In particular, $K_{r,s}$ is not Hamiltonian when $r < s$. Note that deleting the part of size r leaves more than r components. ∎

7.3.2. Proposition. If G has a Hamiltonian cycle, then $\#(G - S) \leq |S|$ whenever $\varnothing \neq S \subseteq V(G)$, where $\#(H)$ denotes the number of components of H.

Proof: Follow a Hamiltonian cycle C. When C leaves a component of $G - S$, it can go only to S. Arrivals in S must be at distinct vertices of S, so S must have as many vertices as $G - S$ has components. ∎

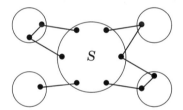

The necessary condition $\#(G - S) \leq |S|$ for all nonempty S sounds like Tutte's 1-Factor Condition, but it is not sufficient for a spanning cycle.

7.3.3. Example. *Necessary but not sufficient.* The graph on the left below fails the condition of Proposition 7.3.2, even though it is bipartite with parts of equal size. Hence it is not Hamiltonian. (Tutte [1971] conjectured that all 3-connected 3-regular bipartite graphs are Hamiltonian; counterexamples were found with 96 vertices (Horton [1982]) and eventually only 50 vertices (Georges [1989]).

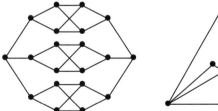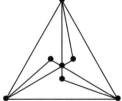

In the graph on the right, the necessary condition holds, but there is no spanning cycle; both edges incident to any vertex with degree 2 must be used, which forces three edges at the central vertex. Exercise 28 generalizes this example. ∎

7.3.4.* Remark. Perhaps strengthening the necessary condition by requiring $|S| \geq t \cdot \#(G - S)$ for some t guarantees a spanning cycle. A connected graph G is *t*-**tough** if $|S| \geq t \cdot \#(G - S)$ for every separating set S, and the **toughness** of G is the maximum such t. By Proposition 7.3.2, Hamiltonian graphs are 1-tough.

For example, the Petersen graph G is 4/3-tough (hence also 1-tough), since

$$\#(G - S) = 2 \Rightarrow |S| \geq 3, \quad \#(G - S) = 3 \Rightarrow |S| \geq 4, \quad \#(G - S) = 4 \Rightarrow |S| = 6,$$

and no larger value of $\#(G - S)$ is possible since $\alpha(G) = 4$. The Petersen graph is not Hamiltonian (Proposition 5.1.18), so 4/3-tough is not sufficient.

Chvátal [1973] conjectured that some t suffices. Many thought $t = 2$ would be enough, but non-Hamiltonian graphs are now known with toughness approaching 9/4 (Bauer–Broersma–Veldman [2000]). For graphs having no induced cycle of length at least 4, Chen–Jacobson–Kézdy–Lehel [1998] showed that toughness 18 suffices, and Kabela–Kaiser [2017] improved that to 10. Enomoto–Jackson–Katerinis–Saito [1985] proved that every k-tough n-vertex graph has a k-factor (Chapter 6) when $n > k$ and nk is even. Broersma [2002] and Bauer–Broersma–Schmeichel [2006] surveyed results on toughness. ∎

For regular graphs with even degree, one can study decompositions into spanning cycles, called **Hamiltonian decompositions** (we write "H-decomposition"). The complete graph K_{2r+1} has H-decompositions (Exercise 5.3.48). Sloane [1969] asked whether every 4-regular graph with an H-decomposition has another.

7.3.5.* Theorem. (Thomason [1978]) The number of Hamiltonian decompositions of any loopless 4-regular multigraph G with at least three vertices is even. Furthermore, for any two edges e and e' in G, the number of Hamiltonian decompositions having e and e' in the same cycle is even.

Proof: Let $R(G)$ be the set of H-decompositions of G. For $e, e' \in E(G)$, let $P_G(e, e')$ and $Q_G(e, e')$ denote the sets of H-decompositions having e and e' in the same or different cycles, respectively. Thus $P_G(e, e') \cup Q_G(e, e') = R(G)$. Let the sizes of these sets be $p_G(e, e')$, $q_G(e, e')$, and $r(G)$, respectively.

Let v be an endpoint of e. In every H-decomposition, the cycle through e contains another edge incident to v. Hence $r(G)$ is even if $p_G(e, e')$ is even whenever e and e' are incident edges. With $r(G)$ even, the stronger statement that $p_G(e, e')$ is even for all $e, e' \in E(G)$ is equivalent to having $q_G(e, e')$ even for all $e, e' \in E(G)$.

We prove the claims by induction on $|V(G)|$. In the only loopless 4-regular 3-vertex multigraph, each vertex pair forms a double edge. When e and e' share both endpoints, $p_G(e, e') = 0$; when they share one endpoint, $p_G(e, e') = 2$.

Now consider larger G. We prove first that $p_G(e, e')$ is even when e and e' have a common endpoint v. Let f and f' be the other two edges incident to v. If e and e' share both endpoints or f and f' share both endpoints, then $p_G(e, e') = 0$. Otherwise, obtain $\hat{G}$ from G by deleting v, adding an edge $\hat{e}$ joining the other endpoints of e and e', and adding an edge $\hat{f}$ joining the other endpoints of f and f'. Note that $\hat{G}$ is a loopless 4-regular multigraph, smaller than G.

The H-decompositions of G in $P_G(e, e')$ put f and f' into the cycle not containing e and e'; hence they correspond to the H-decompositions of $\hat{G}$ in $Q_{\hat{G}}(\hat{e}, \hat{f})$. By the induction hypothesis, $R(\hat{G})$ and $Q_{\hat{G}}(\hat{e}, \hat{f})$ have even size. Hence $p_G(e, e')$ is even, since it equals $q_{\hat{G}}(\hat{e}, \hat{f})$. This holds for all incident e and e', so $r(G)$ is even.

Now consider any $e, e' \in E(G)$. If $R(G) \neq \varnothing$, then G is connected, and some path P contains e and e'. We show by induction on the length of P that $p_G(e, e')$ is even. We have shown it when $E(P) = \{e, e'\}$, the base case.

When e and e' are not incident, let $\hat{e}$ be the edge before e' on P. Note that $Q_G(e, e') = P_G(e, \hat{e}) \triangle P_G(\hat{e}, e')$, since e and e' are in different cycles in a H-decomposition if and only if $\hat{e}$ is in exactly one of those cycles. Since we have already proved that $P_G(\hat{e}, e')$ and $P_G(e, \hat{e})$ (shorter path) have even size, also $q_G(e, e')$ is even. Since we already proved that $r(G)$ is even, also $p_G(e, e')$ is even. ∎

Applications of Theorem 7.3.5 appear in Exercises 32–33. For the number of Hamiltonian cycles, Corollary 5.4.8 proves the result of Smith that a 3-regular graph has an even number of Hamiltonian cycles through any edge. This does not imply that a 3-regular graph has an even number of spanning cycles! The complete graph K_4 and the cartesian product $C_3 \square K_2$ each have exactly three spanning cycles; every edge lies in two of them. Kotzig proved (see Bosák [1967]) that every bipartite 3-regular graph has an even number of spanning cycles.

SUFFICIENT CONDITIONS

The number of edges needed to force a spanning cycle in an n-vertex graph is $\binom{n-1}{2} + 2$ (Exercise 36). Conditions that "spread out" the edges permit a smaller threshold. The idea underlying such results is the edge-switching technique extracted by Ore from an earlier argument of Dirac.

7.3.6. Lemma. (Ore's Lemma; Ore [1960]) Let x and y be distinct nonadjacent vertices of an n-vertex graph G. If $d(x) + d(y) \geq n$, then G is Hamiltonian if and only if $G + xy$ is Hamiltonian.

Proof: If G is Hamiltonian, then so is $G + xy$. If $G + xy$ is Hamiltonian but G is not, then xy lies in each spanning cycle in $G + xy$. Index the vertices from v_1 to v_n along a spanning x, y-path in G.

If some neighbor of x immediately follows a neighbor of y on the path, say $v_{i+1} \in N(x)$ and $v_i \in N(y)$, then omitting $v_i v_{i+1}$ yields the spanning cycle $[x, v_{i+1}, v_{i+2}, \ldots, y, v_i, v_{i-1}, \ldots, v_2]$ in G shown below.

To guarantee such a cycle, we prove that S and T have a common element, where $S = \{i : v_{i+1} \in N(x)\}$ and $T = \{i : v_i \in N(y)\}$. Summing their sizes,

$$|S \cup T| + |S \cap T| = |S| + |T| = d(x) + d(y) \geq n.$$

Neither S nor T contains the index n. Thus $|S \cup T| < n$, and hence $|S \cap T| \geq 1$. We conclude that G has a spanning cycle. ∎

7.3.7. Corollary. (Ore's Theorem; Ore [1960]) Let G be an n-vertex graph. If $n \geq 3$ and $d(x) + d(y) \geq n$ whenever $xy \notin E(G)$, then G is Hamiltonian. (Letting $\sigma_2(G) = \min\{d(x) + d(y) : xy \notin E(G)\}$, **Ore's Condition** is $\sigma_2(G) \geq n$.)

Proof: Since K_2 satisfies Ore's Condition but is not Hamiltonian, we need $n \geq 3$. For $n \geq 3$, we use induction on $|E(\overline{G})|$; the basis step is K_n. When $xy \in E(\overline{G})$, adding xy to G yields a graph G' that also satisfies Ore's Condition. By the induction hypothesis, G' is Hamiltonian, and then by Lemma 7.3.6 so is G. ∎

Ore's Condition is immediately implied by **Dirac's Condition**, which is the stronger hypothesis $\delta(G) \geq n/2$. Thus the sufficiency of Dirac's Condition is a weaker result than the sufficiency of Ore's Condition.

7.3.8. Corollary. (**Dirac's Theorem**; Dirac [1952b]) For $n \geq 3$, an n-vertex graph G with $\delta(G) \geq n/2$ is Hamiltonian, and this threshold is sharp.

Proof: Again $n \geq 3$ is needed, because K_2 satisfies the condition but is not Hamiltonian. For $n \geq 3$, the non-Hamiltonian graph $K_{\lfloor(n-1)/2\rfloor,\lceil(n+1)/2\rceil}$ shows that the bound is sharp. For sufficiency, when $\delta(G) \geq n/2$ we have $d(x) + d(y) \geq n$ whenever $x, y \in V(G)$, and Corollary 7.3.7 applies. ∎

Using Lemma 7.3.6 to add edges, we can test whether G is Hamiltonian by testing whether a larger graph is Hamiltonian.

7.3.9. Definition. A **(Hamiltonian) closure** of an n-vertex graph G, denoted $C(G)$, is obtained from G by iteratively adding edges joining nonadjacent vertices whose degree sum is at least n, until no such pair remains. We say G is **closed** when $C(G) = G$.

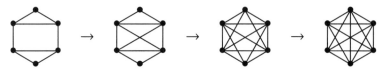

7.3.10. Theorem. (Bondy–Chvátal [1976]) The closure of a graph is unique (does not depend on the order of edge additions), and a graph is Hamiltonian if and only if its closure is Hamiltonian.

Proof: From G, form closures G_1 by adding the edges $e_1, \ldots, e_r$ and G_2 by adding the edges $f_1, \ldots, f_s$. If the degree-sum of two nonadjacent vertices reaches $|V(G)|$, then they must be made adjacent. Thus if $f_1, \ldots, f_{i-1} \in E(G_1)$, then f_i becomes addable and must belong to G_1. Hence neither list contains a first edge omitted by the other, so $G_1 \subseteq G_2$ and $G_2 \subseteq G_1$.

The second statement follows immediately from Lemma 7.3.6. ∎

Bondy–Chvátal [1976] considered many related concepts and closures by varying the threshold for $d(x) + d(y)$. Material on these and other closure concepts includes Ryjáček [1997] and Broersma–Ryjáček–Schiermeyer [2000].

Theorem 7.3.10 characterizes Hamiltonian graphs but does not provide a good algorithm to test whether a graph is Hamiltonian. Nevertheless, it provides a method for proving sufficient conditions. A condition that forces $C(G)$ to be Hamiltonian (as when $C(G) = K_n$) forces a Hamiltonian cycle in G. Chvátal used this method to find the best possible sufficient condition using only vertex degrees. The condition on the degree list in this theorem is called **Chvátal's Condition**.

7.3.11. Theorem. (Chvátal's Theorem; Chvátal [1972]) For $n \geq 3$, let G be a graph with vertex degrees $d_1, \ldots, d_n$ in nondecreasing order. If $d_i > i$ or $d_{n-i} \geq n - i$ whenever $i < n/2$, then G is Hamiltonian.

Proof: By Theorem 7.3.10, it suffices to prove that if G satisfies Chvátal's Condition, then $C(G) = K_n$, since a complete graph is Hamiltonian. Adding edges reduces no entries in the degree list, so Chvátal's Condition is preserved by taking the closure. Thus it suffices to prove the implication in the case where G is already closed. We prove the contrapositive: if G is closed and $G \neq K_n$, then Chvátal's Condition fails. Violation means that for some i less than $n/2$, at least i vertices have degree at most i and at least $n - i$ have degree less than $n - i$.

Since $G \neq K_n$, we may pick a pair $\{u, v\}$ of nonadjacent vertices with largest degree sum. Because G is closed, $uv \notin E(G)$ requires $d(u) + d(v) < n$; label u and v so that $d(u) \leq d(v)$. Since $d(u) + d(v) < n$, we obtain $d(u) < n/2$. Let $i = d(u)$.

Since we chose $\{u, v\}$ with largest degree sum, every nonneighbor of v has degree at most $d(u)$, and there are $n - 1 - d(v)$ of them. Since $n - 1 - d(v) \geq d(u) = i$, at least i vertices have degree at most i.

Similarly, every nonneighbor of u has degree at most $d(v)$, and for u there are $n - 1 - d(u)$ nonneighbors. Since $d(u) \leq d(v)$, also u itself has degree at most $d(v)$. Since $d(v) < n - d(u) = n - i$, these $n - i$ vertices have degree less than $n - i$.

Hence we have proved that $d_i \leq i$ and $d_{n-i} < n - i$ when $i = d(u)$, so Chvátal's Condition does not hold. ∎

7.3.12. Example. *Non-Hamiltonian graphs with "large" vertex degrees.* Theorem 7.3.11 characterizes the degree lists of graphs that force Hamiltonian cycles. If G fails Chvátal's Condition at the value i, then the largest we can make the terms in $d_1, \ldots, d_n$ is $d_j = i$ for $j \leq i$, $d_j = n - i - 1$ for $i + 1 \leq j \leq n - i$, and $d_j = n - 1$ for $j > n - i$. The unique graph with this degree list is not Hamiltonian. The graph is $(\overline{K_i} + K_{n-2i}) \oplus K_i$; it is not Hamiltonian because deleting the i vertices of degree $n - 1$ leaves a subgraph with $i + 1$ components. ∎

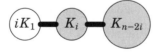

7.3.13. Remark. A **Hamiltonian path** is a spanning path. Results for spanning cycles yield analogous results for paths by a standard transformation: G has a spanning path if and only if $G \oplus K_1$ has a spanning cycle. For example, $\sigma_2(G) \geq n - 1$ yields a spanning path (Ore [1961]); see also Exercise 3 and Exercise 38. ∎

Our next sufficient condition involves connectivity and independence. The proof yields a good algorithm to find a spanning cycle or contradict the hypothesis. The theorem strengthens the degree results in the probabilistic sense (see Chapter 14); the Chvátál–Erdős Condition almost always holds, while the degree conditions do not. Surveys of material related to the Chvátal–Erdős Theorem include Jackson–Ordaz [1990] and Saito [2008].

7.3.14. Theorem. (Chvátal–Erdős Theorem; Chvátal–Erdős [1972]) If $\kappa(G) \geq \alpha(G)$, then G has a Hamiltonian cycle (unless $G = K_2$).

Proof: Let $k = \kappa(G) \geq \alpha(G)$. With $G \neq K_2$, the conditions require $\kappa(G) > 1$, so there is a longest cycle C in G. Since $\delta(G) \geq \kappa(G)$, and since every graph with $\delta(G) \geq 2$ has a cycle of length at least $\delta(G) + 1$ (Exercise 5.3.38), the length of C is at least $k + 1$. Let H be a component of $G - V(C)$. Since $\kappa(G) = k$, at least k vertices of C have neighbors in H.

Let $u_1, \ldots, u_k$ be vertices of C with neighbors in H, indexed in order along C. For each i, let a_i be the vertex following u_i along C. If a_i and a_j are adjacent, then we construct a longer cycle by replacing $u_i a_i$ and $u_j a_j$ with $a_i a_j$ and a u_i, u_j-path through H (see illustration). Similarly, a_i has no neighbor in H. Hence $\{a_1, \ldots, a_l\}$ plus a vertex of H forms an independent set of size greater than k. This contradiction implies that C is a Hamiltonian cycle. ∎

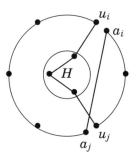

The graph $K_{r,r+1}$ shows that $\kappa(G) \geq \alpha(G) - 1$ is not sufficient.

7.3.15.* Remark. Lu [1994] strengthened the Chvátal–Erdős Theorem by proving that an n-vertex graph G is Hamiltonian when the following inequality holds for every nonempty proper subset S of $V(G)$:

$$\frac{|N(S) \cap \overline{S}|}{|\overline{S}|} \geq \frac{\alpha(G)}{n}.$$

To obtain the Chvátal–Erdős Theorem from Lu's Theorem, we view the latter as lowering the threshold for the size of a separating set. Let S be the vertex set of a component of $G-T$, where T is a smallest separating set. Lu's Condition requires $|N(S) \cap \overline{S}| \geq \frac{\alpha(G)}{n}|S|$. In fact, $N(S) \cap \overline{S} = T$, since T is a minimal separating set. Thus $|T| \geq \alpha(G)(1 - \frac{|S|}{n})$. Since $|T| = \kappa(G)$, the Chvátal–Erdős Condition implies this, so Lu's Theorem implies the Chvátal–Erdős Theorem. This greatly reduces the required size of a separating set T when the components of $G - T$ are large. Cuts that isolate single vertices still must have size at least $\alpha(G)$. ∎

LONG CYCLES (optional)

When a sufficient condition for spanning cycles fails slightly, we still expect a long cycle. The maximum length of a cycle in G is its **circumference**, $c(G)$. We begin with the number of edges to force $c(G) > q$. We also consider long-cycle versions of conditions for spanning cycles.

7.3.16. Theorem. (Erdős–Gallai [1959]) For $q \geq 2$, if an n-vertex graph G has more than $q(n-1)/2$ edges, then $c(G) > q$.

Proof: (Woodall [1972]) Fix q; we use induction on n. When $n = q + 1$, fewer than $(n-1)/2$ edges are missing, so $\delta(G) \geq n/2$ and G is Hamiltonian. Consider $n > q + 1$ and $c(G) \leq q$. If $d(x) \leq q/2$, then $|E(G - x)| > q(n-2)/2$. Applying the induction hypothesis to $G - x$ yields $c(G - x) > q$. Hence we may assume $\delta(G) > q/2$. Using the induction hypothesis we may also assume that G is connected.

Among all longest paths in G, choose P to maximize the degree of the first vertex. Let $P = \langle v_1, \ldots, v_l \rangle$ and $d = d_G(v_1)$. Since G is connected, $v_1 v_l \notin E(G)$ (otherwise an edge from $V(P)$ to $V(G) - V(P)$ would yield a longer path). Let $W = \{v_i : v_{i+1} \in N(v_1)\}$. All neighbors of v_1 lie on P, so $|W| = d$.

For $v_k \in W$, let Q be the path obtained from P by replacing $v_k v_{k+1}$ with $v_1 v_{k+1}$ (see figure below). Since Q has the same length (and vertex set) as P, we have $N(v_k) \subseteq V(P)$, and the choice of P yields $d_G(v_k) \leq d$. Avoiding longer paths and cycles with length greater than q yields $N(v_k) \subseteq Z$, where $Z = \{v_1, \ldots, v_{\min\{l,q\}}\}$.

We bound w, the number of edges incident to W. Since $N(v) \subseteq Z$ for $v \in W$, we have $2w = \sum_{v \in W} d_G(v) + |[W, Z - W]|$. Since $|[W, Z - W]| \leq |W||Z - W|$ and $d_G(v) \leq d = |W|$, we obtain $w \leq \frac{1}{2}(d^2 + d(|Z| - d)) \leq \frac{1}{2}dq$.

Therefore, $G - W$ has $n - d$ vertices and more than $q(n - d - 1)/2$ edges. Using the induction hypothesis, $c(G) \geq c(G - W) > q$.   ∎

Theorem 7.3.16 is sometimes sharp; Kopylov [1977] gave the complete answer. Next, the long-cycle result of Dirac [1952b] will follow from our later results.

7.3.17. Theorem. For G a 2-connected n-vertex graph, $c(G) \geq \min\{n, 2\delta(G)\}$.   ∎

Theorem 7.3.17 improves the observation that $\delta(G) = k$ guarantees a cycle of length at least $k + 1$ (Exercise 5.3.38). Requiring 2-connectedness eliminates the example $K_1 \oplus 2K_{k+1}$, which has circumference $k + 1$.

The long-cycle version of Ore's Theorem (Corollary 7.3.7) came much later. It is implicit in Bondy [1971b] and was explicit in Bermond [1976] and in Linial [1976]. The fundamental argument used in this and many other long-cycle results appears in Bondy [1971b]. Called **Bondy's Lemma**, it strengthens the Ore/Dirac switching argument (Lemma 7.3.6) by considering "gaps".

7.3.18. Lemma. (Bondy [1971b]) If G is a 2-connected n-vertex graph, then $c(G) \geq \min\{n, d(x) + d(y)\}$, where x and y are the ends of a longest path P.

Proof: Suppose $c(G) < n$. Let $P = \langle v_1, \ldots, v_l \rangle$, where $v_1 = x$ and $v_l = y$. For $i, j \in [l]$, let $P_{i,j}$ denote the v_i, v_j-path in P. Since G is connected, l-cycles are forbidden (they yield longer paths or a spanning cycle). In particular, $xy \notin E(G)$.

Case 1: P has a crossover: edges xv_j and $v_i y$ with $i < j$. Adding xv_j and $v_i y$ to $P_{1,i} \cup P_{j,l}$ yields a cycle with length $l - (j - i - 1)$.

Given a crossover $\{xv_j, v_iy\}$ with $j - i$ smallest, x and y have no neighbors between v_i and v_j on P. Also $N(y)$ contains no predecessor on P of a neighbor of x, since l-cycles are forbidden. Hence $N(y)$ lies in $V(P) - \{y\}$ but avoids $\{v_{i+1}, \ldots, v_{j-2}\}$ and $\{v_{r-1}: v_r \in N(x)\}$. Thus $d(y) \leq (l-1) - (j-2-i) - d(x)$, which simplifies to $l - (j - i - 1) \geq d(x) + d(y)$, so the crossover cycle is long enough.

Case 2: *P has no crossover.* We construct a cycle containing x and y and all their neighbors. Since avoiding crossovers requires $|N(x) \cap N(y)| \leq 1$, such a cycle has length exceeding $d(x) + d(y)$.

Let $t_0 = \max\{j: v_j \in N(x)\}$ and $u = \min\{j: v_j \in N(y)\}$. We define paths $Q_1, Q_2, \ldots$. Given t_{i-1}, choose integers s_i and t_i, with $s_i < t_{i-1} < t_i$ and t_i as large as possible, so that G has a path Q_i from v_{s_i} to v_{t_i} that has no internal vertex on P. Such a path exists when $t_{i-1} < l$ since $G - v_{t_{i-1}}$ is connected. If Q_i and Q_j with $i < j$ share an internal vertex, then Q_i would follow Q_j after the intersection to reach v_{t_j}, farther than v_{t_i}. Hence the paths are disjoint. Similarly, $s_{i+1} \geq t_{i-1}$, since otherwise Q_{i+1} would be chosen as Q_i.

Let r be the least index such that $t_r > u$ ($r = 5$ in the figure below). Set

$$a = \min\{j: v_j \in N(x) \text{ and } j > s_1\}, \qquad b = \max\{j: v_j \in N(y) \text{ and } j < t_r\}.$$

Since $s_1 < t_0$ and $t_r > u$, the indices a and b are well-defined. We use the paths Q_i for even i to build one x, y-path; we build another using the paths Q_i for odd i. When r is odd, the two paths are as follows, where $Q : Q'$ denotes the union of paths Q and Q' when the end of Q is the beginning of Q'.

$$xv_a : P_{a,s_2} : Q_2 : P_{t_2,s_4} : Q_4 : \cdots : P_{t_{r-1},b} : v_b y$$

$$P_{1,s_1} : Q_1 : P_{t_1,s_3} : Q_3 : P_{t_3,s_5} : \cdots : Q_r : P_{t_r,l}$$

When r is even, the path starting with xv_a reaches t_r and ends with $P_{t_r,l}$, while the other path reaches v_b and ends with $v_b y$.

Since $s_{i+1} \geq t_{i-1}$,

$$s_1 < a \leq t_0 \leq s_2 < t_1 \leq s_3 < t_2 \cdots < t_{r-1} \leq u \leq b < t_r$$

Thus the two concatenations are x, y-paths forming a cycle. The definitions of a and b yield $N(x) \subseteq V(P_{1,s_1} \cup P_{a,t_0})$ and $N(y) \subseteq V(P_{u,b} \cup P_{t_r,l})$, so the cycle includes x and y and their neighbors and has length exceeding $d(x) + d(y)$. ∎

Bondy's Lemma implies the long-cycle version of Ore's Theorem, which strengthens the long-cycle version of Dirac's Theorem.

7.3.19. Theorem. If G is a 2-connected n-vertex graph and $d(u) + d(v) \geq q$ for every nonadjacent pair $u, v \in V(G)$, then $c(G) \geq \min\{n, q\}$.

Proof: Ore's Theorem provides a spanning cycle if $q \geq n$, so we may assume $q < n$. The endpoints of a longest path are nonadjacent, since G is connected. By hypothesis, their degree sum is at least q, so Bondy's Lemma completes the proof. ∎

Fan [1984] strengthened Theorem 7.3.19 by weakening the degree condition and by requiring it only for some nonadjacent pairs. As shown next, Fan's Theorem (Theorem 7.3.21) yields a sufficient condition for Hamiltonian cycles that does not force the closure to be complete (see Exercise 50 for another).

7.3.20. Example. *A Hamiltonian graph.* With $n = 4r$, let $G_1 = K_{2r}$ and $G_2 = rK_2$, and form G by adding a matching joining $V(G_1)$ and $V(G_2)$. The Hamiltonian closure of G is G itself, so our previous sufficient conditions do not apply. Although G has $n/2$ vertices of degree 2, Fan's Theorem only needs $\max\{d(x), d(y)\} \geq n/2$ when $d_G(x, y) = 2$ and implies that G is Hamiltonian. ∎

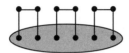

Tian [1988] shortened the proof by applying Bondy's Lemma. Later, Tian [2004] shortened it yet again, obtaining the needed longest path more directly. This improvement uses the switching idea in the proof of Theorem 7.3.16.

7.3.21. Theorem. (**Fan's Theorem**; Fan [1984]) Let G be a 2-connected n-vertex graph. If $d_G(u, v) = 2$ implies $\max\{d(u), d(v)\} \geq q/2$, then $c(G) \geq \min\{n, q\}$.

Proof: (Tian [2004]) Let $U = \{v \in V(G): d(v) \geq q/2\}$. Among all longest paths in G, let Q be one with the most endpoints in U. If both ends are in U, then Bondy's Lemma yields $c(G) \geq \min\{n, q\}$. Hence we may assume that Q starts outside U.

Let x be the last vertex of Q. Among all longest paths in G, let $\mathbf{P}$ be the set of those ending at x. By the choice of Q, all paths in $\mathbf{P}$ start outside U. Let k be the maximum distance along a path in $\mathbf{P}$ from the first vertex to its last neighbor; let P be a path achieving this. Thus P has vertices $v_0, \ldots, v_l$ in order, with v_k being the last neighbor of v_0 in the list. Since $P \in \mathbf{P}$, we have $v_0 \notin U$.

Also $N(v_0) \subseteq W$, where $W = \{v_1, \ldots, v_k\}$. We claim $W \subseteq N(v_0)$. Otherwise, let v_t be the last vertex of W missing from $N(v_0)$. Note that $v_t, \ldots, v_0, v_{t+1}, \ldots, v_l$ in order form a path in $\mathbf{P}$ beginning at v_t, so $v_t \notin U$. This contradicts the hypothesis, since $d_G(v_0, v_t) = 2$ but $v_0, v_t \notin U$. Thus $N(v_0) = \{v_1, \ldots, v_k\}$.

For $1 \leq j \leq k - 1$, the path $v_j, \ldots, v_0, v_{j+1}, \ldots, v_l$ also lies in $\mathbf{P}$. It starts with $v_0, \ldots, v_k$ in some order. Hence the choice of P in $\mathbf{P}$ to maximize the index of the last neighbor of the first vertex yields $N(v_j) \subseteq \{v_0, \ldots, v_k\}$.

We have shown that v_k is the only vertex outside $\{v_0, \ldots, v_{k-1}\}$ having neighbors in this set. Hence $G - v_k$ is disconnected, which is a contradiction since G is 2-connected. We conclude that some longest path has both endpoints in U. ∎

FURTHER DIRECTIONS (optional)

The results of Dirac, Ore, and Chvátal are sharp but only begin the study of cycles. Besides generalizing to long cycles as discussed above, we can consider restricting G or modifying the hypotheses or conclusions in various ways.

7.3.22. Remark. *Restricted classes.* When we restrict attention to spanning sub-graphs of $K_{n/2,n/2}$, the degree threshold cannot be larger; indeed, we hope for an easier threshold. In fact, in this case minimum degree $(n+1)/4$ suffices for a spanning cycle (Moon–Moser [1963]; Exercise 39), and $\delta(G) = n/4$ is not enough. Bagga–Varma [1999] provides a survey.

We can also restrict attention to regular graphs. Every k-regular n-vertex graph with $n \leq 3k$ is Hamiltonian (Jackson [1980]); thus regularity drops the degree threshold from $n/2$ to $n/3$. Only the Petersen graph prevents lowering it to $(n-1)/3$ (Zhu–Liu–Yu [1985], partly simplified in Bondy–Kouider [1988]). The example $K_{r,r+1}$ shows how crucial regularity is.

Spanning cycles are also studied in special classes of graphs. A famous conjecture of Matthews–Sumner [1984] asserts that every 4-connected graph with no four vertices inducing $K_{1,3}$ is Hamiltonian. Connectivity at least 7 is sufficient (Jackson [1989, unpublished], Zhan [1991]). Another proof of this by Ryjáček [1997] reduced the conjecture to the case of line graphs of triangle-free graphs.

The **prism** over a graph G is the cartesian product graph $G \,\square\, K_2$, and G is **prism-Hamiltonian** when $G \,\square\, K_2$ has a spanning cycle. Every graph having a spanning path is prism-Hamiltonian, but spanning paths are not necessary. The concept is interesting because a spanning cycle in $G \,\square\, K_2$ collapses to a 2-**walk** in G, which is a spanning closed walk that visits each vertex at most twice (a weaker version of a spanning cycle). Many classes of graphs are known to be prism-Hamiltonian, including all 3-connected 3-regular graphs (Paulraja [1993]). Kaiser–Ryjáček–Král'–Rosenfeld–Voss [2007] contains a survey.

For graphs on surfaces, see Ellingham [1996] and Holton–Aldred [1999]. ■

7.3.23. Remark. *Stronger conclusions, weaker hypotheses.* A graph with cycles of all lengths (3 to $|V(G)|$) is **pancyclic**. A "meta-conjecture" of Bondy [1971a] asserts that sufficient conditions for spanning cycles usually imply pancyclicity with few exceptions. Bondy [1971a] showed that n-vertex graphs satisfying Ore's Condition are pancyclic except for $K_{n/2,n/2}$ (see Exercise 25). With suitable exceptions, this also holds for Chvátal's Condition (Schmeichel–Hakimi [1974]) and Fan's Condition (Tian–Shi [1986], Benhocine–Wojda [1987]); see Mitchem–Schmeichel [1985] for a survey. Schmeichel–Hakimi [1988] showed that if some spanning cycle has consecutive x and y with $d(x) + d(y) \geq n$, then the graph is pancyclic or bipartite or lacks only cycle length $n-1$ (pancyclic if $d(x) + d(y) \geq n+1$). Using this, Bauer–Schmeichel [1990] gave unified proofs for pancyclicity (with exceptions) under the conditions of Chvátal [1972], Fan [1984], and Bondy [1980].

Strengthening a sufficient condition for spanning cycles often forces more structure. Pósa [1963/4] proved that if the degree sum of nonadjacent vertices is always at least $n+k$, then any set of k edges forming a disjoint union of paths can be extended to a Hamiltonian cycle (Exercise 49). Berge conjectured the same conclusion when $\kappa(G) \geq \alpha(G) + k$, proved in Häggkvist–Thomassen [1982]. Zamani–West [2012] proved the analogue for bipartite graphs. For a survey on disjoint cycles (and paths), see Chiba–Yamashita [2018].

A graph G is **Hamiltonian-connected** if it has a spanning u,v-path whenever $u,v \in V(G)$. Often a slight strengthening of a sufficient condition for Hamiltonian graphs implies that a graph is Hamiltonian-connected (Exercises 44–45).

Ore's Theorem (Corollary 7.3.7) has been strengthened by considering independent sets of size greater than 2 and by considering neighborhood unions instead of degree-sums (see Bondy [1981], Faudree–Gould–Jacobson–Schelp [1989], Faudree–Gould–Jacobson–Lesniak [1992]).   ∎

7.3.24. Remark. *Weakening the conditions for spanning paths.* Spanning paths are usually guaranteed by slight weakenings of sufficient conditions for spanning cycles, since G has a spanning path if and only if $G \ominus K_1$ has a spanning cycle (Remark 7.3.13). Dirac's Condition, Ore's Condition, and the Chvátal–Erdős Condition can be modified in this way to guarantee a spanning path in an n-vertex graph G when $\delta(G) \geq (n-1)/2$, or $\sigma_2(G) \geq n - 1$, or $\alpha(G) \leq \kappa(G) + 1$.

Viewing a spanning path as a restricted spanning tree leads to many generalizations. A spanning path has maximum degree 2, no branch vertices, only two leaves, etc. Weakening the conditions for a spanning path may guarantee similarly weakened spanning trees. Ozeki–Yamashita [2011] provides a thorough survey about such problems, including what we mention here (see also Kouider–Vestergaard [2005]). Say that a spanning tree is k-**bounded** if it has maximum degree at most k, k-**branched** if it has at most k branch vertices (vertices with degree at least 3), and k-**ended** if it has at most k leaves.

For example, Gargano–Hammar–Hell–Stacho–Vaccaro [2004] proved that $\delta(G) \geq (n-1)/3$ guarantees a 1-branched spanning tree. Strengthening earlier conjectures, Ozeki–Yamashita [2011] conjectured that $\delta(G) \geq (n-k)/(k+3)$ guarantees a k-branched spanning tree, which is sharp (Exercise 60); DeBiasio–Lo [2017+] proved the conjecture asymptotically (that is, $\delta(G) \geq (\frac{1}{k+3} + o(1))n$ is sufficient). In fact, the conjecture asserts sufficiency of the Ore-type condition that the sum of the degrees of any $k + 3$ independent vertices is at least $n - k$.

Recall from Remark 7.3.4 on toughness that $\#(G)$ denotes the number of components of G. A k-bounded spanning tree requires $\#(G - S) \leq (k-1)|S| + 1$ for all $S \subseteq V(G)$ (Exercise 61). This necessary condition is almost sufficient: Win [1989] proved that if $\#(G - S) \leq (k - 2)|S| + 2$ for all $S \subseteq V(G)$, then G has a k-bounded spanning tree (Ellingham–Zha [2000] gave a short proof).

In terms of connectivity, $\alpha(G) \leq (k-1)\kappa(G) + 1$ guarantees a k-bounded spanning tree (Neumann-Lara & Rivera-Campo [1991]). Win [1975] (proved below) obtained an Ore-type condition: it suffices that any k independent vertices have degree-sum at least $n - 1$ (implied by $\delta(G) \geq (n-1)/k$). A "long-cycle" analogue for k-bounded subtrees appears in Caro–Krasikov–Roditty [1991].

For k-ended spanning trees, a necessary condition is $\#(G - S) \leq |S| + k - 1$ for all $S \subseteq V(G)$ (Exercise 62). Also the Chvátal–Erdős and Ore conditions were generalized: Win [1979] showed that $\alpha(G) \leq \kappa(G) + k - 1$ suffices, and Broersma–Tuinstra [1998] proved that $\sigma_2(G) \geq (n - k + 1)/2$ suffices.

The notion of Hamiltonian-connected also generalizes to the setting of spanning trees. A graph G with more than k vertices is k-**leaf-connected** if for every set S of k vertices, there is a spanning tree whose set of leaves is S. A graph is Hamiltonian-connected if and only if it is 2-leaf-connected. Gurgel–Wakabayashi [1986] proved that G is k-leaf-connected when $\sigma_2(G) \geq n + k - 1$, and Egawa–Matsuda–Yamashita–Yoshimoto [2008] improved this: $\sigma_2(G) \geq n+1$ suffices when G is $(k + 1)$-connected.   ∎

7.3.25. Theorem. (Win [1975]) For $k \geq 2$, if every independent set of k vertices in a connected n-vertex graph G has degree sum at least $n - 1$, then G has a k-bounded spanning tree (maximum degree at most k), and this is sharp.

Proof: Sharpness is shown by K_{k+1,k^2+1}. In this graph, any k independent vertices have degree sum at least $k^2 + k$, which equals $n - 2$. However, a spanning tree has $k^2 + k + 1$ edges, and in a k-bounded spanning tree only $k^2 + k$ edges can be incident to the small part.

Now assume the degree condition, and let T be a largest k-bounded tree contained in G. If T is not spanning, then some vertex $x \in V(T)$ has a neighbor w outside $V(T)$, since G is connected. Since T cannot be enlarged, $d_T(x) = k$. Let $T_1, \ldots, T_k$ be the components of $T - x$, with t_i the neighbor of x in T_i and $n_i = |V(T_i)|$. Select a leaf p_i in each subtree T_i (with $p_i \neq t_i$ unless $n_i = 1$).

The vertices $p_1, \ldots, p_k$ form an independent set S in G, since an edge $p_i p_j$ would yield a larger k-bounded subtree by replacing xt_i with $p_i p_j$ and xw.

Let Y_j be the set of vertices in T_j having a neighbor in $S - p_j$. If $y \in Y_j$, then $d_T(y) = k$, since otherwise replacing xt_j with xw and an edge yp_i enlarges T.

For $y \in Y_j$, let y' be the neighbor of y on the y, p_j-path in T_j. If there exists $z \in N_T(y) - \{y'\}$ such that $zp_j \in E(G)$, then replacing yz and xt_j with all of xw, zp_j, and some yp_i enlarges T. (If $y = t_j$ and $z = x$, then just replace xy with xw and yp_i.) Hence y has no such neighbor in T.

If z is a common neighbor in T of two vertices $y_1, y_2 \in Y_j$, then $z \in \{y_1', y_2'\}$; otherwise, adding the edges $y_1 z$ and $y_2 z$ to the y_1, p_j-path and y_2, p_j-path completes a cycle in T_j. Thus the sets $N_T(y) - \{y'\}$ are pairwise disjoint for all $y \in Y_j$. For $v \in V(G)$, let $d_j(v) = |N_G(v) \cap V(T_j)|$. From the definition of Y_j,

$$|Y_j| \geq \max_{i \in [k] - \{j\}} d_j(p_i) \geq \frac{1}{k-1} \sum_{i \in [k] - \{j\}} d_j(p_i).$$

We further consider $d_j(p_i)$. The sets $N_T(y) - \{y'\}$ for $y \in Y_j$ are pairwise disjoint and have size $k - 1$. These sets, which have no neighbor of p_i, cannot contain p_j, which is another nonneighbor of p_i. However, x is in one of those sets if $t_j \in Y_j$. Let $\varepsilon = 1$ if $t_j \in Y_j$; otherwise, $\varepsilon = 0$. We obtain

$$d_j(p_i) \leq n_j - 1 - (k-1)|Y_j| + \varepsilon \leq n_j - 1 + \varepsilon - \sum_{r \in [k] - \{j\}} d_j(p_r).$$

Thus $\sum_{r=1}^{k} d_j(p_r) \leq n_j - 1 + \varepsilon$. Since all neighbors of p_r lie in T, summing this inequality over $j \in [k]$ provides an upper bound on D, the sum of the degrees of the k independent vertices in S. If $\varepsilon = 1$ in the inequality for j, then all p_i for $i \neq j$ are not adjacent to x in G, and $D \leq \sum n_j = |V(T)| - 1$. If $\varepsilon = 0$ in each inequality, then vertices of S may be adjacent to x in G; now $D \leq \sum(n_j - 1) + k = |V(T)| - 1$. Since $D \geq n - 1$ is given, T must be a spanning tree. ∎

EXERCISES 7.3

7.3.1. (−) Prove that $K_{n,n}$ has $(n-1)!n!/2$ Hamiltonian cycles when $n \geq 2$.

7.3.2. (−) Prove that every 5-vertex path in the dodecahedron graph extends to a Hamiltonian cycle.

7.3.3. (−) Prove that G has a Hamiltonian path only if for every $S \subseteq V(G)$, the number of components of $G - S$ is at most $|S| + 1$.

7.3.4. (−) Prove that every k-regular k-edge-connected graph is 1-tough. (Comment: Hence every such graph also has a 1-factor.)

7.3.5. (−) Prove or disprove: If an n-vertex graph G has an edge xy with the property that $d(x) + d(y) \geq n + 2$, then G is Hamiltonian if and only if $G - xy$ is Hamiltonian.

7.3.6. (−) Let G be a 3-regular graph whose edge set has exactly one partition into three perfect matchings. Prove that G is Hamiltonian. (Greenwell–Kronk [1973])

7.3.7. (−) Prove or disprove: If G is an n-vertex graph with $n \geq 3$, and G has at least $\alpha(G)$ vertices of degree $n - 1$, then G is Hamiltonian.

7.3.8. A mouse eats its way through a $3 \times 3 \times 3$ cube of cheese by eating all the $1 \times 1 \times 1$ subcubes. If it starts at a corner subcube and always moves on to an adjacent subcube (sharing a face of area 1), can it do this and eat the center subcube last? Give a method or prove impossible. (Ignore gravity.)

7.3.9. (◇) Let x and y be vertices in a Hamiltonian bipartite graph G. Prove that $G - x - y$ has a perfect matching if and only if x and y lie in opposite sets of the bipartition. Use this to conclude that an 8-by-8 chessboard with two missing unit squares can be partitioned into 31 one-by-two rectangles if and only if the two missing squares have opposite colors.

7.3.10. Place n vertices around a circle. Let G_n be the 4-regular graph obtained by making each vertex adjacent to the two nearest vertices in each direction. For $n \geq 5$, prove that G_n is the union of two Hamiltonian cycles.

7.3.11. For $k \geq 3$, let G_k be the graph obtained from two disjoint copies of $K_{k,k-2}$ by adding a matching joining the independent k-sets in the two copies. Determine all values of k such that G_k is Hamiltonian.

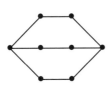 

7.3.12. (◇) For $n \geq 10$ with n even, construct a 3-regular 3-connected graph that is not Hamiltonian. (Hint: Start with the Petersen graph and expand vertices.)

7.3.13. (◇) Prove that the line graph of a graph G is Hamiltonian if and only if G has a closed trail that contains at least one endpoint of each edge (this holds for the Petersen graph). Use the Petersen graph to construct a 3-regular 3-connected graph whose line graph is not Hamiltonian. (Hint: Make an appropriate substitution for each vertex of the Petersen graph.) (Harary–Nash-Williams [1965])

7.3.14. Prove that if n is even, then $K_{n,n}$ does not contain three perfect matchings such that any two together form a spanning cycle. (Hint: Interpret each matching as a permutation.) (Basavaraju–Chandran–Kummini [2010])

7.3.15. (◇) Prove that the cartesian product of two nontrivial graphs with spanning paths fails to have a spanning cycle if and only if both graphs are bipartite and have odd order, in which case the product has a spanning path. (Behzad–Mahmoodian [1969])

7.3.16. ($\Diamond$) On a chessboard, a *knight* can move from one square to another that differs by 1 in one coordinate and by 2 in the other coordinate (some such moves are shown below). A **knight's tour** is a traversal of a rectangular board by knight's moves that visits each square once and returns to the start. Prove that a 4-by-n chessboard cannot have a knight's tour.

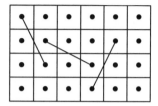

7.3.17. Let G be a claw-free graph (that is, G has no induced $K_{1,3}$). Prove that the maximum t such that G is t-tough is $\kappa(G)/2$. (Matthews–Sumner [1984])

7.3.18. ($\Diamond$) Prove that every 2-connected, claw-free, paw-free graph is Hamiltonian. (The **paw** is obtained from the claw $K_{1,3}$ by adding one edge.) (Goodman–Hedetniemi [1974])

7.3.19. ($\Diamond$) Prove that a graph obtained by deleting one vertex from each part of the bipartite graph $C_{2m} \square K_2$ is Hamiltonian (Stong [2003]). Conclude that deleting one vertex of each parity from the hypercube Q_k with $k \geq 3$ leaves a Hamiltonian graph. (Comment: Locke [2001] asked whether deleting up to $k - 2$ vertices of each parity from Q_k always leaves a Hamiltonian graph. Stong proved that asymptotically $k/2$ of each can be deleted.)

7.3.20. (+) In the k-dimensional hypercube Q_k with $k \geq 2$, prove that every perfect matching extends to a spanning cycle. (Hint: Prove the stronger statement that every perfect matching in the *complete graph* with vertex set $V(Q_k)$ extends to a spanning cycle by adding a perfect matching in Q_k.) (Fink [2007, 2009])

7.3.21. Prove that the k-dimensional hypercube Q_k has a spanning path whose endpoints are complementary if and only if k is odd. Prove that it has a spanning path with endpoints differing in all but one position if and only if k is even.

7.3.22. Let G be an n-vertex Hamiltonian graph. Prove that $G \square K_{1,m}$ is Hamiltonian if and only if $m \leq n$. (Comment: Batagelj–Pisanski [1982] proved that the statement remains true with any tree of maximum degree m in place of $K_{1,m}$.)

7.3.23. (+) For each odd k, construct a $(k - 1)$-connected k-regular bipartite graph that is not Hamiltonian. (Comment: Nash-Williams conjectured that 4-connected 4-regular graphs are Hamiltonian, but the 70-vertex Meredith graph disproves this (duplicate a matching in the Petersen graph to obtain a 4-regular multigraph G, and then form the blow-up graph H as in Example 6.2.15 that seeks a 1-factor in G). Tutte [1971/2] conjectured that every 3-connected 3-regular bipartite graph is Hamiltonian; the first counterexample was the Horton graph with 96 vertices (Bondy–Murty [1976, p. 240]). The smallest known counterexample has 54 vertices (Ellingham–Horton [1983]).)

7.3.24. ($\Diamond$) Determine which complete multipartite graphs are pancyclic. (Recall that an n-vertex graph is **pancyclic** if it has cycles of all lengths from 3 to n.)

7.3.25. (+) Prove that every n-vertex Hamiltonian graph G with more than $n^2/4$ edges is pancyclic. (Bondy [1971a]) (Hint: Use induction on n (Thomassen). Consider separately the cases when G does or does not contain a cycle of length $n - 1$. Comment: Bondy also proved that $n^2/4$ edges are sufficient unless $G = K_{n/2,n/2}$. From this it follows easily that Dirac's Condition and Ore's Condition imply that G is pancyclic unless $G = K_{n/2,n/2}$.)

7.3.26. There is a subgraph of $K_{n,n}$ with $n(n-1)+1$ edges that has no spanning cycle. Prove that every subgraph of $K_{n,n}$ having at least $n(n-1)+2$ edges has cycles of all even lengths from 4 through $2n$. (Entringer–Schmeichel [1988])

7.3.27. Determine whether the graph below has a spanning path.

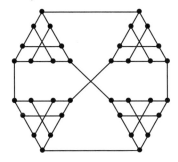

7.3.28. ($\diamond$) The kth **power** of a graph G is the graph G^k with vertex set $V(G)$ and edge set $\{uv: d_G(u,v) \le k\}$.
   (a) Prove that if a vertex x in G has exactly one neighbor in at least three nontrivial components of $G - x$, then G^2 is not Hamiltonian.
   (b) Prove that if G is connected and has at least three vertices, then G^3 is Hamiltonian. (Sekanina [1960], Karaganis [1968]) (Comment: Fleischner [1974] proved that the square of each 2-connected graph is Hamiltonian.)

7.3.29. Let $\mathbf{G}(k,t)$ be the class of connected k-partite graphs in which each part has size t and each subgraph induced by two parts is a matching of size t.
   (a) Show that every graph in $\mathbf{G}(3,t)$ is a cycle and hence is Hamiltonian. (Comment: Also every graph in $\mathbf{G}(k,3)$ is Hamiltonian.)
   (b) For $k \ge 4$ and $t \ge 4$, construct a non-Hamiltonian graph in $\mathbf{G}(k,t)$. (Hint: Start with a cycle spanning three parts, and make the other parts into cliques. Finish the construction so that deleting some three vertices leaves four components.) (Ayel [1982])

7.3.30. ($\diamond$) *The lollipop graph.* (Thomason [1978])
   (a) Let P be a u,z-path in a multigraph G, and let $S = V(P) - \{z\}$. Prove that G has an even number of spanning paths that start along P and end at vertices whose degree in $G - S$ is even. (Hint: Define a graph H on the set of spanning paths beginning along P, making a u,x-path Q and a u,y-path R in $V(H)$ adjacent if and only if $xv \in E(G)$ and $R = Q + xv - vy$, where v is just before y on Q. Which vertices in H have odd degree?)

   (b) Let u and v be vertices in a multigraph G such that every vertex outside $\{u,v\}$ has odd degree. Prove that the number of spanning u,v-paths is even. Conclude that when every vertex has odd degree, the number of spanning cycles through any edge is even. (Comment: This was proved also in Corollary 5.4.8.)

7.3.31. *Hamiltonian cycles through two incident edges.* (Thomason [1978])
   (a) Let uw and wv be edges in a multigraph G with $|V(G)| \ge 4$ such that all vertices outside $\{u,v,w\}$ have odd degree. Use Exercise 7.3.30(a) to prove that if $G-u$ is not Hamiltonian, then the number of Hamiltonian cycles in G using both uw and wv is even.
   (b) For odd k with $k > 1$, prove that every k-regular bipartite graph has an even number of Hamiltonian cycles. (For $k = 3$, this was proved by Kotzig, as noted in Bosák [1967].)

7.3.32. Let G be the union of two spanning cycles C and C'. Say that G is *rearrangeable* if G can be expressed as the union of two other spanning cycles with the same intersection.

(a) Use Theorem 7.3.5 to show that G is rearrangeable if $|E(C) \cap E(C')| = 1$.

(b) For $n \geq 5$, construct an n-vertex example to show that G may not be rearrangeable if $|E(C) \cap E(C')| = 2$. (West [1978])

7.3.33. Let G be the union of m edge-disjoint spanning cycles. Use Theorem 7.3.5 to prove that every edge of G lies in at least $3m - 2$ spanning cycles. Conclude that G has at least $m(3m - 2)$ spanning cycles. (Thomason [1978])

7.3.34. ($\Diamond$) Let G be a graph that is not a forest and has girth at least 5. Use Ore's Theorem to prove that $\overline{G}$ is Hamiltonian. (N. Graham)

7.3.35. ($\Diamond$) Let G be an X, Y-bigraph with $|X| = |Y| = n/2 > 1$. (Moon–Moser [1963])

(a) Given $x \in X$ and $y \in Y$ with $xy \notin E(G)$ and $d(x) + d(y) > n/2$, prove that G is Hamiltonian if and only if $G + xy$ is Hamiltonian (see Lemma 7.3.6).

(b) Prove that $\delta(G) > n/4$ guarantees a spanning cycle in G and that this is sharp.

7.3.36. ($\Diamond$) For $n \geq 2$, give two proofs that the maximum number of edges in a non-Hamiltonian n-vertex graph is $\binom{n-1}{2} + 1$.

(a) Use induction on n. (Ore [1959])

(b) Use Chvátal's Theorem. (Bondy [1972b])

7.3.37. *Edge threshold for spanning cycles in terms of minimum degree.*

(a) Let $f(i) = 2i^2 - i + (n - i)(n - i - 1)$, and suppose $n \geq 6k$. Prove that among i such that $k \leq i \leq n/2$, the maximum value of $f(i)$ is $f(k)$.

(b) Let G be an n-vertex graph with minimum degree k. Use part (a) and Chvátal's condition to prove that if G has at least $6k$ vertices and has more than $\binom{n-k}{2} + k^2$ edges, then G is Hamiltonian. Is this always sharp? (Erdős [1962a])

7.3.38. Let G be a graph with vertex degrees $d_1, \ldots, d_n$ such that $d_1 \leq \cdots \leq d_n$.

(a) Prove that if $d_i \geq i$ or $d_{n+1-i} \geq n - i$ whenever $i < (n + 1)/2$, then G has a spanning path. (Hint: Use Theorem 7.3.11.) Construct an example based on Example 7.3.12 to show that this result is best possible. (Chvátal [1972])

(b) Let $d'_1, \ldots, d'_n$ be the vertex degrees in $\overline{G}$, indexed in nondecreasing order. Use part (a) to prove that if $d_i \geq d'_i$ for $i \leq n/2$, then G has a spanning path. (Hence self-complementary graphs have spanning paths.) (Clapham [1974])

7.3.39. (+) Let G be an X, Y-bigraph having $|X| = |Y| = n/2 > 1$ and vertex degrees $d_1, \ldots, d_n$ in nondecreasing order. Form G' by adding edges to turn Y into a clique.

(a) Prove that G is Hamiltonian if and only if G' is Hamiltonian, and describe the relationship between d and the degree list of G'.

(b) Prove that if $d_k > k$ or $d_{n/2} > n/2 - k$ whenever $k \leq n/4$, then G is Hamiltonian. (Hint: Prove that Chvátal's Condition holds for G'.) (Chvátal [1972])

7.3.40. Prove that Ore's Condition implies Chvátal's Condition. Conclude that Theorem 7.3.11 (Chvátal's Condition is sufficient for a spanning cycle) implies Corollary 7.3.7 (Ore's Condition is sufficient for a spanning cycle).

7.3.41. ($\Diamond$) Use Ore's Lemma to prove that a graph G and its closure $C(G)$ have the same circumference.

7.3.42. For $n \geq 3$, prove that Chvátal's Condition (Theorem 7.3.11) cannot hold for both an n-vertex graph and its complement. (Kostochka–West [2006])

7.3.43. (+) By showing that their hypotheses imply the Chvátal–Erdős Condition, prove that the Chvátal–Erdős Theorem implies each of the following. (Bondy [1978b])

(a) Ore's Theorem (Corollary 7.3.7).

(b) For $k \in \mathbb{N}$, every k-regular graph with $2k + 1$ vertices is Hamiltonian.

7.3.44. Let G be an n-vertex graph such that $\overline{G}$ has m edges. Prove that G is Hamiltonian if $m \leq n - 3$ and G is Hamiltonian-connected if $m \leq n - 4$. (Ore [1963])

7.3.45. Let G be an n-vertex graph with m edges, where $n \geq 4$.
(a) Prove that if G is Hamiltonian-connected, then $m \geq \lceil 3n/2 \rceil$. Use $C_k \,\square\, K_2$ to show that the result is sharp when k is odd. (Moon [1965b])
(b) Prove that G is Hamiltonian-connected if $\sigma_2(G) \geq n + 1$. For $k > 1$, construct a $2k$-vertex graph with minimum degree k that is not Hamiltonian-connected. (Ore [1963])

7.3.46. For a graph G with degrees $d_1, \ldots, d_n$ in nondecreasing order, prove that G is Hamiltonian-connected if $d_i > i + 1$ or $d_{n-i-1} \geq n - i$ whenever $i < (n-1)/2$. (Berge [1973])

7.3.47. For even n, prove that $n - 1$ is the least t such that $\sigma_2(G) \geq t$ forces an n-vertex graph G to have a perfect matching.

7.3.48. For even n, prove that if G has vertex degrees $d_1, \ldots, d_n$ in nondecreasing order, and $d_i \geq i + t - 1$ for $1 \leq i \leq n/2$, then G has t edge-disjoint perfect matchings. (Jahanbekam–West [2016])

7.3.49. ($\diamond$) Let G be an n-vertex graph with $\sigma_2(G) \geq n + k$. Prove that any set of k edges forming a disjoint union of paths in G is contained in a Hamiltonian cycle. Construct an example to show that $\sigma_2(G) \geq n + k - 1$ is not sufficient. (Pósa [1963/4])

7.3.50. (+) *Las Vergnas' Condition.* Let G be a graph having a vertex ordering $v_1, \ldots, v_n$ for which there is no nonadjacent pair v_i, v_j such that $i < j$, $d(v_i) \leq i$, $d(v_j) < j$, $d(v_i) + d(v_j) < n$, and $i + j \geq n$. Las Vergnas [1970] proved that the closure of G is complete (and hence G is Hamiltonian).
(a) Prove that Chvátal's Condition for Hamiltonian cycles implies Las Vergnas' Condition. (Thus Las Vergnas' Theorem strengthens Chvátal's Theorem.)
(b) Prove that each of the graphs below fails Chvátal's Condition but has a complete graph as its Hamiltonian closure. Prove that the smaller graph satisfies Las Vergnas' Condition but the larger one does not.

7.3.51. ($\diamond$) Let G be a connected n-vertex graph with $2 \leq k = \delta(G) < n/2$.
(a) Let P be a maximal path in G (not a subgraph of any longer path). Prove that if P has at most $2k$ vertices, then the induced subgraph $G[V(P)]$ has a spanning cycle (this cycle need not have its vertices in the same order as P).
(b) Use part (a) to prove that G has a path with at least $2k + 1$ vertices. Give an example for each k to show that G may have no cycle with more than $k + 1$ vertices.

7.3.52. Use Theorem 7.3.17 that $c(G) \geq \min\{n, 2\delta(G)\}$ when G is 2-connected to prove that every k-regular graph with $2k + 1$ vertices is Hamiltonian. (Nash-Williams)

7.3.53. ($\diamond$) For $r, k \geq 2$, let G be a k-connected graph with girth at least $2r - 1$. Prove that if G has at least rk vertices, then G has a cycle of length at least rk. (Kostochka)

7.3.54. ($\diamond$) Let G be an n-vertex graph not having $K_{1,t+1}$ as an induced subgraph, where $t \in \mathbb{N}$. Prove that if $\kappa(G) \geq \sqrt{tn}$, then G is Hamiltonian. (Faudree)

7.3.55. Let G be a 4-regular multigraph that is a union of two spanning cycles. Form G' by subdividing one edge from each cycle and adding a double-edge joining the new vertices. Show that G' is also a union of two spanning cycles if $|V(G)| \leq 3$. Conclude for $2 \leq k \leq 4$ that two longest cycles in a k-connected graph have at least k common vertices. (S. Smith)

7.3.56. ($\diamond$) Prove that if $\delta(G) \geq 3k - 1$, or if G is triangle-free and $\delta(G) \geq 2k$, then G contains k disjoint cycles. Show also that lower degree does not suffice. (Thomassen [1988])

7.3.57. ($\diamond$) Use Theorem 7.3.16 to prove that in an n-vertex graph with many edges, one can cover at least half the edges using at most $n/2$ cycles. Conclude that every n-vertex graph decomposes into at most $O(n \log n)$ cycles and edges. (Erdős–Gallai [1959]) (Comment: Erdős and Gallai conjectured that $O(n)$ cycles and edges suffice. Conlon–Fox–Sudakov [2014] reduced the bound to $O(n \log \log n)$.)

7.3.58. Fouquet–Jolivet [1978] conjectured a long-cycle version of the Chvátal–Erdős Theroem: for $k < a$, every n-vertex k-connected graph with independence number a has a cycle of length at least $\frac{k}{a}(n + a - k)$. For $a \geq k \geq 2$ and $r \geq 1$, construct a k-connected graph with $k + ar$ vertices that has no longer cycle, thereby proving sharpness. (Comment: The conjecture was proved in O–West–Wu [2011].)

7.3.59. The **prism** over a graph G is the cartesian product $G \,\square\, K_2$. Prove that the prism over $K_{k,2k+1}$ has no spanning cycle. (Comment: Ellingham–Salehi Nowbandegani [2018] proved that if $\alpha(G) \leq 2\kappa(G)$, then the prism over G is Hamiltonian.)

7.3.60. Ozeki–Yamashita [2011] conjectured that every n-vertex connected graph G with $\delta(G) \geq (n - k)/(k + 3)$ has a spanning tree with at most k branch vertices. Prove sharpness using a connected graph G formed from $(k - 1)K_{t+1} + 2K_{t,t+1}$ by adding one path of length k. (Gargano–Hammar–Hell–Stacho–Vaccaro [2004], DeBiasio–Lo [2017+])

7.3.61. Prove that if G has a k-bounded spanning tree (maximum degree at most k), then $\#(G - S) \leq (k - 1)|S| + 1$ for all $S \subseteq V(G)$. (Ozeki–Yamashita [2011])

7.3.62. ($\diamond$) Prove that if G has a spanning tree with no more than k leaves, then $\#(G - S) \leq |S| + k - 1$ for all $S \subseteq V(G)$. (Salamon–Wiener [2007])

7.3.63. ($\diamond$) Prove that every tournament has a Hamiltonian path and that every strong tournament has a Hamiltonian cycle. (Rédei [1934], Camion [1959])

7.3.64. ($\diamond$) Let T be a strong n-vertex tournament. Suppose $3 \leq k \leq n$.
   (a) For each $u \in V(T)$, prove that u belongs to a cycle of length k in T. (Hint: Use induction on k.)
   (b) Prove that T has at least $n - k + 1$ cycles of length k. Construct an example for each n to show that the bound is sharp for all k, simultaneously. (Moon [1966])

7.3.65. Let T be a 7-vertex tournament in which every vertex has outdegree 3. Prove that T has two disjoint cycles. (Hint: Use Exercise 7.3.64. Comment: Thomassen [1983] proved that every digraph with minimum outdegree at least 3 has two disjoint cycles.)

7.3.66. ($\diamond$) A **bipartite tournament** is an orientation of a complete bipartite graph. Prove that a bipartite tournament has a spanning path if and only if it has a spanning subgraph whose components are all cycles except for possibly one path. (Comment: Gutin [1993] proved this for all complete multipartite graphs.)

7.3.67. Let D be an n-vertex digraph having no loops and having at most one copy of each ordered pair as an edge; such a digraph is **strict**.
   (a) Prove that if $\min\{\delta^-(D), \delta^+(D)\} \geq n/2$, then D has a spanning cycle. (Hint: Consider a longest cycle in D.) (Ghouilà-Houri [1960])
   (b) Construct for each even n an n-vertex digraph D with loops that is not Hamiltonian even though it satisfies all other conditions above.

Chapter 8

Coloring

When the edges of a graph represent vertex conflicts, we often seek the minimum number of conflict-free classes needed to partition the vertices. For example, when scheduling examinations or committee meetings into the fewest time slots, each course or committee is a set of people, and intersecting sets must have different times. The time slots are "colors", and the problem is to use the fewest colors on the vertices of a graph so that adjacent vertices have different colors.

Because labeling to avoid conflicts is such a fundamental concept, coloring theory is enormously broad. Texts and surveys with extensive material on coloring include Berge [1973], Bollobás [1978], Chartrand–Lesniak [1986, etc.], Toft [1995], Jensen–Toft [1995], Molloy–Reed [2002], and Kubale [2004].

8.1. Vertex Coloring

We call vertex labels "colors" because (1) the problem originated with the coloring of regions on maps, and (2) the labels need not be numbers.

8.1.1. Definition. A *k*-**coloring** of G is a vertex labeling $f: V(G) \rightarrow S$, where $|S| = k$. A *k*-coloring f is **proper** if $f(x) \neq f(y)$ whenever $xy \in E(G)$, and G is *k*-**colorable** if it has a proper *k*-coloring. The **chromatic number** $\chi(G)$ is the least k such that G is *k*-colorable. If $\chi(G) = k$, then G is *k*-**chromatic**.

8.1.2. Example. In a proper coloring, each color class is an independent set. Thus G is *k*-colorable if and only if G is *k*-partite. Thus G is 2-colorable if and only if G has no odd cycle. The 5-cycle and the Petersen graph (below) are 3-chromatic.

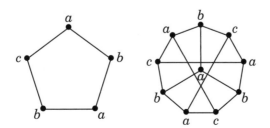

Breadth-first search (Exercise 5.4.70) makes 2-colorability easy to test; a connected graph is bipartite if and only if the sets of vertices at odd and even distance from some vertex x are independent sets. No easily testable characterization of 3-colorable graphs is known; indeed, testing 3-colorability is NP-complete. ∎

We use $\alpha(G)$ for the independence number and $\omega(G)$ for the maximum size of a clique in G, called the **clique number**. As the first and last letters of the Greek alphabet, they suggest the beginning and end of growing a graph.

8.1.3. Remark. *Easy bounds.* When G has n vertices, using distinct colors yields $\chi(G) \leq n$. Since each color class is an independent set, $\chi(G) \geq n/\alpha(G)$. Since adjacent vertices require distinct colors, $\chi(G) \geq \omega(G)$. More generally, a proper coloring of G must properly color its subgraphs, so $\chi(G) \geq \chi(H)$ when $H \subseteq G$.

Equality holds for K_n in these bounds. Each can be very bad. Despite having n vertices, $K_{r,n-r}$ is 2-colorable. Exercise 40(b) shows that $\chi(G) - n/\alpha(G)$ can be large. Theorem 8.1.17 shows that $\chi(G)$ can be large when $\omega(G) = 2$. ∎

8.1.4. Example. *Additivity under joins* (Dirac). For odd cycles, $\chi(C_{2r+1}) = 3$ and $\omega(C_{2r+1}) = 2$. This yields graphs with larger difference via the join operation (Definition 5.1.22). Recall that the join $G \oplus H$ is the graph obtained from the disjoint union $G + H$ by making every vertex of G adjacent to every vertex of H.

The edges that join $V(G)$ and $V(H)$ in $G \oplus H$ force disjoint sets of colors; hence $\chi(G \oplus H) = \chi(G) + \chi(H)$. Also, combining cliques from G and H yields $\omega(G \oplus H) = \omega(G) + \omega(H)$. Thus, $\chi(G) = 3k$ and $\omega(G) = 2k$ when G is the join of k odd cycles. ∎

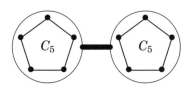

A bound that is rarely sharp begs for improvement. Only complete graphs achieve equality in $\chi(G) \leq |V(G)|$, so we focus first on improving this bound.

UPPER BOUNDS

Most upper bounds on chromatic number arise from coloring algorithms. An n-vertex graph can be properly n-colored without even looking. A natural improvement is to color the vertices in some order and use the first available color.

8.1.5. Algorithm. For a graph G, the **greedy coloring** (or **first-fit coloring**) with respect to a vertex ordering $v_1, \ldots, v_n$ colors vertices in the order $v_1, \ldots, v_n$, giving v_i the least-indexed color not used on its neighbors colored earlier. ∎

8.1.6. Proposition. $\chi(G) \leq \Delta(G) + 1$.

Proof: In a vertex ordering, each vertex has at most $\Delta(G)$ earlier neighbors, so the greedy coloring will not use more than $\Delta(G) + 1$ colors. ∎

Although $\Delta(G) + 1 \leq |V(G)|$, still Proposition 8.1.6 is rarely sharp.

8.1.7. Theorem. (**Brooks' Theorem**; Brooks [1941]) $\chi(G) = \Delta(G) + 1$ only if G has $K_{\Delta(G)+1}$ (or an odd cycle when $\Delta(G) = 2$) as a component.

Cranston–Rabern [2015] surveyed proofs of Brooks' Theorem and its extensions. We will prove it as Corollary 8.2.14 from a stronger result. Meanwhile, since $\chi(G) \leq \Delta(G) + 1$ using any ordering, maybe clever orderings give better greedy colorings. We give an example that yields optimal colorings.

8.1.8. Definition. The **intersection graph** of sets $A_1, \ldots, A_m$ is the graph with vertices $\{v_1, \ldots, v_m\}$ where $v_i v_j$ is an edge if and only if $A_i \cap A_j \neq \varnothing$.

Proper colorings of intersection graphs can model allocation of scarce resources where intersecting sets represent conflicting demands.

8.1.9. Example. An **interval graph** is the intersection graph of a family of intervals on the real line. Computations in a computer program put values of variables into expensive "registers". If two variables are never simultaneously in use, then we can assign them to the same register. Each variable occupies a time interval from its first to its last use. The number of registers needed is the chromatic number of the resulting interval graph. (Other applications include DNA analysis (Benzer [1959]), timing of traffic lights (Roberts [1978]), etc.)

Given intervals representing an interval graph, order the vertices by the left endpoints of the intervals. Under greedy coloring, when x is to be colored, it receives color k only if the left endpoint of the interval for x belongs to intervals already having colors 1 through $k - 1$. If x receives the color k of maximum index, then the neighbors of x already colored at that time induce a k-clique with x. Since $\chi(G) \geq \omega(G)$, the coloring is therefore optimal. For the vertex ordering a, b, c, d, e, f, g of the interval graph below, greedy coloring assigns the colors $1, 2, 3, 1, 1, 2, 3$, respectively, which is optimal. ∎

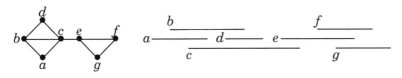

Replacing $\Delta(G)$ with a quantity that is never larger strengthens Proposition 8.1.6. For any vertex ordering, a closer look at the vertex degrees produces the improvement below. This bound is best when we dispose of high-degree vertices first, coloring the vertices in decreasing order of degree.

8.1.10. Proposition. (Welsh–Powell [1967]) If a graph G has vertex degrees $d_1, \ldots, d_n$, then $\chi(G) \leq 1 + \max_i \min\{d_i, i - 1\}$.

Proof: When we color the ith vertex by the greedy algorithm, the number of its neighbors that have already been colored is bounded by d_i and by $i - 1$, so the label used is at most one more than $\min\{d_i, i - 1\}$. ∎

Instead of coloring high-degree vertices first, we can color low-degree vertices last. Applied iteratively to produce an ordering, this generally works better.

8.1.11. Definition. A graph G is **k-degenerate** if every subgraph has a vertex of degree at most k. The **degeneracy** of G is $\max_{H \subseteq G} \delta(H)$; that is, the minimum k such that G is k-degenerate. A **smallest-last ordering** of an n-vertex graph G is constructed iteratively from index n to 1 by letting v_i be a vertex of minimum degree in $G - \{v_{i+1}, \ldots, v_n\}$.

The 1-degenerate graphs are precisely the forests. Some authors call the degeneracy plus 1 the "**coloring number**" of G; this causes some confusion since it generally differs from the chromatic number but provides an upper bound. "Coloring bound" would be a less-confusing term.

8.1.12. Proposition. (Szekeres–Wilf [1968]) If G is a k-degenerate graph, then G is $(k+1)$-colorable. In particular, $\chi(G) \leq 1 + \max_{H \subseteq G} \delta(H)$.

Proof: If G is k-degenerate, then a smallest-last ordering gives each vertex at most k neighbors among the earlier vertices. Hence the greedy coloring for such a vertex ordering is a proper $(k+1)$-coloring. For the second statement, the maximum of $\delta(H)$ is the least k such that G is k-degenerate. ∎

A smallest-last ordering gives the smallest possible greedy coloring bound on $\chi(G)$ in terms of degrees alone, because in every ordering $u_1, \ldots, u_n$ some vertex u_i has at least $\max_{H \subseteq B} \delta(H)$ earlier neighbors (Finck–Sachs [1969]).

Greedy coloring runs fast and constructs a proper coloring even when the graph is shown only one vertex at a time, requiring an immediate and permanent choice of color. With a random vertex ordering on a random graph, the greedy algorithm almost always uses only about twice as many colors as the minimum, although with a bad ordering it may use many colors on a tree (Exercise 23).

Other results bound $\chi(G)$ using orientations of G.

8.1.13. Theorem. (**Minty's Theorem**; Minty [1962]) In a connected graph G that is not a forest, let $\mathbf{C}$ be the set of all cycles, listed in both directions. For an orientation D of G, let $r(D) = \max_{C \in \mathbf{C}} \lceil a/b \rceil$, where a and b are the numbers of edges of D followed forward and backward by C, respectively. Always $\chi(G) \leq 1 + \min_D r(D)$, with equality for some orientation.

Proof: When D has a cycle, $r(D)$ is infinite, and the bound holds. Hence we may assume D is acyclic. We produce a proper coloring of G with $1 + r(D)$ colors. We may assume that G is connected.

Fix a vertex $x \in V(G)$. For a walk W starting at x, let $g(W) = a - b \cdot r(D)$, where a and b are the numbers of steps along W that are forward or backward in D, respectively. For $y \in V(G)$, let $g(y) = \max\{g(W): W \text{ is an } x, y\text{-walk}\}$. By the definition of $r(D)$, traversing a cycle of G (or traversing an edge in both directions) makes no positive contribution to $g(W)$, so $g(y)$ is the maximum over x, y-paths.

If $uv \in E(D)$, then an x, u-walk attaining $g(u)$ extends along uv to yield $g(v) \geq g(u) + 1$. An x, v-walk attaining $g(v)$ extends along uv to yield $g(u) \geq g(v) - r(D)$. Hence $g(u) + 1 \leq g(v) \leq g(u) + r(D)$. We conclude that coloring each vertex y by the congruence class of $g(y)$ modulo $1 + r(D)$ produces a proper $r(D)$-coloring.

For sharpness, consider a proper coloring using colors $1, \ldots, \chi(G)$. Form D^* by orienting each edge from lower color to higher color. Colors increase along every path, so every path has length at most $\chi(G) - 1$. A cycle thus follows at most $\chi(G) - 1$ forward edges before a backward edge. Hence $a/b \leq \chi(G) - 1$ for each cycle. Since $\chi(G)$ is an integer, we obtain $\chi(G) \geq 1 + r(D^*) \geq 1 + \min_D r(D)$.   ∎

Minty's Theorem implies the subsequent better-known Gallai–Roy Theorem, proved in Gallai [1968], Roy [1967], Vitaver [1962], and Hasse [1964/5] (in English, French, German, and Russian, respectively).

8.1.14. Corollary. (**Gallai–Roy Theorem**) $\chi(G) \leq 1 + l(D)$ for every orientation D of G, where $l(D)$ is the length of a longest path in D. Equality holds for some orientation.

Proof: Suppose first that D is acyclic. Since $\chi(G) \leq 1 + r(D)$ by Minty's Theorem, it suffices to prove $r(D) \leq l(D)$. Since $r(D) = \max_C \lceil a/b \rceil$, along a cycle C achieving the maximum we have $a/b > r(D) - 1$. By the Pigeonhole Principle, C has at least $r(D)$ forward edges between some two backward edges. Hence $l(D) \geq r(D)$.

When D is not acyclic, $r(D)$ is infinite. We instead find an acyclic orientation $\widehat{D}$ with $l(\widehat{D}) \leq l(D)$, so $\chi(G) \leq 1 + r(\widehat{D}) \leq 1 + l(\widehat{D}) \leq 1 + l(D)$.

Let D' be a maximal acyclic subdigraph of D. Obtain $\widehat{D}$ from D by reversing all edges of D not in D'. If $\widehat{D}$ contains a cycle C', using reversed edges $e_1, \ldots, e_k$ that create cycles $C_1, \ldots, C_k$ in D when added to D' without reversal, then replacing each e_i in C' with $C_i - e_i$ yields a closed walk in D'. A closed walk in a digraph contains a cycle, but D' is acyclic. Hence $\widehat{D}$ is acyclic.

To prove $l(\widehat{D}) \leq l(D)$, let P be a u, v-path in $\widehat{D}$. If $xy \in E(P)$ was reversed in forming $\widehat{D}$, then D' has an x, y-path. Replacing all such edges of P with paths in D' yields a u, v-walk in D' at least as long as P; it is a path since D' is acyclic.

Equality again is achieved by directing edges from low label to high label with respect to an optimal coloring of G.   ∎

The Gallai–Roy and Minty bounds give the chromatic number exactly on any graph IF we guess the right orientation, as does greedy coloring if we guess the right vertex ordering (Exercise 23).

Tuza [1992] strengthened Theorem 8.1.13 by showing that only some cycles need be considered: if the ratio a/b in the orientation D is at most $k - 1$ for every cycle in G whose length is 1 modulo k, then $\chi(G) \leq k$. Thus if G is not k-colorable, then G has a cycle whose length is congruent to 1 modulo k (Exercise 31).

Many open problems remain for bounds on $\chi(G)$. One of the most famous is Reed's Conjecture (Reed [1998]) that $\chi(G) \leq \lceil \frac{\Delta(G) + \omega(G) + 1}{2} \rceil$ for every graph G, improving the trivial bound $\chi(G) \leq \Delta(G) + 1$.

TRIANGLE-FREE GRAPHS

What forces chromatic number to be large? We know $\chi(G) \geq \omega(G)$, but almost all graphs have large chromatic number but no large clique: when n is large, the values of $\omega(G)$, $\alpha(G)$, and $\chi(G)$ for almost all n-vertex graphs are approximately

$2 \log_2 n$, $2 \log_2 n$, and $n/(2 \log_2 n)$, respectively (see Chapter 14). Hence $\omega(G)$ is usually a weak lower bound on $\chi(G)$, while $n/\alpha(G)$ is usually a good lower bound.

There are many constructions of triangle-free graphs with large chromatic number (see Sachs [1969] for a survey). Our first is perhaps the easiest to describe and best known (see also Exercise 21).

8.1.15. Definition. Given a graph G, **Mycielski's Construction** produces a new graph G' as follows. Letting $V(G) = \{v_1, \ldots, v_n\}$, add a vertex w and an independent set U consisting of vertices $u_1, \ldots, u_n$. For each i, make u_i adjacent to all of $N_G(v_i)$. Finally, let $N_{G'}(w) = U$.

8.1.16. Example. Mycielski [1955] actually considered only the sequence generated from K_2 using Definition 8.1.15, but the construction can be applied to any graph. From K_2, one application yields the 3-chromatic C_5. A second yields the 4-chromatic **Grötzsch graph**, drawn in two ways below. ∎

 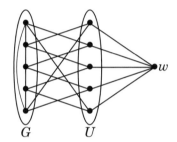

$$G \qquad\qquad U$$

8.1.17. Theorem. Given a graph G, let G' be the graph generated from G by Mycielski's construction. If G is triangle-free, then G' is triangle-free. If G is k-chromatic, then G' is $(k + 1)$-chromatic.

Proof: Name the vertices as in Definition 8.1.15, with $V(G) = \{v_1, \ldots, v_n\}$ and $V(G') = V(G) \cup U \cup \{w\}$. Since U is independent in G', the other vertices of any triangle containing a vertex $u_i \in U$ belong to $V(G)$ and are neighbors of v_i, which would yield a triangle in G. Therefore, Mycielski's Construction preserves the absence of triangles.

A proper k-coloring f of G extends to a proper $(k+1)$-coloring of G' by defining $f(u_i) = f(v_i)$ for each i and $f(w) = k + 1$; hence $\chi(G') \leq \chi(G) + 1$.

To prove that the chromatic number increases, we prove $\chi(G) < \chi(G')$. If G' has a proper k-coloring g, then we may assume $g(w) = k$, which restricts g to $\{1, \ldots, k - 1\}$ on U. Let $A = \{v_i : g(v_i) = k\}$; we change colors on A to obtain a proper $(k - 1)$-coloring of G. For each $v_i \in A$, change the color of v_i to $g(u_i)$.

Since A lies in one color class of g, it is independent. Thus we need only check colors on $\{v_i, v'\}$ with $v_i \in A$ and $v' \in V(G) - A$. If $v'v_i \in E(G)$, then also $v'u_i \in E(G')$, which yields $g(v') \neq g(u_i)$. Hence our alteration does not violate edges within G. We now delete $U \cup \{w\}$ and have a proper $(k - 1)$-coloring of G. ∎

Iterating Mycielski's Construction from $G_2 = K_2$ yields the smallest k-chromatic graphs for $k \leq 4$, but then the graphs grow rapidly: $|V(G_k)| = 2|V(G_{k-1})| + 1$ yields the exponential growth $|V(G_k)| = 3 \cdot 2^{k-2} - 1$. How many vertices are needed?

8.1.18. Proposition. If G is a triangle-free n-vertex graph, then $\chi(G) \le 2\sqrt{n}$. Equivalently, $n \ge k^2/4$ when G is triangle-free and k-chromatic.

Proof: The neighborhood of any vertex in G is an independent set; we can use one color on it. This suggests an algorithm. As long as G has a vertex with at least $\lfloor \sqrt{n} \rfloor$ neighbors not yet colored, use one new color on those vertices.

Because G has only n vertices, this phase uses at most $\sqrt{n}$ colors. Afterwards, the subgraph induced by the remaining vertices has maximum degree less than $\lfloor \sqrt{n} \rfloor$. Thus greedy coloring with any ordering properly colors the remaining subgraph using at most $\sqrt{n}$ additional colors.

Finally, $\chi(G) \le 2\sqrt{n}$ if and only if $n \ge \chi(G)^2/4$.   ∎

The minimum number of vertices needed is surprisingly close to $k^2/4$, but building such small k-chromatic triangle-free graphs is not easy.

8.1.19. Remark. Let $f(k)$ be the smallest order of a triangle-free k-chromatic graph. Using probabilistic (non-constructive) methods (see Theorem 14.2.22), Erdős [1961] proved $f(k) \le c(k \log k)^2$. We now know $c_1 k^2 \log k \le f(k) \le c_2 k^2 \log k$ for some constants c_1 and c_2 (see Chung–Graham [1998, p. 61]). Explicit constructions of relatively small triangle-free k-chromatic graphs appear in Lubotzky–Phillips–Sarnak [1988] and Kriz [1989].

In fact, we can forbid all cycles up to any fixed length and still have large chromatic number. Methods to prove this involve hypergraphs or probabilistic constructions, so we postpone proofs to Chapters 10 and 14.   ∎

EXERCISES 8.1

8.1.1. $(-)$ Compute $\alpha(G)$ and $\chi(G)$ for the graph G below.

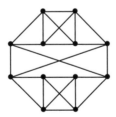

8.1.2. $(-)$ Let the blocks of G be $G_1, \ldots, G_k$. Prove that $\chi(G) = \max_i \chi(G_i)$.

8.1.3. $(-)$ Find an optimal coloring of $C_5 \square C_5$ where sizes of color classes differ by at most 1.

8.1.4. $(-)$ For each $k \in \mathbb{N}$ with $k \ge 2$, construct a k-chromatic graph having no optimal coloring using a color class of size $\alpha(G)$.

8.1.5. $(-)$ Prove or disprove: $\chi(G) \le 1 + a(G)$ for every connected graph G, where $a(G)$ is the average vertex degree in G.

8.1.6. $(-)$ Prove or disprove: If G is an n-vertex graph, and n is sufficiently large, then $\chi(G) \le \omega(G) + n/\alpha(G)$.

8.1.7. (−) Prove that $\chi(G) \geq \frac{n}{n-\delta(G)}$ for every n-vertex graph G. Prove that there are infinitely many graphs (other than complete graphs) where equality holds.

8.1.8. (−) Use the Gallai–Roy Theorem (Corollary 8.1.14) to prove that every tournament has a spanning path.

8.1.9. Let G be a graph without $K_{1,t+1}$ as an induced subgraph. Prove $\chi(G) \geq 1 + \Delta(G)/t$. For $k, t \in \mathbb{N}$, construct a k-chromatic example where equality holds. (Zaker [2011a])

8.1.10. Determine the largest possible minimum degree of an n-vertex k-colorable graph.

8.1.11. (◇) Let G be a graph in which any two odd cycles have a common vertex. Prove $\chi(G) \leq 5$. Construct a graph to show that the bound cannot be improved.

8.1.12. Suppose that every edge of a graph G appears in at most one cycle. Prove that every block of G is an edge or a cycle. Use this to prove $\chi(G) \leq 3$.

8.1.13. (◇) Let G be the **unit-distance graph** in the plane; the vertices are all the points in $\mathbb{R}^2$, adjacent when their Euclidean distance is exactly 1. Prove $4 \leq \chi(G) \leq 7$. (Comment: de Grey [2018] and Heule [2018] proved $\chi(G) \geq 5$; see Soifer [2008, 2019].)

8.1.14. (◇) Given a set of lines in the plane with no three meeting at a point, form a graph G whose vertices are the intersections of the lines, with two vertices adjacent if they appear consecutively on one of the lines. Prove $\chi(G) \leq 3$. (Comment: This can fail when three lines may meet at a point.) (H. Sachs)

8.1.15. Given a spanning tree T in a graph G, let $\chi(G; T)$ denote the least k such that G has a proper coloring using colors in $[k]$ such that vertices adjacent in T have colors differing by at least 2. The **backbone chromatic number** of G is the minimum of $\chi(G; T)$ over all spanning trees T. Prove that every nonbipartite graph has backbone chromatic number at least 4. (The concept originated in Broersma–Fomin–Golovach–Woeginger [2007].)

8.1.16. The **integer simplex** with dimension d and side-length m is the graph T_m^d whose vertices are the nonnegative integer $(d + 1)$-tuples summing to m, with two vertices adjacent when they differ by 1 in two positions and agree elsewhere. Determine $\chi(T_m^d)$.

8.1.17. (◇) The **Kneser graph** $K(n, k)$ has vertex set $\binom{[n]}{k}$, with two vertices are adjacent when they are disjoint k-sets (the Petersen graph is $K(5, 2)$). Prove $\chi(K(n, k)) \leq n - 2k + 2$. Prove that this is optimal when $n = 2k + 1$. (Comment: Lovász [1978] proved Kneser's conjecture that always $\chi(K(n, k)) = n - 2k + 2$; see Theorem 16.2.29.)

8.1.18. Let G be the intersection graph of a family of equal-sized squares in the plane, all having horizontal and vertical sides. Prove that G has a vertex whose neighborhood is covered by two cliques. Conclude that $\chi(G) \leq 2\omega(G) - 1$.

8.1.19. For a graph G containing no 4-cycle, prove $\chi(G) \leq \alpha'(G) + 2$. (Zaker [2011b])

8.1.20. Let G be a 3-regular K_4-free graph with m edges. Prove that G has a bipartite subgraph with at least $\frac{7}{9}m$ edges. (Hint: Use Brooks' Theorem.)

8.1.21. The **achromatic number** of a graph G is the maximum number of colors in a proper coloring of G such that all pairs of colors appear on adjacent vertices.
  (a) Show that the achromatic number is well defined for every graph G.
  (b) For $n \geq 3$, determine the achromatic number of the cycle C_n.

8.1.22. Consider the coloring algorithm that iteratively chooses a maximal independent set and gives it the next color. For $k \in \mathbb{N}$, construct a 2-colorable graph on which this algorithm may use k colors.

8.1.23. ($\diamond$) *Greedy coloring.* Let G be a graph.
(a) Prove that G has a vertex ordering where greedy coloring uses $\chi(G)$ colors.
(b) For $k \in \mathbb{N}$, construct a tree T_k with maximum degree k and an ordering σ of $V(T_k)$ such that greedy coloring with respect to σ uses $k + 1$ colors. (Comment: Thus the performance ratio of greedy coloring to optimal coloring can be $(\Delta(G) + 1)/2$.) (Bean [1976])

8.1.24. Prove that 11 is the least number of vertices in a triangle-free 4-chromatic graph.

8.1.25. Prove that an n-vertex graph with chromatic number k has at most k^{n-k} vertex partitions into k independent sets, with equality only for $K_k + (n - k)K_1$.

8.1.26. The graph below has a proper 3-coloring with each color used twice. Prove that it cannot arise from greedy coloring. (Fon-Der-Flaass)

8.1.27. Let G be an n-vertex graph that does not contain K_{r+1}. Prove $\chi(G) \leq 2^{r-1} n^{1 - 1/2^{r-1}}$. (Comment: This generalizes Proposition 8.1.18.)

8.1.28. ($\diamond$) Prove that if G has no induced $2K_2$, then $\chi(G) \leq \binom{\omega(G)+1}{2}$. (Hint: Cover the vertices using $\binom{\omega(G)}{2} + \omega(G)$ independent sets.) (Wagon [1980])

8.1.29. ($\diamond$) Let G be a k-degenerate graph, and let a and b be positive integers such that $a + b = k - 1$. Prove that $V(G)$ can be partitioned into sets A and B such that $G[A]$ is a-degenerate and $G[B]$ is b-degenerate. Prove that the vertex set of any graph G can be partitioned into $\max_{H \subseteq G} \lceil \frac{\delta(H)+1}{d+1} \rceil$ sets inducing d-degenerate subgraphs. (Comment: Proposition 8.1.12 is the case $d = 0$. The case $d = 1$ is in Chartrand–Kronk [1969].)

8.1.30. Let G be an n-vertex graph. Prove that if $\chi(G) = n - k$, then $\omega(G) \geq n - 2k$. For even k and $n \geq 5k/2$, construct an n-vertex graph with chromatic number $n - k$ and clique number at most $n - 3k/2$.

8.1.31. ($\diamond$) *Chromatic number and cycle lengths.* Let G be a connected graph.
(a) Fix $v \in V(G)$. Choose a spanning tree T to maximize $\sum_{u \in V(G)} d_T(u, v)$. Prove that the endpoints of any edge in G lie along a path in T that starts at v.
(b) Prove that if $\chi(G) > k \geq 2$, then G has a cycle with length congruent to 1 modulo k. (Hint: Define a k-coloring using the tree T of part (a). Comment: This generalizes the characterization of bipartite graphs.) (Tuza [1992])
(c) Prove that a graph having no odd cycle of length more than $2j - 1$ is $2j$-colorable. (Erdős–Hajnal [1966])

8.1.32. ($\diamond$) Prove that a graph is k-colorable if every vertex lies in fewer than $\binom{k}{2}$ odd cycles. (Hint: Use induction on $|V(G)|$.) (Stong [2006])

8.1.33. Prove that $\chi(\overline{H}) = \omega(\overline{H})$ when H is bipartite.

8.1.34. Permutation graphs. Let G be the intersection graph of segments having endpoints on two parallel lines. Prove $\chi(G) = \omega(G)$. (Pnueli–Lempel–Even [1971])

8.1.35. ($\diamond$) Let G and H be graphs. Prove $\chi(G \square H) = \max\{\chi(G), \chi(H)\}$. (Sabidussi [1957], Vizing [1963], Aberth [1964], Behzad–Mahmoodian [1969])

8.1.36. ($\diamond$) Prove that an n-vertex graph G is r-colorable if and only if the cartesian product $G \square K_r$ has an independent set of size n. (Plesnevič–Vizing [1965])

8.1.37. ($\diamond$) Prove that a graph G is 2^k-colorable if and only if G is a union of k bipartite graphs. (Harary–Hsu–Miller [1977])

8.1.38. ($\diamond$) Prove that every k-chromatic graph has at least $\binom{k}{2}$ edges. Conclude that if G is the union of t copies of K_t, then $\chi(G) < 1 + t\sqrt{t-1}$. (Comment: This bound is near tight, but see Conjecture 10.1.33.) (Horák–Tuza [1990])

8.1.39. Let G be an n-vertex graph. Prove $\chi(G) \cdot \chi(\overline{G}) \geq n$ and $\chi(G) + \chi(\overline{G}) \geq 2\sqrt{n}$. Prove that equality can hold when $\sqrt{n}$ is an integer. (Nordhaus–Gaddum [1956], Finck [1968])

8.1.40. ($\diamond$) Let G be an n-vertex graph.
   a) Prove $\chi(G) + \chi(\overline{G}) \leq n + 1$. (Nordhaus–Gaddum [1956])
   b) Using part (a), prove $\chi(G) \cdot \chi(\overline{G}) \leq (n+1)^2/4$. From this, prove $\chi(G) \leq \frac{(n+1)^2}{4\alpha(G)}$. For each odd n, construct G such that $\chi(G) = \frac{(n+1)^2}{4\alpha(G)}$. (Nordhaus–Gaddum [1956], Finck [1968])

8.1.41. For edge-disjoint graphs G and H on the same set of n vertices, let $f(G,H) = \chi(G) + \chi(H) - \chi(G \cup H)$. Determine $\max f(G,H)$ and $\lim_{n \to \infty} \frac{\min f(G,H)}{n}$. (Hint: Use Exercise 8.1.40(a).) (Bloome–Johnson–Saritzky [2012])

8.1.42. ($\diamond$) Use the Gallai–Roy Theorem (Corollary 8.1.14) to prove that every n-vertex digraph D has a path with at least $n/\alpha(D)$ vertices. Conclude the following theorem of Erdős–Szekeres [1935]: every list of $rs + 1$ distinct numbers has an increasing sublist of size $r + 1$ or a decreasing sublist of size $s + 1$. (Schmeichel)

8.1.43. *Paths and chromatic number in digraphs.*
   (a) Let G be the union of graphs F and H. Prove $\chi(G) \leq \chi(F)\chi(H)$.
   (b) Let D be an orientation of a graph G with $\chi(G) > rs$. Assign each $v \in V(D)$ a real number $f(v)$. Use (a) and the Gallai–Roy Theorem to prove that D has a path $\langle u_0, \ldots, u_r \rangle$ with $f(u_0) \leq \cdots \leq f(u_r)$ or a path $\langle v_0, \ldots, v_s \rangle$ with $f(v_0) > \cdots > f(v_s)$.
   (c) Use part (b) to conclude the Erdős–Szekeres Theorem (Exercise 8.1.42).

8.1.44. Prove that every digraph decomposes into two acyclic digraphs.

8.2. Structural Aspects

Since computing the chromatic number is very difficult (testing $\chi(G) \leq k$ is NP-complete when k is a fixed integer at least 3), we study various structural aspects of k-chromatic graphs to understand what makes them hard to color.

COLOR-CRITICAL GRAPHS

8.2.1. Definition. A graph G is **color-critical** if $\chi(H) < \chi(G)$ for every proper subgraph H of G. If also $\chi(G) = k$, then G is k-**critical**.

Every k-chromatic graph contains a k-critical subgraph. The only 1-critical and 2-critical graphs are K_1 and K_2. The 3-critical graphs are the odd cycles. Some structural properties follow easily.

8.2.2. Proposition. If G is a k-critical graph, then $\delta(G) \geq k - 1$.

Proof: If some vertex x has smaller degree, then a proper $(k-1)$-coloring of $G - x$ (guaranteed by k-criticality) extends to G by giving x a color not used on $N(x)$. ∎

Proposition 8.2.2 can be used instead of greedy coloring to prove Propositions 8.1.10–8.1.12 (Exercise 10). Using greedy coloring underscores the algorithmic nature of upper bounds. Properties of k-critical graphs don't help to find optimal colorings, because we have no good algorithm to find k-critical subgraphs. Nevertheless, these properties are useful for proving bounds and for understanding the structure of graphs with large chromatic number.

The technique of proof in Proposition 8.2.2 is typical for proving properties of k-critical graphs. If the desired property of k-critical graphs fails for such a graph G, then we obtain a proper $(k - 1)$-coloring of an appropriate subgraph of G and use the failure of the desired property to produce a proper $(k - 1)$-coloring of G. This contradiction implies that the desired property does hold.

8.2.3. Remark. A nontrivial graph G is color-critical if and only if
    (1) G has no isolated vertex, and
    (2) $\chi(G - e) < \chi(G)$ for every $e \in E(G)$.
Hence to prove that a connected graph is color-critical, we need only consider subgraphs obtained by deleting a single edge. ∎

8.2.4. Proposition. If $v \in V(G)$ and $\chi(G - v) < \chi(G) = k$, then G has a proper k-coloring f having $f(v)$ only on v and the other $k - 1$ colors all on $N(v)$.
    If $e \in E(G)$ and $\chi(G - e) < \chi(G) = k$, then every proper $(k - 1)$-coloring of $G - e$ gives the same color to the endpoints of e.

Proof: Let f be a proper $(k - 1)$-coloring of $G - v$. If any color is not used in $N(v)$, then we can use it on v to complete a proper $(k - 1)$-coloring of G. This contradicts $\chi(G) = k$, so the colors all appear on $N(v)$, and extending f to G by letting $f(v) = k$ completes the desired coloring.

If some proper $(k - 1)$-coloring of $G - e$ gave distinct colors to the endpoints of e, then it would also be a proper $(k - 1)$-coloring of G. ∎

Remark 8.2.3 and Proposition 8.2.4 can be used to show that if G is color-critical, then applying Mycielski's Construction yields a color-critical graph with larger chromatic number (Exercise 16). For k-critical graphs, we strengthen Proposition 8.2.2 to a lower bound on connectivity instead of minimum degree.

8.2.5. Theorem. (Dirac [1953]) Every k-critical graph is $(k - 1)$-edge-connected.

Proof: (Dirac–Sorensen–Toft [1974]) Let G be k-critical, and let $[X, Y]$ be an edge cut. Since G is k-critical, $G[X]$ and $G[Y]$ are $(k - 1)$-colorable. Let $X_1, \ldots, X_{k-1}$ and $Y_1, \ldots, Y_{k-1}$ be the color classes in proper $(k-1)$-colorings of $G[X]$ and $G[Y]$.

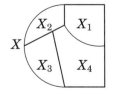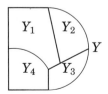

By induction on r, we show that if $\big|[X,Y]\big| < r$ with X and Y partitioned into $X_1,\ldots,X_r$ and $Y_1,\ldots,Y_r$, then $Y_1,\ldots,Y_r$ can be reindexed so that $[X_i,Y_i]$ is empty for all i. This is trivial for $r = 1$. For $r > 1$, some X_i is incident to no edge of $[X,Y]$. If some $e \in [X,Y]$ exists, then let Y_i be the subset of Y incident to e; otherwise any pairing works. The induction hypothesis completes the pairing.

We can thus pair the colors in the proper $(k-1)$-colorings of $G[X]$ and $G[Y]$ to produce a proper $(k-1)$-coloring of G, which contradicts the hypothesis. ∎

A k-critical graph need not be $(k-1)$-connected (Exercise 8 and Exercise 24). Theorem 8.2.5 holds also for graphs where deleting any vertex decreases the chromatic number. This is a larger class (Exercise 5).

LIST COLORING

In graph coloring problems, some colors may be forbidden from certain vertices, representing unavailable resources. Also, when extending partial colorings, the color on a colored vertex is forbidden from use on its neighbors. Such applications lead to the more general model of "list coloring", introduced in Vizing [1976] and Erdős–Rubin–Taylor [1979] and surveyed in Stiebitz–Voigt [2015].

8.2.6. Definition. For a graph G, a **list assignment** L assigns to each vertex $v \in V(G)$ a set $L(v)$ of colors allowed at v. An **L-coloring** is a proper coloring ϕ of G such that $\phi(v) \in L(v)$ for all v. A graph G is **k-choosable** or **list k-colorable** if it has an L-coloring whenever $|L(v)| \geq k$ for all v. The **list chromatic number** or **choice number** or **choosability** $\chi_l(G)$ is the minimum k such that G is k-choosable.

Here the "lists" are actually sets; there is no order or multiplicity for the colors. Nevertheless, the term "list" is thoroughly entrenched for this model, and "set coloring" has other meanings. Exercises 30–39 concern list coloring.

One option for L is assigning all vertices the same list S, in which case an L-coloring exists if and only if $|S| \geq \chi(G)$. Hence $\chi_l(G) \geq \chi(G)$ for every graph G. However, $\chi_l(G)$ may be much larger than $\chi(G)$, even when G is bipartite.

8.2.7. Example. *Choice numbers of C_{2m} and $K_{m,m}$.*
Even cycles are 2-choosable. We may assume that all lists have size 2. If all the lists are the same, then alternate the colors along the cycle. Otherwise, find adjacent vertices x and y such that $L(x)$ has a color c not in $L(y)$. Use c on x, and then color the other vertices in order from x to y. At each new vertex, choose a color from its list that was not used on the previous vertex. Such a choice is always available, and the coloring is proper since the colors on x and y also differ.

As observed in Erdős–Rubin–Taylor [1979], $\chi_l(K_{m,m}) > k$ when $m = \binom{2k-1}{k}$. In each part, let L assign all k-subsets of $[2k-1]$ as lists. In an L-coloring, at least k colors must be used on each part, since otherwise some vertex has no color chosen from the k in its list. Since altogether the lists have only $2k-1$ colors, some color is used on each part. Since $K_{m,m}$ is complete bipartite, we now have adjacent vertices with the same color, so the coloring is not proper. Since $K_{m,m}$ is not L-colorable for this assignment of lists of size k, it is not k-choosable. ∎

Some bounds on chromatic number are strengthened by proving that they hold also for χ_l, such as the Szekeres–Wilf (degeneracy) bound.

8.2.8. Proposition. Every k-degenerate graph is $(k + 1)$-choosable. Thus also
$$\chi_l(G) \leq 1 + \max_{H \subseteq G} \delta(H) \leq 1 + \Delta(G).$$

Proof: When G is k-degenerate, iteratively delete a vertex with at most k neighbors in what remains. In the reverse ordering, σ, each vertex has at most k earlier neighbors. When all lists have size at least $k + 1$, coloring the vertices in the order of σ guarantees that each vertex, when reached, has a color in its list not used on earlier neighbors. ∎

We will prove $\chi_l(G) \leq \Delta(G)$ when G is connected and not a complete graph or an odd cycle. This is stronger than Brooks' Theorem, since $\chi_l(G) \geq \chi(G)$.

8.2.9. Lemma. (Vizing [1976]) Given a connected graph G, let L be a list assignment such that $|L(v)| \geq d(v)$ for all v.
(a) If $|L(y)| > d(y)$ for some vertex y, then G is L-colorable.
(b) If G is 2-connected and some two lists differ, then G is L-colorable.

Proof: (a) Order the vertices by iteratively deleting leaves of a spanning tree rooted at y; each vertex before y has a later neighbor. Choose colors for vertices in this order. At a vertex v earlier than y, we have colored fewer than $d(v)$ neighbors (since v has a later neighbor), so a color remains available in $L(v)$ for v. Although y has no later neighbor, $|L(y)| > d(y)$, so also at y a color is available.

(b) Since the lists are not identical and G is connected, we can find adjacent x and y such that $L(x) - L(y) \neq \varnothing$. Choose $c \in L(x) - L(y)$. Define L' on $V(G - x)$ by $L'(v) = L(v)$ if $v \notin N(x)$ and $L'(v) = L(v) - c$ if $v \in N(x)$. We have $|L'(v)| \geq d_{G-x}(v)$ for all $v \in V(G - x)$, and $|L'(y)| > d_{G-x}(y)$ (since $c \notin L(y)$). Hence part (a) yields an L'-coloring of $G - x$, which extends to an L-coloring of G by using c on x. ∎

8.2.10. Definition. A graph G is f-**choosable** if it is L-colorable whenever $|L(v)| \geq f(v)$ for each vertex v, where $f: V(G) \to \mathbb{N}$. The graph is **degree-choosable** if it is L-colorable whenever $|L(v)| \geq d(v)$ for each vertex v.

8.2.11. Lemma. If a connected graph G has a degree-choosable induced subgraph H, then G is degree-choosable.

Proof: Let L be a list assignment with $|L(u)| \geq d(u)$ for each $u \in V(G)$. Using the same lists, every component of $G - V(H)$ has a vertex y with $d_{G-V(H)}(y) < |L(y)|$, since G is connected. By Lemma 8.2.9a, $G - V(H)$ has an L-coloring f.

Form a list assignment L' for H by deleting from $L(v)$ the colors used by f on neighbors of v. We lose at most one color for each neighbor not in $V(H)$. Since $|L(v)| \geq d_G(v)$, we have $|L'(v)| \geq d_H(v)$ for $v \in V(H)$. Since H is degree-choosable, H has an L'-coloring f'. Together, f and f' form an L-coloring of G. ∎

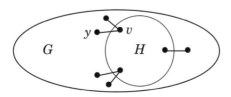

We need an elementary structural result, called **Rubin's Block Theorem** in Entringer [1985]. The proofs by Entringer and Hladký–Král'–Schauz [2010] are shorter than the original, as is the proof here.

8.2.12. Lemma. (Erdős–Rubin–Taylor [1979]) Every 2-connected graph G that is not a complete graph or odd cycle has an even cycle with at most one chord.

Proof: Since G is 2-connected, G contains a cycle. If G contains a triangle, then let Q be a largest complete subgraph of G. Since G is not complete, G has a vertex outside Q. Since G is 2-connected, we may pick a shortest path P joining some vertices u and v of Q via edges not in Q. If P has length at least 3, then its internal vertices have no neighbors in Q, and P forms the desired even cycle with uv or with a u, v-path of length 2 in Q. If P has length 2, then its internal vertex z has some nonneighbor w in $V(Q)$, and the desired cycle is $[u, z, v, w]$.

Hence G has no triangle. A shortest cycle C in G has no chord, so we may assume that C has odd length. Since G is not an odd cycle and is 2-connected, some path joins vertices of C via edges not in C, forming an even cycle with part of C. Hence G has a shortest even cycle C'. Chords of C' create two odd cycles with C'. Any two such chords yield a shorter even cycle (whether they cross or not), since G has no triangle. Thus C' is an even cycle with at most one chord. ∎

The chromatic number of a complete graph or odd cycle exceeds its degree, so these graphs are not degree-choosable.

8.2.13. Theorem. (Borodin [1977], Erdős–Rubin–Taylor [1979]) If graph G is not degree-choosable, then every block of G is a complete graph or an odd cycle.

Proof: By Lemma 8.2.11, it suffices to show that any block B other than a complete graph or odd cycle is degree-choosable. A single edge is a complete graph, so we may assume B is 2-connected. By Lemma 8.2.12, B has a subgraph that is an even cycle with at most one chord. By Lemma 8.2.11, it suffices to show that such a graph H is degree-choosable.

Let L be a list assignment with $|L(v)| \geq d_H(v)$ for all v. If the lists are not identical, then Lemma 8.2.9b applies (H is 2-connected). If they are identical, then ordinary proper coloring suffices, using two colors when H is an even cycle, three colors when H is a cycle with one chord. ∎

Exercise 30 establishes the converse of Theorem 8.2.13; degree-choosability fails when every block is a complete graph or odd cycle.

8.2.14. Corollary. (List Extension of **Brooks' Theorem**) If a connected graph G is not a complete graph or an odd cycle, then $\chi_l(G) \leq \Delta(G)$.

Proof: We prove the contrapositive. Suppose $\chi_l(G) > \Delta(G)$.

By Proposition 8.2.8, G is not $(\Delta(G) - 1)$-degenerate. Hence $\delta(H) \geq \Delta(G)$ for some subgraph H. Since G is connected and vertices with degree $\Delta(G)$ in H have no neighbors outside H, we have $H = G$.

Theorem 8.2.13 now implies that every block of G is a complete graph or an odd cycle. Since G is regular, G must therefore have only one block, since the cut-vertex in a leaf block would have higher degree than the other vertices. Hence G is a complete graph or an odd cycle. ∎

FORCED SUBGRAPHS (optional)

What subgraphs must appear in a k-chromatic graph? We have seen that the chromatic number can be arbitrarily large even when triangles are forbidden. Thus a k-chromatic graph need not contain K_k, but perhaps it must contain some weaker version of a k-clique.

8.2.15. Remark. For a graph F, an F**-subdivision** is a graph obtained from F by successive edge subdivisions. Equivalently, each edge of F is replaced by a path of length at least 1 whose internal vertices are new vertices.

The **Hajós Conjecture** asserts that every k-chromatic graph contains a K_k-subdivision. The thesis of Dirac [1951, p. 5] attributes this question to Hajós in the 1940s. The claim is trivial for $k \leq 3$, since every 3-chromatic graph contains a cycle. As we will show next, Dirac [1952b] proved a stronger version of the conjecture for $k = 4$. Catlin [1979] found counterexamples for $k \geq 7$ (see Exercise 42 for $k \in \{7, 8\}$). For $k = 5$ and $k = 6$, the question is open.

The **Hadwiger Conjecture** (Hadwiger [1943]) is weaker: every k-chromatic graph contains a subgraph reducible to K_k by edge contractions. This is weaker because a K_k-subdivision *is* a graph reducible to K_k by edge contractions. The Hadwiger Conjecture remains open for $k \geq 7$. For $k = 4$, it is equivalent to Hajós' Conjecture (and proved by Hadwiger [1943]). For $k = 5$, it is equivalent to the Four Color Theorem described in the next chapter. For $k = 6$, it was proved by Robertson–Seymour–Thomas [1993] using the Four Color Theorem. Seymour [2016] gave a survey of results on Hadwiger's Conjecture.

In fact, Hajós' Conjecture fails for almost all graphs, while Hadwiger's Conjecture holds for almost all graphs, as discussed in Section 14.4. ∎

Every 4-chromatic graph has a 4-critical subgraph, which has minimum degree at least 3. Hence our next theorem is stronger than guaranteeing K_4-subdivisions in 4-chromatic graphs.

8.2.16. Definition. For $S \subseteq V(G)$, an S**-lobe** of G is a subgraph of G induced by the union of S and the vertices of a component of $G - S$.

8.2.17. Theorem. (Dirac [1952a]) Every graph with minimum degree at least 3 contains a K_4-subdivision.

Proof: Say that a vertex is *weak* if it has degree less than 3. By induction on n, we prove the stronger result that every nontrivial n-vertex graph G with at most one weak vertex contains a K_4-subdivision. We need this stronger statement because we will apply the induction hypothesis to graphs having a weak vertex. For the base step, the only nontrivial graph with at most four vertices that has at most one weak vertex is K_4 itself, which trivially contains a K_4-subdivision.

For $S \subseteq V(G)$, only one S-lobe of G can contain a weak vertex of G not in S, since G has only one weak vertex. All other S-lobes are *good* S-lobes. Let S be a smallest vertex cut. When $|S| \leq 1$, let G' be a good S-lobe. The only possible weak vertex in G' is the vertex of S (when $|S| = 1$). By the induction hypothesis, G' contains a K_4-subdivision, which is also contained in G.

When $|S| = 2$, let $S = \{x, y\}$, and obtain G' from a good S-lobe of G by adding xy if $xy \notin E(G)$ (see the figure below). In G' only x and y can be weak. If at most one is weak, then G' contains a K_4-subdivision H. Replacing xy (if $xy \in E(H)$) with a path through another S-lobe of G yields a K_4-subdivision in G.

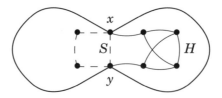

If both x and y are weak in G', then each has exactly one neighbor outside S in H. These neighbors are distinct, since $\kappa(G) \geq 2$. Form G'' from G' by contracting the edge xy. The only weak vertex in G'' is the new vertex z, so G'' contains a K_4-subdivision H. Subdividing an edge incident to z in H (if $z \in V(H)$) yields a K_4-subdivision in G'. Again G contains a K_4-subdivision by replacing xy with a path through another component of $G - S$, if xy is used.

Hence we may assume that G is 3-connected. Select $x \in V(G)$. Since $G - x$ is 2-connected, it has a cycle C of length at least 3. Since G is 3-connected, the Fan Lemma (Lemma 7.2.12) yields an $x, V(C)$-fan of size 3 in G. These three paths combine with C to form a K_4-subdivision in G. ∎

Many forced structures in k-chromatic graphs are in fact forced by minimum degree $k - 1$, which occurs in a k-critical subgraph. For example, large chromatic number forces a subdivision of K_t because large minimum degree is enough. Mader [1967] proved that $\delta(G) \geq 2^{\binom{t}{2}}$ yields a K_t-subdivision in G. Thomassen [1988] extended the argument to F-subdivision for any graph F.

8.2.18. Lemma. (Mader [1967], Thomassen [1988]) If $\delta(G) \geq 2k$, then G contains disjoint subgraphs G' and H such that (1) $\delta(G') \geq k$, (2) each vertex of G' has a neighbor in H, and (3) H is connected.

Proof: We may assume that G is connected. When we contract the edges of a connected subgraph H', the set $V(H')$ becomes a single vertex; let $G \cdot H'$ denote the resulting graph. Consider all connected induced subgraphs H' of G such that $G \cdot H'$ has at least $k|V(G \cdot H')|$ edges. Since $\delta(G) \geq 2k$, every 1-vertex subgraph has this property, so we may let H be a maximal such subgraph. Let G' be the subgraph induced by the vertices of $G - V(H)$ that have neighbors in H.

It remains only to show $\delta(G') \geq k$. Each $x \in V(G')$ has a neighbor y in $V(H)$. In $G \cdot (H \cup xy)$, the edges incident to x in G' collapse onto edges from $V(G')$ to the single vertex representing H in $G \cdot H$. The edge xy also contracts. Other edges of $G \cdot H$ remain in $G \cdot (H \cup xy)$, so contracting the edge xy loses $d_{G'}(x) + 1$ additional edges. The maximality of H implies that contracting xy loses at least $k + 1$ additional edges. Thus $d_{G'}(x) \geq k$. This holds for each x, so $\delta(G') \geq k$. ∎

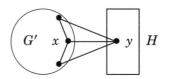

8.2.19. Theorem. (Mader [1967], Thomassen [1988]) If F has m edges and no isolated vertex, and $\delta(G) \geq 2^m$, then G has an F-subdivision.

Proof: We use induction on m; the claim is trivial for $m = 1$. Consider $m \geq 1$. By Lemma 8.2.18, there are disjoint subgraphs H and G' in G such that H is connected, $\delta(G') \geq 2^{m-1}$, and each vertex of G' has a neighbor in H.

If $\delta(F - e) \geq 1$ for some $e \in E(F)$, then the induction hypothesis yields an $(F - e)$-subdivision J in G'. Add a path through H connecting the vertices of J that represent the endpoints of e.

If $\delta(F - e) = 0$ for all $e \in E(F)$, then every edge of F has a leaf endpoint, and F is a forest of stars. If $F = K_{1,m}$, then $F \subseteq G$, since $\delta(G) \geq 2^m \geq m$. Otherwise, deleting the vertex set S of a copy of $K_{1,r}$ in G leaves $\delta(G - S) \geq 2^m - r - 1 \geq 2^{m-r}$ for $r < m$, and the remaining stars can be found iteratively in $G - S$. ∎

8.2.20.* Remark. Better bounds can be proved with more detailed arguments. When F is a tree with m edges, $\delta(G) \geq m$ suffices (Exercise 5.4.17). For $F = K_k$, let $f(k)$ be the smallest $\delta(G)$ that forces a K_k-subdivision in G; Theorem 8.2.19 yields $f(k) \leq 2^{\binom{k}{2}}$. In fact, $\frac{1}{8}k^2 < f(k) < ck^2$. Komlós–Szemerédi [1996] and Bollobás–Thomason [1998] proved the upper bound (the latter shows $c \leq 256$); the lower bound follows from $K_{r,r}$ with r about $\binom{k/2}{2}$ (Exercise 41).

Theorem 8.2.17 yields $f(4) = 3$. Furthermore, $f(5) = 6$. The icosahedron (pictured in Exercise 9.3.32) yields $f(5) \geq 6$, since this graph is 5-regular and has no K_5-subdivision (being planar). Equality holds because every n-vertex graph with at least $3n - 5$ edges contains a K_5-subdivision (conjectured by Dirac [1964] and proved by Mader [1998]). ∎

The chromatic number is unbounded not only for triangle-free graphs, but in fact for the family graphs with any fixed girth, as shown in Chapter 10 and in Chapter 14. Thus the only fixed graphs that must appear in graphs with large chromatic number are forests.

Since G contains every k-vertex tree when $\delta(G) \geq k - 1$ (Exercise 5.4.17), each k-vertex tree appears in every k-critical graph. To strengthen this, we seek a given forest F as an *induced* subgraph. Equivalently, we want to show that forbidding large cliques and induced copies of F bounds the chromatic number.

8.2.21. Definition. A family **G** of graphs is χ-**bounded** if there is a function f such that $\chi(G) \leq f(\omega(G))$ for $G \in$ **G**. A family is **weakly χ-bounded** if there is a bound on the chromatic number of its triangle-free members. A graph is F-**free** if it does not contain the graph F as an induced subgraph. The family of F-free graphs is Forb(F).

If Forb(F) is χ-bounded and F' is an induced subgraph of F, then Forb(F') is also χ-bounded, since Forb(F') $\subseteq$ Forb(F). We have observed that if Forb(F) is χ-bounded, then F must be a forest.

8.2.22. Conjecture. (Gyárfás [1975], Sumner [1981]) For every forest F, Forb(F) is χ-bounded. ∎

For example, Forb($2K_2$) is χ-bounded (Exercise 8.1.28): $G \in$ Forb($2K_2$) implies $\chi(G) \leq \binom{\omega(G)+1}{2}$. In studying Conjecture 8.2.22, it suffices to consider trees, since forests are induced subgraphs of trees. The **broom** $B_{r,s}$ is the tree with $r+s$ vertices consisting of the path P_r plus s edges attached to one end. Gyárfás [1985] proved that $\text{Forb}(B_{r,s})$ is χ-bounded. We present only a weaker earlier result.

8.2.23. Theorem. (Gyárfás–Szemerédi–Tuza [1980]) The class $\text{Forb}(B_{r,s})$ is weakly χ-bounded. In particular, triangle-free $B_{r,s}$-free graphs have chromatic number less than $r + s$.

Proof: It suffices to show that every vertex-color-critical triangle-free graph with chromatic number $r + s$ has $B_{r,s}$ as an induced subgraph. We prove a stronger statement. Let the *end* of the broom $B_{r,s}$ be the vertex farthest from the vertex of degree $s + 1$. Let v be a vertex in a triangle-free connected graph G such that $\chi(G) \geq r+s$. We prove that if all vertices other than v have degree at least $r+s-1$, then G has an induced copy of $B_{r',s}$ with its end at v, for some r' with $r' \geq r$.

We use induction on r. For $r = 2$, the desired broom is a star, and we can take any neighbor of v as the center, since G is triangle-free.

Consider $r > 2$. Let $G' = G - N[v]$. From G', iteratively delete vertices that have degree less than $r+s-2$ in the graph that remains. Let W be the final set of vertices; we claim $\chi(G[W]) \geq r+s-1$. Otherwise, iteratively replace the deleted vertices and color them properly from $[r + s - 2]$ to obtain a proper $(r + s - 2)$-coloring of G', add $N(v)$ with color $r + s - 1$, and assign any color in $[r + s - 2]$ to v. This yields a proper $(r + s - 1)$-coloring of G, a contradiction.

Hence $\chi(H) \geq r + s - 1$ for some component H of $G[W]$. Let v' be the last vertex before $V(H)$ on a shortest path P from v to $V(H)$. Let H' be the subgraph of G induced by $V(H) \cup \{v'\}$. We have $\delta(H) \geq r+s-2$, since vertices with smaller degree would be deleted. By the induction hypothesis, H' has an induced copy of $B_{r',s}$ with end at v' for some $r' \geq r-1$. Adding the rest of P completes the desired induced broom, since on P no vertex before v' has a neighbor in H. ∎

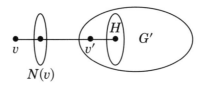

8.2.24.* Remark. Other results toward Conjecture 8.2.22 are harder. Gyárfás [1985] proved that $\text{Forb}(B_{m,n})$ is χ-bounded. Kierstead–Penrice [1994] proved that $\text{Forb}(T)$ is χ-bounded when T has radius 2 (Gyárfás–Szemerédi–Tuza [1980] had proved it weakly χ-bounded). The smallest non-broom tree not having radius 2 is the 7-vertex tree Q obtained by attaching an edge to a central vertex of P_6. Kierstead–Penrice [1990] proved that $\text{Forb}(Q)$ is weakly χ-bounded ($\chi(G) \leq 4$ for every triangle-free graph in $\text{Forb}(Q)$). Kierstead [1992] proved that $\text{Forb}(T)$ is weakly χ-bounded when T is the union of copies of P_4 with a common endpoint.

Kierstead–Zhu [2004] further strengthened these results by proving that if T is a tree of radius 3 obtained from a tree of radius 2 by subdividing every edge incident to a central vertex, then $\text{Forb}(T)$ is χ-bounded. In another direction, Scott [1997] proved that for each tree T, there is a function f_T such that every K_k-free graph with $\chi(G) > f(k)$ contains a subdivision of T as an induced subgraph. Thus $\text{Forb}(T)$ is χ-bounded when T is a subdivision of a star. ∎

EXERCISES 8.2

8.2.1. (−) Let f be a proper k-coloring of a k-chromatic graph G. Prove that each color class under f has a vertex adjacent to vertices of all other colors.

8.2.2. (−) Suppose that $\chi(G - x - y) = \chi(G) - 2$ for all pairs x, y of distinct vertices. Prove that G is a complete graph.

8.2.3. (−) Prove that the Petersen graph can be 2-colored so that the subgraph induced by each color class has no vertex of degree exceeding 1.

8.2.4. (−) Prove $\chi(G) \le 1 + \max_{H \subseteq G} \kappa'(H)$ for every graph G. (Matula [1969])

8.2.5. (−) A graph G is **vertex-color-critical** if $\chi(G - v) < \chi(G)$ for all $v \in V(G)$. Prove that every 3-chromatic vertex-color-critical graph is an odd cycle (and thus color-critical). Prove that the graph on the left below is vertex-color-critical but not color-critical.

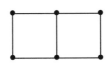

8.2.6. (−) Prove that the graph on the right above is not 2-choosable.

8.2.7. (−) Given that minimum degree 3 forces a K_4-subdivision (Theorem 8.2.17), prove that the maximum number of edges in an n-vertex graph with no K_4-subdivision is $2n - 3$. (Dirac [1964] conjectured and Mader [1998] proved that every n-vertex graph with at least $3n - 5$ edges contains a subdivision of K_5.)

8.2.8. Prove that the graphs below are 4-critical, thereby confirming that k-critical graphs need not be $(k-1)$-connected.

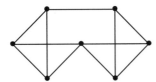
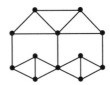

8.2.9. *The smallest k-critical graph other than K_k is $C_5 \oplus K_{k-3}$.*

(a) Prove that $N(x) \subseteq N(y)$ is impossible when x and y are vertices in a k-critical graph G. Conclude that no k-critical graph has $k + 1$ vertices.

(b) Prove that $\chi(G \oplus H) = \chi(G) + \chi(H)$ and that $G \oplus H$ is color-critical if and only if both G and H are color-critical. Conclude that $C_5 \oplus K_{k-3}$ is k-critical. (Comment: In fact, it is the only k-critical graph with $k + 2$ vertices.)

8.2.10. Use Proposition 8.2.2 to prove Propositions 8.1.6, 8.1.10, and 8.1.12.

8.2.11. *Vertex cuts in color-critical graphs.* Let G be a k-critical graph with a vertex cut S. Prove $|S| \ge 2$. Prove that if $S = \{x, y\}$, then $xy \notin E(G)$. Prove also that G has exactly two S-lobes (Definition 8.2.16), which can be named G_1 and G_2 so that $G_1 + xy$ and $G_2 \cdot xy$ (add xy and then contract it) are k-critical.

8.2.12. Determine the least number of edges in an n-vertex k-chromatic connected graph. (Eršov–Kožuhin [1962]; see Bhasker–Samad–West [1994] for higher connectivity.)

8.2.13. For $j < k < n - j$, determine the least number of edges in an n-vertex k-chromatic graph with minimum degree at least j.

8.2.14. For the graph G on the left below, obtain $\chi(G)$ and a $\chi(G)$-critical subgraph.

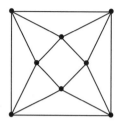 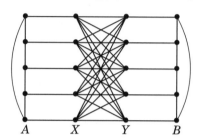

8.2.15. *Dense 4-critical graphs.* For odd m, form G_m with $V(G) = A \cup X \cup Y \cup B$ as follows. Let $G[A] = G[B] = C_m$, let $G[X \cup Y] = K_{m,m}$ (an X, Y-bigraph), and add matchings joining A to X and B to Y (G_5 appears above). Prove that G_m is 4-critical. (Toft [1970])

8.2.16. ($\diamond$) Let G be a color-critical graph. Prove that the graph G' generated from G by Mycielski's Construction (Definition 8.1.15) is also color-critical.

8.2.17. ($\diamond$) A special case of the Erdős–Lovász Tihany Conjecture (see Erdős [1968]) is that if $\chi(G - x - y) = \chi(G) - 2$ for all $xy \in E(G)$ and G is connected, then G is a complete graph. Prove this for $\chi(G) \le 4$. (Comment: For $\chi(G) = 5$, see Stiebitz [1987] and Mozhan [1987].)

8.2.18. Let G be an n-vertex graph with m edges.
 (a) Prove that if $\omega(G) \le r$, then $m \le (1 - 1/r)n^2/2$, and hence $\chi(G) \ge n^2/(n^2 - 2m)$. (Hint: Consider the neighborhood of a vertex of maximum degree.) (Myers–Liu [1972])
 (b) Use part (a) to prove $\alpha(G) \ge \lceil n/(d+1) \rceil$, where d is the average vertex degree of G. (Erdős–Gallai [1961])

8.2.19. ($\diamond$) Let G be a $(k + 1)$-critical graph. Prove that every edge of G lies in at least $(k - 1)!$ cycles whose lengths are congruent to 1 modulo k. (Moore–West [2019])

8.2.20. ($\diamond$) Let G be the 12-vertex graph on the left below. Prove $\chi(G) = 4$. Determine whether G is 4-critical. (Hint: Consider the independent sets. Comment: G is the smallest triangle-free 4-regular 4-chromatic graph.) (Chvátal [1970])

 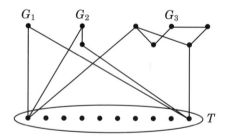

8.2.21. Let $G_1 = K_1$. For $k > 1$, construct G_k by adding to $G_1 + \cdots + G_{k-1}$ a set T of $\prod_{i=1}^{k-1} |V(G_i)|$ additional vertices, such that for each $(v_1, \ldots, v_{k-1}) \in V(G_1) \times \cdots \times V(G_{k-1})$, one vertex of T has neighborhood $\{v_1, \ldots, v_{k-1}\}$. The graph on the right above suggests G_4. Prove that G_k is k-critical and triangle-free. (Zykov [1949], Schäuble [1969])

8.2.22. Let $\{X, Y\}$ be a partition of $V(G)$. Prove that if $G[X]$ and $G[Y]$ are k-colorable, then $\chi(G) \le k + m/k$, where $m = |[X, Y]|$. (Hakimi–Schmeichel [2004])

8.2.23. Let G be a graph, and let $t = \lceil (\Delta(G)+1)/j \rceil$. Prove that $V(G)$ has a partition $V_1, \ldots, V_t$ such that each $G[V_i]$ has no j-edge-connected subgraph. Prove that $\lceil \Delta(G)/j \rceil$ classes do not suffice in the following cases: (1) G is a j-regular j-edge-connected graph, (2) $G = K_n$ with $n \equiv 1 \pmod{j}$, (3) $j = 1$ and G is an odd cycle. (Comment: Matula [1973] generalized Brooks' Theorem by showing that $\lceil \Delta(G)/j \rceil$ suffices except in these cases.)

8.2.24. Hajós construction. For $k \geq 3$, let G and H be k-critical graphs sharing only vertex v, with $vu \in E(G)$ and $vw \in E(H)$. Prove that $(G - vu) \cup (H - vw) \cup uw$ is k-critical. Use this to construct 4-critical n-vertex graphs for $n \geq 6$. (Hajós [1961])

8.2.25. ($\Diamond$) *Generalized Hajós construction.* Let G_1 and G_2 be disjoint k-critical graphs. Choose $w \in v(G_1)$ with $d_{G_1}(w) = k - 1$, and partition $N_{G_1}(w)$ into nonempty sets X and Y. Given $uv \in G_2$, form G from $(G_1 - w) + (G_2 - uv)$ by adding X to $N(u)$ and Y to $N(v)$.
  (a) Prove that G is k-critical.
  (b) For $k \geq 3$, apply this operation once to construct a k-critical n-vertex graph with degree-sum $n(k-1) + k - 3$, where $n = 2k - 1$. (Comment: Every k-critical n-vertex non-complete graph has degree-sum at least $n(k-1) + k - 3$ (Dirac [1957]). Kostochka–Yancey [2014] proved a conjecture of Gallai [1963a] and almost a conjecture of Ore [1967] by showing that every k-critical n-vertex graph has at least $\lceil \frac{(k+1)(k-2)n - k(k-3)}{2(k-1)} \rceil$ edges.)

8.2.26. Let G be a connected graph that is not a complete graph or a cycle whose length is an odd multiple of 3. Prove that every proper $\chi(G)$-coloring of G has two vertices of the same color with a common neighbor. (Tomescu [1990])

8.2.27. ($\Diamond$) *Extension of Brooks' Theorem.* Let G be a graph.
  (a) Given $D_1, \ldots, D_t \in \mathbb{N}_0$ such that $\sum D_i > \Delta(G)$, prove that there is a partition $V_1, \ldots, V_t$ of $V(G)$ such that $\Delta(G[V_i]) < D_i$ for each i. (Lovász [1966])
  (b) Prove that if $\omega(G) = r$, where $3 \leq r \leq \Delta(G)$, then $\chi(G) \leq \lceil \frac{r}{r+1}(\Delta(G)+1) \rceil$. (Borodin–Kostochka [1977], Catlin [1978], Lawrence [1978])

8.2.28. (+) Given a graph G, let $f_1, \ldots, f_k$ be nonnegative integer functions on $V(G)$ such that $\sum_i f_i(v) > d_G(v)$ for all $v \in V(G)$. Prove that G has a coloring $c\colon V(G) \to [k]$ such that for all $v \in V(G)$, fewer than $f_{c(v)}(v)$ neighbors of v have color $c(v)$. (Bernardi [1987])

8.2.29. Gallai's Theorem: *If G is a $(k+1)$-critical graph, then every block of the subgraph of G induced by the vertices of degree k is an odd cycle or a complete graph.* (Gallai [1963a])
  (a) Use Theorem 8.2.13 to prove Gallai's Theorem.
  (b) Gallai's Theorem can be proved without list colorings. Use Gallai's Theorem to prove Brooks' Theorem for chromatic number: Every connected graph G that is not $\Delta(G)$-colorable is a complete graph or an odd cycle.

8.2.30. Prove that if every block of a graph G is a complete graph or an odd cycle, then G is not degree-choosable. (Comment: This is the converse of Theorem 8.2.13.)

8.2.31. ($\Diamond$) Let G be an n-vertex graph. Prove $\max_{H \subseteq G} \delta(H) + \max_{H' \subseteq \overline{G}} \delta(H') \leq n - 1$. Conclude $\chi_l(G) + \chi_l(\overline{G}) \leq n + 1$. (Comment: This strengthens Exercise 8.1.40(a).)

8.2.32. ($\Diamond$) Prove that $K_{k,r}$ is k-choosable if and only if $r < k^k$. (Vizing [1976])

8.2.33. (+) **Small Pot Lemma.** Let $L(X) = \bigcup_{v \in X} L(v)$ for $X \subseteq V(G)$.
  (a) Prove that if G is not k-choosable, then G is not L-colorable for some k-uniform lists L with $|L(V(G))| < |V(G)|$. (Galvin [1998], Kierstead [2000], Reed–Sudakov [2002])
  (b) Let G be a complete k-partite graph. Prove that if one part has size 4, the others have size 2, and k is even, then $\chi_l(G) > k$ (Enomoto–Ohba–Ota–Sakamoto [2002]) (Comment: This graph is a sharpness example for the conjecture of Ohba [2002], proved in Noel–Reed–Wu [2015], that k-colorable graphs with at most $2k+1$ vertices are k-choosable.)

8.2.34. ($\diamond$) Let G be the complete k-partite graph obtained by deleting a perfect matching from K_{2k}. Prove $\chi_l(G) = k$. (Erdős–Rubin–Taylor [1979])

8.2.35. Prove $\chi_l(K_{3,3,1}) = 3$. (Comment: More generally, Gravier–Maffray [1998] proved $\chi_l(G) = \chi(G)$ when $G = K_{3,3,2,\dots,2}$.)

8.2.36. ($\diamond$) Let K_{m*k} denote the complete k-partite graph with m vertices in each part.
    (a) Prove $\chi_l(K_{m*k}) \geq \lceil \frac{(m-1)(2k-1)+1}{m} \rceil$. (Comment: The exact value is $\lceil \frac{4k-1}{3} \rceil$ when $m = 3$ (Kierstead [2000]) and $\lceil \frac{3k-1}{2} \rceil$ when $m = 4$ (Kierstead–Salmon–Wang [2014]).)
    (b) Prove $\chi_l(K_{m*k}) > (k - 1)\frac{\log m}{\log k}$. (Noel–West–Wu–Zhu [2015]) (Hint: Consider m of the form $\binom{kj-1}{(k-1)j}$. Comment: Alon [1992] proved $\chi_l(K_{m*k}) \in \Theta(k \log m)$.)

8.2.37. For a graph G and list assignment L, let $S = \bigcup_{v \in V(G)} L(v)$. Let H be the graph with vertex set $V(G) \cup S$ consisting of G, a complete graph with vertex set S, and the edges $\{vc: v \in V(G), c \in S - L(v)\}$. Prove that G is L-colorable if and only if $\chi(H) \leq |S|$.

8.2.38. ($\diamond$) *Reduction of L-colorability to independence number.* For a graph G and list assignment L, let H be the graph with vertex set $\{v_c: v \in V(G), c \in L(v)\}$ such that u_c and $v_{c'}$ are adjacent if $u = v$ or if $uv \in E(G)$ and $c = c'$. Prove that G is L-colorable if and only if $\alpha(H) = |V(G)|$. (Vizing [1976]) (Comment: This generalizes Exercise 8.1.36. A more general model in which H allows any matching joining the cliques assigned to adjacent vertices u and v in G was introduced in Dvořák–Postle [2018]; it is now called **DP-coloring**.)

8.2.39. (+) *Characterization of 2-choosable graphs.*
    (a) For $n \geq 2$, prove that P_n is not f-choosable when f is 1 at the endpoints and 2 elsewhere. Prove that C_n is not f-choosable when f is 1 at one vertex and 2 elsewhere.
    (b) Let L be a 2-uniform list assignment on a path with lists not all identical. Prove that L-colorings exist having at least three of the four possibilities at the endpoints.
    (c) The graph $\Theta(l_1, \dots, l_k)$ with branch vertices u and v is the union of k pairwise internally-disjoint u, v-paths with lengths $l_1, \dots, l_k$. Prove that $\Theta(k, l, m)$ with $k \leq l \leq m$ is 2-choosable if and only if $(k, l, m) = (2, 2, 2q)$ for some $q \in \mathbb{N}$.
    (d) Prove that a connected bipartite graph is 2-choosable if and only if it has at most one cycle or the edges in its cycles form $\Theta(2, 2, 2q)$ for some q. (Erdős–Rubin–Taylor [1979])

8.2.40. ($\diamond$) Let G be a k-colorable graph, and let P be a set of vertices in G such that $d(x, y) \geq 4$ whenever $x, y \in P$. Prove that every coloring of P with colors from $[k + 1]$ extends to a proper $(k + 1)$-coloring of G. (Albertson [1998])

8.2.41. ($\diamond$) Prove that $K_{r,r}$ contains a subdivision of K_{2k} if and only if $r \geq \binom{k+1}{2}$. Determine the threshold on r for the existence of a K_{2k+1}-subdivision in $K_{r,r}$.

8.2.42. Let G_7 and G_8 denote the graphs on the left and right below. Heavy edges indicate that every vertex inside one circle is adjacent to every vertex inside another. For $k \in \{7, 8\}$, prove that $\chi(G_k) = k$ and that G_k has no K_k-subdivision. (Comment: Thus Hajós' Conjecture is false for $k \in \{7, 8\}$.) (Catlin [1979])

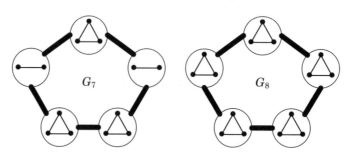

8.2.43. (+) Let F be a forest with m edges. Let G be a graph with at least $|V(F)|$ vertices such that $\delta(G) \geq m$. Prove $F \subseteq G$. (Brandt [1994])

8.2.44. ($\diamond$) Let G be a k-chromatic graph with girth at least 5. Prove that G contains every k-vertex tree as an induced subgraph. (Hint: Reduce to the case $\delta(G) \geq k-1$ and then use induction on k.) (Gyárfás–Szemerédi–Tuza [1980])

8.2.45. (+) *Forcing k-colored subtrees in k-chromatic graphs.* Let f be a proper k-coloring of a graph G, and let T be a tree with vertex set $\{w_1, \ldots, w_k\}$. Prove that there is a map $\phi \colon V(T) \to V(G)$ such that the images of adjacent vertices in T are adjacent in G and $f(\phi(w_i)) = i$ for all i. (Gyárfás–Szemerédi–Tuza [1980])

8.2.46. (+) Let G be a graph having a proper coloring in which no color class has size 1. Prove that G has a proper $\chi(G)$-coloring in which no color class has size 1. (Gallai [1963c])

8.3. Edge-Coloring and Perfection

Just as we partition $V(G)$ into independent sets to avoid conflicts between adjacent vertices, we can also partition $E(G)$ into matchings to avoid conflicts of incident edges. For example, when the vertices are people and the edges are pairs that must meet, we cannot schedule two meetings for one person at the same time. In a sense, this is a better-behaved special case of vertex coloring. Subsequently, we put the results into perspective by discussing the more-better-behaved family of "perfect graphs", where the subject of min-max relations has a natural home.

8.3.1. Definition. A **k-edge-coloring** of a graph is a labeling of its edges from a set of k colors; it is **proper** if incident edges receive distinct colors. A graph is **k-edge-colorable** if it has a proper k-edge-coloring, and the **edge-chromatic number** or **chromatic index** $\chi'(G)$ of G is the minimum k such that G is k-edge-colorable.

These definitions for edge-coloring extend without change to loopless multigraphs. Although multiedges are irrevelant for vertex colorings, they can greatly affect edge-chromatic number. Loops cannot be properly colored.

8.3.2. Remark. Comparing $\chi'(G)$ with $\chi(G)$ is not merely analogy; always $\chi'(G) = \chi(L(G))$. This usage of $'$ parallels $\alpha'(G) = \alpha(L(G))$.

Since the edges at a vertex must have distinct colors, $\chi'(G) \geq \Delta(G)$. Applying greedy coloring to any ordering of the edges of a multigraph (or of the vertices of its line graph) yields $\chi'(G) \leq 2\Delta(G) - 1$. (That is, $\Delta(L(G)) \leq 2(\Delta(G) - 1)$.)

A clique in the line graph $L(G)$ corresponds to pairwise-incident edges in G. In a graph (no loops or multiedges), such edges can have one common endpoint or can form a triangle. Thus for graphs with maximum degree at least 4, the statements $\chi'(G) \geq \Delta(G)$ and $\chi(L(G)) \geq \omega(L(G))$ have the same content. ∎

SPECIAL CLASSES

We have the greedy upper bound $\chi'(G) \leq 2\Delta(G) - 1$. For multigraphs, the bound cannot be reduced below $3\Delta(G)/2$.

8.3.3. Example. *The fat triangle.* In the multigraph below, each edge is incident to all others, so all edges must receive distinct colors, and $\chi'(G) = 3\Delta(G)/2$. ∎

Shannon [1949] proved that Example 8.3.3 is the worst in terms of $\Delta(G)$ (Exercise 26). Vizing [1964] and Gupta [1966] both proved the more detailed bound $\chi'(G) \leq \Delta(G) + \mu(G)$, where $\mu(G)$ is the largest edge-multiplicity. For a graph G, this reduces to $\chi'(G) \leq \Delta(G) + 1$ (*graph* restricts edge-multiplicities to $\{0, 1\}$). There is standard terminology for the two alternatives when G is a graph.

8.3.4. Definition. A graph G is **Class 1** when $\chi'(G) = \Delta(G)$; **Class 2** otherwise. A decomposition of a multigraph into perfect matchings is a 1-**factorization**.

To be Class 1, a regular graph must be colored with the edges of each color forming a perfect matching. Thus regular graphs of odd order are Class 2.

8.3.5. Example. *Edge-coloring of K_{2n}.* In a league with $2n$ teams, we want to schedule games between all pairs, but each team plays at most once a week. Since each team must play $2n - 1$ others, the season lasts at least $2n - 1$ weeks. The games of each week must form a matching. We can schedule the season in $2n - 1$ weeks if and only if K_{2n} has a 1-factorization.

The figure below suggests a solution. Arrange $2n - 1$ vertices cyclically. Let the *length* of an edge be the number of steps between its endpoints along the circle. There are $2n - 1$ edges of each length $1, \dots, n - 1$. In the figure, the bold matching has one edge of each length, plus an edge from the central vertex to one vertex on the circle. Rotating the picture yields others matchings, again with one edge of each length. The $2n-1$ rotations of the figure yield the desired matchings.

The 1-factorization of K_{2n} is related to design theory (see Chapter 13). ∎

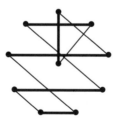

8.3.6. Example. *The cube and the Petersen graph.* The d-dimensional cube is Class 1. A 1-factorization is given by letting the ith 1-factor consist of the edges joining the d-tuples that differ in the ith coordinate.

The Petersen graph is Class 2. Deleting a 1-factor always leaves a 2-factor, and the Petersen graph has no 10-cycle (Proposition 5.1.18). Thus the 2-factor that remains must be $2C_5$, and this has no 1-factor. ∎

There are many proofs that bipartite graphs are Class 1 (see Exercise 18). Perhaps the easiest uses the Marriage Theorem. The proof is also valid when multiedges are allowed; indeed, that simplifies the proof.

8.3.7. Theorem. (König [1916]) $\chi'(G) = \Delta(G)$ when G is a bipartite multigraph.

Proof: Let G be an X, Y-bigraph, and let $k = \Delta(G)$. Constructing a k-regular bipartite supergraph H of G reduces the problem to the regular case, because discarding edges from a proper k-edge-coloring of H yields a proper k-edge-coloring of G. To construct H, add vertices to the smaller of X and Y until both have the same size. Then edges joining vertices of X and Y with degree less than k until all vertices have degree $\Delta(G)$; multiedges may arise.

Hence we may assume G is k-regular. By the Marriage Theorem (Corollary 6.1.6), a k-regular bipartite multigraph has a 1-factor, and hence by induction on k it has a 1-factorization. ∎

Although $\Delta(G) \le \chi'(G) \le \Delta(G) + 1$ for every graph G, deciding between the two options is NP-complete (Holyer [1981]). Thus we seek conditions for Class 1 or Class 2. For example, regular graphs with cut-vertices are Class 2 (Exercise 13). For cartesian products, $G \,\square\, H$ is Class 1 if G or H is Class 1 or if G and H both have 1-factors (Exercise 25).

8.3.8.* Remark. An **overfull subgraph** of a graph G is a subgraph H such that $|E(H)| > \Delta(G)\lfloor |V(H)|/2 \rfloor$. An overfull subgraph forces $\chi'(G) > \Delta(G)$ and can only occur when $|V(H)|$ is odd. When discussing also multigraphs, we add the restriction $\Delta(H) = \Delta(G)$, which holds implicitly for simple graphs.

The **Overfull Conjecture** (Chetwynd–Hilton [1984, 1986]; see also Hilton [1989]) states that an n-vertex graph G with $\Delta(G) > n/3$ is Class 1 if and only if G has no overfull subgraph H. Deleting one vertex from the Petersen graph shows that the condition is not sufficient when $\Delta(G) = n/3$ (Exercise 21).

The Overfull Conjecture implies the following **1-Factorization Conjecture**: If $k \ge m$, or if $k \ge m - 1$ when m is even, then every k-regular graph of order $2m$ is Class 1 (Exercise 22). This is sharp (Exercise 23).

The conclusions of both conjectures hold when the maximum degree is a large enough fraction of the number of vertices (see Chetwynd–Hilton [1989], Niessen–Volkmann [1990], Perkovic–Reed [1997], Plantholt [2004]). ∎

VIZING'S THEOREM AND EXTENSIONS

Vizing [1964, 1965] and Gupta [1966] proved $\chi'(G) \le \Delta(G) + \mu(G)$ for every loopless multigraph G. Commonly known as **Vizing's Theorem**, this has many proofs and many extensions to stronger results.

For example, let $\mu(x, y)$ denote the multiplicity of xy as an edge. In addition, let $\mu(v) = \max_{x \in N(v)} \mu(x, v)$. Ore [1968] proved

$$\chi'(G) \le \max_{v \in V(G)} \{d(v) + \mu(v)\} \le \Delta(G) + \mu(G).$$

Ore [1967] also proved

$$\chi'(G) \le \max \left\{ \Delta(G), \max_{\langle x,y,z \rangle \in \mathbf{P}} \frac{d(x)+d(y)+d(z)}{2} \right\},$$

where $\mathbf{P}$ is the set of all 3-vertex paths. This implies the bound of Shannon [1949]: $\chi'(G) \le \frac{3}{2}\Delta(G)$. Both of Ore's bounds follow immediately from our main result:

8.3.9. Theorem. (Andersen [1977], Goldberg [1977, 1984]) If G is a loopless multigraph, and $\mathbf{P}$ is the set of 3-vertex paths in G, then

$$\chi'(G) \le \max \left\{ \Delta(G), \max_{\langle x,y,z \rangle \in \mathbf{P}} \left\lfloor \frac{d(x) + \mu(x,y) + \mu(y,z) + d(z)}{2} \right\rfloor \right\}. \qquad (*)$$

The proof of the Anderson–Goldberg Theorem, like many proofs of edge-coloring bounds, yields a polynomial-time algorithm to augment a partial proper edge-coloring of G until all edges have been colored using the specified number of colors. The actual value of $\chi'(G)$ may be less than the bound being proved.

The proof here also yields other useful results such as Vizing's Adjacency Lemma (Corollary 8.3.11). The tool that permits extension of partial edge-colorings is the following technical statement. The proofs can be followed for the special case of graphs, where all multiplicities equal 1. In the general case, "uv" indicates a particular edge with endpoints u and v.

8.3.10. Theorem. For $q \in \mathbb{N}$, let G be a multigraph having an edge yw such that $d(y), d(w) \le q$ and $G - yw$ is q-edge-colorable. If

$$d(x) + \mu(x,y) + \mu(y,z) + d(z) \le 2q + 1$$

whenever x and z are distinct neighbors of y, then G is q-edge-colorable.

Proof of Theorem 8.3.9: We use induction on $|E(G)|$, with trivial basis. For the induction step, let q be the bound computed in $(*)$, and let wy be an edge in G. Since deleting edges cannot increase the computation in $(*)$, the induction hypothesis implies that $G - wy$ is q-edge-colorable. Also $d(y), d(w) \le \Delta(G) \le q$.

When x and z are distinct neighbors of y, we have $\langle x, y, z \rangle \in \mathbf{P}$. Also $d(x)+\mu(x,y)+\mu(y,z)+d(z) \le 2q+1$, since otherwise this path $\langle x, y, z \rangle$ contradicts the computation of q in $(*)$. Now Theorem 8.3.10 applies and yields $\chi'(G) \le q$. ∎

Proof of Theorem 8.3.10: (Kostochka [2014]) Let $G' = G - yw$. Let ϕ be a proper q-edge-coloring of G'. For $v \in V(G)$, let $O(v)$ be the set of colors *omitted* at v (used on no edges at v). If w is the only neighbor of y, then any color in $O(w)$ also lies in $O(y)$, and using it on yw extends ϕ to G. Hence we may assume $|N_G(y)| \ge 2$.

A list $v_0, \dots, v_k$ of neighbors of y forms a **color fan** if $v_0 = w$ and $\phi(v_iy) \in O(v_{i-1})$ for $i \ge 1$, where v_iy is some edge joining v_i and y (see w, x, v, z below). If a color α is missing at both y and v_k, then moving color $\phi(v_iy)$ to $v_{i-1}y$ (for $1 \le i \le k$) and using color α on v_ky produces a proper q-edge-coloring of G. Hence $\chi'(G) > q$ requires $O(y) \cap O(z) = \varnothing$ when z is reached by some color fan.

Let X be the set of vertices on color fans. Form an auxiliary digraph H with vertex set X having an edge from u to v for each edge in G with endpoints v and y whose color under ϕ is in $O(u)$. Color fans correspond to paths from w in H.

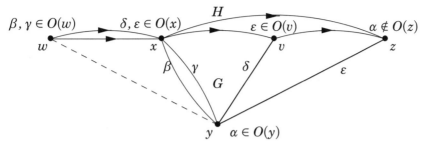

Suppose $\chi'(G) > q$. We will show that the sets $O(v)$ are disjoint for $v \in X \cup \{y\}$. We first show how this yields bounds on $|E(H)|$ leading to a contradiction.

Counting argument: When the sets $O(v)$ are disjoint, each color lies in $O(x)$ for at most one $x \in X$, so each edge of G' with endpoints v and y generates at most one edge of H entering v. Thus counting $E(H)$ by heads yields

$$|E(H)| = \sum_{v \in X} d_H^-(v) \le \left(\sum_{v \in X} \mu(v,y)\right) - 1,$$

where we lose 1 from the sum because the edge yw is not in G'.

Next we count by tails. For $v \in X$, each color in $O(v)$ appears on one edge yz at y, which yields an edge vz in H. Thus $d_H^+(v) = |O(v)| = q - d_{G'}(v)$ for $v \in X$. Again adjusting by 1 since $d_{G'}(w) = d_G(w) - 1$, we have

$$|E(H)| = \sum_{v \in X} d_H^+(v) \ge 1 + \sum_{v \in X}(q - d_G(v)).$$

Combining the two bounds on $|E(H)|$ and grouping by vertices yields

$$\sum_{v \in X}(d_G(v) + \mu(v,y)) \ge q|X| + 2.$$

Note that $q \ge d_G(w)$ yields $d_H^+(w) \ge 1$ and $|X| \ge 2$. By the given hypothesis, the two largest terms on the left sum to at most $2q + 1$, with one of them being at most q. Thus the other $|X| - 2$ terms are also bounded by q, and the total sum is at most $2|X| + 1$. This contradiction yields $\chi'(G) \le q$ if the sets $O(v)$ are disjoint.

Disjointness argument: The definition of color fans yields $O(y) \cap O(z) = \varnothing$ for $z \in X$. If $d(z) = \Delta(G) = q$, then $O(z)$ is empty and disjoint from all sets, so we may consider $\alpha \in O(y)$ and $\beta \in O(z)$. Since α appears at z and β does not, we can follow a maximal path P from z in colors α and β. Let u be the other end of P.

We claim $u = y$. If $u \ne y$, then switching α and β on P changes the omitted set only at z and u. Let Q be a w, z-path in H. The new coloring has a color fan reaching a vertex where α does not appear: u if $u \in V(Q)$ (since $\alpha \notin O(u)$), or z if $u \notin V(Q)$. Again changing colors along the fan extends the q-edge-coloring to G.

If $\beta \in O(v) \cap O(v')$, then the paths from v and v' via α and β both reach y. The vertex where they meet has two edges of the same color, a contradiction.   ∎

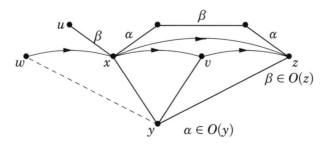

For edge-coloring of simple graphs, a **critical graph** is a graph G with $\chi'(G) = \Delta(G) + 1$ such that $\chi'(H) < \chi'(G)$ for every proper subgraph H. Deleting an edge e would still leave a Class 2 graph if all maximum-degree vertices of G are incident to e. In fact, this cannot happen; every Class 2 graph has at least three maximum-degree vertices. This follows immediately from our next result, given the name **Vizing's Adjacency Lemma** in Fiorini–Wilson [1977], a book that explores early results on edge-coloring. We will apply it in Exercise 9.3.64.

8.3.11.* Corollary. (**Vizing's Adjacency Lemma**; Vizing [1965]) If w and y are adjacent in a critical graph G, then y has at least $\max\{\Delta(G)-d(w)+1, 2\}$ neighbors with degree $\Delta(G)$.

Proof: Since G is critical, $\chi'(G)>\chi'(G-yw)=q$ with $q=\Delta(G)\geq\max\{d(y),d(w)\}$. Since $\chi'(G) > q$, Theorem 8.3.10 yields distinct $x, z \in N(y)$ having the property that $d(x)+\mu(x,y)+\mu(y,z)+d(z) > 2q+1$. Since each multiplicity is 1, we conclude that y has distinct neighbors with degree q.

In addition, since the counting bounds on $E(H)$ in the proof of Theorem 8.3.10 used only the hypothesis $\chi'(G) > q$ and not the assumed inequalities $d(x) + \mu(x,y) + \mu(y,z) + d(z) \leq 2q + 1$, the argument and computation producing $\sum_{v\in X}\mu(v,y) \geq 2 + \sum_{v\in X}(q - d_G(v)$ is also valid here, where $X \subseteq N(y)$.

With multiplicity always 1, the inequality reduces to $|X| \geq 2+\sum_{v\in X}(q-d(v))$. Letting k be the number of vertices in $X - \{w\}$ having degree $\Delta(G)$, we have k terms equal to 0 in the sum on the right, one term $\Delta(G) - d(w)$ for w itself, and $|X| - 1 - k$ terms contributing at least 1 each. Thus the inequality becomes $|X| - 1 \geq 1 + \Delta(G) - d(w) + |X| - 1 - k$, which yields $k \geq \Delta(G) - d(w) + 1$. ∎

LIST EDGE-COLORING

We next consider the list version of edge-coloring, analogous to list coloring. Just as the chromatic number behaves better when restricted to line graphs, where it is equivalent to the edge-chromatic number of the original graph, also the list chromatic number behaves better when restricted to line graphs.

8.3.12. Definition. A **list assignment** L for the edges of a graph G specifies a list $L(e)$ of available colors for each edge e. An **L-edge-coloring** is a proper edge-coloring f with $f(e)$ chosen from $L(e)$ for each e. The **list edge-chromatic number** or **edge-choosability** $\chi_l'(G)$ is the least k such that G is L-edge-colorable whenever $|L(e)| \geq k$ for all $e \in E(G)$. Note that $\chi_l'(G)$ equals $\chi_l(L(G))$, where $L(G)$ is the line graph of G.

Note that $\Delta(L(G)) \leq 2\Delta(G) - 2$. Hence $\chi_l'(G) \leq 2\Delta(G) - 1$ by greedy coloring, as in Proposition 8.2.8. Thus $\chi_l'(G) < 2\chi'(G)$, since $\chi'(G) \geq \Delta(G)$. As in ordinary coloring, the edge version of choosability behaves much better than the vertex version. Even so, the conjectured bound for edge-choosability is surprising. The List Color Conjecture was posed independently by many researchers. It was published in Bollobás–Harris [1985] but was independently formulated earlier by Albertson and Collins in 1981 and by Vizing as early as 1975 (both unpublished).

8.3.13. Conjecture. (**List Color Conjecture**) $\chi'_\ell(G) = \chi'(G)$ for any graph G.

8.3.14.* Remark. Bollobás–Harris [1985] proved $\chi'_\ell(G) < c\Delta(G)$ for each constant c exceeding $11/6$, when $\Delta(G)$ is sufficiently large. This and subsequent improvements used probabilistic methods. Kahn [1996] proved the conjecture asymptotically: $\chi'_\ell(G) \le (1 + o(1))\Delta(G)$. Häggkvist–Janssen [1997] sharpened the error term, proving $\chi'_\ell(G) \le D + 23D^{2/3}\sqrt{\log D}$), where $D = \Delta(G)$. Molloy–Reed [2000] further sharpened it to $\chi'_\ell(G) \le D + O(D^{1/2}(\log D)^4)$.

The special case of the List Color Conjecture for $G = K_{n,n}$ was posed by Dinitz in 1979 (Janssen [1993] proved the LCC for $K_{n,n-1}$). The Dinitz Conjecture became popular in its matrix formulation: If each position of an n-by-n grid contains a set of size n, then one can choose one element from each set so that those chosen in each row are distinct and those chosen in each column are distinct. ∎

Galvin [1995] proved the List Color Conjecture for bipartite multigraphs (generalizing the Dinitz Conjecture). Galvin wrote "The proof is very simple, and uses no new ideas". We start with a fundamental tool for list coloring.

8.3.15. Definition. A **kernel** in a digraph is an independent set S containing a successor of every vertex outside S. A digraph is **kernel-perfect** if every induced subdigraph has a kernel.

In a directed even cycle, either maximum independent set is a kernel. Indeed, a directed even cycle is kernel-perfect, so the next result gives another proof that even cycles are 2-choosable. Recall the definition of f-choosable from Definition 8.2.10. The next proof uses the same idea as Lemma 8.2.11.

8.3.16. Lemma. (Bondy–Boppana–Siegel) If D is a kernel-perfect orientation of G and $f(x) \ge 1 + d_D^+(x)$ for all $x \in V(G)$, then G is f-choosable.

Proof: Let $n = |V(G)|$. We use induction on n, with trivial basis $n = 1$. For $n > 1$, consider a list assignment L with $|L(x)| = f(x)$ for each x. Choose a color α in some list. Let $U = \{v : \alpha \in L(v)\}$. Let S be a kernel of the induced subdigraph $D[U]$, and let $D' = D - S$. Assign color α to all of the independent set S.

Let $L'(v) = L(v) - \{\alpha\}$ and $f'(v) = f(v) - 1$ for each $v \in U - S$, but $L'(v) = L(v)$ and $f'(v) = f(v)$ for $v \notin U$. The assignment L' on $V(D')$ satisfies $f'(x) \ge 1 + d_{D'}^+(x)$, since vertices of $U - S$ that lost a color from their lists also lost a successor in S. By the induction hypothesis, D' is f'-choosable. We complete an L-coloring for G by combining an L'-coloring of D' with color α on S. ∎

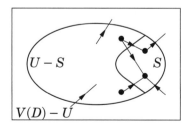

We apply Lemma 8.3.16 to orientations of line graphs of bipartite graphs.

8.3.17. Lemma. (Galvin [1995]) Let G be a bipartite multigraph with parts X and Y. Let c be a proper edge-coloring of G by integers. For $uv, vw \in E(G)$ with $c(uv) < c(vw)$, orient the edge in $L(G)$ as $uv \to vw$ if $v \in X$ and $vw \to uv$ if $v \in Y$. The resulting orientation D_c of $L(G)$ is kernel-perfect.

Proof: Note that multiedges in G become pairs of oppositely-directed edges in D_c. It suffices to prove that the full digraph D_c has a kernel, since the same argument applies to induced subdigraphs, which are obtained by discarding edges of G.

We use induction on $|E(G)|$, with trivial basis for $E(G) = \varnothing$. When $E(G) \neq \varnothing$, let S be the set of edges in G consisting of the highest-colored edge incident to each vertex of X. If S is a matching in G, then S is independent in $L(G)$ and is a kernel in D_c, since every edge $e' \in E(G) - S$ shares an endpoint in X with some edge $e \in S$, and $c(e) > c(e')$, so e is a successor of e' in D_c.

If S is not a matching, then let e and e' be edges in S having a common endpoint $y \in Y$; by symmetry, we may assume $c(e') < c(e)$. Let $G' = G - e$. By the induction hypothesis, the orientation of $L(G')$ obtained by deleting vertex e from D_c has a kernel S'. We claim that S' is also a kernel in D_c; we need only show that e has a successor in S' (see figure on the left below).

Either $e' \in S'$, or e' has a successor $\hat{e}$ in S'. By the choice of S, any successor $\hat{e}$ of e' in D_c shares its endpoint y and has a color less than $c(e')$. Since $c(e) > c(e')$, in either case e shares its endpoint y with a member of S' having smaller color than e, and hence e has a successor in S'. ∎

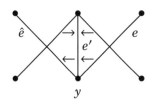

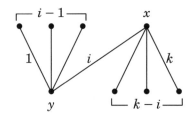

8.3.18. Theorem. (Galvin [1995]) $\chi'_l(G) = \Delta(G)$ for bipartite multigraphs G.

Proof: Let $k = \Delta(G)$; we prove that G is k-edge-choosable. Let c be a fixed proper k-edge-coloring of G using color set $[k]$. Define D_c as in Lemma 8.3.17.

If $c(xy) = i$, then in D_c the vertex xy has at most $i - 1$ successors sharing endpoint y and at most $k - i$ successors sharing endpoint x (see figure on the right above). Hence $d^+_{D_c}(e) \leq k - 1$ for $e \in V(D_c)$. Since Lemma 8.3.17 implies that D_c is kernel-perfect, Lemma 8.3.16 implies that $L(G)$ is k-choosable. ∎

Slivnik [1996] gave another proof. Borodin–Kostochka–Woodall [1997] then strengthened Galvin's result, showing that smaller lists suffice on edges whose endpoints have small degree. Their corollary yielding the best upper bound on $\chi'_l(G)$ in terms of $\Delta(G)$ does not follow directly from Galvin's result.

8.3.19.* Lemma. Let G be an X, Y-bigraph G with $|X| \geq |Y|$ and $\delta(G) \geq 1$. For some nonempty $S \subseteq X$, there is a matching with vertex set $S \cup N(S)$.

Proof: Since $|X| \geq |Y|$, we may choose S to be a minimal nonempty subset of X such that $|S| \geq |N(S)|$. For $x \in S$, the choice of S yields

$$|S| \geq |N(S)| \geq |N(S - x)| \geq |S - x| + 1 = |S|.$$

Hence $|S| = |N(S)|$. Now $|N(T)| > |T|$ for every nonempty proper subset T of S, by the choice of S. Hence the subgraph induced by $S \cup N(S)$ satisfies Hall's Condition and has a perfect matching. ∎

8.3.20.* Theorem. (Borodin–Kostochka–Woodall [1997]) In a bipartite multigraph G, if $f(xy) = \max\{d(x), d(y)\}$ for $xy \in E(G)$, then $L(G)$ is f-choosable.

Proof: Let G be an X, Y-bigraph, named so that $|X| \geq |Y|$ after discarding isolated vertices. By Lemmas 8.3.16–8.3.17, it suffices to produce a proper edge-coloring c (using integers) such that D_c satisfies $d_{D_c}^+(e) \leq \max\{d_G(x), d_G(y)\} - 1$ for each edge e, where e has endpoints x and y. We produce c by induction on $|E(G)|$; when G is just a matching, we use just one color.

For the induction step, let M be a matching (of S to $N(S)$) as given by Lemma 8.3.19. Let $G' = G - M$, and let c' be the edge-coloring of G' guaranteed by the induction hypothesis. Extend c' to a proper edge-coloring c of G by giving one color to M, larger than the colors in c'. Edges not incident to S receive no additional successors in moving from $D_{c'}$ to D_c, so they satisfy the desired bound on $d_{D_c}^+(e)$.

Consider an edge incident to $x \in S$. If $xz \in M$, then $d_{D_c}^+(xz) = d_G(z) - 1 \leq \max\{d_G(x), d_G(z)\} - 1$. If $xy \in E(G')$, then restoring M increases the outdegree: $d_{D_c}^+(xy) = d_{D_{c'}}^+(xy) + 1$, since M is a matching. Nevertheless, both x and y have larger degree in G than in G', since M covers all of $N(S)$. Thus $d_{D_{c'}}^+(xy) \leq \max\{d_{G'}(x), d_{G'}(y)\} - 1$ yields $d_{D_c}^+(xy) \leq \max\{d_G(x), d_G(y)\} - 1$, as desired. ∎

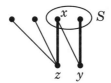

Theorem 8.3.20 implies that the degree bound of Shannon [1949] for edge-chromatic number holds also for edge-choosability. Since the bound is sharp for edge-chromatic number (Example 8.3.3), it is also sharp here. This direct proof of Shannon's bound through list edge-coloring does not require proving Vizing's Theorem or any of its extensions.

8.3.21.* Corollary. (Borodin–Kostochka–Woodall [1997]) If G is a multigraph, then $\chi_l'(G) \leq \frac{3}{2}\Delta(G)$.

Proof: Let L assign lists of size $\left\lfloor \frac{3}{2}\Delta(G) \right\rfloor$. Let H be a largest bipartite subgraph of G, so $d_H(v) \geq \frac{1}{2}d_G(v)$ for $v \in V(G)$ (Theorem 5.2.9). Let $H' = G - E(H)$. Since $\Delta(H') \leq \frac{1}{2}\Delta(G)$, we have $\Delta(L(H')) \leq \Delta(G) - 2$. Since $\frac{3}{2}\Delta(G) \geq \Delta(G) - 1$ when $\Delta(G) \geq 2$, there is an L-edge-coloring for H'.

For $xy \in L(H)$, form $L'(xy)$ by deleting from $L(xy)$ the colors now on edges at x and y in H'. Thus $|L'(xy)| \geq \frac{3}{2}\Delta(G) - d_{H'}(x) - d_{H'}(y)$. Since $d_{H'}(y) \leq \frac{1}{2}\Delta(G)$, we obtain $|L'(xy)| \geq \Delta(G) - d_{H'}(x) \geq d_H(x)$. Similarly, $d_{H'}(x) \leq \frac{1}{2}\Delta(G)$ yields $|L'(xy)| \geq d_H(y)$. By Theorem 8.3.20, H has an L'-edge-coloring that combines with the edge-coloring of H' to complete an L-coloring of G. ∎

PERFECT GRAPHS

Two optimization problems (one maximization and one minimization) are **dual** if every way to meet the constraints for the minimization yields value at least as large as every way to meet the constraints for the maximization. Examples include minimum coloring and maximum clique, minimum edge-coloring and maximum degree, minimum vertex cover and maximum matching, etc.

A **min-max relation** holds for dual integer optimization problems on a family of instances when they have solutions of equal value on those instances. We have proved such relations for coloring and cliques on interval graphs, for edge-coloring and maximum degree on bipartite graphs, etc. Min-max relations are among the most elegant and valuable results in combinatorics. Such a theorem guarantees short proofs of optimality for both optimization problems. Exhibiting a coloring and a clique of the same size proves that neither can be improved, and a min-max relation states that such a coloring and clique can be found.

The general topics of duality and min-max relations belong to the subject of "integer linear programming", which we do not study in this text. Instead, we stay in graph theory and interpret several of the min-max relations we have proved in the context of a family of graphs called "perfect graphs".

8.3.22. Definition. A graph G is **perfect** if $\chi(H) = \omega(H)$ for every induced subgraph H of G.

The induced-subgraph requirement prevents bad local behavior from being hidden by adding a large clique. Claude Berge introduced perfect graphs in 1959; the notion appears first in abstracts of his talks (Berge [1960, 1961]). The concept arose from several families found to be perfect. Many such families are closed under taking induced subgraphs.

8.3.23. Definition. A graph class **G** is **hereditary** if every induced subgraph of a graph in **G** is also in **G**.

8.3.24. Remark. To prove that graphs in a hereditary class **G** are perfect, it suffices to prove $\chi(G) = \omega(G)$ for all $G \in \mathbf{G}$, since the same argument applies to induced subgraphs of G. Note that interval graphs form a hereditary family. ■

8.3.25. Definition. A **hole** or **chordless cycle** in a graph is an induced subgraph that is a cycle of length at least 4 (recall that a chord of a cycle C is an edge not in C with endpoints in C). A **chordal graph** is a graph with no hole.

From the definition, the family of chordal graphs is hereditary (deleting a vertex cannot create a chordless cycle). Using an interval representation, it is easy to see that every interval graph is a chordal graph (see also Exercise 44).

Chordal graphs were first called **triangulated graphs** or **rigid circuit graphs**. They helped suggest the notion of perfect graphs. Chordal graphs are used in problems such as fast numerical inversion of large sparse symmetric matrices (see Golumbic [1978], Andreae [1988]). They have many alternative characterizations; we prove one used in many inductive proofs about them.

8.3.26. Definition. A vertex is **simplicial** if its neighborhood is a clique. A **simplicial elimination ordering** of an n-vertex graph G (also called **perfect elimination ordering**) is a vertex ordering $v_n, \ldots, v_1$ such that each vertex v_i is simplicial in the subgraph induced by $\{v_1, \ldots, v_i\}$.

8.3.27. Theorem. (Dirac [1961]) A graph G is chordal if and only if it has a simplicial elimination ordering.

Proof: *Sufficiency:* If G has a chordless cycle C, then the first vertex deleted from C in an elimination ordering has nonadjacent remaining neighbors. Hence a graph with a chordless cycle has no simplicial elimination ordering.

Necessity: Since the family of chordal graphs is hereditary, it suffices to prove by induction that every chordal graph has a simplicial vertex. To facilitate the induction, we prove the stronger statement that if a chordal graph is not a complete graph, then it has two nonadjacent simplicial vertices.

If G is not complete, then it has a minimal separating set S. By the minimality of S, vertices u and v of S both have neighbors in any components A and B of $G - S$. The union of shortest u, v-paths through A and through B is a cycle of length at least 4 (as on the left below). By the choice of the paths and the absence of edges from A to B, this cycle has no chord other than uv. Since G has no chordless cycle, $uv \in E(G)$. Since $u, v \in S$ were chosen arbitrarily, S is a clique.

Now let G_1 and G_2 be the S-lobes of G containing A and B (Definition 8.2.16), as on the right below. Note that G_i a chordal graph. If G_i is complete, then any vertex of G_i outside S is simplicial. Otherwise, G_i has two nonadjacent simplicial vertices, and since S is a clique at least one of them is outside S. Since a vertex outside S has no neighbors in G outside its S-lobe, vertices outside S that are simplicial in their S-lobes are nonadjacent and simplicial in G. ∎

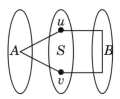 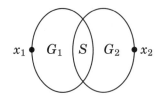

An elimination ordering iteratively deletes vertices. Its reverse is a construction ordering: Every chordal graph can be grown from K_1 by iteratively adding a new vertex whose neighborhood among the existing vertices is a clique.

8.3.28. Theorem. Chordal graphs are perfect.

Proof: (Berge [1960]) Since the chordal graphs form a hereditary family, we need only prove that $\chi(G) = \omega(G)$ when G is chordal.

We color the vertices of G in the reverse order to a simplicial elimination ordering. If v_i has k earlier neighbors, then a color in $[k + 1]$ is available to use on v_i. Since v_i forms a clique with its earlier neighbors, there is a clique whose size is at least the number of the color we use on v_i. After coloring all the vertices, $\omega(G) \geq \chi(G)$. Since always $\chi(G) \geq \omega(G)$, equality holds. ∎

Chordal graphs are also the intersection graphs of families of subtrees of a tree (Exercise 40). This again implies that all interval graphs are chordal graphs.

Many special families of perfect graphs are known. They may have a structural definition, a characterization by a decomposition or construction procedure, a characterization using intersection graphs, fast algorithms for recognition and for optimal coloring, etc. We present some of these families in the exercises.

Meanwhile, we return to the interpretation of our earlier min-max relations in terms of perfection. In discussing independence and coloring, four optimization parameters are natural. A short synonym for "independent set" often used in this context is **stable set**.

8.3.29. Definition. A **clique covering** of a graph G is a partition of $V(G)$ into cliques (a proper coloring of G is a partition of $V(G)$ into stable sets). We consider four parameters:

independence number	$\alpha(G)$	max size of a stable set
clique number	$\omega(G)$	max size of a clique
chromatic number	$\chi(G)$	min size of a coloring
clique covering number	$\theta(G)$	min size of a clique covering

A clique and a stable set share at most one vertex, so always $\chi(G) \geq \omega(G)$ and $\theta(G) \geq \alpha(G)$; covering $V(G)$ requires covering the biggest object of the dual type. Since $\alpha(G) = \omega(\overline{G})$ and $\theta(G) = \chi(\overline{G})$, the statement that $\alpha(H) = \theta(H)$ for every induced subgraph H of G is just the statement that $\overline{G}$ is a perfect graph.

8.3.30. Example. *Bipartite graphs and their line graphs.* Bipartite graphs form a hereditary class, and $\chi(G) = 2 = \omega(G)$ for every nontrivial bipartite graph. Hence bipartite graphs trivially are perfect.

Line graphs of bipartite graphs form another hereditary class (if $G' = L(G)$, then every induced subgraph of G' is the line graph of a subgraph of G'). A stable set in $L(G)$ is a matching in G, and a clique in $L(G)$ is a set of edges in G with a common vertex (or a triangle in G). Hence $\omega(L(G)) = \Delta(G)$ (if G has no triangles), and perfection for line graphs of bipartite graphs is just König's Theorem (Theorem 8.3.7) that $\chi'(G) = \Delta(G)$ when G is bipartite.

Consider the König–Egerváry Theorem ($\alpha'(G) = \beta(G)$ when G is bipartite). We have $\alpha'(G) = \alpha(L(G))$, and cliques in $L(G)$ come from stars in G, so a vertex cover in G provides a clique covering in $L(G)$. Thus the König–Egerváry Theorem says $\alpha(L(G)) = \theta(L(G))$ when G is bipartite. This is another hereditary family, so complements of line graphs of bipartite graphs are perfect.

Since $\omega(G) = 2$, we have $\theta(G) = \beta'(G)$ when G is bipartite with no isolated vertex. Thus König's Other Theorem ($\alpha(G) = \beta'(G)$, Theorem 6.1.16) states that complements of bipartite graphs are perfect. ∎

Bipartite graphs and their line graphs are very special, but these and other families suggest that the complement of every perfect graph is also perfect. Berge conjectured this around 1960; many tried to prove it. Lovász [1972a,b] found the first proof at the age of 22 (Berge–Ramírez-Alfonsín [2001] presents the history). It is now known as the "Perfect Graph Theorem" (PGT). Fulkerson reduced it to a claim he thought was too strong to be true. When Berge told him that Lovász had proved the theorem, within hours he proved the claim, illustrating that a theorem becomes easier to prove when known to be true.

8.3.31. Remark. We proved perfection for each family in Example 8.3.30, but the PGT makes this unnecessary. By the PGT, König's Other Theorem follows from observing that bipartite graphs are perfect. Similarly, the König–Egerváry Theorem and König's Theorem (bipartite graphs are Class 1) imply each other. ■

Lovász proved the PGT twice. Since $\chi(H) \geq \frac{|V(H)|}{\alpha(H)}$, perfection of G (or $\overline{G}$) requires $\omega(H)\alpha(H) \geq |V(H)|$ for each induced subgraph H. Lovász [1972b] proved the converse, making perfection of G and $\overline{G}$ both equivalent to this and hence to each other. Later, Gasparyan [1996] captured the idea in a short proof using linear algebra, which we present. (See Perz–Rolewicz [1990] for another proof.)

8.3.32. Definition. A **minimal imperfect** or **p-critical** graph is an imperfect graph whose proper induced subgraphs are all perfect.

8.3.33. Lemma. If S is a stable set in a p-critical graph G, then $\omega(G) = \omega(G-S)$.

Proof: Since G is p-critical, $\chi(G) > \omega(G)$ and $\chi(G - S) = \omega(G - S)$. Since G can be colored by adding S as one class to an optimal coloring of $G - S$,

$$\chi(G - S) + 1 \geq \chi(G) > \omega(G) \geq \omega(G - S) = \chi(G - S).$$

The two ends of the display differ by 1, so $\chi(G) - 1 = \omega(G) = \omega(G - S)$.   ■

8.3.34. Theorem. (Lovász [1972b]) A graph G is perfect if and only if the inequality $\omega(H)\alpha(H) \geq |V(H)|$ holds for every induced subgraph H.

Proof: (Gasparyan [1996]) Since $\chi(H) \geq \frac{|V(H)|}{\alpha(H)}$, having $\chi(H) = \omega(H)$ requires $\omega(H)\alpha(H) \geq |V(H)|$. Since imperfect graphs have p-critical induced subgraphs, for the converse it suffices to show $\omega(G)\alpha(G) < n$ when G is p-critical with n vertices. Let $a = \alpha(G)$ and $w = \omega(G)$.

Let S_0 be a maximum stable set in G, with $S_0 = \{x_1, \ldots, x_a\}$. By Lemma 8.3.33, each $G - x_r$ is w-colorable. Let $\{S_{(r-1)w+1}, \ldots, S_{rw}\}$ be the stable sets in an optimal coloring of $G - x_r$. For $0 \leq j \leq aw$, Lemma 8.3.33 yields $\omega(G - S_j) = w$, so a maximum clique Q_j in $G - S_j$ is a w-clique in G that avoids S_j.

We claim that Q_j intersects S_i when $i \neq j$. The sets $S_{(r-1)w+1}, \ldots, S_{rw}$ partition $V(G - x_r)$ into w stable sets, and Q_j contains at most one vertex of each (since Q_j is a clique). If $x_r \notin Q_j$, then Q_j must have exactly one vertex from each. If $x_r \in Q_j$, then Q_j has only $w - 1$ vertices other than x_r and hence misses exactly one of the sets in the coloring of $G - x_r$. Since Q_j can only have one vertex in S_0, at most one stable set in the entire list is missed by Q_j. On the other hand, we know that Q_j misses S_j, since Q_j was chosen as a maximum clique in $G - S_j$.

Let A and B be the n-by-$(aw + 1)$ incidence matrices for the set families $\{S_0, \ldots, S_{aw}\}$ and $\{Q_0, \ldots, Q_{aw}\}$, respectively. Since $|S_i \cap Q_j| = 1$ when $i \neq j$, and $|S_j \cap Q_j| = 0$, we obtain $A^T B = J - I$, where J is the all-1 square matrix of order $aw + 1$. Since $J - I$ is nonsingular, both A and B must have rank at least $aw + 1$, and hence $n \geq aw + 1$. We conclude $\omega(G)\alpha(G) < n$.   ■

8.3.35. Corollary. (**Perfect Graph Theorem (PGT)**; Lovász [1972a]) A graph G is perfect if and only if $\overline{G}$ is perfect.

Proof: The condition in Theorem 8.3.34 is the same for both G and $\overline{G}$.   ■

This leads to the question of which graphs are p-critical.

8.3.36. Example. Odd cycles of length at least 5 are imperfect, since $\chi(C_{2k+1}) = 3 > 2 = \omega(C_{2k+1})$. They are p-critical, since proper induced subgraphs are bipartite. By the PGT, $\overline{C}_{2k+1}$ is also p-critical when $k \geq 2$.

Berge [1963] conjectured both the Perfect Graph Theorem and the **Strong Perfect Graph Theorem** (SPGT): *The odd cycles and their complements are the only minimal imperfect graphs.* Since the condition is self-complementary, the SPGT implies the PGT.

The Strong Perfect Graph Conjecture remained open for more than 40 years. It was proved in Chudnovsky–Robertson–Seymour–Thomas [2006] (announced in 2002). The paper is 178 pages long; we will not attempt to describe the proof. ■

Many survey articles and at least four books are devoted to perfect graphs and related topics. Golumbic [1980, 2003] emphasizes algorithmic aspects and special classes. Berge–Chvátal [1984] collects fundamental early papers that developed the theory of perfect graphs. Brandstädt–Le–Spinrad [1999] provides a thorough catalogue of properties of and relationships among nearly 200 classes of perfect graphs. Ramírez-Alfonsín–Reed [2001] discusses classical and more recent aspects, emphasizing the structure of perfect graphs. More recent discussions that emphasize the proof of the SPGT include Roussel–Rusu–Thuillier [2009] and Trotignon [2015].

EXERCISES 8.3

8.3.1. (−) Prove that $L(K_{r,s}) = K_r \square K_s$ and that $\overline{L(K_5)}$ is the Petersen graph.

8.3.2. (−) Prove that if $L(G)$ is connected and regular, then G is regular or is bipartite with equal degree at vertices of the same part. (Ray-Chaudhuri [1967])

8.3.3. (−) Give an explicit edge-coloring to prove that $K_{r,s}$ is Class 1.

8.3.4. (−) Prove that the smallest graph with maximum degree 3 and edge-chromatic number 4 has five vertices.

8.3.5. (−) Determine whether it is possible for a graph G to have more than $\chi'(G)$ pairwise disjoint maximum matchings. Determine whether it is possible to have more than $\chi'(G)$ pairwise disjoint maximal matchings.

8.3.6. (−) Let D be a directed multigraph (loops allowed) such that each indegree and each outdegree is at most d. Prove that D has an edge-coloring with d colors such that edges with a common head or a common tail have distinct colors.

8.3.7. (−) Let G be a Class 1 regular graph of even degree. Prove that $|E(G)|$ is even.

8.3.8. (−) Given integers r and d with $0 \leq r \leq d$, use Vizing's Theorem to prove that every d-regular graph has a spanning subgraph in which every vertex has degree r or $r+1$. (Comment: This is generalized in Tutte [1978].)

8.3.9. (−) Compute $\chi(G)$ and $\omega(G)$ for the complement of an odd cycle.

8.3.10. (−) Show that an n-vertex chordal graph has at most n maximal cliques, with equality if and only if it has no edges. (Fulkerson–Gross [1965])

8.3.11. (−) *Clique identification preserves perfection.* Prove that $G \cup H$ is perfect when G and H are perfect graphs and $G \cap H$ is a complete graph.

8.3.12. Let G be a graph.
(a) Prove that the number of edges in the line graph $L(G)$ is $\sum_{v \in V(G)} \binom{d(v)}{2}$.
(b) Prove that G is isomorphic to $L(G)$ if and only if G is 2-regular.
(c) Determine the graphs G such that $|E(L(G))| < |E(G)|$.

8.3.13. (◊) Prove that every regular graph having a cut-vertex is Class 2.

8.3.14. Use Brooks' Theorem to prove that every graph with maximum degree 3 is 4-edge-colorable. (This is a special case of Vizing's Theorem; do not use Vizing's Theorem.)

8.3.15. (◊) Prove that a 3-regular graph is 3-edge-colorable if and only if it is the union of two even subgraphs.

8.3.16. Prove that every bipartite graph G has a $\Delta(G)$-regular bipartite supergraph (no multiedges).

8.3.17. (◊) Greedy edge-coloring of a graph G produces a proper edge-coloring using at most $2\Delta(G)-1$ colors. For $k \in \mathbb{N}$, construct a tree with maximum degree k and an ordering of its edges such that greedy edge-coloring uses $2k - 1$ colors.

8.3.18. Use induction on $|E(G)|$ to prove König's Theorem that $\chi'(G) = \Delta(G)$ for every bipartite multigraph G. (König [1916])

8.3.19. The d-dimensional **integer simplex** of length m is the graph T_m^d whose vertices are the nonnegative integer $(d + 1)$-tuples summing to m, adjacent when they differ by 1 in two places and agree elsewhere. Compute $\chi'(T_m^d)$ for $m > d$. (Ma–West [2013])

8.3.20. *Density Conditions for Class 2.* Prove that if G is a graph with $2k + 1$ vertices, where $k \in \mathbb{N}$, then $\chi'(G) > \Delta(G)$ under each condition below (with $r \geq 2$).
(a) G has more than $k \cdot \Delta(G)$ edges.
(b) G arises from an r-regular graph by deleting fewer than $r/2$ edges.
(c) G arises from an r-regular graph with $2k$ vertices by subdividing one edge.

8.3.21. Prove that the graph obtained by deleting one edge from the Petersen graph is Class 2 but has no overfull subgraph with maximum degree 3.

8.3.22. (◊) *Overfull Conjecture* $\Rightarrow$ *1-Factorization Conjecture* (Remark 8.3.8).
(a) Prove that in a regular graph of even order, an induced subgraph is overfull if and only if the subgraph induced by the other vertices is overfull.
(b) Let G be a k-regular graph of order $2m$ having an overfull subgraph. Prove $k < m$ for odd m and $k < m - 1$ for even m. Conclude that the Overfull Conjecture implies the 1-Factorization Conjecture.

8.3.23. Let G be the $(m - 1)$-regular connected graph formed from $2K_m$ by applying a 2-switch (Definition 5.2.7). Prove that G has no 1-factorization if m is odd and greater than 3. (Comment: This shows that the 1-Factorization Conjecture in Remark 8.3.8 is sharp.)

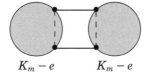

$K_m - e$ \qquad $K_m - e$

8.3.24. (◊) For nontrivial graphs G and H, prove $\chi'(G \square H) = \Delta(G \square H)$ when H satisfies $\chi'(H) = \Delta(H)$. For example, $\chi'(K_3 \square K_2) = 3$. (Mahmoodian [1981])

8.3.25. *Kotzig's Theorem for cartesian product of graphs.*
   (a) Use Vizing's Theorem to prove $\chi'(G \,\square\, K_2) = \Delta(G \,\square\, K_2)$.
   (b) Let G_1 and G_2 be edge-disjoint graphs on V, and let H_1 and H_2 be edge-disjoint graphs on W. Prove $(G_1 \cup G_2) \,\square\, (H_1 \cup H_2) = (G_1 \,\square\, H_2) \cup (G_2 \,\square\, H_1)$.
   (c) Use parts (a) and (b) to prove $\chi'(G \,\square\, H) = \Delta(G \,\square\, H)$ when both G and H have 1-factors. (Kotzig [1979], George [1991])

8.3.26. *Shannon's Theorem.* The edge-coloring theorem of Shannon [1949] states that $\chi'(G) \le \frac{3}{2}\Delta(G)$ for every loopless multigraph G.
   (a) Prove Shannon's Theorem from Vizing's Theorem $\chi'(G) \le \Delta(G) + \mu(G)$.
   (b) Prove Shannon's Theorem without Vizing's Theorem by showing that with $\frac{3}{2}\Delta(G)$ colors available, a partial edge-coloring can be augmented.
   (c) Prove the same bound for even $\Delta(G)$ from Petersen's 2-Factor Theorem.

8.3.27. ($\diamond$) *Equitable edge-colorings.*
   (a) Prove that every graph G with maximum degree $k-1$ has a proper k-edge-coloring with each color used $\lceil |E(G)|/k \rceil$ or $\lfloor |E(G)|/k \rfloor$ times. (de Werra [1971], McDiarmid [1972])
   (b) Prove that every bipartite graph with maximum degree at least k has a k-edge-coloring in which at each vertex v, each color appears $\lceil d(v)/k \rceil$ or $\lfloor d(v)/k \rfloor$ times. (Comment: In particular, bipartite graphs are Class 1. Gupta [1966] notes the case $k = \delta(G)$, which yields a $\delta(G)$-edge-coloring such that every vertex has incident edges of all colors.)

8.3.28. (+) Given an edge-coloring of a graph G, let $c(v)$ be the number of distinct colors on edges at v. An *optimal coloring* maximizes $\sum_{v \in V(G)} c(v)$ among k-edge-colorings of G.
   (a) Suppose that no component of G is an odd cycle. Prove that G has a 2-edge-coloring in which both colors appear at each vertex of degree at least 2.
   (b) Let f be an optimal k-edge-coloring of G in which color a appears at least twice at $u \in V(G)$ and color b does not appear at u. Let H be the subgraph of G consisting of edges colored a or b. Prove that the component of H containing u is an odd cycle.
   (c) Use part (b) to prove Vizing's Theorem for graphs. (Fournier [1973])

8.3.29. Let G_k be the graph consisting of three "parallel" k-cycles $\{x_i\}$, $\{y_i\}$, $\{z_i\}$ and k additional vertices $\{w_i\}$ such that $N(w_i) = \{x_i, y_i, z_i\}$ for each i. Obtain H_k from G_k by deleting $x_k x_1$ and $y_k y_1$ and adding $x_k y_1$ and $y_k x_1$. (Comment: The graphs $\{H_{2j+1} : j \ge 2\}$ are called **flower snarks**. A **snark** is a 2-edge-connected 3-regular Class 2 graph that has girth at least 5 and has no edge cut of size 3 that does not isolate a vertex.)
   (a) Prove that G_k is Class 1.
   (b) Prove that H_k is Class 2 if k is odd. (Isaacs [1975])

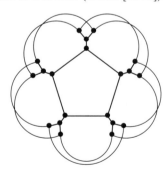

8.3.30. Let M be a maximal k-edge-colorable subgraph of a multigraph G. For $x \in V(G)$, let $\mu(x)$ be the maximum multiplicity of an edge incident to x. Let $S = \{x \in V(G) : d_M(x) \le k - \mu(x)\}$. Puleo [2017] proved that if $x \in S$, then $d_{G[S]}(x) \le d_M(x)$. Prove that Puleo's result implies Ore's strengthening of Vizing's Theorem: $\chi'(G) \le \max_{v \in V(G)} (d_G(v) + \mu(v))$.

8.3.31. The **generalized Petersen graph** $P(n, k)$ is the graph with vertices $\{u_1, \ldots, u_n\}$ and $\{v_1, \ldots, v_n\}$ and edges $\{u_i u_{i+1}\}$, $\{u_i v_i\}$, and $\{v_i v_{i+k}\}$, where addition is modulo n. The usual Petersen graph is $P(5, 2)$, with $\chi' = 4$.

(a) Prove that the subgraph of $P(n, 2)$ induced by k consecutive pairs $\{u_i, v_i\}$ has a spanning cycle if $k \equiv 1 \pmod 3$ and $k \geq 4$. (Comment: The figure below is a proper subgraph of $P(n, 2)$ for $n \geq 10$.)

(b) Use part (a) to prove $\chi'(P(n, 2)) = 3$ for $n \geq 6$.

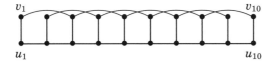

8.3.32. ($\diamond$) Given an edge-coloring of a graph G, for $v \in V(G)$ let $S(v)$ be the set of colors used on edges incident to v. Let $\chi'_k(G)$ be the minimum number of colors in a proper edge-coloring of G such that $|S(x) \cap S(y)| \leq k$ whenever $xy \in E(G)$. Prove $\chi'_k(K_{r,s}) \geq \lceil rs/k \rceil$. (Comment: Borozan et al. [2015] proved $\chi'_k(K_{r,s}) = \lceil rs/k \rceil$, plus $\chi'_2(G) \leq 6$ for $\Delta(G) \leq 3$ and bounds for d-degenerate and complete graphs. **Strong edge-coloring** is the case $k = 1$; Erdős and Nešetřil conjectured $\chi'_1(G) \leq \frac{5}{4}\Delta(G)^2$, improving to $\frac{1}{4}(5\Delta(G)^2 - 2\Delta(G) + 1)$ when $\Delta(G)$ is odd, both achieved by expanding vertices of C_5 into independent sets. The sharp bound of 10 for $\Delta(G) = 3$ was proved in Andersen [1992] and Horák–He–Trotter [1993]; Cranston [2006] proved $\chi'_1(G) \leq 22$ when $\Delta(G) = 4$. See also Molloy–Reed [1997].)

8.3.33. A **total coloring** of G assigns a color to each vertex and each edge so that objects adjacent or incident have different colors. Let $\chi''(G)$ denote the number of colors needed.

(a) Prove $\chi''(G) \leq \chi'_l(G) + 2$ for every graph G.

(b) Prove $\chi''(G) \leq \Delta(G) + 2$ when G is bipartite. (Rosenfeld [1971])

(c) Given the List Color Conjecture, prove $\chi''(G) \leq \Delta(G) + 3$. (Bollobás–Harris [1985]) (Comment: The **Total Coloring Conjecture** is $\chi''(G) \leq \Delta(G)+2$ for all G (Behzad [1965]). It holds for $\Delta(G) \leq 3$ (Rosenfeld [1971], Vijayaditya [1971]), $\Delta(G) = 4$ (Kostochka [1977]), $\Delta(G) = 5$ (Kostochka [1996]), and complete multipartite graphs (Yap [1989]); see Exercise 13.3.5 for $\chi''(K_n)$. Hind–Molloy–Reed [1999] proved $\chi''(G) \leq \Delta(G) + 8 \log^8 \Delta(G)$ when $\Delta(G)$ is large. Geetha–Narayanan–Somasundaram [2018] surveys the topic.)

8.3.34. ($\diamond$) Prove the Total Coloring Conjecture (above) for 4-colorable Class 1 graphs.

8.3.35. (+) An **interval coloring** of G is a proper edge-coloring of G by integers such that at each vertex, the incident colors form an interval of integers.

(a) Let G be an X, Y-bigraph in which every vertex of X has degree 2 and every vertex of Y has degree $2k$ for some $k \in \mathbb{N}$. Prove that G has an interval coloring. (Comment: The statement also holds when vertices of Y have degree $2k + 1$, using the result that a $(2k + 1)$-regular graph with at most k cut-edges has a 2-factor.) (Hansen [1992])

(b) Prove that $K_{r,s}$ has an interval coloring using $r + s - \gcd(r, s)$ colors. (Hint: Combine solutions for the cases $r = s$ and $\gcd(r, s) = 1$. Comment: In fact, fewer colors cannot suffice, but this is harder.) (Hanson–Loten–Toft [1998])

8.3.36. ($\diamond$) A **parity edge-coloring** of a graph G is an edge-coloring such that on every path, some color appears an odd number of times. Prove that G is a subgraph of the hypercube Q_k if and only if G has a parity k-edge-coloring such that on every cycle, every color appears an even number of times. (Havel–Morávek [1972])

8.3.37. With parity edge-coloring defined as above, let $p(G)$ denote the minimum number of colors in a parity edge-coloring of G. Prove $p(K_{2,3}) = 4$ and $p(K_5) = 7$.

8.3.38. ($\diamond$) Give fast algorithms for computing $\chi(G)$, $\alpha(G)$, $\omega(G)$, and $\theta(G)$ when supplied with a simplicial elimination ordering of G.

8.3.39. ($\diamond$) Prove that $\chi_l(G) = \chi(G)$ when G is a chordal graph.

8.3.40. ($\diamond$) *Intersection representation of chordal graphs*.
(a) Prove that pairwise intersecting subtrees of a tree have a common vertex.
(b) Prove that a graph G is chordal if and only if it is an intersection graph of some subtrees of a tree. (Walter [1972, 1978], Gavril [1974], Buneman [1974])
(c) Conclude that a non-complete chordal graph has nonadjacent simplicial vertices.

8.3.41. Show that every intersection graph of subtrees of a tree is an intersection graph of subtrees of a tree with maximum degree 3.

8.3.42. A **clique tree** of a graph G is a tree T whose vertices are the maximal cliques of G, such that for $v \in V(G)$ the cliques containing v induce a subtree of T.
(a) Prove that every tree with fewest vertices in which G has a subtree representation (Exercise 8.3.40) is a clique tree of G.
(b) Let $M(G)$ be the weighted graph whose vertex set is the set $\{Q_1, \ldots, Q_k\}$ of maximal cliques in an n-vertex graph G, with weight $|Q_i \cap Q_j|$ on the edge $Q_i Q_j$. Prove that if T is a spanning tree of $M(G)$, then $w(T) \leq \sum |Q_i| - n$, with equality if and only if T is a clique tree of G. (Acharya–Las Vergnas [1982], McKee [1993])

8.3.43. ($\diamond$) A graph G is a **comparability graph** (see Chapter 12) if and only if it has a **transitive orientation**, which is an orientation F of G such that $xy, yz \in E(F)$ implies $xz \in E(F)$. Prove that comparability graphs are perfect.

8.3.44. (+) This exercise proves that a graph G is an interval graph if and only if G is a chordal graph and $\overline{G}$ is a comparability graph. (Gilmore–Hoffman [1964])
(a) Prove that the condition is necessary.
For sufficiency, henceforth let G be a chordal graph such that $\overline{G}$ is a comparability graph. Let $Q_1, \ldots, Q_k$ be the maximal cliques in G.
(b) Let F be a transitive orientation of $\overline{G}$. Prove that all edges of $\overline{G}$ joining Q_i and Q_j point in the same direction under F.
(c) Define a tournament T on $q_1, \ldots, q_k$ by putting $q_i q_j \in E(T)$ in T when all edges of F joining Q_i and Q_j point from Q_i to Q_j. Prove that T is a transitive tournament.
(d) Prove that $Q_1, \ldots, Q_k$ can be linearly ordered so that the cliques containing any one vertex form a consecutive portion of the list.
(e) Prove that G has an interval representation.

8.3.45. Prove that a connected interval graph G that is not complete has an interval representation whose interval starting leftmost represents a vertex with degree less than $\Delta(G)$. (Comment: Thus G cannot be regular.) (Kostochka–Pelsmajer–West [2003])

8.3.46. Prove that G is an interval graph if and only if G has a vertex ordering $v_1, \ldots, v_n$ such that $v_i v_k \in E(G)$ guarantees $v_j v_k \in E(G)$ whenever $i < j < k$. (Ramalingam–PanduRangan [1988], Jacobson–McMorris–Mulder [1991])

8.3.47. Let Q be a maximal clique in a chordal graph G. Prove that if Q contains no simplicial vertex of G, then $G - Q$ is disconnected. (Voloshin–Gorgos [1982])

8.3.48. The chromatic polynomial $\pi_G(k)$ (Application 4.1.11) counts the proper colorings $f: V(G) \to [k]$. Prove that G is chordal if and only if for every induced subgraph H, $\pi_H(k)$ has the form $k \prod_i (k - j_i)$ with all $j_i \in \mathbb{N}$ (Voloshin [2002]). (Comment: Subdividing one edge of K_6 yields a non-chordal G such that π_G so factors (Read [1975], Dmitriev [1980]).)

8.3.49. Let v be a vertex in a chordal graph G. From the definition (no chordless cycle), prove that there is a simplicial vertex of G among the vertices having maximum distance from v in G. (Voloshin [1982], Farber–Jamison [1986])

8.3.50. Prove $\sum_{r=1}^{\omega(G)} (-1)^{r-1} k_r = 1$ when k_r counts the r-cliques in a connected chordal G.

8.3.51. ($\diamond$) An x, y-**separator** in a graph G is a set S of vertices such that x and y lie in distinct components of $G - S$. A **minimal vertex separator** is a minimal x, y-separator for some pair $\{x, y\}$. Every minimal separating set is a minimal vertex separator. In the graph below, S is a minimal vertex separator (for $\{x, y\}$) but is not a minimal separating set. Prove that a graph is chordal if and only if every minimal vertex separator is a clique.

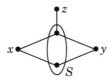

8.3.52. A k-**tree** is a graph grown from K_k by iteratively adding a vertex whose neighborhood is a k-clique. Prove that a connected graph G is a k-tree if and only if (1) G has a k-clique but no $(k + 2)$-clique and (2) Every minimal vertex separator of G is a k-clique.

8.3.53. Let G be an n-vertex chordal graph with clique number r. Prove that G has at most $\binom{r}{j} + \binom{r-1}{j-1}(n - r)$ cliques of size j, with equality only when G is an $(r - 1)$-tree.

8.3.54. A **Meyniel graph** is a graph in which every odd cycle has at least two chords; such graphs are perfect (Meyniel [1976]).
    (a) Prove that every chordal graph is a Meyniel graph.
    (b) Prove that every P_4-free graph is a Meyniel graph.
    (c) Find a comparability graph that is not a Meyniel graph.

8.3.55. ($\diamond$) Let G be a P_4-free graph, meaning that no induced subgraph of G is a 4-vertex path. Prove that every maximal independent set in G intersects every maximal clique in G. Use this to prove that G is perfect.

8.3.56. Let G be a nontrivial connected P_4-free graph. Prove that if Q is a maximal clique in G, then Q contains a neighbor of every vertex of G. Conclude that every connected triangle-free P_4-free graph is a complete bipartite graph.

8.3.57. Let G be a P_4-free graph. Prove that for every vertex ordering σ, the greedy coloring algorithm with respect to σ uses $\chi(G)$ colors (such graphs are called **perfectly orderable**). (Hint: When greedy coloring with respect to σ uses k colors, let i be the least integer such that G has a clique whose vertices receive colors i through k. Prove $i = 1$.)

8.3.58. ($\diamond$) A **complement reducible graph** (or **cograph**), is a graph reducible to a trivial graph by successive complementation within components.
    (a) Prove that a graph G is P_4-free if and only if it is a cograph.
    (b) Use (a) and the PGT to prove P_4-free graphs are perfect. (Seinsche [1974])
    (c) Prove that if G is P_4-free, then G and $\overline{G}$ are comparability graphs.

8.3.59. ($\diamond$) (**Star-Cutset Lemma**) A **star-cutset** of a graph G is a separating set S containing a vertex x adjacent to all of $S - \{x\}$.
    (a) Prove that if G has no stable set intersecting every maximum clique, and every proper induced subgraph of G is $\omega(G)$-colorable, then G has no star-cutset.
    (b) Prove that no p-critical graph has a star-cutset. (Chvátal [1985b])

8.3.60. Find an imperfect graph G having a star-cutset S such that the S-lobes of G are perfect graphs. (Comment: Thus identification at star-cutsets does not preserve perfection, although no p-critical graph has a star-cutset.) (T. Shermer)

8.3.61. A graph G is **weakly chordal** if neither G nor $\overline{G}$ has an induced cycle of length at least 5. Prove that every chordal graph is weakly chordal. (Comment: Hayward [1985] proved that if G and $\overline{G}$ are non-complete weakly chordal graphs, then G or $\overline{G}$ has a star-cutset. Thus weakly chordal graphs are perfect, by the Star-Cutset Lemma.)

8.3.62. ($\diamond$) (**Substitution Lemma**) If $V(G) = \{v_1, \ldots, v_n\}$, and $H_1, \ldots, H_n$ are pairwise disjoint graphs, then the **composition** $G[H_1, \ldots, H_n]$ is the graph $H_1 + \cdots + H_n$ plus the edges $\{xy \colon x \in V(H_i),\ y \in V(H_j),\ v_iv_j \in E(G)\}$. Prove that every composition of perfect graphs is perfect. (Hint: Use the Star-Cutset Lemma, Exercise 8.3.59) (Lovász [1972a])

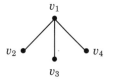

8.3.63. Vertex multiplication is the special case $G[\overline{K}_{h_1}, \ldots, \overline{K}_{h_n}]$ of composition (Exercise 8.3.62), denoted $G \circ h$ (the ith vertex of G "multiplies" into an independent h_i-set). Without using the PGT or Substitution Lemma, prove directly that vertex multiplication preserves perfection of G and perfection of $\overline{G}$.

8.3.64. *Proof of the PGT by vertex multiplication.* The goal: prove that G is perfect when $\overline{G}$ is perfect, using vertex multiplication (Exercise 8.3.63). A minimal counterexample G must be p-critical. Lemma 8.3.33 then implies that every maximal stable set S in G misses some maximum clique $Q(S)$. Let $S_1, \ldots, S_k$ be the list of maximal stable sets in G. Let $h_j = |\{i \colon v_j \in Q(S_i)\}|$. Let $H = G \circ h$.
    (a) By counting vertices of H, prove $\theta(H) \geq k$.
    (b) By counting contributions from cliques, prove $\alpha(H) < k$.
    (c) Obtain a contradiction, using Exercise 8.3.63. (Lovász [1972a])

8.3.65. Let G be a cartesian product of complete graphs. Prove $\alpha(G) = \theta(G)$. Prove that $K_2 \mathbin{\square} K_2 \mathbin{\square} K_3$ is not perfect.

8.3.66. A graph G is a **split graph** if $V(G)$ is the union of a clique and an independent set.
    (a) Prove that if G is a split graph, then G and $\overline{G}$ are chordal.
    (b) Prove that if G and $\overline{G}$ are chordal, then G has no induced C_4, $2K_2$, or C_5.
    (c) Prove that a graph having no induced C_4, $2K_2$, or C_5 is a split graph.

8.3.67. Let $d_1, \ldots, d_n$ be the vertex degrees in a graph G, in nonincreasing order. Let m be the largest index k such that $d_k \geq k - 1$. Prove that G is a split graph if and only if $\sum_{i=1}^{m} d_i = m(m-1) + \sum_{i=m+1}^{n} \min\{m, d_i\}$ (Erdős–Gallai; see Exercise 6.2.43).

8.3.68. ($\diamond$) A graph G is a **threshold graph** if there is a nonnegative weight function w and a threshold t such that $S \subseteq V$ is a stable set if and only if $\sum_{x \in S} w(x) \leq t$; this is a **threshold weighting**. An **edge-threshold weighting** imposes the condition only for $|S| = 2$. Prove A $\Rightarrow$ B $\Rightarrow$ C $\Rightarrow$ D $\Rightarrow$ E $\Rightarrow$ F $\Rightarrow$ A:
    (A) G has a threshold weighting.
    (B) G has an edge-threshold weighting.
    (C) G has no x_1, x_2, x_3, x_4 such that $x_1x_2, x_3x_4 \in E(G)$ and $x_2x_3, x_4x_1 \in E(\overline{G})$.
    (D) G has no 4-vertex induced subgraph isomorphic to C_4, $2K_2$, or P_4.
    (E) G is a P_4-free split graph.
    (F) G arises from K_1 by successively adding vertices made adjacent either to every vertex already present or to no vertex already present.

8.3.69. A graph G is **strongly perfect** if every induced subgraph H has a stable set that intersects all maximal cliques of H. Note that every bipartite graph is strongly perfect.
    (a) Prove that every strongly perfect graph is perfect (justifying the name).
    (b) Construct an example to show that not every perfect graph is strongly perfect.

Chapter 9

Planar Graphs

Topological graph theory studies layouts of graphs on surfaces. In the plane this was stimulated by the famous Four Color Problem, asking whether the (connected) regions of any map on the globe can be colored from four colors so that regions sharing a nontrivial boundary have different colors. More recent motivation comes from circuit layouts on silicon chips. Crossing wires cause problems in layouts, so we want to know which circuits can be laid out without crossings.

Many general texts treat planar graphs and topological graph theory at some length, such as Chartrand–Lesniak [1986], Diestel [1997], and Gross–Yellen [1999] (and more recent editions of all three). Gross–Tucker [1987] and Mohar–Thomassen [2001] are devoted completely to topological graph theory.

9.1. Embeddings and Euler's Formula

A famous puzzle asks for noncrossing paths to enable three hermits to access three utilities. This is equivalent to embedding $K_{3,3}$ in the plane. Postponing definitions, we first explain informally why it cannot be done.

9.1.1. Proposition. K_5 and $K_{3,3}$ are not planar graphs.

Proof: Consider a drawing of K_5 or $K_{3,3}$ in the plane. Let C be a spanning cycle. If the drawing does not have crossing edges, then C is drawn as a closed curve. Chords of C must be drawn inside or outside this curve. Two chords conflict if their endpoints on C occur in alternating order. When two chords conflict, we can draw only one inside C and one outside C.

A 6-cycle in $K_{3,3}$ has three pairwise conflicting chords. We can draw at most one inside and one outside, so the embedding cannot be completed. When C is a 5-cycle in K_5, at most two chords can go inside or outside. There are five chords, so again the embedding cannot be completed. Hence neither graph is planar. ∎

DRAWINGS AND DUALS

We use standard concepts about curves and regions. We treat the formal definitions lightly, since the terms mean what one expects them to mean.

9.1.2. Definition. A **curve** in the plane is the image of a continuous function from the interval $[0, 1]$ into $\mathbb{R}^2$. An **open set** is a set $U \subseteq \mathbb{R}^2$ such that for $p \in U$, all points within some small distance from p belong to U. A **region** is an open set U such that every two points in U lie on a curve contained in U. A curve in the plane is **closed** if its first and last points are the same; it is **simple** if it does not otherwise intersect itself.

Many aspects of planarity need the generality of multigraphs. A **drawing** of a multigraph on a surface maps the vertices into points and the edges into curves, preserving the incidence relation. Edges now have a physical geometric meaning. Since incidence and adjacency are unchanged, we can view these points and curves as the vertices and edges themselves.

By moving edges slightly, we may assume that no three edges share an internal point, that no edge has any vertex as an internal point, and that no two edges are "tangent". We may also assume that no two incident edges cross and that no edge crosses itself, since the change shown below eliminates the crossing. We consider only drawings satisfying these properties.

9.1.3. Definition. In a drawing of a multigraph, a **crossing** of two edges is a common internal point. An **embedding** is a drawing without crossings. A multigraph that embeds in the plane is **planar**. A particular planar embedding of a multigraph is a **plane multigraph**. A subgraph of a plane multigraph inherits its embedding. The **faces** of a plane multigraph are the maximal regions disjoint from the edges.

Every finite plane multigraph has one unbounded face. Computation with faces uses the Jordan Curve Theorem: a simple closed curve cuts the plane into two maximal regions ("inside" and "outside"). In topology this is a deep notion. For finite graphs we may restrict drawings to use only **polygonal curves** (unions of finitely many line segments). In this model the proof is not difficult, and it provides a test for whether a point is inside or outside.

9.1.4. Theorem. (Restricted **Jordan Curve Theorem**) A simple closed polygonal curve C partitions the plane into two faces, each with boundary C.

Proof*: (see Tverberg [1980]) Because C has finitely many segments, nonintersecting segments on C cannot be arbitrarily close. Thus as we follow C, the nearby points on one side all lie in the same face, and similarly for the points on the other side. Every point not on C lies in the same face with at least one of these two sets.

To distinguish the two sets, consider rays in the plane. For a ray ρ from a point p not on C, let $f(\rho)$ be the number of segments of C that ρ intersects. As the direction changes, $f(\rho)$ changes only when ρ visits an endpoint of a segment of C, but the parity is the same before and after that direction. Let the *parity* of p be the parity of $f(\rho)$ for the rays not containing endpoints of segments in C.

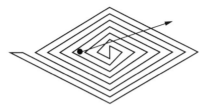

Points x and y in the same face of C are joined by a polygonal x, y-curve P avoiding C. A segment of P lies along a ray. Since the segment does not intersect C, its endpoints have the same parity. Hence any two points in the same face have the same parity. However, the endpoints of a short segment crossing C have opposite parity. Hence there are two faces. The even points and odd points form the outside face and the inside face, respectively.   ∎

A map on the plane or the sphere can be viewed as a plane multigraph. The "boundary-sharing" relation on the faces yields a natural "dual" multigraph.

9.1.5. Definition. The **dual** G^* of a plane multigraph G is a plane multigraph whose vertices correspond to the faces of G and whose edges correspond to the edges of G. The edge $e^* \in E(G^*)$ that corresponds to $e \in E(G)$ joins the vertices corresponding to the faces of G whose boundary contains e.

A canonical embedding of G^* puts the dual vertex x for each face X of G inside X. For each appearance of an edge e in the boundary of X, put a curve within X from x to the midpoint of e; these curves are noncrossing. Each such curve meets another from the other side of e to form the dual edge e^*.

By construction, the dual G^* is a connected plane multigraph, each face of G contains one vertex of G^*, and each edge of G^* is embedded to cross the corresponding edge of G and no other. With this definition, $(G^*)^*$ is isomorphic to G if and only if G is connected (Exercise 15); this motivates using the word "dual".

A plane graph may have loops and multiedges in its dual. A cut-edge of G becomes a loop in G^*; the faces on both sides of it are the same. Multiedges arise in the dual when distinct faces of G have more than one common boundary edge.

9.1.6. Example. Below in bold we show a plane graph with four vertices, four edges, and two faces. The two faces have three common boundary edges, and the cut-edge has the same face on both sides. Hence the dual has a triple-edge and a loop. We show it in solid edges, with four faces, four edges, and two vertices.

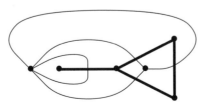

Every planar embedding of K_4 has four faces, and every two of them have a common boundary edge. The dual is another copy of K_4.

Every planar embedding of the cube Q_3 has eight vertices, 12 edges, and six faces. "Opposite" faces have no common boundary; the dual is an embedding of $K_{2,2,2}$, with six vertices, 12 edges, and eight faces.   ■

9.1.7. Example. Two embeddings of a planar graph may have nonisomorphic duals. Each embedding shown below has three faces, so in each case the dual has three vertices. In the embedding on the right, the dual vertex corresponding to the unbounded face has degree 4. In the embedding on the left, no dual vertex has degree 4, so the duals are not isomorphic.

In contrast, every 3-connected planar graph G has essentially only one embedding; the duals of any two drawings of G are isomorphic. This follows from the 2-Isomorphism Theorem of Whitney [1933], which describes when planar graphs have isomorphic duals.   ■

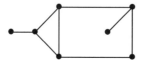

In a connected plane multigraph, the boundary of each face is a closed walk. Otherwise, face boundaries may consist of more than one closed walk.

9.1.8. Definition. The **length** of a face in a plane multigraph G is the total length of the closed walk(s) in G bounding the face.

9.1.9. Example. A cut-edge belongs to the boundary of only one face, and it contributes twice to its length. Each graph in Example 9.1.7 has three faces. In the embedding on the left the lengths are $3, 6, 7$; on the right they are $3, 4, 9$. The sum of the lengths is 16 in each case, which is twice the number of edges.   ■

9.1.10. Proposition. (Dual Degree-Sum Formula) If $l(f_i)$ denotes the length of face f_i in a plane multigraph G with m edges, then $2m = \sum l(f_i)$.

Proof: The face lengths are the degrees of the dual vertices. Since G^* and G have the same number of edges, the statement $2m = \sum l(f_i)$ is just the Degree-Sum Formula for G^*. (Both sums count each edge twice.)   ■

Statements about a connected plane multigraph become statements about its dual when we interchange the roles of vertices and faces. Vertex degrees become face lengths and vice versa, since edges incident to a vertex are exchanged with edges bounding a face.

Geometrically, the Jordan Curve Theorem states that a simple closed curve cuts its interior from its exterior. In plane multigraphs, the duality between curve and cut becomes a duality between cycles and bonds.

9.1.11. Theorem. Edges in a plane multigraph G form a cycle in G if and only if the corresponding dual edges form a bond in G^*.

Proof: Consider $D \subseteq E(G)$, and let D^* denote the set of edges in G^* corresponding to the edges in D. Since a bond is a minimal edge cut, it suffices to show that D^* contains an edge cut if and only if D contains a cycle.

If D is the edge set of a cycle in G, with S being the set of dual vertices corresponding to faces inside the cycle, then the edge cut $[S, \overline{S}]$ consists of all the edges of G^* that cross edges of D; this is the set D^* (by the Jordan Curve Theorem, S and $\overline{S}$ are nonempty).

If D contains no cycle in G, then D encloses no region. The unbounded face of G is reachable from anywhere without crossing D. Hence $G^* - D^*$ is connected, and D^* contains no edge cut. Hence when D is a cycle the dual cut is minimal. ∎

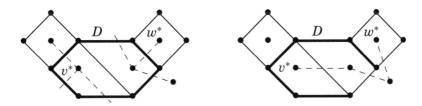

9.1.12. Remark. If a plane graph is not a forest, then every face boundary contains a cycle. ∎

9.1.13. Remark. Deleting a non-cut edge of G has the effect of contracting its dual edge in G^*, since two faces of G combine into one. Contracting a non-loop edge of G has the effect of deleting its dual edge in G^*. Letting G be the central solid graph below, we have $G - e$ on the left and $G \cdot e$ on the right.

To maintain this duality, *when discussing planarity we often keep multiedges and loops that arise from edge contraction.* ∎

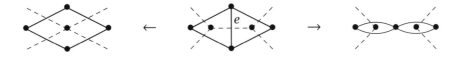

Face boundaries allow us to characterize bipartite planar multigraphs. The result can also be proved by induction (Exercise 17).

9.1.14. Theorem. Equivalent for a plane multigraph G are
  (A) G is bipartite.
  (B) Every face of G has even length.
  (C) The dual G^* is Eulerian.

Proof: A $\Rightarrow$ B. A face boundary consists of closed walks. Every odd closed walk contains an odd cycle. Therefore, in a bipartite plane multigraph every face boundary has even length.

B $\Rightarrow$ A. Let C be a cycle in G. Since G has no crossings, C becomes a simple closed curve in the drawing. Every face of G is wholly inside C or wholly outside C. The total of the face lengths for the faces inside C is even, since each face length is even. The sum counts each edge of C once. It also counts each edge in the region inside C twice, since each such edge belongs twice to faces inside C. Hence the length of C and the full sum have the same parity, which is even.

    B $\Leftrightarrow$ C. The dual G^* is connected, and degrees in G^* are face lengths in G. ∎

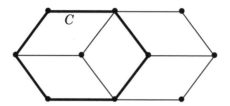

    Many questions we consider for general planar multigraphs can be answered rather easily for graphs in a special class.

9.1.15. Definition. A multigraph is **outerplanar** if it has an embedding with every vertex on the boundary of the unbounded face. An **outerplane multigraph** is such an embedding. A **maximal outerplanar graph** is a (simple) outerplanar graph that is not a proper subgraph of another outerplanar graph on the same vertices.

    By another drawing, the graph in Example 9.1.7 is outerplanar.

9.1.16. Proposition. The boundary of the outer face of a 2-connected outerplane multigraph is a spanning cycle.

Proof: This boundary contains all the vertices. If it is not a cycle, then it visits some vertex more than once; this is a cut-vertex.   ∎

9.1.17. Proposition. K_4 and $K_{2,3}$ are planar but not outerplanar.

Proof: The figure below shows that K_4 and $K_{2,3}$ are planar.

Since they are 2-connected, an outerplane embedding requires a spanning cycle. There is no spanning cycle in $K_{2,3}$ (a bipartite graph has no 5-cycle). There is a spanning cycle in K_4, but the endpoints of the remaining two edges alternate along it. Hence these chords conflict and cannot both be drawn inside.   ∎

    Some facts about outerplane graphs have elegant proofs using duals.

9.1.18. Definition. The **weak dual** of a plane multigraph G is obtained from the dual G^* by deleting the vertex for the unbounded face of G.

9.1.19. Proposition. A plane graph is an outerplane graph if and only if its weak dual is a forest.

Proof: Since components of a plane graph G that are trees do not affect whether G is an outerplane graph, we may assume that all components have cycles. Now components of the weak dual H correspond to components of G. A component of H has a cycle if and only if G has a vertex v surrounded by bounded faces of G, which holds if and only if v is not incident to the unbounded face of G. ∎

9.1.20. Proposition. If G is an outerplanar graph, then $\delta(G) \leq 2$. If also $|V(G)| = n \geq 2$, then G has at most $2n - 3$ edges.

Proof: Let G be an outerplane graph. By Proposition 9.1.19, the weak dual H of G is a forest. If H has no vertices, then G is a forest. If H has an isolated vertex, then G has a component with at most one cycle. In both cases, $\delta(G) \leq 2$.

Otherwise, H has a vertex of degree 1. The corresponding face f in G shares only one edge with a bounded face of G. Since G is simple, f has length at least 3, and the internal vertices of the path along f whose edges are not shared with a bounded face have degree 2 in G.

An outerplanar graph with two vertices has at most one edge. For larger n, deleting a vertex of minimum degree yields the result by induction. ∎

EULER'S FORMULA

Euler's Formula is the basic counting tool relating the numbers of vertices, edges, and faces of multigraphs drawn in the plane.

9.1.21. Theorem. (**Euler's Formula**; Euler [1758]) If a connected plane multigraph G has n vertices, m edges, and r faces, then

$$n - m + r = 2.$$

Proof: We use induction on n. Basis step ($n = 1$): G is a "bouquet" of loops, each a closed curve in the embedding. If $m = 0$, then $r = 1$, and the formula holds. Each added loop passes through a face and cuts it into two faces (by the Jordan Curve Theorem). This augments the edge count and the face count each by 1. Thus the formula holds for $n = 1$.

Induction step ($n > 1$): Since G is connected, it has an edge that is not a loop. Contracting it yields a plane multigraph G' with n' vertices, m' edges, and r' faces. Note that $n' = n - 1$ and $m' = m - 1$. The number of faces does not change (we merely shorten boundaries but may create loops and multiple edges), so $r' = r$. By the induction hypothesis,

$$n - m + r = n' + 1 - (m' + 1) + r' = n' - m' + r' = 2.$$ ∎

$$n = 1 \qquad\qquad n > 1$$

9.1.22. Remark. (1) By Euler's Formula, all planar embeddings of a connected multigraph G have the same number of faces. The dual G^* may depend on the embedding of G, but the number of vertices in G^* does not.

(2) Euler's Formula as stated fails for disconnected multigraphs. If a plane multigraph G has k components, then adding $k-1$ edges to G yields a connected plane multigraph without changing the number of faces. This yields the generalization $n - m + r = k + 1$.

(3) Euler's Formula is familiar for polyhedra as $V - E + F = 2$, where V, E, F are the numbers of vertices, edges, and faces. However, in graph theory V and E are operators returning the sets of vertices and edges, and F is usually a set of edges or a graph. Since we often use n and m for the numbers of vertices and edges, using r to count the faces is consistent and frees f for individual faces. We use r because it is the number of maximal *regions*, that is, faces. ∎

We will see various applications of Euler's Formula.

9.1.23. Theorem. Let G be a planar n-vertex graph with m edges. If $n \geq 3$, then $m \leq 3n - 6$. If also G is triangle-free, then $m \leq 2n - 4$.

Proof: By adding edges, it suffices to consider connected graphs. Let r be the number of faces in some embedding of G, Euler's Formula will relate n and m if we can dispose of r. Proposition 9.1.10 provides an inequality for m and r.

For $n \geq 3$, every face boundary in a connected plane graph has length at least 3. Summing the face lengths yields $2m \geq 3r$. Substituting $r \leq 2m/3$ into $n - m + r = 2$ yields $m \leq 3n - 6$.

When G is triangle-free, all faces have length at least 4. In this case $2m \geq 4r$ and $r \leq m/2$, and we obtain $m \leq 2n - 4$. ∎

9.1.24. Example. Nonplanarity of K_5 and $K_{3,3}$ follows again from Theorem 9.1.23. A planar graph with five vertices has at most nine edges ($3n - 6$), but K_5 has 10 edges. Although $K_{3,3}$ has only nine edges, it is triangle-free, and a triangle-free graph with six vertices has at most eight edges ($2n - 4$). In both cases, there are too many edges to be planar. ∎

9.1.25. Definition. A **maximal planar graph** is a planar graph that is not a spanning subgraph of another planar graph. A **triangulation** is a plane multigraph where every face boundary is a 3-cycle.

9.1.26. Proposition. For an n-vertex plane graph G, equivalent are:
(A) G has $3n - 6$ edges.
(B) G is a triangulation.
(C) G is a maximal plane graph.

Proof: A ⇔ B. For an n-vertex plane graph, the proof of Theorem 9.1.23 shows that having $3n - 6$ edges is equivalent to $2m = 3r$, which occurs if and only if every face has length 3.

B ⇔ C. Some face has length more than 3 if and only if there is a way to add an edge to the drawing and obtain a larger plane graph. ∎

9.1.27. Remark. "Maximal plane multigraph" makes no sense, since edges may have high multiplicity, but an n-vertex triangulation still has $3n - 6$ edges. To obtain a triangulation with multiedges from a maximal plane graph, an edge xy bounding faces A and B can be "widened" into two edges and two vertices inserted as shown below, adding six edges and four faces. Every face still has length 3. In the dual, the edge AB is replaced using four new vertices as shown.   ∎

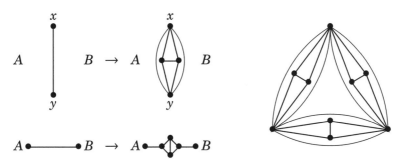

9.1.28. Remark. A graph embeds in the plane if and only if it embeds on a sphere. Given an embedding on a sphere, puncture the sphere inside a face and project the embedding onto a plane tangent to the opposite point. This yields a planar embedding in which the punctured face on the sphere becomes the unbounded face in the plane. The process is reversible.   ∎

9.1.29. Application. A **regular polygon** is a closed simple polygonal curve having sides of equal length and equal angles between successive sides. A **regular polyhedron** is a solid whose faces are isomorphic regular polygons, with the same number of faces meeting at each vertex. When we expand the polyhedron to a sphere and then move the drawing to the plane as in Remark 9.1.28, we obtain a regular plane graph with faces of the same length. Hence the dual also is regular.

Let G be a plane graph with n vertices, m edges, and r faces. If G is k-regular and G^* is l-regular, then $kn = 2m = lr$, using the Degree-Sum Formulas for G and G^*. Substituting for n and r in Euler's Formula yields $m(\frac{2}{k} - 1 + \frac{2}{l}) = 2$. Since m and 2 are positive, the other factor must also be positive, which yields $(2/k) + (2/l) > 1$ and hence $2l + 2k > kl$. We rewrite this as $(k - 2)(l - 2) < 4$. Because the dual of a 2-regular graph is not simple, we require $k, l \geq 3$. The pairs solving $(k - 2)(l - 2) < 4$ are now $(3, 3)$, $(3, 4)$, $(3, 5)$, $(4, 3)$, $(5, 3)$.

Given k and l, there is only one way to lay out the plane graph when we start with any face. Hence the regular polyhedra are only the five Platonic solids listed below, one for each pair (k, l) that satisfies the requirements.   ∎

k	l	$(k-2)(l-2)$	m	n	r	name
3	3	1	6	4	4	tetrahedron
3	4	2	12	8	6	cube
4	3	2	12	6	8	octahedron
3	5	3	30	20	12	dodecahedron
5	3	3	30	12	20	icosahedron

EXERCISES 9.1

9.1.1. (−) Let G be a plane graph with dual G^*. What is the effect on the dual when an edge of G is subdivided? What is the effect when an edge of G is duplicated (that is, when another edge with the same endpoints is added)?

9.1.2. (−) Count the isomorphism classes of planar duals of the graph below.

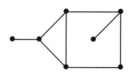

9.1.3. (−) Prove or disprove: If G is a 2-connected plane graph with minimum degree 3 (no multiedges), then its dual G^* has no multiedges.

9.1.4. (−) Let G be a maximal plane graph with $|V(G)| \geq 3$. Prove or disprove each below:
    (a) G must be a chordal graph.
    (b) G must have an even number of faces.
    (c) G^* must be 2-edge-connected and 3-regular.

9.1.5. (−) Let G be a plane graph with an odd number of vertices. Prove or disprove: If G^* is Eulerian, then G cannot be Hamiltonian.

9.1.6. (−) Prove that a 2-edge-connected 3-regular plane graph is 3-connected if and only if no two faces share at least two boundary edges.

9.1.7. (−) Use edge deletion to prove Euler's Formula by induction on the number of edges, with trees in the basis step.

9.1.8. (−) Let S be a set of n points in the plane with no two points in S closer together than 1 inch. Prove that at most $3n - 6$ pairs are exactly 1 inch apart.

9.1.9. (−) Obtain the number of faces in a k-regular plane graph with n vertices.

9.1.10. (−) Prove that every plane graph with minimum degree at least 4 has a face of length 3.

9.1.11. (−) Prove that every plane graph with minimum degree at least 3 has a face with length at most 5.

9.1.12. (−) Determine the number of faces in a plane graph with n vertices, m edges, and k components.

9.1.13. (−) For $n \geq 5$, construct an n-vertex maximal planar graph that does not have three disjoint cycles.

9.1.14. (−) Let G be a maximal planar graph with at least five vertices. Prove that G does not have adjacent vertices of degree 3.

9.1.15. Let G be a plane multigraph. Prove that $(G^*)^* \cong G$ if and only if G is connected.

9.1.16. Contracting an edge of a plane multigraph deletes the corresponding edge from the dual: $(G \cdot e)^* = G^* - e^*$ (Remark 9.1.22(2)). Use this to prove Theorem 9.1.11 inductively: a set $D \subseteq E(G)$ is a cycle in G if and only if the dual set D^* is a bond in G^*.

9.1.17. Prove by induction on the number of faces that a plane multigraph is bipartite if and only if every face has even length.

9.1.18. Prove that every plane n-vertex multigraph isomorphic to its dual has $2n - 2$ edges. For $n \geq 4$, construct an example with no loops or multiedges.

9.1.19. For $k \geq 3$, determine all maximal planar graphs with $k + 2$ vertices such that all vertices have degree 4 except for two vertices of degree k.

9.1.20. Construct a vertex-transitive 3-regular planar graph of diameter 3 with 12 vertices. (Comment: T. Barcume proved the conjecture of Erdős that no 3-regular planar graph with diameter 3 has more than 12 vertices.)

9.1.21. Construct a 16-vertex 5-regular planar graph with diameter 3. (Hint: The unique such graph can be drawn with 4-fold rotational symmetry and contains a 4-regular spanning subgraph with diameter 3. Comment: Preen [2012] proved that this graph is the only 5-regular planar graph with diameter 3 other than the icosahedron.)

9.1.22. Let S be a separating 3-set of the dual of a maximal planar graph G. Prove that $G^* - S$ has two components.

9.1.23. ($\diamond$) Prove that every maximal planar graph having at least four vertices is 3-connected. (Ore [1967, p. 6])

9.1.24. ($\diamond$) A **fullerene** is a 3-regular 3-connected plane graph whose faces have length 6 except for twelve of length 5. For $k \in \mathbb{N}_0$, construct a fullerene with $5k$ faces of length 6 and another with $6k + 2$ faces of length 6. (Comment: Grünbaum–Motzkin [1963] constructed fullerenes with r faces of length 6 for all $r \in \mathbb{N}$ with $r > 1$.)

9.1.25. ($\diamond$) Prove that edges in a connected plane multigraph G form a spanning tree if and only if the duals of the other edges form a spanning tree in G^*. (von Staudt [1847])

9.1.26. ($\diamond$) Let G be a connected plane multigraph. Prove that G is Hamiltonian if and only if $V(G^*)$ can be partitioned into two sets S and T such that $G^*[S]$ and $G^*[T]$ are both trees. (Stein [1970] proved this for triangulations.)

9.1.27. Let G be an Eulerian plane multigraph. Prove that G cannot satisfy either property below. (Comment: These facts can be used in Exercise 9.3.28.)
   (a) One face of G has length 2 or 4 and the rest have length 3.
   (b) Every face of G has length 3 and G has a loop.

9.1.28. Use Euler's Formula to prove that an n-vertex maximal outerplanar graph has $2n - 3$ edges.

9.1.29. *Alternative proof of Proposition 9.1.20.* Prove inductively that every outerplanar graph with at least four vertices has two nonadjacent vertices with degree at most 2.

9.1.30. Let G be an outerplane graph other than K_3 whose bounded faces all have length 3. Let t be the number of faces in G sharing no edge with the unbounded face. Prove that G has exactly $t + 2$ vertices of degree 2.

9.1.31. Let $G_0 = K_3$. For $k \geq 1$, let G_k be the plane graph obtained from G_{k-1} by adding a new vertex v_e for each edge e on the unbounded face of G_{k-1}, adjacent only to the endpoints of e. Note that G_k is outerplanar.
   (a) Determine the degree list of G_k.
   (b) Prove that every outerplanar graph is a subgraph of G_k for some k.

9.1.32. ($\diamond$) A **triangulation** of a convex n-gon is a maximal outerplanar graph obtained by adding chords joining corners of the n-gon. *Flipping* a chord replaces it with the other chord that cuts the union of the two triangles bounding it, as shown below. Flipping a chord yields another triangulation. Prove that any triangulation can be turned into any other using at most $2n - 6$ flips, and prove that at most $2n - 10$ suffice when $n > 12$. (Comment: Sleator–Tarjan–Thurston [1988] used hyperbolic geometry to prove that the bound $2n - 10$ is best possible when n is sufficiently large. Pournin [2014] found a shorter combinatorial proof that $2n - 10$ is sharp when $n > 12$.)

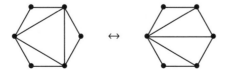

9.1.33. For $G \in \{K_6, K_7\}$, determine whether G decomposes into two outerplanar graphs.

9.1.34. ($\Diamond$) Let G be a nontrivial maximal outerplanar graph with vertex degrees all even. Prove that $|E(G)|$ is divisible by 3. (Jonsson–Propp [2007])

9.1.35. Let l be the length of a longest cycle in a planar triangulation G. Prove that G has cycles of all lengths from 3 through l. (Balister)

9.1.36. ($\Diamond$) *Euler's Formula using linear algebra.* Let A be an n-by-m matrix over a field $\mathbb{F}$; its **nullspace** is $\{x \in \mathbb{F}^m: Ax = 0\}$. The **Rank–Nullity Theorem** of linear algebra states that the rank of A plus the dimension of its nullspace equals the number of columns. Apply this to the incidence matrix (over the field $\mathbb{F}_2$ of size 2) to prove Euler's Formula.

9.1.37. Find the maximum number of edges in a planar subgraph of the hypercube Q_k.

9.1.38. Let G be a 3-regular connected plane graph in which every vertex is incident to one face of length 4, one face of length 6, and one face of length 8. Without drawing G, use Euler's Formula to count the faces of G.

9.1.39. The **rhombicosidodecahedron** is a polyhedron in which every vertex is incident to one triangular face, one pentagonal face, and two (opposite) quadrilateral faces. Determine the number of faces in the rhombicosidodecahedron. (Comment: The toy construction system "Zometool" is based on this polyhedron.)

9.1.40. ($\Diamond$) Use Euler's Formula to count the regions formed by n lines in the plane, assuming that no two are parallel and no three have a common point.

9.1.41. ($\Diamond$) Consider a convex n-gon such that no three segments joining corners have a common internal point. Use Euler's Formula (not induction!) to count the bounded regions in the drawing after adding all $\binom{n}{2}$ segments joining corners of the n-gon. The answers for $3 \le n \le 6$ are 1, 4, 11, and 25.

9.1.42. Let R be a convex region in the plane. Suppose that q chords with distinct endpoints on the boundary of R are drawn and form p points of intersection, with no three chords having a common point. In terms of p and q, compute the number of regions into which the chords cut R. (Alexanderson–Wetzel [1977])

9.1.43. ($\Diamond$) Let f be a proper vertex coloring of an n-vertex triangulation using the colors $\{a, b, c, d\}$. Let t be the number of edges joining colors a and b plus the number of edges joining colors c and d. Determine t. (X. Lv)

9.1.44. Prove that if G is a planar graph with at least 11 vertices, then $\overline{G}$ is nonplanar. Construct a self-complementary planar graph with eight vertices.

9.1.45. ($\Diamond$) Let G be a connected n-vertex planar graph with m edges. Prove that $m \le \frac{g}{g-2}(n-2)$ when G has girth g.

9.1.46. Prove that every n-vertex planar graph with $n + k$ edges has a cycle of length at most $\frac{2(n+k)}{k+2}$. To prove that the bound is sharp, construct an example with girth $\frac{2(n+k)}{k+2}$ whenever $n - 2$ is divisible by $k + 2$.

9.1.47. *Unusual graphs.*

(a) For a planar graph with minimum degree at least 3 and n_i vertices of degree i, prove $3n_3 + 2n_4 + n_5 \ge 12$.

(b) Let G be a 3-connected plane graph in which no three faces have the same length. Prove that G has two faces each of lengths 3, 4, and 5 and no face at all of length at least 7. There are four such graphs (Jorza [2001]); construct one.

9.1.48. ($\Diamond$) Prove that every n-vertex plane graph decomposes into at most $2n - 4$ edges and facial triangles. Determine the graphs for which $2n - 4$ such subgraphs are needed.

9.1.49. ($\Diamond$) Let G be a connected even plane graph. An Eulerian circuit is **noncrossing** if it does not cross itself when viewed as a closed curve. That is, in the embedding around a vertex v, the two edges used on a visit through v are consecutive. Prove the following.

    (a) G has a noncrossing Eulerian circuit. (Abrham–Kotzig [1979], Singmaster [1981])

    (b) If also every bounded face is a triangle, then $|E(G)|$ is divisible by 3.

    (c) If also G is a maximal outerplanar graph, then $|V(G)|$ is divisible by 3.

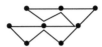

9.1.50. For $k \in \mathbb{N}$, prove that the following exist: an Eulerian planar graph with minimum degree 4 where the distance between degree-4 vertices is at least k, and a planar graph with minimum degree 5 where the distance between degree-5 vertices is at least k.

9.1.51. Prove that every 4-regular plane graph has at least eight triangular faces. For each $k \in \mathbb{N}$, construct a 4-regular plane graph with exactly eight triangular faces and such that any path connecting two of these faces has length at least k.

9.1.52. Determine all n such that there exists a 4-regular planar graph with n vertices. (Chvátal [1969], Owens [1971])

9.1.53. For $k \in \mathbb{N}$, construct a connected 5-regular planar graph with $12k$ vertices.

9.1.54. ($\Diamond$) For $n \geq 4$, prove that every n-vertex planar graph has at least four vertices with degree less than 6. Prove that equality may hold when n is even and at least 8. (Grünbaum–Motzkin [1963])

9.1.55. ($\Diamond$) *Directed plane graphs.* Let G be a plane graph, and let D be an orientation of G. The **dual** D^* of D is an orientation of G^* such that when an edge of D is viewed from tail to head, the dual edge in D^* crosses it from right to left. For example, if the solid edges below are in D, then the dashed edges are in D^*.

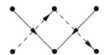

Prove that if D is strongly connected, then D^* has no directed cycle, so $\delta^-(D^*) = \delta^+(D^*) = 0$. Conclude that if D is strong, then D has a face whose boundary is a clockwise cycle and another face whose boundary is a counterclockwise cycle.

9.1.56. Prove that every 2-connected plane graph has an ear decomposition in which ears are successively removed from the boundary of the external face of the remaining graph.

9.1.57. ($\Diamond$) *Pick's Theorem.* A **lattice polygon** is a polygon whose corners are at integer lattice points in the plane.

    (a) Let G be a 2-connected plane graph whose bounded faces are triangles, with B vertices on the unbounded face and I other vertices. In terms of B and I, determine the numbers of edges and faces in G.

    (b) Let T be a lattice triangle. Prove that if no other lattice points lie on the boundary or in the interior of T, then T has area $1/2$. (Hint: Use induction on the area of the smallest lattice rectangle containing T.)

    (c) Let P be a lattice polygon with B lattice points on the perimeter and I lattice points inside. Prove **Pick's Theorem**: the area of the region bounded by P is $I + B/2 - 1$. (DeTemple–Robertson [1974], Gaskell–Klamkin–Watson [1976])

9.2. Structure of Planar Graphs

Which graphs embed in the plane? We know that K_5 and $K_{3,3}$ do not. These two graphs are critical for Kuratowski's characterization of planar graphs. After the publication of Frank Harary's textbook, Kasimir Kuratowski asked Harary about the origin of the notation for K_5 and $K_{3,3}$. Harary replied, "The K in K_5 stands for Kasimir, and the K in $K_{3,3}$ stands for Kuratowski!"

In this section we also discuss the Planar Separator Theorem, a tool for solving computational problems quickly on planar graphs.

KURATOWSKI'S THEOREM

We generalize the notion of subdividing an edge of a graph (Definition 7.2.20).

9.2.1. Definition. A **subdivision** of F or F-**subdivision** is a graph obtained from a graph F via edge subdivisions; edges become paths through new vertices. When $\delta(F) \geq 3$ and H is an F-subdivision, the vertices with degree at least 3 in H are the **branch vertices** of H; these were the vertices of F.

a $K_{3,3}$-subdivision

9.2.2. Proposition. A graph containing a subdivision of K_5 or $K_{3,3}$ is nonplanar.

Proof: Subgraphs of planar graphs are planar, so it suffices to show that subdivisions of K_5 and $K_{3,3}$ are nonplanar. By Proposition 9.1.1, K_5 and $K_{3,3}$ are not planar. Subdividing edges does not affect planarity; the curves in an embedding of a subdivision of G can be used to obtain an embedding of G, and vice versa. ■

By Proposition 9.2.2, avoiding subdivisions of K_5 and $K_{3,3}$ is a necessary condition for being a planar graph. It also is sufficient:

9.2.3. Theorem. (Kuratowski [1930]) A graph is planar if and only if it does not contain a subdivision of K_5 or $K_{3,3}$. ■

Thus K_5 and $K_{3,3}$ are the only *topologically minimal* nonplanar graphs (not a subdivision of another nonplanar graph). Wagner [1937] proved another characterization. Deletions and contractions preserve planarity, so we can seek the minimal nonplanar graphs under these operations. Wagner proved that G is planar if and only if it has no subgraph contractible to K_5 or $K_{3,3}$. This follows easily from Kuratowski's Theorem (see Exercise 13), the proof of which is our next goal.

9.2.4. Definition. A **Kuratowski subgraph** of G is a subgraph of G that is a subdivision of K_5 or $K_{3,3}$. A **minimal nonplanar graph** is a nonplanar graph such that every proper subgraph is planar.

We will show that a smallest nonplanar graph not having a Kuratowski subgraph must be 3-connected. To prove Kuratowski's Theorem, it then suffices to show that 3-connected graphs without Kuratowski subgraphs are planar.

9.2.5. Lemma. If F is the edge set of a face in an embedding of G, then G has an embedding where F is the edge set of the unbounded face.

Proof: Project the embedding onto the sphere, where the edge sets of regions remain the same and all regions are bounded, and then return to the plane by projecting from inside the face bounded by F. ∎

9.2.6. Lemma. Every minimal nonplanar graph is 2-connected.

Proof: Let G be a minimal nonplanar graph. If G is disconnected, then we embed one component of G inside one face of an embedding of the rest.

If G has a cut-vertex v, then let $G_1, \ldots, G_k$ be the $\{v\}$-lobes of G. By the minimality of G, each G_i is planar. By Lemma 9.2.5, we can embed each G_i with v on the outside face. We fit each embedding into an angle smaller than $360/k$ degrees at v and then combine them to obtain an embedding of G. ∎

9.2.7. Lemma. Let $\{x, y\}$ be a separating 2-set of G, let $G_1, \ldots, G_k$ be the $\{x, y\}$-lobes of G, and let $H_i = G_i \cup xy$. If G is nonplanar, then some H_i is nonplanar.

Proof: If H_i is planar, then by Lemma 9.2.5 it embeds with xy on the outside face. Combine such embeddings, with each successive one embedded in a face of the current graph having xy on the boundary. Now delete xy if it is not in $E(G)$. The result is a planar embedding of G. ∎

We next reduce Kuratowski's Theorem to the 3-connected case. The hypothesized graph doesn't exist, but if it did, it would be 3-connected.

9.2.8. Lemma. If G is a graph with fewest edges among all nonplanar graphs without Kuratowski subgraphs, then G is 3-connected.

Proof: Deleting an edge of G cannot create a Kuratowski subgraph in G. Thus deleting one edge produces a planar subgraph, and hence G is a minimal nonplanar graph. By Lemma 9.2.6, G is 2-connected.

Suppose that G has a separating 2-set S, with $S = \{x, y\}$. Since G is nonplanar, the union of xy with some S-lobe is nonplanar (Lemma 9.2.7); let H be such a graph. Since H has fewer edges than G, minimality of G forces H to have a Kuratowski subgraph F. All of F appears in G except possibly the edge xy.

Since S is a minimal vertex cut, both x and y have neighbors in every S-lobe. Thus we can replace xy in F with an x, y-path through another S-lobe to obtain a

Kuratowski subgraph of G. This contradicts the prohibition of Kuratowski subgraphs, so G has no separating 2-set.   ■

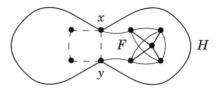

   To prove that 3-connected graphs without Kuratowski subgraphs are planar,
we use induction. For the induction step, we obtain from such a graph a smaller
3-connected graph without Kuratowski subgraphs, by contracting an appropriate edge. We show first that no edge contraction can introduce a Kuratowski
subgraph if none was present before.

9.2.9. Lemma. If $G \cdot xy$ has a Kuratowski subgraph, then G has a Kuratowski
   subgraph.

Proof: Let H be a Kuratowski subgraph of $G \cdot xy$, and let z be the vertex of $G \cdot$
xy obtained by contracting xy. If z is not in H, then H itself is a Kuratowski
subgraph of G. If $z \in V(H)$ but z is not a branch vertex of H, then we obtain a
Kuratowski subgraph of G from H by replacing z with x or y or with the edge xy.
   Similarly, if z is a branch vertex in H and at most one edge at z in H is incident to x in G, then expanding z into xy lengthens that path, and y becomes a
branch vertex for a Kuratowski subgraph in G.
   In the remaining case (shown below), H is a K_5-subdivision, z is a branch
vertex of H, and each of x and y is incident in G to two of the four edges incident
to z in H. Let u_1 and u_2 be the branch vertices of H reached via the paths leaving
z on edges incident in G to x. Similarly let v_1 and v_2 be the branch vertices of H
reached by the paths leaving z on edges incident in G to y. Deleting the u_1, u_2-
path and v_1, v_2-path from H yields a $K_{3,3}$-subdivision in G having y, u_1, u_2 as the
branch vertices for one part and x, v_1, v_2 as the branch vertices for the other.   ■

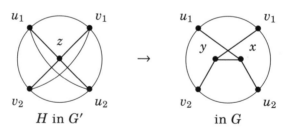

$$H \text{ in } G' \qquad\qquad \text{ in } G$$

   Now we can prove Kuratowski's Theorem. In fact, we prove a stronger property for 3-connected graphs without Kuratowski subgraphs.

9.2.10. Definition. A **convex embedding** of a planar graph is an embedding in
   which each face boundary is a convex polygon (including the unbounded face).

9.2.11. Theorem. (Tutte [1960, 1963]) If G is a 3-connected graph containing no
   subdivision of K_5 or $K_{3,3}$, then G has a convex embedding in the plane with
   no three vertices on a line.

Proof: (Thomassen [1980a, 1981a]) We use induction on n, the order of G. The only 3-connected graph with at most four vertices is K_4, which has such an embedding. For the induction step, consider $n \geq 5$.

Let e with endpoints x and y be an edge such that $G \cdot e$ is 3-connected, guaranteed by Lemma 7.2.18. Let z be the vertex obtained by contracting e, and let $H = G \cdot e$. By Lemma 9.2.9, H has no Kuratowski subgraph. By the induction hypothesis, we obtain a convex embedding of H with no three vertices on a line.

In this embedding, the subgraph obtained by deleting the edges incident to z has a face containing z (perhaps unbounded). Since $H - z$ is 2-connected, the boundary of this face is a cycle C. All neighbors of z lie on C; they may be neighbors in G of x or y or both.

The embedding of H has segments from z to its neighbors. Let $x_1, \ldots, x_k$ be the neighbors of x in cyclic order on C. If all neighbors of y lie in the portion of C from x_i to x_{i+1}, then we obtain a convex embedding of G by putting x at z in H and putting y at a point close to z in the wedge formed by xx_i and xx_{i+1} (slightly moving the segments from z to $N_G(y)$), as shown in Case 0 below.

If this does not occur for any i, then either (1) y shares three neighbors u, v, w with x, or (2) y has neighbors u and v that alternate on C with neighbors x_i and x_{i+1} of x. If (1), then the union of C with xy and the edges from $\{x, y\}$ to $\{u, v, x\}$ is a K_5-subdivision. If (2), then the union of C with the paths uyv, $x_i x x_{i+1}$, and xy is a $K_{3,3}$-subdivision. Since our graph has no Kuratowski subgraph, only Case 0 occurs. ∎

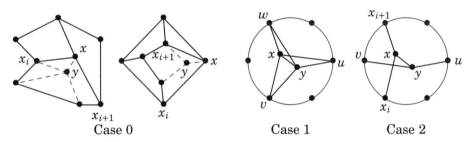

Case 0   Case 1   Case 2

9.2.12.* Remark. Thomassen [1984] proved that the regions in Theorem 9.2.11 can be required to be convex hexagons, but not convex pentagons. Other extensions guarantee special Kuratowski subgraphs in nonplanar graphs. Kelmans [1981] conjectured that every 3-connected nonplanar graph with at least six vertices has a cycle with three pairwise crossing chords. This was proved by Kelmans [1984] and by Thomassen [1984].

A 2-connected planar graph need not have a convex embedding (consider $K_{2,4}$), but every planar graph has an embedding where all edges are straight line segments. Proved by Wagner [1936], Fáry [1948], and Stein [1951], this is known as **Fáry's Theorem** (Exercise 9). ∎

9.2.13.* Remark. Kuratowski's Theorem does not directly yield fast planarity testing: too many subgraphs to check. Demoucron–Malgrange–Pertuiset [1964] obtained a quadratic-time algorithm, and it produces an embedding. The proof of Kuratowski's Theorem by Klotz [1989] provides another algorthm, and it finds a Kuratowski subgraph when the graph is not planar.

Linear-time planarity testing algorithms appear in Hopcroft–Tarjan [1974] and Booth–Luecker [1976]; these are complicated (Gould [1988, pp. 177–185] describes the Hopcroft–Tarjan algorithm). Later linear-time algorithms appear in Boyer–Myrvold [1999] and Shih–Hsu [1999].

For computer display, one seeks a straight-line embedding with the vertices at grid points. Schnyder [1990] proved that a planar n-vertex graph has a straight-line embedding at points of $[n-1] \times [n-1]$, and it can be found in linear time from any embedding of G. We discuss this theorem in Chapter 16. ■

9.2.14.* Remark. Planar graphs have various geometric representations. The **tangency graph** of a family of geometric objects has a vertex for each object, with vertices adjacent when the objects are tangent. Every planar graph is a tangency graph of rectangles in the plane (Thomassen [1986]) and also of circles in the plane (Koebe [1936], Brightwell–Scheinerman [1993], Thurston [1997]).

Many computational problems involve visibility in the plane. The famous **Art Gallery Problem** (Exercises 9.3.25–26) seeks the minimum number of guards needed to watch all locations of interest (see O'Rourke [1987]). Visibility can also be in restricted directions. A graph is a **bar visibility graph** if each vertex can be assigned a bar (a horizontal segment) in the plane so that vertices are adjacent if and only if their bars can "see" each other along an unblocked vertical channel. Wismath [1985] and Tamassia–Tollis [1986] proved that a graph G is a bar visibility graph if and only if G has a planar embedding with all cut-vertices on the unbounded face. Later Hutchinson [2002] gave a short proof. ■

THE SEPARATOR THEOREM (optional)

In "divide-and-conquer" algorithms, we solve a problem recursively by splitting it into smaller pieces. This works well for planar graphs.

9.2.15. Definition. An (m, α)-**separation** of an n-vertex graph G splits $V(G)$ into sets A, B, C such that $|A|, |B| \le \alpha n$, and $|C| \le m$, and no edges join A and B. A graph family **F** has an (f, α)-**separator** if every graph $G \in \mathbf{F}$ has an $(f(|V(G)|), \alpha)$-separation.

An (f, α)-separator may yield good divide-and-conquer algorithms on a hereditary class **F**. For $G \in \mathbf{F}$ with n vertices, the approach is

(1) find an $(f(n), \alpha)$-separation of G, with vertex partition A, B, C,
(2) solve the specified problem on $G[A]$ and $G[B]$, and
(3) combine these answers with C to solve the problem on G.

If $g(n)$ is the worst-case cost for doing steps (1) and (3), then the worst-case cost $h(n)$ for n-vertex graphs in **F** satisfies $h(n) \le 2h(\alpha n) + g(n)$. If $\alpha = 1/2$ and $g(n) = an$, then $h(n) \in O(n \log n)$. When $g(n) = a\sqrt{n}$ with $\alpha = 1/2$, the running time is linear (Exercise 20). With $\alpha = 2/3$, the complexity results are similar.

Trees have a $(1, 2/3)$-separation (Exercise 21). For n-vertex subgraphs of a cartesian product of two paths (**grid graphs**), there is a $(\sqrt{n}, 2/3)$-separation. General planar graphs behave almost as well; the Separator Theorem of Lipton–Tarjan [1979] provides an (f, α)-separator with $f(n) = 2\sqrt{2n}$ and $\alpha = 2/3$.

Planar graphs are sparse, with average degree less than 6, but sparseness alone does not guarantee easy separation. Erdős–Graham–Szemerédi [1976] proved that for $\varepsilon > 0$ there is a positive constant c_ε such that almost all graphs with $(2 + \varepsilon)k$ vertices and $c_\varepsilon k$ edges retain a component with at least k vertices when any k vertices are deleted. (See Chapter 14 for discussion of "almost all".)

Thus the Separator Theorem truly needs the properties of planarity. The original proof (Exercise 24) used weighted vertices and obtained separations from spanning trees of small diameter. We present a later short proof based on connectivity and path-lengths. A **near-triangulation** is a 2-connected plane graph in which every bounded face has length 3.

9.2.16. Lemma. Let G be a near-triangulation with vertices colored red or blue. If the outer cycle C has red points u and v, then G has a red u, v-path or a blue path joining the components of $C - \{u, v\}$.

Proof: Let A be the subset of $V(G)$ reachable from u by paths in red. If G has no red u, v-path, then the edge cut $[A, \mathbf{A}]$ contains a bond B that breaks all u, v-paths. By Theorem 9.1.11, the edges corresponding to B in the dual graph G^* form a cycle B^* with one of $\{u, v\}$ inside and the other outside. The bounded faces in G corresponding to vertices of B^* are triangles having a red vertex and a blue vertex. Deleting A from the union of these triangles (as on the left below) leaves a path through blue vertices joining the two components of $C - \{x, y\}$. ∎

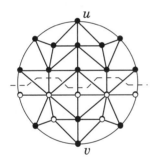
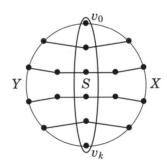

9.2.17. Lemma. Let G be a near-triangulation; outer cycle $C = [v_0, \ldots, v_{2k-1}]$. If $d_G(v_0, v_k) \geq k$, then G has disjoint paths linking v_i to v_{2k-i} for $0 \leq i \leq k$. If also $d_G(v_i, v_j) = d_C(v_i, v_j)$ for all $v_i, v_j \in V(C)$, then $|V(G)| \geq (k+1)^2/2$.

Proof: Let $X = \{v_0, \ldots, v_k\}$ and $Y = \{v_k, \ldots, v_{2k}\}$, where $v_{2k} = v_0$. Let S be a smallest X, Y-barrier. Color S red, and color the remaining vertices blue. Since S is an X, Y-barrier, there is no blue X, Y-path. By Lemma 9.2.16, there is a red v_0, v_k-path. By hypothesis, such paths have length at least k, so $|S| \geq k + 1$.

By Pym's Theorem, G has an X, Y-link of size $|S|$ (shown above). It has $k + 1$ disjoint X, Y-paths. They must link v_i to v_{2k-i} for all i; otherwise edges cross.

If G has no shorter v_i, v_{2k-i}-path than the one along C, then the v_i, v_{2k-i}-path in the set we have obtained has at least $1 + \min\{2i, 2k - 2i\}$ vertices. Hence the total number of vertices in these $k + 1$ paths is at least $1 + 3 + 5 + \cdots + 5 + 3 + 1$, with $k + 1$ terms. The sum of the first j positive odd integers is j^2, so we compute

$$\sum_{i=1}^{\lceil (k+1)/2 \rceil} (2i - 1) + \sum_{i=1}^{\lfloor (k+1)/2 \rfloor} (2i - 1) = \left\lceil \frac{k+1}{2} \right\rceil^2 + \left\lfloor \frac{k+1}{2} \right\rfloor^2 \geq \frac{(k+1)^2}{2}. \qquad \blacksquare$$

9.2.18. Theorem. (**Planar Separator Theorem**; Lipton–Tarjan [1979]). Every n-vertex planar graph G has a $(2\sqrt{2n}, 2/3)$-separation.

Proof: (Alon–Seymour–Thomas [1994]) Since separators of graphs are separators of their spanning subgraphs, we may assume that G is a triangulation. Let $k = \lfloor \sqrt{2n} \rfloor$; since $k > \sqrt{2n} - 1$, we have $n < (k+1)^2/2$. We seek a $(2k, 2/3)$-separation using a cycle as the separating set, separating the inside from the outside. We guarantee this by showing that if a well-chosen cycle fails, then it will meet the conditions of Lemma 9.2.17, yielding $|V(G)| \geq (k+1)^2/2$.

Given a cycle C in G, let c^- and c^+ denote the numbers of vertices in the interior and exterior of C, respectively. Among all cycles of length at most $2k$ such that $c^+ < 2n/3$, let C be one minimizing $c^- - c^+$. Such a cycle exists, since the external triangle has fewer than $2k$ vertices and has no vertices outside.

The $(2\sqrt{2n}, 2/3)$-separation exists if $c^- < 2n/3$; suppose not. Let D be the subgraph on and inside C. For $u, v \in V(C)$, let $c(u,v)$ and $d(u,v)$ be the lengths of shortest u,v-paths in C and D; $C \subseteq D$ implies $d(u,v) \leq c(u,v)$.

It now suffices to prove $d(u,v) = c(u,v)$ for $u,v \in V(C)$ and also $|V(C)| = 2k$. With $d(v_0, v_k) = k$, Lemma 9.2.17 then applies to yield $n \leq |V(D)| \geq (k+1)^2/2$.

Claim 1: $d(u,v) = c(u,v)$ *for* $u,v \in V(C)$. If not, then among the pairs with $d(u,v) < c(u,v)$, choose $\{u,v\}$ with $d(u,v)$ smallest. Let P be a shortest u,v-path in D. By the choice of $\{u,v\}$, no internal vertex of P lies on C. Thus P combines with the two u,v-paths on C to form cycles C_1 and C_2, indexed so $c_1^- \geq c_2^-$.

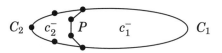

We claim that C_1 contradicts the minimality of C. Since $d(u,v) < c(u,v)$, the cycle C_1 is shorter than C and hence has length at most $2k$. We have $c_1^- \leq c^-$ and $c_1^+ > c^+$, so $c_1^- - c_1^+ < c^- - c^+$. To show $c_1^+ < 2n/3$, use $c_1^- \geq c_2^-$ to compute

$$n - c_1^+ = c_1^- + |V(C_1)| > \tfrac{1}{2}c^- \geq \tfrac{1}{3}n.$$

Thus C_1 is an eligible cycle and would be preferred to C, a contradiction.

Claim 2: C *has* $2k$ *vertices.* By Claim 1, C has no chords; hence every edge of C is one edge of a triangular face with a vertex inside C, since $c^- \geq 2n/3$. If $|V(C)| < 2k$, then obtain C' from C by detouring to absorb one such interior vertex. Since c^+ vertices remain outside C', and fewer than c^- vertices are inside C', the cycle C' contradicts the choice of C. Hence we must have $|V(C)| = 2k$. ∎

Iterating the Separator Theorem reduces α to $1/2$ at the expense of increasing the constant in $f(n)$.

9.2.19. Corollary. (Lipton–Tarjan [1979]). Every n-vertex planar graph G has a $\left(\frac{2\sqrt{2n}}{1-\sqrt{2/3}}, \frac{1}{2} \right)$-separation.

Proof: We build A, B, C gradually; the vertices not yet placed lie in D. Initially $A_0 = B_0 = C_0 = \varnothing$ and $D_0 = V(G)$. Step $i-1$ produces disjoint sets A_{i-1}, B_{i-1}, and D_{i-1} such that no edges join any two of them and $|A_{i-1}| \leq |B_{i-1}| \leq n/2$.

At step i, apply Theorem 9.2.18 to the remaining "large" piece D_{i-1}. This produces C^* separating A^* and B^* within D_{i-1}, labeled so that $|A^*| \le |B^*|$. Combine A^* with A_{i-1}. Since $|A^*| \le |B^*|$ and $|A_{i-1}| \le |B_{i-1}|$, we have $|A_{i-1} \cup A^*| \le n/2$. Also $|B_{i-1}| \le n/2$ is given. Let A_i be the smaller of $A_{i-1} \cup A^*$ and B_{i-1}, and let B_i be the other. Now $|A_i| \le |B_i| \le n/2$. Let $C_i = C_{i-1} \cup C^*$.

We put A^* into the final sets, and B^* becomes D_i to split at the next stage. No edges join D_i to A_i or B_i, so any vertices of D_i can be added to B_i or to A_i.

Since $|D_i| \le \frac{2}{3}|D_{i-1}|$, the bounds on the contributions to $|C|$ form a geometric sum. When $D_k = \varnothing$, we have $|A_k|, |B_k| \le \frac{n}{2}$ and $|C_k| \le \sum_i 2\sqrt{2n(2/3)^i} < \frac{2\sqrt{2n}}{1-\sqrt{2/3}}$. ∎

For $\alpha = 2/3$, the coefficient in Theorem 9.2.18 was reduced successively in five papers from $2\sqrt{2}$ (about 2.828) to reach $\sqrt{2/3} + \sqrt{4/3}$ (about 1.971) in Djidjev–Venkatesan [1997]. For example, Alon–Seymour–Thomas [1994] applied their technique for proving Theorem 9.2.18 more carefully to reduce it to $\frac{3}{2}\sqrt{2}$ (about 2.121). Along the way were also Djidjev [1982, 1987] and Gazit–Miller [1990]. Djidjev [1982] showed that the coefficient cannot be less than $\frac{2}{3}(\pi\sqrt{3})^{1/2}$ (about 1.555). Thus trees of separations of planar graphs down to pieces of constant size need logarithmic depth.

Proofs of the Planar Separator Theorem produce separations in linear time. Due to this, the Planar Separator Theorem yields fast approximation algorithms for computational problems that are difficult on planar graphs. Exercise 23 considers the example of finding a large independent set.

EXERCISES 9.2

9.2.1. (−) Prove that the complement of the 3-dimensional cube Q_3 is nonplanar.

9.2.2. (−) Prove or disprove: The union of any two paths is a planar graph.

9.2.3. (−) For each graph below, prove nonplanarity or give a convex embedding.

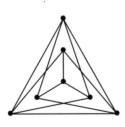 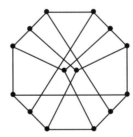

9.2.4. (−) Let M be a matching in G as given below. Determine whether $G - M$ is planar.
   (a) $G = K_{4,4}$ and $|M| = 4$.   c) $G = K_6$ and $|M| = 3$.
   (b) $G = K_{4,4}$ and $|M| = 3$.   d) $G = K_6$ and $|M| = 2$.

9.2.5. (−) Give two proofs that the Petersen graph is nonplanar: one by using Kuratowski's Theorem, and one by using Euler's Formula.

9.2.6. (−) Prove or disprove: Every bipartite graph with minimum degree at least 4 contains $K_{3,3}$ as a subgraph.

9.2.7. When G has degree list $(4, 4, 4, 4, 3, 3, 3, 3)$, also $\overline{G}$ has this degree list. Show that such G and $\overline{G}$ can be both planar, both nonplanar, or one of each type.

9.2.8. ($\diamond$) For $n \geq 5$, let G be the graph whose vertices are n points on a circle, with each vertex adjacent to the four nearest others. Prove that G is planar if and only if n is even.

9.2.9. ($\diamond$) *Fáry's Theorem.* Let R be a planar region bounded by a simple polygon with at most five sides (the edges do not cross). Prove that some point x inside R "sees" all of R, meaning that the segment from x to any point of R does not cross the boundary of R. Conclude inductively that every planar graph has a straight-line embedding.

9.2.10. Find a convex embedding in the plane for the graph below.

9.2.11. ($\diamond$) *Outerplanarity.* (Chartrand–Harary [1967])
(a) Use Kuratowski's Theorem to prove that G is outerplanar if and only if G does not contain a subdivision of K_4 or $K_{2,3}$. (Hint: To *apply* Kuratowski's Theorem, modify G appropriately. This is *much* easier than following the steps of Kuratowski's Theorem.)
(b) Use part (a) to prove that $G \square K_2$ is planar if and only if G is outerplanar.

9.2.12. For G_1 and G_2 connected, with $|V(G_i)| \geq 3$, prove that $G_1 \square G_2$ is planar if and only if one is a path and the other is a cycle or a path. (Behzad–Mahmoodian [1969])

9.2.13. Wagner [1937] proved that a graph G is planar if and only if neither K_5 nor $K_{3,3}$ can be obtained from G by deletions and contractions of edges.
(a) Show that deletion and contraction of edges preserve planarity. Conclude from this that Wagner's condition is necessary.
(b) Use Kuratowski's Theorem to prove that Wagner's condition is sufficient.

9.2.14. ($\diamond$) *The* **Hanani–Tutte Theorem**: *a planarity criterion.*
(a) Prove that in every drawing of K_5 or $K_{3,3}$ in the plane, some two nonincident edges cross an odd number of times. (Hint: Prove that the number of such pairs of edges is always odd.) (Hanani [1934], as Chojnacki)
(b) Prove that a graph is planar if it has a drawing in the plane in which every two nonincident edges cross an even number of times. (Tutte [1970])

9.2.15. ($\diamond$) Prove that every 3-connected graph with at least six vertices that contains a subdivision of K_5 also contains a subdivision of $K_{3,3}$.

9.2.16. ($\diamond$) For a cycle C in a graph G, a C-**fragment** is a component of $G - V(C)$ together with its edges and vertices of attachment to C. Two C-fragments **conflict** if they have three common vertices of attachment or four alternating vertices of attachment on C. The **conflict graph** of C has a vertex for each C-fragment, adjacent when they conflict. Prove that G is planar if and only if the conflict graph of each cycle is bipartite. (Tutte [1958])

9.2.17. Let x and y be vertices in a planar graph G. Prove that G has a planar embedding with x and y on the same face if and only if $G - x - y$ has no cycle C with x and y in conflicting C-fragments in G. (Hint: Use Kuratowski's Theorem. Comment: Tutte proved this without Kuratowski's Theorem and used it to prove Kuratowski's Theorem.)

9.2.18. Let C be a cycle in a 3-connected plane graph G. Prove that C is a face-boundary in G if and only if G has exactly one C-fragment. (Comment: Tutte [1963] used this to prove the result of Whitney [1933] that 3-connected planar graphs have only one planar embedding.)

9.2.19. (+) Let C be a cycle in a graph G, and let H_1 and H_2 be $V(C)$-lobes. Say that H_1 and H_2 are **skew** if C has vertices v_1, v_2, v_3, v_4 in order such that $v_1, v_3 \in V(H_1)$ and $v_2, v_4 \in V(H_2)$. The **skew-overlap graph** of C has a vertex for each C-component and an edge joining every skew pair. Use Kuratowski's Theorem (without Kelman's Conjecture) to prove that G is nonplanar if and only if G contains a cycle whose skew-overlap graph has a cycle of length 3 or 5. (Hint: Use induction on the number of edges.)

9.2.20. Solve the recurrence $f(n) = 2f(n/2) + a\sqrt{n}$, where n is a power of 2, with $f(1) = 0$.

9.2.21. ($\diamond$) *Separator theorems.*
   (a) Prove that every tree has a $(1, 2/3)$-separation.
   (b) Prove that every outerplanar graph has a $(2, 2/3)$-separation.

9.2.22. The naive algorithm to check for K_4 as a subgraph of an n-vertex graph runs in time $O(n^4)$. Use the planar separator theorem to design an algorithm that solves this problem in time $O(n \log n)$ on planar graphs. Assume that the separation guaranteed by the separator theorem can be found in linear time.

9.2.23. Let k be a function of n satisfying $\Omega(\log \log n) \leq k \leq O(\log n)$. Prove that there is an algorithm on n-vertex planar graphs that runs in time $O(n2^k)$ and produces an independent set of size within $O(n/\sqrt{k})$ of the maximum. (Hint: Find a set C of $O(n/\sqrt{k})$ vertices such that every component of $G - C$ has at most k vertices.) (Lipton–Tarjan [1980])

9.2.24. (+) *Lipton–Tarjan [1979] proof of Theorem 9.2.18.* Let G be a planar graph with nonnegative weight $c(v)$ at each vertex v, with $\sum c(v) \leq 1$. An (m, α)-separation is a partition $\{A, B, C\}$ of $V(G)$ where $[A, B] = \varnothing$, with $c(A), c(B) \leq \alpha c(V(G))$ and $|C| \leq m$.
   (a) Prove that a planar graph with a spanning tree of diameter $2r$ has a $(2r + 1, \frac{2}{3})$-separation, with $|C| = 2r + 1$ in such a separation only if C contains the center of T.
   (b) Given a rooted tree T, let T_l denote the set of vertices at distance l from the root. Let T be a tree grown from vertex x in a planar graph G so that $v \in T_l$ if and only if $d_G(x, v) = l$. Let r and s be integers with $r < s$ such that $\sum_{i<r} c(T_i) \leq \frac{2}{3}$ and $\sum_{i>s} c(T_i) \leq \frac{2}{3}$. Prove that G has an $(m, \frac{2}{3})$-separation with $m \leq |T_r| + |T_s| + 2(s - r - 1)$.
   (c) Prove that every weighted n-vertex planar graph has a $(\sqrt{8n}, \frac{2}{3})$-separation.

9.3. Coloring of Planar Graphs

We come now to the notorious Four Color Problem: can the regions of any map be colored from four colors so that adjacent regions have different colors? Equivalently, is every planar graph 4-colorable? The question was in a letter of October 23, 1852 from Augustus de Morgan to Sir William Hamilton. It was asked by de Morgan's student Frederick Guthrie, who heard it from his brother Francis.

Cayley presented the problem to the London Mathematical Society in 1878. Its notoriety stems from its ease of statement (especially the map-coloring form) and from many published faulty proofs. The history is discussed in Ore [1967], Saaty–Kainen [1977, 1986], Aigner [1984, 1987], and Fritsch–Fritsch [1998].

Kempe [1879] published the first "solution", refuted by Heawood [1890]. Tait [1880] published another; we consider both arguments.

EDGE-COLORINGS AND SPANNING CYCLES

Tait reduced the Four Color Problem to edge-coloring of 3-regular planar graphs. It suffices to prove that maximal planar graphs are 4-colorable. It therefore suffices to prove that their duals are 4-face-colorable. The dual of a maximal planar graph is 3-regular and 2-edge-connected.

9.3.1. Theorem. (**Tait's Theorem**; Tait [1880]) A 2-edge-connected 3-regular plane graph is 4-face-colorable if and only if it is 3-edge-colorable.

Proof: Let G be such a graph. Suppose first that G is 4-face-colorable using four colors named as follows: $c_0 = 00$, $c_1 = 01$, $c_2 = 10$, $c_3 = 11$. We define an edge-coloring. To the edge between faces with colors c_i and c_j we assign the color that is the binary coordinate sum of c_i and c_j. Since G is 2-edge-connected, $c_i \neq c_j$, and hence 00 is not obtained. If two edges incident to v receive the same color c, then subtracting the color on their common face from c leaves adjacent faces with the same color, a contradiction. Hence we have produced a proper 3-edge-coloring.

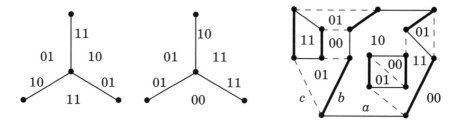

From a proper 3-edge-coloring of G using colors a, b, c on the subgraphs E_a, E_b, E_c, we construct a 4-face-coloring using the four colors as above. Since G is 3-regular, each color appears at every vertex, and the union of two of $\{E_a, E_b, E_c\}$ is a 2-factor of G. Each face in such a subgraph is a union of faces of G. Let $H_1 = E_a \cup E_b$ and $H_2 = E_b \cup E_c$. To a face f, assign the color whose ith coordinate is the parity of the number of cycles in H_i that contain f (0 for even, 1 for odd).

We show that the resulting 4-face-coloring is proper. The faces f and f' bounded by an edge e are distinct, since G is 2-edge-connected. Suppose that e lies on a cycle C in H_i. By the Jordan Curve Theorem, C separates f and f', and no other cycle in H_i separates f and f'. Hence the numbers of cycles enclosing f and f' in H_i have different parity, and the colors we have given to f and f' differ in coordinate i (in both coordinates when e has color b). ∎

Due to Theorem 9.3.1, a proper 3-edge-coloring of a 3-regular graph is called a **Tait coloring**. Proving that all 2-edge-connected 3-regular planar graphs have Tait colorings reduces to the 3-connected case (Exercise 9), and every Hamiltonian 3-regular graph has a Tait coloring. Thus the Four Color Theorem follows if every 3-regular 3-connected planar graph is Hamiltonian. It was noticed that Tait didn't prove that assumption, but no counterexample was found until 1946.

Grinberg [1968] found a simple necessary condition for planar graphs to be Hamiltonian. It yields many 3-regular 3-connected non-Hamiltonian planar graphs, including Tutte's 1946 example and the Grinberg graph of Exercise 13.

9.3.2. Theorem. (Grinberg [1968]) If G is a plane multigraph with a Hamiltonian cycle C, and G has ϕ_i faces of length i inside C and ϕ_i' faces of length i outside C, then $\sum_i (i-2)(\phi_i - \phi_i') = 0$.

Proof: It suffices to prove $\sum_i (i-2)\phi_i = n-2$, since the same argument applies to the regions outside C. Consider the outerplanar graph G' formed by C and the chords inside it, having n vertices, m edges, and r faces.

Note that $\sum_i i\phi_i$ counts the n edges of C once and chords twice, while $\sum_i 2\phi_i$ counts bounded faces twice. Using Euler's Formula,

$$\sum_i (i-2)\phi_i = (2m-n) - 2(r-1) = 2 - 2(n-m+r) + n = n-2. \qquad \blacksquare$$

9.3.3. Example. *The* **Tutte graph**. Tutte [1946] found the 3-connected 3-regular non-Hamiltonian plane graph on the left below. Let H denote each component obtained by deleting the central vertex and the three long edges. Since a Hamiltonian cycle must visit the central vertex, it must traverse one copy of H along a Hamiltonian path joining the other entrances to H, which we call x and y.

We therefore study a graph that has a Hamiltonian cycle if and only if H has a spanning x, y-path. Such a graph H' (on the right below) is obtained by adding an x, y-path of length 2 through a new vertex.

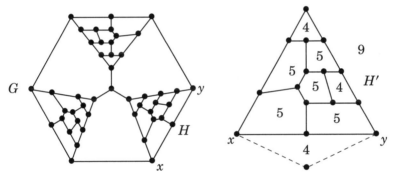

A face of length j is a j-**face**. Considering the faces of each length, Grinberg's condition for H' becomes $2a_4 + 3a_5 + 7a_9 = 0$, where $a_i = \phi_i - \phi_i'$. Since the unbounded face is outside, the equation reduces mod 3 to $2a_4 \equiv 7 \pmod{3}$. Since $\phi_4 + \phi_4' = 3$, we have $a_4 \in \{\pm3, \pm1\}$. The only choice satisfying $2a_4 \equiv 7 \pmod 3$ is $a_4 = -1$, which requires that two 4-faces lie outside the Hamiltonian cycle. However, those having a vertex of degree 2 cannot lie outside the cycle.

A faster contradiction arises by subdividing one edge incident to each vertex of degree 2. This does not change the existence of a spanning cycle. The resulting graph has seven 5-faces, one 4-face, and one 11-face. The required equation becomes $2 \cdot (\pm1) = 9 - 3a_5$, but the left side is not a multiple of 3. $\qquad \blacksquare$

Tutte [1956] proved that all 4-connected planar graphs are Hamiltonian. Thomassen [1983] proved the stronger result (conjectured by Plummer [1975]) that every 4-connected planar graph is Hamiltonian-connected (every vertex pair occurs as the endpoints of a spanning path). By proving that 4-connected triangulations are Hamiltonian, Whitney [1931] reduced the Four Color Problem to the case of Hamiltonian planar graphs (Exercise 10).

5-COLORABLE AND 5-CHOOSABLE

We proved from Euler's Formula that planar n-vertex graphs have at most $3n - 6$ edges (Theorem 9.1.23). Thus planar graphs are 5-degenerate, which implies that they are 6-colorable (recall Proposition 8.1.12). Heawood improved the upper bound from 6 to 5.

9.3.4. Theorem. (**Five Color Theorem**; Heawood [1890]) Every planar graph G is 5-colorable.

Proof: Since subgraphs of G are planar, it suffices to forbid 6-critical planar graphs. If G is 6-critical, then $\delta(G) \geq 5$, so with planarity we may let v be a vertex of degree 5. Let f be a proper 5-coloring of $G - v$. All five colors must appear on $N(v)$. Let v_1, v_2, v_3, v_4, v_5 be the neighbors of v in clockwise order around v, and name the colors so that $f(v_i) = i$.

Let $G_{i,j}$ be the subgraph of $G - v$ induced by the vertices of colors i and j. Switching the colors on a component of $G_{i,j}$ yields another proper 5-coloring of $G - v$. If the component of $G_{i,j}$ containing v_i does not contain v_j, then switching the colors on it removes color i from $N(v)$. Now giving color i to v produces a proper 5-coloring of G. Thus G is 5-colorable unless, for each choice of i and j, the component of $G_{i,j}$ containing v_i also contains v_j. Let $P_{i,j}$ be a path in $G_{i,j}$ from v_i to v_j, shown below for $(i, j) = (1, 3)$.

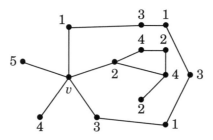

The cycle C completed with $P_{1,3}$ by v separates v_2 from v_4. By the Jordan Curve Theorem, the path $P_{2,4}$ crosses C. Since G is planar, paths cross only at shared vertices. The vertices of $P_{1,3}$ all have color 1 or 3, and those of $P_{2,4}$ all have color 2 or 4, so they have no common vertex.

This contradiction implies that every planar graph is 5-colorable. ∎

After 86 more years, the upper bound was lowered from 5 to 4 in Appel–Haken [1976]. Before we explain the approach to the proof, we describe another way to strengthen the Five Color Theorem.

Vizing asked in 1975 whether every planar graph is 5-choosable; we prove this below. Voigt [1993] found a planar graph with 238 vertices that is not 4-choosable; Mirzakhani [1996] (Exercise 35) reduced this to 63 vertices (both examples generalize to infinite families). Gutner [1996] (Exercise 34) and Voigt–Wirth [1997] showed that in fact some 3-colorable planar graphs are not 4-choosable.

As is common in inductive arguments about plane graphs, the **external vertices** (vertices of the unbounded face) play a special role. Recall that a **near-triangulation** is a 2-connected plane graph in which every bounded face is a triangle. The boundary of the outer face is a cycle.

9.3.5. Theorem. (Thomassen [1994b]) Every planar graph G is 5-choosable.

Proof: Adding edges cannot reduce χ_l, so we may assume that every bounded face is a triangle. By induction on $|V(G)|$, we prove the stronger result that G has an L-coloring even when two adjacent external vertices have distinct lists of size 1 and all other external vertices have lists of size 3. The basis step is a triangle; a color remains available for the third vertex. For $|V(G)| > 3$, let the external cycle C be $[v_1, \ldots, v_p]$, with consecutive vertices v_p and v_1 having fixed colors.

Case 1: C has a chord $v_i v_j$ with $1 \leq i < j - 1 < p$. We apply the induction hypothesis to the subgraph induced by the cycle $[v_1, \ldots, v_i, v_j, \ldots, v_p]$ and the vertices inside it. This selects a proper coloring giving v_i and v_j some fixed colors. Now we apply the induction hypothesis to the subgraph induced by the cycle $[v_i, v_{i+1}, \ldots, v_j]$ and the vertices inside it to complete the L-coloring of G.

Case 2: C has no chord. Let $v_1, u_1, \ldots, u_k, v_3$ be the neighbors of v_2 in order. Since bounded faces are triangles, $\langle v_1, u_1, \ldots, u_k, v_3 \rangle$ is a path. Since C is chordless, $u_1, \ldots, u_k$ are internal vertices, and the outer face of $G - v_2$ is bounded by a cycle C' in which $\langle v_1, u_1, \ldots, u_k, v_3 \rangle$ replaces v_1, v_2, v_3. Let $G' = G - v_2$.

Let c be the color assigned to v_1. Since $|L(v_2)| \geq 3$, there are distinct colors $x, y \in L(v_2) - \{c\}$. We reserve x and y for v_2 by letting $L'(u_i) = L(u_i) - \{x, y\}$. Since $|L(u_i)| \geq 5$, we have $|L'(u_i)| \geq 3$. Hence we can apply the induction hypothesis to G', with $u_1, \ldots, u_k$ having lists of size at least 3 and other lists being the same as in G. Note that v_1 and $u_1, \ldots, u_k$ receive colors outside $\{x, y\}$. We extend the coloring to G by choosing for v_2 a color in $\{x, y\}$ not used on v_3. ∎

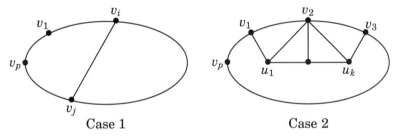

Case 1   Case 2

Building on this method, Grytczuk–Zhu [2020] proved that every planar graph contains a matching whose deletion leaves a 4-choosable graph (Theorem 15.1.45). Bipartite planar graphs are 3-choosable (Exercise 15.1.32). Thomassen [1995a] proved that also planar graphs with girth at least 5 are 3-choosable.

To better understand choosability, a more refined notion was introduced.

9.3.6.* Definition. (Kratochvíl–Tuza–Voigt [1998]) Graph G is (k,d)**-choosable** if G is L-colorable for every list assignment L such that $|L(v)| \geq k$ for all $v \in V(G)$ and $|L(x) \cap L(y)| \leq d$ for all $xy \in E(G)$.

Note that (k, k)-choosable means k-choosable, and for $k \geq 1$ every graph is $(k, 0)$-choosable. Thus when $1 \leq k < \chi_l(G)$ there is a threshold d such that G is (k, d)-choosable but not $(k, d + 1)$-choosable. It is unknown whether all planar graphs are $(4, 2)$-choosable; they are not all $(4, 3)$-choosable (Exercises 34–35).

9.3.7.* Theorem. (Kratochvíl–Tuza–Voigt) Planar graphs are $(4, 1)$-choosable.

Proof: If $|E(H)| \leq d|V(H)|$ for every subgraph H of a graph G, where $d \in \mathbb{N}$, then G has an orientation where every vertex has outdegree at most d (Corollary 6.1.5, applying Hall's Theorem). A planar n-vertex graph G has at most $3n - 6$ edges, so G has an orientation D with maximum outdegree at most 3.

Delete from each list $L(v)$ any color occurring in the list of an outneighbor of v in D. This deletes at most one color for each outneighbor, so a color remains. Now adjacent vertices have disjoint lists, so any coloring chosen from them is proper. ∎

THE FOUR COLOR PROBLEM

In proving the Five Color Theorem, we argued that a smallest non-5-colorable planar graph cannot contain a vertex of degree at most 5. This suggests an approach to the Four Color Problem. We need only consider triangulations, since every subgraph of a 4-colorable graph is 4-colorable.

9.3.8. Definition. In a graph, a **configuration** is a substructure, often a type of subgraph. For the Four Color Problem, a configuration in a triangulation is a separating cycle C (called the **ring**) plus the portion of the graph inside C. A set S of configurations is **unavoidable** for a problem if every instance contains a member of S. A configuration is **reducible** for a property if it cannot occur in a minimal structure lacking that property.

Thus the goal is to find an **unavoidable set of reducible configurations**. For the Four Color Problem, this means showing that every triangulation contains a configuration that is reducible for the property of being 4-colorable.

9.3.9. Example. *An unavoidable set.* We have remarked that $\delta(G) \leq 5$ for every simple planar graph. In a triangulation with at least four vertices, every vertex has degree at least 3. Thus the set of configurations below is unavoidable.

The edges inward from the ring are dashed because a configuration in a triangulation is determined by the interior subgraph and the degrees of its vertices (Exercise 37). Thus we write these as "$\bullet\, 3$", "$\bullet\, 4$", and "$\bullet\, 5$", respectively. ∎

Saying that a configuration C is reducible for 4-colorability of triangulations means that if C appears in G, then G can be changed to a smaller triangulation G' such that every proper 4-coloring of G' leads to a proper 4-coloring of G.

The idea Heawood used in Theorem 9.3.4 came from Kempe's failed proof and is important in proving reducibility of configurations. A path whose colors alternate between two colors is a **Kempe chain**.

9.3.10. Remark. *Kempe's proof.* Let us try to prove the Four Color Theorem by induction using the unavoidable set $\{\bullet\, 3, \bullet\, 4, \bullet\, 5\}$. The approach is similar to Theorem 9.3.4. We can extend a 4-coloring of $G - v$ to complete a 4-coloring of G unless all four colors appear on $N(v)$. Thus "$\bullet\, 3$" is reducible. If $d(v) = 4$, then the Kempe-chain argument works as in Theorem 9.3.4, and "$\bullet\, 4$" is reducible.

Now consider "• 5". When $d(v) = 5$, the repeated color on $N(v)$ in a proper 4-coloring of $G - v$ appears on nonconsecutive neighbors of v, since G is a triangulation. Let v_1, v_2, v_3, v_4, v_5 be the neighbors of v in order. In the 4-coloring f of $G - v$, we may assume by symmetry that $f(v_5) = 2$ and that $f(v_i) = i$ for $1 \le i \le 4$.

Define $G_{i,j}$ and $P_{i,j}$ as in Theorem 9.3.4. We can eliminate color 1 from $N(v)$ unless chains $P_{1,3}$ and $P_{1,4}$ exist from v_1 to v_3 and v_4, respectively, as on the left below. The component H of $G_{2,4}$ containing v_2 is separated from v_4 and v_5 by the cycle through $P_{1,3}$ and v. The component H' of $G_{2,3}$ containing v_5 is separated from v_2 and v_3 by the cycle through $P_{1,4}$ and v. Switching colors 2 and 4 in H and colors 2 and 3 in H' eliminates color 2 from $N(v)$. This was Kempe's final case.

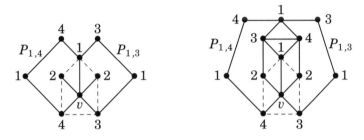

Unfortunately, $P_{1,3}$ and $P_{1,4}$ can intertwine, crossing at a vertex with color 1 as on the right above. Performing the switches in both H and H' now produces a pair of adjacent vertices with color 2. The resulting coloring is not proper.   ■

To replace "• 5", we need to find other reducible configurations.

9.3.11.* Example. All configurations with ring size 3 or 4 are reducible (Exercise 38). Equivalently, no minimal 5-chromatic triangulation has a separating cycle of length at most 4. Birkhoff [1913] proved that configurations with ring size 5 *other than* • 5 are reducible.

Birkhoff also proved reducibility of the configuration below with ring size 6, called the **Birkhoff diamond**. One must consider all proper 4-colorings of the ring. Some cases extend to proper colorings of the interior. In others, Kempe chains are needed to change the coloring into one that extends (Exercise 39).   ■

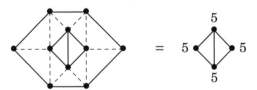

We have barely begun; an enormous amount of detail remains. From 1913 to 1950, enough reducible configurations were found to prove that all planar graphs with at most 36 vertices are 4-colorable. This was slow progress. In the 1960s, Heesch focused attention on the size of the ring, gave heuristics for finding reducible configurations, and developed methods for generating unavoidable sets.

Appel and Haken, working with Koch, improved the heuristics to consider only "promising" configurations. Using 1000 hours of computer time on three computers in 1976, they found an unavoidable set of 1936 reducible configurations with ring size at most 14. A ring of size 13 already has 66430 distinguishable 4-colorings. Proving reducibility requires showing that each leads to a proper 4-coloring of the full graph. Kempe chains and other arguments were used.

9.3.12. Theorem. (**Four Color Theorem**; Appel–Haken–Koch [1977]) Every planar graph is 4-colorable.   ∎

By 1983, refinements led to an unavoidable set of 1258 reducible configurations. Robertson–Sanders–Seymour–Thomas [1997] revisited the proof. Their simplifications yielded an unavoidable set of 633 reducible configurations. They posted their code on the Internet; in 1997, it would prove the Four Color Theorem on a desktop workstation in about three hours. Now it takes just minutes.

We have explained the meaning of reducibility, but how to generate the unavoidable set? Concise terminology for vertices by degrees is very helpful.

9.3.13. Definition. A *j-vertex*, j^+-*vertex*, or j^--*vertex* is a vertex with degree equal to j, at least j, or at most j, respectively. Similarly, a *j-neighbor* of v is a j-vertex adjacent to v. Analogously, *j-face*, j^+-*face*, and j^--*face* refer to the face length, which is the degree in the dual. Let $\overline{d}(G)$ denote the average of the vertex degrees in G.

Every planar graph has a 5^--vertex, but a 5-vertex is not reducible for 4-colorability. We need configurations with small degree on more vertices, such as in Example 9.3.11. Wernicke [1904] guaranteed a 5-vertex having at least one 6^--neighbor. We prove a stronger version to introduce "discharging".

9.3.14. Proposition. (Franklin [1922]) In every planar graph G with minimum degree 5, some 5-vertex has at least two 6^--neighbors.

Proof: We may assume that G is a triangulation, since added edges only increase degrees. Give each vertex initial "charge" equal to its degree. If no 5-vertex has at least two 6^--neighbors, then we modify the charges so that each vertex ends with charge at least 6, without changing the total charge. This contradicts $\overline{d}(G) < 6$, which implies that the desired configuration must occur at some 5-vertex.

Each 5-vertex needs charge; let it take 1/4 from each 7^+-neighbor. When a 5-vertex does not have two 6^--neighbors, it receives charge from at least four neighbors and reaches charge 6. A 6-vertex neither gains nor loses charge.

A 7^+-vertex loses charge 1/4 to each 5-neighbor. Therefore an 8^+-vertex keeps enough charge, since $j - j/4 \geq 6$ when $j \geq 8$. Hence it remains only to show that every 7-vertex ends with charge at least 6. To fall below charge 6, a 7-vertex v must give 1/4 to at least five 5-neighbors. Since G is a triangulation, G has a cycle through $N(v)$. With at least five 5-vertices, this 7-cycle must have three consecutive 5-vertices, yielding a 5-vertex with two 5-neighbors. Hence when the desired configuration does not occur, every vertex ends with charge at least 6. ∎

The condition $\overline{d}(G) < 6$ alone does not guarantee the conclusion. The graph $K_{5,7}$ has no 5-vertex with a 6^--neighbor, despite $\delta(K_{5,7}) = 5$ and $\overline{d}(K_{5,7}) < 6$. We needed planarity to find a cycle through the neighbors of a 7-vertex.

DISCHARGING AND LIGHT EDGES

Discharging arguments yield sparse local structures from global sparseness conditions. For example, $\overline{d}(G) < b$ guarantees some vertex with degree less than b. Adding planarity or stricter bounds on average degree can guarantee more, as in Proposition 9.3.14. Cranston–West [2017] discussed discharging via the approach here, Borodin [2013] surveyed applications to coloring planar graphs, and Jendrol'–Voss [2013] discussed guaranteeing subgraphs with small degree-sum. In order to explain the technique, we first assume only a bound on $\overline{d}(G)$.

Most discharging arguments are motivated by applications to coloring or decomposition problems. We want the configurations forced by discharging to be reducible for the desired property. Thus discharging applications often start by finding reducible configurations and later study what sort of sparseness is needed force their appearance. Combining the reducibility arguments and the discharging arguments to give an inductive proof is called the **Discharging Method**.

To illustrate the method, consider edge-coloring. Just as a vertex with degree less than k is reducible for k-coloring, an edge incident to fewer than k others is reducible for k-edge-coloring. In fact, since we will color this edge after coloring the rest of the graph, the configuration is also reducible for k-edge-choosability.

9.3.15. Definition. When $e = uv \in E(G)$, the **weight** of e is $d_G(u) + d_G(v)$. More generally, the **weight** of a subgraph H in a graph G is $\sum_{v \in V(H)} d_G(v)$.

9.3.16. Lemma. A minimal graph G such that $\chi'_\ell(G) > k$ has no edge of weight at most $k+1$. That is, an edge of weight at most $k+1$ is reducible for $\chi'_\ell(G) \leq k$.

Proof: Let L assign lists of size k to the edges of G. If uv is an edge of weight at most $k+1$, then let $G' = G - uv$. If $\chi'_\ell(G') \leq k$, then G' has an L-coloring ϕ'. Since uv is incident to at most $d_G(u) + d_G(v) - 2$ edges, some color in $L(uv)$ is not used by ϕ' on any edge incident to uv. Using it on uv extends the coloring to G. ∎

Always $\chi'_l(G) \geq \Delta(G)$. For equality, we seek a sparseness condition that guarantees occurrence of our reducible configuration. We measure sparseness by small density, where **density** is usually defined to be $|E(G)| / |V(G)|$. Equivalently, and more pertinently for discharging, we consider graphs with small average degree.

9.3.17. Definition. Degree charging assigns to each vertex in a graph an initial **charge** equal to its degree. **Discharging rules** modify charges assigned to elements of a graph in ways that do not change the total charge.

9.3.18. Remark. *Degree charging and average degree.* We want to show that some set of configurations is unavoidable when $\overline{d}(G)$ is less than some value b. Discharging brings extra charge to vertices with initial charge less than b. Vertices with higher initial charge can give charge but must keep at least b. As in Proposition 9.3.14, the goal is to show that if no desired configuration occurs, then every vertex ends with charge at least b, which requires $\overline{d}(G) \geq b$ and violates the hypothesis. The argument proves the contrapositive of the structural claim.

An edge whose weight is at most some desired value is a **light edge**. More generally, we can speak of light triangles, etc.   ∎

9.3.19. Lemma. If $\delta(G) \geq 2$ and $\overline{d}(G) < \frac{4k}{k+2}$, then G has an edge of weight at most $k+1$ (a light edge). Furthermore, $\overline{d}(G) \leq \frac{4k}{k+2}$ is not sufficient. (In particular, for $k \in \{3, 4, 5, 6\}$, respectively, $\overline{d}(G) < b$ suffices, where $b \in \{\frac{12}{5}, \frac{8}{3}, \frac{20}{7}, 3\}$.)

Proof: The conditions require $k \geq 3$, since $\overline{d}(G) \geq \delta(G)$. We show that if G has no light edge, then $\overline{d}(G) \geq b$, where $b = \frac{4k}{k+2}$. We use degree charging and move charge so that every vertex ends with charge at least b. Here we give the proof for $3 \leq k \leq 6$, leaving $k > 6$ to Exercise 43.

When $k \leq 6$, we have $b \leq 3$, so only 2-vertices need charge (when $k > 6$, also 3-vertices need charge). A 2-vertex v needs $b-2$; let it take half from each neighbor. Since G has no light edge, no neighbor of v needs charge, so v ends with charge b.

We also must show that all vertices still have charge at least b. With no light edges, a 2-vertex has only k^+-neighbors, so only k^+-vertices lose charge. A j-vertex loses $\frac{b-2}{2}$ to each 2-neighbor. Hence its final charge is at least $j - j\frac{b-2}{2}$. With $j \geq k$, it suffices to show $k - k\frac{b-2}{2} \geq b$, which holds when $b \leq \frac{4k}{k+2}$.

To prove sharpness, we construct a graph having average degree $\frac{4k}{k+2}$ and no light edge. Form G by subdividing each edge in a k-regular multigraph H. If $n = |V(H)|$, then G has $n + kn/2$ vertices and kn edges (twice as many as H), for average degree exactly $\frac{4k}{k+2}$. However, every edge has weight $k + 2$. ∎

In applying Lemma 9.3.19 to prove k-edge-choosability, a light edge e allows an L-coloring ϕ of $E(G - e)$ to extend, if ϕ exists. Thus an inductive proof needs the same sparseness condition on subgraphs. This suggests a natural parameter.

9.3.20. Definition. The **maximum average degree** of a graph G, written $\mathrm{Mad}(G)$, is $\max_{H \subseteq G} \overline{d}(H)$; it equals $\max_{H \subseteq G} \frac{2|E(H)|}{|V(H)|}$.

9.3.21. Theorem. If $\mathrm{Mad}(G) < \frac{4\Delta(G)}{\Delta(G)+2}$, then $\chi'_l(G) = \Delta(G)$.

Proof: Always $\chi'_l(G) \geq \Delta(G)$; we prove the upper bound. Deleting edges can reduce the degree, so to permit an inductive proof we prove more generally that $\mathrm{Mad}(G) < \frac{4k}{k+2}$ and $\Delta(G) \leq k$ together imply $\chi'_l(G) \leq k$. Since every subgraph of a graph satisfying these hypotheses also satisfies them, it suffices to show that every graph in this family contains a configuration that is reducible for $\chi'_l(G) \leq k$.

We may ignore isolated vertices. Since $\Delta(G) \leq k$, the edge at a 1-vertex has weight at most $k + 1$. If $\delta(G) \geq 2$, then Lemma 9.3.19 yields an edge of weight at most $k + 1$. By Lemma 9.3.16, this configuration is reducible for $\chi'_l(G) \leq k$. ∎

9.3.22.* Remark. Lemma 9.3.19 is easy to prove when $k \leq 6$: only 2-vertices need charge, and they take what they need from their neighbors. With light edges forbidden, k-vertices are the most in danger of losing too much charge. When 2-vertices take $\frac{b-2}{2}$ from each neighbor to reach b, the final charge will be at least b at each vertex if and only if $k - k\frac{b-2}{2} \geq b$. Hence the proof works if and only if $b \leq \frac{4k}{k+2}$. *We find the statement of the lemma after we have the proof!*

The bound on $\overline{d}(G)$ in the structural result of Lemma 9.3.19 is sharp, but that does not make the coloring application sharp. Our sharpness examples were bipartite, so by Galvin's Theorem (Theorem 8.3.18) they satisfy $\chi'_l(G) = \Delta(G)$. We have not given graphs with average degree $\frac{4k}{k+2}$ and $\chi_l(G) > \Delta(G) = k$. ∎

9.3.23. Remark. Mad(G) *and girth of planar graphs.* By Euler's Formula, an n-vertex plane graph G with all faces of length at least g has at most $\frac{g}{g-2}(n-2)$ edges (Exercise 9.1.45). This holds also for subgraphs, so Mad(G) $< \frac{2g}{g-2}$.

We often get stronger results for planar graphs than for general graphs. If $\frac{2g}{g-2} \le \frac{4k}{k+2}$, then Lemma 9.3.19 guarantees a light edge (weight at most $k+1$) in a planar graph of girth g with minimum degree 2. However, planar graphs with smaller girth, which allow larger average degree, may also force such a light edge. For various k, below we list the bound b on average degree that guarantees a light edge and often find a larger bound on average degree for planar graphs with at least the specified girth where the light edge is also guaranteed. ∎

k	weight	$b = \frac{4k}{k+2}$	girth g	$\frac{2g}{g-2}$	reference
6	7	3	5	10/3	Lemma 9.3.27
5	6	20/7		no gain	Exercise 8
4	5	8/3	7	14/5	Lemma 9.3.28
3	4	12/5	?	?	Exercise 44

This brings us to the special role of discharging on planar graphs. The Discharging Method is particularly effective for planar graphs because the dual is also planar and hence also has bounded average degree. Thus we may also use charge on faces. Euler's Formula yields several ways to exploit this interaction.

9.3.24. Proposition. The following formulas hold for a connected plane graph G, where $F(G)$ denotes the set of faces and $l(f)$ is the length of a face f.

$$\sum_{v \in V(G)} (d(v) - 6) + \sum_{f \in F(G)} (2l(f) - 6) = -12$$

$$\sum_{v \in V(G)} (2d(v) - 6) + \sum_{f \in F(G)} (l(f) - 6) = -12$$

$$\sum_{v \in V(G)} (d(v) - 4) + \sum_{f \in F(G)} (l(f) - 4) = -8$$

Proof: Multiply Euler's Formula by 6 or 4 and split the contribution from the number of edges, obtaining the three formulas below.

$$6n - 2m - 4m + 6p = 12; \quad 6n - 4m - 2m + 6p = 12; \quad 4n - 2m - 2m + 4p = 8.$$

Multiply each equation by -1 and substitute $\frac{1}{2}\sum_{v \in V(G)} d(v)$ for the first occurrence of m in each and $\frac{1}{2}\sum_{f \in F(G)} l(f)$ for the second. Collecting the contributions by vertices and by faces completes the proofs. ∎

9.3.25. Definition. *Three ways to assign charge on planar graphs.*

	charge on $v \in V(G)$	charge on $f \in F(G)$
vertex charging	$d(v) - 6$	$2l(f) - 6$
face charging	$2d(v) - 6$	$l(f) - 6$
balanced charging	$d(v) - 4$	$l(f) - 4$

9.3.26. Remark. For triangulations, vertex charging puts charge 0 on all faces, just a translation of degree charging. Dually, face charging puts charge 0 on all vertices in a 3-regular plane graph and can be useful for plane graphs with large girth. Balanced charging treats vertices and faces symmetrically.

In each case, the total charge is negative. When a hypothesis allows moving charge so that every vertex and face ends with nonnegative charge, the hypothesis must be false. For example, there is no 4-regular bipartite plane graph, since under balanced charging every vertex and face would start with nonnegative charge.

In discharging arguments, we call an element "happy" when it has enough charge. For degree charging, usually "enough charge" means at least the bound on the average degree. For the methods on planar graphs in Definition 9.3.25, it means nonnegative charge. When every vertex and face is happy, the contradiction disproves the assumption that permits the discharging.   ∎

Next we illustrate balanced charging and face charging by proving two of the results mentioned in Remark 9.3.23. Later we illustrate vertex charging.

9.3.27. Lemma. Every planar graph G with girth at least 5 and $\delta(G) \geq 2$ has a 2-vertex with a 5^- neighbor or an edge joining two 3-vertices.

Proof: Suppose that G has no such light edge. Use **balanced charging** (put charge $d(v) - 4$ on each vertex v and $l(f) - 4$ on each face f; all faces have excess charge). The total charge is -8. We obtain a contradiction by showing that each element ends happy (nonnegative charge) under the following discharging rule:

(R1) Each vertex v takes $\frac{4-d(v)}{d(v)}$ from each incident face.

The amount taken is negative when $d(v) \geq 5$, which means that such vertices give charge to faces. The rule immediately makes each vertex happy; they all end with charge 0, having taken or sent charge equally from or to the incident faces.

It remains only to check that each face ends happy. Since light edges are forbidden, a j-face loses charge to at most $\lfloor \frac{j}{2} \rfloor$ vertices. Charge is lost only to 3^--vertices; a 3-vertex takes only $\frac{1}{3}$, but a 2-vertex takes 1.

When a j-face loses charge to a 2-vertex, the next vertex on the face is a 6^+-vertex since G has no light edge. The net loss to these two vertices is only $\frac{2}{3}$. For a 3-vertex and its successor, which is a 4^+-vertex, the loss is only $\frac{1}{3}$.

Hence a j-face ends with charge at least $j - 4 - \frac{2}{3}\lfloor \frac{j}{2} \rfloor$. This is nonnegative when $j \geq 6$. When $j = 5$ it is $-\frac{1}{3}$, but a 5-face with two incident 2-vertices has three incident 6^+-vertices, which provides the extra $\frac{1}{3}$ to make it happy.   ∎

In principle, any method in Proposition 9.3.24 (or others) can be used in a discharging argument on plane graphs, but one method may require more movement of charge (and work) than another.

9.3.28. Lemma. Every planar graph G with girth at least 7 and $\delta(G) \geq 2$ has an edge of weight at most 5.

Proof: Assume that G has no light edge. Use **face charging** (put initial charge $2d(v) - 6$ on each vertex v and $l(f) - 6$ on each face f). Since G has girth at least 7, the only objects with negative initial charge are 2-vertices. Discharging:

(R1) Each 2-vertex takes $\frac{1}{2}$ from each neighboring vertex and each incident face.

To complete the proof, we check that all vertices and faces end happy. The discharging ensures that 2-vertices end with charge 0. Since 3-vertices have no 2-neighbors, their charge remains 0. For $j \geq 4$, a j-vertex may lose $\frac{1}{2}$ along each edge and ends with charge at least $2j - 6 - \frac{j}{2}$, which is nonnegative.

A j-face has at most $\lfloor \frac{j}{2} \rfloor$ incident 2-vertices, since 2-vertices are not adjacent. Hence a j-face has final charge at least $j - 6 - \frac{1}{2}\lfloor \frac{j}{2} \rfloor$, which is nonnegative whenever $j \geq 8$. To help the 7-faces, we add another discharging rule.

(R2) When an edge e joins 4^+-vertices, redirect the charge $\frac{1}{2}$ that each has available for a 2-neighbor so that instead the two faces bounded by e each receive $\frac{1}{2}$.

Now when a 7-face gives away $\frac{3}{2}$ to three 2-vertices, it recovers $\frac{1}{2}$ from the adjacent 4^+-vertices on its boundary and ends with charge 0. ∎

9.3.29. Remark. Lemma 9.3.28 strengthens Lemma 9.3.16 for planar graphs; it can also be proved by balanced charging (Exercise 67). The proof here shows both "redirection" of transmitted charge and the possible need for extra discharging to help elements that lose too much charge.

When $\overline{d}(G) < \frac{12}{5}$ and $\delta(G) = 2$, already Lemma 9.3.19 guarantees an edge of weight 4, meaning adjacent 2-vertices. One may wonder what additional sparseness is guaranteed as the average degree decreases toward 2. Instead of lighter edges, we obtain longer threads, where an ℓ-**thread** is a path in G with ℓ internal vertices having degree 2 in G (see Exercise 53). Long threads are reducible for various properties of colorability or decomposition. ∎

We next use vertex charging to prove a classical result about light edges in a planar graph G. When $\delta(G) = 5$, Franklin's result (Proposition 9.3.14) guarantees a 5-vertex with two 6^--neighbors and thus an edge of weight at most 11. When $\delta(G) = 2$, we cannot guarantee a light edge: every edge in $K_{2,n-2}$ has weight n. The fundamental result of Kotzig [1955] (known as **Kotzig's Theorem**) is that every 3-connected planar graph has an edge of weight at most 13. We prove a stronger version due to Borodin [1989b]. A **normal plane map** is a plane multigraph where every vertex degree and face length is at least 3.

9.3.30.* Lemma. (Borodin [1989b]) Every normal plane map G has an edge of weight at most 11 or a 4-cycle through two 3-vertices and a 10^--vertex (and hence an edge of weight at most 13 at a 3-vertex).

Proof: (Jendrol' [1999], Cranston–West [2017]) Suppose that G has no edge of weight at most 11. If any face f has length more than 3, then adding a chord joining the neighbors on f of a lowest-degree vertex of f does not create one of the desired configurations. Hence any desired subgraph in a triangulation obtained from G must have occurred in G.

We may therefore assume that every face has length 3. Assign charge by **vertex charging** and use this discharging rule (vertices take what they need):

(R1) Every 5^- vertex v takes charge $\frac{6-d(v)}{d(v)}$ from each neighbor.

Faces start and end with charge 0. Since we assume that G has no edge of weight at most 11, only 7^+-vertices lose charge. Hence 5^--vertices become happy and 6-vertices remain happy.

Since G is a triangulation, the neighbors of a vertex v form a cycle C. With $d(v) = j$ and no light edges, at most $\lfloor j/2 \rfloor$ vertices along C take charge from v. A 7-vertex loses charge only to 5-vertices, so it loses at most $3(\frac{1}{5})$. An 8-vertex loses charge only to 4^+-vertices and hence loses at most $4(\frac{1}{2})$. A 9^+-vertex v loses charge at most 1 to at most $\lfloor d(v)/2 \rfloor$ neighbors; this leaves charge at least $\lceil d(v)/2 \rceil - 6$, which is nonnegative when $d(v) \geq 11$. Thus in these cases v remains happy.

This leaves $d(v) \in \{9, 10\}$. If v has no 3-neighbor, then v loses at most $\lfloor \frac{d(v)}{2} \rfloor \frac{1}{2}$, leaving nonnegative charge. Hence already we have proved that G must have an edge of weight at most 13 at a 3-vertex.

In addition, any three consecutive vertices on C form a 4-cycle with v. Thus avoiding the specified 4-cycle separates 3-vertices by at least three edges along C, so v gives charge to at most $\lfloor d(v)/3 \rfloor$ such vertices. If a 9-vertex v has at least three 3-neighbors, then it has exactly three and loses no other charge. Hence a 9-vertex loses at most $\max\{3 \cdot 1, 2 \cdot 1 + 2 \cdot \frac{1}{2}\}$ and ends happy. A 10-vertex has at most five 5^--neighbors, and with three 3-neighbors it can only have one other 5^--neighbor. It loses at most $\max\{4 \cdot 1, 2 \cdot 1 + 3 \cdot \frac{1}{2}\}$ and also ends happy. ∎

9.3.31.* Remark. For sharpness of the guarantee on light edges in Lemma 9.3.30, see Exercise 49. We also mention an application. Proper edge-coloring is decomposition into subgraphs with maximum degree 1. The **arboricity** of a graph is the number of forests needed to decompose it (planar graphs have arboricity at most 3; Exercise 41). A **linear forest** is a disjoint union of paths.

The **linear arboricity**, $\mathrm{la}(G)$, is the minimum number of linear forests needed to decompose G. Since 2-regular graphs contain cycles, $\mathrm{la}(G) \geq (2r + 1)/2$ when G is $2r$-regular. Akiyama–Exoo–Harary [1980, 1981] conjectured that always $\mathrm{la}(G) \leq \lceil (\Delta(G) + 1)/2 \rceil$. Their papers plus Enomoto–Péroche[1984] and Guldan [1986] together proved this for $\Delta(G) \leq 6$ and $\Delta(G) \in \{8, 10\}$. For $\varepsilon > 0$ and $\Delta(G)$ large, Alon [1988] proved $\mathrm{la}(G) \leq (\frac{1}{2} + \varepsilon)\Delta(G)$.

The conjecture has been proved for planar graphs, with $\Delta(G) \geq 9$ in Wu [1999] and $\Delta(G) = 7$ in Wu–Wu [2008]. The proof for planar G with $\Delta(G) \geq 9$ (Exercise 51) uses the fact provided by Lemma 9.3.30 that G has a 4-cycle through two 3-vertices when it has no edge with weight at most 11. ∎

OTHER ASPECTS OF DISCHARGING (optional)

Although the remainder of this section is optional, the proofs illustrate important aspects of discharging arguments, especially Theorem 9.3.32.

We return to the List Color Conjecture $\chi'_l(G) = \chi'(G)$. Thinking that possibly $\chi'_\ell(G) = \Delta(G) + 1$ for some graph G with $\chi'(G) = \Delta(G)$, Vizing conjectured $\chi'_\ell(G) \leq \Delta(G) + 1$. Juvan–Mohar–Skrekovski [1998a, 1999] proved this for $\Delta(G) \leq 4$.

For planar graphs, Borodin [1990] proved $\chi'(G) \leq \Delta(G) + 1$ when $\Delta(G) \geq 9$. We use balanced charging to prove Borodin's result. An interesting tool is a reservoir or "pot" of charge that can flow to or from vertices or faces without regard to location; this enables charge to move long distances. In this proof, the pot allows charge to move from maximum-degree vertices to 3-vertices; we need not name specific recipients. Notions analogous to the pot of charge for long-distance transfer of charge appear in Havet–Sereni [2006] and Borodin–Ivanova [2009]; a general term for such methods is **global discharging**.

9.3.32. Theorem. (Borodin [1990]) $\chi'_\ell(G) \leq \Delta(G) + 1$ for planar G with $\Delta(G) \geq 9$.

Proof: (Cohen–Havet [2010]) To ensure that the family is closed under taking subgraphs, we prove more generally that if $k \geq 9$ and G is planar with $\Delta(G) \leq k$, then $\chi'_\ell(G) \leq k + 1$. Let G be a minimal counterexample, with L assigning lists of size $k + 1$ to edges so that G has no L-edge-coloring. By Lemma 9.3.16, edges with weight at most $k + 2$ are reducible. Hence we may assume that $\delta(G) \geq 3$ and that every edge has weight least $k + 3$. Since $k \geq 9$, edges have weight at least 12.

We use balanced charging and a pot of charge. Initially, the pot has charge 0. Discharging that makes each vertex and face happy and maintains nonnegative charge in the pot contradicts negative total charge. We use two discharging rules.

(R1) The pot takes $\frac{1}{2}$ from every k-vertex and gives 1 to every 3-vertex.

(R2) Each 3-face takes $\frac{1}{2}$ from each incident 8^+-vertex and $\frac{j-4}{j}$ from each incident j-vertex with $j \in \{5, 6, 7\}$.

To ensure positive charge in the pot, we prove $n_k > 2n_3$, where n_j is the number of j-vertices in G. The edges incident to 3-vertices form a bipartite graph H with the 3-vertices as one part and the k-vertices as the other. Any cycle C in H has even length, since H is bipartite. By minimality of the counterexample, $G - E(C)$ has an L-edge-coloring ϕ. Each edge of C has $\Delta(G) - 1$ incident edges colored by ϕ, so it still has at least two available colors. Since even cycles are 2-edge-choosable (Example 8.2.7), the L-edge-coloring extends to G. Thus G being a counterexample requires H to be acyclic, with fewer than $n_3 + n_k$ edges. Since $|E(H)| = 3n_3$, we have $3n_3 < n_3 + n_k$, as desired.

For vertices, (R1) makes 3-vertices happy. A j-vertex v with $j \in \{4, 5, 6, 7\}$ loses altogether at most $j - 4$, its initial charge. Since $k \geq 9$, an 8-vertex gives nothing to the pot and loses at most 4. For $j \geq 9$, possibly sending $\frac{1}{2}$ to the pot, a j-vertex loses at most $\frac{j+1}{2}$ and is happy.

For faces, the 4^+-faces lose no charge and remain happy; we must show that each 3-face f gains at least 1. Let j be the least degree among vertices incident to f. If $j \leq 4$, then f has at least two incident 8^+-vertices, giving $\frac{1}{2}$ each. If $j = 5$, then f has at least two incident 7^+-vertices giving at least $\frac{3}{7}$ each, plus $\frac{1}{5}$ for the 5-vertex. If $j \geq 6$, then each vertex incident to f gives at least $\frac{1}{3}$ to f. ∎

Here again discharging produces an unavoidable set of reducible configurations. The configurations are light edges (weight at most $\Delta(G) + 2$) and cycles alternating between 3-vertices and $\Delta(G)$-vertices.

9.3.33. Remark. Continuing the study of edge-choosability in planar graphs, Bonamy [2015] extended Theorem 9.3.32 to $\Delta(G) \geq 8$ by discharging, but the details are much longer. Cranston [2009] showed that $\Delta(G) \geq 6$ suffices when no two 3-faces share an edge (Exercise 59). For $\Delta(G) = 5$, we have $\chi_l(G) \leq \Delta(G) + 1$ when G is planar with no 3-cycle (Zhang–Wu [2004]), with no 4-cycle (Cranston [2009]), or with no 5-cycle (Wang–Lih [2002]). Higher maximum degree reduces the upper bound; Borodin [1990] proved $\chi_l'(G) = \Delta(G)$ for planar graphs with $\Delta(G) \geq 14$ (Exercise 63), which Borodin–Kostochka–Woodall [1997] reduced to $\Delta(G) \geq 12$. ∎

9.3.34. Remark. Even without planarity, we can study the effect of Mad(G) on edge-coloring. Using the Vizing Adjacency Lemma (Corollary 8.3.11) and degree charging, one can show that when $\Delta(G) \geq 8$ the condition Mad$(G) < 6$ implies $\chi'(G) = \Delta(G)$ (Exercise 64). In fact, Mad$(G) < 6.5$ suffices (Miao–Sun [2010]).

If Mad$(G) < 2$, then G is a forest, and $\chi_l'(G) = \chi'(G) = \Delta(G)$. Hence for both χ' and χ_l' there is a threshold on Mad(G) (in terms of $\Delta(G)$) to guarantee equaling $\Delta(G)$. Vizing [1968] conjectured that Mad$(G) \leq \Delta(G) - 1$ implies $\chi'(G) = \Delta(G)$. In fact, he conjectured $2|E(G)| \geq |V(G)|(\Delta(G) - 1) + 3$ for critical graphs. Fiorini [1975] proved that Mad$(G) < \frac{1}{4}\Delta(G)$ implies $\chi'(G) = \Delta(G)$.

Sanders–Zhao [2002] proved that Mad$(G) < \frac{1}{2}\Delta(G)$ implies $\chi'(G) = \Delta(G)$, and in fact $2|E(G)| \geq \frac{1}{2}|V(G)|(\Delta(G) + \sqrt{2\Delta(G) - 1})$ when G is critical. This result pioneered the interaction between Mad(G) and discharging; they described it as "the first time that the discharging method is applied to a graph theory problem in which Euler's formula is not used and embeddings of graphs in surfaces are not mentioned." Later, Woodall [2007] proved that Mad$(G) < \frac{2}{3}(\Delta(G) + 1)$ suffices.

Woodall [2010] conjectured that Mad$(G) < \Delta(G) - 1$ also implies $\chi_l'(G) = \Delta(G)$, but here the results are much weaker. Using an "iterated discharging" argument as presented in Woodall [2010], Borodin–Kostochka–Woodall [1997] proved that Mad$(G) < \sqrt{2\Delta(G)}$ implies $\chi_l'(G) = \Delta(G)$. ∎

For ordinary proper coloring, the Four Color Theorem begs the question of which planar graphs are 3-colorable. Grötzsch [1959] proved that planar triangle-free graphs are 3-colorable; other proofs (all using discharging) are by Thomassen [1994a], Dvořák–Kawarabayashi–Thomas [2011] (with a linear-time algorithm to find a proper 3-coloring), and Kostochka–Yancey [2014] (see Exercise 31).

9.3.35. Remark. Steinberg's 3-Color Conjecture from 1975 (see Aksionov–Melnikov [1978]) asserts that planar graphs with no 4-cycle or 5-cycle are 3-colorable; triangles are allowed. Erdős asked whether it suffices to forbid cycle lengths 4 through k for some k. This was proved for $k = 11$ (Abbott–Zhou [1991]) and $k = 9$ (Sanders–Zhao [1995] and Borodin [1996a]; see below). With more discharging, Borodin–Glebov–Raspaud–Salavatipour [2005] proved that forbidding j-faces for $4 \leq j \leq 7$ is enough. Many others proved 3-colorability under various restrictions on cycle-lengths. Eventually, Cohen-Addad–Hebdige–Král'–Li–Salgado [2017] showed that Steinberg's Conjecture is false. Their smallest example has 85 vertices.

For choosability, Thomassen [2003] proved $\chi_l(G) \leq 3$ when G is planar with girth at least 5. Dvořák–Postle [2018] (on arXiv in 2015) proved that planar graphs having no cycles of lengths 4 through 8 are 3-choosable. ∎

The traditional proof of the lemma for the result of Sanders–Zhao [1995] and Borodin [1996a] uses balanced charging, but by face charging it is a bit simpler.

9.3.36. Lemma. (Borodin [1996a]) Every 2-connected plane graph G such that $\delta(G) \geq 3$ has two 3-faces with a common edge, or a j-face with $4 \leq j \leq 9$, or a 10-face whose vertices all have degree 3.

Proof: Let G be a plane graph with $\delta(G) \geq 3$ and no such configuration. Use **face charging** ($2d(v) - 6$ on each vertex v and $l(f) - 6$ on each face f). The total charge is -12. With no 4-faces or 5-faces, only triangles start with negative charge, -3.

(R1) Each 3-face takes 1 from each neighboring face.
(R2) Each non-triangular face f takes 1 from each incident 4^+-vertex lying on a triangle sharing an edge with f (to help faces that lose too much).

Here (R1) makes 3-faces happy (since no two 3-faces share an edge), and 3-vertices remain at charge 0. For $j \geq 4$, a j-vertex loses charge at most $\lfloor \frac{j}{2} \rfloor$ (since 3-faces do not share edges) and ends with at least $\lceil \frac{3j}{2} \rceil - 6$, which is nonnegative.

For $j \geq 10$, a j-face f loses 1 for every path along its boundary such that the neighboring faces are triangles and the endpoints have degree 3, as shown on the left below. Face f gives 1 to each of those 3-faces but takes 1 from each intervening vertex, since forbidding 3-faces with shared edges requires the intervening vertices to have degree at least 4. If an endpoint of a maximal such path has degree at least 4, then there is no net loss. Hence the net loss for f is at most $\lfloor \frac{j}{2} \rfloor$, and the final charge is at least $\lceil \frac{j}{2} \rceil - 6$, which is nonnegative when $j \geq 11$.

Hence negative charge occurs only at 10-faces. Losing more than 4 requires losing 1 through five paths. The paths must be single edges sharing no vertices, and all the vertices incident to f must have degree 3, as on the right below. ∎

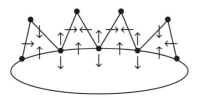 

9.3.37. Theorem. (Borodin [1996a], Sanders–Zhao [1995]) Every plane graph having no j-cycle with $4 \leq j \leq 9$ is 3-colorable.

Proof: Since the family described is hereditary, a smallest counterexample G must be 4-critical, so $\delta(G) \geq 3$ and $\kappa(G) \geq 2$. Since there is no 4-cycle, no two 3-faces share an edge. By Lemma 9.3.36, we may thus assume that G is embedded with at least one 10-face C, whose vertices all have degree 3.

Let ϕ be a proper 3-coloring of $G - V(C)$. If each vertex on C has exactly one neighbor outside C, then two colors remain available at each vertex of C. Otherwise, C has a chord (drawn outside C), and at the endpoints of a chord three colors remain available. Since even cycles with chords are degree-choosable (Theorem 8.2.13), the coloring can be completed. ∎

Discharging can also guarantee *many* desired configurations. For example, the amount by which total charge is negative can force many light edges.

9.3.38. Theorem. (Borodin–Sanders [1994]) For a plane graph G with $\delta(G) = 5$,

$$2e_{5,5} + e_{5,6} + \tfrac{2}{7}e_{5,7} \geq 60,$$

where $e_{i,j}$ is the number of edges joining vertices of degrees i and j. Also, the coefficients in this inequality are sharp.

Proof: Add edges to obtain a triangulation H; still $\delta(H) = 5$. Since $\delta(G) = 5$, no edges at 5-vertices of H were added. Hence $e_{5,j}(H) \leq e_{5,j}(G)$, and it suffices to prove the result for triangulations. We use vertex charging.

Each vertex distributes its charge equally to all incident *edges*, giving $\frac{d(v)-6}{d(v)}$ to each. Thus each vertex ends with 0 and only edges end with nonzero charge. Only 5-vertices take charge ($\frac{1}{5}$ from each edge), so only their incident edges can have negative charge. An 8^+-vertex gives each incident edge at least $\frac{1}{4}$, which exceeds $\frac{1}{5}$. Thus negative charge remains only on edges from 5-vertices to vertices of degrees 5, 6, or 7, with charges $\frac{-2}{5}$, $\frac{-1}{5}$, and $\frac{-2}{35}$, respectively. Since the total charge is -12 and there is no negative charge elsewhere, we obtain $\frac{-2}{5}e_{5,5} + \frac{-1}{5}e_{5,6} + \frac{-2}{35}e_{5,7} \leq -12$. Multiplying by -5 completes the proof.

Equality forbids edges with positive charge, since that would force more edges with negative charge. Hence every sharpness example is a triangulation with maximum degree at most 7. The icosahedron has $e_{5,5} = 30$ and no other edges; thus the coefficient of $e_{5,5}$ cannot be reduced. The graph obtained from the dodecahedron by putting in each face a 5-vertex adjacent to its corners has $e_{5,6} = 60$ with all other edges joining 6-vertices; thus the coefficient of $e_{5,6}$ cannot be reduced. The graph below (three edges wrapping from left to right), with $2e_{5,5} = e_{5,6} = 28$ and $e_{5,7} = 14$, shows that the coefficient of $e_{5,7}$ cannot be reduced. ∎

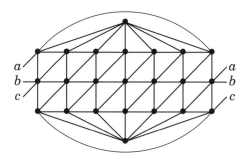

9.3.39. Remark. To explain the method, we have presented discharging arguments with fairly simple details. Many applications are quite intricate. Robertson–Sanders–Seymour–Thomas [1997] used 32 discharging rules to produce their unavoidable set of 633 reducible configurations for the Four Color Theorem.

Discharging was also used for **Tutte's 3-Edge-Coloring Conjecture**: 2-edge-connected 3-regular graphs containing no subdivision of the Petersen graph are 3-edge-colorable. Robertson–Seymour–Thomas [1997, 2019] reduced this to 3-edge-colorability of two types of nearly planar 3-regular graphs: "double-cross graphs" (Edwards–Sanders–Seymour–Thomas [2016]) and "apex graphs" (Sanders–Thomas, unpublished). Exercise 16.1.13 is one step in the proof.

Borodin [1984, 1995] used discharging (without computers) to prove Ringel's **Six Color Conjecture** from 1965: the vertices and faces of a planar graph can be 6-colored so that neighboring objects have different colors. The original discharging argument produced 35 reducible configurations, later cut by about half.

Inductive proofs using discharging to produce reducible configurations also have algorithmic implications. The proof of the Four Color Theorem in Robertson–Sanders–Seymour–Thomas [1997] yields a quadratic-time algorithm that finds a proper 4-coloring; the Appel–Haken proof provides a quartic algorithm. Heawood's proof produces proper 5-colorings in quadratic-time, but Thomassen's proof of 5-choosability (without discharging) yields a linear-time algorithm. ∎

EXERCISES 9.3

9.3.1. (−) Prove that every maximal plane graph other than K_4 is 3-face-colorable.

9.3.2. (−) Exhibit 3-regular graphs with these properties:
(a) planar but not 3-edge-colorable.
(b) 2-connected but not 3-edge-colorable.
(c) planar with connectivity 2, but not Hamiltonian.

9.3.3. (−) Prove that a 2-edge-connected plane graph is 2-face-colorable if and only if it is Eulerian.

9.3.4. (−) Use the Four Color Theorem to prove that every planar graph decomposes into two bipartite graphs. (Hedetniemi [1969], Mabry [1995])

9.3.5. (−) Prove that every graph with maximum average degree at most d is d-degenerate. Prove that every d-degenerate graph has maximum average degree at most $2d$. Give infinitely many examples to show that the second statement is nearly sharp.

9.3.6. (−) Prove or disprove: Every triangle-free planar graph is 4-choosable.

9.3.7. (−) Prove or disprove: For k sufficiently large, a graph G with $\delta(G) = k$ and $\overline{d}(G) < k + 1$ has adjacent k-vertices.

9.3.8. (−) Let G be a planar graph with girth at least 6 and $\delta(G) \geq 2$. By Lemma 9.3.19, G has an edge with weight at most 7. Prove sharpness by constructing a planar graph with girth 6 and minimum degree 2 having no edge with weight at most 6.

9.3.9. *Reduction of Four Color Problem to Tait's conjecture.* Let G be a 3-regular graph with edge-connectivity 2. (Recall that $\kappa(G) = \kappa'(G)$ when G is 3-regular.)
(a) Prove that there exist subgraphs $G_1, G_2 \subseteq G$ and vertices $u_1, v_1 \in V(G_1)$ and $u_2, v_2 \in V(G_2)$ such that $u_1v_1, u_2v_2 \notin E(G)$ and G consists of G_1, G_2, and a "ladder" (of some length) joining them at $\{u_1, v_1, u_2, v_2\}$ as illustrated below.
(b) Prove that $\chi'(G_1 + u_1v_1) = \chi'(G_2 + u_2v_2) = 3$ implies $\chi'(G) = 3$.
(c) Use Tait's Theorem (Theorem 9.3.1) to reduce the Four Color Theorem to **Tait's Conjecture**: "every simple 3-regular 3-connected planar graph is 3-edge-colorable."

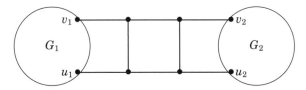

9.3.10. *Reduction of Four Color Problem to Hamiltonian graphs.*

(a) Prove that every triangulation with at least four vertices is 3-connected and that a separating triple in such a graph induces a 3-cycle.

(b) Whitney [1931] proved that every 4-connected planar triangulation is Hamiltonian. Use this to reduce the Four Color Problem to the problem of proving that every Hamiltonian planar graph is 4-colorable. (See also Saaty–Kainen [1977])

9.3.11. Prove Grinberg's Theorem by induction on the number of edges.

9.3.12. (◊) Use Grinberg's Theorem to prove that each graph below is not Hamiltonian.

 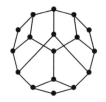 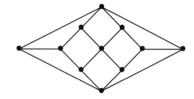

9.3.13. (◊) Use Grinberg's Theorem to prove that the Grinberg graph (on the left below) is not Hamiltonian.

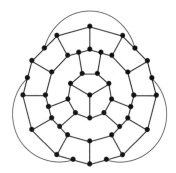 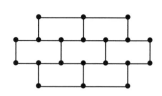

9.3.14. Give a short proof that the graph on the right above is not Hamiltonian. Show that Grinberg's Theorem cannot be used directly to prove this. Show that Grinberg's Theorem does apply to a modification of the graph.

9.3.15. The smallest known 3-regular 3-connected planar graph that is not Hamiltonian has 38 vertices and appears on the left below. Prove that this graph is not Hamiltonian. (Lederberg [1966], Bosák [1967], Barnette)

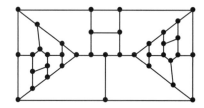

9.3.16. (◊) Let G be the grid $P_r \square P_s$. Let Q be a Hamiltonian path from the upper left corner vertex to the lower right corner vertex, such as that shown on the right above in bold. Note that Q partitions the grid into regions, some opening to the left or downward and others opening to the right or upward. Prove that the total area of the up-right regions (B) equals the total area of the down-left regions (A). (Fisher–Collins–Krompart [1994])

9.3.17. Prove that an outerplanar graph can be 2-colored so that the subgraph induced by each color class is a disjoint union of paths. (Mihók [1983], Akiyama–Era–Gervacio–Watanabe [1989], Goddard [1991])

9.3.18. Prove that a planar graph can be 2-colored so each color class induces an outerplanar graph. (Chartrand–Geller–Hedetniemi [1971], Burštein [1974], Penaud [1975])

9.3.19. Prove that every outerplanar graph can be properly 3-colored so that the union of any two color classes induces a forest. Conclude that the vertices of an outerplanar graph can be partitioned into an independent set and a set inducing a forest so that the independent set has at least 1/3 of the vertices.

9.3.20. ($\Diamond$) *Short proof of the Five Color Theorem*. Let v be a 5-vertex in a graph G. Given that x and y are nonadjacent neighbors of v, let G' be the graph obtained from G by contracting the edges vx and vy. Prove that if G' is 5-colorable, then G is 5-colorable. Use this to prove by induction that every planar graph is 5-colorable. (Kainen [1974])

9.3.21. Albertson–Berman [1979] and Akiyama–Watanabe [1987] conjectured that every n-vertex planar graph has an induced subgraph with at least $\lceil n/2 \rceil$ vertices that is a forest. This would yield an independent set of at least $\lceil n/4 \rceil$ vertices without using the Four Color Theorem. Show that the conjecture is sharp by constructing an n-vertex planar graph having no induced forest with more than $\lceil n/2 \rceil$ vertices. Akiyama–Watanabe [1987] also conjectured that every n-vertex bipartite planar graph has an induced forest with at least $\lceil 5n/8 \rceil$ vertices; show that this conjecture also is sharp.

9.3.22. Without using the Four Color Theorem, prove that every Hamiltonian plane graph is 4-face-colorable (nothing is assumed about the vertex degrees).

9.3.23. ($\Diamond$) Let G be a maximal plane graph. Prove that G^* has a 2-factor and use it to show that $V(G)$ can be 2-colored so that both colors appear on each face. Obtain the same conclusion from the Four Color Theorem. (Burštein [1974], Penaud [1975])

9.3.24. ($\Diamond$) Thomassen [1995b] proved that the vertices of any planar graph can be partitioned into sets inducing a forest and a 3-degenerate graph. To generalize this, prove for $d_1, \ldots, d_k \in \mathbb{N}_0$ and $d \le k - 1 + \sum_{i=1}^{k} d_i$ that the vertices of any d-degenerate graph can be partitioned into sets $V_1, \ldots, V_k$ such that $G[V_i]$ is d_i-degenerate for $1 \le i \le k$.

9.3.25. ($\Diamond$) *The* **Art Gallery Theorem**. (Chvátal [1975], Fisk [1978])
   (a) Prove that outerplanar graphs are 3-colorable (without the Four Color Theorem).
   (b) Prove that an art gallery laid out as an n-gon can be protected by $\lfloor n/3 \rfloor$ guards so that every point of the interior is visible to some guard.
   (c) For each n, construct an art gallery with n outer segments that cannot be watched by fewer than $\lfloor n/3 \rfloor$ guards.

9.3.26. ($\Diamond$) An *art gallery with walls* is a polygon plus nonintersecting chords called "walls" that join vertices. Each interior wall has a tiny "doorway". A guard in a doorway can see along both sides of the wall containing it, but no guard can see through a wall. Determine the least t such that for every walled art gallery with n vertices, one can place t guards so that every interior point is visible to some guard. (Hutchinson [1995], Kündgen [1999])

9.3.27. ($\Diamond$) Heawood [1898] stated that a maximal planar graph is 3-colorable if and only if it is Eulerian. Golovina–Yaglom [1963] proved it (Hutchinson [2001] generalized it).
   (a) Prove that every 3-colorable triangulation (including multigraphs) is Eulerian.
   (b) Let G be an even plane graph with all bounded faces having length 3. By Exercise 9.1.49, G has a noncrossing Eulerian circuit. Prove that every subcircuit has length divisible by 3. Conclude that G is 3-colorable. (Tsai–West [2011])
   (c) Let G be as in part (b). Prove that G is 3-colorable using the fact that the dual is bipartite. (Hint: Use induction on the number of bounded faces.) (Lovász [1993])

9.3.28. (◇) Prove that every 3-colorable planar graph is a subgraph of some 3-colorable triangulation. (Comment: C_5 is not a spanning subgraph of a 3-colorable triangulation; hence it may be necessary to add vertices.) (Król [1972])

9.3.29. Use the Four Color Theorem to prove that every 2-edge-connected 3-regular graph that can be drawn in the plane with only one edge-crossing is 3-edge-colorable.

9.3.30. Grötzsch's Theorem states that every triangle-free n-vertex planar graph G is 3-colorable, so $\alpha(G) \geq n/3$. Tovey and Steinberg showed $\alpha(G) > n/3$. Prove sharpness using the following graphs. Let G_1 be the 5-cycle, with vertices a, x_0, x_1, y_1, z_1 in order. For $k > 1$, obtain G_k from G_{k-1} by adding three vertices x_k, y_k, z_k, the path $\langle x_{k-1}, x_k, y_k, z_k, y_{k-1}\rangle$ and the edge $z_k x_{k-2}$. The graph G_3 appears below. (Fraughnaugh [1985])

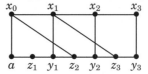

9.3.31. (◇) Kostochka–Yancey [2014] proved (using some discharging) that every 4-critical n-vertex graph has at least $\lceil (5n-2)/3 \rceil$ edges (conjectured by Ore [1967]). Use their result to prove Grötzsch's Theorem: Every triangle-free planar graph is 3-colorable.

9.3.32. It has been conjectured that every planar triangulation has edge-chromatic number $\Delta(G)$, and this has been proved for sufficiently large $\Delta(G)$. Show that $\chi'(G) = \Delta(G)$ for the icosahedron, shown on the left below.

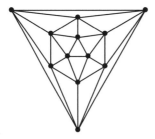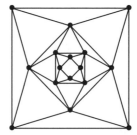

9.3.33. An **equitable coloring** of a graph is a proper coloring whose color classes differ in size by at most 1. Let G_n be the graph formed from n concentric 4-cycles triangulating the region between any two consecutive 4-cycles (each vertex has two consecutive neighbors on the other 4-cycle); G_4 appears above on the right.) Prove that for G_n with n even and for the icosahedron (on the left above), every proper 4-coloring is equitable. (M. Albertson)

9.3.34. (◇) *Non-4-choosable planar graph with 75 vertices.*
(a) Prove that the graph below cannot be properly colored from the given lists; the vertices of degree 5 have lists of size 1 and the others have lists of size 4.
(b) Use part (a) to construct a 3-colorable planar graph with 75 vertices that is not 4-choosable. (Gutner [1996])

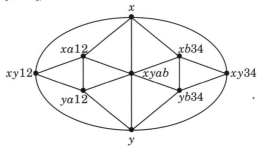

9.3.35. (◊) *Non-4-choosable planar graph with 63 vertices.*

(a) In the list assignments for the graph on the left below, S denotes $[4]$ and i denotes $S - \{i\}$. Prove that this graph has no proper coloring chosen from these lists.

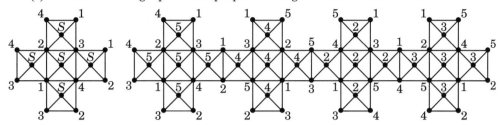

(b) In the list assignments for the graph G on the right above, i denotes $[5] - \{i\}$; each list has size 4. Let G' be the graph obtained from G by adding one vertex with list $[5] - \{1\}$ adjacent to all vertices on the outside face of this drawing of G. Prove that G' has no proper coloring chosen from these lists. (Mirzakhani [1996])

9.3.36. Without the Four Color Theorem, prove that every planar graph with at most 12 vertices is 4-colorable. Conclude that planar graphs with at most 32 edges are 4-colorable.

9.3.37. Let H be a configuration in a planar triangulation (Definition 9.3.8). Let H' be obtained by labeling the neighbors of the ring vertices with their degrees and then deleting the ring vertices. Prove that H can be retrieved from H' when H' has no cut-vertex.

9.3.38. Without using anything not proved in this section, show that a smallest 5-chromatic triangulation has no separating cycle of length 3 or 4. That is, show that any configuration with a ring of length at most 4 is reducible. (Birkhoff [1913])

9.3.39. (+) Prove that the Birkhoff diamond (the configuration on the left below) is reducible for 4-colorability of plane triangulations. (Hint: Prove first that it suffices to consider colorings of the ring obtained by replacing the Birkhoff diamond with the configuration on the right. For each such coloring, extend it to a 4-coloring of the interior diamond, possibly after modifying it via Kempe chain arguments.) (Birkhoff [1913])

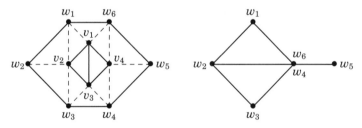

9.3.40. *Sharpness of Steinberg's Conjecture.* Construct two plane graphs with chromatic number 4, one having no 5-cycles and the other having no 4-cycles.

9.3.41. (+) *Arboricity of planar graphs.* Let u be a 5-vertex in a plane triangulation G. Let $N(u) = \{v, w, x, y, z\}$, ordered so $[v, w, x, y, z]$ is a cycle. Form G' from $G - u$ by adding the edges vx and vy. Prove that if G' decomposes into three forests, then also G decomposes into three forests. Conclude that every planar graph decomposes into three forests. (Comment: Balogh–Kochol–Pluhár–Yu [2005] proved that one of the three forests can be required to have maximum degree at most 8. Gonçalves [2009] reduced 8 to 4.)

9.3.42. Every graph G with $\mathrm{Mad}(G) < k + 1$ is k-degenerate. Disprove the converse: for each $k \in \mathbb{N}$ with $k \geq 2$ construct a k-degenerate graph with average degree at least $k + 1$.

9.3.43. Let G be a graph with $\delta(G) \geq 2$ and $\overline{d}(G) < \frac{4k}{k+2}$. Complete the proof of Lemma 9.3.19 by proving when $k > 6$ that G has an edge with weight at most $k + 1$.

9.3.44. ($\diamond$) *Adjacent vertices of minimum degree.*

(a) Determine the largest b such that if $\delta(G) = k$ and $\overline{d}(G) < b$, then G must have two adjacent vertices of degree k.

(b) Determine the least g such that every planar graph G with girth at least g and $\delta(G) \geq 2$ has an edge of weight 4. Explain how this relates to part (a).

9.3.45. Let G be a graph with $\delta(G) = k$. For $0 \leq j < k$, determine the largest ρ such that $\overline{d}(G) < k + \rho$ guarantees in G a k-vertex with more than j neighbors of degree k.

9.3.46. Prove that if $\overline{d}(G) < 4$, then $\delta(G) \leq 2$ or G has a 3-vertex with a 5^--neighbor. Show that when $\delta(G) = 2$ and $\overline{d}(G) < 4$ there is no upper bound on the smallest edge weight.

9.3.47. Let G be a graph with $\delta(G) = 3$ and $\overline{d}(G) < \frac{10}{3}$. Prove that G has a 3-vertex whose neighbors have degree-sum at most 10. (Comment: G. Tardos constructed such a graph with 98 vertices and $\overline{d}(G) < \frac{10}{3}$ having no 3-vertex with three 3-neighbors.)

9.3.48. ($\diamond$) Let G be a connected plane graph such that $\delta(G) \geq 3$ and $\delta(G^*) \geq 3$.

(a) Use balanced charging to prove that G has a 3-vertex on a 5^--face or a 5^--vertex on a 3-face. (Comment: There are five such combinations of degrees and face-lengths; the Platonic solids show that none can be excluded.)

(b) Strengthen part (a) by proving that G contains a 3-vertex on a 5^--face, a 4-vertex on a 3-face, or a 5-vertex with at least four incident 3-faces. (Comment: (Lebesgue [1940] proved this for 3-connected plane graphs.)

9.3.49. Construct a triangulation where all lightest edges join 5-vertices and 6-vertices and another where all lightest edges join 3-vertices and 10-vertices (see Lemma 9.3.30).

9.3.50. Determine whether every planar graph with girth at least 4 and minimum degree 3 has a 3-vertex with a 4^--neighbor. Construct a plane graph G_k with girth 4 and minimum degree 3 where the distance between 3-vertices is at least k. Construct a planar graph H_k with minimum degree 5 where the distance between 5-vertices is at least k.

9.3.51. (+) For $t \geq 5$, prove that every planar graph G with $\Delta(G) < 2t$ decomposes into t linear forests. (Wu [1999])

9.3.52. Vizing's Planar Graph Conjecture states that planar graphs with $\Delta(G) \geq 6$ are Class 1. Prove necessity by finding a graph with maximum degree 6 that is not 6-edge-colorable and a planar graph with maximum degree 5 that is not 5-edge-colorable. (Comment: Vizing [1965] proved the conjecture for $\Delta(G) \geq 8$. Sanders–Zhao [2001] and Zhang [2000] proved it for $\Delta(G) = 7$, but for $\Delta(G) = 6$ it is only known in special classes.)

9.3.53. ($\diamond$) An ℓ-**thread** in a graph G is a path with ℓ internal vertices, all with degree 2 in G. Prove that if $\overline{d}(G) < 2 + \frac{1}{3t-2}$ and G has no 2-regular component, then G has a 1^--vertex or a $(2t-1)$-thread. Construct infinitely many examples with average degree $2 + \frac{1}{3t-2}$ but no $(2t-1)$-thread. (Comment: $(2t-1)$-threads are reducible for C_{2t+1}-coloring, where an H-**coloring** of G is a map $\phi\colon V(G) \to V(H)$ such that $uv \in E(G)$ implies $\phi(u)\phi(v) \in E(H)$.)

9.3.54. ($\diamond$) An I, F-**partition** of a graph G is a partition of $V(G)$ into sets I and F such that $G[F]$ is a forest and the distance between any two vertices of I is at least 3. A **star** k-**coloring** of G is a proper k-coloring such that no 4-vertex path is 2-colored.

(a) Prove that if $\overline{d}(G) < 2 + \frac{1}{2t-1}$ and G has no 2-regular component, then G contains (1) a 1^--vertex, or (2) a 3-vertex with at least $4t - 3$ vertices of degree 2 on its maximal incident threads, or (3) a 4^+-vertex incident to a $(2t-1)$-thread.

(b) Prove that if $\text{Mad}(G) < \frac{7}{3}$, then G has an I, F-partition. (Timmons [2008])

(c) Prove that if G has an I, F-partition, then G has a star 4-coloring.

(d) Construct infinitely many graphs with average degree $\frac{5}{2}$ having no I, F-partition.

(Comment: Brandt–Ferrara–Kumbhat–Loeb–Stolee–Yancey [2016] proved that $\overline{d}(G) < \frac{5}{2}$ implies an I, F-partition of G. Perhaps higher $\text{Mad}(G)$ still guarantees a star 4-coloring.)

9.3.55. Prove that if $\overline{d}(G) < \frac{5}{2}$ and G is connected, then G contains a 3^--vertex with a 1-neighbor, a 4^--vertex with two 2^--neighbors, or a 5^+-vertex v having at least $\frac{d(v)-1}{2}$ neighbors of degree at most 2. (Cranston–Jahanbekam–West [2014])

9.3.56. Let G be a connected graph with at least four vertices. Prove that if $\overline{d}(G) < \frac{5}{2}$ and $\delta(G) \geq 2$, then G contains a 2-thread or a 3-vertex having three 2-neighbors, one of which has a second 3-neighbor. (Cranston–Kim–Yu [2010])

9.3.57. ($\diamond$) Prove that every planar graph with minimum degree 5 contains two 3-faces sharing an edge with weight at most 11. (Borodin [1996b])

9.3.58. Prove that every plane triangulation with minimum degree 5 has two 3-faces sharing an edge such that the non-shared vertices have degree-sum at most 11. (Comment: With this and a 4^--vertex as an unavoidable set, Albertson [1976] proved that $\alpha(G) \geq 2n/9$ when G is an n-vertex planar graph, without using the Four Color Theorem.)

9.3.59. ($\diamond$) Say that a plane graph is *clean* if no two 3-faces share an edge. For $k \geq 7$, prove that a clean plane graph G with $\Delta(G) \leq k$ has an edge with weight at most $k+2$. Conclude that $\chi'_l(G) \leq \Delta(G) + 1$ when G is a clean plane graph with $\Delta(G) \geq 7$. (Comment: Cranston [2009] proved the same conclusion for $\Delta(G) = 6$, strengthening several earlier results.)

9.3.60. ($\diamond$) For $\delta(G) \geq 2$, prove that $\Delta(G) \leq 6$ and $\overline{d}(G) < \frac{7}{2}$ guarantee in G an edge with weight at most 7 or a cycle alternating between 2-vertices and 6-vertices. Conclude that $\Delta(G) \leq 6$ and $\mathrm{Mad}(G) < \frac{7}{2}$ imply $\chi'_l(G) \leq 6$. (Hint: Use degree charging and a pot of charge. Comment: This result strengthens Theorem 9.3.21.)

9.3.61. (+) *Light triangles.* Let G be a planar graph with minimum degree 5. Proposition 9.3.14 guarantees weight at most 17 on some three-vertex path.

(a) Prove that G contains a 3-face with weight at most 17 (Borodin [1989a])

(b) Provide a single example to show that smaller weight cannot be guaranteed. Show that $\delta(G) = 5$ is needed by constructing for each $k \in \mathbb{N}$ a planar graph with minimum degree 4 having no 3-face with weight at most k.

9.3.62. Let G be a plane graph having no 4-cycle and no face length in $\{4, \ldots, k\}$. Prove that the average face length is at least $6 - \frac{18}{k+4}$. Conclude $\mathrm{Mad}(G) < 3 + \frac{9}{2k-1}$. In particular, $\mathrm{Mad}(G) < 4$ when G is a plane graph with no 4-face or 5-face.

9.3.63. (+) *Planar graphs with high maximum degree are Class 1.*

(a) Let G be a plane graph with $\delta(G) \geq 2$. Prove that G contains an edge with weight at most 15 or a cycle that alternates between 2-vertices and higher-degree vertices.

(b) Prove that $\chi'(G) = \Delta(G)$ when G is planar and $\Delta(G) \geq 14$. (Borodin [1990])

9.3.64. Recall that a critical graph is one with $\chi'(G) = \Delta(G) + 1$ such that deleting any edge reduces the edge-chromatic number. The Vizing Adjacency Lemma (VAL, Corollary 8.3.11) says that for an edge yw in a critical graph, y has at least $\max\{\Delta(G) - d(w) + 1, 2\}$ neighbors with degree $\Delta(G)$. Use VAL to prove that if $\Delta(G) \geq 8$ and $\mathrm{Mad}(G) < 6$, then $\chi'(G) = \Delta(G)$. (Vizing [1968], Luo–Zhang [2004])

9.3.65. ($\diamond$) Let G be a planar graph with minimum degree 5. As in Theorem 9.3.38, let $e_{i,j}$ be the number of edges in G with endpoints of degrees i and j. Prove that $e_{5,6} \geq 60$ when $e_{5,5} = 0$, and show that equality can hold. (Grünbaum [1973]) (Comment: More generally, Borodin–Sanders [1994] proved $\frac{7}{3}e_{5,5} + e_{5,6} \geq 60$.)

9.3.66. An **acyclic coloring** is a proper coloring in which the union of any two color classes induces a forest. A graph is **acyclically k-choosable** if an acyclic coloring can be chosen from any lists of size k at the vertices. Prove that if $\mathrm{Mad}(G) < 3$, then G is acyclically 6-choosable. (Comment: Cranston–Yu [2011] proved this also for planar graphs with girth at least 5: the added reducible configuration is a 5-face whose incident vertices are four 3-vertices and a 5^--vertex.)

9.3.67. Use balanced charging to prove that every planar graph with girth at least 7 and minimum degree at least 2 has a 2-vertex adjacent to a 3^--vertex. Prove that the conclusion does not always hold when $\mathrm{Mad}(G) < \frac{14}{5}$ (thus planarity is needed).

9.3.68. A **dynamic coloring** is a proper coloring with each vertex v adjacent to vertices of at least $\min\{2, d(v)\}$ colors (Montgomery [2001]). A graph is **dynamically k-choosable** if a dynamic coloring can be chosen from any vertex lists of size k (Kim–Park [2011]).
   (a) Prove that every cycle of length at least 6 is dynamically 4-choosable.
   (b) Prove that except for C_5, a minimal graph among those that are not dynamically 4-choosable cannot contain an edge of weight at most 5.
   (c) Prove that every planar graph with girth at least 7 is dynamically 4-choosable, but some planar graphs with girth 6 are not dynamically 4-choosable.

9.3.69. ($\diamond$) Prove that every planar graph has a 5^--vertex with at most two 12^+-neighbors. (Comment: Balogh–Kochol–Pluhár–Yu [2005] strengthened 12 to 11; see Exercise 9.3.41. van den Heuvel–McGuinness [2003] applies this to coloring squares of planar graphs.)

9.3.70. ($\diamond$) *Choosability of squares of subcubic graphs.* Let G be a graph.
   (a) Prove that if $\delta(G) = 2$ and $\Delta(G) \le 3$ and $\overline{d}(G) < \frac{14}{5}$, then G has one of the configurations shown below: adjacent 2-vertices, a 3-vertex with two 2-neighbors, adjacent 3-vertices with 2-neighbors, or a 3-vertex with three 3-neighbors having 2-neighbors. (In the last two configurations, the specified 2-vertices need not be distinct.)
   (b) Prove that if $\Delta(G) \le 3$ and $\mathrm{Mad}(G) < \frac{14}{5}$, then $\chi_\ell(G^2) \le 7$. What does this imply for planar graphs? (Cranston–Kim [2008])

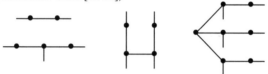

9.3.71. *Choosability of squares.* Let G be a graph.
   (a) Prove that if $2 \le \delta(G) \le \Delta(G) \le 4$ and $\overline{d}(G) < \frac{18}{7}$, then G has adjacent 2-vertices, or a 3-vertex with three 2-neighbors, or a 4-vertex path alternating 2-vertices and 3-vertices.
   (b) Prove that if $\Delta(G) \le 4$ and $\mathrm{Mad}(G) < \frac{18}{7}$, then $\chi_\ell(G^2) \le 7$. What does this imply for planar graphs? (Cranston–Erman–Škrekovski [2014])

9.3.72. *Choosability of denser squares.* Let G be a graph.
   (a) Prove that for $2 \le \delta(G) \le \Delta(G) \le 4$ and $\overline{d}(G) < \frac{10}{3}$, some 2-vertex has a 3^--neighbor, or a 3-vertex has two 3-neighbors, or a 4-vertex has a 2-neighbor and a 3^--neighbor.
   (b) Prove that the result in part (a) is sharp by presenting infinitely many graphs with $\overline{d}(G) = \frac{10}{3}$ and $\Delta(G) = 4$ that contain no such configuration.
   (c) Prove that if $\Delta(G) \le 4$ and $\mathrm{Mad}(G) < \frac{10}{3}$, then $\chi_\ell(G^2) \le 12$. What does this imply for planar graphs? (Cranston–Erman–Škrekovski [2014])

9.3.73. An **injective coloring** gives distinct colors to vertices with a common neighbor, and the **injective chromatic number**, written $\chi_i(G)$, is the minimum number of colors in an injective coloring of G. Injective colorings need not be proper colorings.
   (a) Determine the injective chromatic numbers of C_5 and the Petersen graph.
   (b) Let G be a graph with $\delta(G) = 2$ and $\Delta(G) = 3$. Prove that if $\overline{d}(G) < \frac{36}{13}$ and G has no 2-vertex in a triangle, then G has vertices x and y of degree 2 such that $d(x, y) \le 3$. (Hint: 2-vertices may need more charge than their neighbors can provide.)
   (c) Given $\Delta(G) \le 3$ and $\mathrm{Mad}(G) < \frac{36}{13}$, prove $\chi_i(G) \le 5$. (Cranston–Kim–Yu [2010])
   (d) Use the graph obtained by deleting one vertex from the Heawood graph (Exercise 6.2.27) to prove sharpness for parts (b) and (c).

Chapter 10

Ramsey Theory

Ramsey theory, named for Ramsey's Theorem, is the study of patterns in partitions of large structures. A simple special case of Ramsey's Theorem is the intuitively obvious Pigeonhole Principle. Partitioning theorems in general state that some class in the partition must have special properties.

Motzkin described this informally by saying "Complete disorder is impossible". Stargazers have always found interesting patterns of stars and given them names; Ramsey Theory implies that any given pattern must occur (scaled) if there are enough stars. Similarly, any sufficiently large amount of data (baseball statistics, letters in the Bible, etc.) must have patterns with special structure.

Books on the subject include Graham–Rothschild–Spencer [1980, 1990], Nešetřil–Rödl [1990], Soifer [2011], Prömel [2013], and Landman–Robertson [2014], and recent developments are surveyed in Sudakov [2010] and Conlon–Fox–Sudakov [2015]. Graham–Butler [2015] provides an introduction.

10.1. The Pigeonhole Principle

The Pigeonhole Principle (also called the *Dirichlet Drawer Principle*) is a simple idea with subtle applications. The subtlety often lies in recognizing when it can be used. Its use can eliminate lengthy case analysis.

The Pigeonhole Principle implies that $n + 1$ shoes from a closet containing n pairs of shoes must include a matched pair; they cannot all come from different pairs. We have already used the Pigeonhole Principle in proving the Cycle Lemma (Section 1.3), the existence of a cycle through any k vertices in a k-connected graph (Section 7.2), lower bounds on coloring (Section 8.1), Vizing's Theorem (Section 8.3), etc. The basic idea is that every set of numbers has an element at least as large as the average (also one at least as small).

10.1.1. Proposition. (Pigeonhole Principle) Placing more than kn objects into k classes puts more than n objects into some class.

Proof: With at most n objects per class, there are at most kn objects. ∎

A more general form, which generalizes to Ramsey's Theorem in Section 10.2, allows distinct thresholds (**quotas**) in different classes. In this section, we only need the symmetric form above.

10.1.2. Theorem. (Pigeonhole Principle) If $\left(\sum p_i \right) - k + 1$ objects are put into k classes with quotas $\{p_i\}$, then some class meets its quota.

Proof: If not, then at most $\sum (p_i - 1)$ objects can be accommodated.   ∎

CLASSICAL APPLICATIONS

Rarely is the relevance of the Pigeonhole Principle apparent when a problem is first stated. It may emerge only after the structure of the problem leads to a notion of classes and objects so that having many (or few) of these objects in one class yields a desired property. Often "many" means "more than one".

10.1.3. Example. *The (symmetric) acquaintance relation.* In every set of at least two people, some two people have the same number of acquaintances. Equivalently, in every graph G some two vertices have the same degree. If $|V(G)| = n$, then the degrees lie in $\{0, \ldots, n-1\}$. Degrees $n-1$ and 0 cannot both occur, so at most $n-1$ distinct degrees occur, and hence two must be equal.   ∎

10.1.4. Example. *Midpoints between lattice points.* Given five integer points in the plane, the midpoint of the segment joining some pair also has integer coordinates. The midpoint between (a, b) and (c, d) is $(\frac{a+c}{2}, \frac{b+d}{2})$. This is an integer point if and only if a and c have the same parity and b and d have the same parity.

Hence we group integer points by parity of coordinates: (odd/even, odd/even). With five points, two must be in the same class and yield a lattice midpoint. With four points, we may have one in each class and no lattice midpoint.   ∎

10.1.5. Example. *Covering K_n with bipartite graphs.* We prove that the minimum number of bipartite graphs with union K_n is $\lceil \log_2 n \rceil$. Equivalently, a complete graph covered by k bipartite subgraphs has at most 2^k vertices. (Exercise 8.1.37 presents a generalization; the Pigeonhole Principle is not actually needed.)

Suppose that K_n is covered by bipartite subgraphs $G_1, \ldots, G_k$, with G_i having bipartition $\{X_i, Y_i\}$. We may assume that each G_i contains all the vertices, since adding isolated vertices does not introduce odd cycles.

For each vertex v, define a binary k-tuple a by setting $a_i = 0$ if $v \in X_i$ and $a_i = 1$ if $v \in Y_i$. There are only 2^k binary k-tuples. If $n > 2^k$, then two vertices have the same code. They lie in the same part in each G_i, so the edge between them is not covered. Thus $n \le 2^k$.

To show that $n = 2^k$ is achievable, assign distinct binary k-tuples to the vertices. The ith bipartite subgraph consists of all edges joining vertices with ith coordinate 0 to vertices with ith coordinate 1. Since the k-tuples are distinct, every edge is covered by some such subgraph. Indeed, edges joining complementary codes appear in each subgraph.   ∎

10.1.6. Example. *Forcing divisible pairs.* Erdős [1935] observed that any set of $n + 1$ numbers in $[2n]$ contains a pair such that one divides the other. This is sharp: the n largest numbers in $[2n]$ contain no such pair. To apply the Pigeonhole Principle, we partition $[2n]$ into n classes such that for every two numbers in the same class, one divides the other.

Every natural number is uniquely expressed as an odd number times a power of 2. For fixed k, the set $\{(2k-1)2^{j-1}: j \in \mathbb{N}\}$ has the desired property; the smaller of any two divides the larger. Since only n odd numbers are less than $2n$, we have n such classes. The kth class is $\{m \in [2n]: m = (2k - 1)2^{j-1}$ and $j \in \mathbb{N}\}$. ∎

Like Example 10.1.5, Example 10.1.6 can be viewed as an extremal problem. How many elements can be chosen from $[2n]$ such that none divides another? The Pigeonhole Principle may establish a bound, and a construction can show that the bound is best possible. Nevertheless, presenting a construction and showing that no element can be added does not prohibit larger configurations. For example, we can avoid divisible pairs in Example 10.1.6 by first selecting primes. When $n = 5$, this yields $\{2, 3, 5, 7\}$, but now we cannot add any more elements. Our set is maximal but not maximum-sized: $\{6, 7, 8, 9, 10\}$ is a larger example.

10.1.7. Example. *Inverses modulo n.* If a and n are relatively prime, then $ab \equiv 1 \pmod{n}$ for some $b \in \{1, \ldots, n - 1\}$. Otherwise, $a, 2a, \ldots, (n - 1)a$ lie in $n - 2$ congruence classes other than 1. By the Pigeonhole Principle, two lie in the same class. If they are ia and ja, then $n \mid (i - j)a$. Since a and n are relatively prime, $n \mid (i - j)$, a contradiction. Hence some multiple of a is congruent to 1 modulo n.

Similarly, an element of a group permutes the group elements via composition. Otherwise, it maps two elements into one, contradicting the uniqueness of compositional inverses. We used this in Section 4.2. ∎

10.1.8. Example. *Domino tilings.* A 6-by-6 checkerboard can be partitioned into 18 dominoes consisting of two squares each; this is a **tiling** by dominoes. We prove that every such tiling can be cut between two adjacent rows or adjacent columns without cutting any dominoes. In the example below, the tiling can be cut along the middle horizontal line.

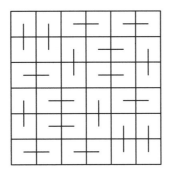

In a tiling, every domino cuts one line between adjacent rows or between adjacent columns. There are 18 dominoes and 10 possible lines to be cut, so the average number of cuts per line is 1.8. Since some line cuts *at most* the average, some line is cut at most once, but perhaps still every line is cut at least once.

It suffices to show that every line is cut by an even number of dominoes, since then a line cut by at most one is not cut at all. Since the number of squares on one side of a line is even, and dominoes pair an even number of those, the number covered by dominoes crossing the line must also be even. ∎

The next theorem implies the result of Graham–Harary [1992] that every spanning tree of Q_k has diameter at least $2k - 1$.

10.1.9. Theorem. (Graham–Entringer–Székely [1994]) If T is a spanning tree of the k-dimensional cube Q_k, then there is an edge of Q_k outside T whose addition to T creates a cycle of length at least $2k$.

Proof: For each vertex v of Q_k, let v' denote the complementary vertex at distance k from v. There is a unique v, v'-path in T; orient its first edge toward v'. Since T has more vertices than edges, the Pigeonhole Principle implies that some edge receives orientations from both endpoints.

Since this edge uv receives an orientation from u and from v, we have v on the u, u'-path and u on the v, v'-path in T. Hence the u, v'-path and v, u'-path in T are disjoint. Each has length at least $k - 1$, since the distance in Q_k between a vertex and its complement is k. Finally, $uv \in E(Q_k)$ implies also $u'v' \in E(Q_k)$, which completes a cycle of length at least $2k$. ∎

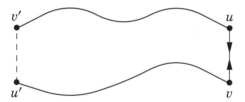

10.1.10.* Theorem. (Two Squares Theorem; Fermat) A natural number is the sum of two squares if and only if all primes congruent to 3 modulo 4 occur with even exponent in its prime factorization.

Proof: *Necessity.* Suppose that $n = x^2 + y^2$, and let p be a prime congruent to 3 modulo 4 that divides n. We will prove that p divides both x and y, which yields $p^2 \mid n$. Dividing the equation by p^2 then expresses n/p^2 as the sum of two squares, and we can use induction on the sum of the exponents in the prime factorization.

By symmetry, it suffices to show that p divides x. If not, then x has a multiplicative inverse modulo p (Example 10.1.7). Since $n \equiv 0 \pmod{p}$, multiplying $x^2 + y^2 = n$ by a number congruent to x^{-2} yields $1 + (x^{-1}y)^2 \equiv 0 \pmod{p}$. However, when $p = 4k + 3$, applying Fermat's Little Theorem (Application 1.3.10) to $1 + a^2 \equiv 0 \pmod{p}$ produces

$$1 \equiv a^{p-1} \equiv a^{4k+2} \equiv 1^k a^2 \equiv -1 \pmod{p}.$$

The contradiction yields $p \mid x$.

Sufficiency (Thue). If $n_1 = x_1^2 + y_1^2$ and $n_2 = x_2^2 + y_2^2$, then $n_1 n_2 = (x_1 x_2 + y_1 y_2)^2 + (x_1 y_2 - x_2 y_1)^2$, so we need only consider $n = 2$, primes congruent to 1 modulo 4, and squares of other primes. Since $2 = 1^2 + 1^2$ and $p^2 = p^2 + 0^2$, it suffices to decompose primes p such that $p \equiv 1 \pmod 4$.

For prime p with $p \equiv 1 \pmod 4$, there is an integer a such that $a^2 + 1 \equiv 0 \pmod p$ (see Exercise 22). Consider $\{ai + j: 0 \le i, j \le \lfloor \sqrt{p} \rfloor\}$. The number of pairs (i, j) indexing this set is $\left(\lfloor \sqrt{p} \rfloor + 1\right)^2$, which exceeds p.

By the Pigeonhole Principle, two of these pairs yield congruent numbers, $ai + j \equiv ai' + j' \pmod p$. Letting $y = j - j'$ and $x = i' - i$, we obtain $y = ax$ with $x, y \not\equiv 0 \pmod p$ and $|x|, |y| \le \lfloor \sqrt{p} \rfloor$. The latter yields $x^2 + y^2 < 2p$. We compute

$$x^2 + y^2 \equiv x^2 + (ax)^2 \equiv x^2(a^2 + 1) \equiv 0 \pmod p.$$

Since $0 < x^2 + y^2 < 2p$, we conclude that $x^2 + y^2 = p$. ∎

10.1.11. Example. *The Chess Player Problem.* A chess player wants to practice for a match for 11 weeks. She wants to play at least one game per day but at most 132 games altogether. No matter how she schedules the games, there is a period of consecutive days (among the 77 days) on which she plays *exactly* 22 games.

We can study totals over consecutive days by considering partial sums. Let a_i be the number of games played on days 1 through i, and set $a_0 = 0$. The number of games played on days $i+1$ through j is $a_j - a_i$. We seek i, j such that $a_i + 22 = a_j$. This suggests considering both $\{a_j: 1 \le j \le 77\}$ and $\{a_i + 22: 0 \le i \le 76\}$. With at least one game each day, each of these sets consists of 77 distinct numbers. Hence repetition among these 154 numbers implies the desired result. Since $a_{77} \le 132$, and $a_{76} + 22 \le 153$, we have 154 numbers in $[153]$, and some number repeats.

Because $a_{76} + 23$ can be as large as 154, this argument does not force a period of consecutive days with exactly k games when $k \ge 23$. ∎

With a total of at most b games on d consecutive days, the partial sum technique in Example 10.1.11 guarantees consecutive days with a total of k games when $k \le 2d - b$. Another trick for $k \in \{23, 24, 25\}$ leads to a general solution. For a circular analogue of the problem, see Clark–Lewis [1989, 1993].

10.1.12.* Theorem. (Hutchinson–Trow [1980]) For $d, b \in \mathbb{N}$ with $b > d$, let k be *forced* if every list of d positive integers summing to at most b has a consecutive segment summing to k. A positive integer k is forced if and only if $r < 2d - b$, where $d = mk + r$ with $0 \le r < k$.

Proof: Call a list of positive integers with sum at most b a *game list*, and call its list of partial sums (increasing and bounded by b) a *sum list*. If a game list has a consecutive segment summing to k (so its sum list has two elements differing by k), then we say that it has *property k*.

A game list alternating $k - 1$ copies of 1 with one copy of $k + 1$ does not have property k. Let $a(k, d)$ be the corresponding sum list of length d. If $d = mk + r$, then each segment of k consecutive terms sums to $2k$, and the final sum $2mk + r$ equals $2d - r$. If $r \ge 2d - b$, then the list has sum at most b and lacks property k.

entry	1	$\cdots$	1	$k+1$	1	$\cdots$	1	$k+1$	1	$\cdots$	1
position	1	$\cdots$	$k-1$	k	$k+1$	$\cdots$	$2k-1$	$2k$	$2k+1$	$\cdots$	d

To prove the converse, we show that $2mk + r$ is the least sum of a list of length d lacking property k. We prove this by induction on d, with $d = 0$ as a trivial basis. If $d > 0$ and d is not a multiple of k, then the sum must exceed the sum of the initial segment of length $d - 1$, which also does not have property k.

Hence we may assume that $d = mk$, so $r = 0$. Let $S = [2mk - 1] - \{k\}$. If a list with length d and sum less than $2mk$ lacks property k, then its partial sums consist of mk numbers among the $2mk - 2$ numbers in S. Partition S into $mk - 1$ pairs by pairing x and $x + k$ whenever $\lfloor x/k \rfloor$ is even, as shown below.

$$
\begin{array}{ccccccc}
1 & \cdots & k-1 & 2k & \cdots & 3k-1 & 4k & \cdots & 2mk-k-1 \\
k+1 & \cdots & 2k-1 & 3k & \cdots & 4k-1 & 5k & \cdots & 2mk-1
\end{array}
$$

By the Pigeonhole Principle, some pair contains two of the partial sums, which implies that the list has property k. ■

We have seen much variety in applications of the Pigeonhole Principle. The classes may have different sizes. Classes are defined so that having many (or few) objects in a class yields the desired outcome. Partial sums may help with numerical problems involving order or sums. An extremal example can suggest classes and objects for applying the Pigeonhole Principle to prove optimality. The principle can be combined with proof by contradiction or with auxiliary results.

MONOTONE SUBLISTS

A classical use of the Pigeonhole Principle guarantees a long monotone sublist in a list of numbers. The result follows from Schensted's Theorem on Young tableaux (Section 4.3), from Dilworth's Theorem on partial orders (Section 12.1), and from other results, but the Pigeonhole Principle gives a short direct proof.

10.1.13. Theorem. (Erdős–Szekeres Theorem; Erdős–Szekeres [1935]) Every list of $n^2 + 1$ distinct real numbers has a monotone sublist with $n+1$ numbers. This is sharp; length n^2 does not suffice.

Proof: (Seidenberg [1959]) For a given list $a_1, \ldots, a_{n^2+1}$, let x_k and y_k be the lengths of a longest increasing and longest decreasing sublist ending at a_k, respectively (see example below). If no monotone sublist has length $n + 1$, then x_k and y_k never exceed n, and there are only n^2 possible pairs (x_k, y_k).

k	1	2	3	4	5	6	7	8	9	10
a_k	7	4	1	8	5	2	9	6	3	10
x_k	1	1	1	2	2	2	3	3	3	4
y_k	1	2	3	1	2	3	1	2	3	1

By the Pigeonhole Principle, some two pairs are the same. However, when $i < j$ and $a_i < a_j$ we can add a_j to a longest increasing sublist ending at a_i. When $i < j$ and $a_i \geq a_j$, we can add a_j to a longest decreasing sublist ending at a_i.

Exercise 28 requests a list proving that the result is sharp. ■

Steele [1995] surveyed seven proofs of Theorem 10.1.13. An extension to paths in digraphs is in Exercise 8.1.43; another generalization is Dilworth's Theorem (Exercise 12.1.15). For a higher-dimensional generalization, we allow repetitions in lists and extend "monotone" to strictly increasing or weakly decreasing: $a_1 < \cdots < a_l$ or $a_1 \geq \cdots \geq a_l$. The proof of Theorem 10.1.13 remains valid.

10.1.14. Definition. A list of d-dimensional vectors is **monotone** if in each coordinate it is monotone.

10.1.15. Theorem. (de Bruijn, unpublished; see Kruskal [1953]) A list of more than l^{2^d} d-dimensional vectors has a monotone sublist with more than l vectors. Length l^{2^d} does not suffice.

Proof: The Erdős–Szekeres Theorem is the case $d = 1$; we use induction on d. For $d > 1$, the given list A has a sublist B of length more than $l^{2^{d-1}}$ in which the last coordinate is monotone. By the induction hypothesis, the $(d-1)$-dimensional list B' formed by the first $d - 1$ coordinates of the vectors in B has a monotone $(d-1)$-dimensional sublist C of length more than l. Since the last coordinate is monotone over all of B, the positions in B corresponding to C form a d-dimensional monotone sublist of A.

Exercise 28 requests a list proving that the result is sharp. ∎

By Theorem 10.1.13, every list of n^2 numbers has a monotone sublist of length at least n. A similar result holds for graphs with an ordering on the edges. Graham–Kleitman [1973] considered the case where the graph is K_n. The idea of the "Pedestrian Algorithm" in the general proof is due to Friedgut, as cited in Winkler [2008].

10.1.16. Theorem. In any n-vertex graph with m edges having distinct numerical labels on the edges, there is an increasing trail of length at least $2m/n$.

Proof: Consider n pedestrians on the graph, starting one at each vertex. Process each edge, in order of the labels, by switching the pedestrians on its endpoints and adding 1 to the number of times each of those two have moved. Always each vertex has one pedestrian. Each pedestrian follows an increasing trail. The total movement is $2m$. By the Pigeonhole Principle, some pedestrian follows a trail of length at least $2m/n$. ∎

PATTERN-AVOIDING PERMUTATIONS (optional)

An increasing sublist of fixed length in a permutation is an example of a "pattern". The Pigeonhole Principle was applied to settle a long-standing open problem about the number of permutations of $[n]$ avoiding a given pattern.

10.1.17. Definition. A permutation σ of $[k]$ is a **pattern** in a permutation π of $[n]$ if the word form of π has a k-element sublist τ such that the element in position i of τ is the σ_ith smallest element in τ. A permutation of $[n]$ is σ-**avoiding** if it does not contain σ as a pattern. Let $S_n(\sigma)$ denote the number of σ-avoiding permutations of $[n]$.

For example, the pattern 4132 occurs in 18256734 at positions $2, 3, 6, 8$; the sublist 8274 has this pattern. An increasing sublist is an occurrence of the identity pattern. The Erdős–Szekeres Theorem states that no permutation of length $k^2 + 1$ avoids both the identity pattern of length $k + 1$ and its reverse.

10.1.18. Example. An inversion is an occurrence of the pattern 21; only identity permutations avoid this. Proposition 3.1.15 gives the generating function for permutations of $[n]$ by the number of occurrences of the pattern 21.

Exercise 2.1.48 shows that $S_n(231)$ is the Catalan number C_n. By combining Schensted's Theorem (Corollary 4.3.20), the Hook-Length Formula (Theorem 4.3.4), and Exercise 4.3.11, it follows that also $S_n(321) = C_n$.

Given a permutation π of $[n]$, form π' by subtracting each entry from $n+1$. A k-element pattern in π' is obtained by subtracting a k-element pattern in π from $k + 1$. Hence $S_n(213) = S_n(231)$. Also the reverse of a σ-avoiding permutation avoids the reverse of σ. We conclude that $S_n(\sigma) = C_n$ for every 3-element σ. ■

The equalities of Example 10.1.18 fail for longer patterns. Bóna [1997] gave the first example of σ and σ' of equal length such that $S_n(\sigma) \neq S_n(\sigma')$, proving $S_n(1234) < S_n(1324)$. (For this and other pattern results, see Bóna [2004].)

One property of 3-element patterns does generalize. Note that $\frac{1}{n+1}\binom{2n}{n} \sim cn^{-3/2}4^n$ for a constant c (Example 2.3.10). Stanley and Wilf conjectured around 1990 that $S_n(\sigma)$ is always bounded by an exponential function of n whose ratio to $n!$ tends to 0. (They conjectured the seemingly stronger statement that $\lim_{n\to\infty}(S_n(\sigma))^{1/n}$ exists, but by Arratia [1999] the two are equivalent.)

Among various partial results, Bóna [1999] proved the conjecture for permutations where each run consists of consecutive elements, such as 456312. Finally, Marcus–Tardos [2004] used the Pigeonhole Principle to prove another conjecture that implies the Stanley–Wilf Conjecture.

10.1.19. Definition. Let A and P be 0, 1-matrices of orders n and k with $n \geq k$. The matrix A **covers** P if A has a submatrix B such that $B_{i,j} = 1$ whenever $P_{i,j} = 1$ (rows and columns from A are indexed in the same order in B). If A does not cover P, then A **avoids** P.

10.1.20. Conjecture. (**Stanley–Wilf Conjecture**) For every pattern σ, the number $S_n(\sigma)$ of σ-avoiding permutations of $[n]$ is at most exponential in n.

(**Füredi–Hajnal Conjecture**; Füredi–Hajnal [1992]) For every permutation matrix P, there is a constant c_P such that the number of 1s in any P-avoiding 0, 1-matrix of order n is at most $c_P n$.

10.1.21. Lemma. (Klazar [2000]) The Füredi–Hajnal Conjecture implies the Stanley–Wilf Conjecture.

Proof: A permutation π avoids a pattern σ if and only if the matrix of π avoids the matrix of σ. Given a permutation matrix P, let T_n be the set of P-avoiding 0, 1-matrices of order n. It suffices to show that $|T_n| \leq c^n$ for some constant c, since the permutation matrices avoiding P lie in T_n.

Consider $A \in T_{2n}$, so A avoids P. Partition A into 2-by-2 blocks. Form an n-by-n matrix A' by replacing each all-0 block with a 0 and each nonzero block with a 1. If A' covers P, then since P only has one 1 in each row and column, we can take any 1 from the corresponding block in A to play the role of the 1 in A' that covers a 1 in P; hence A covers P. By this contradiction, A' avoids P.

There are 15 2-by-2 matrices that can collapse to a 1 in A' and only one that can collapse to a 0. Therefore, each P-avoiding matrix A' arises by collapsing at most $15^{f(n)}$ P-avoiding matrices of order $2n$, where $f(n)$ is the maximum number of 1s in a P-avoiding matrix of order n.

We have proved $|T_{2n}| \leq |T_n|15^{f(n)}$. The Füredi–Hajnal Conjecture is $f(n) \leq c_P n$. For $n = 2^k$, iterating the recurrence yields $|T_n| \leq |T_1|15^{c_P \Sigma_{k=1}^n 2^{k-1}} \leq (15^{c_P})^n$, since $|T_1| = 2$. When n is not a power of 2, let m be the next larger power of 2; note that $m < 2n$. Hence $|T_n| \leq |T_m| \leq (15^{c_P})^{2n}$. With $|T_n| \leq (15^{c_P})^{2n}$ for all n, we obtain the Stanley–Wilf Conjecture with the constant 15^{2c_P}.   ∎

10.1.22. Theorem. (Marcus–Tardos [2004]) Let $f(n)$ be the maximum number of 1s in a P-avoiding $0, 1$-matrix of order n. If P is a permutation matrix of order k, then $f(n) \leq 4k^4\binom{k^2}{k}n$, so the Füredi–Hajnal Conjecture is true.

Proof: Suppose first that n is a multiple of k^2. Let A be a P-avoiding $0, 1$-matrix of order n. Again we use the collapsing technique to obtain a recursive bound. Partition A into blocks of order k^2 by splitting the matrix into n/k^2 sets of consecutive rows and n/k^2 sets of consecutive columns.

Let A' be the matrix of order n/k^2 obtained by collapsing blocks of A into single entries with a block becoming a 0 if and only if it is an all-0 block. The matrix A' avoids P, since otherwise taking any 1 from each block corresponding to a position in A' that represents a 1 in P yields a submatrix of A that covers P.

Let a block be *wide [tall]* if it has 1s in at least k columns [rows]. We claim that each column of blocks has at most $k\binom{k^2}{k}$ wide blocks (and each row has at most $k\binom{k^2}{k}$ tall blocks). Each wide block has k nonzero columns among its k^2 columns, and there are $\binom{k^2}{k}$ possible such sets. With $k\binom{k^2}{k}$ wide blocks in a column of blocks, the some k wide blocks are nonzero in the same set S of k columns. We use the ith highest among these k blocks to cover the 1 in the ith row of P, using the appropriate column in S. That is, in this column of S the block has a 1 in some row, and we use that row from this block. This contradicts that A avoids P.

We thus have at most $\frac{n}{k^2}k\binom{k^2}{k}$ wide blocks and the same bound on the number of tall blocks. Since A' avoids P, at most $f(n/k^2)$ nonzero blocks are neither wide nor tall. Such a block has at most $(k-1)^2$ nonzero entries, and we use the trivial bound of k^4 on the nonzero entries in wide and tall blocks. We now have

$$f(n) \leq 2k^4\frac{n}{k^2}k\binom{k^2}{k} + (k-1)^2 f\left(\frac{n}{k^2}\right).$$

When n is not a multiple of k^2, consider a submatrix whose order is the next smaller multiple of k^2; call it n'. Fewer than $2k^2 n$ positions are lost; at worst they could all be 1. We now prove the desired bound by induction. We compute

$$f(n) < f(n') + 2k^2 n \leq (k-1)^2 f\left(\frac{n'}{k^2}\right) + 2k^3\binom{k^2}{k}n' + 2k^2 n$$

$$\leq (k-1)^2\left[2k^4\binom{k^2}{k}\frac{n'}{k^2}\right] + 2k^3\binom{k^2}{k}n' + 2k^2 n$$

$$\leq 2k^2\left((k-1)^2 + k + 1\right)\binom{k^2}{k}n \leq 2k^4\binom{k^2}{k}n,$$

where the last inequality holds for $k \geq 2$.   ∎

10.1.23. Corollary. If σ is a permutation of $[k]$, then $S_n(\sigma) \le c^n$, where $c = 15^{2k^4\binom{k^2}{k}}$. That is, the Stanley–Wilf Conjecture is true. ∎

The bound on c in Corollary 10.1.23 is quite large; $c = 4$ is adequate when $k = 3$. The best constant is $(k-1)^2$ when σ is the identity permutation (Regev [1981]). Arratia [1999] conjectured that $c = (k-1)^2$ always suffices, but this was disproved at $k = 4$ in Albert et al. [2006] (for $\sigma = 4231$).

Wilf later asked also how many different patterns can occur in a permutation of $[n]$. A trivial upper bound is 2^n. Wilf provided a sequence of examples where the number of patterns grows as fast as the Fibonacci numbers, and Coleman [2004] improved this to $2^{(1-o(1))n}$ (Exercise 37).

LARGE GIRTH AND CHROMATIC NUMBER (optional)

Erdős proved by probabilistic arguments that graphs exist with large girth and large chromatic number. Explicit constructions came later. The construction we present intimately involves the Pigeonhole Principle, applying it iteratively in a way that generates very large structures to ensure the existence of a special substructure. (The proof of Ramsey's Theorem in Section 10.2 will also do this.)

To motivate the hypergraph-based construction, we first construct graphs with girth 6 and large chromatic number. This may be the earliest construction of triangle-free graphs with large chromatic number. It appeared in an undergraduate magazine at Cambridge University. "Blanche Descartes" was a pseudonym used by R.L. Brooks, C.A.B. Smith, A.H. Stone, and W.T. Tutte when undergraduates together at Cambridge; all became mathematicians.

10.1.24. Example. *Graphs with girth 6 and high chromatic number* (Blanche Descartes [1947, 1954], Kelly–Kelly [1954]). Like Mycielski's construction, this is inductive; we start with C_6 for $k = 2$. The construction increases chromatic number but preserves girth 6. Let G be a $(k-1)$-chromatic graph with girth 6.

To construct G', let $r = |V(G)|$ and $N = (r-1)(k-1) + 1$. Begin with a set S of N isolated vertices. Add $\binom{N}{r}$ copies of G, one for each r-subset of S. For $1 \le i \le \binom{N}{r}$, add a matching from the ith copy of G to the ith r-set in S. The resulting graph is G'. A cycle visiting more than one copy of G has at least one interior edge and two departing edges from each copy it visits. Thus cycles in G' not in one copy of G have at least six edges.

A proper $(k-1)$-coloring of G can be extended to a k-coloring of G' by using that $(k-1)$-coloring on each copy of G and using a new color on S. In any $(k-1)$-coloring of G', by the Pigeonhole Principle some r vertices of S have the same color. This color is forbidden from the copy of G corresponding to this r-set. Since this copy of G cannot be properly $(k-2)$-colored, we conclude that $\chi(G') = k$. ∎

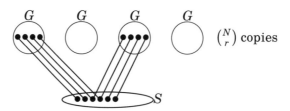

In an early triumph for probabilistic arguments, Erdős [1959] showed that for appropriate sparseness, almost every graph has chromatic number and girth above fixed constants (Theorem 14.2.3). By extending the question to hypergraphs, Lovász [1968b] constructed such graphs explicitly. We present a simplification from Nešetřil–Rödl [1979]. Other constructions appear in Lubotzky–Phillips–Sarnak [1988] and Kriz [1989]; see also Nešetřil [2013] and Alon–Kostochka–Reiniger–West–Zhu [2016] (Exercise 40).

10.1.25. Definition. A **hypergraph** H consists of a vertex set $V(H)$ and an edge set $E(H)$ such that each edge is a subset of $V(H)$. It is r-**uniform** if all edges have size r. In a **complete r-uniform hypergraph**, all r-sets are edges.

A **proper k-coloring** of a hypergraph H labels $V(H)$ from a set of k colors so that no edge with size at least 2 is monochromatic; the least k needed is the **chromatic number** $\chi(H)$. A hypergraph H is k-**chromatic** when $\chi(H) = k$.

Proper coloring of 2-uniform hypergraphs is just proper coloring of graphs. Since the input structures are more general, hypergraph k-coloring is at least as hard as graph k-coloring; it is NP-complete to decide whether a general hypergraph has a proper 2-coloring.

10.1.26. Example. Let F be the 3-uniform 7-vertex hypergraph with edge set $\{124, 235, 346, 457, 561, 672, 713\}$. Three consecutive vertices contain no edge, so $\chi(F) \leq 3$. To show $\chi(F) = 3$, consider a 2-coloring; some four vertices have the same color. Any four vertices not having three consecutive include three vertices of the form $i, i+1, i+3$, forming an edge. A 4-set containing $i, i+1, i+2$ avoids completing an edge only by adding $i + 5$, but then the three vertices with the other color form an edge. Hence every 2-coloring has a monochromatic edge. (This hypergraph is known as the **Fano plane**; see Chapter 13). ∎

10.1.27. Example. *Complete r-uniform N-vertex hypergraph.* An **independent set** in a hypergraph, as in a graph, is a set containing no edge of size at least 2. If $\alpha(H)$ denotes the maximum size of an independent set in H, then $\chi(H) \geq |V(H)| / \alpha(H)$, as for graphs. In the complete r-uniform hypergraph, every r-set is an edge, so $\alpha(H) = \min\{r - 1, N\}$ and $\chi(H) = \lceil N/(r-1) \rceil$. ∎

10.1.28. Definition. For $l \geq 2$, a **cycle** of length l in a hypergraph is a cyclic list $[x_1, e_1, \ldots, x_l, e_l]$ of l distinct vertices and l distinct edges such that x_i and x_{i+1} lie in e_i (indices modulo l). The **girth** of H is the length of a shortest cycle. A hypergraph is N-**partite** if its vertices can be partitioned into N disjoint classes such that each edge has at most one vertex in each class.

Cycles in 2-uniform hypergraphs can be viewed as cycles in graphs. However, when naming a cycle in a graph we normally specify only the edges, because in a 2-uniform hypergraph two vertices can determine only one edge. Hence graphs have no cycles of length 2, though hypergraphs may. Cycles as defined above are often called **Berge cycles**. This is the most general notion of cycle in hypergraphs; there are more restrictive definitions. In multihypergraphs, there may be distinct edges with the same set of vertices; thus three edges that have the same set of vertices in a 3-uniform multihypergraph form a Berge cycle.

We will build an r-uniform k-chromatic hypergraph with girth at least l, using induction on l. Setting $r = 2$ yields the desired graphs. The inductive construction involving N-partite hypergraphs uses hypergraphs with large edges.

10.1.29. Example. *A bigger hypergraph.* From an N-partite hypergraph H and another hypergraph Y, we build a large N-partite hypergraph $f_j(H, Y)$, where $j \in [N]$. We require Y to be s-uniform, where s is the size of the jth class in H.

Let m be the number of edges in Y. We use m copies of H. For $i \in [m]$, let H^i be a copy of H using the vertices of the ith edge in Y as its jth class. The copies of H are disjoint except for the jth class. Below we sketch the construction for $N = 3$ and $m = 5$, with $A_1, \ldots, A_5$ denoting the edges of Y. Each solid ellipse represents a copy of H; the three regions inside represent the three classes.

An edge of Y is not an edge in $f_j(H, Y)$; it is just the jth class in a copy of H contained in $f_j(H, Y)$. The vertex set $V(Y)$ becomes the jth class in $f_j(H, Y)$, and for $k \neq j$ the kth class in $f_j(H, Y)$ is the union of the (disjoint) kth classes in $H^1, \ldots, H^m$. Thus the kth class in $f_j(H, Y)$ is large. If we use the construction again with k in place of j and $f_j(H, Y)$ in place of H, then the new auxiliary Y' will need large edges. ∎

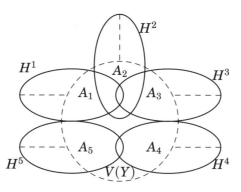

10.1.30. Lemma. (Nešetřil–Rödl [1979]) If H is an N-partite hypergraph with girth at least l and jth class of size s, and Y is an s-uniform hypergraph with girth at least $l/2$, then the hypergraph $f_j(H, Y)$ constructed in Example 10.1.29 is N-partite with girth at least l.

Proof: Let the vertex classes of H^i be $X_1^i, \ldots, X_{j-1}^i, A_i, X_{j+1}^i, \ldots, X_N^i$. Since H is N-partite, no edge of $f_j(H, Y)$ has two vertices in $V(Y)$, so $f_j(H, Y)$ is N-partite with classes $\bigcup_i X_1^i, \ldots, \bigcup_i X_{j-1}^i, V(Y)$, and $\bigcup_i X_{j+1}^i, \ldots, \bigcup_i X_N^i$. Each edge of $f_j(H, Y)$ has the same size as the corresponding edge in H.

Let C be a cycle in $f_j(H, Y)$. If C is contained in one copy of H, then C has length at least l (with equality for some such C). Suppose that C visits more than one copy of H. Since no edge of $f_j(H, Y)$ contains two elements of $V(Y)$, the cycle C cannot have consecutive vertices in $V(Y)$. Also, C can pass from $V(H_i) - V(Y)$ to $V(H_j) - V(Y)$ only through a vertex of $A_i \cap A_j$. Hence the vertices of $V(Y)$ used by C form a cycle in Y. Since C contains at least one vertex not in $V(Y)$ between any two successive vertices in $V(Y)$, the length of each such cycle is at least l. ∎

10.1.31. Theorem. (Lovász [1968b]) For integers $r, k, l \geq 2$, there is an r-uniform k-chromatic hypergraph with girth at least l.

Proof: (Nešetřil–Rödl [1979]) We construct an r-uniform hypergraph $H(r, k, l)$ with girth at least l and chromatic number at least k. Reducing chromatic number to k by deleting edges does not decrease the girth.

We use induction on l. For $r, k \geq 2$, let $N = (r-1)(k-1) + 1$. Let $H(r, k, 2)$ be the complete r-uniform hypergraph with N vertices. When $k = 2$, there is one edge and no cycle, which is treated as infinite girth, and the chromatic number is 2. For $l > 2$, we assume that $H(t, k, u)$ has already been constructed for all $\{t, k, u\}$ with $t \geq 2$ and $2 \leq u < l$.

We construct hypergraphs $M_1, \ldots, M_{N+1}$ and let $H(r, k, l) = M_{N+1}$. Let M_1 be the N-partite r-uniform hypergraph having vertex classes of size $\binom{N-1}{r-1}$ and $\binom{N}{r}$ pairwise disjoint edges. There is one edge for each choice of r classes, consisting of one vertex from each chosen class. The edges are disjoint, so there is no cycle: again infinite girth.

Each successive hypergraph M_i will be r-uniform, N-partite, and have girth at least l, constructed by applying Lemma 10.1.30 with $H = M_{i-1}$. We will argue that every $(k-1)$-coloring of M_{N+1} contains, among its astronomically many copies of M_1, some copy of M_1 in which every class has a single color on its vertices. Since M_1 has $(r-1)(k-1) + 1$ classes, some r of those classes have the same color. By construction, M_1 has some edge consisting of a vertex from each of those r classes. This edge is monochromatic, and hence M_{N+1} is not $(k-1)$-colorable.

Having constructed M_i, let s_i be the size of the ith class in M_i. Let $Y_i = H(s_i, k, l - 1)$. Since Y_i is s_i-uniform, we may define $M_{i+1} = f_i(M_i, Y_i)$ as in Lemma 10.1.30. Note that $s_{i+1} = s_i |E(Y_i)|$, so the construction needs smaller-girth hypergraphs with huge edges. By Lemma 10.1.30, M_{i+1} is r-uniform and N-partite and has girth l.

Let f be a $(k-1)$-coloring of M_{N+1}. Note that M_{N+1} has a copy of M_N for each edge A of Y_N, in which A serves as the Nth class of M_N. Since Y_N is not $(k-1)$-colorable and f uses only $k-1$ colors on $V(Y_N)$, the vertices of some edge A of Y_N (it is not an edge in M_{N+1}) receive the same color under f. The copy of M_N corresponding to A is thus a copy of M_N where f uses one color on the Nth class.

Within this copy L_N of M_N, we apply the argument inductively. In L_N, the $(N-1)$th class is a copy of $V(Y_{N-1})$, and L_N contains a copy of M_{N-1} for each edge of Y_{N-1}. The Nth class of L_N contains the Nth class of each copy of M_{N-1} used in forming it. By the choice of L_N, the Nth class of each copy of M_{N-1} in L_N is monochromatic under f. Since Y_{N-1} is not $(k-1)$-colorable, L_N contains a copy of M_{N-1} in which the $(N-1)$th class is the vertex set of an edge of Y_{N-1} that is monochromatic under f.

At stage i, working back, we have a copy L_i of M_i in which classes i through N are monochromatic under f. Hence they are also monochromatic in each copy of M_{i-1} contained in L_i. Since Y_{i-1} is not $(k-1)$-colorable, in one of these copies the $(i-1)$th part also is monochromatic under f. Iterating through $i = 2$, we obtain a copy of M_1 in which every class is monochromatic under f, yielding a monochromatic edge as discussed earlier. Hence every $(k-1)$-coloring of M_{N+1} contains a monochromatic edge, and we set $H(r, k, l) = M_{N+1}$. ∎

EDGE-COLORING OF HYPERGRAPHS (optional)

Like vertex coloring, proper edge-coloring also generalizes naturally to hypergraphs; the edges containing a given vertex must have distinct colors. We seek an analogue of Vizing's Theorem.

10.1.32. Definition. A hypergraph is **linear** if no two edges share more than one vertex. A **loop** is an edge of size 1. A **simple hypergraph** is a linear hypergraph without loops.

Bounds for edge-coloring of simple hypergraphs have been hard to find. We consider a long-standing conjecture in this area that Erdős described as one of his favorite problems. It originated at a party in 1972 in Boulder, Colorado. The proposers agreed to meet the next day to write out the solution, but they failed. Erdős [1976] offered 50 British pounds for the solution, later raised to $500.

10.1.33. Conjecture. (Erdős–Faber–Lovász Conjecture) Any union of n edge-disjoint copies of K_n has chromatic number n.

10.1.34. Conjecture. Every simple hypergraph with n vertices is properly n-edge-colorable.

The first conjecture sounds a bit artificial, but they are equivalent, and both are known as the Erdős–Faber–Lovász Conjecture. Conjecture 10.1.33 is a way of stating Conjecture 10.1.34 in the language of ordinary graph coloring. The transformation uses hypergraph duality.

10.1.35. Definition. The **dual** of a hypergraph H is the hypergraph H^* whose incidence matrix is the transpose of the incidence matrix of H.

In forming H^* from H, vertices and edges exchange roles, but the incidences remain the same (multiedges arise in H^* when vertices in H have the same incident edges). Proper edge-coloring of H is more restrictive than proper vertex coloring of H^*; it is equivalent to coloring $V(H^*)$ so that in each edge of H^* the vertices all have distinct colors.

10.1.36. Theorem. Conjectures 10.1.33–10.1.34 are equivalent.

Proof: Let G be a union of n pairwise edge-disjoint copies of K_n with vertex sets $Q_1, \ldots, Q_n$. Each Q_i has an element in no other Q_j, since each Q_j has at most one vertex of Q_i. Let Q_i' be the proper subset of Q_i obtained by deleting the elements in no other Q_j. It suffices to show that the union G' of complete graphs on the sets $Q_1', \ldots, Q_n'$ is n-colorable, since the coloring extends to the missing vertices.

View $Q_1', \ldots, Q_n'$ as edges of a hypergraph H. Vertices in an edge are adjacent in the original graph, so we want them to have distinct colors. Hence the vertex coloring problem for G becomes the edge-coloring problem for H^*. The n original cliques in G become the n vertices of H^*.

The original edge-disjointness condition becomes the condition that two edges of H share at most one vertex. Thus H is linear. A hypergraph is linear if and only if its dual is linear, since both mean forbidding a 2-by-2 all-1 submatrix in the incidence matrix. By discarding elements of V that lie in only one clique in G, we have ensured that the dual hypergraph H^* has no edge of size 1. Thus H^* is a simple hypergraph, and G is n-colorable if and only if H^* is n-edge-colorable.

The transformation from G to H^* is reversible. A simple hypergraph H^* with n vertices has each vertex in at most $n - 1$ edges, since no other vertex is in two such edges. Hence the dual hypergraph H is a simple hypergraph with n edges of size at most $n - 1$. Augmenting these edges gives $Q_1, \ldots, Q_n$ and G. ∎

A simple hypergraph whose edges all have size 2 is a graph. An n-vertex graph has maximum degree at most $n - 1$, and thus Vizing's Theorem yields n-edge-colorability in that case. Hindman [1981] proved Conjecture 10.1.34 in the case where the union of the edges with at least three vertices has size at most 10.

Kahn [1997] proved that Conjecture 10.1.34 holds asymptotically. That is, there is a proper $n(1 + o(1))$-edge-coloring (here $o(1)$ represents some function of n that tends to 0 as n tends to ∞).

An upper bound of $2n - 3$ is easy. Color the edges in decreasing order of size, greedily using the least available color. An edge of size k intersects at most $\frac{n-k}{k-1}$ edges of size at least k at each vertex, so the number of previously-colored edges intersecting the current edge is at most $\frac{(n-k)k}{k-1}$. For simple hypergraphs, this is largest at $k = 2$ and always at most $2n - 4$. Chang and Lawler extended this argument to reduce the upper bound.

10.1.37. Theorem. (Chang–Lawler [1988]) The edge-chromatic number of a simple hypergraph on n vertices is at most $\lceil 3n/2 \rceil - 2$.

Proof: Color the edges in non-increasing size order. For an edge of size k, at most $\frac{(n-k)k}{k-1}$ edges intersecting it have already been colored. For $k \geq 3$ this is bounded by $(3n - 9)/2$, which is less than $\lceil 3n/2 \rceil - 3$, so a color is available for the edge. For an edge uv of size 2, if some color avoids both vertices (c avoids v if it has not already been used on an edge containing v), then we can color the edge.

Otherwise, we recolor an edge to free a color for uv. Suppose that a avoiding v and b avoiding u appear on edges uw and vw of size 2, for some vertex w (as shown below). Since each vertex is in at most $n - 1$ edges and uv is uncolored, at most $n - 2$, $n - 2$, and $n - 1$ colors appear at u, v, or w, respectively. Hence at least $n/2$, $n/2$, and $n/2 - 1$ colors avoid u, v, and w, respectively. With only $\lceil 3n/2 \rceil - 2$ colors, some color avoids two of these vertices. No color avoids both u and v, so we may assume that c avoids w and u. We change the color of uw to c and use a on uv.

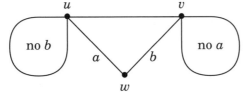

Hence it suffices to find such a vertex w when no color is available for uv. Let A be the set of colors used on edges of size 2 at u and not on larger edges containing v. Similarly, let B be the set of colors used on edges of size 2 at v and not on larger edges at u. Colors in $A - B$ do not appear at v, so edges at u with these colors are candidates for the edge uw. Similarly, colors in $B - A$ do not appear at u. Since colors in $A - B$ specify distinct edges containing u and colors in $B - A$ specify distinct edges containing v, it suffices to prove $|A - B| + |B - A| \geq n - 1$.

We count the $\lceil 3n/2 \rceil - 2$ colors present at u and/or v. A color not in $A \cup B$ appears on an edge of size at least 3 containing u or v. Since each color in A eliminates a potential neighbor of u from such edges and we also exclude $\{u, v\}$, there are at most $(n - |A| - 2)/2$ such edges containing u. Similarly, there are at most $(n - |B| - 2)/2$ such edges containing v. Having counted all the colors, we have

$$|A \cup B| + (n - |A| - 2)/2 + (n - |B| - 2)/2 \geq \lceil 3n/2 \rceil - 2 \geq (3n - 5)/2.$$

This simplifies to $2|A \cup B| - (|A| + |B|) \geq n - 1$. Since the left side equals $|A - B| + |B - A|$, the proof is complete. ∎

EXERCISES 10.1

10.1.1. (–) Prove that every graph with at least three vertices, all of even degree, has at least three vertices with the same degree.

10.1.2. (–) Prove that every set of $2^n + 1$ integer lattice points in $\mathbb{R}^n$ contains a pair of points whose centroid (mean vector) is also an integer lattice point.

10.1.3. (–) Prove that any set of five points in a square of area 1 contains two points separated by distance at most $\sqrt{2}/2$ and that no closer distance can be forced.

10.1.4. (–) Determine the least k such that every k-element subset of $[2n]$ contains two relatively prime elements.

10.1.5. (–) Determine the least k such that every k-element subset of $[3n]$ contains three consecutive numbers.

10.1.6. Given $n, k \in \mathbb{N}$, determine the maximum size of a subset of $[n]$ that has no two numbers differing by k.

10.1.7. (◊) Determine the least n such that every set of n integers contains two elements whose sum or difference is a multiple of k.

10.1.8. Prove that every set of $2k$ integers has elements of opposite parity differing by at least $2k - 1$ or elements of equal parity differing by at least $4k - 2$.

10.1.9. Use the Pigeonhole Principle to prove that for every rational number, the decimal expansion is eventually repeating.

10.1.10. *Generalization of Example 10.1.6.* Show that every set of $(2^m - 1)n + 1$ distinct integers chosen from $\{1, \ldots, 2^m n\}$ contains $m + 1$ distinct integers $a_0, \ldots, a_m$ such that a_{i-1} divides a_i for $1 \leq i \leq m$. (Khare [1989])

10.1.11. Consider the grid $[n] \times [n]$ of n^2 points in the plane. Each point is colored black or white. How large must n be so that every such coloring yields a rectangle with horizontal and vertical sides whose corners have the same color?

10.1.12. An exam consists of k true/false questions. Determine the least n such that whenever n students take the exam, some two students agree in their answers to at least one-third of the questions.

10.1.13. Let T be a complete ternary tree with k levels whose 3^k leaves are 2-colored. Prove that in some complete binary subtree (2^k leaves), all leaves have the same color. (Milans)

10.1.14. In a chess tournament, each person plays every other, scoring 1 for a win, 0 for a draw, and -1 for a loss. Prove that if at least 3/4 of the games are draws, then some two players score the same total. Show that this is asymptotically sharp. (Bóna [2002, p. 50])

10.1.15. In the chess tournament of Exercise 10.1.14, suppose that the players come from two countries (each person plays all others). Prove that some player scores at least as many points against players from the same country as against players from the other country.

10.1.16. ($\diamond$) Determine the least n such that every set of n integers has a nonempty subset whose sum is a multiple of k. (Comment: A more general and stronger statement valid for any group (here $\mathbb{Z}_k$) appears in Moser [1948]. See also Vince [1965] for the special case.)

10.1.17. Let x be a real number. Show that some number in $\{x, 2x, \ldots, (n-1)x\}$ differs by at most $1/n$ from an integer.

10.1.18. ($\diamond$) Let $f(n)$ be the least k such that every set of k elements in $[n]$ has two disjoint subsets with the same sum. Prove $1 + \lfloor \log_2 n \rfloor < f(n) \le \lceil 1 + \log_2 n + \log_2 \log_2 n \rceil$ for $n \ge 3$.

10.1.19. For $n < 2k$, prove that any multiset of k positive integers with sum n has a subset with sum j for all $j \in [n]$, and show that this fails when $n = 2k$. (Brown [1961], Ganter–Teirlinck [1977])

10.1.20. Let $S_1, \ldots, S_m$ be a partition of $[n]$ with at least two sets having size more than m. Prove that there are distinct indices $k, l \in [m]$ such that some difference between two elements of S_k is also a difference between two elements of S_l. (LeSaulnier [2006])

10.1.21. Use Example 10.1.7 to prove the **Chinese Remainder Theorem**: If $n_1, \ldots, n_r$ are pairwise relatively prime natural numbers, and $a_1, \ldots, a_r$ are integers, then one congruence class modulo $\prod n_i$ is congruent to a_i modulo n_i for each i. (Hint: Consider $x = \sum_{i=1}^{r} a_j N_j y_j$, where $N_j = (\prod n_i)/n_j$ and y_j is chosen using multiplicative inverses.)

10.1.22. Given a prime p, let $\mathbb{Z}_p$ be the field of congruence classes modulo p.
(a) The **order** of a nonzero element $a \in \mathbb{Z}_p$ is the least k such that $a^k \equiv 1 \pmod{p}$. Using multiplication by a on $\mathbb{Z}_p$, conclude that $a^{p-1} \equiv 1 \pmod{p}$. This statement is Fermat's Little Theorem, also proved in Application 1.3.10.
(b) Prove that if $a^2 \equiv 1 \pmod{p}$, then $a \equiv \pm 1 \pmod{p}$. Use this and multiplicative inverses to prove **Wilson's Theorem**: $(p-1)! \equiv -1 \pmod{p}$.
(c) Let $r = (p-1)/2$. Use Fermat's Theorem to show that $x^2 \equiv -1 \pmod{p}$ has no solution when $p \equiv 3 \pmod{4}$. Use Wilson's Theorem to show that $x^2 \equiv -1 \pmod{p}$ has $\pm(r!)$ as solutions when $p \equiv 1 \pmod{4}$. (Hint: Modify $(p-1)!$ to express it in terms of r.)

10.1.23. ($\diamond$) For $m \ge 2n$, let S be a set of m points on a circle, no two on a diameter. Say that $x \in S$ is "free" if fewer than n points of $S - x$ lie in the semicircle clockwise from x. Prove that S has at most n free points. (Knuth [1991])

10.1.24. ($\diamond$) Use the technique of Theorem 10.1.9 (not induction!) to prove the following two statements about trees.
(a) The center of a tree T consists of one vertex or two adjacent vertices (see Definition 5.4.10). (Graham–Entringer–Székely [1994])
(b) If $T_1, \ldots, T_k$ are subtrees of T that pairwise intersect, then some vertex of T belongs to all of $T_1, \ldots, T_k$. (Lehel)

10.1.25. ($\diamond$) A club has 90 rooms and 100 members. Keys must be distributed so that any 90 members can fill the 90 rooms without sharing or trading keys. Prove that 990 keys are enough and 989 keys are not enough. (Members may have more than one key, and keys to a room may be given to more than one member.)

10.1.26. Let $x_1, x_2, \cdots$ be a sequence drawn from a finite alphabet. Prove that there is a positive integer k such that $x_{n+k} = x_n$ for infinitely many n.

10.1.27. For $n \in \mathbb{N}$, prove that every list of more than n^3 numbers has a sublist with more than n entries that is strictly increasing or strictly decreasing or constant. Construct an example to show that the result is sharp.

10.1.28. Construct a list of length n^2 having no monotone sublist of length more than n. Extend this by induction to construct a list of l^{2^d} vectors in $\mathbb{R}^d$ having no monotone sublist of length more than l. (Alon–Füredi–Katchalski [1985])

10.1.29. ($\diamond$) Let C_n^k be the $2k$-regular graph obtained from C_n by making any two vertices adjacent if the distance between them in C_n is at most k. For $k \geq 2$, prove that $\chi(C_n^k) > k+2$ if $n = k(k+1)-1$ and $\chi(C_n^k) \leq k+2$ if $n \geq k(k+1)$.

10.1.30. ($\diamond$) A function $f \colon [n] \to [n]$ is **contractive** if $f(i) \leq i$ for all i. A *monotone k-list* for f is a strictly increasing list $a_1, \ldots, a_k$ from $[n]$ such that $f(a_1) \leq \cdots \leq f(a_k)$. Prove that 2^{k-1} is the least n such that for every contractive mapping on $[n]$ there is a monotone k-list. (Hint: For how many a in $[n]$ can the longest monotone list ending with a have j elements? Comment: This result is generalized in West–Trotter–Peck–Shor [1984].)

10.1.31. (+) Prove that Theorem 10.1.16 is sharp for $n \geq 6$ by giving an ordering of $E(K_n)$ such that the maximum length of an increasing trail is $n-1$. (Comment: The construction is easy when n is even.) (Graham–Kleitman [1973])

10.1.32. Prove that at most $(k-1)^{2n}$ permutations of $[n]$ avoid the identity permutation of length k. (Bóna [2004]) (Comment: Although this bound seems loose, in fact there are asymptotically $c \frac{(k-1)^{2n}}{n^{(k^2-2k)/2}}$ such permutations, where c is a known constant.)

10.1.33. Determine the maximum number of 1s in a $0,1$-matrix of order n such that no 1 has a 1 somewhere to its right and somewhere below (see Füredi–Hajnal [1992]).

10.1.34. Prove that the number of permutations of $[n]$ avoiding all three of the patterns 123, 132, and 213 is the adjusted Fibonacci number $\hat{F}_n$. (Simion–Schmidt [1985])

10.1.35. ($\diamond$) Establish a bijection from the 123-avoiding permutations to the 132-avoiding permutations. (Hint: Leave all left-to-right minima fixed.) (Simion–Schmidt [1985])

10.1.36. Generalize Exercise 10.1.35 by proving $S_n(123 \cdots k'k) = S_n(123 \cdots kk')$, where $k' = k-1$ (see Definition 10.1.17). This compares the identity pattern with that switching the last two elements. (Comment: Backelin–West–Xin [2007] proved $S_n(\sigma\tau) = S_n(\sigma\tau')$ whenever σ is a permutation of $[j]$, $\tau = (j+1, \ldots, k)$ in order, and τ' is the reverse of τ.)

10.1.37. *Permutations with many patterns.*
    (a) Prove that $1, n, 2, n-1, \ldots, \lceil n+1 \rceil 2$ contains at least $\hat{F}_n$ patterns. (Wilf)
    (b) Let τ_k be the permutation of k^2 whose runs have size k and step by k. For example, $\tau_3 = 369258147$. Prove that τ_k contains more than $2^{(k-1)^2}$ patterns. (Coleman [2004])

10.1.38. (+) A deck of n cards is a permutation of $[n]$; view it as a word. A **shuffle** splits a word into an initial block and a final block (one may be empty) and merges them, maintaining order within each block. For example, one shuffle can produce 3142 from 1234. Consider permutations of $[n]$ produced from the identity permutation by one shuffle.
    a) Prove that these are the permutations avoiding 321, 2143, and 2413. Count them.
    b) Show that among these, the permutations that can be returned to the identity by one more shuffle are those that split $[n]$ into four segments A, B, C, D with $B, C \neq \emptyset$ and reorder them as A, C, B, D. Count them. Characterize them as the permutations avoiding 321, 2143, 2413, and 3142. (DeSario [2002])

10.1.39. ($\diamond$) Let G be a $(k-1)$-chromatic n-vertex graph with girth l. Let H be a k-chromatic n-uniform hypergraph with girth at least $l/3$, edges $e_1, \ldots, e_m$, and vertices U. Form G' from U and disjoint copies $G_1, \ldots, G_m$ of G by adding a matching of size n from $V(G_i)$ to e_i for each i. Prove that G' has girth l and chromatic number k. (Nešetřil–Rödl)

10.1.40. (+) *Graphs with large girth and chromatic number.* A (d, r, g)-**graph** is a graph with girth at least g formed from a complete d-ary tree by giving each leaf r neighbors (besides its parent) on its path to the root. (A **complete d-ary tree** of height h is a rooted tree where all non-leaves have d children and all leaves have distance h from the root.)

(a) Assume that (d, r, g)-graphs exist. Use a $(k, k + 1, 2g)$-graph to construct a graph with girth at least g that is not k-colorable. (Hint: Use the *canonical edge-coloring* of the underlying tree, which assigns color i to the edge from a non-leaf vertex to its ith child.)

(b) Let $m(d, r, g)$ be the least height of the underlying tree in a (d, r, g)-graph, if one exists. Prove that a (d, r, g)-graph exists for even g by showing (1) $m(d, r, 4) = 2r + 1$, (2) $m(d, r, g + 2) \le 2 + m(d, d^2, g)$, and (3) $m(d, r, g) \le m_1 + m_2 - 1$, where $m_1 = m(d, 1, g)$ and $m_2 = m(d^{m_1}, r - 1, g)$. (Alon–Kostochka–Reiniger–West–Zhu [2016]) (Comment: By setting $m_1 = 2\lfloor m(d, 1, g)/2 \rfloor + 1$ in (3), one can make bipartite (d, r, g)-graphs, which yields additional applications to list coloring.)

10.2. Ramsey's Theorem

The Pigeonhole Principle guarantees that partitioning many objects into classes yields some class with many objects. The famous theorem of F. P. Ramsey [1930] makes a similar statement about partitioning the r-element subsets of the objects. The study of Ramsey Theory often starts with a classic brain-teaser, which appeared on the Putnam examination in 1952.

10.2.1. Example. *Among any six people, there are three mutual acquaintances or three mutual strangers.* Focus on one person, x. Among the remaining five people, x must have at least three acquaintances or at least three non-acquaintances. By symmetry, we may assume that x has at least three acquaintances. If any two of these are acquainted, then they form the desired set with x; if they are pairwise non-acquainted, then we have three mutual strangers. ∎

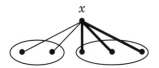

Example 10.2.1 can be modeled as 2-coloring $E(K_6)$. Edges are sets of size 2; more generally, Ramsey's Theorem considers sets of size r. We partition the r-subsets of a set S into k classes and look for p elements of S whose r-sets all lie in the same class. Example 10.2.1 states that when $|S| = 6$, there is always a triple whose 2-sets lie in the same class.

THE MAIN THEOREM

By definition, blocks of a partition are nonempty. The language of coloring allows classes to be empty.

10.2.2. Definition. A k-**coloring** is a function that labels each domain element with one of k *colors*. Let $\binom{S}{r}$ denote the family of r-subsets of a set S. When coloring $\binom{S}{r}$, a set $T \subseteq S$ is **homogeneous** if its r-sets all have the same color.

Roughly speaking, Ramsey's Theorem says that, for $k, r, p \in \mathbb{N}$, every k-coloring of the r-sets of a large enough set has a homogeneous p-set. The precise statement is more detailed, allowing different thresholds in different colors; it facilitates an inductive proof.

10.2.3. Definition. In a k-coloring of $\binom{S}{r}$, a homogeneous set whose r-sets have color i is i-**homogeneous**. For **quotas** $p_1, \ldots, p_k \in \mathbb{N}$, the **Ramsey number** $R(p_1, \ldots, p_k; r)$ is the least $N \in \mathbb{N}$ such that every k-coloring of $\binom{[N]}{r}$ has an i-homogeneous set of size p_i for some i.

Ramsey's Theorem states that $R(p_1, \ldots, p_k; r)$ exists for every choice of the quotas $p_1, \ldots, p_k$. The case $r = 1$ is the quota version of the Pigeonhole Principle (Theorem 10.1.2). To suggest the approach, we first consider the case $r = k = 2$, which includes and extends Example 10.2.1.

10.2.4. Example. When $r = 2$, a partition of $\binom{S}{2}$ is just an edge-coloring of the complete graph with vertex set S. When $k = 2$, by time-honored tradition we make color 1 *red* (solid) and color 2 *blue* (bold). We prove

$$R(p_1, p_2; 2) \le R(p_1 - 1, p_2; 2) + R(p_1, p_2 - 1; 2).$$

Choose a vertex x. The quota version of the Pigeonhole Principle implies that if there are $N + M - 1$ vertices other than x, then x must have N incident red edges or M incident blue edges. Let

$$N = R(p_1 - 1, p_2; 2), \qquad M = R(p_1, p_2 - 1; 2),$$

and consider a 2-coloring of $E(K_{N+M})$. By symmetry, we may assume that x has at least N incident red edges. By the definition of N, the clique induced by the neighbors of x along these edges has a blue p_2-clique or a red $(p_1 - 1)$-clique (which combines with x to form a red p_1-clique). In either case, we obtain an i-homogeneous set of size p_i for some i.

The inequality inductively proves the existence of $R(p_1, p_2; 2)$. We postpone discussion of the resulting bound on $R(p_1, p_2; 2)$. ∎

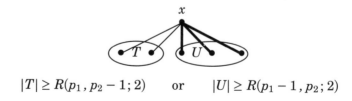

$$|T| \ge R(p_1, p_2 - 1; 2) \qquad \text{or} \qquad |U| \ge R(p_1 - 1, p_2; 2)$$

In Example 10.2.4, we could define an auxiliary coloring on $S - x$ by giving each vertex the color of its edge to x. We would then apply the Pigeonhole Principle to this coloring of the 1-subsets of $S - x$. This is what the induction step of the general theorem reduces to when $r = 2$.

10.2.5. Theorem. (**Ramsey's Theorem**; Ramsey [1930]) For k, r, $p_1, \ldots, p_k \in$ $\mathbb{N}$, the Ramsey number $R(p_1, \ldots, p_k; r)$ exists.

Proof: We use induction on r, proving the induction step by induction on $\sum p_i$. As basis steps, we need the existence of $R(p_1, \ldots, p_k; r)$ for $r = 1$, and we need the existence of $R(p_1, \ldots, p_k; r)$ for $r > 1$ when $\sum p_i$ is small. The case $r = 1$ is the threshold version of the Pigeonhole Principle (Theorem 10.1.2). When $r > 1$ and some quota p_i is less than r, a set of p_i objects has no r-subsets, so vacuously its r-sets all belong to class i. Hence $R(p_1, \ldots, p_k; r) = \min\{p_1, \ldots, p_k\}$ if $\min\{p_1, \ldots, p_k\} < r$.

For clarity, we describe the induction step only for $k = 2$; the general argument is similar (Exercise 12). Write (p, q) for (p_1, p_2). Let

$$p' = R(p - 1, q; r), \quad q' = R(p, q - 1; r), \quad N = 1 + R(p', q'; r - 1).$$

When we consider the triple (p, q, r), the induction hypothesis (induction on $p+q$) for fixed r implies that p' and q' are well-defined. The induction hypothesis for the induction on r then implies that also N is well-defined. Let S be a set of N elements, and choose $x \in S$. Consider a 2-coloring f of $\binom{S}{r}$. We need to show that there is a red-homogeneous p-set or a blue-homogeneous q-set.

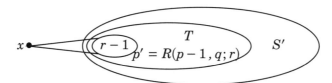

Let $S' = S - \{x\}$. We use f to define a 2-coloring f' of $\binom{S'}{r-1}$. For $A \in \binom{S'}{r-1}$, let $f'(A) = f(A \cup \{x\})$; thus f' is a 2-coloring of the $(r - 1)$-sets in S'. Since $|S'| = R(p', q'; r - 1)$, the induction hypothesis implies that S' contains a red-homogeneous set of size p' or a blue-homogeneous set of size q' under f' (when $r = 2$, this step is the invocation of the Pigeonhole Principle). By symmetry, we may assume that the red quota is met. Let T be a p'-element subset of S' whose $(r - 1)$-sets are red under f'.

We return to the original coloring f on $\binom{T}{r}$. Since $|T| = p' = R(p - 1, q; r)$, within T there is a $(p - 1)$-set that is red-homogeneous under f or a q-set that is blue-homogeneous under f. If there is a blue-homogeneous q-set, then we are done. If there is a red-homogeneous $(p - 1)$-set P, then consider $P \cup \{x\}$. From the definition of T, the $(r - 1)$-sets of P are all red under f', and so their unions with x are red under f. Hence $P \cup \{x\}$ is a red-homogeneous p-set under f. ∎

The structure of this proof is called "double induction"; the proof of the induction step is itself a proof by induction.

APPLICATIONS

Like the Pigeonhole Principle, Ramsey's Theorem has subtle applications and provides elegant existence proofs. Unlike the Pigeonhole Principle, the bounds from Ramsey's Theorem usually are horribly loose.

Our first application was a problem posed to some talented mathematics students in 1932 by their friend Eszther Klein. Paul Erdős, then 19, was in the group. Erdős and Szekeres gave several solutions, including one requiring rediscovery of Ramsey's Theorem, which was unknown to them. Four years later, Klein married Szekeres, so Erdős named this the **Happy End Theorem**.

10.2.6. Theorem. (Erdős–Szekeres [1935]). For $m \in \mathbb{N}$, there is a (least) integer $N(m)$ such that every set of $N(m)$ points in the plane (no three collinear) contains an m-subset forming a convex m-gon.

Proof: We need two facts. *(1) Among any five points in the plane, four determine a convex quadrilateral* (if no three are collinear). Construct the convex hull of the five points. When it is a pentagon or a quadrilateral, the claim is immediate. When it is a triangle, the other two points lie inside. By the Pigeonhole Principle(!), two corners of the triangle are on one side of the line through the points inside. These plus the points inside form a convex quadrilateral.

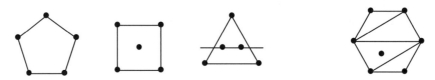

(2) If every 4-subset of m points in the plane forms a convex quadrilateral, then the m points form a convex m-gon. If not, then the convex hull of the m points consists of some t points, where $t < m$. The remaining points lie inside the t-gon. When we triangulate the t-gon, as shown on the right above, a point inside the t-gon lies in one of the triangles. With the vertices of that triangle, it forms a 4-set that does not determine a convex quadrilateral.

In order to combine these two results into a proof of the theorem, let $N = R(m, 5; 4)$. Given N points in a plane, color each 4-set by convexity: red if it determines a convex quadrilateral, blue if it does not. By fact (1), there cannot be 5 points whose 4-subsets are all blue. By Ramsey's Theorem, there must then be m points whose 4-subsets are all red. By fact (2), they form a convex m-gon. Hence $N(m)$ exists and is at most $R(m, 5; 4)$.   ∎

The upper bound $R(m, 5; 4)$ is exact for $m = 4$ by fact (1), but already it is very loose at $m = 5$ (Exercise 6 shows $N(5) = 9$). In general, $N(m) \leq \binom{2m-4}{m-2} + 1$ (Erdős–Szekeres [1935]) and $N(m) > 2^{m-2}$ (Erdős–Szekeres [1960]). In fact, they conjectured $N(m) = 2^{m-2} + 1$ (Exercises 7–9). Erdős offered \$500 for a proof of the conjecture, raised to \$1000 by Graham. After more than 60 years, Chung–Graham [1998] improved the upper bound by 1. About ten papers gave further small improvements, but none improved the leading behavior of 4^m. Finally, Suk [2016] combined various ways of finding a convex polygon to solve the problem asymptotically, proving $N(m) \leq 2^{m+o(m)}$. Other extensions and generalizations are surveyed in Morris–Soltan [2000, 2016].

In the next application, again we need an auxiliary structural lemma and obtain a bound that likely is much larger than needed.

10.2.7. Example. *Table storage and search.* From a large universe of numbers (called "keys"), n numbers arrive to be stored in a table of size n. If they are stored in order, then binary search can test the presence of any key by probing at most $\lceil \lg(n+1) \rceil$ locations, where $\lg$ denotes $\log_2$.

Can another strategy always test membership in fewer probes? It depends on the size of the key space M. Yao [1981] proved that sorted storage and binary search is optimal when M is large.

A *storage strategy* $\mathbf{T}$ assigns each n-set A of keys a storage permutation: $\mathbf{T}(A) = \sigma$ puts the jth smallest element of A in location $\sigma(j)$, for $1 \le j \le n$. A *query* asks whether a key x is present in the table. A *search strategy* $\mathbf{S}$ probes table locations, based on x, $\mathbf{T}$, and the outcome of earlier probes. The answer to a probe is the key stored in that location. The *(worst-case) cost* of a strategy $(\mathbf{S}, \mathbf{T})$ is the maximum, over all table contents A and queries x, of the number of probes used by $\mathbf{S}$ to determine whether $x \in A$ when $\mathbf{T}$ is used to store the contents of the table. For $m = |M|$, let $f(m, n)$ be the minimum cost over all strategies.

When $m = n$, every key is present, and $f(n, n) = 0$. When $m = n + 1$, sorted storage has worst-case search cost 2, since the jth element x may be stored at j or $j - 1$, depending on whether the missing element is higher or lower than x. On the other hand, probing location 1 suffices if we store in location 1 the key that cyclically follows the missing key. Hence $f(n+1, n) = 1$; sorting is not optimal when m is small. Yao [1981] proved that $f(m, n) \le 1$ if and only if $m \le 2n - 2$. ■

The next lemma generalizes the argument that binary search in sorted tables needs $\lg(n + 1)$ probes in the worst case.

10.2.8. Lemma. (Yao [1981]) Let $\mathbf{T}$ be a storage strategy for n-sets from a universe M with $|M| \ge 2$. A set $P \subseteq M$ is *stored consistently* if each n-set in P is stored according to the same permutation. If some set of size $2n - 1$ in M is stored consistently under $\mathbf{T}$, then any strategy using $\mathbf{T}$ for storage costs at least $\lceil \lg(n+1) \rceil$ probes in the worst case.

Proof: Let P be a set of size $2n - 1$ stored consistently under $\mathbf{T}$; there is a permutation σ such that $\mathbf{T}$ stores any n-set from P by putting its jth smallest element in position $\sigma(j)$, for $1 \le j \le n$. Let $P = \{x_1, \ldots, x_{2n-1}\}$ with $x_1 < \cdots < x_{2n-1}$. By reindexing the table locations in the order $\sigma(1), \ldots, \sigma(n)$, we may assume that σ is the identity permutation.

Using induction on n, we prove that an adversary can force the algorithm to use at least $\lceil \lg(n+1) \rceil$ probes to test whether the "central element" x_n of P is present. For $n = 1$, a probe is needed, since $|M| \ge 2$.

For $n \ge 2$, let $n' = \lfloor n/2 \rfloor$. The adversary responds to the first probe to leave a problem of the same type, with size n' and with x_n still the central element. Define P' by eliminating the top $\lceil n/2 \rceil$ and bottom $\lceil n/2 \rceil$ elements of P. Since $2n' - 1 = 2n - 1 - 2\lceil n/2 \rceil$, the size of P' is $2n' - 1$, and the central element is x_n.

If the algorithm first probes location j with $j \le \lceil n/2 \rceil$, then the adversary responds that key x_j is there. The adversary also grants that keys $x_1, \ldots, x_{\lceil n/2 \rceil}$ are present in locations $1, \ldots, \lceil n/2 \rceil$ (in order, since P is consistent) and keys $\{x_{2n-\lceil n/2 \rceil}, \ldots, x_{2n-1}\}$ are not in the table (see exception below for $n \le 3$). Now all is known except which n'-subset of P' appears in locations $\lceil n/2 \rceil + 1, \ldots, n$.

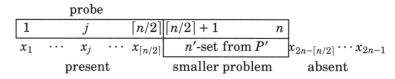

When $n = 2$ or $n = 3$, the adversary instead guarantees the absence of only the top $\lceil n/2 \rceil - 1$ elements. It thus remains uncertain whether x_n is in the last unknown location, and one more probe will be needed.

For $n \geq 4$, consistency of $\mathbf{T}$ on P implies that for n-sets of P containing $x_1, \ldots, x_{\lceil n/2 \rceil}$ and omitting $x_{2n-\lceil n/2 \rceil}, \ldots, x_{2n-1}$, the elements of P' in locations $\lceil n/2 \rceil + 1, \ldots, n$ are in order. Thus the set P' of size $2n' - 1$ with central element x_n is stored consistently under a strategy with n' locations. By the induction hypothesis, the adversary can force at least $\lceil \lg(n' + 1) \rceil$ further probes to test for the presence of x_n. This completes the proof, using $1 + \lceil \lg(\lfloor n/2 \rfloor + 1) \rceil = \lceil \lg(n + 1) \rceil$.

Symmetrically, if the first probe is after $\lceil n/2 \rceil$, then the adversary puts the highest $\lceil n/2 \rceil$ elements in the highest $\lceil n/2 \rceil$ positions and says that the lowest $\lceil n/2 \rceil$ keys in P are absent. Now the induction hypothesis applies with an n'-subset of P' in locations $1, \ldots, n'$. ■

10.2.9. Theorem. (Yao [1981]) Let $f(m, n)$ be the complexity of membership testing when n-sets from a space of m keys are stored in a table of size n. If m is sufficiently large, then $f(m, n) = \lceil \lg(n + 1) \rceil$.

Proof: By Lemma 10.2.8, we need only show that for large m, under every storage strategy $\mathbf{T}$ there is a set P of size at least $2n - 1$ that is stored consistently. This is a job for Ramsey's Theorem. We need P such that every n-subset of P is stored using the same permutation σ, so we color the n-subsets of M using $n!$ colors, with the color on a set being the permutation used to store it. That is, assign color σ to A if $\mathbf{T}$ stores A by putting x_i in location $\sigma(i)$ for all i.

If $m \geq R(2n - 1, \ldots, 2n - 1; n)$ (all $n!$ quotas equal $2n - 1$), then Ramsey's Theorem guarantees a color σ such that some set of $2n - 1$ keys is consistently stored by σ. By Lemma 10.2.8, $\lceil \lg(n + 1) \rceil$ probes are required in the worst case to test the presence of some key in that set. ■

RAMSEY NUMBERS

Ramsey's Theorem defines the Ramsey numbers $R(p_1, \ldots, p_k; r)$. No formula is known, and few values have been computed. To prove $R(p_1, \ldots, p_k; r) = N$, one must guarantee a k-coloring of the r-sets on $N - 1$ points that meets no quota *and* show that *every* coloring on N points meets some quota.

In principle, a computer can check all k-colorings of $\binom{[n]}{r}$ for various n to find the first n such that every coloring meets some quota. Even when $r = k = 2$, the $2^{\binom{n}{2}}$ colorings rapidly become too many to handle. Erdős joked that if an evil spirit threatens to destroy the human race unless we tell it the value of $R(5, 5; 2)$, then we should have all computers in the world look for the answer. If instead it wants $R(6, 6; 2)$, then he advised destroying the evil spirit before it destroys us.

To understand the joke, consider trying to prove $R(5, 5; 2) > 43$. Without making use of symmetry or other reductions of the search, the number of graphs with vertex set [43] is $2^{\binom{43}{2}}$, which equals 2^{903}. We need one having no 5-clique or independent 5-set. The number of particles in the universe is thought to be at most 10^{88}. The age of the universe is thought to be at most 20 billion years, and there are fewer than 32 million seconds per year. Even if every particle in the universe had checked one graph every second since the beginning of time, fewer than 2^{350} graphs would have been checked by now. Hence Erdős was an optimist.

For often-studied subcases, we abbreviate the notation.

10.2.10. Definition. When $r = 2$, we write $R(p_1, \ldots, p_k; r)$ as $R(p_1, \ldots, p_k)$. When $p = p_1 = \cdots = p_k$, we write $R(p_1, \ldots, p_k; r)$ as $R_k(p; r)$ and call it a **diagonal Ramsey number**. In particular, $R_k(3; 2)$ is the least n such that every k-coloring of $E(K_n)$ has a monochromatic triangle.

For $r > 2$, little is known other than $R(4, 4; 3) = 13$ (McKay–Radziszowski [1991]). Even when $r = 2$, only one Ramsey number is known exactly with $k > 2$: $R(3, 3, 3) = 17$ (Greenwood–Gleason [1955]). The table below contains the known values of $R(p, q)$ and the best known bounds for some other values as of August 2009. The current bounds are maintained in Radziszowski [1994–2017] (periodically updated). Computer searches have been slowly increasing the lower bounds.

	3	4	5	6	7	8	9
3	6	9	14	18	23	28	36
4		18	25	36/41	49/61	56/84	73/115
5			43/48	58/87	80/143	101/216	133/316
6				102/165	115/298	134/495	183/780

The values of $R(3, 9)$ (Grinstead–Roberts [1982]), $R(3, 8)$ (McKay–Zhang [1992]), and $R(4, 5)$ (McKay–Radziszowski [1995]) are the most recent. The others were found much earlier (due primarily to Greenwood–Gleason [1955], Kalbfleisch [1967], and Graver–Yackel [1968]).

When $r = k = 2$, we can view the colors on the edges of K_n as "in" and "out". Ramsey's Theorem for this case then becomes: "There exists a smallest integer $R(p, q)$ such that every graph with $R(p, q)$ vertices has a clique of size p or an independent set of size q."

The proof of Ramsey's Theorem yields a recursive upper bound on $R(p, q; r)$. It is very large but gives useful values for small thresholds when $r = 2$. For $r = 2$, it repeats the discussion in Example 10.2.4.

10.2.11. Theorem. $R(p, q) \le R(p - 1, q) + R(p, q - 1)$. If both summands on the right are even, then the inequality is strict.

Proof: If a vertex in an arbitrary graph has $R(p - 1, q)$ neighbors or $R(p, q - 1)$ nonneighbors, then the graph contains K_p or $\overline{K}_q$. With $R(p - 1, q) + R(p, q - 1)$ vertices altogether in the graph, the Pigeonhole Principle guarantees that one of these possibilities occurs.

Equality requires a graph with $R(p - 1, q) + R(p, q - 1) - 1$ vertices having no clique of size p or independent set of size q. If some vertex has at least $R(p - 1, q)$ neighbors or at least $R(p, q - 1)$ nonneighbors, then there is a p-clique or an independent q-set. Otherwise, with exactly $R(p-1,q)+R(p,q-1)-2$ other vertices, every vertex must have degree exactly $R(p-1,q)-1$. When $R(p-1,q)$ and $R(p,q-1)$ are both even, this requires a regular graph of odd degree with an odd number of vertices, which does not exist. ∎

10.2.12. Corollary. $R(p,q) \le \binom{p+q-2}{p-1}$.

Proof: A 2-edge-coloring of K_p has an edge of the second color or doesn't, so $R(p,2) = p$. Hence the upper bound holds with equality when q (or p) is 2. This provides the basis for induction on $p + q$. By Theorem 10.2.11, the induction hypothesis, and Pascal's Formula,

$$R(p,q) \le R(p-1,q) + R(p,q-1) \le \binom{p+q-3}{p-1} + \binom{p+q-3}{p-2} = \binom{p+q-2}{p-1}.$$ ∎

10.2.13. Example. $R(3,3) = 6$ and $R(3,4) = 9$.

For $R(3,3)$, Example 10.2.1 yields 6 as the upper bound. Equality holds because the 5-cycle and its complement have no triangles.

For $R(3,4)$, the inequality in Theorem 10.2.11 is strict. With $R(2,4) = 4$ and $R(3,3) = 6$, we have $R(3,4) < 10$ since both are even. The graph on the right below proves $R(3,4) \ge 9$. Four independent vertices on an 8-cycle must include opposite vertices, but in this graph such vertices are adjacent. ∎

Corollary 10.2.12 yields $R(p+1,p+1) \le \binom{2p}{p} \sim 4^p/\sqrt{\pi p}$. Conlon [2009a] improved the upper bound by putting a superpolynomial function of p in the denominator: $R(p+1,p+1) \le \binom{2p}{p}p^{-c\log p/\log\log p}$ for some constant c.

The constructive lower bounds in Exercises 17–18 are polynomials in p. The best constructive lower bound known grows faster than every polynomial but slower than every exponential (Exercise 19). A survey of constructive bounds appears in Frankl [1990]. An exponential lower bound can be proved by counting.

10.2.14. Theorem. (Erdős [1947]) $R(p,p) > \frac{1}{e\sqrt{2}}p2^{p/2}(1-o(1))$.

Proof: Consider the graphs with vertex set $[n]$. Each possible p-clique occurs in $2^{\binom{n}{2}-\binom{p}{2}}$ of these $2^{\binom{n}{2}}$ graphs. Similarly, each set of p vertices occurs as an independent set in $2^{\binom{n}{2}-\binom{p}{2}}$ of the graphs. Discarding these leaves only graphs with no p-clique or independent p-set. Since there are $\binom{n}{p}$ ways to choose p vertices, the inequality $2\binom{n}{p}2^{\binom{n}{2}-\binom{p}{2}} < 2^{\binom{n}{2}}$ implies $R(p,p) > n$. We seek n such that $\binom{n}{p} < 2^{\binom{p}{2}-1}$.

To obtain the desired bound, we use $\binom{n}{p} < \left(\frac{ne}{p}\right)^p$, proved in Section 14.1. It thus suffices to have $\left(\frac{ne}{p}\right)^p < 2^{\binom{p}{2}-1}$, which simplifies to $n < (1-o(1))\frac{p}{e\sqrt{2}}2^{p/2}$. ∎

We revisit this argument in probabilistic language in Theorem 14.1.3. Determining whether the limit of $R(p,p)^{1/p}$ exists (and its value if it exists) is the foremost open problem in Ramsey theory. By Theorem 10.2.14 and Corollary 10.2.12, the value is between $\sqrt{2}$ and 4.

10.2.15.* Remark. *Fixing one quota.* In contrast to the diagonal case, tighter bounds are known when q is fixed and p is large. Here $R(p,q) \leq cp^{q-1}\frac{\log\log p}{\log p}$ (Graver–Yackel [1968], Chung–Grinstead [1983]). For $q = 3$, the answer is known within a constant factor:

$$\frac{cp^2}{\log p} \leq R(p,3) \leq \frac{c'p^2}{\log p}$$

The upper bound is due to Ajtai–Komlós–Szemerédi [1980], the lower to Kim [1995]. Both use probabilistic methods (see Chapter 14).

This result relates to triangle-free k-chromatic graphs (Remark 8.1.19). Let $n = R(p,3) - 1$. By the meaning of $R(p,3)$, some n-vertex graph has no triangle and no independent p-set. Its chromatic number is thus at least n/p. With $k = n/p$, there is a k-chromatic triangle-free graph with at most n vertices.

We solve for n in terms of k. With $n \geq cp^2/\log p$, we have $p \leq O(\sqrt{n\log n})$. Now $k \approx \sqrt{n/\log n}$. Thus there are k-chromatic triangle-free graphs with at most $O(k^2 \log k)$ vertices. This yields the upper bound in Remark 8.1.19. ∎

Before generalizing the Ramsey problem for $r = 2$, we derive the classical bounds on the diagonal 2-color hypergraph Ramsey numbers. We write simply $R(p;r)$ instead of $R_2(p;r)$. For further discussion of hypergraph Ramsey numbers, see Conlon–Fox–Sudakov [2010, 2013] and Mubayi–Suk [2017].

To introduce the main idea for the upper bound, we consider first an easy explicit upper bound for $R(p;2)$, a bit weaker than Corollary 10.2.12.

10.2.16. Proposition. $R(p;2) \leq 2^{2p}$.

Proof: Consider the elements of $[2^{2p}]$ in order as vertices, and let f be a red/blue coloring of the pairs. Initially, all vertices are "live". When x is the first remaining live vertex, we label x red or blue to agree with the color given by f to the majority of the edges from x to later live vertices. A later live vertex y then remains alive only if $f(xy)$ agrees with the label of x.

After doing this $R(p-1;1)$ times, the Pigeonhole Principle yields $p-1$ live vertices that have the same label, red or blue. Together with one more live vertex, we have a set U of p vertices that were all kept because their edges to earlier vertices in U have that same color.

Since we lose at most a factor of 2 each time we process the first unlabeled live vertex, and $R(p-1;1) = 2p-1$, it suffices to start with $1 + 2^{2p-1}$ vertices. ∎

For larger r, we generalize the idea, considering extensions of $(r-1)$-sets by adding one later vertex.

10.2.17. Theorem. (Erdős–Rado [1952]) $R(p;r) \leq 2^{[R(p-1;r-1)^{r-1}]}$.

Proof: Initially, all vertices are live. Let $N = 2^{[R(p-1;r-1)^{r-1}]}$, consider the elements of $[N]$ in order, and let f be a red/blue coloring of $\binom{[N]}{r}$. Let S be the set of the first $r - 1$ vertices in $[N]$. Label S with the color given by f to the majority of the r-sets obtained by adding a later vertex to S. Keep alive only the later live vertices y such that $f(S \cup \{y\})$ is the same as the label of S.

Consider the situation when x is the first remaining vertex. Iteratively label each $(r - 1)$-set S' consisting of x and $r - 2$ earlier vertices that have been kept alive. The label of S' is the majority color of the extensions of S' to r-sets by adding a later remaining live vertex y. Keep y alive only if $f(S' \cup \{y\})$ is the same as the label on S'.

Each time we label an $(r - 1)$-set, we lose at most half of the remaining live vertices. We want to wind up with $R(p - 1; r - 1)$ vertices whose $(r - 1)$-sets have all been labeled, plus another live vertex. Since $\binom{R(p-1;r-1)}{r-1}$ sets of size $r - 1$ are to be labeled, it suffices to start out with N vertices. By the definition of the Ramsey number, we then have a set U of p vertices in which every $(r - 1)$-set not containing the last vertex has been labeled with the same color, meaning that the r-sets obtained by adding a later vertex of U all have that color. ∎

10.2.18. Corollary. $R(p; r) \leq \mathrm{tow}_{r-1}(c_r p)$, where $\mathrm{tow}_0(x) = x$, $\mathrm{tow}_h(x) = 2^{\mathrm{tow}_{h-1}(x)}$ for $h > 0$, and c_r is a constant that depends on r but is independent of p.

Proof: By Proposition 10.2.16, $R(p; 2) \leq 2^{2p}$, so $c_2 = 2$ suffices. For larger r, Theorem 10.2.17 yields $R(p; r) \leq 2^{[\mathrm{tow}_{r-2}(c_{r-1}(p-1))]^{r-1}} \leq \mathrm{tow}_{r-1}(c_r p)$. The presence of $p - 1$ in the recurrence allows c_r to remain relatively small. ∎

A straightforward extension of the counting argument in Theorem 10.2.14 (counting 3-uniform hypergraphs instead of graphs) yields $R(p; 3) > 2^{cp^2}$. Induction on r leads to a lower bound on $R(p; r)$ that is also a tower, but having height lower by 1. The key is the next lemma, whose proof is due to Erdős and Hajnal but seems to be published only in Graham–Rothschild–Spencer [1980].

10.2.19. Lemma. (Stepping-Up Lemma) If $r \geq 3$, then $R(p; r) > n$ implies $R(2p + r - 4; r + 1) > 2^n$.

Proof*: Let f be a red/blue coloring of $\binom{[n]}{r}$ avoiding homogeneous p-sets. Letting $T = [2^n]$, we form a red/blue coloring of $\binom{T}{r+1}$. For $x \in T$, let $(x_1, \ldots, x_n)$ be the binary expansion of $x - 1$, so that $x = 1 + \sum_{i=1}^{n} x_i 2^{i-1}$.

For $x, x' \in T$, let $\delta(x, x') = \max\{i: x_i \neq x'_i\}$. If $x < x'$ and $\delta(x, x') = j$, then $x_j = 0$ and $x'_j = 1$. Therefore, $x < x' < x''$ implies $\delta(x, x') \neq \delta(x', x'')$; that is, in a monotone list the derived list of high-bit changes has no immediate repeats.

Given $x^1, \ldots, x^t \in T$, indexed in increasing order,

$$\delta(x^1, x^t) = \max_{1 \leq i < t}\{\delta(x^i, x^{i+1})\}. \tag{$*$}$$

The operator δ extends to monotone lists of any length by letting $\delta(x^1, \ldots, x^t) = (\delta(x^1, x^2), \ldots, \delta(x^{t-1}, x^t))$.

Now we define a red/blue coloring f' of $\binom{T}{r+1}$, "stepping up" from f. Index any $(r + 1)$-set Y in increasing order, forming an $(r + 1)$-tuple. If the r-tuple $\delta(Y)$ is strictly monotone, then let $f'(Y) = f(\delta(Y))$. If $\delta(Y)$ has a peak at the second position (value larger than the neighboring values), then let $f'(Y)$ be red. If $\delta(Y)$

has a valley at the second position (value smaller than the neighboring values), then let $f'(Y)$ be blue. Otherwise, $\delta(Y)$ is chosen arbitrarily.

Now let Z be a $(2p + r - 4)$-set in T; we show that Z cannot be homogeneous. Suppose that Z is red-homogeneous (an analogous argument applies for blue). With Z indexed in increasing order, no consecutive values in $\delta(Z)$ are equal.

Case 1: *For some t, the positions t through $t + p - 1$ in $\delta(Z)$ form a strictly monotone p-tuple $\hat{Z}$.* Suppose first that $\hat{Z}$ is decreasing, and let $\delta_s = \delta(Z_s, Z_{s+1})$, so $\hat{Z} = (\delta_t, \ldots, \delta_{t+p-1})$. Since f admits no homogeneous p-tuple, some r entries in $\hat{Z}$ form a set W that is blue under f. Let $i_1, \ldots, i_r$ be the corresponding indices of these positions in Z, and let $Y = (Z_{i_1}, \ldots, Z_{i_r}, Z_{i_r+1})$.

We claim that $f'(Y)$ is blue, contradicting that Z is red-homogeneous. By (∗), the value $\delta(Y_j, Y_{j+1})$ is the largest $\delta(Z_s, Z_{s+1})$ among s such that $i_j \leq s < i_{j+1}$, where $i_{r+1} = i_r + 1$. Since $\delta(Z)$ is monotone decreasing, this largest value is W_j. Thus $\delta(Y) = W$. Since W is a sublist of the monotone decreasing list $\delta(Z)$, also W is monotone decreasing. Hence by definition $f'(Y) = f(W)$, which is blue.

If $\delta(Z)$ is strictly increasing, then an analogous argument applies, with Y consisting of the positions in Z just after all those corresponding to the positions occupied by W, plus the first position occupied by W.

Case 2: *No consecutive p-tuple in $\delta(Z)$ is monotone.* Consider the first $2p - 3$ positions in $\delta(Z)$. None after the first can be a valley, because then Z would contain an $(r + 1)$-tuple assigned blue. Hence $\delta(Z)$ has at most one peak. If it has no peak, then $\delta(Z)$ is monotone. If it has a peak before or after the middle, then the portion on the opposite side of the middle of $\delta(Z)$ is a monotone list of length at least p. In each of these possibilities, Case 1 applies. ∎

10.2.20. Corollary. $R(p; r) > \text{tow}_{r-2}(c_r' p^2)$, where $c_r' = c/4^{2r-6}$ and $R(p; 3) > 2^{cp^2}$.

Proof: Since $R(q; r) > n$ implies $R(2q + r - 4; r + 1) > 2^n$ by the Stepping-Up Lemma, and we may assume $q > r$, also $R(q; r) > n$ implies $R(4q; r + 1) > 2^n$. With $p = 4q$, we have $R(p; r) > 2^{R(p/4; r-1)}$. Thus each iteration increases the height of the tower and decreases the constant by a factor of 4^2. ∎

Thus $\text{tow}_{r-2}(c_r' p^2) < R(p; r) \leq \text{tow}_{r-1}(c_r p)$; the heights differ by 1. Erdős conjectured $R(p; 3) > 2^{2^{cp}}$, yielding $R(p; r) > \text{tow}_{r-1}(c_r' p)$. Erdős and Hajnal proved that $R_4(p; 3)$ is doubly-exponential, so for four colors the towers above and below have the same height. Conlon–Fox–Sudakov [2013] proved $R_3(p; 3) > 2^{p^{c \log p}}$.

GRAPH RAMSEY THEORY

By Ramsey's Theorem, k-coloring the edges of a large enough complete graph forces a monochromatic complete subgraph of order p. It thus also forces a monochromatic copy of every p-vertex graph G. When we seek G and don't need K_p, we may not need $R(p, p)$ vertices. For example, every 2-coloring of the edges of K_3 has a monochromatic P_3, but K_6 is needed to force a monochromatic K_3.

10.2.21. Definition. For graphs $G_1, \ldots, G_k$, the **Ramsey number** $R(G_1, \ldots, G_k)$ is the least n such that every k-coloring of $E(K_n)$ contains a copy of G_i in color i for some i. When $G_i = G$ for all i, we write $R(G_1, \ldots, G_k)$ as $R_k(G)$.

Sometimes $R(G_1, G_2)$ has a simple formula. Again we color red and blue.

10.2.22. Theorem. (Chvátal [1977]). If T is an m-vertex tree, then $R(T, K_n) = (m-1)(n-1)+1$.

Proof: For the lower bound, color $K_{(m-1)(n-1)}$ by letting the red graph consist of $(n-1)K_{m-1}$. Red components of order $m-1$ have no red m-vertex tree. The blue graph is $(n-1)$-partite and hence does not contain K_n.

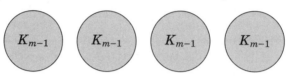

We prove the upper bound by induction on n, using a property of trees proved by induction on m. The basis step is $n = 1$; no edges are needed to obtain K_1.

For $n > 1$, consider a 2-coloring of $E(K_{(m-1)(n-1)+1})$. If some vertex x has more than $(m-1)(n-2)$ incident blue edges, then the induction hypothesis gives a red T or a blue K_{n-1} in $N(x)$, and thus a red T or a blue K_n (with x) in the full graph.

Otherwise, every vertex has at most $(m-1)(n-2)$ incident blue edges and at least $m-1$ incident red edges. Now there is a red T, because every graph G having minimum degree at least $m-1$ contains T.

This claim for trees (Exercise 5.4.17) holds by induction on m; it is trivial for $m = 1$. For $m > 1$, let u be a leaf, with neighbor v, and let $T' = T - u$. By the induction hypothesis, G contains T'. The vertex serving as v has at least $m-1$ neighbors in G, at most $m-2$ of which appear in the copy of T'. Thus v has another neighbor in G to serve as u. ∎

The lower bound holds more generally, with the same coloring. If the largest component of G has m vertices, and H has chromatic number n, then $R(G, H) \geq (m-1)(n-1)+1$ (Chvátal–Harary [1972]). Burr–Erdős [1983] conjectured that $R(G, K_n) = (m-1)(n-1)+1$ when m is sufficiently large relative to n and $\max_{F \subseteq G} \frac{|E(F)|}{|V(F)|}$, but Brandt [1996] disproved this. Nevertheless, Nikiforov [2005] proved $R(C_m, K_n) = (m-1)(n-1)+1$ for $m \geq 4n+2$, partially settling a conjecture of Erdős, Rousseau, and Schelp (see Faudree–Schelp [1978]).

The upper bound for Theorem 10.2.22 depends on color classes in H being single vertices. Otherwise, the lower bound can be weak. For example, the Chvátal–Harary result yields $R(mK_3, mK_3) \geq (3-1)(3-1)+1 = 5$, but the truth is $5m$.

10.2.23. Theorem. (Burr–Erdős–Spencer [1975]). $R(mK_3, mK_3) = 5m$ for $m \geq 2$.

Proof: Let the red graph be $K_{3m-1} + K_{1,2m-1}$, as shown on the left. Every red triangle uses three vertices from the $(3m-1)$-clique, which is not big enough to permit m pairwise disjoint triangles. The complementary blue graph on the right is $(K_{2m-1} + K_1) \oplus \overline{K}_{3m-1}$. Every blue triangle has at least 2 vertices in the $(2m-1)$-clique, which again is not big enough to permit m pairwise disjoint triangles.

For the upper bound, we use induction on m. The basis step $m = 2$ requires case analysis (Exercise 28). If $m \geq 3$, then $5m > R(3,3) = 6$, so every 2-coloring of $E(K_{5m})$ contains a monochromatic triangle. Discarding three vertices, we find more such triangles until fewer than 6 vertices remain. Since $5m - 3m \geq 6$ for $m \geq 3$, we find m disjoint monochromatic triangles. If these have the same color, then we are done; otherwise, we have a triangle in each color.

Let $\{a, b, c\}$ and $\{d, e, f\}$ be the vertex sets of disjoint red and blue triangles. Of the nine edges joining them, we may assume by symmetry that at least five are red. Some two of these have a common endpoint in $\{d, e, f\}$. This yields a red triangle and a blue triangle with a common vertex. If $m > 2$, then the induction hypothesis yields $(m-1)K_3$ in one color among the remaining $5m-5$ vertices. We then add the appropriately colored triangle from the five special vertices. ∎

Using $R(3,3) = 6$ as a basis instead yields $R(mK_3, mK_3) \leq 5m + 1$. A similar but easier result is $R(mK_2, mK_2) = 3m - 1$ (Exercise 27). The main induction argument generalizes to an upper bound on $R(m_1 G_1, \ldots, m_k G_k)$ in terms of $R(G_1, \ldots, G_k)$ (Exercise 29).

Ramsey numbers of graphs (such as K_n) may be exponential in the number of vertices. However, Chvátal–Rödl–Szemerédi–Trotter [1983] proved that for graphs with fixed maximum degree d, the Ramsey number grows only linearly in the number of vertices! That is, $R(G, G) \leq c|V(G)|$, where c is a constant that grows quickly with d but does not depend on $|V(G)|$. We prove this in Section 11.1.

Meanwhile, we mention other directions in graph Ramsey theory. Having generalized the target monochromatic subgraph from complete graphs to any graph, we can also generalize the host graph.

10.2.24. Definition. The notation $H \rightarrow (G_1, \ldots, G_k)$ means that every k-coloring of $E(H)$ has a copy of G_i in color i for some i. When $G_1 = \cdots = G_k$, we write $H \rightarrow_k G$, or just $H \rightarrow G$ when $k = 2$.

Note that $K_n \rightarrow (G_1, \ldots, G_k)$ when $n \geq R(G_1, \ldots, G_k)$, so some such H exists. Classical graph Ramsey numbers minimize $|V(H)|$ such that $H \rightarrow G$, since $K_n \rightarrow G$ whenever some n-vertex graph H suffices. More generally, we can minimize any parameter over the graphs forcing G.

10.2.25. Definition. For a graph parameter ρ, the ρ-**Ramsey number** of a graph G, written $R_\rho(G)$, is $\min\{\rho(H): H \rightarrow G\}$. More generally, $R_\rho(G; s) = \min\{\rho(H): H \rightarrow_s G\}$ and $R_\rho(G_1, \ldots, G_s; s) = \min\{\rho(H): H \rightarrow (G_1, \ldots, G_s)\}$.

We have so far studied vertex Ramsey numbers. There are also the **size Ramsey number** $(\rho(H) = |E(H)|)$, the **clique Ramsey number** $(\rho = \omega)$, the **chromatic Ramsey number** $(\rho = \chi)$, and the **degree Ramsey number** $(\rho = \Delta)$. The initial result on parameter Ramsey numbers was $R_\omega(G) = \omega(G)$ for every graph G (Folkman [1970]), extended to $R_\omega(G; s) = \omega(G)$ by Nešetřil–Rödl [1976].

The size Ramsey number of G, defined in Erdős–Faudree–Rousseau–Schelp [1978], is obviously at most $\binom{R(G)}{2}$; they proved equality when G is a complete graph. Solving a problem for which Erdős offered 100 euros, Beck [1983] proved that the size Ramsey number of P_n is linear in n (improved by Dudek–Prałat

[2015]). The size Ramsey number also grows linearly in n for cycles (Haxell–Kohayakawa–Łuczak [1995], but not for graphs with maximum degree 3 (Rödl–Szemerédi [2000]). The result of Chvátal–Rödl–Szemerédi–Trotter [1983] (Section 11.1) implies a quadratic upper bound in n for graphs with maximum degree at most k, and this was improved to $O(n^2(\frac{\log n}{n})^{1/k})$ by Kohayakawa–Rödl–Szemerédi–Schacht [2011]). Faudree–Schelp [2002] surveyed the early results on the topic, but there are many additional papers on the size Ramsey numbers of special graphs or special families of graphs.

Here we study only the chromatic Ramsey number. This is the primary focus of Burr–Erdős–Lovász [1976], where parameter Ramsey numbers were first proposed. We begin with an old result implying $R_\chi(G) = 2$ when G is bipartite.

10.2.26. Theorem. (Erdős–Rado [1956]) Given p, q, s, there exist m, n such that every s-edge-coloring of $K_{m,n}$ has a monochromatic $K_{p,q}$.

Proof: Let $m = s(p-1)+1$ and $n = s^m(q-1)+1$. Consider an arbitrary s-coloring of $K_{m,n}$ with bipartition X, Y. Let the vertices in X be $x_1, \ldots, x_m$. Label each vertex y_j in Y by a s-ary vector whose ith coordinate is the color of the edge $x_i y_j$. Now we apply the Pigeonhole Principle twice. Since there are s^m such distinct vectors, there must be q vertices in Y with the same vector v as label. Since only s colors are available as entries for the m coordinates of v, v must use some color at least p times. The p vertices of X in those positions and the q vertices of Y with label v induce a monochromatic $K_{p,q}$. ∎

To study chromatic Ramsey numbers of non-bipartite graphs, we apply Theorem 10.2.26 to obtain specially colored complete multipartite subgraphs.

10.2.27. Theorem. (Burr–Erdős–Lovász [1976]) Let $K_r[t]$ denote the complete r-partite graph with parts of size t. For $n, r, s \in \mathbb{N}$, there exists $N \in \mathbb{N}$ such that every s-coloring of $E(K_r[N])$ contains a copy of $K_r[n]$ where any two parts induce a monochromatic copy of $K_{n,n}$.

Proof: By Theorem 10.2.26, sufficiently large M for fixed m guarantees in parts i and j a monochromatic copy of $K_{m,m}$. We want to do this with m large enough so that, even when we keep only m vertices in each part, enough vertices remain to ensure monochromatic copies of $K_{n,n}$ joining all other pairs.

To make this precise, let $b(m)$ be the least M such that s-colorings of $E(K_{M,M})$ force a monochromatic copy of $K_{m,m}$. Let $N_0 = n$, and for $i > 0$ let $N_i = b(N_{i-1})$. Let $N = N_{\binom{r}{2}}$. We claim that $K_r[N]$ has the desired property.

List the pairs of parts in some order. For i from 0 to $\binom{r}{2}$, we show that N is large enough so that for every s-coloring of $E(K_r[N])$, some copy of $K_r[N_{\binom{r}{2}-i}]$ has the property that the edges of the jth pair of parts are colored monochromatically, for each j with $1 \le j \le i$. For $i = 0$, there is nothing to prove; no requirement is imposed on the full graph $K_r[N]$.

After such a copy of $K_r[N_{\binom{r}{2}-i+1}]$ has been found for the first $i-1$ pairs of parts, consider the ith pair. The vertices retained in these two parts induce $K_{M,M}$, where $M = N_{\binom{r}{2}-i+1}$. By the definition of $N_{\binom{r}{2}-i+1}$, the inherited coloring of $E(K_{M,M})$ contains a monochromatic copy of $K_{m,m}$ with $m = N_{\binom{r}{2}-i}$. Keep these m vertices in these two parts and any m vertices in each remaining part. The resulting copy of $K_r[N_{\binom{r}{2}-i}]$ has the desired property for the first i parts. After reaching $i = \binom{r}{2}$, we have the desired copy of $K_r[n]$, since $N_0 = n$. ∎

In Theorem 10.2.27, the colors on monochromatic subgraphs induced by pairs of parts in the resulting $K_r[n]$ may differ. We need two more notions.

10.2.28. Definition. For a family $\mathcal{G}$ of graphs, let $R(\mathcal{G}; s)$ be the minimum number of vertices in a graph H such that every s-coloring of $E(H)$ has a monochromatic copy of some graph in $\mathcal{G}$. Write $R(\mathcal{G})$ for $R(\mathcal{G}; 2)$. A **homomorphism** from a graph G to a graph H is a map $\phi: V(G) \to V(H)$ such that adjacent vertices in G are mapped to adjacent vertices in H.

A proper k-coloring is the same thing as a homomorphism into K_k. Indeed, the notion of homomorphism generalizes graph coloring, and a homomorphism mapping G to H is called an H-**coloring** of G. Graph homomorphism is a vast subject (the book Hell–Nešetřil [2004] is devoted to it), but here we introduce it only to describe the chromatic Ramsey numbers of graphs (see also Exercise 41).

10.2.29. Theorem. (Burr–Erdős–Lovász [1976]) For every graph G, the s-color chromatic Ramsey number equals $R(\mathcal{G}; s)$, where $\mathcal{G}$ is the family of all homomorphic images of G.

Proof: Let $n = |V(G)|$, let $r = R(\mathcal{G}; s)$, and let $H = K_r[N]$, where N is the threshold guaranteed for n, r, s by Theorem 10.2.27. We show that $H \to_s G$, which completes the proof since H has the desired chromatic number.

Consider an s-coloring of $E(H)$. By Theorem 10.2.27, the coloring of some copy H' of $K_r[n]$ is constant on the edges joining any pair of parts. Collapsing each part to a single vertex thus yields an s-coloring of $E(K_r)$. By the definition of $R(\mathcal{G}; s)$, this s-coloring contains a monochromatic copy of some homomorphic image G' of G, say with color c.

Let ϕ be a homomorphism from G to G'. The set mapped to any vertex v in G' by ϕ is an independent set S_v in G, and $|S_v| \le n$. Hence we can return to H' and map each S_v injectively into the part in H' that collapsed to v in G'. If $uv \in E(G')$, then all of S_u is adjacent in H' to all of S_v by edges of color c. Hence H contains a monochromatic copy of G within H'. ∎

There are many variations. In **on-line Ramsey theory**, a Builder presents edges one by one to force Painter (who must color each edge immediately) to make a monochromatic G (see Kierstead–Konjevod [2009], Conlon [2009b/2010], Butterfield et al. [2011], etc.). **Ordered Ramsey theory** was introduced in Fox–Pach–Sudakov–Suk [2012]; with vertex orders on G and H, a monochromatic G respecting the ordering is sought in every coloring of $E(H)$. Theorem 12.4.45 treats such a problem; more recent papers include Balko–Cibulka–Král'–Kynčl [2015], Conlon–Fox–Lee–Sudakov [2017], and Mubayi [2017].

Anti-Ramsey theory asks how many colors are needed on $E(H)$ to force a copy of G with distinct colors on its edges. Erdős–Sós–Simonovits [1975] conjectured that when $H = K_n$ and G is a p-cycle this is $n(\frac{p-2}{2} + \frac{1}{p-1}) + O(1)$, proved by Montellano-Ballesteros & Neumann-Lara [2005]. Kano–Li [2008] and Fujita–Magnant–Ozeki [2010] are early surveys. The area has more than 200 papers.

EXERCISES 10.2

10.2.1. (−) Let D_n be the digraph with vertex set $[n]$ whose edges are the ordered pairs of distinct vertices. A tournament with a linear order on its vertices is **monotone** if every edge points toward its lower endpoint or every edge points toward its higher endpoint. Given k, prove that if n is sufficiently large, then every spanning subdigraph of D_n contains k vertices that induce no edges or a monotone tournament or a copy of D_k.

10.2.2. (−) Let $n = R(k, l)$. Prove that the edges of the graph obtained from K_n by deleting one edge can be 2-colored with no red K_k or blue K_l. (Chvatal [1974])

10.2.3. (−) Use the graph below to prove $R(3, 5) = 14$.

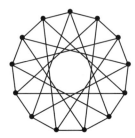

10.2.4. (−) Prove that 2-colorings of $E(K_6)$ have at least two monochromatic triangles.

10.2.5. (◊) For $k \geq 2$, say that a graph G is k-*balanced* if every induced subgraph with $2k$ vertices has independence number k.
    (a) By explicit construction, prove $R(3, k) > 3k − 4$.
    (b) Prove that every k-balanced graph has at most $R(3, k+1)+2$ vertices. (A. Brieden)
    (c) Construct a k-balanced graph with $2k + 2$ vertices. (Comment: Q. Zhu [1989] showed for $k \geq 4$ that no k-balanced graph has more than $2k + 2$ vertices.)

10.2.6. Prove $N(5) = 9$. That is, any nine points in $\mathbb{R}^2$ with no three collinear determine some convex 5-gon, and some set of eight points does not. (Comment: Due first to Makai and Turán in 1935 (unpublished), later proofs include Kalbfleisch–Kalbfleisch–Stanton [1970] and Bonnice [1974]. By computer, Szekeres–Peters [2006] showed $N(6) = 17$.)

10.2.7. (+) In a tournament with weights on the edges, a **nondecreasing [decreasing] path** is a path whose edge weights are successively nondecreasing [decreasing]. Prove that the maximum number of vertices in an edge-weighted tournament having no nondecreasing path of length $p+1$ and no decreasing path of length $q+1$ is $\binom{p+q}{p}$. (Hint: Let (c_{xy}, d_{xy}) denote the maximum length of (nondecreasing, decreasing) paths ending with edge xy. Associate with each vertex the set of maximal labels on entering edges. Show that at most $\binom{p+q}{p}$ distinct sets of labels can occur at vertices.) (Chvátal–Komlós [1971])

10.2.8. A list of points $v_1, \ldots, v_n$, where $v_i = (x_i, y_i)$ with $x_1 < \cdots < x_n$, is *convex* [*concave*] if whenever $i < j$ the segment $v_i v_j$ passes above [below] $\{v_{i+1}, \ldots, v_{j-1}\}$. Use Exercise 10.2.7 to prove that the maximum number of points in the plane (with no three collinear) containing no convex sequence of r points or concave sequence of s points is $\binom{r+s-4}{r-2}$. Conclude that $\binom{2m-4}{m-2} + 1$ points force a convex m-gon.

10.2.9. (+) Find 2^{m-2} points in the plane (no three collinear) that form no convex m-gon. (Hint: Use sets T_i for $0 \leq i \leq m − 2$ (from Exercise 10.2.8) that have $\binom{m-2}{i}$ points but contain no concave $(i + 2)$-sequence or convex $(m − i)$-sequence.) (Erdős–Szekeres [1960])

10.2.10. (◊) Let S be a set of $R(m, m; 3)$ points in the plane with no three collinear. Prove that S contains m points that form a convex m-gon. (Tarsi)

10.2.11. Prove that every graph with 2^k vertices has a clique Q and an independent set S such that $|Q| + |S| = k + 1$. Conclude from this that $R(k, k) \leq 4^{k-1}$.

10.2.12. *Ramsey numbers for $r = 2$ and multiple colors.*
   (a) Let $p = (p_1, \ldots, p_k)$, and obtain q_i from p by subtracting one from p_i but leaving the other coordinates unchanged. Prove $R(p) \leq \sum_{i=1}^k R(q_i) - k + 2$.
   (b) Prove $R(p_1 + 1, \ldots, p_k + 1) \leq \frac{(p_1 + \cdots + p_k)!}{p_1! \cdots p_k!}$.

10.2.13. Prove that every graph G with p vertices, q edges, and automorphism group of size s satisfies $R_k(G) > (sk^{q-1})^{1/p}$.

10.2.14. ($\Diamond$) Let $r_k = R_k(3; 2)$ (see Definition 10.2.10).
   (a) Prove $r_k \leq k(r_{k-1} - 1) + 2$.
   (b) Prove $\lfloor t!e \rfloor = t! \sum_{i=0}^t \frac{1}{i!}$ for $t \in \mathbb{N}$.
   (c) Prove $r_k \leq \lfloor k!e \rfloor + 1$ for $k \geq 2$. (Thus $r_3 \leq 17$; in fact, equality holds).

10.2.15. ($\Diamond$) For the diagonal k-color Ramsey number (Definition 10.2.10), prove the bound $R_k(q; 2) \leq 1 + \sum_{j=0}^{(q-1)k-1} k^j$. (Comment: This implies the simpler-looking but weaker bound $R_k(q; 2) < k^{qk}$, generalized in the next exercise.)

10.2.16. ($\Diamond$) For $k \geq 2$, let $M_k(p; r) = \max\{R(q_1, \ldots, q_k; r): \sum_{i=1}^k q_i = p\}$.
   (a) Prove $M_k(p; r) \leq 1 + M_k(kM_k(p - 1; r); r - 1)$.
   (b) Prove $M_k(p; 2) < k^p$.
   (c) Prove $R_k(q; 3) \leq t(kq + 1, k + 1)$, where $t(1, b) = b$ and $t(h, b) = b^{t(h-1, b)}$ (thus $t(h, b)$ is a tower of h copies of b). (Hint: Prove the stronger bound $M_k(p; 3) < k^{k^{t(p-1, k+1)}}$. Comment: Conlon–Fox–Sudakov [2010] improved the upper bound; in particular, $R_k(6; 3) \leq 2^{2^{(4+o(1))k \log k}}$. The technique of Lemma 10.2.19 yields $R_k(6; 3) > 2^{2^{ck}}$.)

10.2.17. Give a constructive proof of $R(p + 1, p + 1) > \binom{p}{3}$ by considering the graph G whose vertex set is $\binom{[p]}{3}$ with $xy \in E(G)$ if and only if $|x \cap y| = 1$. (Nagy [1972])

10.2.18. ($\Diamond$) The **composition** or **lexicographic product** of graphs G and H is the graph $G[H]$ on $V(G) \times V(H)$ defined by making (u, v) and (u', v') adjacent if and only if (1) uu' is an edge of G, or (2) $u = u'$ and vv' is an edge of H.
   (a) Prove $\alpha(G[H]) \leq \alpha(G)\alpha(H)$.
   (b) Prove that the complement of $G[H]$ is $\overline{G}[\overline{H}]$.
   (c) Use (a) and (b) to prove by construction that
$$R(pq + 1, pq + 1) - 1 \geq [R(p + 1, p + 1) - 1] \cdot [R(q + 1, q + 1) - 1].$$
   (d) Deduce that $R(2^n + 1, 2^n + 1) \geq 5^n + 1$ for $n \geq 0$, and compare this to the non-constructive lower bound for $R(k, k)$ in Theorem 10.2.14. (Abbott [1972])

10.2.19. Frankl–Wilson [1981] constructed explicit n-vertex graphs with clique and independence numbers at most $2^{c\sqrt{\log n \log \log n}}$, where c is a constant. Obtain from this a lower bound for $R(p, p)$ larger than every polynomial in p but smaller than any exponential in p.

10.2.20. (+) Prove $\frac{p}{e} k^{M_1} < R_k(p; r+1) \leq r + k^{M_2}$, where $M_1 = \frac{1}{r+1}\binom{p-1}{r}$ and $M_2 = \binom{R_k(p;r)}{r}$.

10.2.21. ($\Diamond$) Determine $R(K_{1,r_1}, \ldots, K_{1,r_k})$ (in all cases). (Burr–Roberts [1973])

10.2.22. ($\Diamond$) Let T be a tree with m edges, and let n be a multiple of m. Determine the Ramsey number $R(T, K_{1,n+1})$. (Burr [1974])

10.2.23. ($\Diamond$) Let T be a tree with m vertices. Prove that only one 2-coloring of $E(K_{(m-1)(n-1)})$ (up to isomorphism) has no red T or blue K_n. Conclude that if G is formed from $K_{(m-1)(n-1)}$ by adding one vertex adjacent to s vertices of the clique, then all red/blue colorings of G have a red T or a blue K_n if and only if $s > (m - 1)(n - 2)$. (Hook–Isaak [2011])

10.2.24. For m-vertex trees T, show $R(T, K_{n_1},...,K_{n_k}) = (m-1)(R(n_1,...,n_k)-1)+1$. (Burr)

10.2.25. ($\diamond$) Let G and H be graphs with $n = |V(G)|$ and $k = \chi(H)$. Let s be the least size of a color class in any proper k-coloring of H. When $s \leq n$, prove $R(G, H) \geq (n-1)(k-1)+s$. (Rousseau–Sheehan [1978], Burr [1981])

10.2.26. Prove $R(C_4, C_4) = 6$. (Comment: There are many proofs.)

10.2.27. ($\diamond$) Prove $R(mK_2, mK_2) = 3m - 1$.

10.2.28. Complete the proof that $R(2K_3, 2K_3) = 10$.

10.2.29. ($\diamond$) For $1 \leq i \leq k$, let G_i be a graph with p_i vertices, and $m_i \in \mathbb{N}$. Prove $R(m_1 G_1, \ldots, m_k G_k) \leq \sum(m_i - 1)p_i + R(G_1, \ldots, G_k)$.

10.2.30. ($\diamond$) Determine $R(P_3, G)$ for every n-vertex graph G, as a function only of n and the maximum size of a matching in $\overline{G}$.

10.2.31. Let G be an n-vertex graph with m edges. Build a graph H with $n + m$ vertices such that in every red/blue coloring of $E(H)$, the red subgraph has P_3 as an induced subgraph or the blue subgraph has G as an induced subgraph. (Comment: The bound $n + m$ cannot be reduced when the components of G are complete.) (Kostochka–Sheikh [2006])

10.2.32. ($\diamond$) Prove that every k-coloring of $E(K_{s,t})$ has a monochromatic connected subgraph with at least $(s + t)/k$ vertices. Conclude that every k-coloring of $E(K_n)$ has such a subgraph with at least $n/(k-1)$ vertices. For $k = 3$, show that one cannot guarantee more than $n/2$. (Gerencsér–Gyárfás [1967], Gyárfás [1977]) (Comment: Mubayi [2002] and Liu–Morris–Prince [2009] guaranteed double-stars as such subgraphs of $K_{s,t}$. One can use at least $1/k$ of the vertices in each part when $k \leq 3$ (Bucić–Letzter–Sudakov [2018]), but not when $k > 3$ (DeBiasio et al. [2018]). For surveys, see Gyárfás [2011, 2016].)

10.2.33. Let G be an n-vertex graph with m edges.
(a) Prove that if $\sum_{v \in V(G)} \binom{d(v)}{2} > \binom{n}{2}$, then G contains a 4-cycle.
(b) Prove that if $m > \frac{n}{4}(1 + \sqrt{4n-3})$, then G contains a 4-cycle.
(c) Prove $R_k(C_4) \leq k^2 + k + 2$. (Comment: Using difference sets (Section 13.2), there is a lower bound of $k^2 - k + 2$.) (Chung–Graham [1975])

10.2.34. ($\diamond$) Prove that for n sufficiently large, every 2-coloring of $E(K_{n,n})$ has a monochromatic copy of $K_{p,p}$. Show that the analogue fails for tripartite graphs: 2-colorings of $E(K_{n,n,n})$ need not have monochromatic triangles no matter how large n is.

10.2.35. ($\diamond$) Bondy [1971a] proved that if $xy \notin E(G)$ implies $d(x)+d(y) \geq n$ for an n-vertex graph G, then $G = K_{t,t}$ or G has a cycle of each length from 3 to n. Use this to prove $R(C_m, K_{1,r}) = \max\{m, 2r+1\}$, except maybe when m is even and $m \leq 2r$. (Lawrence [1973]).

10.2.36. ($\diamond$) *Ramsey numbers of connected n-vertex graphs.*
(a) Prove that if G is a connected bipartite graph with parts of sizes r and s, with $r \geq s$, then $R(G, G) \geq \max\{2r - 1, r + 2s - 1\}$. (Burr [1974])
(b) Let G be a connected n-vertex graph. Prove $R(G, G) \geq \lfloor (4n-1)/3 \rfloor$.
(c) Prove $R(H, H) \geq \max\{2r - 1, r + 2s - 1\}$, where $H = K_{1,r-1} + K_{1,s}$ and $r > s$.
(d) The formula of part (c) is in fact an upper bound (Rosta). Use this to prove that parts (a) and (b) are sharp by computing $R(G, G)$ when G is the tree obtained from $K_{1,r-1}+K_{1,s-1}$ by adding an edge joining leaves of the two components. (Burr–Erdős [1976]) (Comment: Burr and Erdős conjectured $R(G, G) \leq 2n - 2$ when G is an n-vertex tree.)

10.2.37. (+) For $m \geq n \geq 2$, prove $R(P_m, P_n) = m + \lfloor n/2 \rfloor - 1$. (Gerencsér–Gyárfás [1967])

10.2.38. For $n \geq 3$, prove that every 2-coloring of $E(K_n)$ has a spanning cycle that consists of at most two monochromatic paths. (Gerencsér–Gyárfás [1967]) (Comment: A version for digraphs due to Raynaud [1973] was proved also in Gyárfás [1983].)

10.2.39. (+) *Ramsey numbers for cycles.* Let f be a 2-coloring of $E(K_n)$.

(a) Prove that if f contains a monochromatic copy of C_{2k+1} for some $k \geq 3$, then f also contains a monochromatic copy of C_{2k}.

(b) Prove that if f contains a monochromatic C_{2k} for some $k \geq 3$, then f also contains a monochromatic C_{2k-1} or $2K_k$.

(c) Prove $R(C_m, C_m) \leq 2m - 1$ for $m \geq 5$ ($m = 4$ is Exercise 10.2.26). (Hint: Use the Erdős–Gallai [1959] result that an n-vertex graph with more than $(m-1)(n-1)/2$ edges has a cycle of length at least m (Theorem 7.3.16). There is still one difficult case.)

10.2.40. *3-color Ramsey numbers for cycles.*

(a) Prove $R(C_n, C_n, C_n) \geq 2n$ when n is even.

(b) Prove $R(C_n, C_n, C_n) \geq 4n - 3$ when n is odd.

(Comment: Equality holds in both statements when n is sufficiently large, proved by Benevides–Skokan [2009] for even n. Jenssen–Skokan [2016] proved for odd n that $R_k(C_n) = 2^{k-1}(n-1) + 1$. Both papers use the Regularity Method (Section 11.1).)

10.2.41. (◇) *Chromatic Ramsey number.*

(a) Prove the general lower bound $R_\chi(G; s) > (\chi(G) - 1)^s$. (Comment: Burr–Erdős–Lovász [1976] conjectured that this lower bound is sharp; this was proved by Zhu [2011].)

(b) Let G be a 3-chromatic graph. Prove that $5 \leq R_\chi(G; 2) \leq 6$, with equality in the lower bound if and only if there is a homomorphism from G into C_5.

10.2.42. (◇) In a family of axis-parallel squares in $\mathbb{R}^2$, a set is *homogeneous* if it is pairwise intersecting or pairwise disjoint. Prove that n squares always have a homogeneous set of size at least $\sqrt{n/4}$. Prove for infinitely many n that some family has no homogeneous set of size more than $\sqrt{4n/5}$. (Hagelstein–Herden–Young [2017], Hagelstein–Herden [2018]) (Comment: See Larman–Matoušek–Pach–Törőcsik [1994] for related problems.)

10.2.43. (◇) Given $k, l \in \mathbb{N}$, prove that there exists $N_{k,l}$ such that for $d \geq N_{k,l}$, every k-coloring of the edges of the d-dimensional cube Q_d contains a monochromatic path of length l that is a shortest path joining its endpoints. (Hint: Given the value $N_{k,l}$, obtain an upper bound on $N_{k,2l}$ to prove that it exists.) (Stong [2018])

10.2.44. The **zero-sum Ramsey number** $R(G, \mathbb{Z}_k)$ of a graph G with $|E(G)|$ divisible by k is the least n such that every coloring of $E(K_n)$ by integers has a copy of G on which the sum of the colors is divisible by k. (Caro [1996] surveys such problems.)

(a) Prove that $R(G, \mathbb{Z}_k)$ is well-defined.

(b) Prove $R(K_3, \mathbb{Z}_3) = 11$.

10.3. Further Topics

Ramsey Theory is also called "partition calculus". The idea is that every partition of a large configuration contains a subconfiguration with more structure. We consider coloring integers to obtain arithmetic structure, the infinite version of Ramsey's Theorem and its application to the finite version, and finally the Canonical Ramsey Theorem, which allows infinitely many colors.

VAN DER WAERDEN'S THEOREM

Ramsey's Theorem was not the first result in the partition calculus, but it is the best known and most thoroughly studied. The first result about monochromatic structures in colorings of natural numbers was due to Hilbert [1892] (Exercise 3). We begin with a later result that still pre-dates Ramsey's Theorem and arose from an attempt to prove Fermat's Last Theorem.

10.3.1. Theorem. (Schur's Theorem; Schur [1916]) Given $k > 0$, there is an integer s_k such that every k-coloring of $\{1, \ldots, s_k\}$ yields monochromatic (not necessarily distinct) x, y, z solving $x + y = z$.

Proof: Let $r_k = R_k(3; 2)$. We show that $s_k < r_k$ by showing that every k-coloring f of $[r_k - 1]$ has a monochromatic solution to $x + y = z$. From f, we define a k-coloring f' of $E(K_{r_k})$. Let $V(K_{r_k}) = [r_k]$. Let the color of edge ij in f' be $f(|i - j|)$.

By the choice of r_k, f' yields a monochromatic triangle with some vertices a, b, c. We may assume $a < b < c$. Let $x = b - a$, $y = c - b$, $z = c - a$; we have $f(x) = f(y) = f(z)$. By construction, $x + y = z$. ∎

10.3.2. Example. *Bounds for Schur numbers.* Like Ramsey numbers, few Schur numbers are known exactly. We know $s_1 = 2$, $s_2 = 5$, $s_3 = 14$, and $s_5 = 45$ (by computer). The upper bound of $r_k - 1$ yields $s_k \leq \lfloor k!e \rfloor$ (see Exercise 10.2.14).

Schur proved $s_k \geq (3^k + 1)/2$. Given a k-coloring f of $[n]$ with no monochromatic solution to $x + y = z$, define a $(k + 1)$-coloring f' of $[3n + 1]$ by

$$f'(i) = \begin{cases} f(i) & \text{if } i \leq n, \\ f(i - 2n - 1) & \text{if } i \geq 2n + 2, \\ k + 1 & \text{if } n + 1 \leq i \leq 2n + 1. \end{cases}$$

Checking cases shows that f' produces no monochromatic solution to $x + y = z$ (Exercise 1). The resulting recurrence $s_k \geq 3s_{k-1} - 1$ yields $s_k \geq (3^k + 1)/2$. ∎

Schur's Theorem generalizes considerably. For a homogeneous linear equation $c_1 x_1 + \cdots + c_t x_t = 0$, we may ask whether k-colorings of $[n]$ must contain a monochromatic solution when n is large. Rado (Schur's student) proved that this holds if and only if $\sum_{i \in I} c_i = 0$ for some nonempty index set $I \subseteq [t]$. More generally, he determined the *systems* of homogeneous linear equations such that k-colorings eventually force monochromatic solutions.

10.3.3. Definition. Let $Cx = 0$ be a homogeneous system of linear equations with rational coefficients, where no column of C is 0. The system is **regular** if for all k there is an integer n such that every k-coloring of $[n]$ yields a monochromatic solution to $Cx = 0$. The matrix C satisfies the **Columns Condition** if there is a partition $C_1, \ldots, C_t$ of its columns such that the columns in the first block sum to 0 and in each subsequent block the sum of the columns is a rational linear combination of columns in the preceding blocks.

When C has just one row (with no 0), the Columns Condition reduces to having a nonempty index set I such that $\sum_{i \in I} c_i = 0$. These form the first block, and each later coefficient is a multiple of one in the first block.

10.3.4. Theorem. (Rado's Theorem; Rado [1933]) A linear system $Cx = 0$ is regular if and only if C satisfies the Columns Condition. ∎

Rado's Theorem was extended by Deuber [1973]. We will prove a weaker version of Rado's Theorem for coefficient matrices of the type below.

$$\begin{pmatrix} 1 & -1 & 0 & 0 & 0 & -1 \\ 0 & 1 & -1 & 0 & 0 & -1 \\ 0 & 0 & 1 & -1 & 0 & -1 \\ 0 & 0 & 0 & 1 & -1 & -1 \end{pmatrix}$$

The columns before the last sum to zero, and the last is a linear combination of the others by assigning weight i to column i. Thus the matrix does satisfy the Columns Condition.

To interpret the equations corresponding to the matrix, name the variables $x_1, \ldots, x_l$ and d. The equations then become $x_{i+1} = x_i + d$ for $1 \le i < l$. In other words, $x_1, \ldots, x_l$ must form an l-term arithmetic progression. Under Rado's Theorem, we are guaranteed a monochromatic arithmetic progression where the constant difference d also has the same color. We will prove the weaker statement that does not require d to have the same color as the elements of the progression.

10.3.5. Theorem. (**van der Waerden's Theorem**; van der Waerden [1927])
Given positive integers l and k, there is an integer $w(l, k)$ such that every k-coloring of $[w(l, k)]$ has a monochromatic l-term arithmetic progression.

We present the proof by Graham–Rothschild [1974]. Mills [1983] rephrased it (and that of the subsequent Hales–Jewett Theorem) more compactly. Our presentation follows that of Graham–Rothschild–Spencer [1980]. The details of a small case suggest the ideas in the proof.

10.3.6. Example. We prove $w(3, 2) \le 325$. This is quite loose, since $w(3, 2) = 9$ (Exercise 8). Like the inductive proof of Ramsey's Theorem, this inductive proof gives a ridiculously large upper bound for $w(l, k)$.

Consider a 2-coloring of $[325]$. Partition $[325]$ into 65 blocks of 5 consecutive integers. The coloring gives some red-blue pattern to each block; there are 32 possible patterns. By using 65 blocks, some pattern is repeated by the time we reach the middle block. Thus we can find three equally-spaced blocks such that the first two have the same pattern. Similarly, we use 5-element blocks so that we find a duplicated color by the middle of any block. Thus we can find three equally spaced elements in the block such that the first two have the same color.

For example, when blocks 3 and 30 have the same pattern $RBRBB$, consider block 57 with elements 281-285. We obtain either a blue progression as $\{15, 150, 285\}$ or a red progression as $\{11, 148, 285\}$. ∎

Of the 32 color patterns on blocks, 19 contain 3-term monochromatic progressions. Hence we need only 27 blocks, not 65. We also ignored other ways to obtain 3-term progressions. By considering all possible progressions, Exercise 8 cuts the bound from 325 to the actual minimum, 9.

Nevertheless, block structures lead to a proof of van der Waerden's Theorem by induction on l without seeking a strong bound on $w(l, k)$. The bounds become enormous because when we want a progression of $l - 1$ blocks having the same pattern, the exponentially many patterns behave as distinct colors. With larger

k, we will also need to consider more potential monochromatic progressions. To manage this, we introduce a way to embed a multidimensional grid in $[n]$.

10.3.7. Definition. Let $[0, l]^m$ be the set of integer m-tuples $(x_1, \ldots, x_m)$ such that $0 \le x_i \le l$. The j**th critical class** C_j is defined by

$$C_j = \{x \in [0, l]^m : x_i = l \Leftrightarrow i \le j\}.$$

An m-**dimensional block structure** (of *length l*) in $[n]$ is a choice of positive integers $a, d_1, \ldots, d_n$ that map $[0, l]^m$ to $[n]$ injectively by sending x to $a + \sum x_i d_i$. A block structure is **layered** for a coloring f of $[n]$ if f is constant on the image of each critical class.

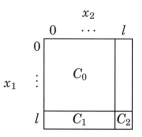

The set $[0, l]^m$ is an m-dimensional grid of lattice points with $l + 1$ points on each side, sketched above for $m = 2$. The critical class C_j consists of the m-tuples whose first j entries equal l and whose last $m - j$ entries are less than l. There are $m + 1$ critical classes, with $|C_j| = l^{m-j}$.

Van der Waerden's Theorem says roughly that for large $[n]$ every k-coloring contains a layered 1-dimensional block structure of length l. The image of C_0 is then the monochromatic arithmetic progression $\{a + rd_1 : 0 \le r \le l - 1\}$.

To prove this by induction on l we will need higher-dimensional layered block structures. In a layered block structure on $[0, l]^m$ using m colors, two of the $m + 1$ critical classes must have the same color. For any two critical classes, there is an l-term arithmetic progression in the image of the larger one that extends to an $(l + 1)$-term progression using an element of the smaller one, as suggested by the figure above. To guarantee progressions of length $l + 1$ we will need to guarantee layered block structures of length l for all dimensions. Hence we again have a double induction, like in Ramsey's Theorem.

10.3.8. Theorem. (van der Waerden [1927], Graham–Rothschild [1974]) For $l, m, t \in \mathbb{N}$, there exists $w(l, m, t)$ such that every t-coloring of $[w(l, m, t)]$ contains a layered block structure on $[0, l]^m$.

Proof: We use induction on l. For each l, we use induction on m. For each l we prove a basis step for $m = 1$ (using the hypothesis for shorter lengths) and an induction step for $m > 1$. Each argument holds for all t.

For the basis step $m = 1$, we bound $w(l + 1, 1, t)$ in terms of $w(l, m, t')$ for all m, t' (focusing the classes in a layered structure). For the induction step $m > 1$, we bound $w(l, m + 1, t)$ in terms of $w(l, k, t')$ for all $k \le m$ and all t' (creating the high-dimensional block structure). Since we have $w(1, 1, t) = 2$ as the overall basis, we do the latter step first.

Step 1: $w(l, m+1, t) \leq MM'$, where $M = w(l, m, t)$ and $M' = w(l, 1, t^M)$. Consider a t-coloring $f: [MM'] \to [t]$. Partition $[MM']$ into M' consecutive blocks of length M. Induce a coloring $f': [M'] \to [t^M]$ by the color patterns used on the blocks. There are t^M possible color patterns for a block of M elements; we have M' such blocks. Since $M' = w(l, 1, t^M)$, there are $l + 1$ equally spaced *special blocks* with the same pattern on the first l of them. Let a' be the index of the first special block, and let d' be the difference between indices of successive special blocks.

Since $M = w(l, m, t)$, the pattern (t-coloring) in the first special block contains a layered block structure on $[0, l]^m$. Stacking the $l + 1$ special blocks increases the dimension of the block structure for the full coloring. Since the pattern is the same for the first l special blocks, the corresponding critical classes have the same color and stack to form a monochromatic critical class l times as large in the $(m + 1)$-dimensional block structure.

To be precise, let the layered block structure in the first special block be $a'', d_1, \ldots, d_m$ when viewing its first position as 1 instead of $1 + (a' - 1)M$. For the block structure in the full coloring, let $d_{m+1} = d'M$ and $a = (a' - 1)M + a''$. The repetition of pattern in the first l special blocks yields monochromatic critical classes l times as large as before. The $(m+1)$th critical class, corresponding to $x = (l, \ldots, l)$, is the last element of the $(l + 1)$th special block. No other element of this block belongs to any critical class, so the colors in this block are irrelevant.

Step 2: $w(l + 1, 1, t) \leq \left\lceil w(l, t, t)\frac{l+1}{l} \right\rceil$. Let $n = \left\lceil w(l, t, t)\frac{l+1}{l} \right\rceil$, and consider $f: [n] \to [t]$. We force the 1-dimensional arithmetic progression by finding a t-dimensional layered block structure. By the choice of n, f is constant on critical classes of some block structure on $[0, l]^t$ in $[w(l, t, t)]$ with parameters $a, d_1, \ldots, d_t$. Two of the $t+1$ monochromatic critical classes must have the same color, say classes r and s with $r < s$. To obtain the desired 1-dimensional block structure, we use special elements in $[0, l]^t$. Let $x^{(i)}$ have the first r coordinates be l, the next $s - r$ coordinates be i, and the remaining coordinates be 0.

The elements $x^{(0)}, \ldots, x^{(l-1)}$ belong to class r, and x^l belongs to class s, but their images have the same color. Set $a' = a + l \sum_{i=1}^{r} d_i$ and $d' = \sum_{i=r+1}^{s} d_i$. These $l + 1$ images form a monochromatic arithmetic progression with constant difference d'; this is the 1-dimensional critical class. Finally, add the element $a' + (l + 1)d'$ as a singleton critical class. This element belongs to $[n]$ due to the factor $(1 + 1/l)$ in the definition of n.   ∎

10.3.9.* Remark. When we color $\mathbb{N}$ with finitely many colors, van der Waerden's Theorem guarantees arbitrarily long arithmetic progressions. "Arbitrarily long" does not mean infinite! Consider 2-coloring $\mathbb{N}$ by the parity of $\lceil \lg n \rceil$. An arithmetic progression in one color has a fixed difference d. No matter where it starts, later there are more than d consecutive integers in the other color. Hence there is no infinite monochromatic arithemetic progression.

We know even less about van der Waerden numbers than about Ramsey numbers: $w(2, 2) = 3$, $w(3, 2) = 9$, $w(4, 2) = 35$, $w(3, 3) = 27$, $w(3, 4) = 76$, $w(5, 2) = 178$, $w(6, 2) = 1132$ (Kouril–Paul [2008]), and $w(4, 3) = 293$ (Kouril [2012]).

The proof of Theorem 10.3.8 yields what Graham–Rothschild–Spencer [1980] called "Eeeeenormous Upper Bounds." For diagonal Ramsey numbers, the induction yields roughly $R(q; r) \leq 2^{R(q; r-1)^{r-1}}$; this builds a "tower" of r exponentiations. The bound from Theorem 10.3.8 grows even faster. Let $w_l(k)$ be the resulting bound on the number required to force monochromatic l-progressions in a k-coloring, viewed as a function of k for each l. Let $v_l(k)$ be defined by letting $v_2(k) = 2k$ and for $l \geq 3$ letting $v_l(k)$ be the result of k iterations of v_{l-1} starting with argument 1. Thus $v_3(k) = 2^k$, $v_4(k)$ is an exponential tower of k 2s, and $v_5(k)$ is already difficult to describe. Even $v_l(2)$, in terms of l, grows unimaginably fast.

Shelah [1988] found a new proof that improves the upper bound. His bound is what logicians call "primitive recursive", whereas the bound from Theorem 10.3.8 is not. Gowers [2001] further improved the bounds.

Another short proof of van der Waerden's Theorem uses the language and properties of "ultrafilters". An **ultrafilter** on a set X is a nonempty family of subsets of X that is upwardly closed, is closed under intersection, omits $\varnothing$, and contains at least one of A and $\overline{A}$ for all $A \subseteq X$. Hindman [1979] proved that results in Ramsey theory that guarantee homogeneous sets correspond to the existence of an ultrafilter on the set being colored. Many results in Ramsey theory have been proved by this method. A thorough survey of the subject, focusing on van der Waerden's Theorem, can be found in Johannson [2007]. ∎

Van der Waerden's Theorem implies Rado's Theorem for systems with one equation, but this still takes some effort. We present an easier application. For a finite set S of integers, let $\mathbf{P}(S)$ be the set of sums of nonempty subsets of S. Folkman and others proved that in every k-coloring of $\mathbb{N}$ there are arbitrarily large sets S such that $\mathbf{P}(S)$ is monochromatic. This follows easily from Rado's Theorem (Exercise 19). To prove it from van der Waerden's Theorem, we need a lemma.

10.3.10.* Lemma. A set $S \subseteq \mathbb{N}$ with sum at most n is *max-governed* under a coloring of $[n]$ if the color of the sum of a subset of S depends only on its maximum element. Given $l, k \in \mathbb{N}$, there exists $M(l, k) \in \mathbb{N}$ such that under every k-coloring of $[M(l, k)]$ some l-set is max-governed.

Proof: We use induction on l; note that $M(1, k) = 1$. Existence of $M(2, k)$ follows using the Schur number s_k. For $l > 2$, let $m = M(l - 1, k)$. Set $n = 2w(m + 1, k)$, and consider a k-coloring f of $[n]$. By van der Waerden's Theorem, the top half of $[n]$ has a monochromatic arithmetic progression T with $m + 1$ terms. Let a_l be its first term and d be its constant difference. Since T fits in an interval of length $n/2$, we have $dm \leq n/2$.

Define a coloring $f' \colon [m] \to [k]$ by $f'(j) = f(dj)$. The induction hypothesis yields an $(l - 1)$-set S' with sum at most m that is max-governed under f'. Let $S = (dS') \cup \{a_l\}$. Since $dm \leq n/2 < a_l$, the largest element of S is a_l. The sum of any subset containing a_l exceeds a_l by a multiple of d that is at most dm, so it belongs to T and has the same color as a_l. ∎

10.3.11.* Theorem. (**Folkman's Theorem**; Folkman [1970], Sanders [1969], Rado [1970]) For $k, t \in \mathbb{N}$, there exists $N \in \mathbb{N}$ such that every k-coloring of $[N]$ yields a t-set with sum at most N whose nonempty subsets have the same color as their sums.

Proof: In Lemma 10.3.10, let $N = M(k(t - 1) + 1, k)$. Given a k-coloring f of $[N]$, we obtain $a_1, \ldots, a_{k(t-1)+1}$ such that the sum of every nonempty subset has the same color as its largest element. The Pigeonhole Principle yields t of them in the same color. These form the desired set. ∎

The block structures in the proof of van der Waerden's Theorem suggest that it concerns sets and arrangements more than integers. The Hales–Jewett Theorem captures the combinatorial essence. The proof is analogous, so we just state the result. The point is that when we color a grid in sufficiently many dimensions, we obtain a monochromatic line. We must define "line" carefully.

Let $S = \{a_0, \ldots, a_{l-1}\}$, and let S^m be the set of m-tuples drawn from S. We designate elements of S^m by the vector of subscripts. Thus S^m can be viewed as C_l^m, the m-dimensional grid with l points on each edge. Only the order of elements in S matters, not their values.

10.3.12. Definition. A **line** in C_l^m is a set of l distinct vectors $x^{(0)}, \ldots, x^{(l-1)}$ determined by a partition B_0, B_1 of $[m]$ and constants c_j for $j \in B_0$ by setting $x_j^{(i)} = c_j$ if $j \in B_0$ and $x_j^{(i)} = i$ if $j \in B_1$ (B_0 may be empty).

In this notion of line, coordinate values vary in the same order on coordinates in B_1, and they are fixed on each coordinate in B_0. Not every geometric line is a line here; $\{210, 111, 012\}$ is not a line in C_3^3.

10.3.13. Theorem. (**Hales–Jewett Theorem**; Hales–Jewett [1963]) Given positive integers l, t, there is an integer $HJ(l, t)$ such that every t-coloring of C_l^m with $m \geq HJ(l, t)$ has a monochromatic line. ∎

The Hales–Jewett Theorem includes van der Waerden's Theorem and implies Folkman's Theorem. We present a geometric application. Two sets $V, W \in \mathbb{R}^m$ are **homothetic** if one arises from the other by a scale change and a translation, meaning a correspondence between $\{v_i\}$ and $\{w_i\}$ such that $w_i = cv_i + z$ with $c \in \mathbb{R} - \{0\}$ and $z \in \mathbb{R}^m$.

10.3.14. Theorem. (**Gallai–Witt Theorem**; see Rado [1933], Witt [1952]) If the points of $\mathbb{R}^m$ are colored with finitely many colors, then for every finite set $V \subset \mathbb{R}^m$ there exists a monochromatic W that is homothetic to V.

Proof: Let $f \colon \mathbb{R}^m \to [t]$ be the coloring. Let $l = |V|$ and $N = HJ(l, t)$. View the elements of the cube C_l^N as lists $(x_1, \ldots, x_N)$ with $x_i \in V$. Define $\phi \colon C_l^N \to \mathbb{R}^m$ by $\phi(x_1, \ldots, x_N) = \sum_{i=1}^N k_i x_i$ for constants $k_1, \ldots, k_N$ chosen to make ϕ injective. We need only avoid finitely many equalities of the form $\sum k_i (x_i - x_i') = 0$ and can choose $k_1, \ldots, k_N$ to be integers.

Once ϕ is injective, define $f' \colon C_l^N \to [t]$ by $f'(x) = f(\phi(x))$. The choice of N guarantees a monochromatic line; let B_0, B_1 be its coordinate partition. The image of the line under ϕ is monochromatic under f. To see that it is homothetic to V, observe that $\phi(x^{(j)}) = cv_j + z$, where $c = \sum_{i \in B_1} k_i$ and $z = \sum_{i \in B_0} k_i x_i^{(j)}$; both c and z are constants. ∎

If the elements of V are positive integer vectors, then we find a monochromatic set homothetic to V in a sufficiently large (i.e., finite) grid. If we choose V to be a cube itself, then Gallai's Theorem guarantees a higher-dimensional version of a monochromatic arithmetic progression.

INFINITE SETS (optional)

Most statements of Ramsey theory can be phrased concisely using the vague-sounding "sufficiently large". A statement about infinite sets would be simpler. That is, when we k-color the r-subsets of $\mathbb{N}$, must there be an infinite homogeneous set? This seems too much to ask but would avoid quotas and Ramsey numbers and would imply the finite version.

Extensions to infinite sets can be dangerous, leading to unsupported leaps of faith and false statements. For example, injections from A to B and B to A must be bijections when A and B are finite, but this fails when A and B are infinite (doubling and negation are injections from $\mathbb{Z}$ to $\mathbb{Z}$, but doubling is not a bijection).

In contrast, Ramsey's Theorem behaves as desired. Like the proof of the finite version, this proof invokes the induction hypothesis for both $r-1$ and 1. However, we ignore quotas and "Ramsey numbers" and don't care which color provides the infinite homogeneous set.

10.3.15. Theorem. (**Infinite Ramsey Theorem**; Ramsey [1930]) For finite r and k, every k-coloring of $\binom{\mathbb{N}}{r}$ has an infinite homogeneous set.

Proof: We use induction on r. The case $r = 1$ is the Pigeonhole Principle: every finite partition of an infinite set has an infinite class. For $r > 1$, consider a k-coloring f of $\binom{\mathbb{N}}{r}$. We produce an infinite sequence of vertices $x_1^*, x_2^*, \cdots$ forming a homogeneous set under f.

First we construct sequences $\langle x \rangle, \langle c \rangle, \langle D \rangle$. The elements of $\langle x \rangle$ are natural numbers (candidates for $\langle x^* \rangle$), those of $\langle c \rangle$ are colors, and those of $\langle D \rangle$ are infinite subsets of $\mathbb{N}$. Begin with $D_0 = \mathbb{N}$. Having constructed D_{n-1}, construct x_n, c_n, and D_n as follows. Let x_n be the least element of D_{n-1}. Induce a k-coloring f' of the $(r-1)$-subsets S of $D_{n-1} - \{x_n\}$ by setting $f'(S) = f(S \cup \{x_n\})$. Since D_{n-1} is infinite, the induction hypothesis guarantees an infinite subset of D_{n-1} that is homogeneous under f' in some color c. Let this set be D_n and this color be c_n.

Since D_n is infinite, the construction proceeds and constructs sequences $\langle x \rangle$ and $\langle c \rangle$. Since there are finitely many colors, some value c^* occurs infinitely often in $\langle c \rangle$, by the Pigeonhole Principle ($r = 1$). Let $\{n_i : i \in \mathbb{N}\}$ be the set of indices such that $c_{n_i} = c^*$. Define $\langle x^* \rangle$ by $x_i^* = x_{n_i}$.

We claim that $f(S) = c^*$ whenever S consists of r terms in $\langle x^* \rangle$. Let the least element of S be x_n. All other elements of S belong to D_n. By the construction, the union of these with the element x_n has color c^* under f. We conclude that $\langle x^* \rangle$ forms the desired infinite set. ∎

Ramsey [1930] proved both versions separately. The infinite version can be used to prove the finite version (but not the other way!) via the "Compactness Principle" (also called **Rado Selection Principle**), which was discovered repeatedly (see Rado [1949], Erdős [1950], Gottschalk [1951], de Bruijn–Erdős [1951]).

The Compactness Principle relates colorings of an infinite hypergraph to colorings of finite subhypergraphs. If A is a subset of the vertex set of H, then the **induced subhypergraph** H_A is the hypergraph with vertex set A whose edges are the edges of H contained in A.

10.3.16. Theorem. (Compactness Principle) Let H be a hypergraph with a countable vertex set and finite edges. If $\chi(H_A) \leq k$ for every finite set of vertices A, then $\chi(H) \leq k$.

Proof: Given $\chi(H_A) \leq k$ for all finite A, we construct a proper k-coloring of H. Rename the vertices so $V(H) = \mathbb{N}$. We develop the coloring iteratively, defining each $f^*(j)$ so that the full coloring $f^*: \mathbb{N} \to [k]$ properly colors H. Let f_n be a proper k-coloring given for $H_{[n]}$ by hypothesis.

Having defined $f^*(1), \ldots, f^*(j-1)$, let $S_{j-1} = \{n \geq j: f_n(i) = f^*(i)$ for $1 \leq i \leq j-1\}$. That is, S_{j-1} is the set of future values n such that f_n agrees with the choices made so far for f^*. Given that S_{j-1} is infinite, we will choose $f^*(j)$ so that S_j is also infinite.

Begin with $S_0 = \mathbb{N}$. Suppose that $f^*(1), \ldots, f^*(j-1)$ have been defined so that S_{j-1} is infinite. For $n \in S_{j-1}$, partition S_{j-1} into k classes according to $f_n(j)$. Since S_{j-1} is infinite, some such class is infinite; call it S_j, and define $f^*(j)$ to agree with $f_n(j)$ for all n in S_j.

This defines a coloring f^*; we check that f^* is a proper k-coloring. Let E be an edge of H, and let m be its highest vertex. Since S_m is infinite, there exists $n \in S_m$ such that f^* agrees with f_n on $[m]$. Since f_n properly colors E, also f^* properly colors E.   ∎

The Compactness Principle yields results on finite structures from results on infinite structures. This can yield shorter existence proofs of theorems on finite sets but gives no information on bounds. In some sense this is not such a great deficiency, since the bounds from inductive proofs are so bad.

The contrapositive is the usual form for applications: If $\chi(H) > k$, then there is a finite $A \subseteq V(H)$ such that $\chi(H_A) > k$. We demonstrate the use of the Compactness Principle on Ramsey's Theorem itself. We will need to restate Ramsey's Theorem in terms of hypergraph coloring.

10.3.17. Corollary. (Finite Ramsey Theorem) For $k, r, q \in \mathbb{N}$, there is $n \in \mathbb{N}$ such that every k-coloring of $\binom{[n]}{r}$ has a homogeneous q-set.

Proof: Let H^q be the hypergraph with vertex set $\binom{\mathbb{N}}{r}$ that has an edge $\binom{S}{r}$ for each q-set $S \subseteq \mathbb{N}$; note that every edge is finite (with size $\binom{q}{r}$). By Ramsey's Theorem (Theorem 10.3.15) every k-coloring of $\binom{\mathbb{N}}{r}$ has an infinite homogeneous set; hence it also has a homogeneous set of size q.

We conclude that H^q is not k-colorable. Every edge of H^q is finite, so the Compactness Principle guarantees a finite set $A \subseteq \mathbb{N}$ such that H_A^q is not k-colorable. Letting n be the largest element of A, we have $H_A^q \subseteq H_{[n]}^q$. We conclude that the desired n exists and is at most $|A|$.   ∎

If we finitely color the r-subsets of an uncountable set, will there be an uncountable homogeneous subset? The answer is yes when $r = 1$, but when $r \geq 2$ we have trouble seeking a set with an order-preserving bijection to $\mathbb{R}$. The **Well-Ordering Principle** states that every set S has a well ordering, meaning a linear ordering with respect to which every nonempty subset of S has a least element. The Well-Ordering Principle is equivalent to the Axiom of Choice.

10.3.18. Theorem. Given the Well-Ordering Principle, there is a 2-coloring of the unordered pairs in $\mathbb{R}$ such that no subset that corresponds to $\mathbb{R}$ under the usual ordering has its pairs monochromatic.

Proof: Let $\prec$ be a well ordering of $\mathbb{R}$, and let $<$ be the usual ordering. For $a, b \in \mathbb{R}$, color $\{a, b\}$ red if a and b have the same order under $<$ and $\prec$; otherwise color it blue. If some red-homogeneous subset corresponds bijectively to $\mathbb{R}$ under $<$, then $<$ agrees with the well ordering $\prec$ on those points, but $\mathbb{R}$ is not well-ordered by $<$. Similarly, such a blue-homogeneous set yields a well ordering of $\mathbb{R}$ under $>$.   ∎

THE CANONICAL RAMSEY THEOREM (optional)

What happens for infinite sets of colors? As usual, we color $\binom{\mathbb{N}}{r}$. When $r = 1$, the situation is simple. If we use infinitely many colors, then we have an infinite set receiving distinct colors. Otherwise, we use only finitely many colors, and then the Pigeonhole Principle guarantees an infinite monochromatic set. When $r \geq 2$, we can no longer guarantee an infinite set whose r-sets all receive distinct colors or all receive the same color, but we *can* guarantee a more general object.

10.3.19. Definition. For $S \subset \mathbb{N}$, a coloring f of $\binom{S}{r}$ is **canonical** if there is a set of indices $I \subseteq [r]$ such that $f(A) = f(B)$ if and only if for $i \in I$ the ith largest elements of A and B are equal. A colored set is **rainbow** (or **polychromatic** or **heterochromatic**) if its elements have distinct colors.

The condition for $f(A) = f(B)$ is *not* that the ith largest elements of A and B are equal if and only if $i \in I$; additional equalities between A and B are allowed. For example, if $I = \varnothing$, then the condition holds for all $\{A, B\}$, so in a canonically colored S with $I = \varnothing$ all r-sets have the same color. On the other hand, if $I = [r]$, then r-tuples get the same color if and only if they agree completely, so $\binom{S}{r}$ is rainbow colored when $I = [r]$, using distinct colors on all r-sets.

When $r = 2$, there are two other possibilities: a canonically-colored S with $I = \{1\}$ has $f(A) = f(B)$ if and only if $\min(A) = \min(B)$, and with $I = \{2\}$ the requirement is $\max(A) = \max(B)$. We state the infinite and finite versions of the general theorem (see Graham–Rothschild–Spencer [1980, 1990, Chapter 5]).

10.3.20. Theorem. (**Canonical Ramsey Theorem**; Erdős–Rado [1950]) Every coloring of $\binom{\mathbb{N}}{r}$ is canonical on some infinite set. For every q, there exists n such that every coloring of $\binom{[n]}{r}$ is canonical on some q-set.   ∎

For the finite version when $r = 1$, see Exercise 11. We prove just the infinite version for $r = 2$. The four possibilities for a canonical coloring are monochromatic, min-coloring, max-coloring, and rainbow.

10.3.21. Theorem. Every coloring f of $\binom{\mathbb{N}}{2}$ is canonical on some infinite set.

Proof: Write a 4-tuple A in $\mathbb{N}$ as $\{a_1, a_2, a_3, a_4\}$ with $a_1 < a_2 < a_3 < a_4$. The behavior of f on A induces a partition of the edges of the complete graph on $\{1, 2, 3, 4\}$ by putting ij and kl in the same class if $f(\{a_i, a_j\}) = f(\{a_k, a_l\})$.

With this fixed vertex set $\{1, 2, 3, 4\}$, there are 203 possible such patterns, corresponding to the 203 distinct partitions of $[6]$ (that is, $\sum_{k=1}^{6} S(6, k) = 203$).

By the Infinite Ramsey Theorem (Theorem 10.3.15) with 203 colors, under f all 4-sets in some infinite set S are colored by the same pattern. In fact, as soon as S has at least seven elements, only a few patterns can hold for all 4-sets in S.

Consider the equalities below, dropping the set brackets within $f(\{i, j\})$.

$$
\begin{array}{ll}
f(a_1, a_2) = f(a_1, a_3) & \qquad f(a_2, a_4) = f(a_3, a_4) \\
f(a_1, a_2) = f(a_1, a_4) & \qquad f(a_1, a_4) = f(a_3, a_4) \\
f(a_1, a_3) = f(a_1, a_4) & \qquad f(a_1, a_4) = f(a_2, a_4) \\
f(a_2, a_3) = f(a_2, a_4) & \qquad f(a_1, a_3) = f(a_2, a_3)
\end{array}
$$

If the pattern on $\{1, 2, 3, 4\}$ satisfies the first column of equalities whenever $\{a_1, a_2, a_3, a_4\} \subset S$ with $a_1 < a_2 < a_3 < a_4$, then the coloring is *min-determined*. This is distinct from min-coloring in that min-determined does not imply that $f(i, j) \neq f(k, l)$ when $\min\{i, j\} \neq \min\{k, l\}$. Similarly, a pattern satisfying the second column is *max-determined*.

Without loss of generality, let $[7]$ be the first seven elements of S, in order. Since the pattern is constant over $\binom{S}{4}$, a single equality yields the four requirements in either column when applied to various 4-sets in $[7]$:

$$
\begin{array}{llll}
\text{if } f(2, 4) = f(2, 6), & & \text{if } f(2, 6) = f(4, 6), & \\
\text{then min-determined} & & \text{then max-determined} & \\
2\ 4\ 6\ 7 & f(a_1, a_2) = f(a_1, a_3) & 1\ 2\ 4\ 6 & f(a_2, a_4) = f(a_3, a_4) \\
2\ 4\ 5\ 6 & f(a_1, a_2) = f(a_1, a_4) & 2\ 3\ 4\ 6 & f(a_1, a_4) = f(a_3, a_4) \\
2\ 3\ 4\ 6 & f(a_1, a_3) = f(a_1, a_4) & 2\ 4\ 5\ 6 & f(a_1, a_4) = f(a_2, a_4) \\
1\ 2\ 4\ 6 & f(a_2, a_3) = f(a_2, a_4) & 2\ 4\ 6\ 7 & f(a_1, a_3) = f(a_2, a_3)
\end{array}
$$

If $f(2, 4) = f(2, 6) = f(4, 6)$, then the pattern puts all edges of K_4 in the same class, and f is monochromatic on $\binom{S}{2}$. To complete the proof, we need to consider the consequences of $f(2, 6) \neq f(4, 6)$ and $f(2, 4) \neq f(2, 6)$. Again we list these in two columns

$$
\begin{array}{llll}
\text{if } f(2, 6) \neq f(4, 6) & & \text{if } f(2, 4) \neq f(2, 6) & \\
2\ 4\ 6\ 7 & f(a_1, a_3) \neq f(a_2, a_3) & 1\ 2\ 4\ 6 & f(a_2, a_3) \neq f(a_2, a_4) \\
2\ 4\ 5\ 6 & f(a_1, a_4) \neq f(a_2, a_4) & 2\ 3\ 4\ 6 & f(a_1, a_3) \neq f(a_1, a_4) \\
2\ 3\ 4\ 6 & f(a_1, a_4) \neq f(a_3, a_4) & 2\ 4\ 5\ 6 & f(a_1, a_2) \neq f(a_1, a_4) \\
1\ 2\ 4\ 6 & f(a_2, a_4) \neq f(a_3, a_4) & 2\ 4\ 6\ 7 & f(a_1, a_2) \neq f(a_1, a_3)
\end{array}
$$

If $f(2, 4) = f(2, 6) \neq f(4, 6)$, then both columns on the left hold. From the first, the coloring is min-determined. The second then ensures that the pattern is a min-coloring. Similarly, $f(2, 4) \neq f(4, 6) = f(2, 6)$ forces max-coloring.

Finally, when $f(2, 4) \neq f(2, 6) \neq f(4, 6)$, we must show that the pattern has no two edges with the same color. The two columns of inequalities establish distinctness for eight pairs among the six edges; there are seven more to check. For example, if $f(a_1, a_2) = f(a_2, a_3)$, then $\{1, 2, 4, 5\}$ and $\{1, 2, 6, 7\}$ yield $f(2, 4) = f(1, 2) = f(2, 6)$, a contradiction. We leave the remaining cases as Exercise 12. ∎

For the finite version with $r = 2$ and $q \geq 7$, it suffices to let n be the Ramsey number such that every 203-coloring of $\binom{[n]}{2}$ has a homogeneous q-set S. The proof above then guarantees that $\binom{S}{2}$ is canonically colored.

The Canonical Ramsey Theorem has interesting applications to graphs. In a graph with vertex set $[n]$, an **increasing path** is a path through successively increasing vertices. The Canonical Ramsey Theorem implies that for large enough n, every coloring of $E(K_{n+1})$ has a monochromatic increasing path of length l or a rainbow increasing path of length k. Let $f(l, k)$ be the least such n. Lefmann–Rödl–Thomas [1992] proved constructively that $f(l, k) \geq l^{k-1}$ and conjectured that equality holds. (Hence also the value of n needed to force a canonical q-set for $r = 2$ is at least $(q + 1)^q$.) Jiang–Mubayi [2000] provided support for the conjecture by proving that $f(l, k)$ is bounded by $(1 + o(1))l^{k-1}$ when $k \in o(\sqrt{l})$.

Another variation ignores the names of colors. For example, when we want a monochromatic copy of G we care not which colors are used. More generally, we may seek certain patterns of contrast among colors.

10.3.22. Definition. An edge-coloring of a graph is **lexical** if the vertices can be linearly ordered so that edges have the same color if and only if they have the same lower endpoint. A **pattern** is a graph G plus a partition of $E(G)$ into color classes. A pattern G *contains* a pattern H if H occurs in G with edges partitioned in G the same as in the given classes in H.

The formally weaker version of the Canonical Ramsey Theorem for $r = 2$ is that for $q \in \mathbb{N}$, there exists $n \in \mathbb{N}$ such that every edge-coloring of K_n contains a monochromatic, a rainbow, or a lexically colored K_q.

10.3.23. Definition. A family $\mathbf{F}$ of patterns is a **Ramsey family** if for some integer n every edge-coloring of K_n contains a pattern in $\mathbf{F}$. The smallest such n is the **pattern Ramsey number** $R(\mathbf{F})$.

10.3.24. Theorem. (Jamison–West [2004]) A family of color patterns is a Ramsey family if and only if it has a monochromatic pattern, a rainbow pattern, and a lexical pattern (not necessarily distinct).

Proof: Let $\mathbf{F}$ be a Ramsey family with pattern Ramsey number n. Every subgraph of a pattern that is monochromatic, rainbow, or lexical inherits that property. By considering monochromatic, rainbow, and lexical colorings of K_n, we see that $\mathbf{F}$ must contain a pattern of each type.

Conversely, when $\mathbf{F}$ contains a monochromatic pattern M, a rainbow pattern R, and a lexical pattern L, let p be the maximum number of vertices among M, R, and L. By the Canonical Ramsey Theorem, there exists n such that every edge-coloring of K_n contains a canonically colored K_p. If an edge-coloring of K_n contains a monochromatic or rainbow K_p, then it contains M or R.

More care is needed when K_n contains a lexical K_p. The lexical coloring is relative to some linear order $<_1$ on $V(K_p)$. Also, the lexical coloring of L assumes a linear order $<_2$ on $V(L)$. Having L as a subpattern requires an order-preserving injection of $V(L)$ into $V(K_p)$. Since both orders are linear and L has at most p vertices, such an injection exists. ∎

Jamison [2011] showed that lexical colorings can be recognized in time that is linear in the number of edges. The same is true for monochromatic and rainbow patterns, so Ramsey families of color patterns can be recognized in linear time.

When $\mathbf{F}$ consists of a monochromatic G and a rainbow $K_{1,t}$, the pattern Ramsey number is the smallest n such that every edge-coloring of K_n with fewer than t colors at each vertex contains a monochromatic G. Introduced in Gyárfás–Lehel–Nešetřil–Rödl–Schelp–Tuza [1987], this is the **local Ramsey number** of G.

When $\mathbf{F}$ consists of a monochromatic and a rainbow copy of one tree T, Chen–Schelp–Wei [2001] showed that the pattern Ramsey number is at most quartic in the order of T, and Jamison–Jiang–Ling [2003] obtained a cubic upper bound. It is conjectured that the truth is quadratic.

EXERCISES 10.3

10.3.1. $(-)$ Confirm that under the $(k+1)$-coloring f' in Example 10.3.2, there is no monochromatic solution to $x + y = z$.

10.3.2. $(-)$ For edge-colorings of graphs (see Definition 10.3.22), prove that the only patterns that are monochromatic and lexical are monochromatic stars, and prove that the only patterns that are rainbow and lexical are rainbow forests. (Jamison–Jiang–Ling [2003])

10.3.3. $(\diamond)$ **Hilbert's Cube Lemma** (Hilbert [1892]). Given $a \in \mathbb{N}_0$ and $d_1, \ldots, d_m \in \mathbb{N}$, the **affine m-cube** $H(a, d_1, \ldots, d_m)$ is the set $\{a + \sum_{i \in I} d_i : I \subseteq [m]\}$. For $m, k \in \mathbb{N}$, prove that there is a least integer $h_{m,k}$ such that every k-coloring of $[h_{m,k}]$ contains a monochromatic affine m-cube. (Hint: Prove $h_{m,k} \leq k^n + n$, where $n = h_{m-1,k}$. Comment: Brown–Erdős–Chung–Graham [1985] proved $h_{2,k} = (1 + o(1))k^2$ and showed $k^{c_1 m} \leq h_{m,k} \leq k^{c_2^m}$ for constants c_1 and c_2. Gunderson–Rödl [1998] gave further refinements.)

10.3.4. $(\diamond)$ Show that for large n, every k-coloring of the nonempty subsets of $[n]$ gives the same color to some two disjoint sets and their union.

10.3.5. $(\diamond)$ Consider the equation $\sum_{i=1}^{m} x_i = x_0$, for $m \geq 1$.
   (a) Prove that $m^2 + m - 1$ is the least n such that every 2-coloring of $[n]$ contains a monochromatic solution (numbers may repeated). (Beutelspacher–Brestovansky [1982])
   (b) For each n, determine the maximum size of a subset of $[n]$ containing no solution.
   (Comment: Given $a_1, \ldots, a_m \in \mathbb{N}$, Guo–Sun [2008] found the least n such that all 2-colorings of $[n]$ have monochromatic solutions of $\sum_{i=1}^{m} a_i x_i = x_0$. It is $rs^2 + s - r$, where $r = \min\{a_1, \ldots, a_m\}$ and $s = \sum_{i=1}^{m} a_i$. Schaal [1993] studied $c + \sum_{i=1}^{m-1} x_i = x_m$. No n suffices when c is odd and m is even, but otherwise the value is known. When $c \geq -(m-2)$, it is $m^2 + (c-1)(m+1)$ if m is odd or c is even. For smaller c it is more complicated.)

10.3.6. $(\diamond)$ Given a homogeneous linear equation with integer coefficients, let $f(k)$ be the least n such that every k-coloring of $[n]$ produces a monochromatic solution to the equation. For each equation below, (1) use Rado's Theorem (Theorem 10.3.4) to determine whether $f(k)$ exists for all k, and (2) determine $f(2)$.

   (a) $x + y = 2z$.   (b) $x + y = 3z$.   (c) $w + x + y = z$.

10.3.7. Given m, prove that if p is prime and sufficiently large, then $x^m + y^m = z^m$ has a solution in integers modulo p. (Dickson [1909], Schur [1916])

10.3.8. Prove $w(3, 2) = 9$. That is, every 2-coloring of $[9]$ has a 3-term monochromatic arithmetic progression, but some 2-coloring of $[8]$ does not. Generalize the latter to prove $w(l, 2) > 2(l - 1)^2$. (Hint: On $[9]$, consider possible colorings of $\{4, 5, 6\}$ or $\{1, 5, 9\}$.)

10.3.9. $(\diamond)$ Let $\langle a \rangle$ and $\langle b \rangle$ be sequences of distinct positive integers, and fix $k \in \mathbb{N}$. Prove that if $\langle a \rangle$ contains arbitrarily long arithmetic progressions, and $|b_n - a_n| \leq k$ for all n, then $\langle b \rangle$ contains arbitrarily long arithmetic progressions. (Rubel [1988])

10.3.10. Prove that every infinite sequence has an infinite nondecreasing subsequence or an infinite decreasing subsequence.

10.3.11. Prove that every finite colored set of at least $(k-1)^2 + 1$ integers has a canonically colored k-set. (Note that the colored objects are 1-sets.)

10.3.12. Finish the proof of the Canonical Ramsey Theorem for $r = 2$ (Theorem 10.3.21).

10.3.13. Prove that for every 2-coloring of the points of $\mathbb{Z}^m$, there is a set of n integer lattice points with the same color whose centroid (mean vector) is an integer lattice point also having that color. (Bòna [1991])

10.3.14. ($\diamond$) Prove that every coloring of $E(K_n)$ has a rainbow triangle or a monochromatic spanning tree. (Thus a graph or its complement is connected.) (Galvin [1975])

10.3.15. ($\diamond$) A **Gallai coloring** is an edge-coloring of a complete graph with no rainbow triangle. (Gyárfás–Simonyi [2004])
    (a) Prove that every Gallai coloring with at least three colors has a color class containing no spanning tree. (Hint: Show that the edges joining two components formed by a color class have the same color.)
    (b) Prove that every Gallai coloring arises from a 2-colored complete graph H by expanding each vertex into a Gallai-colored complete graph. That is, each vertex v_i expands into a vertex set V_i so that each V_i induces a Gallai coloring, and each edge xy with $x \in V_i$ and $y \in V_j$ inherits the color originally on v_iv_j in H.
    (c) Prove that the pattern Ramsey number of the family consisting of a monochromatic $K_{1,m}$ and a rainbow K_3 is the least n such that $\lceil 2n/5 \rceil \geq m$.

10.3.16. ($\diamond$) **Product Ramsey Theorem.** For $r = (r_1, \ldots, r_d) \in \mathbb{N}^d$, an r-**grid** is a set $S_1 \times \cdots \times S_d$, where S_i is a set of r_i natural numbers. Given r, k, and vector thresholds $p_1, \ldots, p_k \in \mathbb{N}^d$, prove that there exists a vector $(n_1, \ldots, n_d) \in \mathbb{N}^d$ such that any k-coloring of the r-grids of $[n_1] \times \cdots \times [n_d]$ yields for some i a p_i-grid whose r-grids all have color i.

10.3.17. Product van der Waerden Theorem. An **arithmetic** m-**grid** of size l is a set of points in $\mathbb{Z}^m$ of the form $\{(a_1 + i_1d_1, \ldots, a_m + i_md_m): 0 \leq i_j < l\}$, where $d_1, \ldots, d_m$ are fixed positive integers.
    (a) Using van der Waerden's Theorem (not the Hales–Jewett Theorem), prove for $r, l, m \in \mathbb{N}$ that there is a number W such that every r-coloring of $[W]^m$ contains a monochromatic arithmetic m-grid of size l.
    (b) For $r, n \in \mathbb{N}$ with $n > 2$, conclude that every r-coloring of $\mathbb{Z}^2$ contains n points with the same color spanning a convex polytope with nonzero 2-dimensional volume whose centroid (mean vector) also has that color. (Comment: This generalizes to $\mathbb{Z}^m$ with $m \geq 2$.)

10.3.18. Determine the least n such that for every k-coloring of $[n]$ there is a pair S of elements for which the color of the sum of every nonempty subset of S is determined by its maximum element (see Lemma 10.3.10).

10.3.19. ($\diamond$) Prove Folkman's Theorem from Rado's Theorem. In particular, show that the system $y_T = \sum_{i \in T} x_i$ for nonempty sets $T \subseteq [t]$ satisfies the Columns Condition.

10.3.20. Consider a positional game with positions $x_1, \ldots, x_n$ and winning sets $\{S_1, \ldots, S_m\}$. (Tic-Tac-Toe has nine positions and eight winning sets.)
    (a) Prove that Player 1 can win Tic-Tac-Toe on the m-dimensional cube C_l^m of side-length l if l is fixed and m is sufficiently large. (Hales–Jewett [1963])
    (b) Suppose that each winning set has size at least a and each position appears in at most b winning sets. Prove that Player 2 can force a draw if $a \geq 2b$.
    (c) In Tic-Tac-Toe on C_l^m, prove that Player 2 can force a draw when $l \geq 3^m - 1$.

Chapter 11

Extremal Problems

In this chapter, we study extremal problems for families of sets and subsets. We seek the largest or smallest structure with certain properties, or extreme values of a parameter over a class of structures.

11.1. Forced Subgraphs

In Ramsey's Theorem, we k-color $\binom{[n]}{t}$ for some large n and find that some q-set is homogeneous. That is, we find a monochromatic copy of the **complete t-uniform hypergraph** with q vertices, written $K_q^{(t)}$. Given n and q with $n \geq q$, we can also force a copy of $K_q^{(t)}$ by having enough edges in a t-uniform hypergraph on n vertices; the maximum number of edges avoiding $K_q^{(t)}$ is **Turán's Problem**.

For $t \geq 3$, the problem is notoriously difficult; Füredi [1991] and Keevash [2011] give extensive surveys. We do not even know the asymptotic answer for $K_4^{(3)}$ (see Exercise 30). The "flag algebra" method pioneered by Razborov [2007, 2010] has improved many of the bounds; Falgas-Ravry–Vaughan [2013] summarizes the results. Hence we focus on $t = 2$, where the central question started the field of extremal graph theory.

TURÁN'S THEOREM

The traditional phrasing asks for the maximum number of edges in an n-vertex graph having no $(r + 1)$-clique. This was solved for $r = 2$ by Mantel [1907] and for general r by Turán [1941]. We already proved the theorem as Theorem 5.2.11; here we take a more quantitative look and give another short proof.

11.1.1. Definition. The **Turán graph** $T_{n,r}$ is the complete r-partite graph with n vertices having b parts of size $a+1$ and $r-b$ parts of size a, where $a = \lfloor n/r \rfloor$ and $b = n - ra$. Let $t_r(n) = |E(T_{n,r})|$.

Turán proved that $T_{n,r}$ is the unique largest n-vertex graph not containing K_{r+1}. The proof by Erdős in Theorem 5.2.11 uses induction on r and the Degree-

Sum Formula, showing that for every graph G not containing K_{r+1}, there is an r-partite graph H with $d_H(v) \geq d_G(v)$ for all $v \in V(G)$. It also uses that $T_{n,r}$ is the unique largest r-partite n-vertex graph (if part-sizes differ by more than 1, shifting a vertex from a larger part to a smallest part gains edges). Here we note the asymptotic value of the number of edges.

11.1.2. Proposition. $t_r(n) = \left(1 - \frac{1}{r}\right)\frac{n^2}{2} - O(n)$.

Proof: With part-sizes $x_1, \ldots, x_r$, there are $\sum_{i<j} x_i x_j$ edges. With $\sum x_i = n$ fixed and all x_i real, the sum is maximized when all x_i equal n/r (averaging any two increases the value). The maximum is thus $\binom{r}{2}\frac{n^2}{r^2}$, which proves $t_r(n) \leq \left(1 - \frac{1}{r}\right)\frac{n^2}{2}$.

For the asymptotic optimality of the bound, note that each part-size exceeds $n/r - 1$. Counting the edges by pairs of parts thus yields

$$t_r(n) > \frac{r(r-1)}{2}\left(\frac{n-r}{r}\right)^2 = \frac{1}{2}\left(1 - \frac{1}{r}\right)\left(n^2 - 2nr + r^2\right) = \left(1 - \frac{1}{r}\right)\frac{n^2}{2} - O(n). \qquad \blacksquare$$

Turán's Theorem has many proofs. Turán's original proof used induction on n (Exercise 7), different from the Erdős proof. The proof in Exercise 8 uses a continuous optimization problem. Theorem 14.1.15 gives a probabilistic proof. Six proofs appear in Aigner [1995], and the five proofs in Aigner–Ziegler [1999] include the proof we present here.

11.1.3. Theorem. (**Turán's Theorem**; Turán [1941]) Among n-vertex graphs with no $(r+1)$-clique, the unique largest graph is $T_{n,r}$.

Proof: (Zykov [1949]) Since $T_{n,r}$ is the unique largest r-partite graph, it suffices to show that any largest graph G not containing K_{r+1} is a complete multipartite graph, which means it does not have $K_2 + K_1$ as an induced subgraph. If G has such a subgraph, with vertex set $\{u, v, w\}$ and $vw \in E(G)$, then we find a larger graph not containing K_{r+1}.

If $d(u) < d(v)$, then we replace u with a new vertex v' having the same neighbors as v, as shown below. Since v' and v are not adjacent, we did not create an $(r+1)$-clique. Since $d(u) < d(v) = d(v')$, the new graph has more edges.

By symmetry in v and w, we may assume $d(u) \geq \max\{d(v), d(w)\}$. Now we replace both v and w by two new copies of u, as shown below. Again we have create no larger clique. We lose only $d(v) + d(w) - 1$ edges, since vw is counted twice, and we gain $2d(u)$ edges, so again the new graph is larger. $\qquad \blacksquare$

In fact, for $2 \leq k \leq r$, the Turán graph $T_{n,r}$ is the unique n-vertex graph without K_{r+1} that has the most k-cliques (Zykov [1949]). This generalization was rediscovered by Erdős [1962b] and by Sauer [1971].

An n-vertex graph with more than $\frac{n^2}{2}\frac{r-2}{r-1}$ edges contains K_r. More generally, how many r-cliques are forced by m edges on n vertices?

11.1.4. Definition. Let $k_r(G)$ denote the number of r-cliques in G.

Let G be an n-vertex graph with m edges. Moon–Moser [1962a] gave a lower bound for $k_r(G)$ in terms of n and m. Goodman [1959] minimized $k_3(G) + k_3(\overline{G})$ in terms of n. Lovász [1972d] obtained both results via inclusion-exclusion. One would expect a *random* n-vertex graph and its complement together to have approximately $2 \cdot \frac{1}{8} \binom{n}{3}$ triangles; surprisingly, the *minimum* asymptotically equals the average. The proof uses a counting formula that has other applications.

11.1.5. Lemma. If a graph G has n vertices and m edges, then $k_3(G) + k_3(\overline{G})$ equals each formula below, where $d(v)$ is the degree of v in G.

(a) $\binom{n}{3} - \frac{1}{2} \sum_{v \in V(G)} d(v)[n - 1 - d(v)]$ (Sauvé [1961])

(b) $\binom{n}{3} - (n-2)m + \sum_{v \in V(G)} \binom{d(v)}{2}$.   (Goodman [1959])

Proof: (Sauvé) View G and $\overline{G}$ as a 2-coloring of $E(K_n)$. Edges of G have one color and edges of $\overline{G}$ the other. We count monochromatic triangles.

Assign each pair of incident edges in K_n weight 2 if they have the same color, weight -1 if not. A vertex triple inducing a monochromatic triangle contributes 6 to the total weight; other triples contribute 0. Hence the sum of the weights of all pairs of incident edges is 6 times the number of monochromatic triangles.

Over the pairs of edges incident to v, the weights sum to

$$2\binom{d(v)}{2} + 2\binom{n-1-d(v)}{2} - d(v)[n - 1 - d(v)].$$

Using $\binom{k}{2} + \binom{m-k}{2} = \binom{m}{2} - k(m - k)$, the contribution from v becomes $2\binom{n-1}{2} - 3d(v)[n - 1 - d(v)]$. Summing over v and dividing by 6 yields (a).

To obtain (b) from (a), replace $n - 1$ with $n - 2 + 1$, apply the Degree-sum Formula, and collect terms. ∎

11.1.6. Theorem. (Goodman [1959]) An n-vertex graph and its complement together have at least $n(n-1)(n-5)/24$ triangles, sharp when $n \equiv 1 \pmod 4$.

Proof: (Sauvé [1961]) To minimize formula (a) of Lemma 11.1.5, we maximize the subtracted terms. This is achieved by setting each $d(v)$ to $(n-1)/2$. The formula becomes $\binom{n}{3} - \frac{n}{2}\frac{(n-1)^2}{4}$, which simplifies to the claimed lower bound.

By Lemma 11.1.5, $k_3(G) + k_3(\overline{G})$ depends only on the vertex degrees, not the choice of edges. Equality in the bound holds if and only if every vertex has degree $(n-1)/2$. This can happen only for n odd and $(n-1)/2$ even, so $n = 4k+1$ and G is $2k$-regular. When $n \equiv 1 \pmod 4$, this is achieved by a regular self-complementary graph, such as the graph obtained by adding one vertex adjacent to the low-degree vertices in the near-regular self-complementary graph $P_4[K_k, \overline{K}_k, \overline{K}_k, K_k]$. For other congruence classes, the bound can be improved slightly (Exercise 27). ∎

We return to the counting of cliques in the graph G itself. For $r = 3$, Lemma 11.1.5 provides a lower bound.

11.1.7. Corollary. (Moon–Moser [1962a]) A graph G with n vertices and m edges has at least $\frac{m}{3n}(4m - n^2)$ triangles.

Proof: Write the formula of Lemma 11.1.5b in terms of $\overline{G}$, letting $m' = |E(\overline{G})|$ and $d'(v) = d_{\overline{G}}(v)$. Note that $3k_3(\overline{G}) \le \sum \binom{d'(v)}{2}$, since incident edges lie in at most one triangle. Subtracting this upper bound on $k_3(\overline{G})$ from Lemma 11.1.5b yields

$$k_3(G) \ge \binom{n}{3} - (n-2)m' + \tfrac{2}{3}\sum\binom{d'(v)}{2}.$$

Replace $\sum\binom{d'(v)}{2}$ with the lower bound $n\binom{2m'/n}{2}$, replace m' with $\binom{n}{2} - m$, and simplify (Exercise 1) to obtain $k_3(G) \ge \frac{m}{3n}(4m - n^2)$. ∎

11.1.8.* Remark. *Lower bounds.* For $q+1 \ge r > p$, the Turán graph $T_{n,q}$ has the fewest r-cliques among n-vertex graphs G with $k_p(G) = k_p(T_{n,q})$. Bollobás [1976b] proved that linear interpolation yields lower bounds on $k_r(G)$ for intermediate values of $k_p(G)$ (see Bollobás [1978, pp. 297–301]). When $p = 2$ and $r = 3$ and $m = |E(G)|$, interpolation improves Corollary 11.1.7 for $m \in [\frac{n^2}{4}, \frac{n^2}{3}]$. Since $K_{n/2,n/2}$ has $n^2/4$ edges and no triangles, while $K_{n/3,n/3,n/3}$ has $n^2/3$ edges and $n^3/27$ triangles, interpolation yields at least $\frac{n}{9}(4m - n^2)$ triangles. This improves Corollary 11.1.7, since $\frac{n}{9} > \frac{m}{3n}$ when $m < n^2/3$.

Just above $t_r(n)$ (with $n^2/4+1$ edges), Turán's Theorem guarantees one triangle, the Moon–Moser bound guarantees $n/3$, and interpolation guarantees $4n/9$. In fact, G has at least $\lfloor n/2 \rfloor$ triangles (Rademacher; see Erdős [1955], in Hebrew). This is sharp, by adding one edge to $T_{n,2}$.

Adding q edges inside one part of $T_{n,2}$ (without forming triangles) creates only $q\lfloor n/2 \rfloor$ triangles. Erdős [1962b] proved that this minimizes $k_3(G)$ when $q < cn$, for some constant c. When $q = n/2$, adding a $(k+1)$-cycle to $K_{k+1,k-1}$ produces only $(n/2)(n/2)-1$ triangles. For large n, Lovász–Simonovits [1976] proved Erdős' conjecture that $c = 1/2$ is best, plus similar results for larger complete graphs.

Mubayi [2010] greatly generalized the results. For an $(r+1)$-chromatic graph F having an edge e such that $\chi(F - e) = r$, there is a constant c_F such that for $1 \le q \le c_F n$, every n-vertex graph with $t_r(n) + q$ edges has at least qs copies of F, where s is the minimum number of copies of F formed by adding one edge to $T_{n,r}$. This is sharp for odd cycles and asymptotically sharp in general. The tool is the Graph Removal Lemma, generalizing the Triangle Removal Lemma (Lemma 11.1.22) obtained from the Szemerédi Regularity Lemma (Theorem 11.1.13).

The problem was also studied for larger numbers of edges: we want to minimize $k_r(G)$ when G is an n-vertex graph with at least γn^2 edges, where $\gamma > \frac{1}{2}(1 - \frac{1}{r-1})$. Lovász–Simonovits [1983] conjectured that the minimizing graph is a complete r-partite graph where all the parts (except one smaller) have the same size. This was proved for $r = 3$ by Razborov [2008], for $r = 4$ by Nikiforov [2011], and in general by Reiher [2016]. ∎

ERDŐS–STONE THEOREM

Just as the Ramsey problem extends by seeking a monochromatic copy of any target graph, so the Turán problem extends by asking how many edges force a given graph. When K_q is forced, every q-vertex graph is forced, but fewer edges may force sparser subgraphs.

11.1.9. Definition. The **Turán number** of F, written $\text{ex}\,(n;F)$, is the maximum number of edges in an n-vertex graph not containing F. Let $G[s]$ denote the graph obtained from G by replacing each vertex v with an independent set S_v of size s, with vertices in S_u and S_v adjacent if and only if $uv \in E(G)$.

Since any edge beyond a maximal triangle-free subgraph must lie in a triangle, n-vertex graphs with $t_2(n)+cn^2$ edges have many triangles when c is a positive constant. Similarly, $t_r(n) + cn^2$ edges force many $(r+1)$-cliques. Erdős and Stone proved that when n is large enough, $t_r(n) + cn^2$ edges force copies of K_{r+1} whose union is a complete $(r+1)$-partite graph with large vertex classes.

11.1.10. Theorem. (**Erdős–Stone Theorem**; Erdős–Stone [1946]) Fix $s \in \mathbb{N}$ and a positive constant c. If n is sufficiently large, then every n-vertex graph with $t_r(n) + cn^2$ edges contains $K_{r+1}[s]$. ∎

This theorem yields the asymptotics of $\text{ex}\,(n;F)$ when $\chi(F) > 2$.

11.1.11. Theorem. (Erdős–Simonovits [1966]) If F is an $(r+1)$-chromatic graph, then $\lim_{n\to\infty} \text{ex}\,(n;F)n^{-2} = \frac{1}{2}(1 - \frac{1}{r})$.

Proof: Since $T_{n,r}$ is r-partite, $F \not\subseteq T_{n,r}$ when $\chi(F) > r$. Thus $\text{ex}\,(n;F) \geq t_r(n) = \frac{1}{2}(1 - \frac{1}{r})n^2 - O(n)$. For the upper bound, let s be the maximum size of a color class in some proper $(r+1)$-coloring of F. In an n-vertex graph with $\frac{1}{2}(1 - \frac{1}{r} + c)n^2$ edges, where n is sufficiently large in terms of c, the Erdős–Stone Theorem guarantees the appearance of $K_{r+1}[s]$ and thus also F. Thus the ratio $\text{ex}\,(n;F)/n^2$ can be brought down as close to $\frac{1}{2}(1 - \frac{1}{r})$ as desired by making n sufficiently large. ∎

Few exact results are known. For odd cycles, in fact $\text{ex}\,(n;C_{2k+1}) = \lfloor n^2/4 \rfloor$ when $n \geq 4k - 2$ (see Füredi–Gunderson [2015], including all n).

For bipartite graphs, Theorem 11.1.11 gives only $\text{ex}\,(n;F) \in o(n^2)$. Counting arguments yield better bounds; in fact, $\text{ex}\,(n;C_4) \in O(n^{3/2})$. The constructions involve designs and projective planes, so we postpone discussion of $\text{ex}\,(n;C_4)$ to Chapter 13. Simonovits [1968] pioneered a method for studying Turán numbers (also called **extremal numbers**). These have been studied for various bipartite graphs and families of graphs, surveyed in Füredi–Simonovits [2013].

Of most interest is $\text{ex}\,(n;C_{2k})$. Bondy–Simonovits [1974] proved $\text{ex}\,(n;C_{2k}) < 90kn^{1+1/k}$ (see Exercise 14.2.11 for a lower bound). The constant was improved by Verstraëte [2000] and then Pikhurko [2012]. More recently, Bukh–Jiang [2017] proved $\text{ex}\,(n;C_{2k}) \leq 80\sqrt{k}\log k\, n^{1+1/k} + 10k^2n$ when $k \geq 2$ and $n \geq (2k)^{8k^2}$.

For the Erdős–Stone Theorem, Exercise 31 requests a direct proof. Here we use an enormously important tool for proving asymptotic results in combinatorics. It is the Szemerédi Regularity Lemma, developed originally to prove that sets of integers with positive density contain arbitrarily long arithmetic progressions.

Roughly speaking, the Regularity Lemma states that every sufficiently large graph contains within it a multipartite subgraph with large parts such that almost all the bipartite subgraphs induced by pairs of parts look fairly random. This is useful when seeking a particular subgraph, because in randomly generated graphs with a fixed density of edges (see Chapter 14) the number of occurences of a particular subgraph is almost always close to its expected value.

The induced subgraphs of fixed size in a "fairly random" bipartite graph have about the same edge density, if they are not too small. Hence we restrict the edge density only for subgraphs with at least a fraction ε of the vertices from each part. Making ε smaller is more restrictive, yielding more "regular" behavior.

11.1.12. Definition. Given disjoint vertex sets A and B in a graph G, the subgraph **generated** by (A, B) is the A, B-bigraph with edge set $[A, B]$. Its **density**, denoted $\rho(A, B)$, is $\frac{\|[A,B]\|}{|A||B|}$. The pair (A, B) is ε-**regular** if $|\rho(A', B') - \rho(A, B)| < \varepsilon$ whenever $A' \subseteq A$ and $B' \subseteq B$ with $|A'| \geq \varepsilon |A|$ and $|B'| \geq \varepsilon |B|$. An **equipartition** of G is a partition of $V(G)$ into parts whose sizes differ by at most 1.

When we partition $V(G)$ into singleton sets, every pair is ε-regular, but applications require larger parts. We need equipartitions with not too many parts such that almost all the pairs of parts are ε-regular. The Regularity Lemma guarantees such a suitable partition in a precise way.

11.1.13. Theorem. (**Regularity Lemma**; Szemerédi [1978]). Given $\varepsilon, l > 0$, there exist constants $M, N \in \mathbb{N}$ such that every graph with at least N vertices has an equipartition with k parts for some $k \in [l, M]$ such that fewer than εk^2 pairs of parts fail to be ε-regular.   ∎

Though Szemerédi [1978] is the seminal paper, already the idea was used in Szemerédi [1975]. Szemerédi [2015] describes a number of variants of the lemma. We will prove the one stated below.

11.1.14. Definition. For an n-vertex graph G, an ε, k-**partition** is a partition $V_0, \ldots, V_k$ of $V(G)$ such that $|V_0| \leq \varepsilon n$ (V_0 may be empty) and $|V_1| = \cdots = |V_k|$; call V_0 the *exceptional part*. An ε, k-partition is an ε-**regular partition** if fewer than εk^2 pairs of nonexceptional parts fail to be ε-regular.

11.1.15. Theorem. (alternate **Regularity Lemma**). For $\varepsilon, l > 0$, there are constants $M, N \in \mathbb{N}$ such that every graph with at least N vertices has an ε-regular ε, k-partition for some k with $l \leq k \leq M$.   ∎

Before proving the Regularity Lemma, we use it to prove the Erdős–Stone Theorem and striking results in graph Ramsey theory and additive combinatorics. Applications of the Regularity Lemma avoid technical detail by capturing the standard technical argument in an "Embedding Lemma". We form a **reduced graph** R from the partition of G provided by the Regularity Lemma. The vertices of R are the (nonexceptional) parts of the equipartition, adjacent when they form an ε-regular pair with density more than some parameter d. With $d > \varepsilon$, the Embedding Lemma implies that any specified subgraph H of $R[s]$ that has small enough maximum degree occurs as a subgraph of G, for suitable s.

For the Erdős–Stone Theorem, the Regularity Lemma provides a reduced graph with enough edges so that Turán's Theorem forces K_{r+1} in R, and then the Embedding Lemma yields $K_{r+1}[s] \subseteq G$. That is, we apply the Embedding Lemma with $R = K_{r+1}$ and $H = R[s]$.

11.1.16. Theorem. (**Embedding Lemma**) Let R be a graph, and fix $m, s \in \mathbb{N}$. Given d, ε with $d > \varepsilon > 0$, let G be a subgraph of $R[m]$ in which each pair of parts corresponding to an edge of R is an ε-regular pair with density at least d. Let H be a subgraph of $R[s]$ with n vertices and maximum degree D. If $\varepsilon \leq \varepsilon'$ and $s - 1 \leq \varepsilon' m$, where $\varepsilon' = (d - \varepsilon)^D/(D + 2)$, then $H \subseteq G$. In fact, G contains more than $(\varepsilon' m)^n$ copies of H (as labeled subgraphs).

Proof: For each vertex $v \in V(R)$, let R_v be the corresponding s-set in $R[s]$, and let A_v be the corresponding m-set in G. Let $V(H) = x_1, \ldots, x_n$. In $V(G)$ we will find representatives $y_1, \ldots, y_n$ of $x_1, \ldots, x_n$ such that $x_i x_j \in E(H)$ implies $y_i y_j \in E(G)$. Furthermore, if $x_t \in R_v$, then $y_t \in A_v$.

We pick $y_1, \ldots, y_n$ in order. At time t, we will choose y_t and update (for $j > t$) the set B_j of candidates for y_j. Initially $B_j = A_v$, where $x_j \in R_v$. The set B_j shrinks when we choose a vertex y_t that y_j needs as a neighbor. For each j with $j \geq t$, let $Y_j = \{y_i: i < t \text{ and } x_i x_j \in E(H)\}$; these already-chosen vertices are required to be neighbors of the vertex we choose as y_j. Since we want $y_j \in A_v$, the set B_j of candidates for y_j is $\{y \in A_v: Y_j \subseteq N_G(y)\}$. Letting $\alpha = d - \varepsilon$, we will make our choices to guarantee that when we are ready to pick y_t,

$$|B_j| \geq \alpha^{|Y_j|} m \qquad \text{for } j \geq t. \qquad (*)$$

When we are choosing y_t, we do not know which vertices will be chosen to represent the neighbors of x_j in $\{x_{t+1}, \ldots, x_{j-1}\}$. The vertex eventually chosen as y_j must be adjacent to all of them. For this reason, we initially preserve many candidates for y_j. As more neighbors of x_j are chosen, the number of candidates we keep decreases. When $|Y_j| = D$, one candidate suffices.

If $x_j \notin N_H(x_t)$, then Y_j and B_j do not change when y_t is chosen. To maintain $(*)$, it suffices to choose y_t from B_t so that y_t has at least $\alpha |B_j|$ neighbors in B_j for each j with $j > t$ and $x_j \in N_H(x_t)$.

For such j, define u and v by $x_t \in R_u$ and $x_j \in R_v$. The pair (A_u, A_v) is ε-regular in G, by definition. We have $B_t \subseteq A_u$ and $B_j \subseteq A_v$. Since the hypotheses guarantee $\alpha^D \geq \varepsilon$, by $(*)$ we have $|B_j| \geq \varepsilon |A_v|$.

Let B be the set of vertices in B_t that do not have at least $\alpha |B_j|$ neighbors in B_j. Thus $\rho(B, B_j) < \alpha$. If $|B| \geq \varepsilon m$, then ε-regularity of (A_u, A_v) guarantees $\rho(B, B_j) \geq d - \varepsilon = \alpha$, a contradiction.

Hence all but εm vertices of B_t have at least $\alpha |B_j|$ neighbors in B_j. Since there are at most D values of j to consider, we discard at most $D\varepsilon m$ vertices of B_t in this way. We also discard vertices of B_t already chosen to represent vertices of H in R_v; there are at most $s - 1$ of these. There remain at least $(\alpha^D - D\varepsilon)m - (s - 1)$ vertices in B_t; choose one to be y_t.

Since at each t there are at least $(\alpha^D - D\varepsilon)m - (s - 1)$ choices for y_t, there are at least $[(\alpha^D - D\varepsilon)m - (s - 1)]^n$ labeled copies of H in G. From $\varepsilon' = \alpha^D/(D + 2)$, we have $\alpha^D = 2\varepsilon' + D\varepsilon' \geq 2\varepsilon' + D\varepsilon$. Thus $2\varepsilon' \leq \alpha^D - D\varepsilon$. Since also $s - 1 \leq \varepsilon' m$, we have $(\alpha^D - D\varepsilon)m - (s - 1) \geq \varepsilon' m$. ∎

The Embedding Lemma is used to find small graphs in a large graph G. (Komlós–Sárközy–Szemerédi [1997] proved a more difficult extension called the **Blow-up Lemma** to find spanning subgraphs.) To complete the proof of the Erdős–Stone Theorem, it now suffices to find a copy of K_{r+1} in the reduced graph that comes from the partition of G provided by the Regularity Lemma.

Proof of Erdős–Stone Theorem (Theorem 11.1.10): Fix $r, s \in \mathbb{N}$ and a positive constant c. We want to prove that if n is sufficiently large, then an n-vertex graph with more than $t_r(n) + cn^2$ edges contains $K_{r+1}[s]$. We may assume $c < \frac{1}{2}$.

Choose l with $l > \max\{r, \frac{1}{2c}\}$. Choose d with $0 < \frac{d}{2} < c - \frac{1}{2l} < \frac{1}{2}$. Choose ε small enough so that $\varepsilon \leq \frac{(d-\varepsilon)^{rs}}{rs+2}$ and $\varepsilon < \frac{1}{3}(c - \frac{d}{2} - \frac{1}{2l})$. Let M and N be the constants in terms of ε and l needed to apply the Regularity Lemma. Choose n so that $n > \max\{N, M(s-1)/\varepsilon'\}$, where $\varepsilon' = \frac{(d-\varepsilon)^{rs}}{rs+2}$.

Let G be an n-vertex graph with more than $t_r(n) + cn^2$ edges. By the alternative form of the Regularity Lemma, G has an ε-regular ε, k-partition $V_0, \ldots, V_k$, where $l \leq k \leq M$. Let $m = |V_1| = \cdots = |V_k|$.

The reduced graph R with vertex set $v_1, \ldots, v_n$ has $v_i v_j \in E(R)$ if and only if the pair (V_i, V_j) is ε-regular with density at least d. Showing that $|E(R)| > t_r(k)$ forces $K_{r+1} \subseteq R$. The Embedding Lemma then implies $K_{r+1}[s] \subseteq G$. Since $\Delta(K_{r+1}[s]) = rs$, applying it requires $\varepsilon \leq \varepsilon'$ and $s - 1 \leq \varepsilon' m$, which were guaranteed by the choice of ε and then n.

The "bad" edges of G are those incident to V_0, within V_i for $i \geq 1$, or joining V_i and V_j when (V_i, V_j) is not ε-regular or is ε-regular with density less than d. To show that R has enough edges, we show that the number of bad edges in G is small enough that more than $t_r(k)$ pairs with density at least d are needed to accommodate the good edges. Note that $n - \varepsilon n \leq km$ yields $n \leq (1 + \varepsilon)km$, and hence $n^2 < 2k^2 m^2$, since $\varepsilon < 1/4$. Of course, $n^2 > k^2 m^2$.

Since $|V_0| \leq \varepsilon n$, the number of edges incident to V_0 is less than εn^2, which is less than $2\varepsilon k^2 m^2$. Since at most εk^2 pairs are not ε-regular, at most $\varepsilon k^2 m^2$ edges of G lie in such pairs. Fewer than $\frac{1}{2}k^2$ pairs have density less than d, and hence at most $\frac{1}{2}k^2 d m^2$ edges lie in such pairs. Each $G[V_i]$ has at most $\frac{1}{2}m^2$ edges, totaling at most $\frac{1}{2}m^2 k$ edges. Finally, each edge of R arises from a pair with density at least d; even when the density is 1, it contributes at most m^2 edges to G. Thus

$$|E(G)| \leq 3\varepsilon k^2 m^2 + \tfrac{1}{2}dk^2 m^2 + \tfrac{1}{2}km^2 + |E(R)| \, m^2.$$

Solving for $|E(R)|$, further substitution yields the desired bound.

$$
\begin{aligned}
|E(R)| &\geq \frac{|E(G)| - 3\varepsilon k^2 m^2 - \frac{1}{2}dk^2 m^2 - \frac{1}{2}km^2}{m^2} \\
&\geq k^2 \left(\frac{t_r(n) + cn^2}{k^2 m^2} - 3\varepsilon - \frac{d}{2} - \frac{1}{2k} \right) \\
&\geq k^2 \left(\frac{t_r(n)}{n^2} + c - 3\varepsilon - \frac{d}{2} - \frac{1}{2l} \right) \\
&> \tfrac{1}{2}k^2(1 - 1/r) \geq t_r(k).
\end{aligned}
$$

Hence $K_{r+1} \subseteq R$, and the Embedding Lemma gives $K_{r+1}[s] \subseteq G$.   ∎

11.1.17.* Remark. As we have noted, the Erdős–Stone Theorem has proofs avoiding these tools. The theorem also can be made more precise. We fixed c and s and showed that $K_{r+1}[s]$ is forced by $t_r(n) + cn^2$ edges when n is large enough. Instead, we can fix c and n and ask how large s can be guaranteed as a function of n when forcing $K_{r+1}[s]$. If it grows with n, then the Erdős–Stone Theorem follows. A lower bound on the growth of s is a quantitative strengthening.

Fixing r and c, one can guarantee s that is logarithmic in n (Bollobás–Erdős [1973]). Several improvements culminated in the result of Ishigami [2002] (via Regularity) that every n-vertex graph with more than $(1 - 1/r + c)n^2/2$ edges contains $K_{r+1}[s]$ with $s \geq \lfloor \frac{\log_{1/c} n}{11r} \rfloor$. Superlogarithmic s is not forced (Bollobás–Erdős–Simonovits [1976]). Bollobás–Kohayakawa [1994] guaranteed that one part has close to linear size. Nikiforov [2008] proved a more general theorem yielding both logarithmic growth of s and near-linear size of one part.   ∎

LINEAR RAMSEY FOR BOUNDED DEGREE

Our second application of the Regularity Lemma is from graph Ramsey theory. The Ramsey number $R(G, G)$ may grow exponentially in the number of vertices of G, such as when $G = K_n$. In some families of relatively sparse graphs, the Ramsey number grows at most linearly in the number of vertices! Most famously, one such family is the family of graphs with bounded maximum degree.

11.1.18. Theorem. (Chvátal–Rödl–Szemerédi–Trotter [1983]) For $d \in \mathbb{N}$, there is a constant c_d such that $R(G, G) \leq c_d |V(G)|$ whenever $\Delta(G) = d$.

Proof: Let $l = 3 \max\{3^d, R(d+1, d+1)\}$, and let $\varepsilon = l^{-1}$. The Regularity Lemma provides constants $M, N \in \mathbb{N}$ such that every graph with at least N vertices has an ε-regular ε, k-partition for some k with $l \leq k \leq M$. Let $c = \max\{N, M/\varepsilon'\}$, where $\varepsilon' = (1/2 - \varepsilon)^d/(d + 2)$. Now fix a graph G with maximum degree d and vertices $x_1, \ldots, x_n$; we prove that $R(G, G) \leq cn$.

Consider a red/blue coloring of $E(K_{cn})$. Let H and $\overline{H}$ be the subgraphs in red and blue, respectively. Via Steps 1 and 2 below, we find $d + 1$ large sets such that every pair of them is ε-regular and has high density in the same color. In Step 3, the Embedding Lemma allows us to find a copy of G among the edges of that color.

Step 1: *Some ε-regular partition of $V(H)$ has at least $l/3$ parts with all pairs ε-regular in H and $\overline{H}$.* For a pair (A, B) of vertex subsets, the densities in H and $\overline{H}$ sum to 1. Hence (A, B) is ε-regular in H if and only if it is ε-regular in $\overline{H}$, so we consider only H. By the choice of c, there is an ε-regular ε, k-partition $A_0, \ldots, A_k$ of H, where $l \leq k \leq M$. All but εk^2 pairs are ε-regular. Let H^* be the graph with vertex set $[k]$ putting i and j adjacent if and only if (A_i, A_j) is ε-regular in H.

By Turán's Theorem, a k-vertex graph with no $(t + 1)$-clique has fewer than $(1 - \frac{1}{t})\frac{k^2}{2}$ edges. Since $|E(H^*)| \geq \binom{k}{2} - \varepsilon k^2$, we find a clique of size at least $\frac{1}{3\varepsilon}$, equal to $\frac{l}{3}$. We thus have $\frac{l}{3}$ parts in $\{A_1, \ldots, A_k\}$ among which every pair is ε-regular.

Step 2: *This ε-regular partition has $d + 1$ parts with all pairs ε-regular of density at least $\frac{1}{2}$ in the same color (H or $\overline{H}$).* We color the edges of the $\frac{l}{3}$-clique in H^* found in Step 1, using red if $\rho_H(A_i, A_j) \geq \frac{1}{2}$ and blue if $\rho_H(A_i, A_j) < \frac{1}{2}$. Since $\frac{l}{3} \geq R(d + 1, d + 1)$, Ramsey's Theorem yields a monochromatic copy of K_{d+1} in H^*; by symmetry, we may assume it is in red. In the original coloring of $E(K_{cn})$, each pair among these parts is ε-regular with density at least $\frac{1}{2}$ in the red graph.

Step 3: *The given 2-coloring of $E(K_{cn})$ has a monochromatic copy of G.* We now have $d + 1$ vertex sets in K_{cn}, all of size at least $n(1 - \varepsilon)/k$, such that each pair is ε-regular with density at least $1/2$ in the red graph, H. Let G' be the subgraph of H induced by these parts. In the language of the Embedding Lemma

(Theorem 11.1.16), the reduced graph R of G' is K_{d+1}. Since $\chi(G) \leq \Delta(G) + 1 \leq d + 1$, the graph G is a subgraph of $R[n]$. If the Embedding Lemma applies, then G' contains G, as desired. It is important that the requirements to apply the Embedding Lemma are only in terms of $\Delta(G)$, not the number of vertices. Since the size of the parts is at most cn/k, we need $\varepsilon \leq \varepsilon'$ and $n - 1 \leq \varepsilon' cn/k$, where $\varepsilon' = (\frac{1}{2} - \varepsilon)^d/(d + 2)$. Since we define ε to be less than $(1/3)^d$ and defined c so that $\varepsilon'c/k \geq \varepsilon'c/M \geq 1$, the hypotheses hold and the proof is complete. ∎

11.1.19.* Remark. Unfortunately, the threshold for N in the Regularity Lemma applies is huge. As a result, the constant c_d from the proof of Theorem 11.1.18 grows like an exponential tower with height d. Using a variant of the Regularity Lemma, Eaton [1998] reduced the constant to doubly-exponential growth ($2^{2^{O(d)}}$).

Graham–Rödl–Ruciński [2000] avoided the Regularity Lemma and reduced c_d to $2^{O(d(\log d)^2)}$ (Conlon–Fox–Sudakov [2012] then reduced it to $2^{O(d \log d)}$). For bipartite graphs, Graham–Rödl–Ruciński [2001] further improved the bound to $8(8d)^d$ and showed that c_d is at least exponential in d. Like Ramsey's Theorem, the Regularity Lemma is a powerful tool to prove the existence of bounds, but accurate bounds generally require more detailed direct arguments.

Theorem 11.1.18 was conjectured by Burr–Erdős [1975], who also conjectured the same claim (with a larger constant) for the larger family of d-degenerate graphs. Kostochka–Rödl [2001, 2004] proved bounds that are quadratic in the number of vertices. Finally, C. Lee [2017] proved the conjecture, showing the existence of a constant c such that every d-degenerate graph H with $\chi(H) = k$ and $|V(H)| \geq 2^{d^2 2^{ck}}$ has Ramsey number at most $2^{d 2^{ck}} |V(H)|$. ∎

ROTH'S THEOREM

We give one more application of the Regularity Lemma. Szemerédi developed the Regularity Lemma to prove the long-standing conjecture of Erdős–Turán [1936] that subsets of the integers with positive density contain long arithmetic progressions. More precisely, subsets of $[n]$ containing no k-term arithmetic progression have size at most $o(n)$ (strengthening van der Waerden's Theorem).

Roth [1953] proved this for $k = 3$; Szemerédi [1970] proved it for $k = 4$. **Szemerédi's Theorem** (Szemerédi [1975]) used a bipartite version of the Regularity Lemma to prove it for all k. A later proof by Gowers [2001] using Fourier analysis avoids regularity and gives much better numerical bounds. We use regularity to prove the case $k = 3$, following Palmer [2015]; see also Szemerédi [2015]. Roth proved that subsets of $[n]$ with size at most $O(n/\log \log n)$ contain a 3-term progression; Behrend [1946] constructed a set of size $n/e^{c\sqrt{\log n}}$ that does not.

We begin with a standard computation, useful also in some of the exercises.

11.1.20. Lemma. (Degree-Density Lemma) In a graph G, let (A, B) be an ε-regular pair with density d. Given a set $Y \subseteq B$ with $|Y| \geq \varepsilon |B|$, the number of vertices in A having at most $(d - \varepsilon)|Y|$ neighbors in Y is less than $\varepsilon |A|$.

Proof: Let X be the set of vertices in A having at most $(d - \varepsilon)|Y|$ neighbors in Y. By direct counting, $\rho(X, Y) \leq d - \varepsilon$. If $|X| \geq \varepsilon |A|$, then ε-regularity of (A, B) requires $\rho(X, Y) > d - \varepsilon$. Hence $|X| < \varepsilon |A|$. ∎

11.1.21. Lemma. (**Triangle-Counting Lemma**) Let A, B, C be a partition of the vertices of a graph G such that each pair is ε-regular. Let α, β, γ be the densities of the pairs (A, B), (B, C), (C, A), respectively. If $\alpha, \beta, \gamma \geq 2\varepsilon$, then the number of triangles having a vertex in each class is at least

$$(1 - 2\varepsilon)(\alpha - \varepsilon)(\beta - \varepsilon)(\gamma - \varepsilon)|A||B||C|.$$

Proof: By Lemma 11.1.20, fewer than $\varepsilon|A|$ vertices in A have at most $(\alpha - \varepsilon)|B|$ neighbors in B, and fewer than $\varepsilon|A|$ vertices in A have at most $(\gamma - \varepsilon)|C|$ neighbors in C. Deleting both such sets from A leaves at least $(1 - 2\varepsilon)|A|$ vertices in A having more than $(\alpha - \varepsilon)|B|$ neighbors in B and $(\gamma - \varepsilon)|C|$ neighbors in C.

For such a vertex x, let $B' = N(x) \cap B$ and $C' = N(x) \cap C$. Since $\beta \geq 2\varepsilon$, by ε-regularity we have $\rho(B', C') \geq \beta - \varepsilon$, so $[B', C'] \geq (\beta - \varepsilon)|B'||C'|$. With at least $(1 - 2\varepsilon)|A|$ choices for x and at least $(\beta - \varepsilon)(\alpha - \varepsilon)(\gamma - \varepsilon)|B||C|$ desired triangles for each such x, the claim follows. ∎

The Triangle-Counting Lemma is a special case of the Embedding Lemma, finding many copies of a fixed subgraph given a partition with all pairs ε-regular. We next show that a graph with few triangles can be made triangle-free by removing few edges. A more recent proof and extension by Fox [2011] gives better bounds and avoids the Regularity Lemma (see also Conlon–Fox [2013]). The more general Graph Removal Lemma states that when F is a k-vertex graph, all copies of F in any n-vertex graph H containing $o(n^k)$ copies of F can be destroyed by deleting $o(n^2)$ edges of H. Rödl–Schacht [2009] generalizes to hypergraphs.

11.1.22. Lemma. (**Triangle-Removal Lemma**; Ruzsa–Szemerédi [1978]) For $\varepsilon > 0$, there exists a positive number δ such that for sufficiently large n, any n-vertex graph G in which at least εn^2 edges must be deleted to break all triangles has at least δn^3 triangles.

Proof: Since a complete graph has only $\binom{n}{2}$ edges, we may assume $\varepsilon < 1/2$. Choose ε' and l with $\varepsilon' < \varepsilon/3$ and $l > 1/\varepsilon'$. By the Regularity Lemma, there exist integers M and N such that for $n > N$ we are guaranteed an equipartition $\{V_1, \ldots, V_k\}$ of $V(G)$ with $l \leq k \leq M$ such that all but $\varepsilon' k^2$ pairs of parts are ε'-regular.

With respect to this partition, we first remove a set of edges that hits all triangles except a certain type. We remove all edges that (1) lie inside parts, (2) join parts that do not form an ε'-regular pair, or (3) join parts forming an ε'-regular pair with density less than $2\varepsilon'$. The number of edges of type (1) is at most $k\binom{\lceil n/k \rceil}{2}$, which is less than $\varepsilon' n^2$ since $k > 1/\varepsilon'$. There are fewer than $\varepsilon' k^2 (n/k)^2$ edges of type (2) and fewer than $\binom{k}{2}2\varepsilon'\lceil n/k \rceil^2$ of type (3), both less than $\varepsilon' n^2$. In total, we have removed fewer than εn^2 edges.

By the hypothesis on G, at least one triangle T remains. By the types of edges removed, T has its vertices in three distinct parts such that each pair among these three is ε'-regular with density at least $2\varepsilon'$. Also, each part has size at least $\lfloor n/k \rfloor$, which is at least n/M. By the Triangle Counting Lemma, among these three parts the number of triangles remaining is at least $(1-2\varepsilon')(\varepsilon')^3(n/M)^3$. Thus the desired conclusion holds with $\delta = (1 - 2\varepsilon')(\varepsilon'/M)^3$. ∎

11.1.23. Theorem. (**Roth's Theorem**; Roth [1953]) A subset of $[n]$ containing no 3-term arithmetic progression has size at most $o(n)$.

Proof: (Ruzsa–Szemerédi [1978]) The meaning of the statement is that for each positive ε, for large enough n every set $S \subseteq [n]$ with $|S| = \varepsilon n$ contains a 3-term arithmetic progression; this is what we must prove.

We construct a graph G with vertex set $A \cup B \cup C$, where $A = \{a_1, \ldots, a_n\}$, $B = \{b_1, \ldots, b_{2n}\}$, and $C = \{c_1, \ldots, c_{3n}\}$. For $s \in S$ and $j \in n$, add to $E(G)$ the triangle with vertices $\{a_j, b_{j+s}, c_{j+2s}\}$. Let $\mathcal{T}$ be this set of triangles.

The εn^2 triangles in $\mathcal{T}$ are edge-disjoint; hence it takes at least εn^2 edges to hit all the triangles in G. By the Triangle-Removal Lemma, G has at least δn^3 triangles (for some constant δ). For large enough n, we have $\delta n^3 > \varepsilon n^2$, and hence G has a triangle T not in $\mathcal{T}$; it uses edges from distinct triangles in $\mathcal{T}$.

We gave G no edges within A, B, or C. Let $V(T) = \{a_j, b_{j+s}, c_{j+t}\}$. By the construction of G, these three edges require $s \in S$ and $t - s \in S$, and also $t = 2s'$ for some $s' \in S$. Also, s and s' are distinct, since $T \notin \mathcal{T}$. Since $2s' - s = t - s \in S$, the numbers $s, s', 2s' - s$ form an arithmetic progression in S. ∎

PROOF OF THE REGULARITY LEMMA (optional)

It remains to prove the Regularity Lemma. For this we will need several lemmas about the behavior of densities of pairs of vertex sets.

11.1.24. Definition. For an A, B-bigraph G with density d, let $f(G) = |A|\,|B|\,d^2$. For a partition Π of $V(G)$ into $V_1, \ldots, V_k$, let $G_{i,j}$ be the subgraph of G generated by (V_i, V_j), and let $f(G, \Pi) = \sum_{i,j} f(G_{i,j})$.

We show first that if Π' refines Π, then $f(G, \Pi) \le f(G, \Pi') \le |E(G)|$. Note that $f(G, \Pi') = |E(G)|$ when the partition consists of singleton sets. The idea is to iteratively refine pairs that are not ε-regular. Each refinement increases f, but the process cannot continue forever, since f is bounded by $|E(G)|$.

11.1.25. Lemma. Let G be an A, B-bigraph with density d, and let $\{A_1, A_2\}$ partition A, with $a_i = |A_i|$. If the subgraph G_i generated by (A_i, B) has density d_i, then $f(G_1) + f(G_2) = f(G) + (a_1/a_2)|A|\,|B|\,(d - d_1)^2$.

Proof: Let $a = |A|$ and $b = |B|$, and let m, m_1, m_2 be the numbers of edges in G, G_1, G_2, respectively. Note that $f(G) = abd^2 = m^2/(ab) = md$. Letting $x = m_1/m$ and $\alpha = a_1/a$, we compute

$$\frac{ba_1a_2}{m^2a}(f(G_1) + f(G_2) - f(G)) = \frac{ba_1a_2}{m^2a}\left(\frac{m_1^2}{a_1b} + \frac{m_2^2}{a_2b} - \frac{m^2}{ab}\right)$$

$$= \left(\frac{m_1}{m}\right)^2\frac{a_2}{a} + \left(\frac{m_2}{m}\right)^2\frac{a_1}{a} - \frac{a_1a_2}{a^2}$$

$$= x^2(1 - \alpha) + (1 - x)^2\alpha - \alpha(1 - \alpha) = x^2 - 2x\alpha + \alpha^2 = (x - \alpha)^2 .$$

Therefore,

$$f(G_1) + f(G_2) - f(G) = \frac{m^2a}{ba_1a_2}(x - \alpha)^2 = \frac{a_1}{a_2}ab\left(\frac{m}{ba_1}\right)^2\left(\frac{a_1}{a} - \frac{m_1}{m}\right)^2$$

$$= \frac{a_1}{a_2}ab\left(\frac{m}{ab} - \frac{m_1}{a_1b}\right)^2 = \frac{a_1}{a_2}ab(d - d_1)^2 . \qquad \blacksquare$$

11.1.26. Lemma. Let G be an A, B-bigraph with density d. Let $\{A_1, A_2\}$ and $\{B_1, B_2\}$ be partitions of A and B, with $|A_1| \geq \varepsilon |A|$ and $|B_1| \geq \varepsilon |B|$. Let $f_{i,j} = f(G[A_i \cup B_j])$. If $|\rho(A_1, B_1) - d| \geq \varepsilon$, then $\sum f_{i,j} \geq f(G) + \varepsilon^4 |A| |B|$.

Proof: Let a, b, a_i, b_j be the sizes of A, B, A_i, B_j, respectively. Also let $d^* = \rho(A_1, B)$ and $d_{i,j} = \rho(A_i, B_j)$. By Lemma 11.1.25,

$$f(G) = f(G[A_1 \cup B]) + f(G[A_2 \cup B]) - \frac{a_1}{a_2} ab(d - d^*)^2$$

$$\leq f_{1,1} + f_{1,2} - \frac{b_1}{b_2} a_1 b(d^* - d_{1,1})^2 + f_{2,1} + f_{2,2} - \frac{a_1}{a_2} ab(d - d^*)^2.$$

Here we have ignored the gain due to the second subsplit.

It now suffices to show $\frac{b_1}{b_2} \frac{a_1}{a}(d^* - d_{1,1})^2 + \frac{a_1}{a_2}(d - d^*)^2 \geq \varepsilon^4$. Since $b_1 \geq \varepsilon b$ and $a_1 \geq \varepsilon a$, we only need $\frac{\varepsilon}{1-\varepsilon}\varepsilon(d^* - d_{1,1})^2 + \frac{\varepsilon}{1-\varepsilon}(d - d^*)^2 \geq \varepsilon^4$. For fixed d and $d_{1,1}$, we minimize the left side by setting $d^* = \frac{d + \varepsilon d_{1,1}}{1+\varepsilon}$, since then

$$\tfrac{\partial}{\partial d^*}[\varepsilon(d^* - d_{1,1})^2 + (d - d^*)^2] = 2\varepsilon(d^* - d_{1,1}) - 2(d - d^*) = 0.$$

With d^* so chosen, $d^* - d_{1,1} = \frac{d - d_{1,1}}{1+\varepsilon}$ and $d - d^* = \frac{\varepsilon}{1+\varepsilon}(d - d_{1,1})$. Using $|d_{1,1} - d| \geq \varepsilon$, we have

$$\frac{\varepsilon}{1-\varepsilon}\varepsilon(d^* - d_{1,1})^2 + \frac{\varepsilon}{1-\varepsilon}(d - d^*)^2 \geq \frac{\varepsilon}{1-\varepsilon}\left[\varepsilon\left(\frac{\varepsilon}{1+\varepsilon}\right)^2 + \left(\frac{\varepsilon^2}{1+\varepsilon}\right)^2\right]$$

$$= \frac{\varepsilon^4}{(1+\varepsilon)^2(1-\varepsilon)}(1+\varepsilon) = \frac{\varepsilon^4}{1-\varepsilon^2} \geq \varepsilon^4. \qquad \blacksquare$$

11.1.27. Lemma. Let G be an n-vertex graph, and let Π be an ε, k-partition of G, where $0 < \varepsilon < \frac{1}{5}$. If Π is not ε-regular and $|V_0| < (\varepsilon - \varepsilon^6)n$, then there is an ε, k'-partition Π' of G with $k \leq k' \leq k2^{k+2}/\varepsilon^6$ such that $\left|V_0'\right| \leq |V_0| + n/2^k$ and $f(G, \Pi') \geq f(G, \Pi) + \varepsilon^5 n^2/2$.

Proof: If Π is not ε-regular, then too many pairs of parts are not ε-regular. Let m be their common size. Each bad pair (V_i, V_j) yields sets $X \subseteq V_i$ and $Y \subseteq V_j$ with $|X|, |Y| \geq \varepsilon m$ such that $|\rho(X, Y) - \rho(V_i, V_j)| \geq \varepsilon$. By Lemma 11.1.26, replacing V_i with $\{X, V_i - X\}$ and V_j with $\{Y, V_j - Y\}$ increases $f(G, \Pi)$ by at least $\varepsilon^4 m^2$.

The idea is to capture this gain simultaneously for all the pairs that are not ε-regular. For each pair involving V_i that is not ε-regular, we obtain a partition $\{X, V_i - X\}$ of V_i. With fewer than k such pairs, the least common refinement of these partitions has fewer than 2^k parts. Apply this to each V_i, obtaining a partition P consisting of V_0 and at most $k2^k$ other parts, with unequal sizes.

To compare $f(G, P)$ with $f(G, \Pi)$, we group the pairs of parts according to the pairs in Π generating them. When (V_i, V_j) is not ε-regular, witnessed by (X, Y), the contributions from subsets of V_i and V_j can be obtained by first splitting V_i and V_j according to X and Y and then continuing the refinement. The gain of at least $\varepsilon^4 m^2$ is followed by further gains we ignore. When computing the contribution for subsets of V_i with subsets of $V_{j'}$, it does not matter that the initial split is different, since we are obtaining a lower bound on the contributions from a set of subpairs disjoint from those arising from $\{V_i, V_j\}$. Since more than εk^2 pairs of parts in Π failed to be ε-regular, we have $f(G, P) \geq f(G, \Pi) + \varepsilon^5 k^2 m^2$.

We still need a partition whose nonexceptional parts have equal size. Let m' be the desired part-size, to be specified later. Break each part in P other than V_0 into blocks of size m', with fewer than m' left over. Combine the leftover vertices with V_0 to form V_0'; this produces Π' with k' parts other than V_0', where $k' \leq n/m'$. Note that $|V_0'| \leq |V_0| + m'k2^k$.

Splitting into blocks of size m' further increases f (we ignore this gain), but combining the leftovers into V_0' reduces f. Combining singletons with V_0 would lose the most. The contributions of a singleton part with all other parts sum to at most the degree of the singleton vertex. Since $m'k2^k$ is an upper bound on $|V_0' - V_0|$, the amount lost by forming V_0' is at most $m'k2^k n$.

Since $km \geq n(1-\varepsilon)$, we now have $f(G, \Pi') \geq f(G, \Pi) + \varepsilon^5 n^2 (1-\varepsilon)^2 - m'k2^k n$. The gain is at least $\varepsilon^5 n^2 - [\varepsilon^5 n^2 (2\varepsilon - \varepsilon^2) + m'k2^k n]$. To ensure net gain at least $\varepsilon^5 n^2/2$, we want $2\varepsilon^6 n + m'k2^k < \varepsilon^5 n/2$. With $\varepsilon < 1/5$, it suffices to choose m' so that $m'k2^k \leq \varepsilon^6 n/2$. Set $m' = \lfloor \varepsilon^6 n/(k2^{k+1}) \rfloor$. Now Π' is an ε, k'-partition with k' bounded by a bit more than $k2^{k+1}/\varepsilon^6$. Doubling this value suffices. Finally, note that the exceptional part has gained at most $\varepsilon^6 n/2$ vertices.   ∎

The simple formula for the gain per split is why we keep equal size for the nonexceptional parts. Later the exceptional part can be distributed to the others.

Proof of the Szemerédi Regularity Lemma (as Theorem 11.1.15): Given positive constants ε and l, we seek positive integers M and N such that every graph with at least N vertices has an ε-regular ε, k-partition for some $k \in [l, M]$.

To apply Lemma 11.1.27, we need $\varepsilon < \frac{1}{5}$. Since an ε-regular ε, k-partition is also an ε'-regular ε', k-partition when $\varepsilon' > \varepsilon$, we may assume $\varepsilon < \frac{1}{5}$. Similarly, if the claim is true for a given value of l, it remains true for smaller values, so we may assume $l > 2/\varepsilon$.

Given such ε and l, define M and N as follows. Let $L_0 = l$ and $L_{i+1} = \lceil L_i 2^{L_i+2}/\varepsilon^6 \rceil$. Let $M = L_{\lceil 1/\varepsilon^5 \rceil}$, and choose N with $N > \max\{M, l/\varepsilon^6\}$. Let G be an n-vertex graph, where $n \geq n$. Since $l > 2/\varepsilon$, we can form an ε, l-partition of G by breaking $V(G)$ arbitrarily into l parts of size $\lfloor n/l \rfloor$, with fewer than $\varepsilon n/2$ vertices left over for the exceptional part.

Since $|E(G)| < n^2/2$, fewer than ε^{-5} iterations of the refinement procedure in Lemma 11.1.27 produce an ε-regular ε, k-partition of G. The initial exceptional set has fewer than $\varepsilon n/2$ vertices, and the exceptional set gains fewer than $\varepsilon^6 n/2$ with each iteration. With fewer than ε^{-5} iterations, the final exceptional set has at most εn vertices. Also, the final partition has at most M parts.   ∎

11.1.28.* Remark. Since the upper bound on the number of parts in Lemma 11.1.27 is $k2^{k+2}/\varepsilon^6$, iterating ε^{-5} times makes N an exponential tower with height ε^{-5}. Thus the Regularity Lemma applies only to enormous graphs.

We have given only a few applications of Regularity; there are many others. Komlós–Simonovits [1996] is an influential early survey; others include Komlós–Shokoufandeh–Simonovits–Szemerédi [2002], Kohayakawa–Rödl [2003], Rödl–Schacht [2010]. Kühn–Osthus [2009] explains the use of the regularity/blow-up method in general. The original lemma gives useful results only for dense graphs; Kohayakawa [1997] developed an analogue for sparse graphs (see also Kohayakawa–Rödl [2003], Gerke–Steger [2005], and Scott [2011]). Frankl–Rödl

[2002] extended the lemma to 3-uniform hypergraphs, and Rödl–Skokan [2004, 2006] extended it to k-uniform hypergraphs. Textbooks discussing Regularity include Bollobás [1998] and Diestel [1997, etc.] (from the 2000 edition onward). There are lecture notes by Lee [2015], Palmer [2015], and Shapira [2016]. See also Martin [2012, Chapters 6–9] and Tao–Vu [2006, Section 10.6].   ∎

EXERCISES 11.1

11.1.1. (−) Complete the computation in Corollary 11.1.7.

11.1.2. (−) Find the minimum size of a maximal triangle-free n-vertex graph.

11.1.3. (−) Determine $\mathrm{ex}\,(n; P_3)$ and $\mathrm{ex}\,(n; P_4)$.

11.1.4. Determine the maximum size of n-vertex graphs of the following types.
(a) Graphs having an independent set of size k.
(b) Graphs having k components.

11.1.5. Let G be a connected graph having neither P_4 nor the 4-vertex "paw" obtained by adding one edge to $K_{1,3}$ as an induced subgraph. Prove that G is complete multipartite.

11.1.6. Prove that every n-vertex graph not containing K_{r+1} has at most $(1 - \frac{1}{r})\frac{n^2}{2}$ edges.

11.1.7. (◇) *Turán's proof of Turán's Theorem.* Recall that $t_r(n) = |E(T_{n,r})|$.
(a) Prove that a maximal graph with no $(r + 1)$-clique has an r-clique.
(b) For $n \geq r$, prove that $t_r(n) = \binom{r}{2} + (n - r)(r - 1) + t_r(n - r)$.
(c) Use parts (a) and (b) to prove Turán's Theorem by induction on n, including the uniqueness of graphs achieving the bound. (Turán [1941])

11.1.8. Let S be the set of nonnegative vectors in $\mathbb{R}^n$ with sum 1. Given a graph G with vertex set $\{v_1, \ldots, v_n\}$, let $f(x) = \sum_{v_i v_j \in E(G)} x_i x_j$ for $x \in S$, and let $\rho = \max_{x \in S} f(x)$.
(a) Prove that f is maximized by some vector whose nonzero coordinates correspond to the vertices of a clique. Conclude $\rho = \frac{1}{2}(1 - 1/\omega(G))$.
(b) Prove that ρ is attained by a vector x with all coordinates nonzero if and only if G is a complete multipartite graph.
(c) Prove Turán's Theorem for K_{r+1}-free n-vertex graphs by induction on $n - \lfloor n/r \rfloor$, using parts (a) and (b) for the base case. (Motzkin–Straus [1965])

11.1.9. (◇) The Turán graph $T_{n,r}$ with size $t_r(n)$ is the complete r-partite graph with b parts of size $a + 1$ and $r - b$ parts of size a, where $a = \lfloor n/r \rfloor$ and $b = n - ra$.
(a) Prove that $t_r(n) = (1 - 1/r)n^2/2 - b(r - b)/(2r)$.
(b) Since $t_r(n)$ is an integer, part (a) yields $t_r(n) \leq \lfloor (1 - 1/r)n^2/2 \rfloor$. Determine the smallest value of r such that strict inequality occurs for some n. For this value of r, determine the values of n such that $t_r(n) < \lfloor (1 - 1/r)n^2/2 \rfloor$.

11.1.10. Let $a = \lfloor n/r \rfloor$. Compare the Turán graph $T_{n,r}$ with the graph $\overline{K}_a + K_{n-a}$ to prove directly that $t_r(n) = \binom{n-a}{2} + (r - 1)\binom{a+1}{2}$.

11.1.11. Given positive integers n and k, let $q = \lfloor n/k \rfloor$, $r = n - qk$, $s = \lfloor n/(k + 1) \rfloor$, and $t = n - s(k + 1)$. Use Turán's Theorem to prove $\binom{q}{2}k + rq \geq \binom{s}{2}(k + 1) + ts$. (Richter [1993])

11.1.12. Let G be a graph with n vertices and m edges. Determine lower bounds on $\alpha(G)$ and $\omega(G)$ in terms of n and m that are sometimes sharp.

11.1.13. Let S be a set of n points in a circular region with radius 1, and consider the distance between any two points in S. Prove that the distance exceeds $\sqrt{2}$ for at most $n^2/3$ of the pairs. (Bondy–Murty [1976, p. 114])

11.1.14. ($\Diamond$) Let G be a graph with n vertices that has $t_r(n) - k$ edges and at least one $(r+1)$-clique, where $k \geq 0$. Prove that G has at least $f_r(n) + 1 - k$ cliques of size $r + 1$, where $f_r(n) = n - \lceil n/r \rceil - r$. (Hint: Prove that a graph with exactly one $(r+1)$-clique has at most $t_r(n) - f_r(n)$ edges.) (Erdős [1964], Moon [1965c])

11.1.15. ($\Diamond$) Let $f_k(n)$ be the minimum size of a family F of subsets of $[n]$ such that every subset of $[n]$ with size at most k is the union of two (possibly equal) members of F. Note that $f_1(n) = f_2(n) = n + 1$. (Comment: Füredi–Katona [2006] determined also $f_4(n)$.)
    (a) Prove $\sqrt{2} \cdot 2^{n/2} - 1/2 \leq f_n(n) \leq 2^{\lceil n/2 \rceil} + 2^{\lfloor n/2 \rfloor} - 1$.
    (b) Prove $f_3(n) = 1 + n + \binom{n}{2} - \lfloor n^2/4 \rfloor$.

11.1.16. *Extensions of Turán's Theorem.* As usual, let $t_r(n) = |E(T_{n,r})|$.
    (a) If $n \geq r + 1 \geq 4$, prove that every n-vertex graph with $t_{r-1}(n) + 1$ edges contains $K_{r+1} - e$ as a subgraph. (Hint: Remove a vertex of minimum degree.)
    (b) If $n \geq n' > r \geq 2$, prove that every n-vertex graph with $t_r(n) + p$ edges has an n'-vertex subgraph with $t_r(n') + p$ edges, where $p \leq 1$.
    (c) If $1 \leq p \leq r - 1$, prove that every n-vertex graph with $t_r(n)$ edges other than $T_{r,n}$ has an $(r + p)$-vertex subgraph with more than $\binom{r+p}{2} - p$ edges. (Dirac [1963])

11.1.17. (+) The Turán graph $T_{n,r}$ is regular when r divides n. Prove that there is a k-regular triangle-free graph on n vertices if and only if n is even and at least $2k$ or n is odd and at least $5k/2$, with k even when n is odd. (Comment: Bauer [1983] determined all triples (r, n, k) such that there is a k-regular K_{r+1}-free graph on n vertices.)

11.1.18. Let G be the Petersen graph. For $n \geq 2$, prove $\mathrm{ex}(n, G) \geq 2n - 3 + \lfloor (n - 2)^2/4 \rfloor$. (Comment: Simonovits [1999] proved that equality holds.)

11.1.19. Prove that the minimum number of edges in a connected n-vertex graph where every edge lies in a triangle is $3(n - 1)/2$ if n is odd, $3n/2 - 1$ if n is even. (Erdős [1988a])

11.1.20. (+) Determine the minimum number of edges in a triangle-free graph on $2n$ vertices whose complement contains no n-clique. (Erdős [1988b])

11.1.21. Determine the maximum number of edges in a 10-vertex graph with no 4-cycle. (Bialostocki–Schönheim [1984])

11.1.22. For $n \geq 6$, prove that the maximum number of edges in an n-vertex graph not having two edge-disjoint cycles is $n + 3$. (Erdős–Pósa [1962])

11.1.23. ($\Diamond$) Prove that if $|V(G)| \geq 6$ and $\delta(G) \geq 4$, then G has two disjoint cycles. Conclude that for $n \geq 6$ the maximum number of edges in an n-vertex graph not having two disjoint cycles is $3n - 6$. (Erdős–Pósa [1962]) (Comment: Corradi–Hajnal [1963] proved more generally that if $|V(G)| \geq 3k$ and $\delta(G) \geq 2k$, then G has k pairwise disjoint cycles.)

11.1.24. Let G be a connected graph with m edges and more than one vertex.
    (a) Prove that if G is P_4-free, then $\overline{G}$ is disconnected. (Seinsche [1974])
    (b) Prove that if $\Delta(G) = D$ and $\overline{G}$ is disconnected, then $m \leq D^2$, with equality only for $K_{D,D}$. (Chung–West [1993])

11.1.25. Let G be a self-complementary n-vertex graph, and let $k = \lfloor n/4 \rfloor$. Conclude from Exercise 27 that G has at least $k(k-1)(4k-2)/3$ triangles if $n = 4k$ and that G has at least $k(k-1)(4k+1)/3$ triangles if $n = 4k + 1$.

11.1.26. ($\Diamond$) Let G be a graph on n vertices such that $\overline{G}$ has no triangles. For $n \geq 6$, prove that the minimum possible number of triangles in G is $\binom{\lfloor n/2 \rfloor}{3} + \binom{\lceil n/2 \rceil}{3}$.

11.1.27. For an n-vertex graph G, prove the bounds below and show that they are sharp.

$$k_3(G) + k_3(\overline{G}) \geq \begin{cases} 2\binom{l}{3} & \text{if } n = 2l \\ \frac{2}{3}k(k-1)(4k+1) & \text{if } n = 4k+1 \\ \frac{2}{3}k(k+1)(4k-1) & \text{if } n = 4k+3 \end{cases}.$$

11.1.28. Let $H_k = K_1 \diamondplus kK_2$. Prove that $\text{ex}(n; H_2) = \lfloor n^2/4 \rfloor + 1$ for $n \geq 5$. (Comment: Erdős–Füredi–Gould–Gunderson [1995] proved $\text{ex}(n; H_k) = \lfloor n^2/4 \rfloor + k^2 - k$ for odd k and $\text{ex}(n; H_k) = \lfloor n^2/4 \rfloor + k^2 - 3k/2$ for even k.)

11.1.29. A graph G with vertex degrees $d_1, \dots, d_n$ is *r-majorizable* if there is an r-partite graph H with vertex degrees $d'_1, \dots, d'_n$ such that $d'_i \geq d_i$ for $1 \leq i \leq n$. By Erdős' proof of Turán's Theorem (Theorem 5.2.11), every K_{r+1}-free graph is r-majorizable.

(a) Construct non-r-majorizable n-vertex graphs containing only one copy of K_{r+1}.

(b) Prove that G is r-majorizable if $\Delta(G) \leq n(1 - 1/r)$.

(c) Consider $n, D, r \in \mathbb{N}$ such that $n - 1 \geq r(n - D) + \binom{r-t}{2}$, where $0 \leq t < r$. Given $n_1, \dots, n_r$ with sum $n - 1$ such that $n - D = n_1 = \cdots = n_t = n_{t+1} < \cdots < n_r$, construct G from $K_{n_1, \dots, n_r}$ by adding one vertex adjacent to all vertices in t smallest parts and one vertex in each other part. Prove that G is not r-majorizable.

(d) The graph of part (c) has $(n - D)^t$ copies of K_{r+1}. For $r = 2$, prove that this is the fewest among non-r-majorizable n-vertex graphs with $\Delta(G) = D$. (Comment: West [1982b] proved this for $r = 2$ and also for $(n, D, r) = (7, 5, 3)$; does it always hold?)

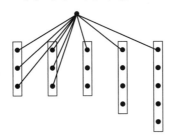

11.1.30. Construct a 3-uniform hypergraph with asympotically $\frac{5}{9}\binom{n}{3}$ edges that does not contain $K_4^{(3)}$, the complete 3-uniform hypergraph with four vertices. (Comment: Turán [1941] conjectured that this is asymptotically the most edges in such an n-vertex hypergraph: that is, $\text{ex}(n; K_4^{(3)})/\binom{n}{3} \to 5/9$. Erdős offered \$1000 for a proof. Chung–Lu [1999] proved an upper bound of .593, and Razborov [2011] lowered it to .561.)

11.1.31. ($\diamond$) *Erdős–Stone Theorem* (Theorem 11.1.10). Fix $\varepsilon \in (0, 1)$, and consider $r, t \in \mathbb{N}$.

(a) Prove that when n is sufficiently large, every n-vertex graph with minimum degree at least $(1 - 1/r + \varepsilon)n$ contains $K_{r+1}[t]$. (Hint: Use induction on r.) (See Lovász [1993])

(b) Prove $\text{ex}(n; K_{r+1}[t]) \leq (1 - 1/r + \varepsilon)n^2/2$ for sufficiently large n. (Erdős–Stone [1946])

11.1.32. An **ordered graph** is a graph with linear order on the vertices. The **interval chromatic number** of an ordered graph is the least number of independent sets of consecutive vertices needed to cover the vertices (ranging from 2 to n for n-vertex ordered paths). An ordered graph G *avoids* an ordered graph H if H does not appear in G via an order-preserving injection on the vertices. Let $\text{ex}(n; H)$ be the maximum number of edges in an n-vertex ordered graph that avoids H. Use the Erdős–Stone Theorem to prove that if H has interval chromatic number k, then $\text{ex}(n; H) = (1 - \frac{1}{k-1})\binom{n}{2} + o(n^2)$. (Pach–Tardos [2006])

11.1.33. A graph G is *F-saturated* if it does not contain F, but adding any edge creates a copy of F. For $n \geq t - 1$, prove that $K_{t-2} \diamondplus \overline{K}_{n-t+2}$ has the fewest edges among all K_t-saturated graphs with n vertices. (Erdős–Hajnal–Moon [1964])

11.1.34. ($\Diamond$) Prove that there is an n-vertex F-saturated graph (see Exercise 11.1.33 for definition) having at most $(\beta(F) - 1)n + \Delta(F)n/2$ edges, where $\beta(F)$ is the vertex cover number of F. (Kászonyi–Tuza [1986]; clarified in Füredi [2007])

11.1.35. Let $\text{sat}(n; F)$ be the minimum number of edges in an F-saturated graph with n vertices (see Exercise 11.1.33 for definition).
  (a) Determine $\text{sat}(n; K_3)$.
  (b) With P being the Petersen graph, show that $\text{sat}(n; P) \leq 4n - 4$.

11.1.36. Given the Regularity Lemma (Theorem 11.1.15), prove that one can instead require an ε-regular partition in which each part belongs to at most εk irregular pairs (instead of a total of εk^2 irregular pairs).

11.1.37. ($\Diamond$) *The* **Ramsey–Turán problem**. (Szemerédi [1972])
  (a) Let G be a graph with $2n$ vertices, partitioned into sets A and B of size n. Show that if $|[A, B]| \geq (1/2 + \varepsilon)n^2$, then G contains K_4 or an independent set of size $\varepsilon n/2$.
  (b) Let G be a graph with $3n$ vertices partitioned into sets A, B, and C, each of size n. Show that if each pair among $\{A, B, C\}$ is ε-regular and has density more than 2ε, then G contains K_4 or an independent set of size at least $\varepsilon^2 n$.
  (c) (+) Let G be an n-vertex graph. Prove that if G does not contain K_4, and $\alpha(G) \in o(n)$, then $|E(G)| < (1/8)n^2 + o(n^2)$.

11.1.38. *The Slicing Lemma.* Let (A, B) be an ε-regular pair with density d, where $0 < \varepsilon \leq \alpha \leq 1/2$. Prove that for any $A' \subseteq A$ and $B' \subseteq B$ with $|A'| \geq \alpha |A|$ and $|B'| \geq \alpha |B|$, the pair (A', B') is ε/α-regular with density greater than $d - \varepsilon$.

11.1.39. ($\Diamond$) Let G be an A, B-bigraph with $|A| = |B| = n$.
  (a) Prove that if G has no independent set consisting of $\delta(G)$ vertices in A and $\delta(G)$ vertices in B, then G has a perfect matching.
  (b) Obtain a perfect matching when (A, B) is an ε-regular pair and $\delta(G) > \varepsilon n$.

11.1.40. ($\Diamond$) Given an ε-regular pair (A, B) with density d and $|A| = |B| = m$, let a vertex v in $A \cup B$ be *good* for the pair if it has at least $(d - \varepsilon)m$ neighbors in the other part.
  (a) Let $x \in A$ and $y \in B$ be fixed good vertices in the pair (A, B) as specified above. Prove that if $d > 5\varepsilon$, then when m is sufficiently large there is an x, y-path of length at least $(1 - \varepsilon - \frac{\varepsilon}{d-\varepsilon})2m$ that avoids any K specified vertices.
  (b) Given $d > 5\varepsilon$, let R be the reduced graph of ε-regular pairs with density at least d resulting from an ε-regular partition of an n-vertex graph G. Prove that if R has k vertices and some component of R has a matching of size t, then G has a cycle of length at least $(1 - \varepsilon - \frac{\varepsilon}{d-\varepsilon})\frac{2t}{k}n$. (See Figaj–Łuczak [2007] for similar results.)

11.1.41. (+) Let G be an A, B-bigraph with $|A| = |B| = n$ and $\delta(G) \geq dn$. Prove that if (A, B) is an ε-regular pair and $d > 3\varepsilon + \sqrt{\varepsilon}$, then G has a spanning cycle. (Haxell [1997])

11.1.42. Let (A, B) be an ε-regular pair with density d and $|A| = |B| = n$.
  (a) For $k \in \mathbb{N}$, consider $Y \subseteq B$ such that $(d - \varepsilon)^{k-1}|Y| \geq \varepsilon|B|$. Prove that the number of k-tuples $(x_1, \ldots, x_k)$ of elements of A having at most $(d - \varepsilon)^k|Y|$ common neighbors in Y is at most $k\varepsilon|A|^k$. (Komlós–Simonovits [1996]) (Comment: Lemma 11.1.20 is $k = 1$.)
  (b) Prove that at least $\frac{1}{4}(1 - 4\varepsilon)(d - \varepsilon)^4 n^4$ 4-cycles alternate between A and B.

11.1.43. ($\Diamond$) **(6,3)-Theorem** (Ruzsa–Szemerédi [1978]).
  (a) Let G be an n-vertex graph with every edge in exactly one triangle. Use the Triangle-Removal Lemma (Lemma 11.1.22) to prove that G has $o(n^2)$ edges.
  (b) Let G be an n-vertex graph that decomposes into at most n induced matchings (vertices of distinct edges in such a matching are not adjacent). Prove that G has $o(n^2)$ edges.
  (c) Let H be a 3-uniform n-vertex hypergraph having no six vertices that induce at least three edges. Prove that H has $o(n^2)$ edges.

11.1.44. ($\diamond$) Fix $c \in (0, 1)$. Prove that for some $n \in \mathbb{N}$, any set of at least cn^2 points in $[n]^2$ contains three distinct points of the form $\{(x, y), (x + a, y), (x, y + a)\}$. (Hint: Use Lemma 11.1.22 on a suitable graph.) (Ajtai–Szemerédi [1974], Solymosi [2003])

11.2. Families of Sets

In this section, we consider various extremal problems for families of subsets of $[n]$. West [1982a], Frankl [1987], and Kleitman [1994] survey these and related topics. See also the books Engel [1997], Jukna [2011], and Gerbner–Patkós [2019]. Frankl–Tokushige [2016] is an authoritative survey. At the end, we discuss entropy as a tool for extremal problems in enumeration.

THE KRUSKAL–KATONA THEOREM

In extensions of the Turán problem, we wanted to choose m edges to have the fewest r-cliques. Similarly, we may ask for m vertex sets of size k that contain the fewest sets of size $k - 1$.

11.2.1. Definition. A **family** of sets or **set system** is a set whose members are sets. A k-**uniform** family is a family of k-sets.

We use "member of family" to avoid confusion with "element of set". A family of subsets of $[n]$ is the edge set of a hypergraph with vertex set $[n]$, and "k-uniform" has the same meaning in both contexts. In Section 10.2 we used hypergraph language to generalize notions from graph theory. Some authors use the term "hypergraph" just for a family of sets without specifying the vertex set.

11.2.2. Definition. The t-**shadow** of a family F is the family of all t-sets contained in members of F. The **shadow** ∂F of a k-uniform family F is its $(k-1)$-shadow. The **shade** is the family of all $(k + 1)$-sets containing members of F.

Among all k-uniform families of size m, we seek the family F with smallest shadow. Since each k-set contributes exactly k sets to the shadow and $|F|$ is fixed, the size of the shadow is minimized when the sets in the shadow arise from many members of F. This can be achieved by confining F to subsets of a small subset of $[n]$. The amazing result is that there exists an ordering of the k-sets such that, for each m, the first m sets in this ordering form an optimal family.

To confine the initial sets to a small subset of $[n]$, we define an ordering of k-sets in which x precedes y if the largest element of x is less than the largest element of y. We use notation like x and y for k-element subsets to emphasize their interpretation as binary incidence vectors.

11.2.3. Definition. The **co-lexicographic** or **colex ordering** on a family of k-sets is obtained by putting $x < y$ if $x_i < y_i$ in the highest coordinate where their binary incidence vectors differ.

11.2.4. Example. *Colex ordering on* $\binom{[5]}{3}$. It is convenient to start the indexing at 0 for both the list of vectors and the coordinates of the vectors.   ∎

index	set	positions of 1s	index	set	positions of 1s
0	11100	012	5	10101	024
1	11010	013	6	01101	124
2	10110	023	7	10011	034
3	01110	123	8	01011	134
4	11001	014	9	00111	234

To facilitate the proof that the first m sets in this order form an optimal family, we study its shadow.

11.2.5. Lemma. If the vector with index m in the colex ordering on $\binom{[n]}{k}$ has 1 in positions $m_1, \ldots, m_k$, with $m_1 < \cdots < m_k$, then $m = \binom{m_k}{k} + \binom{m_{k-1}}{k-1} + \cdots + \binom{m_1}{1}$.

Proof: Let σ be the vector with index m. Indexing starts with 0, so m counts the vectors preceding σ. We count them another way. To reach σ, we must skip all vectors whose kth 1 appears before position m_k. There are $\binom{m_k}{k}$ of these, since the first position is position 0. Some vectors with the last 1 in position m_k also precede σ. These begin with $\binom{m_{k-1}}{k-1}$ vectors having $k-1$ 1s in positions before m_{k-1}. Continuing, the jth term in the summation counts the vectors that appear before σ in the ordering and have their last $j - 1$ 1s in the same positions as σ.   ∎

11.2.6. Corollary. For $m, k \in \mathbb{N}$, there is a unique expression of m in the form $\binom{m_k}{k} + \binom{m_{k-1}}{k-1} + \cdots + \binom{m_i}{i}$ with $m_k > \cdots > m_i \geq i$.

Proof: In the colex ordering on $\binom{[n]}{k}$, appending 0s increases the number of coordinates without changing the position of the first $\binom{n}{k}$ vectors. Lemma 11.2.5 thus applies for any n with $\binom{n}{k} > m$ to obtain an expression for m using the vector with index m in the ordering. Left-justified 1s in the vector yield $m_j = j - 1$ and contribute 0; we drop those terms from the expression in Lemma 11.2.5 to obtain an expression of the desired form. Two such expansions correspond to distinct vectors in the colex ordering and hence distinct values of m.   ∎

11.2.7. Definition. The unique representation of m described in Corollary 11.2.6 is the k-**binomial expansion** of the integer m. For an integer m with k-binomial expansion $m = \binom{m_k}{k} + \binom{m_{k-1}}{k-1} + \cdots + \binom{m_i}{i}$ such that $m_k > m_{k-1} \cdots > m_i \geq i$, let $\partial_k(m) = \binom{m_k}{k-1} + \binom{m_{k-1}}{k-2} + \cdots + \binom{m_i}{i-1}$.

11.2.8. Lemma. The shadow of the first m vectors in the colex order on $\binom{[n]}{k}$ consists of the first $\partial_k(m)$ vectors in the colex order on $\binom{[n]}{k-1}$.

Proof: We count the shadow using Lemma 11.2.5. The family consisting of all elements of $\binom{[n]}{k}$ whose last 1 appears before position m_k has shadow consisting of all vectors of weight $k - 1$ whose last 1 precedes position m_k. In general, the j-th term in the summation for $\partial_k(m)$ considers all vectors before the indexed vector v_m in $\binom{[n]}{k}$ that have their rightmost $j - 1$ ones in the same positions as v_m. It counts the portion of their shadow consisting of sets whose last $j - 1$ ones are in those same positions. This is the portion of the shadow not counted by earlier terms. Thus the full sum counts the entire shadow exactly.   ∎

11.2.9. Definition. On a family F in $\mathbf{2}^{[n]}$, the **shift operator** $\tau_{i,j}$ is defined by

$$\tau_{i,j}(x) = \begin{cases} x - \{j\} + \{i\} & \text{if } j \in x \text{ and } i \notin x \text{ and } x - \{j\} + \{i\} \notin F, \\ x & \text{otherwise.} \end{cases}$$

Let $\tau_{i,j}(F) = \{\tau_{i,j}(x) \colon x \in F\}$.

11.2.10. Lemma. If $F \subseteq \mathbf{2}^{[n]}$ and $i < j$, then $\partial(\tau_{i,j}(F)) \subseteq \tau_{i,j}(\partial F)$.

Proof: Let $G = \partial F$. Let τ_F and τ_G denote $\tau_{i,j}$ on F and G, respectively. It suffices to show $y \in \tau_G(G)$ when $y \in \partial(\tau_F(F))$. For $y \in \partial(\tau_F(F))$, we have $y = x' - s$ for some $s \in x' \in \tau_F(F)$. Note also that $x' = \tau_F(x)$ for some $x \in F$.

$\quad$ **Case 1:** $x' = x$. In this case $y \in G$, and it suffices to show $\tau_G(y) = y$. If not, then $i \notin y$ and $j \in y$. Since $j \in y$ and $y = x - s$, we have $s \neq j$ and $j \in x$. If $s = i$, then $y - j + i = x - j \in G$, so $\tau_G(y) = y$. If $s \neq i$, then $i \notin y$ yields $i \notin x$; now $\tau_F(x) = x$ implies $x - j + i \in F$. Hence $y - j + i \in \partial F = G$, so again $\tau_G(y) = y$.

$\quad$ **Case 2:** $x' \neq x$. Now x contains j but not i, and $x' = x - j + i$. Since $y = x' - s$, we have $j \notin y$, and hence $\tau_G(y) = y$ if $y \in G$. If $i \notin y$, then $i = s$ and $y = x - j \in G$. Hence we may assume $i \in y$ and $y \notin G$. Let $\hat{y} = y + j - i$, so now $\hat{y} \in \partial x \subseteq G$. With $\hat{y} \in G$ and $\hat{y} - j + i = y \notin G$, we have $\tau_G(\hat{y}) = y$, so $y \in \tau_G(G)$ as desired. ∎

11.2.11. Theorem. (Kruskal–Katona Theorem; Kruskal [1963], Katona [1968])
$\quad$ For $F \subseteq \binom{[n]}{k}$ with $|F| = m$, the minimum of $|\partial F|$ occurs when F consists of the first m members of the colex ordering on $\binom{[n]}{k}$, and then $|\partial F| = \partial_k(m)$.

Proof: (Frankl [1984]) By Lemma 11.2.8, the shadow of the specified family has size $\partial_k(m)$, so it suffices to show $|\partial F| \geq \partial_k(m)$ for all $F \subseteq \binom{[n]}{k}$ with $|F| = m$.

$\quad$ We use induction on k; trivially the claim holds for a 1-uniform family. For the induction step, we use induction on m. Note that $m = \binom{k}{k}$ when $m = 1$, and indeed then $|\partial F| = k = \binom{k}{k-1}$. Now consider $m > 1$.

$\quad$ Fix $i \in [n]$. Applying any $\tau_{i,j}$ does not change the number of sets, and it increases the number of sets containing i if it produces a change. From F, application of finitely many such operators therefore leads to a family F^*, invariant under all $\tau_{i,j}$, in which i is a dominant element, meaning $x - \{j\} + \{i\} \in F^*$ for all $j \in x \in F^*$. By Lemma 11.2.10, $|\partial F^*| \leq |\partial F|$. Hence it suffices to prove $|\partial F| \geq \partial_k(m)$ in the case where F has a dominant element i.

$\quad$ From F define $F' = \{x \colon i \notin x \in F\}$ and $F^- = \{x - i \colon i \in x \in F\}$. Note $F' \subseteq \binom{[n]}{k}$, $F^- \subseteq \binom{[n]}{k-1}$, and $|F'| + |F^-| = m$. If $j \in x \in F'$ and $i \notin x$, then $x - j + i \in F$, since i is dominant. Hence $x - j \in F^-$ and $\partial F' \subseteq F^-$. We conclude that ∂F consists of F^- plus the sets formed by adding i to members of the shadow of F^-.

$\quad$ Let $\sum_{j=i}^{k} \binom{m_j}{j}$ be the k-binomial expansion of m, as in Definition 11.2.7. Let $m' = \sum_{j=i}^{k} \binom{m_j - 1}{j}$. If $|F^-| < \partial_k(m')$, then $|F'| = m - |F^-| > m - \partial_k(m') = m'$. Now, since $|F'| < |F| = m$, the induction hypothesis on m yields $|\partial F'| \geq \partial_k(m') > |F^-|$, which contradicts $\partial F' \subseteq F^-$.

$\quad$ Hence we may assume $|F^-| \geq \partial_k(m')$. In this case the induction hypothesis on k yields $|\partial F^-| \geq \partial_{k-1}\partial_k(m') \geq \sum_{j=i}^{k} \binom{m_j - 1}{j-2}$. Finally, we compute

$$|\partial F| \geq |F^-| + |\partial F^-| \geq \partial_k(m') + \sum_{j=i}^{k} \binom{m_j - 1}{j-2} \geq \sum_{j=i}^{k} \binom{m_j}{j-1} = \partial_k(m). \qquad \blacksquare$$

Note that the optimal family depends only on m and k, not n, as long as $\binom{n}{k} \geq m$. The Kruskal–Katona Theorem is a very precise statement. For many applications, a smoothed version due to Lovász [1979] suffices. It treats m as a value of the "choose k" polynomial and can also be proved using shift operators (Exercise 9). Instead, we give a short more recent proof. Exercise 10 requests the characterization of when equality holds.

11.2.12. Theorem. (Lovász [1979, Ex. 13.31(b)]) For $u \geq k$, let $\binom{u}{k} = \frac{u_{(k)}}{k!}$. If $F \subseteq \binom{[n]}{k}$ with $|F| = \binom{u}{k}$, then $|\partial F| \geq \binom{u}{k-1}$.

Proof: (Keevash [2008]) Note that F is contained among the k-sets whose $(k-1)$-subsets are all present in ∂F. Letting $r = k - 1$, it therefore suffices to show that when G is a family of size $\binom{u}{r}$ in $\binom{[n]}{r}$, the number of $(r+1)$-sets whose r-sets are all present in G is at most $\binom{u}{r+1}$.

We use induction on r; the claim is immediate when $r = 1$. For $r > 1$, we treat G as the edge set of a hypergraph. We may assume that each $v \in [n]$ lies in some member of G. Let $d(v)$ be the number of edges of G containing v. Let $G'(v)$ be the family of $(r-1)$-sets obtained by deleting v from edges of G. Let $q(v)$ be the number of $(r+1)$-sets containing v whose r-sets all lie in G.

Note that $S \cup \{v\}$ is counted by $q(v)$ if and only if $S \subseteq G$ and all $(r-1)$-subsets of S lie in $G'(v)$. The first condition implies $q(v) \leq |G| - d(v)$, while the second bounds $q(v)$ by the number of r-sets whose $(r-1)$-subsets all lie in $G'(v)$.

We claim that for all v, at least one of these bounds is at most $\frac{u-r}{r}d(v)$. If $d(v) \geq \binom{u-1}{r-1}$, then $\binom{u}{r} - d(v) \leq \frac{u}{r}\binom{u-1}{r-1} - d(v) \leq \frac{u-r}{r}d(v)$. If $d(v) \leq \binom{u-1}{r-1}$, then define u' by $d(v) = \binom{u'-1}{r-1}$; note that $u' \leq u$. Note also that $|G'(v)| = d(v)$. By the induction hypothesis, at most $\binom{u'-1}{r}$ elements of $\binom{[n]}{r}$ have all their $(r-1)$-sets in $G'(v)$. We compute $\binom{u'-1}{r} = \frac{u'-r}{r}\binom{u'-1}{r-1} \leq \frac{u-r}{r}d(v)$.

Finally, every $(r+1)$-set whose r-sets are all present in G is counted $r+1$ times in $\sum_{v \in [n]} q(v)$. Hence the number of these sets is bounded by $\frac{u-r}{r(r+1)} \sum_{v \in [n]} d(v)$. Since the degree sum is $r|G|$, which equals $r\binom{u}{r}$, the bound simplifies to $\binom{u}{r+1}$. ∎

According to Engel [1997, p. 46], Theorem 11.2.11 was "formulated" by Schützenberger [1959], independently "found" by Kruskal [1963], Harper [1966], Katona [1968], and Clements–Lindström [1969], and reproved later by Hansel [1972], Daykin [1974b], Eckhoff–Wegner [1975], Greene–Kleitman [1978], Hilton [1979], and Frankl [1984]. Clements–Lindström [1969] extended it to multisets with bounded multiplicity. Related exercises include Exercises 7–11.

ANTICHAINS AND INTERSECTING FAMILIES

We began with the Kruskal–Katona Theorem due to its relation to Turán-type problems, but the most fundamental problem in extremal set theory is to maximize the size of a family in which no member contains another. Sperner [1928] gave the answer. The theorem has many proofs, most of which extend to interesting additional results. Chapter 12 presents some of these extensions.

11.2.13. Definition. An **antichain** of sets is a family of sets in which no member contains another.

11.2.14. Theorem. (**Sperner's Theorem**; Sperner [1928]) The maximum size of an antichain of subsets of $[n]$ is $\binom{n}{\lfloor n/2 \rfloor}$, achieved only by the family of $\lfloor n/2 \rfloor$-sets or the family of $\lceil n/2 \rceil$-sets.

Proof: Let k and l be the sizes of the smallest and largest members of a maximum antichain F. Let $A = F \cap \binom{[n]}{l}$. Each l-set has l sets immediately below it, and each $(l-1)$-set lies under $n - l + 1$ sets of size l. Thus $|\partial A| \geq |A| \frac{l}{n-l+1}$. If $l > (n+1)/2$, then replacing the l-sets in F by their shadow yields a larger antichain.

Similarly, the shade of $F \cap \binom{[n]}{k}$ has size at least $|A| \frac{n-k}{k+1}$. If $k < (n-1)/2$, then replacing the k-sets in F by their shade yields a larger antichain. Thus $(n-1)/2 \leq k \leq l \leq (n+1)/2$.

When n is even, we obtain $k = l$. When n is odd, let $A = F \cap \binom{[n]}{(n+1)/2}$. The argument to enlarge F still works unless $|\partial A| = |A|$, which requires all sets above $|\partial A|$ to lie in A. Since one can move from any i-set to any other by alternately adding and deleting elements, $|\partial A| = |A|$ occurs only when A is $\varnothing$ or $\binom{[n]}{(n+1)/2}$. Thus the sets in F all have the same size. ∎

We can also restrict intersections of members of our family.

11.2.15. Definition. An **intersecting family** is a family F such that $A, B \in F$ implies $A \cap B \neq \varnothing$; it is t-**intersecting** if always $|A \cap B| \geq t$. A **star** is a family of sets having a common element; a t-**star** is a family having t common elements.

This definition extends the graph-theoretic notion of star to the hypergraph context. Under various conditions, the large intersecting families are stars.

11.2.16. Example. *Intersecting families.* An intersecting family of subsets of $[n]$ has at most 2^{n-1} members, since a set and its complement cannot both appear. The bound is achieved by the star consisting of all sets containing a fixed element.

Other largest intersecting families consist of all sets with more than half the elements, plus (for even n) one from each complementary pair of sets of size $n/2$. Katona [1964] (Exercise 15) generalized this to determine the largest t-intersecting families (see Ahlswede–Khachatrian [2005] for other proofs).

The family of subsets of $[n]$ containing one element and omitting another is an intersecting family whose complements also form an intersecting family. It is a largest such family, with size 2^{n-2} (Seymour [1973], Schönheim [1974], Anderson [1976], Daykin–Lovasz [1976], Hilton [1976]; see Exercises 28–29.) ∎

We next consider intersecting families of special types. Frankl [1995] surveys many such classical problems. We emphasize the Erdős–Ko–Rado Theorem on t-intersecting families consisting of an antichain of subsets of size at most k.

11.2.17. Definition. An $EKR(k, t)$-**family** is an antichain F that is also a t-intersecting family whose members have size at most k.

We will see that a largest such family uses only k-sets. The k-sets containing t particular elements form a t-star that is an $EKR(k, t)$-family of size $\binom{n-t}{k-t}$. When n is sufficiently large, this is the extremal family. We first consider only the case $t = 1$, using a counting argument. In this case we may express the problem as a Turán-type problem; we seek a largest k-uniform hypergraph not containing the hypergraph consisting of two disjoint edges. Published in 1961, the EKR Theorem was actually proved in 1938, as noted by Erdős [1990].

11.2.18. Theorem. (Erdős–Ko–Rado Theorem $t = 1$; Erdős–Ko–Rado [1961])
For $n \geq 2k$, the maximum size of an $EKR(k, 1)$-family is $\binom{n-1}{k-1}$. For $n > 2k$, equality holds only for stars.

Proof: (Katona [1972b]) If some member of such a family F has size less than k, then replacing all smallest members by their shade strictly enlarges the family (since $k \leq n/2$) while preserving the conditions. Thus we may assume $F \subset \binom{[n]}{k}$.

Given a circular ordering σ of $[n]$, we ask how many members of F occur in σ as a consecutive segment of elements. Write such a member x as $(a_1, \ldots, a_k)$, indexed in order in σ. Every k-element segment that intersects x has a boundary immediately following some a_1 with $1 \leq i < k$. Since $k \leq n/2$, two k-sets having the same boundary within x cannot intersect, as shown below. Hence at most $k - 1$ members of F other than x appear in σ.

Summing over all $(n - 1)!$ circular orderings yields at most $(n - 1)!k$ appearances of members of F. Since each member of F appears in $k!(n - k)!$ such orderings, $|F| \leq \frac{(n-1)!k}{k!(n-k)!} = \binom{n-1}{k-1}$.

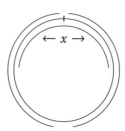

Equality requires having exactly k members of F in every circular ordering of $[n]$. With x occuring in σ, for $1 \leq i < k$ the location between a_i and a_{i+1} must be the boundary of one member x_i of F. If always x_i extends clockwise (or always counterclockwise) from there, then the k appearances successively shift by one element. Otherwise, some x_{i-1} and x_i extend in opposite directions. Since $n > 2k$, they intersect only if they both contain a_i. Now $x_i, \ldots, x_{k-1}, x, x_1, \ldots, x_{i-1}$ successively shift by one element. The argument holds for all orderings.

Let b be the element before a_1 in σ. Consider all circular orderings σ' such that b is followed immediately by $a_1, \ldots, a_{k-1}$ in some order and then a_k. Since $\{b\} \cup \{a_1, \ldots, a_{k-1}\} \notin F$ but $x \in F$, all k-element segments in σ' that contain a_k must lie in F. Using all such σ', every k-set containing a_k belongs to F. ∎

The original proof pushed an $EKR(k, t)$ family toward the claimed extremal family using shift operators.

11.2.19. Lemma. If F is a k-uniform t-intersecting family, then also $\tau_{i,j}(F)$ is t-intersecting, where $\tau_{i,j}$ is the shift operator of Definition 11.2.9.

Proof: View $x, y \in F$ as binary vectors, and let $x' = \tau_{i,j}(x)$ and $y' = \tau_{i,j}(y)$. By symmetry, suppose $i < j$. If x or y has 11 or 00 in these positions, or if they both have 10, then $|x' \cap y'| = |x \cap y|$. If they have 10 and 01, then $x' \cap y' \supseteq x \cap y$.

This leaves the case where x and y both have 01 in these positions. If exactly one of x' and y' changes to 10, then possibly $|x' \cap y'| < t = |x \cap y|$. If $y' = y$ (by symmetry), then $z \in F$, where $z = y - j + i$. Since F is t-intersecting, x and z have at least t common elements outside $\{i, j\}$. Hence this holds also for x' and z and for x' and y', since z and y' agree outside $\{i, j\}$. ∎

11.2.20. Theorem. (**Erdős–Ko–Rado Theorem** $t = 1$, revisited) For $n \geq 2k$, the maximum size of an $EKR(k, 1)$-family is $\binom{n-1}{k-1}$.

Proof: As in Katona's proof, we may let F be a k-uniform intersecting family. We use induction on $n + k$. When $n = 2k$, in F there is at most one member of each complementary pair, and choosing one from each such pair yields an intersecting family. Since $\binom{2k-1}{k-1} = \frac{1}{2}\binom{n}{k}$, the bound holds.

Now suppose $n > 2k$. Since the shift operator $\tau_{i,j}$ does not change the size of a family, by Lemma 11.2.19 we may assume that F is unchanged under all $\tau_{i,j}$ with $i < j$. Partition F into subfamilies F_1 and F_0, with F_1 consisting of those members containing element n. Inductively, $|F_0| \leq \binom{n-2}{k-1}$. Trivially $|F_1| \leq \binom{n-1}{k-1}$; we obtain the desired bound if we can improve this bound to $\binom{n-2}{k-2}$.

Let $F' = \{A - n : A \in F_1\}$. By the induction hypothesis, it suffices to show that F' is an intersecting family. Otherwise, we have $A, B \in F$ with $A \cap B = \{n\}$. Since $n > 2k$, their union omits some other element i. Since F is invariant under $\tau_{i,n}$, we have $A - n + i \in F$. Now B does not intersect $A - n + i$, a contradiction. ∎

Frankl [1987] proved via shift operators that $EKR(k, t)$-families have size at most $\binom{n-t}{k-t}$ when n is sufficiently large. Forbidding t-stars costs a lot; an $EKR(k, 1)$-family with no common element has size at most $\binom{n-1}{k-1} - \binom{n-k-1}{k-1} + 1$ (Hilton–Milner [1967]; see Exercise 14). Bollobás [1986] considered larger t.

11.2.21.* Theorem. (Bollobás [1986]) For $2 \leq t < k$, let F be a t-intersecting k-uniform family of subsets of $[n]$ that is not a t-star. If $n > 2k - t$, then

$$|F| \leq k\binom{n-t-1}{k-t-1} + \sum_{j=1}^{t} \binom{t}{j}\binom{k-t}{j}^2\binom{n-t-j}{k-t-j}.$$

Proof: Let F be a maximal such family. Since $n > 2k - t$, we have $F \neq \binom{[n]}{k}$. If $|A \cap B| > t$ for all $A, B \in F$, then adding a k-set that differs from a member of F by one element enlarges the family. Hence there exist $A, B \in F$ with $|A \cap B| = t$; let $T = A \cap B$. Since F is not a t-star, we find $C \in F$ with $T \not\subseteq C$.

For $0 \leq j \leq t$, let F_j be the subfamily of members of F omitting j elements of T. For $j \geq 1$, a member of F_j must have at least j elements of A and j elements of B outside T. After picking the j elements of T to omit, we thus have

$$|F_j| \leq \binom{t}{j}\binom{k-t}{j}^2\binom{n-t-j}{k-t-j}.$$

For $j = 0$, a member of F_j contains all of T and needs at least one element of C outside T, so $|F_0| \leq k\binom{n-t-1}{k-t-1}$. Since $|F| = \sum_{j=0}^{t} |F_j|$, the bound is complete. ∎

11.2.22.* Theorem. For $2 \le t < k$, let F be a t-intersecting k-uniform family of subsets of $[n]$. If $n \ge 2tk^3$, then $|F| \le \binom{n-t}{k-t}$, with equality only for a t-star.

Proof: The claim is tivial for t-stars. By Theorem 11.2.21, it suffices to show

$$\sum_{j=2}^{t} \binom{t}{j}\binom{k-t}{j}^2\binom{n-t-j}{k-t-j} < \binom{n-t}{k-t} - [k+t(k-t)^2]\binom{n-t-1}{k-t-1}.$$

With $S_j = \binom{t}{j}\binom{k-t}{j}^2\binom{n-t-j}{k-t-j}$, we have $\frac{S_{j+1}}{S_j} = \frac{t-j}{j+1}\left(\frac{k-t-j}{j+1}\right)^2\frac{k-t-j}{n-t-j} < (j+1)^{-3}$. Since $\sum_{j=2}^{\infty}(j+1)^{-3} < 1$, we have $\sum_{j=2}^{t} S_j < 2S_2$, so

$$\sum_{j=2}^{t} S_j < t^2(k-t)^4\binom{n-t-2}{k-t-2} < \frac{t^2(k-t)^5}{n-t}\binom{n-t-1}{k-t-1}.$$

Since $\frac{t^2(k-t)^5}{n-t} < \frac{1}{4}\frac{n-t}{k-t}$ when $n > 2tk^3$, we obtain $\sum_{j=2}^{t} S_j < \frac{1}{4}\frac{n-t}{k-t}\binom{n-t-1}{k-t-1}$, and then $\frac{1}{4}\frac{n-t}{k-t} < \frac{n-t}{k-t} - k - t(k-t)^2$ (for $n > 2tk^3$) finishes the proof. ∎

At first, "sufficiently large" meant $n \ge t + (k-t)\binom{k}{t}^3$; the threshold in Theorem 11.2.22 is better. Frankl [1976] and Wilson [1984] together showed that the t-star of k-sets is optimal when $n \ge (t+1)(k-t+1)$, extending Theorem 11.2.18. For smaller n, other $EKR(k,t)$-families are larger, such as $\binom{[n]}{\lceil(n+1)/2\rceil}$ when $t=1$ and $n < 2k$. Ahlswede–Khachatrian [1997] determined the extreme in all cases (see Exercise 22), as conjectured by Frankl [1978].

The Erdős–Ko–Rado Theorem for $t = 1$ follows easily from the Kruskal–Katona Theorem, as shown by Daykin [1974a]). The earlier result below uses the same ideas and is stronger (simply set $F = G$).

11.2.23.* Theorem. (Kleitman [1968]) Let k, l, n be positive integers such that $k + l \le n$. If $F \subset \binom{[n]}{k}$ and $G \subset \binom{[n]}{l}$ satisfy $x \cap y \ne \emptyset$ whenever $x \in F$ and $y \in G$, then $|F| < \binom{n-1}{k-1}$ or $|G| \le \binom{n-1}{l-1}$.

Proof: Let $F' \subset \binom{[n]}{n-k}$ consist of the complements of members of F. Note that $n - k \ge l$, since $k + l \le n$. The family $G \cup F'$ is an antichain. Thus the l-shadow of F' is disjoint from G. By the Kruskal-Katona Theorem, the size of the l-shadow of F' is at least $\partial_{l+1}\partial_{l+2}\cdots\partial_{n-k}(|F'|)$. Thus $|G| + \partial_{l+1}\partial_{l+2}\cdots\partial_{n-k}(|F'|) \le \binom{n}{l}$.

If $|F| \ge \binom{n-1}{k-1}$, then also $|F'| \ge \binom{n-1}{n-k}$, and the first term in the $(n-k)$-binomial expansion of $|F'|$ is $\binom{n-1}{n-k}$. Regardless of what else is present, iteration of ∂ yields $\binom{n-1}{l}$ as a lower bound on the second term in the inequality above. Thus $|G| \le \binom{n}{l} - \binom{n-1}{l} = \binom{n-1}{l-1}$. ∎

We refer the reader also to Frankl [1978], Wilson [1984], Frankl–Graham [1989], Ahlswede–Khachatrian [1997], Frankl–Füredi [2012], Godsil–Meagher [2016, Chapter 1], and Borg–Meagher [2016] for proofs and extensions of the Erdős–Ko–Rado Theorem. Mubayi–Verstraëte [2016] surveys a related topic.

CHVÁTAL'S CONJECTURE

Every star is an intersecting family. The Erdős–Ko–Rado Theorem implies that in $2^{[n]}$ with $n \geq 2k$, the star F consisting of all sets with size at most k containing 1 is a largest intersecting family of sets with size at most k; it simultaneously achieves equality in the EKR bound for sets of each size. Chvátal conjectured a tantalizing generalization.

11.2.24. Definition. A family I is an **ideal** if $x \in I$ and $y \subseteq x$ imply $y \in I$. The maximal members of an ideal are its **bases**. A family has the **star property** if some largest intersecting subfamily is a star.

11.2.25. Conjecture. (Chvátal [1974]) Every ideal of subsets of $[n]$ has the star property. ∎

The term "ideal" is traditional for this topic and is used in the rest of this chapter. Since "ideal" is used in other ways in other areas of mathematics, "downset" has grown in popularity as a alternative term without alternative meanings.

As we observed before Definition 11.2.24, the Erdős–Ko–Rado Theorem implies the conjecture for the ideal consisting of all subsets of $[n]$ with size at most k, where $k \leq n/2$. Example 11.2.16 proves it for the family of all subsets of $[n]$. Other partial results appear in Chvátal [1974], Kleitman–Magnanti [1974], Berge [1975], and Wang–Wang [1978]; they are surveyed in Kleitman [1979].

We show first that an intersecting family cannot contain more than half of an ideal, because ideals split into pairs of disjoint sets. The proof removes some disjoint pairs from the "top" of the ideal and applies induction to the rest.

11.2.26. Theorem. (Berge [1976]) If I is an ideal of subsets of $[n]$, then I or $I - \{\varnothing\}$ can be partitioned into pairs of disjoint sets.

Proof: (Daykin–Hilton–Miklós [1983], simplified by M. Pelsmajer) We use induction on $|I|$, with a trivial basis for size 0 or 1. For $|I| > 1$, choose a maximal set $T \subseteq [n]$ such that T is a union of disjoint sets $A, B \in I$. Such a set exists, since the empty set qualifies.

Let $H = \{A \subseteq T: A, T - A \in I\}$. The choice of T yields $H \neq \varnothing$. Observe that if $A \in H$ and $A \subseteq C \in I$, then $T - C \subseteq T - A \in I$. Also, if $C \not\subseteq T$, then $C \cup (T - C)$ is larger than T, contradicting the choice of T. Hence $C \in H$, by the definition of H. We conclude that $I - H$ is an ideal.

By construction, H is a subfamily of I partitioned into pairs of disjoint sets (complements within T). Since $I - H$ is an ideal, we complete the desired partition by applying the induction hypothesis to $I - H$. ∎

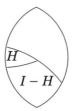

By Theorem 11.2.26, Chvátal's Conjecture holds under any conditions that guarantee a star with half of I, such as when some element belongs to all bases (Exercise 30). Chvátal [1974] proved the conjecture in the case where there is a fixed linear order L on $[n]$ such that $\{a_1, \ldots, a_k\} \in I$ whenever $\{b_1, \ldots, b_k\} \in I$ and $a_i \le b_i$ for $1 \le i \le k$. In this case, the star generated by the initial element in L is a largest intersecting family. It need not contain half of I, as shown when I is generated by singleton sets.

Several survey papers (including one by this author) mistakenly stated that what Chvátal [1974] proved was the stronger result by Snevily that we prove next. Snevily's result is a common generalization of the results of Chvátal and Schönheim. Note the use of $+$ and $-$ for addition and deletion of single elements.

11.2.27. Theorem. (Snevily [1991]) If I is an ideal of subsets of $[n]$, and x is an element of $[n]$ such that $A - a + x \in I$ whenever $a \in A \in I$, then the star generated by x is a largest intersecting family in I.

Proof: Consider a minimal ideal I violating the claim; note that $|I| > 1$. The figure below shows various parts of I that we define. Let $\mathbf{B}$ be the set of bases of I not containing x. Let F be a largest intersecting family in I.

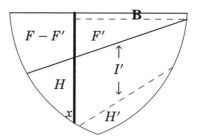

Let a *free base* be a base of I that is not expressible as $B - a + x$ with $a \in B \in \mathbf{B}$. If C is a free base, then let $I^* = I - \{C\}$; we claim that I^* is a smaller ideal satisfying the hypothesis. This holds because if C is needed as $A - a + x$ for some A, then A is below some $B \in \mathbf{B}$, and then C is below $B - a + x$, which is in I. This contradicts that C is a base.

By the induction hypothesis, the star F^* generated by x is a maximum intersecting family in I^*. Since $F^* \subseteq I$, the minimality of I yields $|F| > |F^*|$. Since $F - \{C\} \subseteq I^*$, we obtain $C \in F$ and $|F| = |F^*| + 1$.

If $x \in C$, then I satisfies the claim, since $F^* \cup \{C\}$ is a star with center x in I and has size $|F|$. Hence we may assume that every free base omits x and belongs to F. Thus $\mathbf{B}$ is the set of free bases, and $\mathbf{B} \subseteq F$.

Now let I' be the ideal generated by $\mathbf{B}$. Every member of $I - I'$ lies below a base of the form $B - a + x$ for some $B \in \mathbf{B}$ and must contain x; otherwise it would lie below B and belong to I'.

Let $F' = F \cap I'$. Every member of F' omits x, since it lies in I'. To show that the star generated by x has size at least $|F|$, we will replace F' in F with $|F'|$ sets outside F that contain x. We do this in two steps: first we map the members of F' to their mates in the pairing of I' guaranteed by Theorem 11.2.26, and then we add x to each such mate.

Let H' be the set of mates of members of F', and let $H = \{Y \cup \{x\} : Y \in H'\}$

(it does not matter whether H' is an ideal). Since F' is pairwise intersecting and $Y \in H'$ is disjoint from its mate, $H' \cap F' = \varnothing$. Furthermore, $Y \cup \{x\}$ also is disjoint from the mate of Y in F', so $Y \cup \{x\} \notin F - F'$.

It remains only to show $Y \cup \{x\} \in I$ to obtain $(F - F') \cup H$ as a star generated by x with size $|F|$. Since $\mathbf{B} \subseteq F'$ and $F' \cap H' = \varnothing$, we have $Y \notin \mathbf{B}$. Thus Y is non-maximal in I' and has the form $A - a$ for some $A \in I' \subset I$. The hypothesis then yields $Y \cup \{x\} \in I$. ∎

Subsequent decades have not seen much work on Chvátal's Conjecture. We mention Wang [2002], Borg [2011] (a weighted version), Kamat [2011], and Friedgut–Kahn–Kalai–Keller [2018].

In addition to considering families closed under taking subsets, we can also consider families closed under taking unions. This leads to a conjecture at least as notorious as that of Chvátal. A family of sets is **union-closed** if the union of every two members of the family also belongs to the family.

11.2.28. Conjecture. (Union-Closed Sets Conjecture) If F is a finite union-closed family of finite sets with $|F| \geq 2$, then some element appears in at least half the members of F.

The conjecture is generally attributed to Frankl in 1979 and first appeared in print in 1985. Early special cases were proved by Sarvate–Renaud [1989] (Exercise 32), Poonen [1992], and Johnson–Vaughan [1998]. Bruhn–Schaudt [2015] presents a thorough survey. The fundamental nature of the conjecture is shown by reformulations such as a version in terms of bipartite graphs in Exercise 33.

SUNFLOWERS (optional)

In the problems we have considered thus far, our notion of "star" was a family of sets having a common element. Now we consider a more restricted notion.

11.2.29. Definition. A **Δ-system** or **sunflower** is a family of sets in which any two members have the same intersection.

In a 2-uniform family (a graph), the sunflower condition reduces to a star *or* a family of disjoint edges, where the common intersection and pairwise intersection is always the empty set. The original term is Δ-system, but "Δ" does not evoke the meaning in any way, so we will use the term "sunflower". An early use of this term is in Alon–Frankl–Lovász [1986].

11.2.30. Theorem. (Sunflower Theorem; Erdős–Rado [1960]**)** Every k-uniform family with more than $k!(s-1)^k$ members contains a sunflower of size s, and there is a family of size $(s-1)^k$ having no such sunflower.

Proof: We use induction on k. For $k = 1$, the sets are disjoint, and we have more than $s - 1$ of them.

For $k \geq 2$, let F be the k-uniform family, and let H be a maximal subfamily consisting of disjoint sets. If $|H| \geq s$, then H is the desired sunflower. Otherwise, let B be the union of the sets in H; note that $|B| \le k(s-1)$. Every member of F intersects B. By the Pigeonhole Principle, some element of B lies in at least $|F|/|B|$ members of F. We compute

$$\frac{|F|}{|B|} > \frac{k!(s-1)^k}{k(s-1)} = (k-1)!(s-1)^{k-1}.$$

Deleting x from the sets containing it yields a $(k-1)$-uniform family F' with more than $(k-1)!(s-1)^{k-1}$ members. By the induction hypothesis, F' contains a sunflower of size s. Returning x to its sets yields a sunflower in F of size s.

For the construction, let $X_1, \ldots, X_k$ be disjoint sets of size $s-1$. Let F be the family of all transversals (k-sets consisting of one element from each of $X_1, \ldots, X_k$). If F contains a sunflower $A_1, \ldots, A_s$, then some element x belongs to exactly one of $A_1, \ldots, A_s$. By symmetry, we may assume $x \in A_1 \cap X_1$. Since the pairwise intersections are the same, the sets $A_1, \ldots, A_s$ must intersect X_1 in distinct elements. Since $|X_1| = s-1$, this is impossible. ∎

It is not known whether the bound $k!(s-1)^k$ in Theorem 11.2.30 is sharp.

11.2.31. Conjecture. (Erdős–Rado [1960]) For $k \in \mathbb{N}$ there is a constant c_k such that every k-uniform family F with $|F| > c_k^s$ contains a sunflower of size s.

Kostochka [2000] surveyed results on this and related problems. Abbott–Hanson–Sauer [1972] solved the case $k = 2$. Erdős [1975] offered \$1000 for settling the case $k = 3$. For the case $k = 3$, a more specific conjecture was posed.

11.2.32. Conjecture. (Erdős–Szemerédi [1978]) There is a constant c, less than 2, such that every family of subsets of $[n]$ containing no sunflower of size 3 has at most c^n sets.

By results of Erdős–Szemerédi [1978], Conjecture 11.2.31 implies Conjecture 11.2.32. Using algebraic methods like those in Chapter 15, Naslund–Sawin [2016] proved Conjecture 11.2.32 using a constant c that is just above $3/2^{2/3}$.

The most famous applications of the Sunflower Theorem (and various modifications) are to complexity theory; see Razborov [1985] and Alon–Boppana [1987]. More recently, there is an application to matrix multiplication (Alon–Shpilka–Umans [2013]). Alon–Pudlak [2001] used the method to prove an explicit lower bound for off-diagonal Ramsey numbers: There is a fixed constant c such that for every fixed s and sufficiently large m, the construction produces a graph having neither an s-clique nor an independent m-set. Thus $R(s, m)$ exceeds the order of the constructed graph, which is $m^{c\sqrt{\log s / \log \log s}}$.

ENTROPY (optional)

The notion of entropy is quite different from the rest of this section. We include it here because it can yield extremal results on enumerative problems related to topics studied in this chapter. Our treatment draws on Alon–Spencer [2008] and an excellent tutorial by Galvin [2014]; see also Radhakrishnan [2003].

Imagine an experiment having two possible outcomes, one with probability p and one with probability $1 - p$. If $p = 1/2$, then we can treat the "information" gained by knowing the outcome as one unit, since it takes one bit to specify which outcome occurred. If $p \in \{0, 1\}$, then there is no information in the outcome, since we know before the experiment which will occur.

Now consider three possible outcomes, one with probability $1/2$ and two with probability $1/4$. Using natural binary lists to represent the outcomes, we specify the outcome in one bit with probability $1/2$ and in two bits with probability $1/2$, so the expected information in the outcome is $3/2$. (We will define expectation formally in Chapter 14; here we just use the idea as informal motivation.)

11.2.33. Definition. A finite probability distribution p is a list $(p_1, \ldots, p_n)$ of nonnegative values with sum 1. A **random variable** X with distribution p takes on n possible values, the ith occurring with probability p_i. With lg denoting the base-2 logarithm, the **entropy** of X is a value $H(X)$ defined by

$$H(X) = -\sum_{i=1}^{n} p_i \lg p_i.$$

We think of entropy intuitively as the expected information in the outcome of an experiment with distribution p, motivated by the fact that when the probabilities are powers of $1/2$, the entropy is the expected length of the natural binary word specifying the outcome (Exercise 36). The function can also be obtained as the unique function satisfying certain natural axioms.

The relevance of entropy for extremal problems in enumeration is based on the following lemma.

11.2.34. Lemma. The entropy $H(X)$ of a random variable X with n outcomes is largest only when the outcomes are equally likely, and then $H(X) = \lg n$.

Proof: When the outcomes have the distribution $p_1, \ldots, p_n$, note that $H(X)$ is a weighted average of the values of $\lg(1/p_i)$. For a strictly convex function f, any weighted average of values $f(x_i), \ldots, f(x_n)$ is less than the value at the same weighted average of the arguments $x_1, \ldots, x_n$, unless $x_1 = \cdots = x_n$. (In particular, $\frac{\lambda f(x) + \mu f(y)}{\lambda + \mu} < f(\frac{\lambda x + \mu y}{\lambda + \mu})$ when $x \neq y$. The general statement is known as **Jensen's Inequality**.)

Since lg is strictly convex, when X is not uniform we have

$$H(X) = \sum_{i=1}^{n} p_i \lg(1/p_i) < \lg\left(\sum_{i=1}^{n} p_i(1/p_i)\right) = \lg n.$$

When X is uniform, $H(X) = \sum_{i=1}^{n} p_i \lg n = \lg n$. ∎

The point is this: if we have a random variable X defined over some set S, and we can compute $H(X)$ without knowing $|S|$, then by Lemma 11.2.34 we have $|S| \geq 2^{H(X)}$, with equality if X actually is uniform over S.

We need some additional notions about random variables. When X and Y are random variables, with outcomes respectively in S and T, we denote by (X, Y) the **joint random variable**; it has outcomes in $S \times T$, with $\mathbb{P}(X = i) = \sum_{j \in T} \mathbb{P}((X, Y) = (i, j))$ and $\mathbb{P}(Y = j) = \sum_{i \in S} \mathbb{P}((X, Y) = (i, j))$.

11.2.35. Lemma. If X and Y are random variables, then $H(X) \leq H(X, Y)$.

Proof: The joint distribution breaks events for X into subevents; intuitively, this cannot reduce the expected information in the outcome. In the entropy computation, breaking an outcome with probability $p+p'$ into outcomes with probabilities p and p' replaces $-(p + p')\lg(p + p')$ with $-p\lg p - p'\lg p'$, which is larger since $-\lg$ is a decreasing function. Repeating this leads us from $H(X)$ to $H(X, Y)$. ∎

More importantly, $H(X, Y) \leq H(X) + H(Y)$. Exercise 37 requests a direct proof; we prove this and a generalization via other useful tools (see Galvin [2014]).

11.2.36. Definition. The **conditional probability** of an event F given event E, written $\mathbb{P}(F \mid E)$, is defined by $\mathbb{P}(F \mid E) = \mathbb{P}(F \ \& \ E)/\mathbb{P}(E)$. Write $X \mid E$ to mean the random variable X normalized by the condition that E occurs. Let $p(x \mid E)$ denote the probability of $X = x$ given E; that is, $p(x \mid E) = \mathbb{P}(X = x \ \& \ E)/\mathbb{P}(E)$. This yields a distribution, since $\sum_x p(x \mid E) = 1$. Its entropy is given by $H(X \mid E) = -\sum_x p(x \mid E)\lg p(x \mid E)$. When Y is a random variable and E is the event $Y = y$, taking the weighted average of these distributions over the distribution of Y yields the **conditional entropy**

$$H(X \mid Y) = \mathbb{E}(H(X \mid (Y = y))) = -\sum_y p(Y = y) \sum_x p(x \mid Y = y)\lg p(x \mid Y = y).$$

Intuitively, the information gained in knowing the outcome for the joint variable (X, Y) is the information gained in knowing X plus the information gained in knowing Y given that we already know X. We formalize this.

11.2.37. Lemma. (Chain Rule for entropy) $H(X, Y) = H(X) + H(Y \mid X)$.

Proof: For simplicity, we just write the value for the event that the variable takes that value. Since $\mathbb{P}(Y = y \mid X = x) = \mathbb{P}(X = x \ \& \ Y = y)/\mathbb{P}(X = x)$, we thus write $p(y \mid x) = p(x, y)/p(x)$. Note also $\sum_y p(y \mid x) = 1$. We compute

$$H(X, Y) - H(X) = -\sum_x \sum_y p(x, y)\lg p(x, y) + \sum_x p(x)\lg p(x)$$

$$= -\sum_x p(x) \sum_y p(y \mid x)\lg[p(x)p(y \mid x)] + \sum_x p(x) \sum_y p(y \mid x)\lg p(x)$$

$$= -\sum_x p(x)\Big(\sum_y p(y \mid x)\lg p(x)p(y \mid x) - p(y \mid x)\lg p(x)\Big)$$

$$= -\sum_x p(x)\Big(\sum_y p(y \mid x)\lg p(y \mid x)\Big) = H(X \mid Y).$$ ∎

Another intuitive notion is that the information gained by knowing X is at least that gained by knowing X subject to the restriction of first knowing Y.

11.2.38. Lemma. (Deconditioning) $H(X \mid Y) \leq H(X)$.

Proof: Note that the quantities $p(y)p(x \mid y)$ and $p(x)p(y \mid x)$ both equal $p(x, y)$. Also, since $\sum_y p(y \mid x) = 1$ for fixed x, we can apply Jensen's Inequality (concavity) to the logarithm function to obtain the inequality below. We compute

$$H(X \mid Y) = -\sum_y p(y) \sum_x p(x \mid y) \lg p(x \mid y) = \sum_x p(x) \sum_y p(y \mid x) \lg \frac{1}{p(x \mid y)}$$

$$\le \sum_x p(x) \lg \sum_y \frac{p(y \mid x)}{p(x \mid y)} = \sum_x p(x) \lg \sum_y \frac{p(y)}{p(x)}$$

$$= \sum_x p(x) \lg \frac{1}{p(x)} = H(X). \qquad \blacksquare$$

11.2.39. Lemma. (Subadditivity) $H(X, Y) \le H(X) + H(Y)$.
 More generally, $H(X_1, \ldots, X_k) \le \sum_{i=1}^k H(X_i)$.

Proof: The proof is immediate from the Chain Rule (Lemma 11.2.37) and Deconditioning (Lemma 11.2.38), with induction on k for the generalization. $\qquad \blacksquare$

Already we can use entropy to give a good bound on the number of subsets of $[n]$ with size at most αn. For $0 \le \alpha \le 1$, let $H(\alpha) = -\alpha \lg \alpha - (1 - \alpha) \lg(1 - \alpha)$; note that $H(\alpha)$ is the entropy of a $0, 1$-variable with success probability α.

11.2.40. Theorem. Fix $\alpha \in (0, .5)$. For $n \in \mathbb{N}$,

$$\frac{2^{H(\alpha)n}}{\sqrt{2\pi n \alpha(1 - \alpha)}} \le \sum_{i=0}^{\alpha n} \binom{n}{i} \le 2^{H(\alpha)n} .$$

Proof: (Massey [1974]) The lower bound is the lower bound for the top term from Exercise 2.3.14, using Stirling's Formula. For the upper bound, let F be the family of subsets of $[n]$ with size at most αn. Choose a member of F uniformly at random, generating the random binary n-tuple X where $X_i = 1$ if and only if i is in the chosen set. By Lemma 11.2.34, it suffices to show $H(X) \le H(\alpha)n$.
 By subadditivity and symmetry, $H(X) \le \sum_{i=1}^n H(X_i) = nH(X_1)$. Since X_1 is a $0, 1$-variable, we have $H(X_1) = H(p)$, where $p = \mathbb{P}(X_i = 1)$. An element $i \in [n]$ appears in the fraction k/n of the k-element subsets of $[n]$. Since we choose only from F, we have $k \le \alpha n$ for the restriction to each k, so $p \le \alpha$. Since $H(p)$ is monotone over $[0, .5]$ and $\alpha \le .5$, we have $H(p) \le H(\alpha)$, completing the proof. $\qquad \blacksquare$

Our next application is in extremal set theory, from Kleitman–Shearer–Sturtevant [1981]. We begin with a corollary of subadditivity.

11.2.41. Corollary. Let F be a family of subsets of $[n]$. If p_i is the fraction of sets in F that contain i, for all $i \in [n]$, then $|F| \le 2^{\Sigma_i H(p_i)}$.

Proof: Choose a member of F uniformly at random, generating the random binary n-tuple X where $X_i = 1$ if and only if i is in the chosen set. Note that $\mathbb{P}(X_i = 1) = p_i$ for all i. By Lemma 11.2.39, $H(X) \le \sum_{i=1}^k H(p_i)$. The bound on $|F|$ now follows from Lemma 11.2.34. $\qquad \blacksquare$

11.2.42. Theorem. If F is a family of k-subsets of $[n]$ in which no two pairs of sets have the same intersection, then $|F| \le 2^{.8114k + o(k)}$.

Proof: The condition implies that the intersections of any one set with the others are distinct, and hence $|F| \le 2^k$. To improve the bound, let p_i be the fraction of the sets in F that contain i, and let $m = |F|$. Let G be the family of all pairwise intersections of sets in F. By the condition on F, we have $|G| = \binom{m}{2}$. Element i appears in $\binom{p_i m}{2}$ of the members of G; that is, in a proportion less than p_i^2. By Corollary 11.2.41, $|G| \leq 2^{\Sigma_i H(p_i^2)}$.

Since all sets in F have size k, we have $\sum p_i = k$. Hence dividing by k gives a distribution, and we can apply Jensen's Inequality to the function $H(p^2)/p$ that is concave for $p \in [0, 1]$. We compute

$$\sum H(p_i^2) = k \sum \frac{p_i}{k} \frac{H(p_i^2)}{p_i} \leq k \frac{H(p^2)}{p},$$

where $p = \sum \frac{p_i}{k} p_i$. The maximum of $H(p^2)/p$ occurs at $p \approx .4914$, where the value is about 1.6228. Thus $\binom{m}{2} \leq 2^{1.6228k}$, so roughly $|F| \leq 2^{.8114k+o(k)}$. ∎

The next lemma is a substantial generalization of subadditivity, which is just the special case where F consists of the n singleton sets and $r = 1$. A proof first appeared in Chung–Frankl–Graham–Shearer [1986].

11.2.43. Lemma. (Shearer's Lemma) Let F be a list of subsets of $[n]$. If each $i \in [n]$ appears in at least r members of F, then the joint distribution of random variables $X_1, \ldots, X_n$ satisfies

$$H(X_1, \ldots, X_n) \leq \frac{1}{r} \sum_{\{i_1,\ldots,i_k\}=A\in F} H(X_{i_1}, \ldots, X_{i_k}).$$

Proof: (Radhakrishnan [2003], with Llewellyn) Index the elements of $A \in F$ as $i_1, \ldots, i_k$ in increasing order. Using the Chain Rule and Deconditioning,

$$H(X_{i_1}, \ldots, X_{i_k}) = \sum_{j=1}^{k} H(X_{i_j} | (X_{i_s}: s < j)) \geq \sum_{j=1}^{k} H(X_{i_j} | X_1, \ldots, X_{i_j-1}).$$

Summing over all $A \in F$ yields a lower bound on $\sum_{\{i_1,\ldots,i_k\}=A\in F} H(X_{i_1}, \ldots, X_{i_k})$. In the lower bound, each term of the form $H(X_i | X_1, \ldots, X_{i-1})$ appears at least r times. Hence

$$\sum_{\{i_1,\ldots,i_k\}=A\in F} H(X_{i_1}, \ldots, X_{i_k}) \geq r \sum_{i=1}^{n} H(X_i | X_1, \ldots, X_{i-1}) = rH(X_1, \ldots, X_n),$$

where the final equality is by the Chain Rule. ∎

The next application requires a combinatorial version of Shearer's Lemma.

11.2.44. Lemma. (Combinatorial Shearer's Lemma) Let F be a list of subsets of $[n]$ such that each $i \in [n]$ appears in at least r members of F. Given a family G of subsets of $[n]$, let $t_A(G) = \{B \cap A: B \in G\}$. The following holds:

$$|G| \leq \prod_{A\in F} |t_A(G)|^{1/r}.$$

Proof: Choose X uniformly at random from G, so $|G| \le 2^{H(X)}$. Treat X as its incidence n-tuple $X_1, \ldots, X_n$, with X_i indicating whether $i \in X$. By Shearer's Lemma, $H(X) \le \frac{1}{r} \sum_{A \in F} H(X_A)$, where $X_A = \{X_i : i \in A\}$.

To bound $H(X)$, we use Lemma 11.2.34 to bound $H(X_A)$ by $\lg |S|$, where S is the set of possible values of X_A. This set is the set of subsets of A that can arise when X is chosen from G, which is precisely $t_A(G)$. Thus

$$|G| \le 2^{H(X)} \le 2^{\left(\sum_{A \in F} H(X_A)\right)/r} = 2^{\left(\sum_{A \in F} \lg|t_A(G)|\right)/r} = \prod_{A \in F} |t_A(G)|^{1/r}. \qquad \blacksquare$$

We now consider a restricted family of sets whose pairwise intersections have size at least 3. The sets are restricted to be edge sets of graphs with vertex set $[n]$, and the intersections must contain three edges forming a triangle. By taking all the graphs containing a fixed triangle, a family of size $2^{\binom{n}{2}-3}$ can be formed. Ellis–Filmus–Friedgut [2012] later proved that this family is extremal, as conjectured originally by Simonovits and Sós in 1976. We present the use of Shearer's Lemma in Chung–Frankl–Graham–Shearer [1986] to obtain an easy upper bound.

11.2.45. Theorem. Let $\mathcal{G}$ be a family of graphs with vertex set $[n]$. If any two graphs in $\mathcal{G}$ share a triangle, then $|\mathcal{G}| \le 2^{\binom{n}{2}-2}$.

Proof: View $\mathcal{G}$ as a family of subsets of $\binom{[n]}{2}$. In order to use Lemma 11.2.44 to bound $|\mathcal{G}|$, we introduce a family $\mathcal{F}$ of graphs. We let $\mathcal{F}$ consist of all copies of $K_{\lfloor n/2 \rfloor} + K_{\lceil n/2 \rceil}$ with vertex set $[n]$. The size s of each member of $\mathcal{F}$ is $\binom{\lfloor n/2 \rfloor}{2} + \binom{\lceil n/2 \rceil}{2}$; note that $s \le \frac{1}{2}\binom{n}{2}$.

The total size of all members of $\mathcal{F}$ is $s|\mathcal{F}|$. By symmetry, each vertex pair lies in the same number of members of $\mathcal{F}$, so it occurs in exactly $s|\mathcal{F}|/\binom{n}{2}$ such graphs. This is the value of r in Lemma 11.2.44.

For each $A \in \mathcal{F}$, the family $t_A(\mathcal{G})$ is an intersecting family. In particular, any $B, B' \in \mathcal{G}$ share a triangle, but the complement of A has no triangle, so $B \cap A$ and $B' \cap A$ must share an edge of that triangle.

Since $t_A(\mathcal{G})$ is an intersecting family of subsets of A, and $|A| = s$, we conclude $|t_A(\mathcal{G})| \le 2^{s-1}$ (recall Example 11.2.16). Since $s|\mathcal{F}|/r = \binom{n}{2}$ and $s \le \frac{1}{2}\binom{n}{2}$, using Lemma 11.2.44 yields

$$|\mathcal{G}| \le \prod_{A \in \mathcal{F}} |t_A(G)|^{1/r} \le (2^{s-1})^{|\mathcal{F}|/r} = 2^{\binom{n}{2}-\binom{n}{2}/s} \le 2^{\binom{n}{2}-2}. \qquad \blacksquare$$

We have presented here only the basic ideas of using entropy to prove bounds on extremal combinatorial problems: the relationship between entropy and the number of outcomes, subadditivity, and Shearer's Lemma. Similar applications appear in Exercises 38–41. Meanwhile, Galvin [2014] presents an impressive variety of applications and pointers to applications of entropy. We describe a few below to encourage readers to explore more deeply.

11.2.46. Example. *Antichains of subsets of* $[n]$. By taking families of $n/2$-sets from $[n]$, one can form $2^{\binom{n}{\lfloor n/2 \rfloor}}$ antichains. Kleitman [1969] was the first to show that the logarithm of the total number of antichains is asymptotic to $\binom{n}{\lceil n/2 \rceil}$ (see Section 12.2). Pippenger [1999] showed how to prove this using the properties of entropy discussed here, though his lower-order terms were not as sharp. $\qquad \blacksquare$

11.2.47. Example. *Maximizing the permanent.* The **permanent** per(A) of an n-by-n matrix A is the sum of $\prod_{i=1}^{n} a_{i,\sigma(i)}$ over all permutations σ of $[n]$ (Chapter 15). For the matrix of a bipartite multigraph, this counts the perfect matchings.

An X, Y-subgraph of $K_{n,n}$ trivially has at most $\prod_{x \in X} d(x)$ perfect matchings. In 1963, Minc conjectured that a binary matrix with row-sums $d_1, \dots, d_n$ has permanent at most $\prod_{i=1}^{n} (d_i!)^{1/d_i}$, reducing the bound by a factor almost e^n. When d divides n, equality holds for the graph $(n/d)K_{d,d}$. Brègman [1973] proved the conjecture. Radhakrishnan [1997] gave a proof using subadditivity of entropy. See Alon–Friedland [2008] for the Kahn–Lovász extension to general graphs. ∎

11.2.48. Example. *Embeddings of H into G.* An *embedding* of a graph H into a graph G maps $V(H)$ injectively into $V(G)$ so that all edges of H map into edges of G. Let $f_H(m)$ denote the maximum number of embeddings of H into a graph with m edges. It is easy to show $f_{K_3}(m) \le 2\sqrt{2}m^{3/2}$ (Exercise 42), sharp by $K_{\sqrt{2m}}$.

Alon [1981] proved the existence of a constant c such that $f_H(m) \le cm^{\alpha^*(H)}$ for all H and m. Here $\alpha^*(H)$ is the maximum sum of nonnegative vertex weights such that each edge has total weight at most 1 (note that $\alpha^*(K_3) = 3/2$). Friedgut–Kahn [1998] proved Alon's result using Shearer's Lemma. ∎

EXERCISES 11.2

11.2.1. (–) Fix $m, k \in \mathbb{N}$. When n is sufficiently large, how can a family of m sets of size k be chosen in $[n]$ to maximize the size of the shadow of the family?

11.2.2. (–) Permutations (in word form) *intersect* if they agree in some position. Find the maximum size of an intersecting family of permutations of $[n]$. (Deza–Frankl [1977])

11.2.3. (–) Does every maximal intersecting family in $\mathbf{2}^{[n]}$ have size 2^{n-1}?

11.2.4. (–) Let I be an ideal of subsets of $[n]$. Prove that every element of $[n]$ appears in at most half the members of I.

11.2.5. (–) Use the Sunflower Theorem to prove that every graph with more than $2(k-1)^2$ edges contains a matching of size k or a star of size k.

11.2.6. Determine the maximum size of a family of subsets of $[n]$ such that no element belongs to at least two sets in the family and is omitted by at least two sets in the family.

11.2.7. Use the Kruskal–Katona Theorem to prove that the size of the shade of a family of m elements of $\binom{[n]}{k}$ is minimized by the family consisting of the first m elements of the order on $\binom{[n]}{k}$ in which $x < y$ if $x_i > y_i$ in the highest coordinate where they differ.

11.2.8. (◊) Let G be an n-vertex graph with m edges. For each $k \in \mathbb{N}$, determine the maximum number of k-cliques in G, and construct an example achieving equality. (Bollobás)

11.2.9. Explain how to modify the shift-operator proof of the Kruskal–Katona Theorem (Theorem 11.2.11) to give a direct proof of Theorem 11.2.12 that the shadow of a k-uniform family with size $\binom{u}{k}$ has size at least $\binom{u}{k-1}$. (Frankl [1984])

11.2.10. Let F be a k-uniform family of size $\binom{u}{k}$. Refine the proof of Theorem 11.2.12 to show that the shadow of F has size exactly $\binom{u}{k-1}$ if and only if u is an integer and F consists of all k-sets in a set of size u. (Lovász [1979])

11.2.11. Prove that the list $0, \ldots, 0, f_k, f_{k+1}, \ldots, f_l, 0, \ldots, 0$ is realizable as $f_i = \left| F \cap \binom{[n]}{i} \right|$ for some antichain F of subsets of $[n]$ if and only if

$$f_k + \partial_{k+1}(f_{k+1} + \cdots + \partial_{l-1}(f_{l-1} + \partial_l(f_l))) \le \binom{n}{k}.$$

(Clements [1973], Daykin–Godfrey–Hilton [1974])

11.2.12. Fix $r \ge 2$. For what values of n is it possible to color every square in an n-by-n grid with one of r colors so that, for all i, j, k between 1 and n with $i \ne j$ and $j \ne k$, the square in row i and column j is assigned a different color from the square in row j and column k. (Hint: Use Sperner's Theorem.) (Propp [1998])

11.2.13. Define a **rising antichain** to be an antichain of subsets of $[n]$ whose sizes are distinct. Let two rising antichains be equivalent if one is obtained from the other by complementing all sets and/or relabeling the elements of $[n]$.
  (a) Determine the maximum size of a rising antichain of subsets of $[n]$.
  (b) Prove that when $n \ge 5$, the largest rising antichains are all equivalent to each other. (Bey–Griggs [2002])

11.2.14. For $n > 2k$, construct an intersecting family of k-subsets of $[n]$ with size $\binom{n-1}{k-1} - \binom{n-k-1}{k-1} + 1$ that is not a star. (Comment: Hilton–Milner [1967] proved that none is larger.)

11.2.15. For $n \ge t \ge 1$, let $q = \lceil (n+t)/2 \rceil$. Prove that a largest t-intersecting family in $\mathbf{2}^n$ has size $\sum_{i=q}^{n} \binom{n}{i}$ when $n+t$ is even and $\sum_{i=q}^{n} \binom{n}{i} + \binom{n-1}{q-1}$ when $n+t$ is odd. (Katona [1964])

11.2.16. Let F be a family of 4-sets of $[n]$ that pairwise intersect at most twice. Prove the existence of $S \subset [n]$ with $|S| \ge \lceil (6n-6)^{1/3} \rceil$ that contains no member of F. (Adrian [1991])

11.2.17. Let F be an intersecting family of subsets of $[n]$ whose members have size at most k. Prove that $|F| \le \sum_{i=1}^{k} \binom{n-1}{i-1}$, with equality only for stars if $k < n-1$.

11.2.18. ($\diamond$) Prove that the largest antichain of subsets of $[n]$ consisting of pairs of complementary sets has size $2\binom{n-1}{\lceil n/2 \rceil}$. (Bollobás [1973])

11.2.19. Let π and σ be simple k-words from $[n]$ (lists of k distinct elements). Say that π and σ *intersect* if $\pi_i = \sigma_i$ for some i. Prove that the maximum size of a pairwise intersecting family of simple k-words from $[n]$ is $(n-1)!/(n-k)!$. (see Deza–Frankl [1978], Lovász–Nešetril–Pultr [1980])

11.2.20. Let F be a family of subsets of $[n]$ such that each member has size at least l and each pair of members have at most k common elements, where $l \le k$. Prove that $|F|$ is maximized uniquely by $F = \{A \subseteq [n]: l \le |A| \le k+1\}$.

11.2.21. Let A be an intersecting antichain of subsets of $[n]$ with $|x| \le n/2$ for all $x \in A$. Prove that $\sum_{x \in A} \binom{n-1}{|x|-1}^{-1} \le 1$. (Hint: Assign weight $|x|^{-1}$ to each $x \in A$ and prove that the total weight on members of A appearing as consecutive strings in a single cyclic permutation is at most 1.) (Bollobás [1973], Greene–Katona–Kleitman [1976])

11.2.22. For $0 \le r \le k-t$, let S_r be the family of k-subsets of $[n]$ that contain at least $t+r$ of the smallest $t + 2r$ elements.
  (a) Show that S_r is a t-intersecting family.
  (b) Prove that $|S_r| = \sum_{i=0}^{r} \binom{t+2r}{t+r+i}\binom{n-t-2r}{k-t-r-i}$.
  (c) Prove that $|S_1| > |S_0|$ if and only if $n < (k-t+1)(t+1)$. Thus the t-star S_0 is not a largest $EKR(k, t)$-family when $n < (k-t+1)(t+1)$. (Frankl [1978]) (Comment: Ahlswede–Khachatrian [1997] proved that S_r is a largest $EKR(k, t)$-family when $(k-t+1)(2 + \frac{t-1}{r+1}) \le n < (k-t+1)(2 + \frac{t-1}{r})$. A proof appears in Engel [1997, pp. 50–60].)

11.2.23. ($\Diamond$) Let $A_1, \ldots, A_n$ be r-sets such that $|A_i \cap A_j| \le k$ for distinct $i, j \in [n]$. Prove $\left| \bigcup_{i=1}^n A_i \right| \ge nr^2/[r + (n-1)k]$. Interpret for proper edge-colorings of K_n. (Corrádi [1969])

11.2.24. Let $X_1, \ldots, X_n$ be independent $0, 1$-random variables with $\mathbb{P}(X_i = 1) = p \ge .5$. Let Z be a convex combination of $\{X_i\}$; that is, $Z = \sum \alpha_i X_i$ with $\alpha_i \ge 0$ and $\sum \alpha_i = 1$. Prove that $P(Z \ge .5) \ge p$ when no subset of $\{\alpha_i\}$ sums to $.5$. (Hint: Use the EKR Theorem to bound the number of k-subsets of $\{\alpha_i\}$ whose sum exceeds $.5$. Comment: The claim holds without the restriction that no subset of $\{\alpha_i\}$ sums to $.5$, but eliminating that restriction requires a limit argument.) (Liggett [1977])

11.2.25. Let F and G be ideals in $\mathbf{2} \times \mathbf{3}$. Show that it is not always possible to minimize $|F \cap G|$ by letting F be the first $|F|$ vectors in lexicographic order and letting G be the lexicographically first $|G|$ vectors when the components are read in the opposite order. (Daykin–Kleitman–West)

11.2.26. Find the largest k such that every k-coloring of the subsets of $[n]$ makes all of $\{A, B, A \cup B, A \cap B\}$ the same color for some distinct sets A and B. Solve the same problem for $n = 6$ when $A - B$ and $B - A$ must be nonempty. (Tomescu [1987])

11.2.27. Let I be an ideal of subsets of $[n]$, and let $I' = \{\overline{A} : A \in I\}$. Prove that there exists a bijection $f : I \to I'$ such that $A \subseteq f(A)$ for all $A \in I$. (Erdős–Herzog–Schönheim [1970], Marica [1971], Daykin–Hilton–Miklós [1983])

11.2.28. ($\Diamond$) Use Berge's Theorem (Theorem 11.2.26) to prove the following:
(a) The average size of the sets in any ideal of subsets of $[n]$ is at most $n/2$.
(b) The maximum size of an intersecting family of subsets of $[n]$ whose complements also form an intersecting family is 2^{n-2}.

11.2.29. Let F be an intersecting family of subsets of $[n]$ whose complements also form an intersecting family. Use the shift operators $\tau_{1,j}$ and Theorem 11.2.27 to prove $|F| \le 2^{n-2}$.

11.2.30. *Special cases of Chvátal's Conjecture.* (Schönheim [1976])
(a) Prove the conjecture for ideals whose bases have a common element.
(b) Prove the conjecture for ideals with two bases.
(c) Use Theorem 11.2.27 to give a proof without the Erdős–Ko–Rado Theorem for the ideal consisting of all subsets of $[n]$ with size at most k.

11.2.31. For a set $B \subseteq [n]$ and a family F of subsets of $[n]$, the **translate** of F by B, written $F(B)$, is $\{A \triangle B : A \in F\}$, where $\triangle$ denotes symmetric difference.
(a) Prove that the translate of an ideal by $\{x\}$ has the star property. (B. Reiniger)
(b) Prove that the translate of an ideal I by a set B is an ideal if and only if B is contained in every element of I that is maximal (contained in no other). (P. Wenger)
(c) Let I be an ideal. Prove that every translate of I by a nonempty set B contains a star at least as large as the largest intersecting family in I. (Snevily)

11.2.32. Prove that the Union-Closed Sets Conjecture (Conjecture 11.2.28) is true for families having a member with size at most 2. (Sarvate–Renaud [1989])

11.2.33. ($\Diamond$) Prove each statement below equivalent to the Union-Closed Sets Conjecture.
(a) If a finite family F with $|F| \ge 2$ is closed under taking intersections, then $\bigcup_{A \in F} A$ has an element belonging to at most half of the members of F.
(b) Every nontrivial X, Y-bigraph G contains in both X and Y a vertex that lies in at most half of the maximal independent sets of G. (Bruhn–Charbit–Telle [2013])

11.2.34. Given a graph G, let H be the graph with vertex set $\binom{V(G)}{k}$ such that vertices A and B are adjacent if and only if G has an edge uv with $u \in A - B$ and $v \in B - A$. Suppose that every induced subgraph of G having at most sk vertices has fewer than $\binom{s}{2}$ edges. Use the Sunflower Theorem to prove $\omega(G) \le k!(s-1)^k$.

11.2.35. For $n - s + 1 < k \le n$, prove that the family $\binom{[n]}{k}$ has no sunflower of size s.

11.2.36. Let X be a random variable such that for the ith outcome, the probability p_i that it occurs is a power of $1/2$. Associate distinct binary words with each outcome so that the word associated with the ith outcome has length $- \lg p_i$. Show that the expected length of the word specifying the outcome is the entropy $H(x)$.

11.2.37. For the entropy of the joint variable (X, Y), prove $H(X, Y) \le H(X) + H(Y)$ directly from the definition. (Hint: Express $H(X) + H(Y) - H(X, Y)$ as the double sum $\sum \sum \mathbb{P}(X = i)\mathbb{P}(Y = j)z_{i,j} \lg z_{i,j}$, where $z_{i,j} = \frac{\mathbb{P}[(X,Y)=(i,j)]}{\mathbb{P}(X=i)\mathbb{P}(Y=j)}$.)

11.2.38. Example 10.1.5 used the Pigeonhole Principle to show that $\lg n$ bipartite subgraphs are needed to cover K_n. Use entropy to generalize: The number of k-partite subgraphs needed to cover K_n is at least $(\lg n)/(\lg k)$.

11.2.39. ($\diamond$) A **distinguishing family** for $[n]$ is a family $\mathcal{D}$ of subsets of $[n]$ such that for any two subsets $A, B \subseteq [n]$, there is some $D \in \mathcal{D}$ such that $|A \cap D| \ne |B \cap D|$. Let $f(n)$ be the minimum size of such a family.
    (a) Use entropy to prove $f(n) \ge n/\lg(n + 1)$. (Comment: There is an improvement by an asymptotic factor of 2; see Erdős–Rényi [1963] and Moser [1970]. That matches the upper bound $f(n) \le (1 + o(1))2n/\lg n$ by Lindström [1965] and Cantor–Mills [1966].)
    (b) Given n coins of the same weight, except that some counterfeit coins weigh a fixed smaller amount, prove that $f(n)$ is the minimum number of subsets of the coins whose accurate weighing determines the counterfeit coins. The outcome of a weighing determines how many in the weighed set are light. (Söderberg–Shapiro [1963])

11.2.40. ($\diamond$) Let S be a family of n points in $\mathbb{R}^d$. Let n_j be the number of distinct points obtained by setting the jth coordinates of these points to 0 (projecting S on the hyperplane $x_j = 0$). Prove $n^{d-1} \le \prod_{j=1}^d n_j$, and show that this is sharp for all d. (Hint: Use Lemma 11.2.43, Shearer's Lemma. Comment: Without entropy, Loomis–Whitney [1949] generalized to a measurable body B in $\mathbb{R}^d$. With B_j being the projection of B on the hyperplane $x_j = 0$, they proved $\mathrm{Vol}(B)^{d-1} \le \prod_{j=1}^n (\mathrm{Vol}(B_j))$. The general case can be obtained by approximating B as a union of axis-parallel cubes centered at points of a fine grid.)

11.2.41. ($\diamond$) Let G be a family of subsets of $[n]$ such that any two members of G have k consecutive common elements. Determine the maximum size of G. (Hint: Design a suitable family F in order to apply the combinatorial version of Shearer's Lemma, Lemma 11.2.44.) (Comment: Note that the families in Exercise 11.2.15, where more general k-intersecting families are allowed, are much larger.)

11.2.42. Let G be a graph with m edges. Prove that there are at most $2\sqrt{2}m^{3/2}$ embeddings of K_3 into G (see Example 11.2.48).

11.3. Matroids

    The defining properties of matroids are general enough to occur in many contexts and special enough to yield rich combinatorial structure. Many elegant results from graph theory, linear algebra, and elsewhere generalize in the theory of matroids. These include the greedy algorithm for minimum spanning trees, min-max relations for systems of distinct representatives, dimension properties of vector spaces, and duality properties of planar graphs. When a theorem on a special class holds for all matroids, it immediately yields results for other classes.

Matroids were independently introduced by Whitney [1935] (to study planar graphs) and by Nakasawa [1935, 1936] (to study linear dependence; reprinted and translated in Nishimura–Kuroda [2009]). They were reinvented by van der Waerden [1937] generalizing independence in vector spaces, and they also arose (from a structural viewpoint) in the theory of geometric lattices (MacLane [1936]). The first modern textbook on the subject was Welsh [1976], later Oxley [1992, 2011].

In this brief treatment, we focus on the application of matroids to extremal problems, leaving deeper structural aspects and more subtle matroidal aspects of graphs to a more advanced book. We start from the notions of ideals and antichains defined in the preceding section.

In many mathematical contexts, sets that avoid conflict are called "independent". Inherently, subsets of independent sets (and the empty set) are independent. Thus the family of independent sets is closed under taking subsets. The same family can be specified by its antichain of maximal independent sets, its antichain of minimal nonindependent sets, or other aspects we will describe later.

To obtain the elegant behavior of matroids, we need an additional property. A restriction on the family of independent sets can be translated into a corresponding restriction on some other specification of the system. Because we can specify the system in many ways, we have many equivalent definitions of matroids.

HEREDITARY SYSTEMS AND EXAMPLES

Before discussing the added properties that yield matroids, we begin with the more general notion of hereditary systems, familiar from our study of ideals. Given a finite set E of elements, we use 2^E to denote the family of subsets of E, ordered by inclusion; it has size $2^{|E|}$. We view a nonempty ideal in 2^E and the other ways of specifying it as a single structure. A matroid will then be such a structure that satisfies one of various equivalent additional constraints.

11.3.1. Definition. A **hereditary system** M on E consists of a nonempty ideal $\mathbf{I}_M$ in 2^E and all ways of specifying $\mathbf{I}_M$. The ideal $\mathbf{I}_M$ is the family of **independent sets** of M; the other subsets of E are **dependent**. The **bases** $\mathbf{B}_M$ are the maximal independent sets. The **circuits** $\mathbf{C}_M$ are the minimal dependent sets. The **rank function** $r_M(X)$ of M assigns to each $X \subseteq E$ a value called its **rank**, which is the maximum size of a member of $\mathbf{I}_M$ contained in X.

11.3.2. Example. On the left below we sketch the inclusion order on subsets of E, with the full set E at the top and the empty set at the bottom. We sketch the relationships among the independent sets, bases, circuits, and dependent sets of a hereditary system. The bases are the maximal elements of $\mathbf{I}$, and the circuits are the minimal elements not in $\mathbf{I}$. Always $\varnothing \in \mathbf{I}$. If every set is independent, then there is no circuit, but every hereditary system has at least one base.

On the right we obtain a hereditary system from a multigraph with three edges. The independent sets are the acyclic edge sets. The only dependent sets are $\{1, 2\}$ and $\{1, 2, 3\}$, the only circuit is $\{1, 2\}$, and the bases are $\{1, 3\}$ and $\{2, 3\}$. The rank of an independent set is its size. For the dependent sets, we have $r(\{1, 2\}) = 1$ and $r(\{1, 2, 3\}) = 2$.   ∎

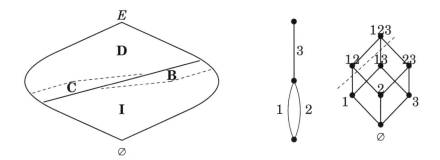

11.3.3. Remark. *Aspects of hereditary systems.* An **aspect** of a hereditary system M is a way of specifying it. For example, M can be specified by any of $\mathbf{I}_M$, $\mathbf{B}_M$, $\mathbf{C}_M$, or r_M, because each of these determines the others. We have expressed $\mathbf{B}_M$, $\mathbf{C}_M$, r_M in terms of $\mathbf{I}_M$. Conversely, if $\mathbf{B}_M$ is given, then $\mathbf{I}_M$ consists of the sets contained in members of $\mathbf{B}_M$. If $\mathbf{C}_M$ is given, then $\mathbf{I}_M$ consists of the sets containing no member of $\mathbf{C}_M$. If r_M is given, then $\mathbf{I}_M$ and the other aspects are found by setting $\mathbf{I}_M = \{X \subseteq E: r_M(X) = |X|\}$. This justifies our view of a hereditary system as a unified structure $(\mathbf{I}, \mathbf{B}, \mathbf{C}, r, \cdots)$. We drop the subscripts on $\mathbf{I}, \mathbf{B}, \mathbf{C}, r, \cdots$ when discussing only one hereditary system. Later we introduce additional aspects that can specify a hereditary system. ∎

Most terminology in matroid theory comes from the motivating contexts that led to the discovery of matroids, particularly graphs and linear algebra. We begin with the fundamental example from graphs.

11.3.4. Definition. The **cycle matroid** $M(G)$ of a multigraph G is the hereditary system on $E(G)$ whose circuits are the edge sets of cycles of G. A hereditary system that can be specified in this way is a **graphic matroid**.

11.3.5. Example. *Bases in a cycle matroid $M(G)$.* The graph K_4^-, with four vertices and five edges, arises by deleting one edge from K_4. Spanning trees in K_4^- have three edges, so every set with more than three edges is dependent. Also the two triangles are dependent. This yields eight dependent sets and 24 independent sets among the subsets of $E(K_4^-)$. There are three minimal dependent sets (the cycles) and eight maximal independent sets (the spanning trees).

For a general multigraph G, the bases of the cycle matroid $M(G)$ are the edge sets of the maximal forests in G. Each contains a spanning tree of each component of G, so they have equal size. Consider $B_1, B_2 \in \mathbf{B}$ and $e \in B_1 - B_2$. Deleting e from B_1 disconnects a component; since B_2 contains a tree spanning the same component of G, some edge $f \in B_2 - B_1$ can be added to $B_1 - e$ to reconnect it.

The **base exchange property** is

If $B_1, B_2 \in \mathbf{B}_M$, then for all $e \in B_1 - B_2$
there exists $f \in B_2 - B_1$ such that $B_1 - e + f \in \mathbf{B}_M$.

Matroids are the hereditary systems satisfying the base exchange property. ∎

11.3.6. Remark. *Notational conventions.* In this subject, we often discuss modifying a set by adding or deleting a single element. For symmetry and simplicity, we use the symbols $+$ and $-$ instead of $\cup$ and $-$ for this, and we drop the set brackets on singleton sets.

Using bold $\mathbf{I}, \mathbf{B}, \mathbf{C}$ for families of subsets of E allows using $I \in \mathbf{I}$, $B \in \mathbf{B}$, $C \in \mathbf{C}$, respectively, to denote membership. We use roman I, B, C, R to denote properties, e, f, x, y for elements of E, and X, Y, F for subsets of E.   ∎

11.3.7. Example. *Rank in cycle matroids.* Let G be an n-vertex graph. For $X \subseteq E(G)$, let G_X be the spanning subgraph of G with edge set X. In $M(G)$, an independent subset of X is the edge set of a forest in G_X. When G_X has k components, the maximum size of such a forest is $n - k$. Hence $r(X) = n - k$. Below we show such a forest Y in X.

If $r(X+e) = r(X)$ for some $e \in E-X$, then the endpoints of e lie in a single component of G_X; adding e does not combine components. If we add two such edges, then again we do not combine components. Therefore, $r(X) = r(X + e) = r(X + f)$ implies $r(X) = r(X + e + f)$.

The **(weak) absorption property** (name suggested by A. Kézdy) is

> If $X \subseteq E$ and $e, f \in E$,
> then $r(X) = r(X + e) = r(X + f)$ implies $r(X + e + f) = r(X)$.

Matroids are the hereditary systems satisfying the absorption property.   ∎

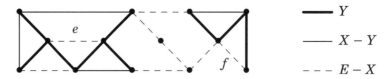

Multigraphs may have loops and multiedges. Analogous terminology in hereditary systems captures the behavior in cycle matroids of elements arising from loops and multiedges in multigraphs.

11.3.8. Definition. In a hereditary system, a **loop** is an element forming a circuit of size 1. Two non-loops are **parallel elements** if they form a circuit. A hereditary system is **simple** if it has no loops or parallel elements.

Another motivating context for matroids is linear independence in vector spaces. Matroids can be obtained from matrices.

11.3.9. Definition. The **vector matroid** on a finite set E of vectors in a vector space (over a field K) is the hereditary system whose independent sets are the linearly independent sets of vectors in E. A matroid expressible in this way is a **linear matroid** (or **representable matroid**) over K. The **column matroid** $M(A)$ of a matrix A is the vector matroid defined on its columns.

11.3.10. Example. *Circuits in vector matroids.* Technically, one vector (or multiples of it) may be used to represent distinct elements of E, just as a matrix may have repeated columns; these yield parallel elements. The circuits are the minimal sets $\{x_1, \ldots, x_k\} \subseteq E$ such that $\sum c_i x_i = 0$ using coefficients not all zero. Minimality forces all $c_i \neq 0$.

Let C_1 and C_2 be distinct circuits containing x. The equations of dependence for C_1 and C_2 let us write x as a linear combination of $C_1 - x$ and of $C_2 - x$. Equating these expressions yields an equation of dependence for $(C_1 \cup C_2) - x$; thus $(C_1 \cup C_2) - x$ contains a circuit.

The **(weak) elimination property** is

> If C_1 and C_2 are distinct circuits and $x \in C_1 \cap C_2$,
> then another member of $\mathbf{C}_M$ is contained in $(C_1 \cup C_2) - x$.

Matroids are the hereditary systems satisfying the weak elimination property. The column matroid of the matrix below is also the cycle matroid of K_4^-.

$$\begin{pmatrix} 0 & 0 & 0 & 1 & 1 \\ 0 & 1 & 1 & 0 & 1 \\ 1 & 1 & 0 & 0 & 0 \end{pmatrix}$$

The weak elimination property implies that in a matroid the relation of being parallel is transitive, but this does not hold for all hereditary systems. ∎

Another class of matroids important for applications arose much later from families of sets. Edmonds–Fulkerson [1965] and Mirsky–Perfect [1967] independently discovered that these are matroids.

11.3.11. Definition. The **transversal matroid** induced by sets $A_1, \ldots, A_m$ with union E is the hereditary system on E in which the independent sets are the systems of distinct representatives (SDR) for subsets of $\{A_1, \ldots, A_m\}$.

11.3.12. Remark. *Transversal matroids and bipartite graphs.* The name "transversal" arises from using this word to mean SDR. However, in the study of optimization problems in hypergraphs, the word "transversal" is used without requiring the representatives to be distinct. We therefore use "SDR" here.

Given $A_1, \ldots, A_m$ with union E, consider the bipartite incidence graph G. The parts are E and $[m]$, with $e \in E$ adjacent to $i \in [m]$ if and only if $e \in A_i$. A set $X \subseteq E$ is independent in the transversal matroid induced by $A_1, \ldots, A_m$ on E if and only if X is covered by some matching in G.

Also, given an E, F-bigraph G, we can associate with $v_i \in F$ the set $N_G(v_i)$. Letting $A_i = N_G(v_i)$ expresses G as the incidence graph of the family of sets. Hence every bipartite graph induces a transversal matroid on each of its parts.

The bipartite graph below yields the transversal matroid of the family $\mathbf{A} = \{\{1, 2\}, \{2, 3, 4\}, \{4, 5\}\}$. This matroid is again $M(K_4^-)$. ∎

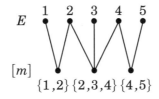

E 1   2   3   4   5

$[m]$ $\{1,2\}$ $\{2,3,4\}$ $\{4,5\}$

11.3.13. Example. *Independent sets in transversal matroids.* The symmetric difference of matchings M and M' in a bipartite graph consists of alternating paths and even cycles. The only components with more edges from M' than from M are the M-augmenting paths. When $|M'| > |M|$, there must be some such path. For an M-augmenting path P, replacing $M \cap P$ with $M' \cap P$ yields a matching of size $|M| + 1$ that covers all vertices of M plus the endpoints of P.

For independent sets I_1 and I_2 in the transversal matroid generated by $A_1, \ldots, A_m$, let M_1 and M_2 be matchings that saturate I_1 and I_2, respectively, in the associated bipartite graph (below, M_1 is bold and M_2 is solid). If $|I_2| > |I_1|$, then the matching obtained from M_1 via an M_1-augmenting path in $M_2 \triangle M_1$ covers I_1 plus an element $e \in I_2 - I_1$; this "augments" I_1.

For a hereditary system on E, the **augmentation property** is

> For distinct $I_1, I_2 \in \mathbf{I}$ with $|I_2| > |I_1|$,
> there exists $e \in I_2 - I_1$ such that $I_1 + e \in \mathbf{I}$.

Matroids are the hereditary systems with the augmentation property.   ∎

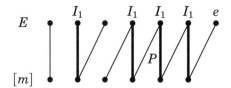

We have defined matroids as hereditary systems with an additional property, but we gave four candidates for this "additional property", claiming that each defines matroids. In fact, the properties are equivalent for hereditary systems; every matroid satisfies them all. After we prove the properties equivalent for hereditary systems, we only need to verify one such property to use them all. To illustrate that the various properties all hold, consider cycle matroids.

11.3.14. Example. *Augmentation in cycle matroids.* Consider $I_1, I_2 \in \mathbf{I}_{M(G)}$ with $|I_2| > |I_1|$. The spanning subgraph G_{I_i} has $n - |I_i|$ components. Therefore, the forest I_2 has some edge with endpoints in two components of G_{I_1}. This edge can be added to I_1 to form a larger forest. Hence the augmentation property holds. ∎

11.3.15. Example. *Weak elimination in cycle matroids.* The circuits of $M(G)$ are the edge sets of cycles in G. Cycles have even degree at each vertex. For $C_1, C_2 \in \mathbf{C}$, the symmetric difference $C_1 \triangle C_2$ also has even degree at each vertex. If $C_1 \neq C_2$, then $C_1 \triangle C_2$ contains a cycle. This is stronger than weak elimination, since $C_1 \triangle C_2 \subseteq C_1 \cup C_2 - x$. In the picture below, C_1 and C_2 are face boundaries of length 9 sharing the dashed edges, and $C_1 \triangle C_2$ is the union of two disjoint cycles.   ∎

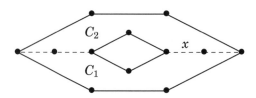

For linear matroids, direct proof of the augmentation or base exchange property uses the fact that k linearly independent vectors cannot all be expressed as linear combinations of a smaller set. On the other hand, since we have verified the weak elimination property for linear matroids, this theorem of linear algebra follows from matroid axiomatics!

11.3.16. Lemma. For the rank function r of a hereditary system on E,
(r1) $r(\varnothing) = 0$.
(r2) $r(X) \le r(X + e) \le r(X) + 1$ whenever $X \subseteq E$ and $e \in E$.

Proof: From the definition $r(X) = \max\{|Y|: Y \subseteq X, Y \in \mathbf{I}\}$, we have $r(\varnothing) = 0$. Because $X + e$ contains every independent subset of X, we have $r(X + e) \ge r(X)$. Because the independent subsets of $X + e$ not contained in X consist of e together with an independent subset of X, we have $r(X + e) \le r(X) + 1$. ∎

11.3.17.* Remark. Every nonempty ideal is the family of independent sets of a hereditary system. A family $\mathbf{B}$ is the set of bases of some hereditary system if and only if it is a nonempty antichain. A family $\mathbf{C}$ is the set of circuits of some hereditary system if and only if it is an antichain of nonempty sets.

Characterizing rank functions of hereditary systems is more subtle. Properties (r1) and (r2) above are used to study matroids but do not suffice to make r the rank function of a hereditary system. (Consider $E = \{1, 2\}$, $r(\varnothing) = r(1) = r(2) = 0$, $r(\{1, 2\}) = 1$; the set $\{1, 2\}$ contains no independent set of size 1, even though it has rank 1.) Another technical condition is needed to characterize rank functions of hereditary systems. Fortunately, we do not need this characterization, because when studying matroids we always start with a hereditary system. ∎

AXIOMATICS OF MATROIDS

We view a matroid as a hereditary system satisfying an additional structural property. A constraint on one aspect of a hereditary system yields corresponding constraints on other aspects; thus we have many equivalent definitions. We can show that a hereditary system is a matroid by verifying any of them, and then we can use them all without additional proof.

Adding an edge to a forest creates at most one cycle. This is the property needed to show that the greedy algorithm works to find minimum spanning trees in weighted graphs. The creation of at most one circuit by adding one element to an independent set is in fact a characterization of matroids, as is the effectiveness of the greedy algorithm itself! Both appear in our list.

Given nonnegative weights on the elements, the **greedy algorithm** iteratively picks a heaviest element whose addition to those already selected yields an independent set. Rado [1957] (see also Borůvka [1926]) proved that matroids are precisely the hereditary systems such that for each nonnegative weight function, the greedy algorithm selects a heaviest independent set.

11.3.18. Definition. A hereditary system M on E is a **matroid** if it satisfies any of the following additional properties, where $\mathbf{I}, \mathbf{B}, \mathbf{C}, r$ are the independent sets, bases, circuits, and rank function of M.

I: **augmentation** - if I_1, $I_2 \in \mathbf{I}$ with $|I_2| > |I_1|$, then $I_1 + e \in \mathbf{I}$ for some $e \in I_2 - I_1$.

B: **base exchange** - if B_1, $B_2 \in \mathbf{B}$, then for all $e \in B_1 - B_2$ there exists $f \in B_2 - B_1$ such that $B_1 - e + f \in \mathbf{B}$.

B': **dual base exchange** - if B_1, $B_2 \in \mathbf{B}$, then for all $e \in B_1 - B_2$ there exists $f \in B_2 - B_1$ such that $B_2 + e - f \in \mathbf{B}$.

A: **weak absorption** - $r(X) = r(X + e) = r(X + f)$ implies $r(X + e + f) = r(X)$ whenever $X \subseteq E$ and $e, f \in E$,

A': **strong absorption** - if $X, Y \subseteq E$, and $r(X + e) = r(X)$ for all $e \in Y$, then $r(X \cup Y) = r(X)$.

U: **uniformity** - for every $X \subseteq E$, the maximal subsets of X belonging to $\mathbf{I}$ have the same size.

R: **submodularity** - $r(X \cap Y) + r(X \cup Y) \leq r(X) + r(Y)$ when $X, Y \subseteq E$.

C: **weak elimination** - if C_1, $C_2 \in \mathbf{C}$ are distinct with $e \in C_1 \cap C_2$, then $(C_1 \cup C_2) - e$ contains a circuit.

J: **induced circuits** - if $I \in \mathbf{I}$, then $I + e$ contains at most one circuit.

G: **greedy algorithm** - for each nonnegative weight function on E, the greedy algorithm selects a maximum-weight independent set.

11.3.19. Remark. *Submodularity.* Submodularity arises naturally in linear algebra. The rank of a set $X \subseteq E$ in a vector matroid is the dimension of the space spanned by X. For subspaces U and V, the result in linear algebra is $\dim U \cap V + \dim W = \dim U + \dim V$, where W is the space spanned by $U \cup V$. When X and Y are spanning sets of vectors in U and V, the dimension of the space spanned by $X \cap Y$ may be less than $\dim U \cap V$. Nevertheless, $\dim U \cap V + \dim W \leq \dim U + \dim V$ is proved naturally for vector spaces using the proof of U $\Rightarrow$ R below. Exercise 31 obtains submodularity directly for cycle matroids. ∎

Various of these properties (plus being a hereditary system) have been used to define matroids. Examples include I (Welsh [1976], Schrijver [2003]), U (Edmonds [1965a,b], Bixby [1981], Nemhauser–Wolsey [1988]), A (Whitney [1935]), C (Tutte [1970]), G (Papadimitriou–Steiglitz [1982]). van der Waerden [1937], Rota [1964], Crapo–Rota [1970], and Aigner [1979] use other conditions.

Many authors list properties of hereditary systems when characterizing some aspect of a matroid. This obscures the essence of the characterization and leads to extra work. Working in the context of hereditary systems is simpler. All properties of hereditary systems are always available; we need not verify them when introducing another aspect. Our chain of implications may seem long, so it is worth noting that augmentation (I) and uniformity (U), for example, are easy to show equivalent, and hence the proof can be given in shorter implication chains.

11.3.20. Theorem. For a hereditary system M, the conditions defining matroids in Definition 11.3.18 are equivalent.

Proof: Property I is often used to show that some hereditary system is a matroid.

I $\Rightarrow$ B. Since a smaller base could be augmented from a larger base, bases have equal size. Now consider B_1, $B_2 \in \mathbf{B}$ with $e \in B_1 - B_2$. Since $B_1 - e \in \mathbf{I}$ and $|B_2| > |B_1| - 1$, the desired f exists.

B $\Rightarrow$ A. Bases cannot differ in size; otherwise, repeated base exchange yields one base inside another. Now let $X' = X + e + f$ with $r(X) = r(X + e) = r(X + f)$

and $r(X') > r(X)$. Among bases containing largest independent sets in X and X', choose B and B' with largest intersection. Since $r(X + e) = r(X + f) = r(X)$, we have $e, f \notin B$. Since $|B' \cap X'| > |B \cap X'|$ and $|B| = |B'|$, there exists $x \in B - B'$ with $x \notin X'$. Base exchange guarantees a base $B - x + x'$ for some $x' \in B' - B$. Since $x \notin X'$, we have $|B - x + x' \cap X| = r(X)$, but $|(B - x + x') \cap B'| > |B \cap B'|$.

A $\Rightarrow$ A'. We use induction on $|Y - X|$. There is nothing to prove when $|Y - X| = 1$. When $|Y - X| > 1$, choose $e, f \in Y - X$ and let $Y' = Y - e - f$. Applying the induction hypothesis using Y' or $Y' + e$ or $Y' + f$ yields $r(X) = r(X \cup Y') = r(X \cup Y' + e) = r(X \cup Y' + f)$. Now weak absorption yields $r(X) = r(X \cup Y)$.

A' $\Rightarrow$ U. If Y is a maximal independent subset of X, then $r(Y + e) = r(Y)$ for all $e \in X - Y$. By strong absorption, $r(X) = r(Y) = |Y|$. Hence all such Y have the same size, $r(X)$.

U $\Rightarrow$ R. Given $X, Y \subseteq E$, choose a maximum independent set I_1 from $X \cap Y$. By uniformity, I_1 can be enlarged to a maximum independent subset of $X \cup Y$; call this I_2. Consider $I_2 \cap X$ and $I_2 \cap Y$; these are independent subsets of X and Y, and each includes I_1. Hence

$$r(X \cap Y) + r(X \cup Y) = |I_1| + |I_2| = |I_2 \cap X| + |I_2 \cap Y| \le r(X) + r(Y).$$

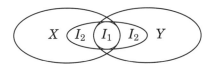

$$U \Rightarrow R$$

R $\Rightarrow$ C. Consider distinct circuits $C_1, C_2 \in \mathbf{C}$ with $e \in C_1 \cap C_2$. We have $r(C_1) = |C_1| - 1$ and $r(C_2) = |C_2| - 1$. Also $r(C_1 \cap C_2) = |C_1 \cap C_2|$, since every proper subset of a circuit is independent. If $(C_1 \cup C_2) - e$ contains no circuit, then its rank is $|C_1 \cup C_2| - 1$, and hence $r(C_1 \cup C_2) \ge |C_1 \cup C_2| - 1$. Applying submodularity to C_1 and C_2 now yields the contradiction

$$|C_1 \cap C_2| + |C_1 \cup C_2| - 1 \le |C_1| + |C_2| - 2.$$

C $\Rightarrow$ J. If $I + e$ contains circuits C_1 and C_2 for some $I \in \mathbf{I}$, then C_1 and C_2 both contain e. Now weak elimination guarantees a circuit in $(C_1 \cup C_2) - e$. However, $(C_1 \cup C_2) - e$ is independent, being contained in I.

J $\Rightarrow$ B'. For distinct $B_1, B_2 \in \mathbf{B}$ and $e \in B_1 - B_2$, by J there is a unique circuit C in $B_2 + e$. Since $C \not\subseteq B_1$, there exists $f \in C - B_1$, and now $B_2 + e - f \in \mathbf{I}$. To prove $B_2 + e - f$ is a base, we show that no two bases have distinct sizes. Otherwise, choose bases B_1' and B_2' with largest intersection such that $|B_1'| < |B_2'|$. Since $B_1' \not\subseteq B_2'$, there is $e \in B_1' - B_2'$. As before, we find $f \in B_2' - B_1'$ with $B_2' + e - f \in \mathbf{I}$. Any base containing $B_2' + e - f$ is as large as B_2' but shares more with B_1' than B_2'.

B' $\Rightarrow$ G. With nonnegative weights, some optimum is a base, and the algorithm chooses a base B. Let B^* be an optimal base having largest intersection with B. If $B^* \ne B$, then let e be the first element chosen for B that is not in B^*. Dual base exchange yields $f \in B^* - B$ such that $B^* + e - f \in \mathbf{B}$. The choice of B^* yields $w(f) > w(e)$. Since the algorithm chose e when f was available, $w(e) \ge w(f)$. Hence $B^* + e - f$ contradicts the choice of B^*, so $B = B^*$.

$G \Rightarrow I$. Given $I_1, I_2 \in \mathbf{I}$ with $k = |I_1| < |I_2|$, we design a weight function so that the greedy algorithm finds the desired augmentation. Let $w(e) = k + 2$ for $e \in I_1$, and let $w(e) = k + 1$ for $e \in I_2 - I_1$. Let $w(e) = 0$ for $e \notin I_1 \cup I_2$. Now $w(I_2) \geq (k+1)^2 > k(k+2) = w(I_1)$, so I_1 does not have maximum weight. However, the greedy algorithm chooses all of I_1 before any element of $I_2 - I_1$. Because it finds a maximum-weight independent set, after absorbing I_1 it adds an element of $I_2 - I_1$ such that $I_1 + e \in \mathbf{I}$.   ∎

11.3.21. Example. *Uniform matroids.* The **uniform matroid** of rank k, denoted $\mathbf{U}_{k,n}$ when $|E| = n$, is defined by $\mathbf{I} = \{X \subseteq E : |X| \leq k\}$. This immediately satisfies the base exchange and augmentation properties. The **free matroid** is the uniform matroid of rank $|E|$. Few uniform matroids are graphic, and few graphic matroids are uniform (Exercise 27). Neither $M(K_4^-)$ nor $M(K_4)$ is uniform.   ∎

11.3.22. Example. *Partition matroids.* The **partition matroid** on E induced by a partition of E into blocks $E_1, \ldots, E_k$ is defined by $\mathbf{I} = \{X \subseteq E : |X \cap E_i| \leq 1$ for all $i\}$. Since $\varnothing \in \mathbf{I}$, and since $X \in \mathbf{I}$ if and only if its elements lie in distinct blocks, $\mathbf{I}$ is a nonempty ideal. Given $I_1, I_2 \in \mathbf{I}$ with $|I_2| > |I_1|$, the set I_2 must intersect more blocks than I_1; an element of I_2 in a block that I_1 misses yields the desired augmentation of I_1. Alternatively, $r(X)$ is the number of blocks having elements in X; this satisfies the absorption property. Every partition matroid is a transversal matroid; consider the $E, [k]$-bigraph consisting of k stars whose leaf sets are the blocks of the partition.

Given a U, V-bigraph G, each part induces a partition matroid with ground set $E(G)$, partitioning the edges by their endpoints in one part. A set $X \subseteq E(G)$ is a matching in G if and only if X is independent both in the partition matroid induced by U *and* in the partition matroid induced by V. This is the motivation for the idea of matroid intersection discussed later in this section.

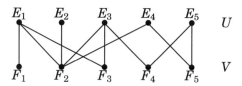

When G has an odd cycle, G has no set of vertices whose incident sets partition $E(G)$; thus $M(K_4^-)$ is *not* a partition matroid, for example. In a digraph, however, each edge has a head and a tail, and we can define the **head partition matroid** and the **tail partition matroid** using the edge partitions induced by incidences with heads and by incidences with tails. For example, the matroid of Example 11.3.2 arises as the partition matroid on E induced by U in the bipartite graph on the left below, as the head partition matroid in the first digraph, and as the tail partition matroid in the second digraph.   ∎

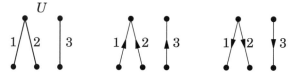

DUALITY AND MINORS

Duality in matroids generalizes duality in planar graphs (Section 9.1). Every connected plane graph G has a natural dual graph G^* such that $(G^*)^* = G$. The dual associates a vertex of G^* to each face of G and an edge of G^* to each edge of G, joining the faces incident to it. Loops and multiedges can arise in G^* (from cut-edges and pairs of faces sharing more than one boundary edge, respectively), so when discussing graphic matroids we allow the generality of multigraphs.

A set of edges forms a spanning tree in a plane graph G if and only if the duals to the remaining edges form a spanning tree in G^* (Exercise 9.1.25). Hence *the bases in the cycle matroid $M(G^*)$ are the complements of the bases in $M(G)$.* We define duality for matroids and for hereditary systems to extend these duality properties of planar graphs.

11.3.23. Definition. The **dual** of a hereditary system M on E is the hereditary system M^* defined by $\mathbf{B}_{M^*} = \{\overline{B}: B \in \mathbf{B}_M\}$. We may write $\mathbf{B}^*$ for $\mathbf{B}_{M^*}$ when M is understood; these are the **cobases** of M. Similarly, the circuits of M^* are the **cocircuits** of M, denoted $\mathbf{C}^*$, etc.

A set in E is **spanning** if it contains a base; $\mathbf{S}_M$ denotes the family of spanning sets. A set is a **hyperplane** if it is a maximal set containing no base; $\mathbf{H}_M$ denotes the family of hyperplanes.

11.3.24. Remark. If $\mathbf{B}$ is a nonempty antichain, then also $\{\overline{B}: B \in \mathbf{B}\}$ is a nonempty antichain (since containment reverses under complementation), so the dual of a hereditary system is a hereditary system. Also it is immediate that $(M^*)^* = M$, which explains the name *dual*.

A set is independent in M^* if and only if it is contained in a cobase of M. Hence $\mathbf{I}_{M^*} = \{\overline{S}: S \in \mathbf{S}_M\}$, and similarly $\mathbf{S}_{M^*} = \{\overline{I}: I \in \mathbf{I}_M\}$.

A set is a circuit in M^* if and only if it is a minimal set contained in no cobase of M. Hence $\mathbf{C}_{M^*} = \{\overline{H}: H \in \mathbf{H}_M\}$ and $\mathbf{H}_{M^*} = \{\overline{C}: C \in \mathbf{C}_M\}$. ∎

Duality is useful because the dual of a matroid is a matroid. This follows from the dual base exchange property (B′). In essence, the statement of the dual base exchange property for M is that of the base exchange property for M^*.

11.3.25. Theorem. (Whitney [1935]) The dual of a matroid M is a matroid.

Proof: We have seen that M^* is a hereditary system; now we prove the base exchange property for M^*. If $\overline{B}_1, \overline{B}_2 \in \mathbf{B}^*$ and $e \in \overline{B}_1 - \overline{B}_2$, then $B_1, B_2 \in \mathbf{B}$, with $e \in B_2 - B_1$. By dual base exchange for M, there exists $f \in B_1 - B_2$ such that $B_1 + e - f \in \mathbf{B}$. Now $\overline{B}_1 - e + f \in \mathbf{B}^*$, proving base exchange for M^*. ∎

Computing the rank function of the dual is easy using an alternative notion of the rank of a set. Instead of viewing $r(X)$ as the maximum size of an independent subset of X, it can be helpful to view it as the maximum size of the intersection of X with a base.

11.3.26. Proposition. The rank function r^* of the dual of a matroid M on E is given by $r^*(X) = |X| - (r(E) - r(\overline{X}))$.

Proof: The rank of a set X is the maximum size of its intersections with bases. Choose $B^* \in \mathbf{B}^*$ so that $r^*(X) = |X \cap B^*|$, and let $B = \overline{B^*}$. Now B is a base having smallest intersection with X and hence largest intersection with $\overline{X}$. Hence $r(\overline{X}) = |B - X|$. Also $|B| = r(E)$, so $r^*(X) = |X| - |X \cap B| = |X| - (r(E) - r(\overline{X}))$. ∎

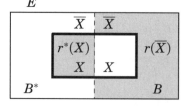

The argument of Proposition 11.3.26 is not valid for arbitrary hereditary systems, because $|B| = r(E)$ uses the uniformity property.

To understand how duality of matroids extends notions of graph theory, we need to know what the cocircuits are in graphic matroids. We begin by characterizing the cocircuits for all matroids.

11.3.27. Proposition. The cocircuits of a matroid are the minimal sets intersecting every base. The bases are the minimal sets intersecting every cocircuit.

Proof: Cocircuits are the minimal sets in no cobase. Since the cobases are the complements of the bases, a set lies in no cobase if and only if it intersects every base. Similarly, cobases are the maximal sets containing no cocircuit, so the complements of the cobases are the minimal sets intersecting every cocircuit. ∎

11.3.28. Corollary. The cocircuits of a cycle matroid $M(G)$ are the bonds of G.

Proof: By Proposition 11.3.27, the cocircuits are the minimal sets intersecting every maximal forest. Hence they are the minimal sets whose deletion increases the number of components; these are the bonds. ∎

11.3.29. Definition. The **bond matroid** or **cocycle matroid** of a graph G is the hereditary system whose circuits are the bonds of G.

By Corollary 11.3.28, the bond matroid of G is the dual of the cycle matroid $M(G)$. By Theorem 9.1.11, a set of edges forms a bond in a plane graph G if and only if the corresponding dual edges form a cycle in G^*. Hence we have shown that the dual of the cycle matroid of G is in fact the cycle matroid of G^*. We will use this to characterize planar graphs. For the proof, we must generalize edge-deletion and edge-contraction to matroids.

Given a graph G, the acyclic subsets of $E(G - e)$ are just the acyclic subsets of $E(G)$ that omit e. To make contraction a dual notion, it makes sense to describe its effect in terms of sets containing bases instead of sets contained in bases. A set $X \subseteq E(G \cdot e)$ contains a spanning tree of each component of $G \cdot e$ if and only if $X + e$ contains a spanning tree of each component of G.

This motivates our notation for hereditary systems. Just as we express the subgraph of G induced by S as both $G[S]$ and $G - \overline{S}$, for matroids we use different notation to emphasize the set of elements remaining or one element eliminated.

11.3.30. Definition. Given a hereditary system M on E and a set $F \subseteq E$, the **restriction** *of M to F*, denoted $M|F$ and obtained by **deleting** $\overline{F}$, is the hereditary system defined by $\mathbf{I}_{M|F} = \{X \subseteq F \colon X \in \mathbf{I}_M\}$. The **contraction** *of M to F*, denoted $M.F$ and obtained by **contracting** $\overline{F}$, is the hereditary system defined by $\mathbf{S}_{M.F} = \{X \subseteq F \colon X \cup \overline{F} \in \mathbf{S}_M\}$. When $F = E - e$, we write $M - e$ for $M|F$ and $M \cdot e$ for $M.F$. The **minors** of a hereditary system M are the hereditary systems arising from M by applying deletions and contractions.

From the definitions, $M|F$ and $M.F$ are hereditary systems. The notations $M|F$ and $M.F$ appear (briefly) in Oxley [1992] (pages 22 and 104, respectively). Note the distinction between "." and "$\cdot$"; the former emphasizes "contracting *to*" a specified set by "contracting *away*" the other elements. We use "$-$" or "$\cdot$" when eliminating one specified element and "|" or "." when specifying the elements that remain. Our notation for $M - e$ and $M \cdot e$ is consistent with our usage for graphs but is nonstandard in the matroid community, where $M \backslash e$ for our $M - e$ and M/e for our $M \cdot e$ are most common.

Defining contraction via spanning sets yields a natural duality.

11.3.31. Proposition. For hereditary systems, restriction and contraction are dual operations: $(M.F)^* = (M^*|F)$ and $(M|F)^* = (M^*.F)$.

Proof: $\mathbf{I}_{(M.F)^*} = \{X \subseteq F \colon F - X \in \mathbf{S}_{M.F}\} = \{X \subseteq F \colon (F - X) \cup \overline{F} \in \mathbf{S}_M\}$
$$= \{X \subseteq F \colon \overline{X} \in \mathbf{S}_M\} = \{X \subseteq F \colon X \in \mathbf{I}_{M^*}\} = \mathbf{I}_{M^*|F}.$$
For the second statement, apply the first to M^* and take duals. ∎

As expected, restrictions and contractions of matroids are matroids.

11.3.32. Theorem. Given $F \subseteq E$ and a matroid M on E, both $M|F$ and $M.F$ are matroids on F. Their bases and rank functions are given by

$$\mathbf{B}_{M|F} = \{B \cap F \colon B \in \mathbf{B}_M \text{ and } |B \cap F| = r_M(F)\}, \quad r_{M|F}(X) = r_M(X),$$
$$\mathbf{B}_{M.F} = \{B \cap F \colon B \in \mathbf{B}_M \text{ and } |B \cap \overline{F}| = r_M(\overline{F})\}, \quad r_{M.F}(X) = r_M(X \cup \overline{F}) - r_M(\overline{F}).$$

Proof: The augmentation property from M applies to sets in $\mathbf{I}_{M|F}$; thus $M|F$ satisfies the augmentation property and is a matroid. Since $M.F = (M^*|F)^*$, duality implies that $M.F$ is also a matroid.

The expressions for the bases and rank function of $M|F$ follow from the definition of $\mathbf{I}_{M|F}$; we do the same for $M.F$ from the definition of $\mathbf{S}_{M.F}$. A base of $M.F$ is a minimal set $\hat{B} \subseteq F$ such that $\hat{B} \cup \overline{F} \in \mathbf{S}_M$. Thus $\hat{B} \cup \overline{F}$ contains a base B of M. The minimality of $\hat{B}$ implies that $\hat{B} \subseteq B$ and that $B \cap \overline{F}$ is a maximal independent subset of $\overline{F}$ (by uniformity on $\hat{B} \cup \overline{F}$). Thus $\mathbf{B}_{M.F}$ consists of the sets in F whose addition to a maximal independent subset of $\overline{F}$ yields a base of M.

To compute $r_{M.F}(X)$, recall that $r_{M.F}(X) = |B' \cap X|$ for some $B' \in \mathbf{B}_{M.F}$; let $Y = B' \cap X$. Also $B' = B \cap F$ for some $B \in \mathbf{B}_M$ such that $|B \cap \overline{F}| = r_M(\overline{F})$; let $Z = B \cap \overline{F}$. Since $B \in \mathbf{B}_M$, we have $Y \cup Z \in \mathbf{I}_M$. If $Y \cup Z$ is not a maximal independent subset of $X \cup \overline{F}$, then it augments to a base in M that contains Z and has larger intersection with X than B', contradicting the choice of B'. Hence uniformity in M yields $|Y \cup Z| = r_M(X \cup \overline{F})$. Now $|Y| = r_M(X \cup \overline{F}) - r_M(\overline{F})$. ∎

11.3.33. Remark. The formula for $r_{M.F}$ yields a description of the independent sets: $X \in \mathbf{I}_{M.F}$ if and only if adding X to $\overline{F}$ increases the rank by $|X|$. Note that when $\overline{F}$ is an independent set $\{e\}$ of size 1, we have $r_{M.F}(X) = r_M(X + e) - 1$. In particular, when we contract a nonloop edge in a graph, the maximum size of a forest among the edges of any set containing that edge decreases by 1.

The duality between deletion and contraction is familiar in plane graphs. Deleting an edge e in a plane graph G contracts the corresponding edge in G^*; contracting e deletes the edge in the dual. The fate of the 4-cycle below illustrates that when a circuit of M intersects F, its intersection with F need not be a circuit of $M.F$, even when only one element has been contracted. In fact, the circuits of $M.F$ are the minimal nonempty sets in $\{C \cap F \colon C \in \mathbf{C}_M\}$ (Exercise 47).

Also, restriction and contraction commute (Exercise 49).   ∎

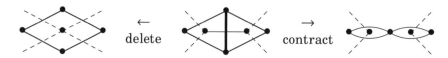

$$\overset{\leftarrow}{\text{delete}} \qquad \qquad \overset{\rightarrow}{\text{contract}}$$

11.3.34. Corollary. The behavior of cycle matroids and bond matroids under deletion or contraction of an edge $e \in E(G)$ is

$$M(G - e) = M(G) - e, \quad M^*(G - e) = M^*(G) \cdot e,$$
$$M(G \cdot e) = M(G) \cdot e, \quad M^*(G \cdot e) = M^*(G) - e.$$

Proof: Matroid deletion and contraction are defined so that the statements in the first column describe the behavior of cycle matroids. Using these and Proposition 11.3.31, for the second column we compute

$$M^*(G - e) = [M(G - e)]^* = [M(G) - e]^* = M^*(G) \cdot e,$$
$$M^*(G \cdot e) = [M(G \cdot e)]^* = [M(G) \cdot e]^* = M^*(G) - e.$$ ∎

Now we characterize planar graphs using matroids. In Theorem 9.1.11, we proved that a set of edges in a plane graph G forms a cycle if and only if the corresponding dual edges form a bond in G^*. Thus under the natural bijection from edges to dual edges, the cycle matroid of a plane graph G is (isomorphic to) the bond matroid of G^*. By Corollary 11.3.28, the bond matroid of a graph H is $[M(H)]^*$. Applying this to G and G^* tells us that the bond matroid of G is (isomorphic to) the cycle matroid of G^*. In particular, the bond matroid of G is graphic. Using Kuratowski's Theorem, we will use this to characterize planarity.

Whitney [1932c] approached this question by defining a non-geometric notion of the dual of a graph. Changing his definition slightly, we say that H is an **abstract dual** of G if there is a bijection $\phi \colon E(G) \to E(H)$ such that $X \subseteq E(G)$ is a bond in G if and only if $\phi(X)$ is the edge set of a cycle in H. With this definition, the statement that G has an abstract dual H is the same as the statement that the bond matroid of G is graphic; the bijection ϕ establishes an isomorphism between $M^*(G)$ and $M(H)$.

11.3.35. Theorem. (Whitney [1932c]) A graph G is planar if and only if its bond matroid $M^*(G)$ is graphic.

Proof: We have observed that planar graphs have abstract duals; this proves necessity. For sufficiency, we first prove that existence of an abstract dual is preserved under deletion and contraction of edges. Suppose that G has an abstract dual H, so that $M(H) \cong M^*(G)$. Let e' be the edge of H corresponding to e under the bijection. To prove that $H \cdot e'$ is an abstract dual of $G - e$ and that $H - e'$ is an abstract dual of $G \cdot e$, we use Corollary 11.3.34 to compute

$$M^*(G - e) = M^*(G) \cdot e \cong M(H) \cdot e' = M(H \cdot e'),$$
$$M^*(G \cdot e) = M^*(G) - e \cong M(H) - e' = M(H - e').$$

By Kuratowski's Theorem, a nonplanar graph contains a subdivision of K_5 or $K_{3,3}$. Hence K_5 or $K_{3,3}$ is a minor of it. Since existence of abstract duals is preserved under deletion and contraction, showing that K_5 and $K_{3,3}$ have no abstract dual implies that every nonplanar graph has no abstract dual.

If H is an abstract dual of G, then also G is an abstract dual of H, since $M^*(G) \cong M(H)$ if and only if $M(G) \cong M^*(H)$. If G has girth g, then bonds of H have size at least g, so $\delta(H) \geq g$. Letting $n = |V(H)|$ and $m = |E(H)|$, we have also $|E(G)| = m$. Thus the Degree-Sum Formula yields $n \leq \lfloor 2m/\delta(H) \rfloor \leq \lfloor 2m/g \rfloor$.

Let H be an abstract dual of K_5. Since K_5 has girth 3, $n \leq \lfloor 20/3 \rfloor = 6$. Since all bonds of K_5 have four or six edges, all cycles of H have four or six edges, and thus H is a simple bipartite graph. However, no simple bipartite graph with at most six vertices has ten edges.

Let H be an abstract dual of $K_{3,3}$. Since $K_{3,3}$ has girth 4, $n \leq \lfloor 18/4 \rfloor = 4$. Since all bonds of $K_{3,3}$ have at least three edges, all cycles of H have at least three edges, and thus H is a simple graph. However, no simple graph with at most four vertices has nine edges. ∎

Planar graphs have no K_5- or $K_{3,3}$-minor, since planarity is preserved under deletion and contraction of edges. Kuratowski's Theorem implies that nonplanar graphs do have such minors. This yields the characterization by Wagner [1937]: K_5 and $K_{3,3}$ are the minimal forbidden minors for planar graphs.

THE SPAN FUNCTION

In addition to the notion of "spanning set" as "set containing a base", another notion of span is suggested by vector spaces. A set S of vectors spans a vector v if v is a linear combination of elements of S, which means that $S + v$ contains a dependent set. This notion defines another aspect of hereditary systems.

11.3.36. Definition. The **span function** of a hereditary system M on E is the function σ_M on 2^E defined by $\sigma_M(X) = X \cup \{e \in E: Y + e \in \mathbf{C}_M \text{ for some } Y \subseteq X\}$. If $e \in \sigma(X)$, then X **spans** e.

A set in a hereditary system is dependent if and only if it contains a circuit, which by Definition 11.3.36 means that $e \in \sigma(X - e)$ for some $e \in X$. Hence we can find the independent sets (and other aspects of M) from the span function via $\mathbf{I}_M = \{X \subseteq E: (e \in X) \Rightarrow (e \notin \sigma_M(X - e))\}$. The properties of σ that we use in studying matroids are (s1, s2, s3) below.

11.3.37. Proposition. If σ is the span function of a hereditary system on E, and $X, Y \subseteq E$, then

    s1) $X \subseteq \sigma(X)$ (σ is **expansive**).

    s2) $Y \subseteq X$ implies $\sigma(Y) \subseteq \sigma(X)$ (σ is **order-preserving**).

    s3) $e \notin \sigma(X)$ and $e \in \sigma(X + f)$ imply $f \in \sigma(X + e)$ (**Steinitz exchange**).

Proof: Definition 11.3.36 implies immediately that σ is expansive and order-preserving. If $e \notin \sigma(X)$, then $e \notin X$. With $e \in \sigma(X + f)$, also e belongs to a circuit C in $X + f + e$. With $e \notin \sigma(X)$, we must have $f \in C$. This circuit yields $f \in \sigma(X + e)$, and hence σ satisfies the Steinitz exchange property. ∎

11.3.38. Example. *Steinitz exchange in cycle matroids.* In the cycle matroid $M(G)$, the meaning of $e \notin \sigma(X)$ is that X (the solid edges below) contains no path between the endpoints of e. If $e \notin \sigma(X)$ but $e \in \sigma(X + f)$, then adding f completes such a path. Adding e to the path completes a cycle, so also $f \in \sigma(X + e)$. ∎

Definition 11.3.36 immediately implies that the circuits of a hereditary system satisfy the **strong dependence property**: $e \in C$ implies $e \in \sigma(C - e)$. The natural notion that an element is spanned by a set X if adding it to X does not increase the rank is valid for all hereditary systems.

11.3.39. Lemma. In a hereditary system, $[r(X+e)=r(X)] \Rightarrow e \in \sigma(X)$.

Proof: Let Y be a maximum independent subset of X. Since $|Y| = r(X) = r(X + e)$, also Y is a maximum independent subset of $X + e$. Hence e completes a circuit with some subset of X contained in Y, and $e \in \sigma(X)$. ∎

In fact, the converse characterizes matroids! We call the converse of Lemma 11.3.39 the "incorporation property".

11.3.40.* Theorem. If M is a hereditary system, then each condition below is necessary and sufficient for M to be a matroid.

P: **incorporation** - $r(\sigma(X)) = r(X)$ for all $X \subseteq E$.

S: **idempotence** - $\sigma^2(X) = \sigma(X)$ for all $X \subseteq E$.

T: **transitivity of dependence** - if $e \in \sigma(X)$ and $X \subseteq \sigma(Y)$, then $e \in \sigma(Y)$.

C': **strong elimination** - whenever $C_1, C_2 \in \mathbf{C}$, $e \in C_1 \cap C_2$, and $f \in C_1 \triangle C_2$, there exists $C \in \mathbf{C}$ such that $f \in C \subseteq (C_1 \cup C_2) - e$.

Proof: See Exercise 50. ∎

Incorporation, idempotence, and transitivity of dependence are well-known properties of matroids. The equivalence of C and C' for hereditary systems was first proved by Lehman [1964]. Brylawski [1986] gave another way to obtain C'. Idempotence is natural for linear matroids; the span of a set of vectors has nothing more in its own span. This suggests another aspect of hereditary systems.

11.3.41. Definition. The **closed sets** of a hereditary system on E are the sets $X \subseteq E$ such that $\sigma(X) = X$ (also called **flats** or **subspaces**).

In a matroid on E, the sets whose span is E are the sets containing bases; hence the term "spanning sets". Similarly, in a matroid the hyperplanes are the maximal proper closed subsets of E. Both fail for general hereditary systems.

The span function of a matroid is also called its **closure function**. A **closure operator** on 2^E is an expansive, order-preserving, idempotent function from 2^E to 2^E. Such an operator is the span function of a matroid if and only if it satisfies Steinitz exchange. Since every hereditary system satisfies Steinitz exchange, our approach to matroids as hereditary systems with an added property is not well suited for studying closure operators. The span function of a hereditary system M is a closure operator if and only if M is a matroid.

Matroids are developed from lattice theory (Section 12.4) in MacLane [1936], Rota [1964], and Aigner [1979]. Brylawski [1986] described the transformations among about a dozen aspects of matroids, calling these maps **cryptomorphisms**.

MATROID INTERSECTION

Matroid theory took a great leap forward with the Matroid Intersection and Union Theorems. They provided a unified context for many well-known min-max relations, which became corollaries. We proved some of these in earlier chapters. Yielding a unified proof for many important theorems, the Matroid Intersection Theorem can be considered among the most beautiful theorems of combinatorics.

The Matroid Intersection Theorem is a min-max relation for the maximum size of common independent sets in two matroids on the same set E. The intersection of two matroids is a hereditary system but generally *not* a matroid. For multiple matroids on a set E, we typically use subscripts to distinguish corresponding aspects, as in $\mathbf{B}_i$ for the bases of M_i, etc. We still use $\overline{X}$ for the complement of X within the full set E.

11.3.42. Definition. The **intersection** of hereditary systems M_1 and M_2 on E is the hereditary system whose independent sets are $\{X \subseteq E: X \in \mathbf{I}_1 \cap \mathbf{I}_2\}$.

11.3.43. Example. Since $\mathbf{I}_1$ and $\mathbf{I}_2$ are closed under taking subsets, the common independent sets in two hereditary systems also form a hereditary family.

In a bipartite graph G with edge set E, each part induces a partition matroid on E. A set of edges forms an independent set in one of these partition matroids if and only if its endpoints in the corresponding part of G are distinct. A set of edges is independent in both matroids if and only if it is a matching in G.

The hereditary system whose independent sets are the matchings in a bipartite graph is not a matroid. The central edge in a path of length 3 forms an independent set, and the two end edges form a larger one, but the smaller set cannot be augmented from the larger, so augmentation fails. This is why the greedy algorithm does not solve maximum matching in bipartite graphs. ∎

The Matroid Intersection Theorem expresses the maximum size of a common independent set in two matroids M_1 and M_2 in terms of their rank functions. We prove it inductively, using restrictions and contractions. Recall that a *loop* is an element forming a circuit of size 1.

11.3.44. Theorem. (**Matroid Intersection Theorem**; Edmonds [1970]) For matroids M_1 and M_2 on E,

$$\max_{I \in \mathbf{I}_1 \cap \mathbf{I}_2} |I| = \min_{X \subseteq E}(r_1(X) + r_2(\overline{X})).$$

Proof: (Woodall; see Seymour [1976]) For the upper bound on $|I|$, consider $I \in \mathbf{I}_1 \cap \mathbf{I}_2$ and $X \subseteq E$. The sets $I \cap X$ and $I \cap \overline{X}$ are also common independent sets, and $|I| = |I \cap X| + |I \cap \overline{X}| \le r_1(X) + r_2(\overline{X})$. Hence $\max\{|I|\} \le \min\{r_1(X) + r_2(\overline{X})\}$.

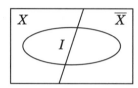

To achieve equality, we use induction on $|E|$; when $|E| = 0$ both sides are 0. If every element of E is a loop in M_1 or in M_2, then $\max |I| = 0 = r_1(X) + r_2(\overline{X})$, where X consists of all loops in M_1. Hence we may assume that $|E| > 0$ and that some $e \in E$ is a non-loop in both matroids. Let $F = E - e$, and consider the matroids $M_1|F$, $M_2|F$, $M_1.F$, $M_2.F$. By Theorem 11.3.32, a set X is independent in $M_1.F$ if and only if $X + e$ is independent in M_1, and similarly for $M_2.F$.

Let $k = \min_{X \subseteq E}\{r_1(X) + r_2(\overline{X})\}$; we seek a common independent k-set in M_1 and M_2. If none exists, then $M_1|F$ and $M_2|F$ have no common independent k-set, and $M_1.F$ and $M_2.F$ have no common independent $(k-1)$-set. By the induction hypothesis and rank formulas (Theorem 11.3.32),

$$r_1(X) + r_2(F - X) \le k - 1 \qquad \text{for some } X \subseteq F,$$
$$r_1(Y + e) - 1 + r_2(F - Y + e) - 1 \le k - 2 \qquad \text{for some } Y \subseteq F.$$

We use $(F - Y) + e = \overline{Y}$ and $F - X = \overline{X + e}$ and sum the two inequalities:

$$r_1(X) + r_2(\overline{X + e}) + r_1(Y + e) + r_2(\overline{Y}) \le 2k - 1.$$

Write $X' = X + e$ and $Y' = Y + e$. We apply submodularity of r_1 to X and Y' and submodularity of r_2 to $\overline{Y}$ and $\overline{X'}$. With the preceding inequality, this yields

$$r_1(X \cup Y') + r_1(X \cap Y') + r_2(\overline{Y} \cup \overline{X'}) + r_2(\overline{Y} \cap \overline{X'}) \le 2k - 1.$$

Since $\overline{Y} \cap \overline{X'} = \overline{X \cup Y'}$ and $\overline{Y} \cup \overline{X'} = \overline{X \cap Y'}$ (see Venn diagram below), the left side sums two instances of $r_1(Z) + r_2(\overline{Z})$. Hence the hypothesis $k \le r_1(Z) + r_2(\overline{Z})$ for all $Z \subseteq E$ yields $2k \le 2k - 1$. This contradiction implies that M_1 and M_2 do have a common independent k-set. ∎

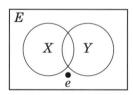

We have already proved various special cases of the Matroid Intersection Theorem, including the König–Egerváry Theorem and the Ford–Fulkerson Theorem

(Theorem 7.2.15). Given any two matroids on the same set, the Matroid Intersection Theorem guarantees a min-max relation for the maximum size of a common independent set, says what the result should be, and provides a proof.

11.3.45. Corollary. (König [1931], Egerváry [1931]) In a bipartite graph, a largest matching and smallest vertex cover have equal size.

Proof: In the partition matroids M_1 and M_2 on $E(G)$ induced by the partite sets U_1 and U_2, the common independent sets are the matchings in G.

For $X \subseteq E$, the number of vertices of U_i incident to edges of X is $r_i(X)$. Hence $r_1(X) + r_2(\overline{X})$ is the size of a vertex cover, using U_1 for X and U_2 for $\overline{X}$. On the other hand, if $T_1 \cup T_2$ is a vertex cover with $T_i \subseteq U_i$, and X is the set of edges incident to T_1, then $r_1(X) + r_2(\overline{X}) \leq |T_1 \cup T_2|$. We conclude that $\min\{r_1(X) + r_2(\overline{X})\}$ is the minimum size of a vertex cover.

With $\alpha'(G)$ and $\beta(G)$ denoting the maximum size of a matching and the minimum size of a vertex cover, we obtain

$$\alpha'(G) = \max\{|I| : I \in \mathbf{I}_1 \cap \mathbf{I}_2\} = \min\{r_1(X) + r_2(\overline{X})\} = \beta(G). \qquad \blacksquare$$

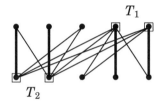

The next application uses the rank function for transversal matroids.

11.3.46. Example. *Transversal matroids* (Definition 11.3.11). The transversal matroid induced by subsets $A_1, \ldots, A_m$ of E has as its independent sets the partial systems of distinct representatives. Equivalently, these are subsets of E that can be matched into $[m]$ in the bipartite incidence graph G with parts E and $[m]$.

If $N(S) < |S|$ for some $S \subseteq X \subseteq E$, then every matching in G leaves at least $|S| - |N(S)|$ elements of X uncovered. Ore's Defect Formula (Corollary 6.1.12) gives $r(X) = \min_{S \subseteq X}\{|X| - (|S| - |N(S)|)\} = \min_{S \subseteq X}\{|N(S)| + |X - S|\}$.

Ore [1955] gave another useful expression for $r(X)$. For $J \subseteq [m]$, let $A(J) = \bigcup_{i \in J} A_i$; in terms of the graph, $A(J) = N(J)$ (see figure below). Since $r(X) = \alpha'(G[X \cup [m]])$, we can write the defect formula in terms of neighborhoods of subsets of $[m]$ instead of subsets of X to obtain

$$r(X) = \min_{J \subseteq [m]}\{|A(J) \cap X| + m - |J|\}. \qquad (*)$$

The upper bound here holds because for any $J \subseteq [m]$, at most $|A(J) \cap X|$ elements of X can be matched into J, and at most $m - |J|$ can be matched into $[m] - J$. The statement that equality holds is equivalent to the König–Egerváry Theorem, since $(A(J) \cap X) \cup ([m] - J)$ is a vertex cover of $G[X \cup [m]]$, and any minimal vertex cover Q can be achieved in this way by setting $J = [m] - Q$. $\blacksquare$

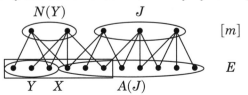

11.3.47. Corollary. (Ford–Fulkerson [1958]) Families $\mathbf{A} = \{A_1, \ldots, A_m\}$ and $\mathbf{B} = \{B_1, \ldots, B_m\}$ have a common system of distinct representatives (CSDR) if and only if, for each $I, J \subseteq [m]$,

$$\left|\left(\bigcup_{i\in I} A_i\right) \cap \left(\bigcup_{j\in J} B_j\right)\right| \geq |I| + |J| - m.$$

Proof: Let E be the union of all the sets in $\mathbf{A} \cup \mathbf{B}$. A common partial transversal is a common independent set in the two transversal matroids M_1, M_2 induced on E by $\mathbf{A}$ and $\mathbf{B}$. A common transversal is a common independent set of size m. We need only restate the condition $r_1(X)+r_2(\overline{X}) \geq m$ to find the appropriate condition on the set systems.

The rank formula $(*)$ from Example 11.3.46 yields

$$r_1(X) + r_2(\overline{X}) = \min_{I\subseteq[m]}\{|A(I) \cap X| - |I| + m\} + \min_{J\subseteq[m]}\{|B(J)\cap \overline{X}| - |J| + m\}.$$

Hence $r_1(X) + r_2(\overline{X}) \geq m$ for all X if and only if

$$|A(I) \cap X| + |B(J)\cap \overline{X}| \geq |I| + |J| - m \text{ for all } X \subseteq E \text{ and } I, J \subseteq [m].$$

Since $|A(I) \cap X| + |B(J)\cap \overline{X}| \geq |A(I) \cap B(J)|$, the Ford–Fulkerson condition is sufficient. To see that it is necessary, let X be a subset of E such that $A(I) - B(J) \subseteq \overline{X}$ and $B(J)-A(I) \subseteq X$. Given I, J, consider the contribution by elements of E to the left side. In this case $|A(I) \cap X| + |B(J)\cap \overline{X}| = |A(I) \cap B(J)|$, so the condition on $r_1(X) + r_2(\overline{X})$ shows that the condition on $A(I)\cap B(J)$ is necessary. ∎

11.3.48.* Remark. The augmenting path approach to maximum bipartite matching generalizes to matroid intersection. The algorithm yields a common independent set I of maximum size and a set X such that $r_1(X) + r_2(\overline{X}) = |I|$ (see Lawler [1976], Edmonds [1979], Faigle [1987]). Finding a maximum common independent set in three matroids is NP-complete (Exercises 25–26).

Maximum matching in general graphs also extends to a matroid context. A common generalization of matroid intersection and general graph matching is known in different phrasings as the **Matroid Matching Problem** or the **Matroid Parity Problem**. It is solvable in polynomial time for linear matroids, in which case it yields a min-max relation (Lovász [1978]). Lovász also proved that it is not solvable in polynomial time for general matroids in the same computational model. The solution technique over linear matroids involves "polymatroids", a polyhedral generalization of matroids. ∎

Next we consider a related min-max relation for "union" of matroids. In this and other applications of matroid intersection, it can be helpful to restrict the range of the minimization.

11.3.49. Corollary. The maximum size of a common independent set in matroids M_1, M_2 on E is the minimum of $r_1(X_1) + r_2(X_2)$ over sets X_1, X_2 such that $X_1 \cup X_2 = E$ and each X_i is closed in M_i.

Proof: The incorporation property implies $r_i(\sigma_i(X)) = r_i(X)$. ∎

MATROID UNION

The intersection of two matroids is seldom a matroid, but a natural concept of union does yield a matroid. The Matroid Union Theorem states this and gives a useful min-max relation for the rank function. The Matroid Intersection and Union Theorems are equivalent; they can be derived from each other. Welsh [1976] proves the Matroid Union Theorem first; here we obtain it from the Matroid Intersection Theorem.

11.3.50. Definition. The **union** $M_1 \cup \cdots \cup M_k$ of hereditary systems $M_1, \ldots, M_k$ on E is the hereditary system M on E defined by $\mathbf{I}_M = \{I_1 \cup \cdots \cup I_k : I_i \in \mathbf{I}_i\}$. The **direct sum** $M_1 \oplus \cdots \oplus M_k$ of hereditary systems $M_1, \ldots, M_k$ on disjoint sets $E_1, \ldots, E_k$ is the hereditary system M on $E_1 \cup \cdots \cup E_k$ defined by $\mathbf{I}_M = \{I_1 \cup \cdots \cup I_k : I_i \in \mathbf{I}_i\}$.

Since $\mathbf{I}_M$ is a nonempty ideal, $M_1 \cup \cdots \cup M_k$ is indeed a hereditary system. The direct sum $M_1 \oplus \cdots \oplus M_k$ on $E_1, \ldots, E_k$ can be expressed as a matroid union. Let $E' = E_1 \cup \cdots \cup E_k$. For $1 \le i \le k$, define M_i' on E' by starting with M_i and addiing each element outside E_i as a loop. Now $M_1 \oplus \cdots \oplus M_k = M_1' \cup \cdots \cup M_k'$.

When each M_i is a uniform matroid, the direct sum is a **generalized partition matroid**. Such a matroid on E consists of a partition $E_1, \ldots, E_k$ of E and positive integers $r_1, \ldots, r_k$ such that $X \in \mathbf{I}$ if $|X \cap E_i| \le r_i$ for all i. The partition matroids defined earlier arise when $r_i = 1$ for all i.

11.3.51. Proposition. Given matroids $M_1, \ldots, M_k$ on disjoint sets $E_1, \ldots, E_k$, the direct sum $M = M_1 \oplus \cdots \oplus M_k$ is a matroid.

Proof: Since the $E_1, \ldots, E_k$ are pairwise disjoint, the intersection of any $I \in \mathbf{I}$ with each E_i is independent in M_i. If $I_1, I_2 \in \mathbf{I}$ with $|I_2| > |I_1|$, then $|I_2 \cap E_i| > |I_1 \cap E_i|$ for some i. Since both sets are independent in M_i, we can augment $I_1 \cap E_i$ from $I_2 \cap E_i$ and therefore I_1 from I_2. Hence $M_1 \oplus \cdots \oplus M_k$ satisfies the augmentation property. ∎

Using a direct sum, we prove that the union of matroids is always a matroid, and we compute the rank function.

11.3.52. Theorem. (**Matroid Union Theorem**; Edmonds–Fulkerson [1965], Nash-Williams [1966]) If $M_1, \ldots, M_k$ are matroids on E with rank functions $r_1, \ldots, r_k$, then the union $M = M_1 \cup \cdots \cup M_k$ is a matroid with rank function $r(X) = \min_{Y \subseteq X}(|X - Y| + \sum r_i(Y))$.

Proof: (following Schrijver [2003]). After proving the formula for the rank function, we will verify the submodularity property to prove that M is a matroid. First we reduce the computation of the rank function to the computation of $r(E)$. In the restriction of the hereditary system M to the set X, we have $\mathbf{I}_{M|X} = \{Y \subseteq X : Y \in \mathbf{I}_M\}$ and $r_{M|X}(Y) = r_M(Y)$ for $Y \subseteq X$. Thus $M|X = \cup_i(M_i|X)$, and applying the formula for the rank of the full union to $M|X$ yields $r_M(X)$.

Consider a k-by-$|E|$ grid of elements E' in which the jth column E_j consists of k copies of the element $e_j \in E$. We define two matroids N_1, N_2 on E' such that the maximum size of a set independent in both N_1 and N_2 equals the maximum size of a set independent in M. We then compute $r_M(E)$ by applying the Matroid Intersection Theorem to N_1 and N_2. Let M_i' be a copy of M_i defined on the elements E^i of row i in E'. Let N_1 be the direct sum matroid $M_1' \oplus \cdots \oplus M_k'$, and let N_2 be the partition matroid induced on E' by the column partition $\{E_j\}$.

Each set $X \in \mathbf{I}_M$ has a decomposition into disjoint subsets $X_i \in \mathbf{I}_i$, since $\mathbf{I}_i$ is an ideal. Given a decomposition $\{X_i\}$ of $X \in \mathbf{I}_M$, let X_i' be the copy of X_i in E^i. Since $\{X_i\}$ are disjoint, $\cup X_i'$ is independent in N_2, and $X_i \in \mathbf{I}_i$ implies that $\cup X_i'$ is also independent in N_1. From $X \in \mathbf{I}_M$, we have constructed $\cup X_i'$ of size $|X|$ in $\mathbf{I}_{N_1} \cap \mathbf{I}_{N_2}$. Conversely, any $X' \in \mathbf{I}_{N_1} \cap \mathbf{I}_{N_2}$ corresponds to a decomposition of a set in $\mathbf{I}_M$ of size $|X'|$ when the sets $X' \cap E^i$ are transferred back to E, because N_2 forbids multiple copies of elements.

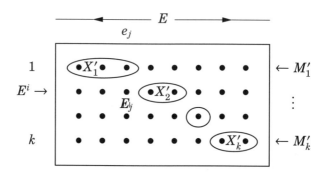

Hence $r(E) = \max\{|I| : I \in \mathbf{I}_{N_1} \cap \mathbf{I}_{N_2}\}$. Let the rank functions of N_1, N_2 be q_1, q_2, and let r_i' be the rank function of the copy M_i' of M_i on E^i. We have $q_1(X') = \sum r_i'(X' \cap E^i)$, and $q_2(X')$ is the number of elements of E that have copies in X'. The Matroid Intersection Theorem yields $r(E) = \min_{X' \subseteq E'}\{q_1(X') + q_2(E' - X')\}$.

By Corollary 11.3.49, the minimum is achieved by a set X' such that $E' - X'$ is closed in N_2. The closed sets in the partition matroid N_2 are the sets containing all or no copies of each element e_j; these are the unions of full columns of E'. Given X' with $E' - X'$ closed in N_2, let $Y \subseteq E$ be the set of elements whose copies comprise X'. Then $q_2(E' - X') = |E - Y|$, and X' contains all copies of the elements of Y, so $q_1(X') = \sum r_i'(X' \cap E^i) = \sum r_i(Y)$. We conclude $r(E) = \min_{Y \subseteq E}\{|E - Y| + \sum r_i(Y)\}$.

To show that M is a matroid, we verify submodularity for r. Given $X, Y \subseteq E$, the formula for r yields $U \subseteq X$ and $V \subseteq Y$ such that

$$r(X) = |X - U| + \sum r_i(U); \qquad r(Y) = |Y - V| + \sum r_i(V).$$

Since $U \cap V \subseteq X \cap Y$ and $U \cup V \subseteq X \cup Y$, we also have

$$r(X \cap Y) \leq |(X \cap Y) - (U \cap V)| + \sum r_i(U \cap V);$$
$$r(X \cup Y) \leq |(X \cup Y) - (U \cup V)| + \sum r_i(U \cup V).$$

After applying the submodularity of each r_i and the diagram below, these inequalities yield $r(X \cap Y) + r(X \cup Y) \leq r(X) + r(Y)$. ∎

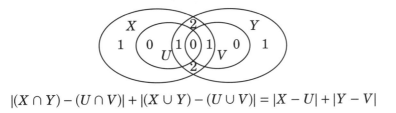

$$|(X \cap Y) - (U \cap V)| + |(X \cup Y) - (U \cup V)| = |X - U| + |Y - V|$$

The Matroid Union Theorem yields short proofs of formulas for packing and covering problems. In each formula below, the optimal subset is closed, because switching from X to $\sigma(X)$ improves the numerator without changing the denominator. The graph corollaries were originally proved by difficult ad hoc arguments.

11.3.53. Corollary. (Matroid Covering Theorem; Edmonds [1965a])
In a loopless matroid M on E, the minimum number of independent sets whose union is E is $\max_{\varnothing \neq X \subseteq E} \left\lceil \frac{|X|}{r(X)} \right\rceil$.

Proof: Let $M_1, \ldots, M_k$ be copies of M on E. The set E is the union of k independent sets in M if and only if E is independent in $M' = M_1 \cup \cdots \cup M_k$. By the Matroid Union Theorem, $r'(E) \geq |E|$ is equivalent to $|E| - |Y| + \sum r_i(Y) \geq |E|$ for all $Y \subseteq E$. Since $r_i(Y) = r(Y)$ for all i, we conclude that E is the union of k independent sets if and only if $kr(Y) \geq |Y|$ for all $Y \subseteq E$. ∎

11.3.54. Corollary. (Nash-Williams [1964]) The number of forests needed to cover the edges of a graph G (its **arboricity**) is $\max_{H \subseteq G: |V(H)| \geq 2} \left\lceil \frac{|E(H)|}{|V(H)| - 1} \right\rceil$.

Proof: (Edmonds [1965a]) This follows immediately by applying Corollary 11.3.53 to $M(G)$. The best lower bound arises from a connected induced subgraph H (corresponding to a closed set in $M(G)$). ∎

11.3.55. Corollary. (Matroid Packing Theorem; Edmonds [1965b])
For a matroid M on E, the maximum number of pairwise disjoint bases equals $\min_{X: r(X) < r(E)} \left\lfloor \frac{|E| - |X|}{r(E) - r(X)} \right\rfloor$.

Proof: There are k disjoint bases if and only if $r'(E) \geq kr(E)$ in the union M' of k matroids $M_1, \ldots, M_k$ that are copies of M on E. By the Matroid Union Theorem, this requires $|E| - |Y| + \sum r_i(Y) \geq kr(E)$ for all $Y \subseteq E$. Since $r_i(Y) = r(Y)$ for all i, there are k disjoint bases if and only if $|E| - |Y| \geq k(r(e) - r(Y))$ for all $Y \subseteq E$. ∎

11.3.56. Corollary. (Nash-Williams [1961], Tutte [1961]) A graph G has k edge-disjoint spanning trees if and only if, for every partition P of $V(G)$, at least $k(|P| - 1)$ edges have endpoints in different parts.

Proof: (Edmonds [1965b]) We may assume G is connected. By Corollary 11.3.55, we need $|E| - |X| \geq k(r(E) - r(X))$ for each closed set X in $M(G)$. The closed sets correspond to partitions of $V(G)$ into sets inducing connected subgraphs. For each such partition $V_1, \ldots, V_p$, the corresponding closed set X is $\bigcup E(G[V_i])$ with rank $n - p$. Since $|E| - |X|$ counts the edges joining parts of the partition and $r(E) - r(X) \geq p - 1$, the graph has k disjoint spanning trees if and only if the condition holds. ∎

Corollary 11.3.56 implies that every $2k$-edge-connected graph has k edge-disjoint spanning trees, and this is sharp (Exercise 63). Nash-Williams and Tutte proved Corollary 11.3.56 independently, by different methods, a few months apart, and the two papers were published in the same volume of the same journal.

EXERCISES 11.3

11.3.1. $(-)$ Let M be the hereditary system on $[4]$ whose bases are $\{1,2\}$ and $\{3,4\}$. Show that all the properties listed in Definition 11.3.18 fail for M.

11.3.2. $(-)$ For each family $\mathbf{C}$ below, determine whether it is the family of circuits of a hereditary system on $[6]$. If it is, determine whether the system is a graphic matroid.
    (a) $\mathbf{C} = \{\{1,2,3\}, \{3,4,5\}, \{5,6,1\}\}$.
    (b) $\mathbf{C} = \{\{1,2,3\}, \{3,4,5\}, \{1,2,4,5\}\}$.

11.3.3. $(-)$ Characterize the G graphs that yield matroids as follows:
    (a) The stable sets of G are the independent sets of a matroid on $V(G)$.
    (b) The matchings in G are the independent sets of a matroid on $E(G)$.

11.3.4. $(-)$ Show that the stable sets of a graph need not be the independent sets of a matroid by finding vertex-weighted graphs where the ratio between the set found by the greedy algorithm and the maximum weight of a stable set is arbitrarily large.

11.3.5. $(-)$ Explain how to obtain the rank function of a hereditary system from the bases.

11.3.6. $(-)$ Let B be a base of a matroid. For $e \in \overline{B}$, prove that $B - f + e$ is a base if and only if f belongs to the circuit formed by adding e to B.

11.3.7. $(-)$ Let e be an element of a circuit C in a matroid. Prove that C is the unique circuit created by adding e to some base.

11.3.8. $(-)$ Prove the following implications directly for hereditary systems.
    (a) The augmentation (I) and uniformity (U) properties are equivalent.
    (b) The uniformity (U) and base exchange (B) properties are equivalent.
    (c) Submodularity (R) implies weak absorption (A).
    (d) Strong absorption (A′) implies base exchange (B).
    (e) Augmentation (I) implies weak elimination (C).

11.3.9. $(-)$ Prove that if $r(X) = r(X \cap Y)$ for some $X, Y \subseteq E$ in a matroid on E, then $r(X \cup Y) = r(Y)$.

11.3.10. $(-)$ A set of $|E| - r(E)$ circuits of a matroid on E form a **fundamental set of circuits** if the elements $e_1, \dots, e_n$ can be ordered so that the last element of C_i is $e_{r(E)+i}$. Prove that every matroid has a fundamental set of circuits. (Whitney [1935])

11.3.11. $(-)$ Describe the circuits of a partition matroid M. Use this description to prove directly that partition matroids satisfy the weak elimination property.

11.3.12. $(-)$ Prove that every partition matroid is a transversal matroid.

11.3.13. $(-)$ Determine whether the cycle matroid of G below is a transversal matroid.

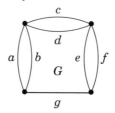

11.3.14. (–) Let B_1 and B_2 be bases of a matroid such that $|B_1 \triangle B_2| = 2$. Prove that there is a unique circuit C such that $B_1 \triangle B_2 \subseteq C \subseteq B_1 \cup B_2$.

11.3.15. (–) Prove that the cycle matroid of a graph G is the column matroid over $\mathbb{Z}_2$ of the vertex-edge incidence matrix of G.

11.3.16. (–) Use matroid duality to prove Euler's Formula for plane graphs.

11.3.17. (–) Let M be the hereditary system on $E(K_n)$ whose independent sets are the edge sets of planar graphs. Determine whether M is a matroid.

11.3.18. (–) Determine whether a set can be a circuit and a cocircuit in the same matroid.

11.3.19. (–) Let C and C^* be a circuit and a cocircuit in a matroid on n elements. Determine the minimum and maximum possible values of $|C| + |C^*|$.

11.3.20. (–) Let M be a matroid on E, and fix $A \subseteq E$. Let $\mathbf{I}'$ be the family of sets $X \subseteq E$ such that $X \in \mathbf{I}$ and $X \cap A = \varnothing$. Prove that $\mathbf{I}'$ is the family of independent sets of a matroid.

11.3.21. (–) Let r and σ be the rank function and span function of a matroid. Prove that $r(X) = \min\{|Y| \colon Y \subseteq X \text{ and } \sigma(Y) = \sigma(X)\}$.

11.3.22. (–) Prove that a matroid of rank r has at least 2^r closed sets. (Lazarson [1957])

11.3.23. (–) Let G be an n-vertex graph, and let $E_1, \ldots, E_{n-1}$ be a partition of $E(G)$ into $n - 1$ sets. Show that matroids can be used to test whether G has a spanning tree with exactly one edge in each subset E_i.

11.3.24. (–) Given matroids $M_1, \ldots, M_k$ on E, the **Matroid Partition Problem** asks whether an input set $X \subseteq E$ partitions into sets $I_1, \ldots, I_k$ with $I_i \in \mathbf{I}_i$. Prove that X is partitionable if and only if $|X - Y| + \sum r_i(Y) \geq |X|$ for all $Y \subseteq X$.

11.3.25. (–) Use HAMILTONIAN PATH in directed graphs to prove that 3-MATROID INTERSECTION is NP-complete.

11.3.26. (–) Use 3-D MATCHING to prove that 3-MATROID INTERSECTION is NP-complete. Given disjoint sets V_1, V_2, V_3 and a family of triples that each consists of one element from each V_i, 3-D MATCHING is the problem of finding the maximum number of triples that together use each element at most once. (In this terminology, ordinary bipartite matching is 2-D MATCHING.)

11.3.27. *Graphic matroids.*
    (a) Determine which graphic matroids are also uniform matroids.
    (b) Determine which graphic matroids are also partition matroids.

11.3.28. (◇) The **Fano matroid** is the vector matroid of the matrix below over the 2-element field $\mathbb{F}_2$. Prove that the Fano matroid is neither a graphic matroid nor a transversal matroid. (Hint: Use the circuits.)

$$1\ 0\ 0\ 1\ 0\ 1\ 1$$
$$0\ 1\ 0\ 1\ 1\ 1\ 0$$
$$0\ 0\ 1\ 0\ 1\ 1\ 1$$

11.3.29. (◇) Let E be the edge set of a graph G. Say that a set $X \subseteq E$ is *weakly acyclic* if the spanning subgraph of G with edge set X has at most one cycle. Prove that the weakly acyclic sets are the independent sets of a matroid on E.

11.3.30. Let s and t be vertices in a digraph D. Let $E = V(D) - \{s, t\}$. For $X \subseteq E$, let $r(X)$ be the number of edges from $s \cup X$ to $\overline{X} \cup t$. Prove that r is submodular.

11.3.31. ($\diamond$) *Submodularity for cycle matroids.* Given a graph G, let $k(X)$ be the number of components of the spanning subgraph G_X with edge set X. Let U and V be the sets of components of G_X and G_Y, respectively. Let H be the U, V-bigraph with U_i adjacent to V_j if U_i in G_X and V_j in G_Y have a common vertex. Relate the numbers of vertices, components, and edges in H to the numbers $k(X)$, $k(Y)$, $k(X \cap Y)$, $k(X \cup Y)$, and conclude directly that the rank function of the cycle matroid of G is submodular. (Aigner [1979])

11.3.32. Without using other characterizations of matroids, prove directly that the base exchange and dual base exchange properties are equivalent.

11.3.33. ($\diamond$) Prove the following implications directly for hereditary systems.
    (a) The base exchange property (B) implies the augmentation property (I).
    (b) The augmentation property (I) implies the absorption property (A).
    (c) The strong absorption property (A′) implies the submodularity property (R) (without using uniformity).

11.3.34. Using only linear dependence, prove that vector matroids satisfy the induced circuit property: adding an element to a linearly independent set of vectors creates at most one minimal dependent set.

11.3.35. ($\diamond$) Prove the following implications directly for hereditary systems.
    (a) The base exchange property (B) implies the induced circuit property (J).
    (b) The induced circuit property (J) implies the augmentation property (I).
    (c) The induced circuit property implies the weak elimination property (C).

11.3.36. Prove that a hereditary system is a matroid if and only if it satisfies the following "ultra-weak" augmentation property: If $I_1, I_2 \in \mathbf{I}$ with $|I_2| > |I_1|$ and $|I_1 - I_2| = 1$, then $I_1 + e \in \mathbf{I}$ for some $e \in I_2 - I_1$. (Chappell)

11.3.37. Given a matroid on a set E, let $C_1, \ldots, C_k$ be circuits such that none is contained in the union of the others. Let X be a subset of E with $|X| < k$. Prove that $\bigcup_{i=1}^{k} C_i - X$ contains a circuit. (Welsh [1979])

11.3.38. ($\diamond$) Let e be an element in a matroid M. Prove that if C is a circuit in $M \cdot e$, then C or $C + e$ is a circuit in M.

11.3.39. A **refinement** of a matroid M is a matroid N on the same elements such that every circuit of M is a circuit of N. Prove that M has a refinement N different from M if and only if no circuit of M has size $r(M) + 1$.

11.3.40. ($\diamond$) Let B_1 and B_2 be bases of a matroid M.
    (a) Prove that for each $e \in B_1$, there exists $f \in B_2$ such that $B_1 - e + f$ and $B_2 - f + e$ are bases of M. (Brualdi [1969])
    (b) Use the cycle matroid $M(K_4)$ to show that there may be no bijection $\pi \colon B_1 \to B_2$ such that setting $f = \pi(e)$ satisfies part (a) for all $e \in B_1$.
    (c) Given $X_1 \subseteq B_1$, prove that for some $X_2 \subseteq B_2$ both $(B_1 - X_1) \cup X_2$ and $(B_2 - X_2) \cup X_1$ are bases. (Brylawski [1973], Greene [1973], Woodall [1974], Greene–Magnanti [1975])

11.3.41. (+) Let B_1 and B_2 be bases of a matroid M. Prove that there is a bijection $\pi \colon B_1 \to B_2$ such that for each $e \in B_1$, the set $B_2 - \pi(e) + e$ is a base of M. (Brualdi [1969]) (Hint: Define a B_1, B_2-bigraph by making $e \in B_1$ and $f \in B_2$ adjacent when $B_2 + e - f \in \mathbf{B}$.)

11.3.42. ($\diamond$) Let M be a matroid on a set E, and let $w \colon E \to \mathbb{N}_0$. Use the greedy algorithm to prove a min-max formula for maximum weighted independent set: $\max_{I \in \mathbf{I}} \sum_{e \in I} w(e) = \min \sum_i r(X_i)$, where the minimum is over all chains (by inclusion) of sets in E such that each element $e \in E$ appears in at least $w(e)$ sets in the chain (sets may repeat in chains).

11.3.43. ($\diamond$) *Circuits and cocircuits.* Consider a matroid M and its dual M^*.

(a) **Dual augmentation property**. Given disjoint sets X and X^* with $X \in \mathbf{I}$ and $X^* \in \mathbf{I}^*$, prove that there are disjoint sets B and B^* with $X \subseteq B \in \mathbf{B}$ and $X^* \subseteq B^* \in \mathbf{B}^*$.

(b) Let e be an element of a base B. Prove that M has exactly one cocircuit disjoint from $B - e$ and that it contains e.

(c) Prove that the cocircuits of M are the minimal nonempty sets C^* such that $|C^* \cap C| \neq 1$ for every $C \in \mathbf{C}$.

(d) For distinct elements x and y of a circuit C, prove that there is a cocircuit C^* such that $C^* \cap C = \{x, y\}$. (Minty [1966])

11.3.44. The k-**truncation** M_k of a matroid M is defined by $\mathbf{I}_{M_k} = \{X \in \mathbf{I}_M : |X| \leq k\}$.

(a) Prove that M_k is a matroid.

(b) Prove that a matroid can be covered by t independent sets of size at most k if and only if $\max_{|X| \geq 1} \frac{|X|}{\min\{k, r(X)\}} \leq t$.

(c) Prove that a matroid of rank at least k with ground set E has t pairwise disjoint independent sets of size k if and only if $\min_{X : r(X) < k} \frac{|E| - |X|}{k - r(X)} \geq t$. (Chen–Lai [1996])

11.3.45. The k-**elongation** of a matroid M is the hereditary system M^k whose bases are the spanning sets of M with size k.

(a) Prove that M^k is a matroid.

(b) Prove that $(M_k)^* = (M^*)^{|E|-k}$ if $k \leq r(M)$, where M_k is the k-truncation of M defined in Exercise 11.3.44. (Welsh [1979])

11.3.46. Prove that a matroid has no circuits of size at most 2 if and only if (1) no element is in every hyperplane, and (2) for any two elements some hyperplane contains exactly one.

11.3.47. Let M be a matroid on E. For $F \subseteq E$, prove that $\mathbf{C}_{M|F} = \{C \subseteq F : C \in \mathbf{C}_M\}$ and that $\mathbf{C}_{M.F}$ is the set of minimal nonempty members of $\{C \cap F : C \in \mathbf{C}_M\}$.

11.3.48. By duality and matroid restriction, prove $r_{M.F}(X) = r_M(X \cup \overline{F}) - r_M(\overline{F})$.

11.3.49. ($\diamond$) Prove that any minor of a matroid obtained by restricting and then contracting can also be obtained by contracting and then restricting. In particular, if M is a matroid on E and $Y \subseteq X \subseteq E$, prove that $(M|X).Y = (M.\overline{X - Y})|Y$ and $(M.X)|Y = (M|\overline{X - Y}).Y$. (Hint: Use the rank function.)

11.3.50. ($\diamond$) Prove that the properties in Theorem 11.3.40 (involving the span function) are equivalent and characterize matroids: incorporation (P), idempotence (S), transitivity of dependence (T), and strong elimination (C'). (Hint: Prove U $\Rightarrow$ P $\Rightarrow$ S $\Rightarrow$ T $\Rightarrow$ C' $\Rightarrow$ C.)

11.3.51. Prove the following properties of closed sets of a matroid.

(a) The closed sets are the complements of the unions of cocircuits.

(b) The intersection of two closed sets is closed.

(c) The span of a set is the intersection of all closed sets containing it.

(d) The union of two closed sets need not be a closed set.

11.3.52. (+) Prove directly that in a hereditary system, the weak elimination property implies the strong elimination property, using induction on $|C_1 \cup C_2|$. (Lehman [1964])

11.3.53. Given a matroid M on E and $e \in E$, the **Shannon switching game** (M, e) played by Spanner and Cutter is as follows. On each round Cutter deletes an element of $E - e$, and then Spanner seizes an element of $E - e$. Spanner wants a set that spans e; Cutter aims to prevent this. Prove that Spanner has a winning strategy when there are disjoint subsets X_1, X_2 of $E - e$ such that $e \in \sigma(X_1) = \sigma(X_2)$. (Lehman [1964]) (Comment: A lengthy proof using the Matroid Union Theorem (Theorem 11.3.52) shows that this sufficient condition is also necessary. A special case using the cycle matroid for a union of two edge-disjoint trees was marketed by Hasbro under the name "Bridg-It"; see West [2001, p.74].)

11.3.54. (+) Given a matroid M, the **base exchange graph** $\beta(M)$ has a vertex for each base of M, with two bases adjacent when their symmetric difference has size 2. Prove that $\beta(M)$ is Hamiltonian when M has at least three bases. (Hint: Prove the stronger statement that $\beta(M)$ has a spanning cycle through any edge.) (Holzmann–Harary [1972])

11.3.55. Use the formula $r(X) = \min_{J\subseteq[m]}\{|A(J)\cap X| + m - |J|\}$ for the rank function of the transversal matroid on E induced by subsets $A_1,\ldots,A_m$ (Example 11.3.46) to prove directly that the rank function satisfies $r(\varnothing) = 0$ and $r(X) \le r(X + e) \le r(X) + 1$.

11.3.56. (◊) Given a bipartite graph G with E being one part, let M be the transversal matroid on E whose independent sets are the subsets of E that can be covered by matchings in G. By Ore's Defect Formula (see Example 11.3.46), $r(X) = \min_{S\subseteq X}\{|N(S)| + |X - S|\}$. Prove directly that r is submodular.

11.3.57. Prove that restrictions and unions of transversal matroids are transversal matroids, but contractions and duals of transversal matroids need not be.

11.3.58. (◊) Let M be the transversal matroid on E induced by subsets $A_1,\ldots,A_m$. Use the Matroid Union Theorem to prove $r_M(X) = \min_{Y\subseteq X}\{|X - Y| + |N(Y)|\}$.

11.3.59. (◊) *Common independent and spanning sets.*
    (a) For matroids M_1 and M_2 on a set E, prove $|I| + |S| = r_1(E) + r_2(E)$, where I is a largest common independent set and S is a smallest common spanning set.
    (b) Let G be an n-vertex bipartite graph with no isolated vertices. Prove $\alpha'(G) + \beta'(G) = n$ (Gallai's Theorem, Exercise 6.1.50).
    (c) Let G be a bipartite graph with no isolated vertices. Without using other results, use part (a) directly to prove $\alpha(G) = \beta'(G)$ (König's Other Theorem, Theorem 6.1.16).

11.3.60. (◊) Use the Matroid Intersection Theorem to prove that in every acyclic digraph, the vertices can be covered by at most k pairwise disjoint paths, where k is the independence number of the underlying graph. (Chappell)

11.3.61. Given matroids M_1 and M_2 whose families of spanning sets are $\mathbf{S}_1$ and $\mathbf{S}_2$, prove that the matroid $(M_1^* \cup M_2^*)^*$ is the hereditary system whose spanning sets are $\{X_1 \cap X_2 : X_1 \in \mathbf{S}_1, X_2 \in \mathbf{S}_2\}$.

11.3.62. *Matroid Intersection from Matroid Union.*
    (a) Without the Matroid Intersection Theorem, prove that the maximum size of a common independent set in matroids M_1 and M_2 on E is $r_{M_1\cup M_2^*}(E) - r_{M_2^*}(E)$.
    (b) Prove the Matroid Intersection Theorem by applying Matroid Union to $M_1 \cup M_2^*$.

11.3.63. (◊) *Connectivity and spanning trees.*
    (a) Let F be a set of at most k edges in a $2k$-edge-connected graph G. Use Corollary 11.3.56 to prove that $G - F$ has k edge-disjoint spanning trees. (Nash-Williams [1961])
    (b) For each k, construct a $(2k - 1)$-edge-connected graph that does not have k edge-disjoint spanning trees.

11.3.64. *Colored trees and b-detachments.*
    (a) Let G be a connected edge-colored graph with color classes $E_1,\ldots,E_k$. Prove that G has a spanning tree with distinct colors if and only if $G - F$ has at most $t + 1$ components whenever F consists of t color classes. (Hint: Use the Matroid Intersection Theorem.)
    (b) A *split* replaces a vertex x with two new vertices x_1 and x_2 whose neighborhoods in the new graph partition $N(x)$. Given $b\colon V(G) \to \mathbb{N}$, a *b-detachment* of a graph G arises by performing splits until there are $b(v)$ copies of each vertex v. Use part (a) to prove that G has a connected b-detachment if and only if for all $U \subseteq V(G)$, graph $G - U$ has at most $f(U) + 1 - b(U)$ components, where $b(U) = \sum_{v\in U} b(v)$ and $f(U)$ is the number of edges incident to U. (Schrijver [2003, p. 704]; see Nash-Williams [1985] for a more general result.)

Chapter 12

Partially Ordered Sets

In a totally ordered set, such as the natural numbers under "$\leq$", any two elements are related. Other orderings are "partial"; examples include the divisibility relation on positive integers and the inclusion relation on subsets of a set. Partially ordered sets can model precedence constraints in scheduling, preferences among objects, partial information about numbers being sorted, etc.

12.1. Structure of Posets

DEFINITIONS AND EXAMPLES

12.1.1. Example. The subsets of a finite set, ordered by inclusion, form a natural poset. We spent most of Chapter 11 studying aspects of it. Elementary understanding of containment yields three natural properties: (1) $A \subseteq A$, (2) $A \subseteq B$ and $B \subseteq A$ together imply $A = B$, and (3) If $A \subseteq B$ and $B \subseteq C$, then $A \subseteq C$.

Similarly, the divisibility relation on the divisors of an integer N satisfies: (1) $x \mid x$, (2) $x \mid y$ and $y \mid x$ imply $x = y$, and (3) If $x \mid y$ and $y \mid z$, then $x \mid z$. ∎

12.1.2. Definition. A **relation** R on a set X is a subset of the cartesian product $X \times X$. We write $(x, y) \in R$ or xRy or say (x, y) *satisfies* R. An **order relation** (or **partial order**) on X is a relation that is

    reflexive (xRx for all x),

    antisymmetric (xRy and yRx imply $x = y$), and

    transitive (xRy and yRz imply xRz).

A **partially ordered set** (or **poset**) is a set P plus an order relation on P. When (x, y) satisfies the relation, we write $x \leq_P y$ or simply $x \leq y$. If $x \leq y$ or $y \leq x$, then x and y are **comparable** (or *related*) in P; otherwise they are **incomparable** (or *unrelated*). All of $\leq, <, \geq, >$ describe ways that elements can be comparable. We write $x \| y$ to mean that x and y are incomparable.

The word "poset" illustrates the evolution of terminology. Originally it was written as "PO-set", emphasizing grammatically that a poset is a set equipped with a partial order.

Like any binary relation, an order relation is specified by a 0, 1-matrix, with 1 in position (x, y) if $x \leq y$. To facilitate study, we seek visual representations. Since an order relation on P is a set of ordered pairs from P, we can view it as a directed graph with vertex set P. Ignoring which way pairs are ordered treats the edges as unordered pairs.

12.1.3. Definition. The **comparability digraph** of a poset P is the digraph whose vertices are the elements of P and whose edges are the ordered pairs xy such that $x \leq_P y$. The **comparability graph** is the graph whose vertices are the elements of P and whose edges are the unordered pairs of distinct vertices that are comparable in P.

12.1.4. Remark. *Comparability graphs and digraphs.* The comparability digraph specifies a poset completely. Since order relations are reflexive, a comparability digraph has a loop at each vertex. Since this is understood, we omit the loops when drawing a comparability digraph and just draw the digraph of the **strict order relation**: the pairs xy with $x < y$.

The comparability graph discards information and does not specify the poset. Below we show distinct posets with the same comparability graph. Reversing all comparable pairs also yields another poset with the same comparability graph.

A comparability graph becomes a comparability digraph by transitively orienting the edges. A **transitive orientation** of a graph G is an orientation such that if xy and yz are (directed) edges, then xz is an edge; that is, x and z are adjacent in G and the orientation directs the edge from x to z. The transitive orientations of G correspond to the posets for which G is the comparability graph.

Exercise 6 gives a necessary condition for comparability graphs that turns out also to be sufficient. Another characterization in Exercise 7 leads to a fast algorithm for testing whether a graph is a comparability graph. Nevertheless, our focus in this chapter is on the order-theoretic properties of posets, so we will not study comparability graphs further.   ■

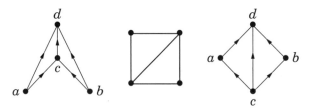

For a useful visual presentation, transitivity of the order relation allows us to describe a poset completely while drawing only some of the comparable pairs.

12.1.5. Definition. If $x < y$ in P and there is no z with $x < z < y$, then y **covers** x (in P), written as $x \lessdot y$ or $y \gtrdot x$. The **cover relation** is the set of pairs (x, y) such that $x \lessdot y$. The **cover digraph** is the digraph on the elements of P whose edge set is $\{xy \colon x \lessdot y\}$.

A **cover diagram** (or **Hasse diagram** or **diagram**) of P is obtained by erasing the directions on edges after drawing the cover digraph in the plane such that each (straight-line) edge points upward. The **cover graph** is the graph on the elements of P whose edge set is $\{xy \colon x \lessdot y$ or $y \lessdot x\}$.

12.1.6. Example. *The diagram.* The comparability digraphs in Remark 12.1.4 direct all edges upward. Erasing the edges implied by transitivity yields the cover digraphs. Since the edges all point upward, erasing the arrowheads then yields the diagrams, which appear on the left below. One of these is the poset of subsets of $\{1, 2\}$, ordered by inclusion.

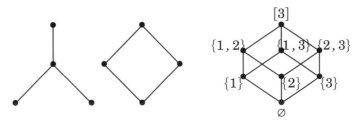

On the right is the poset of subsets of $[3]$, ordered by inclusion. The diagram specifies the poset completely; $x < y$ in P if and only if we can reach y from x in the diagram by moving only upward along edges. Omitting the edges implied by transitivity makes it easier to see the structure. By the convention that the edges for the order relation point upward, the diagram provides the cover digraph and, by transitive closure, the comparability digraph.   ∎

The cover graph (but not the diagram) discards the order information and does not specify the poset. For example, C_4 is the cover graph of two posets (one shown above), and C_6 is the cover graph of seven posets (Exercise 3). In contrast to comparability graphs, testing whether a graph is a cover graph is NP-complete (Nešetřil–Rödl [1987, 1993]; Brightwell [1993] gave a simpler proof).

The simplest posets are totally ordered or totally unordered. Indeed, we say "partial order" as a generalization of total order.

12.1.7. Definition. A **chain** is a poset whose elements are linearly ordered so that $x < y$ if and only if x comes before y in that order. The chain with k elements is denoted **k** and called a k-**chain**. An **antichain** is a poset in which no two elements are comparable.

An element of a poset is **maximal** if no other element is greater than it, **minimal** if no other element is less than it.

The first poset in Remark 12.1.4 has one maximal element (d) and two minimal elements (a and b). A k-chain has one maximal and one minimal element. In an antichain, every element is maximal and minimal. The cover graph of **k** is the path P_k; its comparability graph is K_k.

12.1.8. Definition. Posets P and Q are **isomorphic** if some bijection from the set P to the set Q preserves the order relation. A **subposet** of a poset P is a poset R on a subset of P defined by restricting the comparability relation to R. If Q is isomorphic to a subposet of P, then P **contains** Q or Q **embeds** in P.

The **dual** of P, written P^*, is the poset on the elements of P defined by $y \leq_{P^*} x$ if and only if $x \leq_P y$. A poset isomorphic to its dual is **self-dual**. A poset is **finite** if it has finitely many elements.

The notion of isomorphism allows us to view a diagram of a poset as the poset itself, just as a drawing is a graph. We use **k** for a k-element chain, as an isomorphism class, just as C_k denotes a k-vertex cycle.

12.1.9. Example. *Subposets.* If $n \leq m$, then **n** embeds in **m**, and all n-element subposets of **m** are isomorphic to **n**.

We write $\mathbf{2}^n$ for the inclusion poset on the subsets of $[n]$, rather than $\mathbf{2}^{[n]}$ as in Section 11.3; we will soon explain why. In $\mathbf{2}^3$ (Example 12.1.6) we find six chains of size 4, two antichains of size 3, and 15 subposets isomorphic to $\mathbf{2}^2$ (Exercise 2). Furthermore, $\mathbf{2}^3$ is self-dual, via complementation of subsets of $[3]$.

A chain in a poset is a set of pairwise comparable elements; an antichain is a set of pairwise incomparable elements. They become cliques and independent sets in the comparability graph. When $x_1, \ldots, x_k$ is a chain in P, with $x_1 < \cdots < x_k$, *it need not hold* that x_{i+1} covers x_i. For example, $\mathbf{2}^3$ contains 19 chains of size 2. ∎

12.1.10. Remark. *Subposet* in posets corresponds to *induced subgraph* in graph theory. When G is the comparability graph of P, and Q is a subposet of P, the comparability graph of Q is the subgraph of G induced by the set Q. However, the cover graph of Q need not be a subgraph of the cover graph of P.

A graph consists of a vertex set and an adjacency relation on it. Similarly, a poset consists of a set of elements with an order relation on it. Nevertheless, we generally use the same notation (P) for a partially ordered set and for its set of elements. This abuse of notation works because we treat P as a partially ordered *set* and study as "subposets" only the structures that inherit all comparabilities.

For graphs we manipulate both vertices and edges, so we write $v \in V(G)$ and $e \in E(G)$ instead of $v \in G$ and $e \in G$ to avoid ambiguity. For posets, we just write $x \in P$ when x is in the set of elements, and we use $|P|$ for the **size** of that set. We rarely consider analogues of edge deletion or contraction for posets. ∎

12.1.11. Definition. Let P be a poset. Its **width** $w(P)$ is the size of a largest antichain in P. Its **height** $h(P)$ is the size of a largest chain in P. Its **length** is one less than its height.

12.1.12. Example. The poset $\mathbf{2}^3$ has width 3, height 4, and length 3. An antichain of size $w(P)$ in P is a **maximum antichain**, but a **maximal antichain** (one not contained in a larger antichain) may be smaller. For example, the antichain $\{\{1\}, \{2,3\}\}$ in $\mathbf{2}^3$ is a maximal antichain but not a maximum antichain.

A chain in a poset with finite height is a **maximal chain** if and only if it extends from a minimal element to a maximal element and its successive pairs satisfy the cover relation. For elements, "maximal" and "minimal" refer to the order relation in P, not to containment on sets of elements. ∎

12.1.13. Definition. In a poset P, a **down-set** is a subset D such that $x \in D$ and $y < x$ imply $y \in D$. The complement of a down-set is an **up-set**. The down-set $D[x]$ **generated by** x is $\{y \in P: y \leq x\}$. The down-set $D[A]$ generated by a family A is $\{y: y \leq x$ for some $x \in A\}$. We also write $D(x) = \{y \in P: y < x\}$. Similarly define $U[x]$, $U[A]$, and $U(x)$.

Down-sets have also been called **ideals** or **order ideals**, and up-sets have been called **dual ideals** or **filters**. The terms "down-set" and "up-set" are less sophisticated, but they are explicit and do not conflict with other uses of "ideal" and "filter" in mathematics. In Chapter 11, we used "ideal of sets" for a down-set in 2^n because this is common and is consistent with algebraic notions.

In 2^3, the down-set generated by $\{1, 23\}$ is $\{1, 23, 2, 3, \varnothing\}$. There are exactly 20 antichains and 20 down-sets in 2^3; one of each is empty.

12.1.14. Remark. *There is a one-to-one correspondence between the antichains and the down-sets in any poset.* The natural bijection maps an antichain A to the down-set $D[A]$. The inverse assigns to each down-set the antichain consisting of its maximal elements. The empty antichain corresponds to the empty down-set. ∎

The product operation is a fundamental way to combine posets.

12.1.15. Definition. The **product** $P \times Q$ of two posets P, Q is the poset on $\{(x, y): x \in P, y \in Q\}$ defined by $(x, y) \leq (x', y')$ if and only if $x \leq x'$ and $y \leq y'$. The **sum** $P + Q$ consists of disjoint copies of P and Q with no comparabilities between them.

P

Q

$P \times Q$

12.1.16. Remark. *Products of posets.* The product $P \times Q$ has disjoint copies of Q for each element of P and disjoint copies of P for each element of Q. If the cover graphs of P and Q are G and H, then the cover graph of $P \times Q$ is the cartesian product $G \square H$. However, the comparability graph of $P \times Q$ is not the cartesian product of the comparability graphs of P and Q.

In the sense of isomorphism, the operation is commutative and associative, and we write the product of n copies of P as P^n. ∎

12.1.17. Example. *The subset poset.* The poset **2** is a 2-element chain; call its elements 0 and 1, with $0 < 1$. With each element of $[n]$, associate a copy of **2**. Let P be the poset of subsets of $[n]$, ordered by inclusion. Mapping each member of P to its incidence vector expresses P as 2^n. Thus we denote P as 2^n. (We suggest pronouncing 2^n as "2 sup n".) For the inclusion poset on the subsets of a set X, we write 2^X and call X the **ground set**. The poset 2^n is also called the **Boolean algebra** on n elements, sometimes denoted B_n.

Since 2^n is the product of n copies of **2**, its cover graph is the n-dimensional cube. Note that if x is a k-set and y is an l-set in $[n]$ with $x < y$, then the subposet of 2^n consisting of $\{z: x \leq z \leq y\}$ is isomorphic to 2^{l-k}. Also, 2^n is self-dual (via complementation of subsets of $[n]$). Finally, when writing elements of 2^n as sets or as their incidence vectors, we typically drop set brackets, commas, and parentheses and just write strings of elements. ∎

DILWORTH'S THEOREM AND BEYOND

The archetypal extremal problem for posets is finding a largest antichain. Explicit answers exist only for special classes; we discuss these later. First, we obtain a min-max relation. Since an antichain has no two elements on any chain, the width is bounded by the number of chains needed to cover the elements. Dilworth proved that equality holds.

12.1.18. Theorem. (**Dilworth's Theorem**; Dilworth [1950]) If P is a finite poset, then the maximum size of an antichain in P equals the minimum number of chains needed to cover the elements of P.

Proof: (Perles [1963]) We use induction on $|P|$, with trivial basis. For $|P| > 1$, suppose first that some largest antichain A omits both a maximal element and a minimal element of P. Thus neither the down-set $D[A]$ nor the up-set $U[A]$ generated by A contains all of P, and we can apply the induction hypothesis to each subposet. Also, they share only A.

Let $k = w(P)$. Since both $D[A]$ and $U[A]$ are subposets of P, they have width at most k. Since they both contain A, equality holds. From the induction hypothesis, we obtain k chains covering $D[A]$ and k chains covering $U[A]$, the former with elements of A at the top and the latter with elements of A at the bottom. These combine to form k chains. These chains cover all of P, because the maximality of A implies that $D[A] \cup U[A] = P$. This case is shown on the left below.

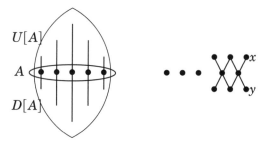

In the remaining case, every maximum antichain of P consists of all maximal elements or all minimal elements. Thus $w(P - \{x, y\}) \le k - 1$ if x is a minimal element and y is a maximal element. Choose a maximal element x and a minimal element y below it (they may be equal). Since $P - \{x, y\}$ is smaller (and has width $k - 1$), we can apply the induction hypothesis to cover $P - \{x, y\}$ with $k - 1$ chains and add the chain consisting of x and y to complete the desired covering.   ∎

Since subposets of chains are chains, we conclude that P is covered by $w(P)$ *pairwise disjoint* chains. A partition of P into chains is a **chain decomposition** of P. In honor of Theorem 12.1.18, a smallest chain decomposition (size $w(P)$) is called a **Dilworth decomposition** of P.

Perles' proof is like Pym's proof of Menger's Theorem. Each uses splicing of paths, with a special argument for a degenerate case. Here we seek a maximum cut (the antichain) instead of a minimum cut. In the language of network flows, the arguments are essentially equivalent.

Meanwhile, we relate Dilworth's Theorem to the min-max relations of graph theory. A poset is **bipartite** if every element is minimal or maximal (or both); equivalently, the comparability graph is bipartite.

12.1.19. Theorem. (Fulkerson [1956]) Dilworth's Theorem is equivalent to the König–Egerváry Theorem on matchings in bipartite graphs.

Proof: To apply Dilworth's Theorem to bipartite matching, view an n-vertex bipartite graph G as a bipartite poset. The vertices of one part become maximal elements; the others are minimal. Chains have one or two elements. Every covering of the poset by $n - k$ chains uses k disjoint 2-chains; these yield a matching of size k in G. Each antichain corresponds to an independent set in G; if it is maximal, then the remaining vertices form a vertex cover. Hence Dilworth's equality between the sizes of a maximum antichain and minimum chain-covering yields a matching and a vertex cover of the same size in G. Since every vertex cover is at least as large as every matching, this proves the König–Egerváry Theorem.

For the converse implication, consider any poset P of size n. We define a bipartite graph $S(P)$, the **split** of P, in which to study matchings. For each element $x \in P$, create two vertices x^- and x^+ (see figure below). The parts of the bipartition of $S(P)$ are $\{x^-: x \in P\}$ and $\{x^+: x \in P\}$. The edge set is $\{x^-y^+: x <_P y\}$.

Every matching in $S(P)$ yields a chain-covering in P as follows: if x^-y^+ is in the matching, then y is immediately above x on a chain in the cover. If x^- or x^+ is unmatched, then x is the top or bottom of its chain, respectively. Since each vertex of $S(P)$ appears in at most one edge of the matching, this defines disjoint chains covering P. If the matching has k edges, then the cover has $n - k$ chains, since each additional edge links the top of one chain with the bottom of another to form a single chain.

Given a vertex cover R of $S(P)$, let $A = \{x \in P: x^-, x^+ \notin R\}$. A relation between elements of A would yield an edge of $S(P)$ uncovered by R, so A is an antichain. No *minimal* vertex cover of $S(P)$ uses both of $\{x^+, x^-\}$, because by transitivity the sets $\{z^-: z \in D(x)\}$ and $\{y^+: y \in U(x)\}$ induce a complete bipartite subgraph in $S(P)$. A vertex cover of a complete bipartite graph must use all of one part. Since x^+ and x^- have no other neighbors, we can drop from R the one of $\{x^+, x^-\}$ that is not in that part.

Thus a minimum vertex cover of size k yields an antichain of size $n - k$. Now the existence of a matching and a vertex cover in $S(P)$ of equal size yields an antichain and a chain-covering of equal size in P. ∎

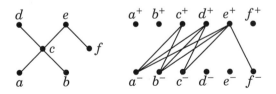

12.1.20. Remark. A simpler min-max relation holds for maximum chains and minimum antichain coverings (Mirsky [1971]). The antichain of maximal elements intersects each maximal chain, so by induction on $h(P)$ we can cover P with $h(P)$ antichains. Via the Perfect Graph Theorem (Corollary 8.3.35), this easy observation *implies* Dilworth's Theorem. ∎

Dilworth's original proof was a complicated induction, cutting and pasting chains. An appropriate generalization is simpler. Here $\alpha(D)$ denotes the independence number of the graph obtained by viewing the edges as unordered pairs.

12.1.21. Theorem. (Gallai–Milgram [1960]) The vertices of an n-vertex digraph D can be covered using at most $\alpha(D)$ disjoint paths.

Proof: Since $V(D)$ is covered by n disjoint paths of length 0, it suffices to prove a stronger claim: If $\mathbf{C}$ is a set of k disjoint paths covering $V(D)$, with $k > \alpha(G)$, and S is the set of sources (initial vertices) of these paths, then $V(D)$ can be covered using fewer than k disjoint paths with sources in S. The statement holds vacuously when $n = 1$; we proceed by induction. The added requirement about sources simplifies the induction step.

Suppose $n > 1$. Since $k > \alpha(D)$, there is an edge xy with $x, y \in S$. Let A and B be the paths in $\mathbf{C}$ starting with x and y. If A has no edge, then moving x to the beginning of B saves one path. Hence A has an edge xz. Deleting x from A yields a cover $\mathbf{C}'$ of $V(D - x)$ by k paths with sources in S', where $S' = S - x + z$. Since $\alpha(D - x) \le \alpha(D)$, the induction hypothesis yields a cover $\mathbf{C}''$ of $V(D - x)$ by fewer than k paths with sources in S'.

All of S' is in S except z. If z is a source in $\mathbf{C}''$, then prepend x to the path starting at z to cover $V(D)$ using fewer than k paths. If z is not a source but y is, then prepend x to the path starting at y. If neither y nor z is a source, then $|S'| = k$ implies that $\mathbf{C}''$ has at most $k - 2$ paths; let x be a path by itself. In all cases, the resulting paths are disjoint and have sources in S. ∎

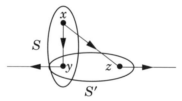

In the special case where D is the comparability digraph of a poset P, the initial covering is the set of $|P|$ trivial paths. Since $\alpha(D) = w(P)$, the paths given by Theorem 12.1.21 form the desired chain-covering, a Dilworth decomposition. The special case where D is acyclic is in Exercise 11.3.60.

Dilworth's Theorem also generalizes beyond antichains.

12.1.22. Definition. A k-**family** in a poset is a subposet containing no $(k + 1)$-chain. The k-**norm** of a set partition $C_1, \ldots, C_m$ is $\sum_{i=1}^{m} \min\{k, |C_i|\}$. A chain decomposition of a poset P is k-**saturated** if its k-norm equals the maximum size of a k-family in P.

Dilworth decompositions are 1-saturated partitions. By Mirsky's observation, k-families are just the unions of (at most) k antichains. Each chain C in a chain partition contributes at most $\min\{k, |C|\}$ to a k-family. Hence the size of a k-family is bounded by the k-norm of a chain partition.

12.1.23. Theorem. (**Greene–Kleitman Theorem**; Greene–Kleitman [1976a])
For every poset P and natural number k, there is a chain decomposition of P that is both k-saturated and $(k + 1)$-saturated. ∎

The original proof of the Greene–Kleitman Theorem was quite long. There are now shorter combinatorial proofs (Saks [1979]) and proofs applying other min-max relations (such as Frank [1980b] via minimum-cost network flow). We leave these to a more advanced book.

For acyclic digraphs, there are extensions of the Greene–Kleitman Theorem due to Linial [1981], Saks [1980], Cameron [1982], and Hoffman [1983]. Berge conjectured an extension for arbitrary digraphs. Let $\mathbf{C}$ be a partition of the vertices of a digraph D using paths. A partition with smallest k-norm is k**-optimal**. On the other hand, a **partial k-coloring** is a union of k independent sets of vertices; view each set as a color class. The number of colors on a path C is bounded by $\min\{k, |V(C)|\}$. We seek a partial k-coloring where equality holds.

12.1.24. Conjecture. (Berge [1982]) For every digraph D, integer k, and k-optimal path partition $\mathbf{C}$ of D, there is a partial k-coloring such that each path C in $\mathbf{C}$ intersects $\min\{k, |V(C)|\}$ color classes. ∎

Berge [1985] noted that the Greene–Kleitman Theorem is the special case for transitive digraphs; he proved it also for $k = 1$ in general. Cameron [1986] proved it for acyclic digraphs, Berger–Hartman [2008] proved it for $k = 2$, and Berger–Hartman [2012] gave a unified proof for various cases using network flows. The general conjecture remains open; Hartman [2006] surveyed partial results.

EXERCISES 12.1

12.1.1. (−) For $n = 3$ and $n = 4$, list the isomorphism classes of posets with n elements. Determine how many are self-dual.

12.1.2. (−) Show that $\mathbf{2}^3$ contains 15 copies of $\mathbf{2}^2$.

12.1.3. (−) Draw the diagrams of the seven posets (isomorphism classes) whose cover graph is a 6-cycle. Determine their widths.

12.1.4. (−) By Remark 12.1.20, every k-family decomposes into k antichains. Use the poset below to show that it may not be possible to obtain a maximum 2-family by adding an antichain to a maximum antichain.

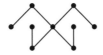

12.1.5. Prove that the graphs below are comparability graphs by exhibiting cover diagrams of posets for which they are the comparability graphs. Use the necessary condition in the next exercise (12.1.6) to prove that their complements are not comparability graphs.

12.1.6. ($\diamond$) For a closed walk $[x_1, \ldots, x_k]$ in a graph, a **triangular chord** is an edge of the form $x_i x_{i+2}$ (with indices modulo k). Prove that in a comparability graph, every closed walk of odd length has a triangular chord. Conclude that the complement of the cycle C_n is not a comparability graph when $n \geq 5$. (Comment: Gilmore–Hoffman [1964] showed that this necessary condition is also sufficient.)

12.1.7. ($\diamond$) For a graph G, the **Ghouilà-Houri graph** G' is the graph defined on the ordered pairs of adjacent vertices of G by putting $(x, y) \leftrightarrow (y, z)$ in G' if and only if $xz \notin E(G)$. Prove that G' is bipartite if and only if every closed odd walk of G has a triangular chord. (Comment: With Exercise 12.1.6, this yields a polynomial-time recognition algorithm for comparability graphs.) (Ghouilà-Houri [1962])

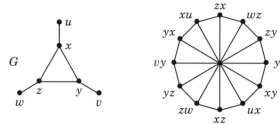

12.1.8. ($\diamond$) Prove that a graph is the cover graph of some poset if and only if it has an acyclic orientation without dependent edges, where a **dependent edge** in a digraph is an edge whose reversal completes a cycle. Conclude that if the chromatic number of a graph is less than its girth, then it is a cover graph.

12.1.9. Use Exercise 12.1.8 to prove that the **Grötzsch graph** below is not a cover graph. (Fisher–Fraughnaugh–Langley–West [1997])

12.1.10. A digraph D is a **cover digraph** if $E(D)$ is the cover relation of some poset. Derive a polynomial-time checkable characterization of cover digraphs. Assume that one can find in polynomial time all vertices reachable from a specified vertex. (Comment: No efficient algorithm is known to check whether an undirected graph is a cover graph.)

12.1.11. Obtain simple formulas (constant number of terms) for
    (a) the number of chains of size 2 and the number of chains of size 3 in $\mathbf{2}^n$.
    (b) the numbers of antichains of size 2 and size 3 in $\mathbf{2}^n$. (Popadić [1970])

12.1.12. A **cutset** in a poset is a subset intersecting every maximal chain. The family $\binom{[n]}{\lceil n/2 \rceil}$ is a minimal cutset in $\mathbf{2}^n$; is it a largest one? (Füredi–Griggs–Kleitman [1989])

12.1.13. There are 16 sentences of the form "In every poset, A B chain intersects C D antichain", where $A, C \in \{\text{some, every}\}$ and $B, D \in \{\text{maximal, largest}\}$. For each such sentence, determine whether it is true or false.

12.1.14. Suppose that the red subgraph in a red/blue-coloring of $E(K_n)$ has a transitive orientation. Prove that the coloring has a monochromatic complete subgraph with at least $\sqrt{n}$ vertices. Show that the bound is sharp.

12.1.15. ($\diamond$) Prove that a poset of size greater than mn has a chain of size greater than m or an antichain of size greater than n. Use this to prove the Erdős–Szekeres Theorem: every list of $mn+1$ distinct integers has an increasing sublist with more than m elements or a decreasing sublist with more than n elements.

12.1.16. ($\diamond$) A family of sets is **union-free** if it has no two distinct members whose union is a third member. Moser asked for $f(n)$, the maximum size of a union-free subfamily that can be guaranteed to exist in any family of n sets.
    (a) Use Dilworth's Theorem to prove $f(n) \geq \sqrt{n}$. (Riddell, Erdős–Komlós [1970])
    (b) Prove $f(n) > \sqrt{2n} - 1$. (Erdős–Shelah [1972], Kleitman [1973])
    (c) Let $A_{i,j}$ consist of the integers from $-i$ to $+j$, and let $F = \{A_{i,j} \colon (i,j) \in [t]^2\}$. Prove that the maximum size of a union-free family in F is $2t-1$. Thus $f(n) \leq 2\sqrt{n} - 1$ when n is a perfect square. (Erdős–Shelah [1972])
    (Comment: Fox–Lee–Sudakov [2012] proved $f(n) = \lfloor \sqrt{4n+1} \rfloor - 1$ for all n.)

12.1.17. Given $n_1, \ldots, n_k \in \mathbb{N}$, let P denote the poset whose elements are the k-tuples x with $x_i \in [n_i]$ for all i, ordered by $x < y$ if and only if $x_i < y_i$ for all i. Prove that every maximal antichain in P is a maximum antichain, and determine its size. (Tsai [2017])

12.1.18. Exercise 6.1.23 used Hall's Theorem to prove that if G is an X, Y-bigraph with no isolated vertices, and $d(x) \geq d(y)$ whenever $xy \in E(G)$ with $x \in X$ and $y \in Y$, then G has a matching that covers X. Use this to show that the family of subsets of $[n]$ decomposes into $\binom{n}{\lceil n/2 \rceil}$ inclusion chains, which therefore form a Dilworth decomposition of $\mathbf{2}^n$. (F. Galvin)

12.1.19. *Another proof of Dilworth's Theorem.* The poset on the right below arises from that on the left by deleting the central covering pair. All other comparabilities remain.
    (a) Suppose that x covers y and z in P. Let Q and R be the posets obtained by deleting (y, x) and (z, x), respectively, from the set of relations in P. Prove that $\min\{w(Q), w(R)\} = w(P)$. (Hint: Take maximum antichains in Q and R, and consider the maximal elements and the minimal elements of the union of these antichains as a subposet of P.)
    (b) Use part (a) to prove Dilworth's Theorem. (Harzheim [1983])

12.1.20. For each poset below and each k, find a chain partition that is both k-saturated and $(k+1)$-saturated. Is some chain partition k-saturated for all k?

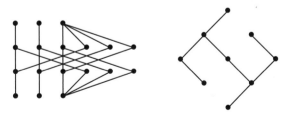

12.1.21. ($\diamond$) Let d_k be the maximum size of a k-family in a poset P, and let $\Delta_k = d_k - d_{k-1}$.
    (a) In a k-saturated chain partition $\mathbf{C}$ of P, let α be the number of chains of size at least k. Prove $\Delta_k \geq \alpha \geq \Delta_{k+1}$, (Comment: The Greene–Kleitman Theorem thus implies $\Delta_1 \geq \cdots \geq \Delta_h$, where P has height h. No direct proof is known.) (Greene–Kleitman [1976a])
    (b) Prove $\alpha = \Delta_k$ when $\mathbf{C}$ is both k-saturated and $(k-1)$-saturated.

12.1.22. A poset P is **polyunsaturated** (West [1986]) if for $1 \le k < l - 1 < h(P)$ no chain partition of P is both k-saturated and l-saturated.

(a) Use the Greene–Kleitman Theorem and Exercise 12.1.21 to prove that $h(P) \le w(P) + 2$ when P is polyunsaturated.

(b) Construct a poset P_k iteratively as follows. For $k = 1$, let $P_k = \mathbf{3}$. For $k > 1$, obtain P_k from P_{k-1} by adding a chain C of $k + 1$ new elements and making the element just below the maximal element on C cover the element just below the maximal element of the (unique) longest chain in P_{k-1}. The diagrams of P_2 and P_3 appear below. Prove that P_k is polyunsaturated. (Comment: Thus the maximum height of a polyunsaturated poset of width k is $k + 2$.) (Chappell [2002])

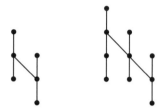

12.2. Symmetric Chains and LYM Orders

Although Dilworth's Theorem applies to all (finite) posets, much of the study of partially ordered sets concerns special families of posets. When suitable constraints are placed on posets, much more can be said about their maximum antichains and other structural aspects.

GRADED POSETS

In this section we consider only finite posets. This property makes the notion of the "rank" of an element well defined.

12.2.1. Definition. In a poset, the **rank** of an element x, written $r(x)$, is the maximum length of a chain having x as its top element. A poset P is **graded** if all its maximal chains have the same length, and its **rank** $r(P)$ is that length.

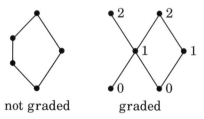

not graded   graded

In a graded poset, $r(x) = r(y) + 1$ when x covers y. The rank of the poset is the rank of its maximal elements. The notion of rank function can be extended to more general posets, but we will study only graded posets in this section.

12.2.2. Definition. If P is graded, then the elements with rank k are the kth **rank** or kth **level** P_k, and we write $N_k(P)$ for the **rank size** $|P_k|$ (also called the kth **Whitney number** of P). A graded poset is **rank-symmetric** if $N_k = N_{r(P)-k}$ for all k. It is **rank-unimodal** if there is a rank k such that $N_i \le N_j$ whenever $i \le j \le k$ or $i \ge j \ge k$. The **rank generating function** is the formal power series $\sum_{k \ge 0} N_k x^k$.

12.2.3. Example. *Subsets.* The poset $\mathbf{2}^n$ is graded, with $r(x) = |x|$ and $r(\mathbf{2}^n) = n$. The kth rank is $\binom{[n]}{k}$. The poset is rank-symmetric and rank-unimodal. Since $N_k(\mathbf{2}^n) = \binom{n}{k}$, the rank generating function is $(1 + x)^n$.

Every maximal chain in $\mathbf{2}^n$ has length n (size $n + 1$). There are maximal antichains of size 1, but for maximum antichains we will again prove Sperner's Theorem (Theorem 11.2.14) that $w(\mathbf{2}^n) = \binom{n}{\lfloor n/2 \rfloor}$. ∎

12.2.4. Example. *Divisors of an integer, or multisets.* The divisors of a positive integer N form a poset $D(N)$ under divisibility. It is graded; the rank of an element is the number of primes (with multiplicity) in its factorization. It is rank-symmetric; mapping x to N/x also shows that $D(N)$ is self-dual.

When N is a product of n distinct primes, $D(N) \cong \mathbf{2}^n$; the subsets select the prime factors. When $N = \prod_{i=1}^n p_i^{e_i-1}$, the divisors of N correspond to integer lists $(a_1, \ldots, a_n)$ with $0 \le a_i < e_i$, ordered by $a \le b$ if and only if $a_i \le b_i$ for all i.

Thus $D(N)$ can be viewed as the containment order on multisets from $[n]$ with at most $e_i - 1$ copies of the ith element, where $A \subseteq B$ if each element appears at least as many times in B as in A. Since $e_1, \ldots, e_n$ determine the multiset poset, we denote it by M^e, where $e = (e_1, \ldots, e_n)$.

The multiset description expresses the poset as a product of chains: $M^e \cong \mathbf{e_1} \times \cdots \times \mathbf{e_n}$. Study of M^e arose from divisibility questions, and some early proofs were phrased using divisibility, but the structure is completely captured by the chain-product description, and we usually view the elements as n-tuples.

For $a \in M^e$, we have $r(a) = \sum a_i$. The rank generating function enumerates multisets, indexed by size: $\prod_{i=1}^n (1 + x + \cdots x^{e_i-1})$. When the ith multiplicity is unbounded, we let $e_i = \infty$ and use $1/(1 - x)$ as the ith factor. ∎

SYMMETRIC CHAIN DECOMPOSITIONS

In light of Dilworth's Theorem, we can determine the width of a poset by exhibiting an antichain and a chain decomposition of the same size. Given a rank-symmetric rank-unimodal poset, we can hope for a special decomposition.

12.2.5. Definition. A chain in a graded poset P is **symmetric** if it has an element of rank $r(P) - k$ whenever it has an element of rank k. A chain is **consecutive** or **skipless** if its elements lie in consecutive ranks. A **symmetric chain decomposition** is a partition into symmetric skipless chains. A poset with a symmetric chain decomposition is a **symmetric chain order**.

A graded poset has the **Sperner property** if a largest-sized rank is a largest antichain. It has the **strong Sperner property** if for all k the union of any k largest ranks form a largest k-family.

12.2.6. Example. Since every chain in a symmetric chain decomposition intersects the middle rank (or both middle ranks), it is a Dilworth decomposition. A symmetric chain order must be rank-symmetric and rank-unimodal. The poset on the left fails, although it is graded and has the Sperner property. The poset on the right is graded but does not have the Sperner property. It has an antichain of size 4, but the maximum rank-size is 3. ∎

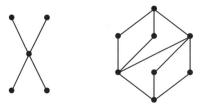

12.2.7. Proposition. Every symmetric chain order satisfies the strong Sperner property.

Proof: In a graded poset P, any k largest ranks form a k-family. Every chain C in a symmetric chain decomposition contributes exactly $\min\{k, |C|\}$ elements to the union of the k largest ranks. No chain C can contribute more than $\min\{k, |C|\}$ elements to a k-family, so these ranks form a maximum k-family. ∎

12.2.8. Theorem. (de Bruijn–Tengbergen–Kruyswijk [1951]) 2^n is a symmetric chain order.

Proof: We use induction on n. When $n = 0$, there is only one element. For $n > 0$, take two copies of a chain decomposition of 2^{n-1}. Add $\{n\}$ to each member of each chain in the second copy. This decomposes 2^n into skipless chains, but they are not symmetric and are too many when n is odd. Alter the two copies of a chain C by transferring the top element of the second copy to the top of the first copy. Since the element moved is the union of $\{n\}$ with the previous top element, the result is still skipless. All chains are now symmetric within the full poset. ∎

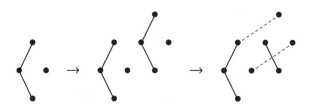

The result was proved more generally for the divisor order, solving a problem posed by the Dutch Wiskundig Genootschap in 1949. Katona observed that the method applies to any product of symmetric chain orders.

12.2.9. Theorem. (Katona [1972a]) Products of symmetric chain orders are symmetric chain orders.

Proof: We need only consider a product of two such orders. If P and Q have symmetric chain decompositions $B_1, \ldots, B_k$ and $D_1, \ldots, D_l$, then the products $B_i \times D_j$

partition $P \times Q$ into "symmetric rectangles". The product $B_i \times D_j$ has elements from ranks k through $r(P \times Q) - k$ for some k, and it has as many elements from rank j as from rank $r(P \times Q) - j$, for each j. Partitioning each such rectangle into symmetric chains provides a symmetric chain decomposition of $P \times Q$.

This reduces the problem to P and Q being chains of sizes $s + 1$ and $t + 1$, with $s \geq t$. The product consists of $\{(i, j): 0 \leq i \leq s$ and $0 \leq j \leq t\}$. Use chains $C_0, \ldots, C_t$, where C_k consists of $(0, k), (1, k), \ldots, (s - k, k), (s - k, k + 1), \ldots, (s - k, t)$. This chain is skipless; it is symmetric because the top and bottom ranks sum to $s + t$. If $i + j \leq s$, then (i, j) belongs only to chain C_j. If $i + j > s$, then (i, j) belongs only to chain C_{s-i}. Thus the chains are disjoint, as shown below. Theorem 12.2.8 is the case $t = 1$. ∎

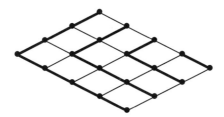

Theorem 12.2.8 is easy but does not explicitly describe the chains. How can we tell whether two given elements lie in the same chain?

12.2.10. Example. *Bracketing decomposition of $\mathbf{2}^n$.* Greene–Kleitman [1976b] and Leeb [unpublished] gave an explicit locally described symmetric chain decomposition of $\mathbf{2}^n$ (see Exercise 12 for multisets); we will see that it is the same as the decomposition in Theorem 12.2.8.

View $S \in \mathbf{2}^n$ as a binary vector x, with $x_i = 1$ if and only if $i \in S$. Encode each 0 as a left bracket and each 1 as a right bracket. Iteratively pair some positions as follows. As long as some unmatched left bracket precedes an unmatched right bracket, there is some closest such pair, say positions l and r with $l < r$. Add (l, r) to the list of paired positions. For any closest unmatched pair, all positions between them are already matched to other positions between them.

The resulting set of position pairs is the **bracketing structure** of x. Having the same bracketing structure is more restrictive than having the same set of matched positions. Below we group the sets in $\mathbf{2}^4$ by their bracketing structure; those in the first column have no paired positions, then three columns have one pair, and finally the two rightmost sets show two ways to pair all four positions.

$$
\begin{array}{llll}
1234 =)))) & & & \\
123 =)))(& 234 = ()) & 134 =)() & 124 =))() \\
12 =))((& 23 = ()(& 13 =)() & 14 =)() & 24 = ()() & 34 = (()) \\
1 =)(((& 2 = ()((& 3 = (()(& 4 = ((() & & \\
\varnothing = ((((& & & \\
\end{array}
$$

When the process ends, all unmatched rights occur before all unmatched lefts. Let C be the family of subsets of $[n]$ with a fixed bracketing structure. The positions with matched rights correspond to elements of $[n]$ belonging to each set in C. Order C by the number of unmatched rights, starting with the set whose unmatched parentheses are all lefts. This expresses C as a chain in $\mathbf{2}^n$. To move up the chain, change the leftmost unmatched left parenthesis to a right parenthesis, changing a 0 to a 1 in the incidence vector. At the bottom of C, all unmatched positions are left parentheses; at the top they are rights.

When the bracketing structure has j matched pairs, the chain extends from rank j to rank $n - j$ (with $n - 2j + 1$ members), since every member contains the elements for the j matched right parentheses and omits those for the j matched left parentheses. Hence the chains are skipless and symmetric. Every member of $\mathbf{2}^n$ occurs on exactly one chain, so this is a symmetric chain decomposition. ∎

12.2.11. Theorem. The inductive (Theorem 12.2.8) and bracketing (Example 12.2.10) decompositions of $\mathbf{2}^n$ are the same.

Proof: We use induction on n. For $n \leq 1$, the poset is a single chain. For $n > 1$, the induction hypothesis states that each chain in the inductive decomposition of $\mathbf{2}^{n-1}$ has a fixed bracketing. It suffices to show that applying the inductive construction to any chain C in the bracketing decomposition of $\mathbf{2}^{n-1}$ yields chains in the bracketing decomposition of $\mathbf{2}^n$.

From C define a chain C' by adding $\{n\}$ to each element of C. In position n, each vector for C gains a left parenthesis, while the corresponding vector for C' gains a right parenthesis. The bracketing structure for C does not change by the added left parenthesis. The top member of C' also has this bracketing structure, since its unmatched positions all have right parentheses, so adding a right at the end does not create another match. Thus this element moves from the top of C' to the top of C in the bracketing decomposition of $\mathbf{2}^n$.

The remaining members of C' all had a rightmost unmatched left parenthesis in the same position. Thus the new right parenthesis matches to the same position to complete the bracketing structure for each member of C' below the top. Hence they lie on the same chain in the bracketing decomposition of $\mathbf{2}^n$.

Furthermore, when the bracketing structure of C consists of j pairs, C starts with $n - 2j$ members. It gains one to reach $n + 1 - 2j$ members (still with j pairs in its bracketing structure), while C' ends with $j + 1$ pairs and $n - 1 - 2j$ members. Hence these are full chains in the bracketing decomposition of $\mathbf{2}^n$.

We have proved that every chain of size at least 2 in the bracketing decomposition of $\mathbf{2}^{n-1}$ becomes two chains in the decomposition of $\mathbf{2}^n$ when the induction step is applied. By induction on n, the two decompositions are thus the same. ∎

The Dilworth decomposition we have described for $\mathbf{2}^n$ arises in many ways. We present a third description, and Exercise 19 has yet a fourth.

12.2.12.* Example. (Aigner [1973]) We use a greedy lexicographic rule. List the k-sets in lexicographic order, from $[k]$ to $[n] - [n - k]$. Match each k-set A in turn to the $(k + 1)$-set that is earliest in lexicographic order on level $k + 1$ among the $(k + 1)$-sets that remain unmatched and contain A. For example, in $\mathbf{2}^4$ we match $\varnothing$ to 1, then $1 \to 12$, $2 \to 23$, $3 \to 13$, and $4 \to 14$, then $12 \to 123$, $13 \to 134$, $14 \to 124$, $23 \to 234$, with none available for 24 and 34, and finally $123 \to 1234$.

The resulting chains (shown below for $n = 4$) are the same as those in Example 12.2.10 (see Exercise 20). It is not even obvious that this produces symmetric chains, and yet we obtain the familiar bracketing decomposition. The construction also extends to multisets.   ∎

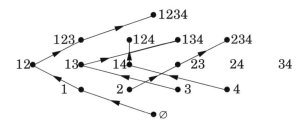

For many classical rank-symmetric rank-unimodal posets, the existence of symmetric chain decompositions remains unknown.

12.2.13. Example. $L(m, n)$: *Bounded integer partitions*. The elements of $L(m, n)$ are the integer lists $(a_1, \ldots, a_m)$ such that $0 \leq a_1 \leq \cdots < a_m \leq n$. These correspond to Ferrers diagrams contained in an m by n rectangle. The order relation puts $a \leq b$ if and only if $a_i \leq b_i$ for all i. Hence $L(m, n)$ is a subposet of $(\mathbf{n+1})^m$.

The complement of a Ferrers diagram in a rectangle fits in that rectangle, so $L(m, n)$ is rank-symmetric (actually, self-dual). The rank generating function (with formal variable q for algebraic reasons), is $\prod_{j=1}^{m} \frac{1-q^{n+j}}{1-q^j}$ (Exercise 3.4.40). Rank-unimodality is difficult; algebraic proofs began with Sylvester (see Proctor [1982]). O'Hara [1990] found an intricate combinatorial proof, presented also in Zeilberger [1989]. Stanley [1982] observed that $L(m, n)$ has the Sperner property, using results of Griggs [1977].

Stanley conjectured that $L(m, n)$ is a symmetric chain order. This is easy for $m \leq 2$ (see Exercises 5–8) and is known for $m \leq 4$ (Riess [1978], Lindström [1980], West [1980]). Solutions for $L(m, n)$ with $m = 5$ and n odd have been rumored. ∎

Finally, we present an application of the bracketing decomposition. We want to count antichains in $\mathbf{2}^n$; this is known as **Dedekind's Problem**. By Remark 12.1.14, antichains correspond to down-sets. There are $2^{\binom{n}{\lceil n/2 \rceil}}$ down-sets whose maximal elements all have size $\lceil n/2 \rceil$; this yields a lower bound. For the upper bound, we view down-sets as **monotone Boolean functions**, defined to be order-preserving functions from $\mathbf{2}^n$ to $\{0, 1\}$.

12.2.14. Theorem. (Hansel [1966]) The number of monotone Boolean functions is at most $3^{\binom{n}{\lceil n/2 \rceil}}$.

Proof: We construct a monotone Boolean function f by specifying its values in order. Consider chains in the bracketing decomposition in increasing order of size. For each new chain, the values of f on some of it are already forced. If x contains a set already assigned 1, then also $f(x) = 1$. Similarly, a subset of a set assigned 0 must be assigned 0. Since there are $\binom{n}{\lceil n/2 \rceil}$ chains, it suffices to show that always there are at most three ways to extend f to the next chain C. To do this, we show that C has at most 2 elements with undetermined labels. When these are x and y with $x > y$, the options $(f(y), f(x))$ will be only $(0, 0)$, $(0, 1)$, and $(1, 1)$.

Consider sets x and y in C, with $x > y$ and $|x - y| > 1$. To show that f is known for x or y, it suffices to construct z between them in $\mathbf{2}^n$ such that $f(z)$ is known. Then $f(z) = 1$ implies $f(x) = 1$, and $f(z) = 0$ implies $f(y) = 0$.

If C has size k, then its bracketing has $n - k + 1$ matched positions. Shorter chains have more matched positions. The unmatched positions for x and y have right parentheses followed by left parentheses, with at least two more rights in x than in y. Construct z from x by changing the next-to-last unmatched right parenthesis in x to a left parenthesis.

Now $y < z < x$. In the bracketing of z, the positions corresponding to the last two unmatched right parentheses in x match, since the positions between them are all matched. The other matches are unchanged, so z has more matched positions and appears on a shorter chain, as desired. ∎

$$z = - - -(- - --) - - - - --$$

$$x = - - -) - - - -) - - - - --$$
$$y = - - -(- - - -(- - - - --$$

By a more detailed argument along the same lines, showing that almost always the new chain contains at most one undetermined element, Kleitman–Markowsky [1975] improved this upper bound to $2^{\binom{n}{\lceil n/2 \rceil}(1+O(\log n/n))}$ (strengthening Kleitman [1969]). Korshunov [1981] obtained a more detailed asymptotic formula, which for even n is

$$2^{\binom{n}{n/2}}e^{\binom{n}{n/2-1}}(2^{-n/2} + n^2 2^{-n-5} - n2^{-n-4}).$$

Exact values for $n \le 7$ are 3, 6, 20, 168, 7581, 7828354, and 2,414,682,040,998, with Korshunov's formula giving 7996118 for $n = 6$.

LYM AND SPERNER PROPERTIES

A short direct proof of Sperner's Theorem counts the maximal chains in $\mathbf{2}^n$ by where they hit an antichain. The proof identifies a property for graded posets that implies the Sperner property and much more.

12.2.15. Theorem. (Sperner [1928]) In $\mathbf{2}^n$, the elements of rank $\lfloor n/2 \rfloor$ form a maximum antichain.

Proof: (Lubell [1966]) A maximal chain has at most one element in an antichain F. We count the chains according to which element of F (if any) they contain. There are $n!$ maximal chains, since they are specified by acquiring the elements of $[n]$ in some order. To pass through x, they must acquire the elements of x first, so exactly $|x|!(n-|x|)!$ maximal chains pass through x. Thus $\sum_{x\in F} |x|!(n-|x|)! \le n!$.

Dividing both sides by $n!$ yields $\sum_{x\in F} 1/N_{r(x)} \le 1$. Replacing $N_{r(x)}$ with the largest binomial coefficient preserves the inequality, and then clearing the fraction yields $|F| \le \binom{n}{\lceil n/2 \rceil}$. ∎

The inequality $\sum_{x\in F} N_{r(x)}^{-1} \le 1$ was discovered independently for antichains in $\mathbf{2}^n$ by Lubell [1966], Yamamoto [1954], and Meshalkin [1963].

12.2.16. Definition. For a family F in a graded poset, the inequality $\sum_{x \in F} N_{r(x)}^{-1} \leq$ 1 is the **LYM inequality**. A graded poset satisfies the **LYM property** if its antichains all satisfy the LYM inequality. Such a poset is an **LYM order**.

By the argument of Theorem 12.2.15, the LYM property implies the Sperner property. An analogue of this argument using circular arrangements yields the strong Sperner property for $\mathbf{2}^n$ (Exercise 28). We will see later that the LYM property implies the strong Sperner property.

We will also see that the argument in Theorem 12.2.15 generalizes; it needs only a list of maximal chains such that, in each rank, each element appears equally often. We introduce a name for such a list of maximal chains in order to use the argument in more generality.

12.2.17. Definition. A nonempty list of maximal chains in a graded poset P is a **regular covering** of P if, for each rank P_k, each element of P_k lies in exactly the fraction $1/|P_k|$ of these chains.

12.2.18. Example. A regular covering of P is a list; chains may be used repeatedly. The list may omit some maximal chains and use some more than others. In $\mathbf{3} \times \mathbf{4}$, shown below, the number of maximal chains containing a specified element of a middle rank is 1, 6, or 3. The full set of maximal chains is not a regular covering. On the other hand, there is a regular covering with 6 chains consisting of two copies of the "outer" indicated chains plus one copy of each "inner" chain. We will prove later that every (finite) product of chains has a regular covering.   ∎

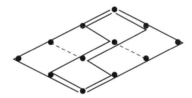

LYM orders are characterized by the existence of a regular covering. To prove this, we use a third and equally important property.

12.2.19. Definition. (Graham–Harper [1969]) A graded poset P has the **normalized matching property** if $|A^*|/N_{k+1} \geq |A|/N_k$ for all k and all $A \subseteq P_k$, where A^* denotes $U[A] \cap P_{k+1}$. The set A^* is called the **shade** of A or its "shadow at the rank above".

12.2.20. Example. $\mathbf{2}^n$ *has the normalized matching property.* For $A \subset \binom{[n]}{k}$, each element of A extends to $n - k$ elements of A^*. Since each element in A^* can lose an element in $k + 1$ ways, $(k+1)|A^*| \geq (n-k)|A|$. Dividing both sides by $(n-k)\binom{n}{k}$ yields $|A|/N_k \leq |A^*|/N_{k+1}$. This was the essence of the argument we used to prove Sperner's Theorem (Theorem 11.2.14).   ∎

The subgraph of the diagram induced by $P_k \cup P_{k+1}$ is a P_k, P_{k+1}-bigraph with $N(A) = A^*$. When $N_{k+1} \geq N_k$, normalized matching is stronger than Hall's condition for a matching that covers P_k (Section 6.1). We use Hall's Theorem to prove equivalence for these properties.

12.2.21. Theorem. (Kleitman [1974]) For a graded poset P, the following statements are equivalent:

(A) P has a regular covering.

(B) P has the LYM property.

(C) P has the normalized matching property.

Proof: A $\Rightarrow$ B. As in the proof of Theorem 12.2.15, counting the chains in a regular covering $\mathbf{C}$ that are hit by the elements of an antichain F yields $\sum_{x \in F} |\mathbf{C}|/N_{r(x)} \le |\mathbf{C}|$ and thus the LYM inequality.

B $\Rightarrow$ C. If $A \subseteq P_k$, then $A \cup (P_{k+1} - A^*)$ is an antichain in P. The LYM inequality yields $|A|/N_k + |P_{k+1} - A^*|/N_{k+1} \le 1$. Since $|P_{k+1}| = N_{k+1}$, this becomes $|A|/N_k \le |A^*|/N_{k+1}$.

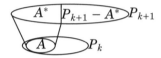

C $\Rightarrow$ A. Letting $M = \prod N_k(P)$, define the **blowup poset** P' with $M/N_{r(x)}$ copies of x for each $x \in P$. A copy of x is less than a copy of y in P' if and only if $x < y$ in P. Use the normalized matching property in P to find perfect matchings joining adjacent ranks in P'; such matchings combine to partition P' into disjoint maximal chains, and these chains collapse to form a regular covering of P with x appearing $M/N_{r(x)}$ times.

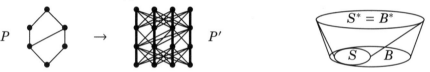

By Hall's Theorem, it suffices to show for $S \subset P'_k$ that $|S^*| \ge |S|$, where S^* is the shade of S (in P'_{k+1}). Let B be the subset of P'_k consisting of all copies of each element of P_k that has at least one copy in S. Now $S^* = B^*$ and $S \subseteq B$, so it suffices to show that $|B^*| \ge |B|$.

Let A be the sets of elements in P that have copies in B. Note that in P' the set B^* consists of all copies of all elements of P_k covering elements of A; that is, all copies of elements of A^*. Note that $|B| = |A| M/N_k$ and $|B^*| = |A^*| M/N_{k+1}$, since all copies of an element appear when any copies appear. The normalized matching property in P now yields $|B^*| \ge |B|$. ∎

12.2.22.* Remark. (1) The LYM inequality and regular coverings are self-dual, so normalized matching is equivalent to requiring $|D[A] \cap P_{k-1}| \ge |A| N_{k-1}/N_k$ when $A \subseteq P_k$ (Exercise 29 requests a direct proof without duality).

(2) Theorem 12.2.21 guarantees short proofs. Having the LYM property can be proved by giving a regular covering. Failing the LYM property can be proved by giving a violation of the normalized matching property (Exercises 26–27).

(3) Since the number of chains is divisible by each rank-size, every regular covering has at least lcm $\{N_k\}$ chains, achieved by blowing up x to lcm $\{N_k\}/N_{r(x)}$ copies (Exercise 4). However, every LYM poset has a regular covering using at most $|P| - r(P)$ *distinct* chains, and this is best possible (Exercise 30).

(4) Graham–Harper [1969] found an efficient method to find a regular covering. We seek nonnegative integer weights on the edges from P_k to P_{k+1} such that

the total weight on edges leaving each element of P_k is N_{k+1} and the total weight on edges entering each element of P_{k+1} is N_k. A solution forms a regular covering for the subposet $P_k \cup P_{k+1}$. The weights give the relative usage of the covering pairs in the full regular covering. Feasibility can be tested by adding a source and sink, modeling the weights with multiedges, and using Menger's Theorem; thus we can test the LYM property in polynomial time.

(5) When the number of elements an element x covers depends only on the rank of x, the set of all maximal chains is a regular covering (Baker [1969]).   ∎

We have seen that the LYM property implies the Sperner property. The LYM property (with its equivalence to regular covering) also implies a more general statement that implies the strong Sperner property.

12.2.23. Theorem. (Greene–Kleitman [1978]) Let $\lambda\colon P \to \mathbb{R}$ be a weight function defined on the elements of P. If P has the LYM property, then for every subset $G \subset P$ and every regular covering $\mathbf{C}$ of P,

$$\sum_{x \in G} \frac{\lambda_x}{N_{r(x)}} \leq \max_{C \in \mathbf{C}} \sum_{y \in C \cap G} \lambda_y \,.$$

Proof: We interpret the inequality probabilistically. Choose a chain from $\mathbf{C}$ uniformly at random, and define a random variable $X = \sum_{y \in C \cap G} \lambda_y$. The expectation of X is $\sum_{x \in G} \lambda_x \mathbb{P}(x \in C)$. Since $\mathbf{C}$ is a regular covering, $\mathbb{P}(x \in C) = 1/N_{r(x)}$. Thus the left side of the desired inequality is the expected value of X, while the right side is its maximum value.   ∎

Suitable choices of λ yield various applications. First let $\lambda_x = N_{r(x)}$.

12.2.24. Corollary. If P is an LYM order with regular covering $\mathbf{C}$ and G is any subset of P, then $|G| \leq \max_{C \in \mathbf{C}} \sum_{y \in C \cap G} N_{r(y)}$.   ∎

Bounds on $|G|$ follow from Corollary 12.2.24 for various chain conditions on $|G|$. Erdős [1945] proved the next corollary for $\mathbf{2}^n$ by generalizing the argument of Sperner [1928] for the Sperner property.

12.2.25. Corollary. (Erdős [1945]) The LYM property implies the strong Sperner property.

Proof: If G is a k-family, then Corollary 12.2.24 limits its size to the sum of the k largest rank sizes.   ∎

Now it is natural to ask whether the LYM property is so strong that it also implies the structural "holy grail" of symmetric chain decomposition. The answer is "yes" when the obvious necessary conditions hold.

12.2.26. Theorem. (Anderson [1976], Griggs [1977]) Every rank-unimodal rank-symmetric LYM poset is a symmetric chain order.

Proof: We use induction on the height of the poset P. For height 1, there is nothing to prove. For even height, the two middle levels have equal size, and normalized matching between them reduces to Hall's Condition. Collapse the two levels to one along the resulting matching. Since normalized matching considers only consecutive levels, the smaller poset P' is still an LYM order, and it also has symmetric and unimodal rank sizes. The guaranteed chain decomposition for P' expands into a symmetric chain decomposition of P using the central matching.

For odd height, let the middle rank be P_{k+1}. If we can match P_k to P_{k+2} through P_{k+1}, then the unused elements of P_{k+1} will become singleton chains. After discarding them, the three levels collapse to one level along the matched edges to obtain a smaller LYM order P'. After obtaining a symmetric chain decomposition of P', expand each element of the middle level into a 3-element chain as the middle of a skipless symmetric chain in P. Together with the discarded singletons, these chains form a symmetric chain decomposition of P.

The covering pairs down from P_{k+2} and up from P_k yield two families $\mathbf{A}$ and $\mathbf{B}$ of subsets of P_{k+1}, each with N_k sets. The two desired matchings exist when there is a common system of distinct representatives for $\mathbf{A}$ and $\mathbf{B}$. For subsets $I, J \subseteq [N_k]$, let $A' = \cup_{i \in I} A_i$ and $B' = \cup_{j \in J} B_j$. By the Ford–Fulkerson Condition (Theorem 7.2.15), $|A' \cap B'| \geq |I| + |J| - N_k$ suffices. Since the dual of an LYM order is an LYM order, both $|A'| \geq |I|(N_{k+1}/N_k)$ and $|B'| \geq |J|(N_{k+1}/N_k)$. Thus

$$|A' \cap B'| = |A'| + |B'| - |A' \cup B'| \geq (|I| + |J|)(N_{k+1}/N_k) - N_{k+1}$$
$$= (|I| + |J| - N_k)(N_{k+1}/N_k) \geq |I| + |J| - N_k. \qquad \blacksquare$$

The LYM property has many consequences. Unfortunately, many graded posets of interest are not LYM orders (see Exercises 25–27). For these, one can still ask whether consequences such as symmetric chain decomposition hold. For the poset $L(m, n)$ (Example 12.2.13), this remains a fascinating open problem.

PRODUCTS OF LYM ORDERS (optional)

So far, our only criteria for the LYM property are the three equivalent defining properties. We have not yet proved that chain-products are symmetric chain orders. We consider the behavior of the LYM property under products. First note the behavior of the rank sizes.

12.2.27. Remark. *Rank of poset products.* If P and Q are graded, then their product is also graded and has rank function given by $r_{P \times Q}(x, y) = r_P(x) + r_Q(y)$. Thus $N_k(P \times Q) = \sum_i N_i(P)N_{k-i}(Q)$. $\qquad \blacksquare$

12.2.28. Example. *The product of LYM posets need not be an LYM poset.* In the example below, the set $\{a1, b1\}$ occupies $\frac{2}{3}$ of its rank, but its shadow at the rank above is only $\frac{1}{2}$ of that rank. $\qquad \blacksquare$

12.2.29. Remark. *Necessary condition for LYM property in a special product.* Let $Q = P \times \mathbf{2}$, where P is an LYM poset. Let $A = \{(x, 1): x \in P_{k-1}\}$. The set A lies in rank k in Q. Note that $A^* \subseteq \{(y, 1): y \in P_k\}$.

Write N_j for $N_j(P)$. Since $\mathbf{2}$ has one element at each rank, Q has $N_{j-1} + N_j$ elements at rank j. Normalized matching thus requires

$$\frac{|A^*|}{N_k + N_{k+1}} \geq \frac{|A|}{N_{k-1} + N_k}.$$

Since $|A| = N_{k-1}$ and $|A^*| = N_k$, we need $N_k^2 \geq N_{k-1}N_{k+1}$. This condition on the sequence of rank sizes is necessary for the product with $\mathbf{2}$ to be an LYM order. We will show that it is sufficient for products in general. ∎

12.2.30. Definition. A sequence $\langle N \rangle$ is **log-concave** if $N_k^2 \geq N_{k-1}N_{k+1}$ for all k.

The condition for log-concavity is that the logarithms of the terms form a concave sequence. Every concave sequence is also log-concave.

12.2.31. Theorem. (Harper [1974], Hsieh–Kleitman [1973]) The family of LYM orders with log-concave rank sizes is closed under taking products.

Proof: We have $r(P_1 \times P_2) = r(P_1) + r(P_2)$ and $N_k(P_1 \times P_2) = \sum_{i=0}^{k} N_i(P_1)N_{k-i}(P_2)$. Log-concavity of the convolution of two log-concave sequences is an exercise in algebraic manipulation (Exercise 31).

To prove the LYM inequality for $P_1 \times P_2$, we generalize Theorem 12.2.23. There we picked a random chain from a regular covering. Here, we pick a random pair C_1, C_2 from regular coverings $\mathbf{C}_1$ and $\mathbf{C}_2$ of P_1 and P_2. This produces a random rectangle in the product.

Given any subset G of $P_1 \times P_2$ and weights λ_x for $x \in P$, let $X = \sum_{x \in G \cap (C_1 \times C_2)} \lambda_x$. The expectation is $\sum_{x \in G} \lambda_x \mathbb{P}(x \in C_1 \times C_2)$. Since $\mathbf{C}_1$ and $\mathbf{C}_2$ are regular coverings, $\mathbb{P}(x \in C_1 \times C_2) = N_{r(x_1)}(P_1)^{-1}N_{r(x_2)}(P_2)^{-1}$, where $x = (x_1, x_2)$. Thus the left side of the inequality below is the expectation of X, while the right side is its maximum.

$$\sum_{x \in G} \frac{\lambda_x}{N_{r(x_1)}(P_1)N_{r(x_2)}(P_2)} \leq \max_{C_i \in \mathbf{C}_i} \sum_{y \in G \cap C_1 \times C_2} \lambda_y$$

Setting $\lambda_x = N_{r(x_1)}(P_1)N_{r(x_2)}(P_2)/N_x$ yields

$$\sum_{x \in G} \frac{1}{N_{r(x)}} \leq \max_{C_i \in \mathbf{C}_i} \sum_{y \in G \cap C_1 \times C_2} \frac{N_{r(y_1)}(P_1)N_{r(y_2)}(P_2)}{N_{r(y)}}.$$

To obtain the LYM inequality, it suffices to show that the right side of this inequality is at most 1 when G is an antichain.

If G lies entirely in the kth rank of $P_1 \times P_2$, then we compute a bound over rank k for arbitrary $C_1 \times C_2$. The elements of rank k in $P_1 \times P_2$ that lie in $C_1 \times C_2$ have the form (y_1, y_2), where y_1 is the element of C_1 with rank i in P_1 and y_2 is the element of C_2 with rank $k - i$ in P_2. There is at most one such element for each $i \in \{0, \ldots, k\}$. Thus

$$\sum_{y \in G \cap C_1 \times C_2} \frac{N_{r(y_1)}(P_1)N_{r(y_2)}(P_2)}{N_{r(y)}} \leq \sum_{i=0}^{k} \frac{N_i(P_1)N_{k-i}(P_2)}{N_k(P_1 \times P_2)} = 1.$$

Suppose that G has elements from more than one rank in $P_1 \times P_2$. We show that pushing those in the lowest occupied rank upward toward the highest occupied rank cannot decrease the specified sum. This suffices, since we have shown that the sum is bounded by 1 when G lies in a single rank.

Let T be a maximal set of "consecutive" elements of G at rank k in $C_1 \times C_2$; "consecutive" means $T = \{(x_i, y_{k-i}): r \leq i \leq s\}$ for some $0 \leq r \leq s$, where x_i is the element of rank i in C_1 and y_j is the element of rank j in C_2. Let T^* be the set of elements covering members of T in $C_1 \times C_2$; this is another consecutive set. Because T is a maximal consecutive set, replacing all of T with T^* does not violate the antichain condition.

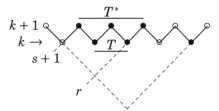

Let $g_k(r, s) = \sum_{i=r}^{s} N_i(P_1)N_{k-i}(P_2)$. The contribution of T to the original sum is $\frac{g_k(r,s)}{g_k(0,\infty)}$. The contribution of T^* to the new sum is $\frac{g_{k+1}(r,s+1)}{g_{k+1}(0,\infty)}$. We need only show that $\frac{g_{k+1}(r,s+1)}{g_{k+1}(0,\infty)} \geq \frac{g_k(r,s)}{g_k(0,\infty)}$. Since this is equivalent to $\frac{g_{k+1}(r,s+1)}{g_k(r,s)} \geq \frac{g_{k+1}(0,\infty)}{g_k(0,\infty)}$, it suffices to show that $\frac{g_{k+1}(r,s+1)}{g_k(r,s)}$ is increasing in r and decreasing in s. See Exercise 32. ∎

12.2.32. Corollary. M^e is an LYM order.

Proof: M^e is a product of chains, each of which is an LYM order with log-concave rank sizes. ∎

Algebraic techniques can yield log-concavity of sequences. When the rank generating function is known, the following result may apply (see Stanley [1989]).

12.2.33. Theorem. If all roots of a polynomial with real coefficients are real, then the sequence of coefficients is log-concave. ∎

12.2.34. Example. The rank generating function for $\mathbf{2}^n$ is $(1 + x)^n$, so Theorem 12.2.33 applies. However, for M^e it is $\prod_{i=1}^{n}(1 + x + \cdots + x^{e_i - 1})$, which has complex roots when $\max e_i > 2$. Log-concavity follows because the convolution of log-concave sequences is log-concave. ∎

EXERCISES 12.2

12.2.1. (−) Use Sperner's Theorem and Dilworth's Theorem to prove (weaker than Theorem 12.2.14) that the number of antichains in $\mathbf{2}^n$ is at most $(n + 1)^{\binom{n}{\lceil n/2 \rceil}}$. (Gilbert [1954])

12.2.2. (−) Let P be a rank-symmetric LYM order of rank n. Let S be a down-set in P. Prove that the elements of S have average rank at most $n/2$.

12.2.3. (−) Show that the sequence of rank sizes of a product poset may be log-concave even though the sequences for the factors are not both log-concave.

12.2.4. (−) Prove that the minimum number of chains in a regular covering of an LYM order is lcm N_k. (Hint: Blow up each element x to $(\text{lcm } N_k)/N_{r(x)}$ copies and use the argument of Theorem 12.2.21.) (West–Harper–Daykin [1983])

12.2.5. (−) Prove that $L(2, n)$ (Example 12.2.13) is a symmetric chain order.

12.2.6. (−) An **automorphism** of a poset is an order-preserving permutation of the elements. Let P be a graded poset such that whenever x and y have the same rank in P, some automorphism maps x to y. Prove that P is an LYM order.

12.2.7. Prove $w(\mathbf{n} \times \mathbf{n} \times \mathbf{n}) = \lfloor (3n^2 + 1)/4 \rfloor$.

12.2.8. (+) Construct a symmetric chain decomposition of $L(3, n)$ for odd n. (West [1980])

12.2.9. Let P be a finite poset whose diagram is connected.
    (a) Show that P may have a maximal chain that is not a longest chain even when every element in P lies in a longest chain of P.
    (b) Suppose that, whenever y covers x in P, some longest chain contains both x and y. Prove that P is a graded poset. (Stanley [1971, p. 19–20])

12.2.10. *k-families in chain products.*
    (a) For $i, j, k \in \mathbb{N}$, find a formula without summations for the maximum size of a k-family in the chain-product $\mathbf{i} \times \mathbf{j}$. (Hint: Consider three cases for $\{i, j, k\}$.)
    (b) Consider three sets of parallel lines in the plane, forming equilateral triangles. Within the sets, the lines need not be equally spaced. Let the sets have sizes r, s, t. Determine the maximum number of points that occur as the intersection of three lines, one from each set. (Matsko–West–Wetzel [2001])

12.2.11. (◊) A **semiantichain** in $P \times Q$ is a subset S having $(u, v) < (u', v')$ for two elements of S only if $u < u'$ and $v < v'$. A product poset has the 2-**part Sperner property** when some single rank is a largest semiantichain.
    (a) Prove that the product of two symmetric chain orders is 2-part Sperner.
    (b) Use part (a) to show that for any 2-coloring of the elements of $[n]$, the largest subposet of $\mathbf{2}^n$ having no related pair with monochromatic difference consists of the middle rank. (Kleitman [1965], Katona [1966])

12.2.12. *Bracketing decomposition of M^e.* Dedicate $e_i - 1$ positions for the ith coordinate. View $a \in M^e$ as a list with $0 \le a_i < e_i$. In the positions for the ith coordinate, put a_i right parentheses followed by $e_i - 1 - a_i$ left parentheses. Match parentheses as in Example 12.2.10. Associate a single chain to the elements having the same matched pairs. Prove that these chains form a symmetric chain decomposition of M^e. (Below is a chain from $M^{(5,4)}$ with one matched pair; it is the second chain in the figure for Theorem 12.2.9.) (Greene–Kleitman [1976b], Leeb [unpublished])

$$\begin{array}{cccccccccccc} ((((\,)((& < &)(((\,)((& < &))((\,)((& < &)))(\,)((& < &)))(\,)((& < &)))(\,))) \\ (0, 1) & & (1, 1) & & (2, 1) & & (3, 1) & & (3, 2) & & (3, 3) \end{array}$$

12.2.13. A finite set of integers is *balanced* if the numbers of even and odd elements differ by at most 1. Let F be a family of subsets of $[n]$ such that whenever $A, B \in F$ with $A \subseteq B$, the set $B - A$ is not balanced. Prove that $|F| \le \binom{n}{\lceil n/2 \rceil}$ and that this bound is best possible. (Greene–Kleitman [1976b])

12.2.14. (◊) A **skew chain order** is a poset having a rank function in which every minimal element has rank 0 and a decomposition into skipless chains starting at rank 0. Describe (with proof) the largest semiantichain (see Exercise 12.2.11) in the product of two skew chain orders. Let P_k be the inclusion order on the set of intervals in $\mathbb{R}$ with endpoints in $[k]$. Give a geometric description of the maximum semiantichain in $P_m \times P_n$. (West–Kleitman [1979])

12.2.15. Use a chain decomposition of $\mathbf{2}^n$ to construct a spanning subgraph of the hypercube graph Q_n that has diameter n and has only $2^n + \binom{n}{\lceil n/2 \rceil} - 2$ edges. (Comment: This subgraph has the same diameter as Q_n while keeping only a vanishing fraction of the edges.) (Graham–Harary [1992])

12.2.16. ($\diamond$) A **universal subset list** on an alphabet S is a word having every subset of S as a consecutive substring. For example, 1231 is such a list on [3], and 123421341 is such a list on [4]. Listing all 2^n subsets successively yields a universal subset list on [n] with length $n2^{n-1}$, since the average size is $n/2$.

(a) For even n, use symmetric chain decompositions of two copies of $\mathbf{2}^{n/2}$ to construct a universal subset list on [n] with length asymptotically at most $(4/\pi)2^n$. (Hint: Use Stirling's formula to approximate $\binom{k}{\lfloor k/2 \rfloor}$.)

(b) Show that the top element on chains in a symmetric chain decomposition of $\mathbf{2}^{[k]}$ has average size $k/2 + O(k^{1/2})$. Use this to reduce the bound in (a). (Lipski [1978])

(c) Prove that these upper bounds are within a factor of $c\sqrt{n}$ of being optimal.

12.2.17. ($\diamond$) *The Littlewood–Offord Problem.* Let $a_1, \dots, a_n$ be vectors in $\mathbb{R}^d$, each having length at least 1. Let $R_1, \dots, R_k$ be regions in $\mathbb{R}^d$, each having diameter less than 1 (i.e., contained in a sphere of diameter 1), and let R be their union. Let $d_k(n) = \sum_{i=r}^{s} \binom{n}{i}$, where $r = \lfloor (n-k+1)/2 \rfloor$ and $s = \lfloor (n+k-1)/2 \rfloor$ (this counts the k middle ranks in $\mathbf{2}^n$).

(a) Prove that $d_k(n) = d_{k+1}(n-1) + d_{k-1}(n-1)$.

(b) Prove that the number of 0, 1-vectors x such that $\sum x_i a_i \in R$ is at most $d_k(n)$. (Hint: To apply part (a) in an inductive proof, one must group these vectors x into sets corresponding to two problems of the same type with $n-1$ vectors, one having $k+1$ regions and one having $k-1$ regions.) (Kleitman [1970])

12.2.18. Let G be an X, Y-bigraph with $X = \{x_1, \dots, x_m\}$ and $Y = \{y_1, \dots, y_n\}$. Define the *greedy matching* of X into Y as follows: having processed each x_i with $i < r$, match x_r to its least-indexed available neighbor in Y, if any is available. Prove that the greedy matchings of X into Y and Y into X are the same.

12.2.19. ($\diamond$) For $X \subseteq [n]$, let $X = \{x_1, \dots, x_k\}$ with $x_1 < \cdots < x_k$, and set $x_0 = 0$. Let t be the largest nonnegative index i minimizing $x_i - 2i$. If $x_t < n$, then form $f(X)$ from X by adding the element $1 + x_t$, but leave $f(X)$ undefined if $x_t = n$. Prove that the pairs $(X, f(X))$ form the chains in the bracketing decomposition of $\mathbf{2}^n$. (For example, applications of f in $\mathbf{2}^4$ starting with $\{3\}$ yield $\{1, 3\}$, then $\{1, 3, 4\}$, then nothing. Other examples are $f(\{1, 3, 4, 7\}) = \{1, 3, 4, 5, 7\}$ with $t = 3$ and $f(\{3, 5, 7, 9\}) = \{1, 3, 5, 7, 9\}$ with $t = 0$.) (White–Williamson [1977])

12.2.20. Prove that Aigner's lexicographically-generated chains (Example 12.2.12) are the same as the chains in Theorem 12.2.8 or the chains in Exercise 12.2.19.

12.2.21. ($\diamond$) Two chain partitions are **orthogonal** if no two elements appear in the same chain in both partitions. (Kleitman–Shearer [1979])

(a) Define $\mathbf{D}$ from the bracketing decomposition $\mathbf{C}$ of $\mathbf{2}^n$ by changing each set to its complement and reversing the chains. For $n \geq 4$, prove that a slight change in $\mathbf{D}$ yields chain partition orthogonal to $\mathbf{C}$.

(b) Construct two orthogonal Dilworth decompositions for each of $\mathbf{2}^2$ and $\mathbf{2}^3$, and construct three pairwise orthogonal Dilworth decompositions for $\mathbf{2}^4$.

(c) Prove that $\mathbf{2}^n$ has at most $\lceil \frac{n+1}{2} \rceil$ pairwise orthogonal Dilworth decompositions.

12.2.22. Prove that $\mathbf{2}^n$ has $n!$ Dilworth decompositions such that no two elements lie on the same chain in more than $n! \lceil (n+1)/2 \rceil^{-1}$ of them. (Kleitman–Shearer [1979])

12.2.23. For x and y chosen from $\mathbf{n} \times \mathbf{m}$ according to some probability distribution p, prove $\mathbb{P}(x \leq y) \geq \frac{n+m}{2nm}$, with equality possible when n divides m. (Hint: For a and b chosen from a distribution over k choices, prove $\mathbb{P}(a = b) \geq 1/k$.) (Kleitman–Shearer [1979])

12.2.24. ($\diamond$) Let G be an X, Y-bigraph with $|Y| \geq |X|$ and no isolated vertices. The *deficiency* $\phi(A)$ of a set $A \subseteq V(G)$ is $|A| - |N(A)|$. Say that G has the **strong Hall property** if $\phi(A) + \phi(B) \leq |Y| - |X|$ when $A, B \subseteq Y$ with $|A| + |B| \leq |Y|$. Let H be a graph formed from $G + G$ by adding a matching joining the copies of Y.

(a) Prove that if G satisfies the strong Hall property, then H has a perfect matching.

(b) Let P be a self-dual rank-symmetric rank-unimodal poset. Prove that if the bipartite graphs joining consecutive levels all have the strong Hall property, then P has a symmetric chain decomposition. (Lu–Wang–Wong [1998])

12.2.25. The **Weak Order** W_n on the permutations of $[n]$ is defined by letting σ cover τ if σ is obtained from τ by transposing two consecutive elements to make them an inversion. The poset is graded: $r(\sigma)$ is the number of inversions in σ, so the rank generating function is $1(1 + x)(1 + x + x^2) \cdots (1 + x + \cdots + x^{n-1})$, with unimodal coefficients (Example 12.2.34).

(a) Show that W_n is a symmetric chain order for $n \leq 4$. (Comment: Also W_5 is a symmetric chain order by using Exercise 12.2.24, but for $n > 5$ the answer is unknown.)

(b) Show that W_4 is not an LYM order by considering the permutation 2143. Generalize the argument to prove that W_{2m} is not an LYM order for $m \geq 2$.

12.2.26. Prove that $L(m, n)$ is an LYM order if and only if $m = n = 3$ or $\min\{m, n\} \leq 2$. (Hint: Find the lowest violation of normalized matching in $L(3, 4)$ and generalize this for $m \geq 3$ and $n \geq 4$. A different violation is needed for $L(4, 4)$; consider 1114.)

12.2.27. ($\diamond$) Let Π_n denote the poset of partitions of $[n]$, with $\sigma \leq \tau$ if σ is a union of partitions of the blocks of τ. Partitions with k blocks have rank $n - k$, so $N_{n-k}(\Pi_n) = S(n, k)$ (the Stirling number). For even n, let A be the set of partitions of $[n]$ into two blocks of size $n/2$. Use A and its shadow to prove that Π_n with n even is not an LYM order when $n \geq 20$. (Spencer [1974]) (Comment: Rota asked whether Π_n always has the Sperner property; Canfield [1978] showed that it doesn't. Shearer [1979] and Jichang–Kleitman [1984] reduced the least n such that Π_n is not Sperner to 4×10^9 and then to 3.4×10^6.)

12.2.28. ($\diamond$) Let F be a k-family in $\mathbf{2}^n$.

(a) Prove that every circular permutation of $[n]$ has at most nk substrings in F.

(b) Use part (a) to prove $\sum_{x \in F} N_{r(x)}^{-1} \leq k$.

(c) Use part (b) to prove the strong Sperner property for $\mathbf{2}^n$. (Füredi–Katona)

12.2.29. Prove directly that the dual of a graded poset satisfying the normalized matching property also satisfies that property. That is, given an X, Y-bigraph satisfying $|N(S)| / |Y| \geq |S| / |X|$ for all $S \subseteq X$, prove $|N(T)| / |X| \geq |T| / |Y|$ for all $T \subseteq Y$.

12.2.30. A **minimal LYM order** is an LYM order such that deleting any covering pair from the order relation destroys the LYM property.

(a) Prove that the relations between adjacent ranks of a minimal LYM order form a forest, connected when the rank sizes are relatively prime. Construct an example to show that the forest may or may not be connected when the rank sizes are not relatively prime. (Hint: Consider the Graham–Harper version of the normalized matching condition.)

(b) Prove that every LYM order P has a regular covering using at most $|P| - r(P)$ distinct chains. Prove that this is sharp. (West–Harper–Daykin [1983])

12.2.31. Let $\langle a \rangle$ and $\langle b \rangle$ be log-concave sequences.

(a) Prove that $a_i a_j \geq a_{i+1} a_{j-1}$ whenever $i \geq j$.

(b) Prove that $\sum_i \sum_j (a_i a_j - a_{i+1} a_{j-1})(b_{k-i} b_{k-j} - b_{k-1-i} b_{k+1-j}) \geq 0$.

(c) (+) Use part (b) to prove that the product of two graded posets with log-concave rank sizes has log-concave rank sizes.

12.2.32. Given log-concave sequences $\langle a \rangle$ and $\langle b \rangle$, let $g_k(r, s) = \sum_{i=r}^{s} a_i b_{k-i}$. Prove that $\frac{g_{k+1}(r, s+1)}{g_k(r, s)}$ is increasing in r and decreasing in s. (This completes Theorem 12.2.31.)

12.2.33. ($\diamond$) Given $s, t \in \mathbb{N}_0$, let $a_n = \sum_{i=s}^{t} \binom{n}{i}$. Prove that $\langle a \rangle$ is log-concave. (Hint: Let $f_n(x) = \sum_{k=s}^{t} \binom{n}{k} x^k$, and study the coefficients of $[f_n(x)]^2 - f_{n-1}(x) f_{n+1}(x)$. (Knuth [2017])

12.2.34. Let $L_n(q)$ be the inclusion order on the subspaces of an n-dimensional vector space over a field with q elements.

(a) Count the bases in a subspace of dimension k.

(b) Prove $N_k(L_n(q)) = \dfrac{(q^n - 1)(q^{n-1} - 1) \cdots (q^{n-k+1} - 1)}{(q^k - 1)(q^{k-1} - 1) \cdots (q^1 - 1)}$.

(c) Prove that elements of rank k cover $\frac{q^k - 1}{q - 1}$ elements. Conclude that $L_n(q)$ is an LYM order. (Comment: $N_k(L_n(q))$ is the **Gaussian polynomial** $\begin{bmatrix} n \\ k \end{bmatrix}_q$ of Exercise 3.4.40.)

12.2.35. A graded poset is **regular** if all elements of rank k are covered by the same number of elements at rank $k + 1$ and cover the same number at rank $k - 1$. A graded poset is **strictly Sperner** if every maximum antichain consists of one rank. A regular covering is **exhaustive** if any two comparable elements both lie on some chain in the covering.

(a) Show that the product of two chains of different lengths is not strictly Sperner (although it is an LYM order).

(b) Prove that a poset with an exhaustive regular covering has the strict Sperner property if and only if for every pair of maximum-sized ranks, the bipartite graph of relations between them is connected. (Broline)

(c) Prove that a regular poset has an exhaustive regular covering, and conclude that 2^n has the strict Sperner property.

12.2.36. *Derived LYM posets.*

(a) Let P be an LYM order of rank r. For $S \subseteq [r]$, let Q be the subposet consisting of all elements whose rank in P belongs to S. Prove that Q is an LYM order.

(b) Let $\langle a_k \rangle$ be a log-concave sequence. Prove that $a_k^2 \geq a_{k+j} a_{k-j}$ for all k, j.

(c) Let $A_1, \ldots, A_r$ be a partition of $[n]$ into r blocks, and let $S_1, \ldots, S_r$ be r arithmetic progressions. Let $P = \{X \subset [n] : |X \cap A_i| \in S_i \text{ for } 1 \leq i \leq r\}$. Prove that the inclusion order on P is an LYM order. (Griggs [1982])

12.2.37. *Bollobás' Inequality.*

(a) Let $A_1, \ldots, A_m$ and $B_1, \ldots, B_m$ be subsets of $[n]$ such that $A_i \cap B_j = \varnothing$ if and only if $i = j$. Prove that $\sum \binom{|A_i| + |B_i|}{|A_i|}^{-1} \leq 1$. (Hint: Consider instances of B_k completely after A_k in permutations of $[n]$.) (Bollobás [1965])

(b) Use part (a) to prove the LYM property for 2^n.

(c) Use part (a) to prove that $\binom{n-k}{\lfloor (n-k)/2 \rfloor}$ is the maximum t such that 2^n contains $(k + 1)$-chains $C_1, \ldots, C_t$ such that every member of each chain is incomparable to all members of all the other chains in the list. (Griggs–Stahl–Trotter [1984])

12.2.38. Let $\mathcal{F}$ be a family of finite sets $A_1, \ldots, A_m$. Let $\tau(\mathcal{F})$ denote the minimum size of a set intersecting each A_i. A family $\mathcal{F}$ is τ-**critical** if $\tau(\mathcal{F} - \{A_i\}) < \tau(\mathcal{F})$ for all i.

(a) Prove $\sum_{i=1}^{m} \binom{|A_i| + s}{s}^{-1} \leq 1$ for τ-critical $\mathcal{F}$ with $\tau(\mathcal{F}) = s + 1$. (Hint: Exercise 12.2.37a.)

(b) Prove that if $\mathcal{F}$ is r-uniform and every $\tau(A) \leq s$ for every subfamily A with $|A| \leq \binom{r+s}{s}$, then $\tau(\mathcal{F}) \leq s$. State the special case for graphs, using the vertex cover number $\beta(G)$.

12.3. Linear Extensions and Dimension

Few partial orders are chains, but chains are useful in understanding more complicated posets. We can describe a poset using chains that are consistent with the order relation, called "linear extensions".

ORDER DIMENSION

When purchasing a new car, a buyer considers many models. Between any two, the buyer may prefer one or be undecided. Assuming that the preference relation is a partial order, let P be the resulting poset. The buyer may try to encode P by rating the cars on criteria such as price, reliability, mileage, roominess, color, styling, etc. Assume that cars are ranked linearly on each scale. The scales "realize" P when car x is preferred to car y in P if and only if x is preferred to y on each scale. We try to realize a poset using a small number of linear orders.

12.3.1. Definition. An **extension** of a poset P is a partial order on the elements of P that contains all the relations of P. A **linear extension** is an extension that is a chain. The **intersection** of partial orders on a given set is the set of relations that appears in each of them.

12.3.2. Proposition. A finite poset is the intersection of its linear extensions.

Proof: A linear extension iteratively lists (and deletes) a minimal unlisted element. All linear extensions arise in this way. For every down-set F in P, this procedure can produce a linear extension that lists all of F before all of $P - F$. If x and y are incomparable elements of P, then $x \notin D[y]$ and $y \notin D[x]$. We thus have a linear extension listing x after y (and all of $D(y)$) and another listing y after x (and all of $D(x)$). Hence $x < y$ in P if and only if x precedes y in every linear extension. Thus P is the intersection of all its linear extensions.　■

In the algorithmic literature, linear extensions are called **topological orderings**. Proposition 12.3.2 is more subtle for infinite posets; one applies Zorn's Lemma to the family of all extensions, ordered by inclusion (Szpilrajn [1930]).

12.3.3. Definition. A **realizer** of a poset P is a set of extensions whose intersection is P. The **(order) dimension** $\dim P$ is the minimum number of linear extensions forming a realizer of P. Let $I(P)$ denote the set of ordered pairs of incomparable elements in P. For $(x, y) \in I(P)$, an extension L **establishes** the pair (x, y) if $x < y$ in L.

A set of extensions realizes P if and only if for every $(x, y) \in I(P)$, some extension in the set has $x < y$. That is, every incomparable pair must be established.

12.3.4. Example. A poset has dimension 1 if and only if it is a chain. Antichains have dimension 2 (list the elements in some order and the reverse order).

The poset below also has dimension 2. Using 123456 as one extension, the other must have $4 < 1$ and $6 < 3$; the extension 412563 completes a realizer. Not every linear extension is part of a realizer of size 2 (see Exercise 3). For example, starting with 124356, we cannot establish all of $4 < 1$, $3 < 4$, and $6 < 3$ with one additional extension, since $6 > 1$.　■

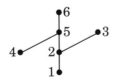

12.3.5. Theorem. (Dushnik–Miller [1941]) A poset P with comparability graph G has dimension at most 2 if and only if $\overline{G}$ is also a comparability graph.

Proof: *Necessity.* Let L_1 and L_2 be two linear extensions of P with intersection P. We have $x \| y$ if and only if x and y appear in opposite order on L_1 and L_2. Hence $x \| y$ if and only if x and y are in the same order on L_1 and L_2', where L_2' is the reverse of L_2. Now $\overline{G}$ is the comparability graph of the intersection of L_1 and L_2'.

Sufficiency. Let F be the comparability digraph of P. Let $\hat{F}$ be a transitive orientation of $\overline{G}$ (edges point up), and let $\hat{F}'$ be its reverse (also transitive). Both $F \cup \hat{F}$ and $F \cup \hat{F}'$ are orientations of K_n, where $n = |P|$. A non-transitive orientation of K_n has a directed 3-cycle, but this would violate the transitivity of F or $\hat{F}$ or $\hat{F}'$. Hence $F \cup \hat{F}$ and $F \cup \hat{F}'$ are transitive orientations of K_n.

A transitive orientation of K_n linearly orders its vertices by outdegree. Hence there are chains L and L' on $V(G)$ whose comparability digraphs are $F \cup \hat{F}$ and $F \cup \hat{F}'$. The intersection of these two chains is P, since two elements are ordered in the same way on L and L' if and only if they are adjacent in F, which is the comparability digraph of P.   ∎

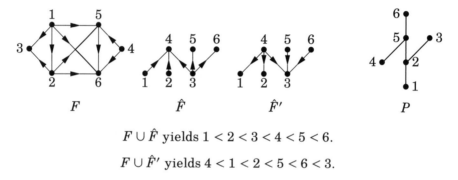

$$F \cup \hat{F} \text{ yields } 1 < 2 < 3 < 4 < 5 < 6.$$

$$F \cup \hat{F}' \text{ yields } 4 < 1 < 2 < 5 < 6 < 3.$$

On the other hand, dimension can be arbitrarily large.

12.3.6. Example. (Dushnik–Miller [1941]) The **standard example** S_n is the subposet of $\mathbf{2}^{[n]}$ induced by the singletons and their complements, denoted by i and $\bar{i}$ for $i \in [n]$ (S_4 appears below). In a realizer of S_n, for each i there must be a linear extension in which i appears above $\bar{i}$, establishing the incomparable pair $(\bar{i}, i)$. This forces the other singletons to appear below this pair and the other sets of size $n - 1$ to appear above them. Thus n distinct extensions are needed. Any n extensions establishing these pairs also establish all others.   ∎

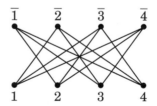

One motivation for studying dimension is compact encoding of n-element posets. A $0, 1$-matrix for the order relation takes n^2 bits, testing $x < y$ in unit time by checking an entry. A realizer of size k uses $kn \log_2 n$ bits. This takes less storage when $\dim P \leq O(n/\log_2 n)$, although testing $x < y$ takes k comparisons.

Encoding each element by its heights on the k extensions of a realizer embeds a poset in $\mathbb{R}^k$ under the product ordering. Indeed, the least such k is another definition of dimension, often attributed to Ore but given earlier by Hiraguchi.

12.3.7. Theorem. (Hiraguchi [1955], Ore [1962]) A partial order P has a realizer of size k if and only if it embeds in the product of k chains.

Proof: Given a realizer, we obtain such an embedding using the heights on the extensions to map each element of P to an element of the product. Conversely, given an embedding that maps $x \in P$ to $(x_1, \ldots, x_k)$, we seek a realizer by placing the elements on the ith extension in the order of their ith coordinates.

These values need not be distinct; there may be "ties" in coordinate i. Let $S = \{x \in P: x_i = t\}$. To break the ties, expand this position on the ith extension into any linear extension of the subposet S. This preserves all the relations of P, so we have obtained linear extensions.

To show that this is a realizer, consider $x, y \in S$ with $x \| y$. Because the encoding embeds P, there are coordinates j and k such that $x_j < y_j$ and $x_k > y_k$. The extensions corresponding to these coordinates have $x < y$ and $y < x$, respectively, which is not affected by breaking ties in another coordinate.   ■

Theorem 12.3.7 yields $\dim \mathbf{2}^{[n]} \leq n$, since $\mathbf{2}^{[n]}$ is a product of n chains. Komm [1948] showed that $\dim \mathbf{2}^{[n]} = n$. To see that $\mathbf{2}^{[n]}$ cannot embed in a product of fewer chains, we use S_n and the next observation.

12.3.8. Corollary. If Q is a subposet of P, then $\dim Q \leq \dim P$.

Proof: A subposet of P embeds wherever P embeds. Also, dropping $P - Q$ from the extensions in a realizer of P yields a realizer of Q.   ■

By Example 12.3.6, Theorem 12.3.7, and Corollary 12.3.8, products of n nontrivial chains have dimension n (Ore [1962]). We next prove a more general result. The upper bound was observed by Hiraguchi [1951]. The sufficient condition for equality was originally proved using completion of lattices; we present a later more explicit proof. A poset is **bounded** if it has a unique minimal element $\hat{0}$ and a unique maximal element $\hat{1}$.

12.3.9. Theorem. (Baker [1961]) If P and Q are posets, then $\dim(P \times Q) \leq \dim P + \dim Q$, with equality if P and Q are bounded posets of size at least 2.

Proof: (Kelly [1981]) Let $f: P \to \mathbb{R}^m$ and $g: Q \to \mathbb{R}^n$ be optimal embeddings. Define $h: P \times Q \to \mathbb{R}^{m+n}$ by letting $h((p,q))$ be the concatenation of $f(p)$ and $g(q)$. This is an embedding of $P \times Q$, so $\dim P \times Q \leq m + n$.

Given bounded posets P and Q with size at least 2, let $L_1, \ldots, L_t$ be a realizer of $P \times Q$, indexed so $L_1, \ldots, L_m$ are the extensions with $(\hat{0}, \hat{1}) < (\hat{1}, \hat{0})$. Using $(\hat{0}, \hat{1}) \| (\hat{1}, \hat{0})$, we force a realizer of $P \times \{\hat{0}\}$ among the m extensions with $(\hat{0}, \hat{1}) < (\hat{1}, \hat{0})$ and a realizer of $\{\hat{0}\} \times Q$ among the $t - m$ extensions with $(\hat{0}, \hat{1}) > (\hat{1}, \hat{0})$. Thus $m \geq \dim P$ and $t - m \geq \dim Q$, so $\dim(P \times Q) = t \geq \dim P + \dim Q$.

If P is a chain, then already $m \geq \dim P$, so we may choose $(a, b) \in I(P)$. Since $(a, \hat{1}) \| (b, \hat{0})$, we must have $(a, \hat{1}) < (b, \hat{0})$ on some extension. Since $(\hat{0}, \hat{1}) < (a, \hat{1})$

and $(b, \hat{0}) < (\hat{1}, \hat{0})$, such an extension will have $(\hat{0}, \hat{1}) < (\hat{1}, \hat{0})$ and must be one of $L_1, \ldots, L_m$. Furthermore, since $(a, \hat{0}) < (a, \hat{1})$, this extension has $(a, \hat{0}) < (b, \hat{0})$. Since $(a, b) \in I(P)$ was arbitrary, $L_1, \ldots, L_m$ contains a realizer of the copy $P \times \{\hat{0}\}$ of P, so $m \geq \dim P$. By analogous reasoning, the extensions with $(\hat{0}, \hat{1}) > (\hat{1}, \hat{0})$ realize $\{\hat{0}\} \times Q$, so there are at least $\dim Q$ of those.   ∎

To see that $\dim P \times Q$ can be less than $\dim P + \dim Q$, consider the simplest product involving a non-bounded poset: $\dim (\bullet\ \bullet) \times (\bullet) = 2 = \dim (\bullet\ \bullet) + \dim (\bullet) - 1$. However, it seems that it cannot be much less.

12.3.10. Conjecture. (Kelly–Trotter [1982]) If P and Q are posets, then $\dim P \times Q \geq \dim P + \dim Q - 2$.   ∎

Little is known about this conjecture. Trotter [1985] proved $\dim S_n \times S_n = 2n-2$, and Reuter [1989a] extended this to $\dim S_m \times S_n = m+n-2$ (he also proved $\dim P \times P \geq 4$ when $\dim P = 3$).

COMPUTATION AND BOUNDS

Since a set of extensions realizes P if and only if each ordered pair $(x, y) \in I(P)$ appears in some extension, computing dimension is equivalent to covering of $I(P)$ by the fewest sets of pairs that can appear in one extension. This expresses dimension as hypergraph coloring (recall that the chromatic number of a hypergraph is the minimum size of a vertex partition into sets containing no edge).

12.3.11. Definition. An **alternating cycle** of incomparable pairs in P is a set $\{(x_i, y_i)\}_{i=1}^k$ in $I(P)$ such that $y_i \leq x_{i+1}$ in P for all i (indices modulo k).

If $x \| y$, then (x, y) and (y, x) together form an alternating cycle.

12.3.12. Lemma. The dimension of P is the minimum number of classes covering $I(P)$ such that no class contains an alternating cycle.

Proof: A single extension cannot establish all pairs in an alternating cycle, because the relations $x_i < y_i$ together with $y_i \leq x_{i+1}$ violate transitivity. Thus the number of classes needed is a lower bound on $\dim P$.

For the opposite inequality, let S be a subset of $I(P)$ containing no alternating cycle. The digraph having $y \to x$ whenever $x < y$ in P or $(x, y) \in S$ is acyclic. A linear extension of the transitive closure of this digraph is a linear extension of P that contains all relations in S. Thus $\dim P$ is at most the number of classes in such a covering.   ∎

Thus $\dim P$ is the chromatic number of a hypergraph where the vertices are $I(P)$, the edges are the alternating cycles, and the colors are the linear extensions. We do not need all of this hypergraph to compute $\dim P$. In computing $\dim S_n$ it was enough to establish the incomparable pairs $(\bar{i}, i)$; the others were then necessarily also established. In general, it suffices to establish the "crucial" incomparable pairs.

12.3.13. Definition. Among ordered incomparable pairs, (a, b) **forces** (c, d) if
$c \leq a$ and $b \leq d$ (every extension with $a < b$ has $c < d$). A pair $(x, y) \in I(P)$
is **unforced** if $x < y$ is not implied by adding any other pair from $I(P)$ to P.
Let $C(P)$ denote the set of unforced pairs.

In the literature, when (x, y) is an unforced pair, the ordered pair (y, x) is
called a **critical pair**. Realizing a poset requires *reversing* the critical pairs,
which is equivalent to *establishing* the unforced pairs.

12.3.14. Proposition. An ordered incomparable pair (x, y) is an unforced pair if
and only if $D(y) \subseteq D(x)$ and $U(x) \subseteq U(y)$.

Proof: The pair (x, y) fails to be unforced if and only if there is an incomparable
pair (a, b) other than (x, y) such that adding the relation $a < b$ forces $x < y$. Such
forcing occurs if and only if $x \leq a$ and $b \leq y$. Hence it will occur if any element of
$U[x]$ is incomparable to any element of $D[y]$. Since x and y must remain incomparable, the forcing fails if and only if all of $U[x]$ is above all of $D[y]$, except for
the pair (x, y) itself. This is equivalent to $U(x) \subseteq U(y)$ and $D(y) \subseteq D(x)$. ∎

12.3.15. Theorem. The dimension of P is the minimum number of linear extensions establishing the unforced pairs of P. This equals the chromatic number
of the hypergraph $H(P)$ with vertex set $C(P)$ whose edges are the minimal alternating cycles consisting of unforced pairs.

Proof: By Definition 12.3.13, extensions that establish $x < y$ for all $(x, y) \in C(P)$
also establish all incomparable pairs. ∎

The hypergraph $H(P)$ is useful for computing dimension in special classes because the lower bound can be established by exhibiting any subgraph of $H(P)$ with
the desired chromatic number. The upper bound is then verified by exhibiting a
realizer that has all the unforced pairs, rather than by verifying that every edge
of $H(P)$ is properly colored. For the standard example, $C(S_n) = \{(\bar{i}, i) \colon 1 \leq i \leq n\}$.
The minimal alternating cycles are precisely the sets of two unforced pairs so
$H(S_n) = K_n$ and $\chi(H(S_n)) = n$.

12.3.16.* Remark. Since computing $\dim P$ is a hypergraph coloring problem,
it is not surprising that testing $\dim P \leq 3$ is NP-complete. Yannakakis [1982]
showed this by constructing from any graph a poset that has dimension 3 if and
only if the graph is 3-colorable. Recognition of 2-dimensional posets runs in time
linear in the number of comparabilities (McConnell–Spinrad [1999]). Structural
descriptions of 2-dimensional posets allow problems that are hard in general to
run quickly on this class (see Möhring [1985], Spinrad [1982, 2003]).

We can also seek a forbidden subposet characterization of d-dimensional
posets. A poset P is **irreducible** if deleting any element reduces its dimension; it
is k-**irreducible** if also $\dim P = k$. The only 2-irreducible poset is the 2-element
antichain. Kelly [1977] and Trotter–Moore [1976b] independently found all 3-
irreducible posets (see the dimension survey in Kelly–Trotter [1982]). The list
includes seven infinite families and ten small examples. ∎

Although exact computation of $\dim P$ is difficult, there are bounds in terms
of other parameters. For S_n, the dimension equals the width and is half the size.

Hiraguchi proved that this is extremal for both parameters. The proof of $\dim P \leq$ $|P|/2$ developed here is due independently to Kimble [1973] and Trotter [1975].

12.3.17. Definition. A linear extension L **puts** Y **over** X if X and Y are disjoint subposets and y is above x in L whenever $x \| y$, $x \in X$, and $y \in Y$. For $Q \subseteq P$, an **upper extension** of Q is a linear extension that puts $P - Q$ over Q; a **lower extension** puts Q over $P - Q$.

12.3.18. Lemma. Every chain in a poset has upper and lower extensions.

Proof: By symmetry, it suffices to find upper extensions. Let C be a chain in a poset P consisting of $x_1, \ldots, x_k$ from bottom to top. Let L be a linear extension of $P - C$. For $1 \leq i \leq k - 1$, let Y_i be the set of elements of $P - C$ that are less than x_{i+1} but not less than x_i; also let $Y_0 = D(x_1)$ and $Y_k = P - D[C]$. Form a linear ordering L' of the elements of P by inserting between x_i and x_{i+1} all elements of Y_i in the order that they have on L. Similarly insert Y_0 before x_1 and Y_k after x_k.

   If $y' < y$ for $y \in Y_i$ and $y' \notin C$, then transitivity yields $y' \in Y_j$ for some $j \leq i$, so L' puts y' and y in the right order. Also x_i and y appear in the right order when they are comparable. If $x_i \| y$, then $y \in Y_j$ for some $j \geq i$. Thus L' is an upper extension of C. ∎

   Not all subposets have upper or lower extensions. Rabinovitch [1978] determined when there is an extension putting Y over X (Exercise 16).

12.3.19. Theorem. (Dilworth [1950], Hiraguchi [1955]) $\dim P \leq w(P)$.

Proof: Start with a Dilworth decomposition of P (a partition into $w(P)$ chains), and take an upper extension of each chain. The resulting extensions form a realizer, since incomparable elements appear on different chains in the original partition **C**. If $x \| y$, then x appears above y on the extension arising from the chain of **C** containing y, and y appears above x on the extension arising from the chain of **C** containing x. ∎

   In addition to the notations $U(x)$ and $D(x)$ for the sets of elements above and below x, we also use $I(x) = \{y \in P : y \| x\}$. Thus $P - x = U(x) \cup D(x) \cup I(x)$.

12.3.20. Theorem. (One-Point Removal Theorem; Hiraguchi [1951]) If x is an element of a poset P, then $\dim P \leq 1 + \dim(P - x)$.

Proof: We construct a realizer of P from a realizer **L** of $P - x$. Extract from P the subsets $U(x)$, $D(x)$, $U(x) \cup I(x)$, and $D(x) \cup I(x)$. Using in each of these subposets the order given by an extension L in **L**, form the two extensions $U(x) \cup I(x) > x > D(x)$ and $U(x) > x > D(x) \cup I(x)$. Replace L in **L** by these two extensions, and insert x anywhere between $U(x)$ and $D(x)$ on the other extensions.

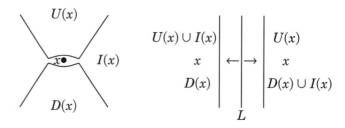

We claim that these $1 + \dim(P - x)$ extensions realize P. The two new extensions establish the incomparable pairs involving x. We must also consider incomparable pairs (y, z) with $y \in I(x)$ and $z \in U(x) \cup D(x)$ that were established by L. Since the two new extensions have $U(x) \cup I(x)$ and $D(x) \cup I(x)$ in order as on L, the pair (y, z) appears as it did in L in one of the new extensions.   ∎

12.3.21. Theorem. If A is an antichain of P, then $\dim P \leq \max\{2, |P - A|\}$.

Proof: As a basis for induction, we want $\dim P \leq 2$ when $|P - A| \leq 2$. This reduces by easy remarks to several cases like the posets shown below, where A is the marked antichain (see Exercise 13).

For $|P - A| > 2$, deleting all but two elements of $P - A$ from P leaves a poset with only two elements outside A. Hence the One-Point Removal Theorem inductively yields $\dim P \leq |P - A|$ when $|P - A| > 2$.   ∎

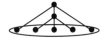  

12.3.22. Corollary. (Hiraguchi's Inequality; Hiraguchi [1955]) If $|P| \geq 4$, then $\dim P \leq |P|/2$.

Proof: To obtain $\dim P \leq |P|/2$ when $w(P) > |P|/2$, apply Theorem 12.3.21; otherwise apply Theorem 12.3.19.   ∎

The technique of Theorem 12.3.20 yields other *removal theorems*, bounding $\dim P$ in terms of the dimension of a subposet Q. Start with a realizer of Q, modify and/or add extensions appropriately, and show that all the incomparable pairs of P are established (Exercises 17–22).

Theorem 12.3.20 and Corollary 12.3.22 suggest a famous conjecture.

12.3.23. Conjecture. (Two-Point Removal Conjecture) Every poset with at least three elements has a **removable pair** of elements; a pair $\{x, y\}$ such that $\dim(P - \{x, y\}) \geq \dim P - 1$.   ∎

Removable pairs are discussed in Exercises 19–26. Tator [1983] proved a weaker statement than Conjecture 12.3.23: always there exist four points in P whose removal decreases the dimension by at most 2 (Exercise 22).

BIPARTITE POSETS

The standard examples S_n are posets whose elements are all maximal or minimal. We consider more general such posets.

12.3.24. Definition. A **bipartite poset** is a poset having no 3-element chains (bipartite comparability graph). A bipartite poset is **normal** if (1) its comparability graph is connected, and (2) when x and y are on the same level, some element other than x is comparable to x but not to y.

12.3.25. Remark. In a normal bipartite poset, the unforced pairs are the pairs $(x, y) \in I(P)$ such that x is maximal and y is minimal.　　　　　■

Various bipartite subposets of $\mathbf{2}^n$ generalize S_n and lead to natural dimension problems. We study those consisting of two ranks.

12.3.26. Definition. Write $\mathbf{2}^n_{l,k}$ for the subposet of $\mathbf{2}^n$ induced by the l-sets and k-sets. Let $d_n(l, k) = \dim \mathbf{2}^n_{l,k}$.

The case of most interest is $d_n(1, k)$. Spencer [1971] showed that $d_n(1, k) \sim c_k \lg \lg n$ when k is constant (we henceforth use lg for $\log_2$ and ln for $\log_e$). Dushnik [1950] computed $d_n(1, k)$ exactly when $k \geq 2\sqrt{n}$; the exact result appears in Theorem 12.3.29. After a slow decline in $d_n(1, k)$ as k decreases from n, the drop becomes rapid for k below $2\sqrt{n}$. The upper and lower bounds when $k \in o(n)$ differ by a factor of $\ln n$.

k	$d_n(1, k)$	reference
$n-1$	n	standard example S_n
$\frac{n-1}{2} \leq k \leq n-2$	$n-1$	Dushnik [1950]
$\frac{n}{3} < k \leq \frac{n}{2} - 1$	$n-2$	Dushnik [1950]
$k = n^{\alpha} \geq 2\sqrt{n}$	$n - n^{1-\alpha} + O(n^{2-3\alpha})$	Dushnik [1950]
general	$< k(k+1)\ln(ne/k)$	Füredi–Kahn [1986]
$2 \leq k \leq 2\sqrt{n} - 4$	$\geq \frac{1}{4}(k+2)^2$	Exercise 29
constant	$\sim c_k \lg \lg n$	Spencer [1971]
2	$\sim \lg \lg n$	Spencer [1971]

These results use different methods. Dushnik's lower bound uses the Pigeonhole Principle, Spencer's bounds are by relating the problem to other questions, and the Füredi–Kahn upper bound is probabilistic.

For bipartite posets in general, the computation of dimension can be reduced to constructing an appropriate set of permutations of the minimal elements. By "permutation", here we mean a linear ordering written out as a list of elements in order; we use "permutation" to distinguish a linear ordering of some of the elements from a linear extension of the full poset. An element of a permutation "comes later than" or "follows" all the elements to its left.

12.3.27. Definition. Given a bipartite poset P, let X and Y be the sets of minimal and maximal elements, respectively. For $y \in Y$, let $S_y = \{x \in X: x < y\}$. A set $\{L_1, \ldots, L_t\}$ of permutations of X is a **suitable set** for P if whenever $y \| x$ with $y \in Y$ and $x \in X$, some L_i puts x later than all of S_y.

12.3.28. Lemma. If P is a normal bipartite poset, then $\dim P$ equals the minimum size of a suitable set for P.

Proof: Let t be the minimum size of a suitable set for P. Given a suitable set of size t, in each permutation we insert each maximal element y immediately after the last element of S_y. The resulting linear orderings can be viewed as linear extensions of P. In fact, they form a realizer of size t, since the unforced pairs are the max-min pairs (y, x) with $x \notin S_y$. Hence $\dim P \leq t$.

For the opposite inequality, consider a smallest realizer of P; this is a set of $\dim P$ linear extensions of P. In each such extension, any maximal element y of P must come later than each element of S_y, since extensions of P preserve the order relation. Hence in an extension where x follows y, also x follows all of S_y. If $x\|y$, then x must follow y in some extension in the realizer. Therefore, deleting the maximal elements from each extension in the realizer yields a set of permutations of the minimal elements that by definition is a suitable set. Hence $t \leq \dim P$.   ∎

In studying $d_n(1,k)$, we thus seek realizers as suitable sets of permutations of $[n]$. Lemma 12.3.28 immediately implies that $d_n(1,k)$ is nondecreasing in k. Dushnik's result below thus yields $d_n(1,k)$ exactly for all k with $k \geq 2\sqrt{n}$.

12.3.29. Theorem. (Dushnik [1950]) If $1 \leq r \leq \sqrt{n}$, then $d_n(1,k) \leq n - r$ if and only if $k \leq n/r + r - 3$.

Proof: We show that a suitable set of $n - r$ permutations of $[n]$ exists for $\mathbf{2}^n_{1,k}$ if and only if $k \leq n/r + r - 3$. Let $t = n - r$.

Necessity. Let $L_1, \ldots, L_t$ be a suitable set for $\mathbf{2}^n_{1,k}$. By symmetry, we may assume that 1 is last in L_1. Thus in L_1, element 1 follows all k-sets omitting 1. Hence in $L_2, \ldots, L_t$ we may move 1 to the beginning. By arguing similarly for $2, \ldots, t$ on $L_2, \ldots, L_t$, we may assume that each L_i ends with i.

Let $R = \{t+1, \ldots, n\}$, so $|R| = r$. For a fixed element $x \in R$, let S be the set of indices $i \in [t]$ such that x appears last among R in L_i. Let $A = (R - \{x\}) \cup S$. Note that x does not follow A_x in any L_i (if $i \in S$, then i follows x in L_i; otherwise, x is not last among $R - \{x\}$ in L_i). Every k-set not containing x must precede x on some L_i, so $|A| > k$. Thus $r - 1 + |S| > k$, so $|S| > k - r + 2$.

On the other hand, the Pigeonhole Principle guarantees that some $x \in R$ appears last among R at most t/r times. Thus $|S| \leq t/r$ for some x. Now $k - r + 2 \leq \min|S| \leq t/r$. Using $t = n - r$, we obtain $k \leq n/r + r - 3$. Hence this inequality is necessary for $d_n(1,k) \leq t$.

Sufficiency. We define permutations $L_1, \ldots, L_t$ of $[n]$ that meet the necessity conditions above and form a suitable set if $k \leq n/r + r - 3$ and $r \leq \sqrt{n}$. The last element of L_i is i, preceded immediately by R in some order. Each element of R is next-to-last in $\lfloor t/r \rfloor$ or $\lceil t/r \rceil$ of the permutations.

Since $r \leq \lfloor \sqrt{n} \rfloor \leq \lfloor n/r \rfloor$, we have $r - 1 \leq \lfloor (n-r)/r \rfloor = \lfloor t/r \rfloor$. Thus for $x \in R$, we can make each element of $R - \{x\}$ appear immediately before x in one of the permutations where x is next-to-last. Thus for distinct x and y in R, there is a permutation L_j that ends y, x, j.

Let A be a set that does not appear before x in any L_j. If x is next-to-last in L_i, then $i \in A$. If x is third-to-last in L_j, followed by z and j, then A must contain z or j (these pairs are disjoint and omit those i where x is next-to-last on L_i). Thus $|A| \geq \lfloor t/r \rfloor + r - 1 > n/r + r - 3 \geq k$. Hence $L_1, \ldots, L_t$ is a suitable set, and $k \leq n/r + r - 3$ yields $d_n(1,k) \leq t$.   ∎

When $k \geq 2\sqrt{n}$, Theorem 12.3.29 gives $d_n(1, k)$ exactly. For smaller k, it gives no better upper bound than $n - \sqrt{n}$. A technique like that in Theorem 12.3.29 yields a lower bound $d_n(1, k) \geq k^2/4$ when $k \leq \sqrt{n}$ (Exercise 29). We present a general upper bound.

12.3.30. Theorem. (Füredi–Kahn [1986]) $d_n(1, k) \leq \lceil k(k + 1) \ln(ne/k) \rceil$.

Proof: Generate permutations $L_1, \ldots, L_t$ of $[n]$ by selecting each at random from all $n!$ orders. We show that if $t \geq k(k+1)\ln(ne/k)$, then with positive probability these form a suitable set of permutations for $\mathbf{2}^n_{1,k}$. Hence some outcome of the experiment is a realizer of the desired size.

For each k-set S and each $x \in [n]$ with $x \notin S$, the probability is $1/(k+1)$ that x follows all of S on L_j. Hence the probability that x follows all of S in none of the random permutations is $(\frac{k}{k+1})^t$. There are $n\binom{n-1}{k}$ such pairs (S, x). Hence we bound the probability that some pair is not established:

$$\mathbb{P}(\text{failure}) \leq n\binom{n-1}{k}\left(\frac{k}{k+1}\right)^t < \binom{n}{k}\left(1 - \frac{1}{k+1}\right)^t$$

$$< \left(\frac{ne}{k}\right)^k e^{-t/(k+1)} \leq \left(\frac{ne}{k}\right)^k e^{-k\ln(ne/k)} = 1 .$$

We used standard inequalities $\binom{n}{k} \leq \left(\frac{ne}{k}\right)^k$ and $1 - x \leq e^{-x}$ (Chapter 14), slightly weakening the Füredi–Kahn bound to simplify computation.   ∎

In the realm of constant k, Spencer's construction of small realizers used special families of sets (he attributed this argument to A. Hajnal).

12.3.31. Definition. A family F of sets is k-**scrambling** if for all choices $S_1, \ldots, S_k$ of k distinct sets in F and all subsets A of the index set $[k]$, the set $(\bigcap_{r \in A} S_r) \cap (\bigcap_{r \notin A} \overline{S}_r)$ is nonempty.

The k-scrambling condition can be stated in several ways. If F is 2-scrambling and $X, Y \in F$, then $X \cap Y$, $X - Y$, $Y - X$, and $\overline{X} \cap \overline{Y}$ are nonempty; for k in general, the condition states that in the Venn diagram on any k sets in F, every cell is nonempty. In terms of elements, the condition is that for $S_1, \ldots, S_k \in F$ and $A \subseteq [k]$, there exists an element x such that $x \in S_r$ for $r \in A$ and $x \notin S_r$ for $r \notin A$. We will use k-scrambling families to prove an upper bound on $d_n(1, k)$.

12.3.32. Example. The four permutations of the elements 0 through 7 shown below form a suitable set for $\mathbf{2}^8_{1,2}$, proving $d_8(1, 2) \leq 4$.

$$L_1: 0, 1, 2, 3, 4, 5, 6, 7$$
$$L_2: 3, 2, 1, 0, 7, 6, 5, 4$$
$$L_3: 5, 4, 7, 6, 1, 0, 3, 2$$
$$L_4: 6, 7, 4, 5, 2, 3, 0, 1$$

It is easy to show directly that this is a suitable set. An element at the end of a permutation follows all pairs among the other elements in that permutation. This takes care of the singletons $1, 2, 4, 7$. The elements $0, 3, 5, 6$ appear next-to-last. When r is next-to-last in a permutation, on that permutation it follows all pairs of other elements except the pairs involving the last element s. In each case, there are two permutations where s immediately precedes r, and each of the other six elements precedes s in one of those permutations.   ∎

These carefully structured permutations arose as a special case of a general construction using k-scrambling sets.

12.3.33. Lemma. Given a k-scrambling family $S_1, \ldots, S_m$ of subsets of $[t]$, there exist permutations $L_1, \ldots, L_t$ of the numbers 0 through $2^m - 1$ such that if $a < b$ and j is the leftmost position where the m-bit binary expansions of a and b differ, then b follows a on L_i if and only if $i \in S_j$.

Proof: Begin with all the m-bit binary integers in one "group". To produce L_i, perform Steps 1 through m in order as follows. On Step j, each current group splits into two smaller groups. In a current group X, let X_r be the numbers whose expansion has r in coordinate j, for $r \in \{0, 1\}$. Replace X with X_1 after X_0 if $i \in S_j$; otherwise put X_0 after X_1.

After Step j, the groups all have size 2^{m-j}; thus an explicit ordering L_i is produced after Step m. Furthermore, the relative ordering between a and b on L_i is determined in Step j, where j is the leftmost position where the expansions of a and b differ (a has 0 there; b has 1). The procedure explicitly puts b in a group after a if and only if $i \in S_j$.   ∎

Example 12.3.32 arises from Lemma 12.3.33 using the family given by $S_1 = \{1, 2\}$, $S_2 = \{1, 3\}$, and $S_3 = \{1, 4\}$. This is a 2-scrambling family of subsets of $[4]$; here $m = 3$. For each L_i, Step 1 splits the elements into the two groups $\{0, 1, 2, 3\}$ and $\{4, 5, 6, 7\}$. The lower group goes first when $i \in S_1$, which holds for $i \in \{1, 2\}$; in L_3 and L_4 the lower group goes last. Step 2 splits the lower group into $\{0, 1\}$ and $\{2, 3\}$ and the upper group into $\{4, 5\}$ and $\{6, 7\}$. Within each group, the lower subgroup goes first when $i \in S_2$, which holds for $i \in \{1, 3\}$. Step 3 finishes the job, deciding which goes first in each pair.

We will use this construction for the upper bound in the next theorem. Given m sets forming a k-scrambling family of subsets of $[t]$, we will obtain $d_n(1, 2) \leq t$ when $n = 2^m$. Hence we seek a large k-scrambling family.

Let $M(t, k)$ be the maximum size of a k-scrambling family in 2^t. We are particularly interested in $k = 2$. The family of all $\lfloor t/2 \rfloor$-sets in $[t]$ that contain the element 1 is 2-scrambling, so $M(t, 2) \geq \binom{t-1}{\lfloor t/2 \rfloor - 1} > 2^t / \sqrt{2\pi t}$. For $k \in \mathbb{N}$, there is a constant c_k such that $M(t, k) \geq c_k^t$ (Exercise 31).

12.3.34. Theorem. (Spencer [1971]) If c is a constant such that $[t]$ has a k-scrambling family of size greater than c^t, then

$$\lg \lg(n - 1) < d_n(1, k) < \tfrac{1}{\lg c} \lg \lg n.$$

Proof: *Upper bound.* We prove $d_n(1, k) \leq t$ for $n = 2^{M(t,k)}$. Since $M(t, k) > c^t$, this yields $d_n(1, k) < \tfrac{1}{\lg c} \lg \lg n$. For convenience, let $m = M(t, k)$. Let $\{S_1, \ldots, S_m\}$ be a largest k-scrambling family of subsets of $[t]$. Lemma 12.3.33 provides orderings $L_1, \ldots, L_t$ of 0 through $n - 1$ such that if $a < b$ and j is the leftmost position where a and b differ as vectors, then b follows a on L_i if and only if $i \in S_j$.

We claim that $\{L_1, \ldots, L_t\}$ is a suitable set for $\mathbf{2}^n_{1,k}$. Viewing the elements as binary m-vectors, consider a vector b and vectors $a_1, \ldots, a_k$ other than b. For $1 \le r \le k$, let j_r be the first coordinate where a_r and b differ. Let $A = \{r : b_{j_r} = 1\}$, so $A \subseteq [k]$. Since $\{S_1, \ldots, S_m\}$ is a k-scrambling family and $\{j_1, \ldots, j_k\}$ is a set of at most k indices, there is a value $i \in [t]$ such that $i \in S_{j_r}$ for $r \in A$ and $i \notin S_{j_r}$ for $r \notin A$. We claim that b occurs after all of $a_1, \ldots, a_k$ on L_i.

In constructing L_i, element b is compared with a_r when processing coordinate j_r. If $b_{j_r} = 1$, then $r \in A$ and $i \in S_{j_r}$. If $b_{j_r} = 0$, then $r \notin A$ and $i \notin S_{j_r}$. In either case, b is placed after a_r at stage j. Hence b follows each of $a_1, \ldots, a_k$ in L_i. We conclude that $L_1, \ldots, L_t$ is a suitable set.

Lower bound. A suitable set of permutations for $\mathbf{2}^n_{1,k}$ is also a suitable set for $\mathbf{2}^n_{1,k-1}$. Hence by Lemma 12.3.28 it suffices to prove the lower bound for $k = 2$. We prove that if $n \ge 2^{2^t} + 1$, then any t orderings of $[n]$ yield a triple that appears monotonically (increasing or decreasing) in each ordering. For such a triple $\{x, y, z\}$ with $x < y < z$, we have y between x and z in each permutation. Hence y never appears after $\{x, z\}$. We conclude that a suitable set must have more than t permutations, so $d_n(1, 2) > t$ if $t \le \lg\lg(n-1)$.

We prove the claim by induction on t using the Erdős–Szekeres Theorem (Theorem 10.1.13): in every list of $m^2 + 1$ distinct numbers some $m + 1$ numbers appear monotonically (Exercise 12.1.15 requests a proof using Dilworth's Theorem). For $t = 1$, five elements suffice to guarantee a monotone triple. For $t > 1$, consider orderings $L_1, \ldots, L_t$ on $[2^{2^t} + 1]$. The Erdős–Szekeres Theorem yields a set S of size $2^{2^{t-1}} + 1$ that appears monotonically in L_t. Within S, the induction hypothesis yields a triple $\{x, y, z\}$ that appears monotonically in each of $L_1, \ldots, L_{t-1}$. By the choice of S, this triple is also monotone in L_t.   ∎

12.3.35. Corollary. (Spencer [1971])

$$\lg\lg n \le d_n(1, 2) < \lg\lg n + \tfrac{1}{2}\lg\lg\lg n + O(1).$$

Proof: We have $M(t, 2) \ge \binom{t-1}{\lfloor t/2 \rfloor - 1} > 2^t/\sqrt{2\pi t}$ (using Stirling's Formula). By the argument in Theorem 12.3.34, $d_n(1, k) \le t$ when $\lg n = M(t, k)$. Solving for t in $\lg n = 2^t/\sqrt{2\pi t}$ yields the more precise upper bound.   ∎

Note the effect of the double exponential. Since $M(t, 2)$ is the number of $\lfloor t/2 \rfloor$-subsets of $[t]$ containing element 1, for $t = 6$ we have $M(t, 2) = 10$. The lower and upper bounds thus yield $4 \le d_{1024}(1, 2) \le 6$.

For $d_n(1, 2)$, Corollary 12.3.35 establishes the asymptotic behavior. We provided the next term of the upper bound because this is in fact sharp. Indeed, there are four natural related problems whose answers in terms of n are $\lg\lg n + (\tfrac{1}{2} + o(1))\lg\lg\lg n$. We discuss three in the rest of this section and the fourth in Exercise 12.4.7.

Meanwhile, we note that Hoşten–Morris [1999] determined exact values of n where $d_n(1, 2)$ increases. Biró–Hamburger–Pór–Trotter [2016] observed that with the result of Kleitman–Markowsky [1975], this determines $d_n(1, 2)$ exactly for almost all n, and within 1 otherwise. In particular, given any positive ε, for sufficiently large n we have $s - \varepsilon < d_n(1, 2) < s + 1 + \varepsilon$, where $s = \lg\lg m + \tfrac{1}{2}\lg\lg\lg n + \tfrac{1}{2}\lg\pi + \tfrac{1}{2}$.

12.3.36. Definition. The **shift graph** G_n is the graph with vertex set $\binom{[n]}{2}$ and edges defined by $ij \leftrightarrow jk$ if and only if $i < j < k$ (disjoint pairs are non-adjacent). The **double shift graph** G'_n is the graph with vertex set $\binom{[n]}{3}$ and edges defined by $ijk \leftrightarrow jkl$ if $i < j < k < l$.

12.3.37. Lemma. (A. Hajnal) The chromatic number of the shift graph G_n is $\lceil \lg n \rceil$; this is the least t such that 2^t has at least n elements.

Proof: Given that 2^t has at least n elements, we properly color G_n using $[t]$ as colors. Let $A_1 < A_2 < \cdots < A_n < \cdots$ be a linear extension of 2^t. For each pair $ij \in V(G_n)$ with $i < j$, color ij with some element of $A_j - A_i$. Since the ordering is an extension of 2^t, such an element exists. Since no element of $A_j - A_i$ can belong to $A_k - A_j$, the coloring is proper.

Conversely, when $\chi(G_n) = t$, we show that $[t]$ has at least n subsets. Consider a proper coloring of G_n using $[t]$ as colors. For each $i \in [n]$, let S_i be the set of colors used on vertices of the form ij with $j > i$. If $S_i = S_j$ with $j > i$, then the color c that appears on ij also appears on jk for some $k > j$. This is impossible in a proper coloring, since $ij \leftrightarrow jk$ in G_n. Hence $S_1, \ldots, S_n$ are distinct. ∎

12.3.38. Theorem. (Füredi–P.Hajnal–Rödl–Trotter [1992]) The chromatic number of the double shift graph G'_n is the least t such that there are at least n antichains in 2^t.

Proof: For the upper bound, we give a proper coloring. Instead of antichains, consider the down-sets they generate. Let $D_1, D_2, \ldots$ be a linear extension of the poset of down-sets in 2^t, ordered by inclusion. Associate with each pair $\{p, q\} \in \binom{[n]}{2}$ having $p < q$ a set $A_{pq} \in D_q - D_p$. Given $i < j < k$, color $ijk \in V(G'_n)$ with an element of $A_{jk} - A_{ij}$. Such an element exists, since otherwise $A_{jk} \subseteq A_{ij}$, but the set A_{jk} not contained in D_j cannot be a subset of the set A_{ij} contained in the down-set D_j. Now suppose that ijk and jkl both have color c. The first requires $c \in A_{jk}$, and the second requires $c \notin A_{jk}$. Thus the coloring is proper.

Conversely, let $t = \chi(G'_n)$ and consider an optimal coloring; we show that 2^t has at least n down-sets. For each pair $i, j \in [n]$, let S_{ij} be the set of colors appearing on vertices of the form ijk with $k > j$, and let $B_i = \{S_{ij} : j > i\}$. Let D_i be the down-set consisting of all subsets of $[t]$ contained in elements of B_i. We must show that these down-sets are distinct. Note that $D_n = \varnothing$ and $D_{n-1} = \{\varnothing\}$; the other down-sets contain non-empty sets.

Consider $D_i = D_j$ with $i < j \leq n - 2$. Since S_{ij} is nonempty and contained in D_i, and D_j is generated by B_j, the condition $D_i = D_j$ requires a set of the form S_{jk} that contains S_{ij}. This contradicts the coloring of G'_n, because the color in S_{ij} that is used on ijk cannot appear in S_{jk}. ∎

12.3.39. Corollary. (Erdős–A.Hajnal) The chromatic number of the double shift graph G'_n is $\lg \lg n + (\frac{1}{2} + o(1)) \lg \lg \lg n$.

Proof: This follows from the Kleitman–Markowsky expression $2^{\binom{n}{\lfloor n/2 \rfloor}(1+o(1))}$ for the number of down-sets in 2^n (see Section 11.3). ∎

12.3.40. Corollary. The dimension of the subposet of 2^n induced by the sets of sizes 1 and 2 is $\lg \lg n + (\frac{1}{2} + o(1)) \lg \lg \lg n$.

Proof: Corollary 12.3.35 establishes the upper bound. We prove that $\chi(G'_n)$ is a lower bound. Let $L_1, \ldots, L_t$ be a realizer. For ijk with $i < j < k$, choose $c \in [t]$ such that the singleton j is above the doubleton ik in L_c. Let c be the color of ijk. We cannot also give color c to jkl with $k < l$, because this would place $ik < j < jl < k < ik$ on L_c. Hence this defines a proper t-coloring of G'_n, and $\chi(G'_n)$ is a lower bound on the dimension.   ∎

Although we have emphasized $d_n(1, k)$, the dimension of the subposet of k-sets and $(n - k)$-sets in 2^n is also of interest. Füredi [1994] proved $d_n(k, n - k) \geq n - 2k + 2$ for $n > 2k$ (Exercise 16.2.18).

EXERCISES 12.3

12.3.1. (–) Prove that the intersection of two order relations on the same set is an order relation.

12.3.2. (–) Prove that a set of elements forms a down-set in a poset P if and only if it is an initial segment of some linear extension of P.

12.3.3. (–) List all linear extensions of the poset in Example 12.3.4. Determine which belong to realizers of size 2.

12.3.4. (–) Describe all pairs of linear extensions realizing $\mathbf{m} + \mathbf{n}$.

12.3.5. (–) Prove that an n-vertex graph G is the comparability graph of a 2-dimensional poset if and only if its vertices can be named $v_1, \ldots, v_n$ so that there is a permutation σ of $[n]$ such that $v_i v_j \in E(G)$ for $i < j$ if and only if $\sigma_i > \sigma_j$.

12.3.6. Prove that the posets below have dimension 3. List four techniques that can be used for the lower bound.

12.3.7. The "fence" poset F_n has minimal elements $x_0, \ldots, x_n$ and maximal elements $y_1, \ldots, y_n$ with comparable pairs given by $x_{i-1}, x_i < y_i$ for $1 \leq i \leq n$. Use an embedding of F_n in $\mathbb{R}^2$ to find all pairs of linear extensions realizing F_n.

12.3.8. Prove that the poset below has dimension 3.

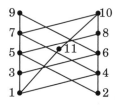

12.3.9. The poset in Exercise 12.3.8 is 3-irreducible (deleting any element leaves a 2-dimensional poset), and it generalizes easily to 3-irreducible posets of all nontrivial heights. Use this and Theorem 12.3.9 to prove the existence of n-irreducible posets of large height.

12.3.10. Prove that a graph G is the complement of a comparability graph if and only if G is the intersection graph of the curves graphing a set of continuous real-valued functions on $[0, 1]$. (Golumbic–Rotem–Urrutia [1983])

12.3.11. Let P be the graded poset with rank sizes $a_0, \ldots, a_r$ such that elements are incomparable if and only if they have the same rank. Prove $\dim P = 2$. Describe the minimal alternating cycles of incomparable pairs and a proper 2-coloring of the hypergraph $H(P)$.

12.3.12. Let $\{(x_i, y_i) : 1 \leq i \leq k\}$ be a minimal alternating cycle of incomparable pairs in a poset P. Prove that $\{x_1, \ldots, x_k\}$ and $\{y_1, \ldots, y_k\}$ are antichains in P.

12.3.13. Complete the proof that $\dim P \leq \max\{2, |P - A|\}$ by proving that $\dim P \leq 2$ if P has an antichain A such that $|P - A| = 2$.

12.3.14. ($\diamond$) The **composition** (also called **lexicographic product**) $Q[P_1, \ldots, P_k]$ is formed from a poset Q of size k by expanding each $x_i \in Q$ to a copy of P_i; elements expanded from x_i and x_j are related as x_i and x_j are related in Q. Prove that if $P = P_0[P_1, \ldots, P_k]$, then $\dim P = \max_{i \geq 0} \dim P_i$. (Hiraguchi [1951])

12.3.15. ($\diamond$) Given a connected graph G, let P be the poset of subsets of $V(G)$ that induce connected subgraphs of G, ordered by inclusion. Prove that $\dim P$ is the number of non-cut vertices in G. (Hint: Use distance from non-cut-vertices to partition the ordered incomparable pairs into classes that avoid alternating cycles.) (Trotter–Moore [1976a])

12.3.16. ($\diamond$) Let X and Y be disjoint subposets of a poset P. Prove that P has a linear extension putting Y over X if and only if P contains no copy of $\mathbf{2} + \mathbf{2}$ with minimal elements in Y and maximal elements in X. (Hint: Use induction on the number of incomparable pairs (x, y) with $x \in X$ and $y \in Y$.) (Rabinovitch [1978])

12.3.17. In a poset P that is not an antichain, let C be a chain, M be the antichain of maximal elements, and A be an antichain. Prove the following inequalities.
  (a) $\dim P \leq 2 + \dim(P - C)$. (Hiraguchi [1951])
  (b) $\dim P \leq 1 + w(P - M)$. (Trotter [1975])
  (c) $\dim P \leq 1 + 2w(P - A)$. (Trotter 1975]; sharpness in Trotter [1974b])

12.3.18. (+) Let C be a chain in a poset P such that each element of $P - C$ is incomparable to at most one element of C. Prove that $\dim P \leq 1 + \dim(P - C)$. (Bogart–Trotter [1973])

12.3.19. ($\diamond$) Suppose that a is a maximal element in P, b is a minimal element in P, and $a \| b$. Prove that $\dim P \leq 1 + \dim(P - \{a, b\})$.

12.3.20. ($\diamond$) Given $a < b$ in a poset P, let $r(a, b)$ count the ordered pairs $(x, y) \in I(P)$ such that $a < x$ and $y < b$. Prove that (a, b) is a removable pair if $r(a, b) \leq \dim P - 3$. (Hint: In a realizer of $P - \{a, b\}$, replace a well-chosen extension with two others.) (Hiraguchi [1951])

12.3.21. ($\diamond$) Given incomparable elements a and b in a poset P, let $r(a, b)$ count the ordered pairs $(x, y) \in I(P)$ such that x is comparable to both a and b and y is incomparable to both a and b. Prove that (a, b) is a removable pair if $r(a, b) \leq \dim P - 3$. (Hint: In a realizer of $P - \{a, b\}$, replace a well-chosen extension with two others.) (Kelly–Trotter [1982])

12.3.22. ($\diamond$) *Four-Point Removal Theorem.* Let P be a poset.
  (a) Let C and D be chains in P such that $x \| y$ for all $x \in C$ and $y \in D$. Prove that P has a linear extension that puts $P - C - D$ over C and puts D over $P - C - D$. (Hint: Partition $P - C - D$ into the set X_1 below the top of C, the set X_2 above the bottom of D, and the remainder X_3. Combine linear extensions of $X_1 \cup C$, $X_2 \cup D$, and X_3.) (Hiraguchi [1955])
  (b) For C and D as in part (a), prove $\dim P \leq 2 + \dim(P - C - D)$. (Hiraguchi [1955])
  (c) For x and y maximal in P, with $D(x) \subseteq D(y)$, prove $\dim P \leq 1 + \dim(P - x - y)$.
  (d) Given $|P| \geq 4$, use parts (b) and (c) to prove that $x, y, z, w \in P$ exist such that $\dim P \leq 2 + \dim(P - \{x, y, z, w\})$. (Tator [1983])

12.3.23. ($\Diamond$) The k-**dimension** $\dim_k(P)$ of a poset P is the minimum t such that P embeds in $\mathbf{k}^t$. Prove that if P decomposes into t chains of size less than k, then $\dim_k P \le t$. (Comment: Thus k-dimension is well defined.) (Trotter [1976])

12.3.24. ($\Diamond$) Prove that the maximum size of a minimal realizer of an n-element antichain is $\lfloor n^2/4 \rfloor$ (for $n \ge 4$). (Maurer–Rabinovitch [1977])

12.3.25. In the poset below, prove that (x, y) is an unforced pair whose removal decreases the dimension by 2. (Reuter [1989b])

12.3.26. For $n \ge 5$, we construct an n-dimensional poset P_n with $4n - 4$ elements and an unforced pair (y, x) such that $\dim(P_n - \{x, y\}) = n - 2$ (see P_5 below). Begin with disjoint copies A and B of S_{n-2}, with sets A_1 and B_1 of minimal elements, A_2 and B_2 of maximal elements. Add four elements x, y, z, w plus covering pairs $B_1 \prec x \prec A_2$ and $A_1 \cup B_1 \prec y$ and $\{z, w\} \prec A_2 \cup B_2$ and $w \prec y$. Prove that $\dim P_n = n$, that (y, x) is an unforced pair, and that $\dim(P_n - \{x, y\}) = n - 2$. Which property fails for $n = 4$? (Kierstead–Trotter [1991])

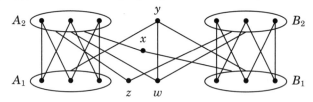

12.3.27. Prove that the n-dimensional standard example S_n is n-irreducible.

12.3.28. The **crown** S_n^k is a bipartite poset with minimal elements i and maximal elements $\bar{i}$ for $i \in [n+k]$, with $i \| \bar{j}$ when $j \in \{i, \ldots, i+k\}$ (modulo $n+k$) and otherwise $i < \bar{j}$.
 (a) Show that $\overline{k+1} < k+1 < \cdots < \bar{1} < 1$ and $\overline{k+2} < 2 < \cdots < \overline{2k+2} < k+2$ together establish the unforced pairs involving $\{1, \ldots, k+2\}$. Conclude $\dim S_n^k \le \lceil \frac{2(n+k)}{k+2} \rceil$.
 (b) Use alternating cycles of length 2 to prove $\dim S_n^k \ge \lceil \frac{2(n+k)}{k+2} \rceil$. (Trotter [1974a])

12.3.29. ($\Diamond$) For $2 \le k \le 2\sqrt{n} - 4$, prove $d_n(1, k) \ge \frac{1}{4}(k+2)^2$. (Hint: Modify the argument for necessity in the proof of Theorem 12.3.29 by letting $R = \{t+1, \ldots, \lceil t + \sqrt{t} \rceil\}$.)

12.3.30. ($\Diamond$) For $k < l$, prove that $d_n(k, l) \ge n - k$ if $l > (n + k - 1)/2$, and $d_n(k, l) \ge l$ if $l \ge 2k$. (Hint: Start with Dushnik's Theorem when $k = 1$.) (Füredi [1994])

12.3.31. ($\Diamond$) k-scrambling sets.
 (a) Prove that the $\lfloor t/2 \rfloor$-sets containing a fixed element form a maximum 2-scrambling family in $\mathbf{2}^t$. (Hint: For odd t, use the Erdős–Ko–Rado Theorem (Theorem 11.2.18).)
 (b) Prove existence of a k-scrambling family in $\mathbf{2}^t$ with size at least $\frac{1}{2}[(1 - 2^{-k})^{-1/k}]^t$. (Hint: Generate m subsets of $[t]$ independently at random, and prove that the probability they are not k-scrambling is less than 1 when m is smaller than desired.) (Spencer [1971])

12.3.32. Let $f(r, k)$ be the maximum size of a k-scrambling family of subsets of $[r]$. Let $\kappa(n, m)$ be the minimum size of a set of vertices in the hypercube Q_n intersecting every m-dimensional subcube. (Graham–Harary–Livingston–Stout [1993])
 (a) Determine $\kappa(4, 2)$.
 (b) Prove $\kappa(n, m) = \min\{r: f(r, n-m) \ge n\}$.

12.4. Special Families of Posets

We begin this section with "chain-like" posets, which maintain some aspects of linear orders as tools of measurement and comparison. Subsequently, we consider posets with special algebraic properties.

SEMIORDERS AND INTERVAL ORDERS

A **ranking** or **weak order** is a partial order whose elements occur in ranks $P_1, \ldots, P_k$ such that elements are incomparable if and only if they belong to the same rank. Rankings are used in voting theory; one seeks a consensus ranking among all voters. Paradoxes abound, and Arrow [1951] proved that no function producing a consensus ranking can satisfy a particular set of four natural axioms.

Rankings are not general enough to model preferences realistically. The big problem is that they require transitivity of indifference. A person given cups of coffee with different amounts of sugar is likely to be indifferent when the amount differs by one or two grains, but a large enough difference yields a preference. A difference of a few dollars in the price of a house won't affect one's attitude toward it, but thousands of dollars will. Luce [1956] introduced a model for "just-noticeable" difference.

12.4.1. Definition. A **semiorder** is a poset representable by a function f and fixed threshold $\delta \geq 0$ so that $x < y$ if and only if $f(y) - f(x) > \delta$ (this is a **semiorder representation**).

The rankings are the posets having semiorder representations with $\delta = 0$. By scaling, we may assume that the threshold δ is 1 when it is nonzero. The terms *weak order* and *semiorder* suggest weakening the conditions for an order relation, but these are quite restricted posets; what they weaken is the condition of total order. We will characterize semiorders as the posets not containing $\mathbf{1 + 3}$ or $\mathbf{2 + 2}$.

There may be uncertainty not only in comparison of elements, but also in assignment of values. For example, the skill of a tennis player may vary from day to day, leading us to represent a player z by an interval $[a_x, b_x]$. We might then conclude that y beats x if the interval for y is wholly above the interval for x. That is, when $a_y > b_x$ we expect y to win.

12.4.2. Definition. An **interval order** is a poset representable by assigning an interval $[a_z, b_z]$ to each element z so that $x < y$ if and only if $b_x < a_y$ (the assignment is an **interval representation**).

12.4.3. Example. Among topics to be discussed in a committee, we set $x < y$ if topic x must be settled before topic y is discussed. On the other hand, we set $x \| y$ if x and y will be available for discussion at the same time. A schedule assigns an interval of time to each topic during which it is available for discussion. ∎

12.4.4. Remark. *Every semiorder is an interval order.* From a semiorder representation f, letting $[a_x, b_x] = [f(x) - \frac{\delta}{2}, f(x) + \frac{\delta}{2}]$ yields an interval representation. The incomparability graph of an interval order is an interval graph, because elements are incomparable when their intervals in an interval representation intersect. The incomparability graph of a semiorder has an interval representation using intervals of length 1.   ■

We will characterize interval orders as the posets not having $\mathbf{2} + \mathbf{2}$ as a subposet. The table below compares the four classes we have discussed.

class	representing function(s)	forbidden subposet
chain	distinct values	$\mathbf{1} + \mathbf{1}$
ranking	$x < y$ if $f(x) < f(y)$	$\mathbf{1} + \mathbf{2}$
semiorder	$x < y$ if $f(x) < f(y) - \delta$	$\mathbf{1} + \mathbf{3}$ and $\mathbf{2} + \mathbf{2}$
interval order	$x < y$ if $f(x) \leq g(x) < f(y) \leq g(y)$	$\mathbf{2} + \mathbf{2}$

In specifying an interval order, the functions f and g above give the left and right endpoints of the corresponding interval, respectively.

The characterizations of interval orders and semiorders by forbidden subposets can be used to construct representations. Semiorders were characterized much earlier, but it is convenient to characterize interval orders first and then characterize semiorders among them. Fishburn–Monjardet [1992] noted that posets without $\mathbf{2} + \mathbf{2}$ were studied as early as Wiener [1914], who gave a characterization in much different terminology.

12.4.5. Theorem. (Fishburn [1970], Mirkin [1972]) A poset is an interval order if and only if it does not contain $\mathbf{2} + \mathbf{2}$ as a subposet.

Proof: (Balof–Bogart [2003]) Suppose that P has an interval representation and that x, y, z, w are four elements inducing $\mathbf{2} + \mathbf{2}$ with $x < y$ and $z < w$. Let $[a_i, b_i]$ be the interval representing $i \in \{x, y, z, w\}$. Because $x < y$ and $z < w$, we have $b_x < a_y$ and $b_z < a_w$. From $x \| w$ and $z \| y$, we have $b_x \geq a_w$ and $b_z \geq a_y$. This yields the contradiction $b_x < a_y \leq b_z < a_w \leq b_x$. Thus the condition is necessary.

For the converse, we use induction on $|P|$ to produce an interval representation for a poset P without $\mathbf{2} + \mathbf{2}$. Choose $x \in P$ to maximize $|D(x)|$; note that x is a maximal element. Let $P' = P - x$. Since P' has no copy of $\mathbf{2} + \mathbf{2}$, the induction hypothesis yields an interval representation of P'.

The key claim is that elements of $I(x)$ are maximal. If z is incomparable to x but not maximal, then choose $w \in U(z)$. Since $|D(x)| \geq |D(w)|$ and $z \in D(w) - D(x)$, there exists $y \in D(x) - D(w)$. Since $z \| x$, also y is incomparable to z and w, and the subposet formed by $\{x, y, z, w\}$ is $\mathbf{2} + \mathbf{2}$.

Conversely, maximal elements are incomparable to x, so $I(x)$ is the set of maximal elements in P other than x. In an interval representation of P', extend the intervals for all maximal elements rightward to a common endpoint a. Since the elements are maximal, this changes no intersections, and a is now the rightmost point in the representation. Among these maximal elements, extend the intervals for those that are also maximal in P further rightward to b. Adding an interval for x that starts at $(a + b)/2$ completes an interval representation of P. ■

The earlier proof of Theorem 12.4.5 in Bogart [1993] produced an explicit interval representation with the minimum number of endpoints (Exercise 5). We build on Theorem 12.4.5 to characterize semiorders, as did Balof–Bogart [2003].

12.4.6. Theorem. (**Scott–Suppes Theorem**; Scott–Suppes [1958]) A poset is a semiorder if and only if neither $\mathbf{2} + \mathbf{2}$ nor $\mathbf{3} + \mathbf{1}$ is a subposet.

Proof: (Bogart–West [1999]) A semiorder is an interval order, so $\mathbf{2} + \mathbf{2}$ is forbidden. Given a semiorder representation f with threshold 1, if $\mathbf{3} + \mathbf{1}$ is a subposet with $x < y < z$ and incomparable element w, then $f(w) \geq f(y)$ contradicts $x \| w$ and $f(w) \leq f(y)$ contradicts $z \| w$. Hence the condition is necessary.

For sufficiency, suppose that P has no $\mathbf{2} + \mathbf{2}$ or $\mathbf{3} + \mathbf{1}$. By Theorem 12.4.5, P is an interval order, representable as in that proof. We convert this representation to one whose intervals have the same length. Since P has no $\mathbf{3} + \mathbf{1}$, there is no pair $x, y \in P$ such that (1) $I_y \subset I_x$ and (2) I_x intersects intervals to the left and right of I_y that do not intersect I_y. This enables us to alter the representation so that no interval properly contains another. If $I_x = [a, b]$ and $I_y = [c, d]$ with $a < c \leq d < b$, then we know that $[a, c]$ or $[d, b]$ contains no endpoint of an interval not intersecting I_y. Hence we can extend I_y past the end of I_x on one end.

Doing this until no more pairs of intervals are related by inclusion yields a proper interval representation. From this we obtain a representation using intervals of length 1: a semiorder representation. When no interval properly contains another, the left ends appear in the same order as the right ends. We process the representation from left to right, adjusting all intervals to have length 1.

Of the remaining unadjusted intervals, let I_x be one with leftmost left endpoint, with $I_x = [a, b]$. If some interval has right end in $[a, b)$, then its left end is before a, and the interval already has length 1. If this occurs, then let α be the largest such right end; otherwise, let $\alpha = a$. In either case, $\alpha \in [a, a + 1)$.

Now, adjust the portion of the representation in $[a, \infty)$ by shrinking or expanding $[\alpha, b]$ to $[\alpha, a + 1]$ and translating $[b, \infty)$ to $[a + 1, \infty)$. The order of endpoints does not change, intervals that begin before a still have length 1, and I_x also now has length 1. Iterating produces the desired representation. ∎

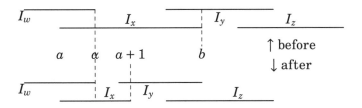

More generally, Fishburn [1984b] proved that the interval orders having interval representations using intervals with lengths in $[n - 2]$ are those not containing the poset $\mathbf{n} + \mathbf{1}$ (see also Fishburn [1985, Theorem 8.3]).

Counting interval orders is hard (Hanlon [1982]); but there are precisely $\frac{1}{n+1}\binom{2n}{n}$ semiorders on n elements, bijectively, using the Catalan numbers (Dean–Keller [1968]; Exercise 9). For semiorders representable using open intervals of length k with integer endpoints, Mitas [1994] obtained a forbidden subposet characterization; the number of forbidden subposets is the $(k + 1)$th Catalan number!

LATTICES

Some special posets admit algebraic operations that generalize the notions of intersection/union for subsets and gcd/lcm for divisibility.

12.4.7. Definition. If $x \leq y$, then x is a **lower bound** for y and y is an **upper bound** for x. A poset L is a **lattice** if for all $x, y \in L$ there is a unique maximal common lower bound (the **meet** $x \wedge y$) and a unique minimal common upper bound (the **join** $x \vee y$). Recall that a poset is **bounded** if it has one minimal element $\hat{0}$ and one maximal element $\hat{1}$.

"Unique maximal common lower bound" means that $x \wedge y$ is an upper bound for every common lower bound of x and y; that is, the subposet of common lower bounds has a unique maximal element. The definition immediately implies that the meet and join operations are commutative and associative.

12.4.8. Example. *Subsets and divisors.* Consider sets $x, y \in 2^n$. A set is contained in both x and y if and only if it is contained in $x \cap y$. Hence $x \cap y$ is the unique maximal common lower bound. Similarly, $x \cup y$ is the unique minimal common upper bound. Hence meets and joins exist, and 2^n is a lattice.

Let $D(n)$ be the divisibility poset on divisors of n. In $D(n)$, meet and join are least common multiple and greatest common divisor, respectively. ∎

12.4.9. Proposition. A product of lattices is a lattice.

Proof: Let $P = L_1 \times \cdots \times L_n$. A lower bound for an element must be a lower bound coordinate by coordinate, and similarly for upper bounds. Thus meets and joins arise componentwise, with $(x \wedge y)_i = x_i \wedge_{L_i} y_i$ and $(x \vee y)_i = x_i \vee_{L_i} y_i$. ∎

The divisibility poset on divisors of an integer and the containment poset on multisets with bounded multiplicities can both be expressed as products of chains. We can write an element a in a product of n chains as $(a_1, \ldots, a_k)$, where a_i is the height of the ith coordinate of a on its chain. The meet and join operations then have simple formulas: $(a \wedge b)_i = \min\{a_i, b_i\}$ and $(a \vee b)_i = \max\{a_i, b_i\}$.

12.4.10. Example. *The **partition lattice** Π_n on partitions of $[n]$.* A **refinement** of a partition replaces each block with a partition of that block. Put $\sigma < \tau$ in Π_n when σ is a refinement of τ (Π_3 and Π_4 appear below).

Rank is given by $r(\sigma) = n - b(\sigma)$, where $b(\sigma)$ is the number of blocks in σ. The meet $\pi \wedge \tau$ is the common refinement of π and τ with the fewest blocks, and $\pi \vee \tau$ is the partition with the most blocks that does not "split" any block of π or τ. ∎

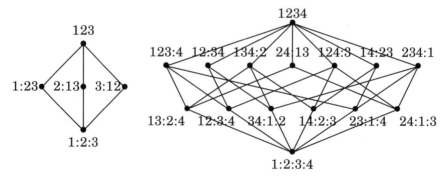

12.4.11. Definition. A subposet P of a lattice L is a **sublattice** of L if $x \wedge_L y$ and $x \vee_L y$ are in P for all $x, y \in P$. Equivalently, a lattice L is a sublattice of a lattice M if there is an embedding $f \colon L \to M$ such that $f(x \wedge_L y) = f(x) \wedge_M f(y)$ and similarly for join.

12.4.12. Example. $J(P)$: *The poset of down-sets.* Let $J(P)$ denote the containment poset on the family of down-sets in P. Each down-set is a subposet of P, so $J(P) \subseteq 2^{|P|}$. Since $J(P)$ is a containment poset, common lower bounds of elements x and y in $J(P)$ are contained in $x \cap y$, and common upper bounds contain $x \cup y$. Since the intersection and union of down-sets in P are down-sets in P, we have $x \wedge y = x \cap y$ and $x \vee y = x \cup y$. Hence meet and join in $J(P)$ agree with meet and join when viewed in all of 2^P. Thus $J(P)$ is a sublattice of $2^{|P|}$. The poset is bounded, with $\hat{1} = P$ and $\hat{0} = \varnothing$. Indeed, $J(P)$ is graded, with $r(I) = |I|$.

Antichains in P correspond to down-sets in P. Consider antichains A and B generating down-sets $D[A]$ and $D[B]$. We have $D[A] \subseteq D[B]$ if and only if, for every $x \in A$, there exists $y \in B$ such that $x \leq y$. This condition just rephrases the containment condition on down-sets, so $J(P)$ and the resulting lattice on antichains are isomorphic. The maximum antichains induce a sublattice (Exercise 26), which was used in the original proof of the Greene–Kleitman Theorem. ∎

12.4.13. Example. $L(m, n)$, *again.* The poset $L(m, n)$ of Example 12.4.13 arises as a lattice of down-sets: $L(m, n) \cong J(\mathbf{m} \times \mathbf{n})$ (Exercise 14). The isomorphism maps $a \in L(m, n)$ to the down-set of $\mathbf{m} \times \mathbf{n}$ generated by $\{(m + 1 - i, a_i) \colon a_i > 0\}$. The down-set in $\mathbf{4} \times \mathbf{5}$ corresponding to $(0, 1, 5, 5) \in L(4, 5)$ is shown below.

Another proof that $L(m, n)$ is a lattice is by applying Definition 12.4.11 to the lattice $(\mathbf{n} + \mathbf{1})^m$ that contains it (Exercise 15). ∎

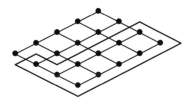

We develop several properties that hold for all lattices.

12.4.14. Lemma. For elements x, y, z of a lattice L,
    (a) $x \wedge y \leq x \leq x \vee z$.
    (b) If $x \leq z$, then $x \wedge y \leq z \wedge y$ and $x \vee y \leq z \vee y$.
    (c) (4-point Lemma) If $z, w \leq x, y$, then $z \vee w \leq x \wedge y$.
    (d) $(x \wedge y) \vee (x \wedge z) \leq x \wedge (y \vee z)$.
    (e) If $z \leq x$, then $x \wedge (y \vee z) \geq (x \wedge y) \vee z$.

Proof: (a): This holds by definition.

(b): Since $x \wedge y$ is a common lower bound for y and z, it lies below the unique greatest lower bound (the second conclusion is symmetric).

(c): Since both z and w are lower bounds for each of x and y, they are lower bounds for $x \wedge y$. Hence $x \wedge y$ is a common upper bound for z and w, which yields $x \wedge y \geq z \vee w$ (see figure below).

(d): By statement (a), x is an upper bound for $x \wedge y$ and $x \wedge z$, and $y \vee z$ is an upper bound for both y and z and hence for $x \wedge y$ and $x \wedge z$. Hence (d) follows from (c) by using x and $y \vee z$ as $\{x, w\}$ in (c) and $x \wedge y$ and $x \wedge z$ as $\{x, y\}$ in (c).

(e): When $z \leq x$, we have $x \wedge z = z$, so (e) follows immediately from (d).   ∎

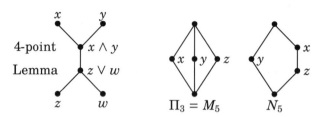

12.4.15. Example. Examples of strict inequality in Lemma 12.4.14(d) and (e) occur in the lattices M_5 and N_5 shown on the right above; we will explain their names later. Note that M_5 is the partition lattice Π_3.   ∎

12.4.16. Definition. For elements $x, y \in P$ with $x \leq y$, the **interval** $[x, y]$ is $\{z \in P \colon x \leq z \leq y\}$. A poset is **locally finite** if every interval is finite.

12.4.17. Example. Every interval $[x, y]$ in $\mathbf{2}^n$ with $r(x) = k$ and $r(y) = l$ is isomorphic to $\mathbf{2}^{l-k}$. Also, every interval in a chain-product is a chain-product. Meets and joins of elements in an interval also lie in the interval.

All intervals in lattices are sublattices, but some sublattices may not be intervals. For example, three of the nine sublattices of $\mathbf{2}^3$ isomorphic to $\mathbf{2}^2$ are not intervals in $\mathbf{2}^3$. Also, a subposet of P that is a lattice need not be a sublattice of P. For example, N_5 is a subposet of $\mathbf{2}^{[3]}$ but is not a sublattice of $\mathbf{2}^{[3]}$.   ∎

The next lemma saves some work in proving that posets are lattices. Given an upper bound (or a lower bound), one need not construct both meets and joins.

12.4.18. Lemma. If P is locally finite, has an upper bound (the element $\hat{1}$), and has a well-defined meet operation, then P is a lattice.

Proof: It suffices to prove that joins exist. Consider $x, y \in P$. The upper bound 1 is a common upper bound for x and y. Since P is locally finite, the interval from $x \wedge y$ to $\hat{1}$ is finite. Thus we can consider the minimal elements among the set of common upper bounds of x and y; they lie in the interval $[x \wedge y, \hat{1}]$.

Let u and v be minimal common upper bounds for x and y. Since x and y are common lower bounds for u and v, we have $x < u \wedge v$ and $y < u \wedge v$. Thus $u \wedge v$ is a common upper bound for x and y. Since u and v are minimal such elements, $u \wedge v \in \{u, v\}$. Thus u and v are comparable. Since they are minimal elements in a subposet of P, they must therefore be equal. Hence there is a unique minimal common upper bound for x and y.   ∎

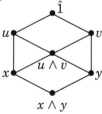

DISTRIBUTIVE LATTICES

Lemma 12.4.14(d) states that $(x \wedge y) \vee (x \wedge z) \leq x \wedge (y \vee z)$ for all x, y, z in any lattice. In the subset lattice, equality holds. We study the class of lattices where equality always holds, seeking characterizations of the class and properties of the subset lattice that extend to such lattices.

12.4.19. Definition. A lattice L is **distributive** if meet distributes over join in L; that is, $x \wedge (y \vee z) = (x \wedge y) \vee (x \wedge z)$ for all $x, y, z \in L$.

Exercise 30 requests a direct proof that a lattice L is distributive if and only if its dual L^* is distributive, by showing that the defining condition for distributivity is equivalent to the property that $x \vee (y \wedge z) = (x \vee y) \wedge (x \vee z)$ for all $x, y, z \in L$. We will also see other ways to prove the equivalence.

12.4.20. Example. *Subset and divisor lattices.* Distributivity for the subset lattice can be seen by marking $x \wedge (y \vee z)$ and $(x \wedge y) \vee (x \wedge z)$ in a Venn diagram.

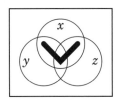

For the divisor lattice, we can argue directly about divisors using gcd and lcm to show distributivity. Alternatively, since every chain is a distributive lattice (min distributes over max for integers), the conclusion that M^e is distributive follows immediately from the next lemma.   ∎

12.4.21. Lemma. A product of lattices is distributive if and only if each factor is distributive.

Proof: The order relation in a product is defined componentwise, so $(x \vee y)_i = x_i \vee y_i$ and $(x \wedge y)_i = x_i \wedge y_i$. Thus distributivity holds for the full lattice if and only if it holds in each factor.   ∎

Recall that P is a sublattice of a lattice L if and only if P is closed under the taking of meets and joins in L.

12.4.22. Lemma. Every sublattice of a distributive lattice is distributive.

Proof: The sublattice inherits the distributivity condition from the full lattice, since meets and joins are computed in the full lattice.   ∎

Lemma 12.4.22 implies that there is a forbidden sublattice characterization of distributive lattices. The two 5-element lattices M_5 and N_5 in Example 12.4.15 are not distributive; they violate the condition with x, y, z as illustrated. Forbidding M_5 and N_5 as sublattices is thus necessary for distributivity. It is also sufficient, but we will not prove this.

12.4.23. Theorem. A lattice is distributive if and only if it does not have M_5 or N_5 of Example 12.4.15 as a sublattice. ∎

Theorem 12.4.23 is quite strong in proving properties of distributive lattices. For example, since M_5 and N_5 are self-dual, Theorem 12.4.23 implies immediately that a lattice L is distributive if and only if L^* is distributive. Thus interchanging meet and join in Definition 12.4.19 yields an equivalent condition.

Lemma 12.4.22 and Example 12.4.20 imply that every sublattice of a subset lattice (or of a chain product) is distributive. For example, since $J(P)$ is a sublattice of $2^{|P|}$, always $J(P)$ is distributive, for any poset P. Our main objective in this discussion of distributive lattices will be a proof that every finite distributive lattice L can be expressed as $J(P)$ for an appropriate poset P.

12.4.24. Definition. An element p of a lattice L is **join-irreducible** if it is non-minimal and is not the join of two other elements; equivalently, $p = x \vee y$ implies $p \in \{x, y\}$. Similarly, p is **meet-irreducible** if it is not maximal and $p = x \wedge y$ implies $p \in \{x, y\}$. In a lattice L, we write $P(L)$ and $Q(L)$ for the subposets formed by the join-irreducible elements and the meet-irreducible elements, respectively.

12.4.25. Example. In a finite lattice, the join-irreducible elements are those covering exactly one element. In the subset lattice these are the 1-sets. In the divisor lattice these are the powers of primes.

In the lattice L on the left below, the elements labeled by single letters are the join-irreducible elements. The resulting subposet $P(L)$ is on the right. The label for each $x \in L$ is the set of join-irreducible elements whose join is x. Uniqueness of such expressions is our next objective. The minimal element of L is not considered join-irreducible, and the maximal element is not meet-irreducible. ∎

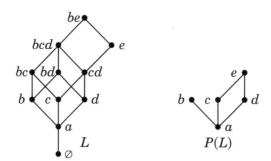

We will see in Theorem 12.4.38 that distributive lattices are the appropriate general setting for many results on 2^n and M^e. It helps to keep the divisor lattice in mind when discussing them. The proof that $L \cong J(P(L))$ when L is distributive generalizes the proof that integers have unique prime factorizations. In this discussion, we abbreviate "join-irreducible" to **irreducible**.

By induction on the size of the set, in a lattice every finite set has a unique least common upper bound, so we write joins of finite sets without parentheses. Induction also shows that meet distributes over a join of a finite set of elements.

12.4.26. Lemma. If $p \leq a_1 \vee \cdots \vee a_k$ for some irreducible element p in a distributive lattice L, then $p \leq a_i$ for some i.

Proof: The definition of meet yields $x \leq y$ if and only if $x \wedge y = x$. With $x = p$ and $y = a_1 \vee \cdots \vee a_k$, we have

$$p = p \wedge (a_1 \vee \cdots \vee a_k) = (p \wedge a_1) \vee \cdots \vee (p \wedge a_k).$$

This expresses p as the join of several elements. Since p is irreducible, it must equal one of them. Thus $p = p \wedge a_i$ for some i, which in turn yields $p \leq a_i$. ∎

12.4.27. Definition. An **irredundant representation** of a is a minimal expression of a as a join of a set of irreducible elements; that is, $a = p_1 \vee \cdots \vee p_k$ and no proper subset of $\{p_1, \ldots, p_k\}$ has join a.

The irreducible elements in an irredundant representation are incomparable, since if $p > q$ then q is redundant in any join of a set containing p. To prove existence and uniqueness of irredundant representations, we need a unique minimal element and a finiteness condition for inductive arguments. For clarity and simplicity, we restrict to finite lattices our discussion of this lemma and its application to characterize distributivity. The results can be extended to infinite lattices satisfying appropriate local finiteness conditions (such as the divisibility order on $\mathbb{N}$), but we will not discuss this.

12.4.28. Lemma. In a finite distributive lattice L having a lower bound $\hat{0}$, every element has a unique irredundant representation.

Proof: We first prove existence. Note that the identity element for join is $\hat{0}$; following our usual convention, $\hat{0}$ is thus the join of the empty set of irreducible elements. Now, if some element has no irredundant representation, then there is a minimal such element x, and x is the join of two lower elements. By minimality, those elements have irredundant representations, and x is the join of the union of those two sets. Deleting the non-maximal elements in the union yields an irredundant representation.

Now let $p_1 \vee \cdots \vee p_k$ and $q_1 \vee \cdots \vee q_l$ be irredundant representations of a. Since $p_i \leq a$, Lemma 12.4.26 implies that each p_r satisfies $p_r \leq q_s$ for some s. Similarly each q_s satisfies $q_s \leq p_t$ for some t. This yields a relation $p_r \leq p_t$, which forces $p_r = q_s = p_t$ since $p_1, \ldots, p_k$ form an antichain. Hence each p_i belongs to $\{q_j\}$, and similarly each q_j belongs to $\{p_i\}$, and the sets are the same. ∎

12.4.29. Definition. The unique irredundant representation of an element x in a finite distributive lattice is the **factorization** of x. For a lattice L, the **ideal map** $\phi \colon L \to J(P(L))$ assigns to each $x \in L$ the down-set of join-irreducibles defined by $\phi(x) = \{p \in P(L) \colon p \leq x\}$.

We use the word "factorization" because for a divisor lattice, the factorization of an element is the set of prime powers in its numerical prime factorization.

12.4.30. Lemma. For x in a finite distributive lattice L, the factorization of x is the antichain of maximal elements in $\phi(x)$, where ϕ is the ideal map on L.

Proof: Let A be the antichain of elements in the factorization of x. Since $x = \bigvee A$ and $A \subseteq P(L)$, we have $A \subseteq \phi(x)$. Hence $\bigvee A \leq \bigvee B$, where B is the set of maximal elements in $\phi(x)$, since every element of A is bounded above by some element of B. Now $\bigvee B \leq x$, since $B \subseteq \phi(x)$ and x is an upper bound for $\phi(x)$. We have proved $x = \bigvee A \leq \bigvee B \leq x$, and hence $\bigvee A = \bigvee B = x$. By Lemma 12.4.28, $A = B$. ∎

12.4.31. Theorem. (Birkhoff [1935]) A finite lattice L is distributive if and only if $L \cong J(P(L))$ (and hence L is a sublattice of $\mathbf{2}^{P(L)}$).

Proof: Because every sublattice of a distributive lattice is distributive, the condition is sufficient. For necessity, suppose that L is distributive. We prove that the ideal map $\phi\colon L \to J(P(L))$ is a lattice isomorphism.

Since the elements in the factorization of x form the antichain of maximal elements in $\phi(x)$, we have $\bigvee \phi(x) = x$. Hence ϕ is injective.

For surjectivity, let D be a down-set in $P(L)$, and let $x = \bigvee D$. Each element of D is bounded above by x, so $D \subseteq \phi(x)$. Since $x = \bigvee D$, we also have $x = \bigvee A$, where A is the antichain of maximal elements in D. An expression of x as a join of an antichain of irreducible elements is irredundant, so Lemma 12.4.28 and Lemma 12.4.30 imply that $D = \phi(x)$.

If $x \leq y$, then $\phi(x) \subseteq \phi(y)$, and $\phi(x) \subseteq \phi(y)$ implies $x = \bigvee \phi(x) \leq \bigvee \phi(y) = y$, so ϕ is a poset isomorphism. For a lattice isomorphism, it remains only to check that ϕ preserves meets and joins. For meets,

$$\phi(x) \wedge_{J(P(L))} \phi(y) = \phi(x) \cap \phi(y) = \{z \in P(L)\colon z \leq x \text{ and } z \leq y\}$$
$$= \{z \in P(L)\colon z \leq x \wedge_L y\} = \phi(x \wedge_L y)$$

The computation for joins is analogous. ∎

12.4.32. Corollary. (1) Every distributive lattice L is graded, with $r(L) = |P(L)|$.
(2) Under ϕ, the irreducible elements of L map to members of $J(P(L))$ generated by single elements.

Proof: (1) $J(P(L))$ is ranked by cardinality of the down-sets in $P(L)$ (for example, compare L and $P(L)$ in the drawing of Example 12.4.25).
(2) The factorization of an irreducible element $p \in P(L)$ is $\{p\}$. ∎

Writing this discussion using meets instead of joins would prove $P(L) \cong Q(L)$; Exercise 33 obtains this statement from Theorem 12.4.31.

12.4.33.* Remark. Like our characterization of semiorders by first characterizing interval orders, Theorem 12.4.23 characterizing distributive lattices by forbidden sublattices is proved by first characterizing a larger class.

A lattice is **modular** if equality always holds in the inequality of Lemma 12.4.14(e). This fails for N_5 with x, y, z as in Example 12.4.15. On the other hand, it holds for M_5. This is the source of the notation: M_5 is modular, and N_5 is non-modular. Using various properties of modular lattices, one can show that a lattice is modular if and only if it does not have N_5 as a sublattice.

Every distributive lattice is modular: when $z \leq x$ in a distributive lattice, $x \wedge (y \vee z) = (x \wedge y) \vee (x \wedge z) = (x \wedge y) \vee z$. Theorem 12.4.23 is proved by showing that a modular lattice is distributive if and only if it does not have M_5 as a sublattice.

Modular lattices lie in a still larger class. A lattice is **semimodular** if $x \vee y$ covers y whenever x covers $x \wedge y$. Semimodular lattices are characterized by having submodular rank functions, meaning $r(x \wedge y) + r(x \vee y) \le r(x) + r(y)$. This links them closely to matroids (Section 11.3); in any matroid, the inclusion order on the family of closed sets is a semimodular lattice.   ∎

CORRELATIONAL INEQUALITIES

A natural probability space arises from linear extensions of posets. We view a poset Q as partial information about an underlying linear order on the elements. This indeed is the setting when we have partially sorted numbers via pairwise comparisons. The set S of possible outcomes is the set of linear extensions of Q.

We assume that each linear extension of Q is equally likely to be the true ordering. An **event** is a subset of the linear extensions. When we specify an event A by a condition (like "$x < y$"), we mean the set of linear extensions in which the condition occurs. The probability $\mathbb{P}(A)$ of an event A is then $|A|/|S|$.

Events A and B are **independent** if $\mathbb{P}(A \cap B) = \mathbb{P}(A)\mathbb{P}(B)$. Events A and B are **positively correlated** if $\mathbb{P}(AB) \ge \mathbb{P}(A)\mathbb{P}(B)$. Our goal is the "XYZ Inequality" (Theorem 12.4.41): for elements x, y, z in any poset Q, the events "$x < y$" and "$x < z$" are positively correlated.

Sampling randomly from a poset raises similar questions. For $F, G \subseteq P$, we ask whether "membership in F" and "membership in G" are positively correlated.

12.4.34. Theorem. (**Kleitman's Inequality**; Kleitman [1966]) If F and G are down-sets in $\mathbf{2}^n$, then $\frac{|F \cap G|}{|P|} > \frac{|F|}{|P|}\frac{|G|}{|P|}$. Equivalently, membership in a down-set and an up-set are negatively correlated.   ∎

Kleitman was motivated by an extremal problem: how large can $|F \cap G|$ be for a down-set F and an up-set G of specified sizes? Later, Anderson [1976] proved that Kleitman's Inequality also holds in the multiset lattice M^e. The proof we present involves Chebyshev's Inequality (a correlational inequality for real numbers) and illustrates an inductive technique for chain-products.

12.4.35. Lemma. (**Chebyshev's Inequality**) If $x_1, \ldots, x_m$ and $y_1, \ldots, y_m$ are both nonincreasing or both nondecreasing sequences, then

$$\frac{\sum x_i y_i}{m} \ge \frac{\sum x_i}{m} \frac{\sum y_i}{m}.$$

Furthermore, if μ is a nonnegative weight function on $[m]$, then

$$\frac{\sum x_i y_i \mu(i)}{\sum \mu(i)} \ge \frac{\sum x_i \mu(i)}{\sum \mu(i)} \frac{\sum y_i \mu(i)}{\sum \mu(i)}.$$

Proof: Form the double sum $\sum_{i,j}(x_i - x_j)(y_i - y_j)$. Since both sequences are monotone, both factors in any term have the same sign (or 0), so each term is nonnegative. Multiplying out and moving the terms expressed with minus signs to the other side yields $2m \sum x_i y_i \ge 2 \sum x_i \sum y_i$. For the weighted version, start with $\sum_{i,j}(x_i - x_j)(y_i - y_j)\mu(i)\mu(j)$ and proceed in the same way.   ∎

We use this to prove Kleitman's Inequality for chain-products.

12.4.36. Theorem. (Anderson [1976]) If F and G are down-sets in L, a product of n chains, then

$$|L||F \cap G| \geq |F||G|.$$

Proof: (Daykin–Kleitman–West [1979]) Let $e_1, \ldots, e_n$ be the chain sizes. We use induction on n. If $n = 1$, then $|F \cap G| = \min\{|F|, |G|\}$ and $|L| \geq \max\{|F|, |G|\}$.

For $n > 1$, partition these sets using the last coordinate. Each $x \in L$ is a vector $(x_1, \ldots, x_n)$ with $0 \leq x_j < e_j$ for $j \in [n]$. For $X \subseteq L$ and $0 \leq i \leq e_n - 1$, let $X_i = \{x \in X: x_n = i\}$. Note that L_i is isomorphic to the product L' of the first $n-1$ chains. If X is a down-set in L, then X_i is a down-set in L_i, and $|X_0| \geq \cdots \geq |X_{e_n-1}|$.

Since F, G, and $F \cap G$ are down-sets, applying this chain of inequalities yields nonincreasing lists for the sizes of their "slices" $|F_i|$, $|G_i|$, and $|(F \cap G)_i|$. Also, $(F \cap G)_i = F_i \cap G_i$. Now we apply Chebyshev's Inequality and then the induction hypothesis (for the subsets of L' obtained by deleting the last coordinate from the elements of F_i, G_i, and $F_i \cap G_i$). The computation is

$$|F||G| = \sum |F_i| \sum |G_i| \leq e_n \sum |F_i|\,|G_i| \leq e_n \sum |L_i||F_i \cap G_i|$$
$$= e_n \prod_{i<n} e_i \sum |(F \cap G)_i| = |L||F \cap G|. \qquad \blacksquare$$

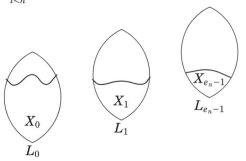

Daykin [1977] proved a more general inequality for more general sets in distributive lattices and thereby characterized distributive lattices (Exercise 43).

As a common extension of Theorem 12.4.36 and Chebyshev's Inequality, we will prove the FKG Inequality for monotone functions on lattices. Discovered jointly by Fortuin, Kasteleyn, and Ginibre, this is the central result about correlational inequalities.

Chebyshev's Inequality can be viewed as a statement about random variables. The weight function $\mu(i)$ gives the probability that the outcome is i. The sequences x and y are monotone functions $f(i)$ and $g(i)$, and we compare the expectation of their product and the product of their expectations. To extend this to distributive lattices, we need a technical condition on the weight function μ.

12.4.37. Definition. A function $f: P \to \mathbb{R}$ is **order-preserving** if $x \leq y$ implies $f(x) \leq f(y)$, **order-reversing** if $x \leq y$ implies $f(x) \geq f(y)$. A real-valued function μ on a lattice is **log-supermodular** if $\mu(x \wedge y)\mu(x \vee y) \geq \mu(x)\mu(y)$.

All weight functions on chains are log-supermodular, since in a chain always $\{x \wedge y, x \vee y\} = \{x, y\}$. Chebyshev's Inequality is the special case of the FKG Inequality where the lattice is a chain.

12.4.38. Theorem. (**FKG Inequality**; Fortuin–Kasteleyn–Ginibre [1971]) In a distributive lattice L, functions f and g that are both order-preserving or both order-reversing are positively correlated with respect to each nonnegative log-supermodular weight function μ, meaning that

$$\sum f(x)g(x)\mu(x) \sum \mu(x) \geq \sum f(x)\mu(x) \sum g(x)\mu(x). \qquad \blacksquare$$

The FKG Inequality yields Kleitman's Inequality for all distributive lattices by setting $\mu = 1$ and letting f and g be the (order-reversing) characteristic functions on the down-sets F and G, where the **characteristic function** of a set A is the function χ_A having value 1 on A and value 0 outside A. Nevertheless, generalizations of Kleitman's Inequality continued to appear long after the FKG Inequality, because the FKG Inequality appeared in the literature of statistical mechanics and discrete mathematicians were unaware of it for years.

Later, Ahlswede and Daykin found a generalization of the FKG Inequality having an easier inductive proof. We give this proof, following the presentation of Graham [1982]. The theorem is known both as the **Ahlswede–Daykin Inequality** and as the **Four Function Inequality**. A result intermediate between the FKG Inequality and the Four Function Inequality appeared in Holley [1974].

12.4.39. Definition. For a set X in a lattice L and a function $f \colon L \to \mathbb{R}$, define $f(X) = \sum_{x \in X} f(x)$. For $X, Y \subseteq L$, define $X \wedge Y = \{x \wedge y \colon x \in X, y \in Y\}$ and $X \vee Y = \{x \vee y \colon x \in X, y \in Y\}$.

12.4.40. Theorem. (Ahlswede–Daykin [1978]) If $\alpha, \beta, \gamma, \delta$ are four nonnegative functions on a distributive lattice L such that

$$\alpha(x)\beta(y) \leq \gamma(x \wedge y)\delta(x \vee y) \text{ for all } x, y \in L,$$

then

$$\alpha(X)\beta(Y) \leq \gamma(X \wedge Y)\delta(X \vee Y) \text{ for all } X, Y \subseteq L.$$

Proof: We first reduce to the case of $\mathbf{2}^n$, using that distributive lattices are sublattices of such posets (Theorem 12.4.31). If the claim holds for $\mathbf{2}^n$, then it holds for a sublattice L of $\mathbf{2}^n$ as follows. Given $\alpha, \beta, \gamma, \delta$ defined on L, extend them to $\mathbf{2}^n$ by giving them value 0 on $\mathbf{2}^n - L$. Whenever $\alpha(x)\beta(y) \neq 0$, we have $x, y \in L$. Since L is a sublattice, L also contains $x \wedge y$ and $x \vee y$, and the hypothesis for $\mathbf{2}^n$ follows from its truth for L. Hence the conclusion holds for any $X, Y \subseteq \mathbf{2}^n$, including when $X, Y \subseteq L$.

The proof for $\mathbf{2}^n$ uses induction on n. For $n = 1$, we check several cases. The elements of L are 0 and 1 (representing $\varnothing$ and $[1]$). The hypothesis gives $\alpha(x)\beta(y) \leq \gamma(\min\{x, y\})\delta(\max\{x, y\})$ for the four possibilities of $x, y \in \{0, 1\}$. The conclusion is easy when $|X| = 1$ or $|Y| = 1$. The case $X = Y = \{0, 1\}$ requires a numerical optimization (Exercise 45).

For $n > 1$, consider fixed $X, Y \subseteq \mathbf{2}^n$. Let $L' = \mathbf{2}^{n-1}$. In terms of X and Y, we define $\alpha', \beta', \gamma', \delta'$ on L' so that the desired inequality $\alpha(X)\beta(Y) \leq \gamma(X \wedge Y)\delta(X \vee Y)$ will become $\alpha'(L')\beta'(L') \leq \gamma'(L' \wedge L')\delta(L' \vee L')$. Note that $L' \wedge L' = L' = L' \vee L'$ in the lattice L'. For $x \in L'$, we define

$$\alpha'(x) = \alpha(x)\chi_X(x) + \alpha(x \cup \{n\})\chi_X(x \cup \{n\})$$
$$\beta'(x) = \beta(x)\chi_Y(x) + \beta(x \cup \{n\})\chi_Y(x \cup \{n\})$$
$$\gamma'(x) = \gamma(x)\chi_{X \wedge Y}(x) + \gamma(x \cup \{n\})\chi_{X \wedge Y}(x \cup \{n\})$$
$$\delta'(x) = \delta(x)\chi_{X \vee Y}(x) + \delta(x \cup \{n\})\chi_{X \vee Y}(x \cup \{n\}).$$

For $t \in \{\alpha, \beta, \gamma, \delta\}$, this definition accumulates in $t'(x)$ the contributions of x and $x \cup \{n\}$ to the "relevant set" Z, which is X, Y, $X \wedge Y$, or $X \vee Y$ when t is α, β, γ, or δ, respectively. Including all contributions, $t'(L') = t(Z)$, and thus the desired inequality becomes $\alpha'(L')\beta'(L') \leq \gamma'(L')\delta(L')$.

To obtain this conclusion from the induction hypothesis, we show that the functions $\alpha', \beta', \gamma', \delta'$ satisfy $\alpha'(x)\beta'(y) \leq \gamma'(x \wedge y)\delta'(x \vee y)$ for all $x, y \in \mathbf{2}^{n-1}$. For fixed x and y, each quantity in this inequality is computed from two elements of L having the form z and $z \cup \{n\}$. Since each of $\{z, z \cup \{n\}\}$ might or might not belong to the relevant set Z, we could complete the proof by checking 16 cases.

We can reduce the verification to four cases by using the induction hypothesis for $n = 1$. We split the contributions to t' by defining new functions on $\mathbf{2}^1$.

$$
\begin{array}{llll}
z = x: & \alpha''(\varnothing) = \alpha(z)\chi_X(z) & \alpha''([1]) = \alpha(z \cup \{n\})\chi_X(z \cup \{n\}) \\
z = y: & \beta''(\varnothing) = \beta(z)\chi_Y(z) & \beta''([1]) = \beta(z \cup \{n\})\chi_Y(z \cup \{n\}) \\
z = x \wedge y: & \gamma''(\varnothing) = \gamma(z)\chi_{X \wedge Y}(z) & \gamma''([1]) = \gamma(z \cup \{n\})\chi_{X \wedge Y}(z \cup \{n\}) \\
z = x \vee y: & \delta''(\varnothing) = \delta(z)\chi_{X \vee Y}(z) & \delta''([1]) = \delta(z \cup \{n\})\chi_{X \vee Y}(z \cup \{n\}).
\end{array}
$$

For $t \in \{\alpha, \beta, \gamma, \delta\}$, we have defined t'' on $\{\varnothing, [1]\}$ so that $t''(\mathbf{2}^1) = t'(z)$, where z is the "relevant element" of $\mathbf{2}^{n-1}$ as listed above.

If $\alpha'', \beta'', \gamma'', \delta''$ satisfy $\alpha''(u)\beta''(v) \leq \gamma''(u \wedge v)\delta''(u \vee v)$ for all $u, v \in \mathbf{2}^1$, then the induction hypothesis for the case $n = 1$ yields $\alpha''(\mathbf{2}^1)\beta''(\mathbf{2}^1) \leq \gamma''(\mathbf{2}^1)\delta''(\mathbf{2}^1)$, which is the needed inequality $\alpha'(x)\beta'(y) \leq \gamma'(x \wedge y)\delta'(x \vee y)$ for elements of $\mathbf{2}^{n-1}$. Using the definition of each t'', the inequalities for $u, v \in \mathbf{2}^1$ are now statements about elements of L, and we can apply the original hypothesis about $\{\alpha, \beta, \gamma, \delta\}$.

Since $\alpha, \beta, \gamma, \delta$ are nonnegative, in checking the hypothesis on $\mathbf{2}^1$ we may assume that the left side of the needed inequality is positive, which requires that the arguments on the left belong to X and Y, respectively. The corresponding arguments to γ and δ then belong to $X \wedge Y$ and $X \vee Y$, respectively, so all the values of the characteristic functions in the definitions of t'' may be assumed to be 1. For $x, y \in L'$, evaluating t'' at the relevant elements yields the four needed inequalities below.

u	v			
$\varnothing$	$\varnothing$	$\alpha(x)\beta(y)$	$\leq$	$\gamma(x \wedge y)\delta(x \vee y)$
$\varnothing$	$[1]$	$\alpha(x)\beta(y \cup \{n\})$	$\leq$	$\gamma(x \wedge y)\delta((x \vee y) \cup \{n\})$
$[1]$	$\varnothing$	$\alpha(x \cup \{n\})\beta(y)$	$\leq$	$\gamma(x \wedge y)\delta((x \vee y) \cup \{n\})$
$[1]$	$[1]$	$\alpha(x \cup \{n\})\beta(y \cup \{n\})$	$\leq$	$\gamma((x \wedge y) \cup \{n\})\delta((x \vee y) \cup \{n\})$

In each nontrivial case, $\alpha''(u)\beta''(v) \leq \gamma''(u \wedge v)\delta''(u \vee v)$ thus reduces to an instance of the hypothesis given on $\mathbf{2}^n$. ∎

The Ahlswede–Daykin Inequality now yields the FKG Inequality.

Proof of FKG Inequality (Theorem 12.4.38): Set $\alpha = \beta = \gamma = \delta = \mu$. By negating the functions if necessary, we may assume that f and g are order-reversing. Also, adding a constant to one of the functions does not affect the inequality, so we may assume that f and g are nonnegative.

We first prove the FKG Inequality when f and g are the characteristic functions of down-sets X and Y. In this case, $\sum_{x \in L} f(x)\mu(x) = \sum_{x \in X} \mu(x) = \mu(X)$, and similarly $\sum_{x \in L} g(x)\mu(x) = \mu(Y)$. The hypothesis that μ is log-supermodular implies the hypothesis for the Ahlswede–Daykin Inequality, and the conclusion of the Ahlswede–Daykin Inequality for X, Y is the desired statement for f and g:

$$\mu(X)\mu(Y) \leq \mu(X \wedge Y)\mu(X \vee Y) \leq \mu(X \cap Y)\mu(L) = \sum f(x)g(x)\mu(x) \sum \mu(x).$$

To complete the proof, we need only show (1) every nonnegative monotone nonincreasing function on L is a linear combination of at most $|L|$ characteristic functions on down-sets, and (2) the FKG Inequality is preserved by taking a positive linear combination of order-reversing functions f_1 and f_2.

(1) follows by induction on the number of nonzero values of f, with a trivial basis. Let ε be the least nonzero value of f, let $S = \{x \in L: f(x) \neq 0\}$, and let χ_S be the characteristic function of S. Now $f - \varepsilon\chi_S$ is a nonnegative nonincreasing function on L with fewer nonzero values than f.

For (2), let f be a positive linear combination of f_1 and f_2, which each satisfy the FKG Inequality with g. Term-by-term linearity of real number arithmetic yields the FKG Inequality for f and g. ∎

The FKG Inequality was notably applied to prove a conjecture by Rival and Sands about positive correlation in random linear extensions of a poset Q. Recall that $\mathbb{P}(A)$ is the fraction of the linear extensions in which event A occurs. For example, when Q is an antichain, $\mathbb{P}(x < y) = \frac{1}{2}$ for any x and y. If $Q = \mathbf{2}+\mathbf{1}$, with $x < z$ and y unrelated to both, then $\mathbb{P}(x < y) = \mathbb{P}(y < z) = \frac{2}{3}$ and $\mathbb{P}(x < z) = 1$.

12.4.41. Theorem. (**XYZ Inequality**; Shepp [1982]) For any elements x, y, z in any poset Q, the events "$x < y$" and "$x < z$" are positively correlated.

Proof: Roughly speaking, we seek a distributive lattice where the events of interest correspond to down-sets and will be positively correlated. Let the elements of Q be $\{q_1, \ldots, q_n\}$, with $(q_1, q_2, q_3) = (x, y, z)$. Henceforth, we use x and y instead as elements of the eventual distributive lattice, as in Theorem 12.4.38.

Let N be a large positive integer, and write $x \in [N]^n$ as $(x_1, \ldots, x_n)$. View x as a map $\lambda: Q \to [N]$ in which $x_i = \lambda(q_i)$. If the coordinates of x are distinct (λ is injective), then x yields an ordering of Q. There are $\binom{N}{n}$ such vectors x for each ordering. When λ is order-preserving, the ordering is a linear extension.

Using $[N]^n$ rather than the extensions themselves makes it easier to define a distributive lattice. We choose N large to make the effect of non-injective mappings negligible. We return to this detail later; for now we view the order-preserving mappings in $[N]^n$ as corresponding to extensions of Q.

Application of the FKG Inequality needs a weight function; let μ be the characteristic function of the set of order-preserving mappings. So that f and

g capture the desired events, let them be the characteristic functions of F and G, where $F = \{x \in [N]^n : x_1 < x_2\}$ and $G = \{x \in [N]^n : x_1 < x_3\}$. Roughly,

$$\sum f(x)\mu(x) / \sum \mu(x) \approx \mathbb{P}(q_1 < q_2),$$
$$\sum g(x)\mu(x) / \sum \mu(x) \approx \mathbb{P}(q_1 < q_3),$$
$$\sum f(x)g(x)\mu(x) / \sum \mu(x) \approx \mathbb{P}(q_1 < q_2 \text{ and } q_1 < q_3).$$

To apply the FKG Inequality, we need a partial order on $[N]^n$ so that f and g are order-preserving, μ is log-supermodular, and the poset is a distributive lattice. For $x, y \in [N]^n$, set $x \leq y$ when $x_1 \geq y_1$ and $x_i - x_1 \leq y_i - y_1$ for $i > 1$.

We have $f(x) = 1$ when $x_1 \leq x_2$. If $x \leq y$, then also $y_1 \leq y_2 - x_2 + x_1 \leq y_2$, so $f(y) = 1$. Thus f is monotone increasing, and similarly for g.

To see that $[N]^n$ is a distributive lattice under this ordering, consider the map $\sigma(x) = (-x_1, x_2 - x_1, \ldots, x_n - x_1)$. We have $x \leq y$ in our ordering if and only if $\sigma(x) \leq \sigma(y)$ in the usual ordering on a product of n chains of size $2N + 1$, indexed from $-N$ to N. Meets and joins are preserved under σ, which follows from

$$(x \wedge y)_i = \min(x_i - x_1, y_i - y_1) + \max(x_1, y_1)$$
$$(x \vee y)_i = \max(x_i - x_1, y_i - y_1) + \min(x_1, y_1).$$

For example, if $i > 1$, then

$$[\sigma(x) \wedge \sigma(y)]_i = \min(x_i - x_1, y_i - y_1) = (x \wedge y)_i - (x \wedge y)_1 = [\sigma(x \wedge y)]_i.$$

Thus our poset is a sublattice of a chain-product and hence a distributive lattice.

To apply the FKG Inequality, we also need μ to be log-supermodular. We defined $\mu(x) = 1$ when x is order-preserving ($x_i \leq x_j$ for $q_i \leq q_j$). We need only check that $x \wedge y$ and $x \vee y$ are order-preserving when x and y are. This is easy from the formulas given above for $x \wedge y$ and $x \vee y$, and it is the reason for subtracting the first coordinate from the others in defining the order relation on $[N]^n$.

From the FKG Inequality, $\sum f(x)g(x)\mu(x) \sum \mu(x) \geq \sum f(x)\mu(x) \sum g(x)\mu(x)$, where the sums run over $x \in [N]^n$. From our definitions of f, g, μ, this is what we want, except that some vectors are not injective. We show that the relative contribution to each term from non-injective x tends to 0 as $N \to \infty$.

Let $A_1 \subset [N]^n$ be the set of injective n-tuples (distinct coordinate values), and let $A_2 = [N]^n - A_1$. For $i \in \{1, 2\}$, define

$$F_i = \sum_{x \in A_i} f(x)\mu(x), \qquad G_i = \sum_{x \in A_i} g(x)\mu(x),$$
$$H_i = \sum_{x \in A_i} f(x)g(x)\mu(x), \quad M_i = \sum_{x \in A_i} \mu(x).$$

Note that M_1 is $\binom{N}{n}$ times the number of extensions of Q. There is at least one extension, so $M_1 \geq \binom{N}{n}$. On the other hand, $M_2 \leq |A_2| = N^n - N_{(n)}$. With n fixed, we have $M_1 \sim N^n/n!$ and $M_2 = O(N^{n-1})$, so $M_2/M_1 \to 0$. Since F_i and G_i and H_i are all bounded by M_i, also F_2/M_1 and G_2/M_1 and H_2/M_1 all tend to 0.

Since each linear extension is represented equally often in A_1, what we want is $H_1 M_1 \geq F_1 G_1$. The FKG Inequality yields

$$(H_1 + H_2)(M_1 + M_2) \geq (F_1 + F_2)(G_1 + G_2).$$

Dividing by M_1^2 and taking the limit as $N \to \infty$ yields $\frac{H_1}{M_1} \geq \frac{F_1}{M_1} \frac{G_1}{M_1}$, which is precisely the desired inequality. ∎

The XYZ Inequality is quite special. Many similar-sounding statements are not true. For example, given $x, y, z, w \in Q$, one might expect that the events $x < y < w$ and $x < z < w$ are positively correlated, but this does not hold for all Q (Exercise 47). Winkler [1986] characterized the very few pairs of subposets that are always positively correlated. Brightwell–Trotter [2002] gives a combinatorial proof of a stronger version of the XYZ Inequality due to Fishburn [1984a].

A PROBLEM IN RAMSEY THEORY (optional)

Here we apply the lattice of down-sets to a problem in ordered Ramsey theory. No aspects of lattice theory are needed, just the concept of the poset of down-sets.

12.4.42. Definition. An **ordered hypergraph** is a hypergraph on a linearly ordered vertex set. An ordered hypergraph G occurs as a **subhypergraph** of an ordered hypergraph H if some order-preserving injection from $V(G)$ to $V(H)$ also preserves edges. We then say that H **contains** (a copy of) G.

Let K_n^r denote the complete r-uniform hypergraph with n vertices. By Ramsey's Theorem, when n is large enough every k-coloring of $E(K_n^r)$ contains a monochromatic copy of K_p^r. Because complete ordered hypergraphs with the same number of vertices are isomorphic, the same statement holds in the ordered sense, and hence Ramsey numbers for ordered hypergraphs are well-defined.

12.4.43. Definition. Let $G_1, \ldots, G_k$ be r-uniform ordered hypergraphs. The **k-color Ramsey number** $R(H_1, \ldots, H_k)$ is the least n such that every k-coloring of an n-vertex complete r-uniform ordered hypergraph contains a copy of H_i in color i for some i. The **r-uniform monotone path** P_t^r is the t-vertex ordered hypergraph whose edges are the sets of r consecutive vertices.

When discussing only ordered hypergraphs, we can use the same notation and terminology as for classical Ramsey numbers. Indeed, when $G_1, \ldots, G_k$ are complete, the Ramsey numbers in the classical and ordered senses are the same.

However, even when $r = 2$ the Ramsey numbers for ordinary paths and monotone paths differ greatly. For a 3-vertex path, the classical k-color Ramsey number is $k + 2$ (with $k + 1$ edges at a vertex, two have the same color), but for the 3-vertex monotone (ordered) path the Ramsey number is $2^k + 1$. This generalizes as follows, using the same idea as the Erdős–Szekeres Theorem (Theorem 10.1.13).

12.4.44. Proposition. For ordered paths, $R(P_{t_1}^2, \ldots, P_{t_k}^2) = 1 + \prod_{i=1}^{k}(t_i - 1)$.

Proof: Fix a k-coloring of $E(K_n)$. For $x \in V(K_n)$ and $i \in [k]$, let x_i be the maximum number of vertices in a monotone increasing path with color i ending at x. Note that $x_i \geq 1$ for all x and i, and avoiding long paths in color i requires $x_i \leq t_i - 1$.

If y comes before z and edge yz has color i, then $z_i > y_i$, so no two k-tuples can be the same. If there is no sufficiently long path, then only $\prod_{i=1}^{k}(t_i - 1)$ vectors are available. Hence with larger n we have a monotone path in color i for some i.

For the lower bound, assign the ℓth vertex to the ℓth element on a linear extension of the product of chains with sizes $t_1 - 1, \ldots, t_k - 1$. When y is before z, we have $y_i < z_i$ for some i; use color i on edge yz. Since coordinate i must increase along any monotone path in color i, no monotone path in color i has t_i vertices. ∎

This proof is the first instance of the induction step in a characterization of $R(P_{t_1}^r, \ldots, P_{t_k}^r)$ as the number of down-sets in an appropriate poset, leading to upper and lower bounds that are exponential towers of height $r-2$ for the Ramsey number. The problem of computing $R(P_{t_1}^r, \ldots, P_{t_k}^r)$ was introduced in Fox–Pach–Sudakov–Suk [2012] with a geometric application, though it had actually been considered by Duffus–Lefmann–Rödl [1995] in the language of shift graphs (Definition 12.3.36) in order to give a lower bound on classical Ramsey numbers. Recall that $J(Q)$ denotes the lattice of down-sets in a poset Q, ordered by inclusion.

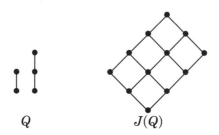

$$Q \qquad\qquad\qquad\qquad J(Q)$$

12.4.45. Theorem. (Moshkovitz–Shapira [2014]) Fix $t_1, \ldots, t_k$ larger than r. Let Q consist of disjoint chains with sizes $t_1 - r, \ldots, t_k - r$. With $Q_1 = Q$ and $Q_i = J(Q_{i-1})$ for $i > 1$, the monotone paths satisfy $R(P_{t_1}^r, \ldots, P_{t_k}^r) = |Q_r| + 1$.

Proof: (Milans–Stolee–West [2015]) For $r = 1$, we use the Pigeonhole Principle. In this case $|Q_r| = \sum_{i=1}^{k}(t_i - 1)$. Any k-coloring of 1-sets chosen from $1 + \sum_{i=1}^{k}(t_i - 1)$ vertices has t_i vertices with color i for some i, forming a monotone copy of $P_{t_i}^1$ in color i. We proceed by induction on r.

Lower Bound. Let $n = |Q_r|$; we construct a k-edge-coloring of K_n^r that avoids $P_{t_i}^r$ in color i for each i. First consider $x, y \in Q_j$, where $2 \leq j \leq r$. The meaning of $x \not\geq y$ is that x does not contain y when they are viewed as down-sets in Q_{j-1}. In this case, let $f(x, y)$ be a fixed element of the family $y - x$ in Q_{j-1}. For $x, y, z \in Q_j$, if $x \not\geq y$ and $y \not\geq z$, then y (in Q_{j-1}) contains $f(x, y)$ but not $f(y, z)$. Since y is a down-set in Q_{j-1}, we obtain $f(x, y) \not\geq f(y, z)$ in Q_{j-1}.

A list $x_1, \ldots, x_s$ of elements in a poset is *descent-free* if $x_i \not\geq x_{i+1}$ for $1 \leq i \leq s - 1$. We extend f to descent-free lists in Q_j by setting $f(x_1, \ldots, x_s) = (f(x_1, x_2), \ldots, f(x_{s-1}, x_s))$. Thus the image under f of a descent-free s-list in Q_j is a descent-free $(s-1)$-list in Q_{j-1}. Let f^0 be the identity map. For a descent-free s-list $x_1, \ldots, x_s$ in Q_j, where $s, j > d > 1$, define $f^d(x_1, \ldots, x_s) = f(f^{d-1}(x_1, \ldots, x_s))$; now $f^d(x_1, \ldots, x_s)$ is a descent-free $(s-d)$-list in Q_{j-d}.

Let $y_1, \ldots, y_n$ be a linear extension of Q_r, so $y_i \leq y_j$ implies $i \leq j$. Each sublist of a linear extension is descent-free. For a sublist $x_1, \ldots, x_r$, note that $f^{r-1}(x_1, \ldots, x_r)$ is a single element in Q. Color $\{x_1, \ldots, x_r\}$ (as an edge in K_n^r) with the index of the chain in Q that contains $f^{r-1}(x_1, \ldots, x_r)$.

Suppose that this coloring has a monotone copy of P_s^r in color i. Let $x_1, \ldots, x_s$ be its vertices, in increasing order. Since $x_1, \ldots, x_s$ is a sublist of a linear extension, it is descent-free. Since each set of r consecutive vertices from the list $x_1, \ldots, x_s$ was given color i, $f^{r-1}(x_1, \ldots, x_s)$ is a descent-free $(s-r+1)$-list in the ith chain of Q. A descent-free list in a chain is strictly increasing (because equality is also considered a descent), so $s - r + 1 \leq t_i - r$. Thus $s < t_i$. We conclude that the coloring avoids $P_{t_i}^r$ in color i for each i, so $R(P_{t_1}^r, \ldots, P_{t_k}^r) > n$.

Upper bound. Given a k-edge-coloring ϕ of $E(K_n^r)$ that avoids $P_{t_i}^r$ in color i for each i, it suffices to define an injection from $[n]$ to Q_r. View each vertex subset $Y \subseteq [n]$ as an increasing list, with $Y^- = Y - \max Y$ and $Y^+ = Y - \min Y$.

For $1 \leq j \leq r < n$, we construct $g_j \colon \binom{[n]}{j} \to Q_{r-j+1}$ such that

$$g_j(Y^-) \not\geq g_j(Y^+) \text{ in } Q_{r-j+1} \text{ when } Y \in \binom{[n]}{j+1}. \tag{$*$}$$

This suffices, since g_1 will then be the desired injection.

We first define g_r. For $X \in \binom{[n]}{r}$, let $i = \phi(X)$, and let $w_1, \ldots, w_{t_i - r}$ be the ith chain in Q. Set $g_r(X) = w_h$, where h is the largest integer such that some copy of P_{h+r-1}^r in color i has last edge X. Note that $h \leq t_i - r$, since ϕ has no copy of $P_{t_i}^r$ in color i. If $\phi(Y^-) = \phi(Y^+)$ for some $Y \in \binom{[n]}{r+1}$, then $g_r(Y^+) > g_r(Y^-)$, and otherwise $g_r(Y^+)$ and $g_r(Y^-)$ are incomparable in Q. In either case, $g_r(Y^-) \not\geq g_r(Y^+)$.

Now consider smaller j, with $g_{j+1} \colon \binom{[n]}{j+1} \to Q_{r-j}$ already defined and satisfying ($*$). For $X \in \binom{[n]}{j}$, let the *precursors* of X be the $(j+1)$-sets obtained from X by adding an element smaller than all of X; that is, $\{Z \in \binom{[n]}{j+1} \colon Z^+ = X\}$. (Note that sets containing the vertex 1 have no precursors.) Define $g_j(X)$ to be the down-set in Q_{r-j} generated by the set of elements $g_{j+1}(Z)$ such that Z is a precursor of X. Since $Q_{r-j+1} = J(Q_{r-j})$, by definition $g_j(X) \in Q_{r-j+1}$.

To check ($*$) for g_j, consider $Y \in \binom{[n]}{j+1}$. Since Y is a precursor of Y^+, the definition of g_j yields $g_{j+1}(Y) \in g_j(Y^+)$. If $g_{j+1}(Y)$ also lies in $g_j(Y^-)$, then by the definition of $g_j(Y^-)$ (using $X = Y^-$), the element $g_{j+1}(Y)$ in Q_{r-j} lies below some element $g_{j+1}(Z)$ such that $Z^+ = Y^-$. Letting $W = Z \cup Y$, we have $Z = W^-$ and $Y = W^+$. Thus $g_{j+1}(W^-) \geq g_{j+1}(W^+)$, which contradicts ($*$) for g_{j+1}.

We conclude $g_{j+1}(Y) \in g_j(Y^+) - g_j(Y^-)$. Hence the down-set $g_j(Y^+)$ in Q_{r-j} does not contain the down-set $g_j(Y^-)$, which means $g_j(Y^-) \not\geq g_j(Y^+)$ in Q_{r-j+1}. $\blacksquare$

Theorem 12.4.45 yields inductive upper and lower bounds for $R_k(P_t^r)$. For $R_k(P_4^3)$, note that $Q_2 = \mathbf{2}^k$, so computing $|Q_3|$ for $R_k(P_4^3)$ is just Dedekind's Problem (see Theorem 12.2.14). For an application, see Exercise 49.

12.4.46. Corollary. (Moshkovitz–Shapira [2014]) Let $\text{tow}_h(x)$ be x when $h = 0$ and $2^{\text{tow}_{h-1}(x)}$ when $h \geq 1$. With $m = t - r + 1$, the monotone paths satisfy

$$\text{tow}_{r-2}(m^{k-1}/2\sqrt{k}) \leq R_k(P_t^r) \leq \text{tow}_{r-2}(2m^{k-1}).$$

Proof: We prove weaker bounds, still exponential towers of the same height.

We bound $|Q_r|$ of Theorem 12.4.45. Since elements of Q_j are subsets of Q_{j-1}, we have $|Q_j| \leq 2^{|Q_{j-1}|}$. With $|Q_2| = m^k$, we obtain $|Q_r| \leq \text{tow}_{r-2}(m^k)$.

For the lower bound, recall that $|J(Q)|$ also counts the antichains in Q. The subsets of an antichain A form $2^{|A|}$ antichains. Since $|Q_1| = k(t-r)$, the down-sets in Q_1 range in size from 0 to $k(t-r)$. Thus Q_2 has fewer than km ranks and has an antichain of size at least m^{k-1}/k. Also, the elements of this antichain have the same size in Q_1. (The chain-product Q_2 is a symmetric chain order, so its middle rank is a largest antichain. Sharper analysis reduces the denominator to $2\sqrt{k}$.)

Given an antichain of size M in Q_j, the subsets of size $M/2$ generate pairwise incomparable down-sets and hence an antichain in Q_{j+1}. The resulting lower bound uses an analogue of exponential towers involving middle binomial coefficients. Let $b_0(x) = x$, and for $h \geq 1$ let $b_h(x) = \binom{b_{h-1}(x)}{\lfloor b_{h-1}(x)/2 \rfloor}$. Since $\binom{n}{\lfloor n/2 \rfloor} \sim 2^n/\sqrt{\pi n/2}$ (Example 2.3.10), inductively we have $b_h(x) \geq \text{tow}_h(x - O(\lg x))$.

Let A_j be a largest antichain in Q_j given by down-sets in Q_{j-1} having the same size; let $a_j = |A_j|$. Thus $|Q_{j+1}| \geq 2^{a_j}$ and $a_{j+1} \geq \binom{a_j}{a_j/2}$. Thus $a_j \geq b_{j-2}(a_2)$. With $a_2 \geq m^{k-1}/k$, we obtain $|Q_r| \geq \text{tow}_{r-2}(m^{k-1}/k - O(k \lg m))$. ∎

EXERCISES 12.4

12.4.1. (−) Prove that a poset is a ranking if and only if no subposet is isomorphic to $\mathbf{2} + \mathbf{1}$.

12.4.2. (−) Prove that a poset is a chain if and only if every subposet is a lattice.

12.4.3. (−) Draw the distributive lattice whose poset of join-irreducible elements is below.

12.4.4. Let P be a poset such that $P - x$ is an interval order. Can relations involving x be added to P to obtain an interval order P' such that $P' - x = P - x$?

12.4.5. *Alternative proof of Theorem 12.4.5.* Let P be a poset without $\mathbf{2} + \mathbf{2}$.
(a) Prove that the "upper holding" sets $U(x)$ for $x \in P$ are linearly ordered by inclusion, and similarly for the sets $D(y)$.
(b) For the $0, 1$-matrix of the order relation, prove that ordering the rows as $x_1, \ldots, x_n$ in decreasing order of $|U(x_i)|$ and the columns as $y_1, \ldots, y_n$ in decreasing order of $|D(y_j)|$ puts the 1s of the matrix in the positions of a Ferrers diagram.
(c) For $x \in P$, let $u(x)$ count distinct sets of the form $U(y)$ that are proper subsets of $U(x)$, and let $d(x)$ count distinct sets of the form $D(y)$ that are proper subsets of $D(x)$. Let h be the total number of distinct nonempty sets of the form $U(x)$. Prove that assigning each $x \in P$ the interval $[d(x), h - u(x)]$ produces an interval representation of P. (Bogart, motivated by Rabinovitch [1977])

12.4.6. (◊) Let $\mathbf{I}_n$ be the poset of nontrivial intervals with integer endpoints in $[n]$, ordered by $[a, b] < [c, d]$ if $b < c$. Prove that $\dim \mathbf{I}_n > k$ when n is sufficiently large. (Hint: Given Ramsey's Theorem, use $n \geq R_k(4; 3)$.) (Rabinovitch [1973])

12.4.7. (◊) Given the interval order $\mathbf{I}_n$ as in Exercise 12.4.6 and the "double-shift graph" G'_n as in Definition 12.3.36, prove $\dim \mathbf{I}_n \geq \chi(G'_n)$. (Comment: Thus $\dim \mathbf{I}_n \geq \lg \lg n + (\frac{1}{2} + o(1)) \lg \lg \lg n$, by Corollary 12.3.39. With some care, Füredi–P.Hajnal–Rödl–Trotter [1992] constructed linear extensions to prove that the lower bound is tight.)

12.4.8. *Representations of interval orders.* Fix $k \in \mathbb{N}$.
(a) Construct an interval order such that every interval representation uses at least k different lengths of intervals. (Fishburn [1983])
(b) Construct an interval order such that every representation has an interval of length more than k times the length of its shortest interval. (Fishburn–Graham [1985])

12.4.9. Establish a one-to-one correspondence between the isomorphism classes of n-element semiorders and the ballot lists of length $2n$ (Definition 1.3.16). (Comment: Thus there are $\frac{1}{n+1}\binom{2n}{n}$ isomorphism classes of semiorders on n elements.) (Dean–Keller [1968])

12.4.10. ($\diamond$) Since a semiorder P has no $\mathbf{2}+\mathbf{2}$, by Exercise 12.3.16 every subposet of P has upper and lower extensions. Let Q be the subposet of P consisting of the elements with even rank. Using upper and lower extensions of Q, prove $\dim P \leq 3$. (Rabinovitch [1978])

12.4.11. ($\diamond$) Although interval orders are more general than chains (or semiorders), prove that the standard example S_n cannot be an intersection of fewer than n interval orders.

12.4.12. ($\diamond$) A general binary relation is modeled by a digraph D, with adjacency matrix $A(D)$. A **biorder representation** of D consists of real-valued functions f and g on $V(D)$ such that $uv \in E(D)$ if and only if $f(u) > g(v)$. For a digraph D, prove that the following five conditions are equivalent.

(A) $A(D)$ has no 2-by-2 submatrix that is a permutation matrix.

(B) The successor sets of D are ordered by inclusion.

(C) The predecessor sets of D are ordered by inclusion.

(D) The rows and columns of $A(D)$ can be permuted independently so that every entry below or to the left of a 1 is a 1.

(E) D has a biorder representation.

(Comment: The equivalences are due to various authors, with a short proof in West [1998]. Such relations are called **biorders**, **Ferrers relations**, or **Ferrers digraphs** (Riguet [1951], Wiener [1914]). If also $f(x) \leq g(x)$ for all x, then D is an interval order.)

12.4.13. ($\diamond$) The *completion* of a graph G with $V(G) = [n]$ is the partition of $[n]$ whose blocks are the vertex sets of components of G. Let $Q(G)$ denote the subposet of Π_n whose elements are the completions of spanning subgraphs of G. The partition of rank 1 with block $\{i, j\}$ lies in $Q(G)$ if and only if $ij \in E(G)$, so these posets are distinct for distinct G.

(a) Show that $Q(G)$ need not be a sublattice of Π_n.

(b) Prove that $Q(G)$ is a sublattice of Π_n isomorphic to $\mathbf{2}^{n-1}$ when G is a tree.

(c) Prove that if P is a copy of $\mathbf{2}^{n-1}$ in Π_n, then $P = Q(G)$ for an n-vertex tree G.

(d) Conclude that Π_n contains n^{n-2} copies of $\mathbf{2}^{n-1}$.

12.4.14. Let $J(P)$ denote the inclusion poset on the down-sets of P. Using the mapping suggested in Example 12.4.12, prove $L(m, n) \cong J(\mathbf{m} \times \mathbf{n}) \cong L(n, m)$.

12.4.15. As a component-wise order on m-tuples from $\{0, \dots, n\}$, the poset $L(m, n)$ is a subposet of $(\mathbf{n+1})^m$. Prove that the meet and join in $(\mathbf{n+1})^m$ of elements of $L(m, n)$ is also in $L(m, n)$, thereby proving by Definition 12.4.11 that $L(m, n)$ is a sublattice of $(\mathbf{n+1})^m$.

12.4.16. ($\diamond$) Write subsets of $[n]$ as strictly decreasing lists, $n \geq a_1 > \cdots > a_k > 0$. Let M_n be the poset of subsets of $[n]$, ordered by $a \leq b$ if and only if $a_i \leq b_i$ for all i, with trailing 0s added to permit comparison, so a subset is never less than a smaller subset.

(a) Describe the rank function and the cover relation in M_n.

(b) Describe M_{n+1} in terms of M_n.

(c) Prove $M_{n+1} \cong J(L(2, n-1))$.

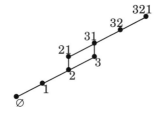

606   Chapter 12: Partially Ordered Sets

12.4.17. ($\diamond$) Let M_n' be the poset of compositions of the integer $n + 1$, ordered by $a \leq b$ if $\sum_{i=1}^{k} a_i \geq \sum_{i=1}^{k} b_i$ for all k (add trailing 0s as needed to test the order relation). Prove $M_n' \cong M_n$, where M_n is the poset of Exercise 12.4.16. (The compositions of 4 are $\{4, 31, 22, 211, 13, 121, 112, 1111\}$.)

12.4.18. Let M_n be the poset of Exercise 12.4.16. Find an explicit symmetric chain decomposition for $n \leq 5$. (Comment: Lindström conjectured that M_n is always a symmetric chain order. Stanley [1982] observed that M_n satisfies the strong Sperner property.)

12.4.19. ($\diamond$) Prove that if (x, y) is an unforced pair in a lattice L, then x is meet-irreducible and y is join-irreducible. Conclude that $\dim L = \dim R$, where R is the subposet of meet-irreducible and join-irreducible elements. (Kelly [1981])

12.4.20. ($\diamond$) *The* **Young lattice**. Let **Y** be the poset of all partitions of all integers, ordered by $a \leq b$ if and only if $a_i \leq b_i$ for all i (trailing zeros appended as needed).
    (a) Prove that **Y** is a lattice.
    (b) Prove that every $a \in \mathbf{Y}$ is covered by one more element than it covers.
    (c) Prove that **Y** is a distributive lattice.
    (d) Describe the join-irreducible elements of **Y**.

12.4.21. ($\diamond$) Prove that the dominance order on the partitions of n (written in decreasing order) is a lattice. Here $\mu \leq \lambda$ if $\sum_{i=1}^{j} \mu_i \leq \sum_{i=1}^{j} \lambda_i$ for all j. (Hint: Apply Lemma 12.4.18.)

12.4.22. ($\diamond$) Obtain a necessary and sufficient condition for λ to cover μ in the dominance order on all integer partitions.

12.4.23. For integers m and k with $0 \leq k \leq m - 1$, let $C(m, k) = \{im + k : i \in \mathbb{Z}\}$; these sets are the **congruence classes**. Let P be the poset of all congruence classes, ordered by inclusion, with the empty set added as a minimal element. Prove that P is a lattice, and describe the meet and join of $C(m, k)$ and $C(n, l)$.

12.4.24. The poset $L_n(q)$ is the containment poset on the set of subspaces of an n-dimensional vector space over a q-element field. Prove that $L_n(q)$ is a graded lattice but in general is not a distributive lattice.

12.4.25. Let Λ_n denote the poset of partitions of the integer n, ordered by refinement. That is, for partitions λ and μ of n, we put $\lambda \leq \mu$ if the multiset of parts in λ is obtained by replacing each integer part in μ with a partition of that part.
    (a) Prove that Λ_n is a graded poset.
    (b) Prove that Λ_7 is not a lattice. (Comment: This holds also for $n > 7$.)

12.4.26. ($\diamond$) Let A and B be maximum antichains in a poset P, and let A' and B' be the down-sets generated by A and B. Prove that the maximal elements of $A' \cup B'$ form a maximum antichain, and similarly for the maximal elements of $A' \cap B'$. Conclude that the maximum-sized antichains in a poset form a sublattice of the lattice of antichains, ordered by $A \leq B$ if and only if for each $x \in A$ there exists $y \in B$ such that $x \leq y$. (Dilworth [1960])

12.4.27. Given the result of Exercise 12.4.26, design a polynomial-time algorithm to construct the maximal element of the lattice of maximum-sized antichains. (Hint: Use the relation of Dilworth's Theorem to maximum matching.)

12.4.28. An **automorphism** of a poset permutes elements but preserves the order relation. Prove that the automorphism group of Π_n is isomorphic to the symmetric group $\mathbb{S}_n$.

12.4.29. Given a connected graph G, let $C(G)$ be the poset of the connected induced subgraphs of G, ordered by inclusion. Prove that $C(G)$ is a lattice if and only if every block of G is a complete graph. (Klavžar–Petkovšek [1988])

12.4.30. For a lattice L, prove directly that $x \vee (y \wedge z) = (x \vee y) \wedge (x \vee z)$ for all $x, y, z \in L$ if and only if $x \wedge (y \vee z) = (x \wedge y) \vee (x \wedge z)$ for all $x, y, z \in L$.

12.4.31. ($\Diamond$) Prove that a lattice L is distributive if and only if $(x \wedge y) \vee (y \wedge z) \vee (z \wedge x) = (x \vee y) \wedge (y \vee z) \wedge (z \vee x)$ for all $x, y, z \in L$. (Comment: There are many proofs.)

12.4.32. Two elements x and z different from y are *y-equivalent* in a lattice L if $x \wedge y = y \wedge z$ and $x \vee y = y \vee z$. Prove that L is distributive if and only if for each $y \in L$ there is no pair of y-equivalent elements. (Hint: Use Exercise 12.4.31 and Theorem 12.4.23.)

12.4.33. ($\Diamond$) The *dual R^** of a poset R is obtained by reversing all relations.
(a) Prove that $J(R^*) = J(R)^*$ for every poset R.
(b) Prove that P and Q are isomorphic if $J(P)$ and $J(Q)$ are isomorphic.
(c) Conclude that in a finite distributive lattice L the posets $P(L)$ and $Q(L)$ (join-irreducibles and meet-irreducibles) are isomorphic.

12.4.34. *Applications of the ideal map (Theorem 12.4.31) on a distributive lattice L.*
(a) For $x, y \in L$, prove that $\{z \in L: x \wedge z = y \wedge z\}$ is a down-set in L. (Birkhoff [1967])
(b) Prove that the number of elements of L that cover exactly k elements equals the number of elements of L covered by exactly k elements. (Hint: Use Exercise 12.4.33.)

12.4.35. ($\Diamond$) *Distributive lattices.* For an element x in a finite lattice L, let $\phi(x) = \{p \in P: p \leq x\}$, where P is the poset of join-irreducible elements of L, and let $\psi(x) = \{q \in Q: q \geq x\}$, where Q is the poset of meet-irreducible elements of L.
(a) Prove that $\phi\colon L \to J(P)$ embeds L as a subposet of $J(P)$.
(b) Prove that the following are equivalent (via A$\Rightarrow$B$\Rightarrow$C$\Rightarrow$D$\Rightarrow$A.)
(A) L is distributive.
(B) L is graded and $|P| = r(L) = |Q|$.
(C) $|P| = |\phi(x)| + |\psi(x)| = |Q|$ for all $x \in L$.
(D) $|\phi(x \vee y)| = |\phi(x) \cup \phi(y)|$ for all $x, y \in L$.

12.4.36. In a lattice, y is a **complement** of x if $x \wedge y = \hat{0}$ and $x \vee y = \hat{1}$.
(a) Using the isomorphism $\phi\colon L \to J(P)$, characterize the elements of a distributive lattice that have complements. Conclude that in any distributive lattice the number of elements having complements is a power of 2.
(b) In a product of chains, which elements have complements?
(c) A lattice is complemented if every element has a unique complement. Show that the only complemented distributive lattices are the subset lattices.

12.4.37. Prove that the order dimension (Section 12.3) of a distributive lattice equals the width of its poset of join-irreducible elements. (Comment: The famous "Dilworth's Theorem" was proved as a lemma for this result.) (Dilworth [1950])

12.4.38. Let L be a distributive lattice. Prove that the smallest dimension of a subset lattice in which L appears as a subposet is the number of join-irreducible elements in L.

12.4.39. A graded lattice L is **semimodular** if its rank function is **submodular**, meaning $r(x \wedge y) + r(x \vee y) \leq r(x) + r(y)$ for all $x, y \in L$. Prove that the product of two semimodular lattices is a semimodular lattice. Conclude that the divisor lattice $D(N)$ is **modular**, meaning that equality always holds in the submodularity inequality.

12.4.40. ($\Diamond$) *Semimodularity of the partition lattice Π_n.*
(a) Prove by induction that $r(\pi \wedge \sigma) + r(\pi \vee \sigma) \leq r(\pi) + r(\sigma)$ for $\pi, \sigma \in \Pi_n$.
(b) Given two partitions π, σ of $[n]$, let $G(\pi, \sigma)$ be the bipartite graph whose parts are the blocks of π and σ, respectively, with vertices adjacent if they intersect as sets. In terms of the number of blocks in $\pi, \sigma, \pi \wedge \sigma, \pi \vee \sigma$, compute the number of vertices, edges, and components of $G(\pi, \sigma)$. Use this to give another proof of part (a). (Aigner [1979])

12.4.41. *Applications of Kleitman's Inequality (Theorem 12.4.34).*
  (a) Prove that 2^{n-2} is the maximum size of an intersecting family whose complements also form an intersecting family. (Seymour [1973], Schönheim [1974], Daykin–Lovász [1976], Hilton [1976], Anderson [1976], Greene–Kleitman [1978])
  (b) Prove that $2^n - 2^{n-k}$ is the maximum size of the union of k intersecting families in $\mathbf{2}^n$. (Kleitman [1966])

12.4.42. *A use of the up-set/down-set form of Kleitman's Inequality (Theorem 12.4.34).*
  (a) Let X be an up-set and Y a down-set in $\mathbf{2}^n$. Let $\overline{X}$ and $\overline{Y}$ be their complements, and let $X - Y$ and $Y - X$ be their differences as sets. Use Kleitman's Inequality to prove $|X - Y| \, |Y - X| \geq |X \cap Y| \, \big|\overline{X} \cap \overline{Y}\big|$, and conclude $|X \cap Y|^{1/2} + \big|\overline{X} \cap \overline{Y}\big|^{1/2} \leq 2^{n/2}$.
  (b) Let F, G be subsets of $\mathbf{2}^n$ such that no member of either contains a member of the other. Apply (a) to prove $|F|^{1/2} + |G|^{1/2} \leq 2^{n/2}$. (Seymour [1973])

12.4.43. ($\diamond$) Using Theorem 12.4.23 and Theorem 12.4.40, prove that a lattice L is distributive if and only if $|F||G| \leq |F \wedge G||F \vee G|$ for all $F, G \subseteq L$ (Definition 12.4.39). Conclude that Kleitman's Inequality holds for distributive lattices. (Daykin [1977])

12.4.44. For $X, Y \subseteq \mathbf{2}^n$, let $X - Y = \{A - B\colon A \in X,\ B \in Y\}$. Apply Exercise 12.4.43 to prove $|X - X| \geq |X|$ for any family X of finite sets. (Comment: Daykin–Lovász [1976] showed in general that every nontrivial boolean function takes at least m distinct values when evaluated over m distinct sets.)

12.4.45. Complete the proofs of the FKG and Ahlswede–Daykin Inequalities by proving the latter for $\mathbf{2}^1$. (Hint: For the case $X = Y = \{\varnothing, [1]\}$, let $w = \alpha(\varnothing)\beta([1])$, $x = \alpha([1])\beta(\varnothing)$, $y = \gamma(\varnothing)\delta([1])$, and $z = \gamma([1])\delta(\varnothing)$. Reduce the problem to inequalities involving w, x, y, z.)

12.4.46. Suppose that the outcome of matches between tennis players is determined by an unknown linear ordering of ability. Suppose that A and B are two teams with two players each, and that initially we know nothing about the relative abilities of the two players on a team. However, we do know $a_2 < b_1$. Determine whether the events $a_1 < b_1$ and $a_2 < b_2$ are positively correlated, assuming that unknown information is random. (Shepp [1980])

12.4.47. Let $Q = \mathbf{2} + \mathbf{4}$. Let z and y be the top and bottom of the 4-chain, and let w and x be the top and bottom of the 2-chain. Prove that the events $x < y < w$ and $x < z < w$ are not postively correlated on Q. (C. Mallows)

12.4.48. ($\diamond$) For $x \in Q$, let the random variable H_x be the height of x on a random linear extension of Q, and let $h(x) = \mathbb{E}(H_x)$. Let y be incomparable to x. Use the XYZ Inequality to prove $h(x|(x > y)) \geq 1 + h(x|(x < y))$. (Winkler [1982])

12.4.49. (+) The **track number** of a graph G, written $\tau(G)$, is the least t such that G is the union of t interval graphs. The **interval number** of G, written $i(G)$, is the least t such that G is the intersection graph of sets that are unions of t intervals in $\mathbb{R}$. Note that always $i(G) \leq t(G)$. We show that $t(G)$ is unbounded even when $i(G) = 2$, as conjectured by Heldt, Knauer, and Ueckerdt (see Milans–Stolee–West [2015]).
  (a) Prove that $i(G) \leq 2$ when G is a line graph.
  (b) Prove that if $n \geq R_k(P')$, then $\tau(L(K_n)) > k$, where P' is the ordered hypergraph with vertices $1, 2, 3, 4, 5, 6$ in order obtained from the monotone path P_6^3 by adding the two edges $\{1, 2, 5\}$ and $\{2, 5, 6\}$.
  (c) Prove that if $n < R_k(P_4^3)$, then $\tau(L(K_n)) \leq k + 2$.
  (d) For the complete 3-uniform hypergraph K_6^3, Erdős–Rado [1952] proved the bound $R_k(K_6^3) \leq 2^{2^{O(k \lg k)}}$ (improved by Conlon–Fox–Sudakov [2010] to $2^{2^{(4+o(1))k \lg k}}$). Use this and the bound $R_k(P_t^r) \geq 2^{m^{k-1}/2\sqrt{k}}$ from Corollary 12.4.46 (where $m = t - r + 1$) to prove

$$\Omega\left(\frac{\lg \lg n}{\lg \lg \lg n}\right) \leq \tau(L(K_n)) \leq \lg \lg n + \tfrac{1}{2} \lg \lg \lg n + O(\lg \lg \lg \lg n).$$

Chapter 13

Combinatorial Designs

In this chapter we study highly structured combinatorial arrangements. Block designs are families of sets (actually, uniform hypergraphs) with special conditions on pairs of elements. They are useful in constructions for extremal problems and in designing statistical experiments.

We study general designs and special configurations such as Latin squares, Hadamard matrices, and projective planes. Tools involve finite fields and difference sets. The final more technical section leads to a constructive disproof of a famous conjecture of Euler.

Many texts and monographs treat classical design theory. For example, Anderson [1997], Cameron–van Lint [1991], Lindner–Rodger [1997, 2009], Stinson [2004], van Lint–Wilson [1992, 2001], and Wallis [1988, 2007] are patient and accessible. Beth–Jungnickel–Lenz [1986, 1999] and Colbourn–Dinitz [1996, 2007] are encyclopedic. Dinitz–Stinson [1992] and Wallis [1996, 2003] provide surveys on special topics.

13.1. Arrangements

For hundreds of years, mathematicians have been fascinated by questions about the arrangements of labels that use all labels or all combinations of labels in equal and symmetric ways. The requirements are quite rigid. For example, solutions to "Sudoku" puzzles are Latin squares satisfying additional constraints.

LATIN SQUARES

We begin with Latin squares and the design of experiments.

13.1.1. Definition. A **Latin square** of *order* n is an n-by-n matrix with n distinct entries such that each label occurs exactly once in each row and each column. Two such squares are **orthogonal** if the n^2 ordered pairs of labels in corresponding positions are distinct. A family of pairwise orthogonal Latin squares is an **orthogonal family**.

13.1.2. Example. *Growing corn.* Suppose we wish to test four different types of corn seeds (A, B, C, D). We want to subdivide our field into plots in order to minimize the effect of the geography of the field on the test results. To guard against the possibility of an east-west or north-south gradient in soil quality, we can divide the field into a four by four grid of subplots and assign seed types to subplots using a Latin square design.

Given also four types of fertilizer to test $(\alpha, \beta, \gamma, \delta)$, we can use another Latin square for those. Ideally, we also want to test each fertilizer with each seed type. We can do this using a Latin square orthogonal to the first square. We can also test levels of irrigation (1,2,3,4) at the same time; the three Latin squares below are *pairwise* orthogonal. There is no larger family of pairwise orthogonal Latin squares of order 4. ∎

$$
\begin{array}{llll}
A\ B\ C\ D & \alpha\ \beta\ \gamma\ \delta & 1\ 2\ 3\ 4 \\
D\ C\ B\ A & \beta\ \alpha\ \delta\ \gamma & 3\ 4\ 1\ 2 \\
B\ A\ D\ C & \gamma\ \delta\ \alpha\ \beta & 4\ 3\ 2\ 1 \\
C\ D\ A\ B & \delta\ \gamma\ \beta\ \alpha & 2\ 1\ 4\ 3
\end{array}
$$

Versatile experimental designs require large orthogonal families. There is a trivial upper bound on the size of such a family.

13.1.3. Lemma. Every orthogonal family of order n has size at most $n-1$.

Proof: By relabeling elements, we can normalize each square so that the entries in each first row are the integers $1, \ldots, n$ in order (see Example 13.1.2). With this normalization, the pairs (i, i) have already all occurred in the first row for all i and every pair of squares in the family.

Since these pairs already occur, in a particular position below the first row the values in all squares in the family must be distinct. They also must differ from the column index, since that value already appears at the top of the column and these are Latin squares. There are only $n - 1$ such values and hence at most $n - 1$ squares in the family. ∎

13.1.4. Definition. An orthogonal family of order n and size $n - 1$ is a **complete family**. A family of k pairwise-orthogonal Latin squares of order n is indicated by MOLS(n, k), read as "mutually orthogonal family of Latin squares".

This use of "mutually" to mean "pairwise" is historical; "pairwise" is more accurate. This differs from the notion of mutual independence in Section 14.2, which is decidedly not pairwise. Dénes–Keedwell [1974, 1991, 2015] provides extensive material on Latin squares.

Complete families arise from finite fields. There is a finite field of size (order) n if and only if n is a power of a prime. The field is unique up to isomorphism and is denoted $\mathbb{F}_n$. The additive identity is 0, the multiplicative identity is 1, and every nonzero element has both an additive and a multiplicative inverse.

13.1.5. Theorem. (Moore [1896], Bose [1939], Stevens [1939]) If n is a power of a prime, then there is a complete family MOLS$(n, n - 1)$.

Proof: Let the elements of $\mathbb{F}_n$ be $x_1, \ldots, x_n$, with x_n the additive identity. The labels in each square are $\{x_i\}$; the field arithmetic determines where to put them. Define the kth square by putting $x_k x_i + x_j$ in position (i, j), for $1 \le k \le n - 1$.

Because addition of a single element or multiplication by a single nonzero element permutes the elements of the field, we have defined $n - 1$ Latin squares. To prove orthogonality, suppose that the pair in position (i, j) of squares k and l is the same as the pair in position (r, s) of squares k and l. That is,

$$x_k x_i + x_j = x_k x_r + x_s \,,$$
$$x_l x_i + x_j = x_l x_r + x_s \,.$$

Subtracting the two equations and canceling $x_k - x_l$ yields $x_r = x_i$, which in turn yields $x_s = x_j$. Thus $(r, s) = (i, j)$. ∎

When n is not a prime power, this construction still yields some orthogonal squares (Exercise 6). Next we give a combining theorem for orthogonal families.

13.1.6. Theorem. (**Moore–MacNeish Theorem**; Moore [1896], MacNeish [1922]) Given h pairwise orthogonal Latin squares of order m and h pairwise orthogonal Latin squares of order n, there are h pairwise orthogonal Latin squares of order mn.

Proof: Let $A_1, \ldots, A_h$ and $B_1, \ldots, B_h$ be the two families. Let $C_1, \ldots, C_h$ be h matrices defined as follows. The elements of C_k are pairs of elements from A_k and B_k. The matrix C_k splits into m^2 square blocks of order n (shown below) such that the first coordinates of the entries in the (i, j)th block all equal $a_{i,j}^{(k)}$ and the second coordinates form a copy of B_k.

Because A_k and B_k are Latin squares, C_k is also. If two pairs of corresponding positions in C_k and C_l have the same entries, then

$$([a_{i,j}^{(k)}, b_{r,s}^{(k)}], [a_{i,j}^{(l)}, b_{r,s}^{(l)}]) = ([a_{p,q}^{(k)}, b_{t,u}^{(k)}], [a_{p,q}^{(l)}, b_{t,u}^{(l)}]).$$

Now the orthogonality of A_k and A_l yields $(i, j) = (p, q)$, and the orthogonality of B_k and B_l yields $(r, s) = (t, u)$. ∎

$$C_k = \begin{array}{|c|c|c|}
\hline
(a_{1,1}^{(k)}, b_{r,s}^{(k)}) & \cdots & (a_{1,m}^{(k)}, b_{r,s}^{(k)}) \\
\hline
\vdots & \ddots & \vdots \\
\hline
(a_{m,1}^{(k)}, b_{r,s}^{(k)}) & \cdots & (a_{m,m}^{(k)}, b_{r,s}^{(k)}) \\
\hline
\end{array}$$

13.1.7. Corollary. If $n = \prod p_i^{e_i}$ with each p_i prime, then there exist $\min_i(p_i^{e_i} - 1)$ pairwise orthogonal Latin squares of order n. ∎

In general, there are larger families of MOLS than guaranteed by Corollary 13.1.7. When n is twice an odd number, Corollary 13.1.7 does not guarantee any pair of orthogonal Latin squares at all.

13.1.8. Example. *Euler's problem.* The problem of finding MOLS$(6, 2)$ was published by Euler in 1779 as the Problem of the 36 Officers. On parade day, there are 36 officers: six ranks, six regiments, one officer of each rank from each regiment. They desire to march with one officer of each rank in each row and column, and one officer of each regiment in each row and column. Thus the appearances of each rank must form a Latin square, as must the appearances of each regiment. Since there is only one officer of a given rank from a given regiment, these Latin squares must be orthogonal.

Euler conjectured impossibility. By normalizing the top rows to the labels 123456 in order, Tarry [1900, 1901] reduced the problem to an exhaustive examination of 9408 pairs, proving Euler right. Stinson [1984] found a much shorter proof (see Anderson [1997, pp. 130–133]). ∎

13.1.9. Remark. Due to Euler's use of Greek and Latin letters for orthogonal squares, orthogonal pairs are also called **Graeco–Latin squares**.

Euler conjectured also that there are no orthogonal n-by-n Latin squares whenever n is an odd multiple of 2, but here he was very wrong. Bose–Shrikhande–Parker [1960] proved that for $n \notin \{1, 2, 6\}$, orthogonal Latin squares of order n exist. We will discuss this in Section 13.3.

It is a long way from 2 to $n - 1$. The next value after 6 that is not a prime power is 10. A long-running supercomputer search concluded that no complete family of order 10 exists (see Lam–Thiel–Swiercz [1989]). Even after this long search, the maximum size of an orthogonal family is not known; in fact, it is not known whether there are 3 pairwise orthogonal 10-by-10 Latin squares. ∎

We will see that complete families of order n are equivalent to various other combinatorial configurations, including projective planes of order n, which can be viewed as a special class of block designs. To indicate more of the scope of design theory, we next introduce the general object.

BLOCK DESIGNS

In the setting of Example 13.1.2, the treatments form a set V. For complete testing, we would like to test each treatment on each experimental unit, but this may be expensive or infeasible. In a more general setting, each unit receives only a subset of V, called a **block**. Several conditions are imposed to balance the usage of treatments and units.

13.1.10. Definition. A (v, k, λ)-**design** is a family of blocks of size k from a set V of size v, such that each element of V appears in the same number of blocks and every two elements of V appear in λ common blocks. Let b denote the number of blocks and r denote the number of blocks containing a given element. When $k < v$, this is a **balanced incomplete block design (BIBD)**.

Having each treatment appear in r blocks ensures that treatments are tested equally often; keeping the blocks the same size ensures that no experimental unit is overused. Fixing λ helps to control for interaction between treatments. These conditions were introduced formally by Yates [1936], but special cases of designs were studied at least as early as 1844, with a seminal paper by Kirkman [1847].

Originally, all five parameters were listed in naming a design (how fast can you say "(v, b, r, k, λ)-design"?). However, two simple counting arguments reduce the number of independent parameters to three (Proposition 13.1.14). Thus we simply call the object a (v, k, λ)-design. (Even the hypothesis that all elements appear in equally many blocks can be dropped; see Exercise 5.)

Historically, the treatments were called **varieties**, and thus the notation reflects the terminology: v counts the varieties, b the blocks, r the "replications" (usage of each element), k the "kardinality" of the blocks, and λ the "linkage" (common appearances) of any two elements. Because we can view the blocks as the edge set of a hypergraph with vertex set V, we usually call V the set of *vertices* rather than the set of *varieties*.

13.1.11. Example. *Complete k-uniform hypergraph.* When $|V| = v$, the family $\binom{V}{k}$ is the set of blocks of a $(v, k, \binom{v-2}{k-2})$-design. There are $\binom{v}{k}$ blocks, each element appears in $\binom{v-1}{k-1}$ blocks, and each pair of elements appears in $\binom{v-2}{k-2}$ blocks. ∎

13.1.12. Example. *Fano plane.* The **Fano plane** is the $(7, 3, 1)$-design with vertex set $[7]$ and blocks $\{124, 235, 346, 457, 561, 672, 713\}$ (for clarity, we delete commas within blocks and drop set brackets on blocks). Viewing the elements as congruence classes modulo 7, the blocks have the form $\{i, i+1, i+3\}$. Every pair $\{r, s\}$ of congruence classes is one apart, two apart, or three apart. It thus occurs uniquely in the block generated by i such that $\{r, s\}$ is $\{i, i+1\}$ or $\{i+1, i+3\}$ or $\{i, i+3\}$, respectively.

We can represent the Fano plane geometrically by letting the blocks correspond to lines through the points; hence the use of "plane" in the name. As shown in the famous diagram below, one line must "bend".

If we delete one of these blocks and delete its points from all other blocks, then we are left with six blocks consisting of all pairs from a set of four points, which is an instance of Example 13.1.11. ∎

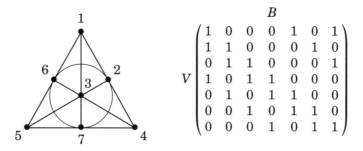

13.1.13. Example. *9-point triple system.* From nine points in a square array, we form four sets of three triples. We use the rows, the columns, the rightward diagonals, and the leftward diagonals. Two points in the same row or column clearly appear together in exactly one block. If they are not in the same row or column, then there is exactly one way to reach one from the other along a diagonal. Hence the 12 triples form a $(9, 3, 1)$-design.   ■

$$\begin{array}{cccc} 123 & 147 & 159 & 168 \\ 456 & 258 & 267 & 249 \\ 789 & 369 & 348 & 357 \end{array}$$

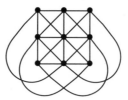

The **incidence matrix** of a design is the matrix A of the incidence relation between the set V of elements and the set B of blocks. Position (i, j) is 1 if the ith element belongs to the jth block, 0 otherwise. The incidence matrix is a fundamental tool in the study of designs, facilitating counting arguments. The first such arguments yield necessary conditions on the parameters and an algebraic interpretation of the definition condition for usage of pairs of vertices.

13.1.14. Proposition. Every (v, k, λ)-design having b blocks and r appearances of each element satisfies the equations below, where A is the incidence matrix of the hypergraph of blocks and J is the all-1 square matrix.
(a) $bk = vr$.
(b) $r(k - 1) = \lambda(v - 1)$.
(c) $AA^T = (r - \lambda)I + \lambda J$.

Proof: The matrix A has k 1s in each column and r 1s in each row, so each side of (a) counts the 1s in A.

For an element x, each side of (b) counts all appearances of elements other than x in blocks containing x, on the left by blocks and on the right by elements.

For (c), the entry in position (i, j) of AA^T is the dot product of rows i and j of A. By definition, this is λ when $i \neq j$ and is r when $i = j$.   ■

Thus $r = \lambda(v - 1)/(k - 1)$ and $b = vr/k$ for a (v, k, λ)-design. Next we use algebraic properties of the incidence matrix to prove a fundamental inequality.

13.1.15. Theorem. (Fisher's Inequality; Fisher [1940]) If $k < v$ in a (v, k, λ)-design, then $b \geq v$.

Proof: Let A denote the incidence matrix. We use Proposition 13.1.14c to show linear independence of the rows of A. If $w^T A = \mathbf{0}$, then associativity of matrix multiplication yields

$$0 = w^T AA^T w = w^T[(r - \lambda)I + \lambda J]w = (r - \lambda)(\sum w_i^2) + \lambda(\sum w_i)^2.$$

Since $r(k - 1) = \lambda(v - 1)$ always, $k < v$ implies $r > \lambda$. Hence the terms on the right are both nonnegative and hence both 0. From $\sum w_i^2 = 0$ we obtain $w = \mathbf{0}$. Hence the rows of A are linearly independent, and therefore A has at least as many columns as rows.   ■

SYMMETRIC DESIGNS

When $b = v$, the incidence matrix is square, and by the proof of Fisher's Inequality it is nonsingular. Such designs have many additional nice properties.

13.1.16. Definition. A (v, k, λ)-design is a **symmetric design** if $b = v$.

By Proposition 13.1.14a, $b = v$ in a symmetric design implies $r = k$; the number of elements in each block is the same as the number of blocks containing each element. We will also show that not only does each pair of elements appear in λ common blocks, but also each pair of blocks has λ common elements.

13.1.17. Example. With 7 blocks and 7 elements, the Fano plane (Example 13.1.12) is a symmetric design. In the Fano plane, we compare two blocks $A = \{i, i + 1, i + 3\}$ and $B = \{j, j + 1, j + 3\}$ by subtracting each element of A from each element of B. The difference $j - i$ appears three times, and the other differences are the six other values from $j - i - 3$ through $j - i + 3$. This means that when $j \neq i$ the difference 0 appears exactly once, and the two blocks have exactly one common element.   ∎

The statement that every two blocks have λ common elements is just $A^T A = (k - \lambda)I + \lambda J$. This statement is nontrivial, whereas the statement $AA^T = (k - \lambda)I + \lambda J$ from Proposition 13.1.14c is essentially by definition. We conclude that for a symmetric design, A and A^T commute.

13.1.18. Theorem. (Bose [1939]) In a symmetric (v, k, λ)-design, every two blocks have λ common elements.

Proof: Let B be a block in the design, and let $x_1, \ldots, x_{v-1}$ be the sizes of the intersections of B with the other blocks. To prove that each x_i equals λ, we will show that $\sum (x_i - \lambda)^2 = 0$.

Since each element of B appears in k blocks, $\sum x_i = k(k - 1) = \lambda(v - 1)$ (applying Proposition 13.1.14). Counting the appearances of pairs of elements from B yields $\binom{k}{2} + \sum \binom{x_i}{2} = \binom{k}{2}\lambda$, which we rewrite as $\sum (x_i^2 - x_i) = k(k - 1)(\lambda - 1) = \lambda(\lambda - 1)(v - 1)$ (applying Proposition 13.1.14). Summing the two identities yields $\sum x_i^2 = \lambda^2(v - 1)$. Now

$$\sum (x_i - \lambda)^2 = \sum x_i^2 - \sum 2\lambda x_i + \sum \lambda^2$$
$$= \lambda^2(v - 1) - (2\lambda)\lambda(v - 1) + (v - 1)\lambda^2 = 0 ,$$

as desired.   ∎

Next we obtain necessary conditions for symmetric (v, k, λ)-designs.

13.1.19. Proposition. If a symmetric (v, k, λ)-design exists, then

$$k(k - 1) = \lambda(v - 1).$$

Proof: Symmetric designs require $r = k$ (Proposition 13.1.14a), so Proposition 13.1.14b becomes $k(k - 1) = \lambda(v - 1)$.   ∎

13.1.20. Example. *Symmetric designs.* Proposition 13.1.19 requires $k > \lambda$ when $k < v$ (we exclude (k, k, k)-designs, which are "complete" designs but can be called trivial designs). Some such triples are easy to achieve.

$(k+1, k, k-1)$: The complete k-uniform hypergraph with $k+1$ vertices. Note that any two k-sets in $[k + 1]$ have $k - 1$ common elements.

$(k(k - 1) + 1, k, 1)$: Symmetric designs with $\lambda = 1$ are called "projective planes", studied in Section 13.2. They exist when $k - 1$ is a power of a prime. The case $(7, 3, 1)$ is the Fano plane of Example 13.1.12. The case $(43, 7, 1)$ will be forbidden by Theorem 13.1.23 in Example 13.1.24.

For $k \leq 7$, the triples with $2 \leq \lambda \leq k - 2$ that satisfy Proposition 13.1.19 are $(7, 4, 2)$, $(11, 5, 2)$, $(11, 6, 3)$, $(16, 6, 2)$, $(15, 7, 3)$, and $(22, 7, 2)$. The last one will be forbidden by the next proposition. The other cases listed here with $\lambda = 2$ exist and are called **biplanes**. ∎

Our remaining necessary conditions use the algebraic statements we have proved about symmetric designs.

13.1.21. Proposition. (Schützenberger [1949], Shrikhande [1950]) If a symmetric (v, k, λ)-design exists with v even, then $k - \lambda$ is a square.

Proof: Let A be the incidence matrix. For a symmetric design, we have $b = v$ and $r = k$. By Proposition 13.1.14c, $AA^T = (k - \lambda)I + \lambda J$. The eigenvalues of λJ are $v\lambda$ with multiplicity 1 and 0 with multiplicity $v - 1$. Adding a multiple of I shifts the eigenvalues, so the eigenvalues of AA^T are $k + (v - 1)\lambda$ with multiplicity 1 and $(k - \lambda)$ with multiplicity $v - 1$.

Since $k(k - 1) = \lambda(v - 1)$ (Proposition 13.1.14b), the largest eigenvalue is k^2. The determinant is the product of the eigenvalues, so $\det(AA^T) = k^2(k - \lambda)^{v-1}$. Since $\det(AA^T) = \det A \det A^T = (\det A)^2$, we have $\det A = k(k-\lambda)^{(v-1)/2}$. Since A is an integer matrix, $\det A$ is an integer, and thus $(k - \lambda)^{(v-1)/2}$ is rational. With v even, this requires $k - \lambda$ to be a square. ∎

Proposition 13.1.21 prohibits $(22, 7, 2)$-designs, since 5 is not a square.

The next theorem is the work of Bruck–Ryser [1949] and Chowla–Ryser [1950]. The proof uses Lagrange's Theorem (every positive integer is a sum of four squares) and a special matrix construction.

13.1.22. Lemma. The matrix below satisfies $HH^T = (b_1^2 + b_2^2 + b_3^3 + b_4^2)I_4$.

$$H = \begin{pmatrix} b_1 & b_2 & b_3 & b_4 \\ -b_2 & b_1 & -b_4 & b_3 \\ -b_3 & b_4 & b_1 & -b_2 \\ -b_4 & -b_3 & b_2 & b_1 \end{pmatrix}$$ ∎

13.1.23. Theorem. (Bruck–Chowla–Ryser Theorem) If a symmetric (v, k, λ)-design exists, then
(a) If v is even, then $k - \lambda$ is the square of an integer.
(b) If v is odd, then $z^2 = (k - \lambda)x^2 + (-1)^{(v-1)/2}\lambda y^2$ has a nonzero solution in integers x, y, z.

Proof: Statement (a) is Proposition 13.1.21. For odd v, let $n = k - \lambda$. Let A be the incidence matrix, and let $\mathbf{x} = (x_1, \ldots, x_v)^T$; this is a column vector of v free variables. Let $(z_1, \ldots, z_v)^T = A\mathbf{x}$. Theorem 13.1.18 yields $A^T A = nI + \lambda J$. Hence

$$\sum_{i=1}^{v} z_i^2 = (\mathbf{x}^T A^T)(A\mathbf{x}) = n \sum_{i=1}^{v} x_i^2 + \lambda \left(\sum_{i=1}^{v} x_i \right)^2 . \tag{*}$$

We will modify this equation to obtain the desired equation satisfied by integers x, y, z. First, Lagrange's Theorem expresses n as the sum of four squares: $n = b_1^2 + b_2^2 + b_3^2 + b_4^2$. Let $(y_1, y_2, y_3, y_4)^T = H^T(x_1, x_2, x_3, x_4)^T$, where H is the matrix of Lemma 13.1.22 defined on b_1, b_2, b_3, b_4. Since $HH^T = nI_4$, we have $\sum_{i=1}^{4} y_i^2 = n \sum_{i=1}^{4} x_i^2$. We call such a replacement *H-substitution*.

Case 1: $v \equiv 1 \pmod 4$. With $w = \sum_{i=1}^{v} x_i$, applying H-substitution to sets of four variables yields

$$\sum_{i=1}^{v} z_i^2 = \sum_{i=1}^{v-1} y_i^2 + nx_v^2 + \lambda w^2 . \tag{**}$$

Since H is invertible and has integer entries, we can write $x_1, \ldots, x_{v-1}$ as rational linear combinations of $\{y_1, \ldots, y_{v-1}\}$. Thus also $z_1, \ldots, z_{v-1}$ are rational linear combinations of $\{y_1, \ldots, y_{v-1}, x_v\}$, treating these as the freely chosen variables. Consider $z_1 = \sum_{i=1}^{v-1} \alpha_i y_i + \alpha_v x_v$. If $\alpha_1 \neq 1$, then restricting the choice of y_1 by requiring $y_1 = \frac{1}{1-\alpha_1} \left[\sum_{i=2}^{v-1} a_i y_i + a_v x_v \right]$ yields $z_1 = y_1$. If $\alpha_1 = 1$, then setting $y_1 = \frac{1}{-1-\alpha_1} \left[\sum_{i=2}^{v-1} a_i y_i + a_v x_v \right]$ yields $z_1 = -y_1$. In either case, we have $z_1^2 = y_1^2$ and can cancel z_1^2 and y_1^2 from (**), leaving an expression in the remaining variables.

With this substitution for y_1, we have expressed z_2 as a rational linear combination of $y_2, \ldots, y_{v-1}$ and x_v. Suitably restricting y_2 in terms of the free variables $y_3, \ldots, y_{v-1}$ and x_v yields $z_2^2 = y_2^2$ and allows us to cancel z_2^2 and y_2^2 from (**).

Repeating this substitution procedure leads to canceling $\sum_{i=1}^{v-1} z_i^2$ and $\sum_{i=1}^{v-1} y_i^2$ from (**) by restricting $y_1, \ldots, y_{v-1}$ as appropriate rational multiples of the remaining free variable x_v, leaving the equation $z_v^2 = nx_v^2 + \lambda w^2$. Since $w = \sum_{i=1}^{v} x_i$ and $(z_1, \ldots, z_v)^T = A\mathbf{x}$, both w and z_v are linear combinations of $x_1, \ldots, x_v$. We expressed $x_1, \ldots, x_{v-1}$ as rational combinations of $y_1, \ldots, y_{v-1}$, which in turn were reduced to rational multiples of x_v. Now multiplying by the least common multiple of all the denominators yields an integer solution to $z^2 = (k-\lambda)x^2 + \lambda y^2$.

Case 2: $v \equiv 3 \pmod 4$. First add nx_{v+1}^2 to both sides of (*), where x_{v+1} is a new free variable. Now the variables $x_1, \ldots, x_{v+1}$ are transformed to $y_1, \ldots, y_{v+1}$ in sets of four by H-substitution. The equation (**) becomes $nx_{v+1}^2 + \sum_{i=1}^{v} z_i^2 = \sum_{i=1}^{v+1} y_i^2 + \lambda w^2$ in terms of the free variables $y_1, \ldots, y_{v+1}$. Iterative restriction and cancellation reduces this to $nx_{v+1}^2 = y_{v+1}^2 + \lambda w^2$ in v steps. Now the clearing of fractions yields an integer solution of $(k-\lambda)x^2 = z^2 + \lambda y^2$, as desired. ∎

13.1.24. Example. Except for $(16, 6, 2)$ and $(43, 7, 1)$, the unresolved triples from Example 13.1.20 have the form $(4m-1, 2m, m)$ or $(4m-1, 2m-1, m-1)$, where m is an integer greater than 1. Both types are allowed by Theorem 13.1.23.

For the first, we need $z^2 = mx^2 - my^2 = m(x-y)(x+y)$; set $x = m+1$, $y = m-1$, and $z = 2m$. For the second, we need $z^2 = mx^2 - (m-1)y^2$; set $x = y = z = 1$. The numerical conditions are only necessary conditions, but these designs do exist for many m (see Proposition 13.1.29).

For $(16, 6, 2)$, since v is even, our only necessary condition is that $6 - 2$ is a square. Such designs exist, but we will not construct them.

However, consider $(43, 7, 1)$. We need a solution to $y^2 + z^2 = 6x^2$. We may assume that x, y, z have no common factor. The sum of two squares is a multiple of 3 only if they are themselves multiples of three. Hence 3 divides both y and z, so 9 divides the sum of their squares. This makes x divisible by 3, contradicting the absence of common factors. (See Example 13.2.9 for a generalization.) ■

HADAMARD MATRICES

Designs with $\lambda = 1$ are studied the most. We pause to discuss special matrices that yield symmetric designs with larger λ. Hadamard [1893] asked for the largest magnitude of the determinant of an n-by-n matrix with entries ± 1. The bound of $n^{n/2}$ is attained only by matrices of the following type (Exercise 13). (Actually, these matrices were first studied by Sylvester [1867].)

13.1.25. Definition. A **Hadamard matrix** of order n is an n-by-n matrix H with entries in $\{1, -1\}$ such that $HH^T = nI$.

With entries ± 1, the dot product of any row with itself is n. The rows (or the columns, equivalently) must be pairwise orthogonal. We will obtain a $(4m - 1, 2m - 1, m - 1)$-design from a Hadamard matrix of order $4m$. First we construct arbitrarily large Hadamard matrices.

13.1.26. Proposition. If A and B are Hadamard matrices of orders m and n, then the matrix H with row indices (i, k), column indices (j, l) and entries $h_{(i,k),(j,l)} = a_{i,j} \cdot b_{k,l}$ is a Hadamard matrix of order mn.

Proof: This is the same construction as in Theorem 13.1.6, except that we record the product of entries instead of the ordered pair. Grouping entries of H by the first coordinate of each index expresses H as a matrix of n-by-n blocks, where the (i, j)th block is $a_{i,j}B$. With b_r denoting row r of B, the computation of the dot product of row (i, k) and row (i', k') of H becomes

$$\sum_j a_{i,j} a_{i',j} (b_k \cdot b_{k'}) = \delta_{k,k'} n \sum_j a_{i,j} a_{i',j} = \delta_{k,k'} \delta_{i,i'} mn. \qquad ■$$

13.1.27. Example. Note that (1) is a 1-by-1 Hadamard matrix. Because $\left(\begin{smallmatrix} 1 & 1 \\ 1 & -1 \end{smallmatrix}\right)$ is a Hadamard matrix, products yield Hadamard matrices for all powers of 2 (observed by Sylvester [1867]). Below is the resulting matrix of order 4.

$$\begin{pmatrix} 1 & 1 & 1 & 1 \\ 1 & -1 & 1 & -1 \\ 1 & 1 & -1 & -1 \\ 1 & -1 & -1 & 1 \end{pmatrix} \qquad ■$$

13.1.28. Definition. A **normalized Hadamard matrix** is a Hadamard matrix in which the first entry of every row and column is 1.

Multiplying a row or column of a matrix by -1 does not change the magnitude of the dot products of any two rows, so a Hadamard matrix of order n yields a normalized Hadamard matrix of order n.

13.1.29. Proposition. (Hadamard [1893]) If $n > 2$ and an n-by-n Hadamard matrix exists, then $m = n/4 \in \mathbb{N}$, and a $(4m - 1, 2m - 1, m - 1)$-design exists.

Proof: First normalize the Hadamard matrix. Since the dot product of the first row with any other is 0, each row other than the first has $n/2$ elements of each sign. Since $n > 2$, there are two other rows. Let m be the number of columns where they both have $+1$. Since they each have $n/2$ elements of each sign, the product is now $2m - 2(n/2 - m)$, as shown below. This must equal 0, so $m = n/4$, and n is divisible by 4.

	$n/2$		$n/2$	
	1		-1	
1	-1	1	-1
m	$n/2 - m$	$n/2 - m$	m

From the normalized Hadamard matrix, discard the first row and column and change each -1 to 0. By the observations above, each row and column has $2m - 1$ 1s, and each pair of rows has 1s in exactly $m - 1$ common columns. ∎

Conversely, a $(4m - 1, 2m - 1, m - 1)$-design yields a Hadamard matrix of order $4m$ (Exercise 3). Hadamard matrices of order $4m$ are also equivalent to the existence of $(4m - 1, 2m, m)$-designs (Exercise 16). Designs with parameters of these two forms are called **Hadamard designs**. The rows of the incidence matrix of a Hadamard design are binary vectors that differ in many positions.

The fundamental problem of coding theory can be described as finding large sets of vectors that pairwise are far apart. For binary vectors, the *length* is the number of positions, the *weight* is the number of positions with 1, and the *distance* between vectors is the number of places where they differ (**Hamming distance**). When the distance between words (vectors) in the code is large, more transmission errors can be tolerated, since they won't change one codeword into another.

Let $A(n, d)$ be the maximum size of a set of binary n-tuples with pairwise distance at least d. Hadamard matrices provide excellent codes, essentially achieving an upper bound on the code size.

13.1.30. Theorem. (Plotkin) If $d > n/2$, then $A(n, d) \le 2d/(2d - n)$.

Proof: Consider a code of size N, and let s be the sum of the distances between all pairs of code words. Since the distance is at least d for each pair, $s \ge d\binom{N}{2}$.

We obtain an upper bound on s by considering the contributions to distances from the n coordinates. A coordinate contributes to the distance between two words if and only if the words differ in that position. If k code words have 1 in position j and $N - k$ have 0 there, then position j contributes $k(N - k)$ to the total of the distances. Counting by positions, we thus have $s \le nN^2/4$.

Together, the bounds on s yield $N \le 2d/(2d - n)$. ∎

13.1.31. Proposition. If $n = 4m - 1$ and there is a Hadamard matrix of order $n + 1$, then $A(n, (n + 1)/2) = n + 1$.

Proof: In a Hadamard matrix of order $n + 1$, every two rows differ in $(n + 1)/2$ places. In a normalized Hadamard matrix, they agree in the first column. We discard the first column and change each -1 to 0. We now have $n + 1$ words of length n with pairwise distance $(n + 1)/2$. This achieves the Plotkin Bound for $d = (n + 1)/2$ and hence is optimal.   ∎

When $d = n/2$, the Plotkin Bound is meaningless. Nevertheless, here again Hadamard matrices give the largest possible codes.

13.1.32. Proposition. If $d = n/2$, then $A(n, d) \leq 2n$, and this is achieveable if there is a Hadamard matrix of order n.

Proof: Consider a code of size N, and let S and T be the sets of code words with first coordinate 1 or 0, respectively. Deleting the first coordinate from every code word in S creates a code of length $n - 1$ with minimum distance d, and similarly for T. Hence $N \leq 2A(n - 1, d) \leq 2n$, by Proposition 13.1.31.

We obtain a code of size N by using the rows of the Hadamard matrix (changing each -1 to 0) *and* the complements of these words. Since the pairs $(1, 1), (1, -1), (-1, 1), (-1, -1)$ occur equally often between two rows, the distance between any two words in this code is $n/2$ or n.   ∎

To illustrate the application of Hadamard matrices and presage similar applications of projective planes, we consider one extremal problem. In applications of designs to extremal problems, there is a bound that arises from a counting argument, and the bound can be attained only when the design-theoretic structure is used in the construction.

13.1.33. Application. *A bipartite Ramsey problem.* (Beineke–Schwenk [1976]) Fix $s, t \in \mathbb{N}$. When n is large, every $0, 1$-matrix of order n has a constant s-by-t submatrix; let $b(s, t)$ be the least such n. In graph-theoretic language, this is a bipartite analogue of Ramsey numbers (Section 10.2). Consider $K_{n,n}$ with bipartition X, Y. A $0, 1$-matrix is a 2-coloring of $E(K_{n,n})$, and we want to force a monochromatic copy of $K_{s,t}$ with the part of size s in X.

Fix $s = 2$. For $n \geq 4t - 3$, we prove that a color used on a subgraph G with at least $n^2/2$ edges of $K_{n,n}$ yields such a copy of $K_{s,t}$. Let $m = |E(G)|$.

Consider the number p of copies of P_3 in G whose center lies in Y. Since any two vertices in X have at most $t - 1$ common neighbors in G, we have $p \leq (t-1)\binom{n}{2}$ if $K_{2,t}$ does not occur in G. On the other hand, $p = \sum_{y \in Y} \binom{d(y)}{2}$, since each $y \in Y$ extends in $\binom{d(y)}{2}$ ways to form P_3. Since $\sum_{y \in Y} d(y) = m$, and the average degree among vertices of Y is m/n, the convexity of quadratic functions yields $\sum_{y \in Y} \binom{d(y)}{2} \geq n\binom{m/n}{2}$. Since $m \geq n^2/2$, we have $m/n \geq n/2$. Thus

$$n\binom{n/2}{2} \leq \sum_{y \in Y} \binom{d(y)}{2} = p \leq (t - 1)\binom{n}{2}.$$

This inequality simplifies to $n\frac{n-2}{n-1} \leq 4t - 4$, which requires $n \leq 4t - 3$ for integer n. Thus if $n > 4t - 3$, then a monochromatic $K_{2,t}$ occurs. However, when n

is odd, the numbers $n^2/2$ and $n/2$ are not integers, which allows us to strengthen the lower bound on p. We leave to Exercise 19 the completion of the proof that also when $n = 4t - 3$, a monochromatic $K_{2,t}$ occurs. Thus $b(2, t) \le 4t - 3$.

Beineke–Schwenk [1976] proved that $b(2, t) = 4t - 3$ when $t \in \{2, 4\}$ and when there is a Hadamard matrix H of order $2t - 2$ (this requires t odd). To prove $b(2, t) > 4t - 4$, we construct from H a matrix of order $4t - 4$ with no constant 2-by-t submatrix. Obtain H' from H by converting each -1 to 0, and let $M = \left(\begin{smallmatrix} H' & H' \\ H' & \overline{H'} \end{smallmatrix}\right)$. Since any two rows of H' have $(t - 1)/2$ common columns with 1s and $(t - 1)/2$ common columns with 0s, any two rows of M have constant 2-by-$(t - 1)$ submatrices in both 0 and 1, but there is no constant 2-by-t submatrix. ∎

Thus far we have constructed Hadamard matrices only of orders that are powers of 2; others are more difficult. As observed in Example 13.1.24, Theorem 13.1.23 does not prohibit $(4m - 1, 2m - 1, m - 1)$-designs for any m. It is conjectured that there is a Hadamard matrix of each order that is a multiple of 4. We will show how to construct Hadamard matrices of order $n = 2^k(q + 1)$, where q is an odd prime power and n is a multiple of 4, and then Proposition 13.1.26 yields matrices of many other orders.

13.1.34.* Example. *The Paley construction.* Below are Hadamard matrices of order 12 from the construction in Theorem 13.1.37 with $q = 11$ and $q = 5$. ∎

```
+|+ + + + + + + + + + +        +|+ + + + +|-|+ + + + +
-|+ + - + + + - - - + -        +|+ + - - +|+|- + - - +
-|- + + - + + + - - - +        +|+ + + - -|+|+ - + - -
-|+ - + + - + + + - - -        +|- + + + -|+|- + - + -
-|- + - + + - + + + - -        +|- - + + +|+|- - + - +
-|- - + - + + - + + + -        +|+ - - + +|+|+ - - + -
-|- - - + - + + - + + +        ---------------------
-|+ - - - + - + + - + +        -|+ + + + +|-|- - - - -
-|+ + - - - + - + + - +        +|- + - - +|-|- - + + -
-|+ + + - - - + - + + -        +|+ - + - -|-|- - - + +
-|- + + + - - - + - + +        +|- + - + -|-|+ - - - +
-|+ - + + + - - - + - +        +|- - + - +|-|+ + - - -
                               +|+ - - + -|-|- + + - -
```

13.1.35.* Definition. A **conference matrix** C is a matrix of order n with diagonal entries 0 and off-diagonal entries ± 1 such that $CC^T = (n - 1)I$; that is, the rows are pairwise orthogonal (Belevitch [1950]).

13.1.36.* Lemma. If C is an antisymmetric conference matrix, then $C + I$ is a Hadamard matrix. If C is a symmetric conference matrix, then $H = \left(\begin{smallmatrix} C+I & C-I \\ C-I & -C-I \end{smallmatrix}\right)$ is a Hadamard matrix.

Proof: When C is antisymmetric, meaning that $C^T = -C$, we have
$$(C + I)(C + I)^T = CC^T + C + C^T + I = (n - 1)I + I = nI.$$

When C is symmetric, computing HH^T again suffices (Exercise 20). ∎

13.1.37.* Theorem. (Paley [1933]) If q is an odd prime power and $n = 2^k(q + 1) \equiv 0 \pmod 4$, then there is a Hadamard matrix of order n.

Proof: By the construction of Proposition 13.1.26, it suffices to consider $n = q+1$ when $q \equiv 3 \pmod 4$ and $n = 2(q+1)$ when $q \equiv 1 \pmod 4$. Using Lemma 13.1.36, it suffices to construct a conference matrix of order $q + 1$, antisymmetric when $q \equiv 3 \pmod 4$ and symmetric when $q \equiv 1 \pmod 4$.

The construction uses the notion of quadratic residues in finite fields. For an element x of $\mathbb{F}_q$, the finite field of order q, we define the **character** $\chi(x)$ to be 1 if x is a nonzero square, -1 if x is not a square, and 0 if $x = 0$.

Observe that $\chi(x)\chi(y) = \chi(xy)$ and that $\sum_{x\in\mathbb{F}_q}\chi(x) = 0$, since half of the nonzero elements are squares. For a nonzero element c of $\mathbb{F}_q$, we have

$$\sum_{x\in\mathbb{F}_q}\chi(x)\chi(x+c) = \sum_{x\neq 0}\chi(x)\chi(x)\chi(1+cx^{-1}) = \sum_{x\neq 0}\chi(1+cx^{-1}) = -1,$$

since the last sum includes characters of all elements except 1.

List $\mathbb{F}_q$ as $x_0, \ldots, x_{q-1}$ with $x_0 = 0$, and define a matrix A by $a_{ij} = \chi(x_i - x_j)$. Since -1 is a square if and only if $q \equiv 1 \pmod 4$, the matrix A is symmetric if $q \equiv 1 \pmod 4$ and antisymmetric if $q \equiv 3 \pmod 4$. We have computed $AA^T = qI - J$.

Now enlarge A to a matrix C of order $q + 1$ by adding a first row and column. The first row is $(0\ 1\ 1\ \cdots\ 1)$. The first column is $(0\ 1\ 1\ \cdots\ 1)^T$ when $q \equiv 1 \pmod 4$ and $(0\ -1\ -1\ \cdots\ -1)^T$ when $q \equiv 3 \pmod 4$. The result is a conference matrix, since each row or column of A sums to 0. The matrix C is symmetric or antisymmetric as desired. ∎

The conference matrices of Theorem 13.1.37 are **Paley matrices**. Scarpis [1898] did this for prime q. Two resulting Hadamard matrices appear in Example 13.1.34. With Proposition 13.1.26, Paley's Theorem guarantees Hadamard matrices of all multiples of 4 through $n = 128$ except 92 and 116 (Exercise 4).

A Hadamard matrix of order 92 was found in Baumert–Golomb–Hall [1962] by computer, based on a method of Williamson [1944] that uses the matrix of Lemma 13.1.22 (see van Lint–Wilson [1992, pp. 177–179]). In 1978, the smallest multiple of 4 for which no Hadamard matrix was known was 268 (Seberry [1978]). In 1992, it was 428 (van Lint–Wilson [1992]). Later, Kharaghani–Tayfeh-Rezaie [2005] constructed one of order 428, leaving 668 as the next open case.

EXERCISES 13.1

13.1.1. (–) Explain how to construct a pair of orthogonal Latin squares of order 15. Include all needed steps, but do not write out the final pair of squares.

13.1.2. (–) Show that there is a (v, k, λ)-design with b blocks and each element appearing in r blocks if and only if there is a $(v, v - k, b + \lambda - 2r)$-design.

13.1.3. (–) Prove that if a $(4m - 1, 2m - 1, m - 1)$-design exists, then there is a Hadamard matrix of order $4m$.

13.1.4. Use Proposition 13.1.26 and Paley's Theorem (Theorem 13.1.37) to show that Hadamard matrices exist of all orders through $n = 128$ that are multiples of 4, except possibly for 92 and 116.

13.1.5. *Generalized block designs.*

(a) Suppose that elements need not appear equally often. Each block has size k and each pair among the v elements appears together λ times, but the ith element appears r_i times. Prove that in fact $r_1 = \cdots = r_v$; no generalization.

(b) Suppose that blocks need not have equal size. Any two elements of $[v]$ appear together λ times, and each element appears in r blocks. Construct an infinite family of examples with $\lambda = 1$ in which the block sizes are not all equal to show that this is really a generalization. Blocks of size 1 are not allowed, since they could be added to any design.

13.1.6. For $n \in \mathbb{N}$ and $1 \le k < n$, define $A^{(k)}$ by $a_{i,j}^k = ki + j \pmod{n}$ (see Theorem 13.1.5).

(a) Prove that $A^{(k)}$ is a Latin square if and only if n and k are relatively prime.

(b) When $A^{(k)}$ and $A^{(l)}$ are Latin squares, prove that they are orthogonal if and only if $k - l$ is relatively prime to n.

(c) Prove that this construction cannot generate any larger families of pairwise orthogonal Latin squares of order n than Corollary 13.1.7.

13.1.7. ($\diamond$) A **transversal** in a Latin square is a set of positions, one in each row and column, containing distinct elements.

(a) Prove that a Latin square of order n belongs to a pair of orthogonal Latin squares if and only if it consists of n disjoint transversals. (Laywine–Mullen [1998, p. 33])

(b) Prove that no Latin square is orthogonal to the Latin square below.

$$\begin{pmatrix} a & b & c & d \\ b & d & a & c \\ c & a & d & b \\ d & c & b & a \end{pmatrix}$$

13.1.8. ($\diamond$) Let M be a Latin square that can be written as $\left(\begin{smallmatrix} X & Y \\ Y & X \end{smallmatrix}\right)$ with X and Y being Latin squares of odd order. Prove that M has no transversal. Use this to prove that there is no Latin square orthogonal to M. (Comment: Maillet [1894] proved that the Cayley table of a group of even order cannot have a transversal.) (Mann [1950])

13.1.9. Determine all n such that there is a Latin square of order n where each row is a rotation of the first and the main diagonal consists of $1, \ldots, n$ in order. (Dályay [2012])

13.1.10. ($\diamond$) For $n \ge 3$, prove that if $\text{MOLS}(n, n - 2)$ exists, then a complete family $\text{MOLS}(n, n - 1)$ exists. (Comment: Shrikhande [1961] proved that $\text{MOLS}(n, n - 3)$ suffices. Bruck proved that if $n > p(n - k)$, where $p(x) = \frac{1}{2}x^4 + x^3 + x^2 + \frac{3}{2}x$, then a family $\text{MOLS}(n, k - 2)$ can be completed to $\text{MOLS}(n, n - 1)$; see Denes–Keedwell [1974, sec. 9.3].)

13.1.11. Prove that if $\text{MOLS}(n, k)$ exists, then there exist $k - 1$ pairwise orthogonal Latin squares whose diagonals all consist of the elements 1 through n in order.

13.1.12. Let S be a k-set of elements in a symmetric (v, k, λ)-design. Prove that if S intersects each block at least λ times, then S is a block in the design. (Hint: Consider the counting argument in Theorem 13.1.18.) (Lander [1981])

13.1.13. Let A be a matrix of order n whose entries are real and have absolute value at most 1. Prove that $|\det A| \le n^{n/2}$, with equality only when A is a Hadamard matrix.

13.1.14. (+) For distinct integers $i, j \in [4]$, let $P(i, j, k, l) = \{s \in \mathbb{Z}_n^4 : s_i = k \text{ and } s_j = l\}$. That is, $P(i, j, k, l)$ is a plane specified by giving coordinates i and j fixed values k and l. Prove that there is a function $\phi \colon \mathbb{Z}_n^4 \to \mathbb{Z}_n^2$ that maps each such plane bijectively onto $\mathbb{Z}_n^2$ if and only if there are two orthogonal Latin squares of order n. (Stong [2007])

13.1.15. Use the Fano plane to prove that $K_{7,7}$ is not 3-choosable.

13.1.16. ($\diamond$) Prove that there is a Hadamard matrix of order $4m$ if and only if there is a $(4m - 1, 2m, m)$-design.

13.1.17. A $[k]$-**pair-covering** of $[n]$ is a list $f_1, \ldots, f_m$ of functions from $[n]$ to $[k]$ such that for every $\{u, v\} \in \binom{[n]}{2}$ and every ordered pair $(r, s) \in [k]^2$, there is some t such that $f_t(u) = r$ and $f_t(v) = s$. Prove that there is a $[k]$-pair-covering of $[n]$ with k^2 functions if and only if there is a family of $n - 2$ pairwise orthogonal Latin squares of order k.

13.1.18. (+) The **product dimension** of a graph is the minimum k such that each vertex can be encoded as an integer k-tuple so that vertices are adjacent if and only if their codes differ in every coordinate (see Definition 15.1.8). Let $G = nK_m$ (the disjoint union of n copies of K_m), with $n \geq 3$. Prove that the product dimension of G is m if and only if there is a family of $(n-1)$ pairwise orthogonal Latin squares of order m. (M. Sohoni [unpublished], Evans–Isaak–Narayan [2000])

13.1.19. ($\diamond$) *Bipartite Ramsey numbers*. Define $b(2, t)$ as in Application 13.1.33.
(a) Strengthen the counting argument in Application 13.1.33 when $n = 4t - 3$ to complete the proof that $b(2, t) \leq 4t - 3$. (Beineke–Schwenk [1976])
(b) Prove that $b(2, t) \geq 4t - 4$ when a Hadamard matrix of order $4t - 4$ exists.

13.1.20. Complete the proof of the construction of a Hadamard matrix of order $2n$ from a symmetric conference matrix of order n (Lemma 13.1.36).

13.2. Projective Planes

Symmetric designs with $\lambda = 1$ have a geometric interpretation. View the elements as **points** and the blocks as **lines**. The condition $\lambda = 1$ says that any two points determine exactly one common line; this suggests the terminology. Symmetrically, any two lines having one common point.

13.2.1. Definition. A **projective plane** is an incidence relation between points and lines such that
(P1) every two points lie in one common line,
(P2) every two lines have one common point, and
(P3) there exist four points of which no three lie in a common line.

Property (P3) avoids degeneracy. (P2) is the non-Euclidean aspect; projective planes have no parallel lines. Definition 13.2.1 has infinite realizations, but we study *finite* projective planes, with finite sets of points and lines. The basic existence question is: for which values of n do projective planes with n points exist?

13.2.2. Example. *Small projective planes.* We will see that the Fano plane of Example 13.1.12 is the smallest projective plane. With 13 points (0 to 9 plus {A,B,C}), the following 13 lines form a projective plane: 0139, 124A, 235B, 346C, 4570, 5681, 6792, 78A3, 89B4, 9AC5, AB06, BC17, C028. ∎

RELATION TO DESIGNS

The postulates in Definition 13.2.1 are quite strong. We will see that the incidence matrix of a projective plane is that of a symmetric design with special parameters and yields a complete family of orthogonal Latin squares.

We begin with point/line duality. Since (P1) and (P2) are dual as stated, we need only show that the dual of (P3) is implied. The duality corresponds to transposing the incidence matrix.

13.2.3. Proposition. In a finite projective plane, there are four distinct lines of which no three contain a common point. Hence any statement about projective planes remains true when the roles of points and lines are interchanged (simply transpose the incidence matrix).

Proof: Let a, b, c, d be four points with no three on a line (guaranteed by (P3)). Let A, B, C, D be the lines guaranteed by (P1) to contain $\{a, b\}$, $\{b, c\}$, $\{c, d\}$, $\{d, a\}$, respectively. By the condition on a, b, c, d, these lines are distinct. If they do not satisfy the claim, then by cyclic symmetry we may assume that A, B, C have a common point x. The condition on a, b, c, d implies that $x \notin \{a, b, c, d\}$, but now A, B have two common points b, x and B, C have two common points c, x, contradicting (P2). ∎

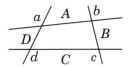

We next prove that the incidence relation of a projective plane has the defining properties of a design, with the points and lines as the elements and blocks of a design. The number of points becomes the number of varieties v, the size of each line becomes the block size k, and $\lambda = 1$.

13.2.4. Theorem. In a finite projective plane, every point lies in the same number of lines, and every line has the same number of points.

Proof: By duality, it suffices to prove the latter, which we do by establishing an injection from one arbitrary line A to another line B. This yields $|B| \geq |A|$, and interchanging them in the argument yields $|A| \geq |B|$.

To define the map, we use a point not on A or B; we first obtain such a point. (P3) provides four points a, b, c, d with no three on a line. If one of these is not in $A \cup B$, then it can be x. Otherwise, A and B each contain exactly two of them, say $a, b \in A$ and $c, d \in B$. Now (P1) provides a line containing $\{a, c\}$ and another line containing $\{b, d\}$, and (P2) guarantees that they have a common point x. If x is in A or B, then we have two points on two distinct lines, violating (P1).

We map each $p \in A$ to a point $f(p) \in B$ using the unique line $L(p)$ containing p and x. Let $f(p)$ be the point where $L(p)$ meets B. If $p \neq q$, then $L(p) \neq L(q)$, else A and $L(p)$ meet twice. This further implies $f(p) \neq f(q)$, else x and $f(p)$ lie on both $L(p)$ and $L(q)$. Hence $|B| \geq |A|$, and by symmetry also $|A| \geq |B|$. ∎

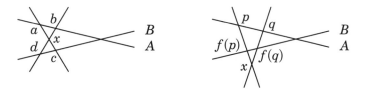

13.2.5. Theorem. In a finite projective plane, the number of lines containing each point is the same as the number of points in each line.

Proof: By (P3), there are a point x and a line L such that x is not in L. By Theorem 13.2.4, it suffices to define a bijection mapping the points in L to the lines through x. For each $y \in L$, there is a unique line containing x, y; let this be $f(y)$. Let L' be an arbitrary line through x. Since L and L' meet exactly once, there is a unique y for which $L' = f(y)$.   ∎

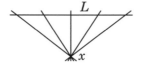

13.2.6. Definition. The **order** of a projective plane is one less than the number of points in each line (or lines through each point).

13.2.7. Theorem. A family of $(q+1)$-sets is the set of lines of a projective plane of order q (≥ 2) if and only if it is the set of blocks of a $(q^2+q+1, q+1, 1)$-design.

Proof: *Necessity.* For a plane with v points and b blocks, Theorems 13.2.4–13.2.5 imply that the incidence relation is that of a $(v, q+1, 1)$-design. Using Proposition 13.1.14, $r = k$ implies $b = v$, and then $r(k-1) = \lambda(v-1)$ implies $v = 1 + (q+1)q$.

Sufficiency. Given a $(q^2 + q + 1, q + 1, 1)$-design, call the blocks lines. Since $(q^2 + q + 1) - 1 = (q+1)q$, by Proposition 13.1.14 the design is symmetric. Hence (P1) and (P2) both hold. When $q = 1$, there are no four points to satisfy property (P3) and hence no projective plane. It suffices to show that if $q \geq 2$, then every $(q^2 + q + 1, q + 1, 1)$-design has four points with no three on a line.

Begin with one line L_1 and distinct points $a, b \in L_1$. With $q^2 + q + 1 > q + 1$, we can choose c outside L_1. Now there are unique lines L_2 and L_3 containing $\{c, a\}$ and $\{c, b\}$, respectively. Since any two lines have exactly one common point, $|L_1 \cup L_2 \cup L_3| = 3(q+1) - 3 = 3q$. Since $q \geq 2$, we have $3q < q^2 + q + 1$. Hence a point d remains outside $L_1 \cup L_2 \cup L_3$, and now $\{a, b, c, d\}$ is the desired 4-set. ∎

13.2.8. Example. Theorem 13.2.7 implies that (P1) and (P2) suffice when lines have more than two elements, and (P3) becomes unnecessary.

No projective plane has order 1. Every projective plane of order 2 is isomorphic (by renaming points) to the Fano plane (Exercise 1). A projective plane of order 3 (Example 13.2.2) has 13 points.   ∎

Theorem 13.2.7 states that a projective plane is just a symmetric design with special parameters. We can thus use the Bruck–Chowla–Ryser Theorem (Theorem 13.1.23) to prohibit some values of q as orders of projective planes.

13.2.9. Theorem. If q is congruent to 1 or 2 modulo 4, and the prime factorization of q has odd exponent on some prime congruent to 3 modulo 4, then there is no projective plane of order q.

Proof: A projective plane of order q is a symmetric $(q^2 + q + 1, q + 1, 1)$-design. If $q \equiv 1 \pmod 4$ or $q \equiv 2 \pmod 4$, then $q^2 + q + 1 \equiv 3 \pmod 4$ and $(v-1)/2$ is odd, where $v = q^2 + q + 1$. The necessary condition for a $(v, q+1, 1)$-design in Theorem 13.1.23b now reduces to having an integer solution to $y^2 + z^2 = qx^2$.

In Theorem 10.1.10 we proved Fermat's Two Squares Theorem: A natural number is the sum of two squares if and only if all primes congruent to 3 modulo 4 occur with even exponent in its factorization. Hence when q is congruent to 1 or 2 modulo 4 the stated condition forbids projective planes of order q.   ∎

As observed by Bruck–Ryser [1949], Theorem 13.2.9 prohibits a projective plane of order q when $q \in \{6, 14, 21, 22, 30, 33, 38, 42, 46, 54, \ldots\}$. For order 10, we seek a $(111, 11, 1)$-design. The equation $y^2 + z^2 = 10x^2$ has the solution $(x, y, z) = (1, 3, 1)$, so Theorem 13.2.9 does not forbid such a design.

Consider prime powers. When $q = p^k$ with $p \equiv 3 \pmod 4$, we have $q \equiv 1 \pmod 4$ if and only if k is even. Hence Theorem 13.2.9 cannot exclude any prime power from being the order of a projective plane. In fact, we will see next that a projective plane of order q exists whenever q is a prime power (this is why we define the order of a plane to be one *less* than the size of the lines). This uses a connection with Latin squares. We proved in Theorem 13.1.5 that there is a complete family $\mathrm{MOLS}(q, q - 1)$ of Latin squares when q is a prime power.

13.2.10. Theorem. (Bose [1939]) If $q \geq 2$, then a projective plane of order q exists if and only if $\mathrm{MOLS}(q, q - 1)$ exists (such as when q is a prime power).

Proof: We first construct an orthogonal family $A^{(1)}, \ldots, A^{(q-1)}$ from a projective plane. In the plane, consider a line L with points $x, y, w_1, \ldots, w_{q-1}$. Let $X_1, \ldots, X_q$ be the other lines through x, let $Y_1, \ldots, Y_q$ be the other lines through y, and let $W_{k,1}, \ldots, W_{k,q}$ be the other lines through w_k.

Each X_i and Y_j have one common point; call it $z_{i,j}$. This uses q^2 points, which are the remaining points of the plane. We use the lines determined by $z_{i,j}$ and points on L to define Latin squares. Let $A^{(k)}_{i,j} = t$, where $W_{k,t}$ is the line determined by $z_{i,j}$ and w_k. This is well-defined, since two points determine one line.

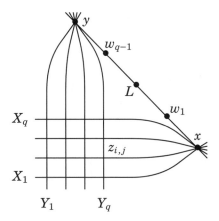

Since $z_{i,j}$ and $z_{i,j'}$ both lie on X_i, which avoids w_k, they cannot both lie on a line with w_k. Thus the entries in row i of $A^{(k)}$ are distinct. Similarly, column entries are distinct, so each $A^{(k)}$ is a Latin square.

For orthogonality, let $(\alpha, \beta) = (A^{(k)}_{i,j}, A^{(l)}_{i,j})$. We have $z_{i,j}$ on $W_{k,\alpha}$ and $W_{l,\beta}$. Since the two lines $W_{k,\alpha}$ and $W_{l,\beta}$ have only one common point, the only position that can yield the pair (α, β) in A^k and A^l is (i, j).

The converse reverses the construction. Begin with squares $A^{(1)}, \ldots, A^{(q-1)}$. Introduce one point w_k for each square A_k. Introduce q^2 points $\{z_{i,j}\}_{i,j=1}^{q}$. Finally, add two points x and y. The line $W_{k,l}$ consists of w_k and all $z_{i,j}$ such that $A_{i,j}^{(k)} = l$. The line X_i consists of x and all $z_{i,j}$ (first index fixed). The line Y_j consists of y and all $z_{i,j}$ (second index fixed). Finally, the line L consists of x, y, and $w_1, \ldots, w_{q-1}$.

We have defined $q^2 + q + 1$ points and $q^2 + q + 1$ lines, each consisting of $q + 1$ points. For $q \geq 2$, points $z_{1,1}, z_{1,2}, z_{2,1}, z_{2,2}$ have no three on a line. It suffices to verify that every two points lie in one common line and every two lines have one common point. We leave this to Exercise 4. ∎

Finite fields of order q exist only when q is a power of a prime, yielding projective planes of such orders. Theorem 13.2.9 forbids projective planes of many non-prime-power orders, but not orders such as 10. In light of Theorem 13.2.10, the computer search mentioned in Remark 13.1.9 also prohibits projective planes of order 10. It remains possible that a complete family of Latin squares (or a projective plane) does not require a finite field of order q. No projective plane of non-prime-power order is known, and none is believed to exist. For $q \in \mathbb{N}$, below we summarize the relationship of $\exists \, \mathrm{MOLS}(q, q - 1)$ to the other two properties.

$$q \text{ is a prime power} \;\Rightarrow\; \exists \, \mathrm{MOLS}(q, q - 1) \;\Leftrightarrow\; \exists \text{ proj. plane of order } q$$

13.2.11.* Example. *Affine planes.* In the picture for Theorem 13.2.10, deleting L and its points leaves q^2 points and leaves $q^2 + q$ lines, each of size q. Any two points that remain lie on exactly one line that remains, but some lines that remain are now "parallel"; they met in a point now deleted. The result is a $(q^2, q, 1)$-design (not symmetric) whose lines group into q "parallel classes", each of which partitions the set of points. Such a design is an **affine plane** (see Example 13.3.23).

The $(9, 3, 1)$-design in Example 13.1.13 is the affine plane that results from deleting a line from the projective plane of order 3. What is the affine plane obtained by deleting a line from the Fano plane? ∎

APPLICATIONS TO EXTREMAL PROBLEMS

The delicate structural properties of projective planes lead to constructions that meet the counting bounds in some extremal problems.

The point-line incidence matrix of a projective plane yields a bipartite graph called the **incidence graph** or **incidence bigraph** of the plane. It is an X, Y-bigraph, where X is the set of points, Y is the set of lines, and x and y are adjacent when x belongs to y. Since any two lines have exactly one common point, this graph has no 4-cycle, even though it has many edges. Since any two lines have a common point, the diameter is only 3, although no vertex has high degree.

To obtain such a graph with diameter 2, we merge points with lines. First we construct projective planes directly from finite fields. Since finite fields yield complete orthogonal families (Theorem 13.1.5), which yield projective planes (Theorem 13.2.10), such a construction must exist. The direct construction is simple and provides a natural way to merge points and lines in the incidence graph.

13.2.12. Theorem. If q is a prime power, there is a projective plane of order q.

Proof: Put triples of elements from the field $\mathbb{F}_q$ into multiplicative classes; The class containing a, b, c is $\{za, zb, zc\colon z \in \mathbb{F}_q - \{0\}\}$. The number of classes of nonzero triples is $(q^3 - 1)/(q - 1)$, which equals $q^2 + q + 1$. We use one copy of these classes as the points and another as the lines, letting (a, b, c) denote the point containing the triple a, b, c and $[a, b, c]$ denote the line containing the triple a, b, c.

To define the incidence relation, put $(x, y, z) \in [a, b, c]$ if and only if $ax + by + cz = 0$. (Geometrically, view $[a, b, c]$ as multiples of a vector normal to a plane through the origin, with the classes (x, y, z) being the points in the plane.) Consider a line $[a, b, c]$; by symmetry, we may assume $c \neq 0$. For each of the $q^2 - 1$ ways to choose x and y not both 0, there is a unique solution $z = -(ax + by)/c$. Solutions equivalent to (x, y, z) arise $q - 1$ times, since multiplying x, y, z by a nonzero field element still yields a solution. Thus each line consists of $q + 1$ points.

The two lines $[a, b, c]$ and $[d, e, f]$ share any point (x, y, z) satisfying both $ax + by + cz = 0$ and $dx + ey + fz = 0$. With $z = -(ax + by)/c$ again, the other equation leaves one multiplicative class as a common solution. Hence two lines have one common point. Since the incidence relation is symmetric in points and lines, also each point is in $q + 1$ lines and every two points lie on one common line.

We thus have a symmetric $(q^2 + q + 1, q + 1, 1)$-design. When $q \geq 2$, it is a projective plane (Theorem 13.2.7). ∎

The construction extends to higher-dimensional *projective geometries* by using multiplicative classes of nonzero $(d + 1)$-tuples.

13.2.13. Definition. A **polarity** of a projective plane is an involution π exchanging points and lines having the property that point p and line l are incident if and only if $\pi(p)$ and $\pi(l)$ are incident. Let V be the set of $q^2 + q + 1$ points of the projective plane of order q constructed above. Given a polarity π, the **polarity graph** has vertex set V, with (a, b, c) and (x, y, z) adjacent if and only if $(x, y, z) \in [a, b, c]$ and $\pi(x, y, z) \neq [a, b, c]$. For the natural polarity that pairs (a, b, c) with $[a, b, c]$, the polarity graph puts (a, b, c) adjacent to (x, y, z) if and only if these vertices lie in distinct multiplicative classes and $ax + by + cz = 0$. We will consider only this polarity and this polarity graph, shown below for orders 2 and 3.

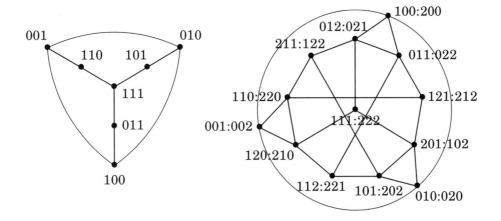

13.2.14. Example. *Diameter 2, small degree, no 4-cycles.* The polarity graph G
collapses the incidence graph by identifying (a, b, c) with $[a, b, c]$ and deleting
loops. Given vertices (a, b, c) and (d, e, f) in G, the system of two homogeneous
equations $ax + by + cz = 0$ and $dx + ey + fz = 0$ always has a nonzero solution
(the line $[x, y, z]$ containing the points (a, b, c) and (d, e, f)). Hence (x, y, z) is a
common neighbor, and diam $G = 2$.

The incidence graph is $(q + 1)$-regular. Since $(x, y, z) \in [a, b, c]$ if and only if
$(a, b, c) \in [x, y, z]$, vertices of G also have degree $q + 1$, except that one edge is lost
when $a^2 + b^2 + c^2 = 0$, since we exclude loops. It is a nontrivial algebraic fact that
in a finite field of order q there are $q + 1$ multiplicative classes of nonzero triples
solving this equation; see Exercises 7–8. Hence G has $q + 1$ vertices of degree
q. The rest have degree $q + 1$, which slightly exceeds $\sqrt{n}$. By the Degree-Sum
Formula, G has $q(q + 1)^2/2$ edges.

The properties of the projective plane forbid distinct multiplicative classes
as common solutions of $u \cdot x = 0$ and $v \cdot x = 0$. Thus no two triples u, v have two
common neighbors, and G has no 4-cycle. ∎

13.2.15. Proposition. Every n-vertex graph G with diameter 2 has maximum
degree at least $\lceil \sqrt{n - 1} \rceil$, with equality for the polarity graph when n equals
$q^2 + q + 1$ for some prime power q.

Proof: Let $k = \Delta(G)$. From one vertex, at most $k + k(k - 1)$ others are reachable
within two steps (see Proposition 5.4.12). Hence diameter 2 requires $n \leq 1 + k^2$.

When $n = q^2 + q + 1$, we have $q < \sqrt{n - 1} < q + 1$. Hence the polarity graph,
which has diameter 2, achieves equality in the bound. ∎

The polarity graph also solves the Turán problem for 4-cycles. In Section 10.1
we defined ex $(n; H)$ to be the maximum number of edges in an n-vertex graph not
containing H and computed ex $(n; K_r)$. The problem of computing ex $(n; C_4)$ was
raised by Erdős in 1938. Perhaps surprisingly, 4-cycles are forced much earlier
than triangles; a simple counting argument shows that ex $(n; C_4) \in O(n^{3/2})$.

13.2.16. Proposition. If $m > \frac{1}{4}n(1 + \sqrt{4n - 3})$, then every simple n-vertex graph
G with m edges contains C_4. When $n = q^2 + q + 1$, the polarity graph has
nearly this many edges.

Proof: Let x count the pairs of incident edges in G. If G has no 4-cycles, then no
two vertices have more than one common neighbor, so $\sum \binom{d_i}{2} = x \leq \binom{n}{2}$. Using the
average degree and the convexity of $\binom{u}{2}$ in terms of u, we have $\sum \binom{d_i}{2} \geq n\binom{2m/n}{2}$.
Combining these inequalities yields $m(2m/n - 1) \leq n(n - 1)/2$. The quadratic
formula yields $m \leq \frac{1}{4}n(1 + \sqrt{4n - 3})$.

When $n = q^2 + q + 1$, the upper bound reduces to $\frac{1}{2}n(q + 1)$. Since the polarity
graph has q^2 vertices of degree $q + 1$ and $q + 1$ vertices of degree q, its number
of edges is just $(q + 1)/2$ less than the bound. The difference is about $\frac{1}{2}\sqrt{n}$ out of
approximately $\frac{1}{2}n^{3/2}$ edges. ∎

Füredi [1996a] improved the upper bound to $m \leq \frac{1}{2}q(q + 1)^2$ when $n = q^2 + q + $
1 and $q \geq 15$. Thus when q is a prime power greater than 13, the polarity graph is

a largest graph not containing C_4. When q is an even prime power, Füredi [1983] proved that only polarity graphs achieve the extreme; the condition for equality is the existence of a polarity with $q+1$ fixed points on the projective plane of order q. This holds also for odd prime powers, but that case is more difficult; Füredi has not published the proof. The case $n = q^2 + q$ for even prime powers is discussed in Firke–Kosek–Nash–Williford [2013].

The incidence graph of a projective plane provides extremal constructions for an analogous bipartite problem. In the context of reduced adjacency matrices, extremal problems for complete bipartite subgraphs of $K_{n,n}$ are natural.

13.2.17. Definition. Zarankiewicz's Problem asks for the maximum number of 1s in an $m \times n$ matrix with no constant $s \times t$ submatrix of 1s. Equivalently, this is the maximum size of an X, Y-bigraph having no copy of $K_{s,t}$ with s vertices in X and t vertices in Y, where $|X| = m$ and $|Y| = n$. Let $z(m, n; s, t)$ denote this value, and put $z(n; t) = z(n, n; t, t)$.

Zarankiewicz asked for the first few values of $z(n; 3)$. Bounds on z apply to other incidence relations. For example, in a set of n points and m circles in the plane (with no three points collinear and no four cocircular), the maximum number of incidences between points and circles is bounded by $z(m, n; 3, 2)$, since any three points determine a unique circle.

The incidence-counting idea in Proposition 13.2.16 also provides an upper bound for $z(m, n; s, t)$. First we observe the connection between $z(n, n; s, t)$ and the corresponding extremal problem for subgraphs of K_n.

13.2.18. Proposition. For $s, t \geq 1$, $\text{ex}(n; K_{s,t}) \leq \frac{1}{2} z(n, n; s, t)$.

Proof: Let G be a simple n-vertex graph with no $K_{s,t}$. We form a subgraph G' of $K_{n,n}$ with no $K_{s,t}$ (or $K_{t,s}$). Let $V(G')$ be the union of copies V_1 and V_2 of $V(G)$. Create edges $x_1 y_2$ and $y_1 x_2$ in G' for each edge xy in G. If G' contains $K_{s,t}$, then since G has no loops the vertices in the two parts of the subgraph arise from distinct vertices of G, and G also contains $K_{s,t}$. ∎

13.2.19. Theorem. (Kővári–Sós–Turán [1954]) For $s, t > 1$,

$$z(m, n; s, t) < (s-1)^{1/t}(n - t + 1)m^{1-1/t} + (t-1)m.$$

$$\text{ex}(n; K_{s,t}) < \frac{1}{2}(s-1)^{1/t} n^{2-1/t} + \frac{t-1}{2} n.$$

Proof: In a $K_{s,t}$-free X, Y-bigraph G with $|X| = m$ and $|Y| = n$, each set of t vertices in Y has at most $s - 1$ common neighbors in X. Hence there are at most $(s-1)\binom{n}{t}$ choices of a vertex $x \in X$ and a t-set $T \subseteq Y$ such that $T \subseteq N(x)$. Since we obtain such a pair for every choice of t vertices from the neighborhood of any $x \in X$, we have $\sum_{x \in X} \binom{d(x)}{t} \leq (s-1)\binom{n}{t}$.

Let $z = |E(G)|$. As in Proposition 13.2.16, we obtain $\sum \binom{d(x)}{t} \geq m\binom{z/m}{t}$ by replacing each degree with the average degree z/m, since $\binom{y}{t}$ is a convex function of y. Multiply both sides by $t!$, let $\alpha = z/m < n$, and use $\frac{\alpha}{n} > \frac{\alpha-1}{n-1} > \cdots > \frac{\alpha-t+1}{n-t+1}$, which holds when $\alpha < n$. This yields $m(\frac{z}{m} - t + 1)^t < (s-1)(n - t + 1)^t$, which simplifies to the bound claimed.

For the bound on $\text{ex}(n; K_{s,t})$, apply Proposition 13.2.18. ∎

The better bound arises by naming the parameters so that $t \leq s$. Znám [1965] improved the leading coefficient when $s = t$, obtaining

$$z(n; t) < (t-1)^{1/t} n^{2-1/t} + (t-1)n/2$$

(see Bollobás [1978, p. 311]). Guy [1968] collected and enhanced early results on the problem. Roman [1975] extended the counting arguments for the upper bound on $z(m, n; s, t)$ to a family of bounds, using a parameter k, each bound optimal for infinitely many pairs (m, n). In particular, for any k at least $t - 1$ he proved $z(m, n; s, t) \leq (s-1)\binom{n}{t}/\binom{k}{t-1} + (k+1)(t-1)m/t$.

Füredi further improved the leading coefficient for $\mathrm{ex}\,(n; K_{s,t})$. This yields asymptotic optimality of the construction for $s = t = 3$ we will describe.

13.2.20. Theorem. (Füredi[1996b]) $\mathrm{ex}\,(n; K_{s,t}) \leq \frac{1}{2}(s-t+1)^{1/t}n^{2-1/t} + o(n^{2-1/t})$.

Equality in the counting bound $\sum_{x \in X} \binom{d(x)}{t} \leq (s-1)\binom{n}{t}$ requires each t-set in Y to have exactly $s - 1$ common neighbors in X and all vertices in X to have the same degree d. When $t = 2$, these are the conditions for the neighborhoods of vertices in X to form an $(n, d, s - 1)$-design on Y. Brown [1966] used finite geometries to show that the counting bound is asymptotically optimal for $z(n; 3)$, but the method does not handle $t > 3$. The upper bound on $\mathrm{ex}\,(n; C_4)$ below was proved earlier by Reiman [1958].

13.2.21. Theorem. (Erdős–Rényi–Sós [1966], Brown [1966]) If q is a power of an odd prime, then $z(q^2 + q + 1; 2) \geq (q^2 + q + 1)(q + 1)$ and $\mathrm{ex}\,(q^2 + q + 1; C_4) \geq \frac{1}{2}q(q+1)^2$. Thus when n is sufficiently large,

$$n^{3/2} - n^{4/3} \leq z(n; 2) \leq \tfrac{1}{2}n(1 + \sqrt{4n-3}) \,,$$

$$\tfrac{1}{2}(n^{3/2} - n^{4/3}) \leq \mathrm{ex}\,(n; C_4) \leq \tfrac{1}{4}n(1 + \sqrt{4n-3}) \,.$$

Proof: The incidence graph and polarity graph of the projective plane yield the lower bounds in terms of q. The second statement uses the number-theoretic result that if m is sufficiently large, then there is a prime between $m - \frac{1}{10}m^{2/3}$ and m. Applying this to $m = \frac{1}{2}(\sqrt{4n-3} - 1)$ and restricting the projective plane construction for this prime to a subset of its vertices yields the bound. ∎

Füredi [1996c] also generalized the polarity construction to $\mathrm{ex}\,(n; K_{2,s})$.

13.2.22.* Example. *Large graphs with no $K_{2,s+1}$* (Füredi [1996c]). When q is a prime power and $n = (q^2 - 1)/s$, we use $\mathbb{F}_q$ to construct an n-vertex graph G with about $\frac{1}{2}\sqrt{s}\,n^{3/2}$ edges and no copy of $K_{2,s+1}$.

The vertices of G are the equivalence classes of a relation on the nonzero ordered pairs of elements of $\mathbb{F}_q$. Since s divides $q^2 - 1$, there is an element ω in $\mathbb{F}_q$ with multiplicative order s. The s elements of $S = \{1, w, \ldots, w^{s-1}\}$ are distinct, and $w^s = 1$. We make (a, b) equivalent to (c, d) if $c = w^i a$ and $d = w^i b$ for some i. This defines an equivalence relation (Exercise 14), and the classes all have size s.

Let $\langle a, b \rangle$ denote the class containing (a, b). Define G by $\langle a, b \rangle \leftrightarrow \langle u, v \rangle$ if $au + bv \in S$; this is symmetric. If $au + bv = w^i$, then $(w^j a)u + (w^j b)v = w^{i+j}$, so the relation is consistent on equivalence classes. The graph $K_{2,s+1}$ is not a subgraph of G, and the vertices of G have degree $q - 1$ or q (Exercise 14). ∎

13.2.23.* Theorem. $\mathrm{ex}\,(n, K_{3,3}) = (\frac{1}{2} + o(1))n^{5/3}$.

Proof: (sketch) The asymptotic statements again use density results about prime powers. The upper bound is from Theorem 13.2.20 (Füredi [1996b]).

The lower bound is due to Brown [1966]. Given a field of order q, where $q \equiv -1 \pmod 4$, we create a graph whose vertices are the triples of field elements. We put $xy \in E(G)$ if and only if $\sum_{i=1}^{3}(x_i - y_i)^2 = 1$.

The idea is that the points at distance 1 from a given point form a sphere. A copy of $K_{3,3}$ would correspond to three spheres having three common points. In such a three-dimensional space, it is not possible for three spheres to have three common points (visualize this in $\mathbb{R}^3$).

For each solution to $\sum_{i=1}^{3} a_i^2 = 1$, one can solve $x_i - y_i = a_i$ in q ways. Since there are $n = q^3$ vertices, it thus suffices to show that there are about $\frac{1}{2}q^2$ solutions in the field to $\sum_{i=1}^{3} a_1^2 = 1$. ∎

The gap between the bounds on $z(n; t)$ is large when $t > 3$. The current best general lower bound arises from a simple non-constructive counting argument. A similar argument gives a lower bound for $\mathrm{ex}\,(n; K_{t,t})$.

13.2.24. Theorem. If $\alpha = \frac{s-1}{st-1}$ and $\beta = \frac{t-1}{st-1}$, then
$$z(m, n; s, t) \geq \left\lfloor (1 - \tfrac{1}{s!t!})m^{1-\alpha}n^{1-\beta} \right\rfloor.$$

Proof: Let $k = \left\lfloor m^{1-\alpha}n^{1-\beta} \right\rfloor$. We count the copies of $K_{s,t}$ over all subgraphs of $K_{m,n}$ with k edges; let the total be T. Some such graph then has at most $T\binom{mn}{k}^{-1}$ copies of $K_{s,t}$. Deleting one edge from each copy of $K_{s,t}$ in this graph yields a $K_{s,t}$-free graph with at least $k - T\binom{mn}{k}^{-1}$ edges.

By summing the number of graphs containing each possible copy of $K_{s,t}$, we obtain $T = \binom{m}{s}\binom{n}{t}\binom{mn-st}{k-st}$. Since $\binom{mn}{k} = \frac{mn(mn-1)\cdots(mn-st+1)}{k(k-1)\cdots(k-t+1)}\binom{mn-st}{k-st}$, the average is bounded by $\binom{m}{s}\binom{n}{t}(\frac{k}{mn})^{st}$, which in turn is bounded by $\frac{1}{s!t!}m^{s-\alpha st}n^{t-\beta st}$. Hence $z(m, n; s, t) \geq e - \frac{1}{s!t!}m^{1-\alpha}n^{1-\beta}$. ∎

For s much larger than t, the exponents in Theorems 13.2.19–13.2.20 are sharp, by explicit constructions. Further improvements are in Alon–Ronyai–Szabó [1999] and Ball–Pepe [2012, 2016].

13.2.25.* Theorem. (Kollár–Rónyai–Szabó [1996]) For $t \geq 4$ and $s > t!$, $\mathrm{ex}\,(n; K_{s,t}) = \Theta(n^{2-1/t})$.

Proof: (sketch) When n has the form q^t with q a prime power, we construct a graph whose vertices are the elements of the field of size q^t. We put x, y adjacent if $N(x+y) = 1$, where $N(z) = z \cdot z^q \cdots z^{q^{t-1}}$. Here $N(z)$ is called the *norm* of z, and the resulting graph G is the **norm graph**.

Except for $N(0) = 0$, $N(z)$ is distributed equally over a subfield with q elements. Thus each vertex in the graph has degree $(q^t - 1)/(q-1)$, except that x has degree one less when $N(2x) = 1$ (since we don't include loops). We thus obtain
$$|E(G)| \geq \frac{1}{2}q^t\left(\frac{q^t - 1}{q - 1} - 1\right) \geq \frac{1}{2}q^{2t-1} = \frac{1}{2}n^{2-1/t}.$$

The norm graph contains $K_{s,t}$ if and only if it has t vertices $z_1, \ldots, z_t$ with s common neighbors. Common neighbors are simultaneous solutions to $N(x + z_i) = 1$ for $1 \leq i \leq t$. Kollár–Rónyai–Szabó [1996] proved that there are at most $t!$ solutions by proving this bound in the more general setting where there are t equations of the form $\prod_{j=1}^{t}(x_j - a_{i,j}) = b_i$ over a field K, and the coefficients are distinct in each column (that is, $a_{i,j} = a_{i',j}$ if and only if $i = i'$).   ∎

DIFFERENCE SETS

Some projective planes and some more general symmetric designs have the simple structure that the blocks are cyclic translates of a single block. A familiar example is the usual description of the Fano plane, in which the blocks are cyclic translates of $\{0, 1, 3\}$ modulo 7.

Here we describe a way of obtaining such a description. It is known that not all symmetric (v, k, λ)-designs arise in this way. In fact, not all projective planes of order 16 can be expressed cyclically.

13.2.26. Definition. A set $\{a_1, \ldots, a_k\}$ is a (v, k, λ)-**difference set** if for every $d \in \mathbb{Z}_v - \{0\}$, there are λ ordered pairs (a_i, a_j) such that $d \equiv a_j - a_i \pmod{v}$.

Since there are $k(k - 1)$ ordered pairs of elements in a k-set, the definition requires $\lambda(v - 1) = k(k - 1)$, which is a familiar necessary condition for symmetric designs (Proposition 13.1.14).

13.2.27. Example. The set $\{0, 1, 3\}$ is a $(7, 3, 1)$-difference set. We have $1 = 1 - 0$, $2 = 3 - 1$, $3 = 3 - 0$, $4 = 0 - 3$, $5 = 1 - 3$, $6 = 0 - 1$. The next result now implies that the Fano plane is a symmetric $(7, 3, 1)$-design.   ∎

13.2.28. Definition. Given a set $D \subseteq \mathbb{Z}_v$, the **translate** $D + x$ (modulo v) is the set $\{a + x : a \in D\}$, with entries reduced modulo v.

13.2.29. Proposition. A set D of size k is a (v, k, λ)-difference set if and only if the translates of D modulo v form (the blocks of) a symmetric (v, k, λ)-design.

Proof: Let D be a (v, k, λ)-difference set. If $a' - a \equiv t \pmod{v}$ for $a, a' \in D$, then the translates of D provide all pairs $r, s \in \mathbb{Z}_v$ such that $s - r \equiv t \pmod{v}$. Furthermore, there one such set of v pairs (r, s) for each pair in D with difference t, so each pair in $\mathbb{Z}_v$ arises exactly λ times.

Conversely, if the translates $D + x$ form a (v, k, λ)-design, then the λ translates in which $\{r, s\}$ occurs yield λ ordered pairs (i, j) such that $a_j - a_i \equiv (s - r) \pmod{v}$. Because $\lambda(v - 1) = k(k - 1)$, this accounts for all ordered pairs from D, and D is a difference set.   ∎

Thus, difference sets correspond to cyclically invariant designs. We next discuss a technique to construct a (v, k, λ)-difference set or show that none exists.

13.2.30. Definition. A **multiplier** of a (v, k, λ)-difference set D is an element t of $\mathbb{Z}_v$ such that tD is a translate of D. If $tD = D$, then t **fixes** D.

13.2.31. Proposition. All translates of D have the same multipliers.

Proof: If t is a multiplier of D, then $tD = D + x$ for some translate $D + x$. For any translate $D + y$, we have $t(D + y) = (tD) + ty = (D + x) + ty$. Since $x + ty = y + (x + (t - 1)y)$, we conclude that $t(D + y)$ is a translate of $D + y$, making t also a multiplier of $D + y$. ∎

13.2.32. Proposition. If $\gcd(v, k) = 1$ and D is a (v, k, λ)-difference set, then some translate of D is fixed by every multiplier of D.

Proof: Let $D = \{a_1, \dots, a_k\}$. Let $\sigma(S)$ denote the sum (modulo v) of a set S. For any translate $D + x$, we have $\sigma(D + x) = \sigma(D) + kx$. Since $\gcd(v, k) = 1$, multiplication by k permutes the nonzero congruence classes, and hence there is exactly one choice of x such that $\sigma(D + x) \equiv 0 \pmod{v}$.

If t is a multiplier of D, then also t is a multiplier of this translate $D + x$, and $t(D + x) = D + y$. With x chosen in this way, $\sigma(D + y) = \sigma(t(D + x)) = t\sigma(D + x) \equiv 0 \pmod{v}$. Since only one translate of D sums to a multiple of v, we have $y = x$, and t fixes $D + x$. ∎

We will show that every prime p that exceeds λ and divides $k - \lambda$ but not v is a multiplier of every (v, k, λ)-difference set. This statement is the **Multiplier Theorem**. Knowing that a number p is a multiplier helps to find a difference set or forbid its existence, because some difference set must consist of complete orbits modulo v under multiplication by p.

13.2.33. Example. *Multipliers and difference sets.* Given a triple (v, k, λ), we seek a (v, k, λ)-difference set in $\mathbb{Z}_v$.

$(7, 3, 1)$. By the Multiplier Theorem, 2 is a multiplier. The orbits under multiplication by 2 are $\{0\}$, $\{1, 2, 4\}$, $\{3, 6, 5\}$. Each of these triples is a difference set. In general, the lines of a projective plane of order q form a $(q^2+q+1, q+1, 1)$-design. If such a design is generated by a difference set, then by the Multiplier Theorem every prime dividing q is a multiplier of that difference set. For example, when $q = 3$ the orbits modulo 13 are $\{0\}$, $\{1, 3, 9\}$, $\{2, 6, 5\}$, $\{4, 12, 10\}$. To form a set of size 4 we combine two orbits: $\{0, 1, 3, 9\}$ is a difference set.

$(11, 5, 2)$. By the Multiplier Theorem, 3 is always a multiplier. The orbits are $\{0\}$, $\{1, 3, 9, 5, 4\}$, $\{2, 6, 7, 10, 8\}$. Sure enough, $\{1, 3, 4, 5, 9\}$ is a difference set, with every difference appearing twice.

$(31, 10, 3)$. By the Multiplier Theorem, 7 is a multiplier of every difference set. Under multiplication by 7, orbits in $\mathbb{Z}_{31}$ have sizes 1, 15, and 15. No combination of these has size 10, so there is no $(31, 10, 3)$-difference set.

$(9, 3, 1)$. By the Multiplier Theorem, 2 is a multiplier of every difference set. The orbits are $\{0\}$, $\{1, 2, 4, 8, 7, 5\}$ and $\{3, 6\}$. The union $\{0, 3, 6\}$ is the only union of size 3, but it is not a difference set. (Of course, here $k(k - 1) = 6 \neq 8 = \lambda(v - 1)$, so this triple already fails the elementary counting condition.) ∎

As shown by the last part of Example 13.2.33, not every union of orbits with the desired size is a difference set (see also Exercise 16).

The remainder of this section is the proof of the Multiplier Theorem and can be skipped without loss of continuity. We note only that although the proof applies only when the specified prime p satisfies $p > \lambda$, it is conjectured that this restriction is unnecessary for the conclusion.

We begin by defining a polynomial from a difference set, using the elements as exponents of monomials. Throughout this discussion, we let $R(x) = \sum_{i=1}^{k} x^{d_i}$ and $Q(x) = \sum_{i=0}^{v-1} x^i$; note that Q records all the congruence classes as exponents.

13.2.34. Lemma. Given $D = \{d_1, \ldots, d_k\}$, let $R(x) = \sum_{i=1}^{k} x^{d_i}$, and let $Q(x) = \sum_{i=0}^{v-1} x^i$. If D is a (v, k, λ)-difference set, then
$$R(x)R(x^{-1}) \equiv k - \lambda + \lambda Q(x) \,(\mathrm{mod}\,(x^v - 1)).$$

Proof: By definition, $R(x)R(x^{-1}) = k + \sum_{i \neq j} x^{d_i - d_j}$. Since D is a difference set, each nonzero congruence class modulo v occurs exactly λ times as $d_i - d_j$. By treating x^v as equivalent to x^0, we thus obtain $R(x)R(x^{-1}) \equiv k + \lambda(\sum_{i=1}^{v-1} x^i) \,(\mathrm{mod}\,(x^v - 1))$. To introduce $Q(x)$, we add and subtract λ. ∎

13.2.35. Lemma. Let $Q(x) = \sum_{i=0}^{v-1} x^i$. If f is a polynomial, then
$$f(x)Q(x) \equiv f(1)Q(x) \,(\mathrm{mod}\,(x^v - 1)).$$

Proof: Working modulo $x^v - 1$, we have $x^v \equiv 1$, so $xQ(x) \equiv Q(x)$. Using induction on j, we have $x^j Q(x) \equiv Q(x) \,(\mathrm{mod}\,(x^v - 1))$. For $f(x) = \sum_{i=0}^{r} c_i x^i$, we compute
$$f(x)Q(x) = \sum_{i=0}^{r} c_i x^i Q(x) \equiv \sum_{i=0}^{r} c_i Q(x) \equiv f(1)Q(x) \,(\mathrm{mod}\,(x^v - 1)). \quad ∎$$

These lemmas yield a sufficient condition for multipliers.

13.2.36. Lemma. Let D be a (v, k, λ)-difference set, and let $n = k - \lambda$. If $R(x^p)R(x^{-1}) \equiv nx^s + \lambda Q(x) \,(\mathrm{mod}\,(x^v - 1))$ for some $p, s \in \mathbb{Z}_v$, then $pD = D + s$.

Proof: All congruence expressions are modulo $x^v - 1$. Multiplying the assumed congruence by $R(x)$ and applying Lemmas 13.2.34–13.2.35 yields
$$R(x^p)(n + \lambda Q(x)) \equiv nx^s R(x) + R(1)\lambda Q(x). \qquad (*)$$

Viewing $R(x^p)$ as a polynomial in x and using $1^p = 1$, Lemma 13.2.35 yields $R(x^p)Q(x) \equiv R(1)Q(x)$. Subtracting $R(1)\lambda Q(x)$ from both sides of $(*)$ now yields $nR(x^p) \equiv nx^s R(x)$. This is just the statement that the congruence classes in pD are the same as the congruence classes in $D + s$. ∎

13.2.37. Theorem. (**Multiplier Theorem**) Let D be a (v, k, λ)-difference set. Any prime p that exceeds λ and divides $k - \lambda$ but not v is a multiplier of D.

Proof*: All congruences are modulo $x^v - 1$; exponents are elements of $\mathbb{Z}_v$. To apply Lemma 13.2.36, we seek s such that $R(x^p)R(x^{-1}) \equiv nx^s + \lambda Q(x)$, where $n = k - \lambda$.

In pD, the differences are p times the differences in D. Since p is relatively prime to v, this multiplication just permutes the congruence classes, and pD is a difference set. By Lemma 13.2.34, $n + \lambda Q(x)$ is thus congruent to both $R(x)R(x^{-1})$ and $R(x^p)R(x^{-p})$. In particular,
$$R(x)R(x^{-1})R(x^p)R(x^{-p}) \equiv [n + \lambda Q(x)]^2. \qquad (1)$$

We will obtain another expression for this product to study $R(x^p)R(x^{-1})$.

Let $m = n/p$. Since $Q(x)$ divides $x^v - 1$, we can write $R(x)R(x^{-1}) \equiv n + \lambda Q(x)$ as $R(x)R(x^{-1}) = pm + Q(x)A(x)$ for some polynomial A. Multiplying by $[R(x)]^{p-1}$ yields $[R(x)]^p R(x^{-1}) = pB(x) + Q(x)C(x)$ for some polynomials B and C.

Since p is prime, p is a factor of all multinomial coefficients in the expansion of $[R(x)]^p$ except when we take the same monomial from each factor. Thus $[R(x)]^p = R(x^p) + pE(x)$ for some polynomial E. We now have

$$R(x^p)R(x^{-1}) = p[B(x) - E(x)R(x^{-1})] + Q(x)C(x).$$

We reduce this modulo $x^v - 1$. Since $x^{-j} \equiv x^{v-j}$, we have $B(x) - E(x)R(x^{-1}) \equiv F(x)$ for some polynomial F. Applying Lemma 13.2.35 to $C(x)$ yields

$$R(x^p)R(x^{-1}) \equiv pF(x) + Q(x)C(1). \qquad (2)$$

When setting $x = 1$, adding a multiple of $x^v - 1$ changes nothing, and congruence becomes equality. Since $R(x)$ has $|D|$ terms with coefficient 1, setting $x = 1$ in (2) thus yields $k^2 = pF(1) + vC(1)$. Since $\lambda(v-1) = k(k-1)$, we have $k^2 = n + \lambda v$. Since p divides n, we conclude that $vC(1) - \lambda v$ is divisible by p. Since v and p are relatively prime, p divides $C(1) - \lambda$.

Thus we can set $C(1) = \lambda + pt$ in (2) for some integer t. Now

$$R(x^p)R(x^{-1}) \equiv pF(x) + ptQ(x) + \lambda Q(x) \equiv pG(x) + \lambda Q(x)$$

for some polynomial G. Setting $x = 1$ yields $k^2 = pG(1) + \lambda v$, and thus $pG(1) = n$. Since $Q(x^{-1}) \equiv Q(x)$, evaluating at x^{-1} yields $R(x^{-p})R(x) \equiv pG(x^{-1}) + \lambda Q(x)$. We have now obtained our second expression for the four-way product:

$$R(x^p)R(x^{-1})R(x^{-p})R(x) \equiv (pG(x) + \lambda Q(x))(pG(x^{-1}) + \lambda Q(x)). \qquad (3)$$

Since $pG(x)$ and $pG(x^{-1})$ are both congruent to polynomials in x, Lemma 13.2.35 yields $pG(x)Q(x) \equiv pG(1)Q(x) = nQ(x)$, and similarly $pG(x^{-1})Q(x) \equiv nQ(x)$. Combining (1) with (3) now yields

$$p^2 G(x)G(x^{-1}) + 2\lambda n Q(x) + \lambda^2 Q^2(x) \equiv n^2 + 2\lambda n Q(x) + \lambda^2 Q^2(x),$$

which simplifies to

$$p^2 G(x)G(x^{-1}) \equiv n^2. \qquad (4)$$

Recall that $R(x^p)R(x^{-1}) \equiv pG(x) + \lambda Q(x)$. Our aim, allowing us to apply Lemma 13.2.36, is to show that $G(x) = ax^s$ for some constants a and s with $s \geq 0$. Since $pG(1) = n$, this will yield $a = m$ and complete the proof.

In the product $R(x^p)R(x^{-1})$, all coefficients are nonnegative. The coefficients do not become negative when we reduce exponents to at most $v - 1$ by applying $x^v \equiv 1$. Do this with both $R(x^p)R(x^{-1})$ and $G(x)$, obtaining $R'(x) = pG'(x) + \lambda Q(x)$ (note that $Q(x)$ does not change). Since each coefficient in $\lambda Q(x)$ is λ and $R'(x) \equiv pG'(x) + \lambda Q(x)$, each coefficient in $R'(x)$ is congruent to λ modulo p. Since $p > \lambda$ and each coefficient is nonnegative, each coefficient is at least λ. Therefore, the polynomial $pG(x)$ congruent to $R(x^p)R(x^{-1}) - \lambda Q(x)$ has nonnegative coefficients.

Since $G(x)$ has nonnegative coefficients, the expression $p^2 G(x)G(x^{-1})$ will have a nonconstant term if $G(x)$ has more than one nonzero term. This would contradict (4). Also (4) implies that G is nonzero. Thus we have shown that G has one monomial term, as desired. ∎

The Multiplier Theorem was proved by Hall [1947] for cyclic difference sets yielding projective planes. Hall–Ryser [1951] extended it for $\lambda > 1$. Extensions appear in McFarland–Mann [1965], Baumert [1971], and Lander [1983].

EXERCISES 13.2

13.2.1. (–) Prove that every projective plane with seven points is isomorphic to the Fano plane, in the sense that renaming elements and permuting lines can turn it into the Fano plane as described in Example 13.1.12.

13.2.2. (–) Prove that the Heawood graph (shown below) is the incidence bigraph of the Fano plane.

13.2.3. (–) Check that $\{1, 2, 5, 15, 17\}$ is a $(21, 5, 1)$-difference set and find a translate that is fixed by every multiplier.

13.2.4. Complete the proof that the construction in Theorem 13.2.10 of points and lines from a complete orthogonal family produces a projective plane.

13.2.5. A hypergraph is k-**colorable** if its vertices can be partitioned into k sets containing no edge. Prove that a hypergraph whose edges form the set of lines of a projective plane of order q is 2-colorable if and only if $q > 2$.

13.2.6. The **transversal number** of a hypergraph is the minimum size of a set of vertices that intersects every edge. Let q be a prime power. Let H be the hypergraph with $q^2 + q + 1$ vertices whose edges are the lines of a projective plane of order q on the vertex set. Let H' be the hypergraph with the same vertex set whose edges are the complements of the edges in H. Determine the transversal numbers of H and H'.

13.2.7. Let q be a prime power congruent to 0 or 1 modulo 4. Prove that $a^2 + b^2 + c^2 = 0$ has exactly q^2 solutions with $a, b, c \in \mathbb{F}_q$, and conclude that there are exactly $q + 1$ multiplicative classes of nonzero solutions. (Hint: When $q \equiv 1 \pmod 4$, show that there exists $j \in \mathbb{F}_q$ with $j^2 = -1$. Use j to reduce the problem to finding solutions of $uv = -c^2$.)

13.2.8. Let q be a prime power with $q \equiv 3 \pmod 4$. Let S_0 and S_1 be the set of nonzero squares and the set of non-squares in $\mathbb{F}_q$. Note that $-1 \in S_1$, and hence $x \in S_0$ if and only if $-x \in S_1$ for $x \neq 0$. For $i, j \in \{0, 1\}$, let $T_{i,j} = \{x \in S_i \colon x + 1 \in S_j\}$ and $t_{i,j} = |T_{i,j}|$.

(a) Prove that $t_{0,0} = t_{1,0} = t_{1,1} = (q - 3)/4$ and $x_{0,1} = (q + 1)/4$. (Hint: Prove $t_{1,0} = t_{1,1}$ by showing that the map taking x to $1/x$ is a bijection from $T_{1,0}$ to $T_{1,1}$.)

(b) For each multiplicative class of nonzero triples solving $a^2 + b^2 + c^2 = 0$, take the representative whose first nonzero coordinate is 1. Prove that each such triple has the form $(1, \pm\sqrt{x}, \pm\sqrt{-x-1})$ for some $x \in T_{0,1}$. Use this to conclude that there are exactly $q + 1$ multiplicative classes of nonzero solutions. (Comment: This approach also yields $q + 1$ classes when $q \equiv 1 \pmod 4$, where the values of $t_{i,j}$ are different.)

13.2.9. ($\diamond$) A **dominating set** in a graph is a vertex subset S such that every vertex outside S has a neighbor in S. Determine the minimum size of a dominating set in the incidence bigraph of a projective plane of order q. (Comment: A **total dominating set** in G is a set that contains a neighbor of every vertex; it is a transversal in the hypergraph whose edges are the neighborhoods in G; see Henning–Yeo [2013b].)

13.2.10. *Zarankiewicz problem for forbidden* $K_{2,t}$. Let G be an n-vertex graph.
(a) Prove that if G is simple and $\sum_{v \in V(G)} \binom{d(v)}{2} > (t-1)\binom{n}{2}$, then G contains $K_{2,t}$.
(b) Prove that $\sum_{v \in V(G)} \binom{d(v)}{2} \geq m(2m/n - 1)$, where G has m edges.
(c) Use (a) and (b) to prove $K_{2,t} \subseteq G$ when $m > \frac{1}{2}(t-1)^{1/2}n^{3/2} + n/4$.
(d) Application: Given n distinct points in the plane, prove that the distance is exactly 1 for at most $\frac{1}{\sqrt{2}}n^{3/2} + n/4$ pairs. (Bondy–Murty [1976, pp. 111–112])

13.2.11. ($\diamond$) Prove that every k-regular graph with girth 6 has at least $2k^2 - 2k + 2$ vertices. For $k \geq 3$, prove that some k-regular graph of girth 6 has $2k^2 - 2k + 2$ vertices if and only if there is a projective plane of order $k-1$. (Karteszi [1960], Singleton [1966])

13.2.12. Use Theorem 13.1.23 to prove that there is no symmetric $(29, 8, 2)$-design. Other triples (v, k, λ) excluded by such arguments are $(43,7,1)$, $(22,7,2)$, $(46,10,2)$, $(67,12,2)$, $(92,14,2)$, $(106,15,2)$, $(137,17,2)$, $(53,13,3)$, $(103,18,3)$, $(34,12,4)$, $(43,15,5)$, $(72,20,5)$.

13.2.13. ($\diamond$) The **degeneracy** $\sigma(G)$ of a graph G is defined by $\sigma(G) = \max_{H \subseteq G} \delta(H)$.
(a) Let $\{G_1, \ldots, G_k\}$ be a decomposition of K_n. Prove $\sum_{i=1}^{k} \sigma(G_i) \leq \sqrt{k}n$. (Hint: Each G_i has a subgraph with minimum degree $\sigma(G_i)$.)
(b) Prove that part (a) is almost sharp, as follows: When q is a power of a prime, $k = q^2 + q + 1$, and $n = mk$ for some integer m, construct a decomposition $\{G_1, \ldots, G_k\}$ such that $\sum_{i=1}^{k} \sigma(G_i) \geq (\sqrt{k} - 1)n$. (Füredi–Kostochka–Škrekovski–Stiebitz–West [2005])

13.2.14. Prove that the construction in Example 13.2.22 produces a $(q^2-1)/s$-vertex graph not containing $K_{2,s+1}$, with all vertices having degree q or $q-1$. (Füredi [1996c])

13.2.15. A (v, k, λ)-**difference family** in $\mathbb{Z}_v$ is a family of k-subsets of $\mathbb{Z}_v$ whose differences cover each nonzero element of $\mathbb{Z}_v$ exactly λ times. Proposition 13.2.29 generalizes immediately to show that the translates modulo v of the sets in a (v, k, λ)-difference family form a (v, k, λ)-design. Show that the translates modulo 41 of $\{0, 1, 4, 11, 29\}$ and $\{0, 5, 14, 20, 22\}$ form a $(41, 5, 1)$-design. (Hanani [1975])

13.2.16. ($\diamond$) *Applications of the Multiplier Theorem.*
(a) Use the Multiplier Theorem to obtain a difference set that generates a projective plane of order 4.
(b) Use the Multiplier Theorem to obtain a difference set that generates a projective plane of order 5. (Comment: Not all unions of orbits of the desired size are difference sets.)

13.2.17. Use the Multiplier Theorem to obtain a $(37, 9, 2)$-difference set and a $(73, 9, 1)$-difference set.

13.2.18. Use multipliers to show that there is no $(56, 11, 2)$-difference set.

13.2.19. Show that every $(n^2 + n + 1, n + 1, 1)$-difference set has n as a multiplier.

13.2.20. ($\diamond$) Prove that there is no $(111, 11, 1)$-difference set. (Hint: Without generating orbits of multipliers, use *two* multipliers to restrict the set of values that can appear in a difference set fixed under all multipliers. Comment: The impossibility of a $(111, 11, 1)$-difference set does not (yet) prohibit a projective plane of order 10.)

13.2.21. Prove that there is no $(n^2 + n + 1, n + 1, 1)$-difference set when n is divisible by 14 or 15 or 21. (Comment: For $n \leq 3600$, it is known that an $(n^2 + n + 1, n + 1, 1)$-difference set exists only when n is a prime power (van Lint–Wilson [1992, p. 348]).)

13.3. Further Constructions

Beyond Latin squares, projective planes, and Hadamard matrices, design theorists study many other structures. Decomposing an object into isomorphic copies of a smaller object has the flavor of and generalizes design theory.

13.3.1.* Remark. *Decomposition into isomorphic copies of a graph or hypergraph* F. A $(v, k, 1)$-design decomposes K_v into copies of K_k. More generally, when does a graph G decompose into copies of a given graph F? Obviously $|E(F)|$ must divide $|E(G)|$, and the greatest common divisor of the vertex degrees in F must divide each vertex degree in G. A graph G is F-**divisible** if these conditions hold.

A deep and difficult theorem of Wilson [1975] states that when v is sufficiently large, F-divisibility of K_v suffices for K_v to decompose into copies of F. When $F = K_k$ these are the familiar necessary conditions $(k-1) \mid (v-1)$ and $k(k-1) \mid v(v-1)$ for a $(v, k, 1)$-design. For $F = K_3$ (triple systems) the necessary conditions are always sufficient, but in general the threshold on v for sufficiency is unknown.

The problem generalizes to hypergraphs. A *t-design* is a k-uniform hypergraph on n points such that each t-set of points appears together in exactly λ edges. Setting $t = 2$ yields the classical (n, k, λ)-designs that we have discussed. For $\lambda = 1$, a t-design is called a **Steiner system** and must have exactly $\binom{n}{t}/\binom{k}{t}$ blocks, since each k-set provides exactly $\binom{k}{t}$ of the t-sets. It was conjectured in the mid-19th century that t-designs exist for all k and t when n is sufficiently large and satisfies the divisibility conditions that $\binom{k-i}{t-i}$ divides $\lambda\binom{n-i}{t-i}$ for $0 \le i < t$.

Wilson's Theorem mentioned above is the case $t = 2$. Its proof was algebraic. In a stunning breakthrough, Keevash [2014+] used "randomized algebraic constructions" to prove the conjecture for general t (an exposition of a special case appears in Keevash [2018]). Stronger and more general results were obtained in Glock–Kühn–Lo–Osthus [2016] by the method of "iterative absorption".

Generalizing beyond decomposition of complete graphs, Nash-Williams [1970] conjectured that when n is sufficiently large, every K_3-divisible graph with minimum degree at least $3n/4$ decomposes into triangles; Exercise 5.3.51 shows that the degree threshold would be sharp. Gustavsson [1991] proved that for every graph F there is a fraction c (perhaps only slightly less than 1) and a threshold n_0 such that when $n \ge n_0$ every n-vertex F-divisible graph with minimum degree at least cn decomposes into copies of F. For $F = K_3$, the value $c = .9$ suffices (Dross [2016]). Results of Barber–Kühn–Lo–Osthus [2016] and those authors with Montgomery in [2017] improved the values of c and n_0 for general graphs and provided a purely combinatorial proof of Wilson's Theorem. ∎

Wilson's Theorem, t-designs, and the general theory of isomorphic decomposition are well beyond the scope of this book. Hence we generally stick with decomposition of complete graphs and $t = 2$ (but see $t = 3$ in Exercise 28). We construct triple systems, we construct decompositions into isomorphic 2-factors, and we construct orthogonal Latin squares to disprove the Euler Conjecture.

STEINER TRIPLE SYSTEMS

For a (v, k, λ)-design with $\lambda = 1$, Fisher's Inequality $b \geq v$ (Theorem 13.1.15) yields $k \leq r$ and thus $k(k-1) \leq v - 1$, by Proposition 13.1.14. Thus projective planes are the designs with the largest possible blocks, given v. At the other end are designs with $k = 3$ (those with $k = 2$ are trivial). A 3-uniform hypergraph is also called a **triple system**; these were the first objects studied in design theory.

13.3.2. Definition. A **Steiner triple system** on v elements, denoted STS(v), is a $(v, 3, 1)$-design.

By Wilson's Theorem, the obvious necessary conditions for existence of $(v, 3, 1)$-designs are sufficient when v is sufficiently large. Fortunately, triple systems are simple enough to be analyzed without this deep result. Steiner triple systems exist for *all* values of v that satisfy the necessary conditions, which require v to be congruent to 1 or 3 modulo 6.

Steiner triple systems were introduced by Woolhouse [1844], who asked for which v they exist. Kirkman [1847] solved the problem, but it seems no one noticed. Steiner [1853] later publicized the notion and conjectured that the obvious necessary conditions are sufficient, but he gave no construction. The constructions by Kirkman [1847] and Moore [1893] were inductive. We present more recent constructions by Bose [1939] and Skolem [1958] of explicit Steiner triple systems for the two congruence classes. Wilson [1974] gave another construction (see Lindner–Rodger [1997, 27–31]). For the history and additional material on triple systems, see Colbourn–Rosa [1999].

The existence problem for $(v, k, 1)$-designs was completely solved in the cases $k = 4$ and $k = 5$ by Hanani [1972]. Wilson's Theorem does not apply for small v; there remain about 30 values of v (between 51 and 801) for which the existence of a $(v, 6, 1)$-design is unknown.

13.3.3. Proposition. If an STS(v) exists, then $v \in \{1, 3\} \pmod 6$.

Proof: Each element in a $(v, 3, 1)$-design appears in $\frac{v-1}{2}$ blocks, and then there are $\frac{1}{3}\binom{v}{2}$ blocks of size 3 (these are just special cases of the usual relationships in Proposition 13.1.14). Since the numbers of occurrences and blocks are integers, v is odd, and v or $v - 1$ is a multiple of 3. ∎

13.3.4. Example. STS($2^l - 1$). The Fano plane is an STS(7). When $v = 2^l - 1$, we can form a Steiner triple system on the set of nonzero binary vectors of length l by letting distinct x and y form a triple with the vector z such that $z = x + y \pmod 2$. Since also $x = y + z$ and $y = x + z$, these triples are well-defined, and each pair belongs to exactly one triple.

This construction can also be described iteratively. The construction for l contains the construction for $l - 1$ on the set of vectors having 0 in the first coordinate. The sum of any two vectors with 1 in the first coordinate is in this set, so the remaining triples consist of one vector starting with 0 and two vectors starting with 1. ∎

Iterative constructions build a design by adding to a smaller design. We will use the iterative approach to disprove the Euler Conjecture. To construct triple systems explicitly, however, we use the algebraic approach. Our presentation is based on Lindner–Rodger [1997, p. 1–14].

13.3.5. Definition. A **quasigroup** is a set Q equipped with a binary operation $\circ$ such that for all $a, b \in Q$, both $a \circ x = b$ and $y \circ a = b$ have unique solutions. A quasigroup is **idempotent** if $a \circ a = a$ for all a.

We consider only finite quasigroups. Applying a via $\circ$ (on the right or the left) permutes Q. Equivalently, the matrix recording $\circ$ by putting $x \circ y$ in the row indexed by x and the column indexed by y is a Latin square with elements Q. Hence quasigroups are equivalent to Latin squares. Quasigroup terminology facilitates algebraic construction and the notation $i \circ j$ for the (i, j)-entry in a Latin square. A quasigroup is commutative if and only if the corresponding Latin square is symmetric. The lemma has a nice application to total coloring of complete graphs (Exercise 5).

13.3.6. Lemma. There is an idempotent commutative quasigroup of order v if and only if v is odd.

Proof: When v is odd, use $\mathbb{Z}_v$ for the elements, and index the rows and colums from 0 to $v - 1$. Since v is odd, the element 2 has a multiplicative inverse. In position (i, j), put the congruence class of $(i + j)/2$. Since adding a constant or dividing by 2 permutes $\mathbb{Z}_j$, the elements are distinct in each row or column. Also the formula is symmetric in i and j, and $(i+i)/2 = i$. Hence we have an idempotent commutative quasigroup; for $v = 5$ this yields the Latin square below.

In the Latin square table of an idempotent commutative quasigroup, idempotence requires each symbol to be used once on the diagonal, and commutativity requires an even number of appearances off the diagonal. Hence each symbol is used an odd number of times, including once in each row, so v is odd. ■

$$
\begin{array}{ccccc}
0 & 3 & 1 & 4 & 2 \\
3 & 1 & 4 & 2 & 0 \\
1 & 4 & 2 & 0 & 3 \\
4 & 2 & 0 & 3 & 1 \\
2 & 0 & 3 & 1 & 4
\end{array}
$$

13.3.7. Example. *The* **Bose Construction.** When $v = 6n + 3$, let the set of elements consist of three copies of the elements of an idempotent commutative quasigroup Q of order $2n + 1$, indicating the three copies by subscripts modulo 3. We use $2n + 1$ Type 1 triples of the form (i_0, i_1, i_2) for $i \in Q$. The Type 2 triples have the form $(i_k, j_k, (i \circ j)_{k+1})$, where $i, j \in Q$ with $i \neq j$ and $k \in \mathbb{Z}_3$. Below we illustrate a Type 1 triple and two Type 2 triples. ■

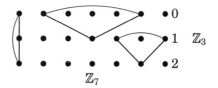

13.3.8. Theorem. (Bose [1939]) For $n \in \mathbb{N}_0$, the triples in the Bose Construction (Example 13.3.7), defined using any idempotent commutative quasigroup of order $2n + 1$, form an STS($6n + 3$).

Proof: We have specified $2n + 1$ Type 1 triples and $3\binom{2n+1}{2}$ Type 2 triples. With $v = 6n + 3$, we compute

$$(2n+1) + 3\binom{2n+1}{2} = (3n+1)(2n+1) = \frac{(6n+2)(6n+3)}{6} = \frac{1}{3}\binom{v}{2}.$$

The number of triples in an STS(v) is $\frac{1}{3}\binom{v}{2}$, since each pair must appear in exactly one triple. Since we have $\frac{1}{3}\binom{v}{2}$ triples, it suffices to show that every pair of elements appears in some triple.

If x and y are copies of the same element of Q, then they appear together in a Type 1 triple (note that idempotence ensures that they appear together in no other triple). If $x = i_k$ and $y = j_k$ with $i \neq j$, then they appear together in the Type 2 triple $(i_k, j_k, (i \circ j)_{k+1})$.

The remaining case is $x = i_k$ and $y = j_l$, with $i \neq j$ and $k \neq l$. Since k and l are distinct in $\mathbb{Z}_3$, we may assume $l = k + 1$. Since Q is a quasigroup, there is a unique $h \in Q$ such that $h \circ i = j$. Now x and y appear in the triple (h_k, i_k, j_l). ∎

When v is odd, there are many idempotent commutative quasigroups (Lemma 13.3.6 presents just one to prove existence). Thus the Bose construction generates many Steiner triple systems of a given order. The original construction of Bose [1939] was more specialized.

When $3 \nmid v$, the Bose construction is not valid, but there is another construction. Again we use three copies of a quasigroup. With $v \equiv 1 \pmod 6$, we designate a special element ∞ and partition the other elements into three copies of a quasigroup of order $2n$. Since commutativity requires each element to appear an even number of times off the diagonal, a commutative quasigroup of even order must also have each element appear an even number of times on the diagonal. We make each element in the "lower half" appear twice.

13.3.9. Definition. A quasigroup on $\mathbb{Z}_{2n}$ is **half-idempotent** if $i \circ i = (n + i) \circ (n + i) = i$ for $0 \leq i \leq n - 1$. Let $\{0, \ldots, n - 1\}$ be the *small* elements of $\mathbb{Z}_{2n}$; the remainder are the *large* elements.

Construction of a half-idempotent quasigroup is similar to Lemma 13.3.6. Note that in $\mathbb{Z}_{2n}$, when a number is a multiple of 2 it is twice a small element *and* twice a large element.

13.3.10. Lemma. For $n \in \mathbb{N}$, there is a half-idempotent commutative quasigroup on $\mathbb{Z}_{2n}$.

Proof: When $i + j$ is even, define $i \circ j$ to be the small element x such that $2x \equiv i + j \pmod{2n}$. When $i + j$ is odd, define $i \circ j$ to be the large element x such that $2x + 1 \equiv i + j \pmod{2n}$. The result for $n = 3$ appears below.

By construction, the diagonals with $i + j$ even are constant in small entries, the diagonals with $i + j$ odd are constant in large entries, and the diagonals cycle. That is, each successive row is a cyclic shift to the left of the previous row. Hence the quasigroup table is a symmetric Latin square, as needed.

Since the positions on the diagonal have even coordinate sum, all diagonal entries are small. In particular, $i + i = 2i \equiv (n + i) + (n + i) \pmod{2n}$, so the quasigroup is half-idempotent. ∎

$$
\begin{array}{cccccc}
0 & 3 & 1 & 4 & 2 & 5 \\
3 & 1 & 4 & 2 & 5 & 0 \\
1 & 4 & 2 & 5 & 0 & 3 \\
4 & 2 & 5 & 0 & 3 & 1 \\
2 & 5 & 0 & 3 & 1 & 4 \\
5 & 0 & 3 & 1 & 4 & 2
\end{array}
$$

13.3.11. Example. *The* **Skolem Construction**. For $v = 6n + 1$, let the elements consist of one special element ∞ plus three copies of a half-idempotent commutative quasigroup Q on $\mathbb{Z}_{2n}$, indicating the three copies by subscripts modulo 3. We use n Type 1 triples of the form (i_0, i_1, i_2) for $0 \le i \le n - 1$. The Type 2 triples have the form $(\infty, (n + i)_k, i_{k+1})$, where $0 \le i \le n - 1$ and $k \in \mathbb{Z}_3$. Finally, Type 3 triples have the form $(i_k, j_k, (i \circ j)_{k+1})$, where $i, j \in Q$ with $i \ne j$ and $k \in \mathbb{Z}_3$. ∎

13.3.12. Theorem. (Skolem [1958]) For $n \in \mathbb{N}$, the triples in the Skolem Construction (Example 13.3.11), defined using any half-idempotent commutative quasigroup on $\mathbb{Z}_{2n}$, form an STS$(6n + 1)$.

Proof: We have specified n Type 1 triples, $3n$ Type 2 triples, and $3\binom{2n}{2}$ Type 3 triples. With $v = 6n + 1$, we compute

$$ n + 3n + 3\tbinom{2n}{2} = 4n + 3n(2n - 1) = n(6n + 1) = \tfrac{1}{3}\tbinom{v}{2}. $$

As in Theorem 13.3.8, since we have $\frac{1}{3}\binom{v}{2}$ triples, it suffices to show that every pair of elements appears in some triple.

First suppose that $\infty \notin \{x, y\}$. If x and y are copies of the same small element of Q, then they appear together in a Type 1 triple. If $x = i_k$ and $y = j_k$ with $i \ne j$, then they appear together in the Type 3 triple $(i_k, j_k, (i \circ j)_{k+1})$.

Hence we may assume that $x = i_k$ and $y = j_l$ with $k \ne l$ and that $\{i, j\}$ consists of two distinct elements or the same large element. Since $k, l \in \mathbb{Z}_3$, by symmetry we may assume $l = k + 1$. If $i = n + j$ with j small, then x and y appear in the Type 2 triple (∞, i_k, j_l). Otherwise, let $h \in Q$ be the unique element such that $h \circ i = j$, which exists since Q is a quasigroup. Since Q is half-idempotent, $i \circ i$ is i when i is small, $i - n$ when i is large; we have already considered those cases. Hence $h \ne i$, so x and y appear in the Type 3 triple (h_k, i_k, j_l).

Finally, consider the pairs involving ∞. This element appears with one small element and one large element in each of the $3n$ Type 2 triples, yielding the desired $6n$ pairs. ∎

There are many half-idempotent commutative quasigroups on $\mathbb{Z}_{2n}$ and hence many Steiner triple systems of order $6n + 1$ from the Skolem construction (the original construction of Skolem [1958] was more specialized). Nevertheless, these constructions lack cyclic invariance.

13.3.13. Definition. A Steiner triple system of order v is **cyclic** if its set of elements is $\mathbb{Z}_v$ and the set of triples is cyclically invariant (that is, adding 1 to each element of any triple yields another triple).

13.3.14. Example. *Cyclic Steiner triple systems.* We generated cyclic projective planes as translations of a single difference set. In a cyclic STS(v), we have more than one class of triples, but each class is invariant under translations modulo v.

Define a **difference triple** in $\mathbb{Z}_v$ to consist of three nonzero elements that sum to 0 modulo v or such that one element is the sum of the other two. Heffter [1896] gave a sufficient condition for a family of difference triples to generate a cyclic STS(v) via its translations, each difference triple generating one cyclically invariant class. For $v = 6n + 1$, the condition is a partition of $\{1, 2, \ldots, 3n\}$ into difference triples. For $v = 6n + 3$, it is a partition of $\{1, 2, \ldots, 3n + 1\} - \{2n + 1\}$ into difference triples.

When $v = 6n + 1$, there are n triples in the partition, which yields nv triples under cyclic translations; this is the desired number of triples, $\frac{1}{3}\binom{v}{2}$. When $v = 6n + 3$, again there are n triples, but we add a "short orbit": the $2n + 1$ translations of $(0, 2n + 1, 4n + 2)$. The resulting number of triples is $(3n + 1)(2n + 1)$, which again is $\frac{1}{3}\binom{v}{2}$. Hence the specified partition yields the needed number of triples. We leave to Exercise 9 the job of checking that every pair appears, thereby proving Heffter's result.

Heffter did not find the needed partitions into difference triples, but Peltesohn [1939] did. The specified partitions exist for all v congruent to 1 or 3 modulo 6 except for $v = 9$. There is no cyclic STS(9) (Exercise 11).   ∎

GRAPHICAL DESIGNS

Thinking about (v, k, λ)-designs only as families of k-sets in a $[v]$-set does not provide much assistance in actually finding designs. It is also difficult to understand the structure of a design just from its incidence matrix, even though a few computational operations on the incidence matrix will confirm that it is a design.

We used algebraic tools in constructing Steiner triple systems above, but there are other ways to find structure that can be exploited. For example, if v has the form $\binom{n}{2}$, then we can view the elements of the design as pairs in an $[n]$-set, or equivalently as edges in K_n. Now k-element blocks become subgraphs with k edges. Confining our attention to such subgraphs with particular structure can help lead us to the desired design. We begin with the example of finding another Steiner triple system on 15 elements (recall that a Steiner triple system on v elements is a $(v, 3, 1)$-design).

13.3.15. Example. *A $(15, 3, 1)$-design.* The graph K_6 has 15 edges; these edges will be the elements of the design. A natural way to form triples of edges is to use the edge-sets of triangles as blocks. There are 20 triangles in K_6; each edge appears in four of them. Although incident edges appear together in exactly one triangle, we do not yet have blocks containing non-incident pairs of edges.

Natural triples containing non-incident pairs are the perfect matchings. There are 15 perfect matchings, every non-incident pair appears in exactly one of them, and no two incident edges appear together in a perfect matching. Thus our set of 35 triples now satisfies the definition of a $(15, 3, 1)$-design. Each edge appears in three perfect matchings and in a total of seven blocks, which confirms the equation $r(k - 1) = \lambda(v - 1)$.   ∎

13.3.16. Example. *An* $(11,6,3)$-*design.* It is hard to think of a natural set with
11 elements. We use the edges of K_5 plus a special element $*$. To keep the goal in
mind, use $r(k-1) = \lambda(v-1)$ and $bk = vr$ to obtain $r = 6$ and $b = 11$; we seek a
symmetric design. We will design blocks that pairwise share three elements; the
transpose of the incidence matrix is the incidence matrix of a design where any
two elements appear together in three blocks.

The edge sets of the five copies of K_4 have six elements, and any two of these
blocks share three edges (a triangle). Add six blocks that consist of $*$ plus a 5-cycle
in K_5. Fixing one 5-cycle, the other five each share two nonconsecutive edges with
it. There is one way to complete each such cycle.

We have ensured that the fixed cycle-block shares three elements with each
other cycle-bloc (two edges plus $*$). Also, since a 5-cycle visits each vertex in K_5,
each cycle-block shares three edges with each clique-block. Note that any two
rotations of the figure on the right below have two common edges (and share $*$).

Thus each of the 11 blocks has size 6, and every two blocks have exactly three
common elements.   ∎

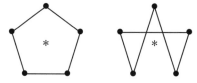

These constructions are examples of "graphical designs". Others can be found
in Exercise 10 and the chapter on graphical designs in Colbourn–Dinitz [1996,
2007]. Building designs from the edges of K_n opens an entire new area of investigation. It is natural to require isomorphic subgraphs. When each subgraph
intersects each other exactly once, the subgraphs are sometimes said to be "orthogonal". The transpose of the incidence matrix then yields a design with $\lambda =$
1. This is then a graphical interpretation of an ordinary design.

13.3.17. Example. *Graphical designs.* Classical $(v,k,1)$-designs can be interpreted as decompositions of K_v into edge-disjoint copies of K_k; *decomposition*
means that the subgraphs are pairwise edge-disjoint. From the graphical viewpoint, it is natural to allow more general subgraphs in seeking decompositions of
K_n into isomorphic subgraphs. For example, Exercise 5.3.48 decomposes K_n into
spanning cycles (for odd n) or spanning paths (for even n).

Ringel [1964] famously conjectured that K_{2m+1} decomposes into $2m+1$ isomorphic copies of any fixed tree T with m edges. Kotzig extended this to say that
some such decomposition is rotationally invariant when $V(K_{2m+1})$ is viewed as the
family of congruence classes modulo $2m+1$ (appearing first in Rosa [1967]). This
became known as the **Ringel–Kotzig Conjecture**, eventually proved for large m
by Montgomery–Pokrovskiy–Sudakov [2020] using probabilistic methods.

A further restriction requires vertex labels 0 through m so that the differences between endpoint labels on the edges are 1 through m. Such a labeling is
called a **graceful labeling**, and the famous **Graceful Tree Conjecture** asserts
that every tree has a graceful labeling. The ever-growing dynamic survey by Gallian [1998] lists hundreds of papers on graceful and related labelings of graphs.
We include just a few exercises about graceful labeling: Exercises 21–24.

Many papers have studied decomposition of K_n into k-cycles when n is odd; here the problem is completely solved. The problem extends to even n by omitting one perfect matching. If the necessary divisibility conditions hold (k divides $\binom{n}{2}$ when n is odd, k divides $\frac{1}{2}n(n-2)$ when n is even), then the desired decomposition exists; this was proved by Alspach–Gavlas [2001] for odd k and by Šajna [2002] for even k (see also an algebraic proof by Buratti [2003] for odd k). Much more generally, Alspach [1981] conjectured that whenever n is odd and $m_1, \ldots, m_t$ are integers at least 3 that sum to $\binom{n}{2}$, the complete graph K_n decomposes into cycles of lengths $m_1, \ldots, m_t$. When n is even, the corresponding conjecture is a decomposition of K_n into a perfect matching plus such cycles, when $\sum m_i = \frac{1}{2}n(n-2)$. Bryant–Horsley–Pettersson [2014] proved these conjectures. ∎

13.3.18. Example. *The* **Oberwolfach Problem.** A conference center in Oberwolfach, Germany, holds small weekly mathematical conferences and seats the participants at round tables of specified sizes for meals. Ideally, each participant should sit next to each other participant exactly once during the week. In 1967, at such a meeting, Ringel asked when it is possible to satisfy this requirement.

This is the *Oberwolfach Problem*: given a partition λ of n, we seek OP(λ), meaning a decomposition of K_n into 2-factors with component lengths $\lambda_1, \ldots, \lambda_t$ (plus a leftover 1-factor if n is even). It is easy to see that OP(3, 3) does not exist, and it is easy to construct OP(3, 4) (Exercise 26). Kotzig–Rosa [1974] showed that OP(3, 3, 3, 3) does not exist; the same is true for OP(4, 5) (Köhler [1977]) and OP(3, 3, 5) (Piotrowski [1979], by computer).

Although the Walecki decomposition is OP(n) with spanning cycles, in general the existence of OP($k, \ldots, k$) (generally written as OP(k^t)) is stronger than the k-cycle decompositions mentioned earlier, because here the k-cycles must be grouped into 2-factors, forming what are called **resolvable cycle systems**. In general, existence of OP(k^t) has been solved: for $k = 3$ by Ray-Chaudhuri–Wilson [1971] (odd n) and Kotzig–Rosa [1974] (even n), for even k by Häggkvist [1985] (when $kt \equiv 2 \pmod 4$) and Alspach–Häggkvist [1985] (when $kt \equiv 0 \pmod 4$), and for odd k by Alspach–Schellenberg–Stinson–Wagner [1989] (when $t \neq 4$) and Hoffman–Schellenberg [1991] (when $t = 4$).

There is also a bipartite version for decomposing $K_{n,n}$; see Piotrowski [1991]. Bryant–Rodger [2007] presents a survey on 2-factorizations.

For $\lambda \in \{(3, 3), (3, 3, 3, 3), (4, 5), (3, 3, 5)\}$, no OP($\lambda$) exists. The general conjecture is that OP(λ) exists in all other cases. The bipartite case has been solved: when all parts in λ are even, n must be even, and we seek a decomposition of K_n minus a perfect matching into 2 factors with cycle lengths given by λ. Here Häggkvist [1985] solved the case $n \equiv 2 \pmod 4$ (see below), and Bryant–Danziger [2011] completed the case $n \equiv 0 \pmod 4$.

For the general case, the conjecture has been confirmed when n is sufficiently large; see Glock–Joos–Kim–Kühn–Osthus [2018]. ∎

13.3.19. Lemma. (Häggkvist [1985]) If G is a path or cycle with m edges, and H is a 2-regular bipartite graph with $2m$ edges, then $G^{(2)}$ decomposes into two copies of H, where $G^{(2)}$ is obtained from G by expanding each vertex into an independent set of size 2 and replacing each edge xy in G with a copy of $K_{2,2}$ whose parts are the vertex pairs expanding x and y.

Proof: Decompose G into paths whose lengths are half the length of the cycles in H. Each such segment P in G with length l expands to $4l$ edges in $G^{(2)}$ and decomposes into two cycles C and C' of length $2l$, as illustrated below. Each end edge of P expands in each of C and C' into two edges at one vertex in the expansion, and each internal edge of P expands in each of C and C' into a 1-factor in the corresponding copy of $K_{2,2}$. Each expansion can be distributed to C and C' in either way. The resulting cycles decompose $G^{(2)}$ into two copies of H.   ∎

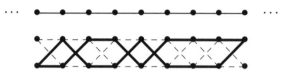

13.3.20. Theorem. (Häggkvist [1985]) If λ is a partition of n and $n \equiv 2 \pmod 4$, then OP(λ) exists.

Proof: Begin with a decomposition of $K_{n/2}$ into $(n-2)/4$ spanning cycles. Double each cycle by replacing each edge with four edges as in Lemma 13.3.19; each doubled cycle then has $2n$ edges. Since each cycle visits each vertex of $K_{n/2}$, we can arrange that in each doubling we omit the *same* perfect matching in K_n. The union of the doubled cycles then decomposes K_n minus the perfect matching. Using Lemma 13.3.19 to decompose each doubled cycle into two copies of the 2-factor with cycle lengths specified by λ then completes the desired decomposition.   ∎

The solution of the case $n \equiv 0 \pmod 4$ in Bryant–Danziger [2011] also uses Lemma 13.3.19.

RESOLVABLE DESIGNS AND OTHER TOOLS

For existence theorems in design theory, one tries to prove that constructions of a desired type exist whenever appropriate necessary conditions are satisfied. In an iterative approach, methods are developed for building larger constructions from smaller ones. We seek enough such combining theorems to obtain a proof that works for all sufficiently large values of the parameters. The Moore–MacNeish Theorem (Theorem 13.1.6) is a combining theorem of this type for orthogonal families of Latin squares. Many constructions for small values may be needed to start the induction.

We use this method to disprove Euler's Conjecture. Euler conjectured that there is no pair of orthogonal Latin squares of order n whenever $n \equiv 2 \pmod 4$. In fact, Euler was completely wrong; such pairs of squares exist if and only if $n \notin \{1, 2, 6\}$. We will develop resolvable designs, pairwise-balanced designs, and orthogonal arrays as tools, transforming the problem from the construction of Latin squares into the construction of certain arrays.

We start with a famous request for a resolvable design.

13.3.21. Example. The **Kirkman Schoolgirls Problem** (Kirkman [1850]) is

Fifteen young ladies in a school walk out three abreast for seven days in succession; it is required to arrange them daily, so that no two shall walk abreast twice.

Each day uses five triples; we want the 35 triples to contain the 105 pairs once each. Thus the triples form a $(15, 3, 1)$-design. The problem asks for more: the 35 triples must group into seven sets of five disjoint triples. ∎

13.3.22. Definition. A **resolvable design** is a (v, k, λ)-design whose blocks group into sets of size v/k such that the blocks in each set cover each element exactly once. The groups are called **parallel classes** or **resolution classes**.

In a resolvable design, v/k must be an integer. Since $v/k = b/r$ it may then be possible to partition the blocks into r parallel classes. The number of blocks is just right for each element to appear once in each parallel class. However, not all designs where $k \mid v$ are resolvable. The triple systems in the Bose construction (Theorem 13.3.8) are not, for example.

13.3.23. Example. *Resolvable designs.* The set of all pairs in $[2n]$ is a $(2n, 2, 1)$-design; the $n(2n - 1)$ blocks are the edges of a complete graph. In Example 8.3.5, we partitioned these edges into perfect matchings. Thus the design is resolvable.

The designs with the largest blocks relative to the number of points are the projective planes. Projective planes are not resolvable, since $q + 1$ does not divide $q^2 + q + 1$, but they lead to resolvable designs. Given a projective plane of order q, delete one line L and all appearances of its $q + 1$ elements. Since every two lines have one common element, the remaining blocks have size q. All $q + 1$ appearances of the remaining elements remain. They appeared together only outside L, so they still appear together once. Thus the family is a $(q^2, q, 1)$-design; it is an **affine plane**. Exercise 13 constructs these explicitly from finite fields.

For each element x of the line L deleted to obtain an affine plane, the remaining n lines through x partition the remaining n^2 points into n sets, forming a parallel class. Each line other than L lies in exactly one of these classes, because it intersects L in one point. Thus an affine plane is a resolvable design.

The blocks of an affine plane obtained from a projective plane of order 3 appear below. The columns in the list of blocks form parallel classes.

$$\left\{ \begin{array}{llll} 123, & 147, & 159, & 168 \\ 456, & 258, & 267, & 249 \\ 789, & 369, & 348, & 357 \end{array} \right\}$$

Like projective planes, affine planes have been used to construct other objects. For example, Fon-Der-Flaass [2002] used them to produce large new families of strongly regular graphs (Definition 15.3.19). ∎

Kirkman did find a construction for the Schoolgirl problem. When supplied with a grouping into parallel classes, resolvable Steiner triple systems become **Kirkman triple systems**. The number of points must be divisible by 3. With that restriction, the full problem was solved by Ray-Chaudhuri–Wilson [1971] using pairwise balanced designs: there is a Kirkman triple system of order n if and only if $n \equiv 3 \pmod 6$ (see Anderson [1997, pp. 198–200]).

The decomposition of K_{2n} into 1-factors in Example 8.3.5 uses one special element and cyclic symmetry among the rest (see also Theorem 13.3.12). A similar idea leads to resolvability for the schoolgirl problem.

13.3.24. Example. *The Schoolgirl design; a resolvable* STS(15). Let the schoolgirls be $\{A, B_1, \ldots, B_7, C_1, \ldots, C_7\}$. On the first day, use blocks

$$
\begin{array}{ccccc}
1 & 2 & 3 & 4 & 5 \\
AB_1C_1 & B_2B_4C_3 & B_3B_7C_5 & B_5B_6C_2 & C_4C_6C_7
\end{array}
$$

The blocks used on the other days are obtained by adding a constant to all the subscripts and treating them modulo 7.

Each such set of five blocks covers each element once. Since there are 35 blocks, it suffices to show that each of the 105 pairs appears. The classes of pairs are covered as follows, with appropriate shifts for i.　■

$$
\begin{array}{ccccc}
1 & 2 & 3 & 4 & 5 \\
B_iC_i & B_iC_{i+1} & B_iC_{i+2} & B_iC_{i+3} & C_iC_{i+1} \\
AB_i & B_iC_{i+6} & B_iC_{i+5} & B_iC_{i+4} & C_iC_{i+2} \\
AC_i & B_iB_{i+2} & B_iB_{i+3} & B_iB_{i+1} & C_iC_{i+3}
\end{array}
$$

The interpretation of affine planes as resolvable designs is reversible.

13.3.25. Example. Every resolvable $(q^2, q, 1)$-design is an affine plane. For each parallel class, we add one element in those blocks, and finally we add one block consisting of the $q+1$ new elements. By construction, the new family of blocks has the properties required to be the set of lines in a projective plane, so the original design was an affine plane.　■

We define more general types of designs.

13.3.26. Definition. For $K \subseteq \mathbb{N}$, a **pairwise balanced design** with parameters (v, K, λ) is a family of blocks on a set of v elements such that the size of each block is in K and any two elements appear together exactly λ times. Let (v, K, λ)-PBD denote such a family.

13.3.27. Example. *Constructions of PBDs.*

(1) By deleting one element from a design wherever it appears, we obtain two possible block sizes, but still each two remaining elements appear together the same number of times. Thus deleting one element from a (v, k, λ)-design yields a $(v - 1, \{k, k - 1\}, \lambda)$-design.

(2) Given any resolvable $(v, k, 1)$-design, pick s of the parallel classes. Add elements $x_1, \ldots, x_s$, with each new element entering all the blocks of one parallel class. Add $\{x_1, \ldots, x_s\}$ as one additional block. The result is a $(v + s, \{s, k + 1, k\}, 1)$-PBD (no blocks of size k remain if s equals the number of parallel classes).

For example, adding one element to each of the seven parallel classes in the schoolgirl design (Example 13.3.24) yields a $(22, \{7, 4\}, 1)$-PBD.　■

We rephrase the orthogonality conditions for Latin squares in order to develop an easier way to generate orthogonal families.

13.3.28. Definition. Two n-by-n matrices (or two vectors of length n^2) are **orthogonal** if the corresponding entries form n^2 distinct ordered pairs. Define special n-by-n matrices R_n and C_n by letting the entry in each position be the row index or the column index, respectively.

Storing a matrix in *row-major order* lists its entries row-by-row, across each row. We can thus interpret each matrix as a vector. In row-major order,

$$R_n = (1, \ldots, 1, 2, \ldots, 2, \cdots, n \ldots, n),$$
$$C_n = (1, \ldots, n, 1, \ldots, n, \cdots, 1 \ldots, n).$$

13.3.29. Definition. An **orthogonal array** $OA(s, n)$ is an s-by-n^2 matrix with entries in $[n]$ in which the rows are pairwise orthogonal.

13.3.30. Proposition. There is an orthogonal family of k Latin squares of order n if and only if there is an orthogonal array $OA(k + 2, n)$.

Proof: An n-by-n matrix M with entries in $[n]$ is a Latin square if and only if it is orthogonal to both R_n and C_n. Therefore, from an orthogonal family, we obtain an orthogonal array by listing R_n, C_n as the first two rows and listing the squares in the family in row-major order as the other rows.

In an orthogonal array, orthogonality and the Pigeonhole Principle require n copies of each entry in each row. By reordering columns, we can convert the first two rows to R_n and C_n without disturbing orthogonality. Now our initial observation implies that the remaining rows are row-major listings of the Latin squares in an orthogonal family. ∎

The book Hedayat–Sloane–Stufken [1999] provides extensive material on orthogonal arrays and their applications.

THE EULER CONJECTURE (optional)

In the remainder of this section, we apply orthogonal arrays and other tools to disprove Euler's Conjecture. The details are somewhat involved.

13.3.31. Conjecture. (Euler Conjecture) When $n \equiv 2 \pmod 4$, there is no pair of orthogonal Latin squares of order n. ∎

Let $N(n)$ be the maximum size of an orthogonal family of n-by-n Latin squares. Since 7 and 4 are prime powers, $N(7) = 6$ and $N(4) = 3$. Applied to the $(22, \{7, 4\}, 1)$-PBD found in Example 13.3.27, the next theorem yields an orthogonal pair of order 22. The value 22 was the first known counterexample to Euler's Conjecture, proved in Bose–Shrikhande [1959].

13.3.32. Theorem. An $(n, \{k_1, \ldots, k_s\}, 1)$-PBD yields $N(n) \geq \min_i\{N(k_i) - 1\}$.

Proof: Let $m = \min_i\{N(k_i)\}$. For each i, take an orthogonal family of m Latin squares of order k_i, written as an orthogonal array $\mathrm{OA}(m + 2, k_i)$.

In $\mathrm{OA}(m + 2, k_i)$, we may assume that R_{k_i} and C_{k_i} are the first two rows. In the other rows, we permute labels so that each row agrees with C_{k_i} in the first k_i columns, having labels $1, \ldots, k_i$ in order. In the remaining $k_i^2 - k_i$ columns, each pair of rows (not using the first row) has all ordered pairs of *distinct* labels from $[k_i]$. Let A_{k_i} be the matrix consisting of the last $m + 1$ rows and last $k_i^2 - k_i$ columns of the orthogonal array written in this form.

For each block B of the given pairwise-balanced design, consider the matrix $A_{|B|}$, and change the entries $1, \ldots, |B|$ in $A_{|B|}$ into the elements of $x_1, \ldots, x_{|B|}$, respectively. Let the resulting matrix be M_B.

Given the blocks $B_1, \ldots, B_b$ of the $(n, \{k_1, \ldots, k_s\}, 1)$-PBD on elements $[n]$, we form a matrix with $m + 1$ rows by concatenating $b + 1$ matrices with $m + 1$ rows each. The first matrix has n columns and has entries $1, \ldots, n$ in each row. The subsequent matrices are $M_{B_1}, \ldots, M_{B_b}$.

Consider rows r and s of this matrix, and consider $x, y \in [n]$. If $x = y$, then x and y appear in a common column in one of the first n columns. If $x \neq y$, then x and y appear together in exactly one block B_t of the design. Thus x and y can appear in a common column only in the block M_{B_t}. By the properties of the matrix $A_{|B_t|}$, rows r and s have x and y (in that order) in a common column exactly once.

Thus every ordered pair of entries occurs once in rows r and s. Hence the matrix has n^2 columns and is an orthogonal array $\mathrm{OA}(m + 1, m)$. By Proposition 13.3.30, there exists an orthogonal family of $m - 1$ Latin squares of order n. ∎

Parker [1959] built orthogonal 10-by-10 squares using orthogonal arrays and a computer. Now Exercise 18 can do this using the general construction below.

13.3.33. Theorem. (Bose–Shrikhande–Parker [1960]) If $N(m) \geq 2$, then also $N(3m + 1) \geq 2$.

Proof: It suffices to construct an $\mathrm{OA}(4, 3m + 1)$. Our $3m + 1$ symbols are the integers $0, \ldots, 2m$ and the labels $x_1, \ldots, x_m$. We concatenate $3 + 2m$ blocks with four rows each. The first block consists of $0, \ldots, 2m$ in order in each row. The second block is an orthogonal array $\mathrm{OA}(4, m)$ using the labels $x_1, \ldots, x_m$. In the remainder we must get the distinct ordered pairs of integers and the pairs (x_i, j) and (j, x_i) as columns in each pair of rows.

First define four row vectors of length m: $C_0 = (0, \ldots, 0)$, $U_0 = (1, \ldots, m)$, $D_0 = (2m, \ldots, m + 1)$, and $X = (x_1, x_2, \ldots, x_m)$ (note that C, U, D stand for "constant", "up", "down", respectively). Obtain C_i, U_i, D_i from C_0, U_0, D_0 by adding i to each coordinate value and reducing modulo $2m + 1$. Define

$$A_i = \begin{pmatrix} C_i & U_i & D_i & X \\ U_i & C_i & X & D_i \\ D_i & X & C_i & U_i \\ X & D_i & U_i & C_i \end{pmatrix}.$$

Complete the array by adding blocks $A_0, \ldots, A_{2m}$ after the first two blocks. Verification that this constructs an $\mathrm{OA}(4, 3m + 1)$ is left to Exercise 17. ∎

13.3.34. Corollary. For each nonnegative integer t, $N(12t + 10) \geq 2$.

Proof: Let $m = 4t + 3$. Since m is odd, the Moore–MacNeish Theorem (Theorem 13.1.6) yields $N(m) \geq 2$. Now apply Theorem 13.3.33.   ∎

To kill the Euler Conjecture for all n, we also strengthen Theorem 13.3.32.

13.3.35. Theorem. (Bose–Shrikhande–Parker [1960]) If an $(n, \{k_1, \ldots, k_m\}, 1)$-PBD exists whose blocks with sizes in $\{k_1, \ldots, k_r\}$ are pairwise disjoint, then

$$N(n) \geq \min\{N(k_1), \ldots, N(K_r), N(k_{r+1}) - 1, \ldots, N(k_m) - 1\}.$$

Proof: Let m be the claimed value plus 1. For $i > r$, construct A_{k_i} with $m+1$ rows as in Theorem 13.3.32. For $i \leq r$, let A_{k_i} be an orthogonal array $OA(m + 1, k_i)$. For each block B, form M_B by substitution in $A_{|B|}$ as in Theorem 13.3.32. Concatenate the matrices M_{B_i} as before, plus a matrix C consisting of one constant column for each element outside all blocks with sizes in $\{k_1, \ldots, k_r\}$.

If $x \neq y$, then x and y appear together in one block B, and $\binom{x}{y}$ occurs as a column in rows r and s in M_B exactly once (and nowhere else). If $x = y$ and x appears in a block B with size in $\{k_1, \ldots, k_r\}$, then $\binom{x}{y}$ appears in rows r and s in M_B; otherwise, it appears in C.   ∎

13.3.36. Example. *Orthogonal pair of order 18.* A projective plane of order 4 is a $(21, 5, 1)$-design. Deleting three points not on a line yields an $(18, \{3, 4, 5\}, 1)$-PBD. Since the three deleted points determined three distinct lines, this has three blocks of size 3. Furthermore, these three blocks are pairwise disjoint, since otherwise two of the original lines would have two common points. Thus we can apply Theorem 13.3.35 to obtain $N(18) \geq \min\{N(3), N(4) - 1, N(5) - 1\} = 2$.

To obtain $N(38) \geq 2$, apply this idea to a $(41, 5, 1)$-design, such as constructed in Exercise 13.2.15.   ∎

We need one more "combining" theorem.

13.3.37. Theorem. If $N(m) \geq k - 1$ and $x < m$, then

$$N(km + x) \geq \min\{N(m), N(x), N(k) - 1, N(k + 1) - 1\}.$$

Proof: Since $N(m) \geq k - 1$, there exists an orthogonal array $OA(k + 1, m)$ with last row R_m. From this we produce a special pairwise balanced design. Since the last row is R_m, in each other row the m distinct entries appear in each successive *group* of m columns. Delete the last row and add $(j - 1)m$ to the entries in row j, for $1 \leq j \leq k$. Row j now consists of entries in the interval $I_j = [1 + (j-1)m, jm]$.

Treat the columns as m^2 blocks of size k on the set $1, \ldots, km$. Each group of m columns contains each element once and is a parallel class. Each block contains one element from each I_j. An ordered pair from I_j and I'_j can only arise from rows j and j'; the orthogonality of the original array yields a unique column in which the desired ordered pair of congruence classes modulo m appeared in these rows.

To complete a pairwise balanced design, we add k further blocks that are precisely the intervals I_j; these form a parallel class of blocks of size m. Thus we have a resolvable $(km, \{k, m\}, 1)$-PBD.

Next we transform this into a $(km + x, \{x, m, k, k + 1\}, 1)$-PBD. We add x new elements. Each new element is added to the blocks of one of the parallel classes of size k, and we add one more block consisting of the new elements. This block of size x and the special class of k blocks of size m together partition the elements. If $m, x \notin \{k, k+1\}$, then the claim now follows from Theorem 13.3.35; otherwise, we drop $N(m)$ or $N(x)$ from the minimization and obtain the same result.

When $x = 1$ there are no pairs involving the new elements, so we don't need the extra block. Now x is not a block size, and again we drop $N(x)$ from the minimization, which we can do by setting $N(1) = \infty$. ∎

13.3.38. Corollary. If $N(m) \geq 3$, $N(x) \geq 2$, and $1 \leq x < m$, then $N(4m + x) \geq 2$.

Proof: Take $k = 4$ in Theorem 13.3.37. ∎

In applying Corollary 13.3.38, it is helpful to take m as a multiple of 4.

13.3.39. Lemma. (see Anderson [1990], p. 129) If $4 \mid m$, then $N(m) \geq 3$.

Proof: The Moore–MacNeish Theorem (Theorem 13.1.6) yields the bound unless m is divisible by 3 but not by 9. Thus m is divisible by 12 but not by 36. We can express m as 12 or 24 times an even power of 2 times an odd number other than 3. By the Moore–MacNeish Theorem, it suffices to prove that $N(12) \geq 3$ and $N(24) \geq 3$. For $n = 12$, there is an explicit family of size 5 (Dulmage–Johnson–Mendelsohn [1961]). For $n = 24$, see Exercise 15. ∎

These combining theorems can yield better lower bounds on $N(m)$ than the Moore–MacNeish Theorem, but we only need $N(m) \geq 2$.

13.3.40. Theorem. (Bose–Shrikhande–Parker [1960]) $N(n) \geq 2$ for all positive integers outside $\{1, 2, 6\}$.

Proof: We need consider only $n \equiv 2 \pmod 4$ with $n > 6$. Corollary 13.3.34 handles the case $n \equiv 10 \pmod{12}$. Since $12t + 6 = 3(4t + 2)$, the Moore–MacNeish Theorem takes care of $n = 12t + 6$ given an orthogonal pair with order $4t + 2$. There is no pair when $n = 6$, so we need an explicit construction for $n = 18$, done in Example 13.3.36. When $18 < 4t + 2 \equiv 6 \pmod{12}$, we reduce by another factor of 3.

This leaves only $n \equiv 2 \pmod{12}$. Explicit pairs are known for $n = 14$ and $n = 26$ (we omit these), and we covered $n = 38$ in Example 13.3.36. We use the Moore–MacNeish Theorem again for $50 = 5 \cdot 10$, $98 = 7 \cdot 14$, $110 = 10 \cdot 11$. We use Corollary 13.3.38 for $62 = 4 \cdot 13 + 10$, $74 = 4 \cdot 16 + 10$, $86 = 4 \cdot 19 + 10$, $122 = 4 \cdot 27 + 14$, $134 = 4 \cdot 27 + 26$. We have now discussed the cases up to $n = 144$.

We express the other values as a multiple of 16 plus one of $\{18, 22, 26, 30\}$. Since $n > 144$, we have $n = 16t + x = 4m + x$, where $4 \div m$ and $m \geq 32$. Thus $x < m$. Also $N(m) \geq 3$, by Lemma 13.3.39. Now $N(n) \geq 2$, by Corollary 13.3.38. ∎

EXERCISES 13.3

13.3.1. (–) For even v, construct a resolvable $(v, 2, 1)$-design.

13.3.2. (–) Prove that K_6 decomposes into three edges and four triangles.

13.3.3. (–) Prove that K_9 decomposes into 4-cycles.

13.3.4. (–) A **caterpillar** is a tree having a single path incident to or containing every edge. Prove that every caterpillar is graceful.

13.3.5. ($\diamond$) A **total coloring** of G color each vertex and edge so that objects adjacent or incident have different colors; $\chi''(G)$ is the number of colors needed. Prove that an idempotent commutative quasigroup of order n (Definition 13.3.9) exists if and only if $\chi''(K_n) = n$. Conclude from Lemma 13.3.10 that $\chi''(K_n)$ is n when n is odd and $n + 1$ when n is even. (Behzad [1965]; see also Yap [1989]) (Comment: Since $\Delta(K_n) = n - 1$, the Total Coloring Conjecture (Exercise 8.3.33) holds for complete graphs.)

13.3.6. Prisoners A, B, and C stand with C in front of B and B in front of A. From k hats of different colors, each receives one hat, selected at random. The prisoners know all the colors but see only the hats in front of them. They guess the colors of their own hats, first A, then B, then C, hearing what is said. If they all guess correctly, they will all be freed. They know the rules and plan a strategy in advance. What strategy maximizes the probability of success? (Hint: Use an idempotent quasigroup.) (Wagon–Zielinski [2018])

13.3.7. View $\mathbb{Z}_v \times \mathbb{Z}_m$ as $\{a_i : a \in \mathbb{Z}_v, i \in [m]\}$ An (i, j)-*difference* is a difference in $\mathbb{Z}_v$ between elements with subscripts i and j. A (v, k, λ, m)-**mixed difference system** is a family $D_1, \ldots, D_t$ of k-subsets of $\mathbb{Z}_v \times \mathbb{Z}_m$ such that each element of $\mathbb{Z}_v$ occurs exactly λ times as an (i, j)-difference for each $(i, j) \in [m]^2$, except that difference 0 does not occur with $i = j$.

(a) Prove that the translates (in $\mathbb{Z}_v$) of the t sets in a (v, k, λ, m)-mixed difference system $D_1, \ldots, D_t$ form a (vm, k, λ)-design.

(b) Let $u = 2n + 1$. Consider sets $\{0_1, 0_2, 0_3\}$ and $\{\{r_i, -r_i, 0_{i+1}\} : i \in \mathbb{Z}_3 \text{ and } 1 \leq r \leq n\}$. Prove that these $3n + 1$ sets form a $(u, 3, 1, 3)$-mixed difference system.

(c) Obtain an STS(v) when $v \equiv 3 \pmod 6$ (see also Example 13.3.7). (Bose [1939])

13.3.8. ($\diamond$) Prove that the complete k-partite graph with parts of size k decomposes into k^2 copies of K_k if and only if there is an affine plane of order k.

13.3.9. Complete the proof that if the small numbers in $\mathbb{Z}_v$ have a partition into difference triples as specified in Example 13.3.14, then there exists a cyclic STS(v).

13.3.10. ($\diamond$) Construct a $(6, 3, 2)$-design. (Hint: Define 10 triangles on six points.)

13.3.11. Prove that there is no cyclic STS(9).

13.3.12. Fisher's Inequality states $b \geq v$ for the number b of blocks in a (v, k, λ)-design. Improve this to $b \geq v + p - 1$ for a resolvable (v, k, λ)-design with p parallel classes.

13.3.13. Let $\mathbb{F}_q$ be the finite field of size q. Let V be the set of ordered pairs of elements from $\mathbb{F}_q$. For $a, b \in \mathbb{F}_q$, let $L_{a,b} = \{(x, y) \in V : y = ax + b\}$, and let $M_a = \{(x, y) \in V : x = a\}$. Prove that the q^2 sets $L_{a,b}$ and the q sets M_a together form a $(q^2, q, 1)$-design.

13.3.14. ($\diamond$) *Affine planes.*

(a) Prove that if B is a block in a $(q^2, q, 1)$-design, and x is an element not in B, then exactly one block in the design contains x and is disjoint from B.

(b) Use part (a) to prove that every $(q^2, q, 1)$-design is resolvable.

(c) Prove that if there is a $(q^2, q, 1)$-design, then there is a $(q^2 + 1, q, q - 1)$-design. (Hint: Make $q - 1$ copies of the blocks in all but one class.) (Rasch–Herrendörfer [1977])

13.3.15. Let q be a prime power. Obtain a $(q^2, q, 1)$-design. Delete one element to obtain a $(q^2 - 1, \{q, q - 1\}, 1)$-PBD in which the blocks of size $q - 1$ are pairwise disjoint. Conclude that $N(q^2 - 1) \geq N(q - 1)$; in particular, $N(24) \geq 3$.

13.3.16. *Pairwise-balanced designs.*

(a) Show that a $(v, \{3, 5\}, 1)$-PBD exists only when v is odd, and construct one for $v = 11$. (Hint: For the construction, start with a resolvable $(6, 2, 1)$-design.)

(b) Show that a $(v, \{4, 5\}, 1)$-PBD exists only when v or $v - 1$ is divisible by 4, and construct one for $v = 17$.

13.3.17. ($\diamond$) Complete the proof of Theorem 13.3.33 that $N(m) \geq 2$ implies $N(3m + 1) \geq 2$.

13.3.18. ($\Diamond$) Use Theorem 13.3.33 to construct an explicit pair of 10-by-10 orthogonal Latin squares. Express it as a "Graeco-Latin" square with the 100 numbers from 00 to 99 in the 100 positions. Explain the steps used (no proof needed).

13.3.19. Without restricting block sizes, a pairwise balanced design on a set E becomes just a family $\mathbf{F}$ of blocks (subsets of E) such that every two elements appear in λ common blocks. Prove that a family $\mathbf{F}$ having no block of size 1 is a pairwise balanced design with $\lambda = 1$ if and only if $\mathbf{F}$ is the set of hyperplanes of a matroid of rank 3 with no circuits of size 1 or 2. (G. Chappell)

13.3.20. ($\Diamond$) Prove that if there exist $n - 1$ pairwise orthogonal Latin squares of order n, then there exists a set of $n(n - 1)$ permutations of $[n]$ such that no two of them have the same ordered pair in any two corresponding positions. The columns of the array below form such a set of permutations when $n = 4$. (Comment: The converse also holds.)

$$\begin{matrix} 1 & 2 & 3 & 4 & 1 & 2 & 3 & 4 & 1 & 2 & 3 & 4 \\ 4 & 3 & 2 & 1 & 2 & 1 & 4 & 3 & 3 & 4 & 1 & 2 \\ 2 & 1 & 4 & 3 & 3 & 4 & 1 & 2 & 4 & 3 & 2 & 1 \\ 3 & 4 & 1 & 2 & 4 & 3 & 2 & 1 & 2 & 1 & 4 & 3 \end{matrix}$$

13.3.21. ($\Diamond$) A **graceful labeling** of a graph G with m edges is an injection $f \colon V(G) \to \{0, \ldots, m\}$ such that $\{|f(u) - f(v)| \colon uv \in E(G)\} = \{1, \ldots, m\}$. Kotzig conjectured that every tree has a graceful labeling (see Ringel [1964]).

(a) Prove that if a graph G with m edges has a graceful labeling, then K_{2m+1} decomposes into $2m + 1$ copies of G. (Rosa [1967])

(b) Prove that if G is a tree and all trees have graceful labelings, then also K_{2m} decomposes into $2m - 1$ copies of G.

13.3.22. An α-**labeling** of a bipartite graph is a graceful labeling where all labels on one part are smaller than all labels on the other. Prove that if a bipartite graph G with m edges has an α-labeling, and $k \in \mathbb{N}$, then K_{2mk+1} decomposes into copies of G. (Rosa [1967])

13.3.23. (+) *Graceful Eulerian graphs.*

(a) Prove that if an Eulerian graph G with m edges is graceful, then $4|m$ or $4|(m+1)$.

(b) Prove that the condition in part (a) is sufficient for cycles. (Hebbare [1976])

13.3.24. (+) Let G be the graph consisting of k 4-cycles with one common vertex. Prove that G is graceful. (Hint: Put 0 at the vertex of degree $2k$.)

13.3.25. Prove that K_9 decomposes into 6-cycles.

13.3.26. For the Oberwolfach Problem, prove that OP(3, 3) does not exist. Using vertex set $\mathbb{Z}_6 \cup \{*\}$, construct OP(3, 4).

13.3.27. ($\Diamond$) *A* (6, 3, 2)-*design.* Let the points be the vertices of K_6. Distinguish one vertex u of K_6. Using five triangles containing u and five triangles not containing v, obtain a (15, 3, 1)-design. Is this design resolvable? (Füredi)

13.3.28. ($\Diamond$) *A* (21, 5, 3)-*design for* $t = 3$. We construct 5-element blocks on 21 elements. The elements are $E(K_7)$. We use the copies of $K_{1,5}$, the 5-cycles, and the spanning subgraphs consisting of a triangle and two isolated edges. Prove that these blocks form a 3-design by showing that every triple of elements appears in exactly three blocks. (Hint: There are five distinguishable ways to arrange three edges in K_7.)

13.3.29. A $(k, \lambda; n)$-**transversal design** is a set $\mathcal{G}$ of k disjoint n-sets ("groups") and a family $\mathcal{F}$ of k-sets ("blocks") such that each block intersects each group and any two elements from different groups appear together in λ blocks. Prove that there exists a $(k, 1; n)$-transversal design if and only if there exists an orthogonal array $OA(k, n)$ (and hence a family of $k - 2$ pairwise orthogonal Latin squares of order n). (Comment: Transversal designs were introduced by Hanani [1961]; see Laywine–Mullen [1998, pp. 28–29].)

Chapter 14

The Probabilistic Method

In its simplest form, the probabilistic method in combinatorics is a nonconstructive method for proving the existence of desired combinatorial objects. A discrete probability space is defined where occurrence of the desired structure is an event. If the event has positive probability, then the desired structure occurs for some point in the space, and hence it exists.

Random variables add considerable power. For example, we proved inequalities for LYM orders in Section 12.2 by comparing average and maximum values of a random variable. Here we will also use random variables to study "threshold functions" in order to make precise statements about "typical behavior" of large structures. Finally, we will also discuss the benefits of showing that random variables are highly concentrated around their expectations.

Several texts and monographs present the probabilistic method in more detail than we can here. Prominent among these is Alon–Spencer [1992, 2000, 2008, 2016]. Others emphasize probabilistic methods in graph theory: Bollobás [1985, 2001], Palmer [1985], Janson–Łuczak–Rucinski [2000], Molloy–Reed [2002], and Frieze–Karoński [2016]. Motwani–Raghavan [1995] discusses probabilistic algorithms. Habib–McDiarmid–Ramirez-Alfonsin–Reed [1998] surveys additional uses of probabilistic methods in algorithmic discrete mathematics.

14.1. Existence and Expectation

In this section we begin with the simplest form of the probabilistic method and then show how to strengthen it by using random variables and expectation. We assume familiarity with the notions of events, probability spaces, and conditional probability as defined in Chapter 0.

Proving that an event occurs with positive probability involves proving an inequality. We use several fundamental tools in proving inequalities, some from elementary calculus. We begin with useful facts about exponentials, where e is the base 2.71828... of the natural logarithm. These inequalities are important in asymptotic arguments, and we use them often, so we distinguish e from the notation for an edge e in a graph by setting the numerical value e in roman font.

14.1.1. Proposition. If $x \in \mathbb{R}$ and $n \in \mathbb{N}$, then $1 + x \le e^x$ and $(1 + \frac{x}{n})^n \le e^x$, with equality in each only when $x = 0$.

Proof: The first inequality is trivial for $x \le -1$ and follows from the series expansion of e^x when $x > -1$. For $x \ne 0$, it yields $(1 + \frac{x}{n})^n < e^{(x/n)n} = e^x$. ∎

We will need bounds on factorials and binomial coefficients. For $\binom{n}{k}$, the bound $n^k/k!$ may suffice for constant k, but often we need a better upper bound.

14.1.2. Proposition. If $k \in \mathbb{N}$, then $\binom{n}{k} < (\frac{ne}{k})^k$.

Proof: Using the factorial expression for the binomial coefficient,

$$\binom{n}{k} = \prod_{i=0}^{k-1} \frac{n-i}{k-i} = \left(\frac{n}{k}\right)^k \prod_{i=0}^{k-1} \frac{1 - i/n}{1 - i/k}.$$

It suffices to prove $\prod_{i=0}^{k-1} \frac{1-i/n}{1-i/k} < e^k$. This is implied by $f(k) < e^k$, where $f(k) = \prod_{i=0}^{k-1} \frac{1}{1-i/k}$. Since $f(0) = 1$ (empty product), the claim follows from

$$\frac{f(k+1)}{f(k)} = \prod_{i=0}^{k} \frac{k+1}{k+1-i} \prod_{i=0}^{k-1} \frac{k-i}{k} = \prod_{i=1}^{k} \frac{k+1}{k} = \left(1 + \frac{1}{k}\right)^k < e,$$

where the last inequality uses Proposition 14.1.1. ∎

Stirling's Formula states $n! \sim n^n e^{-n} \sqrt{2\pi n}$ (Application 2.3.8). It approximates $\binom{n}{k}$ by $\left(\frac{n}{k}\right)^k \left(\frac{n}{n-k}\right)^{n-k} \sqrt{n/[2\pi k(n-k)]}$. The last factor is less than 1 for $1 \le k \le n-1$, and for the middle factor Proposition 14.1.1 yields $\left(\frac{n}{n-k}\right)^{n-k} = \left(1 + \frac{k}{n-k}\right)^{n-k} < e^k$. This is not a rigorous proof of Proposition 14.1.2, because it does not analyze the error in Stirling's Formula and we never derived the $\sqrt{2\pi}$.

THE UNION BOUND

The probabilistic method was popularized by Erdős, who used it in 1947 to obtain lower bounds on Ramsey numbers. Recall that $R(k, k)$ is the least n such that every 2-coloring of $E(K_n)$ has a *homogeneous set* of k vertices, meaning a set whose $\binom{k}{2}$ edges all have the same color. The inductive proof of Ramsey's Theorem yields $R(k, k) \le \binom{2k-2}{k-1} \sim 4^k/\sqrt{\pi k}$. Erdős showed that $\sqrt{2}^k$ vertices do not suffice.

14.1.3. Theorem. (Erdős [1947]) $R(k, k) > \frac{1}{e\sqrt{2}} k 2^{k/2}$.

Proof: Consider a random 2-coloring of $E(K_n)$, with each edge having each color with probability $\frac{1}{2}$, independently. A set of k vertices is homogeneous with probability $2 \cdot 2^{-\binom{k}{2}}$. With $\binom{n}{k}$ such events, the probability that at least one occurs is at most $\binom{n}{k} 2^{1-\binom{k}{2}}$. If this is less than 1, then some outcome of the process has no homogeneous k-set, and hence $R(k, k) > n$. Since $\binom{n}{k} < (\frac{ne}{k})^k$, it suffices (roughly) to have $\frac{ne}{k} \le 2^{(k-1)/2}$, which is equivalent to $n \le \frac{1}{e\sqrt{2}} k 2^{k/2}$. ∎

14.1.4. Remark. There is a large gap between $\sqrt{2}^k$ and 4^k. More sophisticated probabilistic methods have made only small improvements in the lower bound. Nevertheless, the bounds from known constructions are sub-exponential, so this is a triumph for the probabilistic method.

The proof is the same as the counting argument in Theorem 10.2.14, just phrased in probabilistic language. Many probabilistic arguments with finite sample spaces are just weighted counting arguments, but the tools of probability do the job more clearly and efficiently and go well beyond what counting can do.

Existence arguments yield probabilistic constructions. For example, a random 2-coloring of $E(K_{64})$ has a homogeneous 10-set with probability less than $((2^6)^{10}/10!)2^{-44}$, which is about .018. If a randomly generated coloring has a homogeneous 10-set, just try again. The probability of repeated bad outcomes is a product of small numbers and soon becomes incomprehensibly small. ∎

The fundamental probabilistic argument in Theorem 14.1.3 has a name and underlies many easy applications of the probabilistic method. Note in particular that this bound does not require independence; it holds for any events $A_1, \ldots, A_n$.

14.1.5. Proposition. (The **Union Bound**) For any events $A_1, \ldots, A_n$ in a probability space, $\mathbb{P}(\bigcup_{i=1}^n A_i) \leq \sum_{i=1}^n \mathbb{P}(A_i)$. ∎

We next study proper 2-colorings of k-uniform hypergraphs. Recall that the edge set of a k-uniform hypergraph H is a family of k-sets, and a t-coloring of the vertex set $V(H)$ is **proper** if no edge is monochromatic. We can prove that a hypergraph is 2-colorable by showing that a random coloring of the vertex set has positive probability of being proper.

14.1.6. Proposition. (Erdős [1963]) If $f(k)$ denotes the minimum number of edges in a k-uniform hypergraph that is not 2-colorable, then

$$2^{k-1} \leq f(k) \leq \binom{2k-1}{k} \sim \frac{4^k}{2\sqrt{\pi k}}.$$

Proof: A random coloring makes a specified edge monochromatic with probability 2^{1-k}, so with fewer than 2^{k-1} edges the total probability of the bad events is less than 1 and a proper 2-coloring exists.

By the Pigeonhole Principle, the hypergraph with edge set $\binom{[2k-1]}{k}$ is not 2-colorable, since some k vertices must have the same color. Stirling's Formula yields the asymptotics of $\binom{2k-1}{k}$ (see Example 2.3.10). ∎

In this context, 2-colorability was named "Property B" by Erdős in honor of Bernstein. The problem of finding $f(k)$ was posed by Erdős and Hajnal in 1961. Beck [1978] improved the lower bound to $f(k) \geq k^{1/3-o(1)}2^k$ (see Corollary 14.2.7).

A probabilistic argument yields an improved upper bound, logarithmically asymptotic to the lower bound. Instead of seeking existence of a proper coloring, we seek existence of a hypergraph with no proper coloring. Hence we generate a random hypergraph instead of a random coloring. In Proposition 14.1.6 we showed that some outcome of forming a random coloring has no bad edge; now we show that some outcome of forming a random hypergraph has no good coloring.

14.1.7. Theorem. (Erdős [1964]) There exists a k-uniform hypergraph with $(1 + o(1))\frac{e \ln 2}{4} k^2 2^k$ edges that is not 2-colorable.

Proof: We form a k-uniform hypergraph by choosing m edges at random from vertex set $[n]$, where n and m will be chosen later to optimize the resulting bound. Repeated edges cause no problem; indeed, they would improve the bound. We apply the Union Bound to the 2^n events that particular red/blue colorings of the vertex set are proper colorings of the resulting hypergraph.

For a given coloring with r points in one color and s in the other, where $r + s = n$, the probability that a random k-set is monochromatic is $\left[\binom{r}{k} + \binom{s}{k} \right] / \binom{n}{k}$, which is minimized when $r = s$. Let $p = 2\binom{n/2}{k}/\binom{n}{k}$. For a given coloring, the probability that a random k-set is monochromatic is at least p. For m edges chosen independently, the probability that none is monochromatic is at most $(1 - p)^m$.

By the Union Bound, our random hypergraph is 2-colorable with probability at most $2^n(1 - p)^m$. If $2^n(1 - p)^m < 1$, then some k-uniform hypergraph with n vertices and m edges is not 2-colorable. Since $(1 - p) \leq e^{-p}$, it suffices to have $2^n e^{-mp} < 1$. To minimize m such that $2^n < e^{mp}$, we set $m = \left\lceil \frac{n}{p} \ln 2 \right\rceil$.

Now we choose n in terms of k to minimize n/p. Rewrite p as $2^{1-k} \prod_{i=0}^{k-1} \frac{n-2i}{n-i}$. Using $\frac{1}{1-i/n} = 1 + \frac{i}{n} + O((\frac{i}{n})^2)$, we have

$$\frac{n - 2i}{n - i} = \frac{1 - 2i/n}{1 - i/n} = 1 - \frac{i}{n} + O\left(\frac{i^2}{n^2}\right) = e^{-i/n} + O\left(\frac{i^2}{n^2}\right).$$

Using $\sum_{i=0}^{k-1} i = k(k-1)/2$ makes p asymptotic to $2^{1-k} e^{-k^2/(2n)}$. To minimize $n2^{k-1} e^{k^2/(2n)}$, calculus suggests $n = k^2/2$, yielding $m = (1 + o(1))\frac{e \ln 2}{4} k^2 2^k$. ∎

The 2-colorability of k-uniform hypergraphs is closely related to k-choosability of bipartite graphs. Recall that a graph G is *k-choosable* if a proper coloring can be chosen from any color lists of size k assigned to the vertices. The *choice number* $\chi_l(G)$ is the least k such that G is k-choosable. Combinatorial proof of the relationship below is requested in Exercise 15.

14.1.8. Theorem. (Erdős–Rubin–Taylor [1979]) If n_k is the smallest order of a non-k-choosable bipartite graph, and m_k is the smallest size of a non-2-colorable k-uniform hypergraph, then $m_k \leq n_k \leq 2m_k$. ∎

Example 8.2.7 proved $\chi_l(K_{r,r}) > k$ for $r = \binom{2k-1}{k}$, so $\chi_l(K_{r,r}) \geq \left(\frac{1}{2} + o(1) \right) \lg r$. Exercise 16 requests an upper bound about twice that. Both have been improved.

14.1.9. Corollary. $\lg r - (2 + o(1)) \lg \lg r \leq \chi_l(K_{r,r}) \leq \lg r - (\frac{1}{2} - o(1)) \lg \lg r$.

Proof*: The lower bound uses Theorems 14.1.7–14.1.8. Since some hypergraph with at most $O(k^2 2^k)$ edges is not 2-colorable, k-choosability of $K_{r,r}$ requires $r \leq O(k^2 2^k)$. Taking logs and inverting the relationship yields the lower bound.

The improvement to the upper bound combines Theorem 14.1.8 with a lower bound on the number m_k of edges in a non-2-colorable k-uniform hypergraph. Radhakrishnan–Srinivasan [2000] proved $m_k \geq \Omega((k/\log k)^{1/2} 2^k)$ (another proof appears in Kozik–Cherkashin [2015]). For $\chi_l(K_{r,r}) > k$, we have $2r \geq n_k \geq m_k$. Thus there is a constant c such that $r \leq c(k/\lg k)^{1/2} 2^k$ yields $\chi_l(K_{r,r}) \leq k$. Inverting the relationship between r and k yields the upper bound. ∎

Having $\chi_l(G) > k$ requires existence of a k-uniform list assignment from which no proper coloring can be chosen. Hence to apply the existence argument we generate a random list assignment. The subtlety is how to do this so that with positive probability no proper coloring can be chosen. It will suffice to consider a bipartite subgraph with large minimum degree.

14.1.10.* Theorem. (Alon [1993]) For some constant c, every graph G with average degree d has choice number at least $c\frac{\log d}{\log \log d}$.

Proof: It suffices to show that $\frac{d}{4} > \binom{s^4}{s} \log\left(2\binom{s^4}{s}\right)$ implies $\chi_l(G) > s$, since taking logarithms yields $\log d \sim c's \log s$, which is equivalent to $s \sim c\frac{\log d}{\log \log d}$ for some c.

Since G has average degree d, it has a subgraph G' with minimum degree at least $d/2$ (iteratively delete vertices with smaller degree; Exercise 5.2.10). The subgraph G' in turn has a spanning bipartite subgraph H such that $d_H(v) \geq \frac{1}{2}d_{G'}(v)$ for all $v \in V(G')$ (Exercise 27, Theorem 5.2.9), so $\delta(H) \geq d/4$.

We generate a random list assignment L for H with lists of size s (from a set of s^4 colors). We show that with positive probability, H has no L-coloring (Definition 8.2.6). This yields $\chi_l(H) > s$ and hence $\chi_l(G) > s$.

Let A and B be the parts of H, with $|A| \geq |B|$. Let $S = [s^4]$, and let $t = \binom{s^4}{s}$. Each vertex receives a random s-subset of S as a list, with all t such sets equally likely. Say that a vertex of A is *full* if all t possible lists appear on its neighbors. The probability that a particular s-set T fails to appear on the neighbors of x is $(1 - 1/t)^{d_H(x)}$. Since there are t such sets and $d/4 > t \log(2t)$, we obtain

$$\mathbb{P}(x \text{ is not full}) \leq t\left(1 - t^{-1}\right)^{d/4} < te^{-t^{-1}d/4} < te^{-\log(2t)} = \tfrac{1}{2}.$$

Hence the expected number of full vertices is at least $|A|/2$, and there is some outcome of the random list assignment such that at least $|A|/2$ vertices of A are full. Fix such an assignment for B.

We now claim that extending this assignment by randomly chosen lists for A produces with positive probability a list assignment from which no proper coloring can be chosen. Let f be a particular choice of colors from the lists on B (recall that B is independent in H). For a full vertex x in A, since all s-sets appear on its neighbors, at most $s - 1$ colors fail to be chosen on its neighbors. Hence f can be properly extended to x only if $L(x)$ contains one of these missing colors. There are at most $s - 1$ ways to name a usable color, and then $L(x)$ must be filled from the remaining colors, so

$$\mathbb{P}(x \text{ can be colored}) \leq \frac{(s-1)\binom{s^4-1}{s-1}}{t} = \frac{s-1}{s^3} < \frac{1}{s^2}.$$

To extend f to an L-coloring of H, all full vertices must be colored; this has probability at most $(1/s^2)^{|A|/2}$, which equals $s^{-|A|}$. With $s^{|B|}$ choices for f on B from the assignment on B, the probability that some choice f on B extends to an L-coloring is at most $s^{|B|}s^{-|A|}$. Since $|A| \geq |B|$, the bound is less than 1. Hence for some outcome of the random lists on A, no proper coloring can be chosen. ∎

Alon [2000b] later improved Theorem 14.1.10 to $\chi_l(G) \geq \left(\frac{1}{2} - o(1)\right) \ln d$ when G has average degree d. It is conjectured that there is a constant c such that $\chi_l(G) \leq c \log d$ when G is d-regular and bipartite; only $\chi_l(G) \leq O(\frac{d}{\log d})$ is known.

RANDOM VARIABLES

Often we study numerical aspects of points in a probability space. Typically we generate an object randomly, and we consider the value of some parameter of the resulting object. We repeat the relevant definitions from Chapter 0.

14.1.11. Definition. A **random variable** on a discrete probability space S is a function $X \colon S \to \mathbb{R}$. The **expectation** or **expected value** $\mathbb{E}(X)$ of X is $\sum_{a \in S} X(a) \mathbb{P}(a)$ (when this converges). A **discrete random variable** X has countable range, usually $\mathbb{N}_0$. For such X, let $X = k$ denote the event $\{a \in S \colon X(a) = k\}$ and write $\mathbb{E}(X) = \sum_{k=0}^{\infty} k \cdot \mathbb{P}(X = k)$. The **pigeonhole property** of the expectation is the statement that there is an element of the probability space for which the value of X is as large as (or as small as) $\mathbb{E}(X)$.

Using the pigeonhole property requires a value or bound for $\mathbb{E}(X)$. Often we obtain this by expressing X as a sum of simpler random variables. We repeat Lemma 0.13, restricting our attention to finite sums of random variables on discrete probability spaces.

14.1.12. Lemma. (Linearity of expectation) If X and $X_1, \ldots, X_k$ are random variables on the same space such that $X = \sum X_i$, then $\mathbb{E}(X) = \sum \mathbb{E}(X_i)$. Also $\mathbb{E}(cX) = c\mathbb{E}(X)$ for any constant c.

Proof: In a discrete probability space, each sample point contributes the same amount to each side of each of these equations. ∎

14.1.13. Remark. We will apply Lemma 14.1.12 to sums of variables. A **counting variable** X is a sum of variables X_i indicating whether the ith event in some set occurs. An **indicator variable** or **0, 1-variable** takes values in $\{0, 1\}$. When X_i is an indicator variable, $\mathbb{E}(X_i) = \mathbb{P}(X_i = 1)$. Thus the expected number of events that occur is the sum of their probabilities: $\mathbb{E}(X) = \sum \mathbb{P}(X_i = 1)$. ∎

Linearity and pigeonholing capture the existence arguments we have given. For example, to show that k-uniform hypergraphs with fewer than 2^{k-1} edges are 2-colorable, we chose a random vertex partition. Each edge is monochromatic with probability 2^{1-k}. With fewer than 2^{k-1} edges, the expected number of monochromatic edges is less than 1, and hence some coloring has no monochromatic edge.

Use of linearity often simplifies computations of expectation; the exercises present many examples. Also, comparing the expected and maximum values of a random variable yields results such as the next theorem.

14.1.14. Theorem. (Caro [1979], Wei [1981]) For a graph G,

$$\alpha(G) \geq \sum_{v \in V(G)} \frac{1}{d(v) + 1}.$$

Proof: (Alon–Spencer [1992]) In an ordering of $V(G)$, the vertices that appear before all neighbors form an independent set S. In a random ordering, the probability that v precedes all its neighbors is $(d(v) + 1)^{-1}$. Thus the right side of the inequality is the expected size of S. The left side is the maximum size of S. ∎

Theorem 14.1.14 yields a proof of Turán's Theorem (Theorem 11.1.3).

14.1.15. Theorem. Let G be an n-vertex graph with no $(r + 1)$-clique. The number of edges of G is maximized (uniquely) when G is the complete r-partite graph $T_{n,r}$ whose part-sizes differ by at most 1.

Proof: Maximizing $|E(G)|$ such that $K_{r+1} \nsubseteq G$ is equivalent to minimizing $|E(H)|$ such that $\alpha(H) \leq r$, where $H = \overline{G}$. Note that $\overline{T}_{n,r}$ is a disjoint union of complete graphs. Let $m = \left|E(\overline{T}_{n,r})\right|$. It suffices to show that $\overline{T}_{n,r}$ is the only graph with m edges having independence number as small as r, since deleting edges never reduces the independence number and deleting an edge from $\overline{T}_{n,r}$ increases it.

Fix $|E(H)| = m$. The lower bound on $\alpha(H)$ in Theorem 14.1.14 is now minimized only when the degrees differ by at most 1, by convexity of the reciprocal function. Since the degree-sum is $2m$, the only such list is the degree list for $\overline{T}_{n,r}$. Groups of k vertices with degree $k = 1$ contribute 1 to the sum, so the sum is r. We have proved that $\alpha(H) \geq r$ when H has m edges (and n vertices).

To show that equality holds only when $H = \overline{T}_{n,r}$, we observe that equality requires not only the specified degrees, but also equality in the argument of Theorem 14.1.14. To avoid having an independent set larger than $\mathbb{E}(|S|)$, every ordering must generate an independent set of the same size. If there exist $x, y, z \in V(H)$ with $x, z \in N(y)$ and $xz \notin E(H)$, then an ordering σ that starts x, y, z puts $x \in S$ but $y, z \notin S$. For an ordering σ' that agrees with σ except for starting x, z, y, the independent set will be the same except that z will be added. Hence $\alpha(H) > \mathbb{E}(|S|) = r$. A graph with no induced P_3 is a disjoint union of complete graphs. With vertex degrees as in $\overline{T}_{n,r}$, the only such graph is $\overline{T}_{n,r}$. ∎

Applications of expectation and linearity often yield strong results with surprisingly simple proofs. Many occur in the exercises, and another example is in Theorem 16.1.28. By choosing random subgraphs, we show there that when $m \geq 4n$, every drawing in the plane of a graph with n vertices and m edges has at least $m^3/(64n^2)$ crossing pairs of edges.

The pigeonhole property of expectation plays a crucial role in an upper bound on the Ramsey number $R(3, k)$. We seek k that guarantees $\alpha(G) \geq k$ in every triangle-free n-vertex graph G. Weakening the Caro–Wei Theorem (Theorem 14.1.14) yields $\alpha(G) \geq n/(d + 1)$ when $\Delta(G) = d$, which is easily proved directly. The breakthrough by Ajtai–Komlós–Szemerédi [1980] showed $\alpha(G) \geq (cn \lg d)/d$ when G is triangle-free. Simplifications and improvements were due to Shearer [1983] (achieving $c = 1 + o(1)$), Shearer [1995], and Alon [1996]. We present the short proof in Alon–Spencer [2008] based on the earlier papers.

14.1.16. Theorem. If G is a triangle-free n-vertex graph with $\Delta(G) \leq d$, where $d \geq 1$, then $\alpha(G) \geq (n \lg d)/(8d)$.

Proof: If $d \leq 16$, then $\frac{d}{d+1} \geq \frac{1}{2} \geq \frac{\lg d}{8}$, which implies $\frac{n}{d+1} \geq \frac{n \lg d}{8d}$, so the desired bound follows from $\alpha(G) \geq \frac{n}{d+1}$. Hence we may assume $d \geq 16$.

Let W be a random independent set in G, with all independent sets equally likely. To prove $\alpha(G) \geq \frac{n \lg d}{8d}$, it suffices to show $\mathbb{E}(|W|) \geq \frac{n \lg d}{8d}$. To study $\mathbb{E}(|W|)$, we introduce additional random variables. For each vertex v in G, define X_v by

$$X_v = \begin{cases} d & \text{if } v \in W \\ |N(v) \cap W| & \text{if } v \notin W. \end{cases}$$

Let $X = \sum_{v \in V(G)} X_v$. In computing X, each vertex $u \in W$ contributes d to X_u and contributes 1 to X_v for each $v \in N(u)$, since $N(u) \cap W = \varnothing$. Hence the total contribution to X from u is at most $2d$. This yields $X \leq 2d\,|W|$ no matter what W is. Taking expectations, we have $\mathbb{E}(|W|) \geq \mathbb{E}(X)/(2d)$. It therefore suffices to prove $\mathbb{E}(X_v) \geq \frac{1}{4} \lg d$ for each vertex v, taken over all (equally likely) choices of W.

Fix $v \in V(G)$. We will group the choices of W and show that over each group, the expectation of X_v is at least $\frac{1}{4} \lg d$. The desired lower bound on $\mathbb{E}(X_v)$ then follows, regardless of the sizes of the groups, since all W are equally likely.

We group the choices of W according to the subset of W outside the closed neighborhood of v. This set $W - (N_G(v) \cup \{v\})$ is another random variable determined by W; call it S. For a given choice of S, let Y be the set of common nonneighbors of S that lie in $N(v)$, and let $t = |Y|$. The independent set W that generates S may arise by adding v to S (since $S \cap N(v) = \varnothing$) or by adding to S any subset of Y. These $1 + 2^t$ possibilities for W are equally likely, since all W are equally likely. In the first case, the value of X_v is d. Over the other 2^t possibilities the average value of X_v is $t/2$, since the value of X_v is then $|Y \cap W|$. Thus over this group the expected value of X_v is the left side of the desired inequality

$$\frac{d}{2^t + 1} + \frac{(t/2)2^t}{2^t + 1} \geq \tfrac{1}{4} \lg d.$$

If $t = 0$, then the left side simplifies to $d/2$, which is big enough. Hence we may assume $t \geq 1$. If the inequality fails, then clearing fractions and rearranging terms yields $4d - \lg d < 2^t(\lg d - 2t)$. Since the left side is positive, we have $\lg d > 2t \geq 2$. With $t < \frac{1}{2} \lg d$, we have $2^t < \sqrt{d}$. Thus $4d - \lg d < \sqrt{d}(\lg d - 2)$, which fails when $d \geq 16$. ∎

14.1.17. Corollary. (Ajtai–Komlós–Szemerédi [1980]) There is a constant c such that $R(3, k) \leq ck^2 / \lg k$ whenever $k > 1$.

Proof: We prove that $c = 8$ suffices. Consider a graph G with $8k^2/\lg k$ vertices. Suppose that G contains no triangle. If $\Delta(G) \geq k$, then the neighborhood of a maximum-degree vertex contains an independent set of size k. Otherwise, with $\Delta(G) < k$, Theorem 14.1.16 applies to guarantee an independent set of size at least $(|V(G)|\lg k)/8k$. With $|V(G)| = 8k^2/\lg k$, we obtain $\alpha(G) \geq k$. ∎

It turns out that this relatively easy upper bound has the right order. We will obtain the weaker lower bound $R(3, k) \geq \Omega(k^2 / \lg^2 k)$ from Erdős [1961] in Theorem 14.2.22, but Kim [1995] proved $R(3, k) = \Theta(k^2 / \lg k)$.

We close this section with two unusual applications, showing the flexibility and creativity that may be involved in applying probabilistic arguments to combinatorial problems. First we obtain Binet's Formula for the adjusted Fibonacci numbers without directly solving the Fibonacci recurrence.

14.1.18. Theorem. If a_n is the number of 1,2-lists with sum n, then

$$a_n = \frac{1}{\sqrt{5}} \left(\frac{1 + \sqrt{5}}{2} \right)^{n+1} - \frac{1}{\sqrt{5}} \left(\frac{1 - \sqrt{5}}{2} \right)^{n+1}.$$

Proof: (Benjamin–Levin–Mahlburg–Quinn [2000]) Model each $1, 2$-list as a row of squares and dominoes (see Example 2.1.2), forming a tiling. Generate a tiling randomly by iteratively letting the next tile be a square with probability $1/\phi$ and a domino with probability $1/\phi^2$, where $\phi = (1 + \sqrt{5})/2$. Note that $1/\phi + 1/\phi^2 = 1$.

The experiment generates an infinite tiling. The probability that it generates a fixed $1, 2$-list with sum n in the initial positions is $1/\phi^n$ (this is the product of the probabilities for the successive tiles, given what has already happened). Since there are a_n such lists, which are disjoint events, the probability q_n that the experiment produces a break in the tiling after position n is a_n/ϕ^n.

A tiling is unbreakable after length n if and only if it adds a domino following a tiling of length $n - 1$; the probability of this is q_{n-1}/ϕ^2. Thus $q_n = 1 - q_{n-1}/\phi^2$. Since $q_0 = 1$, iteration yields $q_n = 1 - \phi^{-2} + \phi^{-4} - \cdots + (-\phi^{-2})^n$.

With $\alpha = (1 - \sqrt{5})/2$ and $\beta = (1 - \sqrt{5})/2$, we have $\phi\beta = -1$, so $-\phi^{-2} = \beta/\phi$. Evaluating the geometric sum yields

$$q_n = \frac{1 - (-\phi^{-2})^{n+1}}{1 - (-\phi^{-2})} = \frac{1 - (\beta/\phi)^{n+1}}{1 - \beta/\phi}.$$

Now multiplying by ϕ^n yields the same formula as Example 2.2.4:

$$a_n = \phi^n q_n = \frac{\phi^{n+1} - \beta^{n+1}}{\phi - \beta} = \frac{1}{\sqrt{5}}\left[\alpha^{n+1} - \beta^{n+1}\right]. \qquad \blacksquare$$

The next problem, posed by Kearnes–Kiss [1999], succumbs surprisingly easily to a clever probabilistic argument. This time, instead of using $\max X \geq \mathbb{E}(X)$, we use $\mathbb{E}(X) \geq \min X$, but the definition of the random variable is not obvious.

14.1.19. Definition. Let A be the cartesian product of n finite sets $A_1, \ldots, A_n$. A **proper subproduct** of A is a set B of the form $B_1 \times \cdots \times B_n$, where B_i is a proper subset of A_i for $1 \leq i \leq n$. A **proper dissection** of A is a partition of A into proper subproducts; we call these "pieces" of the dissection.

It is not hard to show that proper dissections of 1- and 2-dimensional grids need at least two and four pieces, respectively. Note that the definition of subproduct places no order on the elements of the factors, so the pieces need not be geometrically connected.

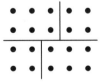

14.1.20. Theorem. (Alon–Bohman–Holzman–Kleitman [2002]) A proper dissection of a cartesian product of n finite sets has at least 2^n pieces.

Proof: Let $B^1, \ldots, B^k$ be the pieces in a proper dissection of $A_1 \times \cdots \times A_n$; recall that $B^j = B_1^j \times \cdots \times B_k^j$. We show that $k \geq 2^n$. For each $i \in [n]$, let R_i be a subset of A_i chosen uniformly at random from among the subsets of A_i with odd size, and let $R = R_1 \times \cdots \times R_n$.

For $j \in [k]$, let X_j be the 0,1-variable that is 1 if $|B^j \cap R|$ is odd. Let $X = \sum_j X_j$. Since $|B^j \cap R|$ is odd if and only if $|B_i^j \cap R_i|$ is odd for all $i \in [n]$, and $\mathbb{P}(|B_i^j \cap R_i|$ is odd$) = \frac{1}{2}$ because B_i^j is a nonempty proper subset of A_i, we have $\mathbb{P}(|B^j \cap R|$ is odd$) = 2^{-n}$. Hence $\mathbb{E}(X_j) = 2^{-n}$ and $\mathbb{E}(X) = k2^{-n}$.

On the other hand, since $B^1 \ldots, B^k$ is a dissection of $A_1 \times \cdots \times A_n$,

$$X = \sum_j X_j \equiv \sum_j |B^j \cap R| \equiv |R| \equiv 1 \pmod 2.$$

Thus always $X \geq 1$. We conclude $k2^{-n} = \mathbb{E}(X) \geq 1$, so $k \geq 2^n$. ∎

EXERCISES 14.1

14.1.1. (–) Choose a random k-subset of $[n]$, with each k-set equally likely. Determine the probability that the element 1 is chosen.

14.1.2. (–) Consider clubs A and B of size n. Each A sends an invitation to a B chosen at random. What is the asymptotic probability that each B receives one?

14.1.3. (–) Construct a random variable and distribution on $\mathbb{N}$ so that the expectation of the random variable is undefined (infinite).

14.1.4. (–) A university study finds that on any given day, permit holders fail to drive to campus with probability .1, independently. The university therefore sells ten permits for a lot with nine spaces and 20 permits for a lot with 18 spaces. Which lot is more likely to turn away the last permit holder who arrives? (adapted from Grimmett–Stirzaker [1992])

14.1.5. (–) Compute the expectations of the following quantities.
 (a) The number of fixed points in a random permutation of $[n]$.
 (b) The number of male/female pairs when $2n$ people consisting of n men and n women are randomly partitioned into n pairs.
 (c) The number of vertices of degree k in a graph with vertex set $[n]$ where each edge occurs with probability p, independently.

14.1.6. (–) Determine the expected number of monochromatic copies of $K_{r,r}$ in a random 2-coloring of the edges of $K_{n,n}$.

14.1.7. (–) Obtain an exponential lower bound on the minimum number of edges in a k-uniform hypergraph that is not t-colorable.

14.1.8. (–) Prove that some n-vertex tournament has at least $n!/2^{n-1}$ spanning paths. (Comment: Alon [1990] proved that the maximum number of Hamiltonian paths in an n-vertex tournament is at most $n!/(2 - o(1))^n$.) (Szele [1943])

14.1.9. Given a graph G with p vertices, q edges, and s automorphisms, let $n = (sk^{q-1})^{1/p}$. Prove that some k-coloring of $E(K_n)$ has no monochromatic G. (Chvátal–Harary [1973])

14.1.10. *Sperner's Theorem by random variables.* Let F be an antichain of subsets of $[n]$. Let X be the number of members of F that occur as initial segments of a random permutation σ of $[n]$. Use X to prove $|F| \leq \binom{n}{\lceil n/2 \rceil}$.

14.1.11. Fix $k, r \in \mathbb{N}$ with $r \geq k$. Determine the value m such that a random k-coloring of the vertices guarantees that every r-uniform hypergraph with at most m edges has a vertex k-coloring such that every color appears in every edge. (Alon–Spencer [2008])

14.1.12. ($\diamond$) Determine the expected number of descents in a random permutation (Definition 3.1.22). (Comment: Since the number of runs in a permutation is one more than the number of descents, this gives an easier proof for Exercise 3.1.36.)

14.1.13. Let S be a finite set of finite binary words such that none is a prefix of another. Prove that $\sum_{a \in S} 2^{-|a|} \le 1$.

14.1.14. An instance of **SATISFIABILITY** is a list of clauses, where each clause is a set of *literals* (a **literal** is a variable or its negation). The instance is *satisfied* when each variable is set to be TRUE or FALSE so that each clause has at least one true literal.
    (a) Prove that an instance with clauses $e_1, \ldots, e_m$ is satisfiable if $\sum_{i=1}^{m} 2^{-|e_i|} < 1$.
    (b) Construct an unsatisfiable instance with 2^k clauses when all clauses have size k.

14.1.15. ($\diamond$) Let n_k be the least number of vertices in a non-k-choosable bipartite graph, and let m_k be the least number of edges in a non-2-colorable k-uniform hypergraph.
    (a) Let H be a k-uniform hypergraph with n edges. Prove that H is 2-colorable if and only if $K_{n,n}$ is L-colorable when each part has $E(H)$ as its color lists. Conclude $n_k \le 2m_k$.
    (b) Let G be a non-k-choosable X, Y-bigraph. Let H be the hypergraph whose edges are the lists in a k-uniform assignment L such that G has no L-coloring. Prove $m_k \le n_k$. (Erdős–Rubin–Taylor [1979])

14.1.16. ($\diamond$) Apply 2-colorability of hypergraphs to prove that if $k > 1 + \lg n$, then every n-vertex bipartite graph is k-choosable. Obtain the same result when $k > \lg n$.

14.1.17. Let H be a hypergraph with no edges $e_1, \ldots, e_k$ and vertex x such that $e_i \cap e_j = \{x\}$ for $i, j \in [k]$. Prove that H is k-choosable. Strengthen the statement to an analogue of degeneracy in graphs. (Hint: No probability needed.) (Berge [1973], from Tomescu [1968])

14.1.18. A **tripartite 3-uniform hypergraph** is a hypergraph whose vertex set partitions into sets X, Y, and Z such that each edge has one vertex in each of those sets. Construct a tripartite 3-uniform hypergraph that is not k-choosable.

14.1.19. The first passenger to board a plane loses her boarding pass and sits in a random seat. Each subsequent passenger sits in his or her own seat if it is available and otherwise chooses a random seat from those that remain. What is the probability that the last passenger sits in his or her own seat? (Winkler [2004])

14.1.20. ($\diamond$) Prove that a tournament cannot contain three 3-cycles on four vertices. Conclude that the maximum number of edges in an n-vertex 3-uniform hypergraph having no set of four vertices containing three edges is at least $\frac{1}{4}\binom{n}{3}$.

14.1.21. A bag has 1000 tiles; on each is printed one of the 26 letters of the alphabet. Player A picks a tile at random and then replaces it. Player B then picks a tile at random. Prove that the probability of the two players picking the same letter is at least $1/26$.

14.1.22. A communication network is configured as a complete k-ary tree with leaves at distance l from the root. At any time, each node is working with probability p, independently of all others. When a node is not working, the subtree rooted at it is inaccessible. What is the expected number of accessible nodes?

14.1.23. ($\diamond$) Let all n^{n-2} trees with vertex set $[n]$ be equally likely. Define the following random variables: X is the number of leaves, Y is the number of vertices of degree 2, and Z is the number of leaves having neighbors of degree 2. Determine the limits as $n \to \infty$ of $\mathbb{E}(X)/n$, $\mathbb{E}(Y)/n$, and $\mathbb{E}(Z)/n$. (Hint: Use an encoding of the trees as n-ary lists.)

14.1.24. ($\diamond$) A pill bottle has m large pills and n small pills. Each day a patient chooses a random pill; a small pill is consumed, but a large pill is split, with half consumed and the other half becoming a small pill. Determine the expected number of small pills after the last large pill is split and the expected day this occurs. (Knuth–McCarthy [1991])

14.1.25. Let n distinct pairs of socks be put into the laundry. After washing, socks are drawn successively at random (without replacement). Compute the expected number of pairs among the first k socks drawn.

14.1.26. For $n, k \in \mathbb{N}_0$ with $0 \le k \le n$, form a permutation of $[n]$ by choosing the first k positions at random and filling the remaining $n - k$ positions in ascending order. Let $E_{n,k}$ be the expected number of left-to-right maxima (for example, $E_{3,1} = 2$ and $E_{3,2} = 11/6$). Compute $E_{n,k}$. Conclude that $E_{n+1,k} - E_{n,k} = 1/(k+1)$. (Deshpande–Deshpande [2008])

14.1.27. ($\diamond$) *Bipartite subgraphs.*
(a) Use a random partition of the vertices to prove that every graph has a bipartite subgraph with at least half its edges.
(b) Improve part (a) by showing that if G has m edges and n vertices, then G has a bipartite subgraph with at least $\frac{m}{2} \frac{\lceil n/2 \rceil}{\lceil n/2 \rceil - 1}$ edges whose part-sizes differ by at most 1.
(c) Prove that every graph with m edges that has a matching with k edges has a bipartite subgraph with at least $(m+k)/2$ edges. (Molloy–Reed [2002, p. 37])

14.1.28. Let G be an n-vertex graph with average degree d and minimum degree k. Prove that G has an induced subgraph with at least $\frac{kn}{d+1}$ vertices that does not contain K_{k+1}.

14.1.29. ($\diamond$) *Bollobás's Inequality.* Let $\{A_i\}_{i=1}^m$ and $\{B_i\}_{i=1}^m$ be subsets of $[n]$, with $|A_i| = a_i$, $|B_i| = b_i$, and $A_i \cap B_j = \emptyset$ if and only if $i = j$. Prove $\sum_{i=1}^m \binom{a_i+b_i}{a_i}^{-1} \le 1$. Conclude that the maximum size of an antichain in $\mathbf{2}^n$ is $\binom{n}{\lceil n/2 \rceil}$ (Sperner's Theorem). (Hint: Define appropriate events over random permutations of $[n]$.) (Bollobás [1965])

14.1.30. ($\diamond$) The **Coupon Collector Problem (Geometric random variable).**
(a) An experiment has success probability p on each trial, independently. Obtain a simple formula (no summation) for the expected number of trials up to the first success.
(b) Every box of a type of candy contains one of n prizes, each with probability $1/n$. The grand prize requires obtaining each of these at least once. Prove that the expected number of the box on which the last prize is obtained is $n \sum_{i=1}^n 1/i$. (Feller [1968, p. 255])

14.1.31. ($\diamond$) Given a distribution of pebbles to the vertices of a graph, a **pebbling move** removes two pebbles from some vertex and adds one pebble to a neighboring vertex. A vertex is *reachable* from a distribution if some (possibly empty) list of pebbling moves results in it having a pebble. Let D be a distribution on the k-cube Q_k from which every vertex is reachable. Moews [1998] proved that D must have at least $(4/3)^k$ pebbles. Here we develop the proof in Bunde–Chambers–Cranston–Milans–West [2008].
(a) Prove $\sum_{t \ge 0} a_{r,t} 2^{-t} \ge 1$ for any $r \in V(Q_k)$, where $a_{r,t}$ is the number of pebbles in D at distance t from r. Thus $\sum_{t \ge 0} \mathbb{E}(a_{r,t}) 2^{-t} \ge 1$ when r is chosen uniformly at random.
(b) Prove $\mathbb{E}(a_{r,t}) = |D| \binom{k}{t} 2^{-k}$, and conclude $|D| \ge (4/3)^k$.

14.1.32. A coloring of a graph G is r-**dynamic** if it is proper and each vertex neighborhood $N(v)$ has at least $\min\{r, d_G(v)\}$ colors. Prove that if $\delta(G) > \frac{r+s}{s+1} r \ln n$, where $n = |V(G)|$, then G has an r-dynamic coloring with at most $\Delta(G) + r + s$ colors. (Hint: Color the vertices properly in the order $v_1, \ldots, v_n$, at random.) (Jahanbekam–Kim–O–West [2016])

14.1.33. Two players alternately flip a coin that has probability p of landing heads up. The winner is the first to obtain heads. What is the probability that Player 1 wins?

14.1.34. ($\diamond$) A standard random walk moves distance 1 to the left or right on the real line with each unit of time, with probability $\frac{1}{2}$ in each direction.
(a) From position 0, prove that the expected number of steps taken to reach position -1 or position n for the first time is n.
(b) Let X_n be the number of steps until the difference between the maximum and minimum positions visited first equals n. Determine $\mathbb{E}(X_n)$. (Palacios–Sandell [1991])

14.1.35. ($\diamond$) A random walk on a graph G starts at a fixed vertex v and at each step moves to a random neighbor of the current vertex, chosen uniformly. When G is a cycle, prove that all vertices other than v are equally likely to be the last vertex visited for the first time. (Comment: This folklore result appeared in Lovász–Winkler [1993], which also showed that cycles and complete graphs are the only graphs with this property.)

14.1.36. An unbiased coin is flipped until k heads occur. Let X be the number of isolated heads in the list, and let Y be the number of runs (when $k = 4$, the list $HHTTHTTH$ yields $X = 2$ and $Y = 5$). Compute $\mathbb{E}(X)$ and $\mathbb{E}(Y)$. (Bhanu–Deshpande [2010])

14.1.37. Consider a committee having both Senators from each of n states. Partition the $2n$ people into pairs at random, with each matching equally likely. All pairs are allowed.
    (a) What is the probability that no Senators from the same state are paired?
    (b) What is the limit of this probability as $n \to \infty$?

14.1.38. ($\diamond$) An urn has n balls, of which k are white. A ball is chosen at random until a white ball is obtained. Determine the expected number of the step on which the first white ball is obtained, under each of the following two scenarios.
    (a) Non-white balls are put back in the urn before the next drawing.
    (b) Non-white balls are discarded before the next drawing.

14.1.39. Let Q be a set of n points chosen independently and uniformly in the unit square. Find the expected number of sets $S \subseteq Q$ such that there is an increasing curve with S on or below the curve and $Q - S$ above it. (Stanley–Steele [1989])

14.1.40. ($\diamond$) Let $v_1, \dots, v_n$ be unit vectors in $\mathbb{R}^n$. Prove that $\varepsilon_1, \dots, \varepsilon_n$ can be chosen in $\{-1, +1\}$ so that $|\sum \varepsilon_i v_i| \geq \sqrt{n}$ and also so that $|\sum \varepsilon_i v_i| \leq \sqrt{n}$. Prove that both results are sharp. (Alon–Spencer [1992])

14.1.41. ($\diamond$) Fix $p_1, \dots, p_n \in [0, 1]$, and let $v_1, \dots, v_n$ be unit vectors in $\mathbb{R}^n$. Let $w = \sum p_i v_i$. Prove that some subset of $v_1, \dots, v_n$ sums to a vector u such that $|w - u| \leq \sqrt{n}/2$. (Alon–Spencer [1992])

14.1.42. ($\diamond$) Let A be an n-set in $\mathbb{Z}_{n^2}$. Prove that there is an n-set B in $\mathbb{Z}_{n^2}$ such that at least half of $\mathbb{Z}_{n^2}$ can be expressed in the form $a + b$ with $a \in A$ and $b \in B$.

14.1.43. (+) A set S is **sum-free** if no distinct $x, y, z \in S$ satisfy $x + y = z$. Prove that every set A of nonzero integers contains a sum-free set S with $|S| > |A|/3$. (Erdős [1965])

14.1.44. Let $A_1, \dots, A_t$ be events in a space of n equally likely outcomes, with $p_i = \mathbb{P}(A_i) = |A_i|/n$ and $s = \sum_{i=1}^{t} p_i$. Given $r \in \mathbb{N}$ with $1 \leq r \leq s$, prove that for some set of r events in $\{A_i\}$, the probability that they all occur is at least $\frac{s_{(r)}}{r!} / \binom{t}{r}$. (Ford [1994])

14.1.45. Determine the probability that n points independently chosen uniformly at random on a circle all lie in a semicircle. (Comment: This result appeared in at least three papers and six books by the 1950s; Jordan [1872/1873] and Wendel [1962] generalized it.)

14.1.46. Fix $n \geq 1$. For each of n independent trials, a ball is placed in some indexed box; it is box j with probability 2^{-j}, for $j \in \mathbb{N}$. Let X denote the number of empty boxes below the highest-indexed occupied box. Compute $\mathbb{E}(X)$. (Comment: Ferguson–Melolidakis [1984] determined the complete distribution of X for each n.)

14.1.47. ($\diamond$) A **binary maze** is a digraph in which every vertex has two exiting edges, one labeled 0 and one labeled 1. A *search list* specifies the label of each successive edge to be followed. A search list b is n-**universal** if for every vertex u in every n-vertex strongly connected binary maze G, following b from u reaches every vertex of G. Prove that the minimum length of an n-universal search list is between 2^{n-1} and $n^2 2^n \ln n$. (Hint: Bound the number of triples (G, u, v) such that G is a maze and a random binary list of length m never reaches v when starting from u.) (Knuth [1999])

14.2. Refinements of Basic Methods

In this section we continue to study existence-type arguments, introducing some refinements that lead to stronger results.

DELETIONS AND ALTERATIONS

When a randomly generated object is close to having a desired property, a slight alteration may produce that property. This technique has been called the **deletion method**, the **alteration principle**, or the **two-step method**. Ramsey numbers provide a classic example.

14.2.1. Theorem. $R(k, k) > n - \binom{n}{k} 2^{1-\binom{k}{2}}$, where $n \in \mathbb{N}$. In particular, $R(k, k) > (1 - o(1))\frac{k}{e} 2^{k/2}$.

Proof: Let X be the number of homogeneous k-sets in a random 2-coloring of $E(K_n)$. Expressing X as a sum of indicator variables for each k-set and applying linearity yields $\mathbb{E}(X) = \binom{n}{k} 2^{1-\binom{k}{2}}$. The pigeonhole property guarantees a coloring with at most this many homogeneous k-sets. By deleting a vertex of each homogeneous k-set, we retain a graph with at least the specified number of vertices but no homogeneous k-set.

Thus $R(k, k) > n - \mathbb{E}(X)$, for each n. Since $\binom{n}{k} < (\frac{ne}{k})^k$, we also have the simpler $n - (\frac{ne}{k})^k 2^{1-k(k-1)/2}$ as a lower bound. We seek n to maximize this bound. Differentiating suggests choosing n to satisfy $1 = k\frac{e}{k}(\frac{ne}{k})^{k-1} 2^{1-k(k-1)/2}$, which requires $n = e^{-1} k 2^{k/2} (2e)^{-1/(k-1)}$. The factor $(2e)^{-1/(k-1)}$ is near 1 when k is large.

For simplicity, we set $n = e^{-1} k 2^{k/2}$ (actually a nearby integer) to obtain the claimed bound. We have

$$n - \left(\frac{ne}{k}\right)^k 2^{1-k(k-1)/2} = \frac{k}{e} 2^{k/2} - 2^{k^2/2} 2^{1-k(k-1)/2} = \frac{1}{e} k 2^{k/2} \left(1 - \frac{2e}{k}\right).$$

Since $2e/k$ tends to 0 for large k, the bound is as claimed.   ■

Deletion arguments can improve basic existence results, although here the improvement from Theorem 14.1.3 is only by a factor of $\sqrt{2}$. Both arguments can also be applied to $R(k, l)$ with $k \neq l$ (Exercise 16).

Next we *add* to a random almost-good-enough structure. In a graph G, a set $S \subseteq V(G)$ is **dominating** if every vertex outside S has a neighbor in S.

14.2.2. Theorem. (Arnautov [1974], Payan [1975]) For $k > 1$, every n-vertex graph with minimum degree k has a dominating set of size at most $n\frac{1+\ln(k+1)}{k+1}$.

Proof: (Alon [1990]) Form a random vertex subset S in such a graph by including each vertex independently with probability $p = \frac{\ln(k+1)}{k+1}$. Given S, let T be the set of vertices outside S having no neighbor in S; adding T to S yields a dominating set. The experiment provides both S and T; we seek the expected size of the union as an upper bound on the minimum size of a dominating set.

Since vertices appear in S with probability p, linearity yields $\mathbb{E}(|S|) = np$. The random variable $|T|$ is the sum of n indicator variables for whether individual vertices belong to T. We have $v \in T$ if and only if v and its neighbors all fail to be in S. This has probability bounded by $(1-p)^{k+1}$, since v has degree at least k. Since $(1-p)^{k+1} < e^{-p(k+1)}$, we have $\mathbb{E}(|S| + |T|) \leq np + ne^{-p(k+1)} = n\frac{1+\ln(k+1)}{k+1}$. The pigeonhole property of the expectation completes the proof. ∎

A greedy algorithm that iteratively adds the vertex dominating the most currently undominated vertices proves the same upper bound constructively (Exercise 8). The coefficient on n in Theorem 14.2.2 is asymptotically sharp; Alon–Wormald [2010] gave a probabilistic construction of k-regular graphs with no dominating set of size less than $(1 - o(1))\frac{\ln k}{k}n$.

The most famous use of the deletion method may be Erdős's proof that graphs with large girth and chromatic number exist. Theorem 10.1.31 presents an explicit construction. Here we present Spencer's simplification of Erdős' proof, with apologies for advance use of Lemma 14.3.8.

14.2.3. Theorem. (Erdős [1959]) Given $k \geq 3$ and $g \geq 3$, there exists a graph with girth at least g and chromatic number at least k.

Proof: We generate graphs with vertex set $[n]$ and edge probability p. Since $\chi(G) \geq \frac{n}{\alpha(G)}$, we choose p large to make large independent sets unlikely. We also choose p small to make the expected number of short cycles (length less than g) small. From a graph satisfying both conditions, we will delete a vertex from each short cycle to obtain the desired graph.

To have few short cycles, let $p = n^{t-1}$, where $t < 1/g$. Each possible cycle of length j occurs with probability p^j. By choosing and cyclically ordering j vertices, $\frac{n_{(j)}}{2j}$ such cycles are possible, so the total number X of cycles of length less than g satisfies

$$\mathbb{E}(X) = \sum_{j=3}^{g-1} \frac{n_{(j)}}{2j} p^j \leq \sum_{j=3}^{g-1} \frac{n^{tj}}{2j} < gn^{(g-1)/g}.$$

With $\mathbb{E}(X) < gn^{1-1/g}$, we have $\mathbb{E}(X)/n \to 0$ as $n \to \infty$. From Markov's Inequality (Lemma 14.3.8), we conclude $\mathbb{P}(X \geq \frac{n}{2}) \to 0$ as $n \to \infty$. For n large, $\mathbb{P}(X \geq \frac{n}{2}) < \frac{1}{2}$.

Since $\alpha(G)$ cannot grow when we delete vertices, at least $\frac{n-X}{\alpha(G)}$ independent sets are needed to color the remaining graph after we delete a vertex of each short cycle. If $X < \frac{n}{2}$ and $\alpha(G) \leq \frac{n}{2k}$, then at least k colors are needed for the graph remaining. By the Union Bound,

$$\mathbb{P}(\alpha(G) \geq r) \leq \binom{n}{r}(1-p)^{\binom{r}{2}} < [ne^{-p(r-1)/2}]^r.$$

The bound tends to 0 as $n \to \infty$ when, for example, $r = \lceil \frac{3}{p} \ln n \rceil$.

With $r = \lceil 3n^{1-t} \ln n \rceil$ and k fixed, we can choose n large enough to obtain $r < \frac{n}{2k}$. If also n is large enough so that $\mathbb{P}(X \geq \frac{n}{2}) < \frac{1}{2}$ and $\mathbb{P}(\alpha(G) \geq r) < \frac{1}{2}$, then there exists an n-vertex graph G such that (1) $\alpha(G) \leq \frac{n}{2k}$ and (2) G has fewer than $\frac{n}{2}$ cycles of length less than g. We delete a vertex from each short cycle and retain a graph with girth at least g and chromatic number at least k.   ∎

The deletion method is also used in the proof of a technique called "dependent random choice". To motivate this, recall that the Turán number $\text{ex}(n; H)$ is the maximum number of edges in an n-vertex graph not containing H. In Theorem 13.2.19, a counting argument showed $\text{ex}(n; K_{r,s}) \leq cn^{2-1/r}$ when $r \leq s$ (Kóvari–Sós–Turán [1954]), which is the right order of growth when $s > r!$ (Theorem 13.2.25, Kollár–Rónyai–Szabó [1996]). Using a counting argument, Füredi [1991] extended this bound on $\text{ex}(n; H)$ to every A, B-bigraph H such that vertices in B have degree at most r. Here we develop a later probabilistic proof.

The elementary counting argument shows that if G has too many edges, then some set of r vertices has s common neighbors, yielding $K_{r,s}$. That is, with nonzero probability a random r-set from $V(G)$ has at least s common neighbors. In the Dependent Random Choice Lemma, we introduce some dependence among the chosen vertices so that r-sets with many common neighbors are more likely to be kept. In this way a smaller number of edges yields nonzero probability of having many r-sets with at least s common neighbors. The comparison with Theorem 14.2.3 is that there we deleted a vertex from each bad cycle (too short), while here we will delete a vertex from each bad r-set (too few common neighbors).

Versions of the lemma were proved in Kostochka–Rödl [2004] and Gowers [1998]. We give roughly the proof by Alon–Krivelevich–Sudakov [2003] (see Alon–Spencer [2008], Lecture 2 of Lee [2015], and Fox–Sudakov [2011]; the latter is an excellent survey presenting a variety of applications of the technique).

14.2.4. Lemma. (**Dependent Random Choice**) Let G be an n-vertex graph with m edges. For $a, b \in \mathbb{N}$, if there is a positive integer t such that

$$\frac{(2m/n)^t}{n^{t-1}} - \binom{n}{r}\left(\frac{b}{n}\right)^t \geq a$$

then G contains a set U of at least a vertices such that every r vertices in U have at least b common neighbors in G.

Proof: We use linearity of expectation repeatedly in the proof.

Let T consist of t vertices chosen independently at random (repetition allowed). Let S be the set of vertices neighboring all of T, and let $X = |S|$. Since $v \in S$ if and only if $T \subseteq N_G(v)$, we have $\mathbb{P}(v \in S) = (d(v)/n)^t$. Therefore $\mathbb{E}(X) = n^{-t}\sum_{v \in V(G)} d(v)^t$. By convexity, $\frac{1}{n}\sum d(v)^t \geq (\sum d(v)/n)^t$, so $\mathbb{E}(X) \geq \frac{(2m/n)^t}{n^{t-1}}$.

Now let Y count the r-subsets of S having fewer than b common neighbors; we want to eliminate such sets. A given r-set R lies in S if and only if each vertex of T lies in $N(R)$. If $|N(R)| < b$, then $\mathbb{P}(R \subseteq S) < \left(\frac{b}{n}\right)^t$. This is where dependence helps; r-sets having more common neighbors are more likely to lie in S. Considering all r-sets, we have $\mathbb{E}(Y) < \binom{n}{r}\left(\frac{b}{n}\right)^r$.

Under the given hypothesis, $\mathbb{E}(X-Y) \geq a$. Hence $X - Y \geq a$ for some outcome T of the experiment. If we delete, from the resulting set S, one vertex of each r-set having fewer than b common neighbors, we will still be left with a set U of at least a vertices in which every r-set has at least b common neighbors.   ∎

14.2.5. Theorem. (Füredi [1991]) If H is an A, B-bigraph such that $d_H(v) \leq r$ for all $v \in B$ then $\operatorname{ex}(n; H) < c_H n^{2-1/r}$ for some constant c_H.

Proof: (Alon–Krivelevich–Sudakov [2003]) Let $a = |A|$ and $b = |V(H)|$, and set $t = r$. Let G be an n-vertex graph with at least $\frac{1}{2} cn^{2-1/t}$ edges, where $c^t \geq \frac{b^t}{r!} + a$. This hypothesis on the constant c yields the hypothesis on G for Lemma 14.2.4.

Lemma 14.2.4 therefore provides a set U of at least a vertices in G such that every r-subset of U has at least b common neighbors. We now find H in G. Use a vertices of U as the vertices of A, assigned arbitrarily. Now, each vertex of B needs to be made adjacent to some set of at most r vertices of A. For each vertex of B in turn, find an r-set of vertices mapped to A containing its desired neighbors, and find a common neighbor of this r-set that has not yet been used. Since the r-set has at least b common neighbors, and $b = |V(H)|$, there is always such a neighbor available until the process is finished.

Hence it suffices to let $c_H = c/2$. ∎

Other applications of dependent random choice appear in Exercises 12–13. Fox–Sudakov [2011] described further enhancements of the method with many impressive applications in additive number theory and extremal graph theory. We mention only a result on Ramsey numbers. For $d \in \mathbb{N}$, Theorem 11.1.18 guarantees a constant c_d such that $R(G, G) \leq c_d |V(G)|$ whenever $\Delta(G) = d$. Proved using the Regularity Lemma, the constant there is a tower of height d. Remark 11.1.19 discusses improvements avoiding the Regularity Lemma. In particular, Fox–Sudakov [2009] used dependent random choice to prove $R(G, G) \leq 8d2^d |V(G)|$ when G is bipartite with $\Delta(G) = d \geq 1$; this is essentially sharp by a construction of Graham–Rödl–Rucinski [2001].

We consider one more alteration method: a second random step. When a random coloring is bad in a few places, we can recolor those elements. For the smallest non-2-colorable k-uniform hypergraph, Beck [1978] used this technique to improve the simple lower bound 2^{k-1} (Proposition 14.1.6) to $\Omega(2^k k^{1/3})$. We present a simplification from Alon–Spencer [1992] that yields a slightly weaker lower bound. A further improvement to $\Omega(2^k k^{1/2}(\ln k)^{-1/2})$ by Radhakrishnan–Srinivasan [2000] appears in Alon–Spencer [2000, 2008], with another proof by Kozik–Cherkashin [2015] in Alon–Spencer [2016].

14.2.6.* Theorem. (Beck [1978]) If $2te^{-pk} + t^2 pe^{pk} < 1$, where $0 \leq p \leq \frac{1}{2}$, then every k-uniform hypergraph with $2^{k-1}t$ edges is 2-colorable.

Proof: (Alon–Spencer [1992]) Given such p and t, consider a k-uniform hypergraph H with $2^{k-1}t$ edges. Let $f: V(H) \to \{0, 1\}$ be a random 2-coloring; each vertex gets each color with probability $\frac{1}{2}$, independently. Call a monochromatic edge a *bad* edge. For each vertex in at least one bad edge under f, change its color with probability p (independently), obtaining a new coloring f'. We show that with positive probability f' is a proper 2-coloring.

First consider the probability that an edge S is bad under both f and f'. Colors under f are the same, and then the vertices in S all keep or all change their color. We compute

$$2 \cdot 2^{-k}(p^k + (1-p)^k) \leq 2^{1-k} \cdot 2(1-p)^k \leq 2^{1-k} \cdot 2e^{-pk}.$$

With $2^{k-1}t$ edges, this occurs for some edge with probability at most $2te^{-pk}$, by the Union Bound.

Now consider an edge S that is bad (in one color) under f' but not f. Since S requires a color change to go bad, some edge T intersecting S has the other color on all its vertices (under f). Given edges S and T that intersect, let $A_{S,T}$ be the event that S becomes monochromatic and the vertices of $S \cap T$ change color.

Let U be the subset of S that changes color in the recoloring step; occurrence of $A_{S,T}$ requires $S \cap T \subseteq U$. The original coloring f gives one color to all of $U \cup T$ and the other to all of $S - U$; this has probability $2(\frac{1}{2})^{2k-r}$, where $r = |S \cap T|$. With $j = |U|$, the probability that the colors on U change in the recoloring step is at most p^j, and the probability of no change on $S - U$ is at most 1.

For specified S, T, U, the probability is thus at most $2^{1-2k+r}p^j$. We sum over U with $S \cap T \subseteq U \subseteq S$ to bound $\mathbb{P}(A_{S,T})$. Letting $i = j - r$ and using the Binomial Theorem and $1 + x < e^x$, the bound becomes

$$\sum_{i=0}^{k-r} \binom{k-r}{i} \left(2^{1-2k+r}p^r\right) p^i = 2^{1-2k+r}p^r(1+p)^{k-r} \le 2^{1-2k}e^{pk}\left(\frac{2p}{1+p}\right)^r.$$

Since $0 \le p \le \frac{1}{2}$, the last expression is maximized when $r = 1$. Thus $\mathbb{P}(A_{S,T}) \le 2^{2-2k}e^{pk}p$ for every intersecting pair (S, T) of edges. Since there are fewer than $(2^{k-1}t)^2$ intersecting ordered pairs of edges, by the Union Bound the probability that some edge is bad under f' but not under f is at most $t^2e^{pk}p$, which is the second term in the hypothesis.

Therefore, when the specified hypotheses hold, the probability that f' is a proper 2-coloring is positive. ∎

14.2.7.* Corollary. For $\varepsilon > 0$, when k is sufficiently large the minimum number of edges in a k-uniform hypergraph that is not 2-colorable is at least $(\sqrt{3} - \varepsilon)(2^{k-1}k^{1/3}(\ln k)^{-1/2})$.

Proof: Choosing $t < (\sqrt{3} - \varepsilon)k^{1/3}(\ln k)^{-1/2}$ and $p = \frac{\ln k}{3k}$ yields the inequality $2te^{-pk} + t^2pe^{pk} < 1$ needed for Theorem 14.2.6. ∎

THE SYMMETRIC LOCAL LEMMA

In simple existence arguments showing that the probability of having some bad event is less than one, parameters are chosen so weakly that even when the events are *disjoint* there remains probability outside their union. In constrast, independent events always have nonzero probability in the complement. We seek a method that can accommodate some lack of independence and still guarantee nonzero probability outside all the events when the events have sufficiently small probability (we will make this precise).

The resulting technique is the **Lovász Local Lemma** or simply the **Local Lemma**. Discussion of the lemma and its applications appears in Alon–Spencer [1992, 2000, 2008, 2016], Molloy–Reed [2002], and Szegedy [2013]. We follow Molloy–Reed by starting with the simpler symmetric version that is easier to understand and suffices for many applications. Subsequently we prove the general version (Theorem 14.2.18) and derive the symmetric version from it.

14.2.8. Definition. Let $A_1, \ldots, A_n$ be events. For disjoint subsets S and T of $[n]$, the **compound event** $A(S, T)$ specifies the occurrence of A_i for $i \in S$ and the non-occurrence of A_j for $j \in T$. An event B is **mutually independent** of $A_1, \ldots, A_n$ if B is independent of each compound event specified by disjoint subsets of $[n]$ (meaning $\mathbb{P}(B \mid A(S, T)) = \mathbb{P}(B)$).

14.2.9. Theorem. (**Symmetric Local Lemma**; Erdős–Lóvasz [1975]) Consider events $A_1, \ldots, A_n$ such that (1) $\mathbb{P}(A_i) \leq p$ for all i, and (2) each A_i is mutually independent of a set of all but d events. If $\mathrm{e}pd \leq 1$, then $\mathbb{P}\left(\bigcap \overline{A_i}\right) > 0$.   ∎

Shearer [1985] proved that replacing the constant e with any smaller constant in the condition $\mathrm{e}pd \leq 1$ no longer guarantees $\mathbb{P}\left(\bigcap \overline{A_i}\right) > 0$.

Most treatments of the Symmetric Local Lemma replace "all but d events" with "all but d other events") but then must require $\mathrm{e}p(d+1) \leq 1$. We phrase the condition using "all but d events" because the formula to count events omitted from the set that is mutually independent of A_i often also counts A_i.

The Local Lemma has many applications. We use it first to improve the lower bound on diagonal Ramsey numbers produced by the elementary existence argument, but only by a factor of 2. Gaining only a factor of 2 is disappointing; the Local Lemma is more effective when dependence is rarer. The improvement for van der Waerden numbers is more dramatic (Exercise 22). The general version of the Local Lemma does provide a significant improvement for off-diagonal Ramsey numbers with one parameter fixed (see Theorem 14.2.22).

14.2.10. Theorem. (Spencer [1975]) $R(k, k) > (1 + o(1)) \frac{\sqrt{2}}{\mathrm{e}} k 2^{k/2}$.

Proof: Form a random 2-coloring of $E(K_n)$, as usual. For each k-set S of vertices, let A_S be the event that S is homogeneous. Knowing the color of all edges outside $\binom{S}{2}$ has no effect on the probability of A_S. Hence we can let d in Theorem 14.2.9 be the number of k-sets in $[n]$ (including S) that share at least two elements with S; the event A_S is mutually independent of the set of all other events.

To obtain an upper bound on d, pick two elements of S and pick $k-2$ elements outside these two. Sets that share more than two elements with S are counted many times. Nevertheless,

$$d < \binom{k}{2}\binom{n-2}{k-2} < \binom{k}{2}\binom{n}{k-2} < \frac{k^2}{2}\left(\frac{n\mathrm{e}}{k-2}\right)^{k-2}.$$

As in Theorem 14.1.3, $\mathbb{P}(A_S) = 2 \cdot 2^{-\binom{k}{2}}$ for all S; let this probability be p.

Fix k, so p is fixed. By making n small, we can make d small. To apply Theorem 14.2.9, it suffices to make n small enough so that

$$\frac{k^2}{2}\left(\frac{n\mathrm{e}}{k-2}\right)^{k-2} < \frac{1}{\mathrm{e}p} = \frac{1}{2\mathrm{e}} 2^{k/2}(2^{k/2})^{k-2}.$$

When this holds, the probability of having no monochromatic copy of K_k is nonzero, and hence $R(k, k) > n$.

Since $2^{k/2} = 2\sqrt{2}^{k-2}$, it suffices to have $n \leq c \frac{\sqrt{2}}{\mathrm{e}} k 2^{k/2}$, where $c = \left(\frac{2}{\mathrm{e}k^2}\right)^{1/(k-2)} \frac{k-2}{k}$. Since $c \to 1$ as $k \to \infty$, the claimed bound holds.   ∎

In the proof of Theorem 14.2.10, we could tighten the bound on d to $\binom{n}{k} - \binom{n-k}{k} - k\binom{n-k}{k-1}$, but this would not asymptotically improve the result.

14.2.11. Remark. *Independence vs. mutual independence.* It is wrong to say "B is independent of $A_1, \ldots, A_n$" when applying the Local Lemma, because it is not generally true that B is mutually independent of $A_1, \ldots, A_n$ when B is independent of each of $A_1, \ldots, A_n$.

For example, in Theorem 14.2.10, each event A_S is (pairwise) independent of all events for sets sharing at most two vertices with it, because when the color of one edge is known, the probability that S is homogeneous remains $(\frac{1}{2})^{\binom{k}{2}-1}$. However, A_S is not mutually independent of that set of events. Let T and U be k-sets such that $S \cap T = \{v, x\}$, $S \cap U = \{v, y\}$, and $T \cap U = \{v, z\}$ (see figure below). The events A_S, A_T, and A_U are pairwise independent, as noted above. However, A_S is not mutually independent of $\{A_T, A_U\}$. If A_T and A_U both occur, then vx and vy have the same color, and $\mathbb{P}(A_S \mid \{A_T, A_U\}) = (\frac{1}{2})^{\binom{k}{2}-2} \neq \mathbb{P}(A_S)$.

An intuitive illustration of pairwise independence not guaranteeing mutual independence is given by the "Borromean rings", shown below on the right. Any two of the three rings are "independent" in the sense that they are not linked. However, the three rings cannot be separated. ∎

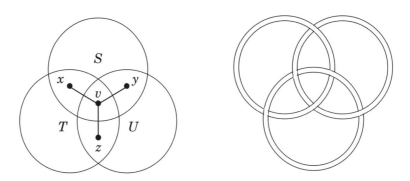

The Local Lemma is effective when the sets on which an event A_i depends are somehow "near" A_i; hence the name "Local Lemma". Often the needed statement about mutual independence uses the following principle, named by Molloy and Reed. It describes a setting under which mutual independence is automatic.

14.2.12. Proposition. (**Mutual Independence Principle**) Let $Z_1, \ldots, Z_m$ be independent experiments and $A_1, \ldots, A_n$ be events such that occurrence of each A_i is determined by a subset S_i of $Z_1, \ldots, Z_m$. If S_i is disjoint from $S_{j_1} \cup \cdots \cup S_{j_k}$, then A_i is mutually independent of $\{A_{j_1}, \ldots, A_{j_k}\}$. ∎

Intuitively, Proposition 14.2.12 is making the observation that no specification of whether each of $A_{j_1}, \ldots, A_{j_k}$ does or does not occur can affect the probability of A_i. In Theorem 14.2.10, $Z_1, \ldots, Z_m$ are the independent "coin-flips" used to determine the colors on the m edges of K_n. Hence Proposition 14.2.12 applies to show that A_S is mutually independent of all sets of events whose members share at most one edge with S.

We next apply the Local Lemma to list coloring. As in Proposition 8.1.12, greedy coloring with vertices in a least-last degree order shows that every k-degenerate graph is $(k + 1)$-choosable. This bound makes no use of the structure of the vertex lists. When the color lists do not overlap much, smaller lists suffice, regardless of the vertex degrees.

14.2.13. Theorem. (Reed [1999]) Let L be a list assignment for a graph G such that each list has size at least k. If every color appears in at most $\frac{k}{2e}$ of the lists in any vertex neighborhood, then G has a proper coloring from the lists.

Proof: We may assume that each list has size exactly k. Form a random coloring by choosing uniformly and independently from each list. For each edge xy and color $c \in L(x) \cap L(y)$, let $A_{xy,c}$ be the event that x and y both receive color c. The probability p of each such event is k^{-2}.

Occurrence of an event $A_{xy,c}$ is determined by the colors chosen for x and y. By the Mutual Independence Principle, $A_{xy,c}$ is mutually independent of the set of all these events indexed by edges not incident to xy. We obtain an upper bound on d by counting the events corresponding to edges incident to xy. Note that there is no bound on the vertex degrees.

For an edge $x'y'$ incident to xy, the event $A_{x'y',c'}$ is defined if and only if c' is in $L(x') \cap L(y')$. By symmetry, suppose that $x' = x$ and $y' \neq y$. The hypothesis allows at most $\frac{k}{2e}$ such events for each color $c' \in L(x)$, one of which is eliminated as $A_{xy,c}$. The same computation holds for edges incident to y. We conclude that each event $A_{xy,c}$ is mutually independent of a set of all but $d - 1$ of the other events, where $d < k^2/e$. Since $epd < 1$, the Local Lemma implies that in some coloring no such event occurs, and hence G has a proper coloring from the lists. ∎

Choosing appropriate events for application of the Local Lemma can be tricky. In Theorem 14.2.13, one might try defining an event A_{xy} for having the same color on x and y, but this fails (Exercise 24). Molloy–Reed [2002] gave various applications of Theorem 14.2.13. By other techniques, Haxell [2001] showed that the conclusion still holds when each color appears in at most $k/2$ lists in each neighborhood. By iterative use of the Local Lemma, Reed–Sudakov [2002] relaxed this to $k(1 - o(1))$. Reed [1999] conjectured that $k - 1$ suffices.

Before proving the general Local Lemma, we apply the symmetric version again. Consider decomposition of graphs into **linear forests**, which are forests whose components are paths. The minimum number of such forests needed is the **linear arboricity** of the graph.

14.2.14. Conjecture. (Akiyama–Exoo–Harary [1981]) Every k-regular graph has linear arboricity at most $\lceil (k + 1)/2 \rceil$. ∎

The bound is sharp, because a k-regular n-vertex graph has $\frac{1}{2}kn$ edges, and each linear forest contains at most $n - 1$. Alon [1988] used the Local Lemma to prove the conjecture asymptotically, showing that for some constant c the linear arboricity is always at most $\frac{1}{2}k + ck^{3/4}(\ln k)^{1/2}$.

We present only the proof of the exact bound for graphs with large girth. As a lemma, we need a variation on the greedy selection that finds independent sets

678   Chapter 14: The Probabilistic Method

of size $\frac{n}{k+1}$. We will be content with smaller independent sets but have some control on the locations of the vertices. (Haxell [2001] strengthened this result: the sets only need to have size at least $2k$.)

14.2.15. Lemma. (Alon [1994]) Given a graph G with maximum degree k and a partition of $V(G)$ into sets of size at least $2ek$, there is an independent set in G consisting of one vertex from each set in the partition.

Proof: Let $a = \lceil 2ek \rceil$, and let $V_1, \ldots, V_r$ be the partition. By discarding vertices from G, we may assume that each V_i has size a. Pick one vertex from each V_i, uniformly and independently. We want the resulting set S to have nonzero probability of being independent.

For each edge xy with endpoints in distinct parts, let A_{xy} be the event that $x, y \in S$. We have $\mathbb{P}(A_{xy}) = \frac{1}{a^2}$. By the Mutual Independence Principle, if $x \in V_i$ and $y \in V_j$, then A_{xy} is mutually independent of the set of events for all edges with both endpoints outside $V_i \cup V_j$. The bound on d for the application of the Local Lemma is thus $2ak - 1$. It suffices to have $epd < 1$. We have $epd < e\frac{2ak}{a^2} < 1$, by the choice of a. ∎

Lemma 14.2.15 implies, for example, that when $11t$ points on a circle are colored, with 11 points in each of t colors, there is a set of t points with distinct colors such that no two are consecutive among the colored points.

Every connected $2k$-regular graph has an Eulerian circuit and hence an orientation in which indegree and outdegree both equal k at every vertex. Such a digraph is called k-**regular**. Decomposing the resulting digraph into $k+1$ forests of directed paths yields the desired decomposition of the original $2k$-regular graph. Thus it suffices to prove the following.

14.2.16. Conjecture. (Nakayama–Peroche [1987]) Every k-regular digraph decomposes into $k + 1$ linear forests. ∎

14.2.17. Theorem. (Alon [1988]) Every k-regular digraph having no cycle of length less than $8ek$ decomposes into $k + 1$ linear forests.

Proof: Let G be such a digraph. Let H be the k-regular bipartite graph obtained by splitting each vertex v of G into two vertices v^- and v^+, inheriting the incoming and outgoing edges at v, respectively. The Marriage Theorem (Corollary 6.1.6) decomposes H into 1-factors, which correspond to a decomposition of G into 1-regular subdigraphs (the argument is the same as that in Theorem 6.2.13).

Each subgraph in the decomposition of G is a spanning union of disjoint cycles. Deleting one edge from each cycle in each factor yields k linear forests. If the deleted edges form a matching, then they form a linear forest to complete the desired decomposition.

Let G' be the underlying undirected graph of G. A matching in G corresponds to an independent set in the line graph H of G'. Since G' is $2k$-regular, H is $(4k - 2)$-regular. Let $V_1, \ldots, V_r$ be the partition of $V(H)$ in which each block is the edge set of one of the cycles in the decomposition of G into 1-regular factors. By the hypothesis, each V_i has size at least $8ek$. By Lemma 14.2.15, H has an independent set with one vertex in each V_i. This independent set forms the desired matching in G. ∎

THE GENERAL LOCAL LEMMA (Optional)

The proof of the general form of the Local Lemma is a subtle but short induction using conditional probability (see Definition 0.10). The set D_i includes indices of all events other than A_i on which A_i is dependent, and possibly more.

14.2.18. Theorem. (**General Lovász Local Lemma**; Erdős–Lovász [1975])
Consider events $A_1, \ldots, A_n$ in a probability space. For each i, let $D_i \subseteq [n]-\{i\}$ be a set such that A_i is mutually independent of $\{A_j: j \notin D_i \cup \{i\}\}$. If there are weights $x_1, \ldots, x_n$ such that $0 \le x_i < 1$ and $\mathbb{P}(A_i) \le x_i \prod_{j \in D_i}(1 - x_j)$ for all i, then $\mathbb{P}\left(\bigcap_{t=1}^n \overline{A}_t\right) > 0$.

Proof: To simplify the notation, let I_S denote $\bigcap_{j \in S} \overline{A}_j$, the event that no event indexed by S occurs. To prove $\mathbb{P}\left(\bigcap_{t=1}^n \overline{A}_t\right) > 0$, it suffices to prove both $\mathbb{P}(I_S) > 0$ and $\mathbb{P}\left(A_t \mid I_S\right) \le x_t$ whenever $t \notin S \subseteq [n]$, since then

$$\mathbb{P}\left(\bigcap_{t=1}^n \overline{A}_t\right) = \prod_{t=1}^n \mathbb{P}\left(\overline{A}_t \mid I_{[t-1]}\right) \ge \prod_{t=1}^n (1 - x_t) > 0.$$

We use induction on $|S|$. For $|S| = 0$, both claims are immediate: $\mathbb{P}(I_\varnothing) = 1$ and $\mathbb{P}(A_t \mid I_\varnothing) = \mathbb{P}(A_t) \le x_t$. Now assume $|S| > 0$.

For the first claim, choose $t \in S$ and let $S' = S - \{t\}$. Using the induction hypothesis, $\mathbb{P}(I_S) = \mathbb{P}(\overline{A}_t \mid I_{S'})\mathbb{P}(I_{S'}) > 0$.

For the second claim, choose $t \notin S$ and let $T = S \cap D_t$. If $T = \varnothing$, then A_t is mutually independent of $\{A_j: j \in S\}$, and hence $\mathbb{P}(A_t \mid I_S) = \mathbb{P}(A_t) \le x_t$. For larger T, since $I_S = I_T \cap I_{S-T}$, the definition of conditional probability yields

$$\mathbb{P}\left(A_t \mid I_S\right) = \frac{\mathbb{P}(A_t \cap I_S)}{\mathbb{P}(I_S)} = \frac{\mathbb{P}(A_t \cap I_T \mid I_{S-T})\mathbb{P}(I_{S-T})}{\mathbb{P}(I_T \mid I_{S-T})\mathbb{P}(I_{S-T})} = \frac{\mathbb{P}(A_t \cap I_T \mid I_{S-T})}{\mathbb{P}(I_T \mid I_{S-T})}. \qquad (*)$$

Intuitively, $A_t \mid I_S$ is the conditional event that results from conditioning on I_T within the probability space obtained by restricting to I_{S-T}.

We seek an upper bound on $\mathbb{P}(A_t \mid I_S)$, expressed as a ratio in $(*)$. The numerator is bounded above by $\mathbb{P}(A_t \mid I_{S-T})$, but this equals $\mathbb{P}(A_t)$ since $S - T \subseteq \overline{D}_t - \{t\}$. With $T(j) = \{i \in T: i < j\}$, the induction hypothesis yields a lower bound on the denominator:

$$\mathbb{P}\left(I_T \mid I_{S-T}\right) = \prod_{j \in T} \mathbb{P}\left(\overline{A}_j \mid I_{T(j)} \cap I_{S-T}\right) \ge \prod_{j \in T}(1 - x_j).$$

With these bounds and the hypothesis $\mathbb{P}(A_t) \le x_t \prod_{j \in D_t}(1 - x_j)$,

$$\mathbb{P}(A_t \mid I_S) = \frac{\mathbb{P}(A_t \cap I_T \mid I_{S-T})}{\mathbb{P}(I_T \mid I_{S-T})} \le \frac{\mathbb{P}(A_t)}{\prod_{j \in T}(1 - x_j)} \le x_t \prod_{j \in D_t - T}(1 - x_j) \le x_t. \qquad \blacksquare$$

When the events are completely independent, automatically $\mathbb{P}\left(\bigcap \overline{A}_i\right) > 0$, which also follows from the Local Lemma (with each D_i being empty) using any weights such that $\mathbb{P}(A_i) < x_i < 1$. Larger sets D_i impose tighter restrictions on $\mathbb{P}(A_i)$ to still guarantee $\mathbb{P}\left(\bigcap \overline{A}_i\right) > 0$. If the events are pairwise disjoint, then $D_i = [n] - \{i\}$ for all i, and finding suitable weights is hard. It becomes impossible when $\bigcup_i A_i$ absorbs enough probability, such as when $n = 2$ and $\mathbb{P}(A_1) = \mathbb{P}(A_2) = .25$ (Exercise 19), even though in this case $\mathbb{P}\left(\bigcap \overline{A_i}\right) = .5$.

The symmetric version is an easy corollary. Note that $(1 - 1/d)^d < e^{-1}$ (Proposition 14.1.1), but $(1 - 1/d)^{d-1} > e^{-1}$ when $d \geq 2$ (Exercise 4). Fortunately, the factor $1 - x_i$ does not appear in the product over D_i in Theorem 14.2.18.

Proof of Theorem 14.2.9 (Symmetric Local Lemma). We are given $A_1, \ldots, A_n$ such that $\mathbb{P}(A_i) \leq p$ and A_i is mutually independent of $\{A_j : j \notin D_i \cup \{i\}\}$, where $D_1, \ldots, D_n$ are subsets of $[n]$ with $|D_i| < d$. We are also given $epd \leq 1$.

To show that $\mathbb{P}\left(\bigcap \overline{A_i}\right) > 0$, we set $x_1 = \cdots = x_n = 1/d$. Let x be this common value $1/d$. We compute

$$\mathbb{P}(A_i) \leq p \leq d^{-1}e^{-1} < d^{-1}(1 - 1/d)^{d-1} \leq x \prod_{j \in D_i}(1 - x_j).$$

Thus Theorem 14.2.18 applies. The inequality $e^{-1} < (1 - 1/d)^{d-1}$ holds because the right side decreases to e^{-1} as $d \to \infty$ (Exercise 4).   ∎

14.2.19.* Remark. The proof given here for the Local Lemma guarantees a point outside all the bad events, called "flawless" in Achlioptas–Iliopoulos [2016]. However, the probability of flawless points may be exponentially small, and we are not given an efficient way to find one. Much effort was devoted to "constructivizing" the proof to obtain such a point in polynomial time.

Work in this direction began with Beck [1991]. Subsequent work obtained randomized algorithms to produce a flawless point in various special cases or under additional restrictions, with a particularly simple example in Moser [2009]. Moser–Tardos [2010] then produced an algorithm that works whenever the sample space can be described as a product of independent choices over a set of variables, as in Proposition 14.2.12.

In particular, if the hypothesis of Theorem 14.2.18 holds for events structured as in Proposition 14.2.12, then one can start with a random point in the probability space and proceed as follows, iteratively. If some bad event A_i occurs, then randomly re-sample the variables in the underlying subset S_i that determine A_i. Moser–Tardos [2010] proved that this leads to a flawless point with expected number of steps at most $\sum_i \frac{x_i}{1-x_i}$ (see Sinclair [2018, Lecture 25] for a proof).

The focus in this area has shifted to efficient local search algorithms described in general settings that go beyond the probabilistic framework of the Local Lemma. For example, Achlioptas–Iliopoulos [2016] replaced the probabilistic formulation by a directed graph on the underlying space and sought a point outside the bad events via a random walk on the digraph.

The original proof in Moser–Tardos [2010] used a technique called "witness trees" to track dependencies and progress toward a flawless point. An alternative approach presented by Moser was named the **Entropy Compression Method** by Tao [2009]. This method has been applied in many situations and sometimes gives stronger results than the Local Lemma (see for example Grytczuk–Kozik–Micek [2013], Esperet–Parreau [2013], and Dujmović–Joret–Kozik–Wood [2016]). Bernshteyn [2017] further strengthened this method with his **Local Cut Lemma**, which captures the sometimes lengthy computations of the Entropy Compression Method and makes it easier to prove the same results.   ∎

A different restriction of the general Local Lemma can be useful when there are several types of events.

14.2.20. Theorem. (**Neighborhood Local Lemma**; Molloy–Reed [2002]) Let $A_1, \ldots, A_n$ be events in a probability space. For each i, let $D_i \subseteq [n] - \{i\}$ be such that A_i is mutually independent of $\{A_j : j \notin D_i \cup \{i\}\}$. If $\mathbb{P}(A_i) \le \frac{1}{4}$ and $\sum_{j \in D_i} \mathbb{P}(A_j) \le \frac{1}{4}$ for all i, then $\mathbb{P}\left(\bigcap_{t=1}^n \overline{A_t}\right) > 0$.

Proof: With $\mathbb{P}(A_i) \le \frac{1}{4}$, we have $0 \le x_i \le \frac{1}{2}$, where $x_i = 2\mathbb{P}(A_i)$. By Theorem 14.2.18, it suffices to prove $\mathbb{P}(A_i) \le x_i \prod_{j \in D_i} (1 - x_j)$.

Let $\alpha = 2 \ln 2$. When $0 \le x \le \frac{1}{2}$, we have $e^{-\alpha x} \le 1 - x$ (equality holds at both ends, and $e^{-\alpha x}$ is convex on the interval). Using $2\mathbb{P}(A_i) = x_i$ and $\frac{1}{2} \ge \sum_{j \in D_i} 2\mathbb{P}(A_j)$,

$$\mathbb{P}(A_i) = 2\mathbb{P}(A_i)e^{-\alpha/2} \le x_i e^{-\alpha \sum_{j \in D_i} 2\mathbb{P}(A_j)} = x_i \prod_{j \in D_i} e^{-\alpha x_j} \le x_i \prod_{j \in D_i} (1 - x_j). \qquad \blacksquare$$

Molloy and Reed call Theorem 14.2.20 the **Asymmetric Local Lemma**. It is useful when events have different probabilities and each "neighborhood" D_i has few high-probability events. Because the Symmetric Local Lemma requires the same upper bound on the probabilities of all the events, that bound must be very small: $\frac{1}{ed}$. Since here we only require $\sum_{j \in D_i} \mathbb{P}(A_j) \le \frac{1}{4}$, the probabilities of individual events can also be near $\frac{1}{4}$. Exercise 26 further underscores the distinction. Exercise 27 develops a generalization.

We illustrate the technique. Hind–Molloy–Reed [1999] used the next result to prove that every graph with maximum degree D has a total coloring (see Exercise 8.3.33) using at most $D + 8 \log^8 D$ colors. A coloring is k-**frugal** if it is a proper coloring and every color appears at most k times in each vertex neighborhood.

14.2.21. Theorem. (Hind–Molloy–Reed [1997]) If $\Delta(G) \ge k^k$, then G has a k-frugal coloring using at most $16(\Delta(G))^{1+1/k}$ colors.

Proof: (Molloy–Reed [2002]) For $k = 1$, a k-frugal coloring is just a proper coloring of G^2, and the claim follows because $\Delta(G^2) \le (\Delta(G))^2$.

For $k \ge 2$, let $q = \lfloor 16(\Delta(G))^{1+1/k} \rfloor$. Color $V(G)$ uniformly at random from $[q]$ (this coloring need not be proper). Define events A_S for all $S \subseteq V(G)$ such that S is a pair of adjacent vertices or a $(k + 1)$-set contained in a vertex neighborhood; the event A_S occurs when S is monochromatic. Note that $\mathbb{P}(A_S) = 1/q^{|S|-1} < \frac{1}{4}$.

By the Mutual Independence Principle, each A_S is mutually independent of the set consisting of all events A_T such that $T \cap S = \varnothing$. That is, let D_S be the set of all T other than S such that $T \cap S \ne \varnothing$ and A_T is an event. By the Neighborhood Local Lemma, it suffices to show $\sum_{T \in D_S} \mathbb{P}(A_T) \le \frac{1}{4}$ for each event A_S.

Every set in D_S is an edge incident to S or a $(k + 1)$-set intersecting S that is contained in a vertex neighborhood. There are at most $|S| \Delta(G)$ sets of the first type and at most $|S| \Delta(G)\binom{\Delta(G)}{k}$ of the second type. Since $|S| \le k + 1$, we compute

$$\sum_{T \in D_S} \mathbb{P}(A_T) \le (k + 1)\Delta(G)\frac{1}{q} + (k + 1)\Delta(G)\binom{\Delta(G)}{k}\frac{1}{q^k}$$

$$< \frac{(k + 1)\Delta(G)}{q} + \frac{(k + 1)(\Delta(G))^{k+1}}{k!q^k} \le \frac{k + 1}{16(\Delta(G))^{1/k}} + \frac{k + 1}{k!16^k} < \frac{1}{4}. \qquad \blacksquare$$

The original proof of Theorem 14.2.21 used the general Local Lemma. The neighborhood version gives a simpler proof, but the symmetric version is not strong enough. It would require a single value p bounding the probabilities of all the events; thus $p \geq 1/q^2$ is needed, where $q = \lfloor 16(\Delta(G))^{1+1/k} \rfloor$. However, in D_S we must include more than $(k+1)\Delta(G)\binom{\Delta(G)}{k}$ events, so we cannot satisfy $epd \leq 1$. The Neighborhood Local Lemma takes advantage of the fact that not many of the events with "high" probability $1/q^2$ occur in each neighborhood.

The Neighborhood Local Lemma is not strong enough for the next application to Ramsey numbers. In Corollary 14.1.17, we proved $R(3,k) \leq 8k^2/\lg k$. The lower bound needs a graph with many vertices having no triangle or independent k-set. The deletion method yields $R(3,k) > k^{3/2+o(1)}$ (Exercise 16).

We will improve the lower bound to $ck^2/(\ln k)^2$. Kim [1995] raised it to $R(3,k) \geq ck^2/\ln k$, matching the growth rate of the upper bound. More recent improvements include Bohman [2009], Bohman–Keevash [2013], and Fiz Pontiveros–Griffiths–Morris [2013], the latter bringing the lower bound asymptotically within a factor of 4 of the upper bound in Shearer [1983]. (For analogous discussion of $R(4,k)$, see Exercise 17.)

The lower bound of $ck^2/(\ln k)^2$ was originally proved by Erdős [1961] using an intricate deletion argument. As noted in Remark 8.1.19, it yields a triangle-free r-chromatic graph with at most $c'(r\ln r)^2$ vertices (Exercise 5). Spencer [1977] gave a simpler proof using the general Local Lemma (see also Spencer [1987]).

14.2.22. Theorem. (Erdős [1961]) $R(3,k) \geq ck^2/\ln^2 k$ for all k when c is a sufficiently small constant.

Proof: (Spencer [1977]) Form a random graph with vertex set $[n]$ by generating each edge with probability p, independently; p will be chosen later. We seek n as large as possible such that the probability of having no triangle or independent k-set is nonzero. Since we can adjust c if needed, we may assume that k is large.

Define events A_S for all $S \subseteq [n]$ with $|S| \in \{3,k\}$; event A_S occurs if S is an independent k-set or induces a triangle. The $\binom{n}{3}$ events A_S with $|S| = 3$ each have probability p^3, and the $\binom{n}{k}$ events A_S with $|S| = k$ each have probability $(1-p)^{\binom{k}{2}}$.

By the Mutual Independence Principle, each A_S is mutually independent of the set consisting of all A_T such that $|T \cap S| \leq 1$. That is, let D_S be the set of all T other than S such that $|T \cap S| \geq 2$.

To simplify the computation, we bound a slightly larger quantity. For each S we include all k-sets in D_S. Taking a closer look at the number of triples, we must include $\binom{|S|}{2}(n-|S|)$ triples in D_S. Since we only have two types of events, to prove $R(3,k) > n$ it suffices to choose y and z in $(0,1)$ such that for some $p \in (0,1)$,

$$p^3 \leq y(1-y)^{3n}(1-z)^{\binom{n}{k}} \quad \text{and} \quad (1-p)^{\binom{k}{2}} \leq z(1-y)^{\binom{k}{2}n}(1-z)^{\binom{n}{k}}.$$

We seek n (in terms of k) such that y, z, and p can be chosen to satisfy these inequalities. We will show that $n = c_0 k^2/(\ln k)^2$ is suitable for some constant c_0. We may assume that k is sufficiently large to make the argument work, since later replacing c_0 by a suitably small constant c will make the conclusion hold also for all smaller k.

For large k, we guarantee the first inequality by setting $y = p^3\alpha$ (where α is a constant larger than 1) and making p and z small enough that $(1-y)^{3n} \to 1$

and $(1-z)^{\binom{n}{k}} \to 1$. By expanding $n \ln(1-x)$, it follows that $(1-x)^n \to 1$ when $nx \to 0$ as $n \to \infty$. Thus the first limit follows from $ny \to 0$, which requires $p = o(n^{-1/3})$; we set $p = c_1 n^{-1/2}$. To obtain $(1-z)^{\binom{n}{k}} \to 1$, we ensure $z\binom{n}{k} \to 0$ by setting $z = \binom{n}{k}^{-\alpha}$.

Because $(1-p)^{\binom{k}{2}} < e^{-p\binom{k}{2}}$ and $(1-z)^{\binom{n}{k}} \to 1$, for the second inequality we study $e^{-p\binom{k}{2}} < z(1-y)^{\binom{k}{2}n}$. To simplify the right side, note that $\binom{k}{2}ny^2 \sim c'\frac{(\ln k)^4}{k^2} \to 0$, which yields $(1-y)^{\binom{k}{2}n} \sim e^{-y\binom{k}{2}n}$ (see Lemma 14.3.11). Note that $\ln z = -\alpha \ln\binom{n}{k}$. By comparing exponents in the desired inequality and canceling $-\binom{k}{2}$, it thus suffices to show $p > \alpha \left[\ln\binom{n}{k}\right] / \binom{k}{2} + ny$.

Using $n^k > \binom{n}{k}$ and $\frac{k^2}{2.5} < \binom{k}{2}$, it suffices to show $p > \alpha 2.5\frac{\ln n}{k} + ny$. Balancing ny and p motivated our choice of p, which yields $ny = pc_1^2\alpha$. Also, $p = c_1 n^{-1/2} = c_1 c_0^{-1/2} \cdot \frac{\ln k}{k}$. Finally, the leading term in $\ln n$ as a function of k is $2\ln k$. If we choose the constants to yield the desired strict inequality in the leading terms, then the lower-order terms can be ignored for sufficiently large k (this also explains why we can ignore the last factor in the second desired inequality). Thus it suffices to have $c_1 > \alpha(5c_0^{1/2} + c_1^3)$. When c_0 and c_1 are made sufficiently small, the inequality holds. $\blacksquare$

The computations in Theorem 14.2.22 are delicate. It would be nice to obtain the result more simply from the Neighborhood Local Lemma, but this fails.

14.2.23.* Remark. In attempting to use the Neighborhood Local Lemma to prove $R(3,k) \geq ck^2/(\ln k)^2$, we define the same events to be avoided as in Theorem 14.2.22, with A_S for all $S \subseteq [n]$ such that $|S| \in \{3,k\}$. Again we generate a random graph using edge-probability p. The formulas for $\mathbb{P}(A_S)$ are as before, and again A_S is mutually independent of $\{A_T : |T \cap S| \leq 1\}$.

By the Neighborhood Local Lemma, choosing p so that $\sum_{T \in D_S} \mathbb{P}(A_T) \leq \frac{1}{4}$ for each A_S suffices. Again we do not lose much by including all k-sets in D_S. Taking a closer look at the number of triples, we must include $\binom{|S|}{2}(n-|S|)$ triples in D_S. Since $k > 3$ and $1-p < e^{-p}$, it suffices to have

$$\binom{k}{2}np^3 + \binom{n}{k}e^{-p\binom{k}{2}} \leq \frac{1}{4}.$$

We want to maximize n in terms of k such that $p \in (0,1)$ can be chosen to make the inequality true. Each term must be less than $\frac{1}{4}$. The first term bounds p from above, and the second bounds it from below. Up to constants c_1 and c_2 (if n is superlinear in k), the bounds are

$$\frac{c_1 \ln n}{k} < p < \frac{c_2}{(k^2 n)^{1/3}}.$$

In order for p to satisfy both inequalities, $n(\ln n)^3 < ck$ is required. Hence this method cannot even yield a linear lower bound! The Symmetric Local Lemma is even worse, providing only a constant lower bound. In this problem the deletion method does better; see Exercise 16. $\blacksquare$

EXERCISES 14.2

14.2.1. (–) Given natural numbers n and t, let $m = n - \binom{n}{t}^2 2^{1-t^2}$. Prove that there is a 2-coloring of the edges of $K_{m,m}$ having no monochromatic copy of $K_{t,t}$.

14.2.2. (–) Let H be a hypergraph in which every edge has size at least k and intersects fewer than $\lfloor 2^{k-1}/e \rfloor$ other edges. Use the Local Lemma to prove that H is 2-colorable.

14.2.3. (–) Let H be a k-uniform hypergraph in which every vertex appears in exactly k edges. Use the Local Lemma to prove that H is 2-colorable if $k \geq 9$.

14.2.4. (–) Prove $(1 - 1/d)^{d-1} > e^{-1}$ for $d \geq 2$.

14.2.5. (–) Given $R(3, k) \geq ck^2/(\ln k)^2$ for some constant c, prove that there is a triangle-free r-chromatic graph with at most $O((r \ln r)^2)$ vertices.

14.2.6. ($\diamond$) Let G be an n-vertex graph with average degree d, where $d \geq 1$. Prove $\alpha(G) \geq n/(2d)$ by first selecting vertices with some probability p and then discarding some. (Comment: This is weaker than Theorem 14.1.14, the Caro–Wei Theorem.)

14.2.7. ($\diamond$) Let G be an n-vertex graph having a proper coloring such that every vertex has neighbors in at most k color classes. Prove that $\alpha(G) \geq n/(k+1)$. (Hint: Modify the argument of Theorem 14.1.14.)

14.2.8. *Dominating set algorithm.* Let $\delta(G) = k$, and let $N[v] = N(v) \cup \{v\}$.
   (a) Given $S \subseteq V(G)$, let $U = V(G) - \bigcup_{v \in S} N[v]$. Prove that some vertex of $G - S$ dominates at least $|U|(k+1)/n$ vertices in U.
   (b) Construct $S \subseteq V(G)$ by iteratively including a vertex that dominates the most vertices yet undominated. Prove that at most $n/(k+1)$ vertices remain undominated after $n \ln(k+1)/(k+1)$ steps, so adding them yields a dominating set of size at most $\lceil n \frac{1+\ln(k+1)}{k+1} \rceil$.

14.2.9. ($\diamond$) Prove that an n-vertex graph with minimum degree k has $\frac{k+1}{2\ln n}$ pairwise disjoint dominating sets. (Feige–Halldórsson–Kortsarz–Srinivasan [2002])

14.2.10. ($\diamond$) A graph is **locally linear** (Fronček [1989]) if every edge lies in exactly one triangle; equivalently, the subgraph induced by each vertex neighborhood is a matching.
   (a) Prove that a union of disjoint triangles is locally linear if and only if it has no two triangles on four vertices and no $(6, 3)$-**configuration** (three triangles on six vertices).
   (b) Use the deletion method to prove that there is a locally linear n-vertex graph with at least $n^{3/2}/c - O(n^{1/2})$ edges, where c is any constant greater than 18. (Comment: Sós–Erdős–Brown [1973] proved existence of $cn^{3/2}$ triangles with no $(6, 3)$-configuration; the "$(6, 3)$-Theorem" of Ruzsa–Szemerédi [1978] showed that a multiple of $n^{2-c/\sqrt{\log n}}$ triangles can occur and that $o(n^2)$ is an upper bound; see Exercise 11.1.43.)

14.2.11. ($\diamond$) Use the deletion method to prove $\mathrm{ex}(n; C_k) \in \Omega(n^{1+1/(k-1)})$ (see Definition 11.1.9). In other words, show that for some constant c_k there is an n-vertex graph with at least $c_k n^{1+1/(k-1)}$ edges that has no k-cycle. (Comment: For even k, counting arguments yield $\mathrm{ex}(n; C_k) \in O(n^{1+2/k})$.) (Bondy–Simonovits [1974])

14.2.12. ($\diamond$) A k-**subdivision** of a graph H is formed by turning each edge of H into a path of length $k+1$. Prove that every n-vertex graph with εn^2 edges contains a 1-subdivision of some complete graph K_a with $a \geq \varepsilon^{3/2}\sqrt{n}$. (Comment: Alon–Krivelevich–Sudakov [2003] showed that $a \geq \varepsilon \sqrt{n}$ can be guaranteed, and Fox–Sudakov [2009] considered 1-subdivision of general graphs. See Fox–Sudakov [2011].)

14.2.13. ($\diamond$) For the Ramsey number of the k-cube Q_k, prove $R(Q_k, Q_k) \leq 2^{3k}$. (Fox–Sudakov [2011]) (Comment: Fox–Sudakov [2009] proved $R(G, G) \leq 8d2^d |V(G)|$ when G is bipartite with $\Delta(G) = d$, which improves this bound to $2^{2k+o(k)}$.)

14.2.14. To show the existence of triangle-free graphs with large chromatic number, consider random n-vertex graphs in which each vertex pair is an edge with probability $n^{-2/3}$, independently. Use these to argue that there is a triangle-free graph with $2n/3$ vertices and chromatic number at least $n^{1/3}/(2\ln n)$. (Comment: This easy result is weaker than what is guaranteed from $R(3, k)$ in Exercise 14.2.5.)

14.2.15. ($\diamond$) Reed [1998] conjectured $\chi(G) \leq \lceil \frac{\Delta(G)+1+\omega(G)}{2} \rceil$ for every graph G. He proved the existence of a positive constant ε such that $\chi(G) \leq \lceil \varepsilon\omega(G) + (1-\varepsilon)(\Delta(G)+1) \rceil$. Reed [1998] showed that $\varepsilon \leq 1/2$ is necessary. Sufficiency remains open.

(a) Generate a random n-vertex graph H with independent edge-probability p. Prove $\mathbb{P}(\alpha(H) < \frac{2+c}{p}\ln n) \to 1$ as $n \to \infty$, where c is a positive constant.

(b) Prove that there is a graph H' with $n - (1/3)n^{3/4}$ vertices having independence number 2 and clique number less than $(2+c)n^{3/4}\ln n$.

(c) Prove that any ε such that Reed's Conjecture holds for all graphs satisfies $\varepsilon \leq 1/2$.

(Comment: As observed by Kostochka, part (c) also follows from the explicit graph G_t obtained by expanding each vertex of C_5 into a t-clique; $\chi(G_t) \geq 5t/2$ while $\omega(G_t) = 2t$ and $\Delta(G) = 3t - 1$. Toward Reed's Conjecture, when $\Delta(G)$ is large it suffices to have $\varepsilon < 1/130{,}000$ (King–Reed [2016]), $\varepsilon < 1/26$ (Bonamy–Perrett–Postle [2018]), or $\varepsilon < 1/13$ (Delcourt–Postle [2017]). The last result also works for the list chromatic number.)

14.2.16. *Off-diagonal Ramsey numbers.* (Note $R(3, k) \geq ck^2/\ln^2 k$ from Theorem 14.2.22.)

(a) Prove that if $\binom{n}{k}p^{\binom{k}{2}} + \binom{n}{l}(1-p)^{\binom{l}{2}} < 1$ for some $p \in (0, 1)$, then $R(k, l) > n$.
Prove also that $R(k, l) > n - \binom{n}{k}p^{\binom{k}{2}} - \binom{n}{l}(1-p)^{\binom{l}{2}}$ for all $n \in \mathbb{N}$ and $p \in (0, 1)$.

(b) Choose n and p in the second part of (a) to prove $R(3, k) > k^{3/2-o(1)}$. What lower bound on $R(3, k)$ can be obtained from the first part of (a)? (Spencer [1977])

14.2.17. (+) *Lower bounds on the Ramsey number $R(4, k)$.*

(a) Use part (a) of Exercise 14.2.16 to show that $R(4, k) \geq \Omega((k/\ln k)^{3/2})$.

(b) Use the Deletion Method to show that $R(4, k) \geq \Omega((k/\ln k)^2)$.

(c) Use the Local Lemma to show that $R(4, k) \geq \Omega((k/\ln k)^{5/2})$. (Spencer [1977])
(Comment: Bohman [2009] proved $R(4, k) \geq \Omega(k^{5/2}/(\ln k)^2)$, and $R(4, k) \leq k^{3+o(1)}$ is known.)

14.2.18. If $n \geq m$, then trivially there is an injective function from $[m]$ to $[n]$. Consider functions where elements of $[m]$ are mapped to uniformly random elements of $[n]$, independently (all n^m functions are equally likely). Show the power of the Local Lemma by comparing the lower bounds on n given by the Union Bound and by the Local Lemma for positive probability of obtaining an injective function. (P. Prałat)

14.2.19. Let A_1 and A_2 be two dependent events, each having probability p. Prove that the hypotheses of the General Local Lemma can hold for A_1 and A_2 if and only if $p \leq .25$.

14.2.20. Let H be a graph with maximum degree k. Prove that if $k \leq s^{3/2}/\sqrt{4e}$, then H has an s-edge-coloring in which no 4-cycle is monochromatic. (Comment: Jiang–Milans–West [2013] more generally considered avoiding monochromatic $K_{p,q}$.)

14.2.21. ($\diamond$) Let G be a graph with $\Delta(G) = k-1$. Let S be a set of disjoint cliques in G, each with size at least ck. Use the Local Lemma to prove that if c is large enough, then G has an independent set consisting of one vertex from each clique in S. Determine how large c must be to allow the argument to work. (Hint: 11/13 is slightly above the threshold.)

14.2.22. ($\diamond$) The **van der Waerden number** $w(l, k)$ is the least n so that every k-coloring of $[n]$ has an l-term arithmetic progression in one color (Theorem 10.3.8).

(a) Use the existence method to prove that $w(l, k) \geq (lk^{l-1})^{1/2}$.

(b) Use the Local Lemma to prove that $w(l, k) \geq (el)^{-1}k^{l-1}$. (Comment: When l is prime, there is a construction for $w(l, 2) > l2^l$ using finite fields.)

14.2.23. ($\diamond$) Let G be a digraph in which every vertex has outdegree k and indegree k. Let $r = \lfloor k/(2.25 + 2\ln k) \rfloor$. Partition $V(G)$ into r nonempty sets $V_1, \ldots, V_r$ by an appropriate experiment. Use the Local Lemma to prove that with positive probability every vertex has a successor in the set containing it. Conclude that every k-regular directed graph has a family of r pairwise disjoint cycles. (Hint: Be careful to ensure that $V_1, \ldots, V_r$ are nonempty. Comment: Alon [1996] proved that $\delta^+(G) \geq k$ guarantees $k/64$ pairwise disjoint cycles (see also Alon–McDiarmid–Molloy [1996]), later improved to $k/18$ (Bucić [2018]). Bermond–Thomassen [1981] conjectured that $k/2$ can be guaranteed, proved for tournaments in Bang-Jensen–Bessy–Thomassé [2014].)

14.2.24. Let L be a k-uniform list assignment for a graph G. Color vertices uniformly at random from their lists, independently. For each edge xy, let A_{xy} be the event that x and y receive the same color. Determine what k is needed for the Symmetric Local Lemma to guarantee a proper coloring from the lists. How useful is this result? (Molloy–Reed [2002])

14.2.25. Let $f(n, r, s)$ be the minimum number of colors on $E(K_n)$ so that every copy of K_r has at least s colors. Let $g(n, r, s)$ be the minimum number of colors on $E(K_{n,n})$ so that every copy of $K_{r,r}$ has at least s colors. Use the Local Lemma to prove the following.

    (a) $f(n, r, s) \leq c_{r,s} n^{(r-2)/[\binom{r}{2}-s+1]}$. (Erdős–Gyárfás [1997])

    (b) $g(n, r, s) \leq c'_{r,s} n^{2(r-1)/[r^2-s+1]}$. (Axenovich–Füredi–Mubayi [2000])

14.2.26. Let H be a hypergraph in which every edge has size at least 3 and each edge of H intersects at most a_r (other) edges of size r. Use the Neighborhood Local Lemma to prove that if $\sum a_r 2^{-r} \leq \frac{1}{8}$, then H is 2-colorable. How would this condition need to change to obtain the conclusion from the Symmetric Local Lemma? (Molloy–Reed [2002])

14.2.27. Weighted Neighborhood Local Lemma (Molloy–Reed [2002]).

    (a) Let $A_1, \ldots, A_n$ be events in a probability space. For each i, let $D_i \subseteq [n]-\{i\}$ be such that A_i is mutually independent of $\{A_j: j \notin D_i \cup \{i\}\}$. Let z and $t_1, \ldots, t_n$ be real numbers such that $0 \leq z \leq \frac{1}{4}$ and $t_i \geq 1$ for all i. Prove that if $\mathbb{P}(A_i) \leq z^{t_i}$ and $\sum_{j\in D_i}(2z)^{t_j} \leq \frac{t_i}{2}$ for all i, then $\mathbb{P}\left(\bigcap_{i=1}^n \overline{A_i}\right) > 0$. (Hint: Mimic the proof of Theorem 14.2.20, with $x_i = (2z)^{t_i}$.)

    (b) Let H be a hypergraph in which every edge has size at least 3 and each vertex of H lies in at most b_i edges of size i. Prove that if $\sum b_i 2^{-i/2} \leq 1/(6\sqrt{2})$, then H is 2-colorable.

14.3. Moments and Thresholds

In this section we develop further techniques for random processes. We focus on random graphs, which have many important applications.

14.3.1. Application. *Melting points.* View a physical solid as a three-dimensional grid of molecules. Neighboring molecules are joined by bonds, as in the cartesian product of three paths. Adding energy excites molecules and breaks bonds, randomly. The fraction of bonds broken increases with temperature. When the graph remains well connected, the material acts solid. Breaking off small pieces doesn't change this, but when all components are small the global nature of the material changes. Small components float freely, like a liquid or gas.

Mathematically, there is a threshold for the number of bonds to be broken (in terms of the size of the grid) such that breaking somewhat fewer bonds almost always leaves a giant component, while breaking somewhat more almost always leaves only tiny components. Below the threshold temperature the material will almost certainly be solid, while above it it will almost certainly not be. ∎

We will need precise notions of "almost always" and "threshold".

14.3.2. Application. *Analysis of algorithms.* Worst-case complexity does not fairly judge algorithms that run quickly on most inputs. *Probabilistic analysis* assumes a probability distribution on the inputs to study the expected running time. Finding a realistic distribution that is easy to analyze can be very hard. Given a distribution over graphs of each order, we study the expected running time as a function of the number of vertices.

Naive algorithms may have good expected behavior. For example, we have no polynomial-time algorithm to find maximum cliques in graphs. If "almost every" graph has clique number about $2 \lg n$, then we can check all vertex subsets up to size $3 \lg n$. If $\omega(G) < 3 \lg n$, then this will compute the clique number (every set of size $\omega(G) + 1$ is not a clique); otherwise it fails. The algorithm rarely fails. There are somewhat too many subsets of size $3 \lg n$ for this to be a polynomial-time algorithm, but it's close. ∎

The probabilistic method was popularized by Erdős beginning in 1947, with systematic study of random graphs initiated by Erdős–Rényi [1959]. There will always be room for new developments, because random structures have different properties under different probability models, and different models may be appropriate for different applications. As in the preceding sections, we aim to convey the main techniques and do not attempt an exhaustive presentation of results.

"ALMOST ALWAYS"

We have proposed studying properties that "almost always" hold.

14.3.3. Definition. Given a sequence Ω_n of probability spaces, let q_n be the probability that property Q holds in Ω_n. Property Q holds **with high probability** if $\lim_{n \to \infty} q_n = 1$.

For example, Ω_n may be a probability distribution over graphs with vertex set $[n]$. A special case of this is our most common model for random graphs, because it leads to the simplest computations.

14.3.4. Definition. Let p be a function of n. Define the probability space $\mathbb{G}(n, p)$ on graphs with vertex set $[n]$ as follows: any two vertices form an edge with probability p, independently. Each graph with m edges has probability $p^m(1 - p)^{\binom{n}{2} - m}$.

The random variable G^p denotes a graph drawn from $\mathbb{G}(n, p)$ for some n. The sequence $\mathbb{G}(n, p)_{n \in \mathbb{N}}$ is known as the **standard** or **binomial random graph model**, because the number of edges has a binomial distribution. It is also called the **Erdős–Rényi model**. "*The* random graph" refers to $\mathbb{G}(n, \frac{1}{2})$.

Computations are much, much simpler in the binomial model (labeled graphs) than in a uniform distribution over isomorphism classes. Since algorithms operate on graphs with specified vertex sets (rather than isomorphism classes) as inputs, this model seems appropriate.

In $\mathbb{G}(n, \frac{1}{2})$, all graphs with vertex set $[n]$ are equally likely. When property Q holds with high probability in $\mathbb{G}(n, \frac{1}{2})$, we say "almost every graph satisfies Q".[1] We often shorten "with high probability" to "whp".

Because we often measure running times of algorithms in terms of the number of vertices and number of edges, we may prefer to control the number of edges. This suggests making the n-vertex labeled graphs with m edges equally likely.

14.3.5. Definition. Let m be an integer-valued function of n. The probability space $\mathbb{G}'(n, m)$ assigns probability $\binom{N}{m}^{-1}$ to each graph with vertex set $[n]$ having m edges, where $N = \binom{n}{2}$. The random variable G^m denotes a graph drawn from $\mathbb{G}'(n, m)$ for some m in the **uniform random graph model**.

Applications seem more natural in the uniform model. For example, we ask "in terms of n, how many edges are needed to make a graph almost surely connected?" rather than "in terms of n, what edge probability makes a graph almost surely connected?" However, calculations are harder in the uniform model.

Fortunately, the binomial model approximates the uniform model when n is large and $m = p\binom{n}{2}$. The proof requires detailed study of the binomial distribution. A graph property Q is **convex** if G satisfies Q whenever $F \subseteq G \subseteq H$ and both F and H satisfy Q.

14.3.6. Theorem. (Bollobás [1985, pp. 34–35]) If Q is a convex property and $p(1-p)\binom{n}{2} \to \infty$, then G^p satisfies Q whp if and only if, for every fixed x, whp G^m satisfies Q, where $m = \lfloor p\binom{n}{2} + x[p(1-p)\binom{n}{2}]^{1/2} \rfloor$. ∎

Many properties of interest are convex; "having an even number of edges" is not. Since Theorem 14.3.6 justifies studying the binomial model to draw conclusions about the uniform model, we will study only $\mathbb{G}(n, p)$. Theorem 14.3.6 also motivates letting p depend on n; to study graphs with a linear number of edges, p must decline like $1/n$. Constant p yields dense graphs.

Proving $\mathbb{P}(Q) \to 1$ is usually easier than computing $\mathbb{P}(Q)$; this distinction is important. Probabilistic analysis uses asymptotic statements; exact computation of probabilities is difficult, unnecessary, and avoided wherever possible.

To compare growth rates of sequences, we use "Oh notation", with the sets $O(f)$, $o(f)$, $\Omega(f)$, and $\omega(f)$ as defined in Chapter 0. We aim to discard lower-order terms that do not affect whether $\lim_{n\to\infty} \mathbb{P}(Q) = 1$. We only need that $\mathbb{P}(\neg Q)$ is *bounded* by something tending to 0. Many arguments are "sloppy" in this sense; it does not matter how loose the bound is as long as it tends to 0. Experience refines one's intuition about what can be discarded without getting into trouble.

[1]In probability theory, "almost always" means "with probability 1". When an event holds with probability tending to 1 in a sequence of spaces, probabilists say it holds "asymptotically almost surely". In discrete spaces, the distinction between "a.a." or "a.s." and "a.a.s" is minor, and the extra word is cumbersome. Erdős and Rényi, and also Alon–Spencer [2000], used "almost always" to mean "with probability tending to 1". A younger generation, interacting more with probabilists, uses "a.a.s." to be more correct. We retreat to "with high probability". Nevertheless, in a sequence of spaces such as $\mathbb{G}(n, \frac{1}{2})$, we continue to use the informal "almost every graph", believing this to be clear and not in conflict with the usage of "almost always" in probability.

14.3.7. Theorem. (Gilbert [1959]) For fixed p, whp G^p is connected.

Proof: We can make G disconnected by picking a vertex bipartition and forbidding edges joining the parts. Occurrence of edges within the parts is irrelevant. We bound the probability q_n that G^p is disconnected by summing $\mathbb{P}([S, \overline{S}] = \varnothing)$ over all S. Graphs with many components are counted many times. When $|S| = k$, there are $k(n-k)$ possible edges in $[S, \overline{S}]$. Each has probability $1-p$ of not appearing, independently, so $\mathbb{P}([S, \overline{S}] = \varnothing) = (1-p)^{k(n-k)}$. Considering all S generates each partition from each side, so $q_n \leq \frac{1}{2} \sum_{k=1}^{n-1} \binom{n}{k}(1-p)^{k(n-k)}$.

With symmetry in k and $n-k$, we have $q_n \leq \sum_{k=1}^{\lfloor n/2 \rfloor} \binom{n}{k}(1-p)^{k(n-k)}$. We loosen the bound to simplify it. Using $\binom{n}{k} \leq n^k$ and $(1-p)^{n-k} \leq (1-p)^{n/2}$ (for $k \leq n/2$) yields $q_n < \sum_{k=1}^{\lfloor n/2 \rfloor} (n(1-p)^{n/2})^k$. Let $x = n(1-p)^{n/2}$. When n is large, $x < 1$, and hence $q_n < \frac{x}{1-x}$. In fact, $x \to 0$ when p is constant, so our bound on q_n approaches 0 as $n \to \infty$.   ∎

Computations simplify further when we introduce **counting variables** (those taking only nonnegative integer values) and use expectation. If X is a counting variable and G^p satisfies Q when $X = 0$, then $\mathbb{E}(X) \to 0$ implies that whp G^p satisfies Q. This is a special case of the next lemma. An analogous statement holds for continuous variables.

14.3.8. Lemma. (**Markov's Inequality**) If X is a counting variable, then $\mathbb{P}(X \geq t) \leq \frac{\mathbb{E}(X)}{t}$. Thus $\mathbb{E}(X) \to 0$ implies $\mathbb{P}(X = 0) \to 1$.

Proof: Letting $p_k = \mathbb{P}(X = k)$, we have

$$\mathbb{E}(X) = \sum_{k \geq 0} k p_k \geq \sum_{k \geq t} k p_k \geq t \sum_{k \geq t} p_k = t \mathbb{P}(X \geq t).$$ ∎

Suppose that $X = 1$ if G is disconnected and $X = 0$ otherwise. In Theorem 14.3.7, we proved directly that $\mathbb{P}(X = 1) \to 0$ (when p is constant) to prove that whp G^p is connected. A different random variable X leads to a simpler proof and stronger result.

14.3.9. Example. *Indicator variables, linearity, and Markov's Inequality.* Let $X = \sum X_i$, where each X_i is an indicator variable for a bad event, and property Q holds when no bad events occur. If the bad events are easy to analyze, then the linearity of expectation and the convenience of $\mathbb{E}(X_i) = \mathbb{P}(X_i = 1)$ for indicator variables simplifies the task of proving $\mathbb{E}(X) \to 0$.

Consider Theorem 14.3.7. The probability that a nontrivial vertex partition is a cut is at most $(1-p)^{n-1}$, since at least $n-1$ pairs of vertices cross the partition. Let X be the number of cuts. There are 2^{n-1} partitions, so $\mathbb{E}(X) \leq [2(1-p)]^{n-1}$. If $p > \frac{1}{2}$, then $\mathbb{E}(X) \to 0$ and the probability of having no cut tends to 1. Thus whp G^p is connected when $p > \frac{1}{2}$.

Other indicator variables yield stronger results. Let X count the pairs of vertices having no common neighbor. When there are no such pairs, G is connected and in fact has diameter 2. Proving $\mathbb{E}(X) \to 0$ in $\mathbb{G}(n, p)$ means that we expect almost no bad pairs, and then Markov's Inequality states that whp G^p has none.

Here X is the sum of $\binom{n}{2}$ indicator variables, say X_i for the ith vertex pair $\{x, y\}$, with $X_i = 1$ when x and y have no common neighbor. Since $n - 2$ vertices must fail, $\mathbb{P}(X_i = 1) = (1 - p^2)^{n-2}$ and $\mathbb{E}(X) = \binom{n}{2}(1 - p^2)^{n-2}$. When p is constant, $\mathbb{E}(X) \to 0$, and whp G^p has diameter 2 (we will see that much smaller p suffices). We generalize further in the next theorem. ∎

14.3.10. Theorem. (Blass–Harary [1979]) Fix s, t, p. Whp, G^p satisfies the property that for every choice of disjoint vertex sets S and T of sizes s and t, there is a vertex $v \notin S \cup T$ adjacent to all of S and none of T.

Proof: Let X be the number of bad pairs (S, T); we need only show that $\mathbb{E}(X) \to 0$. For the ith way to choose disjoint $S, T \subseteq V(G)$, define an indicator variable X_i with value 1 when there is no vertex v with $S \subseteq N(v)$ and $T \subseteq \overline{N}(v)$. That is, (S, T) is bad if every $v \notin S \cup T$ fails to have s specified adjacencies and t specified nonadjacencies. Thus $\mathbb{P}(X_i = 1) = (1 - p^s(1 - p)^t)^{n-s-t}$. The number of variables (choices of (S, T)) is given by a multinomial coefficient. Since $X = \sum X_i$, we obtain

$$\mathbb{E}(X) = \binom{n}{s, t, n - s - t}(1 - p^s(1 - p)^t)^{n-s-t}.$$

For fixed s, t, p, the multinomial coefficient is a polynomial in n. It is bounded by n^{s+t}, while $\mathbb{E}(X_i)$ dies exponentially as $n \to \infty$. The logarithm of the product approaches $-\infty$, and thus $\mathbb{E}(X) \to 0$. ∎

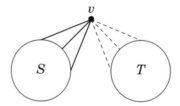

THRESHOLD FUNCTIONS

With constant edge probability, random graphs are dense, with many more edges than needed to be connected. To improve Theorem 14.3.7 and Example 14.3.9, we want to make $p(n)$ smaller while still having G^p connected whp. A closer look at the computation in Example 14.3.9 shows that we can still guarantee $\mathbb{E}(X) \to 0$ when $p(n)$ is much smaller. To make the discussion precise, we need another numerical lemma.

14.3.11. Lemma. If $np^2 \to 0$ as $n \to \infty$, then $(1 - p)^n \sim e^{-np}$.

Proof: Since $np^2 \to 0$ requires $p \to 0$, we can use the series expansion of $\ln(1 - p)$ to write

$$(1 - p)^n = e^{n \ln(1-p)} = e^{n(-p-O(p^2))} = e^{-np}e^{-O(np^2)}.$$

Since $np^2 \to 0$, the second factor tends to 1, and we obtain $(1 - p)^n \sim e^{-np}$. ∎

14.3.12. Example. *Diameter 2, continued.* When X counts the pairs having no common neighbor, $\mathbb{E}(X) = \binom{n}{2}(1 - p^2)^{n-2}$ (Example 14.3.9). If p tends to 0 but not too quickly, then it may still be that $\mathbb{E}(X) \to 0$.

Always $1 - q < e^{-q}$ (Proposition 14.1.1). If $nq^2 \to 0$, then $(1-q)^n \sim e^{-nq}$ (Lemma 14.3.11), and constant powers of $1 - q$ tend to 1. When $p(n)$ is small enough that $np^4 \to 0$, we thus have $\mathbb{E}(X) \sim \frac{1}{2}n^2 e^{-np^2}$. When $p = (\frac{c \ln n}{n})^{1/2}$, the expression simplifies to $\mathbb{E}(X) \sim \frac{1}{2}n^{2-c}$. If $c > 2$, then still $\mathbb{E}(X) \to 0$.

By parametrizing the probability to simplify the asymptotic formula for the expectation, we have shown that G^p has diameter 2 whp for a much smaller edge probability than before (and thus a much smaller number of edges in the uniform model). This strengthens the result. ∎

What happens for smaller edge probability? If $c = 2 - \varepsilon$ in Example 14.3.12, then $\mathbb{E}(X) \to \infty$. This does not yet imply that whp diameter 2 fails, but it suggests the notion of a threshold.

14.3.13. Definition. A **monotone property** is a graph property preserved by adding edges. A **threshold probability function** for a monotone property Q in $\mathbb{G}(n, p)$ is a function $t(n)$ such that when $p(n)$ is sufficiently smaller than $t(n)$ whp G^p fails Q, and when $p(n)$ is sufficiently larger than $t(n)$ whp G^p satisfies Q. The definition of **threshold edge function** for $\mathbb{G}'(n, m)$ is similar.

By Theorem 14.3.6, a threshold edge probability in $\mathbb{G}(n, p)$ generally yields a threshold number of edges for the same property in $\mathbb{G}'(n, m)$.

14.3.14. Remark. Definition 14.3.13 leaves "sufficiently" unspecified. The distinction between above and below the threshold may require $p(n)/t(n)$ to tend to 0 or ∞ (see Theorem 14.3.18). In the candidate for threshold in Example 14.3.12, the distinction is whether the multiplicative constant c exceeds 2. "Sharp" thresholds depend on a constant in a lower-order term (see Theorem 14.3.17).

When our analysis of the model uses a counting variable X such that $X = 0$ implies Q, often the threshold depends on whether $\mathbb{E}(X) \to 0$, and Markov's Inequality does half the job, since $\mathbb{E}(X) \to 0$ implies $\mathbb{P}(Q) \to 1$. In this situation (as in Example 14.3.12), the formula for $\mathbb{E}(X)$ in terms of a parametrized expression for $p(n)$ indicates how sharp the natural threshold is. We obtain candidates for threshold functions by determining which $p(n)$ yield $\mathbb{E}(X) \to 0$. Often we obtain $p(n)$ so that $\mathbb{E}(X) \to 0$ or $\mathbb{E}(X) \to \infty$ depending on the value of a parameter c. We do not aim in advance for a threshold of specified sharpness. ∎

Although $\mathbb{E}(X) \to \infty$ suggests that $\mathbb{P}(X = 0) \to 0$, this is not always true. For example, $\mathbb{E}(X) \to \infty$ when $\mathbb{P}(X = 0) = .5$ and $\mathbb{P}(X = n) = .5$. To obtain $\mathbb{P}(X = 0) \to 0$, we must show that the probability is not dispersed like this.

14.3.15. Definition. The rth **moment** of X is the expectation of X^r. The **variance** of X, written $\mathrm{Var}(X)$, is the quantity $\mathbb{E}\big((X - \mathbb{E}(X))^2\big)$. The **standard deviation** of X is the square root of $\mathrm{Var}(X)$.

An important special case is the variance of a binomial random variable X. With success probability p in n trials, $\mathrm{Var}(X) = np(1 - p)$ (see Example 14.4.3).

14.3.16. Lemma. (Second Moment Method) If X is a counting variable with $\mathbb{E}(X) > 0$, then $\mathbb{P}(X = 0) \leq \frac{\mathbb{E}(X^2) - \mathbb{E}(X)^2}{\mathbb{E}(X)^2}$. Thus $\mathbb{P}(X = 0) \to 0$ when $\frac{\mathbb{E}(X^2)}{\mathbb{E}(X)^2} \to 1$.

Proof: Applied to the variable $(X - \mathbb{E}(X))^2$ and the value t^2, Markov's Inequality yields $\mathbb{P}\big((X - \mathbb{E}(X))^2 \geq t^2\big) \leq \mathbb{E}\big((X - \mathbb{E}(X))^2\big)/t^2$. Written as $\mathbb{P}\big(|X - \mathbb{E}(X)| \geq t\big) \leq \mathrm{Var}(X)/t^2$, this is **Chebyshev's Inequality**. Since

$$\mathbb{E}\big((X - \mathbb{E}(X))^2\big) = \mathbb{E}\big(X^2 - 2X\mathbb{E}(X) + (\mathbb{E}(X))^2\big) = \mathbb{E}(X^2) - (\mathbb{E}(X))^2,$$

the inequality is $\mathbb{P}\big(|X - \mathbb{E}(X)| \geq t\big) \leq \big(\mathbb{E}(X^2) - \mathbb{E}(X)^2\big)/t^2$. Since $X = 0$ only when $|X - \mathbb{E}(X)| \geq \mathbb{E}(X)$, setting $t = \mathbb{E}(X)$ completes the proof. ∎

Intuitively, when the standard deviation grows more slowly than the expectation, all probability is pulled away from 0, and $\mathbb{P}(X = 0) \to 0$.

A connected graph has no isolated vertices, so a connectedness threshold is no smaller than a threshold for no isolated vertices. In fact, the thresholds are the same (Bollobás–Thomason [1985]); later we will suggest intuitively why. Proving the latter threshold is easier, because the indicator variables for isolated vertices have very simple distributions. We need an approximation to $(1 - p)^n$.

14.3.17. Theorem. In the binomial model, $\frac{c \ln n}{n}$ with $c = 1$ is a threshold probability function for the disappearance of isolated vertices (for $\delta(G) \geq 1$). This corresponds to $\frac{c}{2} n \ln n$ in the uniform model.

Proof: Let X be the number of isolated vertices, with X_i indicating whether vertex i is isolated. Now $\mathbb{E}(X) = \sum \mathbb{E}(X_i) = n(1 - p)^{n-1}$. We study the asymptotic behavior of $\mathbb{E}(X)$. As long as $p \in o(1/\sqrt{n})$, we have $np^2 \to 0$ and can write $(1 - p)^n \sim e^{-np}$. Also $(1 - p)^{-1} \sim 1$, so $\mathbb{E}(X) \sim n e^{-np}$.

To simplify further, set $p = \frac{c \ln n}{n}$ to obtain $n e^{-np} = n^{1-c}$, where c is a function of n. When c is constant, $p \in o(1/\sqrt{n})$, as needed to invoke Lemma 14.3.11. When $c > 1$, we have $\mathbb{E}(X) \sim n^{1-c} \to 0$, and we have proved one side of the threshold.

When $c < 1$, we have $\mathbb{E}(X) \to \infty$ and use the Second Moment Method; we need only show that $\mathbb{E}(X^2) \sim \mathbb{E}(X)^2$. Computing $\mathbb{E}(X^2)$ uses another helpful property of indicator variables: $X_i^2 = X_i$. Thus $\mathbb{E}(X^2) = \mathbb{E}(X) + \sum_{i \neq j} \mathbb{E}(X_i X_j)$. Each $X_i X_j$ is an indicator variable, with value 1 only when v_i and v_j are both isolated, which requires forbidding $2(n-2)+1$ edges. Thus $\mathbb{E}(X_i X_j) = (1-p)^{2n-3}$. Again $(1-p)^n \sim e^{-np}$, so $\mathbb{E}(X_i X_j) \sim e^{-2np}$, and

$$\mathbb{E}(X^2) \sim \mathbb{E}(X) + n(n-1)e^{-2np} \sim \mathbb{E}(X) + \mathbb{E}(X)^2.$$

Since $\mathbb{E}(X) \in o(\mathbb{E}(X)^2)$ when $\mathbb{E}(X) \to \infty$, we obtain $\mathbb{E}(X^2) \sim \mathbb{E}(X)^2$. ∎

Next we derive a threshold function for the appearance of fixed subgraphs. A graph is **balanced** if the average vertex degree in every induced subgraph is no larger than the average degree of the entire graph. All regular graphs and all trees are balanced.

14.3.18. Theorem. If H is a balanced graph with k vertices and l edges, then $p = n^{-k/l}$ is a threshold probability function for the appearance of H as a subgraph of G^p.

Proof: Let X count the copies of H that appear. There are $n_{(k)}$ ways to map $V(H)$ into $[n]$. Each copy of H arises under A of these maps, where A is the number of automorphisms of H, so there are $n_{(k)}/A$ possible copies. Note that A depends on H but not on n. Let X_i be the indicator variable for the occurrence of H_i, the ith copy of H, so $X = \sum X_i$. Since H has l edges, $\mathbb{P}(X_i = 1) = p^l$. Because k is fixed, $\mathbb{E}(X) \sim n^k p^l / A$.

Setting $p(n) = c_n n^{-k/l}$ yields $\mathbb{E}(X) \sim c_n^l / A$. Hence $\mathbb{E}(X) \to 0$ when $c_n \to 0$ and $\mathbb{E}(X) \to \infty$ when $c_n \to \infty$. By the Second Moment Method, it remains only to prove that $\mathbb{E}(X^2) \sim \mathbb{E}(X)^2$ when $c_n \to \infty$. Again $\mathbb{E}(X^2) = \mathbb{E}(X) + \sum_{i \neq j} \mathbb{E}(X_i X_j)$. (Note that $\mathbb{E}(X) \in o(\mathbb{E}(X)^2)$ when $\mathbb{E}(X) \to \infty$.)

The summands are not equal; $\mathbb{E}(X_i X_j)$ depends on $H_i \cap H_j$. Group terms by the intersection, H'. Let $E_{H'}$ denote the sum of the contributions to $\sum \mathbb{E}(X_i X_j)$ from all pairs (i, j) with $H_i \cap H_j \cong H'$. When H' has r vertices and s edges, the number of edges needed to create H_i and H_j is $2l - s$, so $\mathbb{E}(X_i X_j) = p^{2l-s}$.

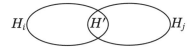

The case $r = s = 0$ occurs when $H_i \cap H_j = \varnothing$. The vertex sets for H_i and H_j can then be chosen in $\frac{n!}{k!k!(n-2k)!}$ ways, and each can be placed on its vertex set in $k!/A$ ways. Since k is constant, there are asymptotically n^{2k}/A^2 disjoint choices of H_i and H_j. Hence the contribution to $\mathbb{E}(X^2)$ from $E_\varnothing$ is asymptotic to $\mathbb{E}(X)^2$. The proof is completed by showing that the total contribution from all other choices of H' has lower order.

When $r \neq 0$, the vertices for pairs (i, j) such that $H_i \cap H_j \cong H'$ can be chosen in $\frac{n!}{r!(k-r)!(k-r)!(n-2k+r)!}$ ways. Since k and r are constant, this is asymptotic to n^{2k-r}. The number of ways to extend H' into copies of H on these sets depends only on H and H', not on n or p; call it $\alpha_{H'}$. Thus $E_{H'}$ is asymptotic to $\alpha_{H'} n^{2k-r} p^{2l-s}$.

Since $2s/r$ is the average degree of H', and H is balanced, we have $r/s \geq k/l$ (including when $s = 0$). Hence $pn^{r/s} \geq pn^{k/l} \to \infty$ when $c_n \to \infty$. Thus $n^{-r} p^{-s} \to 0$, and $E_{H'} \in o(\mathbb{E}(X)^2)$ for $H' \neq \varnothing$. Since the number of such subgraphs H' is constant (independent of n), we obtain $\mathbb{E}(X^2) \sim \mathbb{E}(X) + E_\varnothing \sim \mathbb{E}(X)^2$. ∎

This result generalizes for all H. Let $d(H) = |E(H)| / |V(H)|$; the ratio is the **density** of H. Let $\rho(H) = \max_{F \subseteq H} d(F)$; we call $\rho(H)$ the **maximum density** of H. The density and maximum density are equal when H is balanced, and then $p = n^{-1/\rho(H)}$ is the threshold for appearance of H. Every graph H has a balanced subgraph F such that $d(F) = \rho(H)$. When $pn^{\rho(H)} \to 0$, whp G^p has no copy of F; hence it also has no copy of H. In fact, always $p = n^{-1/\rho(H)}$ is a threshold function for the appearance of H (Exercise 13).

14.3.19. Remark. *Comments on Second Moment computations.*

1. Try to simplify asymptotic expressions. For example, $\binom{n}{k} \sim n^k/k!$ when $k \in o(n^{1/2})$, and $(1 - p)^n \sim e^{-np}$ when $np^2 \to 0$.

2. Analyzing the counting variable by introducing a parameter that governs whether the expectation tends to 0 or ∞ can determine the sharpness of the threshold.

3. The expression $\mathbb{E}(X^2) \sim \mathbb{E}(X)^2$ does not say that the difference approaches 0, only that the difference is in $o(\mathbb{E}(X)^2)$. The notation $\mathbb{E}(X^2) \to c$ makes sense only when c is a constant.

4. Linearity of expectation is valid for sums of random variables, not products. When $X = \sum_{i=1}^{n} X_i$ and each X_i is an indicator variable, $\mathbb{E}(X^2) = \mathbb{E}(X) + \sum_{i \neq j} \mathbb{E}(X_i X_j)$. When they are independent and identically distributed, $\mathbb{E}(X_i X_j) = \mathbb{E}(X_i)\mathbb{E}(X_j)$ and $\mathbb{E}(X^2) = \mathbb{E}(X)^2 + n(n-1)(E(X_i))^2$. In general, analysis of $\mathbb{E}(X_i X_j)$ depends on how X_i and X_j interact.

5. When proving $\mathbb{E}(X^2) \sim \mathbb{E}(X)^2$, isolating the terms that produce $\mathbb{E}(X)^2$ can simplify the treatment of the remainder, because it suffices to prove that the total contribution of the rest has lower order.   ∎

CONVERGENCE OF MOMENTS

Sometimes we can say much more about the distribution of a counting variable X. Beyond the second moment, suppose that we can determine the rth moment $\mathbb{E}(X^r)$ asymptotically. Usually, the sequence of moments determines the distribution (see Exercise 29). Determining the moments asymptotically determines the asymptotic behavior of the distribution, and we can compute $\lim \mathbb{P}(X = k)$.

After developing the method, we present two applications. First is a sharp threshold for the disappearance of isolated vertices. What happens "at" the threshold for a monotone property Q, when $\mathbb{P}(Q)$ does not tend to 0 or 1? When occurrence of Q is equivalent to $X = 0$ for some counting variable X, being *at the threshold* can mean introducing a parametrized lower-order term in the edge probability $p(n)$ to study $\mathbb{P}(X = 0)$ more precisely. We may obtain not only $\mathbb{P}(Q)$ in terms of that parameter, but in fact the entire asymptotic distribution of X.

The second application gives us a way to generate random d-regular graphs. Although whp the minimum and maximum degree of G^p are relatively close, the random graph is decidedly not regular, so we need another method.

We will use two other types of moments.

14.3.20. Definition. The rth **factorial moment** $E_r(X)$ of a random variable X is the expectation of the rth falling factorial $X_{(r)}$. When $X = \sum_{i=1}^{m} X_i$, the rth **binomial moment** $S_r(X)$ is $\sum \mathbb{E}(X_{i_1} \cdots X_{i_r})$, where the sum is taken over all r-subsets of the index set $[m]$.

14.3.21. Lemma. If X is a sum of m indicator variables, then $S_r(X) = \frac{1}{r!} E_r(X)$ and

$$\mathbb{P}(X = k) = \sum_{r=k}^{m} (-1)^{r-k} \binom{r}{k} S_r(X).$$

Proof: The binomial moment $S_r(X)$ is itself the expectation of a sum of indicator variables: $S_r(X) = \sum_{|T|=r} \mathbb{P}(X_i = 1 \text{ for } i \in T)$. The value of X_j for $j \notin T$ is unimportant, so we allow all possibilities for such X_j. Hence $S_r(X) = \sum_{|T|=r} \sum_{S \supseteq T} \mathbb{P}(A_S)$, where A_S is the event that $X_i = 1$ if and only if $i \in S$.

When $|S| = j$, the event A_S appears $\binom{j}{r}$ times in the double sum. Also $\sum_{|S|=j} \mathbb{P}(A_S) = \mathbb{P}(X = j)$. Thus

$$S_r(X) = \sum_{j=r}^{m} \binom{j}{r} \mathbb{P}(X = j) = \frac{1}{r!} \sum_{j} j_{(r)} \mathbb{P}(X = j) = \frac{1}{r!} E_r(X).$$

Finally, given the expression for $S_r(X)$ in terms of $\mathbb{P}(X = j)$, the claimed formula for $\mathbb{P}(X = k)$ follows from the Inclusion-Exclusion Formula for the number of elements belonging to exactly k of the events of the form $X_i = 1$. Both sides of the equality are normalized by the same amount. Alternatively, one can just substitute the formula for $S_r(X)$ into the claimed expression to show that it equals $\mathbb{P}(X = k)$ (Exercise 30). ∎

In applying Lemma 14.3.21, typically $m \to \infty$ as $n \to \infty$. We do not hope to compute $\mathbb{P}(X = k)$ for a given m; we only seek $\lim_{m\to\infty} \mathbb{P}(X = k)$. Fortunately, we can approximate the alternating sum by truncation.

14.3.22. Lemma. (Bonferroni Inequalities) If X is a sum of m indicator variables, and $q \in \mathbb{N}$, then

$$\sum_{r=k}^{k+2q-1} (-1)^{r-k} \binom{r}{k} S_r(X) \leq \mathbb{P}(X = k) \leq \sum_{r=k}^{k+2q} (-1)^{r-k} \binom{r}{k} S_r(X).$$

Proof: Letting $N_i = \mathbb{P}(X = i)$, we have $S_r = \sum_{j=r}^{m} \binom{j}{r} N_j$. Consider a truncation of the sum $N_k = \sum_{r=k}^{m} (-1)^{r-k} \binom{r}{k} S_r$ up to the term $r = t$. Using the expression for S_r, interchanging the order of summation, and using the Subcommittee Identity $\binom{r}{k}\binom{j}{r} = \binom{j}{k}\binom{j-k}{r-k}$ (Theorem 1.2.3(3)) yields

$$\sum_{r=k}^{t} (-1)^{r-k} \binom{r}{k} S_r = \sum_{r=k}^{t} (-1)^{r-k} \binom{r}{k} \sum_{j=r}^{m} \binom{j}{r} N_j$$

$$= \sum_{j=k}^{m} N_j \sum_{r=k}^{t} (-1)^{r-k} \binom{r}{k}\binom{j}{r} = \sum_{j=k}^{m} N_j \binom{j}{k} \sum_{r=k}^{t} (-1)^{r-k} \binom{j-k}{r-k}.$$

The inner sum has the form $s(a, b) = \sum_{l=0}^{a} (-1)^l \binom{b}{l}$, where $l = r - k$, $a = t - k$, and $b = j - k$. Observe that $s(a, 0) = 1$ and $s(a, b) = 0$ for $1 \leq b \leq a$. For $b \geq a$, it follows by induction on b that $s(a, b) = (-1)^a \binom{b-1}{a}$ (Exercise 31). Thus the inner sum is 1 for $j = k$, is 0 for $k < j \leq t$, and is $(-1)^{t-k}\binom{j-k-1}{t-k}$ for $j > t$. For the full sum, this yields

$$\sum_{r=k}^{t} (-1)^{r-k} \binom{r}{k} S_r = N_j + (-1)^{t-k} \sum_{j=t+1}^{m} N_j \binom{j}{k}\binom{j-k-1}{t-k}.$$

The sign of the error to obtain the truncation from N_j depends on the parity of $t - k$, as claimed. ∎

14.3.23. Definition. A discrete random variable X has the **Poisson distribution** with parameter (and mean) μ if $\mathbb{P}(X = k) = e^{-\mu} \mu^k / k!$.

14.3.24. Theorem. Let X be a sum of m indicator variables, where $m \to \infty$. If $S_r(X) \to \mu^r/r!$ for all r, then $\mathbb{P}(X = k) \to e^{-\mu}\frac{\mu^k}{k!}$ (that is, X is asymptotically Poisson distributed with mean μ). In particular, $\mathbb{P}(X = 0) = e^{-\mu}$.

Proof: Letting $m \to \infty$ and using the asymptotic expression for each $S_r(X)$ in the formula we derived for $\mathbb{P}(X = k)$ suggests that

$$\mathbb{P}(X = k) \to \sum_{r \geq k}(-1)^{r-k}\binom{r}{k}\frac{\mu^r}{r!} = \frac{\mu^k}{k!}\sum_{r \geq k}\frac{(-\mu)^{r-k}}{(r-k)!} = \frac{\mu^k}{k!}e^{-\mu}.$$

However, the number of summands is unbounded, so we must be careful about summing asymptotic formulas.

Fix $\varepsilon > 0$. We have observed that $\sum_{r=k}^{k+2q-1}(-1)^{r-k}\binom{r}{k}\frac{\mu^r}{r!} \to \frac{e^{-\mu}\mu^k}{k!}$ as $q \to \infty$. In this series, choose q large enough to make the error at most ε. Since q is now fixed, we can choose M large enough so that for all $r < k + 2q$ simultaneously we have $|S_r(X) - \mu^r/r!| < \frac{\varepsilon}{2q+1}\binom{r}{k}^{-1}$ for $m > M$. The Bonferroni inequalities now imply for $m > M$ that $\mathbb{P}(X = k)$ differs from $\sum_{r=k}^{k+2q-1}(-1)^{r-k}\binom{r}{k}\frac{\mu^r}{r!}$ by at most ε. Now $\left|\mathbb{P}(X = k) - e^{-\mu}\mu^k/k!\right| < 2\varepsilon$ for $m > M$. Finally, let $\varepsilon \to 0$. ∎

The technique used in Theorem 14.3.24 was introduced by V. Brun in 1915; in number theory it is known as **Brun's sieve**. The probabilistic interpretation was initiated by Bonferroni. In probabilistic combinatorics, some call this the **Method of Moments** (Janson–Łuczak–Rucinski [2000, p. 140]) or the **Convergence of Moments Method**. Earlier, Spencer called it the **Poisson Paradigm** (see Alon–Spencer [1992]).

We can now sharpen the threshold for the disappearance of isolated vertices.

14.3.25. Remark. *Proposing a sharp threshold.* We had $p = c\frac{\ln n}{n}$ as a threshold because the probability of having no isolated vertices tends to 0 when $c < 1$ and to 1 when $c > 1$. We are *at the threshold* when $p = (1 + \varepsilon_n)\frac{\ln n}{n}$ with $\varepsilon_n \in o(1)$.

To study the sharpness of a probability threshold for a property determined by $X = 0$ when X is a counting function, examine the computation of $\mathbb{E}(X)$. A function $p(n, x)$ such that $\mathbb{E}(X)$ approaches a nonzero constant μ in terms of x yields a "candidate" for a sharp threshold. We may then be able to prove $\mathbb{P}(X = 0) \to e^{-\mu}$ by finding $\lim E_r(X)$ for $r > 1$.

In counting isolated vertices, we have $\mathbb{E}(X) \sim n^{1-c}$ when $p = \frac{c\ln n}{n}$. Setting $c = 1 + \varepsilon$ with $\varepsilon_n = \frac{x}{\ln n}$ yields $\mathbb{E}(X) \to e^{-x}$. We now invoke Theorem 14.3.24. This needs no familiarity with Poisson distributions. It simply yields $\mathbb{P}(X = k) \to e^{-\mu}\mu^k/k!$ when there is a constant μ such that $S_r(X) \to \mu^r/r!$ for $r \in \mathbb{N}_0$.

If the lower order term is properly chosen, then μ should tend to 0 as $x \to \infty$ (since $\mathbb{P}(X = 0) \to 1$ above the threshold), and μ should tend to ∞ as $x \to -\infty$ (since $\mathbb{P}(X = 0) \to 0$ below the threshold). Note that e^{-x} has this property. ∎

14.3.26. Theorem. If $p = \frac{\ln n}{n} + \frac{x}{n}$, then the number X of isolated vertices in G^p is asymptotically Poisson distributed with expectation $\mu = e^{-x}$. That is, $\mathbb{P}(X = k) \to e^{-\mu}\mu^k/k!$. In particular, $\mathbb{P}(X = 0) \to e^{-e^{-x}}$.

Proof: Since $\mathbb{E}(X) \sim n^{1-c}$ when $p = \frac{c\ln n}{n}$, setting $c = 1 + \frac{x}{\ln n}$ yields $\mathbb{E}(X) \to e^{-x}$. By Theorem 14.3.24, it suffices to prove $S_r(X) \to e^{-\mu}\mu^r/r!$; always $S_1(X) = \mathbb{E}(X)$.

Recall that $X = \sum X_i$, with a variable for each vertex. The term in $S_r(X)$ for a set of variables T is the expectation that the vertices corresponding to T in G^p have no incident edges. Hence $S_r(X) = \binom{n}{r}(1-p)^{r(n-r)+r(r-1)/2}$. For fixed r, we use $p \to 0$ to conclude that constant powers of $(1-p)$ tend to 1, obtaining $S_r(X) \sim (n(1-p)^n)^r/r!$. For constant p, this tends to 0, but with $p = (\ln n + x)/n$, we have $n(1-p)^n \to e^{-x}$. Hence $S_r(X) \to \mu^r/r!$. ∎

Our second application allows us to study random d-regular graphs. A thorough survey of models of random d-regular graphs appears in Wormald [1999b].

14.3.27. Definition. (Bender–Canfield [1978]; Bollobás [1980].) The **pairing model** (or **configuration model**) generates a random d-regular multigraph. Start with n groups of d vertices. Pair the nd vertices at random (we can iteratively pick any unmatched vertex and pair it with another unmatched vertex chosen uniformly). Finally, merge each group into a single vertex.

14.3.28. Remark. The pairing model is useful because it has an asymptotically constant probability p of producing a simple graph. This has two important consequences. First, the probability of failing to obtain a simple graph in t tries is $(1-p)^t$, so we can quickly generate a random d-regular graph.

Second, we can study d-regular graphs by studying the pairing model. If an event Q happens whp in the pairing model, then it also happens whp in the restriction of the model to simple graphs, since for fixed d they occupy an asymptotically constant portion of the space as $n \to \infty$. That is, even if all the failures of Q are for simple graphs, still for simple graphs we have $\mathbb{P}(Q) \to 1$. ∎

14.3.29. Theorem. (Bollobás [1980]) The probability that a d-regular multigraph G generated in the pairing model is simple tends to $e^{-(d^2-1)/4}$.

Proof: The probability that G is simple is the product of the probability that G has no loop and the probability that G has no multiedge given that it has no loop.

Let X be the number of loops in G. The probability of a particular loop arising as an edge joining two specified vertices v_i and v_j in a specified group v is $\frac{1}{dn-1}$, since we can generate the graph by first picking the mate of v_i. The number of these variables is $n\binom{d}{2}$, so $\mathbb{E}(X) = \frac{d-1}{2}\frac{1}{1-1/(dn)}$. Thus $\mathbb{E}(X) \to \frac{d-1}{2}$.

Let $\mu = \frac{d-1}{2}$. If $S_r(X) \to \mu^r/r!$, then X is asymptotically Poisson distributed with mean μ, and the probability of having no loops is $e^{-(d-1)/2}$. Pick r of the indicator variables for loops. The probability that all these loops occur is $\prod_{i=1}^{r} \frac{1}{dn-2i+1}$. The number of choices of k such pairs from distinct groups is $n_{(k)}\binom{d}{2}^k$. There can also be k such pairs with a group contributing more than one loop, but there are only $O(n^{k-1})$ such selections. Thus $S_r(X) \sim (\frac{d-1}{2})^k$, as desired.

Given that no loops occur, let Y count the pairs of edges with the same endpoints that arise. When two vertices are selected from each of two groups, they form a multiedge with probability $2\frac{1}{dn-1}\frac{1}{dn-3}$. There are $\binom{n}{2}\binom{d}{2}^2$ such choices of four vertices, so $\mathbb{E}(Y) = (\frac{d-1}{2})^2\frac{1}{1-1/(dn)}\frac{1}{1-3/(dn)} \to \frac{(d-1)^2}{4}$. Letting $\lambda = \frac{(d-1)^2}{4}$, an argument like that above yields $S_r(Y) \to \lambda^r/r!$. Thus $\mathbb{P}(Y = 0) = e^{-(d-1)^2/4}$.

Finally, $e^{-(d-1)/2}e^{-(d-1)^2/4} = e^{-(d^2-1)/4}$. ∎

Theorem 14.3.29 yields the result of Bender–Canfield [1978] and Wormald [1978] that for fixed d, the number of d-regular graphs with vertex set $[n]$ is asymptotic to $\sqrt{2}e^{(1-d^2)/4}(dn/e)^{dn/2}/(d!)^n$ (Exercise 35). The formula holds as long as $d \leq \sqrt{2\log n} - 1$ (Bollobás [1980]) and also for $d \in o(n^{1/3})$ (McKay [1985]). McKay–Wormald [1991] answered the question for $d \in o(n^{1/2})$, showing that then $\ln(\mathbb{P}(G \text{ is simple})) = (1 - d^2)/4 - d^3/(12n) + O((d^2)/n)$.

GRAPH EVOLUTION

We can generate random graphs with vertex set $[n]$ by iteratively adding an edge that is equally likely to be any edge not already present. This yields the same probability space as the uniform model when we have added m edges.

This procedure is called the **random graph process**, introduced in Erdős–Rényi [1959]. The likely effect of a new edge on the present structure gives intuition about the structure of random graphs, which is usually correct.

A **stage of evolution** is a range for $m(n)$ (or $p(n)$) where the properties of the typical graph do not change much. We describe such stages. Verification uses the techniques we have developed, but the computations can be difficult.

14.3.30. Remark. *Stages of evolution.* With many vertices and few edges, each new edge is likely to be isolated. Later, with small components, each new edge is likely to join two components, thereby avoiding cycles (see Exercise 6).

Let $t_k(n) = n^{-k/(k-1)}$. If $p/t_k \to \infty$ but $p/t_{k+1} \to 0$, then whp every tree on k vertices appears but none on $k + 1$ vertices appears. This uses the fact that if $A_1, \ldots, A_r$ each happen whp, and r is constant, then whp all of $A_1, \ldots, A_r$ happen. For a single tree, Theorem 14.3.18 applies.

When $p = c/n$ with $0 < c < 1$, the number of cycles is asymptotically Poisson distributed with constant mean. With only a few cycles and all components small, we still expect the next edge to join two components or create a cycle in an acyclic component. Whp the largest component has about $\ln n$ vertices, and each component has at most one cycle. Most vertices still lie in acyclic components.

As c reaches and passes 1, the structure of G^p changes radically. There is a **double jump**: the structure is significantly different for $c < 1$, $c \sim 1$, and $c > 1$. At $pn = 1$, the Second Moment Method guarantees that whp G^p has a cycle. Also, the order of the largest component jumps from $\log n$ to $n^{2/3}$. With $c = 1 + O(n^{-2/3})$, whp a giant component appears with a linear number of vertices, the other components are small, and there is a cycle with three crossing chords (nonplanar).

Next, p grows to $\frac{c\ln n}{n}$. For $c < 1$, whp G^p has isolated vertices. When $c > 1$, they disappear. Edges added to a disconnected graph may go within a component or connect two. When all components are small, whp new edges join components. After a giant component emerges, new edges tend to lie within it or join it to another component. The components most likely to be absorbed are the larger ones. Therefore, as c passes through 1 the last remaining small components swallowed by the giant component are isolated vertices, making the threshold for connectedness the same as the threshold for disappearance of isolated vertices.

With $c > 1$, suddenly whp G^p also has a spanning cycle. Minimum degree k (and the spanning cycle when $k = 2$) has a threshold that involves a lower order term: $\frac{\ln n}{n} + (k - 1)\frac{\ln\ln n}{n}$. This is closely related to the result in Exercise 24. ∎

The last stages of evolution are $pn/\ln n \to \infty$ with $p = o(1)$, and then constant p. When $p = \frac{c \ln n}{n}$ with $c \to \infty$, increasing density makes evolutionary viewpoint less valuable. We give less study to probability thresholds and concentrate on the likely value of graph parameters. When $p = 1/2$, properties of the random graph can lead to fast algorithms that whp solve a difficult problem.

14.3.31.* Example. *Vertex degrees and isomorphism testing.* Erdős–Rényi [1966] proved that for $pn/\ln n \to \infty$ and fixed positive ε, whp G^p has all vertex degrees between $(1 - \varepsilon)pn$ and $(1 + \varepsilon)pn$. One expects most degrees to be near pn, but some will be farther away. For $p \le \frac{1}{2}$, Bollobás [1982] showed that the vertex of maximum degree is unique in $\mathbb{G}(n, p)$ whp if and only if $pn/\ln n \to \infty$.

For constant p, whp there are many isolated high degrees before the degrees begin to bunch up. Bollobás [1981b] proved that when $t \in o(n/\ln n)^{1/4}$, whp G^p has different degrees for its t vertices of highest degree, and this is sharp.

Babai–Erdős–Selkow [1980] used this in a quadratic-time algorithm to test isomorphism that succeeds whp. It labels the vertices in a canonical order (usually), so that when it labels vertices as $v_1, \ldots, v_n$ in one graph and $w_1, \ldots, w_n$ in another, the only possible isomorphism is the bijection mapping each v_i to w_i.

Let $r = \lfloor 3 \lg n \rfloor$. The canonical labeling sorts the vertex degrees and lets $v_1, \ldots, v_r$ be those of highest degree in order. Almost always (in $\mathbb{G}(n, \frac{1}{2})$) these are uniquely determined. Almost always the other vertices have distinct neighborhoods within $\{v_1, \ldots, v_r\}$. This is tested by expressing the neighborhoods as binary r-tuples and sorting them!

Almost all graphs pass both tests and have no nontrivial automorphisms. Having canonically labeled two such graphs, we test isomorphism by comparing adjacency matrices (see Exercise 41). Babai–Kučera [1979] presented a refinement with smaller probability of rejection.

Deterministic algorithms take longer but give definite answers. For many years, the best time for testing graph isomorphism for n-vertex graphs was $e^{O((n \log n)^{1/2})}$ (see Babai–Kantor–Luks [1983]). A breakthrough by Babai [2015] showed that it can be tested in "quasipolynomial" time: $e^{(\log n)^{O(1)}}$. ∎

Similarly, some NP-hard problems are trivial for random graphs. Although $\Delta(G) \le \chi'(G) \le \Delta(G)+1$ for every graph G (Vizing [1964]), deciding between these values is NP-hard (Holyer [1981]). Vizing proved that $\chi'(G) = \Delta(G) + 1$ requires at least three vertices of maximum degree. By the uniqueness of the vertex of maximum degree in $G^{1/2}$, Erdős–Wilson [1977] observed that whp $\chi'(G) = \Delta(G)$.

Often the distribution of a graph parameter over $\mathbb{G}(n, p)$ is nontrivial but still is highly concentrated. Given a parameter μ, we want to show $\mu(G^p) \sim f(n)$ whp. We often write $\mu(G^p) = (1 + o(1))f(n)$, meaning that for any positive ε, whp $\mu(G^p)$ is between $(1 - \varepsilon)f(n)$ and $(1 + \varepsilon)f(n)$. Another way of expressing this extends the notion of threshold function from Definition 14.3.13.

14.3.32. Definition. Given a sequence of probability spaces indexed by a parameter n, a **threshold** t for a parameter θ is a function of n such that $\mathbb{P}(\theta \ge k)$ tends to 1 or 0 depending on whether k is above or below the threshold t. As in Remark 14.3.14, "above or below" may mean that the ratio to t tends to ∞ or 0, or it may just mean they differ by a small constant.

The exercises contain examples for various random structures. Here we study connectivity, independence/clique number, and chromatic number in graphs.

14.3.33. Example. *Connectivity of random graphs.* For sparse graphs, the thresholds for a fixed value of the connectivity and the minimum degree are the same. When p is fixed, the expected number of common neighbors for two fixed vertices is $p^2(n-2)$. The techniques in the next section permit showing that whp all pairs of vertices have nearly this many common neighbors, which makes the connectivity linear in n (see Exercise 14.4.3). However, requiring k common neighbors is too restrictive for the best result; Bollobás [1981b] showed for constant p that whp G^p has connectivity equal to minimum degree. ∎

What about clique number? For fixed k, we have derived probability thresholds for the appearance of a k-clique, but for constant p the clique number grows with n. Determining the size of the largest clique in an input graph is NP-complete, but for the random graph we can guess the correct value with high probability without looking at the graph! Amazingly, for constant p, whp G^p has one of two possible values for the clique number (as a function of n). Indeed, for each $k \in \mathbb{N}$ there is a range of n where the clique number is highly likely to be k. The approach is to find bounds on the function $r(n)$ such that whp G^p has an r-clique and whp has no $(r+1)$-clique.

14.3.34. Theorem. (Matula [1972]) For constant p and fixed $\varepsilon > 0$, whp G^p has clique number between $\lfloor d - \varepsilon \rfloor$ and $\lfloor d + \varepsilon \rfloor$, where $d = 2\log_{1/p} n - 2\log_{1/p}\log_{1/p} n + 1 + 2\log_{1/p}(e/2)$.

Proof: (sketch) Let $b = 1/p$. If X_r is the number of r-cliques, then $\mathbb{E}(X_r) = \binom{n}{r}p^{\binom{r}{2}}$. Since $r! \sim (r/e)^r\sqrt{2\pi r}$ (Stirling's Formula),

$$\mathbb{E}(X_r) \sim (2\pi r)^{-1/2}(\tfrac{en}{r}p^{(r-1)/2})^r.$$

If $r \to \infty$ and $(enr^{-1}p^{(r-1)/2}) \le 1$, then $\mathbb{E}(X_r) \to 0$. To obtain such r, take logarithms (base b) in the inequality and solve for r to find

$$r \ge 2\log_b n - 2\log_b r + 1 + 2\log_b e.$$

This is approximately equivalent to $r \ge d(n)$ as defined above. More precisely, if $r > d + \varepsilon$, then whp G^p has no clique of size r.

The lower bound uses the Second Moment Method, as in Theorem 14.3.18, but the dependence of r on n makes the analysis more difficult. The expectation of X_r^2 sums the probability of common occurrence for all ordered pairs of r-cliques. This probability depends only on the number of common vertices, so

$$\mathbb{E}(X_r^2) = \binom{n}{r}\sum_{k=0}^{r}\binom{r}{k}\binom{n-r}{r-k}p^{2\binom{r}{2}-\binom{k}{2}}.$$

We want to show that the term for $k = 0$ dominates. Hence we write $\mathbb{E}(X_r^2)/\mathbb{E}(X_r)^2 = \alpha_n + \beta_n$, where $\alpha_n = \binom{n}{r}^{-1}\binom{n-r}{r}$ and $\beta_n = \binom{n}{r}^{-1}\sum_{k=1}^{r}\binom{r}{k}\binom{n-r}{r-k}b^{\binom{k}{2}}$. We seek $\alpha_n \sim 1$ and $\beta_n \to 0$. When $r \sim 2\log_b n$, an asymptotic formula for $\binom{a}{k}/\binom{b}{k}$ yields $\alpha_n \sim e^{-r^2/(n-r)} \to 1$. The discussion of β_n is harder; Palmer [1985, pp. 75–80] presents further details. ∎

Finally, consider chromatic number for constant p. Since $1 - p$ is also constant, the results on clique number apply. With high probability, no independent set has size more than $(1 + o(1))2 \log_b n$, where $b = 1/(1 - p)$. Hence whp $\chi(G^p) \geq (1 + o(1))\frac{n}{2 \log_b n}$. Achieving this bound requires many disjoint independent sets with near-maximum sizes. For a decade, the best result was an algorithmic guarantee of a coloring with at most twice the number of colors in the lower bound.

Bollobás [1988] proved that the lower bound is achievable by showing that whp, in G^p *every* set having at least $n/(\log_b n)^2$ vertices contains a clique of order at least $2 \log_b n - 5 \log_b \log_b n$. This allows independent sets of near-maximum size to be extracted until too few vertices remain to cause trouble; the remainder can be given distinct colors.

We explore Bollobás' approach in the next section. Here we present the earlier result for its algorithmic interest; whp the greedy algorithm uses at most $(1 + \varepsilon)n/\log_b n$ colors. Thus it usually works as an approximation algorithm in the same sense that the isomorphism algorithm in Example 14.3.31 usually works. Garey–Johnson [1976] showed no fast algorithm uses at most twice the optimum number of colors on *every* graph, unless P = NP. Bollobás' proof does not yield a fast algorithm for coloring almost every graph with an asymptotically optimal number of colors; it is an existence proof only.

14.3.35. Theorem. (Grimmett–McDiarmid [1975]) With edge-probability p, set $b = 1/(1 - p)$. For $\varepsilon > 0$, whp G^p satisfies

$$(\tfrac{1}{2} - \varepsilon)n/\log_b n \leq \chi(G^p) \leq (1 + \varepsilon)n/\log_b n.$$

Proof: For the lower bound, use independent sets as suggested above. For the upper bound, let $h = \lfloor (1 + \varepsilon)n/\log_b n \rfloor$. We show that greedily coloring $v_1, \ldots, v_n$ in order whp uses at most h colors on G^p. Among the n-vertex graphs using more colors, let $\mathbf{B}_m$ be the set such that v_m is the first vertex to use color $h + 1$. We prove $\sum_{m=1}^n \mathbb{P}(\mathbf{B}_m) \to 0$ as $n \to \infty$.

Given G, let $G_m = G[v_1, \ldots, v_{m-1}]$. Before color $h + 1$ is used, color h must be used, so for each $G \in \mathbf{B}_m$ the greedy coloring of G_m uses h colors. Let k_i be the number of uses of color i in this coloring. To require use of color $h + 1$, v_{m+1} must have at least one neighbor of each color $1, \ldots, h$. Given the numbers $\{k_i\}$, the probability of this is $\prod_{i=1}^h [1 - (1 - p)^{k_i}]$.

Bollobás–Erdős [1976] simplified the subsequent computations involving this bound by observing that it is maximized when $k_1 = \cdots = k_h$ (Exercise 40). Thus

$$\prod_{i=1}^h [1 - (1 - p)^{k_i}] \leq [1 - (1 - p)^{(m-1)/h}]^h < [1 - (1 - p)^{n/h}]^h.$$

Let $b_n = [1 - (1 - p)^{n/h}]^h$. Given G_m, the probability that the full graph G belongs to $\mathbf{B}_m$ is at most b_n. Since the bound holds for each G_m, we conclude that $\mathbb{P}(\mathbf{B}_m) < b_n$. This holds for all m, so $\sum_{m=1}^n \mathbb{P}(\mathbf{B}_m) < nb_n$.

Using $1 - x < e^{-x}$, we obtain $nb_n < ne^{-h(1-p)^{n/h}}$. Substituting $h = cn/\log_b n$ yields $(1 - p)^{n/h} = n^{-1/c}$. The logarithm of the bound is $\log n - cn^{1-1/c}/\log_b n$. It tends to $-\infty$ for $c > 1$, so the bound on the probability that the greedy algorithm uses more than h colors tends to 0. ∎

14.3.36. Remark. Some properties known to hold for almost every graph occur in no known examples! For the known lower bound on diagonal Ramsey numbers, there is still no construction of a sequence of graphs such that $\alpha(G) < \log_{\sqrt{2}}(|V(G)|)$ and $\omega(G) < \log_{\sqrt{2}}(|V(G)|)$, even though almost every graph has this property.

The point is that the range from a billion to infinity is much bigger than from 1 to a billion. A statement that is true for almost every graph may not be true for many graphs that are small enough to look at. The probabilistic method gives good results when almost every graph has a desired property, but even then "almost every" is only asymptotic.   ∎

EXERCISES 14.3

14.3.1. (−) Let H be a graph. For constant p, prove that whp G^p contains H as an induced subgraph.

14.3.2. (−) *Generalization of Theorem 14.3.10.*
(a) Fix k, s, t, p. Prove that whp G^p has the following property: for every choice of disjoint vertex sets S and T of sizes s and t, there are at least k vertices that are adjacent to every vertex of S and to no vertex of T.
(b) Conclude that whp G^p is k-connected.
(c) Apply the same argument to random tournaments: whp, for every choice of disjoint vertex sets S and T of sizes s and t, there are at least k vertices with an edge to every vertex of S and from every vertex of T.

14.3.3. (−) Let $\langle \omega \rangle$ be any sequence tending to ∞, fix $p \in (0, 1)$, and let $b = 1/(1-p)$. Prove that whp the first $\log_b(n\omega_n)$ vertices in $[n]$ form a dominating set in the graph G^p.

14.3.4. (−) Determine the smallest connected graph that is not balanced.

14.3.5. For fixed p, and $k \in o(n/\ln n)$, prove that whp G^p is k-connected.

14.3.6. (◊) In $\mathbb{G}(n, p)$ with p depending on n, prove that if $pn \to 0$, then whp G^p has no cycles. Use a *different* random variable to prove that if $pn \to \infty$, then whp G^p has a cycle. (Comment: The result of Theorem 14.3.18 is stronger than this and is not permitted for use here; the desired statements for cycles can be proved with simpler arguments.)

14.3.7. (◊) By Theorem 13.2.19 or Theorem 14.2.5, we know $\mathrm{ex}\,(n; K_{r,r}) < cn^{2-1/r}$ for some constant c. Use the Deletion Method to prove $\mathrm{ex}\,(n; K_{r,r}) > cn^{2-2/(r+1)}$ for some constant c when n is large. (Erdős–Spencer [1974, p. 61])

14.3.8. (◊) With $p = \frac{1}{2}$, whp G^p has a perfect matching when n is even. Let $v_1, \ldots, v_{2k}$ be the vertices. For k rounds, iteratively match the least-indexed vertex that remains to a remaining neighbor, if any exists; otherwise pair it with an arbitrary remaining vertex. Delete these two vertices and continue. Prove that this "**Method of Deferred Decisions**" finds a perfect matching with probability more than $1/3$. (Molloy–Reed [2002, p. 20])

14.3.9. For $p = 1/n$ and fixed $\varepsilon > 0$, show that whp G^p has no component with more than $(1 + \varepsilon)n/2$ vertices. (Hint: Bound the probability of such a component by the probability of another event whose probability tends to 0.)

14.3.10. Suppose that years have t days and that each person's birthday is random among them. Let $\mathbb{P}(s, t)$ be the probability that no two among s given people have the same birthday. Prove that $\mathbb{P}(s, t) \to 0$ as $s \to \infty$ if $t \in o(s^2)$. (Comment: When $t = cs^2$, $\mathbb{P}(s, t)$ tends to a constant depending on c.) (Griggs [1998])

14.3.11. ($\diamond$) Find a small window around $p = 1/2$ such that above it whp G^p has at least $\frac{1}{2}\binom{n}{2}$ edges, below it whp G^p has at most $\frac{1}{2}\binom{n}{2}$ edges, and within it neither statement holds.

14.3.12. Derive a "local" probability threshold in the binomial model for the property "Every edge belongs to a triangle" (the property is not monotone, but within an interval for p there is a threshold). (Hint: Let X be the number of edges of G not in triangles.)

14.3.13. Extend the Second Moment argument of Theorem 14.3.18 to prove that $n^{-1/\rho(H)}$ is a threshold function for the appearance of H as a subgraph of G^p, where $\rho(H)$ is the maximum density of H. (Bollobás [1981a], Rucínski–Vince [1985])

14.3.14. A random tournament T with vertex set $[n]$ is generated by orienting each edge ij as $i \to j$ or $j \to i$ independently with probability $\frac{1}{2}$. A **king** in a tournament is a vertex such that every other vertex can be reached from it by a path of length at most 2. By Proposition 5.2.18, every tournament has a king. Prove that whp in T every vertex is a king. (Palmer [1985]) (Comment: Moon–Moser [1962b] proved that whp T is strongly connected, with probability of failure between $(2n - 1)/2^{n-1}$ and $(2n + 1)/2^{n-1}$.)

14.3.15. ($\diamond$) A k-vertex tournament is **transitive** if it has a vertex ordering $u_1, \ldots, u_k$ such that $u_i \to u_j$ if and only if $i < j$.
    (a) Given a positive constant ε, prove that the fraction of n-vertex tournaments having no transitive subtournament with more than $(1 + \varepsilon) + 2 \lg n$ vertices tends to 1. (Comment: *Every* n-vertex tournament has a transitive subtournament with at least $\lg n$ vertices.)
    (b) Use part (a) to prove $R(k, k) > 2^{(k-2)/2}$ for Ramsey numbers. (Erdős–Hajnal)

14.3.16. For n-vertex tournaments, let Q_k be the property that every set of k vertices has a common successor. Prove that if $k = \lg n - (2 + \varepsilon) \lg \lg n$, where ε is a positive constant, then whp a random n-vertex tournament satisfies Q_k. (Erdős [1963])

14.3.17. ($\diamond$) With $p = (1 - \varepsilon) \ln n / n$, find a large m such that whp G^p has at least m isolated vertices. In particular, what $m(n)$ results from Chebyshev's Inequality?

14.3.18. For fixed p, find a threshold for the maximum k such that whp every k vertices in G^p have a common neighbor.

14.3.19. Let Q_k be the following graph property (in $\mathbb{G}(n, p)$): for every choice of disjoint vertex sets S, T of size k, there is an edge with endpoints in S and T.
    (a) For constant k, obtain a candidate for a threshold probability for property Q_k. Explain what is needed to prove that it is a threshold for occurrence of Q_k. (It may be helpful to write it as a threshold on $1 - p$.)
    (b) For $k = c \lg n$, prove that almost every graph (edge probability $\frac{1}{2}$) has property Q_k if $c > 2$. (Comment: Thus under every vertex ordering, some edge will be "stretched" to length at least $n - 2 \log n$; see "bandwidth" in Section 16.2.)

14.3.20. ($\diamond$) Prove that a longest constant string in a list of n random heads and tails has length $(1 + o(1)) \lg n$. That is, for $\varepsilon > 0$, the probability that a random list has at least $(1 + \varepsilon) \lg n$ consecutive identical flips tends to 0, but the probability that it has at least $(1 - \varepsilon) \lg n$ consecutive identical flips tends to 1.

14.3.21. A **monotone sequence** of vectors in $\mathbb{R}^d$ is a sequence for which the values in each coordinate form a monotone sequence in $\mathbb{R}$ (nondecreasing or nonincreasing). Construct a sequence $v_1, \ldots, v_n$ in $\mathbb{R}^d$ by letting each coordinate be a random permutation of $[n]$. Prove that the expected length of the longest monotone subsequence is bounded by a constant times $n^{1/(d+1)}$. (Comment: By Theorem 10.1.15, any sequence of n vectors in $\mathbb{R}^d$ has a monotone subsequence of length at least $n^{1/2^d}$, and this is sharp (Exercise 10.1.28).)

14.3.22. ($\diamond$) A computer has n messages to send along a channel. Messages are dropped with constant probability p, independently. The computer chooses the next message to transmit uniformly at random from among the n messages. Determine a threshold for the number of transmissions so that whp at least one copy of each message is received.

14.3.23. Each box of cereal contains one of n prizes, equally likely, independently. Obtain a very accurate threshold for the number of boxes that should be opened to obtain at least two copies of each prize. In particular, $n \ln n$ is not sufficiently accurate; adjusting a lower-order term changes the situation from almost-certain failure to almost-certain success.

14.3.24. (+) Let f be a random function from $[m]$ to $[n]$; all n^m functions are equally likely. Let Q be the property that each element of $[n]$ occurs more than k times in the image of f. Prove that $n \ln n + cn \ln \ln n$ with $c = k$ is a threshold on m for Q.

(Hint: Prove that when $p = o(1)$ and $mp \to \infty$ and k is constant, the probability of at most k successes in m coin flips with success probability p is asymptotic to the probability of exactly k successes. Comment: When $k = 0$, this problem extends the Coupon Collector Problem (Exercise 14.1.30) by narrowing the threshold.)

14.3.25. ($\diamond$) Suppose that each trial produces one of n coupons, uniformly and independently. Prove that after $n(\ln n + x)$ trials, the number of coupons that have never been seen has a Poisson distribution with mean e^{-x}.

14.3.26. ($\diamond$) For a real constant c, determine the asymptotic probability that a graph drawn from $\mathbb{G}(n, c/n)$ has no triangle.

14.3.27. For fixed $k \in \mathbb{N}$ and $x \in \mathbb{R}$, consider $\mathbb{G}(n, p)$ with $p = 1 - (\frac{k \ln n + x}{n})^{1/k}$. Determine the asymptotic expected number of dominating sets of size k. What does this suggest about the probability of having a dominating k-set, and what would be needed to prove it?

14.3.28. Obtain a sharp threshold for diameter at most 2. In particular, find $p(n, x)$ such that $\mathbb{P}(\operatorname{diam}(G^p) \le 2) \to e^{-\mu}$, where μ is a function of the parameter x, and $\mu \to 0$ as $x \to \infty$ and $\mu \to \infty$ as $x \to -\infty$. (Hint: Drop lower-order terms when discussing asymptotics of the appropriate random variable.)

14.3.29. The **moment generating function** of a nonnegative integer-valued random variable is the generating function $\sum a_n x^n$, where $a_n = \mathbb{E}(X^n)$. Show that if the moment generating function $\mathbb{E}(e^{tX})$ is finite for some interval of t around 0, then knowing all the moments of a distribution of X is equivalent to knowing the distribution.

14.3.30. For $X = \sum_{i=1}^m X_i$, let $S_r(X)$ be the rth binomial moment of X (Definition 14.3.20). Given $S_r(X) = \sum_{j=r}^m \binom{j}{r} \mathbb{P}(X = j)$ as in Lemma 14.3.21, use direct substitution and computation to prove $\mathbb{P}(X = k) = \sum_{r=k}^m (-1)^{r-k} \binom{r}{k} S_r(X)$.

14.3.31. Prove $\sum_{l=0}^a (-1)^l \binom{b}{l} = \begin{cases} 1 & \text{if } b = 0, \\ 0 & \text{if } 0 < b \le a, \\ (-1)^a \binom{b-1}{a} & \text{if } b \ge a. \end{cases}$

14.3.32. Let $p = c/n$ for some constant c.
  (a) Prove for $s < n(20c^2)$ that whp no set of s vertices in G^p induces at least $2s$ edges.
  (b) Prove that whp $\Delta(G) \le \ln n / \ln \ln n$.
  (c) Prove that whp the vertices of degree at least $\frac{2 \ln n}{3 \ln \ln n}$ form an independent set.

14.3.33. ($\diamond$) Determine the asymptotic probability that a random d-regular graph generated using the pairing model has no triangle.

14.3.34. ($\diamond$) In the pairing model, one can generate the random matching sequentially, picking at each step any desired point to initiate the next pair. Use this with $d = 2$ to prove that in a random 2-regular n-vertex graph the expected number of cycles is asymptotic to $\frac{1}{2} \ln n$. Compare with Theorem 3.1.20. (Alon–Pralat–Wormald [2008])

14.3.35. From Theorem 14.3.29, prove that the number of d-regular graphs on vertex set $[n]$ is asymptotic to $\sqrt{2}e^{(1-d^2)/4}(dn/e)^{dn/2}/(d!)^n$. (Bender–Canfield [1978], Wormald [1978])

14.3.36. ($\diamond$) In the d-regular pairing model on n vertices, prove that the number of vertices in cycles of length at most $\log_{d-1}\frac{n}{\omega_n}$ is $o(n)$, where ω_n is any sequence tending to ∞. (Hint: Explore the graph from a given vertex u to get a bound on the expected number of edges that can yield short cycles through u.)

14.3.37. Consider the model $\mathbb{G}(n,p)$ with p constant. Let $r = c\log_{1/p} n$, where c is constant; in this range, we can approximate $\binom{n}{r}$ by $(ne/r)^r$. For $c > 2$, prove that whp G^p has no r-clique. State what is needed to prove that when $c < 2$, whp G^p has an r-clique.

14.3.38. ($\diamond$) The **strength** of a theorem $A \Rightarrow B$ in a probability space is $\frac{\mathbb{P}(A)}{\mathbb{P}(B)}$. A theorem with strength 0 is useless, since the hypothesis never holds; a theorem with strength 1 characterizes its conclusion. (Palmer [1985, pp. 84–85])

(a) Use the Second Moment Method to prove that almost every graph fails Ore's Condition (Corollary 7.3.7) for spanning cycles, so Ore's Theorem has asymptotic strength 0.

(b) Prove asymptotic strength 1 for the Chvátal–Erdős Theorem (Theorem 7.3.14). (Comment: Chvátal's Condition (Theorem 7.3.11) cannot hold for both G and $\overline{G}$ (Kostochka–West [2006]; Exercise 7.3.42). Thus Theorem 7.3.11 has strength at most $\frac{1}{2}$ for each n. It is stronger than Ore's Theorem, but whether its asymptotic strength is nonzero is unknown.)

14.3.39. The **boxicity** of G is the least number of interval graphs whose intersection is G. The **interval number** of G is the least t such that G is the intersection graph of subsets of $\mathbb{R}$ composed of at most t intervals. Prove that the interval number and boxicity of the random graph ($G^{1/2}$) are each at least $n/(4\lg n)$. (Erdős–West [1985])

14.3.40. Suppose that $0 < p < 1$ and that $k_1, \ldots, k_r$ are nonnegative integers summing to m. Prove that $\prod_{i=1}^{r}[1-(1-p)^{k_i}] \leq [1-(1-p)^{m/r}]^r$.

14.3.41. Given the statements in Example 14.3.31 about the behavior of vertex degrees in $\mathbb{G}(n,\frac{1}{2})$, prove that the isomorphism algorithm there runs in time $O(n^2)$ for n-vertex graphs and works whp. Assume that n numbers can be sorted using $O(n\log n)$ pairwise comparisons. (Babai–Erdős–Selkow [1980])

14.3.42. ($\diamond$) *Perfect matchings in bipartite graphs.* Fix $\varepsilon > 0$. Let G be a random subgraph of $K_{n,n}$, with parts A and B and independent edge probability $(1 + \varepsilon)\frac{\ln n}{n}$. Say S *fails* if $|N(S)| < |S|$. By Hall's Theorem, G has a perfect matching if and only if no set fails.

(a) Prove that if $\varepsilon < 0$, then whp G has no perfect matching.

(b) For a minimal failing set S, prove $|N(S)| = |S| - 1$ and $G[S \cup N(S)]$ is connected.

(c) Prove that if G has no perfect matching, then A or B contains a failing set with at most $\lceil n/2 \rceil$ elements.

(d) If $r, s \geq 1$, then $K_{r,s}$ has $r^{s-1}s^{r-1}$ spanning trees. Use this, part (b), part (c), and Markov's Inequality to prove that if $\varepsilon > 0$, then whp G has a perfect matching. (Hint: A summation in the bound on the expected number of minimal failing sets can be bounded by a geometric series.)

14.3.43. An X, Y-bigraph G with $|X| = |Y| = n$ is an (n, α, β, d)-**expander** if G is regular of degree d and $|N(S)| \geq \beta|S|$ whenever $|S| < \alpha n$. Expanders of constant degree permit rapid widespread communication using only a linear number of edges.

(a) Consider an experiment in which dk-subsets of $[n]$ are chosen at random, and let X be the size of the union. Prove that $\mathbb{P}(X \leq l) \leq \binom{n}{l}(\frac{l}{n})^{kd}$.

(b) Use the probabilistic method to prove that if $\alpha\beta < 1$, then there is a constant d such that, for all n sufficiently large, an (n, α, β, d)-expander exists. (Hint: Using a suitable probability space on d-regular multigraphs, bound the probability that some set fails the expansion property by a geometric series. Then choose d so that the sum will be less than 1. Use $\binom{n}{k} < (ne/k)^k$.)

706 Chapter 14: The Probabilistic Method

14.4. Concentration Inequalities

In the Second Moment Method to obtain threshold functions, we used the second moment to show that the random variable is unlikely to be far from its expectation. We derived the Second Moment Method using the simplest result of this sort: Chebyshev's Inequality, which is obtained by applying Markov's Inequality to the squared deviation from the mean.

The tighter the bound on the "tail probability" (the probability of being at least a specified distance from the mean), the stronger the results that can be proved. The Chernoff Bound strengthens Chebyshev's Inequality when applied to binomial random variables.

A binomial variable is the result of a *sequence* of trials. Sequences of random variables in which the expectation of the next variable is the value of the current one are "martingales". In terms of the amount by which successive variables in a martingale can differ, Azuma's Inequality provides a tight bound on tail probability for the final outcome. Intuitively, a lengthy probabilistic process displays more consistent and predictable global behavior than individual steps do.

The computations required to apply the tail inequalities are often simpler than computing moments, because moment computations are subsumed in the development of these techniques. The theory develops paradigms that can be applied without repeating difficult computations. McDiarmid [1998] surveyed concentration inequalities and their applications in discrete mathematics. Molloy–Reed [2002] discusses many of these results in the context of applications to graph coloring. Dubhashi–Panconesi [2009] also develops the concentration inequalities and applies them to the analysis of randomized algorithms.

CHEBYSHEV AND CHERNOFF BOUNDS

We proved Chebyshev's Inequality to develop the Second Moment Method (Lemma 14.3.16). For a random variable X, Chebyshev's Inequality is $\mathbb{P}(|X - \mathbb{E}(X)| \geq t) \leq \text{Var}(X)/t^2$, where the variance $\text{Var}(X)$ is defined as $\mathbb{E}(X - \mathbb{E}(X))$ and equals $\mathbb{E}(X^2) - \mathbb{E}(X)^2$. We begin by applying Chebyshev's Inequality to a problem in combinatorial number theory (see Alon–Spencer [1992] and Molloy [1998]).

14.4.1. Example. A set A consisting of k positive integers has **distinct sums** if the sums of its 2^k subsets are distinct. By using powers of 2, one can construct a subset of $[n]$ having distinct sums that has size $\lfloor \lg n \rfloor + 1$. Can there be such a subset that is much bigger? Erdős offered \$300 for a determination of whether the maximum size is $\lg n + O(1)$; the question remains open.

The sum of a k-set in $[n]$ is less than kn. If the sums are distinct, then $2^k \leq kn$, which yields $k \leq \lg n + \lg \lg n + 1$. This easy upper bound is not far from the lower bound. Chebyshev's Inequality allows us to bring the upper bound halfway to the lower bound. ∎

14.4.2. Theorem. Every subset of $[n]$ having distinct sums has size at most

$$\lg n + \tfrac{1}{2}\lg\lg n + O(1).$$

Proof: Let $\{a_1,\dots,a_k\}$ be a subset of $[n]$ with distinct sums. Let $s = \sum_{j=1}^{k} a_j$. If k is near the bound $\lg n + \lg\lg n + 1$, then kn is not much bigger than 2^k, which suggests that there must be both small sums and sums near s. Our aim is to show to the contrary that when the sums are distinct, most of them lie in a much smaller interval near the middle of $[s]$. This will improve the upper bound.

Since the 2^k sums are distinct, we can choose uniformly from the sums by choosing a random subset of $a_1,\dots,a_k$. Using each with probability $\tfrac{1}{2}$, independently, let X be the sum of the elements selected. By linearity, $\mathbb{E}(X) = s/2$.

To study the variance, write X as a linear combination of indicator variables: $X = \sum_{i=1}^{k} a_i X_i$. Since $\mathbb{E}(X_i^2) = \tfrac{1}{2}$ and $\mathbb{E}(X_i X_j) = \tfrac{1}{4}$,

$$\mathbb{E}(X^2) = \frac{1}{2}\sum_{i=1}^{k} a_i^2 + \frac{2}{4}\sum_{i<j} a_i a_j.$$

On the other hand, $\mathbb{E}(X) = \tfrac{1}{2}\sum_{i=1}^{k} a_i$, so $\mathbb{E}(X)^2 = \tfrac{1}{4}\sum_{i=1}^{k} a_i^2 + \tfrac{2}{4}\sum_{i<j} a_i a_j$. The difference is $\tfrac{1}{4}\sum_{i=1}^{k} a_i^2$. Since $a_i \le n$ for all i, we have $\mathrm{Var}(X) < \tfrac{1}{4}n^2 k$. Using $t = 2\sqrt{\mathrm{Var}(X)}$ in Chebyshev's Inequality yields

$$\mathbb{P}\left(|X - \mathbb{E}(X)| \ge n\sqrt{k}\right) < \tfrac{1}{4}.$$

Since the 2^k possible values for X are equally likely, we conclude that at least $3/4$ of the possible sums lie in an interval of length $2n\sqrt{k}$ around $s/2$. That is, $\tfrac{3}{4}2^k \le 2n\sqrt{k}$, yielding approximately $k \le \lg n + \tfrac{1}{2}\lg\lg n$. ∎

For binomial random variables, the Chernoff Bound is much tighter on the probability of large deviations from the mean. This leads to results that cannot be proved by Chebyshev's Inequality. First, Chebyshev for binomial variables:

14.4.3. Example. A random variable X has the **binomial distribution** $\mathrm{Bin}(n,p)$ if $\mathbb{P}(X = k) = \binom{n}{k}p^k(1-p)^{n-k}$. This distribution arises when X is the sum of n independent **Bernoulli random variables** $X_1,\dots,X_n$, which are $0,1$-random variables with $\mathbb{P}(X_i = 1) = p$. Since they are independent, $\mathbb{E}(X_i X_j) = \mathbb{E}(X_i)\mathbb{E}(X_j) = p^2$. Also $X_i^2 = X_i$. Thus $\mathbb{E}(X^2) = np + n(n-1)p^2 = np(np+1-p)$, and

$$\mathrm{Var}(X) = \mathbb{E}(X^2) - \mathbb{E}(X)^2 = np(np+1-p) - n^2p^2 = np(1-p).$$

The square root of the variance is the **standard deviation**. Letting $t = \lambda\sqrt{\mathrm{Var}(X)}$, Chebyshev's Inequality yields $\mathbb{P}\left(|X - \mathbb{E}(X)| \ge \lambda\sqrt{\mathrm{Var}(X)}\right) \le \lambda^{-2}$. The number of heads is unlikely to be many multiples of $\sqrt{n}$ from the mean. ∎

Chebyshev's Inequality came from applying Markov's Inequality to a transformed variable, the squared deviation from the mean. This is also the method for the Chernoff Bound, using the variable e^{uX}, where u is a parameter. Note that $\mathbb{E}(e^{uX}) = \mathbb{E}\left(\sum (uX)^k/k!\right) = \sum \mathbb{E}(X^k)u^k/k!$. This is the exponential generating function for the moments of X, called the **moment generating function**.

Using the exponential to improve the tail bound is due to Bernstein in the 1920s. Chernoff [1952] strengthened the analysis for binomial random variables.

14.4.4. Theorem. (**Chernoff Bound**; Chernoff [1952]) If the variable X has the binomial distribution $\mathrm{Bin}(n, p)$ with $0 < p < 1$ and $t > 0$, then

$$\mathbb{P}\big(X - np \geq nt\big) \leq e^{-2nt^2} \qquad \text{and} \qquad \mathbb{P}\big(X - np \leq -nt\big) \leq e^{-2nt^2}.$$

Proof: It suffices to prove the upper tail bound. For the lower, we then apply the bound on $\mathbb{P}(X' - n(1 - p) \geq nt)$, where $X' = n - X$.

Let $q = 1 - p$. If $t \geq q$, then $\mathbb{P}(X - np \geq nt) = 0$, so we may assume $t < q$. Let $m = n(p + t)$. For $u > 0$, Markov's Inequality yields

$$\mathbb{P}(X \geq m) = \mathbb{P}(e^{uX} \geq e^{um}) \leq \mathbb{E}(e^{uX})/e^{um}.$$

To obtain the best bound, we will pick u to minimize $\mathbb{E}(e^{uX})/e^{um}$. Note that $X = \sum_{i=1}^{n} X_i$, where $X_1, \ldots, X_n$ are independent $0, 1$-variables with $\mathbb{P}(X_i = 1) = p$. Using multiplicativity of the expectations of independent variables,

$$\mathbb{E}(e^{uX}) = \mathbb{E}\left(\prod e^{uX_i}\right) = \prod \mathbb{E}\left(e^{uX_i}\right) = (q + pe^u)^n.$$

Thus $\mathbb{E}(e^{uX})/e^{um} = \left(\frac{q+pe^u}{e^{u(p+t)}}\right)^n$. With $x = e^u$, we can choose any x to get a bound; calculus tells us that setting $x = \frac{(p+t)q}{p(q-t)}$ will minimize the function. That is, we set $u = \ln \frac{(p+t)q}{p(q-t)}$. With this choice, careful manipulation (noting $t < q$) yields

$$\frac{q + pe^u}{e^{u(p+t)}} = \left(\frac{p}{p+t}\right)^{p+t} \left(\frac{q}{q-t}\right)^{q-t}.$$

It now suffices to bound this quantity by e^{-2t^2}. Let $f(t) = \ln\left(\left(\frac{p}{p+t}\right)^{p+t} \left(\frac{q}{q-t}\right)^{q-t}\right)$. By computation, $f'(t) = \ln \frac{p(q-t)}{(p+t)q}$ and $f''(t) = \frac{-1}{(p+t)(q-t)} \leq -4$. Since $f(0) = f'(0) = 0$, it follows that $f(t) = \int_0^t \int_0^t f''(s)ds \leq -2t^2$. ∎

14.4.5. Remark. *Other variations.* With $\mu = \mathbb{E}(X)$, Theorem 14.4.4 states

$$\mathbb{P}\big(X \geq \mu + s\big) \leq e^{-2s^2/n} \qquad \text{and} \qquad \mathbb{P}\big(X \leq \mu - s\big) \leq e^{-2s^2/n}.$$

Other variations are more useful when the difference from the mean is larger or when p is small. Another upper tail bound is

$$\mathbb{P}(X \geq \mu + s) \leq e^{-s^2/(2\mu+s)} \qquad \text{or} \qquad \mathbb{P}(X \geq (1 + \delta)\mu) \leq e^{-\delta^2\mu/(2+\delta)}.$$

In the notation of Theorem 14.4.4, this bound is $\mathbb{P}(X - np \geq nt) \leq e^{-nt^2/(2p+t)}$, improving Theorem 14.4.4 when $2p + t < 1/2$. A stronger version reducing the denominator from $2\mu+s$ to $2(\mu+s/3)$ is proved in Janson–Łuczak–Rucinski [2000, pp. 26–28] and Bonato–Prałat [2018]. See also Mitzenmacher–Upfal [2017].

A bound on the lower tail that improves Theorem 14.4.4 when $p < 1/4$ is

$$\mathbb{P}(X \leq \mu - s) \leq e^{-s^2/(2\mu)} \qquad \text{or} \qquad \mathbb{P}(X - np \leq -nt) \leq e^{-nt^2/(2p)}.$$

For small deviations from the mean, combining alternative upper and lower bounds yields $\mathbb{P}\left(|X - \mu| \geq \varepsilon\mu\right) \leq 2e^{-\varepsilon^2\mu/3}$, valid when $\varepsilon \leq 1$. ∎

The Chernoff Bound is much stronger because Chebyshev only uses pairwise independence, while Chernoff uses the full independence of the trials. Exercise 2 compares them. In essence, Chernoff uses all the moments.

Hoeffding [1963] extended the Chernoff Bound to more general sums of independent random variables. The crucial step in the proof of Theorem 14.4.4 required independence, but the computation did not essentially require that $X_1, \ldots, X_n$ be integer-valued, identical, or of limited range.

The proof generalizes immediately when $\mathbb{P}(X_i = 1) = p_i$, with $p = \sum p_i/n$, because the Arithmetic–Geometric Mean Inequality yields $\prod(p_i e^u + 1 - p_i) \leq [\sum(p_i e^u + 1 - p_i)/n]^n = (pe^u + 1 - p)^n$. Extending to variables in the interval $[0, 1]$ uses the convexity of the exponential function, and then arbitrary ranges can be introduced by transforming variables.

14.4.6.* Theorem. (**Chernoff–Hoeffding Bound**) Let $X_1, \ldots, X_n$ be independent random variables. If $0 \leq X_i \leq 1$ and $X = \sum_{i=1}^n X_i$, with $p = \mathbb{E}(X)/n$ and $0 < t < 1$, then

$$\mathbb{P}\big(X - np > nt\big) \leq e^{-2nt^2}.$$

More generally, if $a_i \leq X_i \leq b_i$ and $X = \sum_{i=1}^n X_i$, with $\mu = \mathbb{E}(X)$, then

$$\mathbb{P}(X - \mu \geq s) \leq e^{-2s^2 / \sum (b_k - a_k)^2}.$$

The same bounds apply to $\mathbb{P}\big(X - np < -nt\big)$ and $\mathbb{P}\big(X - \mu < -s\big)$.

Proof*: Again we want to apply Markov's Inequality to the moment generating function, yielding $\mathbb{P}(X \geq m) \leq \mathbb{E}(e^{uX})/e^{um}$, where $m = n(p + t)$.

In comparison to Theorem 14.4.4, the variables are now distinct and are not $0, 1$-random variables. Let $p_i = \mathbb{E}(X_i)$. Independence still yields $\mathbb{E}(e^{uX}) = \prod \mathbb{E}(e^{uX_i})$. For $0 \leq x \leq 1$, we have $e^{ux} \leq (1 - x)e^{0u} + xe^{1u}$, by convexity of the exponential function. Hence the expectation of e^{uX_i} is bounded by the expectation of $1 - X_i + X_i e^u$. Using this and the Arithmetic–Geometric Mean Inequality $\prod_{i=1}^n z_i \leq (\sum_{i=1}^n z_i/n)^n$, we obtain

$$\prod_{i=1}^n \mathbb{E}(e^{uX_i}) \leq \prod_{i=1}^n (1 - p_i + p_i e^u) \leq (1 - p + pe^u)^n.$$

With $q = 1 - p$, we now have $\mathbb{P}(X \geq n(p + t)) \leq \left(\frac{q + pe^u}{e^{u(p+t)}}\right)^n$, which is exactly the bound we had in the proof of Theorem 14.4.4. The argument to show that the right side of this inequality is at most e^{-2nt^2} is exactly the same as there.

Although the distribution need not be symmetric around the mean, letting $X_i' = 1 - X_i$ and applying the bound on the upper tail for $\sum X_i'$ again yields the bound on the lower tail for X.

The extension to arbitrary ranges is requested in Exercise 11.   ∎

As an application of the simple form of the Chernoff Bound, we consider the conjecture of Hajós [1961] that every k-chromatic graph contains a subdivision of K_k (see Section 8.2). The conjecture fails badly, but Chebyshev's Inequality is not strong enough to show this.

14.4.7. Theorem. (Erdős–Fajtlowicz [1981]) In $\mathbb{G}(n, \frac{1}{2})$, almost every graph has chromatic number at least $\frac{n}{2\lg n}$ and has no subdivision of $K_{c\sqrt{n}}$, where $c > 2$.

Proof: Choose G from $\mathbb{G}(n, \frac{1}{2})$. By Theorem 14.3.34, whp $\alpha(G) < 2\lg n$. (The expected number of stable sets of size $\lceil 2\lg n \rceil$ is less than $\frac{1}{n}$, so $\mathbb{P}(\alpha(G) \geq 2\lg n) < \frac{1}{n}$.) Thus whp $\chi(G) > \frac{n}{2\lg n}$.

Now consider a K_r-subdivision H in an n-vertex graph. Let S be the set of r branch vertices. At most $n - r$ paths in H pass through vertices outside S, so S induces at least $\binom{r}{2} - (n-r)$ edges. When $r = c\sqrt{n}$ with $c > 2$, we have $\binom{r}{2} - (n-r) > (\frac{1}{2} + t)\binom{r}{2}$ with $t = \frac{1}{2} - \frac{2}{c^2}$. Thus S induces more than $(\frac{1}{2} + t)\binom{r}{2}$ edges.

The Chernoff Bound shows that this is highly unlikely. The number X of edges induced by a given r-set has the distribution $\mathrm{Bin}(\binom{r}{2}, \frac{1}{2})$. Applying Theorem 14.4.4 yields $\mathbb{P}\left(X \geq (\frac{1}{2} + t)\binom{r}{2}\right) \leq e^{-r(r-1)t^2}$.

We multiply by $\binom{n}{r}$ to consider all S and obtain an upper bound on the probability of a K_r-subdivision. Since $\binom{n}{r} < n^r = e^{r\ln n}$, the exponent in this factor is bounded by $c\sqrt{n}\ln n$, which grows more slowly than $r(r-1)t^2$, which is linear in n. Hence the bound on the probability of having a K_r-subdivision tends to 0. ∎

In the proof of Theorem 14.4.7, Chebyshev's Inequality would not suffice. We need an exponentially small bound on the tail probability because we multiply by $\binom{n}{r}$, the number of experiments. For another consequence of the Chernoff Bound that Chebyshev's Inequality is not strong enough to prove, see Exercise 12.

On the other hand, Chebyshev's Inequality *is* strong enough to prove that a weaker statement than the Hajós Conjecture does hold with high probability.

14.4.8. Remark. *Hadwiger's Conjecture with high probability.* A K_k-subdivision contracts to K_k, by contracting only edges incident to vertices of degree 2. More generally, a graph G has a K_k-**minor** if K_k arises from some subgraph of G by contractions of any edges. Hadwiger conjectured that every k-chromatic graph has a K_k-minor, a statement weaker than the Hajós Conjecture.

Mader [1968] proved the existence of a constant c such that if G has average degree at least $ck\ln k$, then G has a K_k-minor. Kostochka [1982] and Thomason [1984] improved this by showing that G has a K_k-minor when G has a subgraph with minimum degree at least $ck\sqrt{\ln k}$, and Thomason [2001] showed that the best such c is about .319. These are difficult probabilistic results.

The random graph whp has average degree at least $(1 - \varepsilon)n/2$ (using Chebyshev's Inequality on the number of edges). If $k = n/\lg n$, then $n/2$ grows faster than $k\sqrt{\ln k}$. Hence almost every graph has a K_k-minor, but by Theorem 14.3.35 its chromatic number is less than k. Thus whp Hadwiger's Conjecture holds. ∎

The Chernoff Bound also enables us to show that the easy lower bound on the size of bipartite subgraphs proved in Exercise 14.1.27 is surprisingly sharp. Finding the largest bipartite subgraph can be viewed as finding the largest edge cut and is thus also called the **max-cut problem**; determining the maximum size is NP-hard. Every graph has a cut with at least half of its edges. In Exercise 14.1.27, the Existence Method is used to improve the lower bound to $\frac{m}{2} \frac{\lceil n/2 \rceil}{\lceil n/2 \rceil - 1}$ for a **balanced edge cut**, meaning that the two part-sizes differ by at most 1.

The largest cut in the complete graph K_n has size $\lfloor n^2/4 \rfloor$ and is balanced. We show next that the bound from Exercise 14.1.27 is essentially sharp by obtaining graphs where every balanced cut has roughly the guaranteed size.

14.4.9. Theorem. Let d and ε be positive constants with $d \geq (12 \ln 2)/\varepsilon^2$ and $\varepsilon < 1$. If $n > d$ and n is even, then there is an n-vertex graph with between $(1 - \varepsilon)\frac{d(n-1)}{2}$ and $(1 + \varepsilon)\frac{d(n-1)}{2}$ edges such that every balanced edge cut has size between $(1 - \varepsilon)\frac{dn}{4}$ and $(1 + \varepsilon)\frac{dn}{4}$.

Proof: We generate a random graph from $\mathbb{G}(n, p)$, where $p = \frac{d}{n}$. Let Y be the number of edges in the graph, and X be the number of edges in a specified balanced cut, so $\mathbb{E}(Y) = \frac{d(n-1)}{2}$ and $\mathbb{E}(X) = \frac{dn}{4}$. Both variables are binomial with success probability p. In order to study deviation from the mean, we want to bound $\mathbb{P}\big(|Z - \mathbb{E}(Z)| > \varepsilon\mathbb{E}(Z)\big)$ (for $Z \in \{X, Y\}$), so we set $t = \varepsilon p$ in Theorem 14.4.4.

Using the simple form of the (two-tailed) Chernoff Bound (Theorem 14.4.4),

$$\mathbb{P}\big(|Y - \mathbb{E}(Y)|\big) > \varepsilon\frac{d(n-1)}{2}) \leq 2e^{-2\binom{n}{2}\varepsilon^2 d^2/n^2} \approx 2e^{-\varepsilon^2 d^2}.$$

By the two-tailed alternative form $\mathbb{P}\,(|X - \mu| \geq \varepsilon\mu) \leq 2e^{-\varepsilon^2 \mu/3}$ (Remark 14.4.5),

$$\mathbb{P}\big(|X - \mathbb{E}(X)| > \varepsilon\tfrac{dn}{4}\big) \leq 2e^{-\varepsilon^2 dn/12}.$$

The number of balanced partitions is $\frac{1}{2}\binom{n}{n/2}$, which is asymptotic to $\frac{2^n}{\sqrt{2\pi n}}$. It suffices to show that the probabilities of the events for balanced cuts or number of edges outside the desired ranges sum to less than 1. Since $d \geq (12 \ln 2)/\varepsilon^2$, we have $\varepsilon^2 nd/12 > n \ln 2$, so by the Union Bound the probability of some cut failing tends to 0. Also $\varepsilon^2 d^2$ is a constant, so the probability of a large deviation from the expected number of edges is bounded by a constant. Hence for large n some n-vertex graph satisfies the claim. ∎

We next apply the Chernoff Bound to random sampling of multidimensional data. Let P be a set of n points in $\mathbb{R}^d$, such as profiles of n people along d linear attributes. A **query** q specifies an interval in each dimension. The answer is the subset of P lying in the resulting d-dimensional box; such a set of points from P is a "range". The family R of possible ranges from P defines a **range space** (P, R). Using the Chernoff Bound, a relatively small random sample from the population accurately describes the fraction of the population lying in any range. The key is that the number of possible ranges is limited. We give a weaker version of the result of Vapnik–Chervonenkis [1971], as presented in Phillips [2013].

14.4.10.* Theorem. Given a d-dimensional range space (P, R) and positive constants ε and δ, let S randomly sample $(d/\varepsilon^2) \ln(n/\delta)$ points from P. With probability at least $1 - 2\delta^{2d}$, it holds for all $q \in R$ that

$$\left| \frac{q(P)}{|P|} - \frac{q(S)}{|S|} \right| \leq \varepsilon,$$

where $q(T)$ is the number of points in T returned by the query q.

Proof: Let $k = \lceil (d/\varepsilon^2) \ln(n/\delta) \rceil$, and fix a query q. For the ith random choice s_i of a point for S, let X_i be the indicator variable for its membership in the box for

q. Thus $q(S)$ is the sum of k identical Bernoulli random variables with success probability $q(P)/|P|$; the expectation is $kq(P)/|P|$. By the Chernoff Bound,

$$\mathbb{P}\left(\left|q(S) - |S|\frac{q(P)}{|P|}\right| > k\varepsilon\right) \le 2\mathrm{e}^{-2k\varepsilon^2} = 2\delta^{2d}n^{-2d}.$$

To complete the result, we show that at most n^{2d} distinct subsets of P can be returned by queries. A box representing a query can shrink without changing the answer until each endpoint of each of the d intervals defining it agrees in that coordinate with some point of the answer set. Doing this identifies a set of at most $2d$ points in n (in order to indicate the relevant dimensions, with repetition allowed) that generates this box by specifying its projections on the axes. There are fewer than n^{2d} ways to specify this canonical box corresponding to a query, so there are fewer than n^{2d} sets in the range space. Hence with probability at least $1 - 2\delta^{2d}$ all the ranges have sizes approximated within ε.   ∎

Vapnik–Chervonenkis [1971] proved that amazingly only $O(\varepsilon^{-2}d\ln(1/\varepsilon\delta))$ points are needed in the random sample to represent all the ranges this accurately, independent of n. Talagrand [1995] further improved this by showing that only $O(\varepsilon^{-2}(d + \ln(1/\delta)))$ points are needed. This says roughly that the number of samples needed to guarantee accuracy within ε for one query also guarantees the same accuracy for all queries.

MARTINGALES

In the classical random walk on a line, at each step there is probability p of moving one unit left, probability p of moving one unit right, and probability $1 - 2p$ of not moving. Regardless of earlier history, the expected position after t steps (given the earlier history) equals the actual position after $t - 1$ steps. This is the defining property of a martingale.

14.4.11. Definition. A **martingale** is a sequence of random variables $X_0, \dots, X_n$ such that the expectation of X_i given the values of $X_0, \dots, X_{i-1}$ equals X_{i-1}.

The expected position of the random walk after n steps is at the origin. The exact position is the sum of n independent ±1-random variables, so after a simple transformation the Chernoff Bound tells us that the walk is highly unlikely to be very far from the origin, in terms of n.

A martingale $X_1, \dots, X_n$ allows dependence of successive variables. However, if successive positions cannot differ by much, then the final position is still highly concentrated around its expectation. When the technique applies, it makes the detailed computation in the Second Moment Method unnecessary and yields better bounds on tail probabilities. The work is accomplished by Azuma's Inequality, also called the Martingale Tail Inequality. This states that if $|X_i - X_{i-1}| \le 1$, then the probability that $X_n - X_0$ exceeds $\lambda\sqrt{n}$ is bounded by $\mathrm{e}^{-\lambda^2/2}$. We first prove two lemmas.

14.4.12. Lemma. Let Z be a random variable such that $\mathbb{E}(Z) = 0$ and $|Z| \leq 1$. If f is a convex function on $[-1, 1]$, then $\mathbb{E}(f(Z)) \leq \frac{1}{2}[f(-1)+f(1)]$. In particular, $\mathbb{E}(e^{tZ}) \leq \frac{1}{2}(e^t + e^{-t})$ for all $t > 0$.

Proof: Since f is convex and $x = \frac{1-x}{2}(-1) + \frac{1+x}{2}(1)$, the convexity inequality yields $f(x) \leq \frac{1-x}{2}f(-1) + \frac{1+x}{2}f(1)$ for $-1 \leq x \leq 1$. By linearity and $\mathbb{E}(Z) = 0$, we have $\mathbb{E}(f(Z)) \leq \frac{1}{2}[f(-1) + f(1)]$. ∎

We extend the notion of conditional probability (Definition 0.10) to random variables. Recall that the conditional probability of event A *given* event B (when $\mathbb{P}(B) \neq 0$) is obtained by restricting the probability space to B, normalizing by $\mathbb{P}(B)$. Thus $\mathbb{P}(A|B) = \mathbb{P}(A \cap B)/\mathbb{P}(B)$.

Given random variables X and Z, we write $Z \,|\, X$ for the **conditional random variable** "Z given X". We view the slice of the space where $X = a$ as a probability space, after normalizing the resulting distribution for Z by $\mathbb{P}(X=a)$. Thus $Z|X$ becomes an ordinary random variable on a given slice $X=a$. The **conditional expectation** $\mathbb{E}(Z|X)$ is a function of X, defined to be the expectation of Z over the portion of the probability space in which X has the given value.

Thus $\mathbb{E}(Z \,|\, X)$ is itself a random variable, and we can take its expectation. Weighting $\mathbb{E}(Z \,|\, X = a)$ by $\mathbb{P}(X = a)$ and summing over a yields $\mathbb{E}(\mathbb{E}(Z \,|\, X))$. We obtain the expectation for Z over the entire probability space. This removes the effect of conditioning, so $\mathbb{E}(\mathbb{E}(Z \,|\, X)) = \mathbb{E}(Z)$. We make this precise only for discrete random variables.

14.4.13. Lemma. $\mathbb{E}(\mathbb{E}(Z|X)) = \mathbb{E}(Z)$.

Proof: Let $p_{i,j} = \mathbb{P}(X=i \text{ and } Z=j)$, defined for all i such that $\mathbb{P}(X=i) \neq 0$. Since $\mathbb{E}(Z \,|\, X=i) = \Sigma_j j p_{i,j}/\mathbb{P}(X=i)$,

$$\mathbb{E}(\mathbb{E}(Z|X)) = \sum_i \mathbb{E}(Z \,|\, X=i)\mathbb{P}(X=i) = \sum_i \sum_j j p_{i,j} = \mathbb{E}(Z). \qquad \blacksquare$$

The proof for Azuma's Inequality is about the same as that for the Chernoff Bound, with additional care involving conditional expectation.

14.4.14. Theorem. **(Azuma's Inequality)** If $X_0, \ldots, X_n$ is a martingale with $|X_i - X_{i-1}| \leq 1$, then $\mathbb{P}(X_n - X_0 \geq \lambda\sqrt{n}) \leq e^{-\lambda^2/2}$.

Proof: By translation, we reduce to $X_0 = 0$. For $t > 0$, we have $X_n \geq \lambda\sqrt{n}$ if and only if $e^{tX_n} \geq e^{t\lambda\sqrt{n}}$. Hence $\mathbb{P}(X_n \geq \lambda\sqrt{n}) = \mathbb{P}(e^{tX_n} \geq e^{t\lambda\sqrt{n}})$. Markov's Inequality yields $\mathbb{P}(e^{tX_n} \geq e^{t\lambda\sqrt{n}}) \leq \mathbb{E}(e^{tX_n})/e^{\lambda t\sqrt{n}}$. We further simplify (and weaken) the upper bound and then choose t to minimize it.

First we prove by induction on n that $\mathbb{E}(e^{tX_n}) \leq \left[\frac{1}{2}(e^t + e^{-t})\right]^n$. By the induction hypothesis, it suffices to prove $\mathbb{E}(e^{tX_n}) = \frac{1}{2}(e^t + e^{-t})\mathbb{E}(e^{tX_{n-1}})$.

We introduce X_{n-1} to condition on it. Lemma 14.4.13 yields

$$\mathbb{E}(e^{tX_n}) = \mathbb{E}(e^{tX_{n-1}}e^{t(X_n-X_{n-1})}) = \mathbb{E}\left(\mathbb{E}(e^{tX_{n-1}}e^{t(X_n-X_{n-1})} \,|\, X_{n-1})\right).$$

When conditioned on X_{n-1}, in the inner expectation X_{n-1} is constant. Hence we can extract $e^{tX_{n-1}}$ to obtain $\mathbb{E}(e^{tX_n}) = \mathbb{E}\left(e^{tX_{n-1}}\mathbb{E}(e^{tY} \,|\, X_{n-1})\right)$, where $Y = X_n - X_{n-1}$.

Because $\{X_n\}$ is a martingale, $\mathbb{E}(Y) = 0$. By hypothesis, $|Y| \leq 1$. Hence Lemma 14.4.12 applies, yielding $\mathbb{E}(e^{tY} \mid X_{n-1}) \leq \frac{1}{2}(e^t + e^{-t})$. This itself is now a constant, so $\mathbb{E}(e^{tX_n}) = \frac{1}{2}(e^t + e^{-t})\mathbb{E}(e^{tX_{n-1}})$. We weaken the bound to a more useful form using

$$\frac{e^t + e^{-t}}{2} = \sum \frac{t^{2k}}{(2k)!} \leq \sum \frac{t^{2k}}{2^k k!} = e^{t^2/2}.$$

Now we have

$$\mathbb{P}\left(X_n \geq \lambda\sqrt{n}\right) = \mathbb{P}\left(e^{tX_n} \geq e^{t\lambda\sqrt{n}}\right) \leq e^{nt^2/2 - \lambda t\sqrt{n}}$$

for each $t > 0$. We obtain the best bound by minimizing over t. The exponent is quadratic; we minimize it by choosing t so that $tn - \lambda\sqrt{n} = 0$, or $t = \lambda/\sqrt{n}$. The resulting bound is $e^{-\lambda^2/2}$. ∎

Azuma's Inequality is one-sided; it bounds the probability that X_n is much larger than X_0. The conditions are symmetric in sign, so using $\{-X_i\}$ yields the same inequality for the other tail, where X_n is much smaller than X_0.

14.4.15. Example. *The pragmatic gambler*. A gambler can bet up to n times, where n is fixed. Each time he bets, he wins or loses 1 with equal probability. His goal is winning $\lambda\sqrt{n}$, so he stops if he reaches that value. Letting X_i be his winnings after i games, we have $X_i = X_{i-1}$ if $X_{i-1} \geq \lambda\sqrt{n}$, and otherwise $X_i = X_{i-1} \pm 1$, each with probability .5. Hence $\{X_i\}$ is a martingale that changes by at most one at each step, and Azuma's Inequality applies. The probability that the gambler will earn $\lambda\sqrt{n}$ is bounded by $e^{-\lambda^2/2}$. If $\lambda = 1$, then there may be a reasonable chance of success, but if $\lambda = 10$, then there is little hope.

As mentioned earlier, the variable X_n in this martingale is a simple transformation of a binomial random variable. The computation in the proof of Azuma's Inequality reduces to that for the Chernoff Bound, and in fact the bound we obtain on the probability that the gambler earns $\lambda\sqrt{n}$ is the *same* as that provided by the Chernoff Bound when $p = \frac{1}{2}$. However, the Chernoff Bound is better for $\text{Bin}(n, p)$ when $p \neq \frac{1}{2}$. ∎

For combinatorial applications, it is helpful to have a more general notion of martingale to distinguish between an underlying random process and random variables that result from it.

14.4.16. Definition. A sequence $Y_0, \ldots, Y_n$ is a **martingale** *with respect to* a sequence $X_1, \ldots, X_n$ of random variables if (1) each Y_k is a function of $X_1, \ldots, X_k$ and (2) $\mathbb{E}\left(Y_k \mid X_1, \ldots, X_{k-1}\right) = Y_{k-1}$.

The martingales we discussed earlier are the special case $Y_k = X_k$ for all k, where we drop "with respect to".

14.4.17. Lemma. If Y is a martingale with respect to $X_1, \ldots, X_n$, then
(1) $\mathbb{E}\left(Y_{i+j} \mid X_1, \ldots, X_i\right) = Y_i$ for $1 \leq i + j \leq n$.
(2) $\mathbb{E}(Y_i) = Y_0$ for $0 \leq i \leq n$. ∎

14.4.18. Definition. A martingale Y with respect to X satisfies the **Bounded Differences Condition** if there exist constants $c_1, \ldots, c_n$ such that always $|Y_i - Y_{i-1}| \leq c_i$ for all i.

Allowing the bounds on the differences of consecutive values to differ yields a generalization of Azuma's Inequality (Theorem 14.4.14). The resulting inequality is also called the **Hoeffding–Azuma Inequality**.

14.4.19. Theorem. (**Martingale Tail Inequality**; Hoeffding [1963], Azuma [1967]) If a martingale Y with respect to X satisfies the Bounded Differences Condition for $c_1, \ldots, c_n$, then $\mathbb{P}\left(Y_n - Y_0 \geq t\right) \leq e^{-t^2 / \left(2 \sum c_i^2\right)}$ for $t > 0$.   ∎

Applying the argument to $-Y$ yields the same bound on $\mathbb{P}\left(Y_n - Y_0 \leq -t\right)$. Since $\mathbb{E}(Y_n) = Y_0$, this indeed is a bound on the probability of large deviation from the mean. The proof is a relatively straightforward generalization of the proof of Theorem 14.4.14 (see Exercise 19).

Definition 14.4.18 is tailored to the study of parameters on random structures. We "reveal" the structure a little at a time, aiming to show high concentration of a parameter of the structure. For example, we may discover a random graph one edge or vertex at a time. The pieces of the random structure are $X_1, \ldots, X_n$ and the parameter is f.

14.4.20. Definition. Let X denote the list $(X_1, \ldots, X_n)$, where $X_1, \ldots, X_n$ are random variables. For a random variable $f(X)$, the **Doob process** is $(Y_0, \ldots, Y_n)$, where $Y_0 = \mathbb{E}\left(f(X)\right)$ and $Y_i = \mathbb{E}\left(f(X) \mid X_1, \ldots, X_i\right)$ for $i \in [n]$.

14.4.21. Proposition. A Doob process is a martingale with respect to its underlying variables $X_1, \ldots, X_n$.

Proof: Let $X_{(i)}$ denote $(X_1, \ldots, X_i)$. We use the definitions of Y_i and Y_{i-1} at the beginning and end and Lemma 14.4.13 in the middle to "uncondition" on X_i:

$$\mathbb{E}\left(Y_i \mid X_{(i-1)}\right) = \mathbb{E}\left(\mathbb{E}\left(f(X) \mid X_{(i)}\right) \mid X_{(i-1)}\right)$$
$$= \mathbb{E}\left(f(X) \mid X_{(i-1)}\right) = Y_{i-1}. \qquad ∎$$

14.4.22. Example. *Discovering random graphs.* In sampling from the model $\mathbb{G}(n, p)$, a graph is generated, but we may not discover the full graph at once. The **edge-exposure** process looks at one pair of vertices at a time to learn whether they are adjacent. The **vertex-exposure** process includes one vertex at a time to learn its neighbors among the previous vertices. What we discover in one exposure step does not depend on what we learned in previous steps. Random variables $X_1, \ldots, X_n$ record the information discovered.

For a graph parameter f, the random variable $f(X)$ is the value when the full graph is known. In the Doob process, $Y_0 = \mathbb{E}(f(X))$, and Y_n is the actual value on the graph produced by $X_1, \ldots, X_n$. Along the way, Y_i is the expectation of f over the possible completions of the graph, given the knowledge of $X_1, \ldots, X_i$. As observed in Proposition 14.4.21, always Y is a martingale with respect to X.

To apply the martingale tail inequalities, we need the Bounded Differences Condition, which requires bounds on $|Y_i - Y_{i-1}|$. The intuition is that if a small bit of knowledge about the structure (say, X_i) cannot change $f(X)$ by much, then the expectation of $f(X)$ over the possible completions cannot change by much.

For example, let f be the chromatic number, and let X be the vertex-exposure process. Thus X_i tells which edges from v_i to earlier vertices are present. When we know everything about a graph G except X_i, learning that information cannot much affect $\chi(G)$; either $\chi(G) = \chi(G - v_i)$ or $\chi(G) = \chi(G - v_i) + 1$. The next lemma will then yield $|Y_i - Y_{i-1}| \leq 1$.

The figure below suggests a general step in a Doob process. The large oval F_0 is the set of all outcomes for X. For $1 \leq j \leq n$, let F_j be the subset to which we are restricted by knowing $X_1, \ldots, X_i$. Thus F_i is a subset of F_{i-1}.

For example, consider coin flips. The sample points in F_0 are lists of length n, and F_i is the knowledge of the first i flips. Thus $|F_i| = 2^{n-i}$, and always F_i is one block in a partition of F_{i-1} into two sets of equal size.   ∎

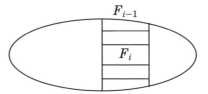

$$F_{i-1}$$
$$F_i$$

14.4.23. Lemma. Let F_0 be the cartesian product space for independent experiments with outcomes $X_1, \ldots, X_n$. Let F_i be the random event defined by $X_1, \ldots, X_i$. Let $Y_0, \ldots, Y_n$ be the Doob process with respect to X for a random variable $f(X)$. Let A be the event defined by $\{X_j : j \neq i\}$. If for each such A the values of f on A differ by at most 1, then $|Y_i - Y_{i-1}| \leq 1$ for all i.

Proof: Consider an instance of F_{i-1}, with $Y_{i-1} = \mathbb{E}(f(X) \mid F_{i-1})$. Note that F_{i-1} is a cartesian product, having all choices for $X_i, \ldots, X_n$, although $X_1, \ldots, X_{i-1}$ are fixed in F_{i-1}. The value of X_i determines a block in the partition of F_{i-1}, represented below by a row. Each column is an event A within F_{i-1} in which all of $X_{i+1}, \ldots, X_n$ are fixed; only X_i varies. By hypothesis, in each column s the minimum and maximum of f (m_s and M_s, respectively), differ by at most 1.

Choices of A ($X_{i+1}, \ldots, X_n$ fixed within column)

Choices
of F_i
(or X_i)

Because X_i and $X_{i+1}, \ldots, X_n$ are specified independently, the probability of the outcome in row r and column s is $q_r p_s$, where q_r is the probability that X_i yields this row and p_s is the probability that $X_{i+1}, \ldots, X_n$ yields this column. The computation of Y_i is expectation across one row:

$$\sum m_s p_s \; \leq \; \mathbb{E}(f(X) \mid F_i) \; \leq \; \sum M_s p_s \; \leq \; 1 + \sum m_s p_s.$$

Since these upper and lower bounds are independent of the row index, taking the expectation over the entire grid to compute Y_{i-1} yields the same inequalities. Hence Y_{i-1} and Y_i lie in one interval of length 1 and differ by at most 1. ∎

When the conditions of Lemma 14.4.23 hold, we conclude immediately that the value of $f(X)$ is highly concentrated around its mean.

14.4.24. Proposition. (Shamir–Spencer [1987]) In $\mathbb{G}(n, p)$, the chromatic number is highly concentrated around its expectation. In particular,

$$\mathbb{P}\big(|\chi(G) - \mathbb{E}(\chi(G))| \geq \lambda \sqrt{n}\big) \; \leq \; 2\mathrm{e}^{-\lambda^2/2}.$$

Proof: Suppose we discover the instance G^p via the vertex martingale. At stage i, we learn the edges from v_i to the previous vertices; this is X_i, and the outcomes of these steps are independent. The event A in which all but X_i are specified is the subgraph $G - v_i$ of the random graph G plus the knowledge of edges from v_i to *later* vertices. Since $\chi(G - v_i) \leq \chi(G) \leq \chi(G - v_i) + 1$, the value of $f(X)$ differs by at most 1 over all possibilities in A. The hypotheses of Lemma 14.4.23 hold, and hence Theorem 14.4.19 applies. Using both tails, the claim follows. ∎

Proposition 14.4.24 says nothing about the value of $\mathbb{E}(\chi(G))$. To approximate this we again use martingales. With constant edge probability p, whp $\omega(G^p)$ is within 1 of $2\log_b n - 2 \log_b \log_b n + 1 + 2 \log_b(e/2)$, where $b = 1/p$ (Theorem 14.3.34). The same holds for $\alpha(G^p)$ using the base $c = 1/(1 - p)$ for the logarithm. To show that the chromatic number of G^p is close to $n/(2 \log_c n)$, Bollobás showed that one can extract independent sets of near-maximum size until few vertices remain.

14.4.25. Theorem. (Bollobás [1988]) For $p = 1 - 1/c$ (constant), whp every induced subgraph of G^p with order at least m has an independent set of size at least r, where $m = \lceil n/\log_c^2 n \rceil$ and $r = 2 \log_c n - 5 \log_c \log_c n$.

Proof: (sketch) Let S be a set of m vertices. We bound the probability that S has no independent r-set by $\mathrm{e}^{-dm^{1+\varepsilon}}$ for some d, ε. This bounds the probability that some m-set has no independent r-set by $\binom{n}{m}\mathrm{e}^{-dm^{1+\varepsilon}}$, which is less than $2^n \mathrm{e}^{-dm^{1+\varepsilon}}$. Since $n = m^{1+o(1)}$, this bound goes to 0, and Markov's Inequality implies that whp G^p has no bad m-set.

It suffices to study the subgraph G induced by $[m]$. Let $f(G)$ be the maximum number of "pair-disjoint" independent r-sets in G; any two of them share at most one vertex. We will show that whp $f(G) \geq 1$. To do this, we show that (1) $f(G)$ is highly concentrated around its mean, and (2) $\mathbb{E}(f(G))$ is unbounded.

We invoke Azuma's Inequality for (1). Consider the edge-exposure martingale. At each step, we learn whether one additional pair of vertices induces an edge. We have $Y_0 = \mathbb{E}(f(X))$ and $Y_{\binom{m}{2}} = f(X)$. The status of one edge slot changes $f(X)$ by at most one, so Lemma 14.4.23 applies, and $\mathbb{P}\big(Y - \mathbb{E}(Y) \leq -\lambda \binom{m}{2}^{1/2}\big) \leq \mathrm{e}^{-\lambda^2/2}$. With $\lambda = \mathbb{E}(Y)/\binom{m}{2}^{1/2}$,

$$\mathbb{P}(Y = 0) \; = \; \mathbb{P}\big(Y - \mathbb{E}(Y) \leq -\mathbb{E}(Y)\big) \; \leq \; \mathrm{e}^{-\mathbb{E}(Y)^2/(m^2 - m)}.$$

Hence it suffices to show that $\mathbb{E}(f(X))/m \to \infty$.

To prove this, we consider another random variable $\hat{f}(X)$, the number of independent r-sets in G sharing no pair with *any* other independent r-set. Such sets are pairwise pair-disjoint, so $f(X) \geq \hat{f}(X)$. We introduced $f(X)$ because the restriction martingale for $\hat{f}(X)$ does not satisfy $\left| \hat{Y}_i - \hat{Y}_{i-1} \right| \leq 1$. In the drawing of $\overline{G}$ below, for example, we have $r = 4$ and seek 4-cliques; if the last (dotted) edge is present in $\overline{G}$, then $\hat{f}(X) = 0$, but if it is absent from $\overline{G}$, then $\hat{f}(X) = 3$.

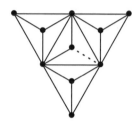

It is easier to compute $\mathbb{E}(\hat{f}(X))$ than $\mathbb{E}(f(X))$. Expressing $\hat{f}(X)$ as the sum of $\binom{m}{r}$ indicator variables, we obtain $\mathbb{E}(\hat{f}(X))$ as $\binom{m}{r}$ times the probability that $[r]$ induces an independent set that is pair-disjoint from all other independent r-sets. This is $(1-p)^{\binom{r}{2}}$ times the conditional probability that $[r]$ does not conflict with other independent r-sets, given the event Z that $[r]$ is in fact independent. Let Y be the number of other independent r-sets overlapping $[r]$ in at least two elements. By Markov's Inequality, $\mathbb{E}(Y \mid Z) \to 0$ implies $\mathbb{P}(Y = 0 \mid Z) \to 1$. Since each set counted shares at least two vertices with $[r]$, we have

$$\mathbb{E}(Y \mid Z) = \sum_{i \geq 2, r-1} \binom{r}{i}\binom{m-r}{r-i}(1-p)^{\binom{r}{2}-\binom{i}{2}}.$$

From the expression for r in terms of m, this can be shown to approach 0 as $m \to \infty$. Hence $\mathbb{E}(\hat{f}(X))$ is asymptotic to $\binom{m}{r}(1-p)^{\binom{r}{2}}$. The expression for r in terms of m yields $\mathbb{E}(\hat{f}(X)) \in \Omega(m^{5/3})$. Thus $\mathbb{E}(Y)/m \to \infty$, as desired. ∎

14.4.26. Corollary. (Bollobás [1988], Matula–Kučera [1990]) For constant edge probability $p = 1 - 1/c$, whp G^p satisfies

$$(1 + \varepsilon)n/(2 \log_c n) \leq \chi(G^p) \leq (1 + \varepsilon')n/(2 \log_c n),$$

where $\varepsilon = \log_c \log_c n / \log_c n$ and $\varepsilon' = 5 \log_c \log_c n / \log_c n$.

Proof: The lower bound holds because whp G^p has no independent set larger than $2 \log_c n - 2 \log_c \log_c n$. The upper bound follows from Theorem 14.4.25, because whp we can select independent sets of size $2 \log_c n - 5 \log_c \log_c n$ until only $n/(\log_c n)^2$ vertices remain. Since $n/(\log_c n)^2 \in o(n/\log_c n)$, we can complete the coloring by using distinct new colors on the remaining vertices. ∎

BOUNDED DIFFERENCES (optional)

We remarked earlier that the constructive version of the Local Lemma was based initially on viewing the underlying probability space as a product space over independent variables. A similar viewpoint allows us to generalize the setting of martingales, although the resulting tail inequalities may be weaker. We follow the development in McDiarmid [1998], beginning by rewriting the Bounded Differences Condition (Definition 14.4.18) in terms of a function.

14.4.27. Definition. A function $f \colon \mathbb{R}^n \to \mathbb{R}$ satisfies the **Bounded Differences Condition** with respect to the variables X if there exist constants $c_1, \ldots, c_n$ such that $\left| \mathbb{E}\big(f(X) \mid X_1, \ldots, X_i \big) - \mathbb{E}\big(f(X) \mid X_1, \ldots, X_{i-1} \big) \right| \le c_i$ for all i.

Applying the Martingale Tail Inequality (Theorem 14.4.19) to the Doob process for a function satisfying this definition yields another tail inequality.

14.4.28. Corollary. If $f \colon \mathbb{R}^n \to \mathbb{R}$ satisfies the Bounded Differences Condition with respect to the variables X using constants $c_1, \ldots, c_n$, then

$$\mathbb{P}\big(f(X) - \mathbb{E}(f(X)) \ge t \big) \le e^{-t^2 / \left(2 \sum c_i^2 \right)} \text{ for } t > 0.$$

14.4.29. Definition. For $c, u, v \in \mathbb{R}^n$, the c-**Hamming distance** $d_c(u, v)$ is defined by $d_c(u, v) = \sum_{i \in I} c_i$, where $I = \{i \in [n] \colon u_i \ne v_i\}$. The **Hamming distance** $d_H(u, v)$ is the case where all $c_i = 1$. We extend Hamming distance by letting the distance $d_H(x, A)$ from x to a finite set A be $\min\{d_H(x, y) \colon y \in A\}$.

Note that the distance $d_c(u, v)$ does not depend on the value of $u_i - v_i$ for the coordinates i where the difference is nonzero, only on which coordinates differ.

14.4.30. Corollary. (Independent Bounded Differences Inequality – IBDI) Let $X = (X_1, \ldots, X_n)$, where $X_1, \ldots, X_n$ are independent random variables. If $|f(u) - f(v)| \le d_c(u, v)$ whenever $u, v \in \mathbb{R}^n$, then

$$\mathbb{P}\big(f(X) - \mathbb{E}(f(X)) \ge t \big) \le e^{-t^2 / \left(2 \sum c_i^2 \right)}$$

for $t > 0$ (with the same bound on lower tails).

Proof: (sketch for the case where the variables X are discrete). It suffices to check the Bounded Differences Condition with respect to X using $c_1, \ldots, c_n$; Corollary 14.4.28 then yields the conclusion. For the details, see Exercise 23. ∎

The condition "$|f(u) - f(v)| \le d_c(u, v)$ for $u, v \in \mathbb{R}^n$" is called a **Lipschitz condition** in analysis. In some contexts the metric $d_c(u, v)$ depends on $|u_i - v_i|$, but here it is only the sum of weights for the coordinates where $u_i \ne v_i$. A function $f \colon \mathbb{R}^n \to \mathbb{R}$ satisfying this condition is a c-**Lipschitz function**.

For a subset A of a product space Ω, let $A_t = \{x \in \Omega \colon d_H(x, A) < t\}$. That is, A_t "fattens" A by adding a boundary of thickness t (differing from points of A in less than t coordinates). When all points in Ω are equally likely, the next result implies that if A is a positive fraction of Ω and $t > \sqrt{n}$, then A_t contains all but an exponentially small part of Ω. McDiarmid [1998] contains a similar result.

14.4.31. Theorem. Let $X = (X_1, \ldots, X_n)$, with $X_1, \ldots, X_n$ independent and each X_k taking values in a probability space Ω_k. If A is a subset of the product space $\prod_i \Omega_i$, then for any $t \in \mathbb{N}$,

$$\mathbb{P}(X \in A)\mathbb{P}(d_H(X, A) \geq t) \leq e^{-t^2/(8n)}.$$

Proof: Since A is fixed, we may let $f(X) = d_H(X, A)$. Let $\mu = \mathbb{E}(f(X))$. Since $d_H(X, A) = 0$ if and only if $X \in A$, the left side of the desired inequality can be written as $\mathbb{P}(f(X) - \mu \leq -\mu)\mathbb{P}(f(X) - \mu \geq t - \mu)$, the product of lower and upper tail probabilities. We use a bound arising from the first factor when $t \leq 2\mu$ and a bound arising from the second factor when $t \geq 2\mu$.

Since $f(X)$ changes by at most 1 when any one coordinate is changed, f satisfies the condition "$|f(u) - f(v)| \leq d_c(u, v)$ whenever $u, v \in \mathbb{R}^n$" in Corollary 14.4.30 with each c_i equal to 1. Hence we can apply the IBDI to both the lower tail and the upper tail. When $t \leq 2\mu$,

$$\mathbb{P}\big(f(X) - \mu \leq -\mu\big) \leq \mathbb{P}\big(f(x) - \mu \leq -t/2\big) \leq e^{-t^2/(8n)}.$$

When $t \geq 2\mu$,

$$\mathbb{P}\big(f(X) - \mu \leq t - \mu\big) \leq \mathbb{P}\big(f(x) - \mu \leq t/2\big) \leq e^{-t^2/(8n)}.$$

With probability at most 1 in the other tail, in each case the claim follows. ∎

14.4.32. Application. A famous result of Harper [1966] concerns the isoperimetric problem on the hypercube. An isoperimetric problem in a metric space asks to minimize the size of the boundary for a subset of specified size. Harper proved that for $A \subseteq 2^{[n]}$ with $|A| \geq \sum_{k=0}^r \binom{n}{k}$, the number of subsets of $[n]$ with Hamming distance at most t from A is at least $\sum_{k=0}^{r+t} \binom{n}{k}$. This result is sharp, since equality holds when A consists of all the subsets with size at most r.

Theorem 14.4.31 leads quickly to an approximate version of Harper's result. Let all subsets X of $[n]$ be equally likely. Given a family A, by Theorem 14.4.31 the probability that a randomly chosen subset of $[n]$ differs in at least t positions from all sets in A is at most $\frac{1}{a}e^{-t^2/(8n)}$, where a is the fraction of all subsets that are in A. Thus A_t occupies a fraction at least $1 - \frac{1}{a}e^{-t^2/(8n)}$ of all sets.

To compare with Harper's result, we must approximate the tail of the binomial distribution. Using the Chernoff Bound $\mathbb{P}(X > \frac{n}{2} + n\tau) \leq e^{-2n\tau^2}$ as in Theorem 14.4.4, when X is distributed as $\mathrm{Bin}(n, \frac{1}{2})$ and $r \geq n/2$ we have $\mathbb{P}(X \geq r+t) \leq e^{-2(r+t-n/2)^2/n}$. With $|A| \geq \sum_{k=0}^r \binom{n}{k}$, we reduce r to $n/2$ to simplify the bound. Harper's result then implies that A_t occupies a fraction at least $1 - \frac{1}{a}e^{-2t^2/n}$ of all sets, similar to the guarantee from Theorem 14.4.31. ∎

Talagrand [1995] greatly strengthened Theorem 14.4.31 to an inequality of a similar form, by introducing a distance function $d_T(X, A)$ that is at least as large as $d_H(X, A)$ and making the upper bound much tighter and independent of n. For a subset A of a product space Ω, let $d_T(x, A)$ be $\sup d_c(x, A)$, where the supremum is taken over all unit vectors c.

14.4.33. Theorem. (Talagrand's Inequality) Under the same conditions as in Theorem 14.4.31, $\mathbb{P}(X \in A)\mathbb{P}(d_T(X, A) \geq t) \leq e^{-t^2/4}$. ∎

The applications of Talagrand's Inequality yield tighter concentration around the median, which also implies tighter concentration around the mean.

For example, let X be the length of a longest increasing subsequence in a random permutation σ of $[n]$. It is known that $\mathbb{E}(X) = 2\sqrt{n}$ (see the discussion in Section 16.3). Before Talagrand's result, it was known that whp $|X - \mathbb{E}(X)| < c\sqrt{n}$. Talagrand's Inequality can be used to show that whp $|X - \mathbb{E}(X)| < cn^{1/4}$.

Further discussion of Talgrand's Inequality and its applications can be found in McDiarmid [1998], Dubhashi–Panconesi [2009], and Alon–Spencer [2016].

EXERCISES 14.4

14.4.1. (−) An algorithm to compute an answer succeeds with probability p, where $p = \frac{3}{4}$. How many times must the algorithm be run to ensure that with probability at least .99 the correct answer is produced at least half of the time?

14.4.2. (−) For a variable X with distribution $\mathrm{Bin}(n, p)$, Chebyshev's Inequality in Example 14.4.3 yields $\mathbb{P}\left(|X - \mathbb{E}(X)| \geq \lambda\sqrt{\mathrm{Var}(X)}\right) \leq \lambda^{-2}$. Determine the improved bound that the simple Chernoff Bound gives for this probability, again independent of n.

14.4.3. (−) Use the Chernoff bound to show for constant p and positive ϵ that whp G^p has connectivity at least $(1 - \epsilon)p^2 n$.

14.4.4. (◇) Use the Chernoff Bound for small p (from Remark 14.4.5) to prove that if $p > \sqrt{c \ln n / n}$ for some constant c, then whp $\mathrm{diam}\,(G^p) \leq 2$. What value of c suffices? (Comment: Exercise 14.3.28 gives the actual threshold.)

14.4.5. (◇) Use the Chernoff Bound to show that almost always the random graph has minimum degree at least $\frac{n}{2} - \sqrt{cn \ln n}$ for some appropriate constant c.

14.4.6. (◇) Use the Chernoff Bound to show that almost always the random graph has connectivity at least $\frac{n}{4} - \sqrt{cn \ln n}$ for some appropriate constant c. (Comment: Bollobás [1981b] showed that the connectivity almost always equals the minimum degree.)

14.4.7. (◇) The expected number of edges joining any two sets of size $n/2$ in the random n-vertex graph is $n^2/8$, and hence the expected maximum size of a bipartite subgraph is at least $n^2/8$. Use the Chernoff Bound and the Union Bound to prove that whp the random graph has no bipartite subgraph with more than $n^2/8 + n^{3/2}$ edges. (Molloy–Reed [2002])

14.4.8. (◇) Place n balls into n boxes, uniformly and independently.

(a) Use the alternative form of the Chernoff Bound (Remark 14.4.5) to prove that whp no box has more than $2 \ln n$ balls, asymptotically.

(b) Use a more direct probability computation to prove that whp no box has more than $O(\frac{\ln n}{\ln \ln n})$ balls. (Comment: Also whp some box has $\Omega(\frac{\ln n}{\ln lnn})$ balls. When the tail is far from the expectation the Chernoff Bound may not be strong enough for optimal results.)

14.4.9. A *ranking* of a tournament T is a linear ordering σ of its vertices. A good ranking agrees with T on many pairs. Let $D_T(\sigma) = a - b$, where σ and T agree on a vertex pairs and disagree on b pairs. Prove that there is an n-vertex tournament T such that $D_T(\sigma) \leq 2n^{3/2}\sqrt{\ln n}$ for every ranking σ. (Comment: De la Vega proved further that for some n-vertex tournament, every ranking agrees with at most $o(n^{3/2})$ pairs.) (Erdős–Moon [1965])

14.4.10. *Polling.* The fraction of the population preferring A to B is p. In a poll of n people, the fraction who prefer A is X. To be accurate, we want $\mathbb{P}(|X - p| \leq \varepsilon p) > 1 - \delta$. Using the simple Chernoff Bound, how large should n be in terms of ε and δ, assuming $p > .25$?

14.4.11. Chernoff–Hoeffding Bound (Hoeffding [1963]) Let $X_1, \ldots, X_n$ be independent random variables, and let $X = \sum_{i=1}^{n} X_i$. Prove the more general version of Theorem 14.4.6: if $a_i \leq X_i \leq b_i$ for $1 \leq i \leq n$, with $\mu = \mathbb{E}(X)$, then $\mathbb{P}(X - \mu \geq s) \leq e^{-2s^2 / \sum (b_i - a_i)^2}$.

14.4.12. ($\Diamond$) Let f color the vertices of a hypergraph using $\{1, -1\}$. The **discrepancy** of f is $\max_{e \in E(H)} \left| \sum_{v \in e} f(v) \right|$. Use the Chernoff Bound to prove that every hypergraph with m edges and n vertices has a $\{1, -1\}$-coloring with discrepancy at most $\sqrt{2n \ln(2m)}$. What weaker bound on discrepancy can be proved using Chebyshev's Inequality? (Comment: For the case $m = n$, Spencer [1985] improved the bound to $5.32\sqrt{n}$; see Alon–Spencer [1992].)

14.4.13. Let H be a hypergraph where every edge has size at least r and intersects at most k other edges. Use the Chernoff Bound and the Local Lemma to prove that H has a coloring with discrepancy at most α if $k \leq \frac{1}{8} e^{\alpha^2 / 6r}$. (Molloy–Reed [2002])

14.4.14. The **discrepancy** of a signing $f \colon E(K_n) \to \{1, -1\}$ is $\max_{S \subseteq V(K_n)} \left| \sum_{\{u,v\} \subseteq S} f(uv) \right|$. Prove that some signing has discrepancy at most $\sqrt{\ln 2}(n^{3/2} + \frac{1}{2} n^{1/2})$. (J.-H. Kim)

14.4.15. In n tosses of a fair coin, let Y be the number of heads minus the number of tails. Compare the bounds on $\mathbb{P}(|Y| > \lambda \sqrt{n})$ using martingales and the simple Chernoff Bound.

14.4.16. Let f be a random function from $[n]$ to $[n]$. Let Y be the number of elements of $[n]$ that are missing from the image of f. Prove $\mathbb{P}(|Y - \mathbb{E}(Y)| \geq t) \leq e^{-t^2/(2n)}$.

14.4.17. ($\Diamond$) *Homogenizing triples.* Let G_0 be an n-vertex graph. At time t, to form G_t from G_{t-1}, first select a random triple S of vertices. If S induces j edges in G_{t-1}, then turn S into a triangle with probability $j/3$ and into an independent set with probability $1 - j/3$.
(a) Determine the probability p that the process ends by turning the graph into K_n.
(b) For $|E(G_0)| = \frac{1}{2}\binom{n}{2}$, prove the expected number of steps is $\Omega(n^4)$. (Wormald [1999a])

14.4.18. ($\Diamond$) Let L_n denote the maximum length of an increasing sublist of a random permutation of $[n]$. Prove $\mathbb{P}(|L_n - \mathbb{E}(L_n)| \geq \lambda \sqrt{n}) \leq 2e^{-\lambda^2/2}$. (Comment: Logan–Shepp [1977] and Versik–Kerov [1977] together proved $\mathbb{E}(L_n) \sim 2\sqrt{n}$; see Section 16.3. Hence the result here is a bit unsatisfying. Frieze [1991] proved a family of stronger bounds, including $\mathbb{P}(|L_n - \mathbb{E}(L_n)| \geq \sqrt{n}) \leq 2e^{-\sqrt{n}}$. Bollobás–Brightwell [1992] generalized.)

14.4.19. (**Martingale Tail Inequality**) Let Y be a martingale with respect to X satisfying the Bounded Differences Condition for $c_1, \ldots, c_n$. Prove Theorem 14.4.19: For $t > 0$,

$$\mathbb{P}(Y_n - Y_0 \geq t) \leq e^{-t^2 / (2 \sum c_i^2)}.$$
(Hoeffding [1963], Azuma [1967])

14.4.20. Generate two random binary lists of length n; the bits are chosen by unbiased coin flips, independently. Let Y_n be the length of a longest common subsequence in the two lists (a common subsequence need not use the same positions in the two lists and need not appear in consecutive positions). Prove $\mathbb{P}(|Y_n - \mathbb{E}(Y_n)| \geq \lambda) \leq 2e^{-\lambda^2/8n}$.

14.4.21. Let X be the number of triangles in the random n-vertex graph. By linearity, $\mathbb{E}(X) = \frac{1}{8}\binom{n}{3}$. Prove $\mathbb{P}(|X - \mathbb{E}(X)| > \lambda n^2) \leq 2e^{-c\lambda^2}$ for some constant c.

14.4.22. ($\Diamond$) *Bin-packing.* The numbers $a_1, \ldots, a_n$ are drawn uniformly and independently from the interval $[0, 1]$. They must be placed in bins, each having total capacity 1. Let X be the number of bins needed. Use IBDI to prove that $\mathbb{P}(|X - \mathbb{E}(X)| \geq \lambda \sqrt{n}) \leq 2e^{-\lambda^2/2}$.

14.4.23. Complete the proof of Corollary 14.4.30 for discrete random variables.

Chapter 15

Linear Algebra

Some combinatorial problems have elegant solutions using algebraic techniques. In this chapter we consider applications of linear algebra. We consider the uses of polynomials and dimension in vector spaces, determinants and permanents of matrices, and eigenvalues of matrices associated with graphs. An important topic that is too large in scope to include here is coding theory.

15.1. Dimension and Polynomials

The dimension of a vector space is an upper bound on the size of a linearly independent set of vectors. This elementary statement in linear algebra is a powerful tool for extremal problems in combinatorics. To apply the method we need a vector space in which the desired set corresponds to linearly independent vectors. Often we use a space of polynomials, viewing a polynomial as a linear combination of monomials. Babai–Frankl [1992] and Matoušek [2010] provide many examples of this and other aspects of linear algebra in combinatorics.

THE DIMENSION ARGUMENT

Two easy examples illustrate the elegance of dimensionality proofs and the process of turning extremal problems into dimension problems.

15.1.1. Example. *Eventown vs. Oddtown.* A town with n people contains many clubs, and every two clubs have an even number of common members. How many clubs can there be if all clubs have even size? How many if all clubs have odd size?

When the clubs have even size ("Eventown"), we can form $2^{\lfloor n/2 \rfloor}$ clubs by grouping the residents into pairs and letting each club be a subset of these pairs. In fact, there is no larger set of clubs (Exercise 4).

When the clubs have odd size ("Oddtown"), we can form n clubs by using clubs of size 1, or clubs of size $n - 1$, for example. In fact, there are between $2^{n(n+2)/8}/(n!)^2$ and $2^{n^2}/n!$ nonisomorphic constructions of size n (Exercise 5). These are much smaller than the Eventown constructions, but nevertheless we show next that there cannot be more than n clubs. ∎

The **incidence vector** of a subset A of $[n]$ is the binary n-tuple u such that $u_i = 1$ if $i \in A$ and otherwise $u_i = 0$. The simple observation that allows us to convert problems about intersections of sets into algebraic problems is that if u and v are the incidence vectors of subsets A and B of $[n]$, then $u \cdot v = |A \cap B|$, where $u \cdot v$ is the ordinary **dot product** of u and v: $u \cdot v = \sum_{i=1}^{n} u_i v_i$. Since subscripts indicate coordinates of vectors, we use superscripts to index distinct vectors.

15.1.2. Theorem. (Berlekamp [1969]) If $\mathcal{F}$ is a family of odd-size subsets of $[n]$ whose pairwise intersections have even size, then $|\mathcal{F}| \leq n$.

Proof: Let $\mathcal{F} = \{A_1, \ldots, A_m\}$. We view the corresponding incidence vectors $u^1, \ldots, u^m$ as elements of the space $\mathbb{F}_2^n$, which is n-dimensional. It suffices to show that $u^1, \ldots, u^m$ are linearly independent, since every n-dimensional vector space has at most n linearly independent vectors. Computations of dot products of vectors in $\mathbb{F}_2^n$ are modulo 2.

The conditions on the sizes of the sets and their intersections require $u^i \cdot u^i \equiv 1$ for $1 \leq i \leq m$ and $u^i \cdot u^j \equiv 0$ for $i \neq j$. To prove linear independence, consider an equation of dependence: $\sum_{i=1}^{m} c_i u^i = \mathbf{0}$. For the dot product of both sides with u^j, the values of $u^i \cdot u^j$ reduce the equation to $c_j u^j \cdot u^j = 0$, and then $c_j = 0$. ∎

The argument using the dot product generalizes. For $u \in K^n$, where K is a field, we can view $u \cdot x$ as a polynomial function of x; that is, $u \cdot x \in K[x_1, \ldots, x_n]$. The polynomial has n variables, with degree 1 in each. The argument applies for functions in a linear (vector) space.

15.1.3. Proposition. (The **Diagonal Criterion**) Let $f_1, \ldots, f_m$ be elements of a linear space of functions. If $v^1, \ldots, v^m$ exist such that $f_i(v^j)$ is nonzero for $i = j$ and zero for $i \neq j$, then $f_1, \ldots, f_m$ are linearly independent.

Proof: Consider $c_1, \ldots, c_m$ such that $\sum_{i=1}^{m} c_i f_i$ is identically zero. Evaluating $\sum_{i=1}^{m} c_i f_i$ at v^j yields $c_j f_j(v^j) = 0$ and hence $c_j = 0$. ∎

Our next application shows the method more fully.

15.1.4. Definition. A k-**distance set** is a set of points such that the distances between points lie in a set of at most k numbers.

For example, a one-distance set in $\mathbb{R}^n$ must lie at the vertices of a simplex, so the size of a 1-distance set in $\mathbb{R}^n$ is at most $n + 1$. For a two-distance set, Exercise 9 requests a construction for a lower bound of $\binom{n+1}{2}$. We prove an upper bound that is not much larger.

15.1.5. Theorem. (Larman–Rogers–Seidel [1977]) Every two-distance set in $\mathbb{R}^n$ has at most $(n + 1)(n + 4)/2$ points.

Proof: Let $\{v^1, \ldots, v^m\}$ be a two-distance set, using distances c and d. To avoid square roots, we compute with squared distances. Write $\|x - y\|^2$ for the square of the distance between x and y; it equals $\sum_{j=1}^{n} (x_j - y_j)^2$.

Define $f_1, \ldots, f_m$ by $f_i(x) = (\|x - v^i\|^2 - c^2)(\|x - v^i\|^2 - d^2)$. Note $f_i(v^i) = c^2 d^2 \neq 0$, and $f_i(v^j) = 0$ for $i \neq j$, since $\|v^j - v^i\| \in \{c, d\}$. By the Diagonal Criterion, $f_1, \ldots, f_m$ are linearly independent.

To obtain a good bound on m, we want to capture $f_1, \ldots, f_m$ within a small subspace of $\mathbb{R}[x_1, \ldots, x_n]$. Written as a polynomial, we have

$$f_i(x) = \Big(\sum_{k=1}^{n} (x_k - v_k^i)^2 - c^2\Big)\Big(\sum_{k=1}^{n} (x_k - v_k^i)^2 - d^2\Big).$$

When expanded completely, the total degree in each monomial term is at most 4. The polynomial is a linear combination of such monomials. The number of ways to distribute total degree at most 4 over n variables, forming such monomials, is less than n^4. Hence $m < n^4$.

To prove a better bound, we capture f_i in the span of fewer monomials. When expanding the product, the terms with degree 4 are all generated by $\big(\sum_{k=1}^{n} x_k^2\big)^2$. Those with degree 3 can be grouped as multiples of $x_j\big(\sum_{k=1}^{n} x_k^2\big)$. Thus f_i is a linear combination of polynomials of the following forms:

$$\Big(\sum_{k=1}^{n} x_k^2\Big)^2, \quad x_j\Big(\sum_{k=1}^{n} x_k^2\Big), \quad x_j x_k, \quad x_j, \quad 1,$$

where $j, k \in [n]$. The number of such polynomials is $1 + n + \binom{n}{2} + n + n + 1$, which simplifies to $(n + 1)(n + 4)/2$. ∎

15.1.6. Remark. The proof of Theorem 15.1.5 outlines the **dimension argument** to show that a set S has size at most m.

(1) Define polynomials associated with the elements of S.
(2) Show that the polynomials are linearly independent.
(3) Show that the polynomials are spanned by a set of size m.

Step 3 shows that the polynomials lie in a space of dimension at most m. Since they are linearly independent, there are at most m of them.

The resulting bound can be improved by adding polynomials if the augmented family is still linearly independent. Blokhuis [1984] did this to improve the bound in Theorem 15.1.5 from $(n + 1)(n + 4)/2$ to $(n + 1)(n + 2)/2$. To the polynomials $f_1, \ldots, f_m$, he added the constant polynomial 1 and the linear polynomials $x_1, \ldots, x_n$ of degree 1. The full set is spanned by the same polynomials as before and is still linearly independent, so the bound $|S| \leq m$ becomes $|S| \leq m - (n + 1)$ (see Exercise 11).

We will apply this augmentation technique in Theorem 15.1.23. ∎

Another criterion for linear independence holds more often than the diagonal criterion; we will use it to study a measure of graph complexity.

15.1.7. Proposition. (The **Triangular Criterion**) Let $f_1, \ldots, f_m$ be functions in a linear space. If $v^1, \ldots, v^m$ exist such that $f_i(v^j)$ is nonzero for $i = j$ and zero for $i > j$, then $f_1, \ldots, f_m$ are linearly independent.

Proof: Consider coefficients $c_1, \ldots, c_m$ such that $\sum_{i=1}^{m} c_i f_i$ is the identically-zero function. Evaluating at v^1 yields $c_1 f_1(v^1) = 0$ and hence $c_1 = 0$. Inductively, if $c_1 = \cdots = c_{j-1} = 0$, then evaluating $\sum_{i=1}^{m} c_i f_i$ at v^j yields $c_j f_j(v^j) = 0$, because the earlier terms have coefficient 0 and the later functions evaluate to 0. Hence $c_j = 0$ for $1 \leq j \leq m$. ∎

15.1.8. Definition. A **product representation** of length d encodes the vertices of a graph G as distinct d-tuples so that $uv \in E(G)$ if and only if the codes for u and v differ in every position. The **product dimension** pdim (G) of G is the minimum length of a product representation.

15.1.9. Example. *Product dimension of K_n, $\overline{K}_n$, and $K_1 + K_{n-1}$*. Trivially, pdim $(K_n) = 1$. Since every two vectors must agree somewhere but the vectors must be distinct, pdim $(\overline{K}_n) \geq 2$; assigning $(0, j)$ to v_j suffices.

For $K_1 + K_{n-1}$, the vectors for the clique must be distinct in each coordinate. The isolated vertex must agree with the others somewhere, but it cannot agree with more than one in each coordinate. Hence at least $n - 1$ coordinates are needed. This suffices, by using $(1, 2, \ldots, n - 1)$ for the isolated vertex and $(i, i, \ldots, i)$ for the ith vertex of the clique. $\blacksquare$

Product dimension is also called **Prague dimension** in honor of the seminal paper of Lovász–Nešetřil–Pultr [1980]. They proved for $n \geq 3$ that pdim $(G) \leq n - 1$ when G has n vertices (Exercise 12). Using linear algebra, they proved a general lower bound whose sharpness is shown by the next construction.

15.1.10. Example. pdim $(\frac{n}{2} K_2) = \lceil \lg n \rceil$. Given k coordinates, the graph that results from using all binary k-tuples as vertex encodings is $2^{k-1} K_2$. Each vector disagrees everywhere only with its complement, so the graph is a matching. If n is not a power of 2, then we can discard some complementary pairs and still encode $\frac{n}{2} K_2$. Thus $\lceil \lg n \rceil$ is an upper bound. Equality follows from the next theorem. $\blacksquare$

15.1.11. Theorem. (Lovász–Nešetřil–Pultr [1980]) If $V(G)$ has vertices $u_1, \ldots, u_r$ and $v_1, \ldots, v_r$ such that $u_i v_i \in E(G)$ for all i and $u_i v_j \notin E(G)$ for $i < j$, then G has product dimension at least $\lceil \lg r \rceil$.

Proof: Given a representation in d coordinates, let $x^1, \ldots, x^r$ and $y^1, \ldots, y^r$ be the encodings of the specified vertices: x^i and y^i differ in each coordinate, but x^i and y^j agree in some coordinate when $i < j$. Thus $\prod_{k=1}^{d} (x_k^i - y_k^j)$ is nonzero when $i = j$ and zero when $i < j$. Using this observation, we construct r linearly independent vectors in $\mathbb{R}^{2^d}$; this yields $r \leq 2^d$ and hence $d \geq \lceil \lg r \rceil$.

Expanding the product $\prod_{k=1}^{d} (w_k - z_k)$ for two vectors $w, z \in \mathbb{R}^d$ yields the sum $\sum_{S \subseteq [d]} \prod_{i \in S} w_i \prod_{j \in \overline{S}} (-z_j)$. We view this as a dot product of vectors in 2^d-dimensional space, with coordinates indexed by subsets of $[d]$. For $w \in \mathbb{R}^d$, we define two vectors $\overline{w}$ and $\hat{w}$ in $\mathbb{R}^{2^d}$ by setting $\overline{w}_S = \prod_{i \in S} w_i$ and $\hat{w}_S = \prod_{i \notin S} (-w_i)$ for each $S \subseteq [d]$. With this definition, $\prod_{k=1}^{d} (w_k - z_k) = \overline{w} \cdot \hat{z}$. The conditions on $x^1, \ldots, x^r$ and $y^1, \ldots, y^r$ thus imply that $\overline{x}^i \cdot \hat{y}^j$ is nonzero if $i = j$ and zero if $i < j$.

The dot product with a fixed vector is a linear function: let $f_i(\hat{y})^j = \overline{x}^i \cdot \hat{y}^j$. With $\hat{y}_1, \ldots, \hat{y}_r$, these $f_1, \ldots, f_r$ satisfy the Triangular Criterion (indexed in reverse). Hence $\overline{x}^1, \ldots, \overline{x}^r$ are independent, yielding $r \leq 2^d$ as desired. $\blacksquare$

Theorem 15.1.11 does not require the two lists to be disjoint, just that v_i is the first vertex in the second list adjacent to u_i. When applied to $\frac{n}{2} K_2$, both lists have all n vertices; each list runs through both halves of the matching to prove pdim $\frac{n}{2} K_2 \geq \lceil \lg n \rceil$, and thus the encoding in Example 15.1.10 is optimal.

RESTRICTED INTERSECTIONS OF SETS (optional)

Restricting the intersections of sets in a family $\mathcal{F}$ restricts $|\mathcal{F}|$. Recall that a family of sets is an **intersecting family** if every two members have a nonempty intersection. For $n \geq 2k$, a special case of the Erdős–Ko–Rado Theorem (Theorem 11.2.18) states that the maximum size of an intersecting family of k-sets in $[n]$ is $\binom{n-1}{k-1}$. We can generalize by specifying all allowed sizes of intersections.

15.1.12. Definition. For $L \subseteq \mathbb{N}_0$, an L-*intersecting family* of sets is a family $\mathcal{F}$ such that $|A \cap B| \in L$ for all $A, B \in \mathcal{F}$.

In this language, the Erdős–Ko–Rado Theorem uses $L = \{1, \ldots, k-1\}$. With $L = \{0, \ldots, k-1\}$, trivially all k-sets can be included, yielding a family of size $\binom{n}{|L|}$. If we do not restrict to a k-uniform family, then we can include all sets of size at most k, yielding size $\sum_{i=0}^{|L|} \binom{n}{i}$. Frankl–Wilson [1981] proved that no L-intersecting family is larger.

To prove the Frankl–Wilson Theorem, we use the Triangular Criterion and introduce another method.

15.1.13. Remark. A polynomial is **multilinear** if each positive exponent in its expression as a sum of monomials is 1. Given a polynomial f in n variables, the **multilinear reduction** of f is the multilinear polynomial $\hat{f}$ whose monomials have the same variables as corresponding monomials in f; that is, each positive exponent in the expression of f is reduced to 1. Because $0^r = 0$ and $1^r = 1$ for $r \in \mathbb{N}$, the values of f and $\hat{f}$ agree on $\{0, 1\}^n$. If $f_1, \ldots, f_m$ are linearly independent because no linear combination of them is 0 on all of $\{0, 1\}^n$, then the same holds for $\hat{f}_1, \ldots, \hat{f}_m$. The idea is to bound m by bounding the number of linearly independent multilinear polynomials. ∎

15.1.14. Theorem. (Frankl–Wilson [1981]) If $\mathcal{F}$ is an L-intersecting family of subsets of $[n]$, where $|L| = s$, then $|\mathcal{F}| \leq \sum_{i=0}^{s} \binom{n}{i}$.

Proof: (Babai [1988]) Let $\mathcal{F} = \{A_1, \ldots, A_m\}$, indexed so $|A_1| \leq \cdots \leq |A_m|$. Let $L = \{l_1, \ldots, l_s\}$. Define polynomials $f_1, \ldots, f_m$ by $f_i(x) = \prod_{k:\, l_k < |A_i|}(x \cdot v^i - l_k)$, where v^i is the incidence vector of A_i. By construction, $f_i(v^j) \neq 0$ for $j = i$. Using the indexing of $\mathcal{F}$ by size, $|A_j \cap A_i| < |A_i|$ for $j < i$, and hence $f_i(v^j) = 0$ for $j < i$. By the Triangular Criterion, $f_1, \ldots, f_m$ are linearly independent.

Since $v^1, \ldots, v^m \in \{0, 1\}^n$, the computations are the same for the multilinear reductions $\hat{f}_1, \ldots, \hat{f}_m$, so those also are linearly independent. We prove the desired bound by capturing *them* in a small space. Each f_i is the product of at most s linear factors, so the total degree of each monomial in the expansion of f_i is at most s. Hence the multilinear reduction of f_i is in the span of the products of at most s distinct variables. There are $\sum_{i=0}^{s} \binom{n}{i}$ such monomials. ∎

The bound in Theorem 15.1.14 can be strengthened when the requirement $|A_i| \notin L$ for $A_i \in \mathcal{F}$ is added. For example, for odd-sized sets and L all even, the Oddtown theorem (Theorem 15.1.2) implies $|\mathcal{F}| \leq n$. This is consistent with the bound $\binom{n}{|L|}$ when we view the intersection sizes as congruence classes modulo 2.

Class 0 contains all intersection sizes, but the sizes of the sets are forbidden from that class. See also Exercise 13.

A variation of Theorem 15.1.14 is proved in Lemma 16.2.40 (by a similar method) and applied in Theorem 16.2.41 to solve a question asked by Borsuk. Meanwhile, here we present a modular version of Theorem 15.1.14. The proof is analogous, and the bound is the same.

15.1.15. Definition. For $L \subseteq \mathbb{Z}_p$ with p prime, write $t \in L \pmod{p}$ when $t \equiv l \pmod{p}$ for some $l \in L$. A family $\mathcal{F} \subseteq 2^{[n]}$ is p-**modular** L-**intersecting** if $|A| \notin L \pmod{p}$ for $A \in \mathcal{F}$ and $|A \cap B| \in L \pmod{p}$ for distinct $A, B \in \mathcal{F}$.

15.1.16. Theorem. (Deza–Frankl–Singhi [1983]) If $\mathcal{F}$ is a p-modular L-intersecting family of subsets of $[n]$, where $|L| = s$ and p is prime, then $|\mathcal{F}| \leq \sum_{i=0}^{s} \binom{n}{i}$.

Proof: (Alon–Babai–Suzuki [1991]) Let $\mathcal{F} = \{A_1, \ldots, A_m\}$, with v^i the incidence vector of A_i. Define polynomials $f_1, \ldots, f_m$ by $f_i(x) = \prod_{l \in L}(x \cdot v^i - l)$. Perform computations over $\mathbb{F}_p$, and write "=" instead of "$\equiv$". Note that $v^j \cdot v^i = |A_i \cap A_j|$. Thus $f_i(v^j) \neq 0$ for $i = j$ (since $|A_i| \notin L \pmod{p}$), while $f_i(v^j) = 0$ for $i \neq j$ (since $|A_i \cap A_j| \in L \pmod{p}$). By the Diagonal Criterion, $f_1, \ldots, f_m$ are independent.

The bound now follows in the same way as in Theorem 15.1.14. The multilinear reductions $\hat{f}_1, \ldots, \hat{f}_m$ are linearly independent by the same criterion as $f_1, \ldots, f_m$, and they lie in a space of dimension $\sum_{i=0}^{s} \binom{n}{i}$. ∎

It may seem that the bound $\sum_{i=0}^{s} \binom{n}{i}$ is much larger than $\binom{n}{s}$, but actually it is not much bigger when s is not too big.

15.1.17. Lemma. If $n \geq 2s$ and $s = n/r$, then
$$\sum_{i=0}^{s} \binom{n}{i} \leq \binom{n}{s}\left(1 + \frac{s}{n - 2s + 1}\right) < \binom{n}{s}\left(1 + \frac{1}{r-2}\right).$$
In particular, if $s \leq n/3$, then $\sum_{i=0}^{s} \binom{n}{i} < 2\binom{n}{s}$.

Proof: By factoring $\binom{n}{s}$ from each term and then enlarging (and extending) the terms to obtain a geometric series,
$$\sum_{i=0}^{s} \binom{n}{i} = \binom{n}{s}\left(1 + \frac{s}{n-s+1} + \frac{s(s-1)}{(n-s+1)(n-s+2)} + \cdots\right)$$
$$\leq \binom{n}{s}\left(1 + \frac{s}{n-s+1} + \frac{s^2}{(n-s+1)^2} + \cdots\right)$$
$$= \binom{n}{s}\frac{1}{1 - \frac{s}{n-s+1}} = \binom{n}{s}\frac{n-s+1}{n-2s+1} = \binom{n}{s}\left(1 + \frac{s}{n-2s+1}\right)$$
$$= \binom{n}{s}\left(1 + \frac{n/r}{n-2n/r+1}\right) < \binom{n}{s}\left(1 + \frac{1}{r-2}\right).$$ ∎

These results yield a constructive superpolynomial lower bound for the Ramsey number $R(t, t)$, weaker than Erdős' nonconstructive exponential bound, but using explicit graphs. Coloring $(t-1)K_{t-1}$ red and $K_{t-1,\ldots,t-1}$ blue yields $R(t, t) >$

$(t-1)^2$. Nagy [1972] increased this to $\binom{t^3}{3}$ (Exercise 18). Frankl [1977] found graphs proving $R(t,t) > t^{\omega(t)}$ using Δ-systems (sunflowers), where $\omega(t) \to \infty$. Frankl–Wilson [1981] obtained similar results from p-modular L-intersecting families. (A vertex set is **homogeneous** if it is a clique or independent set.)

15.1.18. Theorem. (Frankl–Wilson [1981]) Let p be a prime, and choose $n > 2p^2$. Let G be the graph with vertex set $\binom{[n]}{p^2-1}$ defined by $AB \in E(G)$ if and only if $|A \cap B| \not\equiv -1 \pmod p$. The graph G has no homogeneous set with more than $2\binom{n}{p-1}$ vertices. As a consequence, $R(t,t) > t^{(1-\varepsilon)\omega(t)}$, where $\omega(t) = \frac{\ln t}{4\ln\ln t}$.

Proof: If $A_1, \ldots, A_m$ is a clique in G, then it is a p-modular L-intersecting family, where $L = \{0, \ldots, p-2\}$, because $|A_i| = p^2 - 1 \equiv -1 \pmod p$, and $A_iA_j \notin E(G)$ when $|A_i \cap A_j| \equiv -1 \pmod p$. With $|L| = p-1$, Theorem 15.1.16 yields $m \le \sum_{i=0}^{p-1}\binom{n}{i} < 2\binom{n}{p-1}$. If $A_1, \ldots, A_m$ is an independent set, then $|A_i \cap A_j| \in \{p-1, 2p-1, \ldots, p^2-p-1\}$. Here $p-1$ intersection sizes are allowed, so Theorem 15.1.14 yields $m \le \sum_{i=0}^{p-1}\binom{n}{i} < 2\binom{n}{p-1}$.

Fixing t, let p be the largest prime with $2\binom{p^3}{p-1} < t$, and let $n = p^3$. We have proved $R(t,t) > \binom{n}{p^2-1}$. The choice of p yields $p \sim \frac{\ln t}{2\ln\ln t}$, and then $\binom{p^3}{p^2-1} > t^{(1-\varepsilon)\omega(t)}$, where $\omega(t) = \frac{\ln t}{4\ln\ln t}$. That is, we compare roughly $\binom{p^3}{p}$ for t with $\binom{p^3}{p^2}$ for the lower bound on $R(t,t)$. The logarithm of the latter is roughly $p/2$ times the logarithm of the former, so roughly $R(t,t) > t^{p/2}$ (Exercise 16 requests further details). ∎

We present another application. The famous Hadwiger–Nelson problem (Hadwiger [1945]) asks for the fewest colors in a coloring of $\mathbb{R}^n$ such that no two points at distance 1 have the same color. For $n = 2$, the answer is in $\{4,5,6,7\}$ (Exercise 8.1.13). There is an easy general upper bound of $n^{n/2}$ (Exercise 15). Larman–Rogers [1972] presented a quadratic lower bound and an upper bound of $(2\sqrt{2} + o(1))^n$ and conjectured an exponential lower bound. Frankl–Wilson [1981] proved that; we obtain it from a corollary of Theorem 15.1.16.

15.1.19. Corollary. Let p be a prime, and let $\mathcal{F}$ be a $(2p-1)$-uniform family of subsets of $[4p-1]$. If no two members of $\mathcal{F}$ have exactly $p-1$ common elements, then $|\mathcal{F}| \le 2\binom{4p-1}{p-1}$.

Proof: Let $L = \{0, \ldots, p-2\}$. The family $\mathcal{F}$ is p-modular L-intersecting since $\mathcal{F}$ is $(2p-1)$-uniform with $2p-1 \notin L \pmod p$, and the remaining possible intersection sizes $\{p, \ldots, 2p-2\}$ are congruent to elements of L. By $|L| = p-1$, Theorem 15.1.16 and Lemma 15.1.17 yield the bound. ∎

15.1.20. Theorem. (Frankl–Wilson [1981]) For large n, the chromatic number of the unit-distance graph in $\mathbb{R}^n$ is greater than 1.1397^n.

Proof: The graph defined using distance d is isomorphic to the unit-distance graph. Hence it suffices to prove the lower bound for a subgraph of the distance-d graph. We use an appropriate d and a subgraph induced by some binary n-tuples with k nonzero positions.

The squared distance between two points in $\{0,1\}^n$ is the number of coordinates where they differ. As incidence vectors of subsets A and B of n, they differ in $|A \triangle B|$ places. If A and B have size k, then the symmetric difference has size $2(k - |A \cap B|)$. Hence forbidding one distance between the points is equivalent to forbidding one intersection size for the sets. If $k = 2p - 1$, then forbidding intersection size $p - 1$ is equivalent to forbidding squared distance $2p$.

Let p be the largest prime such that $4p - 1 \le n$; we use only $4p - 1$ of the coordinates. Let $d = \sqrt{2p}$. By Corollary 15.1.19, the maximum size of an independent set in the subgraph of the distance-d graph induced by the incidence vectors of the $(2p - 1)$-sets in $[4p - 1]$ is at most $2\binom{4p-1}{p-1}$. Hence the chromatic number is at least $\binom{4p-1}{2p-1}/2\binom{4p-1}{p-1}$, which equals $\binom{4p}{2p}/\binom{4p}{p}$ and simplifies further to $\frac{p!(3p)!}{(2p)!(2p)!}$.

With Stirling's Formula (Application 2.3.8), the ratio is approximately $c(3^{3/4}/2)^{4p}$, where $c = \sqrt{3/4}$. When m is large, there is a prime between m and $m - m^{7/12}$ (Huxley [1973]). Applying this with $m = n/4$ completes the proof, since $3^{3/4}/2 > 1.1397$. (Note: Exercise 17 improves the bound to about $(1.2)^n$.) ∎

Frankl–Füredi [1984] conjectured that the bound in Theorem 15.1.14 on the size of an L-intersecting family improves from $\sum_{i=0}^{s}\binom{n}{s}$ to $\sum_{i=0}^{s}\binom{n-1}{s}$ when $L = [s]$. Ramanan [1997] proved this. Snevily conjectured and proved a substantial generalization, after earlier proving special cases (Snevily [1994, 1999]).

15.1.21. Theorem. (**Snevily's Theorem**; Snevily [2003]) If L is a set of s positive integers, and $\mathcal{F}$ is an L-intersecting family of subsets of $[n]$, then $|\mathcal{F}| \le \sum_{i=0}^{s}\binom{n-1}{i}$. ∎

Fisher's Inequality (Theorem 13.1.15) states that $n \ge v$ when $\mathcal{B}$ is a family of n members of $\binom{[v]}{k}$ in $[v]$ (with $k < v$) such that every two elements of $[v]$ appear in λ common blocks. The dual statement (transposing the incidence matrix) is that an L-intersecting family of subsets of $[n]$ has size at most n when $L = \{\lambda\}$. Since $n = \binom{n-1}{0} + \binom{n-1}{1}$, Fisher's Inequality follows from the case $s = 1$ of Snevily's Theorem, which was proved by Majumdar [1953].

Instead of Snevily's Theorem, we present only a modular version that he proved earlier. The proof uses a refinement of the dimension argument. To the set of polynomials obtained from elements of $\mathcal{F}$, we add more polynomials. If the full set is still linearly independent in the same space, then the bound on $\mathcal{F}$ becomes the dimension of the space minus the number of extra polynomials. This technique is also used in Exercise 11 to improve the bound on two-distance sets.

15.1.22. Lemma. Let $C_1, \ldots, C_t$ be a family of subsets of $[n]$. If polynomials $h_1, \ldots, h_t$ are defined on $\mathbb{R}^n$ by $h_j(x) = \prod_{r \in C_j} x_r$, then $h_1, \ldots, h_t$ are linearly independent on $\{0, 1\}^n$.

Proof: Let w^j be the incidence vector of C_j, indexed so $|C_1| \le \cdots \le |C_t|$. Note that $h_j(w^j) = 1$. If $i > j$, then the indexing guarantees an element $r \in C_i - C_j$. Now h_i has x_r as a factor, but the value in coordinate r of w^j is 0, so $h_i(w^j) = 0$. By the Triangular Criterion, $\{h_1, \ldots, h_t\}$ is linearly independent. ∎

15.1.23. Theorem. (Snevily [1994]) Let p be a prime number. If $\mathcal{F}$ is a p-modular L-intersecting family of subsets of $[n]$ and $|L| = s$, then $|\mathcal{F}| \le \sum_{i=0}^{s}\binom{n-1}{i}$.

Proof: Let $\mathcal{F} = \{A_1, \ldots, A_m\}$, indexed so that $A_1, \ldots, A_q$ omit the element 1 and $A_{q+1}, \ldots, A_m$ contain it. Begin the proof as in Theorem 15.1.16, letting $f_i(x) = \prod_{l \in L}(x \cdot v^i - l)$, where v^i is the incidence vector of A_i. Again the multilinear reductions $\hat{f}_1, \ldots, \hat{f}_m$ are spanned by the $\sum_{i=0}^{s} \binom{n}{i}$ multilinear monomials of degree at most s.

Let $C_1, \ldots, C_t$ be all the sets of size less than s in $[n]$ that omit element 1, indexed so $|C_1| \leq \cdots \leq |C_t|$. Define polynomials h_j and g_j by $h_j(x) = \prod_{r \in C_j} x_r$ and $g_j(x) = (x_1 - 1)h_j(x)$. Each g_j has degree at most s and is multilinear and spanned by the same set of $\sum_{i=0}^{s} \binom{n}{i}$ monomials as $\hat{f}_1, \ldots, \hat{f}_m$. Since $t = \sum_{i=1}^{s} \binom{n-1}{i-1}$, it suffices to prove that $\{\hat{f}_1, \ldots, \hat{f}_m\} \cup \{g_1, \ldots, g_t\}$ is linearly independent.

Let $P = \sum_{i=1}^{m} \alpha_i \hat{f}_i + \sum_{j=1}^{t} \beta_j g_j$, with each α_i and β_j in $\mathbb{F}_p$. Suppose that P is identically 0. Let $A_i' = A_i \cup \{1\}$, and let y^i be the incidence vector of A_i', for $1 \leq i \leq m$. Each y^i has 1 in the first coordinate, so the contribution of the second sum to $P(y^i)$ is always 0.

Note that $A_j' \cap A_i = A_j \cap A_i$ if $i \leq j$. This holds because $A_j' = A_j$ if $j > q$ and $1 \notin A_i$ if $i \leq q$. Thus $f_i(y^j) = f_i(v^j)$ for $i \leq j$. Since $f_i(v^j) = 0$ if and only if $i \neq j$ and $\hat{f}_i(x) = f_i(x)$ when $x \in \{0, 1\}^n$, evaluating P at $y^m, \ldots, y^1$ successively yields $\alpha_m, \ldots, \alpha_1 = 0$.

By Lemma 15.1.22, $h_1, \ldots, h_t$ are linearly independent. Multiplying them all by $x_1 - 1$ leaves $g_1, \ldots, g_t$ independent. Since $\alpha_m, \ldots, \alpha_1 = 0$, making P identically 0 thus also requires $\beta_1, \ldots, \beta_t = 0$. Hence there is no equation of linear dependence for $\{\hat{f}_1, \ldots, \hat{f}_m\} \cup \{g_1, \ldots, g_t\}$. ∎

Taking p sufficiently large in Theorem 15.1.23 yields the bound in Theorem 15.1.21 for the special case of Theorem 15.1.21 where the sizes of members of the intersecting family $\mathcal{F}$ do not lie in L. Extensions of Theorem 15.1.23 to k-wise intersections appear in Grolmusz–Sudakov [2002] and Cao–Hwang–West [2007].

The top two entries in $\sum_{i=0}^{s} \binom{n-1}{i}$ sum to $\binom{n}{s}$. We next show that this smaller value tightens the upper bound on the size of an L-intersecting family $\mathcal{F}$ when we add the restriction that $\mathcal{F}$ is a uniform family, even if 0 is allowed in L. The original proof used linear algebra in a different way; combining the ideas we have presented led to a shorter proof.

15.1.24. Theorem. (Ray-Chaudhuri–Wilson [1975]) If $n \geq 2s$, and L is a set of s nonnegative integers, then every L-intersecting k-uniform family of subsets of $[n]$ has size at most $\binom{n}{s}$.

Proof: (Alon–Babai–Suzuki [1991]) We may assume $k \notin L$. Let $v^1, \ldots, v^m$ be the incidence vectors for the sets in the family. Let $f_i(x) = \prod_{l \in L}(x \cdot v^i - l)$. By the Diagonal Criterion, $f_1, \ldots, f_m$ are linearly independent on $\{0, 1\}^n$. Each f_i has degree s; the multilinear reduction yields polynomials $\hat{f}_1, \ldots, \hat{f}_m$ that are linearly independent on $\{0, 1\}^n$ and spanned by the $\sum_{i=0}^{s} \binom{n}{i}$ multilinear monomials with degree at most s.

As in Theorem 15.1.23, we add polynomials to this set. Let $C_1, \ldots, C_t$ be the sets of size less than s in $[n]$. Define h_j by $h_j(x) = \prod_{r \in C_j} x_r$. By Lemma 15.1.22, $h_1, \ldots, h_t$ are linearly independent over $\{0, 1\}^n$. Define g_j by $g_j(x) = (x \cdot 1_n - k)h_j(x)$. As in Theorem 15.1.23, we have multiplied independent polynomials by one linear factor, and the resulting polynomials are independent.

Let $P = \sum_{i=1}^{m} \alpha_i f_i + \sum_{j=1}^{t} \beta_j g_j$. Consider coefficients $\alpha_1, \ldots, \alpha_m$ and $\beta_1, \ldots, \beta_t$ such that P is identically 0. Since $\mathcal{F}$ is k-uniform, the contribution from the second sum is 0 when evaluated at v^i. Since $f_i(v^j) = 0$ when $j \neq i$, evaluating P at v^i thus yields $\alpha_i = 0$. With each α_i being 0, linear independence of $g_1, \ldots, g_t$ implies also that each β_j is 0.

We conclude that $\{f_1, \ldots, f_m\} \cup \{g_1, \ldots, g_t\}$ is linearly independent. Again we may take the multilinear reduction of g_j since independence was established by evaluation over $\{0, 1\}^n$. The degree of g_j is at most s, so these vectors also lie in the span of the $\sum_{i=0}^{s} \binom{n}{i}$ multilinear monomials with degree at most s. Since $t = \sum_{i=0}^{s-1} \binom{n}{i}$, we conclude $m \leq \binom{n}{s}$. ∎

The bound $\binom{n}{s}$ holds trivially with equality when $L = \{0, \ldots, s-1\}$ and $k = s$. Is it still sharp for larger k? Since $\binom{n}{s} \sim n^s/s!$, the construction below is not so much smaller than the upper bound when s and t are fixed and n is large.

15.1.25. Theorem. For $n \geq 2k^2 \geq 2s^2$ and $L = \{0, \ldots, s-1\}$, some k-uniform L-intersecting family $\mathcal{F}$ satisfies $|\mathcal{F}| > (n/2k)^s$.

Proof: Let p be the largest prime bounded by n/k, so $n/2k < p \leq n/k$. Fix a k-set A contained in $\mathbb{F}_p$; our family $\mathcal{F}$ will consist of p^s k-sets within $A \times \mathbb{F}_p$. Given a polynomial f of degree less than s, let $A_f = \{(a, f(a)): a \in A\}$. There are p^s polynomials over $\mathbb{F}_p$ with degree less than s, so this defines p^s sets of size k. Distinct polynomials of degree d over $\mathbb{F}_p$ agree on at most d points in $\mathbb{F}_p$. Since $k \geq s$, the p^s sets are distinct, and any two of them have fewer than s common elements. ∎

COMBINATORIAL NULLSTELLENSATZ

The Combinatorial Nullstellensatz is a result about zeros of multivariable polynomials over a field. Surprisingly easy to prove, it has many applications in additive number theory, discrete geometry, and graph theory. Applying it is sometimes called the **polynomial method**. The theorem was presented by Noga Alon at a conference in 1995, but the proceedings did not appear until 1999. Alon had already applied the theorem in at least five papers with eight coauthors from 1984 to 1996, proving new results and giving short proofs of old results. Many others have since also used it. Tao [2014] surveyed applications in arithmetic combinatorics and number theory; Clark [2014] explored further extensions.

We need a lemma that generalizes to n variables the familiar statement that a nonzero polynomial of degree d in one variable takes the value 0 at most d times. That statement is proved by induction on d, using the Euclidean algorithm to factor out $x - \alpha$ when α is a root. The discussion is valid in any field. We compute with equalities rather than using congruence notation for finite fields. Also, superscripts return to being exponents.

15.1.26. Lemma. Let f be a polynomial in n variables $x_1, \ldots, x_n$, over a field K. For each i, let the degree of f as a polynomial in x_i be at most d_i, and let S_i be a set of $d_i + 1$ distinct values in K. If $f(x_1, \ldots, x_n) = 0$ for $(x_1, \ldots, x_n) \in \prod_{i=1}^{n} S_i$, then f is identically 0.

Proof: We take the result in one variable as the basis for induction on n. For $n > 1$, we collect terms to write f as a polynomial in x_n. That is, $f = \sum_{j=0}^{d_n} f_j(x_1, \ldots, x_{n-1}) x_n^j$, where each f_j is a polynomial having degree at most d_i in each x_i. For $(x_1, \ldots, x_{n-1}) \in \prod_{i=1}^{n-1} S_i$, evaluating $f_0, \ldots, f_{d_n}$ yields a one-variable polynomial in x_n of degree at most d_n. Furthermore, the hypothesis implies that this polynomial is 0 for $x_n \in S_n$.

By the basis step ($n = 1$), the one-variable polynomial we obtain for a fixed $(x_1, \ldots, x_{n-1}) \in \prod_{i=1}^{n-1} S_i$ is the zero polynomial. Thus each f_i is 0 at all values in $\prod_{i=1}^{n-1} S_i$. By the induction hypothesis, each f_i is identically zero. Thus the coefficients of f are all zero, and f is identically zero. ∎

We want that if the coefficient of a term $\prod x_i^{t_i}$ is nonzero in a polynomial f of degree $\sum t_i$, and $|S_i| > t_i$ for all i, then the polynomial is nonzero at some point in $\prod S_i$. The lemma does not say this, since other terms may have degree larger than t_i in x_i, for some i. Fortunately, it is not hard to overcome this technicality.

The **degree** of a polynomial is the maximum, over all monomials, of the sum of the exponents on the variables. We extract the coefficient of a monomial $\prod_{i=1}^{n} x_i^{t_i}$ in a polynomial $f(x_1, \ldots, x_n)$ using the **coefficient operator** $\left[\prod_{i=1}^{n} x_i^{t_i} \right]$, which for formal power series in one variable was used extensively in Chapter 3.

15.1.27. Theorem. (Combinatorial Nullstellensatz; Alon [1999]) If $\prod_{i=1}^{n} x_i^{t_i}$ is a monomial with nonzero coefficient in a polynomial f having degree $\sum_{i=1}^{n} t_i$ over a field K, and $S_1, \ldots, S_n$ are sets with $|S_i| > t_i$ for $1 \le i \le n$, then $f(x) \ne 0$ for some $x \in \prod S_i$.

Proof: It suffices to prove the statement when $|S_i| = t_i + 1$ for each i. The idea is to change f into another polynomial $\hat{f}$ that agrees with f on $\prod S_i$ but has degree at most t_i as a polynomial in x_i, for each i. Lemma 15.1.26 then implies that $\hat{f}(x) \ne 0$ for some $x \in \prod S_i$. Since $\hat{f}$ agrees with f on $\prod S_i$, also $f(x) \ne 0$.

For each index i, define a polynomial g_i by $g_i(x) = \prod_{s \in S_i} (x_i - s)$; note that g_i depends only on x_i. It has degree $t_i + 1$ in x_i and degree 0 in other variables. Expanding the product yields $g_i(x) = x_i^{t_i+1} - h_i(x)$, where h_i is a polynomial with degree at most t_i in x_i and degree 0 in other variables.

By definition, $g_i(x) = 0$ for $x \in \prod S_i$, since $x_i \in S_i$ in that case. Therefore, $x_i^{t_i+1} = h_i(x)$ for all $x \in \prod S_i$. Thus we can replace each appearance of a variable having too large an exponent with a polynomial having smaller degree in that variable. By making such a replacement as long as the polynomial still has degree greater than t_i in some x_i, we obtain $\hat{f}$ having degree at most t_i in x_i for each i.

We must also show $\left[\prod x_i^{t_i} \right] \hat{f}(x) \ne 0$. Since no exponent is too large, we made no change to that term. Also we did not introduce any terms that could cancel it; since f has degree $\sum t_i$, any monomial containing a variable with too large an exponent has some x_j with exponent less than t_j. Since the substitutions increase no exponents, no substitution can introduce a contribution to $\left[\prod x_i^{t_i} \right]$. ∎

An early application was a short proof of the Cauchy–Davenport Theorem of additive number theory, first proved by Cauchy in 1813 and by Davenport in 1935. Let A and B be subsets of $\mathbb{Z}_n$, with $|A| = a$ and $|B| = b$. How many elements of $\mathbb{Z}_n$ must arise as $x + y$ with $x \in A$ and $y \in B$?

Setting $A = \{0, \dots, a - 1\}$ and $B = \{0, \dots, b - 1\}$ shows that the number can be as small as $\min\{n, a + b - 1\}$. If $a + b > n$, then for any $c \in \mathbb{Z}_n$ the sets A and $\{c - y : y \in B\}$ must intersect, and when an element x is in the intersection we have $x \in A$ and $y \in B$ such that $c = x + y$. Hence the number of sums always equals n if $a + b > n$. Also, when $n = 2k$, taking the "even" classes for both A and B yields only even classes as sums, so here the sum can be as small as $n/2$ even though $a = b = n/2$. This suggests restricting our attention to prime moduli.

15.1.28. Theorem. (Cauchy–Davenport Theorem) If p is prime, and $A, B \subseteq \mathbb{Z}_p$ with $|A| = a$ and $|B| = b$ and $a + b \leq p$, then the smallest possible size of $\{x + y : x \in A, y \in B\}$ is $a + b - 1$.

Proof: It suffices to prove the lower bound; equality holds in the construction presented above. Suppose that there are fewer than $a + b - 1$ sums. Let C be a set of size $a + b - 2$ in $\mathbb{Z}_p$ that contains all the sums. Let $f(x, y) = \prod_{c \in C}(x + y - c)$, over $\mathbb{Z}_p$. We have a polynomial in two variables, and its degree is $a + b - 2$.

We claim that $[x^{a-1}y^{b-1}]f(x, y) = \binom{a+b-2}{a-1} \not\equiv 0 \pmod{p}$. Contributions to this coefficient use x or y in each factor when expanding f, choosing x exactly $a - 1$ times and y exactly $b - 1$ times. The number of ways to do that, each contributing $+1$ to the coefficient, is $\binom{a+b-2}{a-1}$. Finally, that binomial coefficient is nonzero modulo p since $a + b - 2 < p$; there is no factor of p in the numerator and no other way to introduce a factor of p.

Since $|A| = a$ and $|B| = b$, the Combinatorial Nullstellensatz yields $x \in A$ and $y \in B$ such that $f(x, y) \neq 0$. This is a contradiction, since f was constructed to be 0 at all such pairs (x, y). ∎

This short proof illustrates the method for applying the Combinatorial Nullstellensatz. Using the set of sums, we design f that is 0 at (x, y) when $x \in A$ and $y \in B$. If the set of sums is too small, then $A \times B$ is too big for f to be identically 0 there when the appropriate coefficient is nonzero.

When $A = B$, the lower bound in Theorem 15.1.28 is $\min\{2|A| - 1, p\}$. Erdős–Heilbronn [1964] conjectured that almost as much is forced even when ignoring the contributions by adding elements to themselves. Given the ease of proving this from the Combinatorial Nullstellensatz, it is remarkable that the problem was open for 30 years. The original proof used exterior algebra and representation theory of the symmetric group.

15.1.29. Theorem. (Erdős–Heilbronn Conjecture; Dias da Silva & Hamidoune [1994]) If $A \subseteq \mathbb{Z}_p$, where p is prime, and C is the set of sums of distinct elements of A, then $|C| \geq \min\{2|A| - 3, p\}$.

Proof: (Alon–Nathanson–Rusza [1996]) Since there are only p classes, we may assume $2a - 3 < p$, where $a = |A|$. As in the proof of Theorem 15.1.28, we design a polynomial f that is 0 at (x, y) when $x + y \in C$. The polynomial is the same as before, except for including the factor $(x - y)$ to ensure that f is 0 when $x = y$, since $2x$ may not be in C. That is, let $f(x, y) = (x - y)\prod_{c \in C}(x + y - c)$. Note that $\deg(f) = m + 1$, where $m = |C|$.

We study the coefficient of $x^{a-1}y^{m-a+2}$. As before, contributions to the desired coefficient use x or y in each factor. The contributions choosing x in the first factor are positive, and those choosing $-y$ are negative. Thus $\left[x^{a-1}y^{m-a+2}\right]f(x,y) = \binom{m}{a-2} - \binom{m}{a-1} = [1 - \frac{m-a+2}{a-1}]\binom{m}{a-2}$. If $m \le 2a - 4$, then this coefficient is nonzero, and $a > m - a + 2$. Now the Combinatorial Nullstellensatz guarantees $(x, y) \in A^2$ such that $f(x,y) \ne 0$. These are distinct elements of A whose sum is not in C, which is a contradiction. We conclude that $m \ge 2a - 3$. $\blacksquare$

The theorem below extends Theorem 15.1.29 to restricted sums over many variables. See Exercises 23–25 for the proof and applications.

15.1.30. Theorem. (Alon–Nathanson–Rusza [1996]) Let h be a polynomial in k variables over $\mathbb{Z}_p$, where p is prime. Let $A_1, \dots, A_k$ be nonempty subsets of $\mathbb{Z}_p$, with $c_i = |A_i| - 1$ for all i. Let $m = \sum_{i=1}^{k} c_i - \deg(h)$. Let $C = \{\sum_{i=1}^{k} a_i : a_i \in A_i \text{ and } h(a) \ne 0\}$. If $\left[\prod_{i=1}^{k} x_i^{c_i}\right](\sum_{i=1}^{k} x_i)^m h(x) \ne 0$, then $|C| \ge m + 1$. $\blacksquare$

Our next consequence is also number-theoretic but has a geometric application. It was conjectured by Artin in 1934, proved by Chevalley [1935], and extended in Chevalley–Warning [1935]. The proof uses Fermat's Little Theorem (Application 1.3.10), which states that if p is a prime, then $a^{p-1} \equiv 1 \pmod{p}$ for every integer a not divisible by p.

15.1.31. Theorem. (Chevalley–Warning Theorem) Let $P_1, \dots, P_m$ be polynomials over $\mathbb{F}_p$ in n variables, where p is prime. If $\sum_{i=1}^{m} \deg(P_i) < n$ and the polynomials have a common zero, then they have another common zero.

Proof: Let $(c_1, \dots, c_n)$ be a common zero. Let

$$f(x) = \prod_{i=1}^{m}(1 - P_i(x)^{p-1}) - \prod_{j=1}^{n}(1 - (x_j - c_j)^{p-1}).$$

Note that $f(c) = 1 - 1 = 0$. If there is no other common zero, then for $x \in \mathbb{F}_p^n - \{c\}$, there exists i such that $P_i(x) \not\equiv 0 \pmod{p}$. Also there exists j such that $x_j \ne c_j$. By Fermat's Little Theorem, $P_i(x)^{p-1} \equiv 1 \equiv (x_j - c_j)^{p-1} \pmod{p}$. Hence $f(x) = 0$.

The degree of the first term in f is bounded by $(p-1)\sum_{i=1}^{m} \deg(P_i)$, which is less than $(p-1)n$. The degree of the second term is $(p-1)n$, and the coefficient of $\prod x_j^{p-1}$ in f is $(-1)^{n+1}$, which is nonzero modulo p. Since $|\mathbb{F}_p| > p - 1$ and we choose each x_i from $\mathbb{F}_p$, Theorem 15.1.27 guarantees $x \in \mathbb{F}_p^n$ such that $f(x) \ne 0$. By this contradiction, the polynomials must have another common zero. $\blacksquare$

Chevalley proved the case $m = 1$ and Warning proved the general case; both in fact obtained p common zeros. We apply Theorem 15.1.31 to determine the transversal number of a special hypergraph. The vertex set is $\mathbb{F}_p^n$, and for each hyperplane H in $\mathbb{F}_p^n$ we make an edge consisting of all the points in H. These points are the solutions to $a \cdot x = b$, where $a \in \mathbb{F}_p^n - \{0\}$ and $b \in \mathbb{F}_p$ are fixed. Every edge has p^{n-1} vertices. The **transversal number** $\tau(\mathcal{H})$ of a hypergraph $\mathcal{H}$ is the minimum size of a vertex set intersecting all the edges.

15.1.32. Theorem. (Jamison [1977], Brouwer–Schrijver [1978]) For prime p, the transversal number of the hypergraph of all hyperplanes in $\mathbb{F}_p^n$ is $n(p-1)+1$.

Proof: First we produce a transversal of this size. Let B be the set of points in $\mathbb{F}_p^n$ having at most one nonzero coordinate; by construction B has the specified size. To prove that B is a transversal, we use induction on n. For $n = 1$, each point is a hyperplane, and indeed $B = \mathbb{F}_p^1$.

For $n > 1$, hyperplanes of the form $x_n = c$ are hit by the point in B having c in the last coordinate. For other hyperplanes, consider the fixed hyperplane H consisting of $\{x \in \mathbb{F}_p^n : x_n = 0\}$. The hyperplanes of the form $x_n = c$ include H and all hyperplanes disjoint from H. The others intersect H in a hyperplane of $\mathbb{F}_p^{n-1}$ obtained by dropping the last coordinate (0) from the points in the intersection. By the induction hypothesis, these hyperplanes are hit by the points in B that have 0 in the last coordinate.

For the lower bound, let B be any transversal. By applying a translation in each coordinate, we may assume $0 \in B$. Let $A = B - \{0\}$. The set A intersects all hyperplanes not containing 0. This means that for all $x \in \mathbb{F}_p^n - \{0\}$, the equation $x \cdot y = 1$ has a solution $y \in A$.

Let $f(x) = \prod_{a \in A}(x \cdot a - 1)$. Since $x \cdot y = 1$ has a solution in A when $x \neq 0$, we have $f(x) = 0$ for $x \in \mathbb{F}_p^n - \{0\}$ and $f(0) = (-1)^{|A|}$. Given variables $x_i^{(j)}$ for $i \in [n]$ and $j \in [p-1]$, define a polynomial P by

$$P = \left(\sum_{j=1}^{p-1} f(x_1^{(j)}, \ldots, x_n^{(j)})\right) - (-1)^{|A|}(p-1).$$

Since f takes only the values 0 and $(-1)^{|A|}$, the sum has magnitude $p-1$ only when each summand is nonzero, requiring each variable to have value 0. Hence P has value 0 only when all $n(p-1)$ variables have value 0. The contrapositive of the Chevalley–Warning Theorem (for $m = 1$) now yields $n(p-1) \leq \deg P = \deg f = |A| = |B| - 1$. ∎

We turn now to applications in graph theory. Berge and Sauer conjectured that every 4-regular graph has a 3-regular subgraph; Tashkinov [1984] proved this. The claim is false for multigraphs (consider a fat triangle with edges of multiplicity 2), but it becomes true when there is at least one "extra" edge, as seen by setting $p = 3$ in the next theorem. For convenience, when v is a vertex in a graph, let $\Gamma(v)$ denote the set of edges incident to v.

15.1.33. Theorem. (Alon–Friedland–Kalai [1984]) If p is prime, then every loopless multigraph G with average degree greater than $2p - 2$ and maximum degree at most $2p - 1$ contains a p-regular subgraph.

Proof: Suppose that G has n vertices and m edges. We want to design a polynomial f over the field $\mathbb{F}_p$ such that when $f(x) \neq 0$, the point x selects for us a p-regular subgraph. Hence we introduce a variable x_e for each edge e taking value 0 (absent) or 1 (present). To apply the Combinatorial Nullstellensatz, we will want a multilinear monomial term with a nonzero coefficient. Define f by

$$f(x) = \prod_{v \in V(G)}\left[1 - \left(\sum_{e \in \Gamma(v)} x_e\right)^{p-1}\right] - \prod_{e \in E(G)}(1 - x_e).$$

Each factor in the first term has degree $p - 1$, so the degree of the first term is at most $(p-1)n$. This quantity is less than m, since the average degree exceeds

$2p - 2$. Hence the degree is determined by the second term, which has degree m, with $\left[\prod_{e \in E(G)} x_e \right] f(x) = (-1)^{m+1} \neq 0$.

By the Combinatorial Nullstellensatz, $f(\hat{x}) \neq 0$ for some $\hat{x} \in \{0, 1\}^m$. Since $f(0) = 1 - 1 = 0$, this occurs with $\hat{x} \neq 0$. Since $\hat{x} \neq 0$, the second term in $f(\hat{x})$ has a factor that is 0. Hence the first term in $f(\hat{x})$ must be nonzero. By Fermat's Little Theorem, this requires that $\sum_{e \in \Gamma(v)} \hat{x}_e$ is a multiple of p for every vertex v.

Therefore, the degree of each vertex in the subgraph H of G with edge set $\{e \in E(G) : \hat{x}_e = 1\}$ is a multiple of p. Since $\Delta(G) \leq 2p - 1$, the degree is always 0 or p. Since $\hat{x} \neq 0$, it cannot always be 0. Thus H has a nontrivial component, and it is a p-regular subgraph of G. ∎

With no bound on the maximum degree, more edges may be needed to force a k-regular subgraph. Pyber [1985] proved that an n-vertex graph having at least $32k^2 n \ln n$ edges has a k-regular subgraph. The bound is not too far from optimal: Pyber–Rödl–Szemerédi [1995] proved by probabilistic arguments that there are graphs with at least $\Omega(n \log \log n)$ edges that have no 3-regular subgraph (and $O(n \log \Delta(G))$ edges force a 3-regular subgraph).

Instead of fixing the degree at each nonisolated vertex, we can be more flexible. Specify for each $v \in V(G)$ a *bad set* $B(v) \subseteq \{1, \ldots, d_G(v)\}$. We seek a subgraph H with $d_H(v) \notin B(v)$ for $v \in V(H)$. Shirazi–Verstraëte [2008] gave an easy proof from the Combinatorial Nullstellensatz that there is a nontrivial such subgraph H when $\sum_{v \in V(G)} B(v) < |E(G)|$ (Exercise 27), and this inequality is sharp.

They also proved a conjecture of Addario-Berry et al. [2007] that allows 0 to be in the forbidden sets. This was stated originally in terms of allowed degrees, but it is a bit cleaner for forbidden degrees. Consider the design of the polynomial f. We want the multivariate point x with $f(x) \neq 0$ to select the desired subgraph H. Hence we make a variable for each edge, and we restrict it to the values 0 and 1 to model whether the edge is used in H. For each vertex v, we design a factor that is 0 when the constraint at v is violated.

15.1.34. Theorem. (Shirazi–Verstraëte [2008]) For each vertex v in a graph G, specify a *bad set* $B(v) \subseteq \{0, \ldots, d_G(v)\}$. If $|B(v)| \leq \lfloor d(v)/2 \rfloor$ for all v, then G has a subgraph H with $d_H(v) \notin B(v)$ for all v.

Proof: Let $\Gamma(v)$ denote the set of edges in G incident to vertex v. Introduce a variable x_e for each edge e in G, with value 0 or 1, and let $x = (x_1, \ldots, x_{|E(G)|})$. Define a real-valued polynomial f by

$$f(x) = \prod_{v \in V(G)} \prod_{c \in B(v)} \left(\sum_{e \in \Gamma(v)} x_e - c \right).$$

The variables set to 1 yield a subgraph with degree $\sum_{e \in \Gamma(v)} x_e$ at v. The factor for v is 0 if and only if that degree is forbidden. We seek $x \in \{0, 1\}^{|E(G)|}$ with $f(x) \neq 0$.

Since f is a product of linear factors, $\deg(f) \leq \sum_{v \in V(G)} |B(v)|$. By the Combinatorial Nullstellensatz, it suffices to find a monomial with this degree having nonzero coefficient, whose variables all have exponent at most one. Monomials in the product arise by choosing, for each forbidden degree at each vertex, an edge incident to that vertex. We must not choose a given edge from both endpoints.

To avoid repeated selection, we orient G and pick for the monomial at v only variables x_e such that v is then the tail of e. If the orientation has at least $\lfloor d(v)/2 \rfloor$ edges leaving each vertex v, then there are enough such edges to choose distinct ones for the elements of $B(v)$, since $|B(v)| \leq \lfloor d(v)/2 \rfloor$. To form an orientation D such that $d_D^+(v) \geq \lfloor d_G(v)/2 \rfloor$, simply add a vertex w adjacent to all vertices of odd degree in G and orient by following an Eulerian circuit in each component.

We thus obtain a linear monomial. Every contribution to the coefficient of a monomial with degree $\sum_{v \in V(G)} |B(v)|$ is positive, since obtaining that degree requires selecting some x_e (and not c) from each factor.

Finally, when x is the point with $f(x) \neq 0$ guaranteed by the Combinatorial Nullstellensatz, each factor must be nonzero, which means that the number of edges selected at v (via $x_e = 1$) does not lie in $B(v)$.   ∎

Theorem 15.1.34 is sharp; the conclusion may fail when one bad set is a bit too large. Let $G = K_{2r,2r}$ with bipartition X, Y. If $B(x) = \{0, \ldots, r-1\}$ for $x \in X$ and $B(y) = \{r+1, \ldots, 2r\}$ for $y \in Y$, then each $B(v)$ has size $d(v)/2$, and a subgraph is good if and only if it is r-regular. When r is added to one bad set, there is no longer a good subgraph.

THE ALON–TARSI THEOREM

Our final application of the Combinatorial Nullstellensatz to graphs is one of the most famous. The Alon–Tarsi Theorem uses a polynomial associated with a graph G to obtain upper bounds on the list chromatic number $\chi_l(G)$. We first state the result, which can be applied without knowing the algebraic background.

15.1.35. Definition. A **circulation** is a digraph D such that $d_D^+(v) = d_D^-(v)$ for all $v \in V(D)$. The **parity** of a circulation D is the parity of $|E(D)|$. For any digraph D', let $\mathrm{diff}(D')$ denote the absolute difference between the number of even spanning circulations and the number of odd spanning circulations.

In this context, what we call circulations have usually been called "Eulerian subgraphs" (see Alon [1993]). We use "circulation" because there is no connectedness requirement and because the term is used analogously with network flows.

15.1.36. Theorem. (**Alon–Tarsi Theorem**; Alon–Tarsi [1992]) A graph G having an orientation D such that $\mathrm{diff}(D) \neq 0$ is f-choosable, where $f(v) = 1 + d_D^+(v)$ for each $v \in V(D)$.   ∎

15.1.37. Example. Let $G = C_n$. Let D be a cyclic orientation of G. Here $d_D^+(v) = 1$ for all v, and the only spanning circulations in D are the trivial subgraph (no edges) and D itself. If n is even, then $\mathrm{diff}(D) = 2$, and G is 2-choosable. If n is odd, then $\mathrm{diff}(D) = 0$, and D gives us no information.

If we reverse one edge of D to obtain D', then the only circulation is the trivial subgraph, and $\mathrm{diff}(D') = 1$. Since D' has a vertex with outdegree 2, we learn that C_{2k+1} is 3-choosable. The theorem provides only upper bounds; we do not learn that C_{2k+1} is not 2-choosable.

An acyclic digraph contains only one circulation: the trivial subdigraph. Thus Theorem 15.1.36 implies that G is k-choosable if G has an acyclic orientation where every outdegree is less than k. This again proves that k-degenerate graphs are $(k+1)$-choosable.   ∎

Before proving Theorem 15.1.36, we motivate it by describing some applications. A relatively easy application (Alon–Tarsi [1992]) is that every planar bipartite graph is 3-choosable (Exercise 32).

An impressive application is the **Cycle-plus-triangles Theorem**. Consider a 4-regular graph formed from C_{3m} by adding m pairwise disjoint triangles. Du and Hsu conjectured in 1986 that such graphs have independence number m, and in 1987 Erdős conjectured more strongly that they are 3-colorable. Fleischner–Stiebitz [1992] proved this by using Theorem 15.1.36 to prove the stronger result that every such graph is 3-choosable. (Later, Sachs [1993] gave a direct combinatorial proof that such graphs are 3-colorable.)

The analysis of circulations in the Cycle-plus-triangles Theorem is quite lengthy. Instead we present an easier result that also illustrates the use of the Alon–Tarsi Theorem. Recall that C_n^2 is the graph defined on n vertices around a circle by making each vertex adjacent to the four nearest vertices. We have a 4-regular graph. If we can orient it with two edges in and two out at each vertex such that the numbers of spanning circulations of even and odd size differ, then we have 3-choosability. The result arose in the context of "total coloring".

15.1.38. Example. A **total coloring** of a graph colors both the vertices and the edges so that no adjacent or incident objects have the same color. The **total chromatic number** is the number of colors needed; it is the chromatic number of the **total graph**, obtained from a graph G by subdividing every edge and then taking the square (adding edges joining vertices at distance 2 in the subdivision graph). For a cycle, subdividing merely doubles the length, so total coloring of a cycle corresponds to proper coloring of the square of a cycle twice as long.

Note that $3 \mid n$ is already needed for C_n^2 to be 3-colorable; hence it is also needed for 3-choosability. (Exercise 10.1.29 computes $\chi(C_n^k)$ in general.)   ∎

15.1.39. Theorem. (Juvan–Mohar–Škrekovski [1998b]) C_n^2 is 3-choosable if and only if $3 \mid n$. Consequently, C_m is 3-total-choosable if and only if $3 \mid m$.

Proof: As noted above, $3 \mid n$ is necessary. For the converse, we seek a suitable orientation of C_n^2 with maximum outdegree 2. Orient every edge in the clockwise direction as shown above; call this orientation D.

Let S be the set of spanning circulations in D. Let S_-, S_1, S_+ be the subsets of S in which all vertices have outdegree at most 1, all equal 1, or all at least 1, respectively. Obviously, $S_- \cap S_+ = S_1$.

Furthermore, $S = S_- \cup S_+$. A circulation D' in $S - (S_- \cup S_+)$ has a vertex with outdegree 2 and a vertex with outdegree 0. If $d_{D'}^+(v) = 0$ and w is the next vertex following v along the circle, then at most one edge of D' "crosses the gap" between v and w. However, when D' has at most one edge crossing a gap, also D' has at most one edge crossing the next gap. Hence we cannot reach a gap crossed by at least two edges. This forbids existence of a vertex with outdegree 2.

Let t^+ and t^- be the numbers of even spanning circulations and odd spanning circulations, respectively, so $t^+ + t^- = |S|$. To prove $\mathrm{diff}(D) \neq 0$, we prove that $|S| \equiv 2 \pmod 4$ and that t^+ and t^- are both even. The latter statement is easy. For a circulation H, the remaining edges in D also form a circulation H'. Although $H \neq H'$, the numbers of edges in H and H' have the same parity, since D has $2n$ edges. Hence t^+ and t^- are even.

Note that S_1 contains only the outside cycle and the set of "inner" (length 2) edges. Furthermore, complementation of edge sets matches $S_- - S_1$ with $S_+ - S_1$; they have the same size. Hence it suffices to show that $|S_- - S_1|$ is even.

Other than the empty circulation, each circulation in $S_- - S_1$ has exactly one cycle and has a vertex with outdegree 0. Recall that the adjusted Fibonacci number $\hat{F}_i$ is the number of $1, 2$-lists with sum i (Example 2.1.2). For a fixed vertex v, a cycle that omits v corresponds to a $1, 2$-list with sum $n - 2$, while a cycle that visits v corresponds to a $1, 2$-list with sum n. In the latter case, omit the one list having no 2, since the corresponding circulation is in S_1. Replace it with the empty circulation. Now $|S_- - S_1| = \hat{F}_{n-2} + \hat{F}_n = \hat{F}_{n-1} + 2\hat{F}_{n-2}$. Since $\hat{F}_0 = \hat{F}_1 = 1$, both are odd, and hence $\hat{F}_2$ is even. The parity pattern then repeats, with $\hat{F}_r$ even if and only if $r \equiv 2 \pmod 3$ (this is generalized in Exercise 2.1.22). Thus the number of circulations is an odd multiple of 2 if and only if $3 \mid n$. ∎

The key idea here is used also in the proof of the Cycle-plus-triangles Theorem and in other applications of Theorem 15.1.36 to 4-regular graphs: show that the total number of circulations in the specified orientation is an odd multiple of 2. Woodall–Prowse [2003] generalized the application; a special case of their result is that $\chi_l(G) = \chi(G)$ whenever G is a power of a cycle.

To prove the Alon–Tarsi Theorem (Theorem 15.1.36), we view colors as real numbers. We associate with each vertex v_i a variable x_i and define a polynomial that is nonzero precisely when the assigned numbers form a proper coloring.

15.1.40. Definition. Given a graph G with vertex set $v_1, \ldots, v_n$, let $E'(G) = \{(i, j) \colon i < j \text{ and } v_i v_j \in E(G)\}$. The **graph polynomial** p_G of G is defined by $p_G(x_1, \ldots, x_n) = \prod_{(i,j) \in E'(G)} (x_i - x_j)$.

Early applications of the graph polynomial include Petersen [1891], Scheim

[1974], Li–Li [1981]. Indeed, Petersen introduced graphs in order to study such polynomials (see Lützen–Sabidussi–Toft [1992] for the history).

The relation of p_G to orientations is seen by expanding the product.

15.1.41. Definition. Given a fixed ordering $v_1, \dots, v_n$ of the vertices of a graph G, an edge $v_i v_j$ in an orientation of G is *decreasing* if $i > j$. The parity of an orientation is the parity of the number of decreasing edges.

15.1.42. Lemma. Let G be a graph with m edges and vertices $v_1, \dots, v_n$, and let $(d_1, \dots, d_n)$ be a list with sum m. Let S be the set of orientations D of G such that $d_D^+(v_i) = d_i$ for all i. In the graph polynomial p_G, the coefficient of $\prod x_i^{d_i}$ is the number of even minus the number of odd orientations in S.

Proof: The polynomial is homogeneous of degree m, since each factor is homogeneous of degree 1. Each contribution to the expansion is formed by selecting one endpoint of each edge. This corresponds to an orientation by letting the selected vertex be the source of the edge. The resulting contribution to the expansion is $(-1)^t \prod x_i^{d_i}$, where d_i is the outdegree of v_i in the corresponding orientation and t is the number of decreasing edges. For a given list d of outdegrees, the even orientations count $+1$, and the odd orientations count -1. ∎

In order to relate this coefficient to $\text{diff}(D)$ in the statement of Theorem 15.1.36, we establish a bijection from the set of orientations with the same outdegrees as D to the set of spanning circulations in D.

15.1.43. Lemma. For an orientation D of G with $d_i = d_D^+(v_i)$ for each i, the absolute value of the coefficient of $\prod x_i^{d_i}$ in p_G is $\text{diff}(D)$.

Proof: Again the vertex ordering $v_1, \dots, v_n$ is fixed. Let S be the set of orientations D' of G such that $d_{D'}^+(v) = d_D^+(v)$ for all $v \in V(G)$. For $D' \in S$, let $D \oplus D'$ be the spanning subdigraph of D whose edges are the edges of D oriented oppositely in D'. Since $d_D^+(v) = d_{D'}^+(v)$ for all v, the subdigraph $D \oplus D'$ of D is a circulation.

Reversing the orientation on the edges of any circulation in D does not change any outdegree, so it yields a member of S. Thus the map taking D' to $D \oplus D'$ for $D' \in S$ is a bijection from S to the set of circulations contained in D.

For each edge $e \in D \oplus D'$, the reverse edge occurs in D', so exactly one of them is a decreasing edge. Therefore, $D \oplus D'$ has an even number of edges if and only if the numbers of decreasing edges in D and D' have the same parity. That is, the number of members of S with the same parity as D is the number of even circulations in D, and the number with the opposite parity is the number of odd circulations in D. Hence $\text{diff}(D)$ equals the difference between the numbers of even and odd circulations in S, and Lemma 15.1.42 applies. ∎

Proof of Theorem 15.1.36. The Alon–Tarsi Theorem now follows from the Combinatorial Nullstellensatz. We are given a graph G and orientation D. The graph polynomial p_G is homogeneous with degree $|E(G)|$. If $\left[\prod x_i^{d_i}\right] p_G$ is nonzero, then the Combinatorial Nullstellensatz guarantees that $p_G(s) \neq 0$ for some $s \in \prod S_i$ when $|S_i| > d_i$ for each i. By Lemmas 15.1.42–15.1.43, this coefficient is nonzero precisely when there is an orientation D with outdegrees $d_1, \dots, d_n$ such that $\text{diff}(D) \neq 0$. We are given D with these properties. ∎

Schauz [2009, 2010] strengthened the Alon–Tarsi Theorem to obtain the same upper bound on a coloring parameter χ_p called "paint number" that is always at least as large as χ_l. His proof is completely combinatorial, so the Combinatorial Nullstellensatz is no longer needed to prove the Alon–Tarsi Theorem.

After any proof, applying the Alon–Tarsi Theorem is a matter of finding a suitable orientation. This leads to a new parameter that bounds the choosability.

15.1.44.* Definition. An **Alon–Tarsi orientation** or **AT-orientation** of a graph G is an orientation D such that $\text{diff}(D) \neq 0$. The **Alon–Tarsi number** $\text{AT}(G)$ is the least k such that G has an AT-orientation D with $\Delta^+(D) < k$.

By the Alon–Tarsi Theorem, $\chi_l(G) \leq \text{AT}(G)$. By Schauz's result, the paint number is between them. Duraj–Gutowski–Kozik [2016] showed that $\chi_p(K_{n,n})$ exceeds $\chi_l(K_{n,n})$ by $\Theta(\log \log n)$. Hence $\text{AT}(G) \leq k$ suffices for $\chi_l(G) \leq k$ but is not necessary. On the other hand, there are many families where the bound on $\text{AT}(G)$ is sharp for $\chi_l(G) \leq k$, and hence proving the existence of an AT-orientation with the maximum outdegree at most k is stronger than proving $\chi_l(G) \leq k$.

For example, Zhu [2019] proved $\text{AT}(G) \leq 5$ for every planar graph G (we could say they are "5-AT-orientable"). In another direction, Cushing–Kierstead [2010] proved that for any 4-uniform list assignment L of a planar graph G, there is a matching M in G such that $G - M$ is L-colorable. Grytczuk–Zhu [2020] strengthened this in several ways by guaranteeing a matching M such that $\text{AT}(G-M) \leq 4$. Thus one matching chosen in advance works for all L.

A modification of the proof yields an AT-orientation of G with outdegree at most 4 at the tails of edges in M and at most 3 elsewhere, strengthening $\text{AT}(G) \leq 5$. Both Zhu [2019] and Grytczuk–Zhu [2020] use the approach of Thomassen [1994b] (Theorem 9.3.5): inductively prove a stronger statement, using cases depending on whether the outer cycle of the embedded graph has a chord.

15.1.45.* Theorem. (Grytczuk–Zhu [2020]) Every planar graph G contains a matching M such that $\text{AT}(G - M) \leq 4$.

Proof: We prove the following stronger statement. Given an edge e with endpoints v_1 and v_2 on the unbounded face of G, there exists a matching M in G that contains e and an AT-orientation D of $G - M$ such that $d_D^+(v_1) = d_D^+(v_2) = 0$, any other vertex v on the unbounded face satisfies $d_D^+(v) \leq 2 - d_M(v)$, and $d_D^+(v) \leq 3$ for every internal vertex v.

A triangle has such a matching and orientation by letting $M = \{e\}$ and orienting the other two edges in to the endpoints of e. The only circulation is the one with no edges. This provides a basis for induction on the number of vertices.

Exercise 34 implies that if a graph has an AT-orientation D, then every subgraph has an AT-orientation in which each vertex has outdegree at most its outdegree in D. Thus we may assume that G is 2-connected and every bounded face is a triangle. Let C denote the cycle bounding the outer face, with vertices $v_1, \ldots, v_n$ in order. The induction step considers two cases.

Case 1: C has a chord $v_i v_j$ with $i < j$. Let G_1 be the graph enclosed by the cycle through $v_1, \ldots, v_i$ and $v_j, \ldots, v_n$, and let G_2 be the graph enclosed by the cycle through $v_i, \ldots, v_j$. Let $e' = v_i v_j$. Both G_1 and G_2 are 2-connected, so the induction hypothesis yields (M_1, D_1) for (G_1, e) and (M_2, D_2) for (G_2, e'). Let $M =$

$M_1 \cup (M_2 - \{e'\})$ and $D = D_1 \cup D_2$. Note that M is a matching in G and D is an orientation of $G - M$.

Since $d_{D_2}^+(v_i) = d_{D_2}^+(v_j) = 0$, each vertex $v \in V(C) - \{v_1, v_2\}$ inherits $d_D^+(v) \leq 2 - d_M(v)$ from D_1 or D_2, and similarly $d_D^+(v) \leq 3$ for interior vertices.

Since $d_{D_2}^+(v_i) = d_{D_2}^+(v_j) = 0$, every circulation in D is the union of a circulation in D_1 and a circulation in D_2. The parity of the circulation is odd if and only if the parities of the two component circulations differ. Letting $\vec{e}(F)$ and $\vec{o}(F)$ denote the numbers of even and odd circulations in a digraph F, we have $\text{diff}(F) = |\vec{e}(F) - \vec{o}(F)|$. By the computation below, D is an AT-orientation.

$$\vec{e}(D) - \vec{o}(D) = \vec{e}(D_1)\vec{e}(D_2) + \vec{o}(D_1)\vec{o}(D_2) - \vec{e}(D_1)\vec{o}(D_2) - \vec{o}(D_1)\vec{e}(D_2)$$
$$= (\vec{e}(D_1) - \vec{o}(D_1))(\vec{e}(D_2) - \vec{o}(D_2)) \neq 0$$

Case 2: *C has no chord.* Let $G' = G - v_n$; note that G' is 2-connected, with e on its outer cycle C'. Let the neighbors of v_n be $v_1, u_1, \ldots, u_k, v_n$ along C'. All interior vertices of G other than $u_1, \ldots, u_k$ are interior vertices in G'. Let (M', D') be the matching and orientation obtained by applying the induction hypothesis to (G', e). Let $M = M'$ and define a candidate orientation D for $G - M$ by taking D' on $G' - M'$ and orienting the new edges as $v_n v_1$, $v_n v_{n-1}$, and $u_i v_n$ for $1 \leq i \leq k$.

In D, the only vertices that gain outdegree compared to D' are v_n and $u_1, \ldots, u_k$. Since v_n is not covered by M, we have $d_D^+(v_n) = 2 \leq 2 - d_M(v_n)$, and $d_D^+(u_i) = 1 + d_{D'}^+(u_i) \leq 3 - d_M(u_i) \leq 3$, as needed.

Since $d_{D'}^+(v_1) = 0$, any circulation in D that is not contained in D' uses one edge of the form $u_i v_n$ and the edge $v_n v_{n-1}$. Let S_i be the set of circulations in D using u_i and v_n. If each $S_i = \varnothing$, then $\text{diff}(D) = \text{diff}(D') \neq 0$.

If $S_i \neq \varnothing$, then let C_i be a cycle through u_i and v_n. Let $\widehat{D}_i$ be the orientation obtained from D' by reversing $C_i - v_n$. Let $\widehat{S}_i$ be the set of circulations in $\widehat{D}_i$; they do not visit v_n, since $\widehat{D}_i$ does not contain v_n. A circulation in S_i uses some edges of C_i (including $u_i v_n$ and $u_i v_{n-1}$). Deleting those edges and adding the reverse of the edges of C_i that were not used yields a circulation in $\widehat{S}_i$ (it does not visit v_n). Similarly, for any circulation in $\widehat{S}_i$, deleting its edges used from the reverse of C_i and adding instead the edges of C_i whose reverses were not used yields a circulation in S_i. Hence $|S_i| = |\widehat{S}_i|$. The parity of the number of edges changes under this map if and only if C_i has odd length. Hence $|\vec{e}(\widehat{D}_i) - \vec{o}(\widehat{D}_i)|$ is the difference between the numbers of even and odd members of S_i.

If this difference is 0 for all i, then $\text{diff}(D) = \text{diff}(D') \neq 0$, and we are finished. If $|\vec{e}(\widehat{D}_i) - \vec{o}(\widehat{D}_i)| \neq 0$, then obtain an orientation $\widehat{D}$ of $G - M'$ from D by reversing all edges of the cycle C_i except $u_i v_n$. This includes reversing $v_n v_{n-1}$ to become $v_{n-1} v_n$. In $\widehat{D}$ there is no circulation through v_n. Indeed, the circulations in $\widehat{D}$ are simply the circulations in $\widehat{D}_i$. We are in the case $\text{diff}(\widehat{D}_i) \neq 0$, so $\text{diff}(\widehat{D}) = 0$.

In obtaining $\widehat{D}$ from D, outdegree increases only for u_i; it now has two out-neighbors on C_i. Since $d_{D'}^+(u_i) \leq 2 - d_{M'}(u_i)$, we have $d_{\widehat{D}}^+(u_i) \leq 4 - d_{M'}(u_i)$. Since u_i is now internal, this is acceptable if u_i is covered by M'. If not, then we add the edge $u_i v_n$ to obtain M from M' and delete $u_i v_n$ from $\widehat{D}$. Now u_i has the needed outdegree in the orientation of $G - M$, and because $\widehat{D}$ had no circulations using the edge $u_i v_n$ the resulting orientation is still an AT-orientation. ∎

EXERCISES 15.1

15.1.1. (−) Prove that the incidence vectors of the clubs in Oddtown (Example 15.1.1) are linearly independent over the two-element field. (Berlekamp [1969])

15.1.2. (−) Form a digraph D by replacing every edge of the graph C_n with two oppositely directed edges having the same endpoints. Count the circulations in D.

15.1.3. ($\diamond$) Let $A_1, \ldots, A_m$ be a family of even subsets of $[n]$ with odd-sized intersections.
  (a) Prove that $m \le n$, with equality possible when n is odd.
  (b) Prove that $m \le n-1$ when n is even, with equality possible. (Babai–Frankl [1992])

15.1.4. ($\diamond$) Prove that Eventown (Example 15.1.1) has at most $2^{\lfloor n/2 \rfloor}$ clubs. That is, $[n]$ contains at most $2^{\lfloor n/2 \rfloor}$ even sets whose intersections all have even size. (Hint: Prove that the span U of the set of incidence vectors of the clubs is contained in the subspace of vectors orthogonal to all of U.) (Berlekamp [1969])

15.1.5. For even n, prove that there are between $2^{n^2/4}/n!$ and $2^{n^2}/n!$ sets of n odd-sized subsets of $[n]$ such that the intersection of any two has even size. (Hint: Let $n = 2k$. From a k-by-k binary matrix A, form an n-by-n binary matrix $\left(\begin{smallmatrix} A+I_k & A \\ A & A+I_k \end{smallmatrix} \right)$.) (M. Szegedy)

15.1.6. For $0 \le t \le (n-1)/2$, construct a family $\mathcal{F}$ of $n - 2t$ odd-size subsets of $[n]$, whose pairwise intersections have even size, such that no odd-size subset can be added having even intersection with all the current members of $\mathcal{F}$.

15.1.7. ($\diamond$) A town with n people has m sports clubs $A_1, \ldots, A_m$ and m theater clubs $B_1, \ldots, B_m$. Prove that if $|A_i \cap B_i|$ is odd for every i, and $|A_i \cap B_j|$ is even whenever $i > j$, then $m \le n$, and this is sharp. (Babai–Frankl [1992])

15.1.8. Let A be a $2n$-by-$2n$ matrix with entries in $\{1, 0, -1\}$ such that $a_{i,j} = 0$ if and only if $i = j$. Prove that A is nonsingular. Conclude that if $w_1, \ldots, w_{2n+1}$ are real weights such that any $2n$ of them can be partitioned into two n-sets with the same total weight, then all the weights are equal. (Babai–Frankl [1992])

15.1.9. By Theorem 15.1.5, a two-distance set in $\mathbb{R}^n$ has at most $(n + 1)(n + 4)/2$ points. Construct a two-distance set of size $\binom{n+1}{2}$ in $\mathbb{R}^n$. (Hint: Start with such a set in $\mathbb{R}^{n+1}$.)

15.1.10. Prove that any two-distance set of points in the sphere S^{n-1} in $\mathbb{R}^n$ has size at most $n(n + 3)/2$. (Hint: Follow the proof of Theorem 15.1.5, but confine the resulting independent polynomials to a space of dimension $n(n + 3)/2$.) (Delsarte–Goethals–Seidel [1977])

15.1.11. (+) *Improved bound on the size of a two-distance set* $\{v^1, \ldots, v^m\}$ *in* $\mathbb{R}^n$.
  (a) In $\mathbb{R}^n$, the *affine hull* of vectors $\{v^1, \ldots, v^m\}$ is $\{\sum_{i=1}^m c_i v_i \colon \sum_{i=1}^m c_i = 0\}$. Let B be the m-by-$(n + 1)$ matrix whose ith row is v_i plus $\mathbf{1}_m$ as column 0. Prove that if the affine hull is all of $\mathbb{R}^n$, then the columns of B are linearly independent.
  (b) Prove that if the columns of a real m-by-p matrix B are linearly independent, then $B^T B$ is nonsingular.
  (c) Prove that adding $\{1, x_1, \ldots, x_n\}$ to the set of polynomials constructed from a two-distance set in Theorem 15.1.5 yields a linearly independent set. Conclude that the maximum size of a two-distance set in $\mathbb{R}^n$ is at most $\binom{n+2}{2}$. (Blokhuis [1984])

15.1.12. An *equivalence* on G is a spanning subgraph whose components are complete.
  (a) Prove that $\text{pdim}(G)$ is the minimum number of equivalences on $\overline{G}$ whose union is $\overline{G}$ and whose overall intersection is empty.
  (b) Prove that $\text{pdim}(G) \le \chi'(\overline{G})$ when $\chi'(\overline{G}) > 1$, with equality if $\overline{G}$ is triangle-free.
  (c) Use Vizing's Theorem to prove $\text{pdim}(G) \le n - 1$ when $|V(G)| = n \ge 3$.

15.1.13. ($\Diamond$) *L-intersecting families.* (Alon–Babai)

(a) Let A be an m-by-m matrix of integers. Prove that if some prime power divides every off-diagonal entry but no diagonal entry, then A is nonsingular.

(b) Conclude that if the greatest common divisor of the elements of L does not divide k, then L-intersecting families of k-subsets of $[n]$ have size at most n.

(c) Let q be a prime power. Let $A_1, \ldots, A_m$ be subsets of $[n]$ such that $|A_i \cap A_j|$ is divisible by q and $|A_i|$ is not divisible by q whenever $i, j \in [m]$. Prove that $m \le n$. (Comment: When $k = 1$, this becomes an instance of the Deza–Frankl–Singhi Theorem.)

15.1.14. Given $\mathcal{F}_1, \mathcal{F}_2 \subseteq \binom{[n]}{k}$, let $\sigma(\mathcal{F}_2)$ denote the permutation of $\mathcal{F}_2$ obtained by applying the permutation $\sigma \colon [n] \to [n]$.

(a) With all permutations of $[n]$ equally likely, prove $\mathbb{E}\,|\mathcal{F}_1 \cap \sigma(\mathcal{F}_2)| = |\mathcal{F}_1|\,|\mathcal{F}_2|/\binom{n}{k}$.

(b) Suppose that S_1 and S_2 are disjoint, and suppose that $\mathcal{F}_1$ is S_1-intersecting and $\mathcal{F}_2$ is S_2-intersecting. Prove $|\mathcal{F}_1| \cdot |\mathcal{F}_2| \le \binom{n}{k}$. (M. Szegedy)

15.1.15. Prove that the points of $\mathbb{R}^2$ can be colored using asymptotically $n^{n/2}$ colors so that points at distance 1 have different colors.

15.1.16. Use Stirling's Formula to approximate $\ln\binom{p^3}{p^2-1} \big/ \ln\binom{p^3}{p-1}$ for large p.

15.1.17. Choose $n, p \in \mathbb{N}$ with p prime and $n > 2p$. Let $G_{n,p}$ be the graph whose vertices are the incidence vectors of $(2p-1)$-sets in $[n]$, adjacent when their distance in $\mathbb{R}^n$ is $\sqrt{2p}$. Prove that $\chi(G_{n,p}) \ge \binom{n}{2p-1}/\binom{n}{p-1}$. Improve the lower bound on the chromatic number of the unit-distance graph in $\mathbb{R}^n$ by choosing p to maximize the lower bound on $\chi(G_{n,p})$.

15.1.18. ($\Diamond$) Color the edges of the complete graph with vertex set $\binom{[t^3]}{3}$ by making an edge red if its endpoints have one common element and blue otherwise. Conclude $R(t,t) > \binom{t^3}{3}$. (Comment: The graph $(t-1)K_{t-1}$ gives only $R(t,t) > (t-1)^2$.) (Nagy [1972])

15.1.19. Use Snevily's Theorem (Theorem 15.1.21) to prove the Frankl–Wilson Theorem (Theorem 15.1.14).

15.1.20. An **odd representation** of a graph G assigns each vertex a binary k-tuple so that vertices are adjacent if and only if the dot product of their k-tuples is odd. Prove that every n-vertex graph has an odd representation with $k = n - 1$. (Eaton–Grable [1996])

15.1.21. *Cauchy–Davenport in higher dimensions.* The **Hopf–Stiefel function** relative to a prime p is a function $r \circ s$ of positive integers r and s, defined to be the smallest integer n such that $\binom{n}{k}$ is divisible by p whenever $n - r < k < s$. It can be computed recursively from the base-p representations of r and s. Note that $r \circ s = r + s - 1$ when $r + s \le p + 1$.

(a) Prove that if A and B are nonempty subsets of a finite vector space over $\mathbb{F}_p$, with $r = |A|$ and $s = |B|$, then $|A + B| \ge r \circ s$.

(b) Prove that part (a) is sharp for all r and s. (Hint: In one dimension, part (a) reduces to the Cauchy–Davenport Theorem. Prove sharpness by generalizing the sharpness example for that theorem.) (Eliahou–Kervaire [1998])

15.1.22. ($\Diamond$) Let $\mathbb{F}_2^m$ be the vector space of dimension m over $\mathbb{F}_2$. Given nonempty subsets A and B of $\mathbb{F}_2^m$, let $A + B = \{a + b \colon a \in A, b \in B\}$. Use the Combinatorial Nullstellensatz to prove that $|A| + |B| > 2^n$ implies $|A + B| \ge 2^n$. (Comment: The result was proved inductively in Fon-Der-Flaass & Alekseyev [2012]. R. Chapman and T. Viteam independently found proofs (unpublished) using the Combinatorial Nullstellensatz.)

15.1.23. Show that Theorem 15.1.29 is a special case of Theorem 15.1.30. Prove Theorem 15.1.30. (Alon–Nathanson–Rusza [1996])

15.1.24. Let A and B be nonempty subsets of $\mathbb{Z}_p$, where p is prime. Prove that the number of sums $x + y$ such that $x \in A$, $y \in B$, and $xy \neq 1$ is at least $\min\{p, |A| + |B| - 3\}$. (Hint: Use Theorem 15.1.30.) (Alon–Nathanson–Rusza [1995])

15.1.25. *More from Theorem 15.1.30.* (Alon–Nathanson–Rusza [1996]) Let p be a prime.
    (a) For $c_1, \ldots, c_k \in \mathbb{N} \cup \{0\}$ with sum $m + \binom{k}{2}$, where $m \geq 0$, prove

$$\Big[\prod_{i=1}^{k} x_i^{c_i}\Big]\Big(\sum_{i=1}^{k} x_i\Big)^m \prod_{1 \leq i < j \leq k} (x_j - x_i) = \frac{m!}{\prod_{i=1}^{k} c_i!} \prod_{1 \leq i < j \leq k} (c_j - c_i).$$

(Hint: Use the Hook-Length Formula (Theorem 4.3.4) or give a direct proof.)
    (b) Let $A_1, \ldots, A_k$ be nonempty subsets of $\mathbb{Z}_p$. Let $S(A_1, \ldots, A_k)$ be the set of sums of distinct elements $a_1, \ldots, a_k$ such that $a_i \in A_i$ for all i. Prove that if $|A_1|, \ldots, |A_k|$ are distinct and sum to less than $p + \binom{k+1}{2}$, then $|S(A_1, \ldots, A_k)| > \sum_{i=1}^{k} |A_i| - \binom{k+1}{2}$. (Hint: Use part (a) and Theorem 15.1.30.)
    (c) Let $A_1, \ldots, A_k$ be nonempty subsets of $\mathbb{Z}_p$, indexed so that $|A_1| \geq \cdots \geq |A_k|$. Let $b_1 = |A_1|$, and let $b_i = \min\{b_{i-1} - 1, |A_i|\}$ for $2 \leq i \leq k$. Prove that if $b_k > 0$, then $|S(A_1, \ldots, A_k)| \geq \min\{p, \sum_{i=1}^{k} b_i - \binom{k+1}{2} + 1\}$.
    (d) Conclude that if A is a nonempty subset of $\mathbb{Z}_p$, then the number of sums of s distinct elements of A is at least $\min\{p, s|A| - s^2 + 1\}$. (Comment: Theorem 15.1.29 is the special case $s = 2$.) (Dias da Silva–Hamidoune [1994])

15.1.26. Prove that the minimum number of hyperplanes in $\mathbb{R}^n$ that do not contain the origin but together cover all other points of $\{0, 1\}^n$ is n. (Alon–Füredi [1993])

15.1.27. ($\diamond$) For each vertex v in a graph G, specify $B(v) \subseteq \{1, \ldots, d_G(v)\}$.
    (a) Prove that if $\sum_{v \in V(G)} |B(v)| < |E(G)|$, then G has a nontrivial subgraph H such that $d_H(v) \notin B(v)$ for all $v \in V(G)$ (note that degree 0 is allowed, but not at all vertices). (Shirazi–Verstraëte [2008])
    (b) Show that part (a) is sharp by constructing infinitely many examples such that $\sum_{v \in V(G)} |B(v)| = |E(G)|$ and no such subgraph exists.

15.1.28. For odd prime p, let k be an integer with $1 \leq k < p$. Given $a_1, \ldots, a_k \in \mathbb{F}_p$ and distinct elements $b_1, \ldots, b_k \in \mathbb{F}_p$, prove that for some permutation σ of $[k]$ the values $a_i + b_{\sigma(i)}$ are distinct modulo p. (Hint: Use the Vandermonde determinant in an application of the Combinatorial Nullstellensatz.) (Alon [2000a])

15.1.29. Given a permutation σ, let $d_\sigma(i, j) = a - b$, where i and j occupy positions a and b in the word form, respectively. For $i, j \in [k]$ with $i < j$, specify a forbidden distance $f(i, j)$. Prove that there is a permutation $\sigma \in \mathbb{S}_k$ such that $d_\sigma(i, j) \neq f(i, j)$ for $1 \leq i < j \leq k$. Conclude that for $a_1, \ldots, a_k \in \mathbb{Z}_n$ with $2k \leq n + 1$, there is a permutation $\sigma \in \mathbb{S}_k$ such that the elements $a_{\sigma(i)} + i$ for $1 \leq i \leq k$ are distinct modulo n. (Kézdy–Snevily [2002])

15.1.30. A **Kakeya set** in $\mathbb{F}_q^n$ is a set K such that for all $y \in \mathbb{F}_q^n$, there exists $b \in \mathbb{F}^n$ such that $b + ay \in K$ for all $a \in \mathbb{F}$. This means that K contains a line in every direction.
    (a) Prove that if $K \subseteq \mathbb{F}_q^n$ is a Kakeya set with $|K| < \binom{q+n-1}{n}$, then there is a nonzero polynomial $f \in \mathbb{F}_q[x_1, \ldots, x_n]$ of degree less than q that is 0 at all points of K.
    (b) Conclude that for all $y \in \mathbb{F}_q^n$, there exists $b \in \mathbb{F}_q^n$ such that $f(b + ay) = 0$ for all $a \in \mathbb{F}$. As a polynomial in a, conclude that f is identically 0, contradicting (a) and proving that every Kakeya set has size at least $\binom{q+n-1}{n}$. (N. Alon and T. Tao; see Dvir [2017+])

15.1.31. Prove that K_4 is 3-edge-choosable.

15.1.32. ($\diamond$) Hakimi [1965] (Corollary 6.1.5) proved that a graph G has an orientation in which each vertex has outdegree at most d if and only if every subgraph H has at most $d|V(H)|$ edges. Conclude that planar bipartite graphs are 3-choosable. (Alon–Tarsi [1992])

15.1.33. For a plane graph G, let H be the hypergraph with vertex set $V(G)$ whose edges are the vertex sets of faces of G. Use the 5-choosability of G (Theorem 9.3.5) and the 3-choosability of planar bipartite graphs (Exercise 15.1.32) to prove that H is 3-choosable. (Comment: It is conjectured that H is 2-choosable.) (Ramamurthi [2001])

15.1.34. ($\diamond$) Let e be an edge in an AT-orientation D of a graph G. If e lies in no cycle in D, then $D - e$ is an AT-orientation of $G - e$.
    (a) Prove that if e lies in a cycle C in D, then the orientation D' obtained from D by reversing the edges of C is an AT-orientation.
    (b) Prove that $D - e$ or $D' - e'$ is an AT-orientation of G in which every vertex has outdegree at most its outdegree in D (here e' is the reverse of e).

15.2. Matrices

In this section we consider enumerative aspects of matrices. We use determinants to count spanning trees in graphs, permanents to count perfect matchings in bipartite graphs, and matrix inversion to study inversion formulas for functions on partially ordered sets.

DETERMINANTS AND TREES

In Section 4.1, we used signed involutions to express certain enumeration problems as determinant computations. Here we consider another enumeration problem solved by a determinant, but this time the combinatorics is more closely tied to the algebraic aspects of the determinant.

By Cayley's Formula (Theorem 1.3.4), there are n^{n-2} trees with vertex set $[n]$. Viewing them as spanning trees of K_n suggests the general problem of counting the spanning trees in a graph G. Example 2.1.13 develops a recurrence for this number $\tau(G)$, but obtaining $\tau(G)$ from the recurrence generally requires exponential work in the number of edges.

The Matrix Tree Theorem computes $\tau(G)$ using a determinant. Determinants of n by n matrices can be computed using fewer than n^3 operations (for large n), much faster than the recurrence. We delete loops before the computation because they appear in no spanning trees. Multiedges cause no difficulties.

15.2.1. Example. The Matrix Tree Theorem (Theorem 15.2.5) tells us to form the diagonal matrix of vertex degrees, subtract the adjacency matrix, delete a row and a column, and take the determinant. For the graph $K_4 - e$, the vertex degrees are $3, 3, 2, 2$, so we form the matrix on the left below and take the determinant of the matrix in the middle. We get the number of spanning trees. ∎

$$\begin{pmatrix} 3 & -1 & -1 & -1 \\ -1 & 3 & -1 & -1 \\ -1 & -1 & 2 & 0 \\ -1 & -1 & 0 & 2 \end{pmatrix} \quad \rightarrow \quad \begin{pmatrix} 3 & -1 & -1 \\ -1 & 2 & 0 \\ -1 & 0 & 2 \end{pmatrix} \quad \rightarrow \quad 8$$

To state and prove the theorem, we need some basic linear algebra.

748 Chapter 15: Linear Algebra

15.2.2. Definition. The **cofactor** $A_{i,j}$ of position (i, j) in a square matrix A is $(-1)^{i+j}$ times the determinant of the matrix obtained by deleting row i and column j. The **adjugate matrix** $\mathrm{Adj}\, A$ has in position (i, j) the cofactor of position (j, i) in A.

15.2.3. Remark. *Properties of the adjugate matrix.* In linear algebra we obtain $\mathrm{Adj}\,(AB) = \mathrm{Adj}\,(A)\,\mathrm{Adj}\,(B)$ and $(\mathrm{Adj}\, A)A = (\det A)I$. ∎

15.2.4. Lemma. If every row of an n-by-n matrix A has sum 0, then the cofactors of any row are all equal.

Proof: When every row has sum 0, the columns are dependent, so $\mathrm{rank}\,(A) < n$. If $\mathrm{rank}\,(A) < n-1$, then all cofactors are 0. Otherwise, $\mathrm{rank}\,(A) = n-1$ and $\det A = 0$. The equation $A\mathrm{Adj}\, A = 0$ now puts every column of $\mathrm{Adj}\, A$ into the nullspace of A. Every row-sum of A being 0 means that $\mathbf{1}_n$ is in the nullspace. Since $\mathrm{rank}\, A = n-1$, every vector in the nullspace is a multiple of $\mathbf{1}_n$. Hence the columns of $\mathrm{Adj}\, A$ are constant-valued, making the cofactors in each row of A equal. ∎

We also need the **Cauchy–Binet Formula**: If C is the product of an n-by-m matrix A and an m-by-n matrix B, then $\det C = \sum_{S \subseteq [m]} \det A_S \det B_S$, where A_S $[B_S]$ is the n-by-n submatrix of A $[B]$ with columns [rows] indexed by S. We proved this combinatorially in Proposition 4.1.32; also it has a short proof in linear algebra using block multiplication of cleverly-defined matrices (Exercise 11).

15.2.5. Theorem. (**Matrix Tree Theorem**) Given a loopless multigraph G with vertex set $v_1, \ldots, v_n$, let $a_{i,j}$ be the number of edges of the form $v_i v_j$. Let Q be the matrix with entry (i, j) being $-a_{i,j}$ when $i \neq j$ and $d(v_i)$ when $i = j$. If Q^* is the matrix obtained by deleting any row s and column t of Q, then $\tau(G) = (-1)^{s+t} \det Q^*$.

Proof: By Lemma 15.2.4, we need only prove this when $s = t$.

Step 1. *If D is an orientation of G, and M is the incidence matrix of D, then $Q = MM^T$.* With edges $e_1, \ldots, e_m$, we put $m_{i,j} = 1$ when v_i is the tail of e_j, $m_{i,j} = -1$ when v_i is the head of e_j, and $m_{i,j} = 0$ otherwise. Entry i, j in MM^T is the dot product of rows i and j of M. When $i \neq j$, the product counts -1 for every edge of G joining the two vertices; when $i = j$, it counts 1 for every incident edge and yields the degree.

$$M = \begin{array}{c} 1 \\ 2 \\ 3 \\ 4 \end{array}\begin{pmatrix} -1 & 1 & 1 & 0 & 0 \\ 0 & 0 & -1 & -1 & 0 \\ 0 & 0 & 0 & 1 & -1 \\ 1 & -1 & 0 & 0 & 1 \end{pmatrix} \qquad Q = \begin{pmatrix} 3 & -1 & 0 & -2 \\ -1 & 2 & -1 & 0 \\ 0 & -1 & 2 & -1 \\ -2 & 0 & -1 & 3 \end{pmatrix}$$

Step 2. *If B is an $(n-1)$-by-$(n-1)$ submatrix of M, then $\det B = 0$ if the corresponding $n-1$ edges contain a cycle, and $\det B = \pm 1$ if they form a spanning tree of G.* If the edges corresponding to the columns contain a cycle C, then the columns sum to the zero vector when weighted with $+1$ or -1 according as the directed edge is followed forward or backward when following the cycle. This equation of dependence yields $\det B = 0$.

For the other case, we use induction on n. For $n = 1$, by definition a 0-by-0 matrix has determinant 1. For $n > 1$ let T be the spanning tree whose edges are the columns of B. Since T has at least two leaves, B contains a row corresponding to a leaf x of T. This row has only one nonzero entry in B. When computing the determinant by expanding along that row, the only submatrix B' given nonzero weight in the expansion corresponds to the spanning subtree of $G - x$ obtained by deleting x and its incident edge from T. Since B' is an $(n-2)$-by-$(n-2)$ submatrix of the incidence matrix for an orientation of $G - x$, the induction hypothesis implies that det B' is ± 1, and multiplying it by ± 1 gives the same result for B.

Step 3. *Computation of* det Q^*. Let M^* be the matrix obtained by deleting row t of M, so $Q^* = M^*(M^*)^T$. We may assume $m \geq n - 1$, else both sides have determinant 0 and $\tau(G) = 0$. When we use the Cauchy–Binet Formula for det Q^*, the submatrix A_S is an $(n-1)$-by-$(n-1)$ submatrix of M as discussed in Step 2, and $B_S = A_S^T$. Hence the summation counts $1 = (\pm 1)^2$ for each set of $n-1$ edges forming a spanning tree and 0 for each other set of $n-1$ edges. ∎

Rényi proved the Matrix Tree Theorem combinatorially by using generating functions to enumerate trees. Tutte extended the theorem to multidigraphs. To avoid awkwardness, we will use the term "digraph" in the remainder of this section even though multiedges are allowed. Tutte's result yields the Matrix Tree Theorem when the digraph is symmetric (with the same multiplicity for edges xy and yx). Chaiken and Kleitman observed that Rényi's argument also yields Tutte's result and generalized it considerably. First we state Tutte's result.

15.2.6. Definition. A **branching** (or **out-tree**) is an oriented rooted tree with all edges directed away from the root. An **arborescence** is a union of disjoint branchings. An **in-tree** is the reversal of an out-tree. For a digraph G, let $Q^- = D^- - A'$ and $Q^+ = D^+ - A'$, where D^- and D^+ are the diagonal matrices of indegrees and outdegrees, and $A'_{i,j}$ is the number of copies of $v_j v_i$ in $E(G)$.

15.2.7. Theorem. (**Directed Matrix Tree Theorem**; Tutte [1948]) With Q^- and Q^+ defined as above, the number of spanning out-trees [in-trees] rooted at v_i is the value of any cofactor in the ith row of Q^- [ith column of Q^+]. ∎

15.2.8. Example. The digraph below has two out-trees rooted at 1 and two in-trees rooted at 3. The determinants behave as claimed. ∎

$$Q^- = \begin{pmatrix} 0 & 0 & 0 \\ -1 & 1 & 0 \\ -1 & -1 & 2 \end{pmatrix} \qquad Q^+ = \begin{pmatrix} 2 & 0 & 0 \\ -1 & 1 & 0 \\ -1 & -1 & 0 \end{pmatrix}$$

Since the earlier theorems hold for multigraphs, it is natural to generalize using edge weights; non-edges are represented by weight 0. To generalize Theorem 15.2.7, we let the weight of a tree or arborescence in a digraph D be the product of its edge weights. For further generality and a combinatorial proof, we use variables to encode the outdegree of each vertex. This leads to a determinantal formula for the sum of the weights of arborescences with specified roots.

15.2.9. Theorem.(**Matrix Arborescence Theorem**; Chaiken–Kleitman [1978])
   Given real weights $a_{i,j}$, variables $x_1, \ldots, x_n$, and an arborescence A, let $w_A = \prod_{v_j v_i \in E(A)} a_{i,j} x_j$. For $S \subseteq [n]$, let $T(S)$ be the set of all arborescences on $\{v_1, \ldots, v_n\}$ whose roots are $\{v_i : i \in S\}$. Define a matrix Q by $q_{i,j} = -a_{i,j} x_j$ for $i \neq j$ and $q_{i,i} = \sum_{j \neq i} a_{i,j} x_j$. If Q_S denotes the submatrix of Q obtained by omitting all rows and columns indexed by S, then $\sum_{A \in T(S)} w_A = \det Q_S$.

Proof: For $S = \varnothing$, the sum is empty and $\det Q_\varnothing = \det Q = 0$, so the claim holds. Let $k = |S|$ and $V = \{v_1, \ldots, v_n\}$. For nonempty S, we use induction on $n - k$. If $k = n$, then there is one arborescence (no edges) in the sum, with weight 1 (empty product), and the 0-by-0 determinant by convention also equals 1.

Now suppose $k < n$. Both expressions are polynomials in $\{x_j\}$ in which each nonzero term has total degree $n - k$; we write $\det Q_S = f_S(x_1, \ldots, x_n)$ to emphasize this. With total degree $n - k$, each term has degree 0 in at least k variables. Let $\{x_i : i \in S\}$ be "root variables" and the others "non-root variables". To complete the proof, we prove (1) each term in both $\sum w_A$ and f_S has degree 0 in some non-root variable, and (2) the terms in which a particular non-root variable has degree 0 agree in $\sum w_A$ and f_S. The second claim uses the induction hypothesis.

Proof of Claim 1. In w_A, the degree of a variable x_j is the outdegree of v_j in A. When $k < n$, there is a non-root leaf, and it has outdegree 0.

Now consider f_S. Deleting the rows and columns of Q indexed by S does not remove the root variables; they remain in the diagonal terms. Setting the root variables to 0 makes each row of Q_S sum to 0, so $\det Q_S$ and f_S have value 0 regardless of the values of the non-root variables when the root variables are set to 0. Since a polynomial that is 0 at all points has every coefficient 0, each term in f_S has a root variable with positive degree. Since each term has degree 0 in at least k variables, it has degree 0 in some non-root variable.

Proof of Claim 2. Consider terms having degree 0 in a fixed x_t with $t \notin S$. In $\sum w_A$, these arise from arborescences where v_t is a leaf. Such an arborescence A arises from an arborescence A' on $n - 1$ vertices with roots in S by adding some edge $v_j v_t$, yielding $w_A = w_{A'} a_{t,j} x_j$. Let T' be the set of arborescences on $V - \{v_t\}$ with root set S. Since j is arbitrary and we can start with any arborescence in T', the sum of the terms omitting x_t equals $(\sum_{A' \in T'} w_{A'})(\sum_{j \neq t} a_{t,j} x_j)$.

Now consider the terms of f_S that omit x_t; setting $x_t = 0$ in f_S yields their sum. The column for v_t in Q_S becomes 0 except for the diagonal term M, which is $\sum_{j \neq t} a_{t,j} x_j$. Terms involving x_t disappear from other entries in the matrix. Expanding the determinant along column t yields M times the determinant of the matrix Q'_S obtained from Q_S by deleting the tth row and column and setting $x_t = 0$. This is just the determinant defined for the smaller problem with vertex set $V - \{v_t\}$ and the same root set S. By the induction hypothesis, $\det Q'_S = \sum_{A' \in T'} w_{A'}$. Thus the terms omitting x_t are the same in $\sum w_A$ and in f_S. ∎

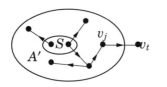

15.2.10. Example. To obtain the Directed Matrix Tree Theorem from the Matrix Arborescence Theorem, let $a_{i,j}$ be the number of edges from v_j to v_i, set each variable to 1, and let S consist of one vertex.

Keeping the variables x_j as indeterminates yields a generating function that enumerates the arborescences rooted at S by their outdegrees. For the digraph in Example 15.2.8, the matrix Q appears below. If $S = \{1\}$, then the determinant is $f(S) = x_1^2 + x_1 x_2$; one branching consists of two edges from v_1 and one is the path $\langle v_1, v_2, v_3 \rangle$. If $S = \{1, 2\}$, then the determinant is $f(S) = x_1 + x_2$, and the choices for the arborescence are the edge $v_1 v_3$ or the edge $v_2 v_3$. ∎

$$\begin{pmatrix} 0 & 0 & 0 \\ -x_1 & x_1 & 0 \\ -x_1 & -x_2 & x_1 + x_2 \end{pmatrix}$$

Spanning in-trees in digraphs are closely related to Eulerian circuits. After distinguishing a vertex v, we will obtain a spanning in-tree to v from each Eulerian circuit by a method that generates each in-tree the same number of times. This enables us to count the circuits (as cyclic lists of edges).

15.2.11. Example. In the digraph below, the edges labeled in order from 1 form an Eulerian circuit C starting along e from v. Each bold edge is the last edge in C that departs from its tail. The bold edges form an in-tree to v. ∎

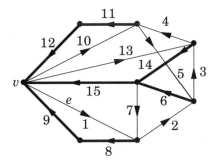

15.2.12. Lemma. For any Eulerian circuit that begins from v along edge e in a digraph G, the set T of edges along which vertices other than v are exited for the last time is the edge set of an in-tree to v.

Proof: Each vertex other than v is left for the last time only once, so $|E(T)| = |V(G)| - 1$. Each vertex other than v has outdegree 1 in T, and v has outdegree 0. A vertex other than v cannot be left for the last time until it is entered for the last time, so T contains no cycle. These properties make T an in-tree to v. ∎

We can reverse the process of obtaining an in-tree from an Eulerian circuit.

15.2.13. Algorithm. (Eulerian circuit in directed graph)
Input: An Eulerian digraph G and a spanning in-tree T rooted at v.
Step 1: For each $u \in V(G)$, specify an ordering of the edges leaving u, such that for $u \neq v$ the edge leaving u in T comes last.
Step 2: Starting at v, construct an Eulerian circuit by always leaving the current vertex u along the next unused edge in the ordering given at u. ∎

15.2.14. Lemma. Algorithm 15.2.13 produces an Eulerian circuit.

Proof: Given the in-tree T to v, Algorithm 15.2.13 constructs a trail. It suffices to show that the trail ends only at v and only after using all edges.

When we enter a vertex u other than v, the edge leaving u in T has not yet been used, since $d^+(u) = d^-(u)$. Thus there is still a way out of u, and hence the trail can end only at v.

We end only after using all edges leaving v. Since $d^-(v) = d^+(v)$, we then also have used all edges entering v, including those in T. Since an edge of T is not used until all other edges leaving its tail are used, and T contains a path from each vertex to v, we cannot end at v until all edges are used at all vertices. ∎

An Eulerian circuit is given by a cyclic ordering of edges; the starting edge does not matter. Hence we may require that Algorithm 15.2.13 always uses a fixed edge e as the first edge leaving v. This associates with each in-tree the same number of circuits. The authors' initials from two relevant papers form an acronym by which the theorem is known; the earlier paper solved the case where each $d_i = 2$.

15.2.15. Theorem. (**BEST Theorem**; van **A**ardenne-**E**hrenfest & de **B**ruijn [1951], **T**utte–**S**mith [1941]) An Eulerian digraph with $d_i = d^+(v_i) = d^-(v_i)$ has exactly $c \prod_i (d_i - 1)!$ Eulerian circuits, where c counts the in-trees to or out-trees from any vertex.

Proof: Since we always start from v and along edge e, the flexibility of choosing the rest of the orderings allows Algorithm 15.2.13 to generate $\prod_{u \in V(G)} (d^+(u) - 1)!$ different Eulerian circuits from each in-tree to v. The last out-edge is fixed by the tree for vertices other than v (after starting along e), so the circuits generated from different trees are different. From the same tree, we reach a difference in the pairs of consecutive edges as soon as we reach a difference in the exit orderings.

Hence we have built $c \prod_i (d_i - 1)!$ distinct Eulerian circuits, and Lemma 15.2.12 ensures that all Eulerian circuits have been found. ∎

Note in this statement that c is independent of the choice of v. If G is Eulerian, then $Q^+ = Q^-$, and the row-sums and column-sums are all 0. Now Lemma 15.2.4 implies that the cofactors are all equal.

CYCLE SPACE AND BOND SPACE

The Matrix Tree Theorem uses the relationship between cycles and the incidence matrix of an oriented graph. In Section 11.3, we used the incidence matrix to represent cycle matroids over $\mathbb{F}_2$. The incidence vectors of cycles formed the null space of the matrix.

15.2.16. Definition. In a graph G with edges $e_1, \ldots, e_m$, the **incidence vector** for a set $F \subseteq E(G)$ has coordinates $a_i = 1$ when $e_i \in F$ and $a_i = 0$ when $e_i \notin F$. The **cycle space** of G is the set $\mathbf{C}(G)$ of incidence vectors of even subgraphs (those with all vertex degrees even). The **bond space** of G is the set $\mathbf{B}(G)$ of incidence vectors of edge cuts.

We view members of $\mathbf{C}(G)$ as even subgraphs or as their incidence vectors; we view members of $\mathbf{B}(G)$ as edge cuts or their incidence vectors.

15.2.17. Theorem. The cycle space and bond space of a connected graph G with n vertices and m edges are binary vector spaces with dimensions $m - n + 1$ and $n - 1$, respectively.

Proof: We have vector spaces over $\mathbb{F}_2$ because the subsets are closed under binary addition: the symmetric difference of two even subgraphs is an even subgraph, and the symmetric difference of two edge cuts is an edge cut (Exercise 12).

Choose a spanning tree T. Each edge of $E(G) - E(T)$ forms a unique cycle with edges in T; these are the **fundamental cycles** relative to T. These cycles are linearly independent in $\mathbf{C}(G)$, since each has a nonzero coordinate outside $E(T)$ that is zero in all other incidence vectors in this set. Hence $\dim \mathbf{C}(G) \geq m - n + 1$.

Choose $n - 1$ vertices in G. Let $v_1, \ldots, v_{n-1}$ be the incidence vectors for the corresponding edge cuts. Let $\sum c_i v_i = 0$ be an equation of dependence. Since each coefficient is 0 or 1, we are just summing the vectors for a set S of vertices. The resulting coordinate for an edge is 0 if and only if the edge has an even number of endpoints in S. Edges in $[S, \overline{S}]$ appear once in the sum, and such coordinates remain nonzero. Since G is connected, $[S, \overline{S}] \neq \varnothing$ unless S or $\overline{S}$ is empty. Hence there is no nontrivial equation of dependence, the vectors are linearly independent, and $\dim \mathbf{B}(G) \geq n - 1$. We have also shown that $[S, \overline{S}] = \sum_{i \in S} v_i$ for every edge cut $[S, \overline{S}]$, so $\dim \mathbf{B}(G) = n - 1$.

The $n - 1$ vectors $v_1, \ldots, v_{n-1}$ are rows in the incidence matrix $M(G)$. By definition, every even subgraph has an even number of edges at each vertex, so the vectors in $\mathbf{C}(G)$ are those satisfying $M(G)x = \mathbf{0}$. Thus $\mathbf{C}(G)$ is the nullspace of $M(G)$, and we have shown that the matrix has rank $n - 1$. The famous **Rank–Nullity Theorem** of linear algebra states that the dimension of the nullspace is the number of columns minus the rank of the matrix (over any field), so $\dim \mathbf{C}(G) = m - n + 1$. (Alternatively, one can show explicitly that the fundamental cycles span the cycle space.) ∎

15.2.18. Definition. A *2-basis* for a graph G is a basis for $\mathbf{C}(G)$ such that each coordinate is nonzero in at most two members of the basis.

15.2.19. Theorem. (MacLane) A graph is planar if and only if it has a 2-basis.

Proof: (Tutte) The facial walks of the bounded faces in a planar embedding form a 2-basis: each cycle is the sum of the bounding walks for the faces inside it.

To show that a nonplanar graph has no 2-basis, it suffices by Kuratowski's Theorem to prove (1) if G has a 2-basis and contains a subdivision of H, then H has a 2-basis, and (2) neither K_5 nor $K_{3,3}$ has a 2-basis.

For $e \in E(G)$, we obtain a 2-basis for $G - e$ from a 2-basis for G. If e is a cut-edge, then $\mathbf{C}(G - e) = \mathbf{C}(G)$. Otherwise, $\dim \mathbf{C}(G - e) = \dim \mathbf{C}(G) - 1$. If e is in one member of the 2-basis for G, delete that member. If e is in two members, replace them by their sum. This leaves $\dim \mathbf{C}(G) - 1$ vectors, all 0 in coordinate e. Deleting coordinate e yields independent elements of $\mathbf{C}(G - e)$ (dependence would make the 2-basis for $\mathbf{C}(G)$ dependent).

Also, a subdivision of a graph H has a 2-basis if and only if H has a 2-basis, because the two edges incident to a vertex of degree 2 in a graph belong to precisely the same even subgraphs.

To show that $K_{3,3}$ and K_5 have no 2-basis, consider a 2-basis $v_1, \ldots, v_k$ for a connected n-vertex graph H with girth g, written as binary m-tuples with $m = |E(G)|$. Let $v_0 = \sum_{i=1}^{k} v_i$. As a nonzero linear combination of basis elements, v_0 is a nonzero element of $\mathbf{C}(G)$. Recall that $k = m - n + 1$.

Let w be the total number of nonzero entries in $v_0, \ldots, v_k$. Since even graphs decompose into cycles, each nonzero vector has at least g such entries. Hence $w \geq g(k + 1)$. Each coordinate contributes at most two such entries, by the definition of 2-basis, so $w \leq 2m$.

Hence $g(m - n + 2) \leq 2m$. Solving for m yields the same bound $m \leq (n - 2)\frac{g}{g-2}$ that holds for planar graphs by Euler's Formula (Exercise 9.1.45). Since both K_5 and $K_{3,3}$ have too many edges to satisfy the bound, each cannot have a 2-basis. ∎

The argument for Theorem 15.2.19 is analogous to that for Whitney's characterization of planarity by matroid duality (Theorem 11.3.35). Also Theorem 15.2.19 can be used to prove Theorem 11.3.35; see Exercise 17.

PERMANENTS AND PLANAR GRAPHS

When we omit the signs on the terms in the permutation expansion of a determinant, the computation still has meaning.

15.2.20. Definition. Let A be a square matrix of order n, with entry $a_{i,j}$ in position (i, j). The **permanent** of A, denoted by $\operatorname{per} A$, is $\sum_{\sigma} \prod_{i=1}^{n} a_{i,\sigma(i)}$, where the sum runs over all permutations of $[n]$.

Terms in the permanent are products of one entry from each row and column. For the biadjacency matrix of a bipartite multigraph G, the permanent counts the perfect matchings. Using row-reduction operations, determinants are easy to compute, but no fast algorithm is known to compute the permanent. Counting the perfect matchings in a bipartite graph belongs to the class of difficult counting problems called #P (Valiant [1979]; see also Jerrum–Sinclair–Vigoda [2001]). In contrast, determining whether there is at least one such matching is easy.

Computing $\operatorname{per} A$ from the definition takes superexponential time in n, as does computing $\det A$ by the permutation expansion. Elementary row operations can compute determinants quickly, but these transformations do not preserve the permanent. When G is planar, however, the permanent of the biadjacency matrix A can be found by computing the determinant of a matrix obtained by negating some entries of A. At first the result is surprising, but the main idea is simple.

Nonzero terms in computing $\operatorname{per} A$ correspond to 1-factors in G. When we negate some entries to form A' and compute $\det A'$, the nonzero terms in the expansion still correspond to the 1-factors in G. The task is to cleverly negate some entries so that all nonzero terms when computing $\det A'$ *have the same sign*. This would yield $\operatorname{per} A = |\det A'|$. The proof is valid for multigraphs, but we phrase it just for graphs. The method was presented in Vazirani–Yannakakis [1989].

15.2.21. Theorem. (Permanent-Determinant Method) Let M be a perfect matching in an X, Y-bigraph G with biadjacency matrix A. A matrix A' obtained from A by negating some entries is **coherent** if for every M-alternating cycle C, the number of edges of C whose positions are negative in A' has opposite parity from k, where $2k$ is the length of C. If A' is a coherent signing of A, then per $A = |\det A'|$.

Proof: Index X and Y so that the edges of M are $\{x_i y_i \colon 1 \le i \le n\}$, where $X = \{x_1, \dots, x_n\}$ and $Y = \{y_1, \dots, y_n\}$; this permutes rows and columns but does not change the permanent or the magnitude of the determinant.

Each perfect matching in G corresponds to a permutation matrix and makes one contribution to per A or det A'. Let M' be a perfect matching in G. The entries in A' for M' are $\{a'_{i,\sigma(i)}\}_{i=1}^n$ for a particular permutation σ of $[n]$. As discussed above, it suffices to show that $\prod_{i=1}^n a'_{i,i} = \operatorname{sgn}(\sigma) \prod_{i=1}^n a'_{i,\sigma(i)}$.

The edges in the symmetric difference $M \triangle M'$ form pairwise disjoint M-alternating cycles. Let C be one such cycle, with length $2k$. We may assume an indexing so that the edges of M in C correspond to the first k positions along the diagonal. Shared edges contribute the same diagonal elements as factors in the product. Since we can make the same argument for each cycle, it suffices to show that $\prod_{i=1}^k a'_{i,i} = \operatorname{sgn}(\sigma') \prod_{i=1}^k a'_{i,\sigma(i)}$, where σ' is the restriction of σ to $[k]$.

The edges of M' on C correspond to off-diagonal positions in the upper-left k-by-k submatrix of A'. The restriction of σ to $[k]$ is a k-cycle. The sign of σ is the product of the signs on the cycles. The sign of a k-cycle is $(-1)^{k-1}$, because it takes $k - 1$ transpositions (or column exchanges) to bring the permutation to the identity (or the entries to the diagonal).

On the other hand, since A' is a coherent signing, the number of elements in $\{a'_{i,i}\}_{i=1}^k \cup \{a'_{i,\sigma(i)}\}_{i=1}^k$ that are negative has the same parity as $k - 1$. That is, $\prod_{i=1}^k a'_{i,i} \prod_{i=1}^k a'_{i,\sigma(i)} = (-1)^{k-1}$. Since each entry is its own multiplicative inverse, the desired equality holds. ∎

The most famous use of the method is the computation by Kasteleyn [1961] of the number of perfect matchings in $P_n \square P_n$. Kasteleyn [1963, 1967] extended the method to all planar bipartite graphs (see also Percus [1969]). We need to show that the biadjacency matrix of any planar bipartite graph has a coherent signing. The argument is based on a special orientation that exists for every planar graph.

15.2.22. Lemma. (Kasteleyn [1963]) Every planar graph G has an orientation such that for every perfect matching M, every M-alternating cycle has an odd number of edges in each direction (forward or backward).

Proof: Consider an embedding of G. Using induction on $|E(G)|$, we first build an orientation such that, when traveling the boundary of any bounded face in the clockwise direction, we traverse an odd number of edges in the forward sense. If G is a tree, then there is nothing to prove.

Otherwise, some edge e along the unbounded face also lies along a bounded face F. Let G' be the orientation of $G - e$ guaranteed by the induction hypothesis. Precisely one choice for the orientation of e yields an odd number of forward edges when F is traversed clockwise. This extends the desired orientation to G.

Let C be an M-alternating cycle, and let t be the number of forward edges in a clockwise traversal of C. For each face F interior to C, let $q(F)$ denote the number of forward edges in a clockwise traversal of F. Since each term in $\sum q(F)$ is odd (summing over faces inside C), the parity of the sum is the same as the number of faces inside. Also, the sum equals t plus the number of edges inside, since each inside edge is forward for the clockwise traversal of exactly one of the faces it bounds. Hence it suffices to show that the numbers of edges and faces inside C have opposite parity.

Let H be the subgraph consisting of C and the vertices and edges inside C. By Euler's Formula, $n - m + p' = 1$, where n, m, p' equal $|V(H)|$, $|E(H)|$, and the number of faces inside C. Since C is M-alternating, the mates of all vertices of C lie in C, and M pairs the points inside C. Hence both n and the number of edges of H not inside C are even. Therefore, by Euler's Formula the number of edges inside C has opposite parity from the number of faces inside. ∎

Given an embedding, which can be found quickly, the orientation can be found quickly. It remains to show that this orientation yields the desired signing of A for application of Theorem 15.2.21.

For an oriented graph G, the adjacency matrix B has $b_{i,j} = 1$ if $v_i \rightarrow v_j$, $b_{i,j} = -1$ if $v_j \rightarrow v_i$, and otherwise $b_{i,j} = 0$; note that B is skew-symmetric. If G is an X, Y-bigraph, then the X, Y-portion of B (rows for X and columns for Y) is a signing of the biadjacency matrix of G.

15.2.23. Theorem. (Kasteleyn [1963]) If G is a planar X, Y-bigraph with biadjacency matrix A, then $\operatorname{per} A = |\det A'|$, where A' is the signing of A given by the orientation from Lemma 15.2.22.

Proof: (Vazirani–Yannakakis [1989]) We show that A' is a coherent signing of A for some (actually, any) perfect matching M. By Lemma 15.2.22, every M-alternating cycle C traverses an odd number of edges of the orientation forward and an odd number backward. Along C, we alternate X-to-Y edges and Y-to-X edges. Only those from X to Y affect A'.

Let $2k$ be the length of C and view C in a fixed direction. The edges of C are of four types; we introduce r and s to count these.

	X to Y	Y to X
backward	r	$k - s$
forward	$k - r$	s

The number of edges of C giving negative entries in A' is $r + s$, since A' records the orientation when an edge is viewed from X to Y. The number of edges along C traversed forward is $k - r + s$. By Lemma 15.2.22, this number is odd. Hence $r + s$ has opposite parity from k. Since this holds for every M-alternating cycle, A' is coherent, and Theorem 15.2.21 applies. ∎

The permanent-determinant method of Theorem 15.2.21 is a special case of the "Hafnian-Pfaffian method", valid also for nonbipartite graphs. Indeed, the original proof of Tutte's 1-Factor Theorem used these ideas. The method is described in Lovász–Plummer [1986, pp. 315–329] and more generally in Kuperberg [1994, 1998]. The combinatorial aspects are outlined in Exercises 25–26.

MÖBIUS INVERSION (optional)

Many inversion formulas can be explained using functions on intervals in a poset. These can be studied via linear algebra when the poset is encoded in a matrix. We obtain a common generalization of Inclusion-Exclusion and many other inversion formulas. Given a function g expressed in terms of another function f, an inversion formula computes f in terms of g. We recall classical examples.

15.2.24. Example. *Summation and difference.* When $g(n) = \sum_{i=1}^{n} f(i)$ for $n \in \mathbb{N}$, we obtain f from g by $f(1) = g(1)$ and $f(n) = g(n) - g(n-1)$ for $n > 1$. This is a discrete version of the Fundamental Theorem of Calculus: just as differentiation is the inverse of integration, so differencing is the inverse of summation. ∎

15.2.25. Example. *Inclusion-Exclusion.* Let $\{A_1, \ldots, A_n\}$ be subsets of a universe U. In Section 4.1, we considered two functions on $2^{[n]}$. For $S \subseteq [n]$, let $g(S) = \left| \bigcap_{i \in S} A_i \right|$, and let $f(S)$ be the number of elements $x \in U$ such that $x \in A_i$ if and only if $i \in S$. An element is counted by $g(T)$ if it lies in all sets indexed by T and perhaps in other sets. Thus $g(T) = \sum_{S \supseteq T} f(S)$. The Inclusion-Exclusion Formula (Theorem 4.1.3) states that $f(T) = \sum_{S \supseteq T} (-1)^{|S|-|T|} g(S)$. ∎

15.2.26. Example. *Möbius inversion in number theory.* Let $f(n)$ be the number of k-ary sequences with period n. Every n-tuple generates a periodic sequence, but $f(n) \neq k^n$, since we must exclude those whose periods divide n. Letting $g(n)$ be the number of k-ary n-tuples, we have $g(n) = k^n$ and $g(n) = \sum_{d|n} f(d)$. The classical Möbius Inversion Formula of number theory inverts such summations:

$$f(n) = \sum_{d|n} \mu\left(\tfrac{n}{d}\right) g(d), \qquad\qquad (*)$$

where $\mu(j) = (-1)^k$ when j is the product of k distinct primes, and $\mu(j) = 0$ when j is divisible by the square of a prime, and $\mu(1) = (-1)^0 = 1$. ∎

We can prove $(*)$ by substituting for $g(d)$ its definition in terms of f and checking that the resulting weighted sum of values of f gives weight 1 to $f(n)$ and weight 0 to all other values of f. The Inclusion-Exclusion Formula can also be proved this way, but we want to prove all such inversion formulas at once.

The functions in these examples are defined on posets: a chain, $2^{[n]}$, or a divisor lattice. In general we want to invert $g(x) = \sum_{y \leq x} f(y)$, where "$\leq$" is the order relation in a poset. We need functions on the intervals of a poset.

15.2.27. Definition. When $x \leq y$ in a poset, the **interval** $[x, y]$ is $\{z: x \leq z \leq y\}$. A poset P is **locally finite** if every interval is finite. Let $\mathrm{Int}\,(P)$ denote the set of intervals in P. An **incidence function** on P is a function $f: \mathrm{Int}\,(P) \to \mathbb{C}$. Write $f(x, y)$ for $f([x, y])$, defined only when $x \leq y$. The **incidence algebra** $\mathbf{A}(P)$ (over $\mathbb{C}$) is the set of incidence functions on P, endowed with operations of **sum** $f + g$, **scalar multiple** cf, and **convolution** (or **product**) fg:

$$(f + g)(x, y) = f(x, y) + g(x, y)$$
$$(cf)(x, y) = c \cdot f(x, y)$$
$$(fg)(x, y) = \sum_{x \leq z \leq y} f(x, z) g(z, y).$$

Restricting to locally finite posets makes each convolution a finite sum. For convenience, we further restrict to finite posets. The set of functions on a set S forms a vector space with S as a canonical basis. Each function is then a linear combination of basis elements. Scalar multiplication and vector addition are coordinate-wise, as in Definition 15.2.27. A vector space becomes an algebra when endowed with a suitable multiplication rule for vectors; here we use convolution.

15.2.28. Remark. *Incidence algebra as matrix multiplication.* Given an ordering $x_1, \ldots, x_n$ of P such that $x_i <_P x_j$ implies $i < j$, we interpret $\mathbf{A}(P)$ another way. Record an incidence function in a matrix A by putting $a_{i,j} = f(x_i, x_j)$ if $x_i \leq_P x_j$ and $a_{i,j} = 0$ otherwise. Under the chosen ordering, A is upper triangular.
The product of such matrices A and B for incidence functions f and g is also upper triangular and is computed by convolutions: $c_{ik} = \sum_j a_{ij} b_{jk}$. If x_i and x_k are incomparable, then there is no x_j between them, and the convolution is 0. For $x_i \leq x_k$, the computation yields $(fg)(x_i, x_k)$. Hence $\mathbf{A}(P)$ is isomorphic to the algebra of upper triangular matrices such that $a_{i,j} = 0$ when $x_i \not\leq x_j$. ∎

Multiplication in $\mathbf{A}(P)$ is not commutative. Nevertheless, $\mathbf{A}(P)$ has a two-sided multiplicative identity corresponding to the identity matrix. This function, called δ or 1, is defined by $1(x, y) = \delta_{x,y}$ (1 if $x = y$ and 0 if $x < y$). Next we seek multiplicative inverses in the incidence algebra.

15.2.29. Lemma. An incidence function f has an inverse if and only if $f(x, x) \neq 0$ for all $x \in P$. If f has an inverse, then f^{-1} is unique and two-sided, and its value on $[x, y]$ depends only on the values of f on subintervals of $[x, y]$.

Proof: A left inverse is a function g such that $gf = 1$. We show first that the existence of a left inverse requires $f(x, x) \neq 0$, and that in this case the equations required by g being a left inverse have a unique solution.
Convolution on $[x, x]$ yields $g(x, x)f(x, x) = 1$, and thus $f(x, x) \neq 0$ and $g(x, x) = f(x, x)^{-1}$. If $f(x, x)$ is nonzero for each x, then the value of $g(x, y)$ for $x < y$ is determined inductively by its values on subintervals. Since $gf = 1$ requires $\sum_{x \leq z \leq y} g(x, z)f(z, y) = 0$ for $x < y$, we must have

$$g(x, y) = -f(y, y)^{-1} \sum_{x \leq z < y} g(x, z)f(z, y).$$

This uniquely determines the left inverse of f.
Given that $gf = 1$, one can verify inductively that $(fg)(x, y) = \delta_{x,y}$, and thus a left inverse g is also a right inverse (Exercise 32). Arguments as above show that the right inverse also is unique. ∎

Since $\mathbf{A}(P)$ is isomorphic to an algebra of upper triangular matrices under matrix multiplication, Lemma 15.2.29 also follows because the invertible triangular matrices are those with nonzero determinant. Triangular matrices have nonzero determinant precisely when the diagonal has no zero.
Various incidence functions measure structural aspects of P.

15.2.30. Definition. (Fundamental incidence functions).
(a) $1(x, y) = \begin{cases} 1 & x = y \\ 0 & x < y \end{cases}$; the identity function.

(b) $\zeta(x, y) = \begin{cases} 1 & x = y \\ 1 & x < y \end{cases}$; the zeta function.

(c) $\lambda(x, y) = \begin{cases} 0 & x = y \\ 1 & x < y \end{cases}$; the "strictly less-than" function; $\lambda = \zeta - 1$.

(d) $\kappa(x, y) = \begin{cases} 1 & x < y \\ 0 & \text{otherwise} \end{cases}$; the cover function.

We can count structures of interest in a poset using these incidence functions and an understanding of the meaning of multiplication. A **weak** x, y**-chain** of length k is a list $x_0, \ldots, x_k$ such that $x = x_0 \leq \cdots \leq x_k = y$; the chain is a **strict chain** if $x_0, \ldots, x_k$ are distinct.

15.2.31. Lemma. Multiplication in $\mathbf{A}(P)$ is associative, and $\prod_{i=1}^{k} f_i$ is specified by $\sum_{x=z_0 \leq \cdots \leq z_k = y} \prod_{i=1}^{k} f_i(z_{i-1}, z_i)$. In particular,
(a) $\zeta^k(x, y)$ is the number of weak x, y-chains of length k, with $\zeta^2(x, y) = \|[x, y]\|$.
(b) $\lambda^k(x, y)$ is the number of strict x, y-chains of length k.
(c) $(1 - \lambda)^{-1}(x, y)$ is the total number of strict x, y-chains.
(d) $\kappa^k(x, y)$ is the number of maximal x, y-chains of length k.
(e) $(1 - \kappa)^{-1}(x, y)$ is the total number of maximal x, y-chains.

Proof: Any order for computing the product yields $\prod_{i=1}^{k} f_i$ as an iterated summation. The distributive law pulls the summations to the front, expressing the formula as one sum. The possible nonzero terms correspond to successive related elements (not necessarily distinct) forming chains from the bottom to the top of the interval. The sum is over all such weak chains of length k.

This yields (a)–(e). If each factor in the term for a weak chain is 1, accomplished by using ζ, then the sum counts the weak chains. To forbid repetitions, use λ instead. Maximal chains are those where no additional element can be inserted; successive elements satisfy the cover relation, and we use κ.

To count all strict chains, form the series $1 + \lambda + \lambda^2 + \cdots$. The function 1 (that is, λ^0) counts chains of length 0; there are none if $x \neq y$. A geometric series of incidence functions behaves like a geometric series of numbers (Exercise 32). We need $1 - \lambda$ to be nonzero on every interval $[x, x]$, which holds by definition. Counting all maximal x, y-chains using $\sum_{k \geq 0} \kappa^k$ is analogous to this. ∎

To solve the inversion problem, we need the inverse of ζ.

15.2.32. Definition. The **Möbius function** of a poset P, denoted by μ, is the incidence function defined by

$$\mu(x, y) = \begin{cases} 1 & \text{if } x = y \\ -\sum_{x \leq z < y} \mu(x, z) & \text{if } x < y. \end{cases}$$

Definition 15.2.32 is just the requirement of Lemma 15.2.29 for an incidence function to be the inverse of ζ. The next theorem extends to posets in which every down-set generated by one element is finite.

15.2.33. Theorem. (**Möbius Inversion Formula**) Let f and g be complex-valued functions defined on a finite poset P. If f and g satisfy $g(x) = \sum_{y \leq x} f(y)$ for all $x \in P$, then $f(x) = \sum_{y \leq x} g(y)\mu(y, x)$ for all $x \in P$.

Proof 1: (Computation) Expand the right side of the desired formula, using the recursive definition of $\mu(x, y)$ and the expression $g(y) = \sum_{z \leq y} f(z)$. The proof is completed by showing that this counts $f(x)$ with weight 1 and other values of f with weight 0 (Exercise 33).

Proof 2: (Incidence functions) Add a unique minimal element $\hat{0}$ to P, yielding P'. Define $f'(\hat{0}, y) = f(y)$, $g'(\hat{0}, y) = g(y)$, and $f'(x, y) = g'(x, y) = 0$ for $x \neq \hat{0}$ (also $f'(\hat{0}, \hat{0}) = g'(\hat{0}, \hat{0}) = 0$). The statement $g(x) = \sum_{y \leq x} f(y)$ becomes $g'(\hat{0}, x) = \sum_{\hat{0} \leq y \leq x} f'(\hat{0}, y) \cdot 1$. This combines with the zero values on other intervals to yield $g' = f'\zeta$ in the incidence algebra. Hence it must hold that $f' = g'\zeta^{-1} = g'\mu$, and the resulting convolution for the value of $f'(\hat{0}, x)$ is the desired formula, because $\mu_{P'}$ restricts to μ_P on $\text{Int}(P)$.

Proof 3: (Matrix multiplication – no incidence functions) After indexing P by a linear extension, the incidence algebra is an algebra of upper triangular matrices under multiplication. View f and g as row vectors, with a value for each element. The statement $g(x) = \sum_{y \leq x} f(y)\zeta(y, x)$ is the statement that g is obtained when f is multiplied on the right by the matrix for ζ. Since the matrix for μ is its inverse, f can be retrieved by multiplying g on the right by the matrix for μ. ∎

Proofs 2 and 3 avoid detailed manipulation and make clear that the validity of $g(x) = \sum_{y \leq x} f(y)$ for all x is not only sufficient but also necessary for $f(x) = \sum_{y \leq x} g(y)\mu(y, x)$ to hold for all x.

Viewing f, g as column vectors instead of row vectors or convolving with μ and ζ on the left instead of the right yields a dual form of the Möbius Inversion Formula, which reduces to the usual Inclusion-Exclusion Formula when $P = 2^n$.

15.2.34. Theorem. (**Dual Möbius Inversion Formula**). Let f, g be complex-valued functions on a finite poset P. The functions f and g satisfy $g(x) = \sum_{y \geq x} f(y)$ for all $x \in P$ if and only if $f(x) = \sum_{y \geq x} \mu(x, y)g(y)$ for all $x \in P$. ∎

To apply the inversion formula, we need the Möbius function.

15.2.35. Example. For the chain formed by $\mathbb{N}$ under the usual ordering, Definition 15.2.32 yields $\mu(j, j) = 1$, $\mu(j, j+1) = -1$, and $\mu(j, k) = 0$ if $k > j + 1$. Hence Theorem 15.2.33 explains why differencing is inverse to summation. ∎

To compute Möbius functions for posets such as $2^{[n]}$ and the divisor poset, we use a combining formula for Möbius functions on products.

15.2.36. Theorem. (**Product Formula**) The Möbius function for the product of locally finite posets P and Q is given by

$$\mu_{P \times Q}((x, y), (x', y')) = \mu_P(x, x')\mu_Q(y, y').$$

Proof: Let $I = [(x, y), (x', y')]$, and let $I_P = [x, x']$ and $I_Q = [y, y']$. We use induction on $|I|$. When $|I| = 1$, both sides equal 1. For $|I| > 1$, the definition of $\mu_{P \times Q}$ yields $\sum_{(x'', y'') \in I} \mu_{P \times Q}((x, y), (x'', y'')) = 0$. Using $I = I_P \times I_Q$ and the definitions of μ_P and μ_Q, we also compute

$$\sum_{(x'', y'') \in I} \mu_P(x, x'') \mu_Q(y, y'') = \sum_{x'' \in I_P} \mu_P(x, x'') \sum_{y'' \in I_Q} \mu_Q(y, y'') = 0.$$

By the induction hypothesis, $\mu_{P \times Q}((x, y), (x'', y'')) = \mu_P(x, x'') \mu_Q(y, y'')$ when $(x, y) \leq (x'', y'') < (x', y')$. Canceling such terms from the two sums that equal 0 leaves the desired equality when $(x'', y'') = (x', y')$. ∎

15.2.37. Example. In the chain product M^e with $e = e_1, \ldots, e_n$, the elements are encoded as lists $(a_1, \ldots, a_n)$ with $0 \leq a_i \leq e_i - 1$. In the ith factor, $\mu(j, j) = 1$, $\mu(j, j+1) = -1$, and $\mu(j, k) = 0$ if $k > j + 1$, as in Example 15.2.35.

In $\mathbf{2}^n$ the factor chains all have size 2. The product formula thus yields $\mu_{\mathbf{2}^n}(T, S) = (-1)^{|S| - |T|}$. Hence the Inclusion-Exclusion Formula (Example 15.2.25) is the special case of Theorem 15.2.34 for $P = \mathbf{2}^n$.

Now view M^e as $D(N)$, the poset of divisors of N. We have $\mu_{D(N)}(k, n) = (-1)^j$ if $\frac{n}{k}$ is the product of j distinct primes, or $\mu_{D(N)}(k, n) = 0$ if $\frac{n}{k}$ is divisible by a square. This special case is the classical Möbius function $\mu(\frac{n}{k})$. ∎

When formulas for both f and g in the Möbius Inversion Formula are known, the formula can be used to find μ. Helped by the Polynomial Principle, this works for the special lattices $L_n(q)$ (Exercise 34) and Π_n.

15.2.38. Example. *Möbius function of the partition lattice Π_n.* Since $\sigma \leq \tau$ when every block of σ lies in a block of τ, $\hat{0}$ consists of singletons and $\hat{1}$ has one block.

If τ has k blocks, then the interval $[\sigma, \tau]$ is the product of partition lattices $\Pi_{n_1}, \ldots, \Pi_{n_k}$, where n_i counts the blocks of σ contained in the ith block of τ. By the product rule, knowing $\mu(\hat{0}, \hat{1})$ for each Π_{n_i} allows us to compute $\mu_{\Pi_n}(\sigma, \tau)$.

Let $g(\sigma)$ count mappings from $[n]$ to a set X of size x that are constant on blocks of σ. If σ has $b(\sigma)$ blocks, then $g(\sigma) = x^{b(\sigma)}$. Let $f(\sigma)$ count the mappings whose inverse images are the blocks of σ. Thus $g(\sigma) = \sum_{\tau \geq \sigma} f(\tau)$, and Möbius inversion guarantees $f(\sigma) = \sum_{\tau \geq \sigma} \mu(\sigma, \tau) x^{b(\tau)}$.

To compute $\mu(\hat{0}, \hat{1})$, consider $f(\hat{0})$. This counts injective mappings, so $f(\hat{0}) = x(x-1) \cdots (x - n + 1)$. Again this is a polynomial equality, since it holds for every positive integer x. Since $\hat{1}$ is the only partition with 1 block, $\mu(\hat{0}, \hat{1})$ is the only contribution to the coefficient of the linear term, so $\mu(\hat{0}, \hat{1}) = (-1)^{n-1}(n-1)!$. ∎

The Möbius function of a lattice yields much structural information; see Aigner [1979] and Stanley [1986]. We briefly mention two related topics.

15.2.39. Remark. *Whitney numbers.* For a graded poset P with lower bound $\hat{0}$, the rank generating function $\sum_k W_k x^k$ has the Whitney numbers (rank sizes) as coefficients; it sums $\zeta(\hat{0}, a)$ over $a \in P$, weighted by $x^{r(a)}$. The **characteristic polynomial** $\sum_k w_k x^k$ sums $\mu(\hat{0}, a)$ over $a \in P$, weighted by $x^{r(P) - r(a)}$. The coefficient w_k is the kth **Whitney number** *of the first kind*, while W_k is the kth **Whitney number** *of the second kind*. For the partition lattice Π_n, we obtain the Stirling numbers: $W_k(\Pi_n) = S_{n, n-k}$ and $w_{n-k}(\Pi_n) = s_{n, k}$. ∎

15.2.40. Remark. Some classical counting problems in terms of k and n are solved by a polynomial f in k of degree n such that the answer to a related problem is $(-1)^n f(-k)$. Stanley [1974] explored this **combinatorial reciprocity**.

For a bounded poset P, the **zeta-polynomial** is defined by $Z_P(k) = \zeta^k(\hat{0}, \hat{1})$. Since $\zeta = 1 + \lambda$ as incidence functions, $Z_P(k) = \sum \binom{k}{i} \lambda^i(\hat{0}, \hat{1})$, so $Z_P(k)$ is a polynomial in k. Recall that $\zeta^k(\hat{0}, \hat{1})$ counts the weak chains $z_0 \leq \cdots \leq z_k$ with $(z_0, z_k) = (\hat{0}, \hat{1})$, while $\lambda^i(\hat{0}, \hat{1})$ counts the strict chains $z_0 < \cdots < z_i$ with $(z_0, z_i) = (\hat{0}, \hat{1})$.

Given an n-element poset P, let $\Omega_P(k)$ be the number of order-preserving maps from P to $\mathbf{k}$, and let $\overline{\Omega}_P(k)$ be the number of strict order-preserving maps from P to $\mathbf{k}$. These are the **order polynomial** and **strict order polynomial** of P, respectively. Each order-preserving map $\sigma \colon P \to \mathbf{k}$ corresponds to a weak chain in $J(P)$ by letting $I_k = \{x \in P \colon \sigma(x) \leq k\}$; we have $\varnothing = I_0 \subseteq I_1 \subseteq \cdots \subseteq I_k = P$. The bijection yields $\Omega_P(k) = Z_{J(P)}(k)$. Thus $\Omega_P(k)$ is a polynomial in k with degree n. Applying Stanley's reciprocity theorem to $Z_{J(P)}$ yields $\overline{\Omega}_P(k) = (-1)^n \Omega_P(-k)$.

For example, when $P = \mathbf{n}$, the order polynomial and strict order polynomial count n-element multisets and subsets from k types, respectively. We saw in Chapters 1 and 3 that there are $(-1)^n \binom{-k}{n}$ multisets.

Another example is the chromatic polynomial $\pi_G(k)$ of a graph G (Application 4.1.11). An acyclic orientation D and a k-coloring σ are **strictly compatible** if $x \to y$ in D implies $\sigma(x) < \sigma(y)$. Thus $\pi_G(k)$ is the number of strictly compatible pairs. Changing the restriction $\sigma(x) < \sigma(y)$ to $\sigma(x) \leq \sigma(y)$ yields **weakly compatible pairs**. Stanley's reciprocity theorem on an appropriate poset yields the number of weakly compatible pairs as $(-1)^n \chi(G; -k)$. As a special case, $(-1)^n \chi(G; -1)$ counts the acyclic orientations of G.   ∎

EXERCISES 15.2

15.2.1. $(-)$ Use the Matrix Tree Theorem to find $\tau(G)$ for the graph below.

15.2.2. $(-)$ In terms of the incidence functions in Lemma 15.2.31, compute
(a) The number of comparable pairs of elements in $[x, y]$.
(b) The number of covering pairs in $[x, y]$.
(c) The number of elements of height k in $[x, y]$ (assume the interval is graded).
(d) The number of elements of co-height k in $[x, y]$.

15.2.3. $(-)$ Let P be the lattice of partitions of 6 with 111111 removed, so that $\hat{0}$ is the partition 21111, and $\hat{1}$ is the partition 6.
(a) Compute $\mu(\hat{0}, x)$ for all x. Suppose that $g(x) = \sum_{y \geq x} f(y)$ and we know that $g(x)$ is given by the rank of x in P. Compute $f(\hat{0})$.
(b) Compute the zeta-polynomial of P. Use it to count the chains $z_0 \leq \cdots \leq z_{10}$ such that $z_0 = \hat{0}$ and $z_{10} = \hat{1}$.

15.2.4. Use the Matrix Tree Theorem to prove Cayley's Formula ($\tau(K_n) = n^{n-2}$).

15.2.5. Use the Matrix Tree Theorem to find $\tau(K_{r,s})$. (Lovász [1979, p. 223])

15.2.6. ($\Diamond$) The **generalized book** $B_{n,k}$ is the graph $K_k \diamond \overline{K}_{n-k}$. Compute $\tau(B_{n,k})$.

15.2.7. A matrix is **totally unimodular** if every square submatrix has determinant in $\{0, 1, -1\}$. Prove that the incidence matrix of every loopless digraph is totally unimodular. Prove that the incidence matrix of a graph is totally unimodular if and only if the graph is bipartite (each column has two +1's).

15.2.8. ($\Diamond$) *de Bruijn cycle* (see Exercise 10.1.28). Prove that the following explicit algorithm generates a binary de Bruijn cycle of length 2^n: start with n copies of 0. Subsequently, append 1 if doing so does not repeat a previous string of length n; otherwise append 0. (Martin [1934], Ungar [1950], and others)

15.2.9. ($\Diamond$) *Tarry's Algorithm* (Tarry [1895], as presented by D.G. Hoffman). A castle has finitely many rooms and corridors. Each corridor has a door into a room at each end. Any room can be reached from any other by traversing corridors and rooms. Initially, no doors have marks. A robot starting in some room explores the castle using the following rules.
   (1) After entering a corridor, traverse it into the room at the other end.
   (2) Upon entering a room with all doors unmarked, mark I on the entry door.
   (3) In a room, mark O on an unmarked door (if any exists), and use it.
   (4) In a room with all doors marked, exit via a door marked I if one exists.
   (5) In a room with all doors marked O, stop.
Prove that the robot stops only after using every corridor exactly twice, once in each direction. (Comment: All decisions are completely local; the robot sees nothing but the current room or corridor. See also König [1936, pp. 35–56] and Fleischner [1983, 1991].)

15.2.10. Let M be the incidence matrix of an orientation of a tree T, and let B be an $(n-1)$-by-$(n-1)$ submatrix obtained by deleting the row of M for a fixed vertex v_0. The columns of B correspond to the edges $\{e_1, \dots, e_{n-1}\}$ of T. Let $b_{i,j}$ be the entry in B^{-1} in the row for vertex v_i and the column for edge e_j. Let P_j be the unique path from v_j to v_0 in T. Prove that $b_{i,j} = 1$ if e_i is an odd edge of P_j, $b_{i,j} = -1$ if e_i is an even edge of P_j, and otherwise $b_{i,j} = 0$. (Branin [1959]; see also Bryant [1967])

15.2.11. Prove the Cauchy–Binet Formula $\det C = \sum_{S \subseteq [m]} \det A_S \det B_S$ by considering the matrix equation $DE = F$ below.

$$\begin{pmatrix} I_m & 0 \\ A & I_n \end{pmatrix} \begin{pmatrix} -I_m & B \\ A & 0 \end{pmatrix} = \begin{pmatrix} -I_m & B \\ 0 & AB \end{pmatrix}$$

15.2.12. Given a graph G, prove combinatorially that the symmetric difference of any two even subgraphs is an even subgraph, and the symmetric difference of any two edge cuts is an edge cut. (Comment: This completes the proof of Theorem 15.2.17.)

15.2.13. ($\Diamond$) Apply the results of this section to obtain the following statements about edge cuts in graphs that earlier were proved combinatorially.
   (a) A set of edges in a graph is an edge cut if and only if it has an even number of edges in every cycle (Exercise 7.1.31).
   (b) Every edge cut is a union of pairwise edge-disjoint bonds (Exercise 7.1.30).

15.2.14. ($\Diamond$) Without using the bond space, show directly that the dimension of the cycle space of a connected graph G is $|E(G)| - |V(G)| + 1$. That is, show that the incidence vectors of the cycles generated by adding one edge to a spanning tree T form a basis.

15.2.15. Let G be a graph with n vertices, m edges, and k components. Determine the dimensions of the cycle space and the bond space of G.

15.2.16. Prove that the cycle space and bond space of a graph are the null space and row space of its incidence matrix.

15.2.17. Prove Theorem 11.3.35 from Theorem 15.2.19.

15.2.18. ($\diamond$) Let W be a subspace of $\mathbb{F}_2^n$. Prove that every coordinate that is not identically 0 in W is 1 in exactly half the elements of W. Conclude that the sum of the sizes of the even subgraphs of a 2-edge-connected graph G with n vertices and m edges is $n2^{m-n}$.

15.2.19. A cycle C in a graph G is **isometric** if $d_C(x,y) = d_G(x,y)$ for all $x, y \in V(C)$. Prove that the set of isometric cycles in G spans the cycle space of G. (see Naatz [2000])

15.2.20. Let Q be a nonempty family of subsets of $[n]$. Prove that the number of subsets of Q using each element of $[n]$ an even number of times equals the number using each element of n an odd number of times. (Beckwith [2001])

15.2.21. (+) Light bulbs $l_1, \dots, l_n$ are controlled by switches $s_1, \dots, s_n$. The ith switch flips the status of the ith light and possibly others, but s_i flips the status of l_j if and only if s_j flips the status of l_i. Initially all lights are off. Prove that it is possible to turn all the lights on. (Lovász [1979], Peled [1992])

15.2.22. Prove that the number of 1-factors in the k-dimensional hypercube Q_k is at least exponential in k. (Clark–George–Porter [1997], Lovász–Plummer [1986])

15.2.23. ($\diamond$) Show that Lemma 15.2.22 can generate the orientation of the grid below. Use it and Theorem 15.2.23 to count the 1-factors in $P_4 \,\square\, P_4$.

15.2.24. *The Permanent Lemma,* from the Combinatorial Nullstellensatz.

(a) Let A be an n-by-n matrix with nonzero permanent over a field K. Prove that for any $b \in K^n$ and sets $S_1, \dots, S_n$ of size 2 in K, there is a vector $x \in \prod_{i=1}^n S_i$ such that Ax differs from b in every coordinate. (Alon–Tarsi [1989])

(b) Let p be a prime. Prove that every list of $2p-1$ members of $\mathbb{Z}_p$ contains p terms that sum to 0 modulo p. (Erdős–Ginzburg–Ziv [1961])

15.2.25. Two permutations are *equivalent* if they differ only in direction on cycles.

(a) Let B be a skew-symmetric matrix. Show that in the permutation expansion of $\det B$, the contribution of the terms from all the permutations containing odd cycles is 0.

(b) Let σ be a permutation in which all cycles have even length. Prove that when B is skew-symmetric the sum of the contributions to $\det B$ over all permutations equivalent to σ is $(-1)^r 2^{r-s} \prod_i b_{i,\sigma(i)}$, where σ consists of r cycles, of which s have length 2.

(c) Given a $2n$-by-$2n$ matrix B and a partition P of $[2n]$ into pairs $\{i_1, j_1\}, \dots \{i_n, j_n\}$, let $b_P = (-1)^n \prod_{r=1}^n b_{i_r, j_r}$, The **Pfaffian** $\mathrm{Pf}\, B$ of a skew-symmetric matrix B is $\sum_P b_P$, summing over all partitions of $[2n]$ into pairs. Establish a bijection that associates a set of terms in the computation of $(\mathrm{Pf}\, B)^2$ with an equivalence class of permutations of $[n]$ whose cycles all have even length. Conclude $\det B = (\mathrm{Pf}\, B)^2$. (Muir [1882, 1906])

15.2.26. For an oriented graph G with adjacency matrix B, prove (A)–(C) equivalent:

(A) $|\text{Pf } B|$ is the number of 1-factors of G (see Exercise 15.2.25 for $\text{Pf } B$).

(B) For every perfect matching M in G, every M-alternating cycle traverses an odd number of edges of G in each direction.

(C) For some perfect matching M in G, every M-alternating cycle traverses an odd number of edges of G in each direction.

15.2.27. For $m \geq n/2$, with n even, determine the multigraph with m edges and n vertices that has the most perfect matchings. (Hajek–Narayanan [1994])

15.2.28. Let M_n be the graph obtained from the cycle C_{2n} by adding edges joining opposite vertices. Count the perfect matchings in M_n and in the graph $C_n \,\square\, K_2$. (Hosoya–Harary [1993], McSorley [1998]) (Comment: Esperet–Kardoš–King–Král'–Norine [2011] proved that the minimum number of perfect matchings in 3-regular bridgeless graphs grows exponentially with the number of vertices.)

15.2.29. Let $\mathbf{G}_n$ be the class of 3-regular connected graphs with n white vertices, n black vertices, and no monochromatic cycles. In a *good orientation* of $G \in \mathbf{G}_n$, white vertices have indegree 2 and black vertices have outdegree 2. (R.E. Prather)

(a) Prove that the numbers of purely white edges and purely black edges are equal, and both color classes induce forests, with white trees $W_1, \ldots, W_k$ and black trees $B_1, \ldots, B_k$.

(b) Let $M(G)$ denote the k-by-k matrix in which entry $m_{i,j}$ is the number of edges from W_i to B_j. Prove that the number of good orientations of G is the permanent of $M(G)$.

(c) Find the least n such that $\mathbf{G}_n$ has a graph with no good orientation.

(d) For even n, construct a member of $\mathbf{G}_n$ with $1 + 3^{n/2}$ good orientations.

15.2.30. A matrix is **doubly stochastic** if all entries are nonnegative and rows and columns all sum to 1. Egorychev [1981] and Falikman [1981] proved the **van der Waerden Conjecture** that the permanent of such a matrix of order n is at least $n!/n^n$, with equality only when every entry is $1/n$. Here are some initial steps.

(a) Prove that a nonnegative matrix of order n has permanent 0 if and only if it has a k-by-$(n + 1 - k)$ submatrix of 0s for some k.

(b) A matrix of order n is **decomposable** if it has a k-by-$(n–k)$ submatrix of 0s, for some k. Prove that a doubly stochastic matrix minimizing the permanent is indecomposable.

(c) A doubly stochastic matrix has 1 as an eigenvalue. Prove that the multiplicity of 1 as an eigenvalue is 1 if and only if the matrix is indecomposable.

15.2.31. (+) *van der Waerden Conjecture for matrices with all entries nonzero.* Doubly stochastic matrices of order n having the smallest permanent are called *optimal*.

(a) Prove that if A is an indecomposable matrix with nonnegative entries (Exercise 15.2.30), then AA^T and $A^T A$ are also indecomposable.

(b) Let $A_{i,j}$ be obtained from A by deleting row i and column j. Prove that if A is an optimal matrix with $a_{i,j} > 0$, then $\text{per } A_{i,j} = \text{per } A$. (Marcus–Newman [1959])

(c) Show that if $\text{per } A_{i,j} = \text{per } A$ for every i, j, then replacing a column by an arbitrary vector with the same sum does not change the permanent.

(d) Conclude that the doubly stochastic matrix with all entries equal has the smallest permanent among doubly stochastic matrices with all entries nonzero.

15.2.32. *Properties of invertible incidence functions.*

(a) If f is an invertible incidence function, prove that the left inverse of f computed in Lemma 15.2.29 is also a right inverse.

(b) If $1 - f$ is an invertible incidence function, prove that $(1 - f)^{-1} = \sum_{k \geq 0} f^k$.

15.2.33. Let f and g be functions on poset P such that $g(x) = \sum_{y \leq x} f(x)$ for all $x \in P$. Using the recursive definition of the Möbius function, prove by algebraic manipulation that $\sum_{y \leq x} g(y)\mu(y, x) = f(x)$. (Comment: This completes Proof 1 of Theorem 15.2.33.)

15.2.34. *Möbius function of the subspace lattice* $L_n(q)$ (subspaces of an n-dimensional vector space V over a q-element field, ordered by inclusion).

(a) Reduce computing the Möbius function to computing $\mu(\hat{0}, \hat{1})$ for all n.

(b) Consider linear functions from V to a vector space X of size x. Let $f(U)$ count such functions with nullspace U. Let $g(U)$ count such functions whose nullspace contains U; thus $g(U) = \sum_{W \supseteq U} f(W)$. Prove that $g(U) = x^{n - \dim U}$.

(c) Prove that $f(\varnothing) = (x-1)(x-q)\cdots(x-q^{n-1})$.

(d) Conclude that $\mu(\hat{0}, \hat{1}) = (-1)^n q^{\binom{n}{2}}$.

15.2.35. Let f and g be incidence functions on a poset P. Let A be the matrix with rows and columns indexed by elements of P whose (x, y)-entry is $\sum_{z \in P} f(z, x) g(z, y)$. Prove that $\det A = \prod_{z \in P} f(z, z) g(z, z)$. (Hint: Generalize the argument for Exercise 4.1.65.) (Altinisik–Sagan–Tuglu [2005])

15.2.36. Let F be a family of finite sets closed under taking subsets (this is the definition of a **simplicial complex**). For $A \in F$, let $U(A)$ denote the family of all members of F that contain A. Fix $k \in \mathbb{N}$. Suppose that $\sum_{B \in U(A)} (-1)^{|B|} = 0$ whenever $|A| < k$. Prove that $|F|$ is divisible by 2^k. (Stanley [2009])

15.3. Eigenvalues

There are important applications of eigenvalues of various combinatorial matrices, such as transition matrices of Markov chains. We restrict our attention to matrices arising from the adjacency matrix of a graph. The classical approach uses the adjacency matrix itself; see Biggs [1974] and a more encyclopedic treatment in Cvetković–Doob–Sachs [1979]. Other texts with further applications and eigenvalues of other graph matrices include Bapat [2010], Brouwer–Haemers [2012], and the broader Godsil–Royle [2001], which studies many algebraic topics in graph theory. We consider first the classical eigenvalues and then briefly discuss a modern alternative.

SPECTRA OF GRAPHS

We assume a basic acquaintance with eigenvalues of matrices and simply state several elementary properties.

15.3.1. Definition. The **eigenvalues** of a graph are the eigenvalues of its adjacency matrix A. These are the roots $\lambda_1, \ldots, \lambda_n$ of the **characteristic polynomial** ϕ given by $\phi(G; \lambda) = \det(\lambda I - A) = \prod_{i=1}^{n}(\lambda - \lambda_i)$. The **spectrum** is the list of distinct eigenvalues with their respective multiplicities $m_1, \ldots, m_t$; we write $\mathrm{Spec}\,(G) = \begin{pmatrix} \lambda_1 \cdots \lambda_t \\ m_1 \cdots m_t \end{pmatrix}$.

15.3.2. Remark. *Elementary properties of eigenvalues.*

(0) The eigenvalues are the values λ for which the square matrix $\lambda I - A$ is singular, which is equivalent to $\det(\lambda I - A) = 0$.

(1) $\sum \lambda_i = \mathrm{Trace}\, A$. The **trace** is the sum of the diagonal elements and is the negative of the coefficient of λ^{n-1} in $\det(\lambda I - A)$. Since $\det(\lambda I - A) = \prod_{i=1}^{n}(\lambda - \lambda_i)$, this also equals $\sum \lambda_i$. For graphs, it is 0.

(2) $\prod \lambda_i = (-1)^n \phi(G; 0) = \det A = \sum_\sigma \text{sign}(\sigma) \prod_{i=1}^n a_{i,\sigma(i)}$, where the sum runs over permutations σ of $[n]$.

(3) For a symmetric real n-by-n matrix A and $\lambda \in \mathbb{R}$, the multiplicity of λ as an eigenvalue of A is $n - \text{rank}(\lambda I - A)$.

(4) Adding c to the diagonal shifts the eigenvalues by c, since $\alpha + c$ is a root of $\det(\lambda I - (cI + A))$ if and only if α is a root of $\det(\lambda I - A)$.

15.3.3. Example. The adjacency matrix of K_n is $J - I$, where J is the matrix of all 1s. Hence the eigenvalues of K_n are 1 less than those of J. Since $\text{Spec } J = \begin{pmatrix} n & 0 \\ 1 & n-1 \end{pmatrix}$, we have $\text{Spec } K_n = \begin{pmatrix} n-1 & -1 \\ 1 & n-1 \end{pmatrix}$.

The adjacency matrix of $K_{m,n}$ has rank 2, so it has two nonzero eigenvalues λ_1, λ_2. The trace is 0, so $\lambda_1 = -\lambda_2$; call this constant b. Hence $\phi(K_{m,n}, \lambda) = \lambda^n - b^2 \lambda^{n-2}$. We compute b using $\phi(G; \lambda) = \det(\lambda I - A)$. Since λ appears only on the diagonal, contributions in the permutation expansion to the coefficient of λ^{n-2} arise only from permutations that use $n-2$ positions on the diagonal. The remaining two positions are $-a_{i,j}$ and $-a_{j,i}$ for some i, j. All mn nonzero contributions of this form are negative. Hence $b^2 = mn$, and $\text{Spec}(K_{m,n}) = \begin{pmatrix} \sqrt{mn} & 0 & -\sqrt{mn} \\ 1 & m+n-2 & 1 \end{pmatrix}$. ∎

We index the coefficients of ϕ so that $\phi(G; \lambda) = \sum_{i=0}^n c_i \lambda^{n-i}$. Since $\phi(G; \lambda) = \det(\lambda I - A)$, always $c_0 = 1$ and $c_1 = -\text{Trace } A = 0$.

The computation of c_2 for $K_{m,n}$ generalizes. Since contributions to $c_2 \lambda^{n-2}$ involve $n - 2$ factors of λ from the diagonal, c_2 is the sum of the principal 2-by-2 subdeterminants of $-A$. For a graph, the off-diagonal elements are both -1 when $v_i v_j \in E(G)$ and both 0 otherwise. Each permutation selecting these positions is odd. Thus summing over all vertex pairs yields $c_2 = -|E(G)|$. The other coefficients of $\phi(G; \lambda)$ can also be described in terms of subgraphs of G (Exercise 7).

Eigenvalues yield characterizations of bipartite graphs.

15.3.4. Proposition. The (i, j)th entry of A^k counts the v_i, v_j-walks of length k. The eigenvalues of A^k are $(\lambda_i)^k$.

Proof: The statement about walks holds easily by induction on k (Exercise 1). For the second statement, $Ax = \lambda x$ implies $A^k x = \lambda^k x$, by repeated multiplication. Since x is any eigenvector, multiplicities of eigenvalues for A and A^k are equal. ∎

15.3.5. Lemma. If G is bipartite and λ is an eigenvalue of G with multiplicity m, then $-\lambda$ is also an eigenvalue with multiplicity m.

Proof: We may assume that the two parts have equal size, since adding isolated vertices merely adds rows and columns of 0s to the adjacency matrix and adds zeros to the list of eigenvalues. This allows us to write the adjacency matrix $A(G)$ in the form $\begin{pmatrix} 0 & B \\ B^T & 0 \end{pmatrix}$, where B is square.

Let λ be an eigenvalue associated with eigenvector v. Write v in the form $\begin{pmatrix} x \\ y \end{pmatrix}$, partitioned via the bipartition of G. Now $\lambda v = Av = \begin{pmatrix} 0 & B \\ B^T & 0 \end{pmatrix}\begin{pmatrix} x \\ y \end{pmatrix} = \begin{pmatrix} By \\ B^T x \end{pmatrix}$. Hence $By = \lambda x$ and $B^T x = \lambda y$.

Let $v' = \begin{pmatrix} x \\ -y \end{pmatrix}$. Compute $Av' = \begin{pmatrix} B(-y) \\ B^T x \end{pmatrix} = \begin{pmatrix} -\lambda x \\ \lambda y \end{pmatrix} = -\lambda v'$. Hence v' is an eigenvector of A with eigenvalue $-\lambda$. Furthermore, m independent eigenvectors for λ yield m independent eigenvectors for $-\lambda$. Since the same argument holds for $-\lambda$, we conclude that $-\lambda$ is an eigenvalue with the same multiplicity as λ. ∎

15.3.6. Theorem. The following are equivalent for a graph G.
(A) G is bipartite.
(B) The nonzero eigenvalues occur in pairs λ_i, λ_j such that $\lambda_i = -\lambda_j$.
(C) $\phi(G; \lambda)$ or $\lambda\phi(G; \lambda)$ is a polynomial in λ^2.
(D) $\sum \lambda_i^{2t-1} = 0$ for every positive integer t.

Proof: We proved A$\Rightarrow$B in the lemma.

B $\Leftrightarrow$ C: We have $(\lambda - \lambda_i)(\lambda - \lambda_j) = (\lambda^2 - a)$ if and only if $\lambda_j = -\lambda_i$.

B $\Rightarrow$ D: If $\lambda_j = -\lambda_i$, then $\lambda_j^{2t-1} = -\lambda_i^{2t-1}$.

D $\Rightarrow$ A: Since $\sum \lambda_i^k$ counts the closed k-walks in the graph (from each vertex), D forbids closed walks of odd length. This forbids odd cycles, since an odd cycle is an odd closed walk, and hence G is bipartite.   ∎

Standard results from linear algebra are useful. For a real symmetric matrix of order n, the **Spectral Theorem** guarantees real eigenvalues and n orthonormal eigenvectors. Also, $x^T A x$ attains its maximum and minimum over unit vectors at eigenvectors of A, where it equals the corresponding eigenvalue. The **Cayley–Hamilton Theorem** states $\phi(A) = 0$. Finally, the **minimum polynomial** of a matrix A is the polynomial ψ of least degree such that $\psi(A) = 0$; it is given by $\psi(A) = \prod_i(\lambda - \mu_i)$, where $\mu_1, \ldots, \mu_r$ are the distinct eigenvalues of A.

EIGENVALUES AND GRAPH PARAMETERS

We first apply the minimum polynomial.

15.3.7. Theorem. The diameter of a graph G is less than the number of distinct eigenvalues of G.

Proof: The adjacency matrix A satisfies a polynomial of degree r if and only if some linear combination of $A^0, \ldots, A^r$ is 0. Since the number of distinct eigenvalues is the degree of the minimum polynomial, we need only show that $A^0, \ldots, A^k$ are linearly independent when $k \leq \text{diam}(G)$.

Finally, for $k \leq \text{diam}(G)$ choose $v_i, v_j \in V(G)$ such that $d(v_i, v_j) = k$. By counting walks, we have $A_{i,j}^k \neq 0$ but $A_{i,j}^t = 0$ for $t < k$. Hence A^k cannot be a linear combination of the smaller powers.   ∎

Let $\lambda_{\max}(G)$ and $\lambda_{\min}(G)$ denote the largest and smallest eigenvalues.

15.3.8. Lemma. If G' is an induced subgraph of G, then
$$\lambda_{\min}(G) \leq \lambda_{\min}(G') \leq \lambda_{\max}(G') \leq \lambda_{\max}(G).$$

Proof: The adjacency matrix A of G is real and symmetric, so $\lambda_{\min}(A) \leq x^T A x \leq \lambda_{\max}(A)$ for every unit vector x. Index $V(G)$ to put the adjacency matrix A' of G' in the upper left of A. Let z' be the unit eigenvector of A' such that $A'z' = \lambda_{\max}(G')z'$. Let z be the unit vector in $\mathbb{R}^n$ obtained by appending zeros to z'. Now
$$\lambda_{\max}(G') = z'^T A' z' = z^T A z \leq \lambda_{\max}(G).$$
Similarly, $\lambda_{\min}(G') \geq \lambda_{\min}(G)$.   ∎

Lemma 15.3.8 is a special case of the **Interlacing Theorem**, which we state without proof (it uses only linear algebra).

15.3.9. Theorem. Indexed in nonincreasing order, the eigenvalues $\lambda_1, \ldots, \lambda_n$ of a graph G and $\mu_1, \ldots, \mu_{n-1}$ of $G - x$ (for $x \in V(G)$) interlace as

$$\lambda_1 \geq \mu_1 \geq \lambda_2 \geq \cdots \geq \mu_{n-1} \geq \lambda_n. \qquad \blacksquare$$

15.3.10. Lemma. For an n-vertex graph G with m edges,

$$\delta(G) \leq \frac{2m}{n} \leq \lambda_{\max}(G) \leq \Delta(G).$$

Proof: Let x be an eigenvector for eigenvalue λ, and let x_j be the largest entry in x. With A being the adjacency matrix, $\lambda \leq \Delta(G)$ follows from

$$\lambda x_j = (Ax)_j = \sum_{v_i \in N(v_j)} x_i \leq d(v_j) x_j \leq \Delta(G) x_j .$$

For the lower bound, evaluate $x^T A x$ by letting x be the unit vector with equal coordinates, written as $\mathbf{1}_n / \sqrt{n}$, where every entry of $\mathbf{1}_n$ is 1. Since the sum of the entries in the adjacency matrix is twice $|E(G)|$,

$$\lambda_{\max} \geq \frac{\mathbf{1}_n^T}{\sqrt{n}} A \frac{\mathbf{1}_n}{\sqrt{n}} = \frac{1}{n} \sum \sum a_{ij} = \frac{2m}{n}. \qquad \blacksquare$$

Lemma 15.3.10 enables us to improve the trivial bound $\chi(G) \leq 1 + \Delta(G)$ given by the greedy coloring algorithm. Replacing $\Delta(G)$ with the average degree is too small; $K_n + K_1$ has chromatic number n and average degree less than $n - 1$. Since $\lambda_{\max}$ is always at least the average degree, $1 + \lambda_{\max}(G)$ has a chance to work and cannot be much improved.

15.3.11. Theorem. (Wilf [1967]) Always $\chi(G) \leq 1 + \lambda_{\max}(G)$.

Proof: Let $k = \chi(G)$. Let H be a minimal induced subgraph of G with chromatic number k. As in Proposition 8.2.2, $\delta(H) \geq k - 1$ (otherwise a proper coloring of a subgraph would extend to H). Since H is an induced subgraph of G, Lemma 15.3.10 and then Lemma 15.3.8 yield

$$k \leq 1 + \delta(H) \leq 1 + \lambda_{\max}(H) \leq 1 + \lambda_{\max}(G) . \qquad \blacksquare$$

The relationship between $\lambda_{\max}$ and the average degree $2m/n$ also aids in developing an eigenvalue proof of the upper bound in Mantel's Theorem, the case of Turán's Theorem (Theorem 5.2.11) for triangle-free graphs. The essence of this argument can be found in Bapat [2010, p. 74].

15.3.12. Theorem. Every n-vertex triangle-free graph has at most $n^2/4$ edges.

Proof: Let G be an n-vertex graph with m edges, t triangles, adjacency matrix A, and eigenvalues $\lambda_1, \ldots, \lambda_n$ in nonincreasing order.

The triangles in a graph are its 3-cycles, which are closed walks. Each cycle of length 3 through vertex v_i is counted twice in the ith diagonal entry of A^3;

hence Trace A^3 counts each triangle six times. The trace of a matrix is the sum of its eigenvalues, and the eigenvalues of A^k are $\lambda_1^k, \ldots, \lambda_n^k$. Hence

$$6t = \text{Trace } A^3 = \sum_{i=1}^n \lambda_i^3 = \lambda_1^3 + \sum_{i=2}^n \lambda_i^3.$$

We want to prove that the number of triangles is positive when $m > n^2/4$. For this it suffices to show that the assumption $m > n^2/4$ implies $\lambda_1^3 > \left| \sum_{i=2}^n \lambda_i^3 \right|$.

If $m > n^2/4$, then $\sqrt{m} > n/2$, and hence $\lambda_1 \geq 2m/n > \sqrt{m}$. Since the ith diagonal element of A^2 counts the edges incident to v_i,

$$2m = \text{Trace} A^2 = \sum_{i=1}^n \lambda_i^2 > m + \sum_{i=2}^n \lambda_i^2.$$

From this and $\lambda_1 > \sqrt{m}$ we obtain $\lambda_1^2 > m > \sum_{i=2}^n \lambda_i^2$. Thus λ_1 is the only eigenvalue of largest absolute value. Now we use the triangle inequality and these facts to compute

$$\left| \sum_{i=2}^n \lambda_i^3 \right| \leq \sum_{i=2}^n \left| \lambda_i^3 \right| = \sum_{i=2}^n |\lambda_i|^3 < \sum_{i=2}^n \lambda_1 |\lambda_i|^2 < \lambda_1 \lambda_1^2 = \lambda_1^3. \qquad \blacksquare$$

15.3.13.* Remark. Mantel's Theorem is just the beginning. Recall that $T_{n,r}$ denotes the r-partite Turán graph with n vertices (Definition 5.2.10), and $t_r(n) = |E(T_{n,r})|$. Nikiforov [2007] proved that if G is an n-vertex graph not containing K_{r+1}, then $\lambda_{\max}(G) \leq \lambda_{\max}(T_{n,r})$, with equality only when $G = T_{n,r}$. Using Lemma 15.3.10 and the fact that $|E(T_{n,r})| = \lfloor n\lambda_{\max}(T_{n,r})/2 \rfloor$, Turán's Theorem that $T_{n,r}$ is the unique such graph with the most edges is a corollary.

Nikiforov went further. Fixing $s \in \mathbb{N}$ and $c > 0$, the Erdős–Stone Theorem (Theorem 11.1.10) states that for sufficiently large n, every n-vertex graph with at least $t_r(n) + cn^2$ edges contains a complete $(r+1)$-partite graph with s vertices in each part. Bollobás–Erdős [1973] showed that exponentially large n suffices, or equivalently that one can guarantee s to be $\Theta(\log n)$.

Nikiforov [2009] proved a stronger version of these results using eigenvalues, showing for $r \geq 3$ that if $s = \lfloor (c/r^r)^r \log n \rfloor \geq 1$ and $t = \lceil n^{1-c^{r-1}} \rceil$, then every n-vertex graph with $\lambda_{\max} \geq (1 - 1/(r-1) + c)n$ contains a complete r-partite graph with $r - 1$ parts of size s and one part of size t. Note that $|E(G)| \geq t_{r-1}(n) + cn^2/2$ implies this condition on $\lambda_{\max}$. $\qquad \blacksquare$

Eigenvalues are also related to the number of complete bipartite graphs needed to decompose a graph. For convenience, here we use **biclique** to mean a complete bipartite graph. Because stars are bicliques, the number needed is at most the vertex cover number $\beta(G)$. Graphs with special structure may have more efficient decompositions. The spectrum provides a lower bound.

15.3.14. Theorem. For every graph G, the number of bicliques needed to partition $E(G)$ is at least $\max\{p, q\}$, where p and q count (including repetition) the positive and negative eigenvalues of the adjacency matrix A.

Proof: When G decomposes into subgraphs $G_1, \ldots, G_t$, we may write $A = \sum_{i=1}^t B_i$, where B_i is the adjacency matrix of the spanning subgraph of G with edge set $E(G_i)$. To write this algebraically, let $\{R_i, S_i\}$ be the bipartition of G_i, and for $U \subseteq V(G)$ let $\mathbf{1}_U$ denote the vector that is 1 in the positions for U and 0 elsewhere. Now $A = \sum_{i=1}^t (\mathbf{1}_{R_i} \mathbf{1}_{S_i}^T + \mathbf{1}_{S_i} \mathbf{1}_{R_i}^T)$.

Let W_0 be the orthogonal complement of the space spanned by $\mathbf{1}_{R_1}, \ldots, \mathbf{1}_{R_t}$; that is, $W_0 = \{x \in \mathbb{R}^n : x^T \mathbf{1}_{R_i} = 0 \text{ for } 1 \le i \le t\}$. Let W_+ be the space spanned by the eigenvectors of A associated with positive eigenvalues. Thus $x^T A x > 0$ for $x \in W_+$. In contrast, for $x \in W_0$,

$$x^T A x = \sum_{i=1}^{t} x^T \mathbf{1}_{R_i} \mathbf{1}_{S_i}^T x + \sum_{i=1}^{t} x^T \mathbf{1}_{S_i} \mathbf{1}_{R_i}^T x = 0.$$

Thus $W_0 \cap W_+ = \{0\}$, which yields $\dim W_0 + \dim W_+ \le n$. Vectors in W_0 satisfy t linear equations, so $\dim W_0 \ge n-t$. By definition, $\dim W_+ = p$. Thus $t \ge p$. Using instead of W_+ the space associated with the negative eigenvalues yields $t \ge q$. ∎

Sylvester's Law of Inertia states that for any symmetric real matrix Q with p positive eigenvalues and q negative eigenvalues, the number of products of linear functionals needed to express $x^t Q x$ is at least $\max\{p, q\}$. The special case above for biclique decomposition is directly tailored to the graph application.

The biclique decomposition problem arose with a network addressing motivation in Graham–Pollak [1971], where the focus was especially on decomposing K_n. They attribute Theorem 15.3.14 to H.S. Witsenhausen. The proof given here is from Gregory & Vander Meulen [1996] (see also Brouwer–Haemers [2012]). For other discussions of the lower bound, see Tverberg [1982], Peck [1984], Reznick–Tiwari–West [1985]. A conjecture of Erdős from the 1980s that a random n-vertex graph decomposes into no fewer than $n - \alpha(G)$ bicliques was disproved in Alon [2015] and more strongly in Alon–Bohman–Huang [2017].

15.3.15. Example. *Biclique decompositions.* When $G = K_n$, there are $n - 1$ negative eigenvalues. Using stars centered at the complement of an independent set always yields an upper bound of $n - \alpha(G)$, so the eigenvalue bound here is sharp. Kratzke–Reznick–West [1988] studies other such graphs. For K_n, no short purely combinatorial proof is known that $n-1$ bicliques are needed. If a smallest decomposition must use a spanning biclique, then the rest follows by induction.

Next consider $G = C_m \,\square\, C_n$. There are simple formulas for the eigenvalues of a cycle (Exercise 12) and for the behavior of eigenvalues under cartesian product (Exercise 13). From these, $C_{(2t+1)n} \,\square\, C_n$ for odd n has $(2t + 1)(n^2 + 1)/2$ positive eigenvalues and $(2t + 1)(n^2 - 1)/2$ negative eigenvalues (0 is not an eigenvalue).

Furthermore, this product decomposes into $(2t + 1)(n^2 + 1)/2$ bicliques, using $(2t + 1)(n - 1)/2$ copies of $K_{2,2}$ and $(2t + 1)(n + 1)/2$ stars (Kratzke–West). All bicliques in $C_m \,\square\, C_n$ are 4-cycles or stars; see $C_{15} \,\square\, C_5$ below. Edges wrap from top to bottom and right to left, and all grid points are vertices. The dots indicate centers of stars in the decomposition; the circles indicate 4-cycles in the decomposition. ∎

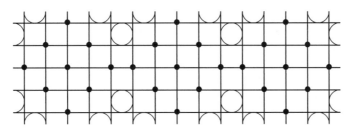

REGULAR AND STRONGLY REGULAR GRAPHS

Like bipartite graphs, regular graphs have spectral characterizations. The n-vector $\mathbf{1}_n$ with all coordinates 1 and the all-1 matrix J play important roles in arguments involving eigenvalues of regular graphs.

15.3.16. Theorem. The eigenvalue of G with largest absolute value is $\Delta(G)$ if and only if some component of G is $\Delta(G)$-regular. The multiplicity of $\Delta(G)$ as an eigenvalue is the number of such components.

Proof: Let A be the adjacency matrix. Since the spectrum of a graph is the union of the spectra of its components, we may assume that G is connected.

The ith entry of $A\mathbf{1}_n$ is $d(v_i)$. When G is k-regular, $A\mathbf{1}_n = k\mathbf{1}_n$, so k is an eigenvalue with eigenvector $\mathbf{1}_n$.

By Lemma 15.3.10, no eigenvalue exceeds $\Delta(G)$. For the lower bound, let x be an eigenvector for eigenvalue λ, with x_j an entry of largest absolute value. Since $-x$ is also an eigenvector for λ, we may assume $x_j \geq 0$. We obtain $|\lambda| \leq \Delta(G)$ from

$$|\lambda x_j| = |(Ax)_j| = \Big| \sum_{v_i \in N(v_j)} x_i \Big| \leq d(v_j)x_j \leq \Delta(G)x_j .$$

Now suppose $\lambda = \Delta(G)$. Removing the absolute value signs, equality requires $d(v_j) = \Delta(G)$ and $x_i = x_j$ for all $v_i \in N(v_j)$. Iterating the argument reaches all coordinates, since G is connected. Hence the eigenvalue associated with x equals $\Delta(G)$ only if G is regular and x is constant. ∎

When G is connected and not regular, eigenvalues of largest absolute value still have multiplicity 1. For non-regular graphs, we will see in the next subsection that the analogous results are much simpler when we modify the adjacency matrix to take the vertex degrees into account. Meanwhile, we study the structure of regular graphs. Powers of the adjacency matrix yield another characterization of regular graphs.

15.3.17. Theorem. (Hoffman [1963]) A graph G with adjacency matrix A is regular and connected if and only if J is a linear combination of powers of A.

Proof: *Sufficiency.* If J can be so expressed, then for each (i,j) we have $(A^k)_{i,j} \neq 0$ for some nonnegative k, which requires a v_i, v_j-walk of length k. Hence G is connected. For regularity, consider JA and AJ. The entry in position (i,j) of AJ is $d(v_i)$ (constant on rows), and in JA it is $d(v_j)$ (constant on columns). Since J is a linear combination of powers of A, each of which commutes with A, we have $JA = AJ$. Thus the entry in position (i,j) is both $d(v_i)$ and $d(v_j)$, and G is regular.

Necessity. Let G be k-regular, so k is an eigenvalue. Hence the minimum polynomial $\psi(G; \lambda)$ is $(\lambda - k)g(\lambda)$ for some polynomial g. Since $\psi(G; A) = 0$, we have $Ag(A) = kg(A)$. Hence each column of $g(A)$ is an eigenvector of A with eigenvalue k. As shown in the proof of Theorem 15.3.16, an eigenvector of a connected regular graph associated with eigenvalue $\Delta(G)$ must have all coordinates equal. However, $g(A)$ is a linear combination of powers of a symmetric matrix and must itself be symmetric. Hence the columns are equal, and $g(A)$ is a multiple of J. ∎

The independence number is bounded in terms of the eigenvalues. Haemers [1980] generalized this to all graphs, proving $\alpha(G) \leq \frac{-\lambda_n}{(\delta(G)^2/\lambda_1) - \lambda_n} n$, but we present only the special case for k-regular graphs, where $\delta(G) = \Delta(G) = \lambda_1 = k$.

15.3.18. Theorem. (A. Hoffman) If G is a k-regular graph with n vertices, then $\alpha(G) \leq \frac{-\lambda_n}{k - \lambda_n} n$. Consequently, $\chi(G) \geq \frac{n}{\alpha(G)} \geq 1 + \frac{k}{-\lambda_n}$.

Proof: The eigenvectors for the adjacency matrix A and for $A - \lambda_n I$ are the same, and the eigenvalues shift by λ_n. In particular, $(A - \lambda_n I)\mathbf{1}_n = (k - \lambda_n)\mathbf{1}_n$. This yields $M\mathbf{1}_n = 0$, where $M = A - \lambda_n I - (k - \lambda_n)J$. The other eigenvalues of A have corresponding eigenvectors that are orthogonal to $\mathbf{1}_n$ and hence in the nullspace of J. Thus the eigenvalues of M are the same as for $A - \lambda_n I$; they are nonnegative.

Let x be the incidence vector of an independent set of size s. Since $x^T A x = \sum_{i,j} x_i a_{i,j} x_j$, we have $x^T A x = 0$. Using the Spectral Theorem,

$$0 \leq x^T M x = x^T A x - \lambda_n x^T I x - \frac{k - \lambda_n}{n} x^T J x = -\lambda_n s - \frac{k - \lambda_n}{n} s^2,$$

which yields the desired inequality. ∎

We study a special class with stronger regularity properties.

15.3.19. Definition. An n-vertex graph G is **strongly regular** with parameters (k, λ, μ) if G is k-regular, every two adjacent vertices have λ common neighbors, and every two nonadjacent vertices have μ common neighbors. We then say that G is (k, λ, μ)-*strongly regular*.

15.3.20. Example. The graph mK_n is a strongly regular graph with parameters $(n - 1, n - 2, 0)$. Its complement, the complete m-partite graph $K_{n,\ldots,n}$, is $(mn - n, mn - 2n, mn - n)$-strongly regular. Indeed, the complement of any strongly regular graph is strongly regular (Exercise 23).

The Petersen graph is $(3, 0, 1)$-strongly regular. The cartesian product $K_n \,\square\, K_n$ is $(2n - 2, n - 2, 2)$-strongly regular. Strongly regular graphs with $\lambda = \mu$ can be obtained from symmetric designs (13). ∎

Counting arguments and then eigenvalue arguments greatly restrict the possibilities for the parameters (k, λ, μ) of a strongly regular graph.

15.3.21. Proposition. If G is a (k, λ, μ)-strongly regular graph with n vertices, then $k(k - \lambda - 1) = \mu(n - k - 1)$.

Proof: We count induced copies of P_3 having v as an endpoint. The middle vertex w can be picked in k ways. For each such w, the third vertex can be any neighbor of w not adjacent to v. With v unavailable, there are $k - \lambda - 1$ ways to pick it. On the other hand, the other endpoint can be picked in $n - k - 1$ ways as a nonneighbor of v, and each such choice has μ common neighbors with v that can serve as w. ∎

15.3.22. Remark. *Every strongly regular graph with parameters $\mu = 0$ or $\lambda = k - 1$ or $k = n - 1$ has the form mK_{k+1}. By Proposition 15.3.21, $\lambda = k - 1$ if and only if $\mu = 0$ or $k = n - 1$. If $\lambda = k - 1$, then every neighbor of v is adjacent to every other, which forces the claimed structure.* ∎

Henceforth, we assume $\mu > 0$ and $\lambda < k-1$. Next we compute the eigenvalues in terms of the parameters.

15.3.23. Lemma. If G is (k, λ, μ)-strongly regular, then the eigenvalues of G are k and $\frac{1}{2}\left(\lambda - \mu \pm \sqrt{(\lambda - \mu)^2 + 4(k - \mu)}\right)$.

Proof: Let A be the adjacency matrix. Since positions in A^2 count walks of length 2, the ijth entry of A^2 is k if $i = j$, is λ if $v_i v_j \in E(G)$, and is μ if $v_i v_j \notin E(G)$. Since $J - I - A$ is the adjacency matrix of $\overline{G}$, we obtain $A^2 = kI + \lambda A + \mu(J - I - A)$. Rearranging terms yields

$$A^2 = (k - \mu)I + (\lambda - \mu)A + \mu J.$$

Since G is k-regular, G has eigenvalue k with eigenvector $\mathbf{1}$. Let x be another eigenvector, with eigenvalue θ. Since $x \cdot \mathbf{1} = 0$, we have $Jx = \mathbf{0}$. Computing $A^2 x$ thus produces $\theta^2 x = (k - \mu)x + (\lambda - \mu)\theta x$.

This yields the quadratic equation $\theta^2 - (\lambda - \mu)\theta - (k - \mu) = 0$ for θ. Its solutions r and s are the only possible values for the eigenvalues other than k. The stated values arise from the Quadratic Formula. ∎

The argument above can also be traced in the opposite direction: a k-regular connected graph G is strongly regular with parameters k, λ, μ if and only if it has exactly three eigenvalues $k > r > s$ and these satisfy $r + s = \lambda - \mu$ and $rs = -(k - \mu)$.

Next we obtain a further necessary condition for the parameters of a strongly regular graph, again using that the multiplicities of the eigenvalues are integers.

15.3.24. Theorem. (Integrality Condition) If G is strongly regular with n vertices and parameters (k, λ, μ), then the following are nonnegative integers:

$$\frac{1}{2}\left(n - 1 \pm \frac{(n - 1)(\mu - \lambda) - 2k}{\sqrt{(\mu - \lambda)^2 + 4(k - \mu)}}\right).$$

Proof: We show that these numbers are the multiplicities a and b of the eigenvalues r and s in Lemma 15.3.23. Remark 15.3.22 describes all cases when $\mu = 0$. Hence we may assume $\mu > 0$, and thus G is connected.

Because G is connected, eigenvalue k has multiplicity 1, and $1 + a + b = n$. Also the eigenvalues sum to 0, so $k + ra + sb = 0$. The solution to these two linear equations is $a = -\frac{k+s(n-1)}{r-s}$ and $b = \frac{k+r(n-1)}{r-s}$. Letting r and s be the numbers computed in Lemma 15.3.23 yields the claimed formulas as the multiplicities. ∎

15.3.25. Example. *Classes of strongly regular graphs.* Consider the two cases $(n - 1)(\mu - \lambda) = 2k$ and $(n - 1)(\mu - \lambda) \neq 2k$.

Excluding trivial values, in the first case $\mu = \lambda + 1$, since $0 < 2k < 2n - 2$, and hence $k = (n - 1)/2$. By Exercise 23, G and $\overline{G}$ are thus strongly regular with the same parameters. Using $\mu = \lambda + 1$ and $k = (n - 1)/2$, Proposition 15.3.21 yields $n = 4\mu + 1$. Furthermore, the eigenvalues r and s have the same multiplicity.

In the second case, rationality requires $(\mu - \lambda)^2 + 4(k - \mu) = d^2$ for some $d \in \mathbb{N}$, and d must divide $(n - 1)(\mu - \lambda) - 2k$. By Lemma 15.3.23, the eigenvalues are integer multiples of $1/2$. For $(\lambda, \mu) = (0, 2)$, three such graphs are known, but it is not known whether the list is finite! The known examples, with parameters (n, k, λ, μ), are C_4 $(4, 2, 0, 2)$, the Clebsch graph $(16, 5, 0, 2)$ (see Exercise 24), and the Gewirtz graph $(56, 10, 0, 2)$ (see Cameron–van Lint [1991], p. 43). Other strongly regular graphs appear in Exercises 25–27. ∎

The most famous application of the Integrality Condition is the nonexistence of certain graphs. Recall that a graph with maximum degree k and diameter d has at most $1 + k \sum_{i=1}^{d} d^{i-1}$ vertices (Proposition 5.4.12); graphs satisfying equality are **Moore graphs**. Damerell [1973] and Bannai–Ito [1973] independently proved that no Moore graph exists with diameter at least 3. The Integrality Condition eliminates almost all possibilities when $d = 2$.

15.3.26. Theorem. (Hoffman–Singleton [1960]) If a k-regular graph G with diameter 2 has $k^2 + 1$ vertices, then $k \in \{2, 3, 7, 57\}$.

Proof: Reaching k^2 other vertices from any one vertex in two steps forbids triangles and 4-cycles. Hence G is $(k, 0, 1)$-strongly regular.

From Theorem 15.3.24, we conclude that $\frac{1}{2}\left(k^2 \pm k(k-2)/\sqrt{4k-3} \right)$ are nonnegative integers. Let $t = \sqrt{4k-3}$. Since $4k - 3$ is a positive integer, also t is a positive integer, or t is irrational. If t is irrational, then $k(k-2)$ must vanish, which is the case where $G = C_5$.

Hence t is a positive integer, and $k = (t^2 + 3)/4$. Since the multiplicities of the eigenvalues are distinct nonnegative integers, their difference is an integer. Its value is $k(k-2)/t$, which equals $(t^4 - 2t^2 - 15)/(16t)$. Letting u be this integer, $t^4 - 2t^2 - 16ut = 15$. Since t divides the left side, it must also divide 15.

Hence $t \in \{1, 3, 5, 15\}$, which leaves for k the possibilities $\{1, 3, 7, 57\}$ (and earlier 2). For $k = 1$, the diameter is 1, and K_2 is a degenerate Moore graph. We restrict to $k \in \{2, 3, 7, 57\}$. The first three give known graphs: C_5, the Petersen graph, and the Hoffman–Singleton graph (Exercise 5.4.59), but it is unknown whether there exists such a graph with degree 57. ∎

Another classical application is the "Friendship Theorem": at any party where every two people have exactly one common acquaintance, some person knows everyone (presumably the host). The resulting graph of the acquaintance relation consists of some number of triangles sharing a vertex.

The symmetry of the condition suggests that G might be regular. The Integrality Condition excludes this; the non-regular case is combinatorial.

15.3.27. Theorem. (**Friendship Theorem**; Higman (see Wilf [1971])) If G is a graph in which any two distinct vertices have exactly one common neighbor, then G has a vertex joined to all others.

Proof: If G is regular, then it is strongly regular with $\lambda = \mu = 1$. By Theorem 15.3.24, $\frac{1}{2}(n - 1 \pm k/\sqrt{k-1})$ are integers. Hence $k/\sqrt{k-1}$ is an integer, which requires $k = 2$. However, K_3 is the only 2-regular graph satisfying the condition, and it does have vertices of degree $n - 1$.

Now suppose that G is not regular. We show first that $v \leftrightarrow w$ requires $d(v) = d(w)$. Since each vertex in $N(v)$ does not equal w, it has a unique neighbor in $N(w)$, since w is in no 4-cycle. Similarly, each vertex of $N(w)$ has a unique neighbor in $N(v)$. Hence $N(v)$ and $N(w)$ are joined by a matching, yielding $|N(v)| = |N(w)|$.

Since G is not regular, it has vertices v and w with $d(w) \neq d(v)$. By the preceding paragraph, $vw \in E(G)$. Let u be their common neighbor. Since u cannot have the same degree as both, we may assume $d(u) \neq d(v)$. If G has a vertex $x \notin N(v)$, then $d(x) = d(v)$, but this requires $w, u \in N(x)$. This creates the 4-cycle v, u, x, w. Hence $d(v) = n - 1$.   ∎

15.3.28. Remark. Such a combinatorial statement should have a short combinatorial proof. An early simple proof was in Longyear–Parsons [1972], and another found in 1975 was not published until Huneke [2002]; we present the latter.

In a k-regular n-vertex graph G where every two vertices have exactly one common neighbor, exactly one walk of length 2 from v reaches each vertex of G, and hence $n = k(k-1) + 1$. Now count closed walks of length p from each vertex. When p is prime, the total number of such walks is divisible by p, since each walk is counted p times (repeated vertices cannot introduce periodicity).

Fixing a vertex v, let $f(j)$ count the closed v, v-walks of length j. Of these, $kf(j-2)$ visit v two steps before the end. The total number of walks of length $j - 2$ from v is k^{j-2}, since G is k-regular. Each one not ending at v extends to a v, v-walk of length j in exactly one way, yielding $k^{j-2} - f(j-2)$ walks. Hence $f(j)$ equals $k^{j-2} + (k-1)f(j-2)$, with $f(0) = 1$.

Letting p be a prime divisor of $k - 1$, we have $f(p) \equiv 1 \pmod{p}$. Summing over v yields $nf(p)$ as the total number of closed walks of length p. Since $n = k(k-1)+1 \equiv 1 \pmod{p}$, we have $nf(p) \equiv 1 \pmod{p}$, but we observed that the total number of these walks is divisible by p.   ∎

LAPLACIAN EIGENVALUES

Since the powers of the adjacency matrix count walks of fixed lengths, eigenvalues are related to distance and diameter. Eigenvalues of regular graphs are also related to vertex degrees. Some of those results extend to non-regular graphs (using essentially the same proofs) when the adjacency matrix is replaced with a matrix that adjusts for the differences in vertex degrees. The "Laplacian matrix" achieves this. A thorough treatment of the resulting "Lapacian eigenvalues" appears in Chung [1997]; see also Grone–Merris–Sunder [1990], Mohar [1991, 1992], Mohar–Poljak [1993], Merris [1994, 1995], and Grone–Merris [1994].

15.3.29. Definition. The **Laplacian matrix** of a graph G is the matrix $D - A$, where A is the adjacency matrix of G and D is the diagonal matrix whose diagonal entry in row i is $d_G(v_i)$. Let Q denote the Laplacian matrix of G. The eigenvalues of Q form the **Laplacian spectrum** of G.

Having used $\lambda_1, \ldots, \lambda_n$ (nonincreasing) for the ordinary eigenvalues, we use $\mu_1, \ldots, \mu_n$ (nondecreasing) for the Laplacian eigenvalues. Since each row of Q has sum 0, the constant vector $\mathbf{1}_n$ is a eigenvector, with eigenvalue 0. That is, 0 is for Laplacian spectra much as $\Delta(G)$ is for ordinary spectra of regular graphs.

15.3.30. Theorem. For the Laplacian spectrum of an n-vertex graph G,
   (a) The smallest Laplacian eigenvalue of G is 0, with multiplicity 1 if and only if G is connected.
   (b) If G is k-regular, then μ is a Laplacian eigenvalue if and only if $k - \mu$ is an ordinary eigenvalue of G, with the same multiplicity.
   (c) If $(0, \mu_2, \ldots, \mu_n)$ is the Laplacian spectrum of G (in nondecreasing order), then $(0, n - \mu_n, \ldots, n - \mu_2\}$ is the Laplacian spectrum of $\overline{G}$.

Proof: Exercise 29.   ∎

Any result involving the Laplacian eigenvalues of a k-regular graph can be restated in terms of its ordinary eigenvalues by subtracting the Laplacian eigenvalues from k, but many results are more natural using the Laplacian.

The Matrix Tree Theorem (Theorem 15.2.5) computes the number of spanning trees $\tau(G)$ of a graph G as the common cofactor of each entry in the Laplacian matrix. Hence it is not surprising that the Matrix Tree Theorem can be restated in terms of the Laplacian eigenvalues. Recall the definition and properties of the adjugate matrix from Definition 15.2.2 and Remark 15.2.3.

15.3.31. Lemma. For a graph G with Laplacian matrix Q, the number of spanning trees is $\det(J + Q)/n^2$.

Proof: The Matrix Tree Theorem states $\mathrm{Adj}\, Q = \tau(G)J$. Since the Laplacian of K_n is $nI - J$, Cayley's Formula yields $\mathrm{Adj}\,(nI - J) = n^{n-2}J$. Since $J^2 = nJ$ and $JQ = 0$, we have $(nI - J)(J + Q) = nQ$. Thus

$$(n^{n-2}J)\mathrm{Adj}\,(J + Q) = \mathrm{Adj}\,(nI - J)\,\mathrm{Adj}\,(J + Q) = \mathrm{Adj}\,[(nI - J)(J + Q)]$$
$$= \mathrm{Adj}\,(nQ) = n^{n-1}\mathrm{Adj}\, Q.$$

Canceling n^{n-2} yields $J\mathrm{Adj}\,(J + Q) = n\tau(G)J$. By Remark 15.2.3, multiplying both sides on the right by $(J + Q)$ yields $J(\det(J + Q)I) = n\tau(G)nJ$. Both sides are multiples of J, so the desired equality holds.   ∎

We now compute $\tau(G)$ from the Laplacian eigenvalues.

15.3.32. Theorem. (Kelmans [1967]; see also Kelmans–Chelnokov [1974]) The number of spanning trees in a graph whose Laplacian eigenvalues are $\mu_1, \ldots, \mu_n$ (in nondecreasing order) is $\frac{1}{n} \prod_{i=2}^{n} \mu_i$.

Proof: As mentioned in Remark 15.3.2(2), the determinant of a matrix is the product of its eigenvalues. In light of Lemma 15.3.31, we seek the eigenvalues of $J + Q$. Since $(J + Q)\mathbf{1}_n = J\mathbf{1}_n + Q\mathbf{1}_n = n\mathbf{1}_n$, we have n as an eigenvalue in place of 0. Since $J + Q$ is real and symmetric, we may take all other eigenvectors to be orthogonal to $\mathbf{1}_n$, just as they are for Q. Letting x be an eigenvector associated with eigenvalue μ_i of Q for $i > 1$, we have $(J + Q)x = Jx + Qx = Qx = \mu_i x$. Hence the eigenvalues of $J + Q$ are the eigenvalues of Q, except for replacing 0 with n. By Lemma 15.3.31, we now have $\tau(G) = \det(J + Q)/n^2 = \frac{1}{n} \prod_{i=2}^{n} \mu_i$.   ∎

15.3.33. Example. The graphs K_n and $K_{p,p}$ are regular, so by Theorem 15.3.30 the Laplacian eigenvalues are the vertex degree minus the ordinary eigenvalues. For K_n, all Laplacian eigenvalues other than the smallest equal n, and Theorem 15.3.32 reduces to Cayley's Formula, $\tau(K_n) = n^{n-2}$. The Laplacian spectrum of $K_{p,p}$ has 0, one copy of $2p$, and $2p-2$ copies of p, so $\tau(K_{p,p}) = \frac{1}{2p}2p^{2p-1} = p^{2p-2}$. ∎

Other methods for counting spanning trees appear in Kelmans [1965/66] and Hartsfield–Kelmans–Shen [1996].

The second smallest Laplacian eigenvalue has important structural aspects. The larger it is, the more highly connected the graph is. We have seen that $\mu_2(G)$ is nonzero if and only if G is connected; this is a special case of $\kappa(G) \geq \mu_2(G)$ (Corollary 15.3.36). Since it bounds the connectivity from below, the second smallest Laplacian eigenvalue is called the **algebraic connectivity**.

15.3.34. Lemma. If G_1 and G_2 are edge-disjoint graphs with the same vertex set, then $\mu_2(G_1) + \mu_2(G_2) \leq \mu_2(G_1 \cup G_2)$.

Proof: Let $G = G_1 \cup G_2$. With Q, Q_1, Q_2 denoting the Laplacian matrices of the three graphs, respectively, we have $Q = Q_1 + Q_2$.

The Laplacian matrix is a real symmetric matrix. By the Spectral Theorem, $\mu_1(G)$ is the minimum of $x^T Q x$ over unit vectors. Its value is 0, using eigenvector **1**. Since a real symmetric matrix has a full set of orthonormal eigenvectors, $\mu_2(G)$ is the minimum of $x^T Q x$ over the set C of unit vectors orthogonal to **1**.

Since $x^T(A + B)x = x^T A x + x^T B x$ and $Q = Q_1 + Q_2$,

$$\mu_2(G) = \min_{x \in C}(x^T Q_1 x + x^T Q_2 x) \geq \min_{x \in C} x^T Q_1 x + \min_{x \in C} x^T Q_2 x = \mu_2(G_1) + \mu_2(G_2). \quad ∎$$

15.3.35. Theorem. If S is a set of vertices in a graph G such that $|S| \leq |V(G)|-2$, then $\mu_2(G - S) \geq \mu_2(G) - |S|$.

Proof: Let $H = G - v$, where $v \in V(G)$. It suffices to show $\mu_2(H) \geq \mu_2(G) - 1$. Let $G' = H \oplus K_1$. Let Q and Q' be the Laplacian matrices of H and G', respectively. Let x be an eigenvector for Q that is associated with eigenvalue $\mu_2(H)$ and satisfies $x^T \mathbf{1} = 0$. With G' having n vertices, block matrix multiplication yields

$$Q'\begin{pmatrix} x \\ 0 \end{pmatrix} = \begin{pmatrix} Q + I & -\mathbf{1} \\ -\mathbf{1}^T & n-1 \end{pmatrix}\begin{pmatrix} x \\ 0 \end{pmatrix} = (\mu_2(H) + 1)\begin{pmatrix} x \\ 0 \end{pmatrix}.$$

Thus $\mu_2(H) + 1$ is a Laplacian eigenvalue of G', which yields $\mu_2(G') \leq \mu_2(H) + 1$. Since all Laplacian eigenvalues are nonnegative, and G' arises from G by adding edges, Lemma 15.3.34 yields $\mu_2(G) \leq \mu_2(G')$. ∎

15.3.36. Corollary. (Fiedler [1973]) For a non-complete graph G, connectivity is bounded below by algebraic connectivity. That is, $\kappa(G) \geq \mu_2(G)$. ∎

The bound fails for K_n, since $\mu_2(K_n) = n$. It holds with equality for non-complete cographs (Exercise 30). de Abreu [2007] surveyed results on $\mu_2(G)$. Meanwhile, the algebraic connectivity also yields an important lower bound for the density of edge cuts.

15.3.37. Theorem. (Alon–Milman [1985]) Let G be an n-vertex graph. If S is a nonempty proper subset of $V(G)$, then

$$\frac{\left|[S,\overline{S}]\right|}{|S|\,|\overline{S}|} \geq \frac{\mu_2(G)}{n}.$$

Proof: The claim is trivial if $\mu_2(G) = 0$, so by Theorem 15.3.30 we may assume that G is connected. With Laplacian Q, the terms of $x^T Q x$ include $-2x_i x_j$ when $v_i v_j \in E(G)$ and x_i^2 with coefficient $d(v_i)$. Hence

$$x^T Q x \;=\; \sum_{v_i v_j \in E(G)} (x_i^2 - 2x_i x_j + x_j^2) \;=\; \sum_{v_i v_j \in E(G)} (x_i - x_j)^2.$$

We choose x so that edges crossing the cut contribute positively. Let $s = |S|$, and set $x_i = -(n - s)$ for $i \in S$ and $x_i = s$ for $i \notin S$. The sum on the right becomes $n^2 \left|[S,\overline{S}]\right|$. Our choice of x yields $\sum x_i = (n - s)s - s(n - s) = 0$, so $x \cdot \mathbf{1} = 0$. Since $\mathbf{1}$ is an eigenvector for $\mu_1(G)$, the Spectral Theorem yields $\frac{x^T Q x}{x^T x} \geq \mu_2(G)$. Hence

$$x^T Q x \;\geq\; \mu_2(G) x^T x \;=\; \mu_2(G)\big(s(n-s)^2 + (n-s)s^2\big) \;=\; \mu_2(G)s(n-s)n.$$

Since $x^T Q x = n^2 \left|[S,\overline{S}]\right|$, we have $\left|[S,\overline{S}]\right| \geq \mu_2(G)s(n-s)/n$. ∎

15.3.38. Corollary. If G is an n-vertex graph with maximum degree k, and S is a nonempty vertex subset with $|S| \leq n/2$, then at least $(\mu_2(G)/2k)|S|$ vertices of $\overline{S}$ have neighbors in S.

Proof: Theorem 15.3.37 yields $\left|[S,\overline{S}]\right| \geq \mu_2(G)|S|\,|\overline{S}|/n$. Each vertex of $\overline{S}$ receives at most k of these edges, so S has at least $\mu_2(G)|S|\,|\overline{S}|/(nk)$ neighbors in $\overline{S}$. Since $|\overline{S}|/n \geq 1/2$, the claim follows. ∎

When applied to k-regular graphs, these results can be stated using ordinary eigenvalues by substituting $k - \lambda_2$ for μ_2. Corollary 15.3.38 states that an n-vertex graph having maximum degree k and second Laplacian eigenvalue μ is an (n, k, c)-magnifier with $c = \mu/(2k)$, as defined below.

15.3.39.* Definition. An (n, k, c)-**magnifier** is an n-vertex graph G such that $\Delta(G) \leq k$ and that $\left|N(S) \cap \overline{S}\right| \geq c \cdot |S|$ for every $S \subseteq V(G)$ with $|S| \leq n/2$. An (n, k, c)-**expander** is an X, Y-bigraph G with $|X| = |Y| = n$ such that $\Delta(G) \leq k$ and that $|N(S)| \geq (1 + c(1 - |S|/n)) \cdot |S|$ for every $S \subseteq X$ with $|S| \leq n/2$.

15.3.40.* Remark. Given an (n, k, c)-magnifier G, construct an X, Y-bigraph H by letting X and Y be copies of $V(G)$, putting $x_i y_j \in E(H)$ if and only if $v_i v_j \in E(G)$, and adding $x_i y_i$ for all i. The result is an $(n, k + 1, c)$-expander (Exercise 5). The expansion condition strengthens Hall's Condition. Expanders appear in the parallel sorting network of Ajtai–Komlós–Szemerédi [1983]. Walters [1996] collects other definitions that have been used to measure expansion properties.

Probabilistic methods (Exercise 31) yield *existence* of large expanders with bounded average degree (Pinsker [1973], Pippenger [1977], Chung [1978b]). Margulis [1973] built explicit examples algebraically (see also Gabber–Galil [1981]).

Random graphs generally are good expanders, but measuring expansion is hard. Tanner [1984] and Alon–Milman [1984, 1985] independently proved good expansion properties for graphs whose two largest ordinary eigenvalues (or two smallest Laplacian eigenvalues) differ greatly. Since eigenvalues are easy to approximate, this yields a good test for expansion.

For k-regular expanders, Alon–Milman [1984, 1985] showed $c \geq \frac{2k-2\lambda_2}{3k-2\lambda_2}$. Alon [1986] proved a partial converse: if a k-regular graph G is an (n, k, c)-magnifier, then the eigenvalue separation $k - \lambda_2$ is at least $\frac{c^2}{4+2c^2}$. Explicit constructions of regular graphs are known with $\lambda_1 - \lambda_2$ nearly as large as possible. The second largest eigenvalue of a k-regular graph with diameter d is at least $2\sqrt{k-1}(1-O(1/d))$ (see Nilli [1991]). Lubotzky–Phillips–Sarnak [1986] and Margulis [1988] constructed infinite families of $(p+1)$-regular graphs, where p is a prime congruent to 1 mod 4, with second largest eigenvalue at most $2\sqrt{p}$. See Alon–Spencer [1992, pp. 119–125]. ∎

EXERCISES 15.3

15.3.1. (–) Let A^k be the kth power of the adjacency matrix of a graph G. Prove that the value in position (i, j) of A^k is the number of v_i, v_j-walks of length k in G.

15.3.2. (–) Prove that if a graph G has s vertices with identical neighborhoods, then 0 is an eigenvalue of G with multiplicity at least $s - 1$. Prove that if G has s vertices with identical closed neighborhoods, then -1 is an eigenvalue with multiplicity at least $s - 1$.

15.3.3. (–) Obtain the eigenvalues of G^2 in terms of the eigenvalues of G, where G^2 is obtained from G by adding the edges xy such that $d_G(x, y) = 2$.

15.3.4. (–) Let F be a cartesian product where any two nonadjacent vertices have exactly two common neighbors. Prove that both factors are complete graphs.

15.3.5. (–) Let G be an (n, k, c)-magnifier with vertices $v_1, \ldots, v_n$. Define an X, Y-bigraph H with $X = \{x_1, \ldots, x_n\}$ and $Y = \{y_1, \ldots, y_n\}$ by $x_i y_j \in E(H)$ if and only if $i = j$ or $v_i v_j \in E(G)$. Prove that H is an $(n, k + 1, c)$-expander.

15.3.6. Let σ_k be the number of k-cycles in G. Let $L_k = \sum \lambda_i^k$ and $D_k = \sum d_i^k$, summing the kth powers of the eigenvalues and the vertex degrees, respectively. Obtain formulas for σ_3 and σ_4 in terms of $\{L_k\}$ and $\{D_k\}$.

15.3.7. (◊) *Coefficients of the characteristic polynomial* $\sum_{i=0}^{n-1} c_i \lambda^{n-i}$.
 (a) Let **H** be the set of spanning subgraphs of a graph G such that every component is an edge or a cycle. Let $k(H)$ and $s(H)$ denote the number of components of H and the number of components of H that are cycles, respectively. Prove that $\det A(G) = \sum_{H \in \mathbf{H}}(-1)^{|V(H)|-k(H)}2^{s(H)}$. (Harary [1962b])
 (b) Prove that $c_i = (-1)^i \sum_{|S|=i} \det A(G[S])$. (Hence $c_3 = -2t$ when G has t triangles.)
 (c) Let $\mathbf{H}_i$ be the set of i-vertex subgraphs of a graph G whose components are edges or cycles. Prove that $c_i = \sum_{H \in \mathbf{H}_i}(-1)^{k(H)}2^{s(H)}$. (Sachs [1967])

15.3.8. Write $\phi(G; \lambda)$ as ϕ_G. Let v and xy denote a fixed vertex and edge in G; let $Z(v)$ and $Z(xy)$ denote the sets of cycles containing v or xy. Use Sach's formula above to:
 (a) Prove $\phi_G = \lambda\phi_{G-v} - \sum_{u \in N(v)} \phi_{G-v-u} - 2\sum_{C \in Z(v)} \phi_{G-V(C)}$.
 (b) Prove $\phi_G = \phi_{G-xy} - \phi_{G-x-y} - 2\sum_{C \in Z(xy)} \phi_{G-V(C)}$.
 (c) Obtain recurrences for the characteristic polynomials of paths and cycles.

15.3.9. Given that $\phi(G; x) = x^8 - 24x^6 - 64x^5 - 48x^4$, determine G.

15.3.10. ($\diamond$) Use Exercise 15.3.7 to prove that the coefficient of λ^{n-2k} in the characteristic polynomial of a tree is $(-1)^k m_k$, where m_k is the number of matchings of size k. Construct two nonisomorphic 8-vertex trees, both having characteristic polynomial $\lambda^8 - 7\lambda^6 + 9\lambda^4$.

15.3.11. (+) Let T be a tree. Prove that $\alpha(T)$ is the number of nonnegative eigenvalues of T. (Cvetkovic–Doob–Sachs [1979, p. 233])

15.3.12. Prove that C_n has eigenvalues $\{2\cos(2\pi j/n): 0 \le j \le n-1\}$.

15.3.13. Let $\lambda_1, \ldots, \lambda_m$ and $\mu_1, \ldots, \mu_n$ be the eigenvalues of G and H, respectively. Show that the mn eigenvalues of the cartesian product $G \square H$ are $\{\lambda_i + \mu_j: i \in [m], j \in [n]\}$. Use this to derive the spectrum of the k-dimensional cube.

15.3.14. Show that the eigenvalues of a graph with n vertices and m edges are bounded by $\sqrt{2m(n-1)/n}$.

15.3.15. For a graph G with adjacency matrix A, prove that the maximum of $x^T A x$ over vectors summing to 1 is $1 - 1/\omega(G)$. Conclude Turán's Theorem. (Motzkin–Straus [1965])

15.3.16. Prove that G is bipartite if G is connected and $\lambda_{\max}(G) = -\lambda_{\min}(G)$.

15.3.17. Prove that a regular graph and its complement have the same eigenvectors. What is the relationship between the corresponding eigenvalues?

15.3.18. Find the spectrum of the complete r-partite graph with m vertices in each part.

15.3.19. ($\diamond$) Let $B_1, \ldots, B_m$ be a biclique decomposition of K_n with vertex set $[n]$, where B_k has parts X_k and Y_k. Let A_k be the 0,1-matrix with 1 in position (i, j) if and only if $i \in X_k$ and $j \in Y_k$. Let $S = \sum_{k=1}^m A_k$. Observe that $S + S^T = J - I$. Prove that every n-by-n matrix satisfying this equation has rank at least $n-1$. Since the rank of the sum of two matrices is at most the sum of their ranks, conclude rank $S \le m$ and hence $m \ge n-1$.

15.3.20. ($\diamond$) For a graph G, a **squashed cube embedding** in dimension t encodes each vertex as a vector in $\{0, 1, *\}^t$ such that $d_G(u, v)$ is the number of coordinates where one of $\{u, v\}$ has 0 and the other has 1. (Graham–Pollak [1972])
   (a) Prove that every n-vertex graph G has a squashed cube embedding. (Comment: Winkler [1983] proved qdim$(G) \le n-1$, where qdim(G) is the least dimension needed.)
   (b) Use a bijection between biclique decompositions and squashed-cube embeddings of K_n to show that qdim K_n is the least size of a biclique decomposition.

15.3.21. ($\diamond$) Let G be a graph with eigenvalues $\lambda_1, \ldots, \lambda_n$ (in nonincreasing order).
   (a) Let H be an induced subgraph of G with eigenvalues $\mu_1, \ldots, \mu_t$. From Theorem 15.3.9 (Interlacing), conclude that $\mu_i \le \lambda_i$ for $1 \le i \le t$ and that $\lambda_2 < 0$ implies $G = K_n$.
   (b) Prove that $\alpha(G) \le n - \max\{p, q\}$, where G has p positive and q negative eigenvalues (including repetition). (Comment: This approach proves the Erdős–Ko–Rado Theorem from the eigenvalues of the Kneser graph $K(n, k)$. See Godsil–Meagher [2016, 2.9–10].)

15.3.22. Given a real symmetric matrix partitioned as $M = \begin{pmatrix} P & Q \\ Q^T & R \end{pmatrix}$ with P, R square, a lemma in linear algebra yields $\lambda_{\max}(M) + \lambda_{\min}(M) \le \lambda_{\max}(P) + \lambda_{\max}(R)$.
   (a) Let A be a real symmetric matrix partitioned into t^2 submatrices $A_{i,j}$ such that each $A_{i,i}$ is square. Prove that $\lambda_{\max}(A) + (t-1)\lambda_{\min}(A) \le \sum_{i=1}^m \lambda_{\max}(A_{i,i})$.
   (b) Prove $\chi(G) \ge 1 + \frac{\lambda_{\max}(G)}{-\lambda_{\min}(G)}$ when G is nontrivial. (Hoffman [1970])
   (c) Prove $\lambda_{\max}(G) + 3\lambda_{\min}(G) \le 0$ when G is planar (use the Four Color Theorem).

15.3.23. ($\diamond$) Let G be strongly regular with n vertices and parameters (k, λ, μ). Prove that $\overline{G}$ is strongly regular, and compute its parameters (k', λ', μ').

15.3.24. ($\diamond$) Let G be a triangle-free n-vertex graph in which any two nonadjacent vertices have exactly two common neighbors.

(a) Prove that G is k-regular, where k satisfies $n = 1 + \binom{k+1}{2}$.

(b) Prove that k is one more than the square of an integer not divisible by 4.

(c) Construct such graphs for $k \in \{1, 2, 5\}$. Let H be the graph for $k = 5$. Prove that $H - N[v]$ is the Petersen graph (for $v \in V(H)$) and that H is obtained from the 4-dimensional hypercube by adding edges joining antipodal vertices. (Comment: A realization for $k = 10$ is known using combinatorial designs.)

15.3.25. Let G be a strongly regular graph containing $K_{n+1} + K_{n,n}$ as an induced subgraph. Use the Interlacing Theorem and the eigenvalue theorem for strongly regular graphs to prove that G must have more than n^2 vertices. (Comment: Vu proved this. Fon-Der-Flaass [2002] proved that every graph with n vertices is an induced subgraph of a strongly regular graph with at most $4n^2$ vertices.)

15.3.26. ($\diamond$) Prove that the Petersen graph is strongly regular and determine its spectrum. Apply the spectrum to show that K_{10} does not decompose into three copies of the Petersen graph. (Hint: Use the spectrum to prove that two copies of the Petersen matrix have a common eigenvector other than the constant vector.) (Schwenk [1983])

15.3.27. The **subconstituents** of a graph G are the induced subgraphs of the form $G[U]$ where $v \in V(G)$ and $U = N(v)$ or $U = \overline{N[v]}$. Vince [1989] defined G to be **superregular** if G has no vertices or if G is regular and every subconstituent of G is superregular. Let $\mathbf{S}$ be the class consisting of $\{aK_b: a, b \geq 0\}$ (disjoint unions of isomorphic cliques), $\{K_m \square K_m: m \geq 0\}$, C_5, and their complements.

(a) Prove that every graph in $\mathbf{S}$ is superregular and that every disconnected superregular graph is in $\mathbf{S}$. (Comment: $\mathbf{S}$ is the set of all superregular graphs, but the full inductive proof requires several pages. (Maddox [1996], West [1996])

(b) Prove that every superregular graph is strongly regular.

15.3.28. A k-regular connected graph of diameter d is **distance-regular** if there exist $c_1, \ldots, c_d$ and $b_0, \ldots, b_{d-1}$ such that whenever $d(x, y) = i$, the neighborhood of y contains exactly c_i vertices with distance $i - 1$ from x and b_i vertices with distance $i + 1$ from x.

(a) Prove that the k-dimensional hypercube Q_k is distance-regular.

(b) The **Odd graph** O_k is the disjointness graph of the k-element subsets of $[2k + 1]$ (the Petersen graph is O_2). Prove that O_k is distance-regular.

(c) Given a distance-regular graph G, let A_i be the matrix with 1 in position (r, s) if $d(v_i, v_j) = i$ and 0 otherwise. Note that $\sum_{i=0}^{d} A_i = J$. Prove that A_i is a polynomial of degree i in the adjacency matrix, that $(A - kI)(\sum_{i=0}^{d} A_i) = 0$, and that $A_0, \ldots, A_d$ is a basis for the space of polynomials in the adjacency matrix.

15.3.29. ($\diamond$) Prove the properties of the Laplacian stated in Theorem 15.3.30. Conclude that the largest Laplacian eigenvalue is $|V(G)|$ when $\overline{G}$ is disconnected. (Kelmans [1967])

15.3.30. ($\diamond$) Let G be a cograph (Exercise 8.3.58), so G and $\overline{G}$ are not both connected.

(a) Prove that the Laplacian eigenvalues of G are all integers. (Merris [1998])

(b) Prove that $\kappa(G) = \mu_2(G)$ (see Corollary 15.3.36). (Abrishami [2019])

15.3.31. (+) *Expanders of linear size.*

(a) Prove $\mathbb{P}(|S| \leq l) \leq \binom{n}{l}(l/n)^{kt}$ when S is the union of k random t-subsets of $[n]$.

(b) For $\alpha\beta < 1$, prove that $k \in \mathbb{N}$ exists such that, when n is large, there exists $G \subseteq K_{n,n}$ with $\Delta(G) \leq k$ such that $|N(S)| \geq \beta|S|$ whenever $|S| \leq \alpha n$. Conclude that (n, k, c)-expanders exist when n is large.

Chapter 16

Geometry and Topology

Many questions about placing objects in the plane with requirements on relative position are combinatorial in nature, such as the characterization of planar graphs. Questions involving specific locations of points become more geometric. We consider a few such questions and later discuss relationships between volume computations and enumeration. We only scratch the surface of these topics. For books on connections between combinatorics and geometry, see Pach–Agarwal [1995], Penrose [2003], Felsner [2004], and Pach–Sharir [2009].

16.1. Graph Drawings

We have observed that every planar graph has a straight-line embedding (Fáry's Theorem). For applications in computer science, we seek a straight-line embedding with the vertices at the integer points in a small grid. The geometric techniques involved in doing this also lead to a characterization of planar graphs in terms of poset dimension. Subsequently, we consider drawings of nonplanar graphs in the plane; the aim is to minimize the number of crossings of edges.

EMBEDDINGS ON GRIDS

Let G be a plane graph with n vertices, where $n \geq 3$. Schnyder [1990] proved that G has a straight-line embedding with its vertices at points of $\{0, \ldots, n-2\} \times \{0, \ldots, n-2\}$. Furthermore, this embedding is computable in linear time from an embedding of G. de Fraysseix–Pach–Pollack [1990] obtained embeddings in an $(n-2)$-by-$(2n-4)$ grid. Our lemmas follow Schnyder [1990], refining Schnyder [1989]. Another approach to embeddings was pioneered by Tutte [1960] (the "rubber band method"), extended in Ribó Mor–Rote–Schulz [2011].

Consider the triangle in space with corners at $\{(1,0,0),(0,1,0),(0,0,1)\}$. The coordinates of any point x in the triangle sum to 1. The coordinates express x as a convex combination of the three corners. Placing the triangle in any plane, these coordinates become the **barycentric coordinates** of x in terms of the three corners. The points with a fixed value of one coordinate lie on a line parallel to the side where that coordinate is 0.

16.1.1. Definition. A **barycentric representation** of G is an injection f from $V(G)$ to $\mathbb{R}^3$ such that (1) the coordinates of $f(v)$ are nonnegative and sum to 1 for all $v \in V(G)$, and (2) if $xy \in E(G)$ and $z \in V(G) - \{x, y\}$, then some coordinate of $f(z)$ is greater than the corresponding coordinates for $f(x)$ and $f(y)$. For all $v \in V(G)$, we write $f(v) = (v_1, v_2, v_3)$.

16.1.2. Lemma. If f is a barycentric representation of G, and α, β, γ are non-collinear points in the plane, then letting $g(v) = v_1\alpha + v_2\beta + v_3\gamma$ yields a straight-line embedding of G in the plane.

Proof: By definition the images are distinct. Suppose that xy and uv are edges having four distinct endpoints. Since each vertex does not belong to the other set, there exist indices $i, j, k, l \in \{1, 2, 3\}$ such that

$$x_i > \max\{u_i, v_i\}, \qquad u_k > \max\{x_k, y_k\},$$
$$y_j > \max\{u_j, v_j\}, \qquad v_l > \max\{x_l, y_l\}.$$

These inequalities require $\{i, j\} \cap \{k, l\} = \varnothing$, but there are only three possible indices. Hence $i = j$ or $k = l$. By symmetry, suppose $i = j$. Now there is a value t with $\min\{x_i, y_i\} > t > \max\{u_i, v_i\}$, and the line of points with i-coordinate t (parallel to the side of the triangle $\alpha\beta\gamma$ not through the ith corner) separates $g(x)$ and $g(y)$ from $g(u)$ and $g(v)$. Hence $g(x)g(y)$ and $g(u)g(v)$ do not intersect.

We must also consider incident edges. Let xy and yz be embedded with one containing the other, say with xy containing yz. In this case z lies on the segment joining x and y, so $f(z)$ is a convex combination of $f(x)$ and $f(y)$, contradicting the existence of a coordinate i such that $z_i > \max\{x_i, y_i\}$. ∎

When the coordinates are rational, multiplying by a constant m yields integer triples summing to m. Letting $\{\alpha, \beta, \gamma\} = \{(1, 0), (0, 1), (0, 0)\}$, we obtain a straight-line plane embedding of G on integer points with coordinates in $[0, m]$.

Hence we seek a rational barycentric representation of G with small denominator. We restrict to triangulations, since a straight-line embedding of a triangulation with vertices at integer points yields such an embedding of every subgraph. Call the bounded faces **cells**. By Euler's Formula, an n-vertex triangulation has $2n - 4$ regions and $2n - 5$ cells. To obtain coordinates for vertex v, we will split the triangulation into three regions meeting at v and count the cells in the three regions. This will yield an embedding using integer coordinates in the range $[0, 2n - 5]$. A refinement to grid-length $n - 2$ counts vertices instead of faces.

The combinatorial structure used to obtain the desired partitions of the cells is a special labeling of the angles of a triangulation. All arithmetic with labels and colors in this discussion will be modulo 3. Since the boundary of each cell is a 3-cycle, we can refer to the cells as triangles and the incidences between vertices and cells as angles (even without straight-line embeddings). The labeling idea was extended to 3-connected planar maps in Felsner [2001] and Miller [2002].

16.1.3. Definition. A **Schnyder labeling** of a triangulation G is a labeling of the angles in cells of G with the labels $1, 2, 3$ such that
(1) Each cell has its angles labeled $1, 2, 3$ in clockwise order.
(2) Each interior vertex has angles with each label, appearing in the clockwise order as all 1s then all 2s then all 3s.

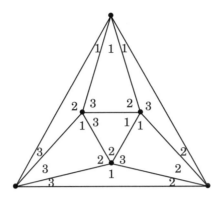

16.1.4. Remark. *Edge-coloring of Schnyder labelings.* A Schnyder labeling leads to a natural labeling and orientation of the internal edges. Each internal vertex has three incident edges where the two adjoining angles have different labels. Letting $\{i, j, k\} = \{1, 2, 3\}$, give color i to the edge between the angles labeled j and k at vertex x, and orient this edge away from x.

An interior edge now has color i if and only if both angles at its head have color i. Hence the edges at an internal vertex appear clockwise in this order: one departing edge with color 1, all entering edges having color 3, one departing edge with color 2, all entering edges having color 1, one departing edge with color 3, all entering edges having color 2. A vertex of degree 3 has no entering edges.

Let T_i be the set of internal edges colored i. We will show that T_i is an in-tree rooted at an external vertex. The paths leaving v in T_1, T_2, T_3 will define the regions where we count cells to obtain a barycentric representation.   ∎

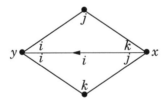

16.1.5. Lemma. In a Schnyder labeling, the external vertices can be named w_1, w_2, w_3 (clockwise) so that all angles at w_i have label i.

Proof: The $3n - 9$ internal edges are oriented by the procedure above. Since each internal vertex is the tail of three of these edges, all internal edges incident to external vertices enter the external vertices. Since the angles at the head of an edge have the same label, the labels at a fixed external vertex are the same. Since each external edge belongs to a cell using two external vertices, the angles at distinct external vertices have different labels, appearing in the order claimed.   ∎

In order to facilitate inductive proofs, we seek an internal edge e whose endpoints have only two common neighbors, which are the other vertices on the cells bounded by e. Contracting e and converting each of the two resulting multiedges to single edges yields a smaller simple triangulation.

16.1.6. Definition. An internal edge of a triangulation is **contractible** if its endpoints have only two common neighbors.

16.1.7. Lemma. (Kampen [1976]) If a is an external vertex of a triangulation G other than K_3, then a has an internal neighbor u such that au is contractible.

Proof: Let $x_0, x_1, \ldots, x_t$ be the neighbors of a in clockwise order, where x_0 and x_t are the other external vertices. Since G is a triangulation, these vertices form a path P. Among the chords of P, choose $x_i x_j$ with $i < j$ to minimize $j - i$; such a chord exists because x_0 and x_t are adjacent. Let u be any x_k with $i < k < j$. Planarity and the minimality of $j - i$ guarantee that the only common neighbors of a and u are x_{k-1} and x_{k+1}. ∎

16.1.8. Theorem. Every triangulation has a Schnyder labeling.

Proof: Let G be an n-vertex triangulation. We use induction on n; the claim is trivial for $n = 3$. For $n > 3$, let a be an external vertex. By Lemma 16.1.7, there is a contractible edge au. Let G' be the smaller triangulation obtained by contracting au into a. Let $N(u)$ consist of a and $x_1, \ldots, x_t$ in clockwise order.

By the induction hypothesis, G' has a Schnyder labeling. By Lemma 16.1.5 and cyclic symmetry, we may assume that all angles at a have label 1. In G, we use the same angle labels as in G', except for the new angles involving u. The angles at u not bordered by au get label 1, and the two angles bordered by au get labels 2 and 3, in clockwise order.

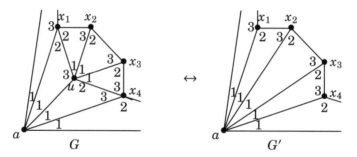

The labeling of a cell in G is inherited from the labeling of G', with u replacing a in the triangles at a having no external edges, and the two triangles involving ua labeled as shown above. The requirements of consecutivity and clockwise order at internal vertices are also inherited from the labeling of G', except for being explicitly enforced at u. ∎

Properties of Schnyder labelings can now be proved inductively if we show that every Schnyder labeling arises as in Theorem 16.1.8 from a labeling of K_3.

16.1.9. Lemma. Given a Schnyder labeling of a triangulation G other than K_3, let a be the external vertex w_i with angles labeled i. There is a contractible edge au such that the internal angles at u all have label i except for the two angles in triangles involving au.

Proof: By symmetry, we may assume $i = 1$. Let $y_0, \ldots, y_t$ be the neighbors of a in clockwise order, where y_0 and y_t are the other external vertices. Since all angles at a have label 1, the angles at y_r and y_{r+1} in their triangle with a have the labels 2 and 3, respectively. Therefore, in the edge-coloring and orientation of Remark 16.1.4, the edge $y_r y_{r+1}$ has color 2 if oriented from y_{r+1} to y_r and color 3 if oriented from y_r to y_{r+1}.

Since y_0 and y_t are external vertices, $y_0 y_1$ has color 2 and $y_{t-1} y_t$ has color 3. Hence there exists k such that $y_{k-1} y_k$ has color 2 and $y_k y_{k+1}$ has color 3. The angles between them (not involving a) have label 1, and the edges between them have color 1 and enter y_k. The only edges leaving y_k go to a, y_{k-1}, and y_{k+1}.

If a and y_k have a common neighbor z outside $\{y_{k-1}, y_{k+1}\}$, then we have shown that zy_k has color 1 and enters y_k. However, za has color 1 and enters a. Since no vertex has two exiting edges colored 1, there is no such vertex z. Hence y_k is the desired vertex u.   ∎

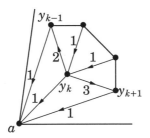

16.1.10. Theorem. (Uniform Angle Lemma) Given a cycle C in a Schnyder labeling of G, and $i \in \{1, 2, 3\}$, there is an *i-uniform* vertex x_i on C (that is, the angles at x_i inside C all have label i). Furthermore, such vertices x_1, x_2, x_3 can be found occurring in clockwise order on C.

Proof: (Lovelace–Kündgen) Consider a Schnyder labeling of a triangulation G with n vertices. We use induction on n. The claim is trivial if C visits all three external vertices, which includes the base case $n = 3$. Otherwise, let a be an external vertex not in C. By symmetry, we may assume $a = w_1$. By Lemma 16.1.9, a has a neighbor u such that contracting au into a yields a Schnyder labeling of a smaller triangulation G' from which the labeling arises as in Theorem 16.1.8.

If $u \notin V(C)$, then C has the same interior in G and G', and the uniform vertices for C from applying the induction hypothesis to G' work also for C in G.

If C visits u, then C becomes a cycle C' through a in G', and u behaves like y_k in the figure above. In particular, since C avoids a, the vertex u is 1-uniform in C, just as a is 1-uniform in C'. The rest of the interior remains the same, so the triple of vertices obtained by applying the induction hypothesis to C' in G' works also for C in G, with a replaced by u if a is the 1-uniform vertex of the triple found in C'.   ∎

16.1.11. Theorem. In the edge-coloring and orientation for a Schnyder labeling, the edges of color i form a tree T_i of paths directed to the external vertex w_i. Also, for each internal vertex v, the paths from v to w_1, w_2, w_3 in T_1, T_2, T_3, respectively, pairwise share only v.

Proof: Each internal vertex has outdegree 1 in T_i; also w_i is a sink in T_i, and the other external vertices have no incident edges of color i in T_i. It thus suffices to show that as a digraph T_i has no cycle. This follows from the Uniform Angle Lemma. A cycle in color i would have an inside angle of label i at the head of each edge and an inside angle of another color at the tail of each edge, so it would have no uniform angle in any label.

For the second statement, let $\{i, j, k\} = \{1, 2, 3\}$. If T_i and T_j each contain v, u-paths P_i and P_j, then $P_i \cup P_j$ is a cycle C in G. However, along P_i inside C is an angle of color i at the head of each edge and an angle of another color at the tail of each edge. The same holds for P_j with color j. Hence the only candidates for uniform vertices are v and u, but the Uniform Angle Lemma requires three such vertices on C.   ∎

Let $P_i(v)$ denote the path from v to the root in the tree T_i. Since no two of $P_1(v)$, $P_2(v)$, and $P_3(v)$ have a common internal vertex, they establish three regions associated with v, as on the left below. Let $R_i(v)$ denote the region (with boundary) enclosed by $P_{i+1}(v)$, $P_{i+2}(v)$, and the external edge not containing w_i. As shown on the right, regions for distinct vertices can be ordered by inclusion; we prove this and use it to obtain the barycentric representation.

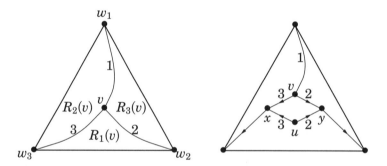

16.1.12. Lemma. If u and v are distinct internal vertices in a Schnyder labeling, and $u \in R_i(v)$, then $R_i(u)$ is properly contained in $R_i(v)$.

Proof: By symmetry, we may assume $i = 1$. We prove first that $P_2(u)$ is confined to $R_1(v)$. Since T_2 is an in-tree, when $P_2(u)$ intersects $P_2(v)$ it thereafter follows $P_2(v)$. Hence if $P_2(u)$ is not confined to $R_1(v)$, then some edge e of T_2 departs $R_1(v)$ from some vertex w of $P_3(v)$. This requires the edge with label 2 that leaves w to be clockwise after the edge with label 3 leaving w and before an edge with label 3 entering w (see figure above), which contradicts Remark 16.1.4. A symmetric argument shows that $P_3(u)$ is confined to $R_1(v)$.

With these paths confined to $R_1(v)$, the region bounded by $P_2(u)$, $P_3(u)$, and the external edge $w_2 w_3$ is contained in $R_1(v)$. Since $u \neq v$, the two regions do not have the same boundary, so $R_1(u)$ omits some cell contained in $R_1(v)$.   ∎

16.1.13. Theorem. Consider a Schnyder labeling of an n-vertex triangulation G. For each internal vertex v of G and each $i \in \{1, 2, 3\}$, let v_i be the number of cells in $R_i(v)$. For the external vertex v at which the angles have label j, let $v_j = 2n - 5$, and let the other coordinates of v be 0. Assigning to each vertex v the triple $\frac{1}{2n-5}(v_1, v_2, v_3)$ yields a barycentric representation of G.

Proof: By construction, the coordinates of a vertex are nonnegative and sum to 1. We must also check that if $xy \in E(G)$ and $z \notin \{x, y\}$, then there exists $k \in \{1, 2, 3\}$ such that $z_k > \max\{x_k, y_k\}$. This holds when z is the external vertex w_k since all other vertices have less 1 in coordinate k. If z is internal, then $x, y \in R_k(z)$ for some k, since the edge xy cannot cross any $P_j(z)$. Again $z_k > \max\{x_k, y_k\}$, since any internal vertex other than z in $R_k(z)$ has smaller value than z in coordinate k, and the external vertices in $R_k(z)$ have 0 in that coordinate. ∎

Theorem 16.1.13 completes the proof that every planar graph with n vertices has a straight-line embedding with the vertices at grid points in the triangle with corners $(2n - 5, 0)$, $(0, 2n - 5)$, and $(0, 0)$.

Using a similar approach, the proof can be modified to obtain a more compact embedding using only the points in a grid with side-length $n - 1$. The trick is to count vertices instead of regions, but we must be more careful about the order relation, and we need a more general notion of barycentric representation.

Let v_i' be the number of vertices in $R_i(v) - P_{i-1}(v)$, so $v_1' + v_2' + v_3' = n - 1$. When v is the jth external vertex, $R_j(v)$ contains all the vertices, $P_j(v)$ contains one vertex, and $P_{j+1}(v)$ contains two vertices. Hence we use the triples $(n-2, 1, 0)$, $(0, n - 2, 1)$, and $(1, 0, n - 2)$ for these vertices. Note that again the coordinates sum to $n - 1$.

Below we list the steps of converting these triples to a straight-line embedding in the grid with side-length $n - 1$, leaving verification of the details as exercises. The arguments are like those using the count of cells.

16.1.14. Definition. In the **lexicographic order** on vectors, $x < y$ if x has smaller value than y in the first coordinate where the vectors differ. A **weak barycentric representation** of G is an injection $f: V(G) \to \mathbb{R}^3$ such that (1) the coordinates of $f(v) = (v_1, v_2, v_3)$ are nonnegative and sum to 1 for all $v \in V(G)$, and (2) if $xy \in E(G)$ and $z \in V(G) - \{x, y\}$, then there exists $k \in \{1, 2, 3\}$ such that (x_k, x_{k+1}) and (y_k, y_{k+1}) are less than (z_k, z_{k+1}) in lexicographic order (with indices taken modulo 3).

16.1.15. Lemma. If f is a weak barycentric representation of G, with $f(v) = (v_1, v_2, v_3)$ for all $v \in V(G)$, and α, β, γ are noncollinear points in the plane, then placing each vertex v at the point $g(v)$ given by $v_1\alpha + v_2\beta + v_3\gamma$ yields a straight-line embedding of G in the plane.

16.1.16. Lemma. If u and v are distinct vertices in a Schnyder labeling and $u \in R_i(v)$, then $(u_i', u_{i+1}') < (v_i', v_{i+1}')$ in lexicographic order.

16.1.17. Theorem. Given a Schnyder labeling of an n-vertex triangulation G, the mapping that sends v to $\frac{1}{n-1}(v_1', v_2', v_3')$ is a weak barycentric representation of G, and the map that sends v to (v_1', v_2') is a straight-line embedding of G on the grid with side-length $n - 2$.

16.1.18. Example. For K_4 and for the octahedron labeled as at the start of this subsection, the resulting embeddings appear below, with the image of the jth external vertex labeled as j. ∎

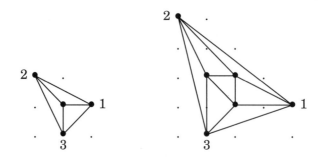

Finally, these results about Schnyder labelings also yield a characterization of planar graphs in terms of dimension of posets.

16.1.19. Definition. The **incidence poset** $P(G)$ of a graph G is the bipartite poset under containment whose minimal elements are the vertices of G and whose maximal elements are the pairs of adjacent vertices.

For an n-vertex graph G, the poset $P(G)$ is a subposet of $\mathbf{2}^n$ using only 1-sets and 2-sets. Thus $\dim P(G) \le \lg \lg n + \frac{1}{2} \lg \lg \lg n + O(1)$ (Corollary 12.3.35).

16.1.20. Theorem. (Schnyder [1989]) A graph G is planar if and only if $\dim P(G) \le 3$.

Proof: By Lemma 12.3.28, $\dim P(G)$ is the minimum number of linear orderings of $V(G)$ such that for each vertex x and edge yz with $x \notin \{y, z\}$, both y and z appear before x in some ordering. Theorem 16.1.13 states that when G is planar, three linear orderings obtained by extending the coordinate rankings defined there have the desired property.

Conversely, suppose that three such linear orderings exist. Assigning each vertex the triple of its heights on them produces a barycentric representation, by Definition 16.1.1. By Lemma 16.1.2, a barycentric representation of a graph yields an embedding of it in the plane.   ∎

For further work on dimension of posets associated with planar graphs, see Brightwell–Trotter [1993, 1997], Felsner [2003], Felsner–Li–Trotter [2010], Barrera-Cruz–Haxell [2011], and Felsner [2014].

CROSSING NUMBER

Every drawing of a nonplanar graph in the plane has edge crossings. A natural objective is to minimize the number of crossings. There are several related parameters, but for now we focus on the most common model. Recall that a crossing is a common internal point of two edges.

16.1.21. Definition. A drawing of G is **proper** if no vertex is an interior point on an edge, no two edges are tangent, no two incident edges cross, and no point is an interior point of more than two edges. The **crossing number** $\mathrm{cr}(G)$ of a graph G is the minimum number of crossings in a proper drawing of G.

The notation $\mathrm{cr}(G)$ is the modern choice; other notations have been used but are now less prevalent. We observed in Section 9.1 that all graphs have proper drawings in the plane; we henceforth consider only proper drawings.

Crossing numbers of small graphs can often be found by extracting maximal planar subgraphs.

16.1.22. Example. $\mathrm{cr}(K_6) = 3$ and $\mathrm{cr}(K_{3,2,2}) = 2$. If H with k edges is a maximal plane subgraph of a drawing of G in the plane, then every edge of G not in H crosses some edge of H, so the drawing has at least $|E(G)| - k$ crossings. If G has n vertices, then $k \leq 3n - 6$. If also G has no triangles, then $k \leq 2n - 4$.

Planar 6-vertex graphs have at most 12 edges, but K_6 has 15. Hence $\mathrm{cr}(K_6) \geq 3$, and the drawing on the left below proves equality.

Since $K_{3,2,2}$ has 16 edges, and planar 7-vertex graphs have at most 15 edges, $\mathrm{cr}(K_{3,2,2}) \geq 1$. The drawing on the right below has two crossings. To strengthen the lower bound, observe that $K_{3,2,2}$ contains $K_{3,4}$. Since $K_{3,4}$ is triangle-free, its planar subgraphs have at most $2 \cdot 7 - 4$ edges, which equals 10, and hence $\mathrm{cr}(K_{3,4}) \geq 2$. Every drawing of $K_{3,2,2}$ contains a drawing of $K_{3,4}$, so $\mathrm{cr}(K_{3,2,2}) \geq \mathrm{cr}(K_{3,4}) \geq 2$. ∎

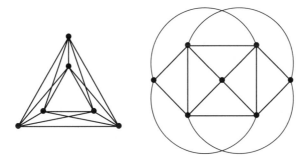

16.1.23. Proposition. Let G be an n-vertex graph with m edges. If k is the maximum number of edges in a planar subgraph of G, then $\mathrm{cr}(G) \geq m - k$. Furthermore, $\mathrm{cr}(G) \geq \frac{m^2}{2k} - \frac{m}{2}$.

Proof: Given a drawing of G in the plane, let H be a maximal subdrawing without crossings. Every edge not in H crosses at least one edge in H; otherwise, it could be added to H. Since H has at most k edges, we have at least $m - k$ crossings between edges of H and edges of $G - E(H)$.

After discarding $E(H)$, we have at least $m - k$ edges remaining. The same argument yields at least $(m - k) - k$ crossings in the drawing of the remaining graph. Iterating the argument yields at least $\sum_{i=1}^{t}(m - ik)$ crossings, where $t = \lfloor m/k \rfloor$. The value of the sum is $mt - kt(t + 1)/2$.

We now write $m = tk + r$, where $0 \leq r \leq k - 1$. Letting $t = (m - r)/k$ in the value of the sum, we simplify to obtain $\mathrm{cr}(G) \geq \frac{m^2}{2k} - \frac{m}{2} + \frac{r(k-r)}{2k}$. ∎

The bound $m - k$ in Proposition 16.1.23 is useful when G is sparse. The computation can be iterated, but for dense graphs the result is too weak. Consider K_n, for example. Proposition 16.1.23 yields $\mathrm{cr}(K_n) \geq \frac{1}{24} n^3 + O(n^2)$, but the growth of $\mathrm{cr}(K_n)$ is quartic, not cubic. The crossing number cannot exceed $\binom{n}{4}$, since we can place the vertices on the circumference of a circle and draw chords. In this drawing of K_n, each set of four vertices contributes one crossing.

In fact, this is the worst straight-line drawing, since four vertices never produce more than one crossing in such a drawing. How much can be saved by a better drawing? We give a recursive lower bound.

16.1.24. Lemma. If a graph G contains a copies of a subgraph H, and each crossing in a drawing of G appears in at most b copies of H, then $\mathrm{cr}(G) \geq \frac{a}{b}\mathrm{cr}(H)$.

Proof: Consider a drawing of G with $\mathrm{cr}(G)$ crossings. Each copy of H in G contributes at least $\mathrm{cr}(H)$ crossings to the drawing. Each crossing in G is counted at most b times. Hence $b\mathrm{cr}(G) \geq a\mathrm{cr}(H)$. ∎

16.1.25. Example. We apply Lemma 16.1.24 with $G = K_n$ and $H = K_{n-1}$. Now $a = n$ and $b = n - 4$, so $\mathrm{cr}(K_n) \geq \frac{n}{n-4}\mathrm{cr}(K_{n-1})$. Dividing by $\binom{n}{4}$ yields

$$\frac{\mathrm{cr}(K_n)}{\binom{n}{4}} \geq \frac{\mathrm{cr}(K_{n-1})}{\binom{n-1}{4}} .$$

Since always $\mathrm{cr}(K_n)/\binom{n}{4} < 1$, the recursive inequality implies that $\mathrm{cr}(K_n)/\binom{n}{4}$ has a limit as $n \to \infty$. However, the value of that limit is unknown. ∎

16.1.26. Theorem. (Guy [1972]) $\frac{1}{80} n^4 + O(n^3) \leq \mathrm{cr}(K_n) \leq \frac{1}{64} n^4 + O(n^3)$.

Proof: Lemma 16.1.24 and Example 16.1.25 yield $\mathrm{cr}(K_n) \geq \frac{1}{5}\binom{n}{4} = \frac{1}{120} n^4 + O(n^3)$ for $n \geq 5$. The denominator of the quartic term can be improved from 120 to 80 by using copies of $K_{6,n-6}$, which has crossing number $6\lfloor \frac{n-6}{2} \rfloor \lfloor \frac{n-7}{2} \rfloor$ (Exercise 15b).

A better drawing lowers the upper bound from $\binom{n}{4}$ to $\frac{1}{64} n^4 + O(n^3)$. Consider $n = 2k$. View the sphere as a tin can. Place k vertices on the top of the can and k vertices on the bottom, drawing chords on the top and bottom for those k-cliques.

The edges from top to bottom fall into k natural classes. The "class number" is the circular separation between the top and bottom endpoints, ranging from $\lceil \frac{-k+1}{2} \rceil$ to $\lceil \frac{k-1}{2} \rceil$. We draw these edges to wind around the can as little as possible in passing from top to bottom, so edges in the same class don't cross. We now twist the can to make the class displacements run from 1 to k. This makes the crossings easier to count but does not change the pairs of edges that cross.

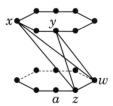

Crossings on the side of the can involve two top vertices and two bottom vertices. For top vertices x, y and bottom vertices z, w, where xz has smaller positive displacement than xw, edges xw and yz cross if and only if the displacements from x to y, z, w are distinct positive values in increasing order. (This holds for x, y, z, w as shown, but not for x, y, z, a, since ya winds around the can.) Hence there are $k\binom{k}{3}$ crossings on the side of the can, so

$$\text{cr}(K_n) \leq 2\binom{k}{4} + k\binom{k}{3} = \tfrac{1}{64}n^4 + O(n^3).$$ ∎

16.1.27. Example. For $\text{cr}(K_{m,n})$, a naive drawing puts the m vertices of one part on one side of a channel and the n vertices of the other part on the other side, with all edges drawn straight across. This produces $\binom{n}{2}\binom{m}{2}$ crossings, but we can do much better. Place the vertices along two perpendicular axes. Put the vertices of the m-vertex part along the x-axis, with $\lceil m/2 \rceil$ vertices on the positive side and $\lfloor m/2 \rfloor$ on the negative side. Similarly split the n-vertex part between the positive and negative portions of the y-axis. Summing the four types of crossings generated by adding straight-line segments for the edges yields $\text{cr}(K_{m,n}) \leq \lfloor \tfrac{m}{2} \rfloor \lfloor \tfrac{m-1}{2} \rfloor \lfloor \tfrac{n}{2} \rfloor \lfloor \tfrac{n-1}{2} \rfloor$ (Zarankiewicz [1954]).

The construction is conjectured optimal (Guy [1969] tells the history). Kleitman [1970] proved this for $\min\{n, m\} \leq 6$. Aided by a computer search, Woodall [1993] brought the smallest unknown cases to $K_{7,11}$ and $K_{9,9}$. Using Kleitman's result, Guy [1970] showed $\text{cr}(K_{m,n}) \geq \tfrac{m(m-1)}{5} \lfloor \tfrac{n}{2} \rfloor \lfloor \tfrac{n-1}{2} \rfloor$ (Exercise 15), not far from the upper bound. ∎

For dense graphs, the inductive argument of Theorem 16.1.26 yields a general lower bound for crossing number conjectured in Erdős–Guy [1973]. There is also an elegant probabilistic proof, which we present here. Stronger results appear in Pach–Tóth [1997].

16.1.28. Theorem. (Ajtai-Chvátal-Newborn-Szemerédi [1982], Leighton [1983])
Let G be an n-vertex graph with m edges. If $m \geq 4n$, then $\text{cr}(G) \geq \tfrac{1}{64}m^3/n^2$.

Proof: Consider a drawing of G with $\text{cr}(G)$ crossings. Take a random induced sub-drawing H by including each vertex independently with probability p. With n' and m' denoting the order and size of H, we have $\mathbb{E}(n') = pn$ and $\mathbb{E}(m') = p^2 m$. Let Y be the number of edge crossings in H. Since all four endpoints of two edges must appear for a crossing in G to appear in H, we have $\mathbb{E}(Y) = p^4\text{cr}(G)$.

Always $Y \geq m' - (3n' - 6)$, by the argument in Proposition 16.1.23. Linearity of expectation yields $p^4\text{cr}(G) = \mathbb{E}(Y) \geq p^2 m - 3pn$, which simplifies to $3n + p^3\text{cr}(G) - pm > 0$. Let $p = 4n/m$, which is allowable since $m \geq 4n$. Now $3n + 64n^3/m^3\text{cr}(G) > 4n$, which yields the desired bound. ∎

16.1.29. Example. The order of magnitude for the lower bound in Theorem 16.1.28 is best possible. Consider $G = \tfrac{n^2}{2m}K_{2m/n}$, where $2m$ is a multiple of n. The total number of vertices is n, and the total number of edges is asymptotic to $\tfrac{n^2}{2m}\tfrac{1}{2}(\tfrac{2m}{n})^2$, which equals m. Since $\text{cr}(K_r) \leq \tfrac{1}{64}r^4$, we have $\text{cr}(G) \leq \tfrac{n^2}{2m}\tfrac{1}{64}(\tfrac{2m}{n})^4 = \tfrac{1}{8}\tfrac{m^3}{n^2}$. This is within a constant factor of the lower bound from Theorem 16.1.28. ∎

We present an application of Theorem 16.1.28 in discrete geometry.

16.1.30. Example. *Unit distances among n points.* Erdős [1946] asked how many unit distances can occur among n points in the plane. In a unit grid, n points can achieve about $2n - O(\sqrt{n})$ unit distances. The points of a finer grid within an appropriate distance from the origin can achieve about $n^{1+c/\log\log n}$ unit distances (Erdős). This growth is superlinear but slower than $n^{1+\varepsilon}$ for all positive ε.

Erdős also proved an upper bound of $O(n^{3/2})$. Let m count the pairs at distance 1 in a given set of n points. Let G be the graph with those pairs forming edges. Since two circles of radius 1 intersect in at most two points, G does not contain $K_{2,3}$, so any two points have at most two common neighbors. Since each vertex v is a common neighbor for its $\binom{d(v)}{2}$ pairs of neighbors, $\sum \binom{d(v)}{2} \leq 2\binom{n}{2}$. Since $2m/n$ is the average vertex degree, convexity yields $\sum \binom{d(v)}{2} \geq n\binom{2m/n}{2}$. Thus $\binom{2m/n}{2} \leq n-1$, which yields the upper bound of $O(n^{3/2})$. (This is the special case for $(s, t) = (3, 2)$ of the argument given for the upper bound on $\mathrm{ex}\,(n; K_{2,3})$ in Theorem 13.2.19.) ∎

Using the Regularity Lemma, Józsa–Szemerédi [1975] improved the upper bound to $o(n^{3/2})$. By number-theoretic arguments about incidences between lines and points, Spencer–Szemerédi–Trotter [1984] improved it to $O(n^{4/3})$. Using Theorem 16.1.28, Székely gave an elegant short proof. A similar argument yields the Szemerédi–Trotter [1983] result that n points and m lines in the plane generate at most $O(n^{2/3}m^{2/3})$ point/line incidences (Exercise 23).

16.1.31. Theorem. (Spencer–Szemerédi–Trotter [1984]) Among a set of n points in the plane, at most $4n^{4/3}$ pairs of points have distance 1.

Proof: (Székely [1997]) The claim is obvious for $n \leq 3$; suppose $n > 3$. By moving points or pairs of points without reducing the number of pairs at distance 1, we can ensure that each point is involved in a unit pair and that no two points have distance 1 only from each other. If any point still is involved in only one unit pair, we can rotate it around its mate until it has distance 1 from another point. Thus we may assume that every point is involved in at least two such pairs.

Let P be an optimal n-point configuration, with q unit distance pairs. We obtain a graph from P, not by using the unit pairs as edges, but by drawing a unit circle around each point. If a point in P has distance 1 from k other points in P, then they partition the circle into k arcs. Each pair of points at distance 1 generates one arc moving clockwise on the circle around each of the two points; altogether we obtain $2q$ arcs. These are the edges of a loopless multigraph G.

Since two points can appear on two (but not three) unit circles, G may have edges of multiplicity 2 but no larger multiplicity. We delete one copy of each duplicated edge to obtain a simple graph G' with at least q edges. We may assume $q \geq 4n$; otherwise the bound already holds.

Because these arcs lie on n circles, they cannot produce many crossings; any two circles cross at most twice. Thus our layout of G' has at most $2\binom{n}{2}$ crossings. By Theorem 16.1.28, G' has at least $\frac{1}{64}q^3/n^2$ crossings. Together, $q \leq 4n^{4/3}$. ∎

16.1.32. Example. *Products of cycles.* After $\mathrm{cr}(K_n)$ and $\mathrm{cr}(K_{m,n})$, the question of computing $\mathrm{cr}(C_m \square C_n)$ arises. For $m \leq n$, a natural drawing proves an upper bound of $(m - 2)n$, which Harary–Kainen–Schwenk [1973] conjectured to be optimal. This has been proved for $m = 3$ (Ringeisen–Beineke [1978]), $m = 4$ (Dean–Richter [1995]), $m = 5$ (Richter–Thomassen [1995] and Klešč–Richter–Stobert [1996], and $m \in \{6, 7\}$ (Anderson–Richter–Rodney [1996, 1997]). Juarez–Salazar [2001] gave a short proof that $\mathrm{cr}(C_m \square C_n) \geq (m - 2)n/2$. ∎

16.1.33. Example. *Variants on crossing number.* In a drawing with the fewest crossings, any two edges cross at most once, since switching the routing of two edges between two points where they cross reduces the number of crossings. However, it may be possible to reduce the number of pairs of edges that cross by allowing edges to cross more than once, using a drawing with crossings at more than $\mathrm{cr}(G)$ points. The minimum number of pairs of edges that cross (among all drawings of G) is the **pair-crossing number** pair-$\mathrm{cr}(G)$. Another variation is the **odd-crossing number** odd-$\mathrm{cr}(G)$, the minimum number of pairs of edges that cross an odd number of times. Since two edges that cross an odd number of times cross at least once, and two edges that cross contribute a crossing,

$$\text{odd-}\mathrm{cr}(G) \leq \text{pair-}\mathrm{cr}(G) \leq \mathrm{cr}(G).$$

It was long unknown whether there exists a graph on which these three parameters are not all equal; Pelsmajer–Schaefer–Stefanovic [2006] proved that odd-cr and cr are not the same parameter. Pach–Tóth [2000a, 2000b] discuss this problem and other variants on crossing numbers.

Determining the crossing number is NP-hard (Garey–Johnson [1983]), even when restricted to 3-regular graphs (Hliněný [2006]). Cabello–Mohar [2013] proved that it is also NP-hard for graphs obtained from a planar graph by adding one edge. Their arguments also give a geometric proof of Hliněný's result. ∎

Another alternative is to restrict the allowable drawings. In particular, given our discussion of straight-line embeddings for planar graphs, it is natural to ask how many crossings are forced in a straight-line drawing of a nonplanar graph.

16.1.34. Definition. The **linear crossing number** $\mathrm{cr}_1(G)$ is the minimum number of crossings in a drawing of G whose edges are drawn as straight-line segments. More generally, the t-**linear crossing number** $\mathrm{cr}_t(G)$ is the minimum number of crossings when the edges are drawn as unions of at most t line segments, allowing $t - 1$ "bends".

The linear crossing number has historically been called the **rectilinear crossing number**, but the t-linear generalization and the permissibility of arbitrary slopes (there is no relation to rectangles or axes) suggest dropping "recti".

Guy [1972] proved that $\mathrm{cr}_1(K_8) > \mathrm{cr}(K_8)$. The smallest graph for which this occurs seems to be $K_{2,2,2,2}$. Bienstock–Dean [1992, 1993] proved many startling results. For graphs with crossing number at most 3, the crossing number and linear crossing number are equal. For $k \geq 4$, there are graphs with crossing number k and linear crossing number arbitrarily high, but $\mathrm{cr}_1(G) \leq c\Delta(G)\mathrm{cr}(G)^2$. Also, allowing one bend in the edges permits a better bound: $\mathrm{cr}_2(G) \leq 2\mathrm{cr}(G)^2$. It is not known whether these bounds are sharp. Bienstock [1991] proved that there is no fixed t such that $\mathrm{cr}_t(G) = \mathrm{cr}(G)$ for all G.

16.1.35. Theorem. (Bienstock–Dean [1993]) For $m \geq k \geq 4$, graphs exist with crossing number k and linear crossing number at least m.

Proof: (sketch). We construct an explicit graph. Begin by forming a graph H as follows: add to $K_{2,4}$ a matching on the larger side, then subdivide each of the eight original edges. There are essentially two ways to embed H in the plane without crossings, shown on the left below.

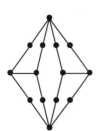 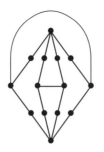 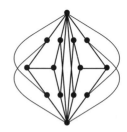

Form H' by adding to H eight more edges, joining each vertex in the original part of size 2 in the bipartition to the four neighbors of the other such vertex in H. Call these *chords* of H'. As shown above, H' has crossing number at most 4.

We complete the construction by replacing each edge of H in H' with a copy of $K_{2,m}$, forming G. Since the crossings in our drawing of H' do not involve edges of H, we also have $\mathrm{cr}(G) \le 4$ (in fact, equality holds).

If $\mathrm{cr}_1(G) < m$, then each copy of $K_{2,m}$ has a "clean" (uncrossed) path joining its "endpoints". Extracting this path yields a drawing of H' in which the edges of H are drawn using two segments, and the edges of $H' - H$ are drawn as straight segments. The contradiction arises from a technical argument showing that this yields a planar embedding of H that is not of the forms on the left above. ∎

16.1.36. Theorem. (Bienstock–Dean [1993]) $\mathrm{cr}_2(G) < 2[\mathrm{cr}(G)]^2$ for any graph G.

Proof: Let D be a drawing of G in the plane with $\mathrm{cr}(G)$ crossings. Obtain G' from D by deleting all edges involved in crossings; G' is a plane graph. From G', form a graph H by placing a vertex inside each face of G' that lost edges. Let v_C be the vertex placed inside a face bounded by a cycle C, and make v_C adjacent to the vertices of C. Note that H is planar.

By Fáry's Theorem, H has a straight-line drawing D'. Deleting v_C from D' still leaves C as the boundary of a face. Finiteness of C allows us to assume that v_C does not lie on a segment joining vertices of C.

To embed G, we replace each deleted edge xy that was removed from the face bounded by C in D. We add it to $D' - \{v_C\}$ as the union of two segments close to xv_C and yv_C, meeting near v_C at a point z on the bisector of the angle made by xv_C and yv_C. We may need to adjust how far z is from v_C.

Let t_C be the number of crossings inside C removed from the original drawing of G; they involved at most $2t_C$ edges, now replaced. We want any two such edges to cross at most once, forming at most $\binom{2t_C}{2}$ crossings by replacing the edges.

One segment of xzy may cross both segments of $x'z'y'$ if x' and y' are on the x, y-path along z. To avoid this, we make z close to v_C when the x, y-path along z is long, farther from v_C when the path is short. The narrower wedge is then captured inside the wider wedge, avoiding crossings. We can make this precise by placing the points near v_C using a linear extension of the containment poset of the paths along C. When the endpoints alternate along C, as in (x, x', y, y'), there is only one crossing, behaving like the original crossing edges.

We have a 2-linear drawing of G. Since the original drawing was optimal, $\sum_C t_C \le \mathrm{cr}(G)$. Hence $\mathrm{cr}_2(G) \le \sum \binom{2t_C}{2} < 2 \sum t_C^2 \le 2[\mathrm{cr}(G)]^2$. ∎

EXERCISES 16.1

16.1.1. $(-)$ Prove that each point in a triangular region has a unique expression as a convex combination of the vertices of the triangle (convex combinations are linear combinations where the coefficients are nonnegative and sum to 1).

16.1.2. Find a Schnyder labeling and small grid embedding of the icosahedron.

16.1.3. Prove Lemmas 16.1.15–16.1.16 and Theorem 16.1.17 to complete the straight-line embeddability of planar graphs in small grids.

16.1.4. Compute the crossing numbers for $K_{4,4}$, $K_{2,2,2,2}$, and the Petersen graph.

16.1.5. Determine $\mathrm{cr}(K_{1,2,2,2})$ and use it to compute $\mathrm{cr}(K_{2,2,2,2})$.

16.1.6. Prove that $K_{3,2,2}$ has no planar subgraph with 15 edges. Use this to give another proof that $\mathrm{cr}(K_{3,2,2}) \geq 2$.

16.1.7. Prove that the crossing number of the graph below is at most 5.

16.1.8. $(\diamond)$ Let M_n be the graph obtained from the cycle C_n by adding chords between vertices that are opposite (if n is even) or nearly opposite (if n is odd). The graph is 3-regular if n is even, 4-regular if n is odd. Determine $\mathrm{cr}(M_n)$. (Guy–Harary [1967])

16.1.9. $(\diamond)$ The graph P_n^k has vertex set $[n]$ and edge set $\{ij \colon |i - j| \leq k\}$. Use a planar embedding of P_n^3 to prove $\mathrm{cr}(P_n^4) = n - 4$. (Harary–Kainen [1993])

16.1.10. Determine the crossing number of the graph below.

16.1.11. $(\diamond)$ For each positive integer k, show constructively that some graph embeddable on the torus requires at least k crossings when drawn in the plane.

16.1.12. $(\diamond)$ A graph is 1-**planar** if it can be drawn in the plane with each edge involved in at most one crossing. Prove that a 1-planar graph with n vertices has at most $4n - 8$ edges, for $n \geq 2$. The bound is sharp for $n \geq 12$; prove sharpness for multiples of 4 at least 8. (Pach–Tóth [1997], Albertson–Mohar [2006], Nagasawa–Noguchi–Suzuki [2018])

16.1.13. $(\diamond)$ Use the Four Color Theorem to prove that every 2-edge-connected 3-regular graph with crossing number 1 is 3-edge-colorable. (Jaeger [1980])

16.1.14. $(\diamond)$ Suppose that n is odd. Prove that in all proper drawings of K_n, the parity of the number of edge crossings is the same. Conclude that $\mathrm{cr}(K_n)$ is even when n is congruent to 1 or 3 modulo 8 and is odd when n is congruent to 5 or 7 modulo 8.

16.1.15. ($\diamond$) Use $\mathrm{cr}(K_{6,n}) = 6\lfloor \frac{n}{2}\rfloor\lfloor \frac{n-1}{2}\rfloor$ (Kleitman [1970]) to prove the following.
   (a) $\mathrm{cr}(K_{m,n}) \geq m\frac{m-1}{5}\lfloor \frac{n}{2}\rfloor\lfloor \frac{n-1}{2}\rfloor$. (Guy [1970])
   (b) $\mathrm{cr}(K_p) \geq \frac{1}{80}p^4 + O(p^3)$.

16.1.16. It is conjectured that $\mathrm{cr}(K_{m,n}) = \lfloor \frac{m}{2}\rfloor\lfloor \frac{m-1}{2}\rfloor\lfloor \frac{n}{2}\rfloor\lfloor \frac{n-1}{2}\rfloor$. Suppose that this conjecture holds for $K_{m,n}$ and that m is odd. Prove that the conjecture then holds also for $K_{m+1,n}$. (Kleitman [1970])

16.1.17. Suppose that m and n are odd. Prove that in all proper drawings of $K_{m,n}$, the parity of the number of edge crossings is the same. Conclude that $\mathrm{cr}(K_{m,n})$ is odd when $m - 3$ and $n - 3$ are divisible by 4 and even otherwise.

16.1.18. Use Exercise 16.1.17 and Kleitman's computation of $\mathrm{cr}(K_{6,n})$ to prove $\mathrm{cr}(K_{7,7}) \in \{77, 79, 81\}$. (Comment: Woodall [1993] proved $\mathrm{cr}(K_{7,7}) = 81$.)

16.1.19. ($\diamond$) Harary–Kainen–Schwenk [1973] conjectured $\mathrm{cr}(C_m \square C_n) = (m-2)n$ for $m \leq n$. This is now known for $m \leq 7$, along with $\mathrm{cr}(K_4 \square C_n) = 3n$. Find plane drawings to establish the upper bounds. (Beineke–Ringeisen [1980]) (Comment:

16.1.20. Prove $\mathrm{cr}(C_3 \square C_3) \geq 2$. (Hint: Find three subdivisions of $K_{3,3}$ that together use each edge exactly twice.)

16.1.21. Let Q_k denote the k-dimensional hypercube. Prove $\mathrm{cr}(Q_4) \leq 8$. (Comment: Faria–de Figueiredo–Sýkora–Vrt'o [2008] proved $\mathrm{cr}(Q_k) \leq \frac{5}{32}4^k - \lfloor \frac{k^2+1}{2}\rfloor 2^{k-2}$, conjectured to hold with equality by Erdős–Guy [1973]. Sýkora–Vrt'o [1993] proved $\mathrm{cr}(Q_k) \geq \frac{1}{20}4^k + O(k^2 2^k)$.)

16.1.22. ($\diamond$) Let $f(n) = \mathrm{cr}(K_{n,n,n})$. Prove that $n^3(n-1)/6 \leq f(n) \leq (9/16)n^4 + O(n^3)$. (Hint: For the lower bound, use Lemma 16.1.24 after proving $\mathrm{cr}(K_{3,3,2}) \geq 5$ and $\mathrm{cr}(K_{3,3,3}) \geq 9$. For the upper bound, embed $K_{l,m,n}$ on a tetrahedron or generalize a drawing of K_n.)

16.1.23. ($\diamond$) In the plane, let P be a set of n points, and let L be a set of m lines, with $m \geq \sqrt{n}$. Each occurrence of a point in P on a line in L is an *incidence*. Prove that there are fewer than $4n^{2/3}m^{2/3}$ incidences. (Szemerédi–Trotter [1983]) (Hint: Use the technique of Theorem 16.1.31. Comment: This short proof is due to Székely [1997].)

16.2. Combinatorial Topology

   Like the probabilistic method, combinatorial topology provides non-constructive existence proofs of natural combinatorial statements. There is a further analogy. Probabilistic arguments can be "derandomized" to obtain constructive procedures. Similarly, the topological theorems generally have discrete algorithmic proofs, which allow these conclusions also to be made constructive. We will not study this aspect of the topic.

   The most famous instance of the topological method in combinatorics may be the use of the Borsuk–Ulam Theorem to prove Kneser's Conjecture on the chromatic number of Kneser graphs. We will derive the topological theorems, but it is important to note that typically one can apply them knowing only the statements.

   We start with easy applications of a simple geometric lemma that turns out to be an intimate part of the topological development.

SPERNER'S LEMMA AND BANDWIDTH

Sperner's Lemma in the plane involves only parity arguments, but its inductive higher-dimensional generalization to simplices takes us toward the applications of simplices considered later.

16.2.1. Definition. A **simplicial subdivision** of a large triangle T is a partition of T into triangular **cells** such that every intersection of two cells is a common edge or corner. We call the corners of cells *nodes*. A **proper labeling** of a simplicial subdivision of T assigns labels from $\{0, 1, 2\}$ to the nodes, avoiding label i on the ith edge of T ($i \in \{0, 1, 2\}$). A **completely labeled cell** has all three labels.

In a proper labeling, each label appears at one corner of T, and label i avoids the edge of T joining the corners not labeled i. The figure below illustrates a simplicial subdivision and the graph we will obtain from it to prove that it has a completely labeled cell.

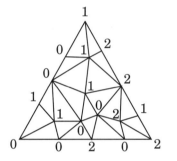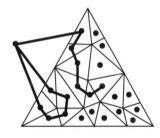

16.2.2. Theorem. (**Sperner's Lemma**; Sperner [1928]) Every properly labeled simplicial subdivision contains a completely labeled cell.

Proof: We prove the stronger result that the number of completely labeled cells is odd. We define a graph G that is a subgraph of the dual of the triangulation. We introduce a vertex for each cell, plus one for the outside region. Vertices of G are adjacent if their regions share a boundary segment with labels 0 and 1. The graph on the right above results from the proper labeling on the left.

A completely labeled cell becomes a vertex of degree 1 in G. A cell with no 0 or no 1 becomes a vertex of degree 0. The other cells are labeled $0, 0, 1$ or $0, 1, 1$ and become vertices of degree 2. Hence the desired cells are the only bounded cells that become vertices of odd degree.

The vertex v for the outside region also has odd degree. In moving from the 0-corner to the 1-corner along the edge of T avoiding label 2, we cross an edge of G involving v for each switch between label 0 and label 1. Since we start with 0 and end with 1, we switch an odd number of times.

Since the number of vertices of odd degree in every graph is even, the number of vertices other than v having odd degree is odd, so there are an odd number of completely labeled cells. ∎

We apply Sperner's Lemma to linear embeddings of triangular grids.

16.2.3. Definition. When the vertices of a graph G are numbered with distinct integers, the **dilation** is the maximum difference between integers assigned to adjacent vertices. The **bandwidth** $B(G)$ of G is the minimum dilation of a numbering of G.

Dilation is minimized when the numbering has no gaps, but gaps can be convenient (Exercise 8). One motivation for studying bandwidth is to minimize the delay between adjacent vertices when the vertices must be processed in a linear order. The name originates from sparse matrix computations; the bandwidth of a matrix is the maximum of $|i - j|$ for a nonzero position (i, j). Computation of bandwidth is NP-hard even for trees with maximum degree 3 (Garey–Graham–Johnson–Knuth [1978]). We present two simple lower bounds.

16.2.4. Lemma. (Local density bound) Always $B(G) \geq \max_{H \subseteq G} \frac{|V(H)|-1}{\operatorname{diam} H}$

Proof: Every numbering of G also numbers each subgraph of G. For every k-vertex subgraph H, the two numbers farthest apart differ by at least $k - 1$. By the Pigeonhole Principle, some step on a path between them has dilation at least $k - 1$ divided by the distance between them. ∎

16.2.5. Definition. For a set S of vertices in a graph G, the **boundary** ∂S is the set of vertices outside S having neighbors in S.

16.2.6. Lemma. (Boundary bound; Harper [1966]**)** For a graph G,
$$B(G) \geq \max_k \min\{|\partial S| : |S| = k\}.$$

Proof: For each value of k, some set S of k vertices must be the first k vertices in an optimal numbering of G. The bandwidth of G must be at least $|\partial S|$, because the vertex among ∂S that has the highest label has an edge of dilation at least $|\partial S|$ to a neighbor in S. ∎

Computation of the bound is nontrivial. For the hypercube Q_k, the value is $\sum_{i=0}^{n-1} \binom{i}{\lfloor i/2 \rfloor}$, studied in Harper [1966]. Nevertheless, one can use the idea of the bound without computing it. To prove $B(G) \geq l$ it suffices to show that an optimal numbering (or every numbering) has some initial segment S with $|\partial S| \geq l$.

16.2.7. Proposition. If $v_1, \ldots, v_n$ is a numbering of the vertices of a graph G that is optimal for bandwidth, then
$$B(G) \geq \max\{|\partial\{v_1, \ldots, v_k\}| : 1 \leq k \leq n\}. \qquad ∎$$

16.2.8. Theorem. (Chvátalová [1975]) $B(P_m \,\square\, P_n) = \min\{m, n\}$.

Proof: Numbering the vertices in successive ranks along the short direction yields maximum difference $\min\{m, n\}$. If $m \geq n \geq 2$, then $P_n \,\square\, P_n$ is a subgraph of $P_m \,\square\, P_n$. Hence it suffices to prove that every numbering of $P_n \,\square\, P_n$ has some initial segment S with at least n neighbors outside S.

Given an ordering $v_1, \ldots, v_n$, let S be the smallest initial segment that has $n - 1$ positions from a single row or column. By symmetry, we may assume this is row r, with position (r, s) being the only element of row r omitted by S. In every column other than s, there is an element of S, but S contains no column, so each such column contains an element of ∂S. Finally, position (r, s) itself lies in ∂S, since it completes row r. Thus $|\partial S| \geq n$, and Proposition 16.2.7 applies.   ∎

The local density bound for $P_n \square P_n$ is only about $(2 - \sqrt{2})n$ (Exercise 3). The boundary bound for a graph may be smaller than the bound in Proposition 16.2.7, but for $P_n \square P_n$ they are equal (Exercise 4).

16.2.9. Example. *Bandwidth of special graphs.* An argument like that in Theorem 16.2.8 shows that $B(C_n \square C_n) = 2n - 1$ (Exercise 5; Li–Tao–Shen [1981]). Also $B(K_n \square K_n) = \lfloor (n^2 + n - 1)/2 \rfloor$ (Balogh–Csirik [2004]).

The **triangular grid** T_l consists of vertices (i, j, k) such that i, j, k are nonnegative integers summing to l, with two vertices adjacent when two corresponding coordinates differ by 1. Below we show T_4. Numbering the vertices by rows produces an upper bound of $l + 1$ for $B(T_l)$. This is optimal, but the local density bound is only about $l/2$.   ∎

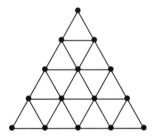

We will use Sperner's Lemma to prove that $B(T_l) = l + 1$.

16.2.10. Lemma. (Hochberg–McDiarmid–Saks [1995]) Let G be a plane graph whose bounded faces are triangles and whose unbounded face is a cycle partitioned into three paths. For any 2-coloring of the vertices, exactly one of the two subgraphs induced by a color class has a **connector**, meaning a component that intersects each of the three bounding paths.

Proof: Call the three bounding paths "sides", as in a triangle; a connector contains a vertex of each side (not necessarily distinct). If neither color contains a connector, then label each vertex v with the least index of a side not reachable from v via vertices of its color. Vertices on the ith side cannot have label i, so the labeling is proper.

By Sperner's Lemma, some cell is completely labeled. Since it has three corners and only two colors, two corners have the same color. Since they are adjacent, they can reach the same vertices via paths in their color, and the least side unreachable from them cannot be different. This contradiction implies that at least one color contains a connector.

A connector in one color splits the other vertices into sets that cannot reach all three sides. Hence connectors cannot exist in both colors.   ∎

Because the vertices of T_l correspond to the nonnegative integer triples (i, j, k) with $i + j + k = l$, and vertices are adjacent in T_l precisely when they agree in one coordinate and differ by 1 in the other two, the sum of the distances from any vertex to the three sides is exactly l. Hence the result $B(T_l) = l + 1$ is immediate from the following theorem.

16.2.11. Theorem. (Hochberg–McDiarmid–Saks [1995]) Let G be a plane graph whose bounded faces are triangles and whose unbounded face is a cycle partitioned into three paths. If l is the minimum over $v \in V(G)$ of the sum of the distances from v to each of the three paths, then $B(G) \geq l + 1$.

Proof: Let the vertices be $v_1, \ldots, v_n$ in an optimal numbering. Let t be the largest index such that the induced subgraph $G[v_1, \ldots, v_t]$ contains no connector. Let $R = \{v_1, \ldots, v_t\}$, let $S = \partial R$, and let $T = V(G) - R - S$. By construction, $v_{t+1} \in S$. Since $R \cup \{v_{t+1}\}$ contains a connector, $R \cup S$ contains a connector and T does not. Since no edge joins R and T, and R contains no connector, also $R \cup T$ contains no connector. Lemma 16.2.10 now implies that S contains a connector.

A connector contains paths from each vertex to each bounding path. By hypothesis, the sum of the lengths of these paths from any vertex is at least l. For some vertex v in S, the paths are disjoint; hence $|S| \geq l + 1$. Since $S = \partial R$, using this argument on each vertex numbering yields $B(G) \geq |\partial R| = |S| \geq l + 1$. ∎

EQUIVALENT TOPOLOGICAL LEMMAS

Sperner's Lemma and Lemma 16.2.10 have higher-dimensional analogues. These statements and several others including the Brouwer Fixed-Point Theorem are equivalent. Our chain of implications was developed by Chris Hartman.

Since its proof is so easy, we begin with Sperner's Lemma. The statement in d dimensions is the same as in two dimensions, but we must define precisely the d-dimensional analogues of the terms.

16.2.12. Definition. A d-**dimensional simplex** in $\mathbb{R}^n$ is the set of convex combinations of $d + 1$ points (its *corners*) not in one affine $(d - 1)$-dimensional space. A **simplicial subdivision** of a region in $\mathbb{R}^n$ decomposes it into cells that are n-dimensional simplices, with the intersection of any two cells being a lower-dimensional simplex. The 0-dimensional cells are called *vertices*.

For a simplex S with corners $w_0, \ldots, w_d$, a **facet** is a $(d - 1)$-dimensional simplex generated by d corners of S. A **proper labeling** of a simplicial subdivision of S is a coloring of the vertices using $\{0, \ldots, d\}$ such that color i does not appear on the facet of S that omits w_i. A **completely labeled cell** is one having all $d + 1$ colors.

In a proper labeling, w_i must have color i, for $0 \leq i \leq d$. The proof in Theorem 16.2.2 is a special case of the general induction step.

16.2.13. Theorem. (**Sperner's Lemma**; Sperner [1928]) Every proper labeling of a simplicial subdivision of a d-dimensional simplex T has a completely labeled cell.

Proof: We use induction on d, proving the stronger statement that the number of completely labeled cells is odd. For $d = 0$ the claim is trivial. For $d > 0$, form a graph G having a vertex for each cell inside T and one vertex w for the "outside". Join two vertices by an edge in G if their intersection is a $(d - 1)$-dimensional simplex whose corners have every label except d.

Every completely labeled cell has degree 1 in G. Every other cell has degree 0 or 2, since having d labels without label d requires a repeated label, and a facet generating an edge is formed by taking either vertex having the repeated label plus all the remaining vertices (if they have distinct colors).

The vertex w has neighbors in G only via the facet T' omitting the corner of T with color d, since no other facet has vertices with all labels but d. Restricting the subdivision to T' yields a properly labeled simplicial subdivision in $d-1$ dimensions. Its completely labeled cells yield the edges incident to w in G.

By the induction hypothesis, w has odd degree in G. Since the number of vertices of odd degree is even and all other vertices of odd degree in G correspond to completely labeled cells, the number of completely labeled cells is odd.   ∎

We note one application: Aharoni–Haxell [2000] used Sperner's Lemma to prove a hypergraph generalization of Hall's Theorem, applied in Aharoni [2001]. We use Sperner's Lemma to obtain a natural generalization of Lemma 16.2.10.

16.2.14. Theorem. (**Connector Lemma**; Hochberg–McDiarmid–Saks [1995])
In a d-coloring of the vertices of a simplicial subdivision of a d-dimensional simplex, some color has a component touching all $(d - 1)$-dimensional facets.

Proof: If there is no such component, then coloring each vertex with the least index of a facet that it cannot reach in its own color (along 1-dimensional faces) yields a proper labeling. By Sperner's Lemma, there is a completely labeled cell. By the Pigeonhole Principle, two vertices of this cell have the same color. Since they are joined by an edge of the subdivision, they cannot have distinct labels. This contradiction implies that the desired component exists.   ∎

16.2.15. Example. *The game of Hex.* In two dimensions, n-by-n **Hex** is played on a grid of n^2 hexagons, shown below on the left for $n = 4$. Two players alternately seize hexagonal cells. One player wants to seize cells forming a path connecting the upper-left border and the lower-right border; the other wants a path connecting the upper-right border and the lower-left border.

The game was invented by the Danish poet and engineer Piet Hein in 1942 and reinvented by John Nash in 1948 as a graduate student at Princeton. Our account comes from Nasar [1998, Chapter 6]. At Princeton the game was known as "Nash" or "John". David Gale suggested the name "Hex" under which Parker Brothers later marketed the game with $n = 11$. Nash proved that, unlike 2-dimensional Tic-Tac-Toe, every game of Hex has a winner. Gale [1979] published the equivalence in $\mathbb{R}^d$ between the extension of this result and the Brouwer Fixed-Point Theorem. Tanaka [2007] showed that the Hex game theorem is also equivalent to Arrow's Impossibility Theorem of social choice theory.

The (weak) dual of the Hex board, for $n = 4$ on the right below, is the square grid graph plus diagonal edges in one direction. In d dimensions, it is the cubical grid plus edges within cubelets for the comparability graph of $\mathbf{2}^n$.   ∎

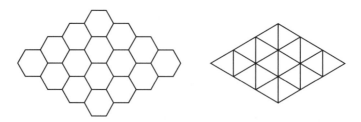

16.2.16. Theorem. (Hex Theorem; Gale [1979]) In a d-coloring of a simplicial
subdivision of a d-dimensional cube such that the points on the opposite faces
in each direction i have color i (but points on more than one face may have
any color), there is for some i a path in color i joining the faces in direction i.

Proof: Add d vertices, one of each color; the new vertex of color i forms a "pyra-
mid" over one face of the cube in direction i. The new points also form a simplex
S with the common point of those d faces. The result can be viewed as a d-colored
simplicial subdivision of a d-dimensional simplex, as illustrated below for $d = 2$.
The vertices of the outer simplex are the d added points plus the common point
of the d faces over which the pyramids were not built.

By the Connector Lemma, in some color c there is a component C that touches
all $(d{-}1)$-dimensional facets. One facet is the $(d{-}1)$-dimensional simplex S formed
by the d added points. In another, all points with color c lie in the face of the cube
opposite the face attached to the point of S with color c. Since C reaches from S
to that facet, C contains a path in color c joining the opposite faces of color c. ■

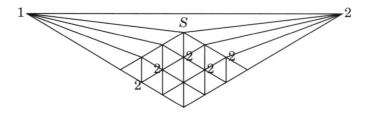

16.2.17. Theorem. (Pouzet's Lemma; see Zaks [1995]) Let A be a cartesian
product of d integer intervals. Points of A are considered adjacent if they
differ by 0 or 1 in every coordinate. Let f assign to each point in A a unit
vector parallel to some axis (positive or negative). If $x + f(x) \in A$ for each
$x \in A$, then there exist adjacent points x and y such that $f(x) = -f(y)$.

Proof: Using A as the vertex set, form a simplicial subdivision of the region en-
compassed by A such that all 1-dimensional simplices join vertices deemed adjacent in A. Color each vertex x with the direction of $f(x)$ (positive and negative
yield the same color) to form a d-coloring of the vertices of the resulting graph.

Associate color i with the extreme faces in coordinate i. By the Hex Theorem,
for some color i a path P in color i joins the extreme faces in color i, say with endpoints u and v. For a point x with color i in an extreme face in coordinate i, there
is only one choice for $f(x)$, since $x + f(x) \in A$. In particular, $f(u) = -f(v)$. Since
there are only two choices for f along P, somewhere along P adjacent vertices x
and y satisfy $f(x) = -f(y)$. ■

A set in $\mathbb{R}^d$ is **compact** if it is closed and bounded. We need the property that every sequence in a compact set has a convergent subsequence. Two sets in $\mathbb{R}^d$ are *homeomorphic* if one can be mapped onto the other by a continuous bijection with a continuous inverse; such a function preserves compactness. Brouwer's Theorem is often stated for the unit ball or simplex, but it holds equivalently for all sets homeomorphic to these, and we use the unit cube in the proof. For extensions of Brouwer's Theorem and its relation to Hex, see Björner–Matoušek–Ziegler [2017].

16.2.18. Theorem. (Brouwer Fixed-Point Theorem; Hadamard [1910], Brouwer [1911]) Let S be a set in $\mathbb{R}^d$ that is homeomorphic to the unit cube. Every continuous function f from S to itself has a fixed point.

Proof: Let Q be the unit cube in $\mathbb{R}^d$. Let $h \colon S \to Q$ be a continuous bijection with a continuous inverse. Now hfh^{-1} is a continuous function from Q to itself. If $x = hfh^{-1}(x)$, then $h^{-1}(x) = f(h^{-1}(x))$. Hence we reduce to the case $S = Q$.

Define g by letting $g(x) = f(x) - x$. If f has no fixed point, then g is never 0. Since Q is compact, $|g(x)|$ is bounded away from 0 (otherwise some sequence converges to a point in Q where g is 0). Choose ε so that $|g(x)| > \varepsilon$ for $x \in Q$.

Place a grid of points on Q using n steps in each direction. At each vertex x of the grid, choose a unit vector v in a coordinate direction closest to the vector $g(x)$. The distance between $g(x)$ and the hyperplane perpendicular to v through the origin is at least $\varepsilon/\sqrt{d}$. Hence the distance between the values of g at vectors assigned opposite coordinate vectors is at least $2\varepsilon/\sqrt{d}$.

Since f is continuous, also g is continuous. Therefore, choosing n large enough yields a grid A within Q such that the distance between values of g at gridpoints that are equal or consecutive in each coordinate is less than $2\varepsilon/\sqrt{d}$. (Actually, this uses uniform continuity of continuous functions on compact sets.)

Scale A so that steps of length $1/n$ in coordinate directions become unit steps. The selection of unit coordinate vectors yields a labeling of the gridpoints as specified in Pouzet's Lemma. Hence some adjacent vertices choose opposite directions. This contradicts the choice of n and implies that f has a fixed point. ∎

16.2.19. Theorem. (Sperner's Lemma; Sperner [1928]) Every proper labeling of a simplicial subdivision of a simplex T has a completely labeled cell.

Proof: (from Brouwer's Theorem). Given a proper labeling of a simplicial subdivision of a d-dimensional simplex T, we define a continuous mapping from T to itself. Map each vertex with label j into the corner of T with label $j + 1 \pmod{d}$. Extend the map into each cell by mapping a convex combination of the vertices of a cell into the same combination of their images. This defines a continuous function f on T. By Brouwer's Theorem, f has a fixed point.

By construction, vertices of the subdivision are not fixed points. Any other point is a convex combination of vertices. Let R be the minimal simplex containing a point x in T. If the vertices of R do not have distinct labels, then $f(x)$ lies in a simplex of smaller dimension, and then $f(x) \neq x$.

If the vertices of R have distinct labels, and S is that set of labels, then R is mapped into a simplex U spanned by the corners of T having labels in $\{j + 1 \colon j \in S\}$. If there is no completely labeled cell, then some label in S is not among those corners. Since the labeling is proper, points with that missing label are forbidden from U. Hence some corner of R is not in U, which yields $x \notin U$, so again $f(x) \neq x$.

Since Brouwer's Theorem requires a fixed point, it cannot happen that there is no completely labeled cell. ∎

By this cycle of implications, the statements of these five theorems are equivalent. Since we proved Sperner's Lemma (Theorem 16.2.13), they are all true.

THE BORSUK–ULAM THEOREM

The Borsuk–Ulam Theorem is one of the most famous theorems of algebraic topology. Some call it just Borsuk's Theorem; Borsuk's paper credited Ulam for conjecturing it. A popular expression of the 2-dimensional case is that at any moment, on the earth's surface there are antipodal locations with the same temperature and wind speed.

The theorem has many proofs, equivalent versions, and applications. In his book *Using the Borsuk–Ulam Theorem*, Matoušek [2003, p.25] notes "hundreds of papers with various new proofs, variations of old proofs, generalizations, and applications have appeared; the most comprehensive survey known to me, Steinlein [1985], lists nearly 500 items in the bibliography."

The Borsuk–Ulam Theorem has many proofs using algebraic topology. Instead, we obtain it from Tucker's Combinatorial Lemma. We then obtain several equivalent forms of the theorem and give combinatorial applications.

16.2.20. Definition. Let $\hat{B}^n = \{x \in \mathbb{R}^n : \sum |x_i| \leq 1\}$ and $\hat{S}^{n-1} = \{x \in \mathbb{R}^n : \sum |x_i| = 1\}$, so $\hat{S}^{n-1}$ is the boundary of $\hat{B}^n$. The set $\hat{B}^n$ has $2n$ extreme points (vertices) obtained by setting one coordinate to ± 1. An **orthant** in $\mathbb{R}^n$ is a simplex having one corner at the origin and at most n other corners reached by unit vectors along distinct coordinate axes. The **octahedral subdivision** is the simplicial subdivision of $\hat{B}^n$ whose simplices are the 2^n full-dimensional orthants. A **special subdivision** of $\hat{B}^n$ is a simplicial subdivision that refines the octahedral subdivision and is centrally symmetric on $\hat{S}^{n-1}$.

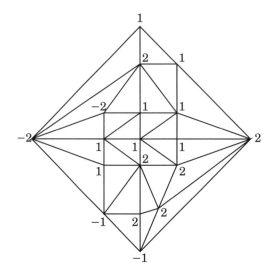

The special subdivision of $\hat{B}^2$ shown above is adapted from Matoušek [2003, pp. 37–40], which also presents the proof we give for Tucker's Lemma. "Refining" the octahedral subdivision means that each orthant is a union of cells in the special subdivision. Labels from $\{\pm 1, \ldots, \pm n\}$ are given to the vertices, with each vertex in $\hat{S}^{n-1}$ given the negation of the label on the antipodal vertex. We say that labels i and $-i$ are *complementary* and that an edge (1-dimensional simplex) with complementary labels is a *complementary edge*. We seek a complementary edge.

Our proof of Tucker's Lemma uses the same tool as Sperner's Lemma: we only need that the number of odd-degree vertices in a special graph is even. In both cases, the resulting proof is algorithmic. That is, following a path in the special graph leads to the desired configuration. Tucker [1946] published a 2-dimensional version. A proof for n dimensions via simplicial topology appears in Lefschetz [1949].

16.2.21. Theorem. (**Tucker's Combinatorial Lemma**) Let T be a special subdivision of $\hat{B}^n$. If the vertices of T are labeled using $\{\pm 1, \ldots, \pm n\}$ so that antipodal points of $\hat{S}^{n-1}$ have complementary labels, then T contains a complementary edge.

Proof: (Freund–Todd [1981]) Let $\mathrm{sgn}(z)$ denote the sign $(0, +1, \text{ or } -1)$ of a real number z; for $x \in \mathbb{R}^n$, let $\mathrm{sgn}(x)$ denote the vector of signs. Sign vectors of points interior to full-dimensional orthants have no zeros, and such sign vectors determine the 2^n orthants. We consider lower-dimensional orthants by letting $C(s)$ be the closure of the set of points in $\hat{B}^n$ with sign vector s.

Say that a simplex σ in the special subdivision T *requires* label i or label $-i$ if the interior points in the smallest orthant containing σ have $+1$ or -1, respectively, in coordinate i of their common sign vector. Also say that σ is *fully labeled* if all labels thus required appear on vertices of σ. Always the 0-dimensional simplex O at the origin is fully labeled, since it has no required label. If a simplex contained in $\hat{S}^{n-1}$ is fully labeled, then so is its antipodal simplex.

Next define a graph G whose vertices are the fully labeled simplices in T. Let fully labeled simplices σ and τ be adjacent if they both lie in $\hat{S}^{n-1}$ and are antipodal or if one is a facet of the other (obtained by deleting one corner) and the smaller has all labels required of the larger.

It suffices to prove the following claims about vertex degrees in G, since they imply that the component of G containing O also contains a vertex having a complementary edge. That component will be a path from O to a simplex containing a complementary edge.

(1) O has degree 1.
(2) A fully labeled simplex containing a complementary edge has degree 1.
(3) All other fully labeled simplices have degree 2.

(In the example below Definition 16.2.20, the path from O to the desired simplex is as shown below. The path has length 9, taking four steps in the top row of the figure. Each step changes the dimension by 1, up or down, except when the adjacency of antipodal simplices in the boundary is used. Such a move occurs here in the next to last step. The horizontal direction is coordinate 1; the vertical is coordinate 2. The graph has another component that is an 8-cycle.)

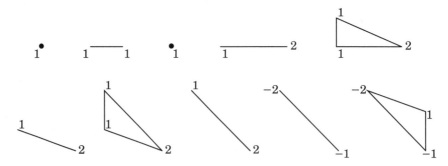

Now we prove the degree claims. Let k be the number of labels required on a fully labeled simplex σ, and let C be the smallest orthant containing σ. Since σ is fully labeled, C has dimension k, and σ has dimension $k - 1$ or k.

Suppose first that σ has dimension $k - 1$. Since it has k vertices and requires labels for k distinct indices, σ has no complementary edge. If $\sigma \not\subset \hat{S}^{n-1}$, then σ is a facet of exactly two k-dimensional simplices in C. They are fully labeled because they have the same smallest orthant as σ and already σ is fully labeled; this gives σ two neighbors in G. Although a facet of σ may be fully labeled (this can happen when the facet lies in a smaller orthant), it cannot be adjacent to σ, because it has only $k - 1$ vertices and thus cannot have all k labels required by σ. If $\sigma \subset \hat{S}^{n-1}$, then by the same argument σ is a facet of one neighbor that is a k-dimensional simplex in C, but the antipodal simplex to σ is also fully labeled and a neighbor. In both cases, σ has degree 2.

Now suppose that σ has dimension k; it has $k + 1$ vertices. Since σ only requires k labels, the $k + 1$ labels on it may (a) have a duplicate label, (b) have a complementary pair, or (c) have an unrequired label $\pm j$ with $s_j = 0$. In case (a), σ has two fully labeled facets with the k required labels, in the same smallest orthant, and any larger fully labeled simplex containing σ and adjacent to σ would force another label on σ; hence σ has degree 2. In case (b), σ contains a complementary edge; since larger neighbors are forbidden by the same argument as above, and only one facet has the required labels, σ has degree 1.

In case (c), form a new sign vector t from the sign vector s associated with σ by setting the jth coordinate to the sign of the extra label on σ. Since $s_j = 0$, orthant $C(s)$ is contained in orthant $C(t)$, and σ is a facet of a fully labeled simplex in $C(t)$ that requires $k + 1$ labels. The other $(k + 1)$-dimensional simplex of which σ is a facet has the opposite sign in the jth coordinate, so even if it is fully labeled it is not adjacent to σ. Meanwhile, as in the paragraph above, σ has exactly one fully labeled facet that has the labels required by σ, so σ has degree 2. The one exception is that when $\sigma = O$ there is no smaller facet, so O has degree 1.   ∎

The central symmetry of the subdivision on the boundary is crucial for giving degree 2 to simplices there. No such symmetry is needed inside, since the adjacency conditions are local. Refinement of the octahedral subdivision allows the proof to use orthants of all dimensions, starting the path from the origin.

Tucker's Combinatorial Lemma sounds somewhat like Pouzet's Lemma (Theorem 16.2.17). A short proof of Tucker's Lemma from Pouzet's Lemma would complete a cycle of implications including the Borsuk–Ulam Theorem with only

Sperner's Lemma proved directly. Tucker's Lemma holds also without requiring refinement of the octahedral subdivision (Exercise 12, using Borsuk's Theorem). To obtain Borsuk's Theorem from Tucker's Lemma any sequence of successively refined subdivisions suffices, so we only need the restricted version proved above.

There are many versions of the Borsuk–Ulam Theorem. We start with one well suited for the use of Tucker's Lemma and one used in our first application.

16.2.22. Definition. Let S^{n-1} denote the $(n-1)$-dimensional sphere (in $\mathbb{R}^n$), consisting of the points having Euclidean distance 1 from the origin; the ball B^n consists of those with distance at most 1. A function f on S^n is **antipodal** if $f(-x) = -f(x)$ for all $x \in S^n$.

16.2.23. Theorem. (Borsuk–Ulam Theorem; Borsuk [1933]) (A) If a continuous function $f\colon B^n \to \mathbb{R}^n$ is antipodal on the boundary S^{n-1}, then $f(x) = 0$ for some $x \in B^n$. (B) If a continuous function $f\colon S^n \to \mathbb{R}^n$ is antipodal on S^n, then there exists $x \in S^n$ with $f(x) = 0$.

Proof: (Tucker [1946]; see Freund–Todd [1981]) Consider (A). To facilitate application of Tucker's Lemma, we may use compositions with continuous functions as in Theorem 16.2.18 (preserving antipodality) to assume instead that f is defined on $\hat{B}^n$ and antipodal on its boundary, which we have called $\hat{S}^{n-1}$.

Suppose that $f(x)$ is never zero. For any special subdivision T of $\hat{B}^n$, label each vertex x with the least index i such that $f_i(x)$ has the largest magnitude among the coordinates of $f(x)$, and let the sign of the label be the sign of $f_i(x)$. Since $f(-x) = -f(x)$ when $x \in \hat{S}^{n-1}$, antipodal points have complementary labels.

The subdivision and labeling satisfy the hypotheses of Theorem 16.2.21, and hence there is a complementary edge. Successively refining the triangulation so that the maximum length of edges tends to 0 yields a sequence of positive ends and a sequence of negative ends of complementary edges. Since $\hat{B}^n$ is compact, the sequence of positive ends has a convergent subsequence. Since the distance between points in the two sequences tends to 0, the corresponding points in the sequence of negative ends converge to the same point x^*, which lies in $\hat{B}^n$. It is approached through points on which the largest nonzero coordinate in f is positive and through points on which it is negative. By continuity, $f(x^*) = 0$.

(A)$\Rightarrow$(B): let $f\colon S^n \to \mathbb{R}^n$ be continuous and antipodal. Note that $S^n \subseteq \mathbb{R}^{n+1}$, so points in S^n have $n+1$ coordinates. For $x \in B^n \subseteq \mathbb{R}^n$, define $x' \in S^n$ by letting $x'_i = x_i$ for $1 \le i \le n$ and $x'_{n+1} = (1 - \sum_{i=1}^n x_i^2)^{1/2}$. Define $g\colon B^n \to \mathbb{R}^n$ by $g(x) = f(x')$. Since f is continuous, also g is continuous. Furthermore, if $x \in S^{n-1}$, then $x' \in S^n$ and $(-x)' = -(x')$. Hence g is antipodal on S^{n-1}, since f is antipodal on S^n. We conclude by the first statement that $g(x) = 0$ for some $x \in B^n$, and hence x' in S^n has the desired property: $f(x') = 0$. (Also (B)$\Rightarrow$(A); Exercise 14.) ∎

16.2.24. Application. *Splitting of necklaces.* Consider a necklace, opened at the clasp, having $2n$ beads chosen from k colors and arranged in some order with $2a_i$ beads of color i. Two jewel thieves want to split the beads so that each receives half of each color, doing this with the fewest possible cuts. When the beads of each color appear consecutively, there must be a cut within each color, so k cuts may be needed. Do k cuts always suffice? The question arose in the context of VLSI circuit design (Bhatt–Leiserson [1981]).

Goldberg–West [1985] answered affirmatively using induction on k, simplicial topology, and a continuous analogue. Alon–West [1986] noted that a suitable continuous analogue was proved 20 years earlier from Borsuk's Theorem.

The continuous analogue allows a k-coloring of the unit interval, where the set of points with a single color is measurable. Without generalizing to measurable sets, just suppose that the density of each color is an integrable function of position on the interval and that the densities sum to 1 at each point. The total amount of each color is the integral of its density function. A **bisection** is a set of points $y_0, \ldots, y_{r+1}$ with $0 = y_0 \leq \cdots \leq y_{r+1} = 1$ such that the union of the intervals $[y_{2i}, y_{2i+1}]$ for all i captures half the amount of each color. We seek a bisection with $r = k$, corresponding to using at most k distinct cuts.

The continuous result implies the discrete result, by modeling a $2n$-bead necklace using $2n$ equal subintervals where the i interval has density 1 in the color of the ith bead and density 0 in the others. A bisection yields the desired split unless cuts occur in the middle of beads. If there is such an internal cut, then there must be another internal cut in the same color, and these two cuts can be moved toward the edges of the subintervals to reduce the number of bad cuts. Similarly, one can capture the ceiling or floor of half the number of beads, as desired, when a color has an odd number of beads. For circular necklaces, see Exercise 13. ∎

The next theorem yields bisections with at most k cuts. Giving the positively signed intervals to one thief and the negatively signed intervals to the other gives each half the measure of each color. When consecutive signs are equal, fewer intervals (and cuts) can be used.

16.2.25. Theorem. (Hobby–Rice [1965]) Given integrable densities $\mu_1, \ldots, \mu_k$ on $[0, 1]$, there are points $y_0, \ldots, y_{k+1}$ with $0 = y_0 \leq \cdots \leq y_{k+1} = 1$ and signs $\varepsilon_0, \ldots, \varepsilon_k \in \{-1, +1\}$ such that $\sum_{j=0}^{k} \varepsilon_j \int_{y_j}^{y_{j+1}} \mu_i(t)dt = 0$ for $1 \leq i \leq k$.

Proof: From the k densities, we define a function $f \colon S^k \to \mathbb{R}^k$. For $x \in S^k$, let $y_0 = 0$ and $y_j = \sum_{i=1}^{j} x_i^2$ for $1 \leq j \leq k + 1$. Note that $y_{k+1} = 1$, since $x \in S^k$; we use $y_1, \ldots, y_k$ as break points on the interval. For $1 \leq i \leq k$, let $\varepsilon_j = \mathrm{sign}(x_j)$ and $f_i(x) = \sum_{j=0}^{k} \varepsilon_j \int_{y_j}^{y_{j+1}} \mu_i(t)dt$. Let $f(x) = (f_1(x), \ldots, f_k(x))$.

The densities are integrable, so f is continuous. For $f(-x)$, subintervals are the same as for $f(x)$, but the coefficients change signs; hence $f(-x) = -f(x)$. By the Borsuk–Ulam Theorem, there exists $x \in S^k$ with $f(x) = 0$. The corresponding values $y_1, \ldots, y_k$ and $\varepsilon_0, \ldots, \varepsilon_k$ are the desired break points and signs. ∎

Alon [1987] generalized the Hobby–Rice Theorem to simultaneous splitting of t continuous measures, using a generalization of the Borsuk–Ulam Theorem in Bárány–Shlosman–Szűcs [1981]. With this, he proved the conjecture of Goldberg and West that an open necklace with k colors of beads can be cut in $k(t-1)$ places so that the resulting segments can be distributed to t thieves with each thief receiving the same amount of each color.

Our second application is more subtle and more important and needs another version of the Borsuk–Ulam Theorem. The original paper (Borsuk [1933]) gave three variants, and Lyusternik–Shnirel'man [1930] earlier gave another. We have proved Statement A below in Theorem 16.2.23; we need Statement E. We only need to work our way down to E; but we include the return trip for complete-ness. See Exercises 14–16 for other equivalent statements. We mostly follow the presentation of Matoušek [2003].

16.2.26. Theorem. The following statements are equivalent.
   (A) For all continuous antipodal $f\colon S^n \to \mathbb{R}^n$, there is $x \in S^n$ with $f(x) = 0$.
   (B) For continuous $f\colon S^n \to \mathbb{R}^n$, there is $x \in S^n$ with $f(x) = f(-x)$.
   (C) There is no continuous antipodal $f\colon S^n \to S^{n-1}$.
   (D) (**Lyusternik–Shnirel'man Theorem**) In a cover of S^n by $n + 1$ closed sets, at least one contains antipodal points.
   (E) In any cover of S^n by $n + 1$ sets such that each of the first n is open or closed, at least one set contains antipodal points.

Proof: A $\Rightarrow$ B. Apply A to g defined by $g(x) = f(x) - f(-x)$.

B $\Rightarrow$ E. Let $U_1, \ldots, U_{n+1}$ be such a cover, and suppose no U_i contains an-tipodal points. Define $f\colon S^n \to \mathbb{R}^n$ by letting $f_i(x)$ be the distance from x to U_i, written $d(x, U_i)$, for $1 \le i \le n$ (the distance from a point to a set is the infimum of distances to points in the set, so this is well-defined also when U_i is open). Note that f is continuous in x.

By Statement B, there exists $x \in S^n$ with $f(x) = f(-x)$. Since U_{n+1} does not contain antipodal points, x or $-x$ belongs to one of the other sets, say U_k. By sym-metry, we may assume $x \in U_k$, and hence $d(x, U_k) = 0$. Hence also $d(-x, U_k) = 0$.

If U_k is closed, then $d(-x, U_k) = 0$ yields $-x \in U_k$, and U_k contains antipodal points. If U_k is open, then $-x$ lies in $\overline{U}_k$, the closure of U_k, which is the unique smallest closed set containing U_k. The complement of an open set is a closed set, so $S^n - (-U_k)$ is a closed set containing U_k, since U_k does not contain antipodal points. We obtain $\overline{U}_k \subseteq S^n - (-U_k)$. Hence $-x \in S^n - (-U_k)$, which yields $-x \notin -U_k$ and $x \notin U_k$, a contradiction.

E $\Rightarrow$ D. Statement D is just a special case of statement E.

D $\Rightarrow$ C. Let σ be a simplex in $\mathbb{R}^n$ containing the origin in its interior. Pro-jecting the facets of σ onto S^{n-1} yields closed sets $F_1, \ldots, F_{n+1}$ that cover S^{n-1} without any of them containing antipodal points. If $f\colon S^n \to S^{n-1}$ is antipodal, then let $G_i = f^{-1}(F_i)$ for $1 \le i \le n + 1$. If G_j contains antipodal points, then f maps them to antipodal points in F_j, which do not exist.

C $\Rightarrow$ A. If $f\colon S^n \to \mathbb{R}^n$ avoids 0, then letting $g(x) = \frac{f(x)}{\|f(x)\|}$ contradicts C. ∎

The special case of Theorem 16.2.26E in which all $n + 1$ sets are open is also called the Lyusternik–Shnirel'man Theorem. The case in which each set is open or closed (slightly weaker than E) was shown by Greene [2002]. Aigner–Ziegler [2014, Chapter 42] observed that no condition is needed on the last set.

KNESER CONJECTURE AND GALE'S LEMMA

The **Kneser graph** $K(n, k)$ (also written as $KG(n, k)$ or $KG_{n,k}$) is the graph with vertex set $\binom{[n]}{k}$ whose edges are the pairs of disjoint k-sets. Note that the special case $K(5, 2)$ is yet another appearance of the Petersen graph. Although not originally posed in graph-theoretic language, the famous conjecture of Kneser [1955] is that $\chi(K(n, k)) = n - 2k + 2$ when $n \ge 2k$. The upper bound is easy.

16.2.27. Proposition. $\chi(K(n, k)) \leq n - 2k + 2$ when $n \geq 2k$.

Proof: We cover $V(K(n, k))$ with $n-2k+2$ independent sets. For $1 \leq i \leq n-2k+1$, the k-sets containing i pairwise intersect and hence form an independent set. The uncovered vertices are the k-sets contained in $\{n - 2k + 2, \ldots, n\}$. This set has size $2k - 1$, so its k-sets pairwise intersect and hence form an independent set. ∎

16.2.28. Remark. The Kneser graph is of interest as an explicit graph with chromatic number much larger than the typical lower bounds. The clique number is only $\lfloor n/k \rfloor$. For example, when $n = 3k - 1$ the graph is triangle-free but has chromatic number $k + 1$.

In addition, $K(n, k)$ has chromatic number much larger than its fractional chromatic number, where the **fractional chromatic number** $\chi_f(G)$ is the infimum of a/b such that every vertex can be covered b times using a total of a independent sets. Always $\chi_f(G) \leq \chi(G)$, since $\chi(G)$ independent sets can cover every vertex once.

Since independent sets have size at most $\alpha(G)$, an n-vertex graph G requires at least $bn/\alpha(G)$ independent sets to cover each vertex b times. Hence $\chi_f(G) \geq n/\alpha(G)$. For a vertex-transitive graph like the Kneser graph, equality holds; in fact, $\chi_f(K(n, k)) = n/k$ (Exercise 17). Usually $n/\alpha(G)$ is a better lower bound than $\omega(G)$, but for the Kneser graph they are essentially equal. ∎

16.2.29. Theorem. (**Kneser Conjecture**) $\chi(K(n, k)) = n - 2k + 2$ when $n \geq 2k$.

Proof: (Greene [2002]) It suffices to prove $\chi(K(n, k)) \geq n - 2k + 2$. Let $d = n - 2k + 1$. Let X be a set of n points in S^d with no more than d on any hyperplane through the origin (this holds for "general position" in $\mathbb{R}^{d+1}$).

Consider a proper d-coloring f of $K(n, k)$. We define subsets of S^d corresponding to color classes under f. For $x \in S^d$ and $1 \leq i \leq d$, put $x \in U_i$ if some k-tuple with color i is contained in the open hemisphere centered at x. Because we use open hemispheres, the sets $U_1, \ldots, U_d$ are open. To complete a cover of S^d, define $U_{d+1} = S^d - \bigcup_{i=1}^{d} U_i$. By Theorem 16.2.26E, some U_i contains antipodal points.

If U_i contains antipodal points, where $i \leq d$, then the open hemispheres around those points are disjoint, and hence we have given color i to disjoint k-sets, violating the choice of f as a proper coloring of $K(n, k)$.

Hence U_{d+1} contains antipodal points x and $-x$. Since these points were not put in another set, the open hemispheres around x and $-x$ each contain at most $k - 1$ points from the set X. This leaves at least $n - 2k + 2$ points lying in the hyperplane separating the two hemispheres. However, $n - 2k + 2 = d + 1$, so this contradicts the choice of X. Hence $K(n, k)$ has no proper d-coloring. ∎

16.2.30.* Remark. Lovász [1978] proved the Kneser Conjecture using topological methods (homotopy). Bárány [1978] gave a short proof from Gale's Lemma (Theorem 16.2.33) and the open set version of the Lyusternik–Shnirel'man Theorem (see Exercise 19). These proofs appear in the book of Matoušek [2003], along with proofs of the Kneser Conjecture by Dol'nikov [1981] and Sarkaria [1990] using other variants of the Borsuk–Ulam Theorem.

Matoušek [2004] found a combinatorial argument for Kneser's Conjecture using Tucker's Lemma without the Borsuk–Ulam Theorem, thereby avoiding topological statements. Ziegler [2002] developed a modular analogue of Tucker's Lemma, extending Matoušek's approach to give combinatorial proofs of related theorems (see also Matoušek–Ziegler [2004]). Although we reached Greene's proof by proving Borsuk's Theorem from Tucker's Lemma, our proof is not fully combinatorial because the proof of Theorem 16.2.26E from Theorem 16.2.26B uses topological properties of open and closed sets.

Exercise 18 presents an application of Theorem 16.2.29. ∎

Schrijver [1978] showed that a small subgraph of $K(n, k)$ already has the full chromatic number $n - 2k + 2$. For the proof we need to develop several tools.

16.2.31. Definition. The **moment curve** in $\mathbb{R}^d$ is $\{\gamma(t) \colon t \in \mathbb{R}\}$, where $\gamma(t) = (t, t^2, \ldots, t^d)$.

16.2.32. Lemma. Every hyperplane in $\mathbb{R}^d$ intersects the moment curve γ in at most d points. Furthermore, if a hyperplane contains d points of γ, then γ crosses it at each intersection point.

Proof: A hyperplane H is the set of solutions to a single linear equation: $\sum_{i=1}^d a_i x_i = b$, where a is a nonzero d-tuple. When $\gamma(t) \in H$, we have $\sum_{i=1}^d a_i t^i = b$. Hence the intersection points are zeros of the polynomial p defined by $p(t) = (\sum_{i=1}^d a_i t^i) - b$. Since p has degree d, there are at most d points of intersection.

Furthermore, if there are d distinct points of intersection, then p has d simple zeros, and the derivative of p is nonzero at all those points. Thus p changes sign at each of those points, which means that γ crosses H at each of them. ∎

The moment curve has many applications in discrete geometry. Here we use it to prove a stronger version of **Gale's Lemma**; the original result of Gale [1956] is the weaker version of the next result obtained by omitting the requirement of cyclically nonconsecutive indices.

16.2.33. Theorem. (Schrijver [1978]) Given $d \geq 0$ and $k \geq 1$, there is a set $x_1, \ldots, x_{2k+d}$ in S^d such that every open hemisphere contains a k-set in $x_1, \ldots, x_{2k+d}$ having no two elements with cyclically consecutive indices.

Proof: (Matoušek–Ziegler [2004]) Open hemispheres correspond to open half-spaces whose boundaries are hyperplanes through the origin. Using projection to S^d via lines through the origin, it thus suffices to find points $x_1, \ldots, x_{2k+d}$ in $\mathbb{R}^{d+1}$ (with distinct projections on S^d) such that every open half-space whose boundary contains the origin contains a k-set in $\{x_1, \ldots, x_{2k+d}\}$ whose indices are cyclically nonconsecutive.

Let H_1 be the hyperplane in $\mathbb{R}^{d+1}$ defined by $x_1 = 1$; points in H_1 have distinct projections on S^d. We move the moment curve into H_1; let $\bar{\gamma}(t) = (1, t, t^2, \ldots, t^d)$. For any hyperplane H in $\mathbb{R}^{d+1}$ not parallel to H_1, dropping the first coordinate of the points in $H \cap H_1$ yields a hyperplane in $\mathbb{R}^d$. Hence Lemma 16.2.32 applies to the intersections with $\bar{\gamma}$ of any hyperplane through the origin not parallel to H_1. Furthermore, there is a one-to-one correspondence between hyperplanes in $\mathbb{R}^d$ and hyperplanes through the origin in $\mathbb{R}^{d+1}$ not parallel to H_1.

Take any $2k + d$ points $w_1, \ldots, w_{2k+d}$ along $\bar{\gamma}$, indexed in order of appearance along $\bar{\gamma}$. Let $W = \{w_1, \ldots, w_{2k+d}\}$. Let H be a hyperplane through the origin in $\mathbb{R}^{d+1}$, bounding open half-spaces H^+ and H^-. Let $x_i = (-1)^i w_i$ and $X = \{x_1, \ldots, x_{2k+d}\}$. We claim that each of H^+ and H^- has at least k points from X, and that their indices are cyclically nonconsecutive within $[2k + d]$.

By Lemma 16.2.32, H contains at most d points from $\bar{\gamma}$. If H contains j points of $\bar{\gamma}$, with $j < d$, then these points plus the origin determine a subspace J of dimension $j + 1$, and we can rotate H around J (always containing J, like a plane rotating around a line in 3-space) until the moving hyperplane acquires an additional point of W. Since we did not pass through any point of W, no point of W moved from one side of H to the other. Hence this process allows us to assume that H contains d points of W.

Let $W_0 = H \cap W$ and $W_1 = W - W_0$. Let the parity of a point in W be the parity of its subscript, so each point of W_1 is even or odd. Color $w \in W_1$ black if exactly one of $\{w$ is even, $w \in H^+\}$ holds; otherwise color w white.

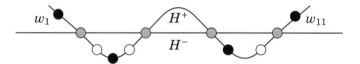

We claim that black and white points alternate along $\bar{\gamma}$. Let j be the number of times that $\bar{\gamma}$ crosses H between w and the next point w' of W_1. The points of crossing all lie in W_0, and by Lemma 16.2.32 the curve crosses through H at each such point. Hence w and w' are on the same side of H if and only if j is even, but they have the same parity in W if and only if j is odd. Hence exactly one of the two properties changes, which gives w and w' opposite colors.

If w_i is odd and lies in H^-, then x_i is in H^+. Thus points of X that lie in H^+ correspond to the white points in W_1, and the points of X that lie in H^- correspond to the black points in W_1. Since the colors alternate on the $2k$ points of W_1 along $\bar{\gamma}$ and $2k$ is even, there are k points of each color, and no two cyclically consecutive points in the indexing lie on the same side of H.   ∎

In [5], the 2-sets having no cyclically consecutive elements are the pairs of the form $\{i, i + 2\}$, computed modulo 5. In the Petersen graph, which is $K(5, 2)$, these vertices induce C_5; already $\chi(C_5) = 3$. Schrijver proved that one can always restrict $K(n, k)$ to such sets without reducing the chromatic number.

16.2.34. Corollary. (Schrijver [1978]) The subgraph of $K(n, k)$ induced by the k-sets with no cyclically consecutive elements has chromatic number $n - 2k + 2$.

Proof: Let $d = n - 2k$, and let X be an n-set in S^d as guaranteed by Theorem 16.2.33, indexed in order. Let G be the specified subgraph of $K(n, k)$; vertices of G correspond to k-sets in X. If G has a proper $(d + 1)$-coloring f, then define $A_1, \ldots, A_{d+1} \subseteq S^d$ by putting x in A_i if some vertex of color i in G corresponds to a k-set contained in the open hemisphere of S^d centered at x.

By Theorem 16.2.33, $A_1, \ldots, A_{d+1}$ are open sets covering S^d. By the Lyusternik–Shnirel'man Theorem (open set version), some A_i contains antipodal points, so f cannot be a proper coloring.   ∎

HAM SANDWICHES AND BISECTIONS

The Ham Sandwich Theorem is a well-known application of the Borsuk–Ulam Theorem. The name arises from a special case: the ham, cheese, and bread of a ham sandwich can be simultaneously bisected by a single knife cut. The theorem was conjectured by Steinhaus and proved by Banach (see Steinlein [1985]).

The theorem holds for general measures, which are beyond our scope to define carefully. Hence we restrict to the case where each object is a nonnegative integrable function; we call it a **mass**. Integrating over any full-dimensional volume gives a portion of the mass; integrating over a lower-dimensional set like a hyperplane gives 0. For a mass μ, we write $\mu(S)$ for the integral over a volume S.

16.2.35. Theorem. (Ham Sandwich Theorem) Given masses $\mu_1, \ldots, \mu_d$ in $\mathbb{R}^d$, some hyperplane cuts $\mathbb{R}^d$ into half-spaces that contain half of each mass.

Proof: (sketch) A half-space H^+ in $\mathbb{R}^d$ is specified by constants $u_0, \ldots, u_d$ (not all zero) such that $H^+ = \{x \in \mathbb{R}^d : \sum_{i=1}^d u_i x_i < u_0\}$. Furthermore, the coefficient vector can be normalized so that $(u_0, \ldots, u_d) \in S^d$. For $u \in S^d$, let $H^+(u)$ denote the half-space specified by this correspondence.

Define $f: S^d \to \mathbb{R}^d$ by $f_i(u) = \mu_i(H^+(u))$. Since $H^+(-u) = \{x \in \mathbb{R}^d : \sum_{i=1}^d u_i x_i > u_0\}$, the half-spaces associated with u and $-u$ are disjoint and omit only the hyperplane defined by $\sum_{i=1}^d u_i x_i = u_0$. If f is continuous, then the Borsuk–Ulam Theorem (Theorem 16.2.26B) provides $u \in S^d$ such that $f(u) = f(-u)$. This cannot occur when $u = (1, 0, \ldots, 0)$, since then $H^+(u) = \mathbb{R}^d$ and $H^+(-u) = \varnothing$. Thus $f(u) = f(-u)$ provides a hyperplane that bisects each mass.

We omit the argument that f is continuous. In more general versions of the theorem this is a matter of some delicacy, but for the integrable functions we use as masses it is intuitive. ∎

16.2.36. Remark. *Equipartitioning.* Using the Ham Sandwich Theorem, it is easy to show that any mass distribution in the plane can be partitioned into four equal parts by two lines (Exercise 23). It is possible (but not easy) to partition a distribution in $\mathbb{R}^3$ into eight equal parts by three planes (for partitioning point sets, see Hadwiger [1966] and Yao–Dobkin–Edelsbrunner–Paterson [1989]).

It is not generally possible to split a set in $\mathbb{R}^d$ into 2^d equal parts by d hyperplanes when $d \geq 5$ (Exercise 23). The question remains open when $d = 4$ (see Exercise 24 for points along the moment curve). What is now called the **Grünbaum–Hadwiger–Ramos problem** asks for the smallest dimension d such that for any m masses in $\mathbb{R}^d$ there are k hyperplanes that cut each mass into 2^k equal pieces. The Ham Sandwich Theorem settles the case $k = 1$, where the answer is m. Ramos [1996] proved the general lower bound $\left\lceil \frac{2^k-1}{k} m \right\rceil$, which is conjectured to be the correct answer. See Blagojević–Frick–Haase–Ziegler [2016, 2018] for a summary and analysis of known results. ∎

Other interesting applications arise when the Ham Sandwich Theorem is applied to bisect finite sets of points. Recall that a set of points in general position in $\mathbb{R}^d$ has no more than d points on any hyperplane.

16.2.37. Theorem. If $A_1, \ldots, A_d$ are disjoint finite sets whose union is in general position in $\mathbb{R}^d$, then there is a hyperplane H in $\mathbb{R}^d$ such that each open half-space with boundary H contains exactly $\left\lfloor \frac{1}{2} |A_i| \right\rfloor$ points from each A_i.

Proof: For each i, proceed as follows. If $|A_i|$ is odd, then let $B_i = A_i$; otherwise, delete any one point from A_i to obtain B_i. Form B_i' by replacing each point in B_i with a ball of radius ε centered at that point, where ε is chosen small enough that each hyperplane intersects at most d of the balls over all i.

Theorem 16.2.35 provides a hyperplane H that bisects all of $B_1', \ldots, B_d'$. Since each $|B_i|$ is odd, H intersects a ball from each B_i', and hence it intersects exactly one ball from each. Furthermore, H passes through the center of each of those balls, and hence H bisects each B_i.

Since H contains one point from each B_i, when $|A_i|$ is even the missing point cannot lie on H. Returning the missing point thus gives one of the half-spaces half of A_i. Let S and T be the sets of points on H from the sets A_i of even size and odd size, respectively. Move each point of S a small distance δ into the half-space containing half of the corresponding set, turning S into a new set S'. Now $S' \cup T$ has size d and determines a unique hyperplane H'. If δ is small enough, then moving from H to H' does not pass the hyperplane through any points in $\bigcup_{i=1}^d A_i$. Now the points of S are on the side of H' that was deficient for each such i, and H' bisects all of the sets. ∎

From this the necklace result follows.

16.2.38. Corollary. Every open necklace with k colors of beads can be cut using at most k cuts, so that the union of odd-indexed intervals captures half the beads of each color, with each color having an odd number of beads rounded up or down as specified arbitrarily.

Proof: Place the beads of the necklace in order as points along the moment curve. Let H be a hyperplane as guaranteed by Theorem 16.2.37, bisecting the k point sets. The cuts of the necklace correspond to where H crosses the moment curve. No color with an even number of beads has a point on H, but each color with an odd number of beads does. The cut made at each such bead can be put before or after the bead depending on whether the half of that color in the odd-indexed intervals should be rounded up or down. ∎

Given n red points and n blue points in $\mathbb{R}^2$, one can find a pairing so that the resulting n segments are disjoint: just take a pairing to minimize the total length of the segments. Like the necklace problem, this has a higher-dimensional generalization that is nontrivial but follows easily from the Ham Sandwich Theorem.

16.2.39. Theorem. (Akiyama–Alon [1989]) If $A_1, \ldots, A_d$ are sets of n points in general position in $\mathbb{R}^d$, then the union splits into n sets, each having one point from each A_i, whose convex hulls are disjoint.

Proof: We use induction on n; for $n = 1$, the union is a single set. For $n > 1$, let H be a hyperplane as guaranteed by Theorem 16.2.37. If n is odd, then the points on H provide one of the sets. Since a half-space contains the convex hull of all its points, it suffices to apply the induction hypothesis separately to each open half-space bounded by H. ∎

BORSUK'S CONJECTURE

The Lyusternik–Shnirel'man version of the Borsuk–Ulam Theorem states that any covering of the sphere S^{d-1} in $\mathbb{R}^d$ by d closed sets puts some pair of antipodal points into the same set. The distance between antipodal points is 2. Since S^{d-1} is a set of diameter 2 in $\mathbb{R}^d$, one way to interpret the theorem is that S^{d-1} has no partition into d sets that have diameter smaller than the full set.

Borsuk [1933] asked whether every bounded set in $\mathbb{R}^d$ (size at least 2) decomposes into $d + 1$ pieces having smaller diameter than the full set. The problem intrigued many people who expected a "yes" answer, so it became known as **Borsuk's Conjecture**, although Borsuk never conjectured an answer. The set of $d + 1$ vertices of a regular simplex already needs $d + 1$ pieces.

Borsuk obtained a positive answer when $d = 2$. Eggleston [1955] proved it for $d = 3$ (see also Heppes–Revész [1956]). For general d, Hadwiger [1946] proved it for bodies with smooth boundaries, Riesling [1971] for centrally symmetric bodies, and Dekster [1995] for bodies of revolution.

Let $f(d)$ be the least k such that every bounded set in $\mathbb{R}^d$ decomposes into k pieces of smaller diameter. It is known that $f(4) \leq 9$. Lassak [1982] proved $f(d) \leq c^d$, and Schramm [1988] proved $f(d) \leq (\sqrt{3/2} + o(1))^d$.

In a 3-page paper, Kahn–Kalai [1993] answered the original question negatively. They produced a finite set that, when $d = 1326$, requires more than $d + 1$ pieces of smaller diameter, and in general yields $f(d) \geq 1.1^{\sqrt{d}}$. In just two pages, Nilli [1994] (pseudonym of Alon) reduced the counterexample to $d = 946$. Further improvements obtained counterexamples with $d = 561$ and $f(d) \geq (1.2255 + o(1))^{\sqrt{d}}$ (Raĭgorodskiĭ [1997]), $d = 321$ (Pikhurko [2002]), $d = 298$ (Hinrichs–Richter [2003]), and $d = 65$ (Bondarenko [2014]). The last shows the existence of a two-distance set (Definition 15.1.4) of 416 points in $\mathbb{R}^{65}$ that cannot be partitioned into 83 sets of smaller diameter.

The idea of Kahn and Kalai is as follows. Create a set of n-tuples such that only small subsets avoid having orthogonal pairs. Use these to form points in $\mathbb{R}^{n^2}$ so that the diameter of the set is achieved precisely by the pairs of points arising from orthogonal pairs in the original set. Since sets of smaller diameter must avoid orthogonal pairs, many sets will be needed.

Recall that a family $\mathcal{F}$ of sets is **L-intersecting** if $|A \cap B| \in L$ for $A, B \in \mathcal{F}$. The lemma about the sets in $\mathbb{R}^n$ uses a variant of the Frankl–Wilson Theorem (Theorem 15.1.14), which states that every L-intersecting family of subsets of $[n]$ has size at most $\sum_{i=0}^{|L|} \binom{n}{i}$. For n-tuples in $\{-1, 1\}^n$, dot product plays a role analogous to intersection for n-tuples in $\{0, 1\}^n$.

16.2.40. Lemma. Let $n = 4p$, where p is an odd prime. Let V be the set of n-tuples in $\{-1, 1\}^n$ such that the first coordinate is positive and the total number of positive coordinates is even. If U is a subset of V containing no two orthogonal vectors, then $|U| \leq \sum_{i=0}^{p-1} \binom{n}{i}$.

Proof: Consider $u, v \in U$. Since u and v agree in the first coordinate, $u \cdot v \neq -n$. If $u \neq v$, then $u \cdot v \neq n$. Since $u, v \in U$, they are not orthogonal, so $u \cdot v \neq 0$.

We claim $u \cdot v \equiv 0 \pmod 4$. Let A and B be the subsets of $[n]$ consisting of the positions where u and v are positive, respectively. Now $u \cdot v = 4p - 2|A \triangle B|$, where

$\triangle$ denotes symmetric difference. Since $|A \triangle B| = |A| + |B| - 2|A \cap B|$, and $|A|$ and $|B|$ are even, the claim follows. Now, since p is odd, $u \cdot v \not\equiv 0 \pmod{p}$ when $u = v$.

To bound $|U|$, we define polynomials over $\mathbb{F}_p$. For $u \in U$, define f_u by $f_u(x) = \prod_{i=1}^{p-1}(u \cdot x - i)$. If $v = u$, then $u \cdot v = 4p$, so $f_u(v) \not\equiv 0 \pmod{p}$. On the other hand, $f_u(v) \equiv 0 \pmod{p}$ when $v \neq u$, since $u \cdot v \neq 0 \pmod{p}$ makes one factor in the definition of f_u divisible by p.

Now we capture $|U|$ linearly independent polynomials in a space of dimension $\sum_{i=0}^{p-1}\binom{n}{i}$. Each polynomial f_u is a product of $p-1$ linear factors and hence has total degree at most $p-1$ in each monomial. Define a new polynomial $\hat{f}_u$ from f_u by reducing the exponents modulo 2 in each monomial. Since $x^{2j} = 1$ and $x^{2j+1} = x$ when $x \in \{-1, 1\}$, we have $\hat{f}_u(x) = f_u(x)$ for all $x \in \{-1, 1\}^n$.

Therefore, $\{\hat{f}_u : u \in U\}$ are polynomials such that $f_u(v) \not\equiv 0 \pmod{p}$ if and only if $v = u$, for $u, v \in U$. By the "Diagonal Criterion" (Section 15.1), these polynomials are linearly independent in the space of all polynomials in n variables over $\mathbb{F}_p$. That is, if $\sum_{u \in U} c_u \hat{f}_u(v) = 0$ for all $v \in U$, then all coefficients must be 0.

Finally, each polynomial $\hat{f}_u$ is a linear combination of monomials that are products of at most $p-1$ variables, each with degree 1. There are $\sum_{i=0}^{p-1}\binom{n}{i}$ such monomials. Hence $|U| \le \sum_{i=0}^{p-1}\binom{n}{i}$. ∎

16.2.41. Theorem. (Kahn–Kalai [1993]) If all bounded sets in $\mathbb{R}^d$ decompose into $f(d)$ sets with smaller diameter, then $f(d) \ge (1.203)^{\sqrt{d}}$ for large enough d.

Proof: (Nilli [1994]) Choose $n = 4p$, where p is a prime, and define $V \subseteq \mathbb{R}^n$ as in Lemma 16.2.40. For $v \in V$, define $p(v) \in \mathbb{R}^{n^2}$ by letting the (i, j)-coordinate of $p(v)$ be $v_i v_j$. Let $P = \{p(v) : v \in V\}$.

For $u, v \in V$, we have
$$p(u) \cdot p(v) = \sum_{i=1}^{n}\sum_{j=1}^{n} u_i u_j v_i v_j = \sum_{i=1}^{n} u_i v_i \sum_{j=1}^{n} u_j v_j = (u \cdot v)^2 \ge 0.$$

Furthermore, each $p(v)$ has length n. Since V does contain orthogonal pairs of vectors and $p(u) \cdot p(v) \ge 0$, it follows that the distance between $p(u)$ and $p(v)$ equals the diameter of P if and only if $u \cdot v = 0$.

Let $m(n) = \sum_{i=0}^{n/4}\binom{n}{i}$. By Lemma 16.2.40, each subset of P with no orthogonal pairs has size at most $m(n)$. Hence the number of sets needed to partition P into pieces with smaller diameter is at least $|P|/m(n)$, which equals $2^{n-2}/m(n)$.

For a slight further improvement, note that P is spanned by a set of $\binom{n}{2}$ vectors (Exercise 26). With $d = \binom{n}{2}$, we have $n \approx \sqrt{2d}$, and $f(d) \ge 2^{n-2}/m(n) > (1.203)^{\sqrt{d}}$ for large d. In the computations, we use $\sum_{i=0}^{n/4}\binom{n}{i} < \frac{3}{2}\binom{n}{n/4}$ (see Theorem 14.4.4) and $\binom{n}{n/4} \approx \sqrt{8/(3\pi n)}(\frac{4}{3^{3/4}})^n$ (by Stirling's Formula). Thus $f(d) \ge c(3^{3/4}/2)^n \approx (1.136)^n$. With $n \approx \sqrt{2d}$, we then raise 1.136 to the $\sqrt{2}$ power. ∎

EXERCISES 16.2

16.2.1. (−) Compute the bandwidths of P_n, K_n, and C_n.

16.2.2. (−) Compute the bandwidth of the complete multipartite graph $K_{n_1,\ldots,n_k}$ with $n_1 \ge \cdots \ge n_k$ and $\sum n_i = n$. (Eitner [1979])

16.2.3. Prove that the local density lower bound for $B(P_n \,\square\, P_n)$ is about $n/2$.

16.2.4. Let S be a set of k vertices in the grid $P_n \,\square\, P_n$.
   (a) Prove that $|\partial S|$ is minimized for $|S| = k$ by a set occupying a Ferrers diagram; that is, the first a_i vertices in row i, with $a_1 \geq \cdots \geq a_n$ and $\sum_{i=1}^n a_i = k$.
   (b) For each k such that $\binom{n}{2} < k < \binom{n+1}{2}$, prove that $|\partial S| \geq n$ whenever $S \subseteq V(G)$ with $|S| = k$. Thus the boundary bound is n. (Chvatalová [1975])

16.2.5. ($\diamond$) Prove that $B(C_n \,\square\, C_n) = 2n - 1$ when n is odd. (Hint: For the lower bound, use the technique of Theorem 16.2.8. Comment: The answer is also $2n - 1$ when n is even, but the argument for the upper bound is messier then.) (Li–Tao–Shen [1981])

16.2.6. Prove that every tree with k leaves is the union of $\lceil k/2 \rceil$ pairwise intersecting paths, and use this to prove that the bandwidth of a tree with k leaves is at most $\lceil k/2 \rceil$. (Ando–Kaneko–Gervacio [1996])

16.2.7. ($\diamond$) Define a graph on the infinite grid $\mathbb{Z}^2$ by making two vertices adjacent if they are consecutive along a line with slope in $\{0, \infty, 1, -1\}$ (thus the graph is 8-regular). Prove that for any set S having vertices from r rows and s columns, the boundary of S is at least $2r + 2s + 4$. (Hint: The lower bound is immediate when S is an r-by-s rectangle. Comment: This lemma was applied in Balogh–Kaul [2007] to random geometric graphs.)

16.2.8. (+) *Bandwidth of caterpillars.* Let G be a caterpillar, and let k be an integer such that $\lceil \frac{|V(H)|-1}{\operatorname{diam} H} \rceil \leq k$ for all $H \subseteq G$. Prove that $B(G) \leq k$. (Hint: Prove that G has a numbering f in which $f(v)$ is a multiple of k whenever v is on the spine and $|f(u) - f(v)| \leq k$ for all $uv \in E(G)$.) (Sysło–Zak [1982], Miller [1981])

16.2.9. Let G be a graph with order n and bandwidth b.
   (a) For $e \in \overline{G}$, prove that $B(G + e) \leq 2b$.
   (b) Prove that if $n \geq 6b$, then $B(G + e)$ can be as large as $2b$.
(Comment: The maximum of $B(G + e)$ is $b + 1$ if $n \leq 3b + 4$ and is $\lceil (n - 1)/3 \rceil$ if $3b + 5 \leq n \leq 6b - 2$.) (Wang–Yao–West [1995])

16.2.10. Prove Brouwer's Theorem (Theorem 16.2.18) from Sperner's Lemma.

16.2.11. Prove Brouwer's Theorem (Theorem 16.2.18) from the Borsuk–Ulam Theorem.

16.2.12. Prove the general form of Tucker's Combinatorial Lemma (Theorem 16.2.21) from the Borsuk–Ulam Theorem (Theorem 16.2.23). That is, if T is a simplicial subdivision of B^n that is antipodally symmetric on the boundary, and the vertices are labeled from $\{\pm 1, \ldots, \pm n\}$ so that labels on antipodal points of the boundary are complementary, then there is a complementary edge.

16.2.13. Consider a circular necklace with beads of k colors. To capture half the beads of each color from such a necklace, an even number of cuts must be made. Show that $2\lceil k/2 \rceil$ cuts suffice, and that when k is odd the first cut can be made between any two beads.

16.2.14. Prove that in Theorem 16.2.23, the second statement implies the first. That is, if every continuous antipodal function from S^n to $\mathbb{R}^n$ is somewhere zero, prove that every continuous function from B^n to $\mathbb{R}^n$ that is antipodal on S^{n-1} is somewhere zero.

16.2.15. Prove that the following statement is equivalent to Theorem 16.2.26C: There is no continuous $f\colon B^n \to S^{n-1}$ that is antipodal on S^{n-1}.

16.2.16. For $\alpha \in \mathbb{R}$ with $0 < \alpha < 2$, the *Borsuk graph* $B(n + 1, \alpha)$ has vertex set S^n, with points adjacent when separated by distance at least α. Prove that the Borsuk–Ulam Theorem is equivalent to the statement that $\chi(B(n + 1, \alpha)) \geq n + 2$ for all n and α.

16.2.17. Prove the following statements.

(a) If G is a vertex-transitive graph, then $\chi_f(G) = |V(G)|/\alpha(G)$.

(b) The Kneser graph is vertex-transitive, and $\chi_f(K(n,k)) = n/k$.

16.2.18. ($\diamond$) Let P be the inclusion poset on the k-sets and $(n-k)$-sets in $[n]$. Apply Theorem 16.2.29 to prove $\dim P \geq n - 2k + 2$ for $n > 2k$. (Hint: Given a realizer $L_1, \ldots, L_t$, consider the unforced pairs consisting of a k-set and its complement.) (Füredi [1994])

16.2.19. *Kneser from Gale.* For $d \geq 0$ and $k \geq 1$, Gale [1956] proved that there is a set X of $2k+d$ points in S^d such that every open hemisphere contains at least k points of X. Use such a set X to give a proof of Theorem 16.2.29. (Hint: Given a proper $(n-2k+1)$-coloring of $K(n,k)$, use X to define appriopriate open sets $A_1, \ldots, A_{d+1}$ in S^d and apply the open set version of the Lyusternik–Shnirel'man Theorem.) (Bárány [1978])

16.2.20. Given a family $\mathcal{F}$ of sets, the *generalized Kneser graph* $K(\mathcal{F})$ has a vertex set $\mathcal{F}$, with vertices adjacent when the corresponding sets are disjoint.

(a) Prove that every graph is a generalized Kneser graph.

(b) Construct families of sets such that $\mathcal{F}$ is the edge set of a hypergraph with chromatic number 2, but $\chi(K(\mathcal{F}))$ is arbitrarily large.

16.2.21. ($\diamond$) Let G be the generalized Kneser graph of a family $\mathcal{F}$, defined in Exercise 16.2.20. Let $c(\mathcal{F})$ be the minimum size of a set $Y \subseteq \bigcup(\mathcal{F})$ such that the hypergraph whose edges are the members of $\mathcal{F}$ disjoint from Y is 2-colorable. Prove $\chi(G) \geq c(\mathcal{F})$. Show $c(\binom{[n]}{k}) = n - 2k + 2$. (Hint: Use the Lyusternik–Schirelman Theorem.)

16.2.22. ($\diamond$) The *Schrijver graph* $S(n,k)$ is the subgraph of the Kneser graph $K(n,k)$ induced by the vertices that are k-sets of $[n]$ whose elements are nonconsecutive, viewed cyclically. Recall that $\chi(S(n,k)) = n - 2k + 2$.

(a) Schrijver proved that $S(n,k)$ is vertex-color-critical (deleting any vertex reduces the chromatic number). Prove that not all Schrijver graphs are edge-critical.

(b) Prove that $S(n,k)$ is guaranteed to be regular only when $k = 2$.

16.2.23. ($\diamond$) Consider a mass distribution in $\mathbb{R}^d$ (no mass at points).

(a) For $d = 2$, use the Ham Sandwich Theorem to show that the distribution can be split into four parts of equal measure by two lines. (Comment: It also holds for $d = 3$ that a distribution can be split into eight equal parts by three planes.)

(b) For $d = 5$, use the moment curve to show that it is not always possible to split a distribution into 32 equal parts using five hyperplanes. (Avis [1984])

16.2.24. Construct a spanning cycle in the 4-dimensional hypercube Q_4 using four edges in each direction. Use this to show that any 16 distinct points on the moment curve in $\mathbb{R}^4$ can be separated using 4 hyperplanes. (Comment: Robinson–Cohn [1981] showed that Q_d has a spanning cycle with $2^d/d$ edges in each direction if and only if d is a power of 2.)

16.2.25. Bisection of a point set A by a hyperplane is defined by requiring $\lfloor |A|/2 \rfloor$ points of A in each open half-space bounded by the hyperplane. Show that if the definition of bisection of A is modified by letting each half-space count "one-half" for each point on the hyperplane and requiring each half-space to capture $|A|/2$, then there are pairs of point-sets in $\mathbb{R}^2$ that cannot both be bisected by one line.

16.2.26. Prove that there is a set S of $\binom{n}{2}$ vectors in $\mathbb{R}^{n^2}$ such that every point in the set P in Theorem 16.2.41 is a linear combination of vectors in S. (Comment: Although $(v_2, \ldots, v_{n-1})$ determines $p(v)$, this dependence is not linear. For example, on the moment curve the first coordinate determines the rest, but any n points on the moment curve in $\mathbb{R}^n$ are independent, by the Vandermonde determinant.)

16.3. Volumes and Containment

In this section we consider combinatorial aspects of volume and position in high-dimensional spaces. We discuss monotone sublists of random permutations, sorting algorithms starting from partial information about an underlying linear order, and geometric representations of partial orders. The common theme is converting problems of counting permutations into problems of computing volumes.

MONOTONE SUBSEQUENCES

Ulam asked for the expected maximum length of a monotone sublist in a random permutation of $[n]$. The Erdős–Szekeres Theorem guarantees always at least $\sqrt{n}$. Hammersley [1972] proved that the expected maximum is asymptotic to $c\sqrt{n}$ for some c. Logan–Shepp [1977] proved $c \geq 2$; Versik–Kerov [1977] proved $c \leq 2$. Pilpel [1990] provided a combinatorial proof of the upper bound.

Here we show only that the expected maximum length is between $(1 - 1/e)\sqrt{n}$ and $e\sqrt{n}$, but we extend to higher dimensions. The proof of the lower bound follows ideas in Winkler [1985] and Bollobás–Winkler [1988]. Recall from Chapter 10 that a d-**dimensional permutation** has a permutation of $[n]$ in each coordinate; we can view it as a d-tuple of permutations of $[n]$. Also, a list of d-dimensional vectors is **monotone** if in each coordinate the list of values is monotone; some coordinates may be strictly increasing and some weakly decreasing.

16.3.1. Theorem. For large n, the expected maximum length $f(n)$ of a monotone sublist of a d-dimensional permutation of $[n]$ satisfies
$$(1 - 1/e)n^{1/(d+1)} < f(n) < en^{1/(d+1)}.$$

Proof: For the upper bound, we consider only $d = 1$, leaving the generalization as Exercise 3. For the lower bound, we treat all dimensions.

Upper bound. Let X be the random variable counting monotone sublists of length t. Each such list occurs in some t positions. For a given set of t positions, the probability that they form a monotone list is $2/t!$. By the linearity of expectation, $\mathbb{E}(X) \leq \binom{n}{t}\frac{2}{t!} \leq 2n^t/(t!)^2$. Using Stirling's Formula to approximate $t!$, the bound becomes $(ne^2/t^2)^t/(\pi t)$. In particular, if $t \geq e\sqrt{n}$, then $\mathbb{E}(X) \to 0$. By Markov's Inequality, whp there is no monotone sublist with length at least $e\sqrt{n}$, so the expected maximum length is less than that.

Lower bound. Instead of studying random d-permutations of $[n]$, we study n random points in a unit cube in $d + 1$ dimensions. Each coordinate of each point is drawn uniformly from $[0, 1]$. Equal values occur with probability 0, so each coordinate gives an ordering. The first d coordinates give permutations by ordering the indices according to the last coordinate. A chain in the product then means points having the values in each coordinate in the same order. This allows us to ignore the order of the generated points; we just look for a chain in the set.

When generating the permutations from n random points, each set of permutations has the same probability. The jth coordinate gives the permutation σ such that $0 \leq a_{\sigma^{-1}(1)} < \cdots < a_{\sigma^{-1}(n)} \leq 1$, where a_i is the jth coordinate of the point whose last coordinate is ith smallest. The n-tuples in the jth coordinate that map to σ form an n-dimensional simplex in $[0, 1]^n$ with volume independent of σ. (We will revisit this simplex in Definition 16.3.7.) The choices of n points in the cube mapping to a particular d-permutation form a cartesian product of d such simplices. Thus when we generate a random element of $([0, 1]^{d+1})^n$, the volume (the probability) that maps to any d-permutation is the same.

We claim that n random points in $[0, 1]^{d+1}$ expect to include a chain of the desired size among chains of a restricted type. For large n, consider $\left\lfloor n^{1/(d+1)} \right\rfloor$ small cubes along the main diagonal in the unit cube; each has side length $1/\left\lfloor n^{1/(d+1)} \right\rfloor$ and volume at least $1/n$. The upper corner of each cube is the lower corner of the next. If the n chosen random points hit X of these small cubes, then picking a point from each of those cubes yields a chain of size X.

It thus suffices to show that $(1 - 1/e)n^{1/(d+1)}$ is a lower bound on $\mathbb{E}(X)$. We have $X = \sum X_i$, where $X_i = 1$ if and only if the ith small cube is nonempty. It is empty only if the n independent points all miss it, which has probability at most $(1 - 1/n)^n$ since the small cube has volume at least $1/n$. This probability is bounded by $1/e$, so $\mathbb{P}(X_i = 1) > 1 - 1/e$. By the linearity of expectation, $\mathbb{E}(X) = \left\lfloor n^{1/(d+1)} \right\rfloor \mathbb{E}(X_i) > n^{1/(d+1)}(1 - 1/e)$, where we have ignored lower-order terms. ∎

Generating n points in a $(d+1)$-dimensional cube is a way to generate random $(d + 1)$-dimensional posets; Theorem 16.3.1 gives bounds for the expected height.

BALANCED COMPARISONS

We return now to partial orders. We view a poset P as partial information about an underlying linear order on its elements, as in Section 12.4. *Sorting* is the problem of finding the true underlying linear order, using comparisons between elements. When P consists of disjoint chains, sorting is called *merging*.

Given no prior information, P is an antichain, and the answer is one of $n!$ linear orders. In general, the answer is a linear extension of P. Each comparison splits the remaining possible orders into two sets, and the result is consistent with one of them. Since an adversary can choose which set remains, in the worst case any algorithm uses at least $\lg e(P)$ comparisons, where we write $e(P)$ for the number of linear extensions of P.

The value $\lg e(P)$ is called the **information-theoretic lower bound**. The bound is achieved if every poset has a comparison that splits the remaining extensions into two equal-sized sets. Fortunately, we do not need perfect balance to obtain performance within a constant factor of the lower bound.

16.3.2. Definition. For $x, y \in P$, the probability of the event $x < y$, written $\mathbb{P}(x < y)$, is the fraction of the linear extensions of P putting x below y. A pair (x, y) of elements is δ-**balanced** if $\delta \leq \mathbb{P}(x < y) \leq 1 - \delta$.

If every poset contains a δ-balanced pair, then sorting from initial information P takes at most $\log_{1/(1-\delta)} e(P)$ comparisons. The closer δ is to $\frac{1}{2}$, the better.

16.3.3. Conjecture. ($\frac{1}{3}, \frac{2}{3}$-**Conjecture**; Kislicyn [1968], Fredman~1975, Linial [1984]) Every poset has a $\frac{1}{3}$-balanced pair. ∎

16.3.4. Example. The poset $\mathbf{1} + \mathbf{2}$ has just three linear extensions, so $\mathbb{P}(x < y) \in \{0, \frac{1}{3}, \frac{2}{3}, 1\}$ for all (x, y). Thus Conjecture 16.3.3 is best possible. ∎

16.3.5. Remark. Conjecture 16.3.3 remains open, but Kahn–Saks [1984] proved that $\delta = \frac{3}{11}$ suffices: in any finite poset, $\frac{3}{11} < \mathbb{P}(x < y) < \frac{8}{11}$ for some pair (x, y). Brightwell–Felsner–Trotter [1995] improved the guarantee to $\delta \geq \frac{1}{2} - \frac{1}{2\sqrt{5}}$ for finite posets and many infinite posets, using the machinery of Kahn–Saks plus the Ahlswede–Daykin Inequality (see also Felsner–Trotter [1993]).

There is reason to think that the conjecture might fail on a semiorder (Definition 12.4.1). Brightwell [1988] constructed infinite semiorders where one cannot guarantee a δ-balanced pair with $\delta > \frac{1}{2} - \frac{1}{2\sqrt{5}}$ (Exercise 8), but Brightwell [1989] proved the $\frac{1}{3}, \frac{2}{3}$-conjecture for finite semiorders (Exercise 9). Brightwell–Wright [1992] proved it for posets in which every element is incomparable to at most five others, Trotter–Gehrlein–Fishburn [1992] for bipartite posets, and Olson–Sagan [2018] for some other classes. Brightwell [1999] provides a survey.

The Kahn–Saks result implies that sorting can be completed using about $2.2 \cdot \lg e(P)$ comparisons. The proof does not produce a balanced comparison in polynomial time, but Kahn–Kim [1995] later addressed this using entropy (Section 11.2), which is easy to compute, unlike $e(P)$ (see Brightwell–Winkler [1991]). Fredman [1976] presented an algorithm using $\lg e(P) + 2|P|$ comparisons.

Kahn and Saks proposed a more detailed conjecture. Let δ_w be the largest δ such that every poset of width w has a δ-balanced pair. They conjectured that δ_w increases with w, approaching $\frac{1}{2}$ from below. Komlós [1990] used functional analysis to prove this for bipartite posets. It is known that $\delta_2 = \frac{1}{3}$ and $\delta_3 > \frac{1}{3}$. The proof below that $\delta_2 = \frac{1}{3}$ is easy; this case includes merging of two lists. ∎

16.3.6. Theorem. (Linial [1984]) Posets of width 2 have $\frac{1}{3}$-balanced comparisons.

Proof: A poset P of width 2 is covered by two chains, by Dilworth's Theorem; let them be $x_1, \ldots, x_n$ and $y_1, \ldots, y_m$. We may assume $x_1 \| y_1$ (otherwise P has a unique minimum and we reduce the problem) and $\mathbb{P}(x_1 < y_1) < \frac{1}{2}$. In terms of i, $\mathbb{P}(x_1 < y_i)$ is strictly increasing. Let j be the largest index with $\mathbb{P}(x_1 < y_j) < \frac{1}{2}$. We have $j < m$ unless P is a chain, because there is at most one extension with $x_1 > y_m$. If $\mathbb{P}(x_1 < y_j) \geq \frac{1}{3}$, then we have a $\frac{1}{3}$-balanced comparison. Otherwise,

$$\tfrac{1}{2} \leq \mathbb{P}(x_1 < y_{j+1}) = \mathbb{P}(x_1 < y_j) + \mathbb{P}(y_j < x_1 < y_{j+1}) \leq 2\mathbb{P}(x_1 < y_j) < \tfrac{2}{3},$$

where the inequality $\mathbb{P}(y_j < x_1 < y_{j+1}) \leq \mathbb{P}(x_1 < y_j)$ holds because $x_1 \| y_j$ allows x_1 and y_j to be switched on any extension with x_1 covering y_j. ∎

The numerical aspects of the Kahn–Saks [1984] $\frac{3}{11}$, $\frac{8}{11}$-result are quite difficult. Instead, we present the later and easier argument of Kahn–Linial [1991]
that there is always a $\frac{1}{2e}$-balanced pair; since $\frac{1}{2e} < \frac{3}{11}$, this is weaker. The idea
is to consider the average height of an element over all linear extensions. It is
easy to show that for some pair of elements this differs by less than 1. The task
then is to show that any such pair is a $\frac{1}{2e}$-balanced pair. We use a geometric object
associated with a linear extension.

16.3.7. Definition. For a poset Q with elements $q_1, \ldots, q_n$, the **order polytope** $\mathbf{O}(Q)$ is the set of vectors $(f(q_1), \ldots, f(q_n))$ in $\mathbb{R}^n$ such that f is an
order-preserving function from Q to $[0, 1]$. For a linear extension σ of Q, the
canonical simplex $\Delta(\sigma)$ in $\mathbf{O}(Q)$ is the set of vectors $(x_1, \ldots, x_n)$ such that
$0 \le x_{\sigma^{-1}(1)} \le \cdots \le x_{\sigma^{-1}(n)} \le 1$. The **centroid** of an n-dimensional region R is
the vector $c(R)$ whose ith coordinate is $(\int_R x_i dV)/(\int_R dV)$.

16.3.8. Lemma. (Stanley [1981]) Let Q be a poset with elements $q_1, \ldots, q_n$, order polytope $\mathbf{O}(Q)$, and canonical simplex $\Delta(\sigma)$ for each linear extension σ.
(a) $\mathbf{O}(Q)$ is the union of $\Delta(\sigma)$ over all linear extensions σ of Q.
(b) $\Delta(\sigma)$ has volume $\frac{1}{n!}$, independent of σ.
(c) $c(\Delta(\sigma))_i = \frac{\sigma(q_i)}{n+1}$, where $c(\Delta(\sigma))$ is the centroid of $\Delta(\sigma)$.
(d) $\mathrm{Vol}(\mathbf{O}(Q)) = \frac{e(P)}{n!}$ and $c(\mathbf{O}(Q))_i = \frac{h_i}{n+1}$, where $h_i = \frac{1}{e(P)} \sum_\sigma \sigma(q_i)$.

Proof: (a) A vector in $\mathbb{R}^n$ satisfying all inequalities imposed by pairs in Q has its
coordinates in some order, which must correspond to a linear extension of Q.

(b) The canonical simplex is seen to be a simplex using the variables $y_0, \ldots, y_n$,
where $y_i = x_{\sigma^{-1}(i+1)} - x_{\sigma^{-1}(i)}$, with x and σ extended by $\sigma(0) = 0$, $\sigma(n+1) = n+1$,
$x_0 = 0$, and $x_{n+1} = 1$. The n-dimensional volume of a "cone" is Bt/n, where B is
the $(n-1)$-dimensional volume of the base and t is the height from base to apex.
The formula for these simplices follows by induction on n, since the height is 1.

(c) This can be evaluated by induction on n (Exercise 4).

(d) The canonical simplices for distinct linear extensions intersect only at
their boundaries (belonging to both requires equality among the coordinates for
which the corresponding elements of Q are ordered differently in the two extensions). Hence the volume of the union $\mathbf{O}(Q)$ is the sum of the individual volumes.

Note that h_i is the average height of q_i on linear extensions. The second
equality is obtained from part (c) by averaging over the linear extensions, since
all canonical simplices have the same volume. ∎

In this setting, the probability of an event A is the fraction of $\mathrm{Vol}(\mathbf{O}(Q))$ in
the canonical simplices for the extensions where A occurs. The event $q_i < q_j$ is
described by the inequality $f(q_i) < f(q_j)$, and each canonical simplex lies on one
side of the hyperplane $f(q_i) = f(q_j)$. The Kahn–Linial proof studies the volume
of this subset of the order polytope using a theorem from convex geometry.

Given a convex body K in $\mathbb{R}^n$, let H^- and H^+ be two parallel hyperplanes that
are both tangent to K, and let H be the hyperplane halfway between them. Let
$\hat{x}$ be a point in H, and let u be the vector from $\hat{x}$ to a point in H^+ along a line perpendicular to H. For $\lambda \in [-1, 1]$, the **cross-section** K_λ of K is the intersection
of K with the hyperplane through $\hat{x} + \lambda u$ parallel to H.

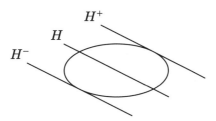

16.3.9. Theorem. (Brunn–Minkowski Theorem) Let K be a convex body in $\mathbb{R}^n$ with cross-sections K_λ for $\lambda \in [-1, 1]$. The function r defined by

$$r(\lambda) = [\text{Vol}_{n-1}(K_\lambda)/V_{n-1}]^{1/(n-1)}$$

is concave on $[-1, 1]$, where V_{n-1} is a normalizing constant equal to the volume of the $(n-1)$-dimensional unit ball.   ■

We are viewing $\sigma(x)$ as the height of x on a linear extension σ of Q, from 1 to $|Q|$. Let $h(x)$ denote the average of $\sigma(x)$ over all linear extensions. Deleting a unique minimum or maximum does not change the probabilities for pairs of other elements. Hence we may assume $1 < h(x) < |Q|$ for $x \in Q$. By the Pigeonhole Principle, some pair (x, y) satisfies $|h(y) - h(x)| < 1$. This requires x below y in some extensions and x above y in others. We show that each case occurs in at least the fraction $\frac{1}{2e}$ of the extensions.

16.3.10. Theorem. (Kahn–Linial [1991]) If in a poset Q the expected heights of q and q' satisfy $h(q) - h(q') > -1$, then $\mathbb{P}(q > q') > \frac{1}{2e}$.

Proof: We write $q_1, \ldots, q_n$ for the elements of Q and take $x, y \in \mathbb{R}^n$. Given $q_i \| q_j$ and $h(q_i) - h(q_j) > -1$, we prove $\mathbb{P}(q_i > q_j) > \frac{1}{2e}$. Using Lemma 16.3.8, we rephrase this in terms of the order polytope; let $K = \mathbf{O}(Q)$. With $K^+ = \{x \in K \colon x_i - x_j \geq 0\}$, we want to prove $\frac{\text{Vol}(K^+)}{\text{Vol}(K)} > \frac{1}{2e}$ when $h(q_i) - h(q_j) > -1$.

We express K^+ using cross-sections of K. Let l be the line in $\mathbb{R}^n$ generated by the vector w with $1/2$ in coordinate i and $-1/2$ in coordinate j and 0 elsewhere. The projection of x on l is λw when $x_i - x_j = \lambda$. The cross-section of K in the hyperplane through λw perpendicular to l is thus $\{x \in K \colon x_i - x_j = \lambda\}$, which we write as K_λ. Note that $K_\lambda \neq \varnothing$ only when $\lambda \in [-1, 1]$. Also, $c(K)_i - c(K)_j > \frac{-1}{n+1}$. We prove the desired inequality for any convex body K satisfying these two facts.

We use Theorem 16.3.9 to transform K into a "double cone" where computations will simplify. We first "symmetrize" K around the line l. Letting $r(\lambda)$ be defined as in Theorem 16.3.9, we replace K_λ with the $(n-1)$-dimensional ball B_λ of radius $r(\lambda)$ centered at λw and contained in the same hyperplane as K_λ. Let $B = \bigcup_{-1 \leq \lambda \leq 1} B_\lambda$. Theorem 16.3.9 implies that B is convex. Also $\text{Vol}(B) = \text{Vol}(K)$ and $\text{Vol}(B^+) = \text{Vol}(K^+)$, where $B^+ = \bigcup_{\lambda \geq 0} B_\lambda$. Furthermore, $c(B)_i - c(B)_j = c(K)_i - c(K)_j > \frac{-1}{n+1}$. Hence it suffices to prove the inequality for B^+ and B.

The symmetrization reduces the problem to two dimensions, parallel and perpendicular to l. These become coordinates y_1 and y_2, with B being the "solid of revolution" of the curve $y_2 = r(y_1)$ around the line l, which is now the y_1 axis (see figure below). In the (y_1, y_2)-plane, let $Q = (0, r(0))$, and choose P on the positive y_1 axis (at $(u, 0)$) so that the volume obtained by revolving PQ equals $\text{Vol}(B^+)$.

Since $r(y_1)$ is concave, $u \geq 1$. Let $S = (-1, 0)$, and choose R on the line PQ so that the double cone D obtained by revolving the triangle PRS has volume $\mathrm{Vol}(B)$.

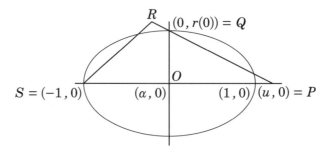

In both the left and the right half-planes, the volume under D is shifted to the right from the volume under B. That is, $\int_{-1}^{t}[r(y_1) - s(y_1)]dy_1 \geq 0$, where $s(y_1)$ is the height of the piecewise linear curve SRP at y_1. This follows from the concavity of r. Hence $c_D(y_1) \geq c_B(y_1)$. The problem thus reduces to showing the following for the double cone:

> If D is an n-dimensional double cone with apexes $(-1, 0, \ldots, 0)$ and $(u, 0, \ldots, 0)$ and centroid $(\alpha, 0, \ldots, 0)$ such that $u \geq 1$ and $\alpha \geq -1/(n+1)$, then $\frac{\mathrm{Vol}(D^{+})}{\mathrm{Vol}(D)} > \frac{1}{2e}$.

Let C_1 and C_2 be the cones comprising D, with common base containing R, apexes $(-1, 0, \ldots, 0)$ and $(u, 0, \ldots, 0)$, and heights h_1, h_2 with $h_1 + h_2 = u + 1$. Because C_1 and C_2 have the same base, we have $\frac{\mathrm{Vol}(C_2)}{\mathrm{Vol}(C_1)} = \frac{h_2}{h_1}$, or $\frac{\mathrm{Vol}(C_2)}{\mathrm{Vol}(D)} = \frac{h_2}{u+1}$. The volume of an n-dimensional cone from a point is proportional to the nth power of its height, so we can rewrite the ratio as

$$\frac{\mathrm{Vol}(D^{+})}{\mathrm{Vol}(D)} = \frac{\mathrm{Vol}(D^{+})}{\mathrm{Vol}(C_2)} \cdot \frac{\mathrm{Vol}(C_2)}{\mathrm{Vol}(D)} = \left(\frac{u}{h_2}\right)^{n}\frac{h_2}{u+1}.$$

Because this formula is minimized by maximizing h_2, which is equivalent to minimizing α when $\mathrm{Vol}(D)$ is fixed, we may assume $\alpha = \frac{-1}{n+1}$.

The distance from the apex to the centroid of an n-dimensional right cone is $\frac{n}{n+1}$ times the height. Hence the centroids of C_1 and C_2 have y_1-coordinates $-1 + h_1\frac{n}{n+1}$ and $u - h_2\frac{n}{n+1}$. Taking the weighted average of the centroids in proportion to their volume yields the centroid of D, so

$$\frac{-1}{n+1} = \frac{1}{u+1}\left[h_1\left(-1 + \frac{h_1 n}{n+1}\right) + h_2\left(u - \frac{h_2 n}{n+1}\right)\right].$$

Using $h_1 + h_2 = u + 1$ and solving for h_2 yields $h_2 = \frac{un}{n-1}$. Substituting this in our expression for the volume ratio and using $u \geq 1$ yields

$$\frac{\mathrm{Vol}(D^{+})}{\mathrm{Vol}(D)} = \left(\frac{n-1}{n}\right)^{n-1}\frac{u}{u+1} > \frac{1}{2e}. \qquad \blacksquare$$

The Kahn–Linial proof was inspired by the proof of Mityagin's Theorem, which says that the fraction of the volume on one side of a hyperplane through the centroid of an n-dimensional convex body is at least $(\frac{n}{n+1})^{n}$ (this is sharp for a simplex and exceeds e^{-1}). In the proof above, the hyperplane $x_i = x_j$ passes through the centroid of the order polytope if $h(q_i) = h(q_j)$. Hence when $h(q_i) = h(q_j)$ we can conclude $e^{-1} < \mathbb{P}(q_i < q_j) < 1 - e^{-1}$ by Mityagin's Theorem.

The Kahn–Saks [1984] proof uses the Alexandrov–Fenchel Inequalities for mixed volumes. We only present a simpler application of these inequalities by Stanley that motivated the Kahn–Saks approach.

16.3.11. Theorem. (Alexandrov–Fenchel Inequalities) Let K_0, K_1 be two convex sets in $\mathbb{R}^n$, and let $K_\lambda = \{(1 - \lambda)x + \lambda y \colon x \in K_0, y \in K_1\}$ for $0 \le \lambda \le 1$. If the dimension of these sets is d, then the volume of K_λ is a homogeneous polynomial in λ of degree d given by

$$\mathrm{Vol}(K_\lambda) = \sum_{i=0}^{d} \binom{d}{i} a_i \lambda^i (1 - \lambda)^{d-i},$$

where the "mixed volumes" a_i form a log-concave sequence. ∎

Here "Inequalities" refers to $a_i^2 \ge a_{i-1} a_{i+1}$, the log-concavity of $\langle a \rangle$. The name "mixed volume" for a_i arises from $a_0 = \mathrm{Vol}(K_0)$ and $a_d = \mathrm{Vol}(K_1)$. A shorthand notation for K_λ is $K_\lambda = (1 - \lambda)K_0 + \lambda K_1$. Stanley applied Theorem 16.3.11 to appropriate subsets of the order polytope to study the height distribution. Since here we view a linear extension of an n-element poset as a map into $[n]$, we say "position k" rather than "height k".

16.3.12. Theorem. (Stanley [1981]) Given a poset Q and an element $x \in Q$, the number of extensions of Q in which x has position k is log-concave in k.

Proof: Let $A = \{f \in \mathbf{O}(Q) \colon y \le x \Rightarrow f(y) = 0\}$, and let $B = \{f \in \mathbf{O}(Q) \colon y \ge x \Rightarrow f(y) = 1\}$. The mixture $K_\lambda = (1 - \lambda)A + \lambda B$ is

$$K_\lambda = \{f \in \mathbf{O}(Q - x) \colon (y \le x \Rightarrow f(y) \le \lambda) \text{ and } (y \ge x \Rightarrow f(y) \ge \lambda)\}.$$

Note that K_λ is $(n - 1)$-dimensional, since $f(x) = \lambda$ for $f \in K_\lambda$.

Theorem 16.3.11 gives one expression for $\mathrm{Vol}(K_\lambda)$; we get another from the decomposition of $\mathbf{O}(Q)$ into canonical simplices. Let σ be a linear extension of Q in which x has position k, and let $S = K_\lambda \cap \Delta(\sigma)$. The vectors in S are the order-preserving functions f respecting σ and satisfying $f(x) = \lambda$. Since these coordinates respect a fixed order and bound, the restriction of S to the first $k - 1$ elements in σ is a $(k - 1)$-dimensional simplex of side-length λ. Similarly, the restriction to the final $n - k$ elements in σ is an $(n - k)$-dimensional simplex of side-length $1 - \lambda$. The vectors in S are all pairings of these, so S is the product of the two simplices. By Lemma 16.3.8 its $(n-1)$-dimensional volume is $\frac{\lambda^{k-1}}{(k-1)!} \frac{(1-\lambda)^{n-k}}{(n-k)!}$.

Since K_λ is composed of its intersections with each of the canonical n-dimensional simplices corresponding to extensions, we have

$$\mathrm{Vol}(K_\lambda) = \sum_{k=1}^{n} e_k(x) \frac{\lambda^{k-1}}{(k-1)!} \frac{(1 - \lambda)^{n-k}}{(n-k)!},$$

where $e_k(x)$ is the number of linear extensions of Q in which x has position k. On the other hand, $\mathrm{Vol}(K_\lambda) = \sum_{k=0}^{n-1} \binom{n-1}{k} a_k \lambda^k (1-\lambda)^{n-1-k}$. Hence $e_k(x)/(n-1)! = a_{k-1}$, and e_k is a log-concave sequence in k. ∎

CONTAINMENT ORDERS

Every family of sets, under inclusion, forms a poset. We have studied such posets for subsets of a finite set, for down-sets in a poset, etc. Specifying a family from which the sets must be drawn yields a "containment class" of posets.

16.3.13. Definition. The **containment order** of sets $S_1, \ldots, S_n$ is the poset P on a fixed set $x_1, \ldots, x_n$ given by $x_i < x_j$ if and only if $S_i \subset S_j$. The sets form a **containment representation** of P. When the sets are constrained to lie in a family Σ, the resulting posets are Σ-**containment posets**.

16.3.14. Example. When Σ is the set of discs in the plane, the Σ-containment posets are the **circle orders**. When Σ is the set of planar "wedges" separated from the rest of the plane by two rays with a common endpoint, the Σ-containment posets are the **angle orders**. Allowing the set of polygonal regions bounded by a path of k segments yields the k-**gon orders**. We may also use boxes in t dimensions, unions of t intervals, etc. When we must use finite sets, we are embedding P in $\mathbf{2}^t$, where t is the size of the union of the sets, and we seek to minimize t. ∎

In a k-dimensional poset, k "criteria" specify the order relation. Intuition suggests that we all k-dimensional posets should be Σ-containment posets if and only if sets in Σ are specified with k "degrees of freedom".

16.3.15. Example. When Σ is the set of k-dimensional boxes having one corner at the origin and another with all coordinates positive, the Σ-containment posets are precisely the k-dimensional posets.

Angles are specified by four numbers: the coordinates of the center and the angles of the rays. As expected, every 4-dimensional poset is an angle order (Exercise 12; Fishburn–Trotter [1985]), but some 5-dimensional posets are not. A k-gon is specified by $2k$ numbers, and every $2k$-dimensional poset is a k-gon order (Exercise 16; Sidney–Sidney–Urrutia [1988]).

Circles in the plane are specified by three numbers: the radius and position of the center. Some 4-dimensional posets are not circle orders. Hurlbert [1988] and Scheinerman–Wierman [1988] proved that some infinite 3-dimensional posets are not circle orders, leaving the finite case open. Finally, Felsner–Fishburn–Trotter [1999] proved the existence of finite 3-dimensional posets that are not containment orders of spheres in any dimension; hence they are not circle orders. ∎

16.3.16. Example. *Containment posets of high dimension.* Recall from Example 12.3.6 the "standard example" S_n, which is the subposet of $\mathbf{2}^{[n]}$ consisting of the sets of size 1 and their complements. We show that S_n is both a circle order and an angle order, as suggested in the figure below.

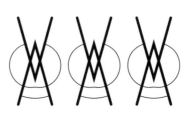

To express S_n as a circle order, assign the 1-set i a tiny disc C_i so that the centers of these are equally spaced on the unit circle, and assign each $(n-1)$-set $\bar{i}$ a large disc so that these centers are equally spaced on a large circle centered at the origin. Arrange them so that $C_{\bar{i}}$ does not contain the unit circle, and the segment joining the centers of C_i and $C_{\bar{i}}$ contains the origin. Choose the radii so that $C_{\bar{i}}$ does not contain C_i but contains all C_j with $j \neq i$.

To express S_n as an angle order, assign the 1-sets thin wedges pointing down from corners on a horizontal line, and assign the $(n-1)$-sets reflex-angle wedges with corners directly below the corners of their complements. Such a wedge contains all the narrow wedges except the one for the complementary set. ∎

Our objective now is a general argument to show that when Σ has k degrees of freedom, not all $(k+1)$-dimensional posets are Σ-containment posets. At the time, many instances of this question for particular Σ were open. To prove the result, we need a lower bound on the number of n-element posets of a fixed dimension and an upper bound on the number of distinct n-element posets that are Σ-containment posets. When the dimension is too large, the former exceeds the latter. Here we take logarithms with respect to any base, since only the ratio of the logarithms matters.

16.3.17. Theorem. (Alon–Scheinerman [1988]) If $p(n,k)$ counts the posets on $[n]$ with dimension at most k, then $\lim_{n\to\infty} \frac{\log p(n,k)}{nk\log n} = 1$.

Proof: Every poset with dimension at most k is the intersection of k linear extensions. We obtain an upper bound by choosing k arbitrary linear orders. Hence $p(n,k) \leq \binom{n!}{k} \leq n^{nk}/k!$, which yields $\log p \leq nk\log n$.

For the lower bound, we construct nearly this many distinct posets. A poset P has dimension at most k if there exists $f: P \to \mathbb{R}^k$ such that $x \leq y$ if and only if $f(x)_t \leq f(y)_t$ for all x, y, t. We construct point-sets $\{f(r) \in \mathbb{R}^k : 1 \leq r \leq n\}$ that yield distinct posets on $[n]$.

Let m be a parameter, which we will later set to $n/\log n$. Define mk boxes in $\mathbb{R}^k$ as follows: for $1 \leq j \leq k$ and $1 \leq i \leq m$, let $B_{i,j}$ be the product of $[i, i+\frac{1}{2}]$ in coordinate j and $[0, m+1]$ in all other coordinates. For fixed j, we have m "parallel" boxes; call this set of boxes M_j. The intersection of a family of boxes consisting of one box from each of $M_1, \ldots, M_k$ is a cube with side-length $\frac{1}{2}$; in this way we form m^k small cubes.

We use these boxes to specify $f(1), \ldots, f(n)$ in $\mathbb{R}^k$. Each box $B_{i,j}$ is determined by two points in $\mathbb{R}^k$. For the first $2mk$ points in $[n]$, let $f(1), \ldots, f(2mk)$ be the determining corners of the boxes $\{B_{i,j}\}$, in a fixed order. Thus the subposet induced by the elements $1, \ldots, 2mk$ in $[n]$ is the same for each poset we construct.

For $2mk+1 \leq r \leq n$, let $f(r)$ be a point inside one of the m^k small cubes. There are $(m^k)^{n-2km}$ ways of making these choices. They determine distinct subposets of relations between $[2mk]$ and $[n]-[2mk]$, because r is between the points assigned to the determining corners of the box $B_{i,j}$ if and only if the small cube containing $f(r)$ is contained in $B_{i,j}$. Hence $p(n,k) \geq (m^k)^{n-2km}$. Setting $m = n/\log n$ yields $\log p \geq n(k - o(1))\log n$. ∎

Next we define "degrees of freedom" more precisely.

16.3.18. Definition. A family of sets Σ is k-**generated** if the sets of Σ are specified by an injection $f: \Sigma \to \mathbb{R}^k$ and there is a finite list of polynomials $\{p_j\}$ in $2k$ variables such that the containment $S \subset T$ for $S, T \in \Sigma$ depends only on the signs of $p_j(f(S), f(T))$.

16.3.19. Example. *Discs are 3-generated.* Let Σ be the set of discs in the plane. For each disc, let f give the radius r and the coordinates (x, y) of the center. For discs C_1 and C_2, we then have $C_1 \subseteq C_2$ if and only if $r_1 - r_2 \leq 0$ and $(x_1 - x_2)^2 + (y_1 - y_2)^2 - (r_1 - r_2)^2 \leq 0$. In this example, only one sign pattern corresponds to containment, but in general there may be more than one. ∎

We show next that when Σ is k-generated, the number of Σ-containment posets on $[n]$ grows more slowly than the number of $(k + 1)$-dimensional posets, and hence not every $(k + 1)$-dimensional poset is a Σ-containment poset. This is an existence proof and does not construct a forbidden $(k + 1)$-dimensional poset. For example, it implies only that some 4-dimensional poset is not representable as a circle order. In fact, $2^{[4]} - \{\varnothing, [4]\}$ is such a poset (Exercise 15).

16.3.20. Theorem. (Alon–Scheinerman [1988]) If Σ is a k-generated family of sets, then some $(k + 1)$-dimensional poset is not a Σ-containment poset.

Proof: Each Σ-containment poset on $[n]$ is specified by nk real numbers, k for each $i \in [n]$ to specify the set $S_i \in \Sigma$ that represents i. Let $p_1, \ldots, p_t$ be the polynomials used to test containment for a pair of sets in Σ. With $n(n - 1)$ ordered pairs of sets, we have $tn(n - 1)$ signs whose pattern specifies the poset. Each polynomial depends on only $2k$ variables, but there are nk variables altogether.

Distinct posets must have distinct signs for some p_j applied to some pair (S_i, S_j). If we could have the full set of $3^{tn(n-1)}$ sign patterns, then we might represent all k-dimensional posets, since the logarithm of $3^{tn(n-1)}$ grows quadratically with n, faster than $nk \log n$. However, the limited number of variables restricts the number of possible sign patterns.

The desired restriction is the result of **Warren's Theorem**. Given r polynomials in l variables with maximum degree d, Warren [1968] proved that the number of plus/minus sign patterns is bounded by $(4edr/l)^l$ if $r \geq l$ and e is the base of the natural logarithms. This can easily be extended to bound the plus/zero/minus patterns by $(8edr/l)^l$ (Exercise 20).

For our application, set $r = tn(n - 1)$ and $l = nk$. Since t, k, d are constants independent of n, the values of r and l yield $(cn)^{nk}$ as an upper bound on the number of sign patterns, for some constant c. This also bounds the number of Σ-containment posets on $[n]$. The logarithm of the bound is asymptotic to $nk \log n$, so by Theorem 16.3.17 not all $(k + 1)$-dimensional posets are representable. ∎

Given a k-generated family Σ, we can now ask whether every finite k-dimensional poset is a Σ-containment order. As a "no" example, the family Σ of vertical rays in the upper half-plane is 2-generated, but all Σ-containment posets are disjoint unions of chains, omitting most 2-dimensional posets (Alon–Scheinerman [1988]). We have noted that the answer is "yes" for k-dimensional boxes, angles, and t-gons, but it is "no" for discs. In general, the question is hard.

EXERCISES 16.3

16.3.1. $(-)$ Show that if Σ is the set of k-dimensional axis-parallel boxes, then the Σ-containment posets are precisely the $2k$-dimensional posets. (Dushnik–Miller [1941] for $k = 1$; Golumbic–Scheinerman [1989] for all k)

16.3.2. Give a formula for the number of linear extensions of $\mathbf{m} \times \mathbf{n}$. (Hint: Model the linear extensions using objects counted in Part I of this book.)

16.3.3. $(\diamond)$ Prove that the expected maximum length of a monotone sublist of a d-dimensional permutation of $[n]$ is at most $en^{1/(d+1)}$. (Hint: See Theorem 16.3.1.)

16.3.4. Let Q be a poset with n elements. Suppose that element $q_i \in Q$ has height k on a linear extension σ. Prove that $c(\Delta(\sigma))_i = \frac{k}{n+1}$ (statement (c) of Lemma 16.3.8).

16.3.5. *Merging many chains.* Let P be a disjoint union of k chains with sizes $n_1, \dots, n_k$.
(a) Determine the information-theoretic lower bound for the number of comparisons needed to sort from the partial information P.
(b) Construct algorithms that come close to this lower bound when all but one chain have size 1 and when all the chains have the same size. (Saks [1985])

16.3.6. Consider arbitrary sets of linear orders equally likely to be the "true underlying linear order". Balanced comparisons are no longer guaranteed. Fix $\delta > 0$. Show that there is a set of linear orders on some finite set admitting no δ-balanced comparison, even when the orders are a subset of the linear extensions of one poset. (Hint: Use orderings obtained from a single ordering by transposing two consecutive elements.)

16.3.7. Kahn–Saks [1984] proved that an incomparable pair is $\frac{3}{11}$-balanced when its expected heights differ by less than 1. For sharpness, Saks [1982] constructed the poset below having a pair (x, y) such that $h(y) - h(x) = 1$ and $\mathbb{P}(y < x) = \frac{3}{11}$. Explain how the proof of Theorem 16.3.6 finds a $\frac{1}{3}$-balanced pair in this poset.

16.3.8. $(\diamond)$ Define a poset Q on $\{x_i : i \in \mathbb{Z}\}$ by $x_i < x_j$ if and only if $i \leq j - 2$. Let Q_n and Q'_n be the subposets consisting of $\{x_1, \dots, x_n\}$ and $\{x_{-n+1}, \dots, x_0, \dots, x_n\}$, respectively.
(a) Determine the number of linear extensions of Q_n. Compute $\lim_{n \to \infty} \mathbb{P}(x_1 > x_2)$.
(b) Within Q'_n, let $\delta_n = \mathbb{P}(x_0 > x_1)$. Compute $\lim_{n \to \infty} \delta_n$. What does this say about balanced pairs in Q? (Brightwell [1988])

16.3.9. $(+)$ In a poset P, say that an element z is *good* for an ordered incomparable pair (x, y) if z covers x and is incomparable to y or if z is covered by y and is incomparable to x.
(a) Prove that if $1 > \mathbb{P}(x < y) > \frac{2}{3}$ and every element $z \in P - \{x, y\}$ satisfies $\mathbb{P}(z < x) \geq \frac{2}{3}$ or $\mathbb{P}(y < z) \geq \frac{2}{3}$, then at least two elements are good for (x, y).
(b) Recall that a semiorder (Definition 12.4.1) is a poset representable by a function f such that $x < y$ if and only if $f(y) - f(x) > 1$. Prove that if a semiorder P has no $\frac{1}{3}$-balanced pair, then $\mathbb{P}(x < y) > \frac{2}{3}$ whenever $f(x) < f(y)$.
(c) Prove that every finite semiorder has a $\frac{1}{3}$-balanced pair. (Brightwell [1989])

16.3.10. Show that every poset is a containment poset of substars of a star. (Golumbic–Scheinerman [1989])

16.3.11. *Interval orders as containment orders.* Let Σ be the family of point-deleted real intervals unbounded below; that is, a set $A(a, b) \in \Sigma$ is determined by two numbers a, b with $a < b$ by setting $A(a, b) = \{x \leq b: x \neq a\}$. Prove that the class of interval orders is properly contained in the set of Σ-containment orders.

16.3.12. Prove that every interval order and every 4-dimensional poset is an angle order. (Fishburn–Trotter [1985])

16.3.13. Prove that every bipartite poset is a sphere containment order for spheres in some dimension. (Scheinerman [1993])

16.3.14. Prove that every 3-dimensional poset is the containment poset of some set of equilateral triangles with the same orientation. (Santoro–Urrutia [1987])

16.3.15. ($\diamond$) $\mathbf{2}^{[4]} - \{\varnothing, [4]\}$ *is not a circle order.* A *function representation* of P assigns each $i \in P$ a continuous function $f_i: [0, 1] \to \mathbb{R}$ such that $i \leq j$ in P if and only if $f_i(x) \leq f_j(x)$ for all $x \in [0, 1]$. The *crossing number* of P is the minimum over all function diagrams of the maximum number of crossings of two curves.

(a) Consider a function representation of the standard example S_n. Prove that there are n different points $x_1, \ldots, x_n$ such that $f_{\bar{i}}(x_i) < f_i(x_i)$ for $1 \leq i \leq n$.

(b) For n even, let T_n be the subposet of $\mathbf{2}^{[n]}$ consisting of the elements of size $1, n/2, n - 1$. Prove that the crossing number of T_n is $n - 1$.

(c) Prove that the crossing number of a circle order is at most 2. (Hint: Show first that any finite circle order has a circle representation in which all the circles have a common interior point. Use such a representation to construct a suitable function representation.) (Sidney–Sidney–Urrutia [1988])

16.3.16. For $k \geq 3$, prove that every $2k$-dimensional poset is the containment order of a family of k-gons, where k-gon is a polygonal region bounded by a closed curve consisting of k segments. (Sidney–Sidney–Urrutia [1988])

16.3.17. Use Theorem 16.3.20 to prove that there are $(2k + 1)$-dimensional posets that are not containment orders of k-gons, where a k-gon is the region enclosed by a simple polygon with k sides. (Hint: Show that if the segment from (x, y) to (x', y') is not vertical, then it intersects the segment from (a, b) to (a', b') if and only if $\left(\frac{b'-y}{a'-x} - \frac{y'-y}{x'-x} \right) \left(\frac{b-y}{a-x} - \frac{y'-y}{x'-x} \right) \leq 0$.)

16.3.18. ($\diamond$) *Interval number of posets.* The *interval inclusion number* of a poset P, written $i(P)$, is the least t such that P is a containment poset of a family of sets in which each set is the union of at most t real intervals. (Madej–West [1991])

(a) Prove that $i(S_n) = 2$, where S_n is the standard example of dimension n.

(b) Prove that $i(P) \leq \lceil r/2 \rceil$ when $\dim P = r$.

(c) Use Theorem 16.3.20 to show that part (b) is sharp for large r.

(d) Prove that $|i(P) - i(P^*)| \leq 1$, where P^* is the dual of P. (Comment: The difference can be 1; $i(P)$ is not a comparability invariant.)

16.3.19. Prove that the poset obtained from the n-dimensional subset lattice $\mathbf{2}^{[n]}$ by deleting the top and bottom element has interval inclusion number $\lceil n/2 \rceil$. (Madej–West [1991])

16.3.20. Use Warren's Theorem (see Theorem 16.3.20) to show that r polynomials in l variables have at most $(8edr/l)^l$ plus/zero/minus sign patterns (Alon–Scheinerman [1988])

Hints to Selected Exercises

1.1.20. Consider the number of runs of odd entries.

1.1.26. Guess an extremal permutation and show that pushing an arbitrary permutation toward it does not decrease the displacement.

1.1.31. Sum the absolute differences of consecutive numbers, in two ways.

1.1.33. Use induction on k.

1.2.18. For the algebraic proof, use the generalized form of Theorem 1.2.3(5).

1.2.22. For part (a), use induction on n. For part (b), first generalize part (a), using induction on m.

1.2.23. Do (b) by induction on m, using (a) to reduce the number of indices.

1.2.26. Count the triples (x, A, B) such that $A, B \subseteq [n]$ and $x \in A \cap B$.

1.2.27. Express each sum as a product.

1.2.39. On the left, group the pairs according to the choice of A. On the right, group the pairs according to the choice of $B \cap N$.

1.2.49. Group the steps in pairs, and encode paths as words of length n in an alphabet of size 4.

1.3.10. Every tree with at least two vertices has a leaf.

1.3.14. Specify two vertices. Delete an edge leaving one on the path to the other.

1.3.16. The term for $k = 0$ may need to be treated separately.

1.3.18. Use Corollary 1.3.7 and the Multinomial Formula.

1.3.26. Use induction on d to add the next term in the sum.

1.3.29. Count the pairs consisting of an ordered tree and one of its leaves, or apply the Narayana numbers.

1.3.32. Change the first peak or valley at even height into the other type.

1.3.38. Consider cyclic arrangements of $a + 1$ 1s and b 0s.

1.3.39. Convert the lists to binary lists, applying Theorem 1.3.17 when n is odd and Theorem 1.3.20 when n is even.

1.3.46. Use the idea of Theorem 1.3.17.

2.1.26. Prove $\sum_{i=1}^{\lfloor n/2 \rfloor} \hat{F}_{n+1-2i} < \hat{F}_n$ for $n \in \mathbb{N}$.

2.1.29. For length at least 2, solve the problem first for $d = 1$.

2.1.39. Interpret $a_{m,n}$ as counting strings in three types of steps, one vertical and two horizontal.

2.1.42. Consider the argument for the derangement recurrence.

2.1.43. Count the permutations satisfying the increase condition. Eliminate the non-derangements by considering the smallest and largest fixed points.

2.1.46. Both sides count the permutations of $[n + k]$ with no fixed points in $[n]$.

2.1.54. Use induction.

2.1.56. With $b_n = a_n/n$, prove that $b_{n+1} = b_n + C_n(pq)^n$.

2.1.57. Find recurrences for both sequences.

2.1.63. One can introduce auxiliary sequences, obtain a system of recurrences, and turn them into a single recurrence for $\langle a \rangle$, or obtain a single recurrence of unbounded order and manipulate it to eliminate summations.

2.1.68. Use row and column operations and then induction.

2.2.27. Manipulate the expression for the sum so that the Binomial Formula can be used to simplify it.

2.2.34. Divide $(1 - x)$ and $(1 - y)$ into the denominator and use the remainder to express the denominator using $(1 - x)$ and $(1 - y)$ as factors. Expanding the remaining factor leads to an infinite sum for the generating function. From the kth term in the sum, extract its coefficient of $x^m y^n$.

2.2.37. With corners $x_0, \ldots, x_{n+1}$, let x_{k+1} be the least indexed vertex such that $x_{k+1}x_{n+1}$ is a chord; note that there may be no such chord.

2.2.39. Develop a system of recurrences for tilings of several types of boards.

2.2.40. Show that $\lfloor \phi n \rfloor$ satisfies the recurrence that defines f.

2.3.16. For part (a), consider the possible locations of the first ball.

2.3.18. For part (c), give a recurrence for f in terms of which seat is filled last, and use this to prove $f(n) = g(n)$ by induction.

3.1.25. For part (a), consider first $\{1, \ldots, r\}$.

3.1.26. Show that the cycle lengths in the canonical cycle representation, from back to front, follow the same process as the production of $(Y_1, \ldots, Y_n)$.

3.1.28. Reduce to the probability that a random element of $\mathbb{S}_n$ has every element in a cycle with one of $\{1, \ldots, k\}$.

3.1.30. Consider canonical cycle representations.

3.1.33. Use $\sum_{n \geq 0} c(k, n)x^n = x^{(k)}$ to find the generating function $\sum_{n \geq 0} F_n(z)x^n$.

3.1.37. Prove that they satisfy the same recurrence.

3.2.5. For (a), be careful about small n. For (b), consider choosing committees.

3.2.19. Begin with a double sum, but evaluate it using two convolutions.

3.2.23. Letting $A_n(x)$ denote the desired sum, find a recurrence in n for $A_n(x)$.

3.2.27. Count the words ending in 0 and the words ending in 1 separately.

3.2.34. For part (a), use induction. For part (b), use part (a). For part (c), use induction and integration without part (a).

3.2.36. First find the generating function, then use partial fractions.

3.2.40. Compare with Example 3.2.17.

3.3.15. For part (a), see Example 3.3.7.

3.3.16. Use the recurrence to obtain a differential equation for the EGF.

3.3.18. Use the correspondence between trees and n-ary lists.

3.3.22. Give a bijective argument using one Stirling number.

3.3.32. For part (a), begin with $\sum_{j=1}^{n} S(n,j)k_{(j)} = k^n$ from Theorem 3.3.13.

3.3.37. For part (a), consider the position of element n in a permutation of $[n]$.

3.3.38. Express the even graphs with vertex set $[n]$ in terms of general graphs with vertex set $[n-1]$.

3.3.43. For part (a), modify the Exponential Formula to require an even number of component structures.

3.3.48. Generalize the proof of Lemma 3.3.33.

3.3.53. Write the left side as a function of y.

3.4.17. Prove that they satisfy the same recurrence.

3.4.21. Interpret both sides using partitions, and then show bijectively that the coefficients are the same.

3.4.34. Count permutations of $[n]$.

3.4.37. For each multiset, choose a canonical path associated with it such that the peak heights occur in nonincreasing order.

3.4.43. Interpret the two multisets using paths along the boundary of the Ferrers diagrams of the partitions.

3.4.45. Build an infinite sum in which each factorization of n contributes $1/n^2$.

3.4.46. Reduce the problem to showing that negative entries can be eliminated.

3.4.49. Express the answer in terms of the greatest integer k such that $\binom{k}{2} < n$.

4.1.15. Group the fractions j/n for $j \in [n]$ by their denominators in lowest terms.

4.1.27. When is a number both an rth power and an sth power?

4.1.30. Count the n-sets in $[2n]$ whose intersection with $[n]$ is $[t]$.

4.1.31. Group $[2n]$ into pairs and consider the subsets of a certain size.

4.1.32. Count a certain type of subsets from a family of k sets of size n.

4.1.33. For the inclusion-exclusion part, count a particular family of $0,1$-lists with no consecutive 1s.

4.1.44. View the squares as 1s in a permutation matrix, and interpret the color of a square using parity.

4.1.46. Show that the left side is the inclusion-exclusion formula to count a set counted more directly by the right side.

4.1.49. Expand the factor $(n - k)^n$ and interchange the order of summation to introduce Stirling numbers, which eliminates most terms. For one nontrivial term that remains, use $k^{-1} = \int_0^1 x^{k-1}$.

4.1.52. Consider the board B' obtained from B by adding x columns of n squares to the left of B.

4.1.57. Say that a partition has Property i if $2i$ is a part or i is a repeated part.

4.1.64. In summing the weights of paths from x_i to y_j, consider how the paths arrive at y_j.

4.1.65. The identity $n = \sum_{d|n} \varphi(d)$ of Exercise 4.1.15 is needed.

4.1.68. Use induction on the number of rows.

4.1.69. For part (a), interpret terms in the product of determinant expansions as pairs of matchings in a weighted bipartite graph.

4.2.12. Encode caterpillars as binary lists or compositions of integers.

4.2.15. Color a crown with p^l positions. Use induction on l.

4.2.18. Use the pattern inventory or Burnside's Lemma with inclusion-exclusion.

4.2.27. Among the graphs with vertex set $[n]$, compare those with even size and odd size left fixed by a given permutation of $[n]$.

4.2.30. In part (b), the answer depends on the parity of d.

4.3.5. How many comparisons are made when an element is processed, in terms of where the shape grows?

4.3.12. Using planarization is shorter than using the bumping procedure.

4.3.13. In part (b), show that both sides satisfy the same recurrence and initial conditions; in particular, $f(\lambda) = 0$ when $\lambda_{i+1} = \lambda_i + 1$.

4.3.16. For part (a), consider the location of element n.

4.3.21. For necessity in part (a), use induction on $n - i - j$. Part (a) is needed to prove in part (b) that ϕ is surjective.

5.1.14. For the impossibility part, consider the usage of the cross edges (copies of the factor K_2) in the two cycles.

5.1.25. When n is divisible by 4, generalize the structure of P_4. When $n - 1$ is divisible by 4, add one vertex to the graph constructed for $n - 1$.

5.1.26. Consider the pairs of consecutive vertices and pairs of opposite vertices along a cycle in σ.

5.1.34. How many independent sets of size 4 or 3 contain a given vertex?

5.1.38. Consider a maximal path. Remember that a construction is needed to prove that the bound cannot be improved.

5.1.43. How many coordinates can change along a cycle of length $2r$?

5.1.45. There may be no 6-cycle. Begin with an edge.

5.1.46. For the last part, use induced subgraphs with $n - 1$ and $n - 2$ vertices.

5.2.12. Define an appropriate graph and use parity of the degrees.

5.2.13. Define a graph to model the movements, and use the even parity of the number of vertices with odd degree.

5.2.15. When n is even, prove that the number of paths of length 2 starting any vertex is odd. Counted in another way, use this to prove that some vertex has odd degree. By studying the triangles at such a vertex, obtain a contradiction.

5.2.16. Use induction on n. In the induction step, be careful to verify the conditions required to apply the induction hypothesis.

5.2.18. Use induction on k.

5.2.20. Follow the method of Theorem 5.2.6.

5.2.27. A short inductive proof uses the statement in the proof of Theorem 5.2.9 about having a bipartite subgraph capturing half the degree at each vertex.

5.2.28. Show that the underlying graph has no triangles.

5.2.29. In part (a), be careful about the base step.

5.2.31. Can G contain K_4?

5.2.34. Show that in every orientation D not having the desired property, there is a path from some vertex x with $d_D^+(x) > k$ to some vertex y with $d_D^+(y) < k$.

5.2.41. What does switching two consecutive vertices in the order do to S?

5.3.28. For the edge v_iv_{i+1} in an odd cycle $[v_1, \ldots, v_k]$ of G, what happens when the edge is not oriented from v_i to v_{i+1}?

5.3.36. Use induction on the number of vertices.

5.3.37. Consider a component that omits some vertex of maximum degree.

5.3.38. Consider a maximal path.

5.3.39. For part (b), use part (a) and the hypothesis on minimum degree to obtain upper and lower bounds on the number of edges joining $V(C)$ and $V(G) - V(C)$. For part (c), construct a graph with vertices grouped into sets of size $k/2$.

5.3.41. Consider a realization with the fewest components.

5.3.42. Use induction or show that there is such a realization among the realizations whose underlying graphs have the fewest edges.

5.3.53. Delete two vertices to apply induction.

5.3.56. For part (a), think about for what kind of graphs it is easy to obtain such an orientation.

5.3.57. For the first part, use an Eulerian circuit in a related even graph.

5.3.60. For part (a), build a list of trails starting at u. For part (b), consider all cycles through v.

5.4.18. Use Exercise 5.4.17 and induction.

5.4.19. For part (a), show that among all the ways to map the vertices of G onto the vertices of H, at least one maps no edge of G onto an edge of H.

5.4.20. Use induction on k. In the induction step, delete one vertex from each component of the subgraph of G induced by the vertices with degree less than k.

5.4.27. Reduce to connected graphs and add edges to a spanning tree.

5.4.29. For part (a), use induction or explicit construction.

5.4.33. For part (a), build a graph H whose vertices are decompositions of G into two subgraphs each having $|V(G)|-1$ edges, such that the desired decompositions into trees with the right degrees are the odd-degree vertices. The even-degree vertices are decompositions $(\hat{S}, \hat{T})$ where the degrees agree with S and T except at two vertices: $\hat{S}$ isolates a fixed leaf w of S and has one cycle through the one other vertex whose degree does not agree with S, and $\hat{T}$ is a spanning tree.

5.4.39. Let $x \in \mathbb{Z}_n$ be a vertex such that $d_G(0, x) \geq 3$. Consider the maximum size of an independent set in the graph with vertex set $\mathbb{Z}_n$ in which vertices are adjacent if and only if they differ by x.

5.4.41. Use induction on the number of leaves or the number of vertices.

5.4.56. Use induction on n. One can prove the stronger statement that an n-vertex tree other than a star occurs in edge-disjoint copies such that each non-leaf vertex appears at distinct vertices in the two copies. Another proof treats stars with one edge subdivided as a special case and then proves the claim for other trees from the induction hypothesis.

5.4.58. Given vertices x and y as in the hint, form G' from $G - x - y$ by adding k disjoint edges joining $N_G(x)$ to $N_G(y)$.

6.1.16. For the second part, generalize the example of C_6 when $k = 2$.

6.1.17. Restrict to a subset of X that is covered by some maximum matching.

6.1.19–20. Look for a perfect matching in an appropriate bipartite graph.

6.1.24. Given n positions and m winning sets, form an X, Y-bigraph with $X = \{x_1, \ldots, x_n\}$ and $Y = \{w_1, \ldots, w_m\} \cup \{w_1', \ldots, w_m'\}$ having edges $x_i w_j$ and $x_i w_j'$ for each instance of position x_i in winning set W_j.

6.1.26. Consider Theorem 6.1.10.

6.1.29. Reduce to the case $\alpha'(G) < n$. Note that it is not possible to prove that $\alpha'(G) \geq 2\delta(G)$ always holds, since that is false when $2\delta(G) > n$.

6.1.31. For part (a), reduce the problem to the case where $r = d - 1$ and the vertices of degree d form an independent set, and then apply Hall's Theorem.

6.1.38. Translate the statements "v belongs to some smallest vertex cover" and "v is covered by every maximum matching" into notation.

6.1.42. When G' has incident edges e_1 and e_2, let S_1 and S_2 be minimal vertex covers of $G' - e_1$ and $G' - e_2$. Add to $S_1 \cap S_2$ an appropriate vertex cover of $G'[(S_1 \triangle S_2) \cup \{x\}]$ to contradict $\beta(G') = \beta(G)$.

6.2.8. Use inclusion-exclusion.

6.2.18. For sufficiency, consider a subgraph with the most vertices among those whose components are nontrivial stars with at most m edges.

6.2.22. If $\alpha'(G) < \lfloor n/2 \rfloor$, then every maximum matching leaves at least two vertices uncovered.

6.2.31. For part (b), consider the edges joining S and $\overline{S}$, where S is the vertex set of a maximum matching. For part (c), use induction on s and the idea of 2-switches (Definition 5.2.7); remember that multiedges are allowed.

6.2.41. For part (a), consider a maximal Tutte set S in $G - \{x, y\}$, and apply Lemma 6.2.7 to $S \cup \{x, y\}$ in G. For part (b), use induction on n.

6.2.42. Design an auxiliary graph H in which the f-factors of G are the vertices of odd degree. The other vertices are spanning subgraphs with the desired number of edges, but they have degree $f(w) - 1$ at a fixed vertex w and degree $f(u) + 1$ at some other vertex u, otherwise agreeing with f.

6.3.7. How do the various edges contribute to the sum of the degrees of the vertices covered by M? Use that relationship to obtain an upper bound on m.

6.3.24. Suppose that the first such occurrence is a rejecting x even though xa is a pair in some stable matching M, and a rejects x for y. Let b be the mate of y in M. What can be deduced about the preferences of these people?

7.1.4. Induction permits a simple proof.

7.1.19. For part (a), argue that some if $|S| \leq k$, then some component of $G - S$ contains a vertex of degree at least d_{n-k}; consider vertices in other components of $G - S$. For part (b), build a graph G with a separating k-set S such that $G - S$ has a component with $n - 1 - d_{n-k}$ vertices.

7.1.20. Consider a minimal subgraph H such that H has at least $2k - 1$ vertices and more than $(2k - 3)(|V(H)| - k + 1)$ edges. Show that H is k-connected.

7.1.27. In part (a), $G \,\square\, H$ has four cyclically overlapping copies of $K_{r+1} \,\square\, K_{r+1}$.

7.1.32. Use induction on n.

7.2.20. Use Menger's Theorem.

7.2.26. For vertices u and v in a k-edge-connected graph G, consider a family of k edge-disjoint u, v-paths with minimum total length.

7.2.27. Make sure that the construction has connectivity k and diameter d.

7.2.30. Define an appropriate digraph from the input graph.

7.2.32. Consider a longest cycle.

7.2.34. Among all the edges and all the k-cuts containing them, choose edge xy and k-cut S so that the resulting subgraph $G - S$ has a smallest possible component, H. Prove that vz is k-contractible, where $z \in S$ and $v \in N(z) \cap V(H)$.

7.2.41. One proof chooses P so that $G - V(P)$ has a component of largest order.

7.2.42. Consider a cycle in $G - u$ whose deletion leaves u in a largest component.

7.2.48. Consider the blocks of $G - v$.

7.2.50. Prove that the cycle C contains at least two such ears.

7.2.53. After Alice chooses u and v, Bob produces a special ear decomposition and a special vertex numbering (in fact, an s, t-numbering with $s = u$ and $t = v$) and uses them to orient at most $2n - 3$ edges in a way that will win.

7.3.16. Find an appropriate set of vertices in the corresponding graph to violate the necessary condition.

7.3.21. Prove both statements in a single induction.

7.3.23. Generalize the construction of the bipartite graph in Example 7.3.3.

7.3.28. For part (a), consider the second graph in Example 7.3.3. For part (b), prove that when G is a tree a stronger statement holds: for any specified edge xy of G, there is a spanning x, y-path in $G^3 - xy$.

7.3.32. For part (b), start with $n \in \{5, 6\}$ and generalize.

7.3.34. Given that Ore's Condition fails for $\overline{G}$, determine the structure of G.

7.3.36. For part (a), delete a vertex with large degree. For part (b), show that if G fails Chvátal's Condition, then $\overline{G}$ has at least $n - 2$ edges.

7.3.43. Consider the location of the vertices of a largest independent set A relative to a smallest separating set S.

7.3.44. Proving the two claims together permits a simpler proof.

7.3.45. In part (b), transform the graph to apply a known result.

7.3.47. Apply Ore's Theorem.

7.3.48. Apply Chvátal's Theorem to a modified graph.

7.3.49. Use induction on k, making an argument like that of Ore's Theorem.

7.3.54. Consider the degree sum for the vertices in a largest independent set, and use the Chvátal–Erdős Theorem.

7.3.56. Use induction on k, after choosing an appropriate cycle in G.

7.3.61. Reduce to trees and use induction on $|S|$.

8.1.13. For the upper bound, present an explicit coloring by regions, with attention to the boundaries.

8.1.17. For the upper bound, cover the vertices with $n - 2k + 2$ independent sets.

8.1.19. Use induction on $|V(G)|$.

8.1.27. Use large neighborhoods and induction on r.

8.1.29. Consider a smallest-last ordering.

8.1.30. Prove that an optimal coloring of G has many color classes of size 1.

8.1.36. Obtain an independent n-set in $G \square K_r$ from a proper r-coloring of G and vice versa.

8.1.40. For part (a), use induction on n or degeneracy.

8.2.17. For $\chi(G) = 4$, consider a shortest odd cycle.

8.2.19. When $G - xy$ is k-colorable and G is not, obtain a cycle through xy with length congruent to 1 modulo k for each cyclic permutation of the k colors.

8.2.27. For part (a), consider a partition $V_1, \ldots, V_t$ minimizing $\sum_i \frac{|E(G_i)|}{D_i}$.

8.2.28. Modify G by adding vertices to make a graph G' where an argument similar to that for part (a) of the previous problem yields the desired result on G.

8.2.30. Construct a special list assignment with list sizes equal to degree.

8.2.33. For part (a), choose a bad assignment L with smallest union, and consider a maximal X such that $|L(X)| < |X|$. For part (b), the union of the bad lists should have size $2k$; force many colors to appear by using the same two disjoint lists on each part of size 2. The construction should restrict to the bad 2-uniform list assignment for $K_{4,2}$.

8.2.34. Reduce to the case where the lists on each part are disjoint, and apply Hall's Theorem to choose a proper coloring from the lists.

8.2.43. Delete one leaf from each nontrivial component of F to obtain F'. Let R be the set of neighbors of the deleted vertices. Map R onto an m-set $X \subseteq V(G)$ chosen to minimize $|E(G[X])|$. Extend X to a copy of F'. Use Hall's Theorem to match X into the remaining vertices.

8.2.46. When an optimal coloring f has a color class of size 1, use g to make an alteration in f.

8.3.6. Use a graph transformation.

8.3.18. If $\chi'(G) > \Delta(G)$, then G has a minimal subgraph G' such that $\chi'(G') > \Delta(G)$. Obtain a proper $\Delta(G)$-edge-coloring of G' from such a coloring of $G' - xy$, by switching colors along a 2-edge-colored path to make a color available for xy.

8.3.22. In part (b), use part (a) to restrict attention to overfull subgraphs with at most half the vertices.

8.3.26. In part (a), first color the subgraph obtained by deleting all copies of edges with multiplicity more than $\Delta(G)/2$. In part (b), for an uncolored edge xy, consider an edge yz having a color that is missing at x. Show that the coloring extends if z has a common missing color with y or x.

8.3.27. Use a graph transformation.

8.3.28. For part (a), use Eulerian circuits. For part (c), consider an optimal $(\Delta(G) + 1)$-edge-coloring of a graph G, and build a color fan at a vertex whose incident colors are not distinct.

8.3.30. Apply Puleo's result with $k = \max_{v \in V(G)}(d_G(v) + \mu(v))$ to obtain $S = V(G)$.

8.3.32. Given a proper edge-coloring with q colors that satisfies the intersection condition, compute upper and lower bounds on the number of ordered pairs of edges having the same color.

8.3.34. Start with a proper vertex coloring and a proper edge-coloring that use the fewest colors, sharing two colors. Uncolor the edges that have the common colors and show that they can be recolored to complete a total coloring.

8.3.35. For part (b), view $K_{r,s}$ as a composition of $K_{m,n}$ and $K_{k,k}$, where m and n are relatively prime.

8.3.36. For sufficiency, use a parity edge-coloring and a spanning tree of G to map the vertices of G into vertices of Q_k.

8.3.47. Use induction on the number of vertices.

8.3.58. Prove that the complement of a connected P_4-free graph is disconnected. (This problem does not allow using that P_4-free graphs are perfectly orderable.)

8.3.63. Reduce to proving $\chi(G') = \omega(G')$ and $\alpha(G') = \theta(G')$ when G and $\overline{G}$ are perfect, respectively, where G' is obtained from G by duplicating one vertex x. The case of $\alpha(G') = \theta(G')$ where x is in no maximum stable set requires looking at a smallest clique cover of G.

8.3.69. For part (b), there is an example with six vertices.

9.1.15. Show that G^* can be drawn so that each edge of G^* intersects only the edge corresponding to it in G.

9.1.23. Show that a separating 2-set must induce two edges.

9.1.26. The solution is easier to write when the characterization used for j-vertex trees is "connected and $j - 1$ edges" rather than "connected and acyclic".

9.1.30. Use induction or duality.

9.1.33. Prove that the complement of any 7-vertex maximal outerplanar graph is not outerplanar.

9.1.34. Show that the edge set decomposes into triangles.

9.1.40. Add a point at infinity or a circle that encloses all intersection points.

9.1.41. Find a simple formula in terms of n for the number of crossings of chords.

9.1.48. Use Euler's Formula or induction on the number of facial triangles.

9.2.12,14b,16. Use Kuratowski's Theorem.

9.2.22. Use the corollary that guarantees a $(c\sqrt{n}, 1/2)$-separation (where $c = (2\sqrt{2})/(1 - \sqrt{2/3})$).

9.3.16. Find a way to use Grinberg's Theorem.

9.3.17. Consider distance from a fixed vertex.

9.3.18. Consider distance from a fixed vertex.

9.3.23. For the first part, use an idea from the proof of Tait's Theorem (Theorem 9.3.1). The proof using the Four Color Theorem is easy.

9.3.25. For part (c), build $\lfloor n/3 \rfloor$ areas that must be watched by different guards.

9.3.27. For the induction step in part (c), delete the edges of a face neighboring the unbounded face.

9.3.31. Prove that a counterexample with fewest vertices has no 4-face, by showing that some two vertices of a 4-face can be merged.

9.3.34. It may help first to construct such a graph with 114 vertices and then one with 86 vertices.

9.3.38. Given a separating cycle C with length at most 4, prove that the C-lobes of G have proper 4-colorings that agree on C. In particular, when C has length 4, show that a C-lobe has a proper 4-coloring where two specified opposite vertices of C receive distinct colors.

9.3.39. Begin by showing that w_4 and w_6 can be assumed not to be adjacent outside the ring.

9.3.40. Force two vertices on the unbounded face to have the same coloring in any proper 3-coloring, and then make them adjacent.

9.3.42. Consider complete bipartite graphs.

9.3.43. Use degree charging, and let each vertex needing charge take what it needs equally from its neighbors.

9.3.44. In part (a), design a discharging argument without knowing b, and determine how small b must be to make it work. Part (b) can be done with face charging or with balanced charging. This exercise completes the bottom line in Remark 9.3.23 in the same way that the first and third lines were completed.

9.3.51. Show that edges of weight at most $2t + 1$ are reducible. When there is no such edge, Lemma 9.3.30 guarantees a 4-cycle through two 3-vertices. Use a decomposition into t linear forests for the graph G' obtained by deleting those two 3-vertices to obtain such a decomposition for G.

9.3.57. Use vertex charging, with 5-vertices taking $\frac{1}{2}$ from incident 4^+-faces and the remaining needed charge from neighbors along edges shared by two triangles.

9.3.58. Use vertex charging; 6-vertices that give charge to 5-neighbors will need charge from 7^+-neighbors.

9.3.64. Use degree charging. Let vertices of degree at most 4 take what they need equally from their neighbors, but let a vertex of degree 5 or 6 take $\frac{1}{4}$ from each 6^+-neighbor. In checking that vertex v ends happy, consider cases depending on $d(v)$, and let j be the least degree among the neighbors of v.

9.3.68. For part (c), use the existence of planar graphs that are not 4-choosable.

10.1.10. Find a set S of size $(m+1)n$ in $[2^m n]$ partitioned into n divisibility chains.

10.1.14. Consider the players with positive scores.

10.1.18. Compare the number of nonempty subsets of k elements with the largest sum of k elements.

10.1.19. Use induction, deleting a largest element of S.

10.1.23. Reduce the problem to the case $m = 2n$.

10.1.25. Use a scheme in which 90 members each get only one key.

10.1.26. Break the sequence into segments whose length is one more than the size of the alphabet.

10.1.31. For odd n, use a decomposition into $(n-1)/2$ isomorphic subgraphs consisting of a 4-cycle plus pendant edges. Use different constructions for $n = 4k-1$ and $n = 4k+1$. In the latter case, use one central vertex and distribute the others evenly over two concentric circles, as $(n-1)/2$ pairs.

10.1.36. Generalize the argument of Exercise 10.1.35 by fixing positions that do not end increasing lists of length $k-1$.

10.2.8. Define a weighted transitive tournament with the points as vertices, where edges point to the right and weights correspond to slopes.

10.2.10. 2-color triples of points so that m points whose triples all have the same color will form a convex m-gon. Possible color criteria include convex/concave shape, indices clockwise or not, and parity of the number of points inside.

10.2.11. Use induction. Consider the neighbors or nonneighbors of one vertex.

10.2.21. Be careful about parity.

10.2.28. Obtain a bow-tie with monochromatic triangles, plus complementary monochromatic 5-cycles, and then use symmetry.

10.2.30. The answer depends on the relationship between n and $\alpha'(\overline{G})$.

10.2.32. For the first part, consider the average of $(x) + d(y)$ over edges xy in a subgraph G of $K_{s,t}$.

10.2.34. Prove a more general statement by induction, or prove that if $\sum_{v\in X}\binom{d(v)}{s} > (t-1)\binom{n}{s}$ for an X, Y-bigraph G contained in $K_{n,n}$, then G contains $K_{s,t}$.

10.2.37. For the upper bound, use induction on m. After finding a red copy of P_{m-1} in a given coloring, consider blue paths that alternate between vertices outside this path and vertices inside it.

10.2.39. For part (a), determine the colors of the edges at displacement 2, 3, and 4 on the monochromatic $(2k+1)$-cycle. For part (b), consider the edges with even displacement on the monochromatic $2k$-cycle.

10.3.4. Take $n \geq R_k(3;2)$, as defined in Definition 10.2.10, and consider the proof of Theorem 10.3.1. One need not use all the nonempty subsets of $[n]$.

10.3.9. Use van der Waerden's Theorem on the indices corresponding to a long arithmetic progression in $\langle a \rangle$.

10.3.11. Consider a monochromatic k-set or a rainbow k-set.

10.3.13. After finding red points $a_1, \ldots, a_n$ whose centroid z is a lattice point, if z is blue consider replacing a_j by a point b_j so that the new set has centroid a_j.

10.3.14. Given a coloring of $E(K_n)$, consider a monochromatic connected subgraph with the most vertices.

11.1.8. For the first part of (a), if $uv \notin E(G)$ and $x_u, x_v \neq 0$, then $f(x') \geq f(x)$ for some x' such that $x'_u x'_v = 0$.

11.1.12. Solve it first for $\omega(G)$, and then use that to give the answer for $\alpha(G)$.

11.1.15. Reduce part (b) to Mantel's Theorem.

11.1.33. Use induction on n. Consider the larger family of graphs in which adding any edge increases the number of t-cliques, regardless of whether the graph already contains a t-clique.

11.1.37. For part (b), first find a vertex in A having at least εn neighbors in each of B and C. For part (c), prove that if G has $(1/8 + \varepsilon)n^2$ edges, then G contains K_4 or an independent set of size at least $(\varepsilon^2/cM)n$ for some appropriate constant c, where M is the bound on the number of classes when the Regularity Lemma is applied with the arguments ε' and l, where $\varepsilon' = \varepsilon/6$ and $l = 1/\varepsilon'$. From the resulting partition with k classes, where $1/\varepsilon' \leq k \leq M$, delete edges except for those in ε'-regular pairs with density more than $2\varepsilon'$. Now applying parts (a) and (b) to the remaining graph shows that with no K_4 and no large independent set, $|E(G)| < (1/8 + \varepsilon)n^2$.

11.1.39. For part (a), first show that $\alpha'(G) > n - \delta(G)$. Then show that a nonperfect maximum matching admits an augmenting path of length 5.

11.1.40. In part (a), the path needs to be grown in two phases. First obtain a path with length at least $(d - 3\varepsilon)2m - 2K$, and then show that a path at least this long can be modified to obtain a longer path until length $(1 - \varepsilon - \frac{\varepsilon}{d-\varepsilon})2m$ is reached.

11.1.41. Let P be a longest path in G having an even number of vertices. Let $2t = |V(P)|$. If $t \leq (1 - \frac{\varepsilon}{d-\varepsilon})n$, then find a detour to obtain a longer such path. If $t > (1 - \frac{\varepsilon}{d-\varepsilon})n$, then find a cycle of length $2t$ through $V(P)$ and then a longer path.

11.1.42. For part (a), use induction on k. In the induction step, consider separately the k-tuples where the first $k - 1$ elements have at most $(d - \varepsilon)^{k-1}|Y|$ common neighbors in Y and those having more such common neighbors.

11.1.43. For part (c), reduce to when every vertex is in at least two edges and no two edges share two vertices. The *link graph* of v has as its edges the pairs that with v form edges of H. Apply part (b) to the union of all the link graphs.

11.2.11. The k-sets in F must avoid the shadows of higher elements.

11.2.15. Use shift operators to compress the family.

11.2.35. Count the elements used in a sunflower of size s.

11.2.39. Express a random subset of $[n]$ in terms of the sizes of its intersections with the members of a distinguishing family.

11.2.41. Let F consist of congruence classes modulo k.

11.3.33. For part (c), use induction on $|X \triangle Y|$.

11.3.35. For part (b), use property (J) and induction on $|I_1 - I_2|$ to obtain the augmentation.

11.3.40. For part (a), use transitivity of dependence or properties of cocircuits. For part (c), using induction on $|X_1|$.

11.3.42. For equality, define a special chain using the weights of the elements.

11.3.53. Devise a strategy using deletions and contractions.

11.3.56. Use the set-counting formula that appears in the proof of the Matroid Union Theorem.

12.1.9. Consider orientations of the ten 4-cycles.

12.1.13. Finding a poset having exactly one largest chain and exactly one largest antichain will show that nine of the statements are false if that chain and antichain are disjoint. One more of the remaining seven statements is also false.

12.1.16. For part (b), when F has no union-free family of size m, how many elements can the inclusion poset on F have at height h?

12.2.8. Consider $(0, i, i) < \cdots < (0, i, n - i) < \cdots (0, 2i, n) < \cdots < (n - 2i, n, n)$.

12.2.31. Rewrite (b) so it is twice the needed inequality in (c).

12.2.32. Compare $g_{k+1}(r, s+1)/g_k(r, s)$ with $g_{k+1}(r, s)/g_k(r, s-1)$ and with $g_{k+1}(r+1, s+1)/g_k(r+1, s)$ by clearing fractions and using log-concavity.

12.2.33. After using the binomial recurrence to express f_{n+1} in terms of f_n, express the coefficients of $[f_n(x)]^2 - f_{n-1}(x)f_{n+1}(x)$ as two sums, and show that the coefficients are nonnegative.

12.3.8. Forbid a 2-realizer or consider the edges of size 2 in the hypergraph of alternating cycles.

12.3.23. Embed P by letting $f_i(x) = j$ if the least element greater than or equal to x on chain i is the jth from the bottom, $f_i(x) = k$ if no such j exists.

12.3.24. For the upper bound, form a directed graph with an edge for each extension in a minimal realizer.

12.3.32. For part (b), use an incidence matrix to establish a bijection between n-tuples of subsets of $[r]$ that are $(n - m)$-scrambling and r-tuples of vertices (as binary vectors) that break all m-cubes in the n-dimensional hypercube.

12.4.31. Use the definition of distributivity, the isomorphism to $J(P(L))$, the embedding of L in a subset lattice, or the forbidden sublattice characterization.

12.4.44. Express $X - X$ as $X \wedge X'$, where X' is the family of complements of members of X in the union of all the sets, and compute $|X \wedge X'|$ and $|X \vee X'|$.

13.1.10. After normalizing all the top rows to be the same, only one choice remains for each other position of the square being added. Show that it works.

13.1.14. Given Latin squares L_1 and L_2, let $\phi(s) = (s_3 - L_1(s_1, s_2), s_4 - L_2(s_1, s_2))$.

13.2.4. Prove that in a complete orthogonal family, any pair of positions not in the same row or column have the same value in exactly one of the squares.

13.2.10. View $K_{2,t}$ as two vertices with t common neighbors.

13.2.13. For part (b), partition $[n]$ into sets $X_1, \ldots, X_k$ of size m, and form pairwise edge-disjoint graphs $H_1, \ldots, H_k$ such that each H_i uses $q + 1$ of the sets $X_1, \ldots, X_k$ as its vertex set.

13.2.18. After obtaining the orbits under a purported multiplier, consider the pairs generating the difference 28.

13.3.6. The operation table of an idempotent quasigroup is a Latin square with element i in position i on the diagonal. For even n, it cannot be commutative, but it can be built from an idempotent commutative quasigroup of order $n - 1$.

13.3.23. Sum the absolute edge differences (mod 2) in two different ways.

13.3.25. Partition the vertices into three copies of $\mathbb{Z}_3$, so the graph has twelve types of edges. Define two 6-cycles that together use one of each type of edge.

14.1.18. Generalize a construction of non-k-choosable bipartite graphs.

14.1.19. What are the possible seats for the last passenger?

14.1.24. This can be done using induction or linearity.

14.1.28. Generalize the Caro–Wei Theorem.

14.1.30. Part (a) can be solved by computing a sum or by a shorter argument. For part (b), apply part (a).

14.1.32. Somewhere use $1 - x \leq e^{-x}$ to simplify a bound.

14.1.34. For part (a), consider the more general situation where the walk starts at a given spot among the $n + 1$ possible positions. Sum the resulting equations.

14.1.35. In order to reach u last, the walk must first visit a neighbor of u and then visit the other neighbor of u before visiting u.

14.1.36. Let t_i be the 0, 1-variable that is 1 if and only if the ith head is immediately preceded by a tail.

14.1.39. Establish a one-to-one correspondence between the selectable sets and the subsets of Q forming chains in the product order on $\mathbb{R}^2$.

14.1.42. Choose B at random and use linearity of expectation.

14.1.43. For a sufficiently large prime p congruent to 2 modulo 3, multiply A by a random element of $\mathbb{Z}_p$ to capture many elements in a sum-free subset consisting of more than 1/3 of $\mathbb{Z}_p$.

14.1.45. Define a suitable event for each point.

14.2.12. Use the Dependent Random Choice Lemma.

14.2.13. Use the Dependent Random Choice Lemma. In coloring K_N with $N = 2^{3k}$, the more plentiful color has at least $2^{-7/3} N^2$ edges. To contain Q_k, we want 2^{k-1} vertices whose k-sets have 2^k common neighbors in this color.

14.2.14. Show that the experiment produces a graph with no large independent set and not many triangles.

14.2.19. Prove first that if there is a solution, then there is a solution where the two weights are equal.

14.2.20. Use the Local Lemma.

14.2.21. Choosing the endpoints of an edge joining members of S is a bad event.

14.3.1. Follow Theorem 14.3.18.

14.3.6. What forces an n-vertex graph to have a cycle? Recall that the variance of a binomial random variable with m trials and success probability p is $mp(1 - p)$.

14.3.7. Consider G^p with p chosen later as a suitable function of n. How many edges remain when the copies of $K_{r,r}$ are broken?

14.3.8. Bound the probability that the pair on the ith round fails to form an edge.

14.3.11. Consider the probability in the tail on the other side of $\frac{1}{2}\binom{n}{2}$ from the expected number of edges.

14.3.12. Design p with a constant parameter c so that the expression for the expected number of edges on no triangle simplifies.

14.3.13. For the Second Moment computation, group the pairs of potential copies of H according to whether or not they share an edge.

14.3.39. Otherwise, the number of possible representations is a vanishing fraction of the total number of graphs.

14.4.6. Show that whp any two vertices in $G^{1/2}$ have many common neighbors.

14.4.7. Obtain an exponentially small bound on the probability of having this many edges across a given bipartition.

14.4.12. For a random coloring f, let X be the number of edges e such that $\left|\sum_{v\in e} f(v)\right| > \sqrt{2n \ln(2m)}$.

14.4.18. Generate a random permutation by choosing a random sequence in the interval $[0, 1]$.

15.1.3. For part (a), one approach applies Theorem 15.1.2 after adding a new resident. For part (b), let M be the incidence matrix of the family. Show that MM^T is nonsingular over $\mathbb{F}_2$ if $m = n$ (when n is even), and hence the incidence vectors are linearly independent over $\mathbb{F}_2$. However, show that they can't be independent.

15.1.6. Use a family of disjoint sets.

15.1.8. Compare $\det A$ with $\det J - I$, and then model the partitioning condition as a matrix equation.

15.1.11. For part (c), consider an equation of dependence for the set consisting of the polynomials $f_1, \ldots, f_m$ from Theorem 15.1.5 plus $\{1, x_1, \ldots, x_n\}$. Take partial derivatives $\partial^2/\partial x_i \partial x_j$ to obtain equations that the coefficients must satisfy. Describe these equations using the matrix B formed as in part (a) from the row vectors $v^1, \ldots, v^m\}$. Use parts (a) and (b) to show that the coefficients on the added polynomials must be 0.

15.1.15. Make use of a cubical grid.

15.1.26. Use the Combinatorial Nullstellensatz.

15.1.31. The claim can be proved using case analysis or the Alon–Tarsi Theorem.

15.2.9. Prove that the process ends in the initial room, that all incident corridors of a reached room are followed both ways, and that every room is reached.

15.2.18. Use induction on $k + n$.

15.2.19. Use induction on the length of C.

15.2.21. Express the desired situation as a matrix equation over $\mathbb{F}_2$.

15.2.31. For part (b), Use the Birkhoff–von Neumann Theorem (Theorem 6.1.8) to argue that optimal matrices are interior points in the space of n-by-n matrices, and then apply Lagrange multipliers.

15.2.36. Although the statement involves the inclusion relation on a family of sets, incidence functions are not needed for the solution.

15.3.9. Use Exercise 15.3.7(b) to interpret the coefficients.

15.3.11. Involve the vertex cover number, the fact that the spectrum of a bipartite graph is symmetric around 0, and Exercise 15.3.9.

15.3.13. Construct eigenvectors for the product from eigenvectors for the factors.

15.3.18. Use the complement.

15.3.29. Recall that a matrix M is positive semidefinite, meaning that all eigenvalues are nonnegative, if and only if $x^T M x \geq 0$ for all real vectors x. Recall also that a real symmetric matrix has an orthonormal basis of eigenvectors.

15.3.30. Use the properties in Theorem 15.3.30, proved in Exercise 8.3.58. For part (b), prove that the largest Laplacian eigenvalue of the complement of a connected cograph G equals the number of vertices in the largest component of $\overline{G}$.

15.3.31. Generate subgraphs of $K_{n,n}$ as the union of k random perfect matchings.

16.1.11. A single toroidal family with arbitrarily large crossing number suffices.

16.1.12. Consider the graph obtained from a maximal such graph by deleting one edge from each crossing. Argue that edges remaining from the crossings do not lie on the same face in the resulting plane graph.

16.1.13. Consider proper 3-edge-colorings of the 3-regular plane graph in which the crossing is replaced with a 4-cycle through four new vertices.

16.1.14. Consider what happens when a vertex moves across an edge. For the second part, consider a drawing where the crossings are easy to count.

16.1.16. Consider the copies of $K_{m,n}$ in a drawing of $K_{m+1,n}$.

16.1.17. Consider what happens when a vertex moves across an edge. For the second part, consider a drawing where the crossings are easy to count.

16.2.6. Choose suitable paths with maximum total length.

16.2.9. For part (b), add a suitable troublesome edge to an appropriate graph.

16.2.10. Use the fact that every infinite sequence of points in a closed and bounded set has a convergent subsequence.

16.2.13. Use the Hobby–Rice Theorem.

16.3.8. For $e(Q_n)$, group the extensions by which element is at the top. For part (b), compare the limit with 1/3 and with other incomparable pairs.

16.3.9. For part (a), begin by mapping extensions having x before y but no element between them that is good for (x, y) injectively into the set of extensions with y before x. Conclude that many extensions with x before y have some element between them that is good for (x, y).

16.3.11. For the second part, consider $\mathbf{2} + \mathbf{2}$.

16.3.14. Use a realizer to specify triangles all containing the origin.

16.3.19. Prove the lower bound by induction on n.

References

The final item in each entry is the number of the citing page in the text.

Abbott H.L., Lower bounds for some Ramsey numbers. *Discr. Math.* **2** (1972), 289–293. [459]

Abbott H.L., D. Hanson, and N. Sauer, Intersection theorems for systems of sets. *J. Combin. Th. A* **12** (1972), 381–389. [504]

Abbott H.L. and B. Zhou, On small faces in 4-critical planar graphs. *Ars Combin.* **32** (1991), 203–207. [414]

Abeledo H. and G. Isaak, A characterization of graphs that ensure the existence of a stable matching. *Math. Soc. Sci.* **22** (1991), 93–96. [288]

Aberth O., On the sum of graphs. *Rev. Fr. Rech. Opér.* **33** (1964), 353–358. [343]

Abrham J. and A. Kotzig, Construction of planar Eulerian multigraphs. In *Proc. 10th Southeastern Intl. Conf. Combin. Graph Th. Comput. (Boca Raton), Congr. Numer.* **23/24** (1979), 123–130. [388]

Abrishami T., *A combinatorial analysis of the eigenvalues of the Laplacian matrices of cographs* (Johns Hopkins Univ., 2019). Masters Thesis. [782]

Acharya B.D. and M. Las Vergnas, Hypergraphs with cyclomatic number zero, triangulated graphs, and an inequality. *J. Combin. Th. B* **33** (1982), 52–56. [374]

Achlioptas D. and F. Iliopoulos, Random walks that find perfect objects and the Lovász local lemma. *J. ACM* **63** (2016), Art. 22, 29. [680]

Addario-Berry L., K. Dalal, C.J.H. McDiarmid, B.A. Reed, and A.G. Thomason, Vertex-colouring edge-weightings. *Combinatorica* **27** (2007), 1–12. [737]

Adrian C., Problem E3459. *Amer. Math. Monthly* **98** (1991), 754. Solution **100** (1993), 593–594. [511]

Aharoni R., Ryser's conjecture for tripartite 3-graphs. *Combinatorica* **21** (2001), 1–4. [803]

Aharoni R. and P. Haxell, Hall's theorem for hypergraphs. *J. Graph Theory* **35** (2000), 83–88. [803]

Ahlswede R. and D.E. Daykin, An inequality for the weights of two families of sets, their unions and intersections. *Z. Wahrsch. Verw. Gebiete* **43** (1978), 183–185. [597]

Ahlswede R. and L.H. Khachatrian, The complete intersection theorem for systems of finite sets. *European J. Combin.* **18** (1997), 125–136. [500, 511]

Ahlswede R. and L.H. Khachatrian, Katona's intersection theorem: four proofs. *Combinatorica* **25** (2005), 105–110. [497]

Ahuja R.K., T.L. Magnanti, and J.B. Orlin, *Network Flows* (Prentice Hall, 1993). [246]

Aigner M., Lexicographic matching in Boolean algebras. *J. Combin. Th. B* **14** (1973), 187–194. [556]

Aigner M., *Combinatorial Theory* (Springer-Verlag, 1979). [123, 520, 529, 538, 607, 761]

Aigner M., *Graphentheorie, Eine Entwicklung aus dem 4-Farben problem* (B.G. Teubner Verlagsgesellschaft, 1984). (English transl. BCS Assoc., 1987). [399]

Aigner M., Turán's graph theorem. *Amer. Math. Monthly* **102** (1995), 808–816. [476]

Aigner M., Lattice paths and determinants. In *Computational discrete mathematics, Lect. Notes Comp. Sci.* **2122** (Springer, 2001), 1–12. [165, 168, 170, 176, 177]

Aigner M. and G.M. Ziegler, *Proofs from The Book* (Springer-Verlag, 1999). Also 2001, 2004, 2009, 2014. [37, 476, 811]

Ajtai M., V. Chvátal, M.M. Newborn, and E. Szemerédi, Crossing-free subgraphs. In *Theory and Practice of Combinatorics*, *Ann. Discr. Math.* **12** (North-Holland, 1982), 9–12.   [793]

Ajtai M., J. Komlós, and E. Szemerédi, A note on Ramsey numbers. *J. Combin. Th. A* **29** (1980), 354–360.   [226, 451, 663, 664]

Ajtai M., J. Komlós, and E. Szemerédi, Sorting in $c \log n$ parallel steps. *Combinatorica* **3** (1983), 1–19.   [779]

Ajtai M. and E. Szemerédi, Sets of lattice points that form no squares. *Stud. Sci. Math. Hungar.* **9** (1974), 9–11 (1975).   [492]

Akin E. and M. Davis, Bulgarian solitaire. *Amer. Math. Monthly* **92** (1985), 237–250.   [143, 144, 152]

Akiyama J. and N. Alon, Disjoint simplices and geometric hypergraphs. In *Combinatorial Mathematics: Proceedings of the Third International Conference (New York, 1985)*, *Ann. New York Acad. Sci.* **555** (New York Acad. Sci., 1989), 1–3.   [816]

Akiyama J., H. Era, S.V. Gervacio, and M. Watanabe, Path chromatic numbers of graphs. *J. Graph Theory* **13** (1989), 569–575.   [418]

Akiyama J., G. Exoo, and F. Harary, Covering and packing in graphs. III. Cyclic and acyclic invariants. *Math. Slovaca* **30** (1980), 405–417.   [412]

Akiyama J., G. Exoo, and F. Harary, Covering and packing in graphs. IV. Linear arboricity. *Networks* **11** (1981), 69–72.   [677]

Akiyama J. and M. Kano, Factors and factorizations of graphs—a survey. *J. Graph Th.* **9** (1985), 1–42.   [268]

Akiyama J. and M. Watanabe, Maximum induced forests of planar graphs. *Graphs Combin.* **3** (1987), 201–202.   [419]

Aksionov V.A. and L.S. Mel'nikov, Essay on the theme: the three-color problem. In *Combinatorics (Proc. 5th Hung. Colloq., Keszthely, 1976), I, Colloq. Math. Soc. János Bolyai* **18** (North-Holland, 1978), 23–34.   [414]

Albert M.H., M. Elder, A. Rechnitzer, P. Westcott, and M. Zabrocki, On the Stanley–Wilf limit of 4231-avoiding permutations and a conjecture of Arratia. *Adv. Appl. Math.* **36** (2006), 96–105.   [433]

Albertson M.O., A lower bound for the independence number of a planar graph. *J. Combin. Th. B* **20** (1976), 84–93.   [423]

Albertson M.O., You can't paint yourself into a corner. *J. Combin. Th. B* **73** (1998), 189–194.   [356]

Albertson M.O. and D.M. Berman, A conjecture on planar graphs. In *Graph Theory and Related Topics* (J.A. Bondy and U.S.R. Murty, eds.) (Academic Press, 1979), 357.   [419]

Albertson M.O. and B. Mohar, Coloring vertices and faces of locally planar graphs. *Graphs Combin.* **22** (2006), 289–295.   [797]

Alexanderson G.L. and J.E. Wetzel, Dissections of a plane oval. *Amer. Math. Monthly* **84** (1977), 442–449.   [388]

Alkan E., Problem 10473. *Amer. Math. Monthly* **102** (1995), 745–746. Solution **104** (1997), 371.   [80]

Alladi K. and V.E. Hoggatt, Jr., Compositions with ones and twos. *Fibonacci Quart.* **13** (1975), 233–239.   [59]

Alon N., On the number of subgraphs of prescribed type of graphs with a given number of edges. *Israel J. Math.* **38** (1981), 116–130.   [510]

Alon N., Eigenvalues and expanders. *Combinatorica* **6** (1986), 83–96.   [780]

Alon N., Splitting necklaces. *Advances in Math.* **63** (1987), 246–253.   [810]

Alon N., The linear arboricity of graphs. *Israel J. Math.* **62** (1988), 311–325.   [412, 677, 678]

Alon N., Transversal numbers of uniform hypergraphs. *Graphs Combin.* **6** (1990), 1–4.   [666, 670]

Alon N., Choice numbers of graphs: a probabilistic approach. *Combin. Probab. Comput.* **1** (1992), 107–114.   [356]

Alon N., Restricted colorings of graphs. In *Surveys in Combinatorics*, *London Math. Soc. Lect. Notes* **187** (Cambridge Univ. Press, 1993), 1–33.   [661, 738]

Alon N., Probabilistic methods in coloring and decomposition problems. *Discr. Math.* **127** (1994), 31–46.   [678]

Alon N., Disjoint directed cycles. *J. Combin. Th. B* **68** (1996), 167–178.   [663, 685]

Alon N., Combinatorial Nullstellensatz. *Combin. Probab. Comput.* **8** (1999), 7–29. [733]

Alon N., Additive Latin transversals. *Israel J. Math.* **117** (2000a), 125–130. [746]

Alon N., Degrees and choice numbers. *Random Structures & Algorithms* **16** (2000b), 364–358. [661]

Alon N., Bipartite decomposition of random graphs. *J. Combin. Th. B* **113** (2015), 220–235. [771]

Alon N., L. Babai, and H. Suzuki, Multilinear polynomials and Frankl–Ray-Chaudhuri–Wilson type intersection theorems. *J. Combin. Th. A* **58** (1991), 165–180. [728, 731]

Alon N., T. Bohman, R. Holzman, and D.J. Kleitman, On partitions of discrete boxes. *Discr. Math.* **257** (2002), 255–258. [665]

Alon N., T. Bohman, and H. Huang, More on the bipartite decomposition of random graphs. *J. Graph Th.* **84** (2017), 45–52. [771]

Alon N. and R.B. Boppana, The monotone circuit complexity of Boolean functions. *Combinatorica* **7** (1987), 1–22. [504]

Alon N., P. Frankl, and L. Lovász, The chromatic number of Kneser hypergraphs. *Trans. Amer. Math. Soc.* **298** (1986), 359–370. [503]

Alon N. and S. Friedland, The maximum number of perfect matchings in graphs with a given degree sequence. *Electron. J. Combin.* **15** (2008), Note 13, 2. [510]

Alon N., S. Friedland, and G. Kalai, Regular subgraphs of almost regular graphs. *J. Combin. Theory B* **37** (1984), 79–91. [736]

Alon N. and Z. Füredi, Covering the cube by affine hyperplanes. *European J. Combin.* **14** (1993), 79–83. [746]

Alon N., Z. Füredi, and M. Katchalski, Separating pairs of points by standard boxes. *European J. Combin.* **6** (1985), 205–210. [442]

Alon N., A. Kostochka, B. Reiniger, D.B. West, and X. Zhu, Coloring, sparseness and girth. *Israel J. Math.* **214** (2016), 315–331. [434, 442]

Alon N., M. Krivelevich, and B. Sudakov, Turán numbers of bipartite graphs and related Ramsey-type questions. **12** (2003), 477–494. [672, 673, 684]

Alon N., C.J.H. McDiarmid, and M. Molloy, Edge-disjoint cycles in regular directed graphs. *J. Graph Theory* **22** (1996), 231–237. [685]

Alon N. and V.D. Milman, Eigenvalues, expanders and superconcentrators. In *Proc. 25th IEEE Symp. Found. Comp. Sci.* (IEEE, 1984), 320–322. [779, 780]

Alon N. and V.D. Milman, λ_1, isoperimetric inequalities for graphs, and superconcentrators. *J. Combin. Th. B* **38** (1985), 73–88. [778, 779]

Alon N., M.B. Nathanson, and I. Rusza, Adding distinct congruence classes modulo a prime. *Amer. Math. Monthly* **102** (1995), 250–255. [746]

Alon N., M.B. Nathanson, and I. Rusza, The polynomial method and restricted sums of congruence classes. *J. Number Theory* **56** (1996), 404–417. [734, 735, 745, 746]

Alon N., P. Prałat, and N. Wormald, Cleaning regular graphs with brushes. *SIAM J. Discrete Math.* **23** (2008/09), 233–250. [704]

Alon N. and P. Pudlák, Constructive lower bounds for off-diagonal Ramsey numbers. *Israel J. Math.* **122** (2001), 243–251. [504]

Alon N., L. Rónyai, and T. Szabó, Norm-graphs: variations and applications. *J. Combin. Th. B* **76** (1999), 280–290. [633]

Alon N. and E.R. Scheinerman, Degrees of freedom versus dimension for containment orders. *Order* **5** (1988), 11–16. [829, 830, 832]

Alon N., P.D. Seymour, and R. Thomas, Planar separators. *SIAM J. Discr. Math.* **7** (1994), 184–193. [396, 397]

Alon N., A. Shpilka, and C. Umans, On sunflowers and matrix multiplication. *Comput. Complexity* **22** (2013), 219–243. [504]

Alon N. and J. Spencer, *The Probabilistic Method* (Wiley, 1992). Also 2000, 2008, 2016. [504, 657, 662, 663, 666, 669, 672–674, 688, 696, 706, 721, 722, 780]

Alon N. and M. Tarsi, A nowhere-zero point in linear mappings. *Combinatorica* **9** (1989), 393–5. [764]

Alon N. and M. Tarsi, Colorings and orientations of graphs. *Combinatorica* **12** (1992), 125–134.
                                                                                                   [738, 739, 746]

Alon N. and D.B. West, The Borsuk–Ulam theorem and bisection of necklaces. *Proc. Amer. Math. Soc.*
    98 (1986), 623–628.   [809]

Alon N. and N. Wormald, High degree graphs contain large-star factors. In *Fete of combinatorics and
    computer science, Bolyai Soc. Math. Stud.* **20** (János Bolyai Math. Soc., 2010), 9–21.   [671]

Alspach B., Research problem 3. *Discr. Math.* **36** (1981), 333.   [646]

Alspach B. and H. Gavlas, Cycle decompositions of K_n and $K_n - I$. *J. Combin. Th. B* **81** (2001), 77–99.
                                                                                                   [646]

Alspach B. and R. Häggkvist, Some observations on the Oberwolfach problem. *J. Graph Th.* **9** (1985),
    177–187.   [647]

Alspach B., P.J. Schellenberg, D.R. Stinson, and D. Wagner, The Oberwolfach problem and factors of
    uniform odd length cycles. *J. Combin. Th. A* **52** (1989), 20–43.   [647]

Altinisik E., B.E. Sagan, and N. Tuglu, GCD matrices, posets, and nonintersecting paths. *Linear Mul-
    tilinear Algebra* **53** (2005), 75–84.   [177, 766]

Amahashi A. and M. Kano, On factors with given components. *Discr. Math.* **42** (1982), 1–6.   [273]

Andersen L.D., On edge-colourings of graphs. *Math. Scand.* **40** (1977), 161–175.   [360]

Andersen L.D., The strong chromatic index of a cubic graph is at most 10. **108** (1992), 231–252. [373]

Anderson I., Perfect matchings of a graph. *J. Combin. Th. B* **10** (1971), 183–186.   [264]

Anderson I., Sufficient conditions for matchings. *Proc. Edinburgh Math. Soc. (2)* **18** (1973), 129–136.
                                                                                                   [275]

Anderson I., Intersection theorems and a lemma of Kleitman. *Discr. Math.* **16** (1976), 181–185.
                                                                                             [497, 561, 595, 607]

Anderson I., *Combinatorial Designs: Construction Methods* (Ellis Horwood, 1990).   [654]

Anderson I., *Combinatorial designs and tournaments, Oxford Lecture Series in Mathematics and its
    Applications* **6** (Oxford Univ. Press, 1997).   [609, 612, 649]

Anderson M., R.B. Richter, and P. Rodney, The crossing number of $C_6 \times C_6$. In *Proc. 27th Southeastern
    Intl. Conf. Combin. Graph Th. Comput. (Boca Raton), Congr. Numer.* **118** (1996), 97–107.   [794]

Anderson M., R.B. Richter, and P. Rodney, The crossing number of $C_7 \times C_7$. In *Proc. 28th Southeastern
    Intl. Conf. Combin. Graph Th. Comput. (Boca Raton), Congr. Numer.* **125** (1997), 97–117.   [794]

Ando K., A. Kaneko, and S.V. Gervacio, The bandwidth of a tree with k leaves is at most $\lceil k/2 \rceil$. *Discr.
    Math.* **150** (1996), 403–406.   [249, 819]

André D., Sur les permutations alternées. *J. Math. Pures et Appliquées* **7** (1881), 167–184.   [208]

André D., Théorème sur les formes quadratiques. *Bull. Soc. Math. France* **15** (1887), 188–192.   [40]

Andreae T., On the unit interval number of a graph. *Discrete Appl. Math.* **22** (1988), 1–7.   [366]

Andrews G. and E. Deutsch, Problem 11908. *Amer. Math. Monthly* **123** (2016), 504.   [148]

Andrews G.E., *The theory of partitions* (Addison-Wesley, 1976).   [140]

Andrews G.E., Problem 10627. *Amer. Math. Monthly* **104** (1997a), 974. Solution **107** (2000), 86. [150]

Andrews G.E., Problem 10628. *Amer. Math. Monthly* **104** (1997b), 974. Solution **107** (2000), 87. [151]

Andrews G.E. and P. Paule, Solution to problem E3376. *Amer. Math. Monthly* **99** (1992), 63–65. Proposed
    97 (1990), 240.   [66]

Andrews P. and E.T. Wang, Problem E3260. *Amer. Math. Monthly* **95** (1988), 350. Solution **97** (1990),
    74–75.   [174]

Appel K. and W. Haken, Every planar map is four colorable. *Bull. Amer. Math. Soc.* **82** (1976), 711–712.
                                                                                                   [402]

Appel K., W. Haken, and J. Koch, Every planar map is four colorable. Part II: Reducibility. *Illinois J.
    Math.* **21** (1977), 491–567.   [406]

Arnautov V.I., Estimation of the exterior stability number of a graph by means of the minimal degree of
    the vertices (Russian). *Prikl. Mat. i Programmirovanie* **11** (1974), 3–8, 126.   [670]

Arratia R., On the Stanley–Wilf conjecture for the number of permutations avoiding a given pattern.
    Electron. J. Combin. **6** (1999), Note, N1, 4 pp. (electronic).   [432, 433]

Arrow K.J., *Social Choice and Individual Values, Cowles Commission Monograph No. 12* (Wiley & Sons; Chapman & Hall, 1951). [585]

Ash P., Problem 11520. *Amer. Math. Monthly* **117** (2010), 649. Solution **119** (2012), 701. [34]

Avis D., Nonpartitionable point sets. *Inform. Process. Lett.* **19** (1984), 125–129. [820]

Axenovich M., Z. Füredi, and D. Mubayi, On generalized Ramsey theory: the bipartite case. *J. Combin. Th. B* **79** (2000), 66–86. [686]

Ayel J., Hamiltonian cycles in particular *k*-partite graphs. *J. Combin. Th. B* **32** (1982), 223–228. [331]

Azuma K., Weighted sums of certain dependent random variables. *Tôhoku Math. J. (2)* **19** (1967), 357–367. [715, 722]

Babai L., A short proof of the nonuniform Ray-Chaudhuri–Wilson inequality. *Combinatorica* **8** (1988), 133–135. [727]

Babai L., Graph isomorphism in quasipolynomial time (2015). (arXiv:1512.03547). [699]

Babai L., P. Erdős, and S.M. Selkow, Random graph isomorphisms. *SIAM J. Comput.* **9** (1980), 628–635. [699, 705]

Babai L. and P. Frankl, *Linear algebra methods in combinatorics with applications to geometry and computer science* (Univ. of Chicago, 1992). [723, 744]

Babai L., W.M. Kantor, and E.M. Luks, Computational complexity and the classification of finite simple groups. In *Proc. 24th IEEE Symp. Found. Comp. Sci.* (1983), 162–171. [699]

Babai L. and L. Kučera, Canonical labelling of graphs in linear average time. In *Proc. 20th IEEE Symp. Found. Comp. Sci.* (IEEE, 1979), 39–46. [699]

Bäbler F., über die Zerlegung regulärer Streckenkomplexe ungerader Ordnung (German). *Comment. Math. Helv.* **10** (1938), 275–287. [267]

Bäbler F., Über eine spezielle Klasse Euler'scher Graphen. *Comment. Math. Helv.* **27** (1953), 81–100. [250]

Bach E., Exact analysis of a priority queue algorithm for random variate generation. In *Proc. 5th ACM-SIAM Symp. Discrete Algorithms (Arlington, 1994)* (ACM, 1994), 48–56. [105]

Bacher R., Problem 10891. *Amer. Math. Monthly* **108** (2001), 668. Solution **110** (2003), 439–440. [66]

Bachmann P., *Die Analytische Zahlentheorie (German)* (Teubner, 1894). [10]

Backelin J., J. West, and G. Xin, Wilf-equivalence for singleton classes. *Adv. in Appl. Math.* **38** (2007), 133–148. [442]

Bagga J.S. and B.N. Varma, Hamiltonian properties in bipartite graphs. *Bull. Inst. Combin. Appl.* **26** (1999), 71–85. [325]

Baker K.A., Dimension, join-independence, and breadth in partially ordered sets (1961). Mimeographed notes. [571]

Baker K.A., A generalization of Sperner's lemma. *J. Combinatorial Theory* **6** (1969), 224–225. [561]

Balko M., J. Cibulka, D. Král', and J. Kynčl, Ramsey numbers of ordered graphs **49** (2015), 419–424. [457]

Ball S. and V. Pepe, Asymptotic improvements to the lower bound of certain bipartite Turán numbers. *Combin. Probab. Comput.* **21** (2012), 323–329. [633]

Ball S. and V. Pepe, Forbidden subgraphs in the norm graph. *Discr. Math.* **339** (2016), 1206–1211. [633]

Ball W.W.R., *Mathematical Recreations and Essays* (McMillan, 1892). [70]

Balof B. and K.P. Bogart, Simple inductive proofs of the Fishburn and Mirkin theorem and the Scott–Suppes theorem. *Order* **20** (2003), 49–51. [586]

Balogh J. and J.A. Csirik, Index assignment for two-channel quantization. *IEEE Trans. Inform. Theory* **50** (2004), 2737–2751. [801]

Balogh J. and H. Kaul, A threshold for random geometric graphs with a hamiltonian cycle (2007). Unpublished manuscript. [819]

Balogh J., M. Kochol, A. Pluhár, and X. Yu, Covering planar graphs with forests. *J. Combin. Th. B* **94** (2005), 147–158. [421, 424]

Banderier C. and S.R. Schwer, Why Delannoy numbers? *J. Statist. Plann. Inference* **135** (2005), 40–54. [28, 41]

Bandlow J., An elementary proof of the hook formula. *Electron. J. Combin.* **15** (2008), Research paper
45, 14.   [190]

Bang C.M. and H. Sharp, Jr., Score vectors of tournaments. *J. Combin. Th. B* **26** (1979), 81–84.   [261]

Bang S.-J., Problem 10490. *Amer. Math. Monthly* **102** (1995), 930. Solution **106** (1999), 588–589.   [32]

Bang-Jensen J., S. Bessy, and S. Thomassé, Disjoint 3-cycles in tournaments: a proof of the Bermond–
Thomassen conjecture for tournaments. *J. Graph Theory* **75** (2014), 284–302.   [685]

Bannai E. and T. Ito, On finite Moore graphs. *J. Fac. Sci. Univ. Tokyo Sect. IA Math.* **20** (1973), 191–
208.   [244, 775]

Bapat R.B., *Graphs and matrices*, Universitext (Springer, 2010).   [766, 769]

Bárány I., A short proof of Kneser's conjecture. *J. Combin. Theory A* **25** (1978), 325–326.   [812, 820]

Bárány I., S.B. Shlosman, and A. Szűcs, On a topological generalization of a theorem of Tverberg. *J.
London Math. Soc. (2)* **23** (1981), 158–164.   [810]

Barát J. and C. Thomassen, Claw-decompositions and Tutte-orientations. *J. Graph Theory* **52** (2006),
135–146.   [238]

Barber B., D. Kühn, A. Lo, R. Montgomery, and D. Osthus, Fractional clique decompositions of dense
graphs and hypergraphs. *J. Combin. Th. B* **127** (2017), 148–186.   [640]

Barber B., D. Kühn, A. Lo, and D. Osthus, Edge-decompositions of graphs with high minimum degree.
Adv. Math. **288** (2016), 337–385.   [640]

Barra M., Editorial comment to solution of problem 10663. *Amer. Math. Monthly* **107** (2000), 370. Proposed **105** (1998), 464.   [62]

Barrera-Cruz F. and P. Haxell, A note on Schnyder's theorem. *Order* **28** (2011), 221–226.   [790]

Basavaraju M., L.S. Chandran, and M. Kummini, *d*-regular graphs of acyclic chromatic index at least
d + 2. *J. Graph Theory* **63** (2010), 226–230.   [329]

Basin S.L. and V.E. Hoggatt, Jr., A primer on the Fibonacci sequence, part II. *Fibonacci Quart.* **1** (1963),
61–68.   [61]

Batagelj V. and T. Pisanski, Hamiltonian cycles in the Cartesian product of a tree and a cycle. *Discr.
Math.* **38** (1982), 311–312.   [330]

Bauer D., Regular K_n-free graphs. *J. Combin. Th. B* **35** (1983), 193–200.   [490]

Bauer D., H.J. Broersma, and E.F. Schmeichel, Toughness in graphs—a survey. *Graphs Combin.* **22**
(2006), 1–35.   [318]

Bauer D., H.J. Broersma, and H.J. Veldman, Not every 2-tough graph is Hamiltonian. In *Proc. 5th
Twente Workshop on Graphs and Combin. Optimization (Enschede, 1997)*, **99** (2000), 317–321.
                                                                                           [318]

Bauer D. and E.F. Schmeichel, Hamiltonian degree conditions which imply a graph is pancyclic. *J. Combin. Th. B* **48** (1990), 111–116.   [326]

Baumert L.D., *Cyclic difference sets*, Lect. Notes Math. **182** (Springer-Verlag, 1971).   [637]

Baumert L.D., S.W. Golomb, and M. Hall, Jr., Discovery of an Hadamard matrix of order 92. *Bull. Amer.
Math. Soc.* **68** (1962), 237–238.   [622]

Bean D.R., Effective coloration. *J. Symbolic Logic* **41** (1976), 469–480.   [342]

Beck J., On 3-chromatic hypergraphs. *Discr. Math.* **24** (1978), 127–137.   [659, 673]

Beck J., On size Ramsey number of paths, trees, and circuits. I. *J. Graph Th.* **7** (1983), 115–129.   [455]

Beck J., An algorithmic approach to the Lovász local lemma. I. *Random Structures Algorithms* **2** (1991),
343–365.   [680]

Beckwith D., Problem 10669. *Amer. Math. Monthly* **105** (1998), 559. Solution **107** (2000), 568–9.   [152]

Beckwith D., Problem 10809. *Amer. Math. Monthly* **107** (2000), 566. Solution **109** (2002), 477–8.   [149]

Beckwith D., Problem 10865. *Amer. Math. Monthly* **108**(2001),371. Solution **109** (2002), 859–60.   [764]

Beckwith D., Problem 11183. *Amer. Math. Monthly* **112** (2005), 839. Solution **114** (2007), 551–2.   [148]

Beckwith D., Problem 11212. *Amer. Math. Monthly* **113**(2006a),268. Solution **115** (2008), 366–7.   [173]

Beckwith D., Problem 11249. *Amer. Math. Monthly* **113**(2006b),760. Solution **115** (2008), 859–60.   [49]

Beckwith D., Problem 11343. *Amer. Math. Monthly* **115** (2008), 166. Solution **116** (2009), 944–5.   [32]

Beckwith D., Problem 11464. *Amer. Math. Monthly* **116** (2009), 845. Solution **118** (2011), 750.   [149]

Beckwith D., Problem 11583. *Amer. Math. Monthly* **118** (2011), 558. Solution **120** (2013), 756–7.   [149]

Beckwith D., Problem 11754. *Amer. Math. Monthly* **121** (2014), 170. Solution **123** (2016), 298.   [59]

Behrend F.A., On sets of integers which contain no three terms in arithmetical progression. *Proc. Nat. Acad. Sci. U. S. A.* **32** (1946), 331–332.   [484]

Behzad M., *Graphs and their chromatic numbers* (Michigan State U., 1965). Ph.D. Thesis.   [373, 654]

Behzad M. and S.E. Mahmoodian, On topological invariants of the product of graphs. *Canad. Math. Bull.* **12** (1969), 157–166.   [329, 343, 398]

Beineke L.W. and R.D. Ringeisen, On the crossing numbers of products of cycles and graphs of order four. *J. Graph Theory* **4** (1980), 145–155.   [798]

Beineke L.W. and A.J. Schwenk, On a bipartite form of the Ramsey problem. In *Proc. 5th British Combin. Conf. (Univ. Aberdeen, 1975), Congr. Numer.* **15** (Utilitas Math., 1976), 17–22.   [620, 621, 624]

Belck H.B., Reguläre Faktoren von Graphen. *J. Reine Angew. Math.* **188** (1950), 228–252.   [268, 276]

Belevitch V., Theory of $2n$-terminal networks with applications to conference telephony. *Electr. Commun.* **27** (1950), 231–244.   [621]

Bender E.A. and E.R. Canfield, The asymptotic number of labeled graphs with given degree sequences. *J. Combin. Th. A* **24** (1978), 296–307.   [697, 704]

Bender E.A. and J.R. Goldman, Enumerative uses of generating functions. *Indiana Univ. Math. J.* **20** (1971), 753–765.   [127]

Bender E.A. and D.E. Knuth, Enumeration of plane partitions. *J. Combin. Th. A* **13** (1972), 40–54.   [207]

Bender E.A., F. Kochman, and D.B. West, Adding up to powers. *Amer. Math. Monthly* **97** (1990), 139–143.   [62]

Benevides F.S. and J. Skokan, The 3-colored Ramsey number of even cycles. *J. Combin. Th. B* **99** (2009), 690–708.   [461]

Benhocine A. and A.P. Wojda, The Geng–Hua Fan conditions for pancyclic or Hamilton-connected graphs. *J. Combin. Th. B* **42** (1987), 167–180.   [326]

Benjamin A.T., G.M. Levin, K. Mahlburg, and J.J. Quinn, Random approaches to Fibonacci identities. *Amer. Math. Monthly* **107** (2000), 511–516.   [664]

Benjamin A.T. and J.J. Quinn, The Fibonacci numbers—exposed more discretely. *Math. Mag.* **76** (2003), 182–192.   [53]

Bennett G., Problem 10216. *Amer. Math. Monthly* **99** (1992), 363. Solution **101** (1994), 917.   [152]

Bentz H.J., Proof of the Bulgarian Solitaire conjectures. *Ars Combin.* **23** (1987), 151–170.   [144, 152]

Benzer S., On the topology of the genetic fine structure. *Proc. Nat. Acad. Sci. USA* **45** (1959), 1607–1620.   [337]

Berge C., Two theorems in graph theory. *Proc. Nat. Acad. Sci. U.S.A.* **43** (1957), 842–844.   [277]

Berge C., Sur le couplage maximum d'un graphe. *C.R. Acad. Sci. Paris* **247** (1958), 258–259.   [266]

Berge C., Les problèmes de coloration en théorie des graphes. *Publ. Inst. Statist. Univ. Paris* **9** (1960), 123–160.   [366, 367]

Berge C., Perfect graphs. In *Six papers on Graph Theory* (Indian Stat. Institute, 1963), 1–21.   [370]

Berge C., Une propriété des graphes k-stables-critiques. In *Combinatorial Structures and Their Applications* (R. Guy, H. Hanani, N.W. Sauer, and J. Schönheim, eds.) (Gordon and Breach, 1970), 7–11.   [222, 263]

Berge C., *Graphs and Hypergraphs* (North-Holland, 1973). (translation and revision of *Graphes et Hypergraphes*, Dunod, 1970).   [267, 333, 335, 667]

Berge C., Nombres de coloration de l'hypergraphe h-parti complet. *Ann. Mat. Pura Appl. (4)* **103** (1975), 3–9.   [501]

Berge C., A theorem related to the Chvátal conjecture. In *Proc. 5th British Combin. Conf. (Univ. Aberdeen, 1975),* **15** (Utilitas Math., 1976), 35–40.   [501]

Berge C., k-optimal partitions of a directed graph. *European J. Combin.* **3** (1982), 97–101.   [549]

Berge C., Path-partitions in directed graphs and posets. In *Graphs and order (Banff, Alta., 1984), NATO Adv. Sci. Inst. C Math. Phys. Sci.* **147** (Reidel, 1985), 447–464.   [549]

Berge C. and V. Chvátal, *Topics on Perfect Graphs, Ann. Discr. Math.* **21** (North-Holland, 1984).   [370]

Berge C. and J.L. Ramírez-Alfonsín, Origins and genesis. In *Perfect graphs*, *Wiley-Intersci. Ser. Discr. Math. Optim.* (Wiley, 2001), 1–12.   [368]

Berger E. and I.B-A. Hartman, Proof of Berge's strong path partition conjecture for $k = 2$. *European J. Combin.* **29** (2008), 179–192.   [549]

Berger E. and I.B-A. Hartman, A unified approach to known and unknown cases of Berge's conjecture. *J. Graph Th.* **71** (2012), 317–330.   [549]

Berlekamp E.R., On subsets with intersections of even cardinality. *Canad. Math. Bull.* **12** (1969), 471–474.   [724, 744]

Berman K.A., Parity results on connected f-factors. *Discr. Math.* **59** (1986), 1–8.   [241, 249]

Bermond J.-C., On Hamiltonian walks. In *Proc. Fifth Brit. Combin. Conf.* (C.St.J.A. Nash-Williams and J. Sheehan, eds.) (Utilitas Math., 1976), 41–51.   [323]

Bermond J.-C. and C. Thomassen, Cycles in digraphs—a survey. *J. Graph Theory* **5** (1981), 1–43.
                                                                                                        [316, 685]

Bernardi C., On a theorem about vertex colorings of graphs. *Discr. Math.* **64** (1987), 95–96.   [355]

Bernshteyn A., The local cut lemma. *European J. Combin.* **63** (2017), 95–114.   [680]

Bertrand J., Solution d'un problème. *Comptes Rendus Acad. Sci., Paris* **105** (1887), 369.   [39]

Beth T., D. Jungnickel, and H. Lenz, *Design Theory* (Cambridge Univ. Press, 1986).   [609]

Beutelspacher A. and W. Brestovansky, Generalized Schur numbers. In *Combinatorial theory (Schloss Rauischholzhausen, 1982)*, *Lecture Notes in Math.* **969** (Springer, 1982), 30–38.   [473]

Bey C. and J.R. Griggs, Problem 10932. *Amer. Math. Monthly* **109** (2002), 298. Solution **111** (2004), 262–263.   [511]

Bhanu K.S. and M.N. Deshpande, Problem 11503. *Amer. Math. Monthly* **117** (2010), 458. Solution **119** (2012), 349–350.   [669]

Bhasker J., T. Samad, and D.B. West, Size, chromatic number, and connectivity. *Graphs Combin.* **10** (1994), 209–213.   [353]

Bhatt S.N. and C.E. Leiserson, Minimizing the longest edge in a VLSI layout (1981). MIT memo.   [809]

Bialostocki A. and J. Schönheim, On some Turán and Ramsey numbers for C_4. In *Graph theory and combinatorics (Cambridge, 1983)* (Academic Press, 1984), 29–33.   [490]

Biedl T., E.D. Demaine, C.A. Duncan, R. Fleischer, and S.G. Kobourov, Tight bounds on maximal and maximum matchings. *Discr. Math.* **285** (2004), 7–15.   [286]

Bienstock D., Some provably hard crossing number problems. *Discrete Comput. Geom.* **6** (1991), 443–459.   [795]

Bienstock D. and N. Dean, New results on rectilinear crossing numbers and plane embeddings. *J. Graph Theory* **16** (1992), 389–398.   [795]

Bienstock D. and N. Dean, Bounds for rectilinear crossing numbers. *J. Graph Theory* **17** (1993), 333–348.   [795, 796]

Biggs N., *Algebraic graph theory* (Cambridge Univ. Press, 1974).   [766]

Biggs N.L., E.K. Lloyd, and R.J. Wilson, *Graph theory: 1736–1936* (Clarendon Press, 1976).   [37, 46]

Binet J.P., Mémoire sur l'intégration des équations linéaires aux différences finies, d'un ordre quelconque, à coefficients variables. *Comptes Rendus hebdomadaires des séances de l'Académie des Sciences* **17** (1843), 559–567.   [67]

Birkhoff G., The reducibility of maps. *Amer. J. Math.* **35** (1913), 114–128.   [405, 421]

Birkhoff G., On the structure of abstract algebras. *Proc. Cambridge Phil. Soc* **31** (1935), 433–454.
                                                                                                          [594]

Birkhoff G., Tres observaciones sobre el algebra lineal. *Rev. Univ. Nac. Tucumán, Series A* **5** (1946), 147–151.   [257]

Birkhoff G., *Lattice theory, Third edition. American Mathematical Society Colloquium Publications, Vol. XXV* (AMS, 1967).   [607]

Biró C., P. Hamburger, A. Pór, and W.T. Trotter, Forcing posets with large dimension to contain large standard examples. *Graphs Combin.* **32** (2016), 861–880.   [580]

Bixby R.E., Matroids and operations research. In *Advanced techniques in practice of operations research* (H.J. Greenberg, F.H. Murphy, and S.H. Shaw, eds.) (North-Holland, 1981), 333–458.   [520]

Björner A., J. Matoušek, and G.M. Ziegler, *Using Brouwer's fixed point theorem*, *A Journey through Discr. Mathematics. A Tribute to JiříMatoušek* (Springer, to appear). (arXiv:1409.7890). [804]

Blagojević P.V.M., F. Frick, A. Haase, and G.M. Ziegler, Hyperplane mass partitions via relative equivariant obstruction theory. *Doc. Math.* **21** (2016), 735–771. [815]

Blagojević P.V.M., F. Frick, A. Haase, and G.M. Ziegler, Topology of the Grünbaum–Hadwiger–Ramos hyperplane mass partition problem. *Trans. Amer. Math. Soc.* **370** (2018), 6795–6824. [815]

Blass A. and F. Harary, Properties of almost all graphs and complexes. *J. Graph Th.* **3** (1979), 225–240. [690]

Blokhuis A., A new upper bound for the cardinality of 2-distance sets in Euclidean space. In *Convexity and graph theory (Jerusalem, 1981)*, *North-Holland Math. Stud.* **87** (1984), 65–66. [725, 744]

Bloom D.M., Problem 10921. *Amer. Math. Monthly* **109** (2002), 201. Solution **110** (2003), 958–9. [33]

Bloome L., P. Johnson, and N. Saritzky, Problem 11625. *Amer. Math. Monthly* **119** (2012), 162. Solution **121** (2014), 273. [344]

Boesch F. and R. Tindell, Robbins's theorem for mixed multigraphs. *Amer. Math. Monthly* **87** (1980), 716–719. [237]

Bogart K.P., *Introductory combinatorics, 2nd ed.* (Harcourt Brace Jovanovich, 1990). Also 1983, 2000. [147]

Bogart K.P., An obvious proof of Fishburn's interval order theorem. *Discr. Math.* **118** (1993), 239–242. [586]

Bogart K.P. and W.T. Trotter, Maximal dimensional partially ordered sets. II. Characterization of 2n-element posets with dimension n. *Discr. Math.* **5** (1973), 33–43. [583]

Bogart K.P. and D.B. West, A short proof that "proper = unit". *Discr. Math.* **201** (1999), 21–23. [587]

Bognár J., J. Mogyoródi, A. Prékopa, A. Rényi, and D. Szász, *Problem Book on Probability (Hungarian)* (Tankönyvkiadó, 1970). [105]

Bohman T., The triangle-free process. *Adv. Math.* **221** (2009), 1653–1677. [682, 685]

Bohman T. and P. Keevash, Dynamic concentration of the triangle-free process. In *7th European Conf. Combin. Graph Th. Appl.*, *CRM Series* **16** (Ed. Norm., 2013), 489–495. [682]

Bollobás B., On generalized graphs. *Acta Math. Acad. Sci. Hungar.* **16** (1965), 447–452. [568, 668]

Bollobás B., Graphs with given diameter and maximal valency and with a minimal number of edges. In *Combinatorial mathematics and its applications (Proc. Conf., Oxford, 1969)* (Academic Press, 1971), 25–37. [244]

Bollobás B., Sperner systems consisting of pairs of complementary subsets. *J. Combin. Th. A* **15** (1973), 363–366. [511]

Bollobás B., On complete subgraphs of different orders. *Math. Proc. Cambridge Philos. Soc.* **79** (1976a), 19–24. [238]

Bollobás B., Relations between sets of complete subgraphs. In *Proc. 5th British Combin. Conf. (Univ. Aberdeen, 1975)*, *Congr. Numer.* **15** (Utilitas Math., 1976b), 79–84. [478]

Bollobás B., *Extremal graph theory*, *London Mathematical Society Monographs* **11** (Academic Press, 1978). [215, 272, 311, 316, 335, 478, 632]

Bollobás B., A probabilistic proof of an asymptotic formula for the number of labelled regular graphs. *European J. Combin.* **1** (1980), 311–316. [697]

Bollobás B., Threshold functions for small subgraphs. *Math. Proc. Camb. Phil. Soc* **90** (1981a), 197–206. [703]

Bollobás B., Degree sequences of random graphs. *Discr. Math.* **33** (1981b), 1–19. [699, 700, 721]

Bollobás B., Vertices of given degree in a random graph. *J. Graph Th.* **6** (1982), 147–155. [699]

Bollobás B., *Random Graphs* (Academic Press, 1985). Also 2001. [657, 688]

Bollobás B., *Combinatorics: Set systems, hypergraphs, families of vectors and combinatorial probability* (Cambridge Univ. Press, 1986). [499]

Bollobás B., The chromatic number of random graphs. *Combinatorica* **8** (1988), 49–55. [701, 717, 718]

Bollobás B., *Modern graph theory*, *Graduate Texts in Mathematics* **184** (Springer-Verlag, 1998). [488]

Bollobás B. and G.R. Brightwell, The height of a random partial order: concentration of measure. *Ann. Appl. Probab.* **2** (1992), 1009–1018. [722]

Bollobás B. and P. Erdős, On the structure of edge graphs. *Bull. London Math. Soc.* **5** (1973), 317–321.
[482, 770]

Bollobás B. and P. Erdős, Cliques in random graphs. *Math. Proc. Camb. Phil. Soc.* **80** (1976), 419–427.
[701]

Bollobás B., P. Erdős, and M. Simonovits, On the structure of edge graphs. II. *J. London Math. Soc. (2)* **12** (1976), 219–224.
[482]

Bollobás B. and A.J. Harris, List-colourings of graphs. *Graphs Combin.* **1** (1985), 115–127.   [362, 373]

Bollobás B. and Y. Kohayakawa, An extension of the Erdős–Stone theorem. *Combinatorica* **14** (1994), 279–286.
[482]

Bollobás B. and E. Szemerédi, Girth of sparse graphs. *J. Graph Theory* **39** (2002), 194–200.   [250]

Bollobás B. and A. Thomason, Random graphs of small order. In *Random graphs '83 (Poznań, 1983)*, *North-Holland Math. Stud.* **118** (North-Holland, 1985), 47–97.
[692]

Bollobás B. and A.G. Thomason, On the girth of Hamiltonian weakly pancyclic graphs. *J. Graph Theory* **26** (1997), 165–173.
[250]

Bollobás B. and A.G. Thomason, Proof of a conjecture of Mader, Erdős and Hajnal on topological complete subgraphs. *Europ. J. Combin.* **19** (1998), 883–887.
[351]

Bollobás B. and P. Winkler, The longest chain among random points in Euclidean space. *Proc. Amer. Math. Soc.* **103** (1988), 347–353.
[821]

Bóna M., Problem E3378. *Amer. Math. Monthly* **97** (1991), 340. Solution **99** (1992), 65-66.   [474]

Bóna M., Permutations avoiding certain patterns: the case of length 4 and some generalizations. *Discr. Math.* **175** (1997), 55–67.
[432]

Bóna M., The solution of a conjecture of Stanley and Wilf for all layered patterns. *J. Combin. Th. A* **85** (1999), 96–104.
[432]

Bóna M., *A walk through combinatorics* (World Scientific Publishing Co., Inc., River Edge, NJ, 2002).
[440]

Bóna M., *Combinatorics of permutations, Discr. Mathematics and its Applications (Boca Raton)* (Chapman & Hall/CRC, 2004).   [101, 102, 106, 133, 158, 432, 442]

Bonamy M., Planar graphs with $\Delta \geq 8$ are $(\Delta+1)$-edge-choosable. *SIAM J. Discr. Math.* **29** (2015), 1735–1763.
[413]

Bonamy M., T. Perrett, and L. Postle, Colouring graphs with sparse neighbourhoods: Bounds and applications (2018). (arXiv:1810.06704).
[685]

Bonato A. and P. Prałat, *Graph searching games and probabilistic methods, Discr. Mathematics and its Applications (Boca Raton)* (CRC Press, 2018).
[708]

Bondarenko A., On Borsuk's conjecture for two-distance sets. *Discrete Comput. Geom.* **51** (2014), 509–515.
[817]

Bondy J.A., Properties of graphs with constraints on degrees. *Stud. Sci. Math. Hung.* **4** (1969), 473–475.
[296]

Bondy J.A., Pancyclic graphs. I. *J. Combin. Th. B* **11** (1971a), 80–84.   [326, 330, 460]

Bondy J.A., Large cycles in graphs. *Discr. Math.* **1** (1971b), 121–132.   [323]

Bondy J.A., Induced subsets. *J. Combin. Th. B* **12** (1972a), 201–202.   [251]

Bondy J.A., Variation on the Hamiltonian theme. *Canad. Math. Bull.* **15** (1972b), 57–62.   [332]

Bondy J.A., A remark on two sufficient conditions for Hamilton cycles. *Discr. Math.* **22** (1978a), 191–193.
[316]

Bondy J.A., Hamilton cycles in graphs and digraphs. In *Proc. 9th Southeastern Intl. Conf. Combin. Graph Th. Comput. (Boca Raton)*, **21** (Utilitas Math., 1978b), 3–28.
[332]

Bondy J.A., *Longest paths and cycles in graphs of high degree, Res. Rep. No. CORR* (Univ. Waterloo, 1980).
[326]

Bondy J.A., Integrity in graph theory. In *The theory and applications of graphs (Kalamazoo, Mich., 1980)* (Wiley, 1981), 117–125.
[327]

Bondy J.A., Basic graph theory: paths and circuits. In *Handbook of combinatorics, Vol. 1, 2* (Elsevier, 1995), 3–110.
[316]

Bondy J.A. and V. Chvátal, A method in graph theory. *Discr. Math.* **15** (1976), 111–136.   [320]

Bondy J.A. and M. Kouider, Hamiltonian cycles in regular 2-connected graphs. *J. Combin. Th. B* **44** (1988), 177–186. [326]

Bondy J.A. and S.C. Locke, Largest bipartite subgraphs in triangle-free graphs with maximum degree three. *J. Graph Theory* **10** (1986), 477–504. [227]

Bondy J.A. and L. Lovász, Cycles through specified vertices of a graph. *Combinatorica* **1** (1981), 117–140. [302]

Bondy J.A. and U.S.R. Murty, *Graph Theory with Applications* (North-Holland, 1976). Also 2007. [215, 228, 330, 489, 639]

Bondy J.A. and U.S.R. Murty, *Graph theory, Grad. Texts Math.* **244** (Springer, 2008). [215]

Bondy J.A. and M. Simonovits, Cycles of even length in graphs. *J. Combin. Th. B* **16** (1974), 97–105. [479, 684]

Bonnice W.E., On convex polygons determined by a finite planar set. *Amer. Math. Monthly* **81** (1974), 749–752. [458]

Booth K.S. and G.S. Luecker, Testing for the consecutive ones property, interval graphs, and graph planarity using PQ-tree algorithms. *J. Comp. Syst. Sci.* **13** (1976), 335–379. [393]

Borchardt C.W., Ueber eine der Interpolation entsprechende Darstellung der Eliminations-Resultante. *J. Reine Angew. Math.* **57** (1860), 111–121. [37]

Borg P., On Chvátal's conjecture and a conjecture on families of signed sets. *European J. Combin.* **32** (2011), 140–145. [503]

Borg P. and K. Meagher, The Katona cycle proof of the Erdős–Ko–Rado theorem and its possibilities. *J. Algebraic Combin.* **43** (2016), 915–939. [500]

Borodin O.V., Criterion of chromaticity of a degree prescription (in Russian). In *Abstracts of IV All-Union Conf. on Theoretical Cybernetics (Novosibirsk)* (1977), 127–128. Details in "Problems of colouring and of covering the vertex set of a graph by induced subgraphs," Ph.D. Thesis (in Russian), Novosibirsk, 1979. [348]

Borodin O.V., Solution of the Ringel problem on vertex-face coloring of planar graphs and coloring of 1-planar graphs. *Metody Diskret. Analiz.* (1984), 12–26, 108. [416]

Borodin O.V., Solving the Kotzig and Grünbaum problems on the separability of a cycle in planar graphs. *Mat. Zametki* **46** (1989a), 9–12, 103. [423]

Borodin O.V., On the total coloring of planar graphs. *J. Reine Angew. Math.* **394** (1989b), 180–185. [411]

Borodin O.V., A generalization of Kotzig's theorem and prescribed edge coloring of planar graphs. *Mat. Zametki* **48** (1990), 22–28, 160. [412, 413, 423]

Borodin O.V., A new proof of the 6 color theorem. *J. Graph Theory* **19** (1995), 507–521. [416]

Borodin O.V., Structural properties of plane graphs without adjacent triangles and an application to 3-colorings. *J. Graph Theory* **21** (1996a), 183–186. [414, 415]

Borodin O.V., Structural theorem on plane graphs with application to the entire coloring number. *J. Graph Theory* **23** (1996b), 233–239. [423]

Borodin O.V., Colorings of plane graphs: A survey. *Discr. Math.* **313** (2013), 517–539. [407]

Borodin O.V., A.N. Glebov, A. Raspaud, and M.R. Salavatipour, Planar graphs without cycles of length from 4 to 7 are 3-colorable. *J. Combin. Th. B* **93** (2005), 303–311. [414]

Borodin O.V. and A.O. Ivanova, 2-distance $(\Delta + 2)$-coloring of planar graphs with girth six and $\Delta \geq 18$. *Discr. Math.* **309** (2009), 6496–6502. [412]

Borodin O.V. and A.V. Kostochka, On an upper bound of the graph's chromatic number depending on the graph's degree and density. *J. Combin. Th. B* **23** (1977), 247–250. [355]

Borodin O.V., A.V. Kostochka, and D.R. Woodall, Total colorings of planar graphs with large maximum degree. *J. Graph Theory* **26** (1997), 53–59. [364, 365, 413, 414]

Borodin O.V. and D.P. Sanders, On light edges and triangles in planar graphs of minimum degree five. *Math. Nachr.* **170** (1994), 19–24. [416, 423]

Borozan V., G.J. Chang, N. Cohen, S. Fujita, N. Narayanan, R. Naserasr, and P. Valicov, From edge-coloring to strong edge-coloring. *Electron. J. Combin.* **22** (2015). [373]

Borsuk K., Drei Sätz über die n-dimensionale euklidische Sphäre (German). *Fundamenta Mathematicae* **20** (1933), 177–190.   [809, 810, 817]

Borůvka O., O jistém problému minimálním (Czech). *Práce mor. přírodevěd. spol. v Brně* **3** (1926), 37–58.   [246, 519]

Bosák J., Hamiltonian lines in cubic graphs. In *Theory of Graphs (Internat. Sympos., Rome, 1966)* (Gordon and Breach; Dunod, 1967), 35–46.   [319, 331, 418]

Bose R.C., On the construction of balanced incomplete block designs. *Ann. Eugenics* **9** (1939), 353–399.
                                                                                  [610, 615, 627, 641–643, 655]

Bose R.C. and S.S. Shrikhande, On the falsity of Euler's conjecture about the non-existence of two orthogonal Latin squares of order $4t + 2$. *Proc. Nat. Acad. Sci. U.S.A.* **45** (1959), 734–737.   [651]

Bose R.C., S.S. Shrikhande, and E.T. Parker, Further results on the construction of mutually orthogonal Latin squares and the falsity of Euler's conjecture. *Canad. J. Math.* **12** (1960), 189–203.
                                                                                              [612, 652–654]

Bouchet A. and J.L. Fouquet, Trois types de décompositions d'un graphe en chaînes. In *Combinatorial mathematics (Marseille-Luminy, 1981), North-Holland Math. Stud.* **75** (North-Holland, 1983), 131–141.   [273]

Boyer J. and W. Myrvold, Stop minding your P's and Q's: a simplified $O(n)$ planar embedding algorithm. In *Proc. 10th ACM-SIAM Symp. Discrete Algs. (Baltimore)* (Assoc. Comput. Mach., 1999), 140–146.   [393]

Brandstädt A., V.B. Le, and J.P. Spinrad, *Graph classes: a survey, SIAM Monographs on Discr. Mathematics and Applications* (Society for Industrial and Applied Mathematics (SIAM), 1999).
                                                                                                            [370]

Brandt A., M. Ferrara, M. Kumbhat, S. Loeb, D. Stolee, and M. Yancey, I,F-partitions of sparse graphs. *European J. Combin.* **57** (2016), 1–12.   [422]

Brandt J., Cycles of partitions. *Proc. Amer. Math. Soc.* **85** (1982), 483–486.   [143–145]

Brandt S., Subtrees and subforests of graphs. *J. Combin. Th. B* **61** (1994), 63–70.   [262, 356]

Brandt S., Expanding graphs and Ramsey numbers (1996). Bielefeld preprint server, No. A 96-24.
                                                                                                            [454]

Branin F.H., Jr., The relation between Kron's method and the classical methods of network analysis. In *IRE Wescon Convention Record, II* (1959), 3–28.   [763]

Brègman L.M., Certain properties of nonnegative matrices and their permanents. *Dokl. Akad. Nauk SSSR* **211** (1973), 27–30.   [510]

Brightwell G.R., Linear extensions of infinite posets. *Discr. Math.* **70** (1988), 113–136.   [823, 831]

Brightwell G.R., Semiorders and the $\frac{1}{3}-\frac{2}{3}$ conjecture. *Order* **5** (1989), 369–380.   [823, 831]

Brightwell G.R., On the complexity of diagram testing. *Order* **10** (1993), 297–303.   [543]

Brightwell G.R., Balanced pairs in partial orders. *Discr. Math.* **201** (1999), 25–52.   [823]

Brightwell G.R., S. Felsner, and W.T. Trotter, Balancing pairs and the cross product conjecture. *Order* **12** (1995), 327–349.   [823]

Brightwell G.R. and E.R. Scheinerman, Representations of planar graphs. *SIAM J. Discr. Math.* **6** (1993), 214–229.   [394]

Brightwell G.R. and W.T. Trotter, The order dimension of convex polytopes. *SIAM J. Discr. Math.* **6** (1993), 230–245.   [790]

Brightwell G.R. and W.T. Trotter, The order dimension of planar maps. *SIAM J. Discr. Math.* **10** (1997), 515–528.   [790]

Brightwell G.R. and W.T. Trotter, A combinatorial approach to correlation inequalities. *Discr. Math.* **257** (2002), 311–327.   [600]

Brightwell G.R. and P. Winkler, Counting linear extensions. *Order* **8** (1991), 225–242.   [823]

Brightwell G.R. and C. Wright, The 1/3–/3 conjecture for 5-thin posets. *SIAM J. Discr. Math.* **5** (1992), 467–474.   [823]

Broersma H., Z. Ryjáček, and I. Schiermeyer, Closure concepts: a survey. *Graphs Combin.* **16** (2000), 17–48.   [320]

Broersma H. and H. Tuinstra, Independence trees and Hamilton cycles. *J. Graph Th.* **29** (1998), 227–237. [327]

Broersma H.J., On some intriguing problems in Hamiltonian graph theory—a survey. *Discr. Math.* **251** (2002), 47–69. [318]

Broersma H.J., F.V. Fomin, P.A. Golovach, and G.J. Woeginger, Backbone colorings for graphs: tree and path backbones. *J. Graph Theory* **55** (2007), 137–152. [342]

Broline D.M., Renumbering of the faces of dice. *Math. Mag.* **52** (1979), 312–314. [103]

Broline D.M. and D.E. Loeb, The combinatorics of Mancala-type games: Ayo, Tchoukaillon, and $1/\pi$. *UMAP J.* **16** (1995), 21–36. [144]

Brooks R.L., On colouring the nodes of a network. *Proc. Cambridge Phil. Soc.* **37** (1941), 194–197. [336]

Brouwer A.E. and W.H. Haemers, *Spectra of graphs*, *Universitext* (Springer, 2012). [766, 771]

Brouwer A.E. and A. Schrijver, The blocking number of an affine space. *J. Combin. Th. A* **24** (1978), 251–253. [735]

Brouwer L.E.J., Über Abbildung von Mannigfaltigkeiten. *Math. Ann.* **71** (1911), 97–115. [805]

Brown J.L., Note on complete sequences of integers. *Amer. Math. Monthly* **68** (1961), 557–560. [441]

Brown T.C., P. Erdős, F.R.K. Chung, and R.L. Graham, Quantitative forms of a theorem of Hilbert. *J. Combin. Th. A* **38** (1985), 210–216. [473]

Brown W.G., On graphs that do not contain a Thomsen graph. *Canad. Math. Bull.* **9** (1966), 281–285. [632, 633]

Brown W.G. and F. Harary, Extremal digraphs (1970), 135–198. [228]

Brualdi R.A., Comments on bases in dependence structures. *Bull. Austral. Math. Soc.* **1** (1969), 161–167. [538]

Brualdi R.A., *Introductory combinatorics (5th ed.)* (Pearson Prentice Hall, 2010). First ed. North–Holland, 1977. [174]

Brualdi R.A. and S. Kirkland, Aztec diamonds and digraphs, and Hankel determinants of Schröder numbers. *J. Combin. Th. B* **94** (2005), 334–351. [34]

Bruck R.H. and H.J. Ryser, The nonexistence of certain finite projective planes. *Canadian J. Math.* **1** (1949), 88–93. [616, 627]

Bruhn H., P. Charbit, and J.A. Telle, The graph formulation of the union-closed sets conjecture. In *The Seventh European Conference on Combinatorics, Graph Theory and Applications*, CRM Series **16** (Ed. Norm., 2013), 73–78. [512]

Bruhn H. and O. Schaudt, The journey of the union-closed sets conjecture. *Graphs Combin.* **31** (2015), 2043–2074. [503]

Bryant D. and P. Danziger, On bipartite 2-factorizations of $K_n - I$ and the Oberwolfach problem. *J. Graph Th.* **68** (2011), 22–37. [647, 648]

Bryant D., D. Horsley, and W. Pettersson, Cycle decompositions V: Complete graphs into cycles of arbitrary lengths. *Proc. Lond. Math. Soc. (3)* **108** (2014), 1153–1192. [646]

Bryant D. and C.A. Rodger, Cycle decompositions. In *The CRC Handbook of Combinatorial Designs (2nd ed.)* (Colbourn, Charles J. and Dinitz, Jeffrey H., eds.) (CRC Press, 2007), 373–382. [647]

Bryant P.R., Graph theory applied to electrical networks. In *Graph Theory and Theoretical Physics* (Academic Press, 1967), 111–137. [763]

Brylawski T.H., Some properties of basic families of subsets. *Discr. Math.* **6** (1973), 333–341. [538]

Brylawski T.H., Appendix of matroid cryptomorphisms. In *Theory of matroids* (N. White, ed.), *Encyc. Math. Appl.* **26** (Cambridge Univ. Press, 1986), 298–316. [528, 529]

Bucić M., An improved bound for disjoint directed cycles. *Discr. Math.* **341** (2018), 2231–2236. [685]

Bucić M., S. Letzter, and B. Sudakov, Three colour bipartite Ramsey number of cycles and paths (2018). (arXiv:1803.03689). [460]

Buckley F. and M. Lewinter, *A Friendly Introduction to Graph Theory* (Prentice Hall, 2002). [215]

Bukh B. and Z. Jiang, A bound on the number of edges in graphs without an even cycle. *Combin. Probab. Comput.* **26** (2017), 1–15. [479]

Bunde D.P., E.W. Chambers, D.W. Cranston, K.G. Milans, and D.B. West, Pebbling and optimal pebbling in graphs. *J. Graph Theory* **57** (2008), 215–238.   [668]

Buneman P., A characterization of rigid circuit graphs. *Discr. Math.* **9** (1974), 205–212.   [373]

Buneman P. and L. Levy, The Towers of Hanoi problem. *Info. Process. Lett.* **10** (1980), 243–244.   [80]

Buratti M., Rotational k-cycle systems of order $v < 3k$; another proof of the existence of odd cycle systems. *J. Combin. Des.* **11** (2003), 433–441.   [646]

Burns D. and S. Schuster, Embedding $(p, p-1)$ graphs in their complements. *Israel J. Math.* **30** (1978), 313–320.   [251]

Burr S.A., Generalized Ramsey theory for graphs—a survey. In *Graphs and Combinatorics, Proc. Capital Conf. (Washington, 1973), Lect. Notes Math.* **486** (Springer, 1974), 52–75.   [459, 460]

Burr S.A., Ramsey numbers involving graphs with long suspended paths. *J. London Math. Soc. (2)* **24** (1981), 405–413.   [459]

Burr S.A. and P. Erdős, On the magnitude of generalized Ramsey numbers for graphs. In *Infinite and finite sets (Colloq., Keszthely, 1973), I, Colloq. Math. Soc. János Bolyai* **10** (North-Holland, 1975), 215–240.   [484]

Burr S.A. and P. Erdős, Extremal Ramsey theory for graphs. *Utilitas Math.* **9** (1976), 247–258.   [460]

Burr S.A. and P. Erdős, Generalizations of a Ramsey-theoretic result of Chvátal. *J. Graph Th.* **7** (1983), 39–51.   [454]

Burr S.A., P. Erdős, and L. Lovász, On graphs of Ramsey type. *Ars Combinatoria* **1** (1976), 167–190.
                                                                                          [456, 457, 461]

Burr S.A., P. Erdős, and J. Spencer, Ramsey theorems for multiple copies of graphs. *Trans. Amer. Math. Soc.* **209** (1975), 87–99.   [454]

Burr S.A. and J.A. Roberts, On Ramsey numbers for stars. *Utilitas Math.* **4** (1973), 217–220.   [459]

Buršteĭn M.I., An upper bound for the chromatic number of hypergraphs. *Sakharth. SSR Mecn. Akad. Moambe* **75** (1974), 37–40.   [418, 419]

Burungale A., Problem 11262. *Amer. Math. Monthly* **113** (2006), 940. Solution **115** (2008), 862.   [238]

Butler S. and J. Mao, Problem 11265. *Amer. Math. Monthly* **114** (2007), 77. Solution **116** (2009), 181.
                                                                                              [237]

Butterfield J., T. Grauman, W.B. Kinnersley, K.G. Milans, C. Stocker, and D.B. West, On-line Ramsey theory for bounded degree graphs. *Electron. J. Combin.* **18** (2011), Paper 136, 17.   [457]

Cabello S. and B. Mohar, Adding one edge to planar graphs makes crossing number and 1-planarity hard. *SIAM J. Comput.* **42** (2013), 1803–1829.   [795]

Callan D., Problem 10643. *Amer. Math. Monthly* **105** (1998), 175. Solution **107** (2000), 278–279.   [63]

Callan D., Solution to problem 10596. *Amer. Math. Monthly* **106** (1999), 367. Proposed **104** (1997), 456.
                                                                                              [174]

Callan D., Solution to problem 10878. *Amer. Math. Monthly* **110** (2003a), 342–343. Proposed **108** (2001), 470.   [35]

Callan D., Solution to problem 10894. *Amer. Math. Monthly* **110** (2003b), 443–444. Proposed **108** (2001), 770.   [64]

Callan D., Problem 11013. *Amer. Math. Monthly* **110** (2003c), 438. Solution **112** (2005), 184.   [81]

Callan D., Problem 11091. *Amer. Math. Monthly* **111** (2004), 534. Solution **113** (2006), 462–463.   [49]

Callan D., Problem 11362. *Amer. Math. Monthly* **105** (2008), 461. Solution **117** (2010), 187.   [59]

Callan D., Problem 11567. *Amer. Math. Monthly* **118** (2011), 371. Solution **120** (2013), 370.   [134]

Callan D. and E. Deutsch, The run transform. *Discr. Math.* **312** (2012), 2927–2937.   [151]

Cameron K. and J. Edmonds, Some graphic uses of an even number of odd nodes. *Ann. Inst. Fourier (Grenoble)* **49** (1999), 815–827.   [241, 242, 249, 276]

Cameron K.B., *Polyhedral and Algorithmic Ramifications of Antichains* (ProQuest, 1982). Ph.D. Thesis, Univ. Waterloo.   [549]

Cameron K.B., On k-optimum dipath partitions and partial k-colourings of acyclic digraphs. *European J. Combin.* **7** (1986), 115–118.   [549]

Cameron P.J. and J.H. van Lint, *Designs, graphs, codes and their links, London Math. Soc. Student Texts* **22** (Cambridge Univ Press, 1991).   [609, 774]

Cameron P.J. and I.M. Wanless, Covering radius for sets of permutations. *Discr. Math.* **293** (2005), 91–109. [261]

Camion P., Chemins et circuits hamiltoniens des graphes complets. *C. R. Acad. Sci. Paris* **249** (1959), 2151–2152. [334]

Campbell C. and W. Staton, On extremal regular graphs with given odd girth. In *Proc. 22nd Southeastern Intl. Conf. Combin. Graph Th. Comput. (Baton Rouge)*, *Congr. Num.* **81** (1991), 157–159. [238]

Canfield E.R., On a problem of Rota. *Adv. in Math.* **29** (1978), 1–10. [567]

Cannings C. and J. Haigh, Montreal solitaire. *J. Combin. Th. A* **60** (1992), 50–66. [144]

Cantor D.G. and W.H. Mills, Determination of a subset from certain combinatorial properties. *Canadian J. Math.* **18** (1966), 42–48. [513]

Cao W., K.W. Hwang, and D.B. West, Improved bounds on families under *k*-wise set-intersection constraints. *Graphs Combin.* **23** (2007), 381–386. [731]

Carlitz L., Eulerian numbers and polynomials. *Math. Mag.* **32** (1958/1959), 247–260. [101]

Carlitz L., D.P. Roselle, and R.A. Scoville, Some remarks on ballot-type sequences of positive integers. *J. Combinatorial Theory A* **11** (1971), 258–271. [177]

Caro N. and C. Pohoata, Solution to problem 11403 (solved independently). *Amer. Math. Monthly* **118** (2011), 276–277. Proposed **115** (2008), 949. [174]

Caro Y., New results on the independence number. Tech. Rep. 05-79, Tel-Aviv University (1979). [263, 662]

Caro Y., Zero-sum problems—a survey. *Discr. Math.* **152** (1996), 93–113. [461]

Caro Y., I. Krasikov, and Y. Roditty, On the largest tree of given maximum degree in a connected graph. *J. Graph Th.* **15** (1991), 7–13. [327]

Caro Y. and Y. Roditty, Connected colorings of graphs. *Ars Combin.* **70** (2004), 191–196. [236]

Caro Y. and Z. Tuza, Improved lower bounds on *k*-independence. *J. Graph Theory* **15** (1991), 99–107. [263]

Cartwright D.P. and F. Harary, Structural balance: a generalization of Heider's theory. *Psychological Review* **63** (1956), 277–293. [237]

Catalan E., Note sur une équation aux différences finies. *J. Math. Pures Appl.* **3** (1838), 508–516. [41]

Catlin P.A., A bound on the chromatic number of a graph. *Discr. Math.* **22** (1978), 81–83. [355]

Catlin P.A., Hajós' graph-coloring conjecture: variations and counterexamples. *J. Combin. Th. B* **26** (1979), 268–274. [349, 356]

Cayley A., A theorem on trees. *Quart. J. Math.* **23** (1889), 376–378. [36, 37, 47]

Chaiken S. and D.J. Kleitman, Matrix tree theorems. *J. Combin. Th. A* **24** (1978), 377–381. [749]

Chandrasekharan K., *Arithmetical functions*, Die Grundlehren der mathematischen Wissenschaften, Band 167 (Springer-Verlag, 1970). [140]

Chang W.I. and E.L. Lawler, Edge coloring of hypergraphs and a conjecture of Erdős, Faber, Lovász. *Combinatorica* **8** (1988), 293–295. [439]

Chappell G.G., Polyunsaturated posets and graphs and the Greene–Kleitman theorem. *Discr. Math.* **257** (2002), 329–340. [551]

Charalambides C.A., *Enumerative combinatorics*, CRC Press Series on Discr. Mathematics and its Applications (Chapman & Hall/CRC, 2002). [101]

Chartrand G., D. Geller, and S. Hedetniemi, Graphs with forbidden subgraphs. *J. Combinatorial Theory B* **10** (1971), 12–41. [418]

Chartrand G. and F. Harary, Planar permutation graphs. *Ann. Inst. H. Poincaré Sect. B (N.S.)* **3** (1967), 433–438. [398]

Chartrand G. and F. Harary, Graphs with prescribed connectivities. In *Theory of Graphs (Tihany, 1966)* (P. Erdős and G. Katona, eds.) (Academic Press, 1968), 61–63. [295]

Chartrand G., S.F. Kapoor, L.M. Lesniak, and S. Schuster, Near 1-factors in graphs. In *Proc. 2nd West Coast Conf. Combin. Graph Th. Comput. (Eugene, OR, 1983)*, *Congr. Numer.* **41** (1984), 131–147. [273]

Chartrand G. and H.V. Kronk, The point-arboricity of planar graphs. *J. London Math. Soc.* **44** (1969), 612–616. [343]

Chartrand G. and L.M. Lesniak, *Graphs and Digraphs* (2nd ed.) (Wadsworth, 1986). Also 1996, 2005, 2011.   [215, 250, 315, 377]

Chartrand G., A.D. Polimeni, and M.J. Stewart, The existence of 1-factors in line graphs, sqaures, and total graphs. *Indagationes Math.* **35** (1973), 228–232.   [272]

Chein M., Graphe régulièrement décomposable. *Rev. Fran. Info. Rech. Opér.* **2** (1968), 27–42.   [312]

Chen G., R.J. Gould, and X. Yu, Graph connectivity after path removal. *Combinatorica* **23** (2003), 185–203.   [314]

Chen G., M.S. Jacobson, A.E. Kézdy, and J. Lehel, Tough enough chordal graphs are Hamiltonian. *Networks* **31** (1998), 29–38.   [318]

Chen G., J. Lehel, M.S. Jacobson, and W.E. Shreve, Note on graphs without repeated cycle lengths. *J. Graph Theory* **29** (1998), 11–15.   [249]

Chen G., R.H. Schelp, and B. Wei, Monochromatic-rainbow Ramsey numbers (2001). Presented 14th Cumberland Conference.   [473]

Chen Z.H. and H.J. Lai, The higher-order edge toughness of a graph and truncated uniformly dense matroids. *J. Combin. Math. Combin. Comput.* **22** (1996), 157–160.   [539]

Cherkashin D.D. and J. Kozik, A note on random greedy coloring of uniform hypergraphs. *Random Structures Algorithms* **47** (2015), 407–413.   [660, 673]

Chernoff H., A measure of asymptotic efficiency for tests of a hypothesis based on the sum of observations. *Ann. Math. Statistics* **23** (1952), 493–507.   [707]

Chetwynd A.G. and A.J.W. Hilton, Partial edge-colourings of complete graphs or of graphs which are nearly complete. In *Graph theory and combinatorics (Cambridge, 1983)* (Academic Press, London, 1984), 81–97.   [359]

Chetwynd A.G. and A.J.W. Hilton, Star multigraphs with three vertices of maximum degree. *Math. Proc. Cambridge Math. Soc.* **100** (1986), 303–317.   [359]

Chetwynd A.G. and A.J.W. Hilton, 1-factorizing regular graphs of high degree—an improved bound. In *Graph Theory and Combinatorics (Cambridge, 1988)*, *Discr. Math.* **75** (1989), 103–112.   [359]

Chevalley C., Démonstration d'une hypothèse de M. Artin. *Abh. Math. Sem. Univ. Hamburg* **11** (1935), 73–75.   [735]

Chevalley H. and E. Warning, Bemerkung zur vorstehenden Arbeit. *Abh. Math. Sem. Univ. Hamburg* **11** (1935), 76–83.   [735]

Chiba S. and T. Yamashita, Degree conditions for the existence of vertex-disjoint cycles and paths: a survey. *Graphs Combin.* **34** (2018), 1–83.   [326]

Chiue W.S. and B.S. Shieh, On connectivity of the Cartesian product of two graphs. *Appl. Math. Comput.* **102** (1999), 129–137.   [292]

Choi J.O., L. Özkahya, and D.B. West, Degree-splittability of multigraphs and caterpillars. In *Proc. 41st Southeastern Intl. Conf. Combin. Graph Th. Comput. (Boca Raton)*, *Congr. Numer.* **202** (2010), 137–147.   [239]

Chow T., A short proof of the rook reciprocity theorem. *Electron. J. Combin.* **3** (1996), R10.   [175]

Chowla S. and H.J. Ryser, Combinatorial problems. *Canadian J. Math.* **2** (1950), 93–99.   [616]

Chudnovsky M., N. Robertson, P.D. Seymour, and R. Thomas, The strong perfect graph theorem. *Ann. of Math. (2)* **164** (2006), 51–229.   [370]

Chung F.R.K., On partitions of graphs into trees. *Discr. Math.* **23** (1978a), 23–30.   [249]

Chung F.R.K., On concentrators, superconcentrators, generalizers and nonblocking networks. *Bell Syst. Tech. J.* (1978b), 1765–1777.   [779]

Chung F.R.K., *Spectral graph theory*, CBMS Conf. Series **92** (Amer. Math. Soc., 1997).   [776]

Chung F.R.K. and R.L. Graham, On multicolor Ramsey numbers for complete bipartite graphs. *J. Combin. Th. B* **18** (1975), 164–169.   [460]

Chung F.R.K. and R.L. Graham, *Erdős on graphs* (A K Peters, 1998).   [341, 446]

Chung F.R.K., R.L. Graham, P. Frankl, and J.B. Shearer, Some intersection theorems for ordered sets and graphs. *J. Combin. Th. A* **43** (1986), 23–37.   [508, 509]

Chung F.R.K. and C.M. Grinstead, A survey of bounds for classical Ramsey numbers. *J. Graph Th.* **7** (1983), 25–37.   [451]

Chung F.R.K. and L. Lu, An upper bound for the Turán number $t_3(n, 4)$. *J. Combin. Theory A* **87** (1999), 381–389. [491]

Chung K.L. and W. Feller, On fluctuations in coin-tossing. *Proc. Nat. Acad. Sci. U. S. A.* **35** (1949), 605–608. [50]

Chung M.-S. and D.B. West, Large P_4-free graphs with bounded degree. *J. Graph Th.* **17** (1993), 109–116. [490]

Chvátal V., Planarity of graphs with given degrees of vertices. *Nieuw Arch. Wisk. (3)* **17** (1969), 47–60. [389]

Chvátal V., The smallest triangle-free 4-chromatic 4-regular graph. *J. Combin. Th.* **9** (1970), 93–94. [354]

Chvátal V., On Hamilton's ideals. *J. Combin. Th. B* **12** (1972), 163–168. [320, 326, 332]

Chvátal V., Tough graphs and Hamiltonian circuits. *Discr. Math.* **2** (1973), 215–223. [318]

Chvátal V., Intersecting families of edges in hypergraphs having the hereditary property. In *Hypergraph Seminar (Proc. First Working Sem., Ohio State Univ., 1972; dedicated to Arnold Ross), Lecture Notes in Math.* **411** (Springer, 1974), 61–66. [458, 501, 502]

Chvátal V., A combinatorial theorem in plane geometry. *J. Combin. Th. B* **18** (1975), 39–41. [419]

Chvátal V., Tree-complete graph Ramsey numbers. *J. Graph Th.* **1** (1977), 93. [453]

Chvátal V., Star-cutsets and perfect graphs. *J. Combin. Th. B* **39** (1985b), 138–154. [375]

Chvátal V. and P. Erdős, A note on Hamiltonian circuits. *Discr. Math.* **2** (1972), 111–113. [321]

Chvátal V. and F. Harary, Generalized Ramsey theory for graphs, III. Small Off-diagonal Numbers. *Pac. J. Math.* **41** (1972), 335–345. [454]

Chvátal V. and F. Harary, Generalized Ramsey theory for graphs, I. Diagonal numbers. *Period. Math. Hungar.* **3** (1973), 115–124. [666]

Chvátal V. and J. Komlós, Some combinatorial theorems on monotonicity. *Canad. Math. Bull.* **14** (1971), 151–157. [458]

Chvátal V. and L. Lovász, Every directed graph has a semi-kernel. In *Hypergraph Seminar (Columbus, 1972), Lect. Notes Math.* **411** (Springer, 1974), 175. [229]

Chvátal V., V. Rödl, E. Szemerédi, and W.T. Trotter, The Ramsey numbers of a graph with bounded maximum degree. *J. Combin. Th. B* **34** (1983), 239–243. [455, 483]

Chvátalová J., Optimal labelling of a product of two paths. *Discr. Math.* **11** (1975), 249–253. [800, 818]

Cigler J., Some remarks on Catalan families. *European J. Combin.* **8** (1987), 261–267. [50]

Clapham C.R.J., Hamiltonian arcs in self-complementary graphs. *Discr. Math.* **8** (1974), 251–255. [332]

Clark D.S. and J.T. Lewis, Avoiding-sequences with minimum sum. *Discrete Appl. Math.* **22** (1989), 103–108. [429]

Clark D.S. and J.T. Lewis, Circular avoiding sequences with prescribed sum. *Discrete Appl. Math.* **43** (1993), 27–36. [429]

Clark L.H., J.C. George, and T.D. Porter, On the number of 1-factors in the n-cube. In *Proc. 28th Southeastern Intl. Conf. Combin. Graph Th. Comput. (Boca Raton), Congr. Numer.* **127** (1997), 67–69. [764]

Clark P.L., The Combinatorial Nullstellensätze revisited. *Electron. J. Combin.* **21** (2014), Paper 4.15, 17. [732]

Clements G.F., A minimization problem concerning subsets of a finite set. *Discr. Math.* **4** (1973), 123–128. [511]

Clements G.F. and B. Lindström, A generalization of a combinatorial theorem of Macaulay. *J. Combinatorial Theory* **7** (1969), 230–238. [496]

Cohen N. and F. Havet, Planar graphs with maximum degree $\Delta \geq 9$ are $(\Delta + 1)$-edge-choosable—a short proof. *Discr. Math.* **310** (2010), 3049–3051. [413]

Cohen-Addad V., M. Hebdige, D. Král', Z. Li, and E. Salgado, Steinberg's conjecture is false. *J. Combin. Th. B* **122** (2017), 452–456. [414]

Colbourn C.J. and J.H. Dinitz (eds.), *The CRC handbook of combinatorial designs, CRC Press Series on Discr. Math. Appl.* (CRC Press, 1996). Also 2007. [609, 646]

Colbourn C.J. and A. Rosa, *Triple systems*, *Oxford Mathematical Monographs* (Clarendon Press, 1999).
[641]

Coleman M., An answer to a question by Wilf on packing distinct patterns in a permutation. *Electron. J. Combin.* **11** (2004), Note 8, 4 pp.   [434, 442]

Comtet L., *Analyse combinatoire. Tomes I, II*, Collection SUP: "Le Mathématicien", *4* **5** (Presses Universitaires de France, 1970).   [65]

Comtet L., *Advanced combinatorics* (D. Reidel Publishing Co., 1974).   [65]

Conlon D., A new upper bound for diagonal Ramsey numbers. *Ann. of Math.* (2) **170** (2009a), 941–960.
[450]

Conlon D., On-line Ramsey numbers. *SIAM J. Discr. Math.* **23** (2009b/10), 1954–1963.   [457]

Conlon D. and J. Fox, Graph removal lemmas. In *Surveys in combinatorics 2013, London Math. Soc. Lecture Note Ser.* **409** (Cambridge Univ. Press, 2013), 1–49.   [485]

Conlon D., J. Fox, C. Lee, and B. Sudakov, Ordered Ramsey numbers. *J. Combin. Th. B* **122** (2017), 353–383.   [457]

Conlon D., J. Fox, and B. Sudakov, Hypergraph Ramsey numbers. *J. Amer. Math. Soc.* **23** (2010), 247–266.   [451, 459, 608]

Conlon D., J. Fox, and B. Sudakov, On two problems in graph Ramsey theory. *Combinatorica* **32** (2012), 513–535.   [484]

Conlon D., J. Fox, and B. Sudakov, An improved bound for the stepping-up lemma. *Discrete Appl. Math.* **161** (2013), 1191–1196.   [451, 453]

Conlon D., J. Fox, and B. Sudakov, Cycle packing. *Rand. Struct. Alg.* **45** (2014), 608–626.   [334]

Conlon D., J. Fox, and B. Sudakov, Recent developments in graph Ramsey theory. In *Surveys in combinatorics 2015, London Math. Soc. Lecture Note Ser.* **424** (Cambridge Univ. Press, 2015), 49–118.
[425]

Cook S., The complexity of theorem proving procedures. In *Proc. 3rd ACM Symposium on Theory of Computing* (1971), 151–158.   [11]

Corrádi K., Problem at Schweitzer competition. *Mat. Lapo* **20** (1969), 159–162.   [511]

Corradi K. and A. Hajnal, On the maximal number of independent circuits in a graph. *Acta Math. Acad. Sci. Hungar.* **14** (1963), 423–439.   [490]

Cox D.A. and U. Thieu, Problem 11862. *Amer. Math. Monthly* **122** (2015), 802. Solution **124** (2017), 473.
[173]

Cranston D.W., Strong edge-coloring of graphs with maximum degree 4 using 22 colors. *Discr. Math.* **306** (2006), 2772–2778.   [373]

Cranston D.W., Edge-choosability and total-choosability of planar graphs with no adjacent 3-cycles. *Discuss. Math. Graph Theory* **29** (2009), 163–178.   [413, 423]

Cranston D.W., R. Erman, and R. Škrekovski, Choosability of the square of a planar graph with maximum degree four. *Australas. J. Combin.* **59** (2014), 86–97.   [424]

Cranston D.W., S. Jahanbekam, and D.B. West, The 1, 2, 3-conjecture and 1, 2-conjecture for sparse graphs. *Discuss. Math. Graph Theory* **34** (2014), 769–799.   [422]

Cranston D.W. and S.J. Kim, List-coloring the square of a subcubic graph. *J. Graph Theory* **57** (2008), 65–87.   [424]

Cranston D.W., S.J. Kim, and G. Yu, Injective colorings of sparse graphs. *Discr. Math.* **310** (2010), 2965–2973.   [423, 424]

Cranston D.W. and L. Rabern, Brooks' theorem and beyond. *J. Graph Th.* **80** (2015), 199–225.   [336]

Cranston D.W. and D.B. West, Problem 11712. *Amer. Math. Monthly* **120** (2013), 569. Solution **122** (2015), 505–506.   [152]

Cranston D.W. and D.B. West, An introduction to the discharging method via graph coloring. *Discr. Math.* **340** (2017), 766–793.   [407, 411]

Cranston D.W. and G. Yu, Linear choosability of sparse graphs. *Discr. Math.* **311** (2011), 1910–1917.
[423]

Crapo H.H. and G.C. Rota, *On the Foundations of Combinatorial Theory: Combinatorial Geometries* (MIT Press, 1970).   [520]

Cruse A.B., A note on i-factors in certain regular multigraphs. *Discr. Math.* **18** (1977), 213–216.　　[274]

Cull P., Tours of graphs, digraphs, and sequential machines. *IEEE Trans. Comp.* **C29** (1980), 50–54.
　　[237]

Cull P. and E.F. Ecklund, On the Towers of Hanoi and generalized Towers of Hanoi problems. In *Proc. 13th Southeastern Intl. Conf. Combin. Graph Th. comput. (Boca Raton), Congr. Numer.* **35** (1982), 229–238.　　[80]

Curran S., O. Lee, and X. Yu, Finding four independent trees. *SIAM J. Comput.* **35** (2006), 1023–1058.
　　[315]

Cushing W. and H.A. Kierstead, Planar graphs are 1-relaxed, 4-choosable. *European J. Combin.* **31** (2010), 1385–1397.　　[742]

Cvetković D.M., M. Doob, and H. Sachs, *Spectra of graphs: Theory and applications*, Pure and Applied *Mathematics* **87** (Academic Press, 1979). Also 1982, 1995.　　[766, 781]

Czipzer J., Solution to problem 127 (Hungarian). *Mat. Lapok* **14** (1963), 373–374.　　[216]

Dályay P.P., Problem 11631. *Amer. Math. Monthly* **119** (2012), 247–8. Solution **121** (2014), 367.　　[623]

Dályay P.P., Problem 11897. *Amer. Math. Monthly* **123** (2016), 297. Solution **125** (2018), 86.　　[115]

Damerell R.M., On Moore graphs. *Proc. Cambridge Philos. Soc.* **74** (1973), 227–236.　　[244, 775]

David K., Solution to problem 11144. *Amer. Math. Monthly* **114** (2007), 262–264. Proposed **112** (2005), 274.　　[47]

Daykin D.E., Erdős–Ko–Rado from Kruskal–Katona. *J. Combin. Th. A* **17** (1974a), 254–255.　　[500]

Daykin D.E., A simple proof of the Kruskal–Katona theorem. *J. Combin. Th. A* **17** (1974b), 252–253.
　　[496]

Daykin D.E., A lattice is distributive if and only if $|A||B| \leq |A \vee B||A \wedge B|$. *Nanta Math.* **10** (1977), 58–60.　　[596, 608]

Daykin D.E., J. Godfrey, and A.J.W. Hilton, Existence theorems for Sperner families. *J. Combinatorial Theory A* **17** (1974), 245–251.　　[511]

Daykin D.E., A.J.W. Hilton, and D. Miklós, Pairings from down-sets and up-sets in distributive lattices. *J. Combin. Th. A* **34** (1983), 215–230.　　[501, 512]

Daykin D.E., D.J. Kleitman, and D.B. West, The number of meets between two subsets of a lattice. *J. Combin. Th. A* **26** (1979), 135–156.　　[596]

Daykin D.E. and L. Lovász, The number of values of a Boolean function. *J. London Math. Soc. (2)* **12** (1976), 225–230.　　[497, 607, 608]

de Abreu N.M.M., Old and new results on algebraic connectivity of graphs. *Linear Algebra Appl.* **423** (2007), 53–73.　　[778]

de Bruijn N.G., A combinatorial problem. *Nederl. Akad. Wetensch., Proc.* **49** (1946), 758–764; Indagationes Math. 8, 461–467 (1946).　　[239]

de Bruijn N.G. and P. Erdős, A colour problem for infinite graphs and a problem in the theory of relations. *Nederl. Akad. Wetensch. Proc. A.* **54** = *Indagationes Math.* **13** (1951), 369–373.　　[468]

de Bruijn N.G. and T. van Aardenne-Ehrenfest, Circuits and trees in oriented linear graphs. *Simon Stevin* **28** (1951), 203–217.　　[752]

de Bruijn N.G., E. van Tengbergen, and D. Kruyswijk, On the set of divisors of a number. *Nieuw Arch. Wisk. (2)* **23** (1951), 191–193.　　[554]

de Fraysseix H., J. Pach, and R. Pollack, How to draw a planar graph on a grid. *Combinatorica* **10** (1990), 41–51.　　[783]

de Grey A., The chromatic number of the plane is at least 5. *Geombinatorics* **28** (2018), 18–31. (Also arXiv:1804.02385)　　[342]

de Moivre A., *The Doctrine of Chances* (Pearson, 1718).　　[153]

de Moivre A., *Miscellanea Analytica* (London, 1730).　　[67]

de Werra D., Equitable colorations of graphs. *Rev. Française Informat. Recherche Opérationnelle* **5** (1971), 3–8.　　[372]

Dean A.M. and R.B. Richter, The crossing number of $C_4 \times C_4$. *J. Graph Theory* **19** (1995), 125–129.
　　[794]

Dean N., *Contractible Edges and Conjectures about Path and Cycle Numbers* (ProQuest, 1987). Ph.D. Thesis, Vanderbilt University.   [249]

Dean R.A. and G. Keller, Natural partial orders. *Canad. J. Math.* **20** (1968), 535–554.   [587, 604]

DeBiasio L., A. Gyárfás, R.A. Krueger, M. Ruszinkó, and G.N. Sárközy, Monochromatic balanced componenets, matchings, and paths in multicolored complete bipartite graphs (2018). (arXiv:1804.04195).   [460]

DeBiasio L. and A. Lo, Spanning trees with few branch vertices (2017). (arXiv:1709.04937).   [327, 334]

Dekster B.V., The Borsuk conjecture holds for bodies of revolution. *J. Geom.* **52** (1995), 64–73.   [817]

Delannoy H., Emploi de l'échiquier pour la résolution de divers problèmes de probabilité. *Assoc. Franç. Paris* **18** (1889), 43–52.   [28, 29]

Delcourt M. and L. Postle, On the list coloring version of Reed's conjecture. *Elec. Notes Discr. Math.* **61** (2017), 343–349.   [685]

Delsarte P., J.M. Goethals, and J.J. Seidel, Spherical codes and designs. *Geometriae Dedicata* **6** (1977), 363–388.   [744]

Demoucron G., Y. Malgrange, and R. Pertuiset, Graphes planaires: reconnaissance et construction des représentations planaires topologiques. *Rev. Française Recherche Opérationnelle* **8** (1964), 33–47.   [393]

Dénes J. and A.D. Keedwell, *Latin squares and their applications* (Academic Press, 1974). Also 2015 (Elsevier/North-Holland).   [610, 623]

Dénes J. and A.D. Keedwell, *Latin squares, Annals Discr. Math.* **46** (North-Holland, 1991).   [610]

Derbyshire J., *Prime obsession* (Joseph Henry Press, 2003).   [10]

DeSario R., Problem 10931. *Amer. Math. Monthly* **109** (2002), 298. Solution **111** (2004), 169–70.   [442]

Descartes B., A three colour problem. *Eureka* (1947). Solution 1948.   [434]

Deshpande B. and M.N. Deshpande, Problem 11350. *Amer. Math. Monthly* **115** (2008), 262. Solution **117** (2010), 89.   [668]

Deshpande B. and M.N. Deshpande, Problem 11500. *Amer. Math. Monthly* **117** (2010), 371. Solution **119** (2012), 348.   [92]

Deshpande M.N. and K. Laghate, Problem 11042. *Amer. Math. Monthly* **110** (2003), 843–842. Solution **112** (2005), 574.   [104]

Deshpande M.N. and R.M. Welukar, Problem 11033. *Amer. Math. Monthly* **110** (2003), 742. Solution **112** (2005), 470–472.   [115]

DeTemple D. and J.M. Robertson, The equivalence of Euler's and Pick's theorems. *Math. Teacher* **67** (1974), 222–226.   [389]

Deuber W., Partitionen und lineare Gleichungssysteme. *Math. Z.* **133** (1973), 109–123.   [462]

Deutsch E., Problem 10649. *Amer. Math. Monthly* **105** (1998), 271. Solution **107** (2000), 279–280.   [62]

Deutsch E., Problem 10751. *Amer. Math. Monthly* **106** (1999), 686. Solution **108** (2001), 872–873.   [49]

Deutsch E., Problem 10795. *Amer. Math. Monthly* **107** (2000), 367. Solution **108** (2001), 980.   [49]

Deutsch E., Problem 10877. *Amer. Math. Monthly* **108** (2001), 470. Solution **110** (2003), 245–246.   [82]

Deutsch E., Problem 10902. *Amer. Math. Monthly* **108** (2001a), 871. Solution **110** (2003), 639–640.
                                                                                          [22, 62]

Deutsch E., Problem 11071. *Amer. Math. Monthly* **111** (2004a), 259. Solution **113** (2006), 460–1.   [50]

Deutsch E., Problem 11108. *Amer. Math. Monthly* **111** (2004b), 725. Solution **113** (2006), 466–7.   [59]

Deutsch E., Problem 11150. *Amer. Math. Monthly* **112** (2005), 367. Solution **114** (2007), 264–265.
                                                                                          [48, 164]

Deutsch E., Problem 11237. *Amer. Math. Monthly* **113** (2006), 655. Solution **115** (2008), 666–7.   [149]

Deutsch E., Problem 11373. *Amer. Math. Monthly* **115** (2008), 568. Solution **117** (2010), 462.   [105]

Deutsch E., Problem 11424. *Amer. Math. Monthly* **116** (2009), 277. Solution **118** (2011), 376.   [21]

Deza M. and P. Frankl, Problem. In *Combinatorics (1976), Coll. Math. Soc. J. Bolyai* **18** (North-Holland, 1978), 1193.   [511]

Deza M., P. Frankl, and N.M. Singhi, On functions of strength *t*. *Combinatorica* **3** (1983), 331–339.
                                                                                            [728]

Dias da Silva J.A. and Y.O. Hamidoune, Cyclic spaces for Grassman derivatives and additive theory. *Bull. London Math. Soc.* **26** (1994), 140–146. [734, 746]

Díaz-Barrero J.L., Problem 11164. *Amer. Math. Monthly* **112** (2005), 568. Solution **114** (2007), 364–365. [32]

Dickson L.E., *College Algebra* (Wiley & Sons, 1902). [48]

Dickson L.E., Lower limit for the number of sets of solutions of $x^e + y^e + z^e \equiv 0 \pmod{p}$. *J. Reine Angew. Math.* **135** (1909), 181–188. [473]

Diestel R., *Graph theory, Graduate Texts in Mathematics* **173** (Springer-Verlag, 1997). Also 2000, 2006, 2010, 2016. [215, 377, 488]

Dijkstra E.W., A note on two problems in connexion with graphs. *Numer. Math.* **1** (1959), 269–271. [246, 252]

Dilcher K., Some q-series identities related to divisor functions. *Discr. Math.* **145** (1995), 83–93. [32]

Dilworth R.P., A decomposition theorem for partially ordered sets. *Ann. of Math. (2)* **51** (1950), 161–166. [546, 574, 607]

Dilworth R.P., Some combinatorial problems on partially ordered sets. In *Proc. Sympos. Appl. Math.* **10** (Amer. Math. Soc., 1960), 85–90. [606]

Dinitz J.H. and D.R. Stinson, A brief introduction to design theory. In *Contemporary design theory, Wiley-Intersci. Ser. Discr. Math. Optim.* (Wiley, 1992), 1–12. [609]

Dirac G.A., *On the Colouring of Graphs: Combinatorial Topology of Linear Complexes* (Univ. London, 1951). Ph.D. Thesis. [349]

Dirac G.A., A property of 4-chromatic graphs and some remarks on critical graphs. *J. London Math. Soc.* **27** (1952a), 85–92. [349]

Dirac G.A., Some theorems on abstract graphs. *Proc. London Math. Soc. (3)* **2** (1952b), 69–81. [320, 323, 349]

Dirac G.A., The structure of k-chromatic graphs. *Fund. Math.* **40** (1953), 42–55. [345]

Dirac G.A., A theorem of R. L. Brooks and a conjecture of H. Hadwiger. *Proc. London Math. Soc. (3)* **7** (1957), 161–195. [355]

Dirac G.A., In abstrakten Graphen vorhandene vollständige 4-graphen und ihre Unterteilungen. *Math. Nachr.* **22** (1960), 61–85. [301, 302]

Dirac G.A., On rigid circuit graphs. *Abh. Math. Sem. Univ. Hamburg* **25** (1961), 71–76. [367]

Dirac G.A., Extension of Turán's theorem on graphs. *Acta Math. Acad. Sci. Hungar.* **14** (1963), 417–422. [490]

Dirac G.A., Homomorphism theorems for graphs. *Math. Ann.* **153** (1964), 69–80. [351, 353]

Dirac G.A., Minimally 2-connected graphs. *J. Reine Angew. Math.* **228** (1967), 204–216. [315, 316]

Dirac G.A., B.A. Sørensen, and B. Toft, An extremal result for graphs with an application to their colourings. *J. Reine Angew. Math.* **268/269** (1974), 216–221. [345]

Djidjev H.N., A linear algorithm for partitioning graphs. *C. R. Acad. Bulgare Sci.* **35** (1982), 1053–1056. [397]

Djidjev H.N., On the constants of separator theorems. *C. R. Acad. Bulgare Sci.* **40** (1987), 31–34. [397]

Djidjev H.N. and S.M. Venkatesan, Reduced constants for simple cycle graph separation. *Acta Inform.* **34** (1997), 231–243. [397]

Djokovic D., Problem E3465. *Amer. Math. Monthly* **98** (1991), 852. Solution **100** (1993), 800. [32]

Dmitriev I.G., Weakly cyclic graphs with integral chromatic spectra. *Metody Diskret. Analiz.* (1980), 3–7, 100. [374]

Dobiński G., Summirung der Reihe $\sum \frac{n^m}{n!}$ für $m = 1, 2, 3, 4, 5, \ldots$. *Grunert's Archiv* **61** (1877), 333–336. [134]

Dodgson C.L., Condensation of determinants. *Proc. Roy. Soc. A* **15** (1866), 150–155. [170, 178]

Dol'nikov V.L., Transversals of families of sets. In *Studies in the theory of functions of several real variables (Russian)* (Yaroslav. Gos. Univ., 1981), 30–36, 109. [812]

Došlić T. and D. Rautenbach, Factor-critical graphs with the minimum number of near-perfect matchings. *Discr. Math.* **338** (2015), 2318–2319. [275]

Doster D., Problem E3391. *Amer. Math. Monthly* **97** (1990), 528. Solution **98** (1991), 860–861. [82]

Doster D., Problem 10403. *Amer. Math. Monthly* **101** (1994), 792. Solution **104** (1997), 368.   [91]

Doubilet P., G.C. Rota, and R.P. Stanley, On the foundations of combinatorial theory. VI. The idea of generating function. In *Proceedings of the Sixth Berkeley Symposium on Mathematical Statistics and Probability (Univ. California, Berkeley, Calif., 1970/1971), Vol. II: Probability theory* (Univ. California Press, 1972), 267–318.   [127]

Dross F., Fractional triangle decompositions in graphs with large minimum degree. *SIAM J. Discr. Math.* **30** (2016), 36–42.   [640]

Dubhashi D.P. and A. Panconesi, *Concentration of measure for the analysis of randomized algorithms* (Cambridge Univ. Press, 2009).   [706, 721]

Dudek A. and P. Prałat, An alternative proof of the linearity of the size-Ramsey number of paths. *Combin. Probab. Comput.* **24** (2015), 551–555.   [455]

Dudek A. and J.R. Schmitt, On the size and structure of graphs with a constant number of 1-factors. *Discr. Math.* **312** (2012), 1807–1811.   [276]

Duffus D., H. Lefmann, and V. Rödl, Shift graphs and lower bounds on Ramsey numbers $r_k(l; r)$. *Discr. Math.* **137** (1995), 177–187.   [601]

Dujmović V., G. Joret, J. Kozik, and D.R. Wood, Nonrepetitive colouring via entropy compression. *Combinatorica* **36** (2016), 661–686.   [680]

Duraj L., G. Gutowski, and J. Kozik, Chip games and paintability. *Electron. J. Combin.* **23** (2016), Paper 3.3, 12.   [742]

Dushnik B., Concerning a certain set of arrangements. *Proc. Amer. Math. Soc.* **1** (1950), 788–796.   [576, 577]

Dushnik B. and E.W. Miller, Partially ordered sets. *Amer. J. Math.* **63** (1941), 600–610.   [569, 570, 831]

Dvir Z., On the size of Kakeya sets in finite fields (to appear).   [746]

Dvořák Z., K.I. Kawarabayashi, and R. Thomas, Three-coloring triangle-free planar graphs in linear time. *ACM Trans. Algorithms* **7** (2011), Art. 41, 14.   [414]

Dvořák Z. and L. Postle, Correspondence coloring and its application to list-coloring planar graphs without cycles of lengths 4 to 8. *J. Combin. Th. B* **129** (2018), 38–54.   [356, 414]

Dvoretzky A. and T.S. Motzkin, A problem of arrangements. *Duke Math. J.* **14** (1947), 305–313.   [42]

Dzhumadildaeva A.A., Problem 11406. *Amer. Math. Monthly* **116** (2009), 82. Solution **117** (2010), 935.   [115]

Dziobek O., Eine Formel der Substitutionstheorie. *Sitzungsber. Berl. Math. G.* **17** (1917), 64–67.   [46]

Eaton N., Ramsey numbers for sparse graphs. *Discr. Math.* **185** (1998), 63–75.   [484]

Eaton N. and D.A. Grable, Set intersection representations for almost all graphs. *J. Graph Theory* **23** (1996), 309–320.   [745]

Eckhoff J. and G. Wegner, Über einen Satz von Kruskal. *Period. Math. Hungar.* **6** (1975), 137–142.   [496]

Edelman P. and C. Greene, Balanced tableaux. *Adv. in Math.* **63** (1987), 42–99.   [208]

Edmonds J., Existence of k-edge connected ordinary graphs with prescribed degrees. *J. Res. Nat. Bur. Standards Sect. B* **68B** (1964), 73–74.   [297]

Edmonds J., Minimum partition of a matroid into independent subsets. *J. Res. Nat. Bur. Standards Sect. B* **69B** (1965a), 67–72.   [251, 520, 535]

Edmonds J., Lehman's switching game and a theorem of Tutte and Nash-Williams. *J. Res. Nat. Bur. Standards Sect. B* **69B** (1965b), 73–77.   [251, 520, 535]

Edmonds J., Paths, trees, and flowers. *Canad. J. Math.* **17** (1965c), 449–467.   [264, 275, 278]

Edmonds J., Maximum matchings and a polyhedron with 0,1-vertices. *J. Res. Nat. Bur. Standards* **69B** (1965d), 125–130.   [278, 282]

Edmonds J., Submodular functions, matroids and certain polyhedra. In *Combinatorial Structures and Their Applications (Calgary, 1969)* (Gordon and Breach, 1970), 69–87.   [529]

Edmonds J., Matroid intersection. In *Discrete Optimization I* (P.L. Hammer, E.L. Johnson, and B.H. Korte, eds.), *Ann. Discr. Math.* **4** (1979), 39–49.   [532]

Edmonds J. and D.R. Fulkerson, Transversals and matroid partition. *J. Res. Nat. Bur. Standards Sect. B* **69B** (1965), 147–153.   [517, 533]

Edwards K., D.P. Sanders, P. Seymour, and R. Thomas, Three-edge-colouring doublecross cubic graphs. *J. Combin. Th. B* **119** (2016), 66–95.   [416]

Egawa Y., H. Matsuda, T. Yamashita, and K. Yoshimoto, On a spanning tree with specified leaves. *Graphs Combin.* **24** (2008), 13–18.   [327]

Eğecioğlu Ö. and J.B. Remmel, Bijections for Cayley trees, spanning trees, and their q-analogues. *J. Combin. Th. A* **42** (1986), 15–30.   [36]

Egerváry E., On combinatorial properties of matrices (Hungarian with German summary). *Mat. Lapok* **38** (1931), 16–28.   [258, 530]

Eggleston H.G., Covering a three-dimensional set with sets of smaller diameter. *J. London Math. Soc.* **30** (1955), 11–24.   [817]

Egorychev G.P., The solution of van der Waerden's problem for permanents. *Adv. in Math.* **42** (1981), 299–305.   [256, 765]

Eitner P.G., The bandwidth of the complete multipartite graph (1979). Presentation at Toledo Symposium on Applications of Graph Theory.   [818]

Eliahou S. and M. Kervaire, Sumsets in vector spaces over finite fields. *J. Number Theory* **71** (1998), 12–39.   [745]

Elkies N., G. Kuperberg, M. Larsen, and J.G. Propp, Alternating-sign matrices and domino tilings. I. *J. Algebraic Combin.* **1** (1992), 111–132.   [34]

Ellingham M. and P. Salehi Nowbandegani, The Chvátal–Erdős condition for prism-Hamiltonicity (2018). Unpublished preprint.   [334]

Ellingham M.N., Spanning paths, cycles, trees and walks for graphs on surfaces. *Congr. Numer.* **115** (1996), 55–90.   [326]

Ellingham M.N. and J.D. Horton, Non-Hamiltonian 3-connected cubic bipartite graphs. *J. Combin. Th. B* **34** (1983), 350–353.   [330]

Ellingham M.N. and X. Zha, Toughness, trees, and walks. *J. Graph Th.* **33** (2000), 125–137.   [327]

Ellis D., Y. Filmus, and E. Friedgut, Triangle-intersecting families of graphs. *J. Eur. Math. Soc. (JEMS)* **14** (2012), 841–885.   [509]

Enchev O., Problem 10390. *Amer. Math. Monthly* **101** (1994), 574. Solution **104** (1997), 367–368.   [261]

Engel A., *Problem-Solving Strategies* (Springer, 1998).   [226]

Engel K., *Sperner theory, Encyclopedia of Mathematics and its Applications* **65** (Cambridge Univ. Press, 1997).   [493, 496, 511]

Enomoto H., B. Jackson, P. Katerinis, and A. Saito, Toughness and the existence of k-factors. *J. Graph Th.* **9** (1985), 87–95.   [318]

Enomoto H., K. Ohba, K. Ota, and J. Sakamoto, Choice number of some complete multi-partite graphs. *Discr. Math.* **244** (2002), 55–66.   [355]

Enomoto H. and B. Péroche, The linear arboricity of some regular graphs. *J. Graph Theory* **8** (1984), 309–324.   [412]

Entringer R.C., A short proof of Rubin's block theorem. In *Cycles in graphs (Burnaby, B.C., 1982), North-Holland Math. Stud.* **115** (North-Holland, 1985), 367–368.   [347]

Entringer R.C. and E.F. Schmeichel, Edge conditions and cycle structure in bipartite graphs. *Ars Combin.* **26** (1988), 229–232.   [330]

Erdős P., Problem 3739. *Amer. Math. Monthly* **42** (1935), 396. Solution **44** (1937), 120.   [426]

Erdős P., On a lemma of Littlewood and Offord. *Bull. Amer. Math. Soc.* **51** (1945), 898–902.   [561]

Erdős P., On sets of distances of n points. *Amer. Math. Monthly* **53** (1946), 248–250.   [793]

Erdős P., Some remarks on the theory of graphs. *Bull. Amer. Math. Soc.* **53** (1947), 292–294.   [450, 658]

Erdős P., Some remarks on set theory. *Proc. Amer. Math. Soc.* **1** (1950), 127–141.   [468]

Erdős P., Some theorems on graphs. *Riveon Lematematika* **9** (1955), 13–17.   [478]

Erdős P., Graph theory and probability. *Can. J. Math.* **11** (1959), 34–38.   [434, 671]

Erdős P., Graph theory and probability, II. *Canad. J. Math.* **13** (1961), 346–352.   [341, 664, 682]

Erdős P., Remarks on a paper of Pósa. *Magyar Tud. Akad. Mat. Kut. Int. Közl.* **7** (1962a), 227–229.
                                                                                                       [332]

Erdős P., On the number of complete subgraphs contained in certain graphs. *Magy. Tud. Akad. Mat. Kutató Int. Közl.* **7** (1962b), 459–464.   [476, 478]

Erdős P., On a combinatorial problem. *Nord. Mat. Tidskr.* **11** (1963), 5–10.   [659, 703]

Erdős P., Extremal problems in graph theory. In *Theory of Graphs and Its Applications* (Academic Press, 1964), 29–36.   [489, 659]

Erdős P., Extremal problems in number theory. In *Proc. Sympos. Pure Math., Vol. VIII* (AMS, 1965), 181–189.   [669]

Erdős P., Problem 2. In *Theory of Graphs (Tihany, 1966)* (P. Erdős and G. Katona, eds.) (Academic Press, 1968).   [354]

Erdős P., On the graph theorem of Turán. *Mat. Lapok* **21** (1970), 249–251 (1971).   [223]

Erdős P., Problems and results on finite and infinite combinatorial analysis. In *Infinite and finite sets (Colloq., Keszthely, 1973), I, Colloq. Math. Soc. János Bolyai* **10** (North-Holland, 1975), 403–424.
                                                                                                         [504]

Erdős P., Problems and results in graph theory and combinatorial analysis. In *Proceedings of the Fifth British Combinatorial Conference (Aberdeen, 1975), Congr. Numer.* **15** (Utilitas Math., 1976), 169–192.   [438]

Erdős P., Problem E3284. *Amer. Math. Monthly* **95** (1988a), 762. Solution **97** (1990), 248–249.   [490]

Erdős P., Problem E3255. *Amer. Math. Monthly* **95** (1988b), 259. Solution **97** (1990), 848–849.   [490]

Erdős P., Some of my favourite unsolved problems. In *A tribute to Paul Erdős* (Cambridge Univ. Press, 1990), 467–478.   [497]

Erdős P. and S. Fajtlowicz, On the conjecture of Hajós. *Combinatorica* **1** (1981), 141–143.   [709]

Erdős P., R.J. Faudree, C.C. Rousseau, and R.H. Schelp, The size Ramsey number. *Period. Math. Hungar.* **9** (1978), 145–161.   [455]

Erdős P., Z. Füredi, R.J. Gould, and D.S. Gunderson, Extremal graphs for intersecting triangles. *J. Combin. Th. B* **64** (1995), 89–100.   [491]

Erdős P. and T. Gallai, On maximal paths and circuits of graphs. *Acta Math. Acad. Sci. Hung.* **10** (1959), 337–356.   [322, 334, 460]

Erdős P. and T. Gallai, Graphs with prescribed degrees of vertices (Hungarian). *Mat. Lapok* **11** (1960), 264–274.   [269, 276]

Erdős P. and T. Gallai, On the minimal number of vertices representing the edges of a graph. *Publ. Math. Inst. Hung. Acad. Sci.* **6** (1961), 181–203.   [273, 354]

Erdős P., A. Ginzburg, and A. Ziv, Theorem in the additive number theory. *Bull. Res. Council Israel* **10** (1961).   [764]

Erdős P., A.W. Goodman, and L. Pósa, The representation of graphs by set intersections. *Canad. J. Math.* **18** (1966), 106–112.   [238]

Erdős P., R.L. Graham, and E. Szemeredi, On sparse graphs with dense long paths. In *Computers and mathematics with applications* (Pergamon, 1976), 365–369.   [394]

Erdős P. and R.K. Guy, Crossing number problems. *Amer. Math. Monthly* **80** (1973), 52–58.   [793, 798]

Erdős P. and A. Gyárfás, A variant of the classical Ramsey problem. *Combinatorica* **17** (1997), 459–467.
                                                                                                         [686]

Erdős P. and A. Hajnal, On chromatic numbers of graphs and set systems. *Acta Math. Acad. Sci. Hung.* **17** (1966), 61–99.   [343]

Erdős P., A. Hajnal, and J.W. Moon, A problem in graph theory. *Amer. Math. Monthly* **71** (1964), 1107–1110.   [491]

Erdős P. and H. Heilbronn, On the addition of residue classes mod *p*. *Acta Arith.* **9** (1964), 149–159.
                                                                                                         [734]

Erdős P., M. Herzog, and J. Schönheim, An extremal problem on the set of noncoprime divisors of a number. *Israel J. Math.* **8** (1970), 408–412.   [512]

Erdős P., C. Ko, and R. Rado, Intersection theorems for systems of finite sets. *Quart. J. Math. Oxford (2)* **12** (1961), 313–320.   [498]

Erdős P. and J. Komlós, On a problem of Moser. In *Combinatorial theory and its applications, I (Proc. Colloq., Balatonfüred, 1969)* (North-Holland, 1970), 365–367.   [551]

Erdős P. and L. Lovász, Problems and results on 3-chromatic hypergraphs and some related questions. In *Infinite and finite sets (Colloq., Keszthely, 1973), II, Colloq. Math. Soc. János Bolyai* **10** (North-Holland, 1975), 609–627. [675, 679]

Erdős P. and J.W. Moon, On sets of consistent arcs in a tournament. *Canad. Math. Bull.* **8** (1965), 269–271. [721]

Erdős P. and L. Pósa, On the maximal number of disjoint circuits of a graph. *Publ. Math. Debrecen* **9** (1962), 3–12. [273, 490]

Erdős P. and R. Rado, A combinatorial theorem. *J. London Math. Soc.* **25** (1950), 249–255. [470]

Erdős P. and R. Rado, Combinatorial theorems on classifications of subsets of a given set. *Proc. London Math. Soc. (3)* **2** (1952), 417–439. [451, 608]

Erdős P. and R. Rado, A partition calculus in set theory. *Bull. Amer. Math. Soc.* **62** (1956), 427–489. [456]

Erdős P. and R. Rado, Intersection theorems for systems of sets. *J. London Math. Soc.* **35** (1960), 85–90. [503, 504]

Erdős P. and A. Rényi, On random graphs, I. *Publ. Math. Debrecen* **6** (1959), 290–297. [687, 698]

Erdős P. and A. Rényi, On a problem in the theory of graphs. *Magyar Tud. Akad. Mat. Kutató Int. Közl.* **7** (1962), 623–641. [252]

Erdős P. and A. Rényi, On two problems of information theory. *Magyar Tud. Akad. Mat. Kutató Int. Közl.* **8** (1963), 229–243. [513]

Erdős P. and A. Rényi, On the existence of a factor of degree one of a connected random graph. *Acta Math. Acad. Sci. Hung.* **17** (1966), 359–368. [699]

Erdős P., A. Rényi, and V.T. Sós, On a problem of graph theory. *Studia Sci. Math. Hungar.* **1** (1966), 215–235. [244, 632]

Erdős P., A.L. Rubin, and H. Taylor, Choosability in graphs. In *Proc. West Coast Conf. Combin. Graph Th. Comput. (Arcata), Congr. Numer.* **26** (Utilitas Math., 1979), 125–157. [346, 348, 356, 660, 667]

Erdős P. and H. Sachs, Reguläre Graphen gegebener Taillenweite mit minimaler Knotenzahl. *Wiss. Z. Martin-Luther-Univ. Halle-Wittenberg Math.-Natur. Reihe* **12** (1963), 251–257. [219, 251, 274]

Erdős P. and S. Shelah, Separability properties of almost-disjoint families of sets. *Israel J. Math.* **12** (1972), 207–214. [551]

Erdős P. and M. Simonovits, A limit theorem in graph theory. *Studia Sci. Math. Hungar.* **1** (1966), 51–57. [479]

Erdős P., M. Simonovits, and V.T. Sós, Anti-Ramsey theorems **10** (1975), 633–643. Colloq. Math. Soc. János Bolyai. [457]

Erdős P. and J. Spencer, *Probabilistic methods in combinatorics* (Academic Press, 1974). [702]

Erdős P. and A.H. Stone, On the structure of linear graphs. *Bull. Amer. Math. Soc.* **52** (1946), 1087–1091. [479, 491]

Erdős P. and G. Szekeres, A combinatorial problem in geometry. *Composito Math* **2** (1935), 464–470. [344, 430, 446]

Erdős P. and G. Szekeres, On some extremum problems in elementary geometry. *Ann. Univ. Sci. Budapest. Eötvös Sect. Math.* **3–4** (1960/1961), 53–62. [446, 458]

Erdős P. and E. Szemerédi, Combinatorial properties of systems of sets. *J. Combin. Th. A* **24** (1978), 308–313. [504]

Erdős P. and P. Turán, On Some Sequences of Integers. *J. London Math. Soc.* **S1-11** (1936), 261. [484]

Erdős P. and D.B. West, A note on the interval number of a graph. *Discr. Math.* **55** (1985), 129–133. [705]

Erdős P. and R.J. Wilson, On the chromatic index of almost all graphs. *J. Combin. Th. B* **23** (1977), 255–257. [699]

Erdős P.L. and L.A. Székely, Applications of antilexicographic order. I. An enumerative theory of trees. *Adv. in Appl. Math.* **10** (1989), 488–496. [47, 135]

Eršov A.P. and G.I. Kožuhin, Estimates of the chromatic number of connected graphs (Russian). *Dokl. Akad. Nauk. SSSR* **142** (1962), 270–273. [353]

Esperet L., F. Kardoš, A.D. King, D. Král', and S. Norine, Exponentially many perfect matchings in cubic graphs. *Adv. Math.* **227** (2011), 1646–1664.   [765]

Esperet L. and A. Parreau, Acyclic edge-coloring using entropy compression. *European J. Combin.* **34** (2013), 1019–1027.   [680]

Etienne G., Tableaux de Young et solitaire bulgare. *J. Combin. Th. A* **58** (1991), 181–197.   [144, 152]

Euler L., Solutio problematis ad geometriam situs pertinentis (Latin). *Comment. Acad. Sci. U. Petrop* **8** (1736), 128–140.   [234]

Euler L., *Introductio in analysin infinitorum (Latin)* (Culture et civilisation (Bruxelles), 1748). Reprinted in 2000 (Sociedad Andaluza de Educación Matemática "Thales").   [141, 148, 151]

Euler L., Demonstratio nonnullarum insignium proprietatum, quibus solida hedris planis inclusa sunt praedita (Latin). *Novi Commentarii acad. sci. Petropolitanae* **4** (1758), 140–160.   [383]

Evans A.B., G. Isaak, and D.A. Narayan, Representations of graphs modulo *n*. *Discr. Math.* **223** (2000), 109–123.   [624]

Everman D., A.E. Danese, and K. Venkannayah, Problem e1396. *Amer. Math. Monthly* **67** (1960), 63–64. Solution **67** (1960), 694.   [61]

Faà di Bruno F., Sullo sviluppo delle funzioni (Italian). *Ann. di Sci. Math. e Fisiche* **6** (1855), 479–480.   [127]

Faà di Bruno F., Note sur une nouvelle formule de dalcul differentiel (French). *Quart. J. Pure and Appl. Math.* **1** (1857), 359–360.   [127]

Faigle U., Matroids in combinatorial optimization. In *Combinatorial Geometries* (N. White, ed.) (Cambridge Univ. Press, 1987), 161–210.   [532]

Falgas-Ravry V. and E.R. Vaughan, Applications of the semi-definite method to the Turán density problem for 3-graphs. *Combin. Probab. Comput.* **22** (2013), 21–54.   [475]

Falikman D.I., Proof of the van der Waerden conjecture on the permanent of a doubly stochastic matrix. *Mat. Zametki* **29** (1981), 931–938, 957.   [256, 765]

Fan G., New sufficient conditions for cycles in graphs. *J. Combin. Th. B* **37** (1984), 221–227.   [324–326]

Farber M. and R.E. Jamison, Convexity in graphs and hypergraphs. *SIAM J. Algeb. Disc. Meth.* **7** (1986), 433–444.   [374]

Faria L., C.M.H. de Figueiredo, O. Sýkora, and I. Vrt'o, An improved upper bound on the crossing number of the hypercube. *J. Graph Theory* **59** (2008), 145–161.   [798]

Fáry I., On the straight line representations of planar graphs. *Acta Sci. Math.* **11** (1948), 229–233.
                                                                    [393]

Farzad B., M. Mahdian, E.S. Mahmoodian, A. Saberi, and B. Sadri, Forced orientation of graphs. *Bull. Iranian Math. Soc.* **32** (2006), 79–89, 98.   [237]

Fasenmyer M.C., *Some Generalized Hypergeometric Polynomials* (ProQuest, 1946). Ph.D. Thesis, Univ. Michigan.   [90]

Faudree R.J., Forbidden subgraphs and Hamiltonian properties: a survey. *Congr. Numer.* **116** (1996), 33–52.   [316]

Faudree R.J., Survey of results on *k*-ordered graphs. *Discr. Math.* **229** (2001), 73–87.   [316]

Faudree R.J., R.J. Gould, M.S. Jacobson, and L.M. Lesniak, Neighborhood unions and a generalization of Dirac's theorem. *Discr. Math.* **105** (1992), 61–71.   [327]

Faudree R.J., R.J. Gould, M.S. Jacobson, and R.H. Schelp, Neighborhood unions and Hamiltonian properties in graphs. *J. Combin. Th. B* **47** (1989), 1–9.   [327]

Faudree R.J., A. Gyárfás, R.H. Schelp, and Z. Tuza, Induced matchings in bipartite graphs. *Discr. Math.* **78** (1989), 83–87.   [262]

Faudree R.J. and R.H. Schelp, Various length paths in graphs. In *Theory and applications of graphs (Proc. Internat. Conf., Western Mich. Univ., Kalamazoo, Mich., 1976), Lecture Notes in Math.* **642** (Springer, 1978), 160–173.   [454]

Faudree R.J. and R.H. Schelp, A survey of results on the size Ramsey number. In *Paul Erdős and his mathematics, II (Budapest, 1999), Bolyai Soc. Math. Stud.* **11** (János Bolyai Math. Soc., 2002), 291–309.   [455]

Favaron O., F. Genest, and M. Kouider, Regular path decompositions of odd regular graphs. *J. Graph Theory* **63** (2010), 114–128. [273]

Feige U., M.M. Halldórsson, G. Kortsarz, and A. Srinivasan, Approximating the domatic number. *SIAM J. Comput.* **32** (2002), 172–195. [684]

Feller W., *An introduction to probability theory and its applications, I, Third edition* (Wiley & Sons, 1968). [40, 668]

Felsner S., Convex drawings of planar graphs and the order dimension of 3-polytopes. *Order* **18** (2001), 19–37. [784]

Felsner S., Geodesic embeddings and planar graphs. *Order* **20** (2003), 135–150. [790]

Felsner S., *Geometric graphs and arrangements, Advanced Lectures in Mathematics* (Friedr. Vieweg & Sohn, 2004). [783]

Felsner S., The order dimension of planar maps revisited. *SIAM J. Discr. Math.* **28** (2014), 1093–1101. [790]

Felsner S., P.C. Fishburn, and W.T. Trotter, Finite three-dimensional partial orders which are not sphere orders. *Discr. Math.* **201** (1999), 101–132. [828]

Felsner S., C.M. Li, and W.T. Trotter, Adjacency posets of planar graphs. *Discr. Math.* **310** (2010), 1097–1104. [790]

Felsner S. and W.T. Trotter, Balancing pairs in partially ordered sets. In *Combinatorics, Paul Erdős is eighty, Vol. 1, Bolyai Soc. Math. Stud.* (János Bolyai Math. Soc., 1993), 145–157. [823]

Feng W., Bounds on maximum *b*-matchings. *Discr. Math.* **309** (2009), 4162–4165. [286]

Ferguson T. and C. Melolidakis, Problem E3061. *Amer. Math. Monthly* **91** (1984), 580. Solution **94** (1987), 189. [669]

Fiedler M., Algebraic connectivity of graphs. *Czech. Math. J.* **23** (1973), 298–305. [778]

Fielder D.C., Fibonacci numbers in tree counts for sector and related graphs. *Fibonacci Quart.* **12** (1974), 355–359. [65]

Figaj A. and T. Łuczak, The Ramsey number for a triple of long even cycles. *J. Combin. Th. B* **97** (2007), 584–596. [492]

Finck H.-J., On the chromatic numbers of a graph and its complement. In *Theory of Graphs (Tihany, 1966)* (P. Erdős and G. Katona, eds.) (Academic Press, 1968), 99–113. [344]

Finck H.-J. and H. Sachs, Über eine von H. S. Wilf angegebene Schranke für die chromatische Zahl endlicher Graphen. *Math. Nachr.* **39** (1969), 373–386. [338]

Fink J., Perfect matchings extend to Hamilton cycles in hypercubes. *J. Combin. Th. B* **97** (2007), 1074–1076. [330]

Fink J., Matching graphs of hypercubes and complete bipartite graphs. *European J. Combin.* **30** (2009), 1624–1629. [330]

Fiorini S., Some remarks on a paper by Vizing on critical graphs ("Critical graphs with a given chromatic class" (Russian), Diskret. Analiz. No. 5 (1965), 9-17). *Math. Proc. Cambridge Philos. Soc.* **77** (1975), 475–483. [414]

Fiorini S. and R.J. Wilson, *Edge-colourings of graphs* (Pitman; distrib. Fearon-Pitman, 1977). [361]

Firke F.A., P.M. Kosek, E.D. Nash, and J. Williford, Extremal graphs without 4-cycles. *J. Combin. Th. B* **103** (2013), 327–336. [630]

Fishburn P. and B. Monjardet, Norbert Wiener on the theory of measurement (1914, 1915, 1921). *J. Math. Psych.* **36** (1992), 165–184. [586]

Fishburn P.C., Intransitive indifference with unequal indifference intervals. *J. Mathematical Psychology* **7** (1970), 144–149. [586]

Fishburn P.C., Interval lengths for interval orders: a minimization problem. *Discr. Math.* **47** (1983), 63–82. [604]

Fishburn P.C., A correlational inequality for linear extensions of a poset. *Order* **1** (1984a), 127–137. [600]

Fishburn P.C., Numbers of lengths for representations of interval orders. In *Progress in combinatorial optimization (Waterloo, 1982)* (Academic Press, 1984b), 131–146. [587]

Fishburn P.C., *Interval orders and interval graphs*, Wiley-Interscience Series in Discr. Mathematics (Wiley & Sons, 1985).   [587]

Fishburn P.C. and R.L. Graham, Classes of interval graphs under expanding length restrictions. *J. Graph Theory* **9** (1985), 459–472.   [604]

Fishburn P.C. and W.T. Trotter, Angle orders. *Order* **1** (1985), 333–343.   [828, 832]

Fisher D.C., K.L. Collins, and L.B. Krompart, Problem 10406. *Amer. Math. Monthly* **101** (1994), 793. Solution **104** (1997), 572–573.   [418]

Fisher D.C., K. Fraughnaugh, L. Langley, and D.B. West, The number of dependent arcs in an acyclic orientation. *J. Combin. Th. B* **71** (1997), 73–78.   [550]

Fisher R.A., An examination of the different possible solutions of a problem in incomplete blocks. *Ann. Eugenics* **10** (1940), 52–75.   [614]

Fisk S., A short proof of Chvátal's watchman theorem. *J. Combin. Th. B* **24** (1978), 374.   [419]

Fiz Pontiveros G., S. Griffiths, and R. Morris, The triangle-free process and $r(3,k)$ (2013). (arXiv:1302.6279).   [682]

Fleischner H., The square of every two-connected graph is hamiltonian. *J. Combin. Th. B* **16** (1974), 29–34.   [331]

Fleischner H., Eulerian graphs. In *Selected Topics in Graph Theory, 2* (L.W. Beineke and R.J. Wilson, eds.) (Academic Press, 1983), 17–54.   [763]

Fleischner H. and M. Stiebitz, A solution to a colouring problem of P. Erdős. *Discr. Math.* **101** (1992), 39–48.   [739]

Floyd R.W., Problem E3399. *Amer. Math. Monthly* **97** (1990), 611–612.   [257]

Foata D. and J. Riordan, Mappings of acyclic and parking functions. *Aequationes Math.* **10** (1974), 10–22.   [50]

Foata D. and M.P. Schützenberger, *Théorie géométrique des polynômes eulériens, Lecture Notes in Mathematics, Vol. 138* (Springer-Verlag, 1970).   [99, 101, 106, 127, 177]

Folkman J., Graphs with monochromatic complete subgraphs in every edge coloring. *SIAM J. Appl. Math.* **18** (1970), 19–24.   [455, 466]

Fon-Der-Flaass D.G., New prolific constructions of strongly regular graphs. *Adv. Geom.* **2** (2002), 301–306.   [649, 782]

Ford G.W. and G.E. Uhlenbeck, Combinatorial problems in the theory of graphs. I and III. *Proc. Nat. Acad. Sci. U.S.A.* **42** (1956), 122–128, 529–535.   [135]

Ford K., Problem 10383. *Amer. Math. Monthly* **101** (1994), 473. Solution **104** (1997), 457.   [669]

Ford L.R., Jr. and D.R. Fulkerson, Maximal flow through a network. *Canad. J. Math.* **8** (1956), 399–404.   [300]

Ford L.R., Jr. and D.R. Fulkerson, Network flows and systems of representatives. *Canad. J. Math.* **10** (1958), 78–85.   [302, 531]

Fortuin C.M., P.W. Kasteleyn, and J. Ginibre, Correlation inequalities on some partially ordered sets. *Comm. Math. Phys.* **22** (1971), 89–103.   [596]

Foulkes H.O., Enumeration of permutations with prescribed up-down and inversion sequences. *Discr. Math.* **15** (1976), 235–252.   [197]

Fouquet J.L. and J.-L. Jolivet, In "problémes". In *Problémes combinatoires et théorie des graphes (Orsay, 1976), Colloq. Internat. CNRS,* **260** (1978), 437–443.   [334]

Fournier J.-C., Colorations des arêtes d'un graphe. In *Colloque Théorie des Graphes (Bruxelles, 1973), Cahiers Ctr. Étud. Rech. Opér* **15** (1973), 311–314.   [372]

Fox J., A new proof of the graph removal lemma. *Ann. of Math. (2)* **174** (2011), 561–579.   [485]

Fox J., C. Lee, and B. Sudakov, Maximum union-free subfamilies. *Israel J. Math.* **191** (2012), 959–971.   [551]

Fox J., J. Pach, B. Sudakov, and A. Suk, Erdős–Szekeres-type theorems for monotone paths and convex bodies. *Proc. Lond. Math. Soc. (3)* **105** (2012), 953–982.   [457, 601]

Fox J. and B. Sudakov, Density theorems for bipartite graphs and related Ramsey-type results. *Combinatorica* **29** (2009), 153–196.   [673, 684]

Fox J. and B. Sudakov, Dependent random choice. *Random Structures Algorithms* **38** (2011), 68–99.
                                                                                                    [672, 673, 684]

Frame J.S., G.de B. Robinson, and R.M. Thrall, The hook graphs of the symmetric groups. *Canadian J. Math.* **6** (1954), 316–324.   [190]

Frank A., On the orientation of graphs. *J. Combin. Th. B* **28** (1980a), 251–261.   [310]

Frank A., On chain and antichain families of a partially ordered set. *J. Combin. Th. B* **29** (1980b), 176–184.   [548]

Frank A., Applications of submodular functions. In *Surveys in Combinatorics, 1993* (K. Walker, ed.), *Lond. Math. Soc. Lect. Notes* **187** (Cambridge Univ. Press, 1993), 85–136.   [307, 310]

Frank A. and E. Tardos, Matroids from crossing families. In *Finite and infinite sets (Eger, 1981), I–II*, *Colloq. Math. Soc. János Bolyai* **37** (North-Holland, 1984), 295–304.   [310]

Frankl P., Families of finite sets satisfying an intersection condition. *Bull. Austral. Math. Soc.* **15** (1976), 73–79.   [500]

Frankl P., A constructive lower bound for some Ramsey numbers. *Ars Combinatoria* **3** (1977), 297–302.
                                                                                                    [728]

Frankl P., The Erdős–Ko–Rado theorem is true for $n = ckt$. In *Combinatorics (Proc. 5th Hung. Colloq., Keszthely, 1976), I, Colloq. Math. Soc. János Bolyai* **18** (North-Holland, 1978), 365–375.
                                                                                                    [500, 511]

Frankl P., A new short proof for the Kruskal–Katona theorem. *Discr. Math.* **48** (1984), 327–329.
                                                                                                    [495, 496, 510]

Frankl P., The shifting technique in extremal set theory. In *Surveys in combinatorics 1987 (New Cross, 1987), London Math. Soc. Lecture Note Ser.* **123** (Cambridge Univ. Press, 1987), 81–110.
                                                                                                    [493, 499]

Frankl P., Constructive Ramsey bounds and intersection theorems for sets. In *Mathematics of Ramsey theory, Algorithms Combin.* **5** (Springer, 1990), 53–56.   [450]

Frankl P., Extremal set systems. In *Handbook of combinatorics, Vol. 1, 2* (Elsevier Sci. B. V., Amsterdam, 1995), 1293–1329.   [497]

Frankl P. and M. Deza, On the maximum number of permutations with given maximal or minimal distance. *J. Combin. Th. A* **22** (1977), 352–360.   [510]

Frankl P. and Z. Füredi, Families of finite sets with missing intersections. In *Finite and infinite sets (Eger, 1981), I–II, Colloq. Math. Soc. János Bolyai* **37** (North-Holland, 1984), 305–318.   [730]

Frankl P. and Z. Füredi, A new short proof of the EKR theorem. *J. Combin. Th. A* **119** (2012), 1388–1390.   [500]

Frankl P. and R.L. Graham, Old and new proofs of the Erdős–Ko–Rado theorem. *Sichuan Daxue Xuebao* **26** (1989), 112–122.   [500]

Frankl P. and V. Rödl, Extremal problems on set systems. *Random Structures Algorithms* **20** (2002), 131–164.   [488]

Frankl P. and N. Tokushige, Invitation to intersection problems for finite sets. *J. Combin. Th. A* **144** (2016), 157–211.   [493]

Frankl P. and R.M. Wilson, Intersection theorems with geometric consequences. *Combinatorica* **1** (1981), 357–368.   [459, 727–729]

Franklin F., On Newton's Method of Approximation. *Amer. J. Math.* **4** (1881), 275–276.   [151]

Franklin P., The Four Color Problem. *Amer. J. Math.* **44** (1922), 225–236.   [406]

Franzblau D.S. and D. Zeilberger, A bijective proof of the hook-length formula. *J. Algorithms* **3** (1982), 317–343.   [192]

Fraughnaugh (Jones) K., Minimum independence graphs with maximum degree four. In *Graphs and applications (Boulder, 1982), Wiley-Intersci. Publ.* (Wiley, 1985), 221–230.   [420]

Fredman M.L., How good is the information theory bound in sorting? *Theoret. Comput. Sci.* **1** (1976), 355–361.   [823]

Fredman M.L. and D.E. Knuth, Recurrence relations based on minimization. *J. Math. Anal. Appl.* **48** (1974), 534–559.   [92]

Freund R.M. and M.J. Todd, A constructive proof of Tucker's combinatorial lemma. *J. Combin. Th. A* **30** (1981), 321–325.   [807, 809]

Friedgut E. and J. Kahn, On the number of copies of one hypergraph in another. *Israel J. Math.* **105** (1998), 251–256.   [510]

Friedgut E., J. Kahn, G. Kalai, and N. Keller, Chvátal's conjecture and correlation inequalities. *J. Combin. Th. A* **156** (2018), 22–43.   [503]

Frieze A., On the length of the longest monotone subsequence in a random permutation. *Ann. Appl. Probab.* **1** (1991), 301–305.   [722]

Frieze A. and M. Karoński, *Introduction to Random Graphs* (Cambridge, 2016).   [657]

Fritsch R. and G. Fritsch, *The Four-Color Theorem* (Springer, 1998). (published in German by F.A. Brockhaus, 1994).   [399]

Frobenius G., Ueber die Congruenz nach einem aus zwei endlichen Gruppen gebildeten Doppelmodul. *J. Reine Angew. Math.* **101** (1887), 273–299.   [181]

Frobenius G., über die Charaktere der symmetriscen Gruppe (German). *Sitzungberichte der Königlich Preussischen Akademie der Wissenschaften zu Berlin* (1900), 516–534.   [189, 190, 207]

Frobenius G., Über zerlegbare Determinanten. *Sitzungsber. König. Preuss. Adad. Wiss.* **XVIII** (1917), 274–277.   [256]

Fronček D., Locally linear graphs. *Math. Slovaca* **39** (1989), 3–6.   [684]

Fujita S., C. Magnant, and K. Ozeki, Rainbow generalizations of Ramsey theory: a survey. *Graphs Combin.* **26** (2010), 1–30.   [457]

Fulkerson D.R., Note on Dilworth's decomposition theorem for partially ordered sets. *Proc. Amer. Math. Soc.* **7** (1956), 701–702.   [547]

Fulkerson D.R. and O.A. Gross, Incidence matrices and interval graphs. *Pac. J. Math.* **15** (1965), 835–855.   [370]

Fulkerson D.R., A.J. Hoffman, and M.H. McAndrew, Some properties of graphs with multiple edges. *Canad. J. Math.* **17** (1965), 166–177.   [222]

Fulton W., *Young tableaux, London Mathematical Society Student Texts* **35** (Cambridge Univ. Press, 1997).   [189]

Füredi Z., Graphs without quadrilaterals. *J. Combin. Th. B* **34** (1983), 187–190.   [630]

Füredi Z., Turán type problems. In *Surveys in combinatorics, 1991 (Guildford, 1991), London Math. Soc. Lecture Note Ser.* **166** (Cambridge Univ. Press, 1991), 253–300.   [475, 672]

Füredi Z., The order dimension of two levels of the Boolean lattice. *Order* **11** (1994), 15–28.
                                                                                      [582, 584, 819]

Füredi Z., On the number of edges of quadrilateral-free graphs. *J. Combin. Th. B* **68** (1996a), 1–6.
                                                                                      [630]

Füredi Z., New asymptotics for bipartite Turán numbers. *J. Combin. Th. A* **75** (1996b), 141–144.   [633]

Füredi Z., An upper bound on Zarankiewicz' problem. *Combin. Probab. Comput.* **5** (1996c), 29–33.
                                                                                      [632, 639]

Füredi Z., *A proof on the stability of extremal graphs, Workshop on Extremal Graphs and Hypergraphs (Carnegie-Mellon)* (2007).   [491]

Füredi Z., A proof of the stability of extremal graphs, Simonovits' stability from Szemerédi's regularity. *J. Combin. Th. B* **115** (2015), 66–71.   [228]

Füredi Z., J.R. Griggs, and D.J. Kleitman, A minimal cutset of the Boolean lattice with almost all members. *Graphs Combin.* **5** (1989), 327–332.   [550]

Füredi Z. and D.S. Gunderson, Extremal numbers for odd cycles. *Combin. Probab. Comput.* **24** (2015), 641–645.   [479]

Füredi Z. and P. Hajnal, Davenport–Schinzel theory of matrices. *Discr. Math.* **103** (1992), 233–251.
                                                                                      [432, 442]

Füredi Z., P. Hajnal, V. Rödl, and W.T. Trotter, Interval orders and shift graphs. In *Sets, graphs and numbers (Budapest, 1991), Colloq. Math. Soc. János Bolyai* **60** (North-Holland, 1992), 297–313.
                                                                                      [581, 604]

Füredi Z. and J. Kahn, On the dimensions of ordered sets of bounded degree. *Order* **3** (1986), 15–20.
[576, 578]

Füredi Z. and G.O.H. Katona, 2-Bases of quadruples. *Combin. Probab. Comput.* **15** (2006), 131–141.
[490]

Füredi Z., A.V. Kostochka, R. Škrekovski, M. Stiebitz, and D.B. West, Nordhaus–Gaddum-type theorems for decompositions into many parts. *J. Graph Theory* **50** (2005), 273–292.   [239, 639]

Füredi Z. and M. Simonovits, The history of degenerate (bipartite) extremal graph problems. In *Erdős centennial, Bolyai Soc. Math. Stud.* **25** (János Bolyai Math. Soc., 2013), 169–264.   [479]

Fuss N., Solutio quaestionis, quot modis polygonum n laterum in polygona m laterum, per diagonales resolvi queat. *Nova Acta Academiae Sci. Petropolitanae* **9** (1791), 243–251.   [42]

Gabber O. and Z. Galil, Explicit construction of linear-sized superconcentrators. *J. Comput. Systems Sci.* **22** (1981), 407–420.   [779]

Gabow H.N., An efficient implementation of Edmonds' algorithm for maximum matchings on graphs. *J. Assoc. Comput. Mach.* **23** (1975), 221–234.   [282]

Gabow H.N., Data structures for weighted matching and nearest common ancestors with linking. In *Proc. 1st ACM-SIAM Symp. Disc. Algs (San Francisco)* (SIAM, 1990), 434–443.   [282]

Gabow H.N., Z. Galil, T. Spencer, and R.E. Tarjan, Efficient algorithms for finding minimum spanning trees in undirected and directed graphs. *Combinatorica* **6** (1986), 109–122.   [246]

Gabow H.N. and R.E. Tarjan, Faster scaling algorithms for general graph matching problems. Tech. Rep. CU-CS-432-89, Dept. Comp. Sci., Univ. Colorado–Boulder (1989).   [282]

Gale D., Neighboring vertices on a convex polyhedron. In *Linear inequalities and related system, Annals of Mathematics Studies, no. 38* (Princeton Univ. Press, 1956), 255–263.   [813, 820]

Gale D., The game of Hex and the Brouwer fixed-point theorem. *Amer. Math. Monthly* **86** (1979), 818–827.   [803, 804]

Gale D. and L.S. Shapley, College admissions and the stability of marriage. *Amer. Math. Monthly* **69** (1962), 9–15.   [285, 288]

Gallai T., On factorisation of graphs. *Acta Math. Acad. Sci. Hungar.* **1** (1950), 133–153.   [264]

Gallai T., Über extreme Punkt- und Kantenmengen. *Ann. Univ. Sci. Budapest, Eötvös Sect. Math.* **2** (1959), 133–138.   [260, 263]

Gallai T., Kritische Graphen I. *Magyar Tud. Akad. Mat. Kut. Int. Közl.* **8** (1963a), 165–192.   [355]

Gallai T., Neuer Beweis eines Tutte'schen Satzes. *Magyar Tud. Akad. Mat. Kut. Int. Közl.* **8** (1963b), 135–139.   [275]

Gallai T., Kritische Graphen II. *Magyar Tud. Akad. Mat. Kut. Int. Közl.* **8** (1963c), 373–395.   [357]

Gallai T., On directed paths and circuits. In *Theory of Graphs (Tihany, 1966)* (P. Erdős and G. Katona, eds.) (Academic Press, 1968), 115–118.   [339]

Gallai T. and A.N. Milgram, Verallgemeinerung eines graphentheoretischen Satzes von Rédei. *Acta Sci. Math. Szeged* **21** (1960), 181–186.   [548]

Gallian J.A., A dynamic survey of graph labeling. *Electron. J. Combin.* **5** (1998), Dynamic Survey 6, 43.
[646]

Gallian J.A. and D.J. Rusin, Cyclotomic polynomials and nonstandard dice. *Discr. Math.* **27** (1979), 245–259.   [103]

Galperin G. and H. Gauchman, Problem 11103. *Amer. Math. Monthly* **111** (2004), 724. Solution **113** (2006), 465–466.   [61]

Galvin D., Three tutorial lectures on entropy and counting (2014). (arXiv:1406.7872).   [504, 506, 509]

Galvin F., Problem 6034. *Amer. Math. Monthly* **82** (1975), 592. Solution **84** (1977), 224.   [474]

Galvin F., The list chromatic index of a bipartite multigraph. *J. Combin. Th. B* **63** (1995), 153–158.
[363, 364]

Galvin F., Problem 10701. *Amer. Math. Monthly* **105** (1998), 956. Solution **108** (2001), 79–80.   [355]

Galvin F., Problem 10761. *Amer. Math. Monthly* **106** (1999), 864. Solution **108** (2001), 773–774.   [228]

Ganter B. and L. Teirlinck, A combinatorial lemma. *Math. Z.* **154** (1977), 153–156.   [441]

Gardner M., Mathematical games (a.k.a. Bulgarian Solitaire and other seemingly endless tasks). *Sci. Amer.* **249** (1983), 8–13.   [143]

Garey M.R., R.L. Graham, D.S. Johnson, and D.E. Knuth, Complexity results for bandwidth minimization. *SIAM J. Appl. Math.* **34** (1978), 477–495.   [800]

Garey M.R. and D.S. Johnson, The complexity of near-optimal graph coloring. *J. Assoc. Comp. Mach.* **23** (1976), 43–49.   [701]

Garey M.R. and D.S. Johnson, *Computers and intractability* (W. H. Freeman and Co., 1979). A guide to the theory of NP-completeness, A Series of Books in the Mathematical Sciences.   [11]

Garey M.R. and D.S. Johnson, Crossing number is NP-complete. *SIAM J. Algebraic Discrete Methods* **4** (1983), 312–316.   [795]

Gargano L., M. Hammar, P. Hell, L. Stacho, and U. Vaccaro, Spanning spiders and light-splitting switches. *Discr. Math.* **285** (2004), 83–95.   [327, 334]

Gaskell R.W., M.S. Klamkin, and P. Watson, Triangulations and Pick's theorem. *Math. Mag.* **49** (1976), 35–37.   [389]

Gasparyan G.S., Minimal imperfect graphs: a simple approach. *Combinatorica* **16** (1996), 209–212.
                                                                                                                                   [369]

Gavril F., The intersection graphs of subtrees in trees are exactly the chordal graphs. *J. Combin. Th. B* **16** (1974), 47–56.   [373]

Gazit H. and G.L. Miller, Planar separators and the Euclidean norm. In *Algorithms (Tokyo, 1990), Lecture Notes in Comput. Sci.* **450** (Springer, Berlin, 1990), 338–347.   [397]

GCHQ Problem Solving Group, Solution to problem 11883. *Amer. Math. Monthly* **125** (2018), 82–83. Proposed **123** (2016), 97.   [149]

Geetha J., N. Narayanan, and K. Somasundaram, Total colorings-A survey (2018). (arXiv:1812.05833v1).
                                                                                                                                   [373]

George J.C., *1-Factorizations of tensor products of graphs* (ProQuest, 1991). Ph.D. Thesis, Univ. Illinois at Urbana–Champaign.   [273, 371]

Georges J.P., Non-Hamiltonian bicubic graphs. *J. Combin. Th. B* **46** (1989), 121–124.   [317]

Gerbner D. and B. Patkós, *Extremal Finite Set Theory* (CRC Press, 2019).   [493]

Gerencsér L. and A. Gyárfás, On Ramsey-type problems. *Ann. Univ. Sci. Budapest. Eötvös Sect. Math.* **10** (1967), 167–170.   [460]

Gerke S. and A. Steger, The sparse regularity lemma and its applications. In *Surveys in combinatorics 2005, London Math. Soc. Lecture Note Ser.* **327** (Cambridge Univ. Press, 2005), 227–258.   [488]

Gessel I.M., *Generating Functions and Enumeration of Sequences* (ProQuest LLC, 1977). Ph.D. Thesis, Massachusetts Inst. Tech.   [102]

Gessel I.M., Problem 10357. *Amer. Math. Monthly* **101** (1994), 75. Solution **104** (1997), 177–178.   [115]

Gessel I.M., Problem 10424. *Amer. Math. Monthly* **102** (1995), 70. Solution **104** (1997), 466–467.   [117]

Gessel I.M., The Smith College diploma problem. *Amer. Math. Monthly* **108** (2001), 55–57.   [106]

Gessel I.M., Lagrange inversion. *J. Combin. Th. A* **144** (2016), 212–249.   [129]

Gessel I.M. and R.P. Stanley, Stirling polynomials. *J. Combin. Th. A* **24** (1978), 24–33.   [102]

Gessel I.M. and G. Viennot, Binomial determinants, paths, and hook length formulae. *Adv. in Math.* **58** (1985), 300–321.   [165, 167, 168]

Gessel I.M. and G. Viennot, Determinants, paths, and plane partitions (1989). Available online at http://contscience.xavierviennot.org/xavier/articles_files/determinant_89.pdf
                                                                                                                                   [169]

Getz M. and D. Jones, Problem 11005. *Amer. Math. Monthly* **110** (2003), 340. Solution **112** (2005), 89–90.   [61]

Ghouila-Houri A., Une condition suffisante d'existence d'un circuit Hamiltonien. *C. R. Adac. Sci. Paris* **251** (1960), 495–497.   [334]

Ghouila-Houri A., Caractérisation des graphes non orientés dont on peut orienter les arêtes de manière à obtenir le graphe d'une relation d'ordre. *C. R. Acad. Sci. Paris* **254** (1962), 1370–1371.   [550]

Gilbert E.N., Lattice theoretic properties of frontal switching functions. *J. Math. Physics* **33** (1954), 57–67.   [564]

Gilbert E.N., Random graphs. *Ann. Math. Stat.* **30** (1959), 1141–1144.   [688]

Gilmore P.C. and A.J. Hoffman, A characterization of comparability graphs and of interval graphs. *Canad. J. Math.* **16** (1964), 539–548. [374, 549]

Glaisher J., A theorem in partitions. *Messenger of Math.* **12** (1883), 158–170. [142, 148]

Glock S., F. Joos, J. Kim, D. Kühn, and D. Osthus, Resolution of the Oberwolfach problem (2018). (arXiv:1806.04644). [647]

Glock S., D. Kühn, A. Lo, and D. Osthus, The existence of designs via iterative absorption (2016). (arXiv:1611.06827). [640]

Goddard W., Acyclic colorings of planar graphs. *Discr. Math.* **91** (1991), 91–94. [418]

Godsil C. and K. Meagher, *Erdős–Ko–Rado theorems: algebraic approaches, Cambridge Studies in Advanced Mathematics* **149** (Cambridge Univ. Press, 2016). [500, 781]

Godsil C. and G. Royle, *Algebraic graph theory, Graduate Texts in Mathematics* **207** (Springer-Verlag, 2001). [766]

Goldberg C.H. and D.B. West, Bisection of circle colorings. *SIAM J. Algebraic Discrete Methods* **6** (1985), 93–106. [809]

Goldberg M.K., Structure of multigraphs with restrictions on the chromatic class (Russian). *Metody Diskret. Analiz.* **30** (1977), 3–12. [360]

Goldman J.R., J.T. Joichi, and D.E. White, Rook theory. I. Rook equivalence of Ferrers boards. *Proc. Amer. Math. Soc.* **52** (1975), 485–492. [175]

Golomb S.W. and L.R. Welch, Perfect codes in the Lee metric and the packing of polyominoes. *SIAM J. Appl. Math.* **18** (1970), 302–317. [29]

Golovina L.I. and I.M. Yaglom, *Introduction in Geometry, Topics in mathematics* (Heath, 1963). [419]

Golumbic M.C., Trivially perfect graphs. *Discr. Math.* **24** (1978), 105–107. [366]

Golumbic M.C., *Algorithmic Graph Theory and Perfect Graphs* (Academic Press, 1980, 2004). [370]

Golumbic M.C., D. Rotem, and J. Urrutia, Comparability graphs and intersection graphs. *Discr. Math.* **43** (1983), 37–46. [582]

Golumbic M.C. and E.R. Scheinerman, Containment graphs, posets, and related classes of graphs. In *Combinatorial Mathematics: Proceedings of the Third International Conference (New York, 1985), Ann. New York Acad. Sci.* **555** (New York Acad. Sci., 1989), 192–204. [831, 832]

Gonçalves D., Covering planar graphs with forests, one having bounded maximum degree. *J. Combin. Th. B* **99** (2009), 314–322. [421]

Gondran M. and M. Minoux, *Graphs and algorithms, Wiley-Interscience Series in Discr. Mathematics* (Wiley & Sons, 1984). [287]

Good I.J., Normal recurring decimals. *J. Lond. Math. Soc.* **21** (1946), 167–169. [239]

Goodman A.W., On sets of acquaintances and strangers at any party. *Amer. Math. Monthly* **66** (1959), 778–783. [477]

Goodman S. and S. Hedetniemi, Sufficient conditions for a graph to be Hamiltonian. *J. Combin. Th. B* **16** (1974), 175–180. [330]

Gosper R.W., Jr., Decision procedure for indefinite hypergeometric summation. *Proc. Nat. Acad. Sci. U.S.A.* **75** (1978), 40–42. [89]

Gottschalk W.H., Choice functions and Tychonoff's theorem. *Proc. Am. Math. Soc.* **2** (1951), 172. [468]

Gould H.W., Explicit formulas for Bernoulli numbers. *Amer. Math. Monthly* **79** (1972), 44–51. [23, 30, 65]

Gould R.J., *Graph Theory* (Benjamin/Cummings, 1988). Also 2012. [215, 393]

Gould R.J., Updating the Hamiltonian problem—a survey. *J. Graph Theory* **15** (1991), 121–157. [316]

Gould R.J., Advances on the Hamiltonian problem—a survey. *Graphs Combin.* **19** (2003), 7–52. [316]

Gould R.J., Recent advances on the Hamiltonian problem: Survey III. *Graphs Combin.* **30** (2014), 1–46. [316]

Goulden I.P. and L.G. Serrano, Maintaining the spirit of the reflection principle when the boundary has arbitrary integer slope. *J. Combin. Th. A* **104** (2003), 317–326. [42]

Govorčin J. and R. Škrekovski, On the connectivity of Cartesian product of graphs. *Ars Math. Contemp.* **7** (2014), 293–297. [297]

Gowers W.T., A new proof of Szemerédi's theorem for arithmetic progressions of length four. *Geom. Funct. Anal.* **8** (1998), 529–551.   [672]

Gowers W.T., A new proof of Szemerédi's theorem. *Geom. Funct. Anal.* **11** (2001), 465–588.   [466, 484]

Graham N., R.C. Entringer, and L.A. Székely, New tricks for old trees: maps and the pigeonhole principle. *Amer. Math. Monthly* **101** (1994), 664–667.   [428, 441]

Graham N. and F. Harary, Changing and unchanging the diameter of a hypercube. *Discrete Appl. Math.* **37-38** (1992), 265–274.   [428, 565]

Graham N., F. Harary, M. Livingston, and Q.F. Stout, Subcube fault-tolerance in hypercubes. *Inform. and Comput.* **102** (1993), 280–314.   [584]

Graham R.L., Linear extensions of partial orders and the FKG inequality. In *Ordered sets (Banff, Alta., 1981), NATO Adv. Study Inst. C: Math. Phys. Sci.* **83** (Reidel, 1982), 213–236.   [597]

Graham R.L. and S. Butler, *Rudiments of Ramsey theory (2nd ed.), CBMS Regional Conference Series in Mathematics* **123** (AMS, 2015).   [425]

Graham R.L. and L.H. Harper, Some results on matching in bipartite graphs. *SIAM J. Appl. Math.* **17** (1969), 1017–1022.   [559, 560]

Graham R.L. and D.J. Kleitman, Increasing paths in edge ordered graphs. *Period. Math. Hungar.* **3** (1973), 141–148.   [431, 442]

Graham R.L., D.E. Knuth, and O. Patashnik, *Concrete mathematics* (Addison-Wesley, 1989). 2nd edition 1994.   [24, 33, 41, 61, 89, 101]

Graham R.L. and H.O. Pollak, On the addressing problem for loop switching. *Bell Sys. Tech. J.* **50** (1971), 2495–2519.   [771]

Graham R.L. and H.O. Pollak, On embedding graphs in squashed cubes. In *Graph Th. and Appl. (Proc. Conf., Western Michigan Univ., Mich., 1972; dedicated to the memory of J. W. T. Youngs), Lect. Notes Math.* **303** (Springer, 1972), 99–110.   [781]

Graham R.L., V. Rödl, and A. Ruciński, On graphs with linear Ramsey numbers. *J. Graph Theory* **35** (2000), 176–192.   [484]

Graham R.L., V. Rödl, and A. Ruciński, On bipartite graphs with linear Ramsey numbers. *Combinatorica* **21** (2001), 199–209.   [484, 673]

Graham R.L. and B.L. Rothschild, A short proof of van der Waerden's theorem on arithmetic progressions. *Proc. Amer. Math. Soc.* **42** (1974), 385–386.   [463, 464]

Graham R.L., B.L. Rothschild, and J. Spencer, *Ramsey Theory* (Wiley, 1980). Also 1990.
                                                                                   [425, 452, 463, 465, 470]

Graver J.E. and J. Yackel, Some graph theoretic results associated with Ramsey's Theorem. *J. Combin. Th.* **4** (1968), 125–175.   [449, 451]

Gravier S. and F. Maffray, Graphs whose choice number is equal to their chromatic number. *J. Graph Theory* **27** (1998), 87–97.   [356]

Greene C., A multiple exchange property for bases. *Proc. Amer. Math. Soc.* **39** (1973), 45–50.   [538]

Greene C., An extension of Schensted's theorem. *Advances in Math.* **14** (1974), 254–265.   [205]

Greene C., G.O.H. Katona, and D.J. Kleitman, Extensions of the Erdős–Ko–Rado theorem. *Studies in Appl. Math.* **55** (1976), 1–8.   [511]

Greene C. and D.J. Kleitman, The structure of Sperner *k*-families. *J. Combin. Th. A* **20** (1976a), 41–68.
                                                                                                   [548, 551]

Greene C. and D.J. Kleitman, Strong versions of Sperner's theorem. *J. Combin. Th. A* **20** (1976b), 80–88.   [555, 565]

Greene C. and D.J. Kleitman, Proof techniques in the theory of finite sets. In *Studies in combinatorics, MAA Stud. Math.* **17** (Math. Assoc. Amer., 1978), 22–79.   [496, 561, 607]

Greene C. and T.L. Magnanti, Some abstract pivot algorithms. *SIAM J. Appl. Math.* **29** (1975), 530–539.
                                                                                                   [538]

Greene C., A. Nijenhuis, and H.S. Wilf, A probabilistic proof of a formula for the number of Young tableaux of a given shape. *Adv. in Math.* **31** (1979), 104–109.   [190]

Greene J.E., A new short proof of Kneser's conjecture. *Amer. Math. Monthly* **109** (2002), 918–920.
                                                                                                   [811, 812]

Greenwell D.L. and H.V. Kronk, Uniquely line colorable graphs. *Canad. Math. Bull.* **16** (1973), 525–529. [329]

Greenwood R.E. and A.M. Gleason, Combinatorial relations and chromatic graphs. *Canad. J. Math.* **7** (1955), 1–7. [449]

Gregory D.A. and K.N. Vander Meulen, Sharp bounds for decompositions of graphs into complete r-partite subgraphs. *J. Graph Theory* **21** (1996), 393–400. [771]

Griggs J.R., Sufficient conditions for a symmetric chain order. *SIAM J. Appl. Math.* **32** (1977), 807–809. [557, 561]

Griggs J.R., Collections of subsets with the Sperner property. *Trans. AMS* **269** (1982), 575–591. [568]

Griggs J.R., Problem 10665. *Amer. Math. Monthly* **105** (1998), 464. Solution **107** (2000), 653–4. [702]

Griggs J.R. and C.C. Ho, The cycling of partitions and compositions under repeated shifts. *Adv. in Appl. Math.* **21** (1998), 205–227. [144, 145, 152]

Griggs J.R., J. Stahl, and W.T. Trotter, A Sperner theorem on unrelated chains of subsets. *J. Combin. Th. A* **36** (1984), 124–127. [568]

Grimmett G.R. and C.J.H. McDiarmid, On colouring random graphs. *Math. Proc. Camb. Phil. Soc.* **77** (1975), 313–324. [701]

Grimmett G.R. and D.R. Stirzaker, *Probability and random processes, 2nd ed.* (Oxford Univ. Press, 1992). Also 1982, 2001. [666]

Grinberg E.J., Plane homogeneous graphs of degree three without hamiltonian circuits. *Latvian Math. Yearbook* **5** (1968), 51–58. [400]

Grinstead C.M. and S.M. Roberts, On the Ramsey numbers $R(3,8)$ and $R(3,9)$. *J. Combin. Th. B* **33** (1982), 27–51. [449]

Grolmusz V. and B. Sudakov, On k-wise set-intersections and k-wise Hamming-distances. *J. Combin. Th. A* **99** (2002), 180–190. [731]

Grone R. and R. Merris, The Laplacian spectrum of a graph. II. *SIAM J. Discr. Math.* **7** (1994), 221–229. [776]

Grone R., R. Merris, and V.S. Sunder, The Laplacian spectrum of a graph. *SIAM J. Matrix Anal. Appl.* **11** (1990), 218–238. [776]

Gross J.L. and T.W. Tucker, *Topological graph theory, Wiley-Interscience Series in Discr. Mathematics and Optimization* (Wiley & Sons, 1987). [377]

Gross J.L. and J. Yellen, *Graph Theory* (CRC Press, 1999). [377]

Grötzsch H., Ein Dreifarbensatz für dreikreisfreie Netze auf der Kugel. *Wiss. Z. Martin-Luther-U., Halle-Wittenberg, Math.-Nat. Reihe* **8** (1959), 109–120. [414]

Grünbaum B., Acyclic colorings of planar graphs. *Israel J. Math.* **14** (1973), 390–408. [423]

Grünbaum B. and T.S. Motzkin, The number of hexagons and the simplicity of geodesics on certain polyhedra. *Canad. J. Math.* **15** (1963), 744–751. [387, 389]

Grytczuk J., J. Kozik, and P. Micek, New approach to nonrepetitive sequences. *Random Structures Algorithms* **42** (2013), 214–225. [680]

Grytczuk J. and X. Zhu, The Alon–Tarsi number of a planar graph minus a matching (2020). (arXiv:1811.12012). [403, 742]

Guan M., Graphic programming using odd and even points. *Chinese Math.* **1** (1962), 273–277. [287]

Guldan F., The linear arboricity of 10-regular graphs. *Math. Slovaca* **36** (1986), 225–228. [412]

Gunderson D.S. and V. Rödl, Extremal problems for affine cubes of integers. *Combin. Probab. Comput.* **7** (1998), 65–79. [473]

Guo S. and Z.W. Sun, Determination of the two-color Rado number for $a_1 x_1 + \cdots + a_m x_m = x_0$. *J. Combin. Th. A* **115** (2008), 345–353. [473]

Gupta H., Combinatorial proof of a theorem on partitions into an even or odd number of parts. *J. Combin. Th. A* **21** (1976), 100–103. [176]

Gupta R.P., The chromatic index and the degree of a graph (Abstract 66T-429). *Notices Amer. Math. Soc.* **13** (1966), 719. [358, 359, 372]

Gurgel M.A. and Y. Wakabayashi, On k-leaf-connected graphs. *J. Combin. Th. B* **41** (1986), 1–16. [327]

Gusfield D. and R.W. Irving, *The Stable Marriage Problem: Structure and Algorithms* (MIT Press, 1989). [286]

Gustavsson T., *Decompositions of Large Graphs and Digraphs with High Minimum Degree* (Univ. of Stockholm, 1991). Ph.D. Thesis. [640]

Gutin G., Finding a longest path in a complete multipartite digraph. *SIAM J. Discr. Math.* **6** (1993), 270–273. [334]

Gutner S., The complexity of planar graph choosability. *Discr. Math.* **159** (1996), 119–130. [402, 420]

Guy R.K., A problem of Zarankiewicz. In *Theory of Graphs (Proc. Colloq., Tihany, 1966)* (Academic Press, 1968), 119–150. [632]

Guy R.K., The decline and fall of Zarankiewicz's Theorem. In *Proof Techniques in Graph Theory* (F. Harary, ed.) (Academic Press, 1969), 63–69. [793]

Guy R.K., Sequences associated with a problem of Turán and other problems. In *Combin. Conf. Balatonfüred, 1969* (North-Holland, 1970), 553–569. [793, 797]

Guy R.K., Crossing numbers of graphs. In *Graph Theory and Applications* (Y. Alavi et al, ed.), *Lect. Notes Math.* **303** (Springer, 1972), 111–124. [792, 795]

Guy R.K., The strong law of small numbers. *Amer. Math. Monthly* **95** (1988), 697–712. [105]

Guy R.K., Catwalks, sandsteps, and Pascal pyramids. *J. Integer Seq.* **3** (2000), Article 00.1.6 (electronic). [64]

Guy R.K. and F. Harary, On the Möbius ladders. *Canad. Math. Bull.* **10** (1967), 493–496. [797]

Gyárfás A., On Ramsey covering-numbers. In *Infinite and finite sets (Colloq., Keszthely, 1973), II, Colloq. Math. Soc. János Bolyai* **10** (North-Holland, 1975), 801–816. [351]

Gyárfás A., Particiófedések és lefogóhalmazok hipergráfokban. *Tanulmányok-MTA Számitástechn. Automat. Kutató Int. Budapest* (1977), 62. [460]

Gyárfás A., Vertex coverings by monochromatic paths and cycles. *J. Graph Th.* **7** (1983), 131–135. [460]

Gyárfás A., Problems from the world surrounding perfect graphs. *Tanulmányok—MTA Számitástech. Automat. Kutató Int. Budapest* (1985), 53. [351, 352]

Gyárfás A., Large monochromatic components in edge colorings of graphs: a survey. In *Ramsey theory*, *Progr. Math.* **285** (Birkhäuser/Springer, 2011), 77–96. [460]

Gyárfás A., Vertex covers by monochromatic pieces—a survey of results and problems. *Discr. Math.* **339** (2016), 1970–1977. [460]

Gyárfás A., J. Lehel, J. Nešetřil, V. Rödl, R.H. Schelp, and Z. Tuza, Local k-colorings of graphs and hypergraphs. *J. Combin. Th. B* **43** (1987), 127–139. [472]

Gyárfás A. and G. Simonyi, Edge colorings of complete graphs without tricolored triangles. *J. Graph Theory* **46** (2004), 211–216. [474]

Gyárfás A., E. Szemerédi, and Z. Tuza, Induced subtrees in graphs of large chromatic number. *Discr. Math.* **30** (1980), 235–244. [352, 357]

Győri E., On division of graphs to connected subgraphs. In *Combinatorics (Proc. 5th Hung. Colloq., Keszthely, 1976), I, Colloq. Math. Soc. János Bolyai* **18** (North-Holland, 1978), 485–494. [312]

Habib M., C.J.H. McDiarmid, J. Ramirez-Alfonsin, and B.A. Reed (eds.), *Probabilistic methods for algorithmic discrete mathematics*, *Algorithms and Combin.* **16** (Springer-Verlag, 1998). [657]

Hadamard J., Résolution d'une question relative aux déterminants (French). *Bull. Sci. Math.* **17** (1893), 240–246. [618, 619]

Hadamard J., Note sur quelques applications de l'indice de Kronecker in Jules Tannery. In *Introduction à la théorie des fonctions d'une variable* **2** (Hermann & Fils, 1910), 737–477. [805]

Hadwiger H., Über eine Klassifikation der Streckenkomplexe. *Vierteljschr. Naturforsch. Ges. Zürich* **88** (1943), 133–142. [349]

Hadwiger H., Ueberdeckung des Euklidischen Raumes durch kongruente Mengen. *Portugaliae Math.* **4** (1945), 238–242. [729]

Hadwiger H., Mitteilung betreffend meine Note: Überdeckung einer Menge durch Mengen kleineren Durchmessers. *Comment. Math. Helv.* **19** (1946), 72–73. [817]

Hadwiger H., Simultane Vierteilung zweier Körper. *Arch. Math. (Basel)* **17** (1966), 274–278. [815]

Haemers W.H., *Eigenvalue techniques in design and graph theory*, Mathematical Centre Tracts **121** (Mathematisch Centrum, 1980). [772]

Hagelstein P. and D. Herden, Problem 12071. *Amer. Math. Monthly* **125** (2018), 851. Solution **127** (2020), 464. [461]

Hagelstein P., D. Herden, and D. Young, Ramsey-type theorems for sets satisfying a geometric regularity condition. *J. Math. Anal. Appl.* **447** (2017), 951–956. [461]

Häggkvist R., A lemma on cycle decompositions. In *Cycles in graphs (Burnaby, B.C., 1982)*, North-Holland Math. Stud. **115** (North-Holland, 1985), 227–232. [647, 648]

Häggkvist R. and J.C.M. Janssen, New bounds on the list-chromatic index of the complete graph and other simple graphs. *Combin. Probab. Comput.* **6** (1997), 295–313. [362]

Häggkvist R. and C. Thomassen, Circuits through specified edges. *Discr. Math.* **41** (1982), 29–34. [326]

Hajek B. and B. Narayanan, Multigraphs with the most edge covers. *Inst. Math. Appl. Preprint Series* (1994). [765]

Hajós G., Über eine Konstruktion nicht n-färbbarer Graphen. *Wiss. Z. Martin-Luther-Univ. Halle-Wittenberg Math.-Nat. Reihe* **10** (1961), 116–117. [355, 709]

Hakimi S.L., On the realizability of a set of integers as degrees of the vertices of a graph. *SIAM J. Appl. Math.* **10** (1962), 496–506. [221, 238]

Hakimi S.L., On the degrees of the vertices of a directed graph. *J. Franklin Inst.* **279** (1965), 290–308. [255, 261, 746]

Hakimi S.L. and E.F. Schmeichel, Improved bounds for the chromatic number of a graph. *J. Graph Theory* **47** (2004), 217–225. [354]

Hales A.W. and R.I. Jewett, Regularity and positional games. *Trans. Amer. Math. Soc.* **106** (1963), 222–229. [467, 474]

Halin R., A theorem on n-connected graphs. *J. Combin. Th.* **7** (1969), 150–4. [310]

Hall M., Distinct representatives of subsets. *Bull. Amer. Math. Soc.* **54** (1948), 922. [254, 255, 262]

Hall M. and H.J. Ryser, Cyclic incidence matrices. *Canadian J. Math.* **3** (1951), 495–502. [637]

Hall M., Jr., Cyclic projective planes. *Duke Math. J.* **14** (1947), 1079–1090. [637]

Hall P., On representatives of subsets. *J. Lond. Math. Soc.* **10** (1935), 26–30. [254]

Halmos P.R. and H.E. Vaughan, The marriage problem. *Amer. J. Math* **72** (1950), 214–215. [254]

Hammersley J.M., A few seedlings of research. In *Proceedings of the Sixth Berkeley Symposium on Mathematical Statistics and Probability (Univ. California, Berkeley, Calif., 1970/1971), Vol. I: Theory of statistics* (Univ. California Press, 1972), 345–394. [821]

Hanani H., Über wesentlich unplättbare Kurven im drei-dimensionalen Raume (German). *Fund. Math.* **23** (1934), 135–142. As Chaim Chojnacki. [398]

Hanani H., The existence and construction of balanced incomplete block designs. *Ann. Math. Statist.* **32** (1961), 361–386. [656]

Hanani H., On balanced incomplete block designs with blocks having five elements. *J. Combin. Th. A* **12** (1972), 184–201. [641]

Hanani H., Balanced incomplete block designs and related designs. *Discr. Math.* **11** (1975), 255–369. [639]

Hanlon P., Counting interval graphs. *Trans. Amer. Math. Soc.* **272** (1982), 383–426. [587]

Hansel G., Sur le nombre des fonctions booléennes monotones de n variables. *C. R. Acad. Sci. Paris Sér. A-B* **262** (1966), A1088–A1090. [557]

Hansel G., Complexes et décompositions binomiales. *J. Combinatorial Theory A* **12** (1972), 167–183. [496]

Hansen H.M., *Scheduling with minimum waiting periods (Danish)* (Odense Univ., 1992). Master Thesis. [373]

Hanson D., C.O.M. Loten, and B. Toft, On interval colourings of bi-regular bipartite graphs. *Ars Combin.* **50** (1998), 23–32. [273, 373]

Harary F., On the notion of balance of a signed graph. *Michigan Math. J.* **2** (1953–54), 143–146. [237]

Harary F., The number of linear, directed, rooted, and connected graphs. *Trans. Amer. Math. Soc.* **78** (1955a), 445–463.   [224]

Harary F., Note on the Pólya and Otter formulas for enumerating trees. *Michigan Math. J.* **3** (1955b), 109–112.   [184]

Harary F., The maximum connectivity of a graph. *Proc. Nat. Acad. Sci. U.S.A.* **48** (1962a), 1142–1146.   [290]

Harary F., The determinant of the adjacency matrix of a graph. *SIAM Review* **4** (1962b), 202–210.   [780]

Harary F., *Graph Theory* (Addison-Wesley, 1969).   [215]

Harary F., Reviews: The Polya Picture Album–Encounters of a Mathematician. *Amer. Math. Monthly* **96** (1989), 750–753.   [224]

Harary F., D.F. Hsu, and Z. Miller, The biparticity of a graph. *J. Graph Th.* **1** (1977), 131–133.   [343]

Harary F. and P.C. Kainen, The cube of a path is maximal planar. *Bull. Inst. Combin. Appl.* **7** (1993), 55–56.   [797]

Harary F., P.C. Kainen, and A.J. Schwenk, Toroidal graphs with arbitrarily high crossing numbers. *Nanta Math.* **6** (1973), 58–67.   [794, 798]

Harary F. and P. Kovács, The smallest graphs with prescribed odd and even girth. *Caribbean J. Math.* **1** (1982), 24–26.   [216]

Harary F. and C.St.J.A. Nash-Williams, On eulerian and hamiltonian graphs and line graphs. *Canad. Math. Bull.* **8** (1965), 701–709.   [329]

Harary F. and E.M. Palmer, *Graphical enumeration* (Academic Press, 1973).   [135]

Harary F. and G. Prins, The block-cutpoint-tree of a graph. *Publ. Math. Debrecen* **13** (1966), 103–107.   [297]

Harary F. and A.J. Schwenk, The number of caterpillars. *Discr. Math.* **6** (1973), 359–365.   [187]

Hardy G.H. and S. Ramanujan, Asymptotic formulæ in combinatory analysis. *Proc. London Math. Soc.* *(2)* **17** (1918), 75–115.   [140]

Harper L.H., Optimal numberings and isoperimetric problems on graphs. *J. Combin. Th.* **1** (1966), 385–393.   [496, 720, 800]

Harper L.H., The morphology of partially ordered sets. *J. Combinatorial Theory A* **17** (1974), 44–58.   [563]

Hartke S.G., D. Stolee, D.B. West, and M. Yancey, Extremal graphs with a given number of perfect matchings. *J. Graph Theory* **73** (2013), 449–468.   [276]

Hartman I.B-A., Berge's conjecture on directed path partitions—a survey. *Discr. Math.* **306** (2006), 2498–2514.   [549]

Hartsfield N., A.K. Kelmans, and Y.Q. Shen, On the Laplacian polynomial of a K-cube extension. In *Proc. 27th Southeastern Intl. Conf. Combin. Graph Th. Comput. (Baton Rouge), Congr. Numer.* **119** (1996), 73–77.   [778]

Harzheim E., Remarks on Dilworth's decomposition theorem. *Ars Combin.* **16** (1983), 27–31.   [551]

Hasse M., Zur algebraischen Begründung der Graphentheorie. I. *Math. Nachr.* **28** (1964/1965), 275–290.   [339]

Havel I. and J. Morávek, b-valuations of graphs. *Czechoslovak Math. J.* **22** (1972), 338–351.   [373]

Havel V., A remark on the existence of finite graphs (Czech). *Časopis Pěst. Mat* **80** (1955), 477–480.   [221]

Havet F. and J.S. Sereni, Improper choosability of graphs and maximum average degree. *J. Graph Theory* **52** (2006), 181–199.   [412]

Haxell P.E., Partitioning complete bipartite graphs by monochromatic cycles. *J. Combin. Th. B* **69** (1997), 210–218.   [492]

Haxell P.E., A note on vertex list colouring. *Combin. Probab. Comput.* **10** (2001), 345–347.   [677]

Haxell P.E., Y. Kohayakawa, and T. Łuczak, The induced size-Ramsey number of cycles. *Combin. Probab. Comput.* **4** (1995), 217–239.   [455]

Hayes P.J., A note on the Towers of Hanoi problem. *Computer J.* **20** (1977), 282–285.   [80]

Hayward R.B., Weakly triangulated graphs. *J. Combin. Th. B* **39** (1985), 200–208.   [375]

Heawood P.J., Map-colour theorem. *Q. J. Math.* **24** (1890), 332–339. [399, 402]

Heawood P.J., On the four-colour map theorem. *Q. J. Math.* **29** (1898), 270–85. [419]

Hebbare S.P.R., Graceful cycles. *Utilitas Math.* **10** (1976), 307–317. [656]

Hedayat A.S., N.J.A. Sloane, and J. Stufken, *Orthogonal arrays, Springer Series in Statistics* (Springer-Verlag, 1999). [651]

Hedetniemi S., On partitioning planar graphs. *Canad. Math. Bull.* **11** (1969), 203–210. [417]

Heffter L., Ueber gemeinsame Vielfache linearer Differentialausdrücke und lineare Differentialgleichungen derselben Klasse. *J. Reine Angew. Math.* (1896), 157–166. [645]

Hell P. and J. Nešetřil, *Graphs and homomorphisms, Oxford Lecture Series in Mathematics and its Applications* **28** (Oxford Univ. Press, 2004). [457]

Henning M.A. and A. Yeo, Tight lower bounds on the size of a maximum matching in a regular graph. *Graphs Combin.* **23** (2007), 647–657. [274]

Henning M.A. and A. Yeo, Hypergraphs with large transversal number. *Discr. Math.* **313** (2013a), 959–966. [263]

Henning M.A. and A. Yeo, *Total domination in graphs, Springer Monographs in Mathematics* (Springer, 2013b). [638]

Heppes A. and P. Révész, A splitting problem of Borsuk. *Mat. Lapok* **7** (1956), 108–111. [817]

Herrendörfer G. and D. Rasch, Complete block designs. II. Analysis of partially balanced designs. *Biometrical J.* **19** (1977), 455–461. [655]

Heule M.J., Computing small unit-distance graphs with chromatic number 5. *Geombinatorics* **28** (2018), 32–50. (Also arXiv:1805.12181) [342]

Hierholzer C., Über die Möglichkeit, einen Linienzug ohne Wiederholung und ohne Unterbrechung zu umfahren. *Math. Ann.* **6** (1873), 30–32. [234]

Hilbert D., Ueber die Irreducibilität ganzer rationaler Functionen mit ganzzahligen Coefficienten. *J. Reine Angew. Math.* **110** (1892), 104–129. [461, 473]

Hill C., Solution to problem 10992. *Amer. Math. Monthly* **111** (2004), 827–829. Proposed **110** (2003), 155. [34]

Hilton A.J.W., The number of spanning trees of labeled wheels, fans and baskets. In *Combinatorics (Proc. Conf. Combin. Math., Math. Inst., Oxford, 1972)* (Inst. Math. Appl., 1972), 203–206. [65]

Hilton A.J.W., A theorem on finite sets. *Quart. J. Math. Oxford (2)* **27** (1976), 33–36. [497, 607]

Hilton A.J.W., A simple proof of the Kruskal–Katona theorem and of some associated binomial inequalities. *Period. Math. Hungar.* **10** (1979), 25–30. [496]

Hilton A.J.W., Two conjectures on edge colouring. *Discr. Math.* **74** (1989), 61–64. [359]

Hilton A.J.W. and E.C. Milner, Some intersection theorems for systems of finite sets. *Quart. J. Math. Oxford Ser. (2)* **18** (1967), 369–384. [499, 511]

Hilton P. and J. Pederson, Catalan numbers, their generalization, and their uses. *Math. Intel.* **13** (1991), 64–75. [41]

Hind H., M. Molloy, and B.A. Reed, Colouring a graph frugally. *Combinatorica* **17** (1997), 469–482. [681]

Hind H., M. Molloy, and B.A. Reed, Total coloring with Δ+poly(log Δ) colors. *SIAM J. Comput.* **28** (1999), 816–821. [373, 681]

Hindman N., Ultrafilters and combinatorial number theory. In *Number theory, Carbondale 1979 (Proc. Southern Illinois Conf., Carbondale, 1979), Lecture Notes in Math.* **751** (Springer, 1979), 119–184. [466]

Hindman N., On a conjecture of Erdős, Faber, and Lovász about n-colorings. *Canad. J. Math.* **33** (1981), 563–570. [439]

Hinrichs A. and C. Richter, New sets with large Borsuk numbers. *Discr. Math.* **270** (2003), 137–147. [817]

Hiraguchi T., On the dimension of partially ordered sets. *Sci. Rep. Kanazawa Univ.* **1** (1951), 77–94. [571, 574, 583]

Hiraguchi T., On the dimension of orders. *Sci. Reports Kanazawa Univ.* **4** (1955), 1–20. [571, 574, 575, 583]

Hladký J., D. Král', and U. Schauz, Brooks' theorem via the Alon–Tarsi theorem. *Discr. Math.* **310** (2010), 3426–3428.   [347]

Hliněný P., Crossing number is hard for cubic graphs. *J. Combin. Th. B* **96** (2006), 455–471.   [795]

Hoare A.H.M., An involution of blocks in the partitions of *n*. *Amer. Math. Monthly* **93** (1986), 475–476.
[152]

Hobby C.R. and J.R. Rice, A moment problem in L_1 approximation. *Proc. Amer. Math. Soc.* **16** (1965), 665–670.   [810]

Hochberg R., C.J.H. McDiarmid, and M.E. Saks, On the bandwidth of triangulated triangles. In *14th Brit. Combin. Conf. (Keele, 1993)* **138** (Discr. Math., 1995), 261–5.   [801–803]

Hoeffding W., Probability inequalities for sums of bounded random variables. *J. Amer. Statist. Assoc.* **58** (1963), 13–30.   [708, 715, 722]

Hoffman A.J., On the polynomial of a graph. *Amer. Math. Monthly* **70** (1963), 30–36.   [772]

Hoffman A.J., On eigenvalues and colorings of graphs. In *Graph Theory and Its Applications* (B. Harries, ed.) (Academic Press, 1970), 79–91.   [781]

Hoffman A.J., Extending Greene's theorem to directed graphs. *J. Combin. Theory A* **34** (1983), 102–107.
[549]

Hoffman A.J. and R.R. Singleton, On Moore graphs with diameters 2 and 3. *IBM J. Res. Develop.* **4** (1960), 497–504.   [244, 775]

Hoffman D.G. and P.J. Schellenberg, The existence of C_k-factorizations of $K_{2n}-F$. *Discr. Math.* **97** (1991), 243–250.   [647]

Holland F., Problem 11798. *Amer. Math. Monthly* **121** (2014), 798. Solution **124** (2017), 370.   [23, 117]

Holley R., Remarks on the FKG inequalities. *Comm. Math. Phys.* **36** (1974), 227–231.   [597]

Hollingsworth S., Packing trees into complete bipartite graphs. *Discr. Math.* **313** (2013), 945–948.
[248]

Holton D. and R.E.L. Aldred, Planar graphs, regular graphs, bipartite graphs and Hamiltonicity. *Australas. J. Combin.* **20** (1999), 111–131.   [326]

Holton D.A. and J. Sheehan, *The Petersen Graph* (Cambridge Univ. Press, 1993).   [213]

Holyer I., The NP-completeness of edge-coloring. *SIAM J. Comput.* **10** (1981), 718–720.   [359, 699]

Holzmann C.A. and F. Harary, On the tree graph of a matroid. *SIAM J. Appl. Math.* **22** (1972), 187–193.
[540]

Hook J. and G. Isaak, Star-critical Ramsey numbers. *Discrete Appl. Math.* **159** (2011), 328–334.   [459]

Hopcroft J.E. and R.M. Karp, An $n^{5/2}$ algorithm for maximum matchings in bipartite graphs. *SIAM J. Comput.* **2** (1973), 225–231.   [283, 284]

Hopcroft J.E. and R.E. Tarjan, Efficient planarity testing. *J. Assoc. Comput. Mach.* **21** (1974), 549–568.
[393]

Horadam A.F., A generalized Fibonacci sequence. *Amer. Math. Monthly* **68** (1961), 455–459.   [61]

Horák P., Q. He, and W.T. Trotter, Induced matchings in cubic graphs. *J. Graph Th.* **17** (1993), 151–160.
[373]

Horák P. and Z. Tuza, A coloring problem related to the Erdős–Faber–Lovász conjecture. *J. Combin. Th. B* **50** (1990), 321–322.   [344]

Horton J.D., On two-factors of bipartite regular graphs. *Discr. Math.* **41** (1982), 35–41.   [317]

Hosoya H. and F. Harary, On the matching properties of three fence graphs. In *Appl. Graph Th. and Discr. Math. Chem. (Saskatoon, 1991)* **12** (1993), 211–218.   [765]

Hoşten S. and W.D. Morris, Jr., The order dimension of the complete graph. *Discr. Math.* **201** (1999), 133–139.   [580]

Howard F.T., The number of multinomial coefficients divisible by a fixed power of a prime. *Pacific J. Math.* **50** (1974), 99–108.   [48]

Hsieh W.N. and D.J. Kleitman, Normalized matching in direct products of partial orders. *Studies in Appl. Math.* **52** (1973), 285–289.   [563]

Huang D. and D. Scully, Problem 11052. *Amer. Math. Monthly* **110** (2003), 957. Solution **113** (2006), 183.
[152]

Huneke C., The friendship theorem. *Amer. Math. Monthly* **109** (2002), 192–194.   [776]

Hurlbert G.H., A short proof that $\mathbf{n}^3$ is not a circle containment order. *J. Order* **5** (1988), 235–237.
[828]

Hutchinson J.P., Problem 10478. *Amer. Math. Monthly* **102** (1995), 746. Solution **105** (1998), 274–275.
[419]

Hutchinson J.P., Three- and four-coloring nearly triangulated surfaces. In *Proc. 32nd Southeastern Intl. Conf. Combin. Graph Th. Comput. (Boca Raton)*, **150** (2001), 129–143.
[419]

Hutchinson J.P., Arc- and circle-visibility graphs. *Australas. J. Combin.* **25** (2002), 241–262.
[394]

Hutchinson J.P. and P.B. Trow, Some pigeonhole principle results extended. *Amer. Math. Monthly* **87** (1980), 648–651.
[429]

Huxley M., The difference between consecutive primes. In *Analytic Number Theory (St. Louis Univ., 1972), Proc. Sympos. Pure Math* **24** (Amer. Math. Soc., 1973), 141–145.
[730]

Igusa K., Solution of the Bulgarian solitaire conjecture. *Math. Mag.* **58** (1985), 259–271.
[144, 152]

Ionin Y., Solution to problem 11678. *Amer. Math. Monthly* **121** (2014), 952–953. Proposed **119** (2012), 880.
[178]

Isaacs R., Infinite families of nontrivial trivalent graphs which are not Tait colorable. *Amer. Math. Monthly* **82** (1975), 221–239.
[372]

Ishigami Y., Proof of a conjecture of Bollobás and Kohayakawa on the Erdős–Stone theorem. *J. Combin. Th. B* **85** (2002), 222–254.
[482]

Itai A. and M. Rodeh, Covering a graph by circuits. In *Automata, Langs. and Prog. (Udine, 1978), Lect. Notes Comp. Sci.* **62** (Springer-Verlag, 1978), 289–299.
[249]

Jackson B., Hamilton cycles in regular 2-connected graphs. *J. Combin. Th. B* **29** (1980), 27–46.
[326]

Jackson B., Hamilton cycles in 7-connected line graphs (1989). Unpublished preprint.
[326]

Jackson B. and O. Ordaz, Chvátal–Erdős conditions for paths and cycles in graphs and digraphs. A survey. *Discr. Math.* **84** (1990), 241–254.
[321]

Jacobi C.G.J., De resolutione aequationum per series infinitas. *J. Reine Angew. Math.* **6** (1830), 257–286.
[129]

Jacobson M.S., F.R. McMorris, and H.M. Mulder, Tolerance intersection graphs. In *Graph Theory, Combinatorics, and Applications (Kalamazoo, 1988)* (Y. Alavi, G. Chartrand, O.R. Oellerman and A.J. Schwenk, eds.) (Wiley, 1991), 705–724.
[374]

Jaeger F., Tait's theorem for graphs with crossing number at most one. *Ars Combin.* **9** (1980), 283–287.
[797]

Jahanbekam S., J. Kim, S. O, and D.B. West, On r-dynamic coloring of graphs. *Discrete Appl. Math.* **206** (2016), 65–72.
[668]

Jahanbekam S. and D.B. West, New lower bounds for matching numbers of general and bipartite graphs. *Congr. Numer.* **218** (2013), 57–59.
[262, 273]

Jahanbekam S. and D.B. West, Anti-Ramsey problems for t edge-disjoint rainbow spanning subgraphs: cycles, matchings, or trees. *J. Graph Th.* **82** (2016), 75–89.
[333]

Jamison R.E., Covering finite fields with cosets of subspaces. *J. Combin. Th. A* **22** (1977), 253–266.
[735]

Jamison R.E., Orientable edge colorings of graphs. *Discrete Appl. Math.* **159** (2011), 595–604.
[472]

Jamison R.E., T. Jiang, and A.C.H. Ling, Constrained Ramsey numbers of graphs. *J. Graph Theory* **42** (2003), 1–16.
[473]

Jamison R.E. and D.B. West, On pattern Ramsey numbers of graphs. *Graphs Combin.* **20** (2004), 333–339.
[472]

Janson S., T. Łuczak, and A. Ruciński, *Random Graphs* (Wiley-Interscience, 2000).
[657, 696, 708]

Janssen J.C.M., The Dinitz problem solved for rectangles. *Bull. Amer. Math. Soc. (N.S.)* **29** (1993), 243–249.
[363]

Jarník V., O jistém problému minimálnim. *Acta Societatis Scientiarum Natur. Moravicae* **6** (1930), 57–63.
[246, 252]

Jendrol' S., A short proof of Kotzig's theorem on minimal edge weights of convex 3-polytopes (1999), 35–38.
[411]

Jendrol' S. and H.-J. Voss, Light subgraphs of graphs embedded in the plane—a survey. *Discr. Math.* **313** (2013), 406–421.   [407]

Jensen T.R. and B. Toft, *Graph coloring problems*, Wiley-Interscience Series in Discr. Mathematics and Optimization (Wiley & Sons, 1995).   [335]

Jenssen M. and J. Skokan, Exact Ramsey numbers of odd cycles via nonlinear optimisation (2016). (arXiv:1608.05705).   [461]

Jerrum M., A. Sinclair, and E. Vigoda, A polynomial-time approximation algorithm for the permanent of a matrix with non-negative entries. In *Proceedings of the Thirty-Third Annual ACM Symposium on Theory of Computing* (ACM, 2001), 712–721.   [754]

Jiang T., K.G. Milans, and D.B. West, Degree Ramsey numbers for cycles and blowups of trees. *European J. Combin.* **34** (2013), 414–423.   [685]

Jiang T. and D. Mubayi, New upper bounds for a canonical Ramsey problem. *Combinatorica* **20** (2000), 141–146.   [471]

Jichang S. and D.J. Kleitman, Superantichains in the lattice of partitions of a set. *Studies in applied mathematics* **71** (1984), 207–241.   [567]

Johannson K.R., *Variations on a theorem by van der Waerden* (Univ. Manitoba, 2007). Masters Thesis. Available online at http://mspace.lib.umanitoba.ca/bitstream/handle/1993/321/thesismain.pdf   [466]

Johnson D.M., A.L. Dulmage, and N.S. Mendelsohn, Orthomorphisms of groups and orthogonal latin squares. I. *Canad. J. Math.* **13** (1961), 356–372.   [654]

Johnson N.L. and S. Kotz, *Urn models and their application* (Wiley, 1977).   [106]

Johnson R.T. and T.P. Vaughan, On union-closed families. I. *J. Combin. Th. A* **84** (1998), 242–249.   [503]

Jonsson J. and J.G. Propp, Problem 11298. *Amer. Math. Monthly* **114** (2007), 547. Solution **116** (2009), 371–372.   [387]

Jordan C., Sur les assemblages de lignes. *J. Reine Angew. Math.* **70** (1869), 185–190.   [243, 251]

Jordan C., Questions de probabilités. *Bull. Soc. Math. France* **1** (1872), 256–258.   [669]

Jorza A., Problem 10856. *Amer. Math. Monthly* **108** (2001), 172. Solution **113** (2006), 180–183.   [388]

Joyal A., Une théorie combinatoire des séries formelles. *Adv. in Math.* **42** (1981), 1–82.   [37, 127]

Józsa S. and E. Szemerédi, The number of unit distances on the plane. In *Infinite and finite sets (Colloq., Keszthely, 1973), II, Colloq. Math. Soc. Jáanos Bolyai* **10** (North-Holland, 1975), 939–950.   [794]

Juarez H.A. and G. Salazar, Drawings of $C_m \times C_n$ with one disjoint family. II. *J. Combin. Th. B* **82** (2001), 161–165.   [794]

Jukna S., *Extremal combinatorics (2nd ed.)*, Texts in Theoretical Computer Science. An EATCS Series (Springer, 2011).   [493]

Juvan M., B. Mohar, and R. Škrekovski, On list edge-colorings of subcubic graphs. *Discr. Math.* **187** (1998a), 137–149.   [412]

Juvan M., B. Mohar, and R. Škrekovski, List total colourings of graphs. *Combin. Probab. Comput.* **7** (1998b), 181–188.   [60, 739]

Juvan M., B. Mohar, and R. Škrekovski, Graphs of degree 4 are 5-edge-choosable. *J. Graph Theory* **32** (1999), 250–264.   [412]

Kabela A. and T. Kaiser, 10-tough chordal graphs are Hamiltonian. *J. Combin. Th. B* **122** (2017), 417–427.   [318]

Kahn J., Asymptotically good list-colorings. *J. Combin. Th. A* **73** (1996), 1–59.   [362]

Kahn J., On some hypergraph problems of Paul Erdős and the asymptotics of matchings, covers and colorings. In *The mathematics of Paul Erdős, I, Algorithms Combin.* **13** (Springer, 1997), 345–371.   [439]

Kahn J. and G. Kalai, A counterexample to Borsuk's conjecture. *Bull. Amer. Math. Soc. (N.S.)* **29** (1993), 60–62.   [817, 818]

Kahn J. and J.H. Kim, Entropy and sorting. *J. Comput. System Sci.* **51** (1995), 390–399.   [823]

Kahn J. and N. Linial, Balancing extensions via Brunn–Minkowski. *Combinatorica* **11** (1991), 363–368. [824, 825]

Kahn J. and M.E. Saks, Balancing poset extensions. *Order* **1** (1984), 113–126. [823, 824, 826, 831]

Kainen P.C., A generalization of the 5-color theorem. *Proc. Amer. Math. Soc.* **45** (1974), 450–453. [419]

Kaiser T., E. Rollová, and R. Lukot'ka, Nowhere-zero flows in signed graphs: A survey. *Lect. Notes. Seminar. Interdiscip. Matematica* **14** (2017), 85–104. [237]

Kaiser T., Z. Ryjáček, D. Král', M. Rosenfeld, and H.J. Voss, Hamilton cycles in prisms. *J. Graph Th.* **56** (2007), 249–269. [326]

Kalbfleisch J.D., J.G. Kalbfleisch, and R. Stanton, A combinatorial problem on convex *n*-gons. In *Proc. Louisiana Conf. on Combinatorics, Graph Theory and Computing (Louisiana State Univ., 1970)(RC Mullin, KB Reid, and DP Roselle, eds.)* (1970), 180–188. [458]

Kalbfleisch J.G., Upper bounds for some Ramsey numbers. *J. Combin. Th.* **2** (1967), 35–42. [449]

Kamat V.M., *Erdős–Ko–Rado Theorems: New Generalizations, Stability Analysis and Chvatal's Conjecture* (ProQuest LLC, 2011). Thesis (Ph.D.)–Arizona State University. [503]

Kampen G.R., Orienting planar graphs. *Discr. Math.* **14** (1976), 337–341. [786]

Kano M. and X. Li, Monochromatic and heterochromatic subgraphs in edge-colored graphs—a survey. *Graphs Combin.* **24** (2008), 237–263. [457]

Kantrowitz M., Problem E3130. *Amer. Math. Monthly* **93** (1986), 131. Solution **95** (1988), 555–6. [61]

Kapoor S.F., A.D. Polimeni, and C.E. Wall, Degree sets for graphs. *Fund. Math.* **95** (1977), 189–194. [227]

Karaganis J.J., On the cube of a graph. *Canad. Math. Bull.* **11** (1968), 295–296. [331]

Karaivanov B. and T.S. Vassilev, Solution to problem 11798. *Amer. Math. Monthly* **124** (2017). Proposed **121** (2014). [35]

Karger D.R., P.N. Klein, and R.E. Tarjan, A randomized linear-time algorithm to find minimum spanning trees. *J. Assoc. Comput. Mach.* **42** (1995), 321–328. [246]

Karp R.M., Reducibility among combinatorial problems. In *Complexity of computer computations (Proc. Sympos., IBM Thomas J. Watson Res. Center, Yorktown Heights, N.Y., 1972)* (Plenum, 1972), 85–103. [11]

Kárteszi F., Piani finiti ciclici come risoluzioni di un certo problema di minimo. (Italian). *Boll. Un. Mat. Ital. (3)* **15** (1960), 522–528. [639]

Kasteleyn P.W., The statistics of dimers on a lattice, I. The number of dimer arrangements on a quadratic lattice. *Physica* **27** (1961), 1209–1225. [755]

Kasteleyn P.W., Dimer statistics and phase transitions. *J. Math. Phys.* **4** (1963), 287–293. [755, 756]

Kasteleyn P.W., Graph theory and crystal physics. In *Graph Theory and Theoretical Physics* (Academic Press, 1967), 43–110. [755]

Kászonyi L. and Z. Tuza, Saturated graphs with minimal number of edges. *J. Graph Theory* **10** (1986), 203–210. [491]

Katerinis P., Some conditions for the existence of *f*-factors. *J. Graph Theory* **9** (1985), 513–521. [276]

Katona G.O.H., Intersection theorems for systems of finite sets. *Acta Math. Acad. Sci. Hungar.* **15** (1964), 329–337. [497, 511]

Katona G.O.H., On a conjecture of Erdős and a stronger form of Sperner's theorem. *Studia Sci. Math. Hungar.* **1** (1966), 59–63. [565]

Katona G.O.H., A theorem of finite sets. In *Theory of graphs (Proc. Colloq., Tihany, 1966)* (Academic Press, 1968), 187–207. [495, 496]

Katona G.O.H., A generalization of some generalizations of Sperner's theorem. *J. Combinatorial Theory B* **12** (1972a), 72–81. [554]

Katona G.O.H., A simple proof of the Erdős–Chao Ko–Rado theorem. *J. Combin. Th. B* **13** (1972b), 183–184. [498]

Kauers M., Problem 11545. *Amer. Math. Monthly* **118** (2011), 84. Solution **119** (2012), 885–886. [174]

Kawarabayashi K.i., A survey on Hamiltonian cycles. In *Proc. Workshop on Graph. Th. and Related Topics (Sendai, 1999)*, **7** (2001), 25–39. [316]

Kearnes K.A. and E.W. Kiss, Finite algebras of finite complexity. *Discr. Math.* **207** (1999), 89–135.
[665]

Keedwell A.D. and J. Dénes, *Latin squares and their applications* (Elsevier/North-Holland, 2015), second edn.   [610]

Keevash P., Shadows and intersections: stability and new proofs. *Adv. Math.* **218** (2008), 1685–1703.
[496]

Keevash P., Hypergraph Turán problems. In *Surveys in combinatorics 2011, London Math. Soc. Lecture Note Ser.* **392** (Cambridge Univ. Press, 2011), 83–139.   [475]

Keevash P., The existence of designs (2014). (arXiv:1401.3665).   [640]

Keevash P., Counting designs. *J. Eur. Math. Soc. (JEMS)* **20** (2018), 903–927.   [640]

Kellogg A., Problem 10585. *Amer. Math. Monthly* **104** (1997), 361. Solution **106** (1999), 170–171.   [177]

Kelly D., The 3-irreducible partially ordered sets. *Canad. J. Math.* **29** (1977), 367–383.   [573]

Kelly D., On the dimension of partially ordered sets. *Discr. Math.* **35** (1981), 135–156.   [571, 606]

Kelly D. and W.T. Trotter, Dimension theory for ordered sets. In *Ordered sets (Banff, Alta., 1981), NATO Adv. Study Inst. C: Math. Phys. Sci.* **83** (Reidel, 1982), 171–211.   [572, 573, 583]

Kelly J.B. and L.M. Kelly, Paths and circuits in critical graphs. *Amer. J. Math.* **76** (1954), 786–792.
[434]

Kelmans A.K., The number of trees in a graph, I and II. *Automat. Remote Control* **26/27** (1965/66), 2118–2129, 233–241.   [778]

Kelmans A.K., The properties of the characteristic polynomial of a graph (Russian). *Cybernetics* **4** (Izdat. "Énergija", 1967), 27–41.   [777, 782]

Kelmans A.K., A new planarity criterion for 3-connected graphs. *J. Graph Th.* **5** (1981), 259–267.
[393]

Kelmans A.K., A strengthening of the Kuratowski planarity criterion for 3-connected graphs. *Discr. Math.* **51** (1984), 215–220.   [393]

Kelmans A.K., On Hamiltonicity of {claw, net}-free graphs. *Discr. Math.* **306** (2006), 2755–2761.   [297]

Kelmans A.K. and V.M. Chelnokov, A certain polynomial of a graph and graphs with an extremal number of trees. *J. Combin. Th. B* **16** (1974), 197–214.   [777]

Kempe A.B., On the geographical problem of four colours. *Amer. J. Math.* **2** (1879), 193–200.   [399]

Kempe A.B., A memoir on the theory of mathematical form. In *Philosophical Transactions of the Royal Society of London* **177** (1886), 1–70.   [213]

Kendall M.G. and B.B. Smith, On the method of paired comparisons. *Biometrika* **31** (1940), 324–345.
[228]

Kerimov A., Problem 11454. *Amer. Math. Monthly* **116** (2009), 746. Solution **118** (2011), 659.   [315]

Keselman G., Solution to problem 11274. *Amer. Math. Monthly* **115** (2008). Proposed **114** (2007), 165.
[116]

Kézdy A.E. and H.St.C. Snevily, Distinct sums modulo n and tree embeddings. *Combin. Probab. Comput.* **11** (2002), 35–42.   [746]

Khan M.A., Problem E3451. *Amer. Math. Monthly* **98** (1991), 645. Solution **100** (1993), 303.   [133]

Kharaghani H. and B. Tayfeh-Rezaie, A Hadamard matrix of order 428. *J. Combin. Des.* **13** (2005), 435–440.   [622]

Khare C.B., Problem E3315. *Amer. Math. Monthly* **96** (1989), 253. Solution **98** (1991), 366–367.   [440]

Kierstead H.A., Long stars specify χ-bounded classes. In *Sets, graphs and numbers (Budapest, 1991), Colloq. Math. Soc. János Bolyai* **60** (North-Holland, 1992), 421–428.   [352]

Kierstead H.A., On the choosability of complete multipartite graphs with part size three. *Discr. Math.* **211** (2000), 255–259.   [355, 356]

Kierstead H.A. and G. Konjevod, Coloring number and on-line Ramsey theory for graphs and hypergraphs. *Combinatorica* **29** (2009), 49–64.   [457]

Kierstead H.A. and A.V. Kostochka, Efficient graph packing via game colouring. *Combin. Probab. Comput.* **18** (2009), 765–774.   [248]

Kierstead H.A. and S.G. Penrice, Recent results on a conjecture of Gyárfás. In *Proc. 21st Southeastern Intl. Conf. Combin. Graph Th. Comput. (Boca Raton)*, **79** (1990), 182–186.   [352]

Kierstead H.A. and S.G. Penrice, Radius two trees specify χ-bounded classes. *J. Graph Th.* **18** (1994), 119–129.   [352]

Kierstead H.A., A. Salmon, and R. Wang, On the choice number of complete multipartite graphs with part size four (2014). (arXiv:1407.3817v1).   [356]

Kierstead H.A. and W.T. Trotter, Explicit matchings in the middle levels of the Boolean lattice. *Order* **5** (1988), 163–171.   [50]

Kierstead H.A. and W.T. Trotter, A note on removable pairs. In *Graph theory, combinatorics, and applications. Vol. 2 (Kalamazoo, MI, 1988)*, Wiley-Intersci. Publ. (Wiley, 1991), 739–742.   [584]

Kierstead H.A. and Y. Zhu, Radius three trees in graphs with large chromatic number. *SIAM J. Discr. Math.* **17** (2004), 571–581.   [352]

Kim J.H., The Ramsey number $R(3, t)$ has order of magnitude $t^2 / \log t$. *Random Structures Algorithms* **7** (1995), 173–207.   [451, 664, 682]

Kim S.J. and W.J. Park, List dynamic coloring of sparse graphs. In *Combinatorial optimization and applications, Lecture Notes in Comput. Sci.* **6831** (Springer, 2011), 156–162.   [424]

Kimble R.J., Jr, *Extremal Problems in Dimension–Theory for Partially-Ordered Sets* (ProQuest, 1973). Ph.D. Thesis, Massachusetts Inst. Tech.   [573]

Kimble R.J., Jr. and A.J. Schwenk, On universal caterpillars. In *The theory and applications of graphs (Kalamazoo, 1980)* (Wiley, 1981), 437–447.   [187]

King A.D. and B.A. Reed, A short proof that χ can be bounded ϵ away from $\Delta + 1$ toward ω. *J. Graph Th.* **81** (2016), 30–34.   [685]

Kirchhoff G., über die Auflösung der Gleichungen, auf welche man bei der Untersuchung der linearen Verteilung galvanischer Ströme geführt wird. *Ann. Phys. Chem.* **72** (1847), 497–508.   [37]

Kirdar M.S. and T.H.R. Skyrme, On an identity relating to partitions and repetitions of parts. *Canad. J. Math.* **34** (1982), 194–195.   [152]

Kirkman T.P., On a problem in cominations. *Cambridge and Dublin Math. J.* **2** (1847), 191–204.   [612, 641]

Kirkman T.P. In *The Lady's and Gentleman's Diary, (J. Greenhill, 1850)* (1850).   [649]

Kislicyn S.S., Finite partially ordered sets and their corresponding permutation sets. *Mat. Zametki* **4** (1968), 511–518.   [823]

Klavžar S., U. Milutinović, and C. Petr, On the Frame–Stewart algorithm for the multi-peg Tower of Hanoi problem. *Discrete Appl. Math.* **120** (2002), 141–157.   [70]

Klavžar S. and M. Petkovšek, Problem E3281. *Amer. Math. Monthly* **95** (1988), 655. Solution **97** (1990), 924–925.   [606]

Klazar M., The Füredi–Hajnal conjecture implies the Stanley–Wilf conjecture. In *Formal power series and algebraic combinatorics (Moscow, 2000)* (Springer, 2000), 250–255.   [432]

Kleitman D., On Dedekind's problem: The number of monotone Boolean functions. *Proc. Amer. Math. Soc.* **21** (1969), 677–682.   [509, 558]

Kleitman D., Review of "On a problem of Moser". *MathSciNet Review* (1973). MR0297582 (45 #6636)   [551]

Kleitman D.J., On a lemma of Littlewood and Offord on the distribution of certain sums. *Math. Z.* **90** (1965), 251–259.   [565]

Kleitman D.J., Families of non-disjoint subsets. *J. Combin. Th.* **1** (1966), 153–155.   [595, 608]

Kleitman D.J., On families of subsets of a finite set containing no two disjoint sets and their union. *J. Combinatorial Theory* **5** (1968), 235–237.   [500]

Kleitman D.J., The crossing number of $K_{5,n}$. *J. Combin. Th.* **9** (1970), 315–323.   [566, 793, 797]

Kleitman D.J., On an extremal property of antichains in partial orders. The LYM property and some of its implications and applications. In *Combinatorics (Proc. NATO Advanced Study Inst., Breukelen, 1974), Part 2: Graph theory; foundations, partitions and combinatorial geometry* (Math. Centrum, 1974), 77–90. Math. Centre Tracts, No. 56.   [559]

Kleitman D.J., A note on some subset identities. *Studies in Appl. Math.* **54** (1975), 289–292.   [40]

Kleitman D.J., Extremal hypergraph problems. In *Surveys in combinatorics (Proc. Seventh British Combinatorial Conf., 1979)*, London Math. Soc. Lecture Note Ser. **38** (Cambridge Univ. Press, 1979), 44–65.   [501]

Kleitman D.J., Extremal problems on hypergraphs. In *Extremal problems for finite sets (Visegrád, 1991)*, Bolyai Soc. Math. Stud. **3** (János Bolyai Math. Soc., 1994), 355–374.   [493]

Kleitman D.J. and T.L. Magnanti, On the number of latent subsets of intersecting collections. *J. Combin. Th. A* **16** (1974), 215–220.   [501]

Kleitman D.J. and G. Markowsky, On Dedekind's problem: the number of isotone Boolean functions. II. *Trans. Amer. Math. Soc.* **213** (1975), 373–390.   [558, 580]

Kleitman D.J., J. Shearer, and D. Sturtevant, Intersections of k-element sets. *Combinatorica* **1** (1981), 381–384.   [507]

Klešč M., R.B. Richter, and I. Stobert, The crossing number of $C_5 \times C_n$. *J. Graph Theory* **22** (1996), 239–243.   [794]

Klotz W., A constructive proof of Kuratowski's Theorem. *Ars Combinatoria* **28** (1989), 51–54.   [393]

Klove T., Problem 10460. *Amer. Math. Monthly* **102** (1995), 553. Solution **105** (1998), 69.   [22]

Kneser M., Aufgabe 300 (German). *Jahresber Deutsch. Math.-Verein* **58** (1955), 27.   [251, 811]

Knuth D., Problem 12055. *Amer. Math. Monthly* **125** (2018), 660. Solution **127** (2020), 186.   [152]

Knuth D.E., *The Art of Computer Programming, Vol. 1: Fundamental Algorithms* (Addison-Wesley, 1968). Also 1973, 1997.   [99, 101, 122]

Knuth D.E., Permutations, matrices, and generalized Young tableaux. *Pacific J. Math.* **34** (1970), 709–727.   [193, 195, 198, 208]

Knuth D.E., *The Art of Computer Programming. Vol. 3: Sorting and searching* (Addison-Wesley, 1973).
                                                                             [50, 101, 190, 193]

Knuth D.E., Big omicron and big omega and big theta. *ACM SIGACT News* **8** (1976a), 18–24.   [10]

Knuth D.E., *Mariages Stables* (Les Presses de l'Univ. de Montréal, 1976b). English trans. AMS, 1997.
                                                                             [286]

Knuth D.E., Problem E3463. *Amer. Math. Monthly* **98** (1991), 852. Solution **100** (1993), 693–4.   [441]

Knuth D.E., Problem 10298. *Amer. Math. Monthly* **100** (1993), 400. Solution **103** (1996), 80–81.   [133]

Knuth D.E., Problem 10546. *Amer. Math. Monthly* **103** (1996), 695. Solution **105** (1998), 867–8.   [176]

Knuth D.E., Problem 10720. *Amer. Math. Monthly* **106** (1999), 264. Solution **110** (2003), 60–61.   [669]

Knuth D.E., Problem 11151. *Amer. Math. Monthly* **112** (2005), 367. Solution **114** (2007), 265–266.   [81]

Knuth D.E., Problem 11274. *Amer. Math. Monthly* **114** (2007), 165. Solution **116** (2009), 548–549.
                                                                             [35, 116]

Knuth D.E., Problem 11452. *Amer. Math. Monthly* **116** (2009), 648. Solution **118** (2011), 657.   [65]

Knuth D.E., Problem 11985. *Amer. Math. Monthly* **124** (2017), 275. Solution **126** (2019), March.   [567]

Knuth D.E. and T.J. Buckholtz, Computation of tangent, Euler, and Bernoulli numbers. *Math. Comp.* **21** (1967), 663–688.   [65]

Knuth D.E. and J. McCarthy, Problem E3429. *Amer. Math. Monthly* **98** (1991), 264. Solution **99** (1992), 684.   [667]

Koebe P., Contaktprobleme der konformen Abbildung (German). *Berichte Über die Verhandlungen der Sächsischen Akad. Wissen. Leipzig, Math.-Phys. Klasse* **88** (1936), 141–164.   [394]

Kohayakawa Y., Szemerédi's regularity lemma for sparse graphs. In *Foundations of computational mathematics (Rio de Janeiro, 1997)* (Springer, 1997), 216–230.   [488]

Kohayakawa Y. and V. Rödl, Szemerédi's regularity lemma and quasi-randomness. In *Recent advances in algorithms and combinatorics*, CMS Books Math./Ouvrages Math. SMC **11** (Springer, 2003), 289–351.   [488]

Kohayakawa Y., V. Rödl, M. Schacht, and E. Szemerédi, Sparse partition universal graphs for graphs of bounded degree. *Adv. Math.* **226** (2011), 5041–5065.   [455]

Köhler E., über das Oberwolfacher Problem. In *Beiträge zur geometrischen Algebra (Proc. Sympos., Duisburg, 1976)*, Lehrbücher Monograph. Geb. Exakten Wissensch., Math. Reihe **21** (Birkhäuser, Basel, 1977), 189–201.   [647]

Kollár J., L. Rónyai, and T. Szabó, Norm-graphs and bipartite Turán numbers. *Combinatorica* **16** (1996), 399–406. [633, 672]

Komlós J., A strange pigeonhole principle. *Order* **7** (1990), 107–113. [823]

Komlós J., G.N. Sárközy, and E. Szemerédi, Blow-up lemma. *Combinatorica* **17**(1997),109–123. [481]

Komlós J., A. Shokoufandeh, M. Simonovits, and E. Szemerédi, The regularity lemma and its applications in graph theory. In *Theoretical aspects of computer science (Tehran, 2000)*, *Lect. Notes Comp. Sci.* **2292** (Springer, 2002), 84–112. [488]

Komlós J. and M. Simonovits, Szemerédi's regularity lemma and its applications in graph theory. In *Combinatorics, Paul Erdős is eighty, Vol. 2 (Keszthely, 1993)*, Bolyai Soc. Math. Stud. **2** (János Bolyai Math. Soc., 1996), 295–352. [488, 492]

Komlós J. and E. Szemerédi, Topological cliques in graphs II. *Combin. Probab. Comput.* **5** (1996), 79–90. [351]

Komm H., On the dimension of partially ordered sets. *Amer. J. Math.* **70** (1948), 507–520. [571]

Konheim A.G. and B. Weiss, An occupancy discipline and applications. *SIAM J. Applied Math.* **14** (1966), 1266–1274. [50]

König D., Über Graphen und ihre Anwendung auf Determinantentheorie und Mengenlehre. *Math. Ann.* **77** (1916), 453–465. [256, 259, 359, 371]

König D., Graphen und Matrizen. *Mat. Lapok* **38** (1931), 116–119. [258, 530]

König D., *Theorie der endlichen und unendlichen Graphen* (Akademische Verlagsgesellschaft, 1936). Also Chelsea, 1950; Teubner, 1986. [763]

Kopylov G.N., Maximal paths and cycles in a graph. *Dokl. Akad. Nauk SSSR* **234** (1977), 19–21. [323]

Korshunov A.D., The number of monotone Boolean functions. *Problemy Kibernet.* (1981), 5–108, 272. [558]

Koshy T., *Fibonacci and Lucas numbers with applications*, Pure and Applied Mathematics (New York) (Wiley-Interscience, 2001). [53]

Kostochka A.V., The total coloring of a multigraph with maximal degree 4. *Discr. Math.* **17** (1977), 161–163. [373]

Kostochka A.V., The minimum Hadwiger number for graphs with a given mean degree of vertices. *Metody Diskret. Analiz.* (1982), 37–58. [710]

Kostochka A.V., The total chromatic number of any multigraph with maximum degree five is at most seven. *Discr. Math.* **162** (1996), 199–214. [373]

Kostochka A.V., Extremal problems on Δ-systems. In *Numbers, information and complexity (Bielefeld, 1998)* (Kluwer Acad. Publ., 2000), 143–150. [504]

Kostochka A.V., A new tool for proving Vizing's theorem. *Discr. Math.* **326** (2014), 1–3. [360]

Kostochka A.V., M.J. Pelsmajer, and D.B. West, A list analogue of equitable coloring. *J. Graph Theory* **44** (2003), 166–177. [374]

Kostochka A.V. and V. Rödl, On graphs with small Ramsey numbers. *J. Graph Th.* **37** (2001), 198–204. [484]

Kostochka A.V. and V. Rödl, On graphs with small Ramsey numbers. II. *Combinatorica* **24** (2004), 389–401. [672]

Kostochka A.V. and N. Sheikh, On the induced Ramsey number $IR(P_3, H)$. In *Topics in discrete mathematics*, *Algorithms Combin.* **26** (Springer, 2006), 155–167. [460]

Kostochka A.V. and D.B. West, Chvátal's condition cannot hold for both a graph and its complement. *Discuss. Math. Graph Theory* **26** (2006), 73–76. [332, 705]

Kostochka A.V. and M. Yancey, Ore's conjecture for $k = 4$ and Grötzsch's theorem. *Combinatorica* **34** (2014), 323–329. Generalized in "Ore's conjecture on color-critical graphs is almost true," *J. Combin. Th. B* (2014). [355, 414, 420]

Kotzig A., Contribution to the theory of Eulerian polyhedra. *Mat.-Fyz. Časopis. Slovensk. Akad. Vied* **5** (1955), 101–113. [411]

Kotzig A., Aus der Theorie der endlichen regulären Graphen dritten und vierten Grades. *Časopis Pěst. Mat.* **82** (1957), 76–92. [273]

Kotzig A., On the theory of finite graphs with a linear factor. I, II, III. *Mat.-Fyz. Časopis. Slovensk. Akad. Vied.* **9/10** (1959), 73–91, 136–159; 205–215.   [264, 276]

Kotzig A., On even regular graphs of the third degree. *Mat.-Fyz. Časopis Sloven. Akad. Vied* **16** (1966a), 72–75.   [227]

Kotzig A., 1-factorizations of Cartesian products of regular graphs. *J. Graph Th.* **3**(1979),23–34.   [371]

Kotzig A. and A. Rosa, Nearly Kirkman systems. In *Proc. 5th Southeastern Intl. Conf. Combin. Graph Th. Comput. (Boca Raton), Congr. Numer.* **10** (1974), 607–614.   [647]

Kouider M. and P.D. Vestergaard, Connected factors in graphs—a survey. *Graphs Combin.* **21** (2005), 1–26.   [327]

Kouril M., Computing the van der Waerden number $W(3, 4) = 293$. *Integers* **12** (2012), Paper No. A46, 13.   [465]

Kouril M. and J.L. Paul, The van der Waerden number $W(2, 6)$ is 1132. *Experiment. Math.* **17** (2008), 53–61.   [465]

Kovář P., M. Kubesa, and M. Meszka, Factorizations of complete graphs into brooms. *Discr. Math.* **312** (2012), 1084–1093.   [239]

Kövari T., V.T. Sós, and P. Turán, On a problem of K. Zarankiewicz. *Colloquium Math.* **3** (1954), 50–57.   [631, 672]

Kraitchik M., The Gambler's Ruin. *Mathematical Recreations* (W. W. Norton, 1942), 140.   [66]

Kratochvíl J., Z. Tuza, and M. Voigt, Brooks-type theorems for choosability with separation. *J. Graph Theory* **27** (1998), 43–49.   [403]

Krattenthaler C., Bijective proofs of the hook formulas for the number of standard Young tableaux, ordinary and shifted. *Electron. J. Combin.* **2** (1995), Research Paper 13.   [192]

Krattenthaler C., The enumeration of lattice paths with respect to their number of turns. In *Advances in combinatorial methods and applications to probability and statistics, Stat. Ind. Technol.* (Birkhäuser Boston, 1997), 29–58.   [50]

Krattenthaler C., Advanced determinant calculus. *Sém. Lothar. Combin.* **42** (1999), Art. B42q, 67 pp. (electronic). The Andrews Festschrift (Maratea, 1998).   [170]

Kratzke T., B. Reznick, and D.B. West, Eigensharp graphs: decomposition into complete bipartite subgraphs. *Trans. Amer. Math. Soc.* **308** (1988), 637–653.   [771]

Kriz I., A hypergraph-free construction of highly chromatic graphs without short cycles. *Combinatorica* **9** (1989), 227–229.   [341, 434]

Król M., On a sufficient and necessary condition of 3-colorableness for the planar graphs. I, II. *Prace Nauk. Inst. Mat. Fiz. Teoret. Politechn. Wrocław. Ser. Studia i Materiały* (1972), 37–40.   [419]

Kruskal J.B., On the shortest spanning subtree of a graph and the traveling salesman problem. *Proc. Am. Math. Soc.* **7** (1956), 48–50.   [245]

Kruskal J.B., The number of simplices in a complex. In *Mathematical optimization techniques* (Univ. California Press, 1963), 251–278.   [495, 496]

Kruskal J.B., Jr., Monotonic subsequences. *Proc. Amer. Math. Soc.* **4** (1953), 264–274.   [431]

Kubale M. (ed.), *Graph colorings, Contemporary Mathematics* **352** (AMS, 2004).   [335]

Kuczma M.S., A multi-well problem for phase transformations. In *The mathematics of finite elements and applications, X, MAFELAP 1999 (Uxbridge)* (Elsevier, 2000), 271–282.   [24]

Kühn D. and D. Osthus, Embedding large subgraphs into dense graphs. In *Surveys in combinatorics 2009, London Math. Soc. Lecture Note Ser.* **365** (Cambridge Univ. Press, 2009), 137–167.   [488]

Kühn D. and D. Osthus, Hamilton cycles in graphs and hypergraphs: an extremal perspective. In *Proc. of the Int. Congr. Math. IV (Seoul, 2014)* (Kyung Moon Sa, 2014), 381–406.   [316]

Kuhn H.W., The Hungarian method for the assignment problem. *Naval Research Logistics Quarterly* **2** (1955), 83–97.   [280]

Kündgen A., Art galleries with interior walls. *Discrete Comput. Geom.* **22** (1999), 249–258.   [419]

Kündgen A. and R. Ramamurthi, Coloring face-hypergraphs of graphs on surfaces. *J. Combin. Th. B* **85** (2002), 307–337.   [287]

Kuo E.H., Applications of graphical condensation for enumerating matchings and tilings. *Theoret. Comput. Sci.* **319** (2004), 29–57.   [34]

Kuperberg G., Symmetries of plane partitions and the permanent-determinant method. *J. Combin. Th. A* **68** (1994), 115–151. [756]

Kuperberg G., An exploration of the permanent-determinant method. *Electron. J. Combin.* **5** (1998), Research Paper 46. [756]

Kuperberg G., Kasteleyn cokernels. *Electron. J. Combin.* **9** (2002), Research Paper 29, 30. [34]

Kupka J., Problem E3402. *Amer. Math. Monthly* **97** (1990), 612. Solution **99** (1992), 367. [49]

Kuratowski K., Sur le problème des courbes gauches en topologie. *Fund. Math.* **15** (1930), 271–283. [390]

Labelle G., Une nouvelle démonstration combinatoire des formules d'inversion de Lagrange. *Adv. in Math.* **42** (1981), 217–247. [129]

Lagrange J.L., Nouvelle méthode pour résoudre des équations littérales par le moyen des séries. *Mém. Acad. Roy. des Sci. et Belles-Lettres de Berlin* **24** (1770). [129]

Lam C.W.H., L. Thiel, and S. Swiercz, The nonexistence of finite projective planes of order 10. *Canad. J. Math.* **41** (1989), 1117–1123. [612]

Lamé G., Note sur la limite du nombre des divisions dans la recherche du plus grand commun diviseur entre deux nombres entiers. *C. R. Acad. Sci. Paris* **19** (1844), 867–870. [67]

Landau E., *Handbuch der Lehre von der Verteilung der Primzahlen (German)* (Teubner, 1909). [10]

Landau H.G., On dominance relations and the structure of animal societies, III: The condition for score structure. *Bull. Math. Biophys.* **15** (1953), 143–148. [225, 228, 261]

Lander E.S., Symmetric designs and self-dual codes. *J. London Math. Soc. (2)* **24** (1981), 193–204. [623]

Lander E.S., *Symmetric designs: an algebraic approach*, London Mathematical Society Lecture Note Series **74** (Cambridge Univ. Press, 1983). [637]

Landman B.M. and A. Robertson, *Ramsey theory on the integers (2nd ed.)*, Student Mathematical Library **73** (AMS, 2014). [425]

Larman D.G., J. Matoušek, J. Pach, and J. Törőcsik, A Ramsey-type result for convex sets. *Bull. London Math. Soc.* **26** (1994), 132–136. [461]

Larman D.G. and C.A. Rogers, The realization of distances within sets in Euclidean space. *Mathematika* **19** (1972), 1–24. [729]

Larman D.G., C.A. Rogers, and J.J. Seidel, On two-distance sets in Euclidean space. *Bull. London Math. Soc.* **9** (1977), 261–267. [724]

Las Vergnas M., Sur l'existence de cycles hamiltoniens dans un graphe. *C. R. Acad. Sci. Paris Sér. A-B* **270** (1970), A1361–A1364. [333]

Las Vergnas M., A note on matchings in graphs. *Cahiers Centre Études Recherche Opér.* **17** (1975), 257–260. [272]

Lassak M., An estimate concerning Borsuk partition problem. *Bull. Acad. Polon. Sci. Sér. Sci. Math.* **30** (1982), 449–451 (1983). [817]

Lavallée I., Note sur le problème des Tours de Hanoï. *Acta Math. Vietnam.* **7** (1982), 131–137 (1984). [80]

Lawler E.L., *Combinatorial Optimization: Networks and Matroids* (Holt, Rinehart, and Winston, 1976). [282, 532]

Lawrence J., Covering the vertex set of a graph with subgraphs of smaller degree. *Discr. Math.* **21** (1978), 61–68. [355]

Lawrence S.L., Cycle-star Ramsey numbers. *Notices Amer. Math. Soc.* **20** (1973), A–420 (Notice #73T-157). [460]

Laywine C.F. and G.L. Mullen, *Discrete mathematics using Latin squares*, Wiley-Interscience Series in Discr. Mathematics and Optimization (Wiley & Sons, 1998). [623, 656]

Lazarson T., *Independence functions in algebra* (U. London, 1957). Thesis. [537]

Lebesgue H., Quelques conséquences simples de la formule d'Euler. *J. Math. Pures Appl.* **19** (1940), 27–43. [422]

Lederberg J., Systematics of organic molecules, graph topology and Hamiltonian circuits (Instrumentation Res. Lab. Rept.). Tech. Rep. 1040, Stanford Univ. (1966). [418]

Lee C., Lecture notes for Extremal Combinatorics (Spring 2015 course at mit) (2015). Available online
    at http://math.mit.edu/~cb_lee/18.318/materials.html.   [488, 672]

Lee C., Ramsey numbers of degenerate graphs (2017). (arXiv:1505.04773).   [484]

Lefmann H., V. Rödl, and R. Thomas, Monochromatic vs multicolored paths. *Graphs Combin.* **8** (1992),
    323–332.   [471]

Lefschetz S., *Introduction to Topology, Princeton Mathematical Series, vol. 11* (Princeton Univ. Press,
    1949).   [807]

Lehman A., A solution of the Shannon switching game. *J. Soc. Indust. Appl. Math.* **12** (1964), 687–725.
                                                                                     [528, 539]

Lehmer D.H., Lacunary recurrence formulas for the numbers of Bernoulli and Euler. *Ann. of Math. (2)*
    36 (1935), 637–649.   [65]

Leighton F.T., *Complexity Issues in VLSI: optimal layouts for the shuffle-exchange graph and other net-
    works, Foundations of Computing* (MIT Press, 1983).   [793]

Lekkerkerker C.G., Representation of natural numbers as a sum of Fibonacci numbers. *Simon Stevin*
    29 (1952), 190–195.   [61]

LeSaulnier T., Finding a repeated difference (solution to problem 11084). *Amer. Math. Monthly* **113**
    (2006), 371–372. Proposed **111** (2004), 440.   [441]

Lesniak L.M., Hamiltonicity in some special classes of graphs. *Congr. Numer.* **116** (1996), 53–70.
                                                                                          [316]

Leuck D.H., Solution to problem E3057. *Amer. Math. Monthly* **94** (1987), 187–188. Proposed **91** (1984),
    515.   [175]

Li H., Generalizations of Dirac's theorem in Hamiltonian graph theory—a survey. *Discr. Math.* **313**
    (2013), 2034–2053.   [316]

Li Q., M.Q. Tao, and Y.Q. Shen, The bandwidth of the discrete tori $C_m \times C_n$. *J. China Univ. Sci. Tech.*
    11 (1981), 1–16.   [801, 819]

Li S.Y.R. and W.C.W. Li, Independence numbers of graphs and generators of ideals. *Combinatorica* **1**
    (1981), 55–61.   [740]

Li X.L., The connectivity of path graphs. In *Combinatorics, graph theory, algorithms and applications
    (Beijing, 1993)* (World Sci. Publ., 1994), 187–192.   [312]

Lick D.R., Characterizations of n-connected and n-line-connected graphs. *J. Combin. Th. B* **14** (1973),
    122–124.   [312]

Liggett T.M., Extensions of the Erdős–Ko–Rado theorem and a statistical application. *J. Combinatorial
    Th. A* **23** (1977), 15–21.   [511]

Lih K.W., C.Y. Lin, and L.D. Tong, On an interpolation property of outerplanar graphs. *Discrete Appl.
    Math.* **154** (2006), 166–172.   [287]

Lindner C.C. and C.A. Rodger, *Design Theory* (CRC Press, 1997). Also 2009.   [609, 641]

Lindström B., On a combinatorial problem in number theory. *Canad. Math. Bull.* **8** (1965), 477–490.
                                                                                          [513]

Lindström B., On the vector representations of induced matroids. *Bull. London Math. Soc.* **5** (1973), 85–
    90.   [165, 167, 169]

Lindström B., A partition of $L(3, n)$ into saturated symmetric chains. *European J. Combin.* **1** (1980),
    61–63.   [557]

Linial N., A lower bound for the circumference of a graph. *Discr. Math.* **15** (1976), 297–300.   [323]

Linial N., Extending the Greene–Kleitman theorem to directed graphs. *J. Combin. Th. A* **30** (1981), 331–
    334.   [549]

Linial N., A new derivation of the counting formula for Young tableaux. *J. Combin. Theory A* **33** (1982),
    340–342.   [207]

Linial N., The information-theoretic bound is good for merging. *SIAM J. Comput.* **13** (1984), 795–801.
                                                                                          [823]

Liouville B., Sur la connectivité des produits de graphes. *C. R. Acad. Sci. Paris Sér. A-B* **286** (1978),
    A363–A365.   [292]

Lipski W., Jr., On strings containing all subsets as substrings. *Discr. Math.* **21** (1978), 253–259.   [566]

Lipton R.J. and R.E. Tarjan, A separator theorem for planar graphs. *SIAM J. Appl. Math.* **36** (1979), 177–189.   [394–396, 399]

Lipton R.J. and R.E. Tarjan, Applications of a planar separator theorem. *SIAM J. Comput.* **9** (1980), 615–627.   [399]

Littlewood D.E., *The Theory of Group Characters and Matrix Representations of Groups* (Oxford Univ. Press, 1940). Also 1950.   [207]

Liu B., The theorem on partition of connected graph and its applications in graphical enumeration. *J. South China Normal Univ. (Natural Science)* **1** (1985), 51–56.   [236]

Liu C.L., *Topics in combinatorial mathematics* (MAA, 1972). MAA Summer Seminar lect. notes.   [17]

Liu H., R. Morris, and N. Prince, Highly connected monochromatic subgraphs of multicolored graphs. *J. Graph Th.* **61** (2009), 22–44.   [460]

Liu J. and H. Zhou, Maximum induced matchings in graphs. *Discr. Math.* **170** (1997), 277–281.   [262]

Locke S.C., Problem 10447. *Amer. Math. Monthly* **102** (1995), 360. Solution **104** (1997), 976.   [229]

Locke S.C., Problem 10892. *Amer. Math. Monthly* **108** (2001), 668. Solution **110** (2003), 440–1.   [330]

Logan B.F. and L.A. Shepp, A variational problem for random Young tableaux. *Advances in Math.* **26** (1977), 206–222.   [722, 821]

Long C.T., On the Moessner theorem on integral powers. *Am. Math. Monthly* **73** (1966), 846–851.   [62]

Long C.T., Strike it out: Add it up. *Math. Gazette* **66** (1982), 273–277.   [62]

Long C.T., A note on Moessner's process. *Fibonacci Quart.* **24** (1986), 349–355.   [62]

Longyear J.Q. and T.D. Parsons, The friendship theorem. *Nederl. Akad. Wetensch. Proc. A* **75**=*Indag. Math.* **34** (1972), 257–262.   [776]

Loomis L.H. and H. Whitney, An inequality related to the isoperimetric inequality. *Bull. Amer. Math. Soc* **55** (1949), 961–962.   [513]

Lossers O.P. and R.S. Pinkham, Noncrossing trees, solution I to problem e3170. *Amer. Math. Monthly* **96** (1989), 359–361. Proposed by Howard University Group, **93** (1986), 650.   [129]

Lovász L., On graphs not containing independent circuits. *Mat. Lapok* **16** (1965), 289–299.   [272]

Lovász L., On decomposition of graphs. *Studia Sci. Math. Hungar.* **1** (1966), 237–238.   [249, 262, 355]

Lovász L., On covering of graphs. In *Theory of Graphs (Tihany, 1966)* (P. Erdős and G. Katona, eds.) (Academic Press, 1968a), 231–236.   [232]

Lovász L., On chromatic number of finite set-systems. *Acta Math. Acad. Sci. Hung.* **19** (1968b), 59–67.   [434, 436]

Lovász L., Normal hypergraphs and the perfect graph conjecture. *Discr. Math.* **2** (1972a), 253–267.   [368, 369, 375, 376]

Lovász L., A characterization of perfect graphs. *J. Combin. Th. B* **13** (1972b), 95–98.   [369]

Lovász L., On the structure of factorizable graphs. *Acta Math. Hungarica* **23** (1972c), 179–195.   [276]

Lovász L., On the sieve formula. *Mat. Lapok* **23** (1972d), 53–69 (1973).   [477]

Lovász L., Problem 5. *Period. Math. Hungar.* **4** (1974), 82.   [308, 310]

Lovász L., Three short proofs in graph theory. *J. Combin. Th. B* **19** (1975), 269–271.   [263, 264]

Lovász L., A homology theory for spanning trees of a graph. *Acta Math. Acad. Sci. Hungar.* **30** (1977), 241–251.   [312]

Lovász L., Kneser's conjecture, chromatic number, and homotopy. *J. Combin. Th. A* **25** (1978), 319–324.   [342, 532, 812]

Lovász L., *Combinatorial problems and exercises* (North-Holland, 1979).
                                                                                       [47, 104, 495, 496, 510, 762, 764]

Lovász L., *Combinatorial problems and exercises, 2nd ed.* (North-Holland, 1993).   [419, 491]

Lovász L., J. Nešetřil, and A. Pultr, On a product dimension of graphs. *J. Combin. Th. B* **28** (1980), 47–67.   [511, 726]

Lovász L. and M.D. Plummer, *Matching Theory, Ann. Discrete Math.* **29** (Akademiai Kiado and North-Holland, 1986).   [262, 275, 756, 764]

Lovász L. and M. Simonovits, On the number of complete subgraphs of a graph. In *Proc. 5th British Combin. Conf. (Univ. Aberdeen, 1975)*, **15** (Utilitas Math., 1976), 431–441.   [478]

Lovász L. and M. Simonovits, On the number of complete subgraphs of a graph. II. In *Studies in pure mathematics* (Birkhäuser, 1983), 459–495.   [478]

Lovász L. and P. Winkler, A note on the last new vertex visited by a random walk. *J. Graph Theory* **17** (1993), 593–596.   [668]

Lu X., A Chvátal–Erdős type condition for Hamiltonian graphs. *J. Graph Th.* **18** (1994), 791–800.
                                                                                                  [322]

Lu X., D.W. Wang, and C.K. Wong, On avoidable and unavoidable claws. *Discr. Math.* **184** (1998), 259–265.   [566]

Lubell D., A short proof of Sperner's lemma. *J. Combinatorial Theory* **1** (1966), 299.   [558]

Lubell D., Problem 10992. *Amer. Math. Monthly* **110** (2003), 155. Solution **111** (2004), 827–829.   [34]

Lubotzky A., R. Phillips, and P. Sarnak, Explicit expanders and the Ramanujan conjectures. In *Proc. 18th ACM Symp. Th. Comp.* (Assoc. Comput. Mach., 1986), 240–246.   [780]

Lubotzky A., R. Phillips, and P. Sarnak, Ramanujan graphs. *Combinatorica* **8** (1988), 261–277.
                                                                                             [341, 434]

Lucas F.E.A., Sur les congruences des nombres eulériens et les coefficients différentiels des functions trigonométriques suivant un module premier. *Bull. Soc. Math. France* **6** (1878), 49–54.   [47]

Lucas F.E.A., Note sur les intersections de trois quadriques. *Bull. Soc. Math. France* **19** (1891), 118–119.
                                                                                                  [162]

Lucas F.E.A., Sur les polygones inscrits dans les coniques. *Bull. Soc. Math. France* **20** (1892), 33–34.
                                                                                                  [238]

Luce R.D., Semiorders and a theory of utility discrimination. *Econometrica* **24** (1956), 178–191.   [585]

Luo R. and C.Q. Zhang, Edge coloring of graphs with small average degrees. *Discr. Math.* **275** (2004), 207–218.   [423]

Lützen J., G. Sabidussi, and B. Toft, Julius Petersen 1839–1910. A biography. *Discr. Math.* **100** (1992), 5–82.   [740]

Lyusternik L.A. and L. Shnirel'man, *Topological Methods in the Calculus of Variations (Russian)* (Moscow, 1930).   [810]

Ma M. and D.B. West, Problem 11731. *Amer. Math. Monthly* **120** (2013), 755. Solution **122** (2015), 807.
                                                                                                  [371]

Mabry R., Bipartite graphs and the Four-color Theorem. *Bull. ICA* **14** (1995), 119–112.   [417]

MacLane S., Some interpretations of abstract linear dependence in terms of projective geometry. *Amer. J. Math.* **58** (1936), 236–240.   [514, 529]

MacMahon P.A., *Combinatory analysis, I and II* (Cambridge Univ. Press, 1916). Reprinted Chelsea, 1960; Dover, 2004.   [45, 149, 169, 178, 207]

MacNeish H.F., Euler squares. *Ann. of Math. (2)* **23** (1922), 221–227.   [611]

Maddox R., The superregular graphs (solution to Problem 6617). *Amer. Math. Monthly* **103** (1996), 600–603. Proposed **96** (1989), 942.   [782]

Madej T. and D.B. West, The interval inclusion number of a partially ordered set. *Discr. Math.* **88** (1991), 259–277.   [832]

Mader W., Homomorphieeigenschaften und mittlere Kantendichte von Graphen. *Math. Ann.* **174** (1967), 265–268.   [350]

Mader W., Homomorphiesätze für Graphen. *Math. Ann.* **178** (1968), 154–168.   [710]

Mader W., Minimale *n*-fach kantenzusammenhängende Graphen. *Math. Ann.* **191** (1971), 21–28.
                                                                                                  [309]

Mader W., Existenz *n*-fach zusammenhängender Teilgraphen in Graphen genügend grosser Kantendichte. *Abh. Math. Sem. Univ. Hamburg* **37** (1972), 86–97.   [296, 310, 311, 315, 316]

Mader W., 1-Faktoren von Graphen. *Math. Ann.* **201** (1973), 269–282.   [264]

Mader W., A reduction method for edge-connectivity in graphs. *Ann. Discr. Math.* **3** (1978), 145–164.
                                                                                             [308, 310]

Mader W., Zur Struktur minimal *n*-fach zusammenhängender Graphen. *Abh. Math. Sem. Univ. Hamburg* **49** (1979), 49–69.   [311, 316]

Mader W., On vertices of degree n in minimally n-connected graphs and digraphs. In *Combinatorics, Paul Erdős is eighty (Keszthely, 1993)*, Bolyai Soc. Math. Stud. (János Bolyai Math. Soc., 1996), 423–449.   [311]

Mader W., $3n - 5$ edges do force a subdivision of K_5. *Combinatorica* **18** (1998), 569–595.   [351, 353]

Mahmoodian S.E., On edge-colorability of Cartesian products of graphs. *Canad. Math. Bull.* **24** (1981), 107–108.   [371]

Maillet E., Contributions à la théorie des groupes. *TIM (9)* **6** (1894), 258–280.   [623]

Majumdar K.N., On some theorems in combinatorics relating to incomplete block designs. *Ann. Math. Statistics* **24** (1953), 377–389.   [730]

Mann H.B., On orthogonal Latin squares. *Bull. Amer. Math. Soc.* **50** (1950), 249–257.   [623]

Mann H.B. and H.J. Ryser, Systems of distinct representatives. *Amer. Math. Monthly* **60** (1953), 397–401.   [254]

Mantel W., Problem 28, soln. by H. Gouwentak, W. Mantel, J. Teixeira de Mattes, F. Schuh and W.A. Wythoff. *Wiskundige Opgaven* **10** (1907), 60–61.   [224, 475]

Marcus A. and G. Tardos, Excluded permutation matrices and the Stanley–Wilf conjecture. *J. Combin. Th. A* **107** (2004), 153–160.   [432, 433]

Marcus M. and M. Newman, On the minimum of the permanent of a doubly stochastic matrix. *Duke Math. J.* **26** (1959), 61–72.   [765]

Marcus M. and R. Ree, Diagonals of doubly stochastic matrices. *Quart. J. Math.* **2** (1959), 295–302.
                                                                                                      [257]

Margulis G.A., Explicit constructions of concentrators. *Problems of Information Transmission* **9** (1973), 325–332.   [779]

Margulis G.A., Explicit group-theoretic constructions of combinatorial schemes and their applications in the construction of expanders and concentrators (Russian). *Problems of Information Transmission* **24** (1988), 39–46.   [780]

Marica J., Orthogonal families of sets. *Canad. Math. Bull.* **14** (1971), 573.   [512]

Martin M.H., A problem in arrangements. *Bull. Amer. Math. Soc.* **40** (1934), 859–864.   [763]

Martin R., Notes on Extremal Graph Theory (2012), 101. Available online at
    http://orion.math.iastate.edu/rymartin/ISU608EGT/EGTbook.pdf.   [488]

Martinov N., Uncontractible 4-connected graphs. *J. Graph Th.* **6** (1982), 343–344.   [313]

Massey J.L., On the fractional weight of distinct binary n-tuples. *IEEE Trans. Information Theory* **IT-20** (1974), 131.   [507]

Matoušek J., *Using the Borsuk–Ulam theorem*, Universitext (Springer-Verlag, 2003).   [806, 810, 812]

Matoušek J., A combinatorial proof of Kneser's conjecture. *Combinatorica* **24** (2004), 163–170.   [812]

Matoušek J., *Thirty-three miniatures*, Student Mathematical Library **53** (AMS, 2010).   [723]

Matoušek J. and G.M. Ziegler, Topological lower bounds for the chromatic number: a hierarchy. *Jahresber. Deutsch. Math.-Verein.* **106** (2004), 71–90.   [812, 813]

Matsko V.J., D.B. West, and J.E. Wetzel, Trifold arrangements and cevian dissections. *J. Geom.* **72** (2001), 115–127.   [565]

Matthews M.M. and D.P. Sumner, Hamiltonian results in $K_{1,3}$-free graphs. *J. Graph Th.* **8** (1984), 139–146.   [326, 330]

Matula D.W., The cohesive strength of graphs. In *The Many Facets of Graph Theory (Proc. Conf., Western Mich. Univ., Kalamazoo, Mich., 1968)* (Springer, 1969), 215–221.   [353]

Matula D.W., The employee party problem. *Notices Amer. Math. Soc.* **19** (1972), A–382.   [700]

Matula D.W., An extension of Brooks' Theorem. Tech. Rep. 69, Center for Numerical Analysis, University of Texas–Austin (1973).   [354]

Matula D.W., Ramsey theory for graph connectivity. *J. Graph Theory* **7** (1983), 95–103.   [297]

Matula D.W. and L. Kučera, An expose-and-merge algorithm and the chromatic number of a random graph. In *Random graphs '87 (Poznań, 1987)* (Wiley, 1990), 175–187.   [718]

Maunsell F.G., A note on Tutte's paper "The factorization of linear graphs.". *J. London Math. Soc.* **27** (1952), 127–128.   [264]

Maurer R., E2404. *Amer. Math. Monthly* **80** (1973), 316. Proposed **81** (1974), 287.   [106]

Maurer S., The king chicken theorems. *Math. Mag.* **53** (1980), 67–80.   [228]

Maurer S. and I. Rabinovitch, Large minimal realizers of a partial order. *Proc. Amer. Math. Soc.* **66** (1977), 211–216.   [584]

Maurer S., I. Rabinovitch, and W.T. Trotter, Large minimal realizers of a partial order II. *Discr. Math.* **31** (1980), 297–314.   [228]

McConnell R.M. and J.P. Spinrad, Modular decomposition and transitive orientation. *Discr. Math.* **201** (1999), 189–241.   [573]

McDiarmid C.J.H., The solution of a timetabling problem. *J. Inst. Math. Appl.* **9** (1972), 23–34.   [372]

McDiarmid C.J.H., Concentration. In *Probabilistic methods for algorithmic discrete mathematics*, *Algorithms Combin.* **16** (Springer, 1998), 195–248.   [706, 719, 721]

McFarland R. and H.B. Mann, On multipliers of difference sets. *Canad. J. Math.* **17** (1965), 541–542.   [637]

McKay B.D., Asymptotics for symmetric 0-1 matrices with prescribed row sums. *Ars Combin.* **19** (1985), 15–25.   [697]

McKay B.D. and S.P. Radziszowski, The first classical Ramsey number for hypergraphs is computed. In *2nd Symp. Disc. Alg. (San Francisco)*, ACM-SIAM (1991), 304–308.   [449]

McKay B.D. and S.P. Radziszowski, $R(4, 5) = 25$. *J. Graph Th.* **19** (1995), 309–322.   [449]

McKay B.D. and N.C. Wormald, Asymptotic enumeration by degree sequence of graphs with degrees $o(n^{1/2})$. *Combinatorica* **11** (1991), 369–382.   [697]

McKay B.D. and K.M. Zhang, The value of the Ramsey number $R(3, 8)$. *J. Graph Th.* **16** (1992), 99–105.   [449]

McKee T.A., Recharacterizing Eulerian: intimations of new duality. *Discr. Math.* **51** (1984), 327–242.   [239]

McKee T.A., How chordal graphs work. *Bull. ICA* **9** (1993), 27–39.   [374]

McSorley J.P., Counting structures in the Möbius ladder. *Discr. Math.* **184** (1998), 137–164.   [765]

Melham R.S. and A.G. Shannon, A generalization of the Catalan identity and some consequences. *Fibonacci Quart.* **33** (1995), 82–84.   [60]

Menger K., Zur allgemeinen Kurventheorie. *Fund. Math.* **10** (1927), 95–115.   [298, 300]

Merca M., Problem 11767. *Amer. Math. Monthly* **121** (2014a), 267. Solution **123** (2016), 505–506.   [150]

Merca M., Problem 11772. *Amer. Math. Monthly* **121** (2014b), 366. Solution **123** (2016), 614.   [148]

Merris R., Laplacian matrices of graphs: a survey. **197/198** (1994), 143–176.   [776]

Merris R., A survey of graph Laplacians. *Linear and Multilinear Algebra* **39** (1995), 19–31.   [776]

Merris R., Laplacian graph eigenvectors. *Linear Algebra Appl.* **278** (1998), 221–236.   [782]

Meshalkin L.D., Generalization of Sperner's theorem on the number of subsets of a finite set (Russian). *Th. Prob. Appl.* **8** (1963), 203–204.   [558]

Meyniel H., On the perfect graph conjecture. *Discr. Math.* **16** (1976), 339–342.   [375]

Miao L. and Q. Sun, On the size of critical graphs with maximum degree 8. *Discr. Math.* **310** (2010), 2215–2218.   [414]

Micali S. and V.V. Vazirani, An $O(\sqrt{|V|} \cdot |E|)$ algorithm for finding maximum matching in general graphs. In *Proc. 21th IEEE Symp. Found. Comp. Sci.* (Assoc. Comput. Mach., 1980), 17–27.   [284]

Mihók P., On vertex partition numbers of graphs. In *Graphs and Other Combin. Topics (Prague, 1982)*, *Teubner-Texte Math.* **59** (Teubner, 1983), 183–8.   [418]

Mikola M., The Lucas number as the number of spanning trees. *Práce Štúd. Vysokej Školy Doprav. Žiline Sér. Mat.-Fyz.* **2** (1980), 69–77.   [65]

Milans K.G., D. Stolee, and D.B. West, Ordered Ramsey theory and track representations of graphs. *J. Comb.* **6** (2015), 445–456.   [602, 608]

Miller E., Planar graphs as minimal resolutions of trivariate monomial ideals. *Doc. Math.* **7** (2002), 43–90.   [784]

Miller Z., The bandwidth of caterpillar graphs. In *Proc. 12th Southeastern Intl. Conf. Combin. Graph Th. Comput. (Baton Rouge)*, *Congr. Numer.* **33** (1981), 235–252.   [819]

Mills G., A quintessential proof of van der Waerden's theorem on arithmetic progressions. *Discr. Math.* **47** (1983), 117–120.   [463]

Minty G.J., A theorem on n-coloring the points of a linear graph. *Amer. Math. Monthly* **69** (1962), 623–624.   [338]

Minty G.J., On the axiomatic foundations of the theories of directed linear graphs, electrical networks and network-programming. *J. Math. Mech.* **15** (1966), 485–520.   [539]

Mirkin B.G., Description of some relations on the set of real-line intervals. *J. Mathematical Psychology* **9** (1972), 243–252.   [586]

Mirsky L., *Transversal theory. Mathematics in Science and Engineering* **75** (Academic Press, 1971).   [547]

Mirsky L. and H. Perfect, Applications of the notion of independence to combinatorial analysis. *J. Combin. Th.* **2** (1967), 327–357.   [517]

Mirzakhani M., A small non-4-choosable planar graph. *Bull. Inst. Combin. Appl.* **17** (1996), 15–18.   [402, 421]

Mitas J., Minimal representation of semiorders with intervals of same length. In *Orders, algorithms, and applications (Lyon, 1994)*, *Lect. Notes Comp. Sci.* **831** (Springer, 1994), 162–175.   [587]

Mitchem J. and E.F. Schmeichel, Pancyclic and bipancyclic graphs—a survey. In *Graphs and applications (Boulder, Colo., 1982)*, *Wiley-Intersci. Publ.* (Wiley, 1985), 271–278.   [326]

Mitzenmacher M. and E. Upfal, *Probability and computing* (Cambridge University Press, Cambridge, 2017), second edn.   [708]

Moessner A., Eine Bemerkung über die Potenzen der natürlichen Zahlen. *S.-B. Math.-Nat. Kl. Bayer. Akad. Wiss.* **1951** (1951), 29 (1952).   [62]

Moews D., Optimally pebbling hypercubes and powers. *Discr. Math.* **190** (1998), 271–276.   [668]

Mohanty S.G., *Lattice path counting and applications* (Academic Press, 1979).   [50]

Mohar B., The Laplacian spectrum of graphs. In *Graph theory, combinatorics, and applications. Vol. 2 (Kalamazoo, MI, 1988)*, *Wiley-Intersci. Publ.* (Wiley, 1991), 871–898.   [776]

Mohar B., Laplace eigenvalues of graphs—a survey. **109** (1992), 171–183.   [776]

Mohar B. and S. Poljak, Eigenvalues in combinatorial optimization. In *Combinatorial and graph-theoretical problems in linear algebra (Minneapolis, MN, 1991)*, *IMA Vol. Math. Appl.* **50** (Springer, 1993), 107–151.   [776]

Mohar B. and C. Thomassen, *Graphs on surfaces*, *Johns Hopkins Studies in the Mathematical Sciences* (Johns Hopkins Univ. Press, 2001).   [377]

Möhring R.H., Algorithmic aspects of comparability graphs and interval graphs. In *Graphs and order (Banff, 1984)*, *NATO Adv. Sci. Inst. C Math. Phys. Sci.* **147** (Reidel, 1985), 41–101.   [573]

Molloy M., The probabilistic method. In *Probabilistic methods for algorithmic discrete mathematics*, *Algorithms Combin.* **16** (Springer, 1998), 1–35.   [706]

Molloy M. and B.A. Reed, A bound on the strong chromatic index of a graph. *J. Combin. Th. B* **69** (1997), 103–109.   [373]

Molloy M. and B.A. Reed, Near-optimal list colorings. In *Proceedings of the Ninth International Conference "Random Structures and Algorithms" (Poznan, 1999)*, **17** (2000), 376–402.   [362]

Molloy M. and B.A. Reed, *Graph colouring and the probabilistic method*, *Algorithms and Combinatorics* **23** (Springer-Verlag, 2002).   [335, 657, 668, 674, 677, 681, 686, 702, 706, 721, 722]

Montágh B., A simple proof and a generalization of an old result of Chung and Feller. *Discr. Math.* **87** (1991), 105–108.   [50]

Montellano-Ballesteros J.J. and V. Neumann-Lara, An anti-Ramsey theorem on cycles. *Graphs Combin.* **21** (2005), 343–354.   [457]

Montgomery B., *Dynamic coloring of graphs* (ProQuest LLC, 2001). Ph.D. Thesis, West Virginia University.   [424]

Montgomery R., A. Pokrovskiy, and B. Sudakov, A proof of Ringel's conjecture (2020). (arXiv:2001.02665v2).   [646]

Montmort P.R., *Essay d'Analyse sur les Jeux de Hazard* (Paris, 1708).   [162]

Moon J.W., On the diameter of a graph. *Michigan Math. J.* **12** (1965a), 349–351.   [251]

Moon J.W., On a problem of Ore. *Math. Gaz.* **49** (1965b), 40–41.   [333]

Moon J.W., On the number of complete subgraphs of a graph. *Canad. Math. Bull.* **8** (1965c), 831–834.
[489]

Moon J.W., On subtournaments of a tournament. *Canad. Math. Bull.* **9** (1966), 297–301.   [334]

Moon J.W., Various proofs of Cayley's formula for counting trees. In *A seminar on Graph Theory* (Holt,
Rinehart and Winston, 1967), 70–78.   [37, 47]

Moon J.W., *Counting Labeled Trees* (Canad. Math. Congress, 1970).   [37]

Moon J.W. and L. Moser, On a problem of Turán. *Magyar Tud. Akad. Mat. Kutató Int. Közl.* **7** (1962a),
283–286.   [477]

Moon J.W. and L. Moser, Almost all tournaments are irreducible. *Canad. Math. Bull.* **5** (1962b), 61–65.
[703]

Moon J.W. and L. Moser, On Hamiltonian bipartite graphs. *Israel J. Math.* **1** (1963), 163–165.
[325, 332]

Moore B.R. and D.B. West, Cycles in color-critical graphs (2019). (arXiv:1912.03754v2).   [354]

Moore E.H., Concerning triple systems. *Math. Ann.* **43** (1893), 271–285.   [641]

Moore E.H., Tactical memoranda I–III. *Am. J. Math.* **18** (1896), 264–303.   [610, 611]

Morris W. and V. Soltan, The Erdős–Szekeres problem. In *Open problems in mathematics* (Springer,
[Cham], 2016), 351–375.   [446]

Morris W.D., Jr. and V. Soltan, The Erdős–Szekeres problem on points in convex position—a survey.
Bull. Amer. Math. Soc. (N.S.) **37** (2000), 437–458.   [446]

Moser L., Problem 4300. *Amer. Math. Monthly* **55** (1948), 369. Solution **57** (1950), 47.   [441]

Moser L., Problem B-6: Some reflections. *Fibonacci Quarterly* **1** (1963), 75–76.   [59]

Moser L., The second moment method in combinatorial analysis. In *Combin. Struct. Appl. (Proc. Calgary
Internat. Conf., 1969)* (Gordon and Breach, 1970), 283–284.   [513]

Moser R.A., A constructive proof of the Lovász local lemma. In *STOC'09—Proceedings of the 2009 ACM
International Symposium on Theory of Computing* (ACM, 2009), 343–350.   [680]

Moser R.A. and G. Tardos, A constructive proof of the general Lovász local lemma. *J. ACM* **57** (2010),
Art. 11, 15.   [680]

Moshkovitz G. and A. Shapira, Ramsey theory, integer partitions and a new proof of the Erdős–Szekeres
theorem. *Adv. Math.* **262** (2014), 1107–1129.   [602, 603]

Motwani R. and P. Raghavan, *Randomized algorithms* (Cambridge University Press, 1995).   [657]

Motzkin T.S. and E.G. Straus, Maxima for graphs and a new proof of a theorem of Turán. *Canad. J.
Math.* **17** (1965), 533–540.   [489, 781]

Mozhan N.N., Twice critical graphs with chromatic number five. *Metody Diskret. Analiz.* (1987), 50–59,
73.   [354]

Mubayi D., Generalizing the Ramsey problem through diameter. *Electron. J. Combin.* **9** (2002), Research
Paper 42, 10.   [460]

Mubayi D., Counting substructures I: color critical graphs. *Adv. Math.* **225** (2010), 2731–2740.   [478]

Mubayi D., Variants of the Erdős–Szekeres and Erdős–Hajnal Ramsey problems. *European J. Combin.*
62 (2017), 197–205.   [457]

Mubayi D. and A. Suk, Off-diagonal hypergraph Ramsey numbers. *J. Combin. Th. B* **125** (2017), 168–
177.   [451]

Mubayi D. and J. Verstraëte, A survey of Turán problems for expansions. In *Recent trends in combinatorics, IMA Vol. Math. Appl.* **159** (Springer, [Cham], 2016), 117–143.   [500]

Muir T., *A Treatise on the Theory of Determinants* (Macmillan, 1882). Also 1906.   [764]

Mulder H.M., Julius Petersen's theory of regular graphs. *Discr. Math.* **100** (1992), 157–175.   [253, 274]

Mullin R. and G.C. Rota, On the foundations of combinatorial theory. III. Theory of binomial enumeration. In *Graph Theory and its Applications (Proc. Advanced Sem., Madison, WI, 1969)* (Academic
Press, 1970), 167–213.   [123]

Munkres J., Algorithms for the assignment and transportation problems. *J. Soc. Indust. Appl. Math.* **5**
(1957), 32–38.   [280]

Mycielski J., Sur le coloriage des graphes. *Coll. Math.* **3** (1955), 161–162.   [340]

Myers B.R., On spanning trees, weighted compositions, Fibonacci numbers, and resistor networks. *SIAM Rev.* **17** (1975), 465–474.   [65]

Myers B.R. and R. Liu, A lower bound on the chromatic number of a graph. *Networks* **1** (1972), 273–277.
                                                                                                [354]

Naatz M., A connectivity lemma. In *6th International Conference on Graph Theory (Marseille, 2000)*, *Electron. Notes Discr. Math.* **5** (Elsevier, 2000), 4.   [312, 764]

Nagasawa T., K. Noguchi, and Y. Suzuki, Optimal 1-embedded graphs on the projective plane which triangulate other surfaces. *J. Nonlinear Convex Anal.* **19** (2018), 1759–1770.   [797]

Nagy Z., A certain constructive estimate of the Ramsey number. *Mat. Lapok* **23** (1972), 301–302.
                                                                                        [459, 728, 745]

Nakasawa T., Zur Axiomatik der linearen Abhängigkeit. I, II, III (in German). *Sci. Rep. Tokyo Bunrika Daigaku Sect. A* **2** (1935), 129–149. II and III in **3** (1936), 17–41 and 77–90.   [514]

Nakayama A. and B. Péroche, Linear arboricity of digraphs. *Networks* **17** (1987), 39–53.   [678]

Narayana T.V., A combinatorial problem and its application to probability theory. I. *J. Indian Soc. Agric. Statist.* **7** (1955), 169–178.   [45]

Nasar S., *A beautiful mind* (Simon & Schuster, 1998). Also 2001.   [803]

Nash-Williams C.St.J.A., On orientations, connectivity and odd-vertex-pairings in finite graphs. *Canad. J. Math.* **12** (1960), 555–567.   [307, 309]

Nash-Williams C.St.J.A., Edge-disjoint spanning trees in finite graphs. *J. Lond. Math. Soc.* **36** (1961), 445–450.   [251, 535, 540]

Nash-Williams C.St.J.A., Decomposition of finite graphs into forests. *J. Lond. Math. Soc.* **39** (1964), 12.
                                                                                            [251, 535]

Nash-Williams C.St.J.A., An application of matroids to graph theory. In *Theory of Graphs, Intl. Sympos., Rome* (Dunod, 1966), 263–265.   [533]

Nash-Williams C.St.J.A., An unsolved problem concerning decomposition of graphs into triangles. *Combinatorial Theory and Its Applications III* (1970), 1179–1183.   [238, 640]

Nash-Williams C.St.J.A., Connected detachments of graphs and generalized Euler trails. *J. London Math. Soc. (2)* **31** (1985), 17–29.   [540]

Naslund E. and W.F. Sawin, Upper bounds for sunflower-free sets (2016). (arXiv:1606.09575).   [504]

Nemhauser G.L. and L.A. Wolsey, *Integer and combinatorial optimization* (Wiley, 1988).   [520]

Nešetřil J., A combinatorial classic—sparse graphs with high chromatic number. In *Erdős Centennial*, *Bolyai Soc. Math. Stud.* **25** (János Bolyai Math. Soc., Budapest, 2013), 383–407.   [434]

Nešetřil J. and V. Rödl, The Ramsey property for graphs with forbidden complete subgraphs. *J. Combin. Th. B* **20** (1976), 243–249.   [455]

Nešetřil J. and V. Rödl, A short proof of the existence of highly chromatic hypergraphs without short cycles. *J. Combin. Th. B* **27** (1979), 225–227.   [434, 436, 437]

Nešetřil J. and V. Rödl, Complexity of diagrams. *Order* **3** (1987), 321–330. Corrigendum: *Order* **10** (1993), 393.   [543]

Nešetřil J. and V. Rödl (eds.), *Mathematics of Ramsey theory*, *Algorithms and Combinatorics* **5** (Springer-Verlag, 1990).   [425]

Netto E., *Lehrbuch der Combinatorik (German)* (Teubner, 1901). Also 1927.   [41]

Neumann-Lara V. and E. Rivera-Campo, Spanning trees with bounded degrees. *Combinatorica* **11** (1991), 55–61.   [327]

Nicoaescu L.I., Problem E3157. *Amer. Math. Monthly* **93** (1986), 482. Solution **95** (1988), 557. [22]

Niessen T. and B. Randerath, Regular factors of simple regular graphs and factor-spectra. *Discr. Math.* **185** (1998), 89–103.   [273]

Niessen T. and L. Volkmann, Class 1 conditions depending on the minimum degree and the number of vertices of maximum degree. *J. Graph Th.* **14** (1990), 225–246.   [359]

Nikiforov V., The cycle-complete graph Ramsey numbers. *Combin. Probab. Comput.* **14** (2005), 349–370.
                                                                                                  [454]

Nikiforov V., Bounds on graph eigenvalues. II. *Linear Algebra Appl.* **427** (2007), 183–189.   [770]

Nikiforov V., Graphs with many r-cliques have large complete r-partite subgraphs. *Bull. Lond. Math. Soc.* **40** (2008), 23–25.   [482]

Nikiforov V., A spectral Erdős–Stone–Bollobás theorem. *Combin. Probab. Comput.* **18** (2009), 455–458.   [770]

Nikiforov V., The number of cliques in graphs of given order and size. *Trans. Amer. Math. Soc.* **363** (2011), 1599–1618.   [478]

Nikšić F., Are surjections $[n] \to [k]$ more common than injections $[k] \to [n]$? Available online at www.mathoverflow.net/questions/268544 (2017). Solution by M. Wildon, July 2, 2017.   [133]

Nilli A., On the second eigenvalue of a graph. *Discr. Math.* **91** (1991), 207–210.   [780]

Nilli A., On Borsuk's problem. In *Jerusalem combinatorics '93, Contemp. Math.* **178** (AMS, 1994), 209–210.   [817, 818]

Nishimura H. and S. Kuroda, *A Lost Mathematician, Takeo Nakasawa: The Forgotten Father of Matroid Theory, Mathematics and Statistics* (Mirkhäuser, 2009).   [514]

Niu Y.Y. and B.W. Zhu, Connectivities of Cartesian products of graphs. In *Combinatorics, graph theory, algorithms and applications (Beijing, 1993)* (World Sci. Publ., 1994), 301–305.   [292]

Niven I., Formal power series. *Amer. Math. Monthly* **76** (1969), 871–889.   [107]

Noel J.A., B.A. Reed, and H. Wu, A proof of a conjecture of Ohba. *J. Graph Theory* **79** (2015), 86–102.   [355]

Noel J.A., D.B. West, H. Wu, and X. Zhu, Beyond Ohba's conjecture: a bound on the choice number of k-chromatic graphs with n vertices. *European J. Combin.* **43** (2015), 295–305.   [356]

Nordhaus E.A. and J.W. Gaddum, On complementary graphs. *Amer. Math. Monthly* **63** (1956), 175–177.   [344]

Norman R.Z. and M. Rabin, Algorithm for a minimal cover of a graph. *Proc. Amer. Math. Soc.* **10** (1959), 315–319.   [263]

Novelli J.C., I. Pak, and A.V. Stoyanovskii, A direct bijective proof of the hook-length formula. *Discr. Math. Theor. Comput. Sci.* **1** (1997), 53–67.   [192]

Nyblom M., Problem 11117. *Amer. Math. Monthly* **111** (2004), 915. Solution **113** (2006), 762.   [172]

O S. and D.B. West, Balloons, cut-edges, matchings, and total domination in regular graphs of odd degree. *J. Graph Theory* **64** (2010), 116–131.   [237, 274]

O S., D.B. West, and H. Wu, Longest cycles in k-connected graphs with given independence number. *J. Combin. Th. B* **101** (2011), 480–485.   [334]

O'Hara K.M., Unimodality of Gaussian coefficients: a constructive proof. *J. Combin. Th. A* **53** (1990), 29–52.   [557]

Ohba K., On chromatic-choosable graphs. *J. Graph Theory* **40** (2002), 130–135.   [355]

Ohtsuka H. and R. Tauraso, Problem 12016. *Amer. Math. Monthly* **125** (2018), 81. Solution **126** (2019), 665.   [33]

Olmsted C., Problem E3175. *Amer. Math. Monthly* **93** (1986), 732. Solution **96** (1989), 59–60.   [176]

Olson E.J. and B.E. Sagan, On the 1/3–2/3 conjecture. *Order* **35** (2018), 581–596.   [823]

Ore O., A problem regarding the tracing of graphs. *Elemente der Math.* **6** (1951), 49–53.   [250]

Ore O., Graphs and matching theorems. *Duke Math. J.* **22** (1955), 625–639.   [258, 531]

Ore O., Graphs and subgraphs. II. *Trans. Amer. Math. Soc.* **93** (1959), 185–204.   [332]

Ore O., Note on Hamilton circuits. *Amer. Math. Monthly* **67** (1960), 55.   [319]

Ore O., Arc coverings of graphs. *Ann. Mat. Pura Appl.* **55** (1961), 315–321.   [321]

Ore O., *Theory of graphs, Amer. Math. Soc. Colloquium Publications* **38** (Amer. Math. Soc., 1962).   [260, 571]

Ore O., Hamiltonian connected graphs. *J. Math. Pures Appl.* **42** (1963), 21–7.   [332, 333]

Ore O., *The four-colour problem* (Academic Press, 1967).   [355, 360, 387, 399, 420]

Ore O., Diameters in graphs. *J. Combinatorial Theory* **5** (1968), 75–81.   [249, 359]

O'Rourke J., *Art gallery theorems and algorithms, International Series of Monographs on Computer Science* (Oxford Univ. Press, 1987).   [394]

Owens A.B., On the planarity of regular incidence sequences. *J. Combinatorial Theory B* **11** (1971), 201–212.   [389]

Oxley J.G., *Matroid theory*, *Oxford Science Publications* (Oxford Univ. Press, 1992). 2nd edition 2011.
[514, 525]

Ozeki K. and T. Yamashita, Spanning trees: a survey. *Graphs Combin.* **27** (2011), 1–26.   [327, 334]

Paasche I., Eine Verallgemeinerung des Moessnerschen Satzes. *Compositio Math.* **12** (1956), 263–270.
[62]

Pach J. and P.K. Agarwal, *Combinatorial geometry*, *Wiley-Interscience Ser. Discr. Math. Optim.* (Wiley &
Sons, 1995).   [783]

Pach J. and M. Sharir, *Combinatorial geometry and its algorithmic applications*, *Mathematical Surveys
and Monographs* **152** (AMS, 2009).   [783]

Pach J. and G. Tardos, Forbidden paths and cycles in ordered graphs and matrices. *Israel J. Math.* **155**
(2006), 359–380.   [491]

Pach J. and G. Tóth, Graphs drawn with few crossings per edge. *Combinatorica* **17** (1997), 427–439.
[793, 797]

Pach J. and G. Tóth, Thirteen problems on crossing numbers. *Geombinatorics* **9** (2000a), 194–207.
[795]

Pach J. and G. Tóth, Which crossing number is it anyway? *J. Combin. Th. B* **80** (2000b), 225–246.
[795]

Pak I., Partition bijections, a survey. *Ramanujan J.* **12** (2006), 5–75.   [141, 142, 152]

Palacios J.L., Problem 11672. *Amer. Math. Monthly* **119** (2012), 800. Solutions by P. Condon and R.
Strong **121** (2014), 948–949.   [66]

Palacios J.L. and D.P. Sandell, Problem 6665. *Amer. Math. Monthly* **98** (1991), 655. Solution **100** (1993),
405–407.   [668]

Paley R.E.A.C., On orthogonal matrices. *J. Math. Phys.* **12** (1933), 311–320.   [621]

Palmer C., *Extremal Combinatorics* (2015). Lecture notes.   [484, 488]

Palmer E.M., *Graphical Evolution: An Introduction to the Theory of Random Graphs* (Wiley, 1985).
[657, 700, 703, 705]

Palumbíny D., On decompositions of complete graphs into factors with equal diameters. *Boll. Un. Mat.
Ital.(4)* **7** (1973), 420–428.   [249]

Papadimitriou C.H. and K. Steiglitz, *Combinatorial Optimization: Algorithms and Complexity* (Prentice
Hall, 1982). Also Dover, 1998.   [520]

Parker E.T., Orthogonal latin squares. *Proc. Nat. Acad. Sci. U.S.A.* **45** (1959), 859–862.   [652]

Paulraja P., A characterization of Hamiltonian prisms. *J. Graph Th.* **17** (1993), 161–171.   [326]

Payan C., Sur le nombre d'absorption d'un graphe simple. In *Colloque sur la Théorie des Graphes (Paris,
1974), Cahiers Centre Études Recherche Opér.* **17** (1975), 307–317.   [670]

Peart P. and W.J. Woan, A bijective proof of the Delannoy recurrence. In *Proc. 33rd Southeastern Intl.
Conf. Combin. Graph Th. Comput. (Boca Raton), Congr. Numer.* **158** (2002), 29–33.   [64]

Peck G.W., A new proof of a theorem of Graham and Pollak. *Discr. Math.* **49** (1984), 327–328.   [771]

Peck G.W., Noncrossing trees, solution II to problem e3170. *Amer. Math. Monthly* **96** (1989), 359–361.
Proposed by Howard University Group, **93** (1986), 650.   [42]

Peled U., Problem 10197. *Amer. Math. Monthly* **99** (1992), 162. Solution **100** (1993), 806–807.   [764]

Pelsmajer M.J., M. Schaefer, and D. Štefanovič, Odd crossing number is not crossing number. In *Graph
drawing, Lect. Notes Comp. Sci.* **3843** (Springer, 2006), 386–396.   [795]

Peltesohn R., Eine Lösung der beiden Heffterschen Differenzenprobleme. *Compositio Math.* **6** (1939),
251–257.   [645]

Penaud J.G., Une propriété de bicoloration des hypergraphes planaires. In *Colloque sur la Théorie des
Graphes (Paris, 1974), Cahiers Centre Études Recherche Opér.* **17** (1975), 345–349.   [418, 419]

Penrose M., *Random geometric graphs*, *Oxford Studies in Probability* **5** (Oxford Univ. Press, 2003).
[783]

Percus J.K., *Combinatorial methods*, *Notes recorded by Ora Engelberg Percus* (Courant Inst. Math. Sci.,
New York Univ., 1969).   [755]

Perkovic L. and B.A. Reed, Edge coloring regular graphs of high degree. In *Graphs Combinatorics (Marseille, 1995)* **165/166** (Discr. Math., 1997), 567–578.   [359]

Perles M.A., A proof of Dilworth's decomposition theorem for partially ordered sets. *Israel J. Math.* **1** (1963), 105–107.   [546]

Perron O., Beweis des Moessnerschen Satzes. *S.-B. Math.-Nat. Kl. Bayer. Akad. Wiss.* **1951** (1951), 31–34 (1952).   [62]

Perz S. and S. Rolewicz, Norms and perfect graphs. *Z. Oper. Res.* **34** (1990), 13–27.   [369]

Petersen J., Die Theorie der regulären Graphen. *Acta Math.* **15** (1891), 193–220.
                                                                 [213, 267, 268, 273, 274, 277, 740]

Petersen T.K., *Eulerian numbers, Birkhäuser Advanced Texts: Basler Lehrbücher*. (Birkhäuser/Springer, 2015).   [101]

Petkovsek M., *Finding closed-form solutions of difference equations by symbolic methods* (ProQuest LLC, 1991). Ph.D. Thesis, Carnegie Mellon Univ.   [90]

Petkovšek M., H.S. Wilf, and D. Zeilberger, $A = B$ (A K Peters, 1996).   [87, 89, 90]

Phillips J., Chernoff–Hoeffding inequality and applications (2013). (arXiv:1209.63946).   [711]

Pikhurko O., Borsuk's conjecture fails in dimensions 321 and 322 (2002). (arXiv:math/0202112).   [817]

Pikhurko O., A note on the Turán function of even cycles. *Proc. Amer. Math. Soc.* **140** (2012), 3687–3692.
                                                                                                    [479]

Pilpel S., Descending subsequences of random permutations. *J. Combin. Th. A* **53** (1990), 96–116.
                                                                                                    [821]

Pinsker M., On the complexity of a concentrator. *7th International Teletraffic Conference* (1973), 318/1–318/4.   [779]

Piotrowski W.L., Untersuchungen über das Oberwolfacher Problem (1979). Unpublished manuscript.
                                                                                                    [647]

Piotrowski W.-L., The solution of the bipartite analogue of the Oberwolfach problem. *Discr. Math.* **97** (1991), 339–356.   [647]

Pippenger N., Superconcentrators. *SIAM J. Comput.* **6** (1977), 298–304.   [779]

Pippenger N., Entropy and enumeration of Boolean functions. *IEEE Trans. Inform. Theory* **45** (1999), 2096–2100.   [509]

Pité E., Problem 11957. *Amer. Math. Monthly* **124** (2017), 179. Solution **125** (2018), November.   [133]

Pitman J., Coalescent random forests. *J. Combin. Th. A* **85** (1999), 165–193.   [37]

Plantholt M., Overfull conjecture for graphs with high minimum degree. *J. Graph Theory* **47** (2004), 73–80.   [359]

Plaza A. and S. Falcón, Problem 11920. *Amer. Math. Monthly* **123** (2016), 296. Solution **125** (2018), 374–375.   [80]

Plesnevič P.S. and V.G. Vizing, On the problem of the minimal coloring of the vertices of a graph (Russian). *Sibirsk. Mat. Zh.* **6** (1965), 234–236.   [343]

Plesník J., Connectivity of regular graphs and the existence of 1-factors. *Mat. Časopis Sloven. Akad. Vied* **22** (1972), 310–318.   [267]

Plesník J., Critical graphs of given diameter. *Acta Fac. Rerum Natur. Univ. Comenian. Math.* **30** (1975), 71–93.   [294]

Plesník J. and Š. Znám, On equality of edge-connectivity and minimum degree of a graph. *Arch. Math. (Brno)* **25** (1989), 19–25.   [296]

Plummer M.D., On minimal blocks. *Trans. Amer. Math. Soc.* **134** (1968), 85–94.   [315, 316]

Plummer M.D., On the cyclic connectivity of planar graphs. In *Graph theory and applications (Proc. Conf., Western Michigan Univ., 1972; dedicated to the memory of J. W. T. Youngs)*, Lecture Notes in Math. **303** (Springer, 1972), 235–242.   [297]

Plummer M.D., Problem. In *Infinite and finite sets* (A. Haynal, R. Rado, and V.T. Sós, eds.) **10** (North-Holland, 1975), 1549–1550.   [401]

Plummer M.D., Matching theory—a sampler: from Dénes König to the present. *Discr. Math.* **100** (1992), 177–219.   [253]

Plummer M.D., Graph factors and factorization: 1985–2003: a survey. *Discr. Math.* **307** (2007), 791–821.
                                                                                                    [268]

Pnueli A., A. Lempel, and S. Even, Transitive orientation of graphs and identification of permutation graphs. *Canad. J. Math.* **23** (1971), 160–175. [343]

Pólya G., Kombinatorische Anzahlbestimmungen für Gruppen, Graphen und chemische Verbindungen. *Acta Math.* **68** (1937), 145–254. [37, 128, 184]

Poonen B., Union-closed families. *J. Combin. Th. A* **59** (1992), 253–268. [503]

Popadić M.S., On the number of antichains of finite power sets. *Mat. Vesnik* **7 (22)** (1970), 199–203. [550]

Popescu C., Problem 10770. *Amer. Math. Monthly* **106** (1999), 963. Solution **108** (2001), 979–80. [172]

Pósa L., On the circuits of finite graphs. *Magyar Tud. Akad. Mat. Kutató Int. Közl.* **8** (1963), 355–361 (1964). [326, 333]

Pournin L., The diameter of associahedra. *Adv. Math.* **259** (2014), 13–42. [387]

Pratt R., Problem 11573. *Amer. Math. Monthly* **118** (2011), 463. Solution **120** (2013), 372. [22]

Preen J., A census of all 5-regular planar graphs with diameter 3. *Ars Combin.* **106** (2012), 129–135. [386]

Prim R.C., Shortest connection networks and some generalizations. *Bell Syst. Tech. J.* **36** (1957), 1389–1401. [246, 252]

Proctor R.A., Solution of two difficult combinatorial problems with linear algebra. *Amer. Math. Monthly* **89** (1982), 721–734. [557]

Proctor R.A., Let's expand Rota's twelvefold way for counting partitions! (2006). (arXiv:math/0606404). [147]

Prodinger H., A correspondence between ordered trees and noncrossing partitions. *Discr. Math.* **46** (1983), 205–206. [49, 64]

Prömel H.J., *Ramsey theory for discrete structures* (Springer, 2013). [425]

Propp J.G., Problem 10679. *Amer. Math. Monthly* **105** (1998), 666. Solution **107** (2000), 374. [511]

Prowse A. and D.R. Woodall, Choosability of powers of circuits. *Graphs Combin.* **19** (2003), 137–144. [740]

Prüfer H., Neuer beweis eines satzes über permutationen (German). *Arch. Math. Phys.* **27** (1918), 742–744. [37]

Pudaite P.R., Problem 10801. *Amer. Math. Monthly* **107** (2000), 368. Solution **109** (2002), 394–5. [66]

Puleo G.J., Maximal k-edge-colorable subgraphs, Vizing's theorem, and Tuza's conjecture. *Discr. Math.* **340** (2017), 1573–1580. [372]

Pyber L., An extension of a Frankl–Füredi theorem. *Discr. Math.* **52** (1984), 253–268. [249]

Pyber L., Regular subgraphs of dense graphs. *Combinatorica* **5** (1985), 347–349. [737]

Pyber L., V. Rödl, and E. Szemerédi, Dense graphs without 3-regular subgraphs. *J. Combin. Th. B* **63** (1995), 41–54. [737]

Pyke R., The supremum and infimum of the Poisson process. *Ann. Math. Statist.* **30** (1959), 568–576. [50]

Pym J.S., The linking of sets in graphs. *J. London Math. Soc.* **44** (1969), 542–550. [299]

Qiao P. and X. Zhan, The largest graphs with given order and diameter: A simple proof. *Graphs and Combinatorics* **36** (2019), 1715–1716. [249]

Rabinovitch I., *The Dimension-Theory of Semiorders and Interval-Orders* (ProQuest LLC, 1973). Ph.D. Thesis, Dartmouth College. [604]

Rabinovitch I., The Scott–Suppes theorem on semiorders. *J. Mathematical Psychology* **15** (1977), 209–212. [604]

Rabinovitch I., The dimension of semiorders. *J. Combin. Th. A* **25** (1978), 50–61. [574, 583, 604]

Rademacher H., On the partition function $p(n)$. *Proc. London Math. Soc.* **43** (1937), 241–254. [140]

Radhakrishnan J., An entropy proof of Brégman's theorem. *J. Combin. Th. A* **77** (1997), 161–164. [510]

Radhakrishnan J., Entropy and counting. In *Computational mathematics, modeling and algorithms* (J. C. Misra, ed.) (Narosa, 2003), 146–148. [504, 508]

Radhakrishnan J. and A. Srinivasan, Improved bounds and algorithms for hypergraph 2-coloring. *Random Structures Algorithms* **16** (2000), 4–32. [660, 673]

Rado R., Studien zur Kombinatorik. *Math. Z.* **36** (1933), 424–470.   [462, 467]

Rado R., Axiomatic treatment of rank in infinite sets. *Canadian J. Math.* **1** (1949), 337–343.   [468]

Rado R., Note on independence functions. *Proc. Lond. Math. Soc.* **7** (1957), 300–320.   [519]

Rado R., Some partition theorems. In *Combinatorial Theory and its Applications, III (Proc. Colloq., 1969)* (North-Holland, 1970), 929–936.   [466]

Radoux C., Nombres de Catalan généralisés. *Bull. Belg. Math. Soc. Simon Stevin* **4** (1997), 289–292.   [177]

Radziszowski S.P., Small Ramsey numbers. *Electronic J. Combin.* (1994), Dynamic Survey 1.   [449]

Raĭgorodskiĭ A.M., On the dimension in Borsuk's problem (Russian). *Uspekhi Mat. Nauk* **52** (1997), 181–182.   [817]

Ramalingam G. and C. PanduRangan, A unified approach to domination problems on interval graphs. *Inform. Process. Lett.* **27** (1988), 271–274.   [374]

Ramamurthi R., *Coloring problems on graphs and hypergraphs* (ProQuest, 2001). Ph.D. Thesis, Univ. Illinois, Urbana-Champaign.   [746]

Ramanan G.V., Proof of a conjecture of Frankl and Füredi. *J. Combin. Th. A* **79** (1997), 53–67.   [730]

Ramírez-Alfonsín J.L. and B.A. Reed (eds.), *Perfect graphs*, Wiley-Interscience Series in Discr. Mathematics and Optimization (Wiley & Sons, 2001).   [370]

Ramos E.A., Equipartition of mass distributions by hyperplanes. *Discr. Comput. Geom.* **15** (1996), 147–167.   [815]

Ramras M., Minimum cutsets in hypercubes. *Discr. Math.* **289** (2004), 193–198.   [295]

Ramsey F.P., On a Problem of Formal Logic. *Proc. Lond. Math. Soc.* **30** (1930), 264–286.   [443, 444, 468]

Raney G.N., Functional composition patterns and power series reversion. *Trans. Amer. Math. Soc.* **94** (1960), 441–451.   [50, 129]

Ray-Chaudhuri D.K., Characterization of line graphs. *J. Combin. Th.* **3** (1967), 201–214.   [370]

Ray-Chaudhuri D.K. and R.M. Wilson, Solution of Kirkman's schoolgirl problem (1971), 187–203.   [647, 649]

Ray-Chaudhuri D.K. and R.M. Wilson, On *t*-designs. *Osaka J. Math.* **12** (1975), 737–744.   [731]

Raynaud H., Sur le circuit hamiltonien bi-coloré dans les graphes orientés. *Period. Math. Hungar.* **3** (1973), 289–297.   [460]

Razborov A.A., Lower bounds on the monotone complexity of some Boolean functions. *Dokl. Akad. Nauk SSSR* **281** (1985), 798–801.   [504]

Razborov A.A., Flag algebras. *J. Symbolic Logic* **72** (2007), 1239–1282.   [475]

Razborov A.A., On the minimal density of triangles in graphs. *Combin. Probab. Comput.* **17** (2008), 603–618.   [478]

Razborov A.A., On 3-hypergraphs with forbidden 4-vertex configurations. *SIAM J. Discr. Math.* **24** (2010), 946–963.   [475]

Razborov A.A., On the Fon-der-Flaass interpretation of extremal examples for Turán's (3, 4)-problem. *Tr. Mat. Inst. Steklova* **274** (2011), 269–290.   [491]

Read R.C., Review of "die chromatischen Polynome unterringfreier Graphen". *MathSciNet Review* (1975). MR0354428 (50 #6906)   [374]

Rédei L., Ein kombinatorischer Satz. *Acta Litt. Szeged* **7** (1934), 39–43.   [334]

Redfield J.H., The Theory of Group-Reduced Distributions. *Amer. J. Math.* **49** (1927), 433–455.   [184]

Reed B.A., ω, δ, and χ. *J. Graph Th.* **27** (1998), 177–212.   [339, 685]

Reed B.A., A strengthening of Brooks' Theorem. *J. Combin. Th. B* **76** (1999), 136–149.   [677]

Reed B.A. and B. Sudakov, Asymptotically the list colouring constants are 1. *J. Combin. Th. B* **86** (2002), 27–37.   [355, 677]

Rees D., Note on a paper by I.J. Good. *J. Lond. Math. Soc.* **21** (1946), 169–172.   [239]

Regev A., Asymptotic values for degrees associated with strips of Young diagrams. *Adv. in Math.* **41** (1981), 115–136.   [433]

Reiher C., The clique density theorem. *Ann. of Math. (2)* **184** (2016), 683–707.   [478]

Reiman I., Über ein Problem von K. Zarankiewicz. *Acta. Math. Acad. Sci. Hungar.* **9** (1958), 269–273.
[632]

Remmel J.B., Bijective proofs of formulae for the number of standard Young tableaux. *Linear and Multilinear Algebra* **11** (1982), 45–100.   [192]

Renault M., Lost (and found) in translation: André's actual method and its application to the generalized ballot problem. *Amer. Math. Monthly* **115** (2008), 358–363.   [40]

Rényi A., Some remarks on the theory of trees. *Magyar Tud. Akad. Mat. Kut. Int. Közl.* **4** (1959), 73–85.
[133]

Rényi A., Théorie des éléments saillants d'une suite d'observations (French). *Ann. Fac. Sci. Univ. Clermont-Ferrand No.* **8** (1962), 7–13.   [99]

Rényi A., On the enumeration of trees. In *Combinatorial Structures and their Applications (Proc. Calgary Internat. Conf., 1969)* (Gordon and Breach, 1970), 355–360.   [37]

Reuter K., On the dimension of the Cartesian product of relations and orders. *Order* **6** (1989a), 277–293.
[572]

Reuter K., Removing critical pairs. *Order* **6** (1989b), 107–118.   [584]

Rey J.G., Problem 10615. *Amer. Math. Monthly* **104** (1997), 767. Solution **106** (1999), 692–693.   [47]

Reznick B., P. Tiwari, and D.B. West, Decomposition of product graphs into complete bipartite subgraphs. *Discr. Math.* **57** (1985), 179–183.   [771]

Ribó Mor A., G. Rote, and A. Schulz, Small grid embeddings of 3-polytopes. *Discrete Comput. Geom.* **45** (2011), 65–87.   [783]

Richter R.B., Problem 10330. *Amer. Math. Monthly* **100** (1993), 796. Solution **103** (1996), 700–1.   [489]

Richter R.B. and C. Thomassen, Intersections of curve systems and the crossing number of $C_5 \times C_5$. *Discrete Comput. Geom.* **13** (1995), 149–159.   [794]

Riddell R.J., *Contributions to the theory of condensation* (Univ. Michigan, Ann Arbor, 1951). Ph.D. Thesis.
[135]

Riddell R.J., Jr. and G.E. Uhlenbeck, On the theory of the virial development of the equation of state of mono-atomic gases. *J. Chem. Phys.* **21** (1953), 2056–2064.   [127]

Riesling A.S., Boruk's problem in three-dimensional spaces of constant curvature. *Ukr. Geom. Sborni* **11** (1971), 78–83.   [817]

Rieß W., Zwei Optimierungsprobleme auf Ordnungen. *Arbeitsber. Inst. Math. Masch. Datenverarb. (Inform.)* **11** (1978), 59.   [557]

Riguet J., Les relations de Ferrers. *C. R. Acad. Sci. Paris* **232** (1951), 1729–1730.   [605]

Ringeisen R.D. and L.W. Beineke, The crossing number of $C_3 \times C_n$. *J. Combin. Th. B* **24** (1978), 134–136.   [794]

Ringel G., Selbstkomplementare graphen. *Arch. Math.* **14** (1963), 354–358.   [218]

Ringel G., Problem 25. In *Theory of Graphs and Its Applications (Smolenice, 1963)* (Czech. Acad. Sci., 1964), 162.   [646, 656]

Riordan J., *An introduction to combinatorial analysis* (Wiley; Chapman & Hall, 1958).   [153, 161]

Riordan J., Inverse relations and combinatorial identities. *Amer. Math. Monthly* **71** (1964), 485–498.
[133]

Riordan J., *Combinatorial identities* (Wiley & Sons, 1968). Also 1979.   [47, 65]

Robbins H., A theorem on graphs, with an application to a problem in traffic control. *Amer. Math. Monthly* **46** (1939), 281–283.   [306]

Roberts F.S., *Graph Theory and Its Applications to the Problems of Society*, CBMS-NSF Monograph **29** (SIAM Publications, 1978).   [337]

Robertson N., D. Sanders, P. Seymour, and R. Thomas, The four-colour theorem. *J. Combin. Th. B* **70** (1997), 2–44.   [406, 416, 417]

Robertson N., P. Seymour, and R. Thomas, Excluded minors in cubic graphs. *J. Combin. Th. B* **138** (2019), 219–285.   [416]

Robertson N., P.D. Seymour, and R. Thomas, Hadwiger's conjecture for K_6-free graphs. *Combinatorica* **13** (1993), 279–361.   [349]

Robertson N., P.D. Seymour, and R. Thomas, Tutte's edge-colouring conjecture. *J. Combin. Th. B* **70** (1997), 166–183.   [416]

Robinson G.de B., On the representations of the symmetric group. *Amer. J. Math.* **60** (1938), 745–760.   [193, 195]

Robinson J.P. and M. Cohn, Counting sequences. *IEEE Trans. Comput.* **30** (1981), 17–23.   [820]

Rödl V. and M. Schacht, Generalizations of the removal lemma. *Combinatorica* **29** (2009), 467–501.   [485]

Rödl V. and M. Schacht, Regularity lemmas for graphs. In *Fete of combinatorics and computer science*, *Bolyai Soc. Math. Stud.* **20** (János Bolyai Math. Soc., 2010), 287–325.   [488]

Rödl V. and J. Skokan, Regularity lemma for *k*-uniform hypergraphs. *Random Structures Algorithms* **25** (2004), 1–42.   [488]

Rödl V. and E. Szemerédi, On size Ramsey numbers of graphs with bounded degree. *Combinatorica* **20** (2000), 257–262.   [455]

Roman S.M., *On a Problem of Zarankiewicz Concerning Matrices of Zeros and Ones* (ProQuest, 1975). Ph.D. Thesis, Univ. Washington.   [632]

Roman S.M., *The umbral calculus*, *Pure and Applied Mathematics* **111** (Academic Press, 1984).   [123]

Roman S.M. and G.C. Rota, The umbral calculus. *Advances in Math.* **27** (1978), 95–188.   [123]

Rosa A., On certain valuations of the vertices of a graph. In *Theory of Graphs (Rome, 1966)*, *Intl. Symp.* (Gordon and Breach; Dunod, 1967), 349–355.   [646, 656]

Rosenfeld M., On the total coloring of certain graphs. *Israel J. Math.* **9** (1971), 396–402.   [373]

Rota G.C., On the foundations of combinatorial theory I. *Z. Wahrsch. Verw. Gebiete* **2** (1964), 340–368.   [520, 529]

Roth K.F., On certain sets of integers. *J. London Math. Soc.* **28** (1953), 104–109.   [484, 485]

Rotman J.J., Problem E3462. *Amer. Math. Monthly* **98** (1991), 755. Solution **100** (1993), 594.   [238]

Rousseau C.C. and J. Sheehan, A class of Ramsey problems involving trees. *J. London Math. Soc. (2)* **18** (1978), 392–396.   [459]

Roussel F., I. Rusu, and H. Thuillier, The strong perfect graph conjecture: 40 years of attempts, and its resolution. *Discr. Math.* **309** (2009), 6092–6113.   [370]

Roy B., Nombre chromatique et plus longs chemins d'un graphe. *Rev. Française Automat. Informat. Recherche Opérationelle sér. Rouge* **1** (1967), 127–132.   [339]

Royle G., Graphs where every two vertices have odd number of mutual neighbours (2010). Available online at https://mathoverflow.net/questions/17809/graphs-where-every-two-vertices-have-odd-number-of-mutual-neighbours.   [226]

Rubel L.A., Problem 6565. *Amer. Math. Monthly* **95** (1988), 60. Solution **97** (1990), 80–81.   [473]

Ruciński A. and A. Vince, Balanced graphs and the problem of subgraphs of random graphs. In *Proc. 16th Southeastern Intl. Conf. Combin. Graph Th. Comput. (Boca Raton)*, **49** (1985), 181–190.   [703]

Rupp C.A., Problem 3468. *Amer. Math. Monthly* **37** (1930), 552. Solution **38** (1931), 355.   [178]

Ruzsa I.Z. and E. Szemerédi, Triple systems with no six points carrying three triangles. In *Combinatorics (Proc. Fifth Hungarian Colloq., Keszthely, 1976), II*, *Colloq. Math. Soc. János Bolyai* **18** (North-Holland, 1978), 939–945.   [485, 492, 684]

Ryjáček Z., On a closure concept in claw-free graphs. *J. Combin. Th. B* **70** (1997), 217–224.   [320, 326]

Rymer N.W., Projects, problems and patience. *Math. Gazette* **63** (1979), 1–7.   [66]

Ryser H.J., Combinatorial properties of matrices of zeros and ones. *Canad. J. Math.* **9** (1957), 371–377.   [227]

Ryser H.J., *Combinatorial mathematics*, *The Carus Mathematical Monographs, No. 14* (Math. Assoc. Amer.; distributed by Wiley, 1963).   [314]

Ryser H.J., Matrices of zeros and ones in combinatorial mathematics. In *Recent Advances in Matrix Theory (Madison, 1963)* (Univ. Wisc. Press, 1964), 103–124.   [228]

Saaty T.L. and P.C. Kainen, *The four-color problem* (McGraw-Hill, 1977).   [399, 418]

Sabidussi G., Graphs with given group and given graph-theoretical properties. *Canad. J. Math.* **9** (1957), 515–525.   [343]

Sachs H., Über Teiler, Faktoren und charakteristische Polynome von Graphen II. *Wiss. Z. Techn. Hochsch. Ilmenau* **13** (1967), 405–412. [780]

Sachs H., Finite graphs (Investigations and generalizations concerning the construction of finite graphs having given chromatic number and no triangles). In *Recent Progress in Combinatorics (Proc. 3rd Waterloo Conf. Combin., 1968)* (Academic Press, 1969), 175–184. [339]

Sachs H., Elementary proof of the cycle-plus-triangles theorem. In *Combinatorics, Paul Erdős is eighty, Vol. 1, Bolyai Soc. Math. Stud.* (János Bolyai Math. Soc., 1993), 347–359. [739]

Sachs H. and H. Zernitz, Remark on the dimer problem. *Discrete Appl. Math.* **51** (1994), 171–179. [34]

Sagan B., On selecting a random shifted Young tableau. *J. Algorithms* **1** (1980), 213–234. [207]

Sagan B.E., *The symmetric group, The Wadsworth & Brooks/Cole Mathematics Series* (Wadsworth & Brooks, 1991). [189]

Saito A., Chvátal–Erdős theorem: old theorem with new aspects. In *Computational geometry and graph theory, Lecture Notes in Comput. Sci.* **4535** (Springer, Berlin, 2008), 191–200. [321]

Šajna M., Cycle decompositions. III. Complete graphs and fixed length cycles. *J. Combin. Des.* **10** (2002), 27–78. [646]

Saks M.E., A short proof of the existence of k-saturated partitions of partially ordered sets. *Adv. in Math.* **33** (1979), 207–211. [548]

Saks M.E., *Duality Properties of Finite Set Systems* (ProQuest, 1980). Ph.D. Thesis, Mass. Inst. Tech. [549]

Saks M.E., A class of perfect graphs associated with planar rectilinear regions. *SIAM J. Algebraic Discrete Methods* **3** (1982), 330–342. [831]

Saks M.E., The information theoretic bound for problems on ordered sets and graphs. In *Graphs and order (Banff, Alta., 1984), NATO Adv. Sci. Inst. Ser. C Math. Phys. Sci.* **147** (Reidel, 1985), 137–168. [831]

Salamon G. and G. Wiener, Leaves of spanning trees and vulnerability. In *5th Hungarian–Japanese Sym. Discr. Math. and Appl.* (2007), 225–235. [334]

Sanders D.P. and Y. Zhao, A note on the three color problem. *Graphs Combin.* **11** (1995), 91–94. [414, 415]

Sanders D.P. and Y. Zhao, Planar graphs of maximum degree seven are class I. *J. Combin. Th. B* **83** (2001), 201–212. [422]

Sanders D.P. and Y. Zhao, On the size of edge chromatic critical graphs. *J. Combin. Th. B* **86** (2002), 408–412. [414]

Sanders P.R., The central automorphisms of a finite group. *J. London Math. Soc.* **44** (1969), 225–228. [466]

Sands A.D., On generalised Catalan numbers. *Discr. Math.* **21** (1978), 219–221. [50]

Santoro N. and J. Urrutia, Angle orders, regular n-gon orders and the crossing number. *Order* **4** (1987), 209–220. [832]

Sarkaria K.S., A generalized Kneser conjecture. *J. Combin. Theory B* **49** (1990), 236–240. [812]

Sarvate D.G. and J.-C. Renaud, On the union-closed sets conjecture. *Ars Combin.* **27** (1989), 149–153. [503, 512]

Sauer N., A generalization of a theorem of Turán. *J. Combinatorial Th. B* **10** (1971), 109–112. [476]

Sauer N. and J. Spencer, Edge disjoint placement of graphs. *J. Combin. Theory B* **25** (1978), 295–302. [248]

Sauvé L., On chromatic graphs. *Amer. Math. Monthly* **68** (1961), 107–111. [477]

Savage C.D. and C.Q. Zhang, The connectivity of acyclic orientation graphs. *Discr. Math.* **184** (1998), 281–287. [312]

Scarpis U., Sui determinanti di valore massimo (Italian). *Rendiconti della R. Istituto Lombardo di Sci. e Lettere* **31** (1898), 1441–1446. [622]

Schaal D., On generalized Schur numbers. In *Proc. 24th Southeastern Intl. Conf. Combin. Graph Th. Comput. (Boca Raton)*, **98** (1993), 178–187. [473]

Schäuble M., Bemerkungen zur Kounstruktion dreikreisfreier k-chromatischer Graphen. *Wiss. Zeitschrift TH Ilmenau* **15** (1969), 59–63. [354]

Schauz U., Mr. Paint and Mrs. Correct. *Electron. J. Combin.* **16** (2009), Research Paper 77, 18.   [741]

Scheim D.E., The number of edge 3-colorings of a planar cubic graph as a permanent. *Discr. Math.* **8** (1974), 377–382.   [740]

Scheinerman E.R., A note on graphs and sphere orders. *J. Graph Theory* **17** (1993), 283–289.   [832]

Scheinerman E.R. and J.C. Wierman, On circle containment orders. *Order* **4** (1988), 315–318.   [828]

Schensted C., Longest increasing and decreasing subsequences. *Canad. J. Math.* **13** (1961), 179–191.   [193, 195, 198, 206]

Schläfli L., *Theorie der vielfachen Kontinuität (German,* completed 1852*)* (George & Co., 1901). Reprinted in *Collected mathematical works Vol. I* (Verlag Birkhäuser, 1949).   [48]

Schmeichel E.F. and S.L. Hakimi, Pancyclic graphs and a conjecture of Bondy and Chvátal. *J. Combin. Th. B* **17** (1974), 22–34.   [326]

Schmeichel E.F. and S.L. Hakimi, A cycle structure theorem for Hamiltonian graphs. *J. Combin. Theory B* **45** (1988), 99–107.   [326]

Schmidt F.W., Problem 10285. *Amer. Math. Monthly* **100** (1993), 185.   [188]

Schmidt F.W., Problem 10364. *Amer. Math. Monthly* **101** (1994), 177. Solution **104** (1997), 179.   [105]

Schmidt F.W., Problem 10481. *Amer. Math. Monthly* **102** (1995), 840. Solution **104** (1997), 877–878.   [134]

Schmidt F.W., Problem 10629. *Amer. Math. Monthly* **104** (1997), 974. Solution **107** (2000), 87–8.   [152]

Schmuland B., Solution to problem 11590. *Amer. Math. Monthly* **120** (2013), 760–761. Proposed by Bibak **118** (2011), 653.   [176]

Schnyder W., Planar graphs and poset dimension. *Order* **5** (1989), 323–343.   [783, 790]

Schnyder W., Embedding planar graphs on the grid. In *Proc. 1st Annual. ACM–SIAM Symp. Discrete. Algorithms, SODA* (SIAM, 1990), 138–148.   [394, 783]

Schönberger T., Ein Beweis des Petersenschen Graphensatzes. *Acta Scientia Mathematica Szeged* **7** (1934), 51–57.   [267]

Schönheim J., On a problem of Daykin concerning intersecting families of sets. In *Combinatorics (Proc. British Combinatorial Conf. Univ. Coll. Wales, 1973), London Math. Soc. Lecture Note Ser.* **13** (Cambridge Univ. Press, 1974), 139–140.   [497, 607]

Schönheim J., Hereditary systems and Chvátal's conjecture. In *Proc. 5th British Combin. Conf. (Aberdeen), Congr. Numer.* **15** (Utilitas Math., 1976), 537–539.   [512]

Schramm O., Illuminating sets of constant width. *Mathematika* **35** (1988), 180–189.   [817]

Schrijver A., Vertex-critical subgraphs of Kneser graphs. *Nieuw Arch. Wisk. (3)* **26** (1978), 454–461.   [813, 814]

Schrijver A., *Combinatorial optimization. Polyhedra and efficiency. Vol. A, B, C, Algorithms and Combinatorics* **24** (Springer-Verlag, 2003).   [277, 520, 533, 540]

Schröder E., Vier combinatorische Probleme. *Zeit. fur Math.* **15** (1870), 361–376.   [47, 64, 81, 135]

Schur I., Über die Kongruenz $x^m + y^m \equiv z^m (\bmod p)$. *Jber. Deutsch. Math.-Verein.* **25** (1916), 114–116.   [462, 473]

Schützenberger M.P., A non-existence theorem for an infinite family of symmetrical block designs. *Ann. Eugenics* **14** (1949), 286–287.   [616]

Schützenberger M.P., A characteristic property of certain polynomials of E.F. Moore and C.E. Shannon. In *RLE Quarterly Progress Report* **55** (1959), 117–118.   [496]

Schützenberger M.P., Quelques remarques sur une construction de Schensted. *Math. Scand.* **12** (1963), 117–128.   [197, 198, 206, 208]

Schützenberger M.P., Sur un théorème de G. de B. Robinson. *C. R. Acad. Sci. Paris Sér. A-B* **272** (1971), A420–A421.   [129]

Schützenberger M.P., La correspondance de Robinson. In *Combinatoire et représentation du groupe symétrique (Strasbourg, 1976), Lect. Notes Math.* **579** (Springer, 1977), 59–113.   [203]

Schwenk A.J., Problem 6434. *Amer. Math. Monthly* **90** (1983), 403. Solution **94** (1987), 885–887.   [782]

Schwenk A.J., Problem E3143. *Amer. Math. Monthly* **93** (1986), 299. Solution **95** (1988), 352.   [206]

Schwer S.R., S-arrangements avec répétitions. *C. R. Math. Acad. Sci. Paris* **334** (2002), 261–266.   [31]

Scott A., Szemerédi's regularity lemma for matrices and sparse graphs. *Combin. Probab. Comput.* **20** (2011), 455–466. [488]

Scott A.D., Induced trees in graphs of large chromatic number. *J. Graph Th.* **24** (1997), 297–311. [352]

Scott D. and P. Suppes, Foundational aspects of theories of measurement. *J. Symb. Logic* **23** (1958), 113–128. [587]

Seberry J., A computer listing of Hadamard matrices. In *Combinatorial mathematics (Proc. Internat. Conf. Combinatorial Theory, Australian Nat. Univ., 1977), Lect. Notes Math.* **686** (Springer, 1978), 275–281. [622]

Sedláček J., On the skeletons of a graph or digraph. In *Proc. Calgary Intl. Conf. Combin. Structures and Their Applications (Univ. Calgary, 1969)* (Gordon & Breach, 1970), 387–391. [65]

Segner J.A., Enumeratio modorum quibus figurae planae (Latin). *Novi Comment. Acad. Sci. Imp. Petropol.* **7** (1759), 203–210. [41]

Seidenberg A., A simple proof of a theorem of Erdős and Szekeres. *J. London Math. Soc.* **34** (1959), 352. [430]

Seinsche D., On a property of the class of n-colorable graphs. *J. Combin. Th. B* **16** (1974), 191–193. [375, 490]

Sekanina M., On an ordering of the set of vertices of a connected graph. *Spisy Přírod. Fak. Univ. Brno* **1960** (1960), 137–141. [331]

Servedio R. and Y.N. Yeh, A bijective proof on circular compositions. *Bull. Inst. Math. Acad. Sinica* **23** (1995), 283–293. [144]

Seymour P., Hadwiger's conjecture. In *Open problems in mathematics* (Springer, 2016), 417–437. [349]

Seymour P.D., On incomparable collections of sets. *Mathematika* **20** (1973), 208–209. [497, 607, 608]

Seymour P.D., A short proof of the matroid intersection theorem (1976). Unpublished note. [530]

Seymour P.D., Disjoint paths in graphs. *Discr. Math.* **29** (1980), 293–309. [314]

Shamir E. and J. Spencer, Sharp concentration of the chromatic number on random graphs $G_{n,p}$. *Combinatorica* **7** (1987), 121–129. [717]

Shannon A.G., Fibonacci and Lucas numbers and the complexity of a graph. *Fibonacci Quart.* **16** (1978), 1–4. [65]

Shannon C.E., A theorem on coloring the lines of a network. *J. Math. Phys.* **28** (1949), 148–151. [358, 360, 365, 372]

Shapira A., Extremal Graph Theory (2016). Lecture notes scribed by Guy Rutenberg. Available online at http://www.math.tau.ac.il/~asafico/ext-graph-theory/notes.pdf. [488]

Shapiro L.W., A short proof of an identity of Touchard's concerning Catalan numbers. *J. Combin. Th. A* **20** (1976), 375–376. [82]

Shapiro L.W., Problem 10753. *Amer. Math. Monthly* **106** (1999), 777. Solution **108** (2001), 873–4. [48]

Shapiro L.W. and W. Hamilton, The Catalan numbers visit the world series. *Math. Mag.* **66** (1993), 20–22. [65]

Shapiro L.W. and D. Rogers, Problem E3343. *Amer. Math. Monthly* **96** (1989), 734–735. Solution **97** (1991), 368. [64]

Shearer J.B., A simple counterexample to a conjecture of Rota. *Discr. Math.* **28** (1979), 327–330. [567]

Shearer J.B., A note on the independence number of triangle-free graphs. *Discr. Math.* **46** (1983), 83–87. [663, 682]

Shearer J.B., On a problem of Spencer. *Combinatorica* **5** (1985), 241–245. [675]

Shearer J.B., On the independence number of sparse graphs. *Random Structures Algorithms* **7** (1995), 269–271. [663]

Shearer J.B. and D.J. Kleitman, Probabilities of independent choices being ordered. *Stud. Appl. Math.* **60** (1979), 271–276. [566]

Shelah S., Primitive recursive bounds for van der Waerden numbers. *J. Amer. Math. Soc.* **1** (1988), 683–697. [466]

Shepp L.A., The FKG inequality and some monotonicity properties of partial orders. *SIAM J. Algebraic Discrete Methods* **1** (1980), 295–299. [608]

Shepp L.A., The *XYZ* conjecture and the FKG inequality. *Ann. Probab.* **10** (1982), 824–827.   [599]

Shih W.K. and W.L. Hsu, A new planarity test. *Theoret. Comput. Sci.* **223** (1999), 179–191.   [393]

Shirazi H. and J. Verstraëte, A note on polynomials and *f*-factors of graphs. *Electron. J. Combin.* **15** (2008), Note 22, 5.   [737, 746]

Shrikhande S.S., The impossibility of certain symmetrical balanced incomplete block designs. *Ann. Math. Statistics* **21** (1950), 106–111.   [616]

Shrikhande S.S., A note on mutually orthogonal Latin squares. *Sankhyā Ser. A* **23** (1961), 115–116.   [623]

Sidney J.B., S.J. Sidney, and J. Urrutia, Circle orders, *n*-gon orders and the crossing number. *Order* **5** (1988), 1–10.   [828, 832]

Silwal S., Problem 12039. *Amer. Math. Monthly* **125** (2018), 371. Solution **127** (2020), 86.   [226]

Simion R. and F.W. Schmidt, Restricted permutations. *European J. Combin.* **6** (1985), 383–406.   [442]

Simonovits M., A method for solving extremal problems in graph theory, stability problems. In *Theory of Graphs (Proc. Colloq., Tihany, 1966)* (Academic Press, 1968), 279–319.   [479]

Simonovits M., How to solve a Turán type extremal graph problem? (linear decomposition). In *Contemporary trends in discrete mathematics (Štiřín Castle, 1997)*, DIMACS Ser. Discr. Math. Theoret. Comput. Sci. **49** (Amer. Math. Soc., 1999), 283–305.   [490]

Sinclair A., *Lecture notes for Randomness and Computation (Spring 2018)* (U California–Berkeley, 2018). Available at https://people.eecs.berkeley.edu/~sinclair/cs271/s18.html.   [680]

Singleton R.R., On minimal graphs of maximum even girth. *J. Combinatorial Theory* **1** (1966), 306–332.   [639]

Singmaster D., Problem E2897. *Amer. Math. Monthly* **88** (1981), 537. Solution by J.W. Grossman **90** (1983), 287–288.   [388]

Singmaster D., Reviews: After Math. Puzzles and Brainteasers // The Chicken from Minsk // New Mathematical Diversions. *Amer. Math. Monthly* **105** (1998), 579–587.   [70]

Skolem T., Some remarks on the triple systems of Steiner. *Math. Scand.* **6** (1958), 273–280.   [641, 644]

Sleator D.D., R.E. Tarjan, and W.P. Thurston, Rotation distance, triangulations, and hyperbolic geometry. *J. Amer. Math. Soc.* **1** (1988), 647–681.   [387]

Slivnik T., Short proof of Galvin's theorem on the list-chromatic index of a bipartite multigraph. *Combin. Probab. Comput.* **5** (1996), 91–94.   [364]

Sloane N.J.A., Hamiltonian cycles in a graph of degree 4. *J. Combinatorial Theory* **6** (1969), 311–312.   [318]

Smith H.J.S., On the value of a certain arithmetical determinant. *J. Reine Angew. Math.* **251** (1876), 100–109.   [177]

Smolenskii E.A., A method for the linear recording of graphs (English translation). *U.S.S.R. Comput. Math. and Math. Phys.* **2** (1962), 396–397.   [250]

Smoot N., Solution to problem 11908. *Amer. Math. Monthly* **125** (2018). Proposed **123** (2015), 504.   [148]

Snevily H.St.C., *Combinatorics of finite sets* (ProQuest LLC, 1991). Ph.D. Thesis, Univ. Illinois, Urbana–Champaign.   [502]

Snevily H.St.C., On generalizations of the de Bruijn–Erdős theorem. *J. Combin. Th. A* **68** (1994), 232–238.   [730]

Snevily H.St.C., A generalization of Fisher's inequality. *J. Combin. Th. A* **85** (1999), 120–125.   [730]

Snevily H.St.C., A sharp bound for the number of sets that pairwise intersect at *k* positive values. *Combinatorica* **23** (2003), 527–533.   [730]

Snevily H.St.C. and D.B. West, The bricklayer problem and the strong cycle lemma. *Amer. Math. Monthly* **105** (1998), 131–143.   [50]

Soderberg S. and H.S. Shapiro, A Combinatory Detection Problem. *Amer. Math. Monthly* **70** (1963), 1066–1070.   [513]

Sofair I., Problem 11775. *Amer. Math. Monthly* **121** (2014), 455. Solution **123** (2016), 508.   [174]

Soifer A., *The Mathematical Coloring Book: Mathematics of Coloring and the Colorful Life of its Creators* (Springer, 2008).   [342]

Soifer A., *Ramsey Theory: Yesterday, Today, and Tomorrow* **285** (Birkhauser, 2011). [425]

Soifer A., Progress in my favorite open problem of mathematics, chromatic number of the plane: An étude in five movements. *Geombinatorics* **28** (2019), 206–210. [342]

Solymosi J., Note on a generalization of Roth's theorem. In *Discrete and Computational Geometry: The Goodman–pollack Festschrift* (B. Aronov, S. Basu, J. Pach, and M. Sharir, eds.), *Algorithms and Combinatorics* **25** (Springer, Berlin, 2003), 825–827. [492]

Sondow J., Problem 11026. *Amer. Math. Monthly* **110** (2003), 636. Solution **112** (2005), 367–369. [32]

Soneoka T., H. Nakada, M. Imase, and C. Peyrat, Sufficient conditions for maximally connected dense graphs. *Discr. Math.* **63** (1987), 53–66. [296]

Sorel J., Problem 11899. *Amer. Math. Monthly* **123** (2016), 297. Solution **125** (2018), 88. [33]

Sós V.T., P. Erdős, and W.G. Brown, On the existence of triangulated spheres in 3-graphs, and related problems. *Period. Math. Hungar.* **3** (1973), 221–228. [684]

Špacapan S., Connectivity of Cartesian products of graphs. *Appl. Math. Lett.* **21** (2008), 682–685. [292, 297]

Spencer J., Minimal scrambling sets of simple orders. *Acta Math. Acad. Sci. Hungar.* **22** (1971/72), 349–353. [576, 579, 580, 584]

Spencer J., A generalized Rota conjecture for partitions. *Studies in Appl. Math.* **53** (1974), 239–241. [567]

Spencer J., Ramsey's theorem—a new lower bound. *J. Combin. Th. A* **18** (1975), 108–115. [675]

Spencer J., Asymptotic lower bounds for Ramsey functions. *Discr. Math.* **20** (1977), 69–76. [682, 685]

Spencer J., Six standard deviations suffice. *Trans. Amer. Math. Soc.* **289** (1985), 679–706. [722]

Spencer J., *Ten lectures on the probabilistic method*, CBMS-NSF Regional Conference Series in Applied Mathematics **52** (SIAM, 1987). [682]

Spencer J., E. Szemerédi, and W. Trotter, Unit distances in the Euclidean plane. In *Graph theory and combinatorics (Cambridge, 1983)* (Academic Press, London, 1984), 293–303. [794]

Sperner E., Neuer Beweis für die Invarianz der Dimensionszahl und des Gebietes. *Hamburger Abhand.* **6** (1928), 265–272. [496, 558, 561, 799, 802, 805]

Spinrad J.P., *Two-dimensional partial orders* (ProQuest LLC, 1982). Ph.D. Thesis, Princeton Univ. [573]

Spitzer F., A combinatorial lemma and its application to probability theory. *Trans. Amer. Math. Soc.* **82** (1956), 323–339. [50]

Stanley R.P., Problem E2315. *Amer. Math. Monthly* **78** (1971), 904. Solution **79** (1972), 908–910. [81, 565]

Stanley R.P., *Ordered structures and partitions* (AMS, 1972). [207]

Stanley R.P., Combinatorial reciprocity theorems. *Advances in Math.* **14** (1974), 194–253. [761]

Stanley R.P., Generating functions. In *Studies in Combinatorics*, MAA Stud. Math. **17** (Math. Assoc. America, 1978), 100–141. [127, 132, 135]

Stanley R.P., Two combinatorial applications of the Aleksandrov–Fenchel inequalities. *J. Combin. Th. A* **31** (1981), 56–65. [824, 827]

Stanley R.P., Some aspects of groups acting on finite posets. *J. Combin. Th. A* **32** (1982), 132–161. [557, 606]

Stanley R.P., On the number of reduced decompositions of elements of Coxeter groups. *European J. Combin.* **5** (1984), 359–372. [208]

Stanley R.P., *Enumerative combinatorics. Vol. I*, The Wadsworth & Brooks/Cole Mathematics Series (Wadsworth & Brooks, 1986). [48, 76, 93, 145, 150, 177, 761]

Stanley R.P., Log-concave and unimodal sequences in algebra, combinatorics, and geometry. In *Graph theory and its applications: East and West (Jinan, 1986)*, Ann. New York Acad. Sci. **576** (New York Acad. Sci., 1989), 500–535. [564]

Stanley R.P., Problem 10199. *Amer. Math. Monthly* **99** (1992), 162. Solution **101** (1994), 278–279. [61]

Stanley R.P., *Enumerative combinatorics. Vol. 2*, Cambridge Studies in Advanced Mathematics **62** (Cambridge Univ. Press, 1999). [42, 127, 129, 135, 136, 189]

Stanley R.P., Problem 11453. *Amer. Math. Monthly* **116** (2009), 746. Solution **118** (2011), 658–9. [766]

Stanley R.P., A survey of alternating permutations. In *Combinatorics and graphs*, *Contemp. Math.* **531** (AMS, 2010), 165–196.   [207]

Stanley R.P., Problem 11610. *Amer. Math. Monthly* **118** (2011), 937. Solution **120** (2013), 943.   [116]

Stanley R.P., Problem 11762. *Amer. Math. Monthly* **121** (2014), 266. Solution **123** (2016).   [150, 206]

Stanley R.P., *Catalan Numbers* (Cambridge Univ. Press, 2015a).   [41, 42]

Stanley R.P., Problem 11838. *Amer. Math. Monthly* **122** (2015b), 500.   [261]

Stanley R.P. and J.M. Steele, Problem E3344. *Amer. Math. Monthly* **96** (1989), 734. Solution **98** (1991), 649.   [669]

Stanton D. and D.E. White, *Constructive combinatorics*, *Undergraduate Texts in Mathematics* (Springer-Verlag, 1986).   [100]

Stathopoulos D., Problem 11668. *Amer. Math. Monthly* **119** (2012), 700. Solution **121** (2014), 743–744.   [62, 135]

Steele J.M., Variations on the monotone subsequence theme of Erdős and Szekeres. In *Discrete probability and algorithms (Minneapolis, 1993)*, *IMA Vol. Math. Appl.* **72** (Springer, 1995), 111–131.   [430]

Stein S.K., Convex maps. *Proc. Amer. Math. Soc.* **2** (1951), 464–466.   [393]

Stein S.K., *b*-sets and coloring problems. *Bull. Amer. Math. Soc.* **76** (1970), 805–806.   [387]

Steiner J., Einige Gesetze über die Theilung der Ebene und des Raumes. *J. Reine Angew. Math.* **1** (1826), 349–364.   [48]

Steiner J., Combinatorische aufgabe. *J. Reine Angew. Math.* **45** (1853), 181–182.   [641]

Steinlein H., Borsuk's antipodal theorem and its generalizations and applications: a survey. In *Topological methods in nonlinear analysis*, *Sém. Math. Sup.* **95** (Presses Univ. Montréal, 1985), 166–235.   [806, 815]

Stevens W.L., Solution to a geometrical problem in probability. *Ann. Eugenics* **9** (1939), 315–320.   [610]

Stiebitz M., K_5 is the only double-critical 5-chromatic graph. *Discr. Math.* **64** (1987), 91–93.   [354]

Stiebitz M. and M. Voigt, List-colourings. In *Topics in chromatic graph theory*, *Encyclopedia Math. Appl.* **156** (Cambridge Univ. Press, 2015), 114–136.   [346]

Stinson D.R., A short proof of the nonexistence of a pair of orthogonal Latin squares of order six. *J. Combin. Th. A* **36** (1984), 373–376.   [612]

Stinson D.R., *Combinatorial designs* (Springer-Verlag, 2004).   [609]

Stockmeyer P.K., Solution to problem 10663. *Amer. Math. Monthly* **107** (2000), 370. Proposed **105** (1998), 464.   [92]

Stong R., Solution to problem 10892. *Amer. Math. Monthly* **110** (2003), 440–441. Proposed **108** (2001), 668.   [330]

Stong R., Solution to problem 11086. *Amer. Math. Monthly* **113** (2006), 372. Proposed **111** (2004), 440.   [343]

Stong R., Solution to problem 1192. *Amer. Math. Monthly* **114** (2007), 839–840. Proposed **112** (2005), 930.   [623]

Stong R., Solution to problem 11931. *Amer. Math. Monthly* **125** (2018). Proposed **123** (2016), 831.   [461]

Strehl V., Binomial identities—combinatorial and algorithmic aspects. *Discr. Math.* **136** (1994), 309–346.   [31]

Subba Rao K., Some properties of Fibonacci numbers. *Amer. Math. Monthly* **60** (1953), 680–684.   [60]

Sudakov B., Recent developments in extremal combinatorics: Ramsey and Turán type problems. In *Proc. Inter. Congr. Math IV* (Hindustan Book Agency, 2010), 2579–2606.   [425]

Suk A., On the Erdős–Szekeres convex polygon problem (2016). (arXiv:1604.08657v1).   [446]

Sulanke R.A., Bijective recurrences concerning Schröder paths. *Electron. J. Combin.* **5** (1998), Research Paper 47, 11.   [64]

Sulanke R.A., Problem 10894. *Amer. Math. Monthly* **108** (2001), 770. Solution **110** (2003), 443–444.   [62]

Sulanke R.A., Objects counted by the central Delannoy numbers. *J. Integer Seq.* **6** (2003), Article 03.1.5, 19 pages.   [28, 29, 34]

Sumner D.P., Graphs with 1-factors. *Proc. Amer. Math. Soc.* **42** (1974), 8–12.   [272]

Sumner D.P., 1-factors and antifactor sets. *J. London Math. Soc. (2)* **13** (1976), 351–359.   [273]

Sumner D.P., Subtrees of a graph and the chromatic number. In *The theory and applications of graphs (Kalamazoo, Mich., 1980)* (Wiley, 1981), 557–576.   [351]

Sved M., Counting and recounting: the aftermath. *Math. Intelligencer* **6** (1984), 44–45.   [40]

Sýkora O. and I. Vrt'o, On crossing numbers of hypercubes and cube connected cycles. *BIT* **33** (1993), 232–237.   [798]

Sylvester J.J., On the change of systems of independent variables. *Quart. J. Math.* **1** (1857), 42–56.
                                                                                         [37]

Sylvester J.J., Thoughts on orthogonal matrices, simultaneous sign-successions, and tessellated pavements in two or more colours, with applications to Newton's rule, ornamental tile-work, and the theory of numbers. *Phil. Mag.* **34** (1867), 461–475.   [618]

Sylvester J.J. and F. Franklin, A Constructive Theory of Partitions, Arranged in Three Acts, an Interact and an Exodion. *Amer. J. Math.* **5** (1882), 251–330.   [141, 142]

Sysło M.M. and J. Zak, The bandwidth problem: critical subgraphs and the solution for caterpillars. In *Bonn Workshop on Combin. Opt. (Bonn, 1980)* (North-Holland, 1982), 281–286.   [819]

Szegedy M., The Lovász local lemma—a survey. In *Computer science—theory and applications, Lecture Notes in Comput. Sci.* **7913** (Springer, 2013), 1–11.   [674]

Székely L.A., Crossing numbers and hard Erdős problems in discrete geometry. *Combin. Probab. Comput.* **6** (1997), 353–358.   [794, 798]

Szekeres G. and L. Peters, Computer solution to the 17-point Erdős–Szekeres problem. *ANZIAM J.* **48** (2006), 151–164.   [458]

Szekeres G. and H.S. Wilf, An inequality for the chromatic number of a graph. *J. Combin. Th.* **4** (1968), 1–3.   [338]

Szele T., Combinatorial investigations concerning complete directed graphs (Hungarian). *Mat. Fiz. Lapok* **50** (1943), 223–236.   [666]

Szemerédi E., On sets of integers containing no four elements in arithmetic progression. In *Number Theory (János Bolyai Math. Soc., Debrecen, 1968)* (North-Holland, 1970), 197–204.   [484]

Szemerédi E., On graphs containing no complete subgraph with 4 vertices. *Mat. Lapok* **23** (1972), 113–116.   [492]

Szemerédi E., On sets of integers containing no k elements in arithmetic progression. *Acta Arith.* **27** (1975), 199–245.   [480, 484]

Szemerédi E., Regular partitions of graphs. In *Problémes combinatoires et théorie des graphes (Orsay, 1976), Colloq. Internat. CNRS,* **260** (1978), 399–401.   [480]

Szemerédi E., Arithmetic progressions, different regularity lemmas and removal lemmas. *Commun. Math. Stat.* **3** (2015), 315–328.   [480, 484]

Szemerédi E. and W.T. Trotter, Extremal problems in discrete geometry. *Combinatorica* **3** (1983), 381–392.   [794, 798]

Szpilrajn E., Sur l'extension de l'ordre partiel (French). *Fundamenta Mathematicae* **16** (1930), 386–389.
                                                                                         [569]

't Woord A.N., Solution to problem 10490. *Amer. Math. Monthly* **106** (1999), 589. Proposed **102** (1995), 930.   [32]

Tagiuri A., On some recurrent sequences with positive integer terms (Italian). *Periodico di Mat. (2)* **3** (1900), 1–12.   [61]

Tait P.G., Remarks on the colourings of maps. *Proc. R. Soc. Edinburgh* **10** (1880), 729.   [399, 400]

Talagrand M., Concentration of measure and isoperimetric inequalities in product spaces. *Inst. Hautes Études Sci. Publ. Math.* (1995), 73–205.   [712, 720]

Tamassia R. and I.G. Tollis, A unified approach to visibility representations of planar graphs. *Discrete Comput. Geom.* **1** (1986), 321–341.   [394]

Tanaka Y., Equivalence of the HEX game theorem and the Arrow impossibility theorem. *Appl. Math. Comput.* **186** (2007), 509–515.   [803]

Tanner R.M., Explicit construction of concentrators from generalized n-gons. *SIAM J. Algebr. Discrete Meth.* **5** (1984), 287–293.   [779]

Tao T., Moser's entropy compression argument (blog post) (2009). Available at http://terrytao. wordpress.com/2009/08/05/mosers-entropy-compression-argument.   [680]

Tao T., Algebraic combinatorial geometry: the polynomial method in arithmetic combinatorics, incidence combinatorics, and number theory. *EMS Surv. Math. Sci.* **1** (2014), 1–46.   [732]

Tao T. and V. Vu, *Additive combinatorics*, Cambridge Studies in Advanced Mathematics **105** (Cambridge Univ. Press, 2006).   [488]

Tarjan R.E., A simple version of Karzanov's blocking flow algorithm. *Oper. Res. Letters* **2** (1984), 265–268.   [246]

Tarry G., Le problème des labyrinthes (French). *Nouv. Ann. Math.* **14** (1895), 187–190.   [763]

Tarry G., Le problème de 36 officiers. *Compte Rendu de l'Assoc. Français Avanc. Sci. Naturel* **1** (1900), 122–123.   [612]

Tarry G., Le problème de 36 officiers. *Compte Rendu de l'Assoc. Français Avanc. Sci. Naturel* **2** (1901), 170–203.   [612]

Tashkinov V.A., 3-regular subgraphs of 4-regular graphs. *Mat. Zametki* **36** (1984), 239–259.   [736]

Tator C., *On the Dimension of Ordered Sets* (Technische Hochschule Darmstadt, 1983). Thesis.   [575, 583]

Tauraso R., Problem 11241. *Amer. Math. Monthly* **113** (2006), 656. Solution **115** (2008), 858–859.   [81]

Taylor H., Problem E3448. *Amer. Math. Monthly* **98** (1991), 553. Solution **100** (1993), 298–300.   [91]

Thomas R. and P. Wollan, An improved linear edge bound for graph linkages. *European J. Combin.* **26** (2005), 309–324.   [314]

Thomason A.G., Hamiltonian cycles and uniquely edge colourable graphs. *Ann. Discr. Math.* **3** (1978), 259–268.   [242, 249, 318, 331, 332]

Thomason A.G., An extremal function for contractions of graphs. *Math. Proc. Cambridge Philos. Soc.* **95** (1984), 261–265.   [710]

Thomason A.G., The extremal function for complete minors. *J. Combin. Th. B* **81** (2001), 318–338.   [710]

Thomassen C., Planarity and duality of finite and infinite graphs. *J. Combin. Th. B* **29** (1980a), 244–271.   [304, 392]

Thomassen C., 2-linked graphs. *European J. Combin.* **1** (1980b), 371–378.   [314]

Thomassen C., Kuratowski's Theorem. *J. Graph Th.* **5** (1981a), 225–241.   [392]

Thomassen C., Nonseparating cycles in k-connected graphs. *J. Graph Theory* **5** (1981b), 351–354.   [305, 313]

Thomassen C., A remark on the factor theorems of Lovász and Tutte. *J. Graph Theory* **5** (1981c), 441–442.   [262]

Thomassen C., A theorem on paths in planar graphs. *J. Graph Th.* **7** (1983), 169–176.   [334, 401]

Thomassen C., A refinement of Kuratowski's Theorem. *J. Combin. Th. B* **37** (1984), 245–253.   [393]

Thomassen C., Interval representations of planar graphs. *J. Combin. Th. B* **40** (1986), 9–20.   [394]

Thomassen C., Paths, circuits and subdivisions. In *Selected topics in graph theory, 3* (Academic Press, 1988), 97–131.   [333, 350]

Thomassen C., Grötzsch's 3-Color Theorem. *J. Combin. Th. B* **62** (1994a), 268–279.   [414]

Thomassen C., Every planar graph is 5-choosable. *J. Combin. Th. B* **62** (1994b), 180–181.   [402, 742]

Thomassen C., 3-list-coloring planar graphs of girth 5. *J. Combin. Th. B* **64** (1995a), 101–107.   [403]

Thomassen C., Decomposing a planar graph into degenerate graphs. *J. Combin. Th. B* **65** (1995b), 305–314.   [419]

Thomassen C., A short list color proof of Grötzsch's theorem. *J. Combin. Th. B* **88** (2003), 189–192.   [414]

Thrall R.M., A combinatorial problem. *Michigan Math. J.* **1** (1952), 81–88.   [192, 207]

Thurston W.P., *Three-dimensional geometry and topology. Vol. 1*, *Princeton Mathematical Series* **35** (Princeton Univ. Press, 1997). Also 1991.   [394]

Tian F., A short proof of a theorem about the circumference of a graph. *J. Combin. Th. B* **45** (1988), 373–375.   [325]

Tian F., A short proof of Fan's theorem. *Discr. Math.* **286** (2004), 285–286.   [325]

Tian F. and R.H. Shi, A new class of pancyclic graphs. *J. Systems Sci. Math. Sci.* **6** (1986), 258–262.   [326]

Timmons C., Star coloring high girth planar graphs. *Electron. J. Combin.* **15** (2008), Research Paper 124, 17.   [422]

Toft B., On the maximal number of edges of critical *k*-chromatic graphs. *Studia Sci. Math. Hungar.* **5** (1970), 461–470.   [354]

Toft B., Colouring, stable sets and perfect graphs. In *Handbook of combinatorics, Vol. 1, 2* (Elsevier, 1995), 233–288.   [335]

Toida S., Properties of an Euler graph. *J. Franklin Inst.* **295** (1973), 343–5.   [239]

Tomescu I., Sur le problème du coloriage des graphes généralisés. *C. R. Acad. Sci. Paris Sér. A-B* **267** (1968), A250–A252.   [667]

Tomescu I., Problem E3188. *Amer. Math. Monthly* **94** (1987), 72. Solution **95** (1988), 876–877. [512]

Tomescu I., Problem E3409. *Amer. Math. Monthly* **97** (1990), 916. Solution **99** (1991), 860–861. [355]

Touchard J., Sur les cycles des substitutions. *Acta Math.* **70** (1939), 243–297.   [127]

Tracy P., Problem 10811. *Amer. Math. Monthly* **107** (2000), 566. Solution **109** (2002), 478.   [249]

Trotignon N., Perfect graphs. In *Topics in chromatic graph theory*, *Encyclopedia Math. Appl.* **156** (Cambridge Univ. Press, 2015), 137–160.   [370]

Trotter W.T., Dimension of the crown S_n^k. *Discr. Math.* **8** (1974a), 85–103.   [584]

Trotter W.T., Irreducible posets with large height exist. *J. Combin. Th. A* **17** (1974b), 337–344. [583]

Trotter W.T., Inequalities in dimension theory for posets. *Proc. Amer. Math. Soc.* **47** (1975), 311–316.   [573, 583]

Trotter W.T., A generalization of Hiraguchi's: inequality for posets. *J. Combin. Th. A* **20** (1976), 114–123.   [583]

Trotter W.T., The dimension of the Cartesian product of partial orders. *Discr. Math.* **53** (1985), 255–263.   [572]

Trotter W.T., W.G. Gehrlein, and P.C. Fishburn, Balance theorems for height-2 posets. *Order* **9** (1992), 43–53.   [823]

Trotter W.T. and J.I. Moore, Jr., Some theorems on graphs and posets. *Discr. Math.* **15** (1976a), 79–84.   [583]

Trotter W.T. and J.I. Moore, Jr., Characterization problems for graphs, partially ordered sets, lattices, and families of sets. *Discr. Math.* **16** (1976b), 361–381.   [573]

Tsai M.T. and D.B. West, A new proof of 3-colorability of Eulerian triangulations. *Ars Math. Contemp.* **4** (2011), 73–77.   [419]

Tsai S.F., Problem 11987. *Amer. Math. Monthly* **124** (2017), 563. Solution **126** (2019), March. [551]

Tucker A.W., Some topological properties of disk and sphere. In *Proc. First Canadian Math. Congress, Montreal, 1945* (Univ. Toronto Press, 1946), 285–309.   [807, 809]

Turán P., Eine Extremalaufgabe aus der Graphentheorie. *Mat. Fiz Lapook* **48** (1941), 436–452.   [223, 475, 476, 489, 491]

Tutte W.T., On Hamiltonian circuits. *J. Lond. Math. Soc.* **21** (1946), 98–101.   [242, 401]

Tutte W.T., The factorization of linear graphs. *J. Lond. Math. Soc.* **22** (1947), 107–111.   [264]

Tutte W.T., The dissection of equilateral triangles into equilateral triangles. *Proc. Cambridge Philos. Soc.* **44** (1948), 463–482.   [749]

Tutte W.T., The factors of graphs. *Canad. J. Math.* **4** (1952), 314–328.   [268, 270, 276]

Tutte W.T., A short proof of the factor theorem for finite graphs. *Canad. J. Math.* **6** (1954), 347–352.   [269, 270, 276]

Tutte W.T., A theorem on planar graphs. *Trans. Amer. Math. Soc.* **82** (1956), 99–116.   [401]

Tutte W.T., A homotopy theorem for matroids, I, II. *Trans. Amer. Math. Soc.* **88** (1958), 144–174. [398]

Tutte W.T., Convex representations of graphs. *Proc. Lond. Math. Soc.* **10** (1960), 304–320.   [392, 783]

Tutte W.T., A theory of 3-connected graphs. *Indag. Math.* **23** (1961), 441–55.   [304, 305, 314, 535]

Tutte W.T., On the problem of decomposing a graph into n connected factors. *J. Lond. Math. Soc.* **36** (1961a), 221–230.   [251]

Tutte W.T., How to draw a graph. *Proc. London Math. Soc. (3)* **13** (1963), 743–767.   [398]

Tutte W.T., *Connectivity in Graphs* (Toronto Univ. Press, 1966).   [314]

Tutte W.T., *Introduction to the Theory of Matroids* (Amer. Elsevier, 1970).   [398, 520]

Tutte W.T., On the 2-factors of bicubic graphs. *Discr. Math.* **1** (1971), 203–208.   [317, 330]

Tutte W.T., The subgraph problem. *Ann. Discr. Math.* **3** (1978), 289–295.   [370]

Tutte W.T. and C.A.B. Smith, On Unicursal Paths in a Network of Degree 4. *Amer. Math. Monthly* **48** (1941), 233–237.   [752]

Tuza Z., Graph coloring in linear time. *J. Combin. Th. B* **55** (1992), 236–243.   [339, 343]

Tverberg H., A proof of the Jordan Curve Theorem. *Bull. Lond. Math. Soc.* **12** (1980), 34–38.   [378]

Tverberg H., On the decomposition of K_n into complete bipartite subraphs. *J. Graph Th.* **6** (1982), 493–494.   [771]

Ungar P., Problem 4385. *Amer. Math. Monthly* **57** (1950), 189. Solution **58** (1951), 573.   [763]

Ungar P., Problem E3052. *Amer. Math. Monthly* **91** (1984), 438. Solution **94** (1987), 185–186.   [175]

Valencia-Pabon M. and J.C. Vera, On the diameter of Kneser graphs. *Discr. Math.* **305** (2005), 383–385.   [251]

Valiant L.G., The complexity of enumeration and reliability problems. *SIAM J. Comput.* **8** (1979), 410–421.   [754]

van den Heuvel J. and S. McGuinness, Coloring the square of a planar graph. *J. Graph Theory* **42** (2003), 110–124.   [424]

van der Waerden B.L., Beweis einer Baudetschen Vermutung (in German). *Niew. Arch. Wisk.* **15** (1927), 212–216.   [463, 464]

van der Waerden B.L., *Moderne Algebra Vol. 1* (2nd ed.) (Springer-Verlag, 1937). (Many editions.)   [514, 520]

van Lint J.H., *Combinatorial Theory Seminar, Eindhoven University of Technology, Lect. Notes in Math.* **382** (Springer-Verlag, 1974).   [139]

van Lint J.H. and R.M. Wilson, *A course in combinatorics* (Cambridge Univ. Press, 1992).   [609, 622, 639]

Vanden Eynden C., Problem E3435. *Amer. Math. Monthly* **98** (1991), 365. Solution **99** (1992), 881–882.   [92]

Vandenbussche J. and D.B. West, Matching extendability in hypercubes. *SIAM J. Discr. Math.* **23** (2009), 1539–1547.   [260]

Vandermonde A.-T., Mémoire sur des irrationanelles de différents ordres avec une application au cercle (French). *Acad. des Sci.* (1772).   [26]

Vapnik V.N. and A.Y. Chervonenkis, Theory of uniform convergence of frequencies of events to their probabilities and problems of search for an optimal solution from empirical data. *Avtomat. i Telemeh.* (1971), 42–53.   [711, 712]

Vassilev M. and K. Atanassov, On Delanoy numbers. *Annuaire Univ. Sofia Fac. Math. Inform.* **81** (1994), 153–162.   [29]

Vazirani V.V., A theory of alternating paths and blossoms for proving correctness of the $O(|V^{1/2}\|E|)$ general graph matching algorithm. *Combinatorica* **14** (1994), 71–91.   [284]

Vazirani V.V. and M. Yannakakis, Pfaffian orientations, 0-1 permanents, and even cycles in directed graphs. *Discrete Appl. Math.* **25** (1989), 179–190.   [754, 756]

Veršik A.M. and S.V. Kerov, Asymptotic behavior of the Plancherel measure of the symmetric group and the limit form of Young tableaux. *Dokl. Akad. Nauk SSSR* **233** (1977), 1024–1027.   [722, 821]

Verstraëte J., On arithmetic progressions of cycle lengths in graphs. *Combin. Probab. Comput.* **9** (2000), 369–373.   [479]

Viennot G., Une forme géométrique de la correspondance de Robinson–Schensted. In *Combinatoire et représentation du groupe symétrique (Actes Table Ronde CNRS, Univ. Louis-Pasteur Strasbourg, 1976), Lect. Notes Math.* **579** (Springer, 1977), 29–58.   [199, 200]

Vijayaditya N., On total chromatic number of a graph. *J. London Math. Soc. (2)* **3** (1971), 405–408.   [373]

Vince A., Problem E1771. *Amer. Math. Monthly* **72** (1965), 316. Solution **73** (1966), 543.   [441]

Vince A., Problem 6617. *Amer. Math. Monthly* **96** (1989), 642.   [782]

Vitaver L.M., Determination of minimal coloring of vertices of a graph by means of Boolean powers of the incidence matrix (Russian). *Dokl. Akad. Nauk. SSSR* **147** (1962), 758–759.   [339]

Vizing V.G., The Cartesian product of graphs. *Vyč. Sis.* **9** (1963), 30–43.   [343]

Vizing V.G., On an estimate of the chromatic class of a *p*-graph. *Diskret. Analiz.* **3** (1964), 25–30.   [358, 359, 699]

Vizing V.G., Critical graphs with a given chromatic class (Russian). *Diskret. Analiz.* **5** (1965), 9–17.   [362, 422]

Vizing V.G., Some unsolved problems in graph theory. *Uspekhi Mat. Nauk (Russian Math. Surveys)* **23** (1968), 9117–134.   [414, 423]

Vizing V.G., Coloring the vertices of a graph in prescribed colors (Russian). *Diskret. Analiz.* **29** (1976), 3–10.   [346, 347, 355, 356]

Vizing V.G. and M.K. Goldberg, The length of a circuit of a strongly connected graph. *Kibernetika (Kiev)* (1969), 79–82.   [237]

Voigt M., List colourings of planar graphs. *Discr. Math.* **120** (1993), 215–219.   [402]

Voigt M. and B. Wirth, On 3-colorable non-4-choosable planar graphs. *J. Graph Th.* **24** (1997), 233–235.   [402]

Volkmann L., Bemerkungen zum *p*-fachen Kantenzusammenhang von Graphen. *An. Univ. Bucureşti Mat.* **37** (1988), 75–79.   [296]

Volkmann L., *Graphen und Digraphen* (Springer-Verlag, 1991).   [215]

Volkmann L., Regular graphs, regular factors, and the impact of Petersen's theorems. *Jahresber. Deutsch. Math.-Verein.* **97** (1995), 19–42.   [268]

Volkmann L., *Fundamente der Graphentheorie* (Springer-Verlag, 1996).   [215]

Voloshin V.I., Properties of triangulated graphs (Russian). In *Oper. Research and Progr.* (B. A. Shcherbakov, ed.) (Shtiintsa, 1982), 24–32.   [374]

Voloshin V.I., Problem 10976. *Amer. Math. Monthly* **109** (2002), 855. Solution **111** (2004), 444–445.   [174, 374]

Voloshin V.I. and I.M. Gorgos, Some properties of 1-simply connected hypergraphs and their applications. *Mat. Issled.* (1982), 30–33, 191.   [374]

von Neumann J., A certain zero-sum two-person game equivalent to the optimal assignment problem. In *Contributions to the theory of games, vol. 2, Annals of Mathematics Studies, no. 28* (Princeton Univ. Press, 1953), 5–12.   [257]

von Staudt K.G.C., *Geometrie de Lage, Verlag von Bauer and Rapse 25* (Julius Merz, 1847).   [387]

Wagner K., Bemerkungen zum Vierfarbenproblem. *Jber. Deutsch. Math. Verein.* **46** (1936), 21–22.   [393]

Wagner K., Über eine Eigenschaft der ebenen Komplexe. *Math. Ann.* **114** (1937), 570–590.   [390, 398, 527]

Wagon S., A bound on the chromatic number of graphs without certain induced subgraphs. *J. Combin. Th. B* **29** (1980), 245–246.   [343]

Wagon S., Fourteen proofs of a result about tiling a rectangle. *Amer. Math. Monthly* **94** (1987), 601–617.   [221]

Wagon S. and P. Zielinski, Problem 12082. *Amer. Math. Monthly* **125** (2018), 945. Solution **127** (2020), 571.   [655]

Wallis W.D., *Combinatorial designs, Monographs and Textbooks in Pure and Applied Mathematics* **118** (Marcel Dekker, 1988). Also 2007.   [609]

Wallis W.D. (ed.), *Computational and constructive design theory*, Mathematics and its Applications **368** (Kluwer Acad., 1996).   [609]

Wallis W.D., *A beginner's guide to graph theory* (Birkhäuser Boston, 2000).   [215]

Wallis W.D. (ed.), *Designs 2002*, Mathematics and its Applications **563** (Kluwer Acad., 2003).   [609]

Walsh T.R., The towers of Hanoi revisited: moving the rings by counting the moves. *Inform. Process. Lett.* **15** (1982), 64–67.   [80]

Walsh T.R., Iteration strikes back—at the cyclic Towers of Hanoi. *Inform. Process. Lett.* **16** (1983), 91–93.   [80]

Walter J.R., *Representations of rigid cycle graphs* (Wayne State Univ., 1972). Ph.D. Thesis.   [373]

Walters I.C., Jr., The ever expanding expander coefficients. *Bull. Inst. Combin. Appl.* (1996), 97.   [779]

Wang D.L. and D.J. Kleitman, On the existence of n-connected graphs with prescribed degrees ($n \geq 2$). *Networks* **3** (1973), 225–239.   [226]

Wang D.L. and P. Wang, Some results about the Chvátal conjecture. *Discr. Math.* **24** (1978), 95–101.   [501]

Wang J.F., D.B. West, and B. Yao, Maximum bandwidth under edge addition. *J. Graph Theory* **20** (1995), 87–90.   [819]

Wang W. and K.W. Lih, Choosability and edge choosability of planar graphs without five cycles. *Appl. Math. Lett.* **15** (2002), 561–565.   [413]

Wang Y., Notes on Chvátal's conjecture. *Discr. Math.* **247** (2002), 255–259.   [503]

Wardlaw W.P., Problem E3358. *Amer. Math. Monthly* **96** (1989), 928. Solution **98** (1991), 650.
                                                                                                        [116, 173]

Warren H.E., Lower bounds for approximation by nonlinear manifolds. *Trans. Amer. Math. Soc.* **133** (1968), 167–178.   [830]

Watkins M.E., A lower bound for the number of vertices of a graph. *Amer. Math. Monthly* **74** (1967), 297.   [313]

Watkins M.E., On the existence of certain disjoint arcs in graphs. *Duke Math. J.* **35** (1968), 231–246.   [314]

Weaver W., Questions, Discussions, and Notes: Lewis Carroll and a Geometrical Paradox. *Amer. Math. Monthly* **45** (1938), 234–236.   [60]

Wei V.K., A Lower Bound on the Stability Number of a Simple Graph. Tech. Rep. TM 81-11217-9, Bell Laboratories (1981).   [263, 662]

Weinstein J.M., On the number of disjoint edges in a graph. *Canad. J. Math.* **15** (1963), 106–111.
                                                                                                            [263]

Welsh D.J.A., *Matroid Theory* (Academic Press, 1976).   [514, 520, 532]

Welsh D.J.A., Colouring problems and matroids. In *Surveys in combinatorics (Proc. 7th British Combi. Conf., 1979)*, London Math. Soc. Lecture Note Ser. **38** (Cambridge Univ. Press, 1979), 229–257.
                                                                                                      [538, 539]

Welsh D.J.A. and M.B. Powell, An upper bound for the chromatic number of a graph and its application to timetabling problems. *Computer J.* **10** (1967), 85–87.   [337]

Wendel J.G., A problem in geometric probability. *Math. Scand.* **11** (1962), 109–111.   [669]

Wernicke P., Über den kartographischen Vierfarbensatz. *Math. Ann.* **58** (1904), 413–426.   [406]

West D.B., Pairs of adjacent Hamiltonian circuits with small intersection. *Stud. Appl. Math.* **59** (1978), 245–248.   [331]

West D.B., A symmetric chain decomposition of $L(4, n)$. *European J. Combin.* **1** (1980), 379–383.
                                                                                                      [557, 565]

West D.B., Extremal problems in partially ordered sets. In *Ordered sets (Banff, Alta., 1981)*, NATO Adv. Study Inst. C: Math. Phys. Sci. **83** (Reidel, 1982a), 473–521.   [493]

West D.B., The number of complete subgraphs in graphs with non-majorizable degree sequences. In *Proc. Silver Jubilee Conf. (Waterloo, 1982), Progress in Graph Theory* (Academic Press, 1982b), 509–521.   [491]

West D.B., "Poly-unsaturated" posets: the Greene–Kleitman theorem is best possible. *J. Combin. Th. A* **41** (1986), 105–116.   [551]

West D.B., *Introduction to graph theory* (Prentice Hall, 1996). Also 2001.   [215]

West D.B., The superregular graphs. *J. Graph Th.* **23** (1996), 289–295.   [782]

West D.B., Short proofs for interval digraphs. *Discr. Math.* **178** (1998), 287–292.   [605]

West D.B., *Introduction to graph theory, 2nd ed.* (Prentice Hall, 2001). Also 1996.   [539]

West D.B., A short proof of the Berge–Tutte formula and the Gallai–Edmonds structure theorem. *European J. Combin.* **32** (2011), 674–676.   [275]

West D.B., L.H. Harper, and D.E. Daykin, Some remarks on normalized matching. *J. Combin. Th. A* **35** (1983), 301–308.   [564, 567]

West D.B. and D.J. Kleitman, Skew chain orders and sets of rectangles. *Discr. Math.* **27** (1979), 99–102.   [565]

West D.B., W.T. Trotter, G.W. Peck, and P. Shor, Regressions and monotone chains: a Ramsey-type extremal problem for partial orders. *Combinatorica* **4** (1984), 117–119.   [442]

West D.B. and W.H. Wiedemann, Problem E3290. *Amer. Math. Monthly* **95** (1988), 872. Solution **97** (1990), 428–429.   [176]

West Don, Solution to problem E2404. *Amer. Math. Monthly* **81** (1974), 287. Proposed **80** (1973), 316.   [106]

White D.E. and S.G. Williamson, Recursive matching algorithms and linear orders on the subset lattice. *J. Combinatorial Theory A* **23** (1977), 117–127.   [566]

Whiting P.D. and J.A. Hillier, A method for finding the shortest route through a road network. *Operations Research Quart.* **11** (1960), 37–40.   [246]

Whitney H., A theorem on graphs. *Ann. of Math.* (2) **32** (1931), 378–390.   [401, 418]

Whitney H., Congruent graphs and the connectivity of graphs. *Amer. J. Math.* **54** (1932a), 150–168.   [293, 300, 306, 314]

Whitney H., A logical expansion in Mathematics. *Bull. Amer. Math. Soc.* **38** (1932b), 572–579.   [158]

Whitney H., Non-separable and planar graphs. *Trans. Amer. Math. Soc.* **34** (1932c), 339–362.   [526]

Whitney H., 2-isomorphic graphs. *Amer. J. Math.* **55** (1933), 245–254.   [380, 398]

Whitney H., On the abstract properties of linear dependence. *Amer. J. Math.* **57** (1935), 509–533.
                                                                                        [514, 520, 523, 536]

Whitworth W.A., *Choice and Chance (2nd ed.)* (Deighton, Bell & Co., 1897). First edition 1867.
                                                                                          [33, 40, 46, 115]

Wiener N., A contribution to the theory of relative position. *Proc. Cambridge Philos. Soc.* **17** (1914), 441–449.   [586, 605]

Wilf H.S., The eigenvalues of a graph and its chromatic number. *J. Lond. Math. Soc.* **42** (1967), 330–332.   [769]

Wilf H.S., The friendship theorem. In *Combinatorial Mathematics and Its Applications (Oxford, 1969)* (Academic Press, 1971), 307–309.   [775]

Wilf H.S., *generatingfunctionology* (Academic Press, 1990).   [87, 112, 117, 129, 135, 136, 140]

Wilf H.S., Problem 10578. *Amer. Math. Monthly* **104** (1997), 270. Solution **106** (1999), 169.   [91]

Will T.G. and H. Hulett, Parsimonious multigraphs. *SIAM J. Discr. Math.* **18** (2004), 241–245.   [238]

Williamson J., Hadamard's determinant theorem and the sum of four squares. *Duke Math. J.* **11** (1944), 65–81.   [622]

Wilson R.M., Constructions and uses of pairwise balanced designs. In *Combin. (Proc. NATO Advanced Study Inst., 1974) Part 1: Theory of designs, finite geometry and coding theory, Centre Tracts* **55** (Math. Centrum, 1974), 18–41.   [641]

Wilson R.M., An existence theory for pairwise balanced designs, II–III. *J. Combin. Th. A* **18** (1975), 71–79.   [640]

Wilson R.M., The exact bound in the Erdős–Ko–Rado theorem. *Combinatorica* **4**(1984),247–257.
                                                                                                    [500]

Win S., Existenz von Gerüsten mit vorgeschriebenem Maximalgrad in Graphen. *Abh. Math. Sem. Univ. Hamburg* **43** (1975), 263–267.   [327, 328]

Win S., On a conjecture of Las Vergnas concerning certain spanning trees in graphs. *Resultate Math.* **2** (1979), 215–224.   [327]

Win Z., On the windy postman problem on Eulerian graphs. *Math. Programming* **44** (1989), 97–112.
                                                                                                    [327]

Winkler P., Random orders. *Order* **1** (1985), 317–331.   [821]

Winkler P.M., Average height in a partially ordered set. *Discr. Math.* **39** (1982), 337–341.   [608]

Winkler P.M., Proof of the squashed cube conjecture. *Combinatorica* **3** (1983), 135–139.   [781]

Winkler P.M., Correlation and order. In *Combinatorics and ordered sets (Arcata, 1985)*, Contemp. Math. **57** (Amer. Math. Soc., 1986), 151–174.   [600]

Winkler P.M., *Mathematical puzzles: a connoisseur's collection* (A K Peters, 2004).   [221, 667]

Winkler P.M., Puzzled: Solutions and sources. *Commun. ACM* **51** (2008), 118–118.   [431]

Wismath S.K., Characterizing bar line-of-sight graphs. In *Proc. 1st Symp. Comput. Geo. (Baltimore, 1985)* (ACM, 1985), 147–152.   [394]

Witt E., Ein kombinatorischer Satz der Elementargeometrie (in German). *Math. Nachr.* **6** (1952), 261–262.   [467]

Wood D., The towers of Brahma and Hanoi revisited. *J. Recreational Math.* **14** (1981/82), 17–24.   [80]

Woodall D.R., Sufficient conditions for circuits in graphs. *Proc. Lond. Math. Soc.* **24** (1972), 739–755.
                                                                                                    [323]

Woodall D.R., An exchange theorem for bases of matroids. *J. Combin. Th. B* **16** (1974), 227–228.   [538]

Woodall D.R., Cyclic-order graphs and Zarankiewicz's crossing-number conjecture. *J. Graph Th.* **17** (1993), 657–671.   [793, 798]

Woodall D.R., The average degree of an edge-chromatic critical graph. II. *J. Graph Th.* **56** (2007), 194–218.   [414]

Woodall D.R., The average degree of a multigraph critical with respect to edge or total choosability. *Discr. Math.* **310** (2010), 1167–1171.   [414]

Woolhouse W.S.B., Prize question 1733. *Lady's and Gentleman's Diary* (1844).   [641]

Wormald N.C., *Some problems in the enumeration of labelled graphs* (Univ. Newcastle, 1978). Ph.D. Thesis.   [697, 704]

Wormald N.C., The differential equation method for random graph processes and greedy algorithms. In *Lectures on Approximation and Randomized Algorithms* (M. Karonski and H.J. Proemel, eds.) (PWN, Warsaw, 1999a), 73–155.   [722]

Wormald N.C., Models of random regular graphs. In *Surveys in combinatorics, 1999 (Canterbury)*, London Math. Soc. Lecture Note Ser. **267** (Cambridge Univ. Press, 1999b), 239–298.   [697]

Worpitzky J., Studien über die *bernoulli*schen und *euler*schen Zahlen (German). *J. Reine Angewandte* **94** (1883), 203–232.   [101]

Wu J.L., On the linear arboricity of planar graphs. *J. Graph Theory* **31** (1999), 129–134.   [412, 422]

Wu J.L. and Y.W. Wu, The linear arboricity of planar graphs of maximum degree seven is four. *J. Graph Theory* **58** (2008), 210–220.   [412]

Xiao S., Solution to problem 11509. *Amer. Math. Monthly* **119** (2012), 430. Proposed **117** (2010), 558.
                                                                                                     [24]

Xiao Y. and H. Zhao, New method for counting the number of spanning trees in a two-tree network. *Phys. A* **392** (2013), 4576–4583.   [65]

Yaglom A.M. and I.M. Yaglom, *Challenging mathematical problems with elementary solutions. Vol. I: Combinatorial analysis and probability theory*, Translated by James McCawley, Jr.; revised and edited by Basil Gordon (Holden-Day, 1964).   [24]

Yamamoto K., Logarithmic order of free distributive lattice. *J. Math. Soc. Japan* **6** (1954), 343–353.
                                                                                                    [558]

Yamashita T., Degree sum conditions and Chvátal–Erdős conditions for cycles. In *The COE Seminar on Mathematical Sciences 2004*, Sem. Math. Sci. **31** (Keio Univ., 2004), 185–199.   [316]

Yan C.H., Parking functions. In *Handbook of enumerative combinatorics*, Discr. Math. Appl. (Boca Raton) (CRC Press, 2015), 835–893.   [50]

Yannakakis M., The complexity of the partial order dimension problem. *SIAM J. Algebraic Discrete Methods* **3** (1982), 351–358.   [573]

Yao A.C.C., Should tables be sorted? *J. Assoc. Comput. Mach.* **28** (1981), 615–628.   [447, 448]

Yao F.F., D.P. Dobkin, H. Edelsbrunner, and M.S. Paterson, Partitioning space for range queries. *SIAM J. Comput.* **18** (1989), 371–384. [815]

Yap H.P., Total colourings of graphs. *Bull. London Math. Soc.* **21** (1989), 159–163. [373, 654]

Yates F., A new method of arranging field trials involving a large number of varieties. *J. Agric. Sci. Comb.* **26** (1936), 424–455. [612]

Yeh Y.N., A remarkable endofunction involving compositions. *Stud. Appl. Math.* **95** (1995), 419–432. [144]

Young A., Quantitative substitutional analysis I. *Proc. London Math. Soc.* **33** (1901), 97–146. [189]

Young A., Quantitative substitutional analysis II. *Proc. London Math. Soc.* **35** (1902), 361–397. [190, 207]

Yu X., Problem 10575. *Amer. Math. Monthly* **104** (1997), 168–169. Solution **106** (1999), 266–268. [134]

Zaker M., On lower bounds for the chromatic number in terms of vertex degree. *Discr. Math.* **311** (2011a), 1365–1370. [341]

Zaker M., Bounds for chromatic number in terms of even-girth and booksize. *Discr. Math.* **311** (2011b), 197–204. [342]

Zaks J., Towards a simpler proof of the Brouwer fixed point theorem. *Geombinatorics* **5** (1995), 35–37. [804]

Zamani R. and D.B. West, Spanning cycles through specified edges in bipartite graphs. *J. Graph Theory* **71** (2012), 1–17. [326]

Zarankiewicz K., On a problem of P. Turán concerning graphs. *Fund. Math.* **41** (1954), 137–145. [793]

Zaslavsky T., Glossary and bibliography of signed and gain graphs and allied areas. *Electron. J. Combin.* **5** (1998), Dynamic Surveys 8, 124 and 9, 41. [237]

Zeckendorf E., A generalized Fibonacci numeration. *Fibonacci Quart.* **10** (1972), 365–372. [61]

Zehavi A. and A. Itai, Three tree-paths. *J. Graph Theory* **13** (1989), 175–188. [315]

Zeilberger D., Sister Celine's technique and its generalizations. *J. Math. Anal. Appl.* **85** (1982), 114–145. [90]

Zeilberger D., A short hook-lengths bijection inspired by the Greene–Nijenhuis–Wilf proof. *Discr. Math.* **51** (1984), 101–108. [192]

Zeilberger D., Kathy O'Hara's constructive proof of the unimodality of the Gaussian polynomials. *Amer. Math. Monthly* **96** (1989), 590–602. [557]

Zeilberger D., A fast algorithm for proving terminating hypergeometric identities. *Discr. Math.* **80** (1990), 207–211. [90]

Zeilberger D., The method of creative telescoping. *J. Symbolic Comput.* **11** (1991), 195–204. [90]

Zeilberger D., Reverend Charles to the aid of Major Percy and Fields medalist Enrico. *Amer. Math. Monthly* **103** (1996), 501–502. [178]

Zeilberger D., Dodgson's determinant-evaluation rule proved by two-timing men and women. *Electron. J. Combin.* **4** (1997), Research Paper 22. [178]

Zhan S., On Hamiltonian line graphs and connectivity. *Discr. Math.* **89** (1991), 89–95. [326]

Zhang K.M. and Z.M. Song, Cycles in digraphs—a survey. *Nanjing Daxue Xuebao Ziran Kexue Ban* **27** (1991), 188–215. [316]

Zhang L., Every planar graph with maximum degree 7 is of class 1. *Graphs Combin.* **16** (2000), 467–495. [422]

Zhang L. and B. Wu, Edge choosability of planar graphs without small cycles. *Discr. Math.* **283** (2004), 289–293. [413]

Zhou B., A note on the Erdős–Sós conjecture. *Acta Math. Sci. (English Ed.)* **4** (1984), 287–289. [248]

Zhou L., Problem 11187. *Amer. Math. Monthly* **112** (2005), 929. Solution **114** (2007), 554. [66]

Zhu Q.C., The structure of α-critical graphs with $|V(G)|-2\alpha(G) = 3$. In *Graph theory and its applications: East and West (Jinan, 1986)*, *Ann. New York Acad. Sci.* **576** (1989), 716–722. [458]

Zhu X., The fractional version of Hedetniemi's conjecture is true. *European J. Combin.* **32** (2011), 1168–1175. [461]

Zhu X., The Alon–Tarsi number of planar graphs. *J. Combin. Th. B* **134** (2019), 354–358. [742]

Zhu Y.J., Z.H. Liu, and Z.G. Yu, An improvement of Jackson's result on Hamilton cycles in 2-connected regular graphs. In *Cycles in Graphs (Burnaby, 1982)* (B. Alspach and C. Godsil, eds.) (North-Holland, 1985), 237–247.   [326]

Ziegler G.M., Generalized Kneser coloring theorems with combinatorial proofs. *Invent. Math.* **147** (2002), 671–691.   [812]

Zito J., The structure and maximum number of maximum independent sets in trees. *J. Graph Theory* **15** (1991), 207–221.   [264]

Znám Š., Two improvements of a result concerning a problem of K. Zarankiewicz. *Colloq. Math.* **13** (1965), 255–258.   [631]

Zykov A.A., On some properties of linear complexes (Russian). *Mat. Sbornik* **24** (1949), 163–188.
                                                                                [354, 476]

Author Index

Glossary of Notation

Relations, operators, positional notation

$\{\}$ - set

$\cap, \cup, \in, \triangle$ - intersection, union, membership, symmetric difference

$\lfloor x \rfloor, \lceil x \rceil$ - floor, ceiling (nearest integer at most or at least x)

$[k]$ - $\{1, \ldots, k\}$

$k\text{-}object$ - "object" with parameter value k

$[a, b]$ - $\{a, \ldots, b\}$, if a, b are integers

$[x, y]$ - $\{z : x \le z \le y\}$, if x, y are real numbers or poset elements

$[x^k]$ - coefficient of x^k in a formal power series in x

$[S, T]$ - cut consisting of all edges from S to T

$\langle v_1, \ldots, v_n \rangle$ - path with vertices $v_1, \ldots, v_n$ in order

$[v_1, \ldots, v_n]$ - cycle with vertices $v_1, \ldots, v_n$ in order

$n! - \prod_{i=1}^{n} i$

$n!! - \prod_{i=0}^{\lfloor (n-1)/2 \rfloor} (n - 2i)$

$n_{(r)} - n(n - 1) \cdots (n - r + 1)$

$n^{(r)} - n(n + 1) \cdots (n + r - 1)$

$d \mid n$ - d divides n

$d \nmid n$ - d does not divide n

$|S|$ - size of set S

$\#\{i : i \in S\}$ - size of described set

$\#(G)$ - number of components of a graph

A^T - transpose of a matrix A

$\overline{S}$ - complement of set S (within a given universe)

$\overline{G}$ - complement of graph G

G^*, P^*, M^* - dual (of a planar graph, poset, or matroid)

$\leftrightarrow$ - adjacency relation on vertices of graph

$\rightarrow$ - succession relation on nodes (vertices) of digraph

$\parallel$ - incomparability relation on elements of poset

$\prec, \succ$ - cover relation on elements of poset

$a_n \sim b_n$ - asymptotic to (ratio approaching 1)

$\lambda \vdash n$ - λ partitioning the integer n, with $\lambda_1 \ge \cdots \ge \lambda_k$ and $\sum \lambda_i = n$

$G - v$, $G - S$, $G - e$, $G - F$ - deletion of vertex, vertex set, edge, edge set

$G \cdot e$, $M \cdot e$ - contraction of edge or matroid element

G^- - graph formed by deleting one edge from an edge-transitive graph G

G^+ - graph formed by adding one edge to a graph G with edge-transitive $\overline{G}$

$M|F$, $M.F$ - restriction or contraction of a matroid to the elements F

$\mathbf{0}$, $\mathbf{1}$ - unique lower and upper bounds of a poset

$\mathbf{1}_n$ - column vector of n 1s

$\mathbf{2}$, $\mathbf{n}$ - two-element, n-element chains in posets

$\mathbf{2}^n$ - subset poset of order n

$\mathbf{2}^E$ - subsets of E, ordered by inclusion

$\binom{n}{k}$ - binomial coefficient counting k-subsets of the set $[n]$

$\binom{n}{k_1, \ldots, k_m}$ - multinomial coefficient

$\binom{S}{k}$ - the collection of k-subsets of set S

$G + H$ - disjoint union of graphs or posets

mG - disjoint union of m copies of graph or poset

$G \oplus H$ - join (disjoint union plus all edges between)

$G[S]$ - subgraph of G induced by S

$G[H_1, \ldots, H_n]$, $G[H]$ - lexicographic product, composition (graphs or posets)

$G \square H$ - Cartesian product of graphs

$G \boxtimes H$ - strong product

$G * H$ - weak product, tensor product

$P \times Q$ - Cartesian product of sets or posets

P^k - Cartesian product of k copies of poset P

G^k - graph on $V(G)$ with $u \leftrightarrow v$ if $d_G(u, v) \le k$

P_k - kth rank of a ranked poset P

$x \vee y$ - least upper bound in poset (join)

$x \wedge y$ - greatest lower bound in poset (meet)

X^-, X^+ - down-set and up-set generated by set X in poset

Usage of Roman alphabet

$A(G)$ - adjacency matrix of a graph G

$\mathbf{A}(P)$ - incidence algebra of a poset P

A_n - n-element antichain

$A(n, k)$ - Eulerian number counting permutations of $[n]$ with k runs

$A(x)$, $B(x)$ - typical generating functions with formal variable x

AP - arithmetic progression

box G - boxicity of a graph G

$B(G)$ - bandwidth of a graph G, block graph of G

B_n - Bell number counting partitions of $[n]$

$\mathbf{B}(M)$ - bases of a matroid M

Bin (n, p) - binomial distribution (n trials, success probability p)

BIBD - balanced incomplete block design

$c(G)$ - circumference of a graph G

$c(n, k)$ - number of permutations of $[n]$ with k cycles

ch(G) - choosability (list chromatic number) of a graph G

$\mathrm{cr}(G)$ - crossing number of a graph G

$\mathrm{cr}'(G)$ - (recti)linear crossing number of a graph G

$\mathrm{cr}_k(G)$ - crossing number when each edge is at most k segments

$\mathbb{C}$ - complex numbers

C_n - cycle of length n, nth Catalan number

$C(P)$ - set of critical (unforced) incomparable pairs of a poset P

$\mathbf{C}(M)$ - circuits of a matroid M

$d(v), d_G(v)$ - degree, valence (of a vertex v in G)

$\overline{d}(G)$ - average vertex degree in a graph G

$d(u,v), d_G(u,v)$ - distance between u and v (in a graph G)

$d_1, \ldots, d_n$ - degree list

$d_k(P)$ - size of maximum k-family in a poset P

$\hat{d}_k(P)$ - size of maximum k-cofamily in a poset P

$d_{m,n}$ - Delannoy number counting up/right/diagonal paths from $(0,0)$ to (m,n)

$\mathrm{def}(S)$ - deficiency of a vertex set S (given by $o(G-S) - |S|$)

$\mathrm{df}(S)$ - defect of a set S in a bipartite graph (given by $|S| - |N(S)|$)

det - determinant of a matrix

$\mathrm{diam}\,(G)$ - diameter of a graph G

$\dim\,(P)$ - order dimension of a poset P

$\partial_k(m)$ - numerical shadow of k-binomial expansion of an integer m

$\partial(F)$ - shadow of a set-family F

$D[S]$ - down-set (ideal) of poset, generated by set S, often $D[x]$

D_n - number of derangements of $[n]$

D_N - divisors of N, ordered by divisibility

e - base of natural logarithm, 2.71828...

$e(P)$ - number of linear extensions of a poset P

$e_k(x)$ - number of linear extensions with x at height k

$\mathrm{ex}\,(n; H)$ - maximum number of edges in n-vertex graph not containing H

E - typical set of elements of a matroid

$E(G)$ - set of edges in a (hyper)graph G

$\mathbb{E}(X)$ - expectation of random variable X

EGF - exponential generating function

f_k - rank parameters of family F in a ranked poset

f_λ - number of Young tableaux with shape λ

$f(T)$ - in inclusion-exclusion, #elements whose properties are indexed by T

F_n - nth classical Fibonacci number

$\hat{F}_n$ - nth adjusted Fibonacci number (equal to F_{n+1})

$\mathbb{F}_q$ - q-element field

F, G, H - typical graphs, digraphs, or hypergraphs

$g(G)$ - girth

G^p - random graph in Model A (edge-probability model)

G^m - random graph in Model B (fixed-size model)

$\mathbb{G}(n, p)$ - random graph model with edge-probability p

$G(P)$ - comparability graph of a poset P

$h(P)$ - height of a poset P

$h(x)$ - height of poset element x, or expected height of element

H_n - nth harmonic number, $\sum_{i=1}^{n} 1/i$

$H_{k,n}$ - k-connected Harary graph with n vertices

$i(G)$ - interval number of a graph G

$\mathbf{I}(M)$ - independent sets of a matroid M

$I(P)$ - set of incomparable pairs of poset P

$\text{Int}(P)$ - family of intervals in a poset P

$I(x)$ - set of elements incomparable to x

I - identity matrix (diagonal matrix with 1s on diagonal),

J - square matrix with every entry equal to 1

$J(P)$ - lattice of order ideals of a poset P, ordered by inclusion

K_n - complete graph of order n

$K_{m,n}$ - complete bipartite graph with parts of sizes m and n

$K_{S,T}$ - complete bipartite graph with parts S and T

$K(n, k)$ - Kneser graph with vertex set $\binom{[n]}{k}$

$l(P)$ - length of a poset P (length of longest chain)

lg - logarithm in base 2

ln - natural logarithm

$L(G)$ - line graph of a graph G

$L(v)$ - list of colors at a vertex in a list assignment

$L(m, n)$ - poset of m-tuples a with $0 \leq a_1 \leq \cdots \leq a_m$, componentwise ordered

$L_n(q)$ - lattice of subspaces of finite vector space of dimension n over $GF(q)$

m - often the number of edges in a graph

$m_k(\mathbf{C}), m_k(a)$ - k-norm of chain partition or sequence $[= \sum_i \min\{k, a_i\}]$

M - a matching, a matroid, etc.

M^e - chain product $\underline{e_1} \times \cdots \times \underline{e_n}$

$M(G)$ - cycle matroid of a graph G

$\text{Mad}(G)$ - maximum average degree among subgraphs of a graph G

$\text{MOLS}(n, k)$ - family of k pairwise (mutually) othogonal Latin squares of order n

M_5 - the 5-element modular non-distributive lattice

n - typically the number of vertices or elements in a (hyper)graph or poset

N_5 - the 5-element non-modular non-distributive lattice

$N_k(P)$ - size of kth rank of a poset P

$N(x), N_G(x)$ - (open) neighborhood of x in a graph G

$N[x]$ - $N(x) \cup \{x\}$ (closed neighborhood)

$N(S)$ - $\cup_{x \in S} N(x)$

$N^+(x)$ - out-neighborhood (successor set) of x in a digraph

$N^-(x)$ - in-neighborhood (predecessor set) of x in a digraph

$N(m)$ - maximum number of pairwise orthogonal Latin squares of order m

$\mathbb{N}$ - set of natural numbers

$\mathbb{N}_0$ - $\mathbb{N} \cup \{0\}$

$o(G)$ - number of odd components

$o(f(n))$ - functions whose ratio to $f(n)$ approach 0

$O(f(n))$ - functions bounded (for large n) by a constant multiple of $f(n)$

O_k - Odd graph (disjointness graph on k-sets and $(k + 1)$-sets in $[2k + 1]$)

$\mathbf{O}(Q)$ - order polytope of a poset Q

OGF - ordinary generating function

OA - orthogonal array

$p(n, \varepsilon)$ - threshold probability function

per - permanent of a matrix

$\mathbb{P}(A)$ - probability of event A

P_n - path with n vertices

$P(n, k)$ - generalized Petersen graph

PIE - inclusion-exclusion principle

$P(L)$ - poset of join-irreducibles of a lattice L

PBD - pairwise balanced design

$P(\sigma)$ - P-symbol of a permutation σ

$Q(\sigma)$ - Q-symbol of a permutation σ

$\mathbb{Q}$ - rational numbers

Q_k - hypercube of dimension k (as a graph)

$r(M)$, $r(P)$ - rank of a matroid or poset

$r(x)$, $r(X)$ - rank of an element in a poset, rank of a set in a matroid

r_M - rank function of matroid M

$\mathbb{R}$ - real numbers

$R_B(x)$ - rook polynomial of the board B

$R(p_1, \ldots, p_k; r)$ - Ramsey number for k-coloring r-sets

$R_k(p; r)$ - Ramsey number for k-coloring r-sets with common threshold p

$R(p, q)$ - Ramsey number for 2-coloring 2-sets

$R(G_1, \ldots, G_k)$ - graph Ramsey number

$\mathbb{S}_n$ - symmetric group on n elements (set of permutations of $[n]$)

S_g - surface with g handles

$S(n, k)$, $s(n, k)$ - Stirling numbers of second and first kinds

SDR - system of distinct representatives

STS(v) - Steiner triple system of order v

T - typical tree or tournament

$T_{n,r}$ - Turán graph (equipartite complete r-partite n-vertex graph)

$U[S]$ - up-set in poset generated by set S, typically $U[x]$

$V(G)$ - set of vertices of a graph or hypergraph G

$w(l, k)$ - van der Waerden number (guaranteeing l-term AP in k-coloring)

$w(P)$ - width of a poset P

$W_k(P)$ - size of kth rank of a poset (Whitney numbers)

X - typical random variable

X^+, X^- - positive and negative parts in a signed involution

X, Y - typical bipartition of a graph into two independent sets

Y - Young lattice of all partitions of integers, ordered componentwise

$\mathbb{Z}$ - integers

Usage of Greek alphabet

$\alpha(G)$ - maximum size of independent set in a graph G

$\alpha'(G)$ - maximum size of a matching in G (pairwise independent edges)

$\beta(G)$ - minimum number of vertices covering all edges in G

$\beta'(G)$ - minimum number of edges covering all vertices in G

$\gamma(G)$ - genus, domination number of a graph G

$\delta_{i,j}$ - Kronecker delta (1 if $i = j$, 0 if $i \neq j$)

$\delta(G)$ - minimum vertex degree of a graph G

$\Delta(G)$ - maximum vertex degree of a graph G

Δ - a simplex

$\Delta_k(P)$ - $d_k(P) - d_{k-1}(P)$ for a poset P

$\hat{\Delta}_k(P)$ - $\hat{d}_k(P) - \hat{d}_{k-1}(P)$ for a poset P

$\varepsilon(v)$ - eccentricity of a vertex v

$\zeta, \eta, \mu, \kappa, \lambda$ - incidence functions on posets

$\theta(G)$ - minimum number of cliques to cover all vertices in a graph G

$\kappa(G)$ - connectivity of a graph G

$\kappa'(G)$ - edge-connectivity of a graph G

$\kappa(x, y)$ - local connectivity for vertices in a graph

$\kappa'(x, y)$ - local edge-connectivity for vertices in a graph

$\lambda(x, y)$ - maximum number of independent x, y-paths

$\lambda'(x, y)$ - maximum number of edge-disjoint x, y-paths

$\lambda_1, \ldots, \lambda_n$ - eigenvalues (of adjacency matrix)

Λ_n - partitions of integer n, ordered by refinement

$\mu(n)$ - Möbius function of an integer n

$\mu(x, y)$ - Möbius function on interval $[x, y]$

$\mu_1, \ldots, \mu_n$ - Laplacian eigenvalues of a graph

π - circumference of circle with unit diameter, 3.1415926...

π, σ, τ - typical permutations

Π_n - lattice of partitions of an n-set, ordered by refinement

ρ - density

$\sigma(X)$ - span of a set X in a matroid

Σ - universe of sets for intersection classes, surface

$\sum$ - summation

$\tau(G)$ - number of spanning trees in a graph G

$\tau(G)$ - minimum partition into complete bipartite subgraphs

$\Upsilon(G)$ - arboricity of a graph G

ϕ - a mapping, often an isomorphism, coloring, or homomorphism

$\varphi(m)$ - Euler totient function (numbers in $[m]$ relatively prime to m)

$\chi(G)$ - chromatic number of a graph G

$\chi'(G)$ - edge-chromatic number of G

$\chi''(G)$ - total chromatic number of G

$\chi_\ell(G)$ - list-chromatic number of G

$\chi'_\ell(G)$ - list-edge-chromatic number of G

$\omega(G)$ - clique number of a graph G

ω_n - unbounded sequence

$\Omega(\mathbf{A})$ - intersection graph of collection $\mathbf{A}$

$\Omega(\Sigma)$ - intersection class for subsets of Σ

$\Omega(f(n))$ - functions at least a constant multiple of $f(n)$ (for large n)

Subject Index

A page number in italics indicates a definition. A single listing in italics may indicate the definition for a concept so prevalent (such as "graph") that it would not be productive to list its occurrences. An item that appears on few pages may have none italicized for the definition.

Page ranges in bold indicate material such as the proof of a major result or the main treatment of the concept; this may also include a definition. Pages ranges may also include isolated pages where the term does not appear.

Parenthetic clarifiers act as subheadings. Terms consisting of a word modified by prefatory notation are alphabetized according to the root word.